To my son, Christopher

BIOLOGY PUBLISHER: Jack C. Carey

EDITORIAL ASSISTANTS: Kathryn Shea, Laura Jersild

DEVELOPMENTAL EDITOR: Mary Arbogast

PRODUCTION AND ART SUPERVISION:
Hal Lockwood, Christi Payne Fryday/Bookman
Productions

PRODUCTION COORDINATOR: Harold Humphrey

PRINT BUYER: Karen Hunt

PERMISSIONS EDITOR: Robert Kauser

COPY EDITOR: Elizabeth F. Gehman

PAGE MAKE-UP: Judith Levinson

TECHNICAL ILLUSTRATORS: Darwen and Vally
Hennings, John and Judy Waller, Illustrious, Inc.,
Academy ArtWorks, Inc., Precision Graphics

CHAPTER-OPENING ILLUSTRATORS: Chapters 1, 4, 5,
7, 9, 10, 12, 13, 14, 18, 20, 21, 24, 25, and 27 designed and
rendered by Darwen and Vally Hennings; Chapters 6, 8,
11, 19, 22, and 26 designed by Darwen Hennings and
rendered by Teresa Snyder; Chapters 2, 3, 15, 16, 17, and
23 designed and rendered by John and Judy Waller

COMPOSITOR: Syntax International

COLOR SEPARATOR: H&S Graphics, Inc.

PRINTER: R. R. Donnelley and Sons Company/Willard

COVER DESIGN: Stephen Rapley

COVER ILLUSTRATION: Tomo Narashima, based on a
design by Darwen and Vally Hennings

2345678910   96 95 94 93

Library of Congress Cataloging-in-Publication Data

Wolfe, Stephen L.
    Molecular and cellular biology / Stephen L. Wolfe.
        p.   cm.
    Includes bibliographical references and index
    ISBN 0–534–12408–9
    1. Cytology.   2. Molecular biology.   I. Title.
QH581.2.W646 1993
574.87--dc20   92–38636

BOOKS IN THE WADSWORTH BIOLOGY SERIES

Environmental Science, 3rd, Miller

Resource Conservation and Management, Miller

Biology: Concepts and Applications, Starr

Biology: The Unity and Diversity of Life, 6th, Starr and
    Taggart

Biology of the Cell, 2nd, Wolfe

Cell Ultrastructure, Wolfe

Dimensions of Cancer, Kupchella

Evolution: Process and Product, 3rd, Dodson and Dodson

Introduction to Cell Biology, Wolfe

Oceanography: An Introduction, 4th, Ingmanson and
    Wallace

Plant Physiology, 4th, Devlin and Witham

Exercises in Plant Physiology, Witham et al.

Plant Physiology, 4th, Salisbury and Ross

Plant Physiology Laboratory Manual, Ross

Plants: An Evolutionary Survey, 2nd, Scagel et al.

Psychobiology: The Neuron and Behavior, Hoyenga and
    Hoyenga

Sex, Evolution, and Behavior, 2nd, Daly and Wilson

MOLECULAR AND CELLULAR BIOLOGY

STEPHEN L. WOLFE
University of California at Davis

Wadsworth Publishing Company
Belmont, California
A Division of Wadsworth, Inc.

PREFACE

The aim of this book is to integrate molecular biology, biochemistry, and cell biology into a unified course of study. Until now, molecular biology texts have concentrated on gene structure and activity to the near exclusion of more "traditional" biochemistry and cell biology. This reflects a past tendency to throw out the baby with the bath water—to regard many of the findings and conclusions developed by traditional biochemical and cell biological investigations as unreliable and uninteresting because the research was conducted before the molecular revolution. However, the emphasis in molecular biology has now shifted from a concentration on genes for their own sake to the application of molecular genetic studies to all areas of cell biology and biochemistry. As a result, more traditional research areas—ion transport and the control of cell division come to mind as two examples—have become the subjects of highly productive investigations in which the powerful methods of molecular biology are producing many new findings and pushing these areas to the forefront of research.

With these developments in mind, this book offers a balanced view of contemporary molecular biology, biochemistry, and cell biology not currently available in other texts. The central topics of molecular biology are included, among them DNA structure, messenger RNA gene structure and activity, and the molecular methods for studying these genes. However, the molecular biology of genes encoding the other major RNA types is also emphasized, along with their regulation and the biochemistry of their transcription. These topics are integrated with the cell biology of the nucleus, including the structure of nucleosomes, chromatin, the nucleolus, and the nuclear envelope; the changes occurring in nucleosomes and chromatin during the shift between inactivity and gene transcription; and the role of the nuclear envelope in transport between the nucleus and cytoplasm. The organization of genes into genomes and the techniques of genetic engineering are also included in this integration. Similarly, the coverage of cell division includes an integrated view of the molecular biology and biochemistry of cell cycle genes and their regulation, DNA replication and repair, and the changes in nucleosome and chromatin structure that accompany cell division. The molecules, forces, and structures separating the chromosomes during mitotic cell division are also considered, as well as the molecular biology, biochemistry, and cell biology of genetic recombination during meiosis, the formation of gametes, and their functions in fertilization.

Among the more traditional topics of cell biology and biochemistry that are integrated with molecular biology and emphasized in this text are protein synthesis and the modification and distribution of newly synthesized proteins by the endoplasmic reticulum and Golgi complex; membrane structure and transport, and the role of ion transport in the generation and conduction of nerve impulses; the activities of mitochondria and chloroplasts in cellular metabolism, and the mechanisms governing genetic inheritance in these organelles; the roles of microtubules and microfilaments in the generation of motility and, with intermediate filaments, in cytoskeletal support; and the structure and function of the cell surface and extracellular matrix, along with the cellular regulatory mechanisms linked to receptors at the cell surface.

These topics are distributed between the chapters of text, which cover subjects that are central to most or all courses, and chapter Supplements, which add important specialized or peripheral information. This distribution keeps the chapters as direct and to the point as possible. Also included in the chapters are Information Boxes, presenting short but essential items of background information. By selecting among the material presented in the chapters and Supplements, an instructor can tailor the text to suit the aims of the course and its students.

To make the book as clear and graphic as possible, every topic that merits explanation by pictures as well as words is illustrated by a diagram or micrograph. The many light and electron micrographs, in particular, emphasize that molecular and biochemical processes have a structural basis and help relate these processes to cell biology. In addition, each chapter opens with a major piece of art that focuses interest and illustrates a central topic explained in that chapter.

Where possible, molecular and cellular biology and biochemistry are brought home to students by integrating examples from human biology, especially medicine. These examples show that the topics described in the book, as well as having scientific and academic importance, touch directly on human affairs. The regulation of cell division, for example, is illustrated by a description of the cell cycle controls that go awry in the transformation of normal cells into cancer cells; the organization of DNA sequences in genomes is elaborated by a discussion of genetic engineering and the genetic rearrangements underlying the production of antibodies, the immune response, and the rejection of transplanted tissues.

The findings, conclusions, and principles described in the text are presented in terms of experimental evidence drawn from work with prokaryotes, fungi, animals, and plants. Controversies as well as conclusions are presented in order to show that the body of scientific information in molecular and cellular biology is not fixed, and that many significant questions remain to be answered. This experimental foundation is bolstered by essays contributed by original investigators, presenting their experiences in the conception and execution of classic and contemporary experiments that have produced key contributions in molecular biology, biochemistry, and cell biology. The essays add a personal element to the information presented in the book and emphasize that this information is the result of experimental work by individual scientists. The essayists include both established researchers, who are essentially household names among biologists, and relative newcomers, who are likely to be among the next generation of famous names. Hopefully, in addition to piquing students' interest, the essays may help turn them toward the possibility of research as a career.

The spate of multiple-author textbooks in molecular and cellular biology has led to a belief that the information in this field has become so extensive that it lies beyond the reach of a single author. I hope this book will make it apparent that the opposite is true, and that coverage by a single author brings a degree of organization, integration, and unity that lies beyond even the most carefully prepared multiple-author text.

Although this is a single-author book, I must acknowledge my debt to the many people who have contributed to it. The book could not have been written without the help of colleagues and friends who reviewed the text and offered suggestions for improving its accuracy and content. I am also deeply indebted to the many investigators who generously supplied micrographs, diagrams, and tables. I must also acknowledge my debt to my biology editor, Jack Carey, who provided endless encouragement and guidance in the preparation of the text; to my developmental editor, Mary Arbogast, who provided expert editorial help and unscrambled more glitches than I care to remember; and to Hal Humphrey, Hal Lockwood, and Christi Fryday, who coordinated production at Wadsworth and Bookman Productions, and solved the innumerable problems attending publication of a book of this size and complexity. I am also indebted to Darwen and Vally Hennings, and John and Judy Waller, who prepared original art and redrew many of the diagrams sent by other authors. Darwen Hennings in particular was instrumental in the design and preparation of the more complex illustrations in the book, including the majority of drawings that serve as chapter openers.

REVIEWERS

THE EXPERIMENTAL PROCESS ESSAYISTS

BRIEF CONTENTS

DETAILED CONTENTS

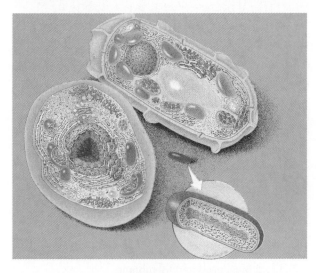

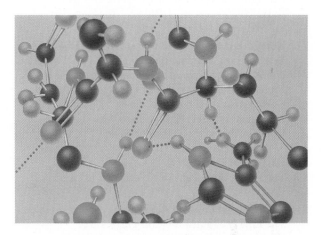

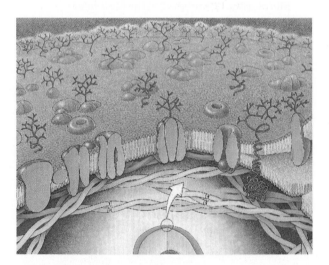

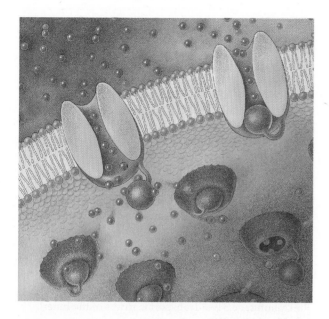

CHAPTER 9

ENERGY FOR CELL ACTIVITIES: CELLULAR OXIDATIONS AND THE MITOCHONDRION

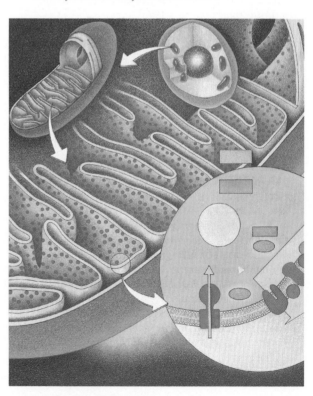

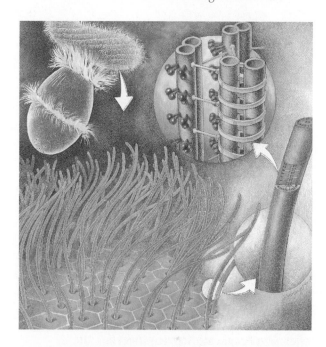

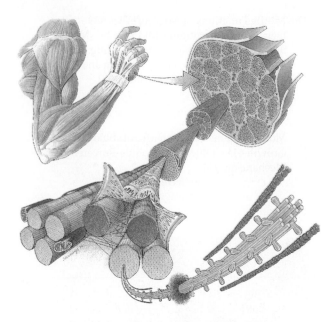

CHAPTER 13

MICROTUBULES, MICROFILAMENTS, AND INTERMEDIATE FILAMENTS IN THE CYTOSKELETON

CHAPTER 14

THE NUCLEUS AND ITS MOLECULAR CONSTITUENTS

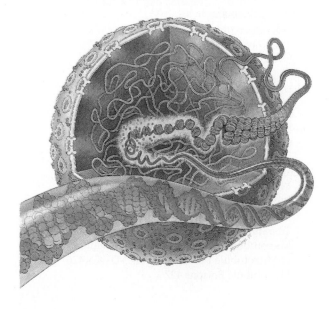

CHAPTER 17

TRANSCRIPTIONAL AND TRANSLATIONAL REGULATION

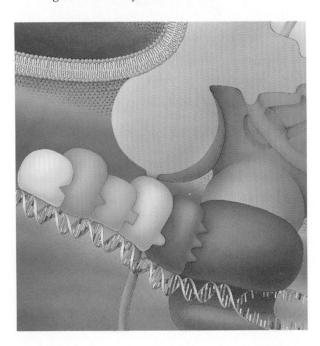

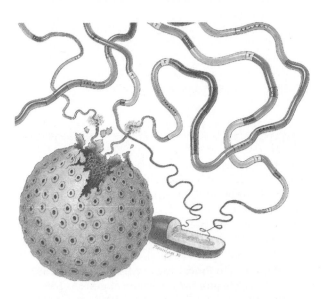

CHAPTER 18

ORGANIZATION OF THE GENOME AND GENETIC REARRANGEMENTS

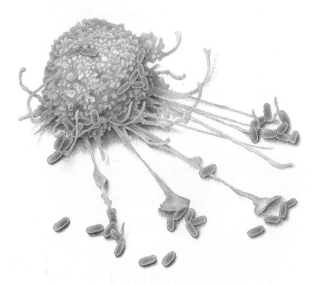

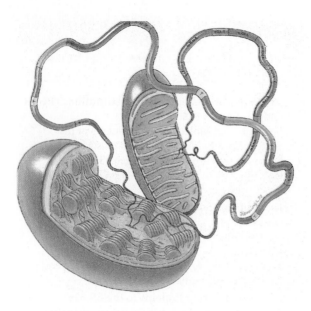

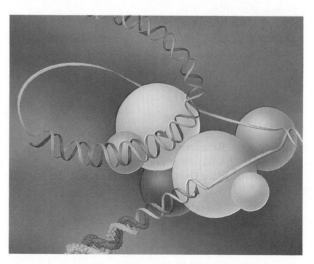

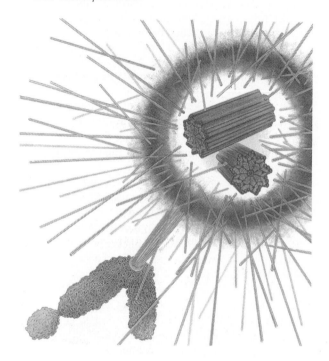

CHAPTER 24

NUCLEAR AND CYTOPLASMIC DIVISION

CHAPTER 25

MEIOSIS AND GENETIC RECOMBINATION

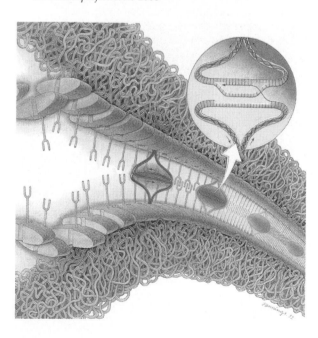

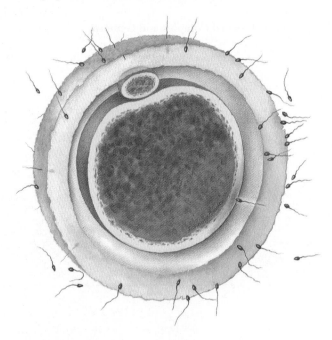

INTRODUCTION TO CELL
AND MOLECULAR BIOLOGY

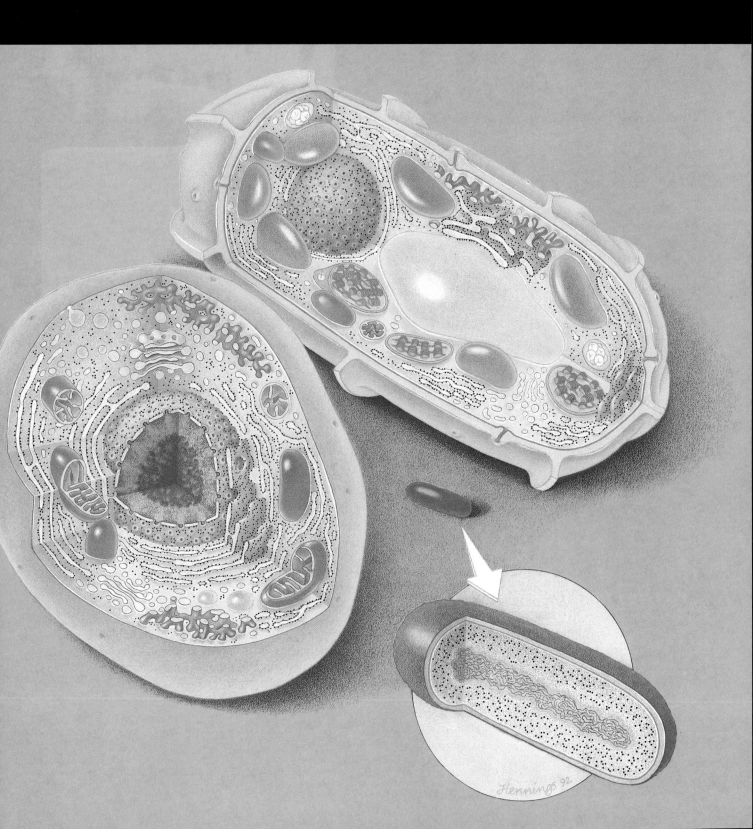

Prokaryotic cell structure ▪ *Eukaryotic cell structure* ▪ *Viruses* ▪ *Viroids and prions* ▪ *History of cell and molecular biology*

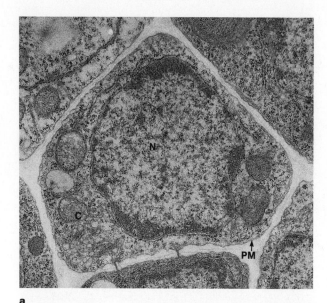

a

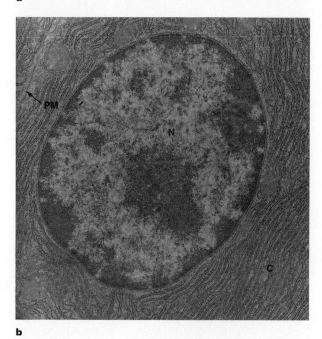

b

Figure 1-1 Electron micrographs of plant and animal cells. N, nucleus; C, cytoplasm; PM, plasma membrane. **(a)** An embryonic plant cell of *Sorghum bicolor*, a type of grass. × 12,000. Courtesy of Chin Ho Lin. **(b)** A cell from the pancreas of the rat. × 11,500. Photograph by the author.

C ells are the fundamental structural and functional units of all living organisms. They contain highly organized molecular and biochemical systems that are capable of storing information, translating the information into the synthesis of cellular molecules, and using energy sources to power these activities. In addition, cells are capable of movement and can compensate for environmental fluctuations by altering their internal biochemical reactions. Cells can also duplicate and pass on their hereditary information, as well as their major biochemical and molecular systems, as part of cellular reproduction. All these activities are packed into structural units that represent the ultimate in miniaturization—in most cases the cells of living organisms are of microscopic dimensions, either invisible or barely visible to the naked eye.

In some groups of organisms, such as bacteria and protozoa, cells and individuals are one and the same. Each cell of these organisms is functionally independent and capable of carrying out all the activities of life. In more complex multicellular organisms major life activities are divided among groups of specialized cells. However, even the cells of multicellular organisms are potentially capable of independent activity. Many of these cells, if removed and cultured under the proper conditions, retain all the qualities of life, including the ability to grow and reproduce.

Many of the inner structures of cells can also be maintained in the test tube in fully functional form. However, once cells are broken, the quality of life is lost: the interior structures, although capable of limited biochemical or molecular activities, are unable to grow, reproduce, or respond to outside stimuli in a coordinated, potentially independent fashion. Therefore, life as we know it does not exist in units more simple than individual cells.

INTRODUCTION TO CELL STRUCTURE

Cells take highly varied forms in different plants, animals, and microorganisms (Figs. 1-1 to 1-5 give some idea of the wide range of cell types). They may exist singly, as in bacteria and protozoa, or packed together by the millions or billions, as in larger plants and animals. In size cells range from bacteria, which with diameters of about 0.5 micrometer (μm) are barely visible in the light microscope, to units as large as a hen's egg: the yolk of a hen's egg is a single cell, several centimeters in diameter. (Table 1-1 explains the micrometer and other units of measurement used in cell and molecular biology.) In a multicellular animal cells range from about 10 to 30 μm in diameter; in plants cells range from about 10 μm to as much as several hundred micrometers. Cells may be roughly spherical in shape, flattened, cuboidal, or columnar; some, like

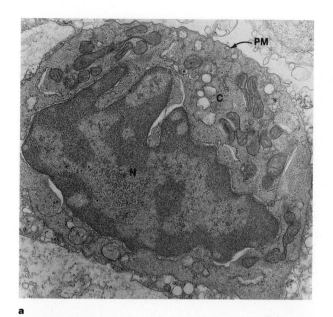

a

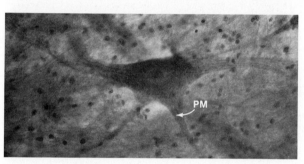

b

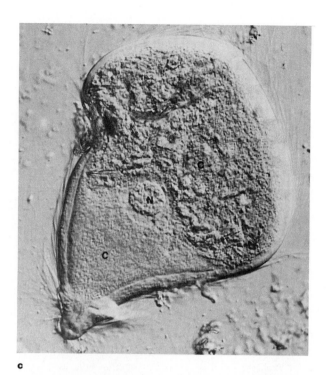

c

Figure 1-2   Additional eukaryotic cell types. N, nucleus; C, cytoplasm; PM, plasma membrane. **(a)** An electron micrograph of a human leucocyte, or white blood cell. ×13,000. Courtesy of S. Brecher. **(b)** A light micrograph of a nerve cell from bovine spinal cord. ×1,600. Photograph by the author. **(c)** A light micrograph of a protozoan cell from the gut of a termite. ×330. Courtesy of T. K. Golder.

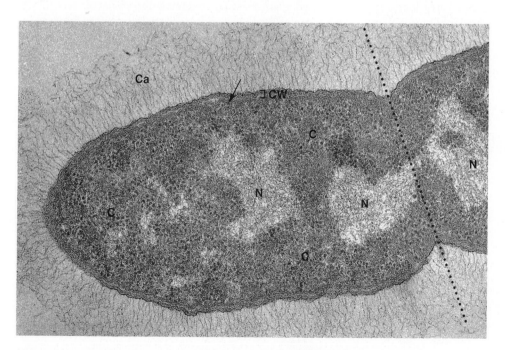

Figure 1-3   A prokaryotic cell, the bacterium *Klebsiella pneumoniae*. The nucleoid (N), the prokaryotic equivalent of a nucleus, occupies the center of the cell. The cytoplasm (C) surrounding the nucleoid is packed with ribosomes. The cell is surrounded by a cell wall (CW); the plasma membrane (arrow) lies just beneath the cell wall. The capsule (Ca) is visible just outside the cell wall. The cell is dividing; the plane of division is shown by the dotted line. ×52,000. Courtesy of E. N. Schmid, from *J. Ultrastr. Res.* 75: 41 (1981).

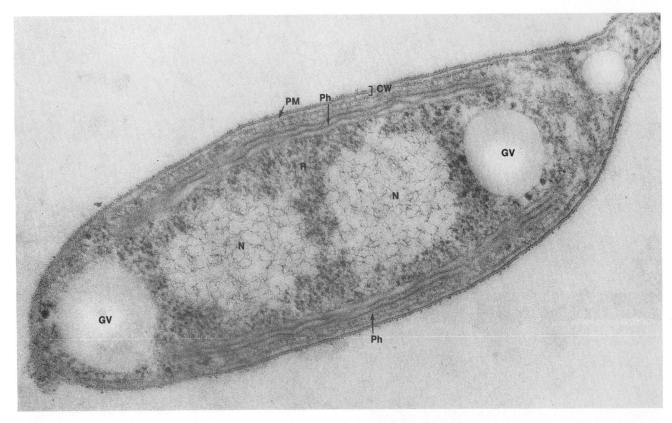

Figure 1-4 A photosynthetic bacterium. N, nucleoid; PM, plasma membrane; Ph, photosynthetic membrane; R, ribosomes; GV, gas vacuole; CW, cell wall. ×75,000. Courtesy of W. C. Trentini and the American Society for Microbiology, from *J. Bact.* 93:1699 (1967).

the nerve cells of larger animals, may carry long extensions that are of microscopic diameter but more than a meter in length.

In spite of their varied sizes, shapes, and activities, all cells are divided into two major internal regions that reflect a fundamental division of labor in cell function.

The *nuclear region* contains *deoxyribonucleic acid (DNA)* molecules that store hereditary information required for cell growth and reproduction. In addition, the nuclear region contains enzymatic systems that can copy the hereditary information for cell reproduction and for directing the synthesis of proteins. The second region,

Table 1-1	Units of Measure Used in Cell and Molecular Biology			
		Equivalent in:		
Unit	Angstroms	Nanometers	Micrometers	Millimeters
Angstrom (Å)	1	0.1	0.0001	0.0000001
Nanometer (nm)	10	1	0.001	0.000001
Micrometer (µm)	10,000	1000	1	0.001
Millimeter (mm)	10,000,000	1,000,000	1000	1

Units on a log scale:

1 Å	10 Å	100 Å	1000 Å	10,000 Å	10,000,000 Å
	1 nm	10 nm	100 nm	1000 nm	1,000,000 nm
				1 µm	1000 µm
					1 mm

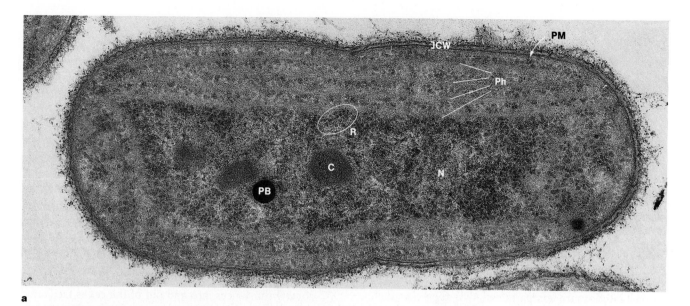

a

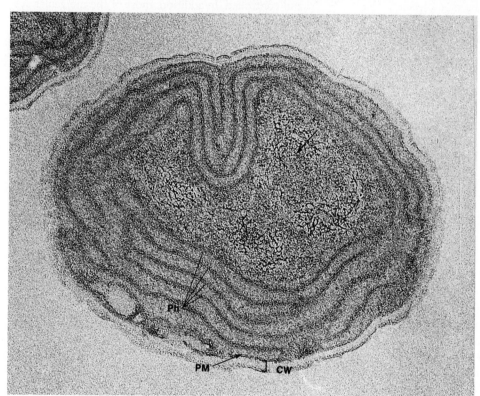

b

Figure 1-5 Two representative cyanobacteria. **(a)** *Agmenellum quadruplicatum.* N, nucleoid; R, ribosomes; PM, plasma membrane; Ph, photosynthetic membranes; PB, polyphosphate body; C, carboxysome, a body containing semicrystalline deposits of the enzyme active in carbon dioxide fixation; CW, cell wall. ×60,000. Courtesy of S. A. Nierzwicki-Bauer, D. L. Balkwill, and S. E. Stevens, Jr., from *J. Ultrastr. Res.* 84:73 (1983). **(b)** *Synechococcus lividus.* The center of the cell is occupied by the nucleoid, in which DNA fibers are clearly visible (arrow). The cytoplasm is packed with photosynthetic membranes (Ph); the cell is surrounded by the plasma membrane (PM) and the cell wall (CW). ×72,000. Courtesy of M. R. Edwards, New York State Department of Health and *Journal of Phycology.*

the *cytoplasm,* makes proteins according to directions copied in the nucleus and also synthesizes most of the other molecules required for growth and reproduction (some are made in the nuclear region). The cytoplasm carries out several additional vital functions. Among the most important of these are the conversion of fuel substances into forms of chemical energy that can be used by the cell, the conduction of stimulatory signals from the outside to the cell interior, the transport of materials to and from the cell, and cell motility. The

total living matter of cells, including both the nuclear region and the cytoplasm, is collectively called *protoplasm.*

Cells are maintained as distinct environments and collections of matter by exceedingly thin layers of lipid and protein molecules called *membranes.* These molecular layers, not much more than 7 to 8 nanometers (nm) in thickness, set up continuous outer boundaries that effectively separate the cell contents from the exterior. In many cell types membranes also subdivide

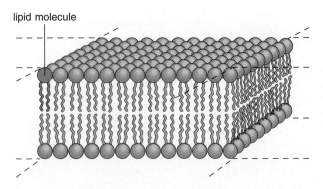

lipid molecule

Figure 1-6 The arrangement of lipid molecules in a layer two molecules in thickness. The double layer provides the framework of all biological membranes.

the cell interior into compartments with their own distinct environments and collections of molecules. The lipid part of membranes, consisting of a double layer of molecules (Fig. 1-6), provides a structural framework and sets up a barrier to the passage of water-soluble substances. Membrane proteins, which are suspended individually or in groups in the lipid framework, form channels allowing selected water-soluble molecules to pass from one side of the membrane to the other. The selectivity of the protein channels allows membranes to control the movement of molecules in and out of cells, and between membrane-bound compartments within cells. Most other functions of membranes, such as the recognition and binding of molecules at the membrane surfaces, are also carried out by the protein component. Thus, the structural framework of membranes depends primarily on lipids, and the functions of membranes primarily on proteins.

The cells of all organisms fall into one of two major divisions according to the number and arrangement of cellular membranes and the complexity of the nuclear region. The smaller and more primitive division, the *prokaryotes* (from *pro* = before and *karyon* = nucleus), includes only two major groups—bacteria and *cyanobacteria* (or *blue-green algae*). The membrane systems of prokaryotes are limited to the surface, or *plasma membrane*, and a relatively simple collection of inner membranes in the cytoplasm. The nuclear region of prokaryotes, called the *nucleoid*, is suspended directly in the cytoplasm with no system of boundary membranes setting it off as a separate compartment.

The second major division of living organisms, the *eukaryotes* (from *eu* = typical and *karyon* = nucleus), includes all the remaining organisms of the earth— animals, plants, fungi, and protozoa. A plasma membrane covers the surface of the cells of these organisms as in the prokaryotes. In addition, several distinct internal membrane systems divide eukaryotic cells into interior compartments with specialized functions.

PROKARYOTIC CELLS

Prokaryotic cells (see Figs. 1-3 to 1-5) are comparatively small, usually not much more than a few micrometers long and a micrometer or slightly less in width. (Fig. 1-7 compares the dimensions of several molecules and cell structures in prokaryotes and eukaryotes.) In almost all prokaryotes the boundary membrane of the cell, the plasma membrane, is surrounded by a rigid external layer of material, the *cell wall*. The cell wall may range in thickness from 15 to 100 nm or more and may itself be coated with a thick, jellylike *capsule* (see Fig. 1-3). The cell wall provides rigidity to prokaryotic cells and, with the capsule, protects the cell within the wall.

The plasma membrane carries out a variety of vital functions in prokaryotes. Most important of these functions is *transport*, in which the movement of water-soluble substances into and out of the cell is facilitated and controlled. In addition, most of the molecular systems that break down fuel substances to release energy for cell activities are housed in the plasma membrane in prokaryotes. In photosynthetic bacteria and cyanobacteria the molecules absorbing light and converting it to chemical energy are associated with the plasma membrane and its interior extensions or derivatives. The plasma membrane also contains proteins that can act as *receptors* by recognizing and binding specific molecules that penetrate through the wall from the surrounding medium. Binding the external molecules triggers internal reactions that allow prokaryotic cells to respond to their environment. The prokaryotic plasma membrane may also play a part in replication and division of the nuclear material.

The nucleoid of a prokaryotic cell, suspended directly in the cytoplasm without boundary membranes, appears as a structure of irregular outline that contains masses of very fine fibers 3 to 5 nm in thickness (see Fig. 1-3). When isolated, the nucleoid of bacteria proves to contain a single, large DNA molecule in the form of a closed circle. In *Escherichia coli*, the best-studied bacterium, the nucleoid circle contains 1360 μm of DNA. Other bacteria have circles ranging from about 250 μm to a maximum of about 1500 μm of included DNA. All this DNA is packed into cells that are only 1 to 2 μm long.

Comparatively little is known about the structure of DNA in cyanobacteria. Considerably more DNA, up to several times the amount in bacteria such as *E. coli*, is present in these prokaryotes. Whether the DNA of cyanobacteria also forms closed circles is as yet unknown. Relatively few proteins can be detected in association with the DNA in either bacteria or cyanobacteria.

The cytoplasm surrounding the nucleoid usually appears densely stained in electron micrographs. Most of this density is due to the presence of large numbers

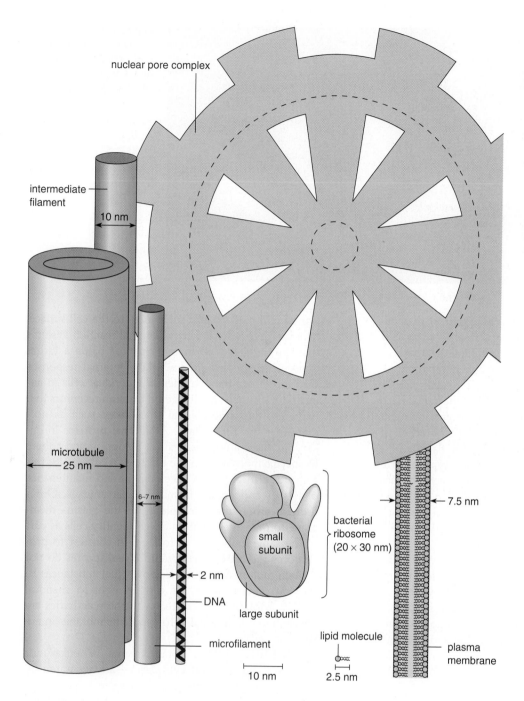

intermediate filament

nuclear pore complex

10 nm

microtubule
25 nm

6–7 nm

2 nm

DNA

microfilament

10 nm

small subunit

large subunit

bacterial ribosome
(20 × 30 nm)

lipid molecule

2.5 nm

7.5 nm

plasma membrane

Figure 1-7 The dimensions of some cell structures in prokaryotes and eukaryotes. In the diagram of a nuclear envelope pore complex the annulus that fills the pore is shown in outline as it might appear from the outside of the nuclear envelope. The smaller dotted circle at the center is the approximate location of the channel extending through the annulus; the larger dotted circle is the diameter of the opening in the nuclear envelope membranes that is filled by the annulus. The diagram shown for the pore complex is based on a model developed by C. W. Akey; from *J. Cell Biol.* 109:955 (1989).

of *ribosomes*, which are small, roughly spherical particles about 20 to 30 nm in diameter (see Fig. 1-7). In bacteria ribosomes contain more than 50 different proteins in combination with several types of *ribonucleic acid (RNA).* These small but complex spherical bodies are the sites where amino acids are assembled into proteins.

Other structures may be present in the cytoplasm of more complex prokaryotes. In some bacteria and cyanobacteria the cytoplasm contains numerous bag-like, closed sacs, collectively called *vesicles* or *vacuoles,* with walls formed by a single, continuous membrane.

Molecules carrying out photosynthesis are associated with some of these internal sacs in the photosynthetic bacteria (see Fig. 1-4) and cyanobacteria (see Fig. 1-5). Many photosynthetic bacteria and cyanobacteria also contain sacs called *gas vacuoles.* The gas vacuoles evidently provide buoyancy, holding the cells near liquid surfaces or sources of light. The cytoplasm of prokaryotic cells may also contain deposits of lipids, polysaccharides, or inorganic phosphates that appear as small, very dense spherical bodies scattered in the cytoplasm (see Fig. 1-5a). The deposits of inorganic phosphates,

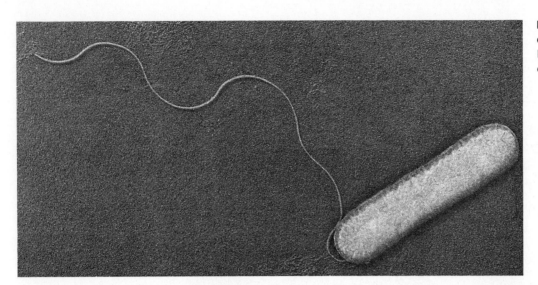

Figure 1-8 A bacterial cell with a single flagellum. × 30,000. Courtesy of J. Pangborn.

called *polyphosphate bodies*, may provide a store of phosphate groups acting as an energy reserve or as raw materials for nucleic acid or phospholipid synthesis.

Many types of bacteria are capable of rapid movement generated by the action of long, threadlike *flagella* (singular = *flagellum*) that extend from the cell surface (Fig. 1-8). Bacterial flagella are formed from long corkscrew-shaped chains of protein molecules attached at one end to the cell surface. Usually, the flagella of a single bacterial species contain a single type of protein called a *flagellin*. Bacterial flagella, which produce motion by rotating like a propeller (see p. 447 and Fig. 11-33), are fundamentally different from the much larger and more complex flagella of eukaryotic cells (see below).

The apparent simplicity of prokaryotic cells is deceptive. Most bacteria and cyanobacteria contain complex molecular systems, can use a wide variety of substances as energy sources, and are able to synthesize all their required organic molecules from simple starting substances, such as water, carbon dioxide, and inorganic sources of nitrogen, phosphorus, and sulfur. In many respects, in fact, prokaryotes are more versatile in their biochemical activities than eukaryotes.

EUKARYOTIC CELLS

The complex membrane systems of eukaryotic cells (Figs. 1-9 to 1-11) define the nucleus and separate the cytoplasm into distinct compartments called *organelles*. The plasma membrane surrounding a eukaryotic cell, like that of prokaryotes, carries out a variety of functions. Most significant of these functions is transport, provided by proteins forming channels in the membrane. Eukaryotic plasma membranes also contain a variety of proteins acting as receptors, which can recognize and bind specific substances to the cell surface. Most receptors are linked to internal reaction systems that are triggered when the receptor binds a target molecule at the cell surface. Many internal responses that coordinate the activities of individual cells in animals, for example, are triggered through receptors that recognize and bind "signal" molecules, such as hormones, at the cell surface. Additional proteins in eukaryotic plasma membranes can recognize and adhere to specific molecules on the surfaces of other cells; this ability is critical to the development and maintenance of tissues and organs in multicellular animals. Other protein-carbohydrate complexes in the plasma membrane form markers that, in effect, identify cells as part of the same individual or foreign. In contrast to prokaryotes the plasma membranes of eukaryotes do not contain molecules that break down fuel substances to provide cellular energy or convert light to chemical energy. Molecules associated with these activities are concentrated in internal cytoplasmic membranes in eukaryotic cells.

The two distinct compartments of eukaryotic cells—the nucleus and cytoplasm—are separated by the *nuclear envelope*. This membrane system, which covers the nucleus, consists of two concentric membranes, one layered just inside the other. The outer membrane faces the cytoplasm, and the inner membrane, separated from the outer membrane by a narrow space, faces the nucleoplasm. The nuclear envelope is perforated by *pores* that form openings through both membranes and serve as channels conducting substances between the nucleus and cytoplasm (arrows, Fig. 1-9; see also Figs. 1-7 and 14-27 to 14-33). At the margins of the pores, the outer nuclear membrane folds inward and becomes continuous with the inner membrane, forming an opening in the nuclear envelope about 70 to 90 nm in diameter. The pores are filled by a ringlike plug of dense material known as the *annulus*, which contains a

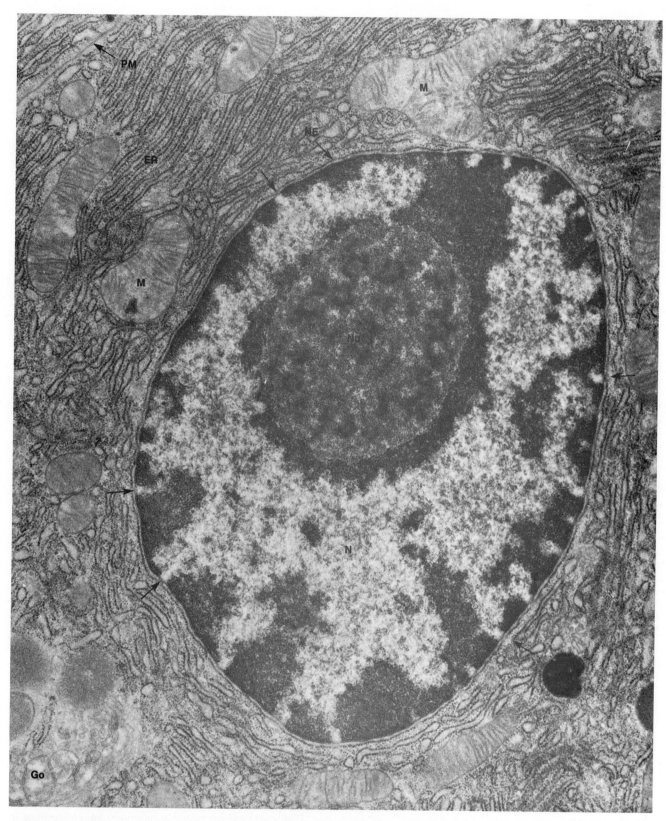

Figure 1-9 A eukaryotic cell from rat pancreas. N, nucleus; Nu, nucleolus; M, mitochondrion; ER, endoplasmic reticulum; Go, Golgi complex; PM, plasma membrane; NE, nuclear envelope; arrows, pore complexes. ×24,000. Photograph by the author.

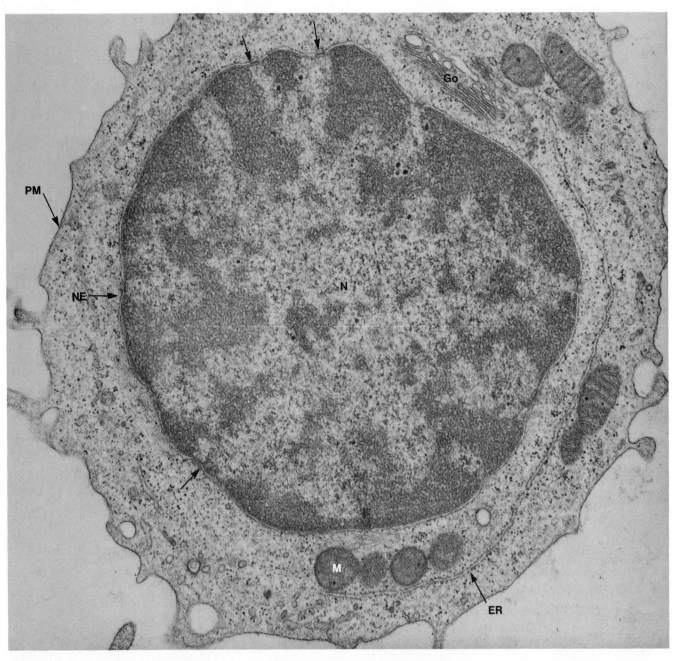

Figure 1-10 A human lymphocyte, a type of white blood cell. N, nucleus; NE, nuclear envelope; M, mitochondrion; ER, endoplasmic reticulum; Go, Golgi complex; PM, plasma membrane; arrows, nuclear pore complexes. Courtesy of W. R. Hargreaves.

narrow aperture at its center. The annulus acts in some as yet unknown way to control the movement of larger molecules, such as RNA and proteins, through the nuclear envelope. The pore and the annulus together form the *pore complex.*

The Nucleus

Most of the space inside the nucleus is occupied by masses of very fine, irregularly folded fibers. The fibers, which range from about 10 to 30 nm in thickness, contain the DNA of the nucleus, along with two types of protein specifically associated with DNA in eukaryotes, the *histone* and *nonhistone chromosomal proteins.* The two proteins occur in relatively large quantities in eukaryotic nuclei, together amounting to approximately twice the DNA quantity by weight. This is in distinct contrast to bacteria, in which the ratio of protein to DNA in nucleoids is less than 1. Histones are primarily structural molecules that pack DNA into chro-

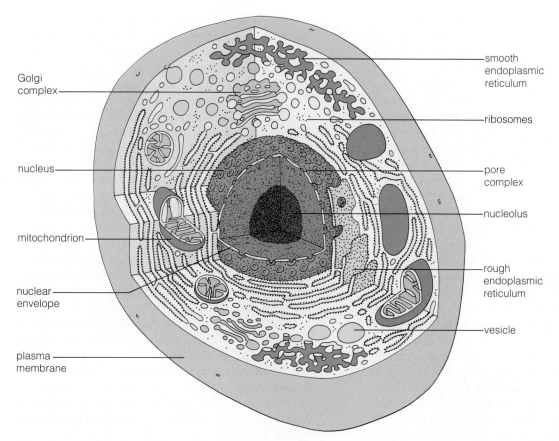

Golgi complex

nucleus

mitochondrion

nuclear envelope

plasma membrane

smooth endoplasmic reticulum

ribosomes

pore complex

nucleolus

rough endoplasmic reticulum

vesicle

Figure 1-11 Structures visible in thin-sectioned eukaryotic cells (see text).

matin fibers. Nonhistones include a group of proteins that carry out one of the most important cellular functions, the regulation of gene activity.

Instead of being concentrated in a single circular molecule as in bacteria, the hereditary information of a eukaryotic nucleus is subdivided among several to many linear DNA molecules. Each individual DNA molecule, with its associated proteins, is a *chromosome.* A human nucleus, for example, contains 46 chromosomes, each a single linear DNA molecule with its associated histone and nonhistone chromosomal proteins. The total collection of DNA molecules, in combination with the histone and nonhistone chromosomal proteins, is called the *chromatin* of a eukaryotic nucleus.

Eukaryotic nuclei contain much greater quantities of DNA than do prokaryotic nucleoids. The entire complement of chromosomes in a human cell, for example, includes a total DNA length of about a meter. The nuclei of some eukaryotes contain even more—a cell nucleus of an amphibian, such as a frog or salamander, for example, contains about 10 meters of DNA!

Suspended within the chromatin of the nucleus are one or more irregularly shaped bodies, the *nucleoli* (sin-

gular = *nucleolus*). Nucleoli are so densely structured that their outlines are clearly visible, even though no membranes separate them from the surrounding chromatin. Chromatin fibers and two other major components, *nucleolar fibers* and *nucleolar granules*, are visible inside the nucleolus. The nucleolar fibers are indistinct and difficult to trace in electron micrographs; generally they appear to be somewhat thinner than chromatin fibers. The nucleolar granules, in contrast, are distinctly visible as small, spherical particles about half the size of cytoplasmic ribosomes.

Nucleoli are actually subparts of the chromatin specialized for the synthesis and assembly of ribosomal subunits. The nucleolar fibers and granules are structural forms taken by successive stages in this synthesis and assembly. The overall size and shape of the nucleolus and the distribution of fibers and granules inside it change as cells go through their cycles of growth and division, reflecting greater or lesser requirements for newly synthesized ribosomes.

The nucleus is the ultimate control center for cell activities. Within the chromatin the information required to synthesize cellular proteins is coded into the

DNA. Each DNA segment containing the information for making a single protein molecule constitutes a *gene.* The information in a protein-encoding gene is copied into a *messenger RNA (mRNA)* molecule that moves to the cytoplasm through the pore complexes of the nuclear envelope. In the cytoplasm mRNA molecules are used by ribosomes as directions for the assembly of proteins. The DNA of the entire nuclear region contains the codes for many thousands of different proteins.

Other DNA regions store the information for making additional RNA types that carry out accessory roles in protein synthesis and other functions in the nucleus and cytoplasm. One of the additional RNAs, *ribosomal RNA (rRNA),* is encoded in DNA regions forming parts of the nucleolus. This RNA type forms a part of ribosomes. Another RNA, *transfer RNA (tRNA),* binds to amino acids during protein synthesis and provides a necessary link between the nucleic acid code and the amino acid sequence of proteins. Other RNAs are involved in the processing of mRNA, rRNA, and tRNA molecules from initial to finished forms. Each DNA segment encoding an rRNA, tRNA, or other accessory RNA type is also known as a gene. The process in which any of the genes is copied into an mRNA, rRNA, tRNA, or other RNA equivalent is called *transcription.*

A second major function of the nucleus involves duplication of the chromatin as a part of cell reproduction. Just before cell division all the components of chromatin, including both DNA and chromosomal proteins, are precisely doubled. The duplication of DNA in the chromatin is known as *replication.* During cell division the two copies of each duplicated chromosome are separated and exactly divided so that the two cells resulting from the division each receive a complete set of genes.

The Cytoplasm

Eukaryotic cytoplasm is packed with ribosomes, vesicles, and a variety of membrane-bound organelles. The boundary membranes set off the organelles as separate chemical and molecular environments that are specialized to carry out different functions of the cytoplasm.

Ribosomes and Protein Synthesis   The reactions assembling proteins take place entirely in the cytoplasm. Following their transcription in the nucleus, mRNAs and tRNAs pass through the nuclear envelope and enter the cytoplasm as individual molecules; rRNAs enter in the form of ribosomal subunits. In the cytoplasm the ribosomal subunits join by twos with mRNA molecules to form complete ribosomes active in protein synthesis. Eukaryotic ribosomes, at diameters of about 25 to 35 nm, are somewhat larger than the ribosomes of prokaryotes and contain more protein and RNA molecules.

In protein synthesis, called *translation,* a ribosome starts at the beginning of the coding portion of an mRNA molecule and moves along the mRNA, reading the code for protein assembly as it goes. As it moves, the ribosome assembles amino acids into a gradually lengthening protein chain. At the end of the encoded message translation stops, the ribosomal subunits separate and detach from the mRNA, and the completed protein is released. At any instant several ribosomes may be at different places on a single mRNA molecule, engaged in reading the message and assembling protein chains. Ribosomes and mRNA may recycle through the mechanism many times. In this way each mRNA molecule may serve as a template for hundreds of protein molecules.

The tRNA molecules function as the "dictionary" in the translation mechanism. Each kind of tRNA corresponds to 1 of the 20 amino acids used in protein synthesis. Individual tRNAs are attached to their respective amino acids by enzymes that can recognize both a particular amino acid and the tRNA (or tRNAs) corresponding to that amino acid. As a result of the activity of these enzymes, each amino acid is linked to a specific kind of tRNA. The tRNA, in turn, is capable of recognizing and binding the nucleic acid code word (called a *codon*) specifying its attached amino acid in an mRNA molecule.

The interaction of tRNAs with mRNA codons takes place on ribosomes (Fig. 1-12). As a ribosome encounters an mRNA codon specifying a given amino acid (Fig. 1-12*a*), the tRNA carrying that amino acid recognizes and binds to the codon (Fig. 1-12*b*). This binding places the amino acid in its correct location in the growing protein chain. The ribosome then moves to the next mRNA codon, causing the next tRNA-amino acid complex specified by the code to bind (Fig. 1-12*c* and *d*). As each successive amino acid arrives at the ribosome, it is split from its tRNA and linked into the gradually lengthening protein chain. The process repeats until the ribosome reaches the end of the message and completes assembly of the protein.

Ribosomes, Endoplasmic Reticulum, and the Golgi Complex   Ribosomes active in protein synthesis may be either freely suspended in the cytoplasm or attached to the surface of membranous sacs (Fig. 1-13). The freely suspended ribosomes make proteins that primarily become part of the soluble background substance of the cytoplasm or form important structural or motile cytoplasmic elements. The ribosomes attached to membranous sacs synthesize proteins that become a part of membranes or are packaged into vesicles for storage in the cytoplasm or export to the cell exterior.

The membranous sacs with their attached ribosomes interconnect to form an extensive system of cytoplasmic channels known as the *rough endoplasmic*

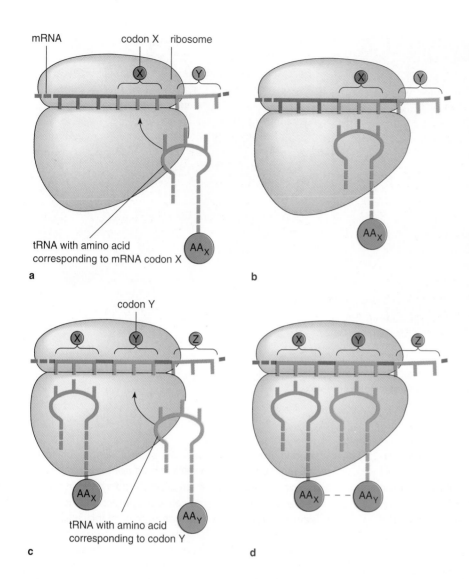

mRNA   codon X   ribosome

tRNA with amino acid
corresponding to mRNA codon X

a

b

codon Y

tRNA with amino acid
corresponding to codon Y

c

d

Figure 1-12 An outline of the overall mechanism by which amino acids are assembled into proteins on ribosomes. As a ribosome encounters an mRNA codon specifying a given amino acid (codon X in **a**), the tRNA carrying that amino acid recognizes and binds to the codon **(b)**. This binding places the amino acid in its correct location in the growing protein chain. The ribosome then moves to the next mRNA codon (codon Y in **c**), causing the next tRNA-amino acid complex specified by the code to bind **(d)**. As each successive amino acid arrives at the ribosome, it is split from its tRNA and linked into the gradually lengthening protein chain. The process repeats until the ribosome reaches the end of the message and completes assembly of the protein.

Figure 1-13 A mitochondrion from bat pancreas, surrounded by cytoplasm containing rough endoplasmic reticulum (rough ER). Cristae (arrows) extend into the interior of the mitochondrion as folds from the inner boundary membrane. The darkly stained granules (G) are believed to be calcium deposits. A segment of the rough ER is circled; ribosomes (arrows within circle) are clearly visible on the surfaces of the ER membranes. ×50,000. Courtesy of K. R. Porter.

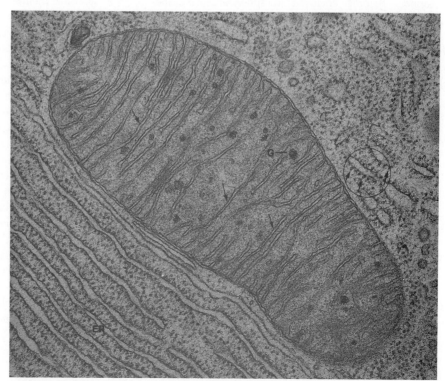

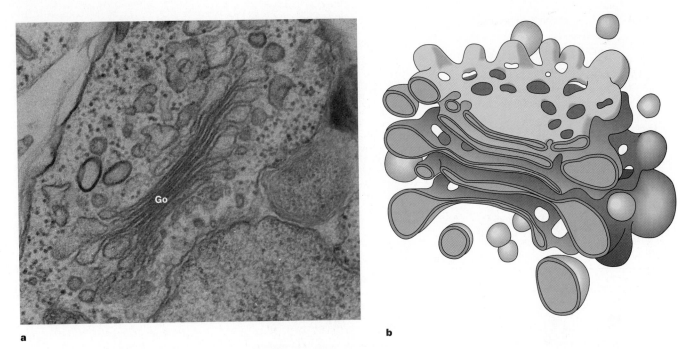

a   b

Figure 1-14   The Golgi complex. **(a)** A Golgi complex (Go) in a plant cell. Courtesy of
W. A. Jensen. **(b)** A three-dimensional reconstruction of a Golgi complex.

reticulum, or *rough ER* (see Figs. 1-9 and 1-10; see also
Fig. 1-13 and Figs. 20-1 to 20-3). Proteins synthesized
on the ribosomes penetrate into the membranes of the
rough ER, forming a part of the membrane structure,
or pass entirely through the membranes to enter the
enclosed ER channels. Most of the proteins that become
part of the rough ER membranes eventually move from
the ER to other cellular membranes, such as the plasma
membrane. The proteins penetrating into the rough ER
channels become enclosed in spherical membrane-
bound vesicles that pinch off from the ER. These proteins are subsequently processed into chemically altered
forms and distributed in vesicles to various sites in the
cytoplasm or released to the cell exterior.

The reactions processing and distributing the proteins released in vesicles from the rough ER occur
primarily in another system of membranous sacs, the
Golgi complex. This system, named for its discoverer,
Camillo Golgi, often takes the form of closely stacked,
flattened sacs (Fig. 1-14; see also Figs. 20-5 to 20-8).
The small vesicles pinching off from the rough ER fuse
with the Golgi membranes and release their proteins
into the interior of a Golgi sac. Within the Golgi complex the proteins are modified by the attachment of
other chemical groups, which, depending on the protein
type, may include small groups such as sulfates, or
larger chemical structures such as lipid or sugar units.
Following modification, the proteins are sorted into
small vesicles that pinch off from the Golgi membranes.
The protein-containing sacs may either remain sus-

pended in the cytoplasm as *storage vesicles* or form
secretory vesicles, which release their contents to the cell
exterior. Assembly of sugar units into precursors of
cellulose and other major cell wall components has been
identified with the Golgi complex in plant cells.

The release of the contents of secretory vesicles
to the cell exterior occurs by a process known as *exocytosis* (Fig. 1-15). A secretory vesicle moves to the cell
boundary (Fig. 1-15*a*); as it makes contact with the
plasma membrane, the vesicle membrane fuses with
the plasma membrane (Fig. 1-15*b*). The fusion makes
the secretory vesicle membrane continuous with the
plasma membrane and spills the vesicle contents to
the cell exterior (Fig. 1-15*c*). The proteins and lipids of
the vesicle membrane become a temporary or permanent part of the plasma membrane.

Eukaryotic cells regularly take up particulate matter
or large molecules such as proteins by a mechanism
called *endocytosis* (see Fig. 1-15), which essentially reverses exocytosis. In endocytosis the molecule or particle is attached to the external surface of the cell, usually
by receptor molecules forming part of the plasma membrane (Fig. 1-15*d*). The plasma membrane then invaginates as a pocket (Fig. 1-15*e*), which subsequently
pinches off (Fig. 1-15*f*) and sinks into deeper layers of
the cytoplasm as an *endocytotic vesicle.* Once in the
cytoplasm, the vesicle contents are routed to various
locations in the cell. One of the major destinations is
the Golgi complex, where proteins taken in by endocytosis are sorted and placed into vesicles for routing

Exocytosis

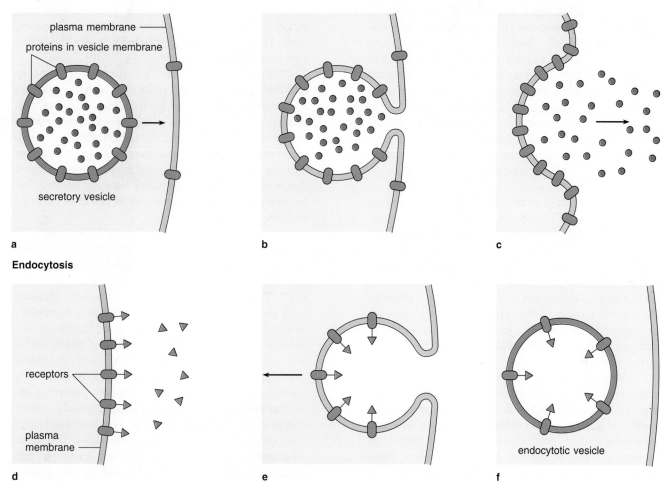

Endocytosis

Figure 1-15 Exocytosis (**a** to **c**) and endocytosis (**d** to **f**). In exocytosis a secretory vesicle moves to the cell boundary and makes contact with the plasma membrane (**a**). As contact is made, the vesicle membrane fuses with the plasma membrane (**b**), making the two membranes continuous and spilling the vesicle contents to the cell exterior (**c**). The proteins and lipids of the vesicle membrane become a temporary or permanent part of the plasma membrane. In endocytosis a molecule or particle is attached to the external surface of the cell, usually by receptor molecules forming part of the plasma membrane (**d**). The plasma membrane then invaginates as a pocket (**e**) that subsequently pinches off (**f**) and sinks into deeper layers of the cytoplasm as an endocytotic vesicle.

to other locations. The Golgi thus serves as a major sorting station for materials traveling both into and out of the cell.

Not all the interconnected membranous vesicles collectively identified as endoplasmic reticulum are associated with ribosomes. The ribosome-free membranes, known as *smooth endoplasmic reticulum,* or *smooth ER* (see Fig. 20-1), have various functions in the cytoplasm. One is to break down fats as a first step in the series of reactions using these substances as an energy source. Some segments of smooth ER have also been identified with the synthesis of lipids or with

reactions that break down toxic substances absorbed by cells.

Lysosomes One type of protein-containing sac produced by the activity of the rough ER and Golgi complex, the *lysosome* (see Fig. 20-35), is especially important in animal cells. Lysosomes contain enzymes that, in aggregate, are capable of breaking down all the major classes of biological molecules. In many cell types molecules taken in by endocytosis are frequently delivered in vesicles to lysosomes, where the extracellular molecules are digested by the lysosomal enzymes. Lysosomes may also be released to the cell exterior by exocytosis to break down molecules on the outside. In some situations lysosomes rupture to release their enzymes inside the cell, where the enzymes break down the molecules of the cell itself and cause cell death. Self-destruction of this type may be a part of pathological conditions or may form part of normal developmental processes. Loss of the tail as tadpoles develop into adult frogs or disappearance of the webbing between fingers and toes in human development, for example, is mediated by lysosome-caused cell death occurring as a normal part of embryogenesis.

Mitochondria   Most of the chemical energy required for the activities of ribosomes, the ER, the Golgi complex, and, in fact, all other functions of eukaryotic cells is derived from reactions taking place in a complex membrane-bound cytoplasmic organelle known as the *mitochondrion* (plural = *mitochondria;* from *mitos* = thread and *chondros* = grain). The name given to the organelle refers to the fact that mitochondria may be long and filamentous or compact and granulelike in different cell types, or may change between these forms in the same cells.

Mitochondria are surrounded by two separate membrane systems, one enclosed within the other (see Fig. 1-13 and Figs. 9-7 and 9-8). The *outer boundary membrane* is smooth and continuous. The *inner boundary membrane* extends into the mitochondrial interior in numerous folds or tubular projections called *cristae* (singular = *crista*). The innermost mitochondrial compartment, surrounded by both membranes, is the *matrix.*

Mitochondria are frequently called the "power-houses" of the cell because they carry out most of the oxidative reactions that release energy for cellular ac-

tivities. Fuel for these reactions is provided by chemical derivatives of all the major classes of cellular molecules, including carbohydrates, fats, proteins, and nucleic acids. The reactions are distributed between the cristae membranes and the matrix.

Mitochondria are partially autonomous; they contain their own DNA and ribosomes and the enzymes and other factors required for transcription and protein synthesis, all concentrated in the matrix. Many features of mitochondrial DNA and the transcription and translation mechanisms resemble the equivalent systems of bacteria. The similarities to bacterial systems suggest that mitochondria probably evolved from bacteria that became established as symbionts in cell lines destined to form the eukaryotes.

Microbodies   Eukaryotic oxidative and other reactions also are localized in another group of organelles collectively known as *microbodies* (Fig. 1-16; also Fig. 9-37). These are relatively simple structures enclosed by a single boundary membrane. Inside a microbody is a *matrix* consisting primarily of a solution or suspension of enzymes. Often a crystalline *core* is visible in the matrix (Fig. 1-16). Rather than directly providing chemical energy for cellular activities, microbodies carry out reactions that link major oxidative pathways occurring elsewhere in the cytoplasm. Typically, microbodies make hydrogen peroxide as a final product of their oxidative reactions and break down this toxic substance by an enzyme that converts hydrogen peroxide to water and oxygen. Because of this characteristic reaction, microbodies are frequently termed *peroxisomes.* In plants a group of microbodies called *glyoxisomes* converts products of lipid breakdown into sugars.

Figure 1-16   A microbody (M) in the cytoplasm of a tobacco leaf cell. Ma, matrix of the microbody; C, crystalline core; Ch, chloroplast; Mt, mitochondrion. × 49,000. Courtesy of S. E. Frederick, from *Ann. N. Y. Acad. Sci.* 386:228 (1982).

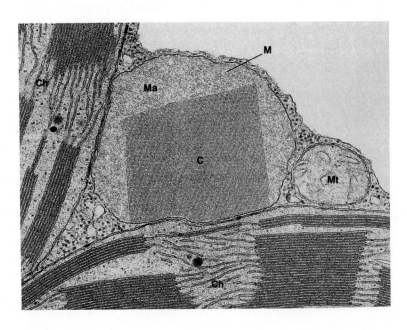

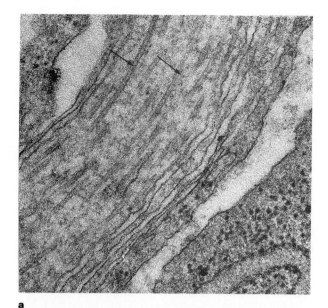

a

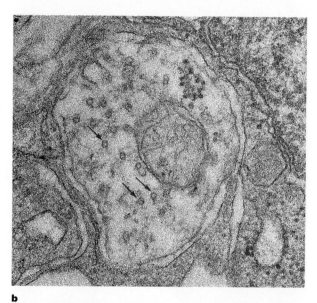

b

Figure 1-17 Microtubules (arrows) in longitudinal section **(a)** and cross section **(b)**. ×65,000. Courtesy of M. P. Daniels and the New York Academy of Sciences.

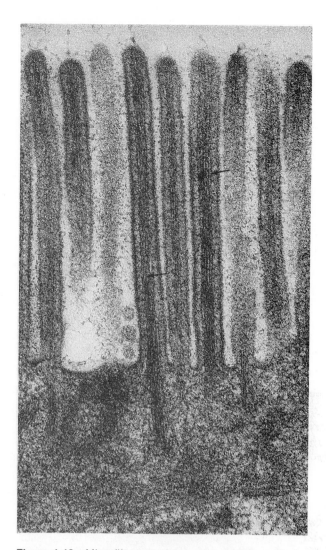

Figure 1-18 Microfilaments (arrows) inside fingerlike projections extending from the surface of a chick intestinal cell. The surface projections are called microvilli. ×95,000. Courtesy of C. Chambers.

Microtubules, Microfilaments, and Cell Movement
Almost all cell movements in eukaryotes are generated by the activities of two cytoplasmic structures—*microtubules* and *microfilaments*—acting singly or in coordination. Microtubules (Fig. 1-17; see also Fig. 1-7) are fine tubelike structures about 25 nm in diameter, with walls about 5 nm thick enclosing a central channel about 15 nm in diameter. Microfilaments (Fig. 1-18; see also Fig. 1-7) are solid fibers 5 to 7 nm in diameter, not much thicker than the wall of a microtubule. Each structure is assembled from subunits of a different protein—microtubules from *tubulin* and microfilaments from *actin*.

Both microtubules and microfilaments produce motion by an active sliding mechanism. The sliding force is developed by protein crossbridges extending from the microtubule or microfilament surfaces. One end of a crossbridge is firmly attached to the surface of a microtubule or microfilament. The opposite end has a reactive site that may attach to another microtubule or microfilament or to other cell structures. The crossbridge makes an attachment at its reactive end, swivels a short distance, and then releases, working much like the oar of a boat (see Figs. 11-18 and 12-12). Each of the two motile elements has distinct proteins forming the swiveling crossbridges. Microfilament crossbridges consist of the protein *myosin*. Several different proteins, among them *dynein* and *kinesin,* operate as crossbridges in microtubule-powered motile systems.

Dynein crossbridges may extend between microtubules, causing one microtubule to slide forcibly over another. Either dynein or kinesin crossbridges may make attachments between microtubules and other structures, such as mitochondria or pigment granules, thereby imparting movement to these attached elements. The myosin crossbridges of microfilaments may produce forcible sliding between microfilaments or between microfilaments and a variety of cell structures such as mitochondria or chloroplasts. None of the known crossbridges form links between microtubules and microfilaments. Although the two motile systems may coordinate their activities to produce cellular movements, they remain structurally independent.

Some cellular movements, such as the beating of flagella, depend exclusively on the activity of microtubules. Microfilaments are solely responsible for other types of movements, including the active, flowing motion of cytoplasm called cytoplasmic streaming and the contraction of muscle cells. Cell division in animals involves the coordinated activities of both microtubules and microfilaments—the chromosomes are divided by microtubules and the cytoplasm by microfilaments.

The flagella of eukaryotic cells are fundamentally different from the much smaller and simpler flagella of bacteria. Within a eukaryotic flagellum is a remarkably complex system of microtubules (Fig. 1-19) consisting of a circle of nine peripheral double microtubules (the *doublets*) surrounding a central pair of single microtubules (the *central singlets*). With rare exceptions the same 9 + 2 arrangement of microtubules is found inside eukaryotic flagella of all types, including the tails of plant and animal sperm cells; flagella found on many animal cells, such as the cells lining the respiratory tract of mammals; and flagella of protozoa and algae. In all these systems flagellar movements are powered by dynein crossbridges.

The Cytoskeleton In addition to their functions in cell motility, both microtubules and microfilaments form supportive networks in the cytoplasm of eukaryotic cells (Fig. 1-20). In some eukaryotic cells, most notably

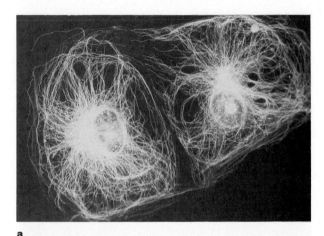

a

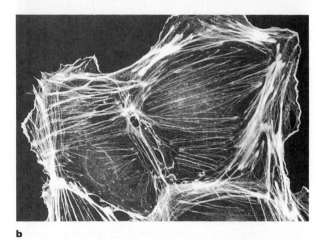

b

Figure 1-20 The cytoskeleton in cultured rat kangaroo cells. The staining technique used to prepare these micrographs makes the cytoskeletal elements appear white against a dark background. **(a)** The microtubule network of the cytoskeleton; **(b)** the microfilament network. ×550. Reprinted from M. Osborn, W. W. Franke, and K. Weber, *Proc. Nat. Acad. Sci.* 74:2490 (1977).

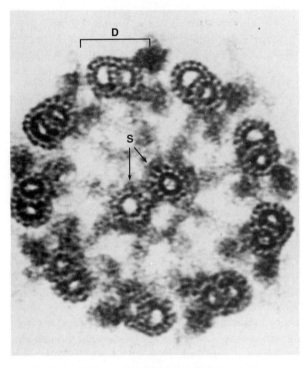

Figure 1-19 The 9 + 2 system of microtubules inside a eukaryotic flagellum. D, doublet; S, central singlet. Subunits in the microtubule walls have been made visible by tannic acid staining. Courtesy of K. Fujiwara and the New York Academy of Science.

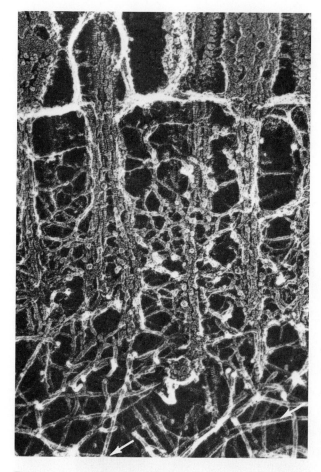

Figure 1-21 Intermediate filaments (arrows) in a mouse intestinal cell prepared by the freeze-fracture technique (see p. 120). ×90,000. Courtesy of J. E. Hauser; reproduced from *J. Cell Biol.* 91:399 (1981) by permission of the Rockefeller University Press.

in animals, *intermediate filaments* (Fig. 1-21; see also Figs. 13-1 and 13-2) also provide support to the cytoplasm. These fibers, with diameters averaging about 10 nm, are built up from a diverse but interrelated group of proteins that is completely distinct from the tubulins and actins.

One of the intermediate filament proteins most widely distributed in higher animals is *desmin*, which occurs in muscle cells. Muscle cells are also reinforced by intermediate filaments constructed from a protein called *vimentin*. Vimentin-based intermediate filaments are also found in a variety of additional cell types. The epithelial cells covering body surfaces in animals are reinforced by intermediate filaments containing different combinations of a large group of proteins collectively known as *cytokeratins*; these proteins also are the major constituent of the nails, claws, horns, and hair of animals. Other intermediate filament proteins occur in cells of the nervous system. Proteins related to the other intermediate filament types, the *lamins*, reinforce the inside of the nuclear envelope. All intermediate filament proteins are related to a greater or lesser extent in amino acid sequence and functional properties, suggesting that they probably evolved from a single ancestral protein type.

The supportive networks set up by microtubules, microfilaments, and intermediate filaments are known collectively as the *cytoskeleton*. Different cytoskeletal systems support the plasma membrane, cell extensions, regions of the cytoplasm between the plasma membrane and the nucleus, and the nuclear envelope.

Except for their arrangement as supportive elements, the microtubules and microfilaments of the cytoskeleton are structured essentially exactly as they are in motile systems. However, the dynein or myosin crosslinks responsible for motility of these elements may be greatly reduced or absent in cytoskeletal structures.

Specialized Cytoplasmic Structures of Plant Cells

All the nuclear and cytoplasmic structures described up to this point, with the possible exception of lysosomes and possibly intermediate filaments, occur in both plant and animal cells. Plant cells also have several organelles and components not found in animal cells (Figs. 1-22 and 1-23). The most conspicuous of these are *plastids*, large, specialized membrane-bound sacs called *central vacuoles*, and the *cell wall*.

Plastids are a family of membrane-bound organelles with various functions in plants. In green plant tissues the characteristic plastid is the *chloroplast* (Fig. 1-24; see also Figs. 10-2 and 10-3), an organelle built up from three membrane systems. A smooth *outer boundary membrane* completely covers the surface of the organelle. A highly folded and convoluted *inner boundary membrane* lines the outer membrane. These two boundary membranes enclose an inner compartment, the *stroma*, equivalent in location to the mitochondrial matrix. Within the stroma is the third membrane system, consisting of flattened, closed sacs called *thylakoids*.

Thylakoid membranes house photosynthetic pigments and molecules that absorb light energy and convert it to chemical energy. The chemical energy is used by enzyme systems suspended in the stroma to drive the assembly of carbohydrates and other complex organic molecules from water, carbon dioxide, and other simple inorganic precursors.

The chloroplast stroma, like the mitochondrial matrix, also contains DNA, ribosomes, and all the enzymes and other factors required for transcription and protein synthesis. As in mitochondria, the DNA and other elements are believed to be derived from the biochemical systems of ancient prokaryotes resembling cyanobacteria. The ancient prokaryotes became established as permanent residents and evolved into chloroplasts

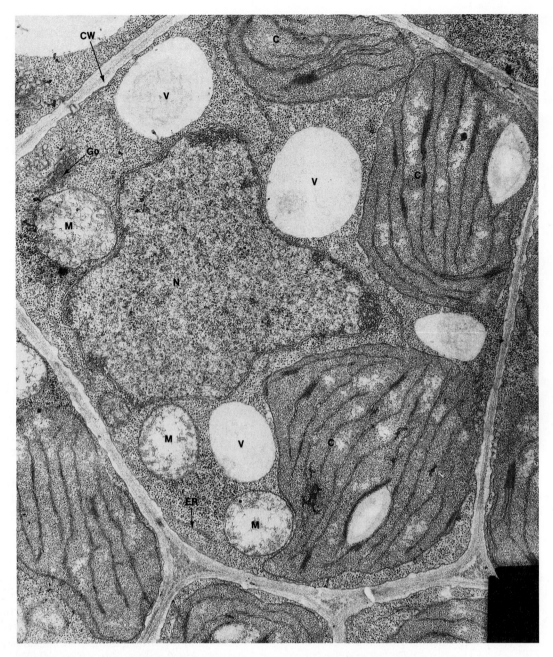

Figure 1-22 A plant cell from a bean seedling *(Phaseolus vulgaris)* showing the major structures occurring in plant but not animal cells: the cell wall (CW), chloroplasts (C), and a series of large vacuoles (V). Structures common to both plant and animal eukaryotes are also visible: the nucleus (N), surrounded by the nuclear envelope; mitochondria (M); endoplasmic reticulum (ER); and the Golgi complex (Go). In this embryonic plant cell the vacuoles are small in size and have not yet coalesced into a single, large central vacuole. Courtesy of Chin Ho Lin.

in the cytoplasm of cell lines destined to found the modern eukaryotic algae and plants.

The plastids of embryonic plant tissues are small and contain only a few inner membranes. These plastids, called *proplastids,* develop into chloroplasts if the tissue is exposed to light. Other plastids, packed with stored lipid, protein, or starch rather than photosynthetic mem-

branes, are called *leucoplasts* (from *leukos* = colorless or white). Ripening fruit or leaves displaying fall colors contain *chromoplasts,* colored plastids in which red and yellow lipid pigments predominate.

The various types of plastids represent different developmental fates of proplastids. Although many variations occur, the most common developmental se-

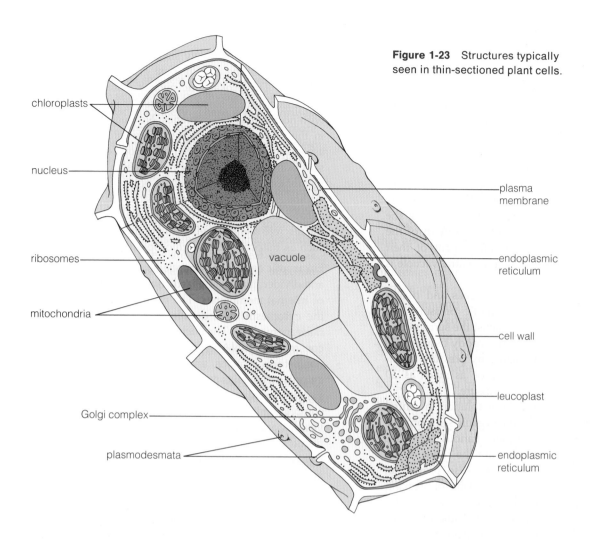

Figure 1-23 Structures typically seen in thin-sectioned plant cells.

chloroplasts

nucleus

ribosomes

mitochondria

Golgi complex

plasmodesmata

plasma membrane

endoplasmic reticulum

cell wall

leucoplast

endoplasmic reticulum

vacuole

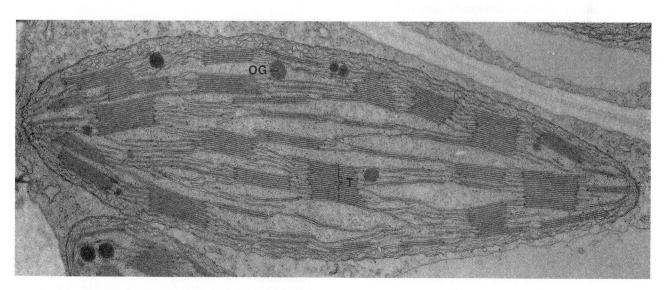

Figure 1-24 Chloroplast in a tobacco leaf cell. The reactions of photosynthesis converting light into chemical energy are concentrated in the thylakoid membranes (T) inside the chloroplast. The osmiophilic granule (OG) is a darkly staining granule commonly observed inside chloroplasts. Courtesy of W. M. Laetsch.

Eukaryotic Cells   21

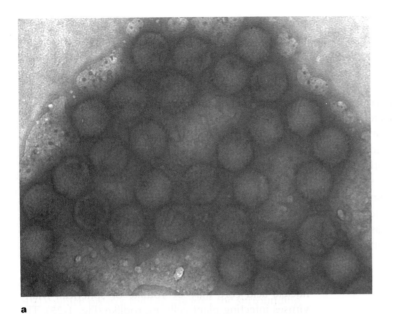
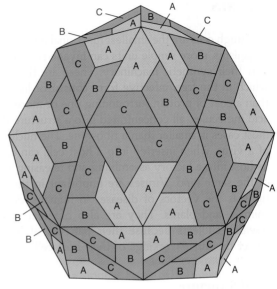

a

b

Figure 1-27 Icosahedral viruses. **(a)** Examples of the adenovirus, a polyhedral virus taking the icosahedral form. × 102,000. Courtesy of G. M. Beards. **(b)** The arrangement of three protein subunits in the icosahedral coat of the southern bean mosaic virus. In this virus the subunits A, B, and C are the same protein arranged in slightly different folding patterns. In other icosahedral viruses with the same arrangement of subunits, A, B, and C are different proteins. Some icosahedral viruses have more than three subunits in a triangular facet.

30 edges and 12 corners (Fig. 1-27). The triangular facets of the smallest icosahedral viruses are assembled from 3 protein subunits, giving a total of 60 protein subunits in the entire capsid. Larger viruses have additional protein subunits in each facet, giving totals that are multiples of 60, such as 120, 180, and so on. The three or more proteins of each facet may be identical or different. In some of the icosahedral viruses, *spike proteins* providing host recognition extend from the points or corners of the capsid. Some spherical viruses are covered by a membranous envelope derived from their host cells (Fig. 1-28).

Some of the viruses infecting bacteria, called *bacteriophages*, are among the most complex spherical viruses. Some bacteriophages have a *tail* (Fig. 1-29) that functions in host cell recognition and attachment and in injection of the nucleic acid core. The tail, which extends from the capsid, consists of a *collar* at the point of attachment of the tail to the head, a cylindrical *sheath* extending from the collar, and a hexagonal *baseplate* at the end of the sheath. The baseplate carries six long, hairlike extensions, the *tail fibers*, which function in host cell recognition and binding.

Viral Infective Cycles

In the free *virion* form viral particles are incapable of independent movement. They are moved about by

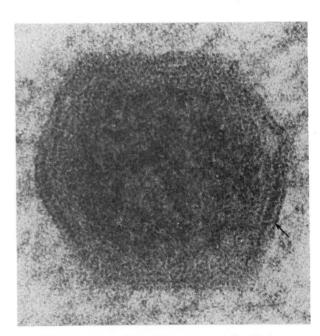

Figure 1-28 An animal virus covered by a membrane (arrow) derived from a host cell. The virus shown causes lymphocytosis in fishes. × 260,000. Courtesy of L. Berthiaume and Academic Press, from *Virology* 135:10 (1984).

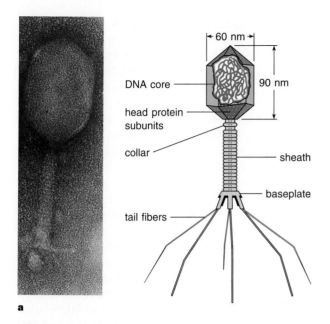

DNA core

head protein
subunits

collar

tail fibers

60 nm

90 nm

sheath

baseplate

Figure 1-29 Tailed bacteriophages. **(a)** A particle of a tailed bacteriophage that infects *E. coli* cells. × 340,000. Courtesy of the Perkin–Elmer Corporation. **(b)** The structures of a tailed bacteriophage.

random molecular collisions until contact is made with the surface of an appropriate host cell. After contact is made, a series of automated events allows either the entire virus or the nucleic acid core to enter the host cell. Once inside the host, the genes encoded in the viral nucleic acid direct the host cell to make and release additional viral particles of the same type. In most cases the cycle of viral infection and release damages or kills the host cell.

Infection of Bacterial Cells   In the infection of a bacterial cell by a tailed bacteriophage, for example, random collision of a viral particle with the host is followed immediately by tight binding between recognition proteins in the viral tail fibers and molecules of the bacterial cell wall (Fig. 1-30a). The head and tail sheath of the virus then contract and inject the DNA core into the cell (Fig. 1-30b). The proteins of the virus remain outside.

Once inside the bacterial cell, a part of the bacteriophage DNA is immediately transcribed into mRNAs by bacterial enzymes. These first mRNAs direct the bacterial ribosomes to synthesize several viral pro-

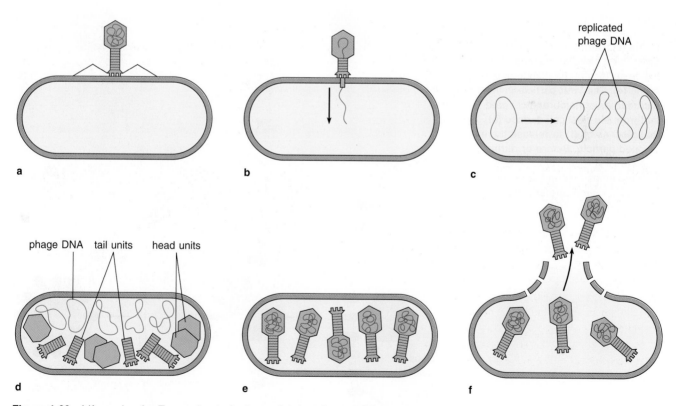

Figure 1-30 Life cycle of a T-even bacteriophage. **(a)** Attachment of bacteriophage to a bacterial cell; **(b)** injection of viral DNA; **(c)** replication of viral DNA; **(d)** synthesis of head and tail proteins; **(e)** assembly of DNA, head, and tail units into finished virus particles; and **(f)** release of completed bacteriophage particles by rupture of the infected cell.

teins, among them enzymes necessary for replication of the viral DNA. The bacterial cell, now converted to the production of viral DNA, continues replication until as many as a thousand new bacteriophage DNA molecules are made (Fig. 1-30c). At some time after viral DNA replication begins, other segments of the viral DNA are transcribed into mRNAs. These late viral messengers direct the bacterial cell to synthesize structural proteins of the bacteriophage and assemble these proteins into heads and tails (Fig. 1-30d). As the head and tail segments accumulate in the bacterial cytoplasm, the newly synthesized bacteriophage DNA packs into the heads, and the heads and tails assemble into completed particles (Fig. 1-30e). After assembly of the particles is complete, a final viral protein is synthesized that causes the bacterial cell wall to rupture (Fig. 1-30f), releasing the newly completed viral particles to the surrounding medium. Chance collisions may lead to another cycle of infection and release of virus particles if a correct host cell is encountered.

Infection of Eukaryotic Cells Infection of eukaryotic cells follows a similar pattern except that both the core and capsid enter the host cell. In the infection of animal cells, viral particles without membrane envelopes are bound tightly to the outer surface of the plasma membrane by their recognition proteins. The binding stimulates the host cell to take in the viral particles by endocytosis. Enveloped viruses enter by a fusion of their surface membranes with the host cell plasma membrane (Fig. 1-31a to c). The fusion inserts the virus inside the host cell with its protein coat but without the surface membrane. In plants viral particles usually enter initially through cellular regions exposed by wounds or abrasions in the cell wall. Within the plant infective particles are passed from infected to healthy cells via plasmodesmata and the vascular system.

Once inside the eukaryotic host cell, the viral capsids rupture or disassemble to release nucleic acid cores. The nucleic acid directs the assembly of additional viral nucleic acids and capsid proteins as in the bacterial

Figure 1-31 Entry (**a** to **c**) and release (**d** to **f**) of enveloped viruses. During entry the virus binds to the plasma membrane of the host cell (**a**). Binding is followed by fusion of the viral envelope with the host cell plasma membrane (**b**), which strips the membrane from the viral particle and inserts the particle into the cytoplasm (**c**). During release a newly assembled particle attaches to the inside surface of the plasma membrane (**d**). The viral particle is then extruded as a membrane-covered bud from the host cell surface (**e**) that pinches off (**f**) to release the enveloped particle. Before or during release viral recognition proteins become embedded in the segments of plasma membrane enveloping the viral particles.

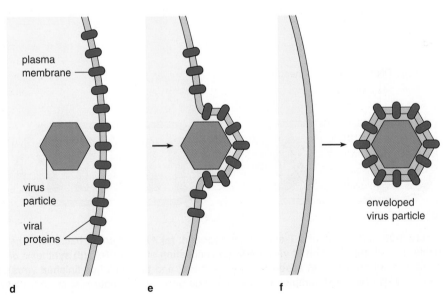

Viral Entry

plasma membrane

enveloped virus

virus particle free in cytoplasm

a　　　　　b　　　　　c

Viral Release

plasma membrane

virus particle

viral proteins

enveloped virus particle

d　　　　　e　　　　　f

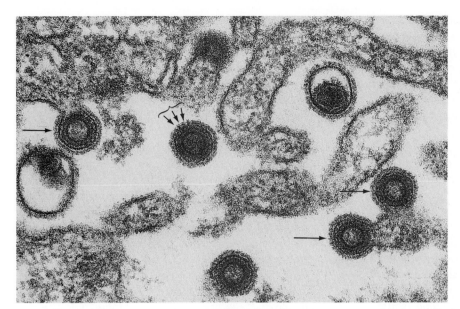

Figure 1-32 Particles of a mouse mammary tumor virus (arrows) in the process of acquiring membranous coats during release from a host cell. Viral proteins (bracketed arrows) are closely packed in the membrane segments. × 140,000. Courtesy of N. H. Sarkar.

viruses. When synthesis of these components is complete, the new viral particles are assembled and released from the infected cell.

The enveloped viruses (see Fig. 1-28) derive their membranous outer coats from the host cell plasma membrane during release. After a new viral particle has been assembled in the host cell cytoplasm, it attaches to the inside surface of the plasma membrane (Fig. 1-31d) and is extruded as a membrane-covered bud from the host cell surface (Fig. 1-31e). The bud then pinches off (Fig. 1-31f) to release the virus in free form. As a result the released virus is covered by a segment of plasma membrane derived from the host cell. During extrusion of the virus some of its proteins are inserted in the covering membrane to extend as recognition spikes from the particle surface. This pattern of viral extrusion, since it does not break the plasma membrane, may proceed without rupture of the host cell. Figure 1-32 shows particles of an enveloped virus being released from an infected animal cell.

Infective Pathways of RNA and DNA Viruses Most viruses that contain RNA rather than DNA cores follow a pathway of infection in which the viral RNA molecule serves as a template for assembly of additional RNA copies. The copying is done by a viral enzyme that can use RNA as a template for the assembly of RNA molecules, in contrast to the typical cellular pattern in which only DNA is used as a template for the assembly of RNA copies. The enzyme, called an *RNA replicase,* is included among coat proteins in some of the RNA viruses. In others the enzyme is assembled by ribosomes of the host cell, using the RNA of the infecting virus as an mRNA for the replicase. Some RNA copies made by the viral replicase are used by ribosomes as mRNAs

giving directions for making viral coat proteins; others become packed into the coat proteins as the cores of newly synthesized infective particles.

One group of RNA-containing viruses infecting animal cells, the *retroviruses,* undergoes an intermediate step in which viral RNA is copied into DNA. The enzyme accomplishing this task, a *reverse transcriptase,* reverses the process of transcription, producing a DNA copy from an RNA template. The DNA copy is subsequently inserted into the DNA of the host cell. In this form the integrated viral DNA is known as a *provirus.* At some point the DNA is copied into mRNAs that direct the synthesis of viral coat proteins and, eventually, into RNA copies that become the cores of finished infective particles. Among the retroviruses are a number that have been implicated in the changes that alter normal cells into cancer cells in humans and other mammals.

Many viruses with a DNA core also enter a stage in which the viral DNA is inserted into the host cell DNA as a provirus. During much of the time in which the DNA of these viruses is integrated into the host DNA, it may remain inactive in directing rounds of self-replication or directing the synthesis of coat proteins. In this form, called the *lysogenic phase* of the infective cycle, the inactive viral DNA is replicated and passed on with the host cell DNA to all descendants of the infected cell. At some later time, frequently during environmental conditions that are stressful to descendants of the original host, the viral DNA becomes active. Among the first viral proteins synthesized in response are enzymes that cut the viral DNA from the host cell DNA. The viral DNA then directs its replication and the production of capsid proteins. This active stage, culminating in the release of infective viral particles, is called the *lytic phase* of the infective cycle. The switch of a virus from the lysogenic to the lytic phase during periods of environmental stress has been likened to a rat deserting a sinking ship. Among the viruses entering lytic phases in which their DNA is integrated into the host cell DNA are some of the bacteriophages and the herpes viruses infecting humans.

Origin and Significance of Viruses

Viral particles in free form carry out none of the activities of life and are inert except for the capacity to attach to their host cells. Many can be purified and crystallized in this form, and stored indefinitely without change or damage. The viral nucleic acid molecule carries only the information required to direct the host cell machinery to make more viral particles, and is active in this function only when inside a host cell. Thus, a virion is probably best classified as nonliving matter.

Although the origin of viruses is obscure, these particles probably represent nothing more or less than fragments of a nucleoid or chromosome derived from a once-living cell, or an RNA copy of such a fragment, surrounded by a layer of protein with protective and cell recognition functions. The information encoded in the core of the virus is reduced to a set of directions for maintenance of the virus in a host cell, and production of more virus particles of the same kind.

The scientific, medical, and economic importance of viruses can hardly be overestimated. The molecular revolution in biology has primary roots in experiments carried out with viruses, particularly bacteriophages. These experiments were among the first to reveal how genes are structured, regulated, and duplicated.

Viruses are responsible for human disabilities ranging from relatively mild irritations such as the common cold, through more serious infections such as influenza, chicken pox, and polio, to deadly diseases such as AIDS and some forms of cancer. Viruses also cause diseases of domestic animals and infect plants in both agricultural and natural environments. With a few notable exceptions, such as polio and influenza, which are controllable by vaccines, most viral diseases are difficult or impossible to treat and simply must be allowed to run their course. Some viral diseases, such as oral and genital herpes, are presently lifetime infections with little possibility for a permanent cure.

Although the polio and smallpox vaccines have remained highly effective, some viral infections that can be treated by vaccines present recurring problems because the viruses mutate rapidly to forms for which a previously successful vaccine is ineffective. The influenza virus, for example, constantly appears in new forms that cannot be treated by previously effective vaccines. The fact that the virus responsible for AIDS also mutates rapidly will undoubtedly cause problems for the treatment of this disease by vaccines.

Some viruses that usually cause only moderately incapacitating diseases can also mutate into more deadly forms. The influenza virus, for example, infects millions but normally causes only about 20,000 deaths in the United States each year, primarily among infants and the aged. However, in 1918 the virus mutated into a highly dangerous form that killed between 20 and 30 million people of all ages in Europe and America, so many that few families survived without the death of at least one close relative from influenza. Fortunately, the virus has since mutated into less virulent forms.

On the plus side the viruses infecting bacterial cells probably benefit humanity through the elimination of potentially dangerous bacteria. Some idea of the extent of bacterial destruction that may result from bacteriophages can be gained from the fact that the concentration of these viruses in natural waters often approaches levels as high as 100 million particles per milliliter.

Viroids and Prions

Two additional infective agents with properties remotely similar to viruses have been identified in eukaryotic host cells. *Viroids* are small, circular RNA molecules that can induce their own duplication. No known viroids develop capsids or occur in a free form. Viroids are responsible for several diseases in plants, among them, the *cadang-cadang disease* of coconut palms, which has caused extensive losses in the coconut palm industry of countries such as the Philippines.

The RNA molecules of viroids are the smallest molecular agents known to be capable of infection. Whether viroids are a form of virus without a capsid or a distinct infective type unrelated to the viruses is uncertain. It is also uncertain how viroids are transmitted from one host to another in nature or how they cause disease. It is considered possible that viroid RNAs enter the nucleus and interfere with gene regulation or transcription, perhaps by interacting with the host cell DNA or proteins regulating gene activity.

Prions are agents responsible for *scrapie* in goats and sheep, and for several relatively rare human neurological ailments, including *Creutzfeldt-Jacob disease* and *kuru*. Scrapie is so called because sheep with this brain disease rub against fences or trees until they scrape off most of their wool. Humans afflicted by prion-related disorders exhibit lack of coordination, general physical debilitation, and dementia. Although the effects are initially mild and typically progress slowly, the prion-based diseases are eventually fatal.

Prion-based diseases can be transmitted experimentally from one individual or cell culture to another, indicating that they are caused by an infective agent. Attempts to identify the agent, however, have revealed only a relatively small protein with no detectable traces of nucleic acids—hence, the name *prion*, which stands for *proteinaceous infective particle.*

Recent information indicates that the prion protein is a modified product of a cellular gene present in both normal and infected individuals. Infective prions somehow alter transcription or processing of the gene's mRNA product toward production of the prion, or induce the production of other enzymes that alter reserves of the normal protein to the disease-causing, infective prion form. It is also possible that prions contain an as yet undetected nucleic acid component responsible for transmission of the disease. Although the function of the protein that is altered to the prion is uncertain, recent research by G. D. Fischbach suggests that it may regulate production of receptors that receive signals in nerve cells. If so, this activity may account for the neurological disorders produced in individuals with the prion form of the protein. Both the normal and mutant forms of the prion protein occur on the surfaces of neurons.

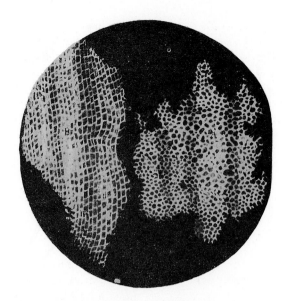

Figure 1-33 Hooke's drawing of cell walls in the cork tissue of a plant, published in his *Micrographia* in 1665.

HISTORICAL ORIGINS OF CELL AND MOLECULAR BIOLOGY

Cell and molecular biology developed gradually from the first description of cells in the seventeenth century. During the earliest period investigations in cell biology consisted almost entirely of morphological descriptions of cell structure. In the eighteenth and nineteenth centuries the study of cell chemistry and physiology began, largely as an effort that proceeded independently from the morphological studies. As a result, the early investigations into cell chemistry and physiology had little effect on the scientists studying cell structure and vice versa. Cell structure on one hand, and cell chemistry and physiology on the other, continued as separate fields of experimentation until the beginning of the twentieth century, when the rapidly developing field of biochemistry began to influence cell biology. At the same time genetics became established as a new field of study.

The integration of cell biology, genetics, physiology, and biochemistry began in earnest in the 1930s, and research in cell biology started its shift from primarily morphological investigations to biochemical and molecular studies of cell function. In recent years the biochemical and molecular approach has dominated the study of cells.

The Discovery of Cells

The earliest developments in cell biology were closely tied to the invention and gradual improvement of the light microscope. In 1665, soon after the microscope was invented, the English scientist Robert Hooke published *Micrographia,* which included the first de-

scriptions of cells. Hooke reported seeing small, compartmentlike units in the woody tissues of plants, which he named "pores" or "cells" (Fig. 1-33). It is frequently claimed that Hooke did not actually observe living cells because in some of the tissues he examined, such as the cork cells shown in Figure 1-33, the cells were simply empty spaces outlined by residual cell walls. But in other plant tissues, such as the "inner pulp or pith of an Elder, or almost any other tree," Hooke "plainly enough discover'd these cells or Pores fill'd with juices." Thus, he actually saw living cells.

Hooke's observations of cells were extended by several investigators toward the end of the 1600s, most notably by Anton van Leeuwenhoek, a Dutch amateur microscopist. Leeuwenhoek made remarkably accurate observations of the microscopic structure of protozoa, blood, sperm, and a variety of other "animalcules," as he called them (Fig. 1-34). He reported his findings to the British Royal Society in some 200 letters written over a period of about 50 years.

After Leeuwenhoek's death in the early 1700s, cell biology entered a period of relative quiescence that lasted until well into the nineteenth century. This lag in the evolution of the field was largely because of imperfections in the lenses of the early light microscopes, which were serious enough to prevent investigators from seeing the inner details of cells. The correction of some of the more serious imperfections in the early nineteenth century led to a burst of new discoveries in cell biology.

The early observations of Hooke and others, made primarily in plant tissues, gave the impression that cells were units of living matter outlined by conspicuous cell walls. Because the cells of animal tissues lack distinct walls, structures equivalent to plant cells could not at

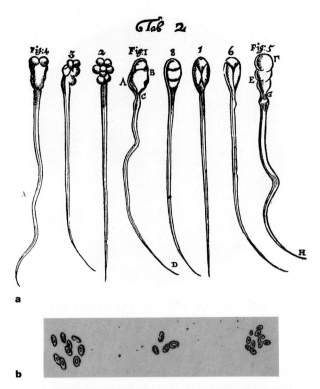

a

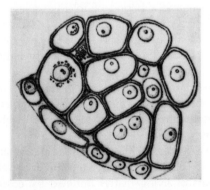

b

Figure 1-34 Leeuwenhoek's drawings of sperm cells **(a)** and red blood cells from fishes **(b)**. The nuclei shown in the blood cells, which Leeuwenhoek described as "a little clear sort of light in the middle" of the cells, are the first published drawings of cell nuclei.

first be seen in animals. As a result, the basic similarities in microscopic structure between plants and animals escaped notice.

The parallels in structure between animal and plant tissues were finally drawn in 1839 when the German biologist Theodor Schwann observed that animal cartilage contains a microscopic structure that "exactly resembles . . . [the] cellular tissues of plants" (Fig. 1-35). Schwann's work was aided by the fact that extracellular material in cartilage occupies a position analogous to the cell wall in plants and clearly outlines the cells. This enabled Schwann to recognize the cellular nature of animal tissue for the first time. Schwann also remarked

Figure 1-35 Schwann's drawing of cartilage cells in an amphibian larva.

on the presence of nuclei in the cartilage cells, but was not as impressed by the cell contents as by the structures he thought were walls.

Continued improvements in light optics soon allowed microscopists to recognize the much thinner boundaries of cells in other types of animal tissues. As the details of structure in the cell interior became discernible, interest gradually shifted from the walls to the contents, and the term *cell* began to take on its modern connotations. At this time investigators recognized that the fluid contents of the cell are the primary substance of living organisms. A physiologist at a Polish university, Jan Purkinje, proposed the term *protoplasm* for this substance in 1840.

Development of the Cell Theory

In 1833, a few years before the definition of protoplasm by Purkinje, Robert Brown published a paper in England describing the microscopic structure of the reproductive organs of plants. In this paper, which subsequently received wide notice, Brown drew attention to the nucleus as a constant feature of plant cells. Brown's study established that nucleated cells are the units of living tissue in plants and laid the foundations for the concept that cells with nuclei are the fundamental units of all living organisms. This idea is part of what is now known as the *cell theory.*

The cell theory was developed by Schwann and an eminent German botanist, Matthias Schleiden. During his work with cartilage in the 1830s, Schwann's attention was drawn to the cell nucleus by Schleiden, who had developed a series of hypotheses emphasizing the importance of the nucleus in cell reproduction. Although Schleiden's ideas about cell reproduction were erroneous, his preoccupation with the cell nucleus led Schwann to recognize the universality of nucleated cells in living matter. This work led to two of the three postulates of the cell theory: (1) that all living organisms are composed of one or more nucleated cells and (2) that cells are the minimum functional units of living organisms. Historians usually attribute these two hypotheses jointly to Schwann and Schleiden.

Their conclusions were soon supplemented by a third postulate that completed the cell theory. This idea was developed by scientists investigating cell origins, who observed that cells arise in both plants and animals by the division of a parent cell into two daughter cells. By 1855 this work had progressed far enough for Rudolf Virchow, a German pathologist, to affirm that all cells arise only from preexisting cells by a process of division. Virchow's famous statement[1] of this concept, "*Omnis cellula e cellula,*" completed the cell theory:

[1] Although Virchow's statement became famous, the expression "*omnis cellula e cellula*" originated much earlier, in a work written in 1825 by the French investigator Francois-Vincent Raspail.

1. All living organisms are composed of nucleated cells.

2. Cells are the functional units of life.

3. Cells arise only from preexisting cells by a process of division.

Further work established that the nucleus is the repository of hereditary information and that the essential feature of cell division is transmission of hereditary information to daughter cells. The physical continuity of the nuclear material through cycles of division was confirmed by the German scientists Eduard Strasburger and Walther Flemming, who discovered that the chromatin is transformed into compact rodlets, the chromosomes (Fig. 1-36), during cell division. When fully formed, chromosomes are clearly seen to be double. As division progresses, the two halves of each chromosome split apart and pass into separate daughter cells, where they form the daughter nuclei. Later work by Flemming and others showed that the number of chromosomes remains constant for all members of a species.

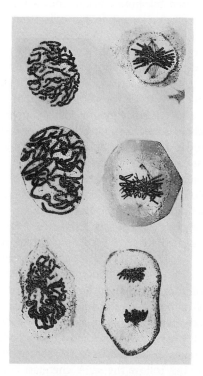

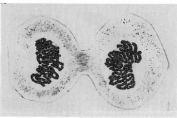

Figure 1-36 Flemming's drawings of division in cells of a salamander larva, clearly showing the chromosomes.

With the discovery that chromosomes are duplicated and passed on in constant numbers during cell division, enthusiasm grew for the hypothesis that cell heredity is probably controlled by the cell nucleus. In 1884 Strasburger declared that the physical basis of heredity resides in the nucleus; in 1885 the German zoologist August Weismann concluded that " the complex mechanism for cell division exists practically for the sole purpose of dividing the chromatin, and . . . thus the [chromatin] is without doubt the most important part of the nucleus."

Some years before these developments took place, an Austrian monk, Gregor Mendel, had studied the inheritance of traits such as seed shape and color in garden peas. Through his mathematical analysis of the distribution of these traits in parents and offspring, Mendel discovered genes and their patterns of inheritance.

The significance of Mendel's work was not appreciated at the time of its publication in 1865. One reason for its lack of impact was that the biologists of the time were unaccustomed to thinking about cellular processes in abstract, mathematical terms. A second was that investigations had not progressed far enough to allow Mendel's mathematical conclusions to be related to physical units in the cell. A third reason was that Mendel's conclusions ran counter to the beliefs of prominent biologists at that time, who generally held that inheritance depended on an averaging or mixing of parental traits rather than being determined by discrete hereditary units. As a result, Mendel's findings were not taken seriously and were forgotten until just after the turn of the century when the same conclusions were reached independently by a Dutch plant physiologist, Hugo de Vries, and the German and Austrian botanists Carl Correns and Erich von Tschermak. These investigators, on searching the earlier literature, were surprised to find that they had been scooped by a monk some 40 years earlier!

By the time Mendel's results were rediscovered, the behavior of chromosomes in cell division was sufficiently well known for the correlation between Mendel's genes and chromosomes to become apparent. In 1903 the American cell biologists W. A. Cannon, Edmund Wilson, and Wilson's student Walter Sutton pointed out the precise equivalence between the patterns of inheritance of genes and chromosomes in organisms that reproduce sexually, that is, by the union of eggs and sperm. Both genes and chromosomes occur in pairs in which one member of the pair is inherited from the male and the other from the female parent of an individual. Both gene and chromosome pairs are separated in the formation of gametes and reunited in fertilization. In total these findings and conclusions established the concept that heredity is controlled by discrete physical units, the genes, that genes are carried on the chromosomes, and therefore that the nucleus and its contents store and transmit hereditary information. Research with genes and their patterns of transmission to offspring expanded greatly during the early decades of this century, particularly in the laboratory of Thomas Hunt Morgan at Columbia University. Morgan's work with the fruit fly, *Drosophila melanogaster*, led directly to an understanding of the linear order of genes on chromosomes, and of the new patterns into which genes are mixed during the formation of gametes in sexually reproducing organisms.

Early Chemical Investigations

The application of physics and chemistry to cell biology began in 1772, when the English chemist Joseph Priestley discovered that green plants release oxygen when they are exposed to light. At almost the same time, Antoine Lavoisier, a Frenchman who is considered the father of modern chemistry, recognized that "respiration is a . . . combustion, slow, it is true, but otherwise perfectly similar to that of charcoal." The fundamental importance of these discoveries was not appreciated at the time because it was still widely believed that the substances and processes in living organisms were different from those of the inorganic world. Therefore, the chemical and physical techniques used to study inanimate objects were not believed to be applicable to life.

The earliest significant movement away from these attitudes came in 1828, when the German chemist Friedrich Wöhler achieved the first artificial synthesis of organic molecules from inorganic precursors. Wöhler converted the inorganic chemical ammonium cyanate to urea, an organic substance commonly excreted by animals. Wöhler also synthesized oxalic acid, an organic chemical found in plant tissues. Many additional organic molecules were subsequently synthesized by others, and it gradually became clear that the same elements occur in both living and inanimate objects and are governed by the same chemical and physical laws. By the end of the nineteenth century investigators had isolated, identified, and synthesized many organic substances found in plants and animals. Most successful in this work was the German organic chemist Emil Fischer, who extracted, degraded, and resynthesized many substances from living organisms and laid the foundation for the chemical description of amino acids, proteins, fats, and sugars. In 1902 Fischer and another German chemist, F. Hofmeister, independently described the structure and formation of the peptide bond, a chemical linkage that ties amino acids together in proteins. The structure of nucleotides, the chemical building blocks of the nucleic acids, was also worked out by the turn of the century.

This chemical work was complemented in the early nineteenth century by the first functional biochemical studies. Before this time the substances and reactions of living systems were generally thought to be moved by a mysterious "vital force." The discovery of chemical catalysts at the beginning of the nineteenth century provided clues to the real nature of the vital force underlying chemical interactions in living organisms. In 1836 Jons Jakob Berzelius, a Swedish chemist, wrote that it is "justifiable . . . to suppose that, in living plants and animals, thousands of catalytic processes take place . . . and result in the formation of the great number of dissimilar chemical compounds, for whose formation out of the common raw material . . . no probable cause could be assigned."

Berzelius's intuitions about the role of catalysts in living organisms were later proved correct by investigations into the nature of alcoholic fermentation. In the 1850s the French scientist Louis Pasteur began his efforts to determine the cause of fermentation. Pasteur found that fermentation occurs in nature only if living microorganisms are present—if the microorganisms are eliminated or killed, sugar is not converted into alcohol. Pasteur's work also confirmed that microorganisms and cells in general can arise only from preexisting cells, and eliminated the possibility of spontaneous generation—the production of living cells from nonliving matter.

The catalytic nature of fermentation was discovered through the research of Eduard Buchner in Germany. In 1897 Buchner was attempting to isolate and preserve yeast extracts for medicinal purposes. To preserve the extracts, made by grinding and pressing yeast cells, Buchner added a sugar solution, commonly used then as today for preserving food. To his surprise, the sugar was rapidly fermented by the cell-free extract. Buchner subsequently demonstrated, in yeast extracts from which all intact cells were carefully removed, that the fermentation was catalyzed by protein-based catalysts. The protein-based catalysts were given the name *enzymes*, coined from a Greek word that means "in yeast."

This work culminated in general acceptance of the conclusion that living organisms contain the same elements as nonliving matter and are motivated by no vital forces other than enzymes and the reactions they catalyze. Living systems could now be studied by chemical and physical techniques with the confidence that their activities follow the same chemical and physical laws as do nonliving systems.

Integration of Chemical and Morphological Studies

Significantly, investigations into the chemical nature of the nucleus were among the first efforts to integrate chemical and morphological approaches to the study of cellular life. This work stemmed directly from the growing realization in the latter half of the nineteenth century that the nucleus is of central importance in cell function and heredity.

Johann Friedrich Miescher, a Swiss physician and physiological chemist, became interested in the chemical composition of cell nuclei and developed a method for isolating nuclei in quantity for analysis. From his preparations he isolated a previously unknown substance with properties then considered unusual for organic matter, including high phosphorus content and a strongly acidic reaction. Miescher, in announcing his discovery in 1871, called the new substance *nuclein.* Soon afterward, Flemming concluded that if the chromatin of the nucleus and nuclein were not one and the same substance, "one carries the other." Nuclein was later called a nucleic acid.

A method for purifying nucleic acids was worked out by R. Altman in 1889, and their chemical subunits, the nucleotides, were identified. One type of nucleic acid, DNA, subsequently was established to be characteristic of all cell nuclei. Work in the 1920s and 1930s confirmed that DNA is located in the chromosomes, and many cell biologists began to suspect direct involvement of this substance in heredity. Finally, in the 1940s and 1950s, a series of experiments with bacteria by Oswald Avery and his colleagues, and with viruses by Alfred D. Hershey and Martha Chase, confirmed that DNA is the hereditary molecule.

Avery and his colleagues Colin MacLeod and MacLyn McCarty, working at the Rockefeller Institute, took advantage of two forms of a bacterium, *Streptococcus pneumoniae,* that causes a form of pneumonia in humans and mice. Cells of a virulent, infective form that can cause pneumonia are surrounded by capsules and produce smooth, gellike colonies when grown in culture dishes. Cells of a nonvirulent type of the same bacterium have no capsules and form colonies that appear lusterless or rough. It had been found previously that killed cells of the virulent form could transform rough, nonvirulent cells into fully infective cells with capsules. The change was permanent and was passed on to descendants of the transformed cells.

Avery and his colleagues were interested in identifying the agent in killed virulent cells that could transform nonvirulent cells to the virulent form. They treated extracts of killed cells with enzymes that catalyze the breakdown of DNA, RNA, or proteins, and exposed nonvirulent cells to the treated extracts (Fig. 1-37). Only the enzyme breaking down DNA destroyed the capacity of the extract to transform nonvirulent cells into the infective form. From these findings Avery, MacLeod, and McCarty proposed in 1944 that DNA is the substance responsible for transforming the noninfective cells. Because DNA could carry genetic information in this pattern, it was considered likely to be the normal carrier of genetic information in the cell nucleus.

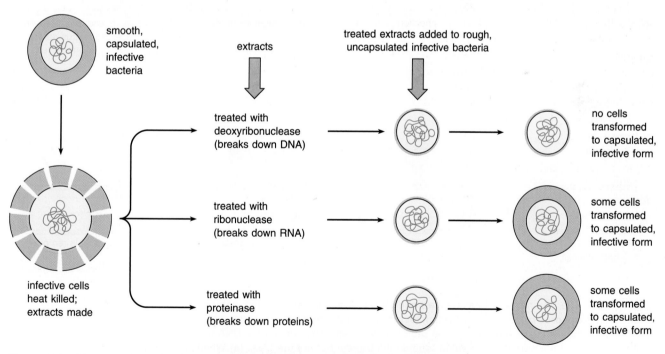

Figure 1-37 The experiment by Avery and his colleagues demonstrating that DNA is the factor able to transform nonvirulent *Streptococcus pneumoniae* into the virulent, infective form. Heat-killed extracts of the infective form were treated with deoxyribonuclease, ribonuclease, or a proteinase. Only the deoxyribonuclease, which breaks down DNA, destroyed the ability of the extracts to transform nonvirulent cells into the virulent form. On this basis the transforming agent was concluded to be DNA.

Their conclusion was directly supported and extended by the experiments of Hershey and Chase in 1952 with bacterial viruses at the Carnegie Laboratory of Genetics. It was known that the viruses, which consist only of a DNA core surrounded by a protein coat, cause bacterial cells to cease production of their own molecules and to make instead the DNA and protein of new viral particles of the same kind as the infecting virus. It was assumed that during the infection of a bacterium either the DNA or the protein of the virus enters the bacterial cell to alter its genetic and synthetic activity.

Hershey and Chase used radioactive isotopes of phosphorus and sulfur to label differentially the protein or DNA of the virus. This was possible because the protein of the virus contains sulfur but no phosphorus, and DNA contains phosphorus but no sulfur. The labeled viruses were used to infect bacterial cells. Radioactivity was found inside the infected cells only if the virus contained labeled DNA (Fig. 1-38a). If bacteria were infected by viruses with labeled proteins, the radioactivity remained outside the cell and could be dislodged by shaking (Fig. 1-38b). The experiments showed that only the DNA of the virus entered the bacterial cells and, therefore, that it must contain the genetic information that converts the cells to the production of viral DNA and protein. The Hershey and Chase and Avery experiments, taken together, established that DNA is the hereditary molecule.

Insights into the type of genetic information encoded in DNA were obtained during the same period by George Beadle and Edward Tatum. Previous work indicated that genes control individual steps in biochemical reactions. Beadle and Tatum set out to identify the basis for this control by investigating mutants in the bread mold *Neurospora crassa*. They found that mutations in individual genes frequently caused the mold to grow poorly. Growth could be restored to normal

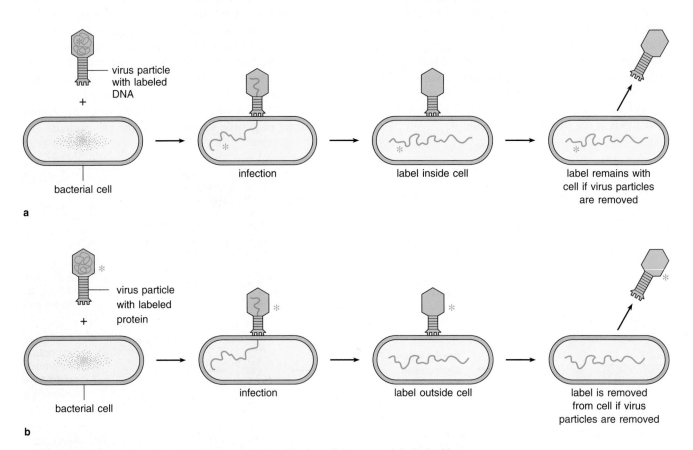

Figure 1-38 The Hershey and Chase experiment. Viral particles were labeled with radioactive phosphorus or sulfur by growing infected cells in the presence of compounds containing either of these substances. The radioactive phosphorus labeled the DNA but not the protein of the particles; the sulfur labeled the protein but not the DNA. **(a)** When viral particles containing labeled DNA were used to infect bacteria, the radioactivity entered the cells. **(b)** When particles containing labeled protein were used for infection, the radioactivity remained outside the cells and could be removed by shaking. The experiment demonstrated that the bacteriophage DNA enters the cell and delivers the genetic information necessary to convert the cell to production of new viral particles.

levels, however, by the addition of single substances, such as individual amino acids, to the growth medium. Beadle and Tatum concluded that a mutant gene encodes a faulty form of an enzyme necessary to produce a substance needed for normal growth. On this basis they proposed that each gene identified in their study codes for a single enzyme. This proposal became famous as the "one gene—one enzyme" hypothesis.

Research with cytoplasmic organelles led to similar integrations of morphology and biochemistry in the cytoplasm, beginning just before the turn of the century with investigations into the functions of chloroplasts and mitochondria. Work in this area proceeded slowly until the 1930s, when Albert Claude developed a technique for isolating and purifying cell parts by centrifugation. In centrifugation, centrifugal force is used to separate structures or molecules into distinct groups according to size or density (for details, see p. 124). Claude and his associates George Palade, Keith Porter, and Christian de Duve soon isolated and identified ribosomes, mitochondria, lysosomes, peroxisomes, and the Golgi complex in cell fractions by this method, and made possible the biochemical analysis of the fractions. By the 1950s it had even become possible to prepare and analyze subfractions of mitochondria and chloroplasts, producing separate outer membrane, inner membrane, and matrix or stroma preparations.

Application of the electron microscope to biological materials, which began in the late 1940s, greatly extended the details visible in cells. Membranes and their disposition could be directly observed, and features as small as ribosomes and even individual protein molecules could be seen in the microscope. Palade and Porter pioneered application of the electron microscope to cell biology, and worked out the ultrastructure of many of the organelles studied biochemically in cell fractions.

As this work progressed, research with cell extracts, particularly from *E. coli*, yeast, and green algae, gradually identified the molecules and enzymes participating in individual steps of cellular oxidation and photosynthesis. Fractionation of eukaryotic cells by centrifugation soon allowed these reaction sequences to be placed in their correct locations in mitochondria and chloroplasts or in the cytoplasm surrounding these organelles. Similar work with ribosomes led to the development of cell-free systems able to carry out all of the reactions of protein synthesis.

In more recent years centrifugation has been supplemented by gel electrophoresis, which uses an electrical field instead of centrifugal force to drive molecules through highly viscous plastic gels (for details, see p. 126). Even small molecules are separated readily and rapidly according to size, density, and charge by the technique. Application of gel electrophoresis has allowed many thousands of protein, nucleic acid, and other cellular molecules to be individually isolated and identified.

The Molecular Revolution

The experiments of Avery and Hershey and Chase sparked an intensive effort to work out the structure of DNA. Although its chemical components had been identified, the three-dimensional structure of DNA was still unknown when the molecule was established as the hereditary material. An intense competition among scientists culminated in the discovery of DNA structure at Cambridge University in 1953 by an American, James D. Watson, and an Englishman, Francis H. C. Crick (Fig. 1-39). Their model for DNA structure was based primarily on data obtained by X-ray diffraction (see p. 132) of DNA samples by Maurice Wilkins and Rosalind Franklin. The DNA structure worked out by Watson and Crick, perhaps the most significant single accomplishment in the history of biology, sparked an effort to determine the molecular structure of genes and their modes of action. This effort was greatly facilitated by rapid methods for nucleic acid sequencing developed by Fred Sanger, Allan M. Maxam, and Walter Gilbert.

Figure 1-39 James D. Watson (left) and Francis H. C. Crick, demonstrating their model for DNA structure deduced from X-ray diffraction data obtained by Wilkins and Franklin. Courtesy of Cold Spring Harbor Laboratory Archives.

By the 1970s cell biology and most other biological disciplines were dominated completely by molecular biology.

Many genes, and in some organisms the entire DNA complement, have been completely sequenced. Through the gene sequences, the amino acid sequences of many proteins have been deduced, including in some organisms those of every protein encoded in the DNA. The comparisons of gene and protein structure in normal and mutant forms made possible by these accomplishments have provided fundamental insights into the molecular functions of genes and proteins, and how genes are regulated and controlled. The aims of this approach have been furthered by the development of genetic engineering, which allows genes and their controls to be altered in specific, predetermined ways. Experimentally induced mutations have revealed the functions not only of specific regions of genes and their controls but also of proteins encoded in the genes.

The molecular approach has produced an explosion in our knowledge of cell structure and function, and a revolution in the methods and attainments of almost all of the related fields of biology: genetics, biochemistry, developmental biology, physiology, medicine, and even taxonomy, systematics, evolution, and behavior. The importance of the molecular approach has made an understanding of cell and molecular biology a prerequisite for basic comprehension and advanced study in any field of biology. It has also created a new responsibility for nonscientist citizens and their governments, who must now understand and cope with the achievements and conclusions of molecular biology to vote and govern wisely and effectively.

The contemporary field of cell biology has developed through an integration of structural and biochemical studies, recently revolutionized by the molecular studies of gene structure and function and of the primary encoded products of the genes—the proteins. The unity of cellular structure and function revealed by the integration of morphological and biochemical studies, and the molecular revolution and its accomplishments are the subjects of this book.

Suggestions for Further Reading

Allen, M. M. 1984. Cyanobacterial cell inclusions. *Ann. Rev. Microbiol.* 38:1–25.

Bracegirdle, B. 1989. Microscopy and comprehension: the development of understanding of the nature of the cell. *Trends Biochem. Sci.* 14:464–68.

Brock, T. D. 1988. The bacterial nucleus: a history. *Microbiol. Rev.* 52:397–411.

Carlson, G. A.; Hsiao, K.; Oesch, B.; Westaway, D.; and Prusiner, S. B. 1991. Genetics of prion infections. *Trends Genet.* 7:61–65.

Claude, A. 1975. The coming of age of the cell. *Science* 189:433–35.

De Duve, C. 1975. Exploring cells with a centrifuge. *Science* 189:186–94.

———, and Beaufay, H. 1981. A short history of tissue fractionation. *J. Cell Biol.* 91:293s–99s.

Fruton, J. S. 1976. The emergence of biochemistry. *Science* 192:327–34.

Hughes, A. 1959. *A history of cytology.* New York: Abelard-Schuman.

van Iterson, W. 1984. *Inner structures of bacteria.* New York: Van Nostrand Reinhold.

May, Y.; Pattison, B. W. J.; and Pattison, J. R., eds. 1984. *The microbe, 1984, Part 1: Viruses.* 36th Symposium of the Society for General Microbiology. New York: Cambridge University Press.

Quastel, J. H. 1984. The development of biochemistry in the 20th century. *Canad. J. Cell Biol.* 62:1103–10.

Rossman, M. G., and Johnson, J. E. 1989. Icosahedral virus structure. *Ann. Rev. Biochem.* 58:533–73.

Schenk, F. 1988. Early nucleic acid chemistry. *Trends Biochem. Sci.* 13:67–69.

Watson, J. D. 1968. *The double helix.* New York: Atheneum.

Weissman, C. 1989. Prions: sheep disease in human clothing. *Nature* 338:298–99.

———. 1991. A "unified theory" of prion propagation. *Nature* 352:679–83.

Review Questions

1. Define and outline the overall functions of the nuclear region and cytoplasm.

2. What are the major differences between eukaryotic and prokaryotic cells? What organisms fall into the two divisions?

3. What cytoplasmic structures occur in prokaryotic cells?

4. Compare the structures of prokaryotic nucleoids and eukaryotic nuclei.

5. What major differences exist in the DNA of eukaryotic and prokaryotic cells?

6. Compare the flagella of prokaryotes and eukaryotes.

7. What is an organelle?

8. Define transcription and translation.

9. Outline the major roles of mRNA, rRNA, and tRNA in protein synthesis.

10. Outline the structure and primary functions of mitochondria, the endoplasmic reticulum, the Golgi complex, lysosomes, and microbodies.

11. Contrast microtubules and microfilaments, and outline their major cellular roles.

12. What are intermediate filaments?

13. What cellular and extracellular structures occur in plant but not animal cells? Briefly outline the roles of these structures.

14. Outline the structure of a rod-shaped and spherical virus particle.

15. How do coated viruses obtain their surface membrane?

16. Describe the pattern of infection followed by bacterial viruses. What does this pattern suggest about the living or nonliving nature of viruses?

17. Contrast the lytic and lysogenic phases of viral infective cycles.

18. List the major hypotheses forming parts of the cell theory. Why are the first and second hypotheses listed as separate ideas?

19. What was the significance of Wöhler's synthesis of an organic molecule to the development of the chemical approach to cell biology?

20. Why was Mendel's discovery of genes largely unnoticed when he published his results? What parallels between the behavior of genes and chromosomes convinced investigators that genes are carried on the chromosomes?

21. What experiments established that DNA, and not protein, is the hereditary molecule? Outline the steps carried out in these experiments.

22. What major techniques and discoveries led to the molecular revolution in biology?

2

CHEMICAL BONDS AND
BIOLOGICAL MOLECULES

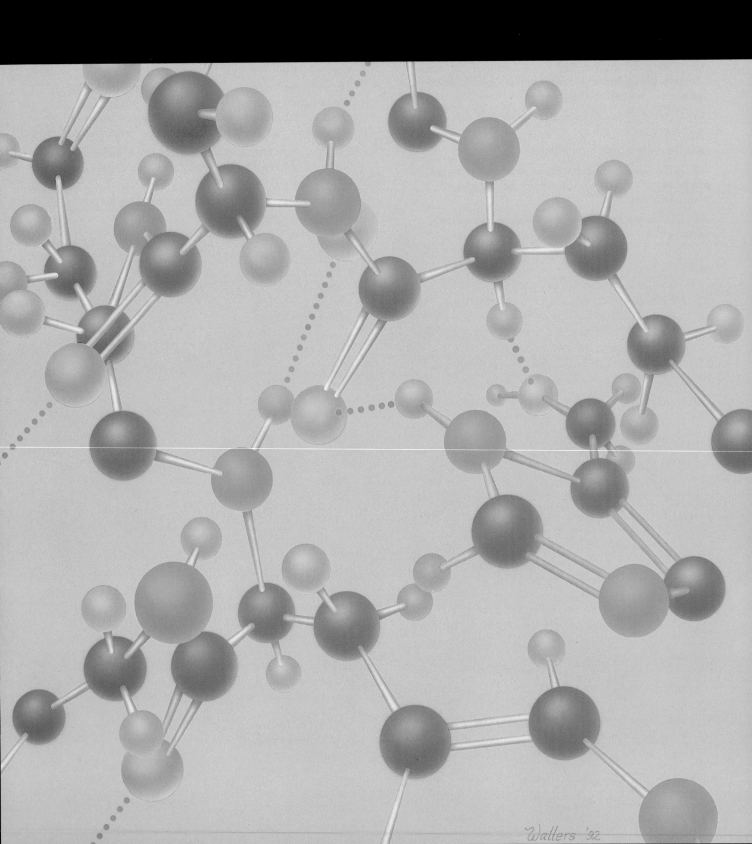

Walters '92

C ell biology has been transformed into a science that is now primarily biochemical and molecular, and molecular biology is by nature chemical and physical in approach. As a result, it is impossible to understand the conclusions of cell and molecular biology without an introduction to the forces holding atoms together in molecules, molecular structure, chemical reactions, and the major types of biological molecules carrying out the activities of life.

Although many of the molecules of living systems are highly complex, most fall into one of four classes—carbohydrates, lipids, proteins, and nucleic acids—or contain substructures belonging to one or more of these classes. Along with water these four classes of organic molecules form almost the entire substance of living organisms. These molecules are held together and interact through chemical bonds and forces of several different types: ionic, covalent, and hydrogen bonds; nonpolar associations; and van der Waals forces.

For those already familiar with chemical bonds and the major classes of biological molecules, this chapter will serve for reference and review. For others the chapter will provide a foundation for the discussion of molecular structures, functions, and interactions in this book.

CHEMICAL BONDS IMPORTANT IN BIOLOGICAL MOLECULES

The chemical properties of an atom are determined largely by the number of electrons in its outermost energy level, or shell. (Supplement 2-1 to this chapter reviews atomic structure.) Most significant is the difference between the number in the outermost shell and the stable numbers of either two electrons for atoms with only one shell or eight electrons for atoms with two or more shells. Helium, with two electrons in its single shell, and atoms such as neon and argon, with eight electrons in their outer shells, are stable and essentially inert chemically. Atoms with outer shells containing electrons near these numbers tend to gain or lose electrons to approximate these stable configurations. For example, sodium has two electrons in its first shell, eight in the second, and one in the third and outermost shell. This atom readily loses its single outer electron to leave a stable second shell with eight elec-

trons. Chlorine, with seven electrons in its outermost shell, tends to attract an electron from another atom to attain the stable number of eight. Atoms differing from the stable configuration by more than one electron are inclined to *share* electrons with other atoms rather than to gain or lose electrons completely. Among the atoms forming biological molecules, electron sharing is most characteristic of carbon, which has four outer electrons and thus falls at the midpoint between the tendency to gain or lose electrons. Oxygen, with six electrons in its outer shell, nitrogen, with five electrons, and hydrogen, with a single electron, also share electrons readily to complete their outer shells. The relative tendency to gain, share, or lose electrons underlies the chemical bonds and forces holding the atoms of molecules together.

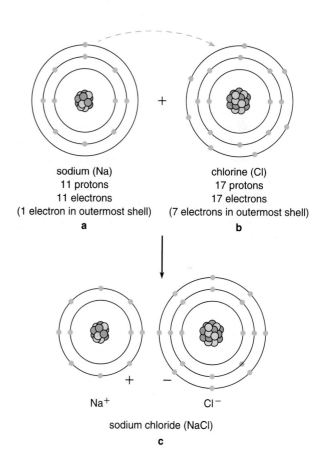

sodium (Na)
11 protons
11 electrons
(1 electron in outermost shell)
a

chlorine (Cl)
17 protons
17 electrons
(7 electrons in outermost shell)
b

Na⁺ Cl⁻

sodium chloride (NaCl)
c

Figure 2-1 Formation of ionic, or electrostatic, bonds. **(a)** Sodium, with one electron in its outermost shell or energy level, readily loses an electron to attain a stable state in which its second shell, with eight electrons, becomes the outer shell. **(b)** Chlorine, with seven electrons in its outer shell, readily gains an electron to attain the stable number of eight electrons in its outer shell. **(c)** Transfer of an electron from sodium to chlorine creates stable outer shells in both atoms and creates the ions Na^+ and Cl^-. The attractive force holding the oppositely charged ions together is the ionic, or electrostatic, bond.

Table 2-2 van der Waals Contact Radii	
Atom	Radius (angstroms)
C	2.0
H	1.2
O	1.4
N	1.5
S	1.8
P	1.9

radius (Table 2-2), establishes the diameter of the spheres representing atoms in a *space-filling model*, such as those shown in Figure 2-10.

An attractive van der Waals force at the optimum separation between two atoms requires only about 1 kcal/mol for breakage (see Table 2-1). At this level the force is only about one-half to one-fifth as strong as a hydrogen bond. However, the collective effects of van der Waals forces can be enough to stabilize the three-dimensional shape of a molecule or a combination between interacting molecules. The van der Waals attractions are also significant as forces holding nonpolar substances excluded by water in tightly packed masses. In many intra- and intermolecular reactions, the collective van der Waals attractions add to stabilizing forces provided by hydrogen bonds and ionic attractions.

Covalent linkages and their patterns of electron sharing are, therefore, responsible for several interatomic and intermolecular attractions and repulsions. Besides holding atoms together in molecular structures, the patterns of electron sharing in covalent bonds may produce polarity, polar and nonpolar associations, hydrogen bonds, and van der Waals forces. All of these bonding forces are critical to the structure and interactions of biological molecules.

FUNCTIONAL GROUPS IN BIOLOGICAL MOLECULES

Although covalent bonds are inherently stable, some arrangements of atoms place bonds in positions in which they are disturbed more readily. The atoms of such groups can enter into reactions in which their existing covalent bonds are broken and new ones are formed. These reactive arrangements, called *functional groups*, serve as reaction centers converting one type of organic subunit to another or forming linkages that bind subunits into larger molecular assemblies. Although the complete list of such groups is long, several stand out as important reactive subparts of biological

molecules: *hydroxyl* (—OH), *carbonyl* (C=O), *carboxyl* (—COOH), *amino* (—NH$_2$), *phosphate* (—PO$_4$), and *sulfhydryl* (—SH) groups (Table 2-3).

Depending on their positions, these functional groups take on characteristic patterns of interaction, giving the organic molecules containing them properties of *alcohols, aldehydes, ketones*, or *acids*. Each of these substances figures importantly in the molecular interactions that occur in cells, and often serves as a building block for the assembly of more complex biological molecules. The four major classes of biological molecules—carbohydrates, lipids, proteins, and nucleic acids—are each based primarily on different combinations of building blocks consisting of alcohols, aldehydes, ketones, and organic acids.

The Hydroxyl Group and Alcohols

Hydroxyl (—OH) groups are split readily from their molecules or enter into interactions that link subunits into larger molecular assemblies. Among the most frequent of these interactions are *condensation*, in which the elements of a water molecule are released as subunits assemble, and *hydrolysis*, in which the elements of a water molecule are incorporated as subunits are split from a larger molecule. (Condensation and hydrolysis are defined further in Information Box 2-1.) Hydroxyl groups contribute to the polarity of molecules because of unequal electron sharing between the oxygen and hydrogen atoms in the structure. Hydroxyl groups are so strongly polar that a single —OH can make an entire molecule or a molecular subregion polar.

A hydroxyl group forms the reactive part of the alcohol group, in which a hydrogen is also linked to the same carbon atom:

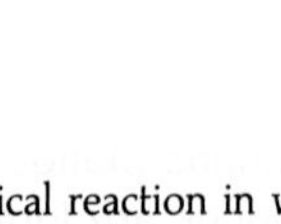

The alcohol group is readily oxidized[2] to form aldehydes or ketones and because of its hydroxyl segment easily forms linkages. Alcohol groups are important functional groups in many organic molecules including carbohydrates.

[2] An *oxidation* is a chemical reaction in which electrons are removed from a molecule or reactive group. The opposite reaction, in which electrons are added to a molecule or reactive group, is a *reduction*. Because electrons have an associated energy, represented by the velocity at which they travel through space, oxidations remove energy from the molecule oxidized, and reductions add energy to the molecule reduced. (For further details, see Ch. 9 and Information Box 9-1.)

Table 2-3 Common Reactive Groups of Organic Molecules

Functional Group	Structural Formula	Reactivity
Hydroxyl	—OH	Strongly polar; highly reactive with other groups in formation of covalent bonds, particularly in condensations.
Alcohol	H │ —C—OH │	Strongly polar; readily reduced to aldehydes and ketones; also reacts with organic acids through —OH group to form many biological substances; part of many carbohydrate molecules.
Carbonyl	│ —C═O	Weakly polar; highly reactive; enters into linkages, particularly with bases; primary reactive group of aldehydes and ketones.
Aldehyde	H │ —C═O	Similar in reactivity because of carbonyl group but aldehydes more reactive than ketones; oxidized to form acids or reduced to form alcohols; components of carbohydrates, fats, and intermediates formed in synthesis and oxidative breakdown of fats and carbohydrates.
Ketone	│ C═O │	
Carboxyl	O ∥ —C ＼OH	Strongly polar; ionizes in solution to release H^+ and thus acts as acid; part of amino acids, fatty acids, and many additional organic acids that form fuel substances and intermediates of a variety of biological reactions.
Amino	H ／ —N ＼H	Polar; basic because it combines with H^+ in solution to produce $—NH_3^+$ group, thereby reducing H^+ concentration; enters reactions with other groups forming covalent linkages; component of amino acids and other biological molecules.
Phosphate	O^- │ —O—P—O^- ∥ O	Acidic and polar; forms linking bridge between major organic groups in complex biological molecules; part of molecules storing and releasing chemical energy; serves as control element regulating molecular activity.
Sulfhydryl	—S—H	Readily oxidized; reactive group that forms linkages between major organic groups; two —SH groups enter into disulfide linkages that stabilize biological molecules including proteins.

The Carbonyl Group, Aldehydes, and Ketones

The oxygen atom of the carbonyl (—C═O) group is highly susceptible to chemical attack by other reactive groups, particularly bases. (Information Box 2-2 defines acids, bases, and pH.) The carbonyl group can be oxidized or reduced, depending on the presence of other active groups. Although only slightly polar, the carbonyl group may form a part of other functional groups, such as the carboxyl (—COOH) group, which are strongly polar.

Carbonyl groups figure as part of aldehydes and ketones, which are particularly important as building blocks of carbohydrates and fats. In an aldehyde the carbonyl group occurs with a hydrogen at the end of a carbon chain:

$$-C{\overset{H}{\underset{\text{O}}{\diagdown}}}$$

In a ketone the carbonyl group occurs in the interior of a carbon chain:

$$-\underset{\underset{\text{O}}{\parallel}}{C}-$$

Although it reacts similarly in aldehydes and ketones, the carbonyl oxygen is more exposed in the aldehyde group, making this the more reactive of the two. Aldehydes and ketones serve as potential linkage sites and can be oxidized to form acids or reduced to form alcohols. Their susceptibility to oxidation, particularly

Condensation and Hydrolysis

Many types of biological reactions involve addition or removal of the components of a molecule of water. As two glucose molecules interact to form a disaccharide, for example, a molecule of water appears as an additional product (see part **a** of the figure in this box; the atoms contributing to the formation of water are in blue). In this type of reaction, called a *condensation*, the components of a molecule of water, H^+ and OH^-, split from reacting groups of the combining molecules. Most biological reactions in which complex molecules are assembled from smaller subunits are condensations. The interactions assembling amino acids into proteins and

nucleotides into nucleic acids, for example, are typical condensation reactions.

The disassembly of biological molecules into subunits usually involves *hydrolysis*, a reaction that is the reverse of condensation. In this process the components of a molecule of water are *added* as a covalent bond is broken (see part **b** of the figure in this box). The reactions of digestion, which break down proteins, fats, carbohydrates, and nucleic acids into smaller chemical subunits that can be absorbed in the small intestine, are typical hydrolysis reactions.

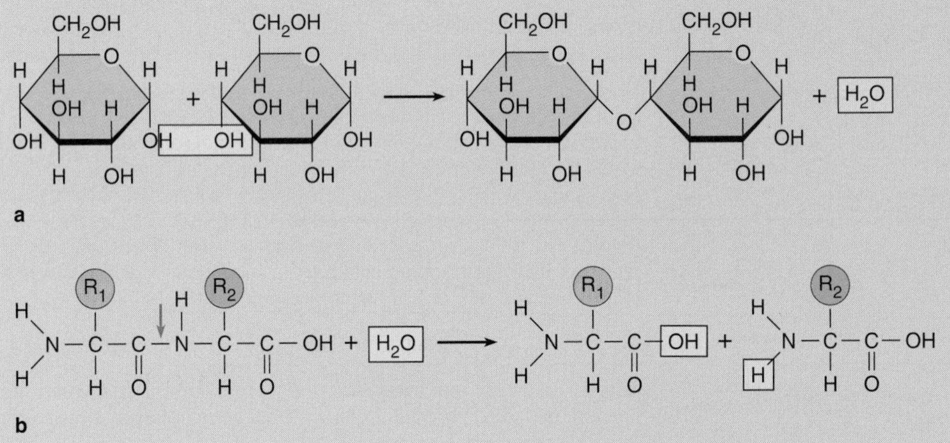

a

b

the aldehydes, makes them important intermediates in cellular energy metabolism, as well as building blocks for reactions assembling macromolecules.

The Carboxyl Group and Organic Acids

Carbonyl and hydroxyl groups combine to form the carboxyl group, the characteristic functional group of organic acids:

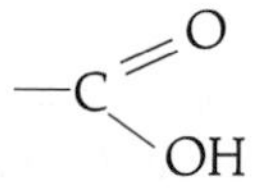

The reactivity of the carboxyl group depends primarily on its hydroxyl segment, which, because of the chemical influence of the nearby carbonyl oxygen, readily dissociates to release its hydrogen as H^+:

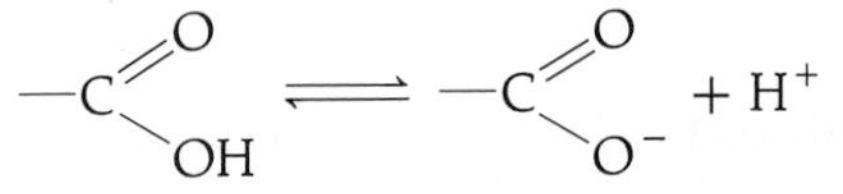

The strongly polar carboxyl group is characteristic of amino acids, fatty acids, and a long list of other organic acids. Many organic acids, such as citric, pyruvic, and acetic acid, are oxidized readily and serve as important intermediates in energy metabolism. Carboxyl groups also enter into linkages combining organic subunits into larger assemblies.

The Amino Group

Amino ($-NH_2$) groups are significant for both their chemical reactivity and their ability to act as a base in organic molecules. They are particularly important in the amino acids, which contain an amino group at one end and a carboxyl group at the other. A condensation reaction between the amino group of one amino acid and the carboxyl group of another is responsible for the formation of a *peptide linkage*, the bond that ties amino acids into the backbone chains of proteins (see Fig. 2-21).

An amino group reacts as a base because of its ability to take up a hydrogen ion:

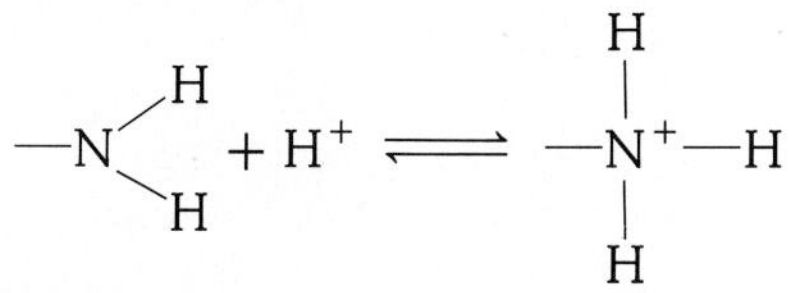

$$-N\overset{H}{\underset{H}{\diagup}} + H^+ \rightleftharpoons -\overset{H}{\underset{H}{\overset{|}{N^+}}}-H$$

Removal of the H^+ effectively raises the pH of the surrounding medium.

The Phosphate Group

The phosphate ($-PO_4$) group

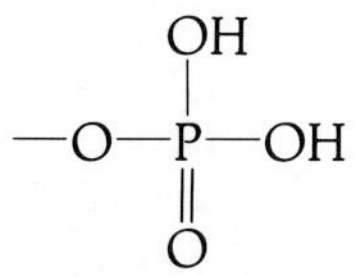

$$-O-\overset{OH}{\underset{O}{\overset{|}{\underset{\|}{P}}}}-OH$$

plays several important roles in biological molecules. One of the most significant is its ability to form chemical bridges linking organic building blocks into larger molecules. These bridges form because the oxygens held by single bonds are available to form covalent bonds on either side of the central phosphorus atom. Most notable of the phosphate-bridged superstructures are the nucleic acids DNA and RNA (see Fig. 2-33) and the phospholipids (see Fig. 2-13), which provide the structural framework of biological membranes. Phosphate groups are also important in molecules that conserve and release energy, and in metabolic regulation. The activity of many proteins, for example, is turned on or off by the addition or removal of phosphate groups.

Phosphate groups may dissociate to release H^+, thus acting as an acid:

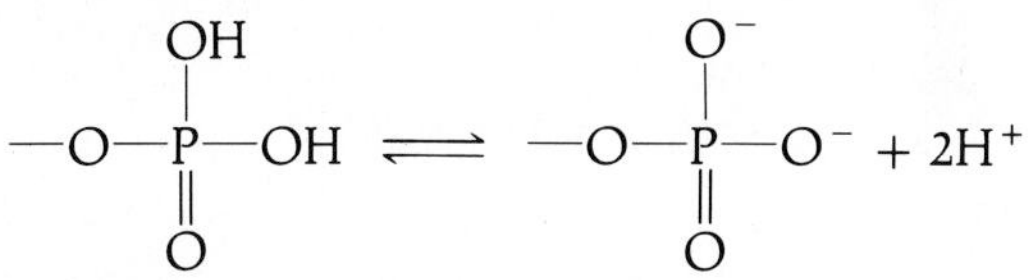

$$-O-\overset{OH}{\underset{O}{\overset{|}{\underset{\|}{P}}}}-OH \rightleftharpoons -O-\overset{O^-}{\underset{O}{\overset{|}{\underset{\|}{P}}}}-O^- + 2H^+$$

The strongly acidic reaction of the nucleic acids, for example, results from the release of H^+ from the many phosphate groups linking the subunits of these molecules.

The Sulfhydryl Group

The sulfhydryl ($-SH$) group serves as a reactive site that is easily oxidized or converted into a covalent linkage. One of the most important of these linkages occurs when two sulfhydryl groups interact within a protein to form a *disulfide* ($-S-S-$) *linkage* (see Fig. 2-29):

$$-SH + HS- \longrightarrow$$
$$-S-S- + 2H^+ + 2\ electrons$$

The reaction is an oxidation, in which two electrons and two hydrogen nuclei (or protons) are removed from the interacting $-SH$ groups. The disulfide linkage is a molecular "button" that stabilizes protein structures by holding subparts of a protein together.

These functional groups and organic building blocks combine in various ways to produce the four major classes of biological molecules—carbohydrates, lipids, proteins, and nucleic acids. The particular combinations give the classes individual chemical properties that allow them to serve vital roles in living systems: carbohydrates and lipids as structural and fuel molecules; proteins as enzymes, structural, transport, motile, and recognition molecules and among other functions; and nucleic acids as informational molecules.

MAJOR CLASSES OF BIOLOGICAL MOLECULES: CARBOHYDRATES

Carbohydrates serve many functions. Along with fats they provide the primary molecules oxidized to supply chemical energy for cell activities. Carbohydrate fuels are stored in cells as *starches*, molecules consisting of long chains of repeating carbohydrate units linked end to end. Molecules assembled from chains of carbohydrate subunits also form structural molecules such as *cellulose*, one of the primary constituents of plant cell walls. Carbohydrates link with lipids to form *glycolipids*, and with proteins to form *glycoproteins*. Both glycolipids and glycoproteins appear in quantity at cell surfaces, where they act in such roles as receptors and cell-cell recognition molecules. Glycoproteins are abundant among the molecules released by cells as secretions.

Carbohydrate Structure

Carbohydrates are named for their characteristic content of carbon, hydrogen, and oxygen atoms, which occur in a 1C:2H:1O ratio or in numbers that closely approximate this ratio. Short chains containing from three to seven carbons form the *monosaccharides*, the individual building blocks of the carbohydrate family (Fig. 2-6). Of these, *trioses* (three carbons), *pentoses* (five

Acids, Bases, and pH

Many inorganic and organic molecules act as either acids or bases in water solution. Acids and bases are substances that affect the relative concentrations of hydrogen ions (H^+) and hydroxyl ions (OH^-) in water. Water always contains both ions because a proportion of the water molecules in liquid water dissociates to produce H^+ and OH^-:

$$H_2O \rightleftharpoons H^+ + OH^-$$

When dissolved in water, acids release additional hydrogen ions, increasing the relative H^+ concentration. Bases bind H^+ or release additional OH^- when dissolved in water, thereby reducing the relative H^+ concentration.

The relative concentrations of H^+ and OH^- in a water solution determine the *acidity* of the solution. The degree of acidity alters the chemical reactivity of many organic and inorganic substances dissolved in water and modifies the folding conformations of protein molecules.

Acidity is expressed quantitatively as *pH*, on a number scale ranging from 0 to 14 (see the figure in this box). The number scale reflects a constant relationship between the relative concentrations of H^+ and OH^- in water solutions. At a temperature of 25°C the product of the concentrations of H^+ and OH^- has a constant value of 1×10^{-14}:

(concentration of H^+) × (concentration of OH^-)
$$= 1 \times 10^{-14}$$

These concentrations are given in moles per liter, abbreviated M. The pH of a solution is defined as the negative logarithm (to the base 10) of the H^+ concentration:

$$pH = -\log_{10}[H^+]$$

where the brackets [ ] indicate concentration in moles per liter. For example, if the H^+ concentration is $0.0000001\ M$ ($1 \times 10^{-7}\ M$), the $\log_{10}$ of this concentration is -7. The negative of the logarithm -7 is 7. Thus, a water solution with an H^+ concentration of $1 \times 10^{-7}\ M$ is said to have a pH of 7.

At pH 7 the concentrations of H^+ and OH^- are equal:

$$(1 \times 10^{-7}\ M\ H^+)(1 \times 10^{-7}\ M\ OH^-) = 1 \times 10^{-14}$$

At this concentration the solution is said to be neutral. Solutions with pH higher than 7 have OH^- in excess and are basic or alkaline; solutions with pH less than 7 have H^+ in excess and are acid.

The product of the concentrations of H^+ and OH^- remains equal to 1×10^{-14}. Thus, if the H^+ concentration rises from 1×10^{-7} to 1×10^{-5}, the OH^- concentration falls so that the product remains equal to 1×10^{-14}:

$$\underset{\text{(H}^+\text{ concentration)}}{(1 \times 10^{-5})} \times \underset{\text{(OH}^-\text{ concentration)}}{(1 \times 10^{-9})} = 1 \times 10^{-14}$$

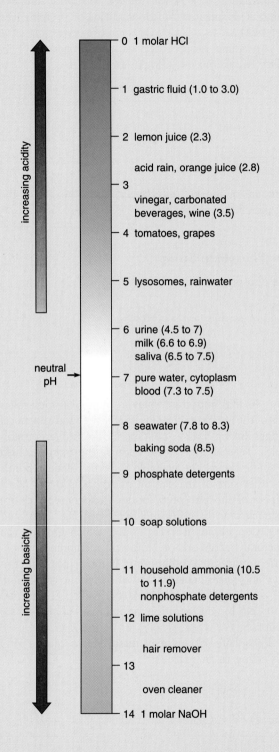

Because pH values are logs to the base 10 of the concentration, a change of one pH unit represents a concentration difference of ten times. Thus, a solution of pH 8 has ten times as many OH^- ions as a solution at pH 7. Cells typically have a neutral cytoplasmic pH or a slightly basic pH in the range 7.0 to 7.2.

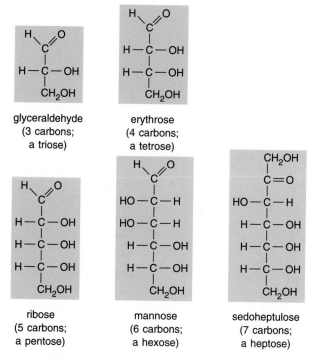

glyceraldehyde
(3 carbons;
a triose)

erythrose
(4 carbons;
a tetrose)

ribose
(5 carbons;
a pentose)

mannose
(6 carbons;
a hexose)

sedoheptulose
(7 carbons;
a heptose)

Figure 2-6 Some representative carbohydrates containing three to seven carbons.

carbons), and *hexoses* (six carbons) are most common in cells. (Table 2-4 lists a number of monosaccharides and gives some of their functions in living organisms.)

All monosaccharides can occur in linear form. In such form (Fig. 2-6) each carbon atom in the chain, except one, carries an —OH group. The remaining carbon carries a —C=O (carbonyl) group. In monosaccharides all the other available binding sites of the carbons are occupied by hydrogen atoms. The carbonyl oxygen of a linear sugar may be located at the end of

the carbon chain as an aldehyde group or inside the chain as a ketone group (see Table 2-3).

Monosaccharides with five or more carbons can exist in ring as well as linear conformations. The rings are formed through a reaction between two functional groups in the same molecule. In the six-carbon monosaccharide *glucose*, for example, a covalent bond can form through a reaction between the aldehyde at the 1-carbon and the hydroxyl at the 5-carbon. This reaction produces either of two closely related *glucopyranose* ring structures (Fig. 2-7a and b). The aldehyde at the 5-carbon can also react with the hydroxyl at the 4-carbon to produce a *glucofuranose* ring (Fig. 2-7c). The glucopyranose ring is the most common form of glucose in living cells.

Monosaccharide rings, such as that of glucose, are frequently depicted as a *Haworth projection* (Fig. 2-7d), a convenient diagram that suggests the three-dimensional orientation of the attached —H, —OH, and —CH₂OH groups in relation to the ring. However, the glucopyranose ring does not actually lie in a flat plane as suggested by this projection. Instead, the ends of the ring are bent up or down, most frequently in the *"chair" conformation* shown in Figure 2-7e. Moreover, the side groups attached to the ring extend at the various angles shown in Figure 2-7e, not at the right angles depicted in the Haworth projection. A space-filling model of the glucopyranose ring, showing the volumes occupied by its atoms, appears in Figure 2-7f.

The Haworth projection clearly depicts the two forms of the glucopyranose ring and other monosaccharide ring structures, which differ only in the direction pointed by the —OH group (see Fig. 2-7b). The —OH group points downward in the *alpha* (α) form of the sugar, as in α-glucose. The other form, in which the

Number of Carbons	Type	Examples	Major Activity
		Table 2-4 Carbohydrate Units Found in Nature	
3	Triose	Glyceraldehyde, dihydroxyacetone	Intermediates in energy-yielding reactions and photosynthesis.
4	Tetrose	Erythrose	Intermediate in photosynthesis.
5	Pentose	Ribose, deoxyribose, ribulose	Intermediates in photosynthesis; components of molecules carrying energy; components of the informational nucleic acids DNA and RNA; structural molecules in cell walls of plants.
6	Hexose	Glucose, fructose, galactose, mannose	Fuel substances; products of photosynthesis; building blocks of starches, cellulose, and carbohydrate units of glycolipids and glycoproteins.
7	Heptose	Sedoheptulose	Intermediate in photosynthesis.

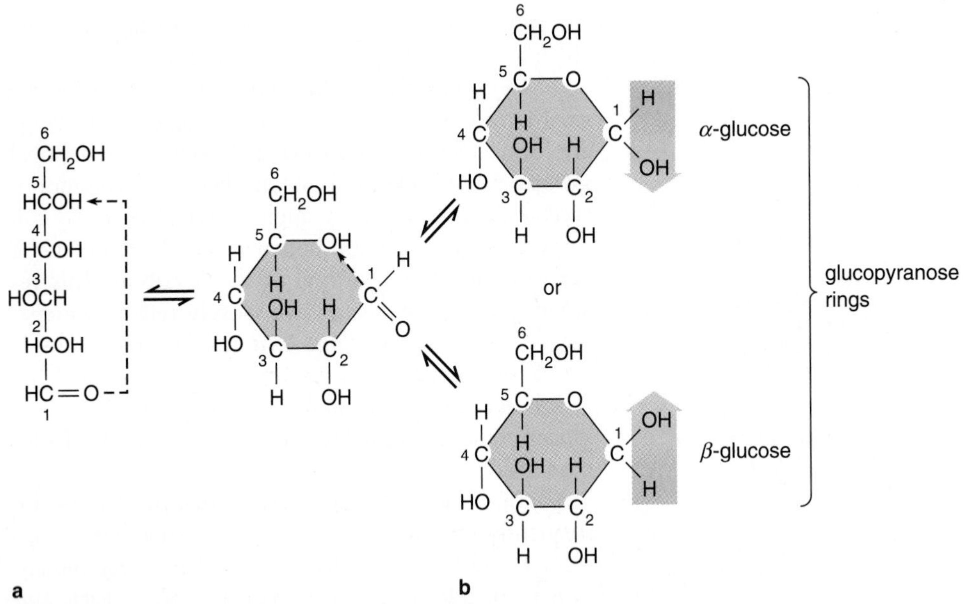

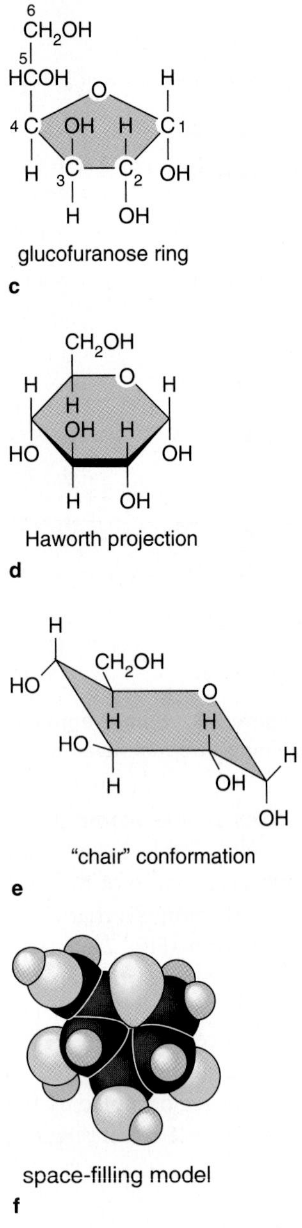

Figure 2-7 Various states and representations of the glucose molecule. **(a)** Glucose in linear form. The carbons of glucose and other monosaccharides are numbered in order with the carbon of the aldehyde (—CHO) group as number 1. **(b)** Pyranose ring formation by glucose, in which the aldehyde group at the 1-carbon reacts with the hydroxyl group at the 5-carbon. The reaction produces the two alternate conformations of the glucopyranose ring, α and β, determined by the orientation of the hydroxyl (—OH) group linked to the 1-carbon (shaded). **(c)** The glucofuranose ring, formed by reaction of the aldehyde at the 1-carbon with the hydroxyl group at the 4-carbon. The glucofuranose ring, relatively rare in nature, also exists in two forms equivalent to α- and β-glucopyranose. **(d)** A Haworth projection of glucose, in which the letter C designating carbon is omitted and the covalent bonds of the ring are shown as thicker lines along one side. This depiction indicates that the ring lies in a plane perpendicular to the page, with the thickest edge closest to the viewer. The various groups attached to the ring (the —H, —OH, and —CH$_2$OH groups) are considered to extend at right angles to the ring, in a plane parallel to the surface of the page. **(e)** The "chair" conformation, a more accurate depiction of the most common three-dimensional arrangement of the glucose ring. The side groups extend at various angles from the ring as shown. **(f)** A space-filling model of glucose, showing the volumes and arrangement of the atoms according to their van der Waals radii. Carbon atoms are shown in black, oxygen in red, and hydrogen in blue.

—OH group points upward from the ring, is the *beta* (β) form of the sugar, as in β-glucose.

Although the difference between the two ring forms might seem trivial, it has great significance for the chemical properties of polysaccharides assembled from monosaccharide rings. For example, starches, which are assembled from α-glucose, are soluble and easily digested. Cellulose, assembled from β-glucose units, is insoluble and cannot be digested as a food source by most animals.

The directions "pointed" by other groups attached to the carbon chain of carbohydrates also affect the properties and interactions of monosaccharides. The alternative forms of a molecule in which atoms or groups attached to the carbon chain point in different directions are termed *stereoisomers*. By convention stereoisomers are identified as D- or L-, depending on the directions pointed by atoms attached to the carbon chain (see Information Box 2-3).

With several important exceptions the different stereoisomers of the monosaccharides have identical chemical and physical properties. Most important of the exceptions for biological substances is that stereoisomers react at widely different rates with the D- and L-forms of other biological molecules that either are stereoisomers themselves or contain subunits that are stereoisomers. Enzymes contain subunits that are stereoisomers and, as a result, react much more effi-

Stereoisomers

Many internal carbon atoms of monosaccharides and other organic molecules are *asymmetric* in their linkages to other atoms. This means that each of their four covalent bonds links to a different atom or chemical group. The middle carbon of the triose *glyceraldehyde*, for example, is asymmetric: it shares electrons in covalent bonds with —H, —OH, —CHO, and —CH_2OH:

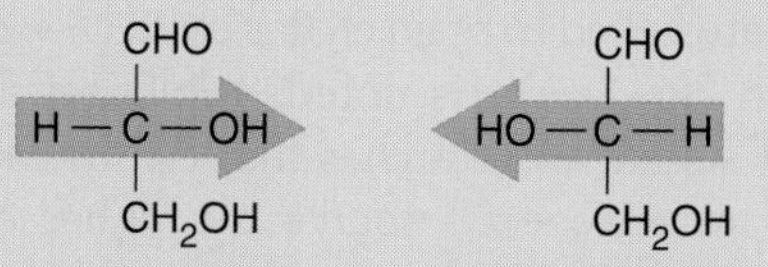

D-glyceraldehyde   L-glyceraldehyde

The groups attached to asymmetric carbon atoms can take up either of two fixed positions with respect to other carbon atoms in a carbon chain. In glyceraldehyde (see above) the —OH group can extend either to the left or to the right of the carbon chain with reference to the —CHO and —CH_2OH groups. These two possible forms are called *stereoisomers*. By convention the stereoisomer in which the —OH extends to the right is called *D-glyceraldehyde* (from the Latin *dexter* = right), and the stereoisomer in which the —OH extends to the left is *L-glyceraldehyde* (from the Latin *laevus* = left).

All monosaccharides with chains from three to seven carbons long have different stereoisomeric forms. The longer-chain carbohydrates have several asymmetric carbons, giving more than two possible types, and so naming them becomes complicated. However, by convention, the longer chains are designated as D- or L-isomers *according to the direction pointed by the —OH group on the next-to-last carbon at the end of the chain opposite the carbonyl (C=O) oxygen:*

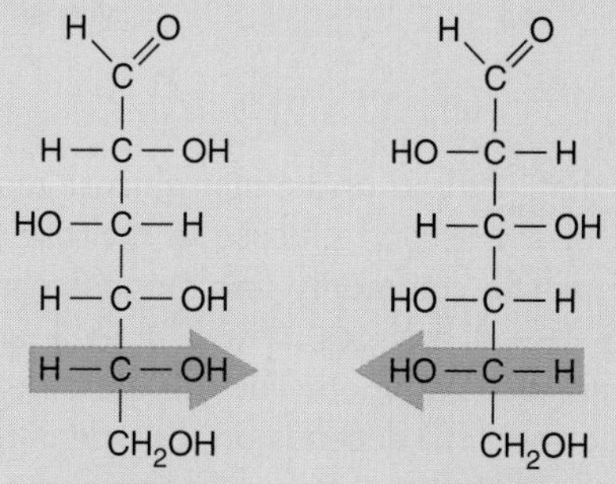

D-glucose   L-glucose

Many other kinds of biological molecules are stereoisomers, including the amino acids. These are named by comparing them to the "handedness" of glyceraldehyde. The amino acids, for example, are named as D- and L-forms by noting the direction pointed by the NH_2 group attached to the central asymmetric carbon:

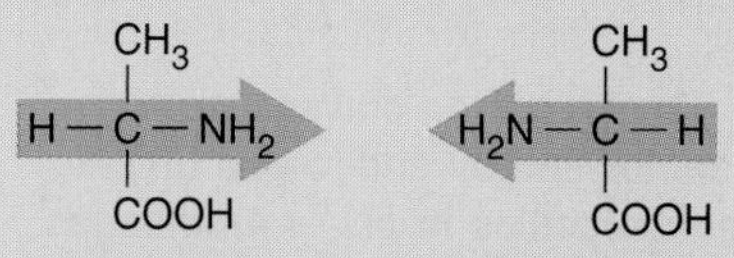

D-alanine   L-alanine

The different isomers of a molecule generally have the same chemical and physical properties. However, they frequently interact at significantly different rates with other substances that are also stereoisomers or contain subunits that are stereoisomers, such as enzymes. The rates are often different enough to limit use effectively to one stereoisomer. Proteins, for example, are assembled by the enzymatic activities of ribosomes exclusively from L-isomers of the amino acids.

ciently with only one of the two stereoisomers of monosaccharides. In most cases this is the D-form, which accounts for the fact that D-forms are much more common among cellular carbohydrates than L-forms.

Linkage of Monosaccharides into Disaccharides and Polysaccharides

The ring forms of monosaccharides can link together by twos to form *disaccharides* or, in greater numbers, to form *polysaccharides*. Some polysaccharides, such as plant starch and cellulose, consist of monosaccharides linked in unbranched chains. Others, such as glycogen, contain chains with forks or branches. Some polysaccharides, such as cellulose, contain only a single type of monosaccharide building block. Others, including the polysaccharide units of glycolipids and glycoproteins, are built up from a variety of different units.

Linkage of two glucose molecules to form the disaccharide *maltose* illustrates the general pattern of the reactions that assemble disaccharides and polysaccharides (Fig. 2-8). The linkage shown in Figure 2-8 extends

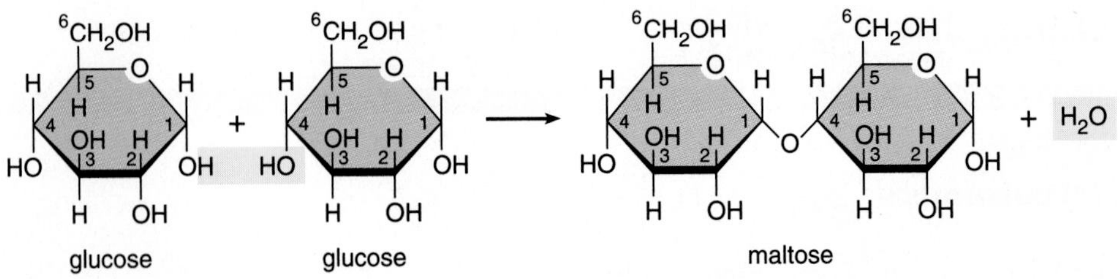

Figure 2-8 Combination of two glucose molecules to form the disaccharide maltose. In the reaction the elements of a water molecule are removed from the monosaccharides (shown in blue).

glucose   glucose   maltose

between the 1-carbon of the first glucose unit and the 4-carbon of the second glucose in maltose. Bonds of this type, which commonly link monosaccharides, are known as $1 \rightarrow 4$ linkages; $1 \rightarrow 2$, $1 \rightarrow 3$, and $1 \rightarrow 6$ linkages are also common. The linkage designation as alpha (α) or beta (β) depends on the orientation of the —OH group at the 1-carbon forming the bond. In maltose the —OH group is in the α-position. Therefore, the link between the two glucose subunits of maltose is an $\alpha(1 \rightarrow 4)$ linkage.

A series of reactions of the type shown in Figure 2-8 forms cellulose (Fig. 2-9a), except that the bonds holding the glucose subunits together are $\beta(1 \rightarrow 4)$ linkages. The glucose subunits of plant starch molecules (Fig. 2-9b) are held together by $\alpha(1 \rightarrow 4)$ linkages. Glycogen, the primary storage polysaccharide of animal cells, also contains glucose subunits held in chains by $\alpha(1 \rightarrow 4)$ linkages. In contrast to plant starch, glycogen molecules form branched structures in which the side branches are linked to main chains by $\alpha(1 \rightarrow 6)$ linkages (shaded in Fig. 2-9c). As noted, α-linkages are characteristic of sugars and starches that can be metabolized and used as an energy source by higher organisms; β-linkages occur in structural molecules such as cellulose that cannot be digested by most eukaryotic organisms.

Linkage of glucose units into chains in cellulose or starch molecules illustrates the general pattern of a *polymerization reaction*. This reaction type underlies as-

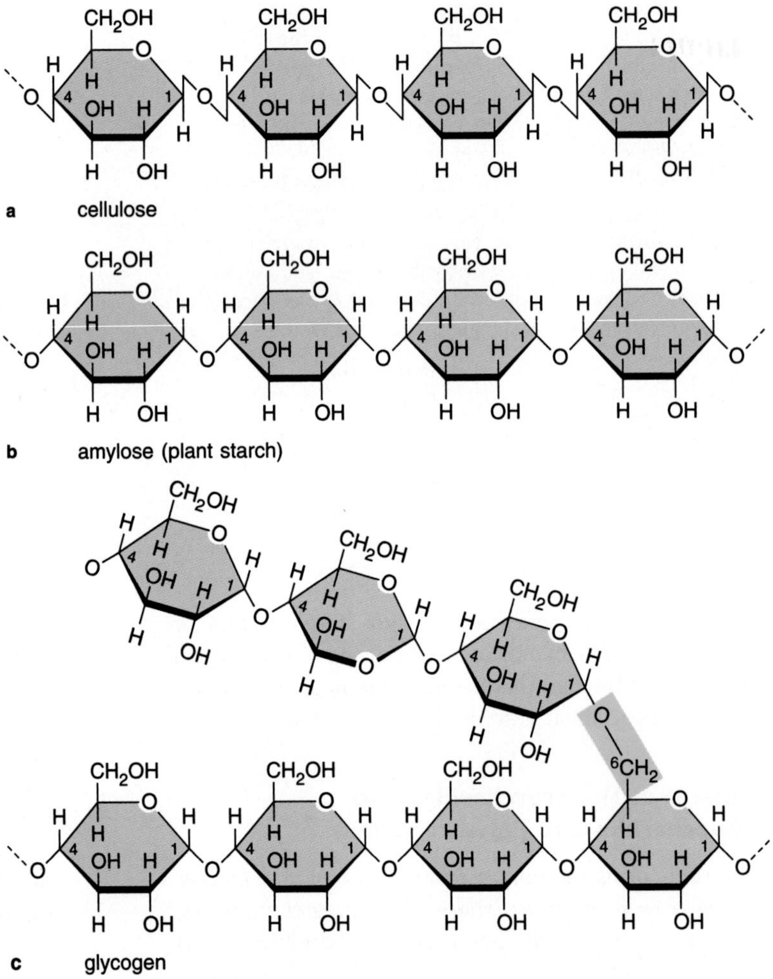

Figure 2-9 Three comon polysaccharides. **(a)** Cellulose, formed by end-to-end binding of β-glucose molecules in $\beta(1 \rightarrow 4)$ linkages. **(b)** Amylose (plant starch), formed by end-to-end binding of α-glucose molecules in $\alpha(1 \rightarrow 4)$ linkages. **(c)** Glycogen, a branched polysaccharide formed by α-glucose units joined by $\alpha(1 \rightarrow 4)$ and $\alpha(1 \rightarrow 6)$ linkages; the $\alpha(1 \rightarrow 6)$ linkage, which occurs at branches, is shown in blue in **(c)**.

a   cellulose

b   amylose (plant starch)

c   glycogen

sembly mechanisms in which many identical or nearly identical subunits, called *monomers,* join together like links in a chain to form a *polymer.* Thus, in cellulose formation, the individual glucose molecules are monomers and the finished cellulose molecule is the polymer. Enzymes that catalyze polymerization reactions are generally called *polymerases.*

Much of the earth's organic matter consists of (1) carbohydrates in the form of monosaccharides, (2) polysaccharides containing glucose units as the sole building block, or (3) polysaccharides containing glucose in combination with other hexoses. One of the polysaccharides containing glucose, cellulose, is probably the most abundant organic molecule on the earth. Most polysaccharides serve as a reservoir of stored chemical energy or provide structural support for cells. Other polysaccharides are linked to proteins in glycoproteins, which take part in an almost endless variety of reactions and function as enzymes and recognition, adhesion, receptor, and structural molecules. Figure 5-12 shows the branched or antennalike polysaccharide structure typical of many membrane glycoproteins. Similar carbohydrate "antennas" attached to lipids in glycolipids (see Fig. 2-17) occur commonly in cellular membranes, including the plasma membrane, chloroplasts, and mitochondria.

LIPIDS

Lipids are so diverse that they defy a simple, all-inclusive definition. Perhaps the most workable one is based on the solubility of lipids in nonpolar solvents. By this criterion lipids are biological substances that dissolve more readily in nonpolar solvents, such as acetone, ether, chloroform, and benzene, than in water. This solubility property reflects the fact that nonpolar groups make up all or nearly all segments of lipid structures, giving these molecules a hydrophobic character. This characteristic is of extreme importance in cells because lipids tend to associate into nonpolar groups and barriers, as in the cell membranes that form boundaries between and within cells. The hydrophobic regions established by lipids in membranes also provide a locale for interactions that proceed most favorably in a nonaqueous environment, such as some of the reactions of photosynthesis and cellular oxidations.

Besides having structural and functional roles in membranes, lipids are an important energy source. Other lipids form parts of cellular regulatory mechanisms. Lipids link covalently with carbohydrates to form glycolipids and with proteins to form *lipoproteins.* Both of these composite molecular groups play a wide variety of structural and functional roles in cells.

The lipid molecules carrying out these varied roles occur in one of three major forms: (1) *neutral lipids,* (2) *phospholipids,* and (3) *steroids.* Each of these lipid types is present in quantity in different cells and tissues.

Neutral Lipids

The neutral lipids, commonly found in cells as storage *fats* and *oils,* are so called because at cellular pH they bear no charged groups. Generally, they are completely nonpolar with no affinity for water or aqueous solutions. Almost all neutral lipids are a combination of *fatty acids* with the alcohol *glycerol.*

A fatty acid is a long, unbranched chain of carbon atoms with attached hydrogens and other groups (Fig. 2-10*a* and *b*). A carboxyl (—COOH) group at one end

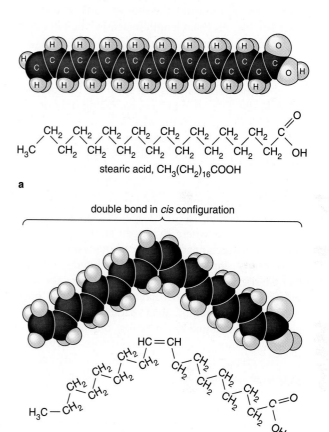

stearic acid, $CH_3(CH_2)_{16}COOH$

a

double bond in *cis* configuration

oleic acid, $CH_3(CH_2)_7CH=CH(CH_2)_7COOH$

b

double bond in *trans* configuration

c

glycerol, $HOCH_2CH(OH)CH_2OH$

d

Figure 2-10 The components of a neutral lipid or triglyceride, which include fatty acids and glycerol. **(a)** Stearic acid, a saturated fatty acid. **(b)** Oleic acid, an unsaturated fatty acid. The space-filling molecular model shows the "kink" introduced in the fatty acid chain by a double bond in the *cis* configuration. **(c)** The *trans* configuration of a double bond in a fatty acid, in which the chain is not conspicuously kinked. **(d)** Glycerol. The —OH groups (blocked in color) react with the —COOH groups of fatty acids to form triglycerides.

gives the molecule its acidic properties. Most naturally occurring fatty acids contain an even number of carbon atoms in their backbone chains. Although a few with odd numbers are found in all organisms, these make up only a minor fraction of the total.

The carbon chains of fatty acids vary in length from as few as 4 carbons to much longer structures containing 24 or more. Most of the fatty acids linked into neutral lipids have even-numbered backbone chains with 14 to 22 carbons; those with either 16 or 18 carbons occur most frequently. The polar —COOH group is enough to make the shortest fatty acid chains water soluble. As chain length increases, the fatty acid types become progressively less water soluble and take on oily or fatty characteristics.

Within a fatty acid chain hydrogen atoms are bound to the carbon atoms that form the backbone of the molecule. This combination, consisting only of carbons and hydrogens, forms what is known as a *hydrocarbon* (the prefix *hydro* refers to hydrogen, not water). If the maximum possible number of hydrogen atoms binds to the carbons in a fatty acid chain, the fatty acid is *saturated* (Fig. 2-10a). If hydrogen atoms are missing, so that the total number of hydrogens is less than the possible maximum, the fatty acid is *unsaturated* (Fig. 2-10b). At points where hydrogen atoms are missing from adjacent carbon atoms, the carbons share a double instead of a single bond. If double bonds occur at multiple sites (up to a maximum of about six), the fatty acid is *polyunsaturated*. Unsaturated fatty acids have lower melting points than saturated fatty acids and are more abundant in living organisms. (Table 2-5 lists common saturated and unsaturated fatty acids.)

The carbon chain of a fully saturated fatty acid is more or less straight, without major bends or kinks (see Fig. 2-10a). An unsaturated fatty acid may take one of two forms at a double bond. In the *cis* form the chain bends at an angle of about 30°, producing a definite kink (see Fig. 2-10b). In the *trans* form the chain con-

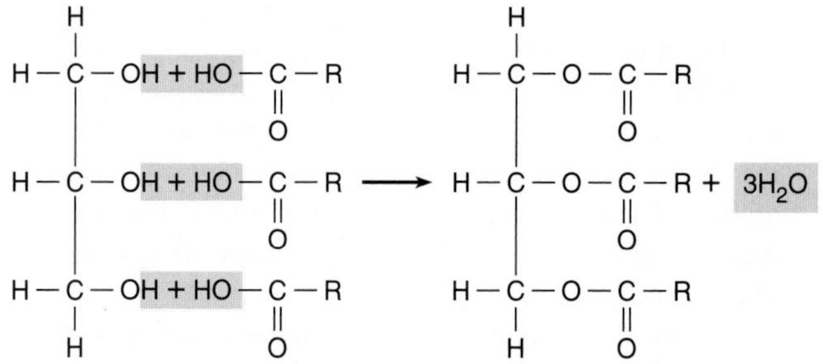

Figure 2-11 Combination of glycerol with three fatty acids to form a triglyceride. The R groups represent the carbon chains of the fatty acids. The components of a water molecule (shown in blue) are removed from the glycerol and fatty acids in each of the three bonds formed.

tinues in the same direction after the double bond (Fig. 2-10c). The pronounced kink of the *cis* form affects the packing of unsaturated fatty acid chains in cell structures, making the chains more disordered and consequently more fluid at biological temperatures. Unsaturated fatty acids in the *cis* form are much more common in cells than the *trans* form.

Besides fatty acids the other major component of neutral lipids—glycerol—has three —OH groups at which fatty acids may attach (shaded in Fig. 2-10d). If a fatty acid binds to each of the three sites, the resulting compound is a *triglyceride*. Most neutral lipids in living systems are triglycerides.

In the formation of triglycerides each of the three —OH sites of a glycerol molecule reacts with the —COOH group of a fatty acid (Fig. 2-11). The additions are condensation reactions (see Information Box 2-1), in which one water molecule is released as each linkage forms. (Fatty acids and other organic subunits, such as amino acids, are called *residues* after their linkage into larger molecules.) The three fatty acids linking to glycerol may occur in any combination: two or all three may be the same, as in the example shown in Figure

Table 2-5	Saturated and Unsaturated Fatty Acids		
Saturated Fatty Acids		Unsaturated Fatty Acids	
Butyric acid	$CH_3(CH_2)_2CO_2H$	Crotonic acid	$CH_3CH{=}CHCO_2H$
Caproic acid	$CH_3(CH_2)_4CO_2H$	Palmitoleic acid*	$CH_3(CH_2)_5CH{=}CH(CH_2)_7CO_2H$
Caprylic acid	$CH_3(CH_2)_6CO_2H$	Oleic acid*	$CH_3(CH_2)_7CH{=}CH(CH_2)_7CO_2H$
Capric acid	$CH_3(CH_2)_8CO_2H$	Linoleic acid*	$CH_3(CH_2)_3(CH_2CH{=}CH)_2(CH_2)_7CO_2H$
Lauric acid	$CH_3(CH_2)_{10}CO_2H$	Linolenic acid	$CH_3(CH_2CH{=}CH)_3(CH_2)_7CO_2H$
Myristic acid	$CH_3(CH_2)_{12}CO_2H$	Arachidonic acid	$CH_3(CH_2)_3(CH_2CH{=}CH)_4(CH_2)_3CO_2H$
Palmitic acid	$CH_3(CH_2)_{14}CO_2H$	Nervonic acid	$CH_3(CH_2)_7CH{=}CH(CH_2)_{13}CO_2H$
Stearic acid	$CH_3(CH_2)_{16}CO_2H$		
Arachidic acid	$CH_3(CH_2)_{18}CO_2H$		
Lignoceric acid	$CH_3(CH_2)_{22}CO_2H$		

* Occur as major fatty acids in human storage fats.

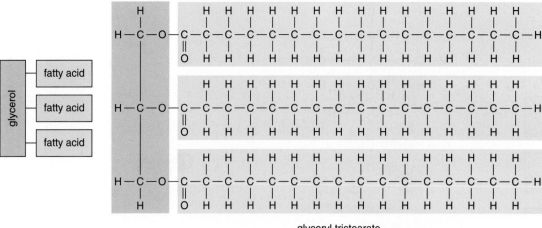

glyceryl tristearate

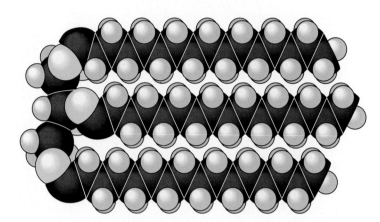

Figure 2-12 A triglyceride, formed by the reaction between glycerol and three fatty acids.

2-12, or they may be all different. As a result, many varieties of the basic triglyceride format are possible. In fishes, for example, at least 20 different fatty acids may link in various combinations with glycerol, giving more than 1500 possible triglycerides. Different species usually have distinctive combinations of fatty acid residues in their triglycerides, and the combinations may change somewhat depending on diet.

Generally, as with fatty acids, triglycerides become less fluid as the length of their fatty acid chains increases. Triglycerides that are liquid at biological temperatures are oils; those that are semisolid or solid are called fats. Most of the oils and fats inside cells are stored in droplets as energy reserves. *Waxes*, which contain very long-chain fatty acids in combination with glycerol or other alcohols, occur most frequently on the exterior surfaces of cells or cell walls, where they form a protective coating that resists water loss or invasion by infective agents.

Phospholipids

The primary lipids of biological membranes are phospholipids, a group of phosphate-containing molecules with structures related to the triglycerides. In the most common phospholipids, called *phosphoglycerides*, glycerol forms the backbone of the molecule but only two of its binding sites link to fatty acid residues. The third site links instead to a bridging phosphate group (Fig. 2-13). The carbon linked to the phosphate group is called the 3-carbon; the carbons attached to fatty acid residues are the 1- and 2-carbons. The other end of the phosphate bridge links to another organic subunit, most commonly a nitrogen-containing alcohol. Other organic subunits that may link at this position include the amino acids serine and threonine and a sugar, inositol (Fig. 2-14).

The phosphoglyceride in Figure 2-13, *phosphatidyl choline*, is a major lipid component of cellular membranes. Because different fatty acids may bind at the 1- and 2-carbons of the glycerol residue in phospholipids of this type, phosphatidyl choline is actually a family of closely related molecules differing in the particular fatty acids present. Other phospholipids (see pp. 154–158) differ from phosphoglycerides to a greater or lesser extent. All, however, contain a phosphate bridge linking a more or less complex polar unit to a segment containing nonpolar hydrocarbon chains.

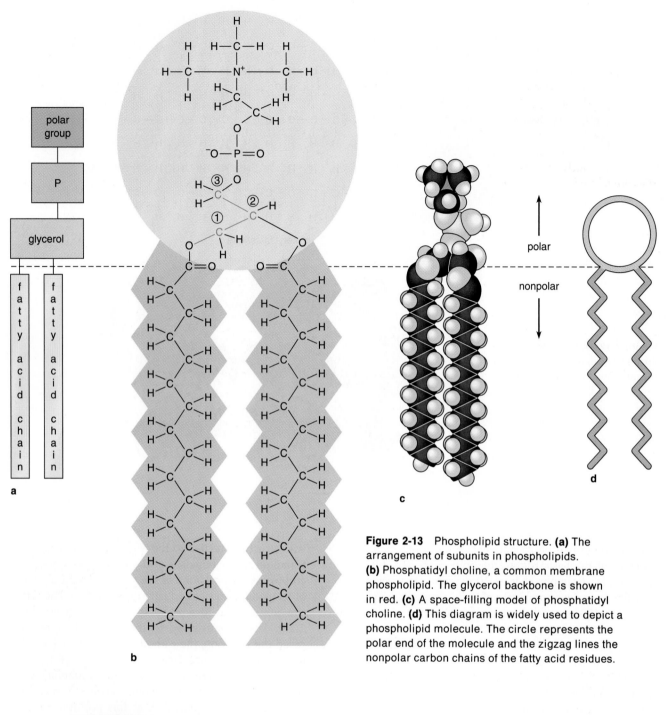

Figure 2-13 Phospholipid structure. **(a)** The arrangement of subunits in phospholipids. **(b)** Phosphatidyl choline, a common membrane phospholipid. The glycerol backbone is shown in red. **(c)** A space-filling model of phosphatidyl choline. **(d)** This diagram is widely used to depict a phospholipid molecule. The circle represents the polar end of the molecule and the zigzag lines the nonpolar carbon chains of the fatty acid residues.

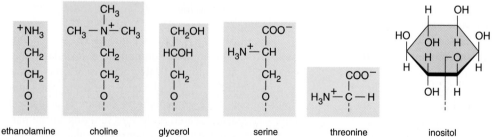

Figure 2-14 Organic subunits commonly linked to glycerol by a phosphate group in phospholipids. The site at which the subunit links to glycerol via a phosphate group is indicated by the dashed line.

Phospholipids have dual-solubility properties; that is, one end of the molecule is hydrophilic, the other hydrophobic. The end with the phosphate bridge and the nitrogenous alcohol or other group linked to it is polar and hydrophilic. Depending on the alcohol or other group present, the polar segment may carry a net charge at cellular pH. The end with long hydrocarbon chains, in contrast, is strongly nonpolar and hydrophobic. Such molecules with dual-solubility properties are termed *amphipathic* or *amphiphilic*.

Phospholipids take up arrangements in polar or nonpolar environments that satisfy their dual-solubility properties. When introduced into the interface formed when a nonpolar solvent such as benzene is layered on top of water, phospholipids orient so that their nonpolar hydrocarbon chains extend into the benzene and their polar groups extend into the water (Fig. 2-15a).

Phospholipids placed under the surface of water (or any strongly polar liquid) satisfy their dual-solubility requirements in an interesting and highly significant way, by forming layers just two molecules thick called *bilayers*. In a bilayer the phospholipid molecules orient so that the hydrocarbon chains associate in a nonpolar, hydrophobic region in the interior of the layer. The polar phosphate-alcohol groups face the surrounding water molecules (Fig. 2-15b).

A phospholipid bilayer is stable in water because the hydrophilic and hydrophobic parts of its molecules are suspended in environments with like solubility properties. Displacement of the phospholipids would expose their nonpolar regions to the surrounding aqueous medium or bury their polar segments in the nonpolar interior of the bilayer. Phospholipid bilayers, held together by polar and nonpolar associations, are the primary structural framework of cellular membranes. (Ch. 5 further discusses phospholipid types and the arrangement of phospholipids in cellular membranes.)

Steroids

Steroids comprise a class of lipids based on a framework of four interconnected carbon rings (Fig. 2-16a). The

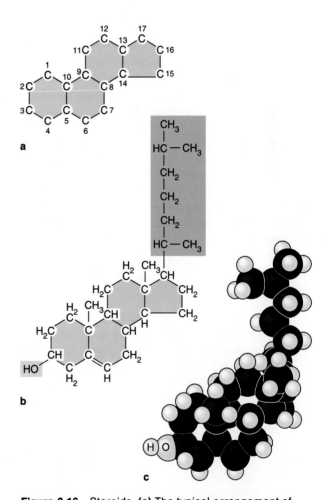

Figure 2-16 Steroids. **(a)** The typical arrangement of four carbon rings in a steroid molecule. **(b** and **c)** A sterol, cholesterol. Sterols have a long nonpolar side chain linked to the 17-carbon at one end and a single —OH group at the 3-carbon at the other end (boxed in red in **b**). The —OH group makes the end of the cholesterol molecule that carries it slightly polar. The remaining parts of the molecule are nonpolar.

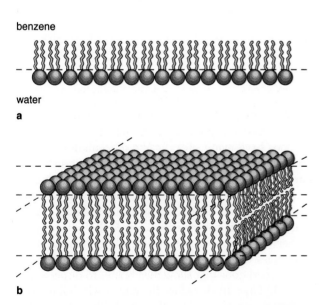

Figure 2-15 Arrangements satisfying the dual solubility property of phospholipids. **(a)** Phospholipid molecules introduced into the interface between benzene and water layers orient with their nonpolar tails extending into the benzene and their polar heads facing the water. **(b)** Phospholipid molecules completely surrounded by water assemble into bilayers with their nonpolar tails associated in the bilayer interior and their polar heads facing the surrounding water molecules.

various steroids differ in the position and number of double bonds in the rings and in the side groups attached to the rings. The most abundant group of steroids, the *sterols,* has a hydroxyl group linked to one end and a complex, nonpolar carbon chain at the opposite end of the ring structure (blocked groups, Fig. 2-16*b*). Although sterols are almost completely hydrophobic, the single hydroxyl group gives one end of the molecules a slightly polar, hydrophilic character. As a result the sterols are slightly amphipathic, and tend to orient in arrangements that satisfy their dual-solubility properties.

Cholesterol (Fig. 2-16*b* and *c*) is an important component of the plasma membrane in all animal cells; similar sterols occur in plant plasma membranes. When present in membranes, cholesterol loosens the packing of membrane phospholipids. Among other effects its interference with phospholipid packing maintains the fluidity of membranes at low temperatures. Deposits derived from cholesterol also form a major part of the material collecting inside arteries in the disease *atherosclerosis* (hardening of the arteries).

Other steroids are important as *hormones* in animals. Although steroid hormones occur in only trace amounts in animal tissues, they have regulatory effects far out of proportion to their concentrations. The male and female sex hormones of humans and other animals are steroid hormones; so are the hormones of the adrenal cortex that regulate cell growth and activity. (Details of steroid hormone regulation appear in Ch. 18.) Some steroids are toxic as, for example, those that occur in the venoms of toads and other animals.

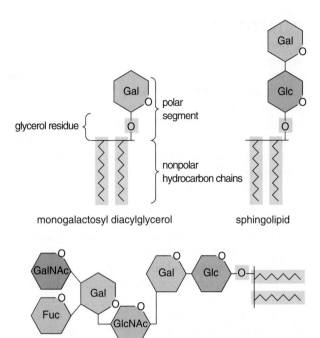

Figure 2-17 Some representative glycolipids in diagrammatic form. Monogalactosyl diacylglycerol and the sphingolipids occur in both plant and animal membranes; sphingolipids are especially abundant in brain and nerve cells in higher vertebrates. Glc, glucose; Gal, galactose; Fuc, fucose; GlcNAc, glucosamine; GalNAc, galactosamine. The structures of these sugars appear in Figure 5-5.

Glycolipids

Glycolipids are amphipathic lipids with one or more carbohydrate groups linked to their polar segments. The carbohydrate groups may be relatively simple and include only one or two monosaccharide building blocks. Or they may be complex, branched structures containing a variety of carbohydrate units linked in different patterns. (Fig. 2-17 diagrams several representative glycolipids.)

Glycolipids occur in the highest proportions in bacterial plasma membranes and in the thylakoid membranes of chloroplasts. Animal plasma membranes, depending on the species and cell type, contain glycolipids in varying amounts. The plasma membranes of most mammalian cells contain low concentrations of glycolipids with carbohydrate units ranging from very simple structures with only a few sugar units to complex "antennae" built up from as many as 30 different sugars. A few mammalian cell types, including cells of the brain and intestinal lining, contain glycolipids in relatively higher concentrations.

Glycolipids lend varied structural and functional properties to membranes. Because their carbohydrate groups have many sites capable of forming hydrogen bonds, glycolipids reinforce bilayer structure and occur in quantity in membranes subject to disruptive physical stress or chemical agents. For example, the glycolipids in the intestinal lining of mammals stabilize plasma membranes against the effect of bile salts, which would otherwise disrupt membrane structure. (Bile salts are examples of detergents, which generally create solubility conditions that allow membranes to disperse; see p. 163.) Glycolipids reduce the rate at which ions penetrate membranes. This effect probably accounts for their elevated concentrations in the plasma membranes of bacteria living in highly saline environments. Glycolipids also function as recognition markers at the surfaces of some mammalian cells. The ABO blood groups of humans, for example, depend on small differences in the carbohydrate groups linked to glycolipids or glycoproteins of red blood cell plasma membranes. (Fig. 2-17 shows a red blood cell glycolipid.)

PROTEINS

Proteins are large, complex molecules that carry out many vital functions in living organisms. As structural molecules they provide much of the cytoskeletal framework of cells; as enzymes they act as biological catalysts that speed the rate of cellular reactions; and as motile molecules they impart movement to cells and cell structures. They stabilize and control the activity of the two nucleic acids DNA and RNA, including the DNA of genes; form active parts of ribosomes; and provide the transport and recognition functions of cellular membranes. Many cells secrete proteins to the exterior in large quantities. Some secreted proteins, such as the collagen of animal cells and the wall proteins of plant cells, form parts of extracellular supportive frameworks. Others, including hormones and growth factors, transmit control signals from one cell to another; still others act as antibodies or digestive enzymes. Many toxins or venoms are based on secreted proteins.

The protein molecules carrying out these diverse functions are fundamentally similar in structure. All consist of one or more long, unbranched chains of subunits, the amino acids. Although only 20 different amino acids link initially in different combinations to make proteins, these may be modified to other forms after proteins are synthesized. As a result, as many as 150 different amino acids may occur in finished proteins. The smallest proteins contain as few as 3 to 10 amino acids, while the largest contain more than 50,000. The most common proteins are assembled from between about 50 and 1000 amino acids.

An almost endless variety of proteins is possible. At any internal point in the amino acid chains of different proteins, any one of the 20 amino acids may be present as a link in either modified or unmodified form. In a relatively small protein with only 50 amino acids this allows 20^{50} different sequences without modifying individual amino acids. This huge number is equivalent to one unique protein for every gram of matter in the universe! Modification of individual amino acids to other forms allows additional possibilities, so that the potential number of different proteins is essentially infinite. For any given protein, however, the amino acid sequence shows little or no variation; this sequence determines the properties and chemical activity of the protein type.

The Amino Acid Subunits of Proteins

With one exception the 20 amino acids assembled into proteins are based on the same structural plan (Fig. 2-18). These amino acids have a central carbon atom, the *alpha* (α) carbon, to which are attached, one on either side, an amino ($-NH_2$) and a carboxyl ($-COOH$) group. One of the two remaining bonds

of the central carbon is linked to a hydrogen atom, giving the

$$NH_2-\overset{\displaystyle |}{\underset{\displaystyle |}{C}}-COOH$$
$$H$$

structure common to these amino acids. The fourth bond of the α-carbon may be attached to any one of 19 different side chains, ranging in complexity from a single hydrogen atom in the simplest amino acid, *glycine*, to longer carbon chains or rings in other amino acids. Some of the more complex side chains contain oxygen, nitrogen, or sulfur in addition to carbon and hydrogen. The remaining amino acid, *proline*, an exception to these general rules, is based on a ring structure that includes the central carbon atom (see Fig. 2-18). However, proline has a $-COOH$ group at one side and an *imino* ($-NH-$) group forming part of the ring at the other, which reacts in essentially the same pattern as the $-NH_2$ groups of the other amino acids in the formation of proteins.

The types and configurations of atoms in the side chains give the individual amino acids distinct chemical properties. Some of the side chains include functional groups, such as $-NH_2$, $-OH$, $-COOH$, and $-SH$, which contribute to their chemical properties. These functional groups may interact with atoms, molecules, and ions located outside the protein or elsewhere on the amino acid chains of the same or different proteins. Many side groups have atoms in positions that can form hydrogen bonds. The various side groups differ in polarity, and some of the polar ones are charged or capable of acting as acids or bases at pH ranges characteristic of living cells.

Although each amino acid is distinctive, several fall into groups with similar properties. Glycine, alanine, valine, methionine, leucine, isoleucine, proline, phenylalanine, and tryptophan have side chains with no groups readily capable of entering into chemical reactions or forming hydrogen bonds; these amino acids are generally nonpolar and relatively inert chemically. Although cysteine is usually included among the nonpolar amino acids, its $-SH$ group gives it some capacity to form hydrogen bonds and actually places it on the polar/nonpolar borderline. Its $-SH$ group can also form disulfide ($-S-S-$) linkages by reaction with the $-SH$ group of another cysteine (Fig. 2-19). Serine, threonine, tyrosine, asparagine, and glutamine have groups capable of forming hydrogen bonds and are polar and uncharged. The $-OH$ groups of serine, threonine, and tyrosine are targets of enzymes that add phosphate groups as a part of major mechanisms regulating the activity of proteins in cells. Aspartic acid and glutamic acid both have terminal $-COOH$ groups

sequence also determines the number and position of disulfide linkages between cysteine residues, the location of proline residues, and the position at which other substances, such as metallic ions, lipids, and carbohydrates, may bind. Also important, and equally dependent on the amino acid sequence, are the positions of hydrogen bonds, attractions between positively and negatively charged side groups, van der Waals forces, and polar and nonpolar associations. The net effect of these factors is to establish a distinct three-dimensional shape for each protein of a unique sequence. This three-dimensional shape is the *tertiary structure* of a protein. (Fig. 2-25 shows the tertiary structure of several proteins.)

The three-dimensional shape of a protein represents a compromise between many opposing factors, such as regions at the surface or interior occupied by polar and nonpolar amino acid side groups, maximum possible packing to the limits of van der Waals radii, and opportunities for hydrogen bonding. On the average globular proteins fold to about 75% of the minimum volume expected from their van der Waals radii. This degree of packing makes a protein a solid structure, equivalent in rigidity to most organic solids. The folding of globular proteins typically moves a high percentage of the nonpolar groups to the interior in aqueous solutions. For example, in the much-studied ribonuclease T_1 molecule, whose three-dimensional structure was worked out in the laboratory of W. Saenger, the final folding conformation buries 87% of the nonpolar amino acids in the interior.

The compromise represented by the three-dimensional folding pattern is a state in which the smallest possible amount of energy is required to maintain the folded structure. Departures from this *minimum energy state* require an input of energy from the surrounding medium, such as an increase in temperature. Because the minimum energy state is a balance of opposing factors, the three-dimensional shape of many proteins remains sensitive to disturbance and may change readily.

In addition to temperature changes, alterations in the pH or salt concentration of the medium may contribute to changes in protein shape. Molecular collisions or the binding of ions or chemical groups may also cause changes in tertiary structure. These changes may involve alterations in the angle of rotation of atoms around a single bond, movements of segments of the amino acid chain, or vibrational or "breathing" movements of the entire protein. Any of these structural alterations is known as a *conformational change*.

The flexibility of protein structure contributes to the function of many proteins. Conformational changes are vital to protein functions such as enzymatic catalysis and motility. For example, in the molecular movements responsible for muscle contraction, a major segment of the protein *myosin* undergoes conformational changes that cycle it through oarlike motions (see Fig. 12-12).

The oarlike movement, pushing against other proteins forming parts of muscle, results in the muscle contractions responsible for voluntary and involuntary movements in animals.

Conformational changes can be detrimental under extreme conditions. Extreme changes in temperature or pH can seriously disrupt the internal bonds and attractions that hold proteins in their tertiary conformations, causing the molecules to unfold or refold into random shapes. Excessive heat increases the motions of amino acid side groups until the relatively weak attractions of hydrogen bonds and van der Waals forces can no longer hold them in place. The resultant unfolding is one reason why few living organisms can tolerate temperatures above 45° to 55°C. Changes in pH alter the charge of amino acid side groups, destroying or weakening ionic bonds that hold polypeptide chains in their three-dimensional form. If the unfolding is extensive enough to seriously alter the functional activity of the protein, the effects are called *denaturation*.

Some proteins can return to their native folded form if the temperature or pH returns to natural values, so that denaturation is temporary. In others the denaturation is permanent. The most familiar example of permanent denaturation is the cooking of an egg white, in which a clear, soluble solution of the protein albumin is hardened into an insoluble white mass.

Disulfide linkages are particularly important in limiting the degree of protein denaturation by heat and other factors. These linkages prevent amino acid chains from completely unfolding, and frequently hold them in intermediate conformations that greatly facilitate refolding to a native conformation. The extra stability conferred by disulfide linkages probably accounts for their characteristic occurrence in proteins that are released to the more disruptive conditions in the cell exterior, such as the collagen proteins forming parts of extracellular supports in some types of animal cells.

Frequently, the tertiary structure of proteins contains large blocks of amino acids folded into distinct subregions called *domains* (Fig. 2-27). A *cleft* or *crevice* spanned only by a segment of random coil usually separates domains. In many enzymatic proteins the regions catalyzing chemical reactions are located in clefts or crevices between domains. The clefts and crevices provide unique environments containing charged, acidic, or basic amino acid side groups, or patterns of hydrogen bonding. These environments promote reactions that would proceed only very slowly in free and open solutions.

In proteins with multiple functions, such as an enzyme capable of catalyzing several different reactions, the individual functions are often associated with different domains (as in the protein segment diagrammed in Fig. 2-27). Thus, protein domains often represent subdivisions of function as well as structure. Multifunctional proteins that have some of their functions

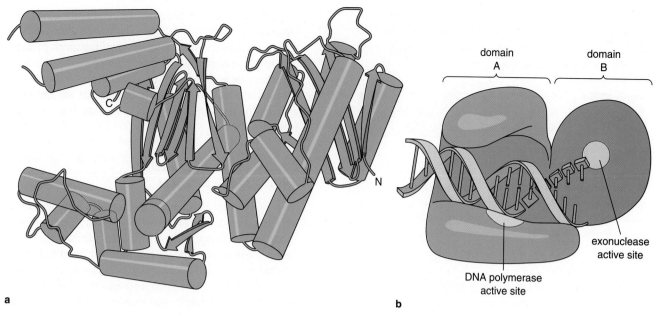

a

b

Figure 2-27 The domain structure of a part of the *E. coli* DNA polymerase enzyme. **(a)** The domains, with cylinders representing alpha-helical segments and arrows representing beta strands. **(b)** A three-dimensional drawing from the same view showing only the domain surfaces. This part of the enzyme catalyzes both the polymerization of DNA molecules from their nucleotide subunits (called *DNA polymerase activity*) and the reverse reaction, which removes nucleotides from the ends of a nucleotide chain (called *DNA exonuclease activity*). The DNA polymerase activity is centered in the large horizontal groove in domain A; the exonuclease activity occurs in domain B. Redrawn from originals courtesy of T. A. Steitz, from *Trends Biochem. Sci.* 12:288 (1987).

in common frequently share domains associated with the similar functions. The shared domains suggest that these related but different proteins may have evolved through a mechanism that mixed existing domains into new combinations. (A discussion of the mechanism by which this pattern of evolution may have taken place appears on p. 594.)

Quaternary Structure: Associations of Multiple Polypeptide Chains

Many complex proteins, such as hemoglobin and antibody molecules, are built up from several polypeptide chains. The same bonds and forces folding individual amino acid chains into tertiary structures hold the chains together. (See Fig. 2-19c, which shows the disulfide linkages stabilizing the quaternary structure of antibody molecules.) Hemoglobin and antibody molecules (Fig. 2-28) both consist of 4 individual polypeptide chains; some proteins have as many as 10 or 12. The pattern in which the individual chains of a multichain protein are held together is called its *quaternary structure*.

Fibrous proteins frequently contain specialized structures conferring high elasticity and tensile strength

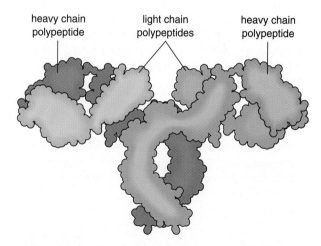

Figure 2-28 Quaternary structure in the hemoglobin molecule, assembled from four polypeptide chains—two α-chains and two β-chains. The four chains are stabilized and held in place by the pattern of disulfide linkages shown in Figure 2-19c.

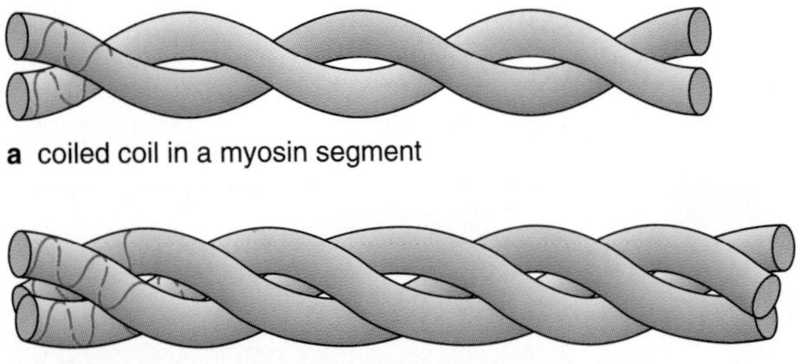

a coiled coil in a myosin segment

b triple helix in a collagen segment

Figure 2-29 Coiled structures of fibrous proteins.
(a) The coiled-coil structure in a segment of the myosin molecule in which two polypeptide chains wind in a double helix. **(b)** The triple helix of polypeptide chains in a collagen molecule.

that are intimate combinations of tertiary and quaternary structure. One of these is the *coiled coil,* a structure in which two polypeptide subunits consisting of long alpha-helical segments wind around each other in a double spiral (Fig. 2-29*a* and Information Box 2-4). Coiled coils occur, for example, in several of the fibrous proteins of muscle tissue. A *triple helix* of three alpha-helical polypeptide segments makes up much of the structure of collagen, the primary fibrous protein of the extracellular matrix in animals (Fig. 2-29*b*; for further details, see Fig. 8-1 and p. 274).

Proteins thus have four levels of structure (summarized in Table 2-6). The sequence of amino acids in a protein constitutes its primary structure. The arrangement of segments in the amino acid chain into the alpha helix, beta strand, or random coil defines the secondary structure of a protein. Folding of the chain upon itself produces the tertiary or three-dimensional structure of a protein. Quaternary structure refers to the number and arrangement of individual polypeptide chains within a complex protein containing more than one chain.

Many types of proteins associate noncovalently by twos as *dimers,* by threes as *trimers,* by fours as *tetramers,* and so on. Many of the transport proteins of the plasma membrane, for example, take the form of dimers or higher levels of association. In these cases the activity of the protein is dependent on the association; the proteins are inactive, or largely so, in single form.

Combinations of Proteins with Other Substances

The many reactive groups on the side chains of amino acid residues allow proteins to enter a wide range of biochemical interactions with other substances. Covalent linkage of proteins with carbohydrate or lipid units produces composite molecules with great importance to cellular functions. Glycoproteins—formed by covalent binding of protein and carbohydrate units—function as enzymes, antibodies, recognition and receptor molecules at the cell surface, and as parts of extracellular supports such as collagen. All the known proteins located at the cell surface or in the spaces between cells, in fact, are glycoproteins. The attached carbohydrate groups, because of their many sites capable of forming hydrogen bonds, increase conformational stability. The hydrogen-bonding capacity also gives carbohydrate groups the ability to take up and partially immobilize water molecules in large quantity. This property makes some glycoproteins highly effective as extracellular cushions or lubricants. The carbohydrate groups of glycoproteins increase resistance to enzymatic attack; in some glycoproteins carbohydrates provide groups recognized by receptors at the cell surface.

Lipoproteins—formed by covalent linkage of proteins to lipid units—are widely distributed among prokaryotic and eukaryotic cells. At least 50 known eukaryotic proteins have covalent linkages to lipids, including individual fatty acids or entire phospholipids.

Table 2-6	Structural Nomenclature of Proteins	
Structural Level	Definition	Primary Bonds of Structural Level
Primary structure	Sequence of amino acids.	Covalent bonds of peptide linkages
Secondary structure	Folding of amino acid chain into conformations such as the alpha helix, beta strand, and random coil.	Hydrogen bonds
Tertiary structure	Complete three-dimensional folding pattern of a protein containing a single polypeptide chain.	Hydrogen bonds, disulfide linkages, polar and nonpolar associations, van der Waals forces
Quaternary structure	Complete three-dimensional folding pattern of a protein containing two or more polypeptide chains.	Same as tertiary structure

The Coiled Coil

The coiled coil, which occurs in fibrous protein elements such as the long, taillike extension of the myosin molecule, consists of two alpha-helical amino acid chains twisted around each other to form a double helix. This highly stable structure depends on a regularly repeated sequence of seven amino acids that is characteristic of proteins of this type. In the repeated sequence, designated *a-b-c-d-e-f-g*, amino acids *a* and *d* in the sequence are nonpolar, *g* is positively charged, and *e* is negatively charged (see part *a* of the figure in this box). The twist of the coiled-coil helix brings the hydrophobic amino acid groups together in the region of contact between the two alpha helices, creating a hydrophobic association that stabilizes the twist (see part *b* of the figure in this box). The association is further stabilized by attractions between the positively and negatively charged amino acid groups, which are also held in a position favoring an electrostatic attraction between these residues. The outermost surfaces of the coiled coil, formed by the amino acids at the *b*, *c*, and *f* positions in the repeated sequences, are hydrophilic and stabilize the coiled coil in its interactions with the surrounding aqueous medium.

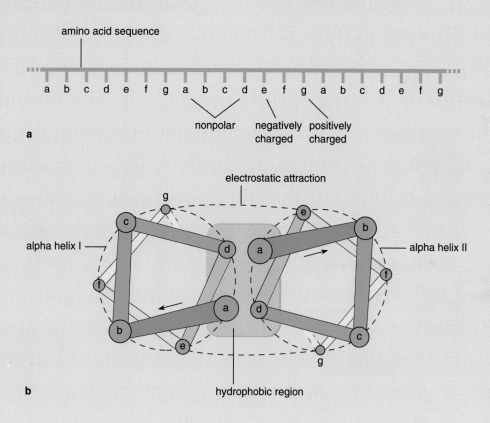

Some have covalent linkages to glycolipids, producing a complex, three-way structure that includes covalently linked protein, lipid, and carbohydrate units. In most of these complexes the lipid units link to either the N- or C-terminal end of the protein chain (see Fig. 5-28). Most lipoproteins are in cellular membranes, particularly in the plasma membrane, where the lipid segment appears to provide a covalently bound anchor tying the protein to the membrane.

Proteins interact with inorganic ions, such as sodium, potassium, zinc, copper, iron, iodine, magnesium, and calcium, through electrostatic attractions set up by charged amino acid residues. The presence of one or more of these ions is necessary for full activity of many proteins, particularly enzymes. Individual amino acid side chains covalently bind inorganic and organic groups, including hydroxyl, methyl ($-CH_3$), acetyl ($-COCH_3$), and phosphate groups. Addition of these groups alters amino acids to other forms, and often modifies or regulates the activity of the entire protein. The addition and removal of calcium ions or phosphate groups are particularly important in modifying the activity of proteins, and form part of pathways regulating cellular activities as diverse as cell division and muscle contraction. The reversible addition of phosphate groups to serine, threonine, and tyrosine side groups is a particularly important mechanism regulating the activity of proteins.

Proteins, in addition, combine either covalently or noncovalently with larger, more complex organic structures such as the pyrrole ring or nucleotides (Fig. 2-30). In most instances the organic structure adds a vital function to the protein, such as the ability to accept and release electrons in oxidative reactions.

Linked ions or other substances that contribute to the function of an enzyme or other protein are called *cofactors*. Large, complex organic structures are often called *coenzymes* or *prosthetic groups*. Many complex coenzymes or prosthetic groups of enzymatic proteins are derived from vitamins.

NUCLEOTIDES AND NUCLEIC ACIDS

The two nucleic acids, *deoxyribonucleic acid (DNA)* and *ribonucleic acid (RNA)*, are the informational molecules of all living organisms. Besides storing or transmitting information, RNA forms structural parts of units such as the ribosome, and in some systems has a catalytic function. Both DNA and RNA are long, chainlike polymers assembled from repeating subunits, the *nucleotides.* The sequence of nucleotides in informational nucleic acid molecules makes up a code that stores and transmits the directions required for assembling all types of proteins.

Individual nucleotides carry out a variety of biological functions. Many nucleotides or molecules built on nucleotides transport chemical energy in the form of phosphate groups or electrons from one reaction system to another. (The prosthetic groups FMN and

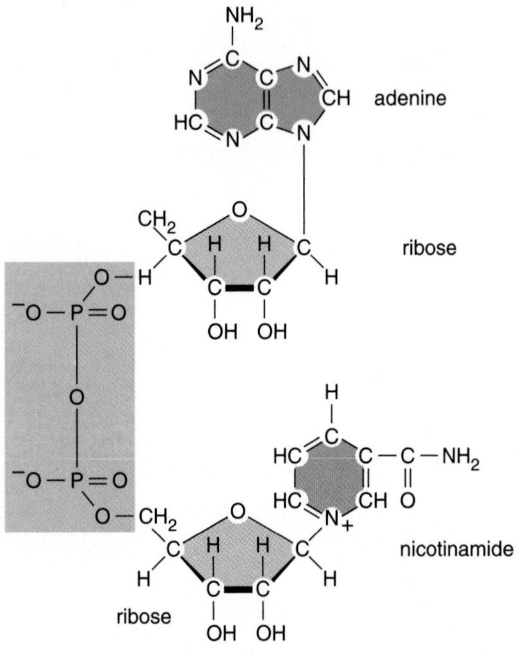

nicotinamide adenine dinucleotide (NAD)

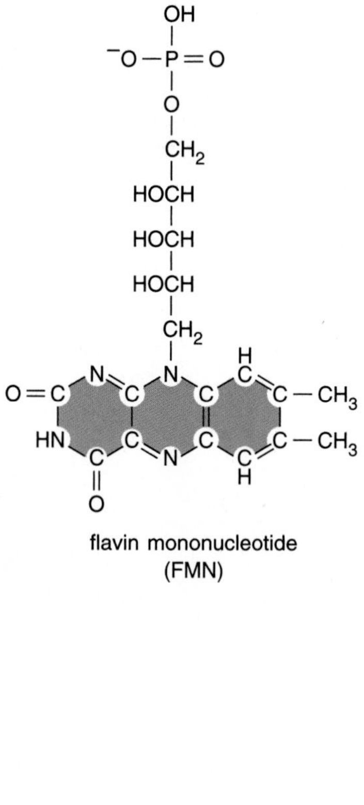

flavin mononucleotide
(FMN)

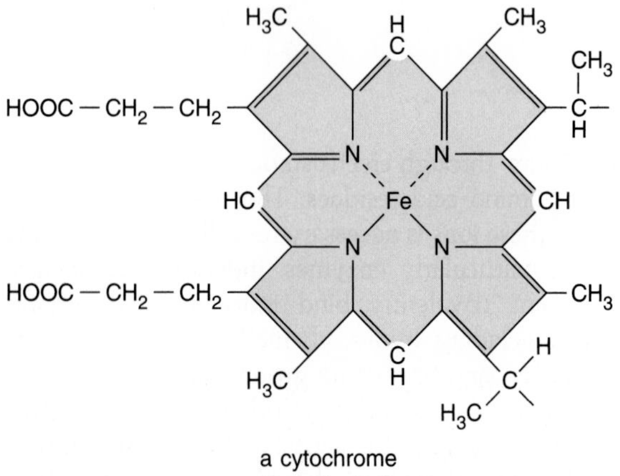

a cytochrome
(pyrrole ring)

Figure 2-30 Some complex organic groups that link covalently to proteins as coenzymes or prosthetic groups. Each of these groups functions as an electron carrier in biological oxidative reactions.

NAD shown in Fig. 2-30 are nucleotides.) Others carry metabolites, such as acetyl groups, between reactions. Still other nucleotides in cyclic form (see Fig. 7-5) act as important regulatory substances in cells.

The Nucleotides

A nucleotide (Fig. 2-31) consists of three parts: (1) a nitrogen-containing base, (2) a five-carbon sugar, and (3) one or more phosphate groups, all linked together

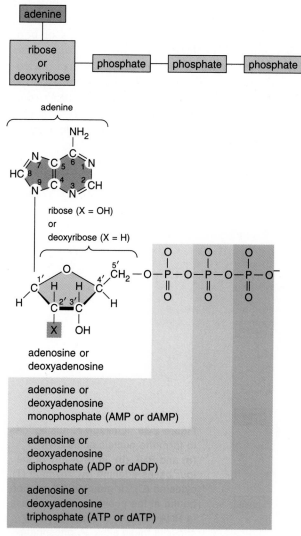

Other nucleotides:
containing guanine: guanosine or deoxyguanosine mono-, di-, or triphosphate
containing cytosine: cytidine or deoxycytidine mono-, di-, or triphosphate
containing thymine: thymidine mono-, di-, or triphosphate*
containing uracil: uridine mono-, di-, or triphosphate*

*Thymidine occurs only in DNA in the deoxyribose form; uridine occurs only in RNA in the ribose form.

Figure 2-31 Structural plan of the nucleotides (see text).

by covalent bonds. The nitrogenous bases in naturally occurring nucleotides, *pyrimidines* and *purines* (Fig. 2-32), are ring-shaped molecules containing both carbon and nitrogen atoms. Pyrimidines contain one carbon-nitrogen ring, and purines contain two rings. Three pyrimidine bases, *uracil (U)*, *thymine (T)*, and *cytosine (C)*, and two purine bases, *adenine (A)* and *guanine (G)*, are assembled into nucleic acids in cells.

Nitrogenous bases link covalently to one of two five-carbon sugars, either *ribose* or *deoxyribose*, in nucleotides. The two sugars differ only in the chemical group bound to their 2'-carbons (blue box in Fig. 2-31; the carbons in the sugars are written with primes—1', 2', 3', and so on—to distinguish them from the carbons of the bases, which are written without primes). Ribose has an —OH group at this position and deoxyribose a single hydrogen. A chain of one, two, or three phosphates links to the ribose or deoxyribose sugar at its 5'-carbon to complete the mono-, di-, or triphosphate form of a nucleotide.

The names used for nucleotides can be confusing. The term nucleo*tide* refers to a complete unit containing all three subunits: a nitrogenous base, a five-carbon sugar, and one or more phosphates. A unit consisting

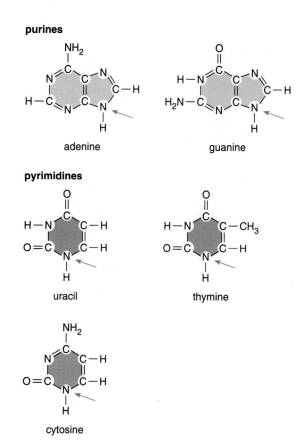

Figure 2-32 The purine and pyrimidine bases of nucleic acids and nucleotides. The arrows indicate where the base links to ribose or deoxyribose sugars in the formation of nucleotides.

of only the base and sugar without phosphates is called a nucleo*side*, and is named according to its nitrogenous base. For example, the base-sugar complex containing adenine and ribose is called *adenosine*; if deoxyribose is the sugar in the complex, it is called *deoxyadenosine*. To allow the phosphate groups to be numbered, complete nucleotides are usually named by considering them as nucleosides with added phosphates. The adenine-ribose complex with one phosphate, for example, is called *adenosine monophosphate*, or *AMP*; with two phosphates, *adenosine diphosphate*, or *ADP*; with three, *adenosine triphosphate*, or *ATP*. All the remaining nucleoside mono-, di-, and triphosphates use equivalent names and abbreviated forms (Fig. 2-31). In abbreviations, a lowercase *d* prefix indicates that the deoxyribose form of the five-carbon sugar is present, as in dATP or dGTP.

DNA and RNA

Nucleotides link together into long *nucleotide chains* to form the two nucleic acids DNA and RNA (Fig. 2-33). The chains consist of nucleotides held together by a bridging phosphate extending between the 5'-carbon of one sugar and the 3'-carbon of the next sugar in line. This arrangement produces a backbone chain of alternating sugar and phosphate groups.

The nucleotides of DNA chains each contain the sugar deoxyribose and one of the four bases adenine, thymine, guanine, or cytosine. Each nucleotide in an RNA chain contains the sugar ribose and one of the four bases adenine, uracil, guanine, or cytosine. Thus, DNA and RNA differ in the sugar present, either ribose or deoxyribose, and the presence of either thymine in DNA or uracil in RNA. (Thymine and uracil differ only in a methyl group linked to the ring in thymine but absent in uracil; see Fig. 2-32.)

The fully processed and finished forms of both DNA and RNA contain a number of *modified bases* formed by chemical alteration of the original nucleotides to other types. (Fig. 15-4 shows some of the modified bases.) Modified bases are particularly common in some types of RNA. As many as 10% to 15% of the bases in tRNA molecules, for example, may be modified to other forms.

DNA exists in cells as a *double helix* containing two intertwined chains of nucleotides (Fig. 2-34). In the DNA double helix, as first proposed by J. D. Watson

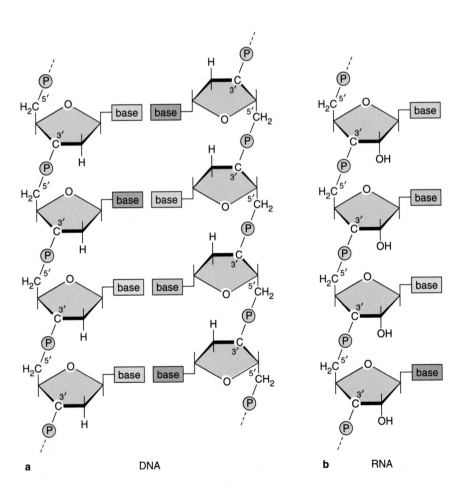

Figure 2-33 Linkage of nucleotides to form the nucleic acids DNA **(a)** and RNA **(b)**. In DNA any of the four bases adenine (A), thymine (T), cytosine (C), or guanine (G) may be bound at the positions marked *base*. In RNA A, G, C, or uracil (U) may occur at these sites. P, phosphate group.

a   DNA   b   RNA

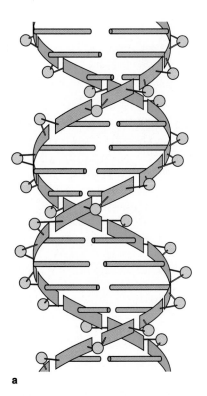

a

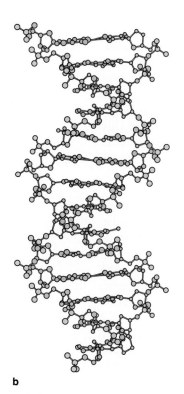

b

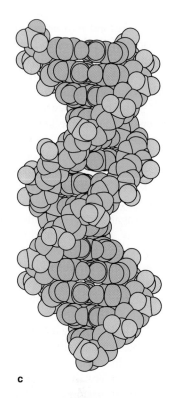

c

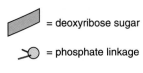

= deoxyribose sugar

= phosphate linkage

= base pair

Figure 2-34 The DNA double helix. **(a)** The arrangement of sugars, phosphate groups, and bases in the DNA double helix. **(b)** Positions of atoms and bonds in the DNA double helix. The paired bases, which lie in flat planes, are seen on edge in this view. **(c)** Space-filling diagram of the DNA double helix in the B conformation. **(b** and **c)** Redrawn from originals, courtesy of W. Saenger and Springer-Verlag, from *Principles of Nucleic Acid Structure*, 1984.

and F. H. C. Crick in 1953, the sugar-phosphate backbones of the two chains twist together in a right-handed direction to form the double spiral. The backbone chains, which are located at the surface of the double helix, are separated from each other by a regular distance that is filled in by the nitrogenous bases. The bases extend inward from the sugars toward the axis of the helix as *base pairs*, which lie stacked in flat planes roughly perpendicular to the long axis of the helix. Figure 2-34*b* shows these planes from the side (the horizontal pairs). Figure 2-35 shows the base pairs as viewed from one end of the molecule. Each complete turn of the double helix includes ten base pairs.

The space separating the sugar-phosphate backbones of a DNA double helix is just wide enough to accommodate a purine-pyrimidine base pair. Purine-purine pairs are too wide, and pyrimidine-pyrimidine pairs too narrow to fit this space exactly. The shapes of the bases and the locations of groups capable of forming stabilizing hydrogen bonds impose further restrictions on base pairing. Note from Figure 2-35 that three hydrogen bonds can form between guanine and

adenine   thymine

guanine   cystosine

Figure 2-35 The A-T (adenine-thymine) and G-C (guanine-cytosine) base pairs of DNA. Dotted lines designate hydrogen bonds.

cytosine, and two between adenine and thymine. The shapes of the bases and their sites available for hydrogen bonding effectively restrict the pairing to the guanine-cytosine (G-C) and adenine-thymine (A-T) pairs shown in Figure 2-35.

The two nucleotide chains run in opposite directions in the double helix and are thus *antiparallel.* This arrangement is most easily understood if the two chains are unwound and laid out flat, as illustrated in Figure 2-33a. The phosphate linkages in the chain on the left, if traced from the bottom to the top, extend from the 5'-carbon of the sugar below to the 3'-carbon of the sugar above each linkage. On the other chain the 5' → 3' linkages run in the opposite direction, from top to bottom. This feature of DNA structure has great significance for both DNA replication and RNA transcription, since a new DNA or RNA chain being copied must run in the opposite direction from its template (see Fig. 15-2).

Three primary intermolecular forces hold the DNA double helix together. One is hydrogen bonding between the base pairs in the interior of the molecule. Cumulatively, the hydrogen bonds form a stable structure if the helix includes at least 10 base pairs. Attractive van der Waals forces between the closely packed atoms of the double helix, particularly between atoms in the tightly stacked base pairs, provide the second stabilizing force. The third stabilizing force results from hydrophobic associations among base pairs in the interior of the helix. The nitrogenous bases, which are primarily nonpolar, pack tightly enough to exclude water and form a stable, primarily nonpolar environment in the helix interior.

The double-helical structure deduced by Watson and Crick, known as the *B conformation* (see Fig. 2-34c), is one of several possible conformations taken by DNA. The double helix in four of its other conformations, known as the *A, C, D, and E conformations*, also winds to the right but differs in details such as total diameter, the distance separating backbone chains, and the spacing and packing of base pairs (Fig. 14-5 shows the A conformation). In a fifth, the *Z conformation,* the double helix winds to the left (see Fig. 14-6). Of these only the A, B, and Z conformations are common in cells.

RNA exists largely as single, rather than double, nucleotide chains in living cells. However, segments of RNA molecules may pair temporarily in double-helical form or may fold back on themselves to set up extensive double-helical regions (Fig. 2-36). These fold-back double helices and their arrangement are often more important to the functions of RNA molecules than their nucleotide sequence, particularly in noncoding RNAs such as those of ribosomes. RNA molecules also wind temporarily with DNA to form hybrid double helices consisting of one DNA and one RNA chain. Such

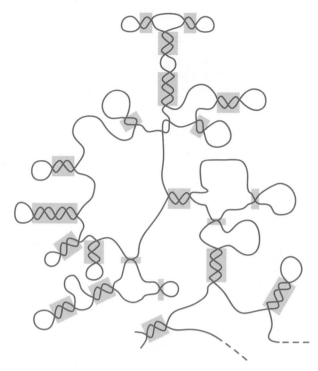

Figure 2-36 Pattern of fold-back double helices in one segment of an rRNA molecule from the bacterium *E. coli.* The double-helical regions take the A conformation typical of RNA molecules.

hybrid helices are set up, for example, during transcription of RNA molecules from DNA templates (see Fig. 15-2). Both the fold-back RNA double helices and the hybrid RNA-DNA double helices take the A conformation.

Like the three-dimensional structures of proteins, the DNA and RNA double helices are sensitive to disturbance by conditions in the surrounding medium. Particularly important as a disturbing factor is elevated temperature. Rising temperatures progressively disturb hydrogen bonds and other weak forces holding the backbones together until the two chains come apart and unwind. This denaturation, or *melting* as it is called, takes place in most DNA molecules at temperatures of about 50° to 60°C. Because an additional hydrogen bond stabilizes G-C base pairs, DNA molecules or segments containing higher proportions of G-C pairs are more stable and melt at higher temperatures than those rich in A-T pairs. In many cases DNA or RNA molecules can wind back into native form if temperatures return to normal values. Unwinding and rewinding DNA and RNA molecules by adjustments in temperature is a frequent experimental technique in cell and molecular biology (for details, see p. 528).

For Further Information

Carbohydrates
    in cell walls, *Ch. 8*
    formation in photosynthesis, *Ch. 10*
    oxidation, *Ch. 9*
    as subunits of membrane glycoproteins and glycolipids, *Chs. 5 and 7*
Glycolipids and glycoproteins, *Chs. 5 and 7*
Lipids
    in chloroplasts, *Ch. 10*
    in membranes, *Ch. 5*
    in mitochondria, *Ch. 9*
    oxidation, *Ch. 9*
Lipoproteins, *Chs. 5, 7, and 8*
Nucleic acids
    in the cell nucleus, *Ch. 14*
    in mitochondria and chloroplasts, *Ch. 21*
    in protein synthesis, *Ch. 16*
    replication, of DNA, *Ch. 23*
    transcription, of RNA, *Ch. 15*
Nucleotides in photosynthesis and oxidation, *Chs. 9 and 10*
Proteins
    in cell membranes, *Ch. 5*
    and enzymatic catalysis, *Ch. 3*
    in membrane transport, *Ch. 6*
    in motility, *Chs. 11 and 12*
    synthesis, *Ch. 16*

Suggestions for Further Reading

General Books and Articles

Brown, T. L., and Lemay, H. E. 1985. *Chemistry: The central science.* 3rd ed. New York: Prentice-Hall.

Eisenberg, D., and Kauzmann, T. 1969. *The structure and properties of water.* New York: Oxford University Press.

Lehninger, A. L. 1982. *Principles of biochemistry.* New York: Worth.

Pauling, L. 1960. *The nature of the chemical bond.* 3rd ed. New York: Cornell Univ. Press.

Stillinger, F. H. 1980. Water revisited. *Science* 209:451–57.

Stryer, L. 1988. *Biochemistry.* 3rd ed. New York: Worth.

Zubay, G. 1983. *Biochemistry.* Reading, Mass.: Addison-Wesley.

Carbohydrates

Ginsburg, V., and Robbins, P., eds. 1984. *Biology of carbohydrates.* New York: Wiley.

Sharon, N. 1980. Carbohydrates. *Sci. Amer.* 243:90–114 (November).

Lipids

Quinn, P. J. 1976. *The molecular biology of cell membranes.* London: University Park Press.

Proteins

Blake, C. C. F. 1984. Protein structure. *Trends Biochem. Sci.* 10:421–425.

Branden, C., and Tooze, J. 1991. *Introduction to protein structure.* New York: Garland.

Chothia, C., and Lesk, A. 1985. Helix movements in proteins. *Trends Biochem. Sci.* 10:116–118.

————, and Finkelstein, A. V. 1990. The classification and origins of protein folding patterns. *Ann. Rev. Biochem.* 59:1007–930.

Creighton, T. E. 1984. *Proteins, structures and molecular properties.* New York: W. H. Freeman.

————. 1988. Disulfide bonds and protein stability. *Bioess.* 8:57–63.

Dickerson, R. E., and Geis, I. 1969. *The structure and action of proteins.* New York: Harper & Row.

Doolittle, R. F. 1985. Proteins. *Sci. Amer.* 253:88–99 (October).

Frauenfelder, H.; Sligar, S. G.; and Wolynes, P. G. 1991. The energy landscapes and motions of proteins. *Science* 254:1596–1603.

Karplus, M., and McCammon, J. A. M. 1986. The dynamics of proteins. *Sci. Amer.* 254:42–51 (April).

Pace, C. N. 1990. Conformational stability of proteins. *Trends Biochem. Sci.* 15:14–17.

Schultz, G. E., and Schirmer, R. H. 1979. *Principles of protein structure.* New York: Springer-Verlag.

Nucleic Acids

Adams, R. L. P.; Burdon, R. H.; Campbell, A. M.; Leader, D. P.; and Smellie, R. M. S. 1981. *The biochemistry of the nucleic acids.* 9th ed. New York: Chapman and Hall.

Dickerson, R. E. 1983. The DNA double helix and how it is read. *Sci. Amer.* 249:94–111 (October).

Felsenfeld, G. 1985. DNA. *Sci. Amer.* 253:58–66 (October).

Rich, A., and Kim, S.-H. 1978. The three-dimensional structure of a transfer RNA. *Sci. Amer.* 238:52–62 (January).

Saenger, W. 1984. *Principles of nucleic acid structure.* New York: Springer-Verlag.

Review Questions

1. What determines the chemical activity of an atom?

2. What is an ionic, or electrostatic, bond? How is this bond formed?

3. What is a covalent bond? How are covalent bonds formed? How are covalent bonds represented in molecular diagrams?

4. What are polar and nonpolar molecules? What conditions produce polarity?

5. What do the terms hydrophobic and hydrophilic mean? How is the property of being hydrophobic or hydrophilic related to polarity?

6. How is polarity important in molecular and cellular structure?

7. What is a hydrogen bond? What conditions are necessary for formation of a hydrogen bond? In what ways are hydrogen bonds related to polarity?

8. What are van der Waals forces? How can these forces be either attractive or repulsive?

9. Diagram a hydroxyl, carbonyl, alcohol, carboxyl, aldehyde, amino, sulfhydryl, and phosphate group. What are the overall chemical properties of each group? What is an aldehyde? A ketone?

10. What structural features do carbohydrates have in common? What is a monosaccharide? A disaccharide? A polysaccharide?

11. What is a polymerization? List several polymerization reactions of importance in biological systems.

12. What is the difference between the α- and β-forms of glucose? Does this difference have any biological significance?

13. What are isomers? Do the different isometric forms of molecules such as sugars and amino acids have any biological significance?

14. What are the primary differences between cellulose, plant starches, and glycogen?

15. Define a lipid.

16. Diagram a saturated and unsaturated fatty acid. What are the characteristics of cellular fatty acids?

17. What is the structural plan of a neutral lipid? Why are neutral lipids nonpolar?

18. Diagram a phospholipid. Why do phospholipids have dual solubility properties?

19. What is a bilayer? How is bilayer formation related to the properties of phospholipids? Could a neutral lipid form a bilayer? Why are bilayers stable?

20. What is a steroid? A sterol? In what way are the solubility properties of steroids and phospholipids similar?

21. What is a glycolipid? Where do glycolipids occur in cells?

22. What functions do proteins carry out in cells?

23. What is the structural plan of an amino acid? List the acidic, basic, polar, and nonpolar amino acids. How does proline differ from the other amino acids?

24. How do amino acids link together to form proteins? Where does this reaction take place in cells?

25. Define the primary, secondary, tertiary, and quaternary structure of proteins. What determines the three-dimensional structure of proteins?

26. What is an alpha helix? A beta strand? A random coil? How do these folding arrangements affect protein structure?

27. What are conformational changes in proteins? How are they related to the functions of proteins? What is denaturation?

28. What is a protein domain? What is the significance of domains for protein structure and function?

29. What are glycoproteins and lipoproteins? Coenzymes? Prosthetic groups?

30. What are the organic subunits of nucleotides? How are nucleotides named? What is a purine? A pyrimidine?

31. What is the difference between ATP and dATP?

32. List the structural similarities and differences between DNA and RNA.

33. What is the DNA double helix? What are base pairs? Can RNA exist in double helical form? What does the term *antiparallel* mean with reference to nucleic acids?

34. List the primary functions of nucleotides and nucleic acids.

Supplement 2-1
Atomic Structure

Molecules are formed from atoms linked together in definite numbers and ratios by chemical bonds. Although nearly a hundred different kinds of atoms occur naturally on the earth and link in various ways to form the molecules of both living and nonliving systems, all are basically similar in structure. Each atom consists of an atomic *nucleus* surrounded by one or more smaller, fast-moving particles, the *electrons* (Fig. 2-37). Most of the space occupied by an atom contains the electrons; the nucleus represents only about 1/10,000 of the total volume of an atom. The nucleus, however, makes up more than 99% of the total *mass* of an atom.[3]

The Atomic Nucleus

All atomic nuclei contain one or more positively charged particles called *protons.* The number of protons in the nucleus of a particular type of atom is always the same. Hydrogen, the smallest atom, has a single proton in its nucleus; the nucleus of uranium, the largest naturally occurring atom, has 92 protons. Since the atoms of a given type always contain the same number of protons, this number, called the *atomic number,* specifically identifies an atom. Hydrogen, for example, has the atomic number 1. Similarly, nitrogen, with seven protons, and oxygen, with eight, have atomic numbers of 7 and 8, respectively (Fig. 2-38).

Each type of atom, except hydrogen, also has a number of uncharged particles called *neutrons* in its nucleus. Neutrons occur in atomic nuclei in variable

[3] Mass is defined as the tendency of a body or particle to resist an accelerating force. Applying an accelerating force by kicking a soccer ball filled with air and one filled with lead would provide an unforgettable demonstration of mass and its meaning.

numbers approximately equal to the number of protons. For example, all carbon atoms have nuclei with six protons. Most carbon atoms also have six neutrons. About 1% of naturally occurring carbon atoms have six protons and seven neutrons in their nuclei, and an even smaller percentage have six protons and eight neutrons. These different forms of an atom, all with the same proton number but varying numbers of neutrons, are called *isotopes.* The different isotopes of an atom have essentially the same chemical properties but differ in mass and other physical characteristics.

A neutron and a proton have almost the same mass. This mass is given an arbitrary value of 1, and atoms are assigned a *mass number* based on the total number of protons and neutrons in the atomic nucleus. (The mass of electrons is so small by comparison that it can be ignored in determinations of atomic mass.) The three isotopes of carbon with mass numbers of 12 (six protons plus six neutrons), 13 (six protons plus seven neutrons), and 14 (six protons plus eight neutrons) are identified as ^{12}C, ^{13}C, and ^{14}C, respectively.

Some isotopes of an atom are unstable and change into other atoms through radioactive decay. For example, the unstable carbon isotope ^{14}C slowly breaks down through a mechanism in which one neutron in the ^{14}C nucleus splits into a proton and an electron. The proton remains in the nucleus, and the electron is ejected. (Electrons ejected from decaying nuclei in this way are termed *beta particles.*) The new total of seven protons and seven neutrons in the nucleus is characteristic of the most common isotope of nitrogen. Another pattern of radioactive decay ejects *alpha particles.* These consist of two protons and two neutrons or the equivalent of the nucleus of a helium atom. The emission of alpha or beta particles can be detected as *radioactivity.*

The radioactive isotopes of hydrogen, carbon, sulfur, and phosphorus have been of great value in bio-

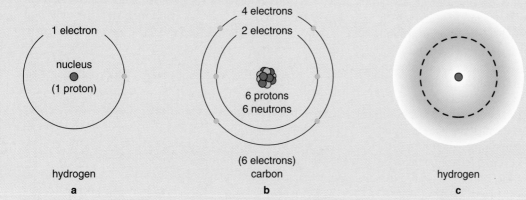

Figure 2-37 Atoms consist of a nucleus surrounded by fast-moving electrons. **(a)** In hydrogen, the simplest atom, the orbital surrounding the nucleus has a single electron. **(b)** Carbon, a more complex atom, has a nucleus surrounded by electrons at two energy levels. **(c)** The orbital (dashed line over region of deepest color) represents the location most frequently occupied by an electron in its travel around the hydrogen nucleus. Regions of lighter color indicate less frequently occupied locations.

Figure 2-38 The atomic number is equivalent to the number of protons in the nucleus of an atom. **(a)** Hydrogen and helium with one and two protons, respectively, have atomic numbers of 1 and 2. **(b)** Atoms with atomic numbers from 7 to 10; carbon, with an atomic number of 6, appears in Figure 2-37*b*.

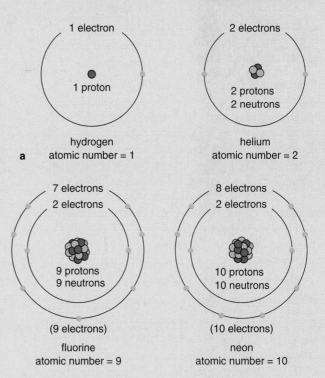

a

hydrogen
atomic number = 1

helium
atomic number = 2

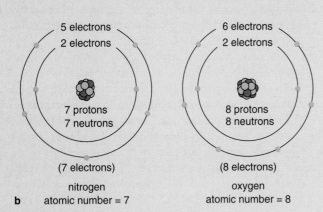

b

nitrogen
atomic number = 7

oxygen
atomic number = 8

fluorine
atomic number = 9

neon
atomic number = 10

logical research because molecules containing them can easily be traced by their radioactivity as they go through chemical reactions in living organisms. A number of stable, nonradioactive isotopes, such as ^{15}N (heavy nitrogen), can be detected by their mass differences and have also proved to be valuable as tracers in biological experiments.

The Electrons of an Atom

The electrons surrounding an atomic nucleus carry a negative charge and occur in numbers equal to the number of protons in the atomic nucleus. Electrons do not contribute significantly to the mass of an atom, because the mass of an electron is equivalent to only 1/1800 of the mass of a proton.

Electrons are in constant, rapid motion around the nucleus of an atom. Until the 1920s electrons were believed to follow definite paths around a nucleus, much as the planets follow a narrowly defined orbit around the sun. More accurate analysis of the behavior of electrons has shown that an orbiting electron may actually be found in almost any location, ranging from the immediate vicinity of an atomic nucleus to practically infinite space. In moving through these locations, an electron travels so fast that it can almost be regarded as being at all points at the same time. However, if its movements could be tracked, an electron would be found to pass through some locations much more frequently than others. The most probable locations surround the atomic nucleus in layers of different shapes called *orbitals*. In the hydrogen atom shown in Figure 2-37*c*, a shaded region depicts the orbital of the single electron, with the electron's most probable locations shown as areas of deepest shade. Either one or two

electrons may occupy a given orbital. However, the most stable and balanced condition occurs when an orbital contains a *pair* of electrons. Orbitals often have complex shapes rather than the spherical form depicted in Figure 2-37*c*.

The electrons surrounding an atomic nucleus travel at velocities or energy levels that increase in discrete steps as the distance of the electron from the nucleus increases. The discrete difference in energy between one level and the next is termed a *quantum*. As a matter of convenience the successive energy levels are often termed *shells*. However, they do not actually have the rigid shapes or physical form suggested by the word shell. The lowest energy level, or shell, may be occupied by a maximum of two electrons and is thus filled by a single orbital. Hydrogen has one electron in this orbital; helium has two (see Fig. 2-38*a*). Atoms with electrons numbering between three (lithium) and ten (neon) have two shells. The innermost contains two electrons in a single orbital. The second shell may contain as many as eight electrons traveling in four orbitals. As this shell is filled, the atoms from lithium to neon form, including beryllium (with four electrons), boron (with five), carbon (six), nitrogen (seven), oxygen (eight), fluorine (nine), and neon (with ten electrons). (Carbon appears in Fig. 2-37; the series from nitrogen to neon in Fig. 2-38*b*.) Larger atoms have more shells; although the outermost typically contains from 1 to 8 electrons, intermediate levels may contain as many as 32. However, no matter how many electrons occur in the intermediate shells, the most stable atoms chemically are those with eight electrons in the outermost shell. For hydrogen and helium, which have electrons in only one shell, the most stable is helium, with a complete orbital containing two electrons.

3

ENERGY, ENZYMES, AND BIOLOGICAL REACTIONS

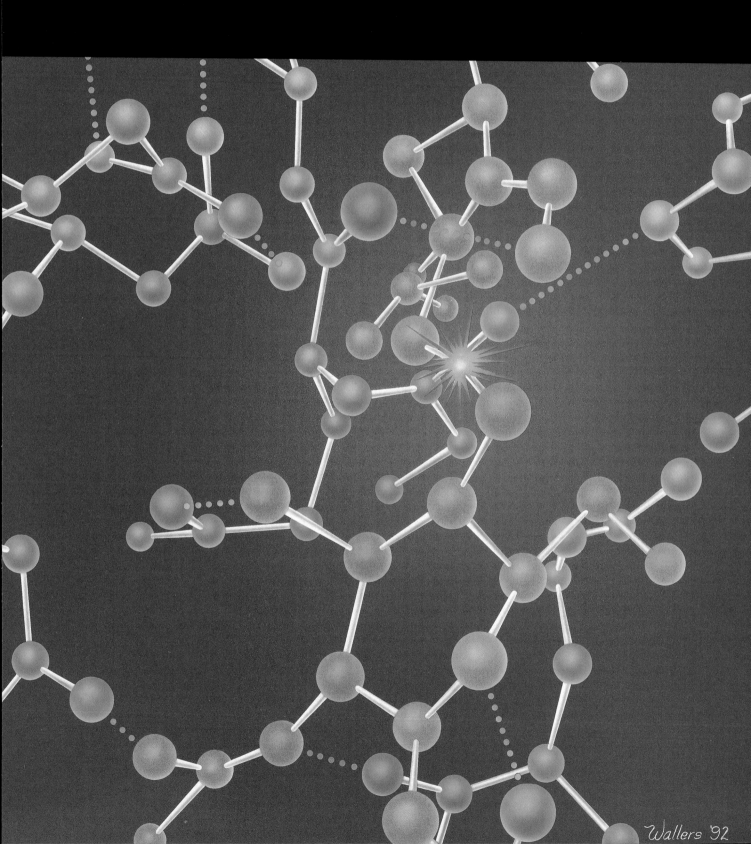

Wallers '92

At the cellular level literally thousands of individual biochemical reactions accomplish the special activities we associate with life—growth, reproduction, movement, and the ability to respond to stimuli. The progress of these cellular reactions depends on enzymes: enzymes speed their rate, making practical chemical interactions that would otherwise take place far too slowly at the temperatures characteristic of cellular life.

Control of the reactions occurring at any time or location is accomplished in cells by the synthesis of the enzymes required for a series of reactions or by the regulation of enzymes already present. Regulation activates only those enzymes needed for the cellular activities taking place; it leaves others in inactive form. As a consequence of controlled synthesis and regulation of enzymes, the right enzymes appear in active form at the right place and time to increase the rate of the biochemical interactions necessary for a particular life activity.

The chemical reactions making up the activities of life are not unique to cells or to the living condition. They can proceed in a test tube with no cells present if the necessary physical and chemical conditions exist for the reactants. Not even the enzymes normally catalyzing the reactions are necessary: the role of enzymes is simply *to increase the rate of reactions that could still proceed, however slowly, without enzymes.* This is true because biochemical and all other reactions obey the same chemical and physical laws operating anywhere in the universe, whether inside cells or in the outside world.

Therefore, understanding whether biological reactions will take place and predicting their direction depend on a knowledge of the basic chemical and physical laws that govern chemical interactions. Understanding the rate at which such reactions take place inside cells depends on an understanding of enzymes and how they work. These fundamental relationships are the subjects of this chapter.

CHEMICAL AND PHYSICAL LAWS GOVERNING REACTIONS

The possibility and direction of reactions of any kind depend on physical laws deduced through the study of *thermodynamics.* Thermodynamics examines all chemical and physical changes and considers the relationship of energy to these changes.

Reactions that are possible are called *spontaneous* reactions in thermodynamics. This usage of the word *spontaneous* does not imply that the change is instantaneous. In thermodynamics spontaneous reactions are simply reactions that will "go"; they may take place at any rate from unmeasurably slow to practically instantaneous.

Groups of reacting molecules are called *systems* in thermodynamics. A system can be defined to include any reacting molecules of interest. Everything outside the defined system is called the *surroundings.* In undergoing any type of change, such as a chemical reaction, systems go from an *initial state* to a *final state.* If the total amount of matter in the system remains the same during the change, the system is *closed.* If matter is transferred between the system and its surroundings during the change, the system is *open.* Living organisms are open systems because they constantly exchange matter with their surroundings. However, many of the individual reactions within living organisms may operate as closed systems.

Spontaneous Reactions and the Laws of Thermodynamics

The study of thermodynamics has produced two laws that together can be used to predict whether reactions are possible and will occur spontaneously. These laws have been of inestimable value in predicting the rate and direction of all chemical and physical reactions, including those taking place in living organisms.

The First Law of Thermodynamics The most fundamental law of thermodynamics was developed partly through a study of the many unsuccessful attempts to construct perpetual motion machines. All such machines, no matter how ingenious, eventually run down and stop. The initial push of energy used to start the machines is inevitably lost to the surroundings as heat resulting from friction between the working parts. To keep the machines running, the lost energy must constantly be replaced by energy added from the surroundings.

Careful measurement of the heat flowing from such machines shows that as they come to a stop, the total amount of heat lost is exactly equivalent to the amount of energy used to start them. These and similar observations in other reacting systems led to the deduction of the *first law of thermodynamics.* This law states that in any nonnuclear process involving an energy change (as in the change from mechanical to heat energy in a machine) the *total amount of energy in a system and its surroundings remains constant.* In other words, in such nonnuclear changes *energy can neither be created nor destroyed.*

Because the systems of interest in cell and molecular biology consist of molecules taking part in chemical reactions, it is useful to consider what total energy means when applied to collections of molecules. One part of the energy content of molecules at temperatures above absolute zero ($-273°C$) is *kinetic energy*, reflected in the constant rotation, vibration, and lateral movement of the molecules. The second part of the energy content, *potential energy*, depends on the energy contained in the arrangement of atoms and chemical bonds in the molecules.

Kinetic and potential energy are interchangeable. For example, an unlit match at room temperature contains considerable potential energy but only moderate kinetic energy. The potential energy content reflects the arrangement of atoms and their bonds in the wood of the match and the phosphorus at its tip. Once the match is struck, much of the potential energy is transformed into kinetic energy, reflected in increased motion of molecules in the match and its immediate surroundings. Some energy also flows from the burning match as heat and light.

The energy content of the reactants in a chemical reaction is usually larger than that of the products. This can be determined directly by igniting equal quantities of reactants and products separately in a bomb calorimeter and comparing the heat given off by the combustion. Such observations support the conclusion that spontaneous reactions generally progress to a state in which the reactants have *minimum energy content.* The difference between the energy content of reactants and products is lost to the surroundings as heat. Reactions that release heat to the surroundings are termed *exothermic* reactions.

Although most spontaneous reactions are exothermic and proceed to a state of minimum energy content, there are exceptions. For example, as ice melts spontaneously at temperatures above 0°C, the energy content of the meltwater is higher than that of the ice. The energy required for melting the ice is absorbed from the surroundings. Reactions that absorb heat from the surroundings as they proceed are called *endothermic* reactions.

The Second Law of Thermodynamics Clearly, knowing the energy content of reactants and products is not enough to predict whether a process or chemical reaction will take place spontaneously. Thermodynamic investigation into a phenomenon that is part of our common experience supplied the missing element. As any type of change takes place, the things involved generally tend to get out of order rather than spontaneously assuming more ordered arrangements (your room is probably the best example of this phenomenon). The scientific study of this phenomenon led to

deduction of the *second law of thermodynamics*: in any process involving a spontaneous change from an initial to a final state, *the total disorder or randomness of the system plus its surroundings always increases.* In thermodynamics disorder is termed *entropy*. If the system is defined as the entire universe, the second law means that as changes take place anywhere, the total disorder, or entropy, of the universe constantly increases.

Using the First and Second Laws to Predict the Rate and Direction of Chemical Reactions

Combining the first and second laws of thermodynamics provides the complete criteria needed to determine whether a reaction is possible and spontaneous. *Reactions in which the system goes toward a condition of minimum energy and maximum entropy* (maximum disorder) are most favored and will tend to proceed on their own.

Energy released by reactions moving spontaneously to a final state is termed *free energy*. This energy is available to do work. In living organisms the primary work accomplished by free energy is the chemical and physical work involved in activities such as the synthesis of cellular molecules, movement, and reproduction.

It is possible for reactions to proceed spontaneously when only one of the two conditions is met. A reaction may take place in which, although the energy content of the system decreases, entropy also decreases. Or, conversely, although entropy increases, the energy content of the system may rise between the initial and the final state. In such reactions the change in energy content and entropy in the system act as opposing tendencies, one tending to favor the reaction and the other to oppose it. In these interactions the *balance* of the opposing tendencies determines whether the reaction will proceed spontaneously.

For example, when ice melts, the energy content of the system increases, which is an unfavorable condition for a spontaneous reaction. However, there is a very large increase in entropy, reflecting the change in the water molecules from the highly ordered arrangement in ice crystals to the more random and disordered condition in liquid water. The balance between the two opposing factors is sufficiently favorable to make the change proceed spontaneously. This change does not violate the first law of thermodynamics. Because the required energy is absorbed from the surroundings, the total energy content of the system and its surroundings remains the same.

The reverse process, the conversion of liquid water to ice, provides a convenient example of a change in which a large loss in energy content compensates for a relatively smaller decrease in entropy. At temperatures below 0°C water freezes spontaneously, even though the arrangement of water molecules in ice crystals is more ordered than that in liquid water. Although entropy decreases in this reaction, the very large drop in energy content as water freezes is sufficient to balance the entropy change.

Some biological systems seem to follow a pathway in which the energy content increases and entropy decreases, so that the energy and entropy changes fail to strike a favorable balance. For example, as a fertilized egg develops into an adult animal, there is an obvious increase in energy content and a decrease in entropy. As it grows, the animal synthesizes more and more organic molecules from less complex substances, greatly increasing its energy content and decreasing its entropy. However, the increase in energy content and decrease in entropy are only apparent in this and similar biological situations because the system has not been defined to include all reactants and products.

In order to include all reactants and products, the system in the initial state—fertilization—must include the complex organic molecules such as glucose that the developing animal will use as an energy source. At the final state when development is complete, the system must include the waste products excreted by the animal, all of them simple, less complex substances than the organic molecules used as fuels. With the inclusion of these necessary reactants and waste products, the total change is toward both minimum energy and maximum entropy.

Reversible Reactions

In many chemical reactions the conversion of reactants to products does not proceed entirely to completion. In these reacting systems complete conversion into an arrangement containing all products and no reactants would represent a very high degree of order (greatly reduced entropy) with respect to concentrations, so high that it cannot be balanced by the energy and entropy changes in the reaction itself.

Reactions of this type proceed to a point at which the tendency of the reaction to go completely to products with minimum energy and maximum entropy is exactly balanced by the tendency to move backward toward a more disordered state with equal numbers of reactants and products. (Equal numbers of reactants and products would produce minimum energy and maximum entropy with respect to concentration differences.) The balance achieved is called the *equilibrium point* for the reaction. At the equilibrium point equal numbers of molecules undergo conversion from reactants to products and from products to reactants. If more reactants are added, the reaction will proceed toward products until an equilibrium is again established. Conversely, if more products are added, the reaction will move backward toward reactants until the system

again balances. Such reactions are termed *reversible* and are written with a double arrow:

$$\text{reactants} \rightleftharpoons \text{products} \qquad (3\text{-}1)$$

Most biological reactions are reversible.

The equilibrium point of a reversible reaction depends on the difference in energy content and entropy between the reactants and products. If this difference is large, so that the products contain considerably less energy and much greater entropy than the reactants, the equilibrium point will lie far to the right, and most of the reaction mixture will consist of products. If the energy and entropy difference between reactants and products is small, the equilibrium point will lie farther to the left, and a greater proportion of reactants will be present in the equilibrium mixture.

The ability of reversible reactions to proceed in either direction allows them to be written in either direction, so that the names given to reactants and products are interchangeable. For example, in one important biological reaction the amino acid glutamine is converted to glutamic acid by the removal of an amino group:

$$\text{glutamine} + H_2O \longrightarrow \text{glutamic acid} + NH_3 \qquad (3\text{-}2)$$

The reaction proceeds far in the direction of products and releases free energy.

Because the reaction is reversible, it can be written in the opposite direction:

$$\text{glutamic acid} + NH_3 \longrightarrow \text{glutamine} + H_2O \qquad (3\text{-}3)$$

The reaction will actually proceed spontaneously in this direction if the solution initially contains only glutamic acid and NH_3. However, very little glutamine and water are expected as products because the equilibrium point, when the reaction is written in this way, now lies far to the left.

How Living Organisms Push
Reversible Reactions Uphill

Cells rely for their functioning on many reactions such as Reaction 3-3, even though the equilibrium point lies far to the left. How do cells circumvent this problem, so that the products can be made in useful quantities? The answer is by a biological "trick" in which the desired "uphill" reaction is joined, or *coupled*, with another reaction that proceeds "downhill" with a large free energy release. The total changes toward minimum energy content and maximum entropy for the combined reactions produce a new equilibrium point that lies far to the right.

One of the most interesting features of the mechanism coupling uphill and downhill reactions in living cells is that the primary coupling agent, the nucleotide ATP, is the same in all organisms from bacteria to humans. ATP (*adenosine triphosphate*; see Fig. 3-1*a*; see also p. 310) consists of a nitrogenous base, *adenine*, linked to a five-carbon sugar, *ribose*. The ribose sugar links in turn to a chain of three phosphate groups.

Much of the energy available in the ATP molecule is associated with the arrangement of the three phosphate groups. The oxygens of the phosphate groups carry a strongly negative charge at cellular ranges of acidity. The three phosphate groups line up side by side in ATP in highly ordered positions that bring the strongly negative charges close together. Removing one or two of the phosphates produces a large drop in energy content and greatly increases entropy. As a consequence, free energy is released in large quantity:

$$ATP + H_2O \longrightarrow ADP + P_i \qquad (3\text{-}4)$$

$$ATP + H_2O \longrightarrow AMP + 2P_i \qquad (3\text{-}5)$$

The products of these reactions are *adenosine diphosphate* (ADP) and *adenosine monophosphate* (AMP; see Fig. 3-1*b*). P_i represents the removed phosphate group in these reactions (i = inorganic).

Cells use ATP breakdown to drive uphill reactions by coupling Reaction 3-4 or 3-5 to the reactions requiring energy. For example, the reaction synthesizing glutamine from glutamic acid is run far to the right by coupling Reactions 3-3 and 3-4:

$$\text{glutamic acid} + NH_3 + ATP \longrightarrow$$
$$\text{glutamine} + ADP + P_i \quad (3\text{-}6)$$

The total products—glutamine, ADP, and phosphate—have considerably less energy content and much greater entropy than the total reactants, which include glutamic acid, ATP, and NH_3. As a result, the total reaction runs spontaneously far to the right and releases free energy.

In coupled reactions, therefore, an uphill reaction requiring energy is coupled to the breakdown of ATP to produce an overall reaction that is spontaneous and releases energy. Almost all the uphill chemical, electrical, and mechanical work of cells is made energetically favorable in this way, including the energy-requiring activities of growth, reproduction, movement, and responses to stimuli (Fig. 3-2).

The Standard Free Energy Change

The amount of energy released by ATP breakdown, or required for the synthesis of glutamine from glutamic acid, is often shown in biological reactions in calories

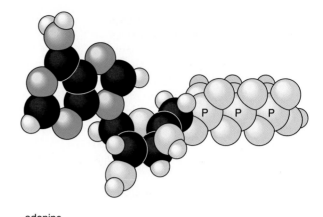

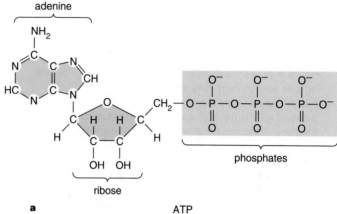

a ATP

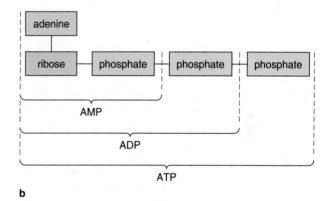

b

Figure 3-1 Adenosine triphosphate (ATP) and its derivatives adenosine diphosphate (ADP) and adenosine monophosphate (AMP). **(a)** Space-filling model and chemical structure of the ATP molecule. P, the phosphorus atoms of the three phosphate groups. **(b)** Sequential removal of the phosphate groups produces ADP and AMP.

per mole,[1] as determined from reactions run under standard conditions. (Standard conditions include a temperature of 25°C, constant pressure at 1 atmosphere, and the reaction initiated with the reactants and products in equal quantities of 1 mole each.) Under these

[1] Defined in footnote 1 on p. 40.

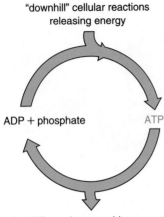

"downhill" cellular reactions
releasing energy

ADP + phosphate   ATP

"uphill" reactions requiring energy
in growth, movement, reproduction,
responsiveness, and other cell activities

Figure 3-2 ATP and ADP cycle between energy-releasing and energy-requiring activities in the cell. Reactions that release energy are indirectly coupled to the synthesis of ATP from ADP and phosphate (top half of diagram). Reactions that require energy are made favorable by coupling them to the breakdown of ATP to ADP and phosphate, which releases enough energy for the reactions to proceed (bottom half of diagram). Some energy-requiring activities remove two phosphates from ATP, producing AMP and phosphate, which return to the energy-releasing mechanisms to be converted back to ATP. Most of the "downhill" cellular reactions releasing energy for eukaryotic ATP synthesis are oxidations that take place inside mitochondria.

Table 3-1	Standard Free Energy Changes at pH 7 and 25°C for Some Chemical Reactions of Biological Interest	
	Reaction	$\Delta G'$ (cal/mol)
Oxidation		
	glucose + $6O_2 \rightarrow 6CO_2 + 6H_2O$	−686,000
	lactic acid + $3O_2 \rightarrow 3CO_2 + 3H_2O$	−326,000
	palmitic acid + $23O_2 \rightarrow$	
	$16CO_2 + 16H_2O$	−2,338,000
Hydrolysis		
	sucrose + $H_2O \rightarrow$ glucose + fructose	−5,500
	glucose-6-phosphate + $H_2O \rightarrow$	
	glucose + H_3PO_4	−3,300
	glycylglycine + $H_2O \rightarrow$ 2 glycine	−4,600
Rearrangement		
	glucose-1-phosphate $\rightarrow$	
	glucose-6-phosphate	−1,745
	fructose-6-phosphate $\rightarrow$	
	glucose-6-phosphate	−400
Ionization		
	$CH_3COOH + H_2O \rightarrow$	
	$H_3O^+ + CH_3COO^-$	+6,310
Elimination		
	malate $\rightarrow$ fumarate + H_2O	+750

SOURCE: From *Bioenergetics*, 2nd ed., by Albert L. Lehninger (Menlo Park, CA: Benjamin/Cummings Publishing Company, 1971), p. 32. Reprinted by permission.
NOTE: $\Delta G' = \Delta G°$ at pH 7.

conditions the reaction will proceed from the initial conditions to equilibrium and give off or absorb free energy as it proceeds. The free energy released or absorbed under these conditions is termed the *standard free energy change*, designated $\Delta G°$. (The G is in honor of J. W. Gibbs, an American physicist who was a major contributor to the science of thermodynamics.)

The standard free energy change $\Delta G°$ indicates directly whether a reaction is spontaneous. It also provides a way to compare different reactions on the same scale and to estimate in relative terms the amount of energy released or required in running them. (The standard free energy changes for a number of reactions of biological interest appear in Table 3-1.)

For example, running Reaction 3-3 to the right requires an input of 3400 calories for each mole of glutamic acid fully converted to glutamine under standard conditions. By convention, the standard free energy change is written as a positive number if a reaction requires the addition of energy:

$$\text{glutamic acid} + NH_3 \longrightarrow \text{glutamine} + H_2O \quad (3\text{-}7)$$

$$\Delta G° = +3400 \text{ cal/mol}$$

A positive value for the standard free energy change indicates that a reaction is nonspontaneous.

Running Reaction 3-4 to the right releases 7000 calories for each mole of ATP converted into ADP + P_i. The standard free energy change is written as a negative number if a reaction releases free energy:

$$\text{ATP} + H_2O \longrightarrow \text{ADP} + P_i \quad (3\text{-}8)$$

$$\Delta G° = -7000 \text{ cal/mol}$$

A negative value for the standard free energy change indicates that a reaction is spontaneous.

This information can be used to estimate the free energy released when Reactions 3-7 and 3-8 are coupled to drive the synthesis of glutamine under standard conditions. Adding +3400 cal/mol and −7000 cal/mol gives an expected free energy change of −3600 calories for each mole of glutamic acid converted to glutamine by the coupled reaction:

$$\text{glutamic acid} + NH_3 + \text{ATP} \longrightarrow$$
$$\text{glutamine} + \text{ADP} + P_i \quad (3\text{-}9)$$

$$\Delta G° = -3600 \text{ cal/mol}$$

Because the value for the standard free energy change is negative, the reaction is expected to be spontaneous and release energy.

Keep in mind that $\Delta G°$ is valuable only for making comparisons between reactions. The actual energy released or required for a reaction taking place inside cells depends on the initial concentrations of reactants and products, temperature, and pH, which under biological conditions are likely to differ from the standard values used in calculating $\Delta G°$. In particular the breakdown of ATP to ADP + P_i under cellular conditions is likely to release considerably more than 7000 cal/mol.

The ATP broken down in coupled reactions such as 3-9 is replenished through energy released by cellular oxidations. This energy is used to add phosphate groups to AMP or ADP:

$$AMP + P_i \longrightarrow ADP + H_2O \qquad (3\text{-}10)$$

$$ADP + P_i \longrightarrow ATP + H_2O \qquad (3\text{-}11)$$

The effect of adding or removing phosphate groups in the ATP/ADP/AMP system is much like compressing or releasing a spring. Adding phosphate groups, up to a limit of three, compresses the chemical spring and stores energy. Removing one or both terminal phosphates releases the spring and makes energy available for useful work.

THE ROLE OF ENZYMES IN BIOLOGICAL REACTIONS

Enzymes and Enzymatic Catalysis

The laws of thermodynamics apply to chemical reactions anywhere in the universe. No reactions can violate the rule that they must proceed to a level of minimum energy and maximum entropy or to a favorable balance between the two factors. What effects do enzymes have, then, on biochemical reactions? The answer is that enzymes simply increase the *rate* at which spontaneous reactions take place.[2] Enzymes cannot make a reaction occur that would not already proceed spontaneously without the enzyme.

The same principles apply equally to reversible reactions. Enzymes do not alter the equilibrium point of a reversible reaction. The same equilibrium is established whether an enzyme is present or not—an enzyme simply increases the rate at which the reaction reaches equilibrium.

[2] *Rate* in this sense means the number of reactant molecules converted to products per unit of time under constant temperature and pressure.

Most biological reactions, although spontaneous, would proceed so slowly at the temperatures characteristic of living organisms that their rate without enzymes would be essentially zero. For example, the oxidation of glucose to CO_2 and H_2O is spontaneous and proceeds almost completely in the direction of the products. However, without enzymes glucose oxidation occurs so slowly at physiological temperatures as to be unmeasurable. Within cells enzymes speed the oxidation of glucose in a number of substeps. Even though the overall breakdown pathway is complex, with enzymes the oxidation of glucose takes only minutes at the relatively low temperatures characteristic of living organisms.

The increases in rate achieved by enzymes, depending on the enzyme and reaction, range from a minimum of about a million to as much as a trillion times faster than the uncatalyzed reaction at equivalent concentrations and temperatures (Table 3-2). In some instances, these rates are as much as a billion times faster than the rates achieved by inorganic catalysts used in the chemical industry.

Enzymes and Activation Energy

How do enzymes accomplish these feats? Their most basic and fundamental effect is to reduce the *activation energy* required for a chemical interaction to proceed. Activation energy is the energy barrier over which the molecules in a system must be raised for a reaction to take place (Fig. 3-3a). This condition is analogous to a rock resting in a depression at the top of a hill (Fig. 3-3b). As long as the rock remains undisturbed, it will not spontaneously roll downhill, even though the total "reaction"—the progression of the rock downward—is energetically favorable. In this physical example the effort required to raise the rock over the lip of the depression is the activation energy.

The requirement for activation energy raises the question of why reactions proceed spontaneously at all. What is the source of the energy required to push molecules over the energy barrier? Movement over the

Table 3-2 Enzymatic Increases in Reaction Rates

Enzyme	Increase in Reaction Rate over Uncatalyzed Rate
Carbonic anhydrase	1.1×10^8
Creatine kinase	4×10^8
Hexokinase	8×10^{10}
Lysozyme	1×10^7 to 1×10^9
Phosphorylase	9×10^{11}
Serine proteases	1×10^5 to 1×10^{10}
Urease	1×10^{14}

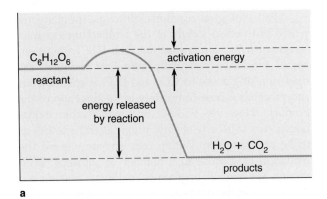

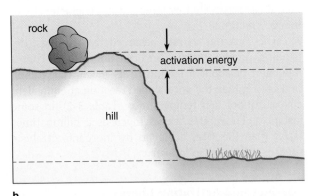

a

b

Figure 3-3 Activation energy. **(a)** The activation energy for the oxidation of glucose, an energy barrier over which glucose molecules must pass before they can oxidize to H_2O and CO_2. **(b)** An analogous physical situation in which a rock rests in a depression at the top of a hill. The rock will not move downward unless enough "activation energy" is added to raise it over the lip of the depression.

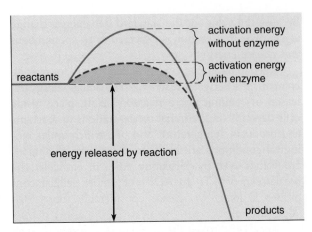

Figure 3-4 Enzymes increase the rate of a spontaneous reaction by reducing the activation energy. This reduction allows biological reactions to proceed rapidly at the relatively low temperatures that living organisms can tolerate. Enzymes combine briefly with reactants as part of the mechanism reducing the activation energy.

heat to keep the reaction temperature high enough to push the remaining reactants over the barrier. Glucose oxidation may be pushed into a self-sustaining reaction, for example, by igniting glucose over an open flame.

Ignition is obviously not a satisfactory approach for pushing reacting molecules over the activation barrier in biological systems. Rather than igniting the reactants, enzymes speed reactions by *lowering the activation energy required for a reaction to proceed* (Fig. 3-4). Lowering the activation energy increases the possibility that molecules will gain enough energy to pass over the energy barrier at normal temperatures. The mechanism by which enzymes accomplish this feat is still not totally understood.

Enzymes are protein molecules that are tailored to recognize specific reactants and speed their conversion into products. These proteins are responsible for increasing the rates of practically all of the many thousands of reactions taking place inside cells. Although enzymes are the primary biological catalysts, several types of RNA molecules known as *ribozymes* can also speed certain reactions by lowering activation energy.

Characteristics of Protein-Based Enzymes

All enzymatic proteins have several characteristics in common (Table 3-3). Enzymes combine briefly with the reacting molecules during catalysis and are released unchanged when the reaction is complete. Depending on the enzyme, the combination may occur through the temporary formation of any type of chemical bond, association, or attraction—ionic, covalent, or hydrogen bonds, polar or nonpolar associations, or van der Waals forces. Because enzymes are released unchanged after

barrier is possible because molecules, unlike the rock in Figure 3-3*b*, are in constant motion at temperatures above absolute zero. Although the average amount of movement, or kinetic energy, is below the amount required for activation, some molecular collisions may raise a number of molecules to the energy level required for the reaction to proceed. A high activating barrier indicates that collisions of this type are not very likely to occur, and few molecules will proceed over the energy barrier per unit of time.

A rise in temperature increases the probability that molecules in the reacting system will pass over the energy barrier. As the temperature rises, both the speed of individual molecules and the frequency of collisions increase, making it more likely that sufficiently forceful collisions will occur at the correct angle and place. Once large numbers of molecules begin to pass over the barrier, the reaction is frequently self-sustaining. The molecules being converted to products release enough

Table 3-3   Characteristics of Enzymatic Proteins

1. Enzymes combine briefly with reactants during an enzyme-catalyzed reaction.

2. Enzymes are released unchanged after catalyzing the conversion of reactants to products.

3. Enzymes are specific in their activity: each enzyme catalyzes the reaction of a single type of molecule or a group of closely related molecules.

4. Enzymes are saturated by high substrate concentrations.

5. Many enzymes contain nonprotein groups called cofactors, which contribute to their activity. Inorganic cofactors are all metallic ions. Organic cofactors, called coenzymes, are complex groups derived from vitamins.

Table 3-4   Turnover Numbers for Some Enzymes

Enzyme	Turnover Number*
Carbonic anhydrase	600,000
Ketosteroid isomerase	285,000
Catalase	93,000
Amylase	18,000
Penicillinase	2,000
Lactic acid dehydrogenase	1,000
Galactosidase	200
Chymotrypsin	100
Succinic acid dehydrogenase	20
Tryptophan synthetase	2

* Number of substrate molecules converted to products by one enzyme molecule in one second.

a reaction, a single enzyme molecule may combine repeatedly with reactants and release products. (It is worth noting that enzymes undergo changes *during* a reaction, even in some cases to the extent of covalent bonding with the substrate; however, these changes are transient, so at the close of the reaction the enzyme returns to its original state.) The rate of combination and release, known as the *turnover number*, lies near 1000 per second for most enzymes. However, some enzymes have turnover numbers as small as 100 per second or as large as 10 million per second. (Table 3-4 gives some examples.) As a result of enzyme turnover, a relatively small number of enzyme molecules can catalyze a large number of reactant molecules.

Enzymatic proteins catalyze only a single type of biochemical reaction and combine only with a single type of molecule or a group of closely related molecules. This characteristic of enzymatic proteins is known as *specificity*. The specific molecule or molecular group whose reaction is catalyzed is known as the *substrate* of the enzyme.

Literally thousands of different enzymes have been detected and described. These enzymatic proteins vary from relatively small molecules, with single polypep-

tide chains containing as few as 100 amino acids, to large complexes with several to many polypeptide chains totaling thousands of amino acids. Many enzymes include an inorganic ion or a nonprotein organic group that contributes to their catalytic function. These groups, called *cofactors*, may link to the enzyme by covalent, ionic, or hydrogen bonds or other attractions.

The inorganic cofactors, all metallic ions, include iron, copper, magnesium, zinc, potassium, manganese, molybdenum, and cobalt. When present as cofactors, these ions contribute directly to the reduction of activation energy by the enzyme. The organic cofactors, called *coenzymes*, are all complex chemical groups of various kinds, many of them derived in higher animals from vitamins (Table 3-5; Figs. 9-5 and 9-11 show representative coenzymes). Coenzymes frequently act as carriers of chemical groups, atoms, or electrons removed from substrates during reactions. When tightly linked to an enzyme by covalent bonds, coenzymes are sometimes called *prosthetic groups.*

Enzymatic proteins are named and placed in one of six major classes according to their substrates and the type of reaction they catalyze (Table 3-6). Within

Table 3-5   Some Coenzymes and Their Vitamin Sources

Coenzyme	Vitamin Source	Units Carried
NAD (nicotinamide adenine dinucleotide)	Nicotinic acid	Electrons and hydrogen
CoA (coenzyme A)	Pantothenic acid	Acetyl groups
FMN (flavin mononucleotide)	Riboflavin	Electrons and hydrogen
FAD (flavin adenine dinucleotide)	Riboflavin	Electrons and hydrogen
TPP (thiamin pyrophosphate)	Thiamin	Aldehyde groups

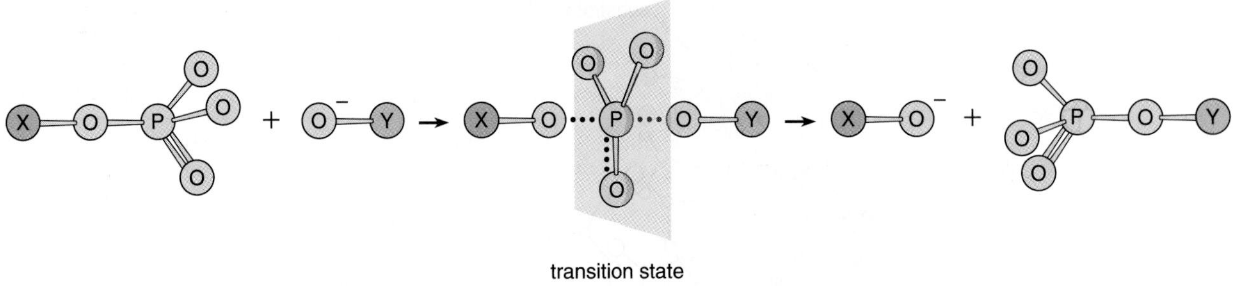

transition state

Figure 3-6 Formation of the transition state in the transfer of a phosphate group between two molecules designated as X and Y. The transition state is unstable and easily pushed in the direction of either reactants or products.

site of an enzyme, rather than fitting only the reactant molecules, has additional conformations that also fit the transition state and the products. Of these conformations the tightest binding, and the most precise fit, is to the transition state. Somewhat later, in 1958, D. E. Koshland added the idea that, after attaching to the enzyme through a relatively loose fit to the active site, reactant molecules are distorted or "warped" toward the transition state by a conformational change in the enzyme that provides tightest binding to the active site. Once in this state the reacting molecules require only a small energy input to move in the direction of products. The energy may be supplied by collisions between the reacting molecules in the active site or between the enzyme-substrate complex and other molecules in the surrounding solution. The energy of these collisions, transmitted through the flexible structure of the enzyme molecule, is easily large enough to push the unstable transition state in the direction of products.

Experiments using molecules that resemble the transition state more closely than either the reactants or products of an enzyme-catalyzed reaction directly support these ideas. Such molecules, called *transition state analogs*, bind readily and tightly to the enzyme, more tightly than either the reactants or products, as expected if the active site fits the transition state most closely. In addition, some naturally occurring antibiotics that work by binding and inactivating enzymes were found to contain molecular structures that resemble the transition state for the reactions catalyzed by the enzymes.

Further support for Pauling's hypothesis came from interesting and highly significant research with antibodies. Some years after Pauling advanced his idea that enzymes bind most tightly to the transition state, W. P. Jencks realized that antibodies tailored to fit the transition state could test the hypothesis. Antibodies are protein molecules generated in higher animals in response to exposure to a foreign substance called an *antigen.* Antibody molecules contain a binding site tailored exactly to fit their antigen. Combination of an antibody with its antigen leads to removal of the invading substance from the body (see Ch. 19 for details).

To test Pauling's hypothesis, Jencks proposed that transition state analogs be injected into animals as antigens. The antibodies developed in response should fit the analogs exactly. If binding the transition state does indeed achieve enzymatic catalysis, the antibodies fitting the transition state should also be able to act as enzymes. In 1986 two groups working independently under the direction of R. A. Lerner and P. G. Schultz showed that antibodies can indeed be induced to act as enzymes by modeling them to the transition state.

The Lerner group studied the enzymatic catalysis of a typical biological reaction called an *acyl transfer* in which an ester bond is hydrolyzed (Fig. 3-7a). The transition state for this reaction is mimicked by another group of stable molecules called *phosphonate esters* (Fig.

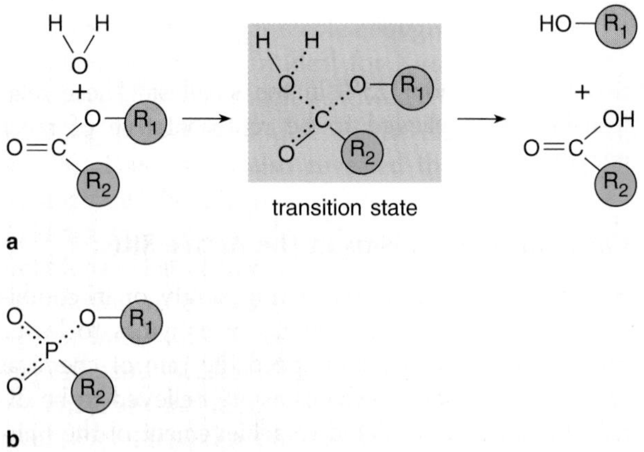

transition state

a

b

Figure 3-7 The acyl transfer reaction used by R. A. Lerner and his coworkers to demonstrate binding of the transition state by the active site of enzymes. **(a)** Formation of the transition state in an acyl transfer reaction. R_1 and R_2 designate organic groups. **(b)** A phosphonate ester, a stable structure that closely resembles the transition state of acyl transfer reactions.

3-7b). Using the phosphonate esters as antigens, the Lerner group induced the formation of antibodies tailored to fit the transition state for the acyl transfer reaction. The resulting antibodies, as predicted by the Pauling–Jencks hypotheses, were able to act as enzymes (or *abzymes,* as they are now called) and speed the rate of the acyl transfer reaction. The Schultz laboratory obtained similar results using a different reaction system (see The Experimental Process box on p. 92).

These results directly support the proposal that achievement of the transition state is an important part of enzymatic catalysis. The technique of converting antibodies to enzymes by tailoring them to fit the transition state also opened possibilities for the production of "designer enzymes"—enzymes designed specifically to catalyze a reaction desired in research, medicine, or industry. (A second highly promising technique for producing designer enzymes for these purposes is *site-directed mutagenesis,* in which amino acids are deliberately introduced, substituted, or deleted in active sites by induced mutations.)

As a result of these experiments, tight binding by the active site is now considered to bend substrate atoms to positions approximating the transition state. Tight binding may also increase the distance between substrate atoms, thereby decreasing the attraction of the bonds and making them easier to break.

Mechanisms Contributing to the Transition State

A number of mechanisms operating at the active site probably contribute to formation of the transition state or help to push reacting molecules over the activation barrier once they are in the transition state. These include:

1. Bringing substrate molecules into close proximity.

2. Orienting substrate molecules in positions favoring reaction.

3. Placing proton (H^+) donors or acceptors in positions promoting acid/base reactions.

4. Exposing substrates to altered environments.

These mechanisms, which may operate singly or in combination in different enzymes, supplement the warpage or strain of reacting molecules toward the transition state by their combination with the active site.

Bringing Substrate Molecules into Close Proximity
Many reactions involve the combination or interaction of two or more reactant molecules. In order for the reaction to take place, the substrate molecules must collide. The required collisions may be rare in solutions, particularly if the substrate molecules are present in low concentrations. Binding at the active site of an enzyme brings the reactants close together, effectively raising their local concentration to many times the concentration in the surrounding solution. The increase in effective concentration brought about by this close proximity may be sufficient in itself to elevate the rate of a reaction by as much as 10,000 to 100 million times.

Orienting Substrate Molecules at the Active Site
Besides raising the effective concentration, alignment of reactant molecules at the active site may also bring substrate molecules into an arrangement in which they can collide at exactly the correct positions and angles required for achievement of the transition state. For example, in the reaction shown in Figure 3-6, orientation in the active site could place the phosphorus atom adjacent to substance Y, directed toward Y in such a way that the cloud of electron orbitals surrounding the phosphorus atom could penetrate and alter the clouds surrounding the reactive oxygen atom of substance Y. This penetration would achieve the transition state and greatly facilitate formation of the new covalent bond accomplishing the phosphate transfer. Orientation of reactant molecules at the active site may also restrict the rotation of substructures of the reactants around single bonds, thereby confining them to the most favorable positions for reaction. Orientation of reactants at the active site is estimated by itself to be able to enhance reaction rates by as much as 10,000 times.

Promoting Acid/Base Reactions
In many biological reactions, particularly those involving hydrolysis, a substrate gains or loses a proton (hydrogen ion). In some reactions a series of transfers takes place in which protons are sequentially added and removed. Addition of a proton in these reactions induces a rearrangement of shared electrons that reduces the stability of covalent bonds and lowers the energy required for their breakage. Several amino acid side groups can act as proton donors or acceptors—those with a basic reaction, including arginine and lysine, can act as proton acceptors, and those with an acidic reaction, including aspartic and glutamic acids, can act as proton donors. Histidine can act as either a proton donor or acceptor (see p. 48), making it especially effective in catalysis involving acid/base reactions. Under certain conditions even groups normally considered as either proton donors or acceptors can work in both capacities—the —COOH group in the undissociated form, for example, can act as a proton donor; in the dissociated —COO⁻ state, the group can act as a proton acceptor. The alignment of acidic or basic amino acid side groups at the active site positions them precisely to donate or accept protons from interacting segments of substrate molecules.

The Experimental Process

Catalytic Antibodies: Tailor-Made Catalysts

KEVAN M. SHOKAT AND PETER G. SCHULTZ

KEVAN M. SHOKAT (left) did his undergraduate work at Reed College in Portland, Oregon, and graduate work at the University of California at Berkeley with Peter Schultz, completing his Ph.D. in 1991. He is currently a postdoctoral fellow of the Life Sciences Research Foundation at Stanford University in the Howard Hughes Medical Institute. Peter G. Schultz (right) did his undergraduate and graduate work at Caltech. After postdoctoral studies at MIT, he joined the faculty of the University of California at Berkeley in 1985, where he is currently a Professor of Chemistry. Schultz has received a number of awards, including the NSF Waterman Award and ACS Pure Chemistry and Eli Lilly Awards. He is also a founder of Affymax Research Institute, a novel pharmaceutical venture.

Enzymes are currently used in many applications such as the treatment of disease (e.g., tissue plasminogen activator), the manipulation of biopolymers such as DNA (e.g., restriction enzymes), and industrial scale organic syntheses (e.g., lipases for chiral resolutions). For many applications, however, naturally occurring enzymes cannot be used because of their limited substrate specificity. As a result, one of the most exciting areas of enzymology has been the design and generation of novel synthetic biological catalysts. While site-directed mutagenesis has produced a limited number of enzymes with novel structures and functions, most attempts to alter enzyme specificity have resulted in a loss in the function or efficiency of these enzymes. In 1986 two research groups, one at the Scripps Institute in San Diego and our group at the University of California at Berkeley, developed an alternative approach to the problem. Rather than alter the specificity of existing catalysts, we sought to introduce catalytic function into a class of highly evolved *ligand binding proteins*, antibodies. In contrast to enzymes, whose specificity evolves over millions of years, antibodies can be generated with virtually any specificity within a matter of weeks.

The immune system has evolved to recognize, bind, and clear foreign pathogens from the body. It possesses the ability to generate up to 10^{12} immunoglobulins (antibodies) which can bind almost any target protein, carbohydrate, or small organic molecule of interest. The immune system accomplishes this through a remarkable system of cells, soluble factors, and complex positive and negative selection processes. The challenge then is to devise strategies for endowing antibodies with catalytic activity. That is, can we design a ligand (hapten) which can induce antibodies capable of catalyzing a target reaction rather than simply binding the substrate? In theory, once a strategy exists for generating an antibody with a specific catalytic activity, such as ester bond hydrolysis, esterases with any substrate specificity can be generated by immunizing with a corresponding hapten.

Not only have antibodies with novel (nonenzymatic) substrate specificities been generated, but reactions not catalyzed by any known enzymes have also been catalyzed. Antibodies have been generated that catalyze ester, amide, and phosphate hydrolysis, Claisen rearrangements, Diels—Alder reactions, elimination reactions, photochemical thymine dimer cleavage, redox reactions, lactonization reactions and bimolecular amide, ester, and imine formation, decarboxylations, porphyrin metallation, and porphyrin-catalyzed peroxidation, and others. Additionally, catalytic antibodies have been useful for testing fundamental notions of enzymatic reaction mechanisms such as transition state stabilization, ground state strain, and proximity effects.

In order to illustrate the thought process that goes into the generation of a catalytic antibody we will consider a model reaction involving proton abstraction and flouride elimination from substrate **1** to form product **2**:

①

O_2N ... $+ HF$

②

A number of considerations must be taken into account when designing a hapten to elicit a catalytic antibody, including reaction mechanism, the rate-limiting step of the reaction, substrate binding, product dissociation, and *in vivo* hapten stability. In this reaction the rate limiting step is proton abstraction. Enzymes that catalyze proton abstraction reactions utilize amino acids such as Glu, Asp, and His in their substrate-binding pockets to serve as general base catalysts. Consequently, one would like to generate an antibody-combining site which contains a general base. What hapten does one design to induce antibodies that contain an appropriately positioned active site containing Asp or Glu? From early studies of the complementarity between haptens and antibodies it was known that positively charged ammonium ion-containing haptens elicited antibody-combining sites with complementary negatively charged Asp or Glu residues. We therefore decided to synthesize the following hapten:

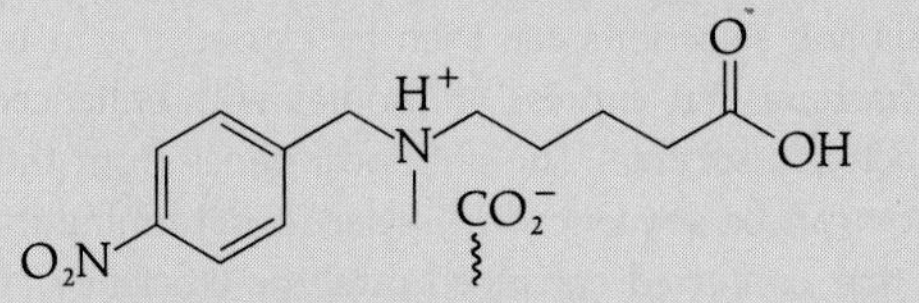

to induce an antibody-combining site containing an appropriately positioned basic amino acid.[1] The position of the shaded group in this hapten corresponds to the position of the abstractable proton in substrate 1 and should elicit a complementary catalytic carboxylate within bonding distance. A *p*-nitrophenyl ring was included to serve as a common recognition element between hapten and substrate. Moreover, replacement of hapten by substrate in the antibody-combining site should increase the pK_a of the catalytic carboxylate group (making it a better base) since a stabilizing salt bridge interaction is lost. Finally, the substrate and product are structurally quite different because of the formation of a double bond in the product, suggesting that a substratelike hapten should avoid any potential problem with product inhibition.

Standard synthetic organic methods were used to synthesize the hapten and to covalently couple it to the surface of carrier proteins. Carrier proteins are necessary in order to make small molecules capable of generating antibodies. The most common strategy for protein coupling is to form an amide bond between the side-chain amino groups of surface lysine residues and a carboxylate group in the hapten. Typically this carboxylate group is linked to the hapten via a chemical group long enough to ensure that antibodies can bind the hapten without interference with the surface of the protein carrier. Following a 30–60 day immunization schedule in mice, a serum enzyme-linked immunosorbent assay (ELISA; the same test used to test for antibodies against HIV and for CG in pregnancy) was used to test for hapten-specific antibodies. Once a high titre of hapten-specific antibodies had been achieved, the antibody-producing cells of the immune mice were fused with myeloma cells, thus rendering them immortal [see p. 805].

A total of six monoclonal antibodies specific for the hapten were produced in this manner. With the antibodies in hand, an analytical technique was necessary in order to monitor the conversion of 1 to 2. In this case the product alkene possesses an extended π-system of electrons not present in substrate 1. This results in a red-shift of the UV/Vis spectrum of 2. Consequently, by monitoring the increase in absorbance at 330 nm the conversion of 1 to 2 could be followed. Four of the six antibodies accelerated the conversion of 1 to 2, and were completely inhibitable by hapten. Hapten inhibition demonstrates that catalysis is occurring in the binding site of the antibody.

In general the K_M values [see p. 105] of antibody catalysts are in the same range as those of enzymes. The rate acceleration by antibody compared to the background rate with acetate ion was 8.8×10^4, reflecting the contribution of proximity of substrate and a catalytic group in a protein-binding site to rate enhancement. By comparison the rate accelerations achieved by peptidases are as high as 10^{13}. The antibody discriminated between substrate 1 and a related analog by a factor of ten in overall rate. Chemical modification and affinity labeling studies as well as the pH dependence of catalysis indicate Glu 46 of the heavy chain as the base responsible for catalysis by the antibody. Following the successful demonstration of antibody-hapten charge complementarity a number of other important reactions have been catalyzed using the same strategy, including *cis-trans* alkene isomerization (similar to the reactions of vision) and a dehydration reaction (similar to reactions in glycolysis).

As the field of catalytic antibodies develops, experiments tend to focus on reactions of biological, medical, or synthetic interest. For example, can a key biosynthetic reaction—one required for the survival of the an organism—be carried out by an antibody? One such reaction—the conversion of chorismate, 3, into prephenate, 5—is a step in the aromatic amino acid biosynthetic pathway of plants and bacteria:

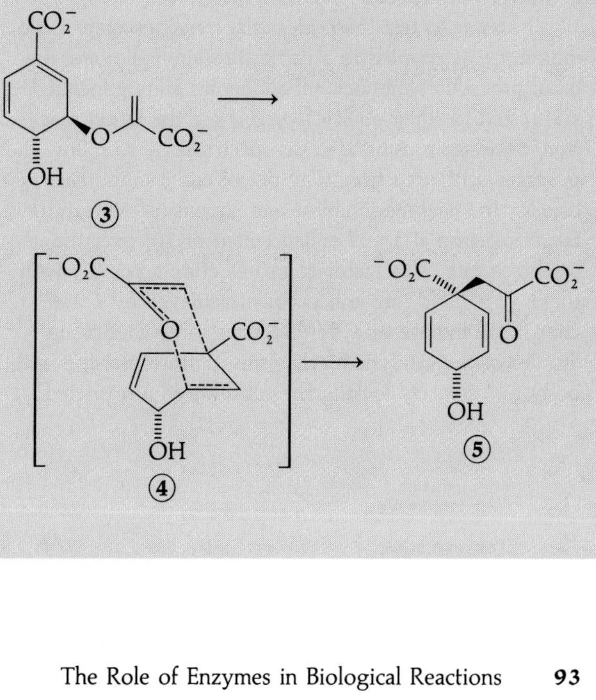

This reaction involves the concerted intramolecular cleavage of a carbon-oxygen bond and the formation of a carbon-carbon bond. The reaction is formally termed a Claisen rearrangement and is the only known example of such a reaction catalyzed by an enzyme. In order to induce the immune system to produce antibodies that catalyze this reaction we chose the following compound, a known inhibitor of the enzyme chorismate mutase[2]:

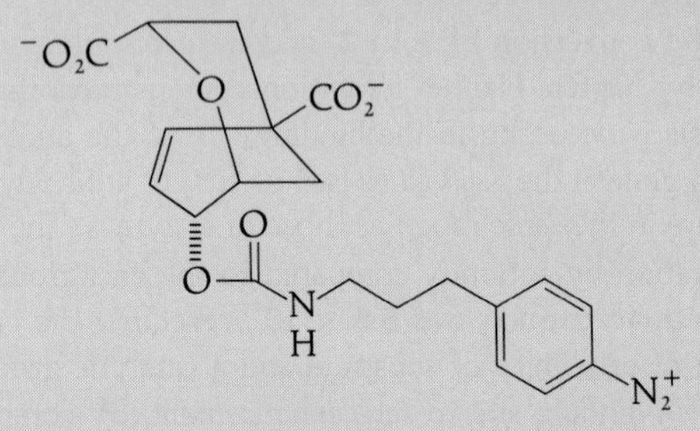

This compound belongs to a class of inhibitors termed transition state analogs. The transition state of a given reaction is the highest energy species encountered in the conversion of reactant(s) to product(s). The transition state for the conversion of chorismate to prephenate is the chairlike species **4**. Linus Pauling in 1948 pointed out that enzymes function by virtue of their ability to stabilize the transition state structure of a given reaction (in contrast, antibodies classically function by stabilizing ground states). In other words the enzyme active site is most electronically and configurationally complementary to the transition state of a reaction. Because the lifetime of a transition state is on the order of a bond vibration, 10^{-13} sec, transition states cannot be isolated and thus cannot be used as an enzyme inhibitor or hapten. Consequently, the key to the design of tight-binding enzyme inhibitors is to create stable analogs of the putative transition state species. As early as 1969 William Jencks suggested that if antibodies could be elicited against such transition state analogs they should function as enzymes and accelerate the corresponding reaction.

In order to test these ideas the transition state analog inhibitor was coupled to a carrier protein. Following the usual procedures monoclonal antibodies were generated and tested for their ability to accelerate the target reaction, once again using UV/Vis spectroscopy to follow the progress of the reaction. One out of eight antibodies specific for the enzyme inhibitor was shown to catalyze the target reaction at a rate enhancement of 10^4 over the uncatalyzed rate. This factor compares quite favorably with the 3×10^6-fold rate enhancement achieved by *E. coli* chorismate mutase assayed under the same conditions. Studies of the catalytic mechanisms indicate that this antibody functions by locking the substrate in a restricted conformation; that is, the antibody reduces the entropic barrier to the reaction to zero! This antibody was also shown to be highly stereoselective with a 30:1 preference for the (−) over the (+) isomer of chorismate. The demonstration that catalytic antibodies function as stereoselective catalysts may lead to industrial applications of this research.

An exciting prospect for the development of highly efficient catalytic antibodies is the use of biological selection to enhance the activity of existing catalysts. In the case of an antibody that catalyzes a reaction in a biological pathway, a complementation experiment can be carried out. A strain of bacteria is obtained which is deficient in the enzyme that catalyzes the relevant reaction. This bacterial strain is then transformed with a plasmid containing the gene encoding the catalytic antibody. Random mutagenesis and selections can then be carried out in which only bacteria that express antibodies with enhanced catalytic activity survive. The antibody genes from these bacteria can be sequenced in order to determine the mutations that conferred enhanced catalytic efficiency. The initial experimental requirements in the development of such biological selections have recently been satisfied using a chorismate mutase catalytic antibody.[3]

Where do we go from here? In general the rate accelerations achieved by antibodies have not yet reached those of enzymes. Investigators are working on approaches to increase the rate accelerations of existing antibodies as well as designing new strategies for generating more efficient antibody catalysts. These approaches include the use of combinatorial antibody libraries, generating antibodies that contain catalytic dyads, and the use of biological selection coupled with random mutagenesis. General strategies are also being developed for the catalysis of many biologically or chemically important reactions such as selective peptide bond hydrolysis, carbohydrate hydrolysis, glycosyl transfer, adenosine deamination, and cocaine hydrolysis. The detailed mechanisms of antibody-catalyzed reactions are also being studied in an effort to better understand how both antibodies and enzymes carry out chemical reactions. For a more comprehensive survey of catalytic antibodies a recent review is available.[4]

References

[1]Shokat, K. M.; Leumann, C. J.; Sugasawara, R.; and Schultz, P. G. *Nature (London)* 338:269–71 (1989).

[2]Jackson, D. Y.; Jacobs, J. W.; Reich, S.; Sugasawara, R.; Bartlett, P. A.; and Schultz, P. G. *J. Am. Chem. Soc.* 110:4841–42 (1988).

[3]Tang, Y.; Hicks, J. B.; and Hilvert, D. *Proc. Natl. Acad. Sci. USA* 88:8784–86 (1991).

[4]Lerner, R. A.; Benkovic, S. J.; and Schultz, P. G. *Science* 252:659–67 (1991).

Exposing Substrates to Altered Environments Active sites that effectively reduce the concentration of water molecules, or even exclude water entirely, may favor reaction of substrate molecules. The water may be excluded by tight binding of substrate molecules to pockets or crevices lined with hydrophobic amino acid side groups within the enzyme. The nonpolar environment created in the active site can greatly enhance achievement of the transition state for biochemical reactions, such as the addition of a carboxyl group, that take place more easily in nonpolar solvents. Placing such reacting systems in a nonpolar environment may reduce the activation energy by as much as 500,000 times as compared to the highly polar cellular medium.

The Active Site, Transition State, and Reversible Reactions

The contemporary view of the active site as fitting most tightly to the transition state provides an explanation for the ability of enzymes to catalyze reversible reactions in either direction. According to this viewpoint, the active site, while binding most strongly to the transition state, also has conformations that fit the reactants and the products. This multiple binding capacity is not as difficult to achieve as it might at first appear because the reactants, transition state, and products of enzymatically catalyzed reactions generally have similar overall structures. Whether reactants or products bind to the active site depends on their relative concentrations in the solution surrounding the enzyme. If the reactants are present in highest concentration, collisions between the enzyme and reactant molecules will be more likely and the reactants will bind most frequently. If the products are present in highest concentration, the reverse will be true. Once the molecules bind to the enzyme, they enter the tightly bound and unstable transition state from which they can easily be pushed in either direction, toward reactants or products.

This concept of enzymatic activity is fundamentally different from an earlier idea proposed by Emil Fischer in 1884. Fischer viewed the active site as a rigid arrangement of amino acid side groups that precisely matches and complements chemical groups on the substrate molecules (Fig. 3-8). He proposed that the fixed arrangement of side groups in the active site formed a "lock" that only a specific substrate "key" could fit. Release of the products of the reaction was believed to be automatic because the products could not fit the active site. Although now superseded, Fischer's lock-and-key hypothesis, which held sway from the late 1800s until the 1960s, was an important step on the pathway to the current understanding of enzymatic mechanisms at the active site.

FACTORS AFFECTING ENZYME ACTIVITY

A number of external factors affect the activity of enzymes in speeding the conversion of reactants to products. These factors, including variations in the concentration of substrate molecules, temperature, and pH, speed or slow enzymatic activity in highly characteristic patterns. The patterns afford many insights into enzymatic mechanisms as well as providing a means to determine whether an enzyme catalyzes a process or if a process resembles enzymatic catalysis. Many features of the transport of substances across cellular membranes, for example, follow the same patterns as enzymes in response to changes in substrate concentration. These parallels in behavior gave some of the first clues that transport molecules are proteins with many functional similarities to enzymes.

The Effects of Substrate Concentration on Enzyme Activity

Enzyme Saturation Enzymes react to alterations in the concentration of substrate molecules in highly characteristic fashion. At very low substrate concentrations

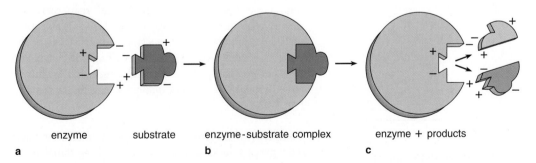

Figure 3-8 The lock-and-key model, now outmoded, for the active site. **(a)** The active site in this model contains charged, polar, and nonpolar groups shaped into a conformation that exactly matches and complements the substrate. Binding the substrate **(b)** converts it to products that no longer fit the active site. **(c)** The enzyme consequently releases the products.

collisions between the enzyme and substrate molecules are infrequent and the reaction proceeds slowly. As the concentration of substrate molecules increases, the reaction rate initially increases proportionately as collisions between enzyme molecules and the reactants become more frequent (Fig. 3-9). When the enzymes begin to approach the maximum rate at which they can combine with reactants and release products, however, the effects of increasing substrate concentration diminish. At the point at which the enzymes are cycling as rapidly as possible, further increases in substrate concentration have no effect on the reaction rate. At this point the enzyme is said to be *saturated,* and the reaction remains at the saturation level represented by the horizontal dashed line in Figure 3-9.

The characteristic saturation curve in Figure 3-9 provides a valuable biochemical tool for determining whether an enzyme speeds a given reaction or process in a biological system. To determine whether an enzymatic catalysis is involved, the concentration of reactants is increased experimentally and the rate of the reaction followed. If the reaction reaches a point at which further increases in reactants have no effect in increasing its rate, indications are good that an enzyme catalyzes the reaction. Uncatalyzed reactions, in contrast, increase in rate almost indefinitely as the concentration of reactants increases. (Supplement 3-1 presents a quantitative description of the responses of enzymes to increasing substrate concentration—the *Michaelis–Menten equation.*)

Enzyme Inhibition   The fact that enzymes combine briefly with their reactants makes them susceptible to *inhibition* by nonreactant molecules that resemble the substrate. The inhibiting molecules can combine with the active site of the enzyme, but tend to remain bound to the active site without change and block access by the normal substrate. As a result, the rate of the reaction slows. If the concentration of the inhibitor becomes high enough, the reaction may stop completely.

Some inhibitors also interfere with enzyme-catalyzed reactions by combining with enzymes at locations outside the active site. These inhibitors act by reducing the ability of the enzyme to lower the activation energy rather than by interfering with enzyme-substrate combination.

Enzyme inhibition is a normal regulatory mechanism for some cellular interactions. But it may also interfere with normal enzymatic activities, as in the case of poisons or toxins that act as enzyme inhibitors. For example, the action of cyanide and carbon monoxide as poisons depends on their ability to inhibit enzymes important in the utilization of oxygen in cellular respiration.

Most inhibitors acting as natural regulators in biological reactions combine reversibly with an enzyme, so that changes in their concentrations result in adjustments in the amount of the inhibitor combined with the enzyme. As a consequence, changes in the concentration of these inhibitors produce precise and sensitive adjustments in enzyme activity. Poisons and toxins, in contrast, typically act irreversibly by combining so strongly with enzymes that the inhibition is essentially permanent.

Reversible inhibition through direct combination of the inhibitor with the active site is called *competitive inhibition* because the inhibitor competes directly with the substrate molecule for its binding site on the enzyme. If a reversible inhibitor slows enzyme action by binding to a site elsewhere on the enzyme molecule, so that substrate molecules can still combine at normal rates with the active site, the pattern of inhibition is called *noncompetitive.* (Supplement 3-1 shows how the Michaelis–Menten equation can be used to determine whether inhibition is competitive or noncompetitive.)

Both types of reversible inhibition are common regulatory mechanisms in biological reactions. Frequently, the substance acting as an inhibitor is a product of the enzyme-catalyzed reaction. As an excess of the product accumulates, the enzymatic reaction producing it is automatically slowed, producing what is known as *end-product,* or *feedback, inhibition* (Fig. 3-10; see also Fig. 3-15).

Temperature and pH   Interactions between amino acid side groups hold proteins in their three-dimensional conformations (see p. 62). The pattern and strength of three of these interactions are highly sensitive to changes in the conditions of the solution surrounding the enzyme: hydrogen bonds and van der Waals forces

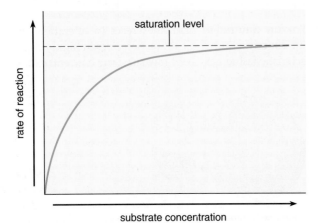

Figure 3-9   The effect of increases in substrate concentration on the rate of an enzyme-catalyzed reaction. At saturation (horizontal dashed line) further increases in substrate concentration do not increase the rate of the reaction.

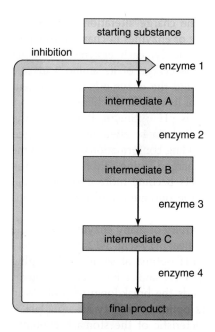

Figure 3-10 End-product, or feedback, inhibition. The final product of a multistep pathway acts as an inhibitor of the enzyme catalyzing an early step in the pathway. If the final product accumulates in excess, inhibition of the enzyme catalyzing the early step by the product turns off the entire pathway (see also Fig. 3-15).

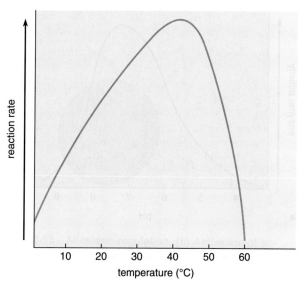

Figure 3-11 The effect of temperature on enzymatic activity. As the temperature rises, the rate of the catalyzed reaction increases proportionally until the temperature reaches the point at which the enzyme begins to denature. The rate drops off steeply as denaturation becomes complete.

to increases in temperature, and electrostatic linkages to alterations in pH.

As the temperature rises, the kinetic motions of both the amino acid chains and side groups forming an enzyme molecule increase; the strength and frequency of collisions between enzymes and surrounding molecules also increase. At elevated temperatures these disturbances become strong enough to overcome the attraction of hydrogen bonds and van der Waals forces, which are individually relatively weak. As these bonds and attractions break, an enzyme gradually unfolds and loses its native three-dimensional folding conformation.

These changes affect enzymatic activity in a characteristic way (Fig. 3-11). Over the range from 0°C to about 40°C increases in temperature have little significant effect on enzyme structure and increase enzyme activity along the lines followed by all chemical reactions: each 10°C rise in temperature approximately doubles the reaction rate. This effect results from increases in the force and frequency of collisions between enzymes and reactant molecules, reflecting the heightened kinetic motion of all molecules in the solution. As the temperature rises above 40°C, the increase in rate begins to fall off as collisions become violent enough to break hydrogen bonds and unfold the enzymes. The drop in activity becomes steep at 55°C and falls to zero at 60°C, when the disturbance in hydrogen bonding causes the enzyme to unfold into

a completely inactive, denatured form (see p. 66). Once complete, denaturation totally counteracts the positive effects of increased kinetic motion at elevated temperatures.

As a result of the two opposing effects of a rise in temperature, all enzymes have an optimum temperature at which kinetic motion is greatest but no significant unfolding has yet occurred. For most enzymes this optimum lies between 40°C and 50°C. Some enzymes have lower temperature optima. The enzymes of maize pollen, for example, have optima near 30°C and become inactive above 32°C. As a result, extremely hot weather seriously inhibits fertilization in maize. Among the enzymes with the lowest known temperature optima are those of arctic snow fleas, which are most active at −10°C. At the other extreme a few organisms, such as the bacteria living in hot springs, possess enzymes with structures so resistant to disturbance that they remain active at temperatures of 85°C or more.

Changes in pH affect enzyme structure and activity primarily by altering the charge of amino acid residues with charged groups. These amino acid residues change between charged and uncharged forms at a characteristic pH. Charge alterations of this type affect the strength of electrostatic bonds holding enzymes in their final three-dimensional shape, and also the activity of proton donors and acceptors acting as catalytic groups within the active site.

4. What does energy content mean with reference to a collection of molecules? Define kinetic and potential energy.

5. What is entropy? How do energy content and entropy interact to determine whether chemical reactions will proceed to completion?

6. What is a reversible reaction? What happens at the equilibrium point of a reversible reaction? What happens if more reactants are added to a reaction at equilibrium? If more products are added?

7. How do biological organisms run reactions that require energy? What does coupling mean?

8. It is sometimes claimed that organisms violate the second law of thermodynamics because they become more complex as they develop spontaneously from a seed or fertilized egg to an adult. Why is this statement incorrect?

9. How does ATP act as the primary agent coupling uphill and downhill reactions in living organisms?

10. Draw the structures of ATP, ADP, and AMP.

11. What are calories? Moles?

12. What effects do enzymes have on spontaneous reactions?

13. What is the activation energy of a reaction? What effects do enzymes have on the activation energy?

14. What are the characteristics of protein-based enzymes? What is the active site of an enzyme?

15. Define the transition state and outline its importance to enzyme-catalyzed reactions. What mechanisms may possibly push molecules toward the transition state at the active site of enzymes? How does attainment of the transition state affect the activation energy for a reaction?

16. Why is it likely that the active site of an enzyme can fit both the reactants and products of a reaction?

17. Outline one experiment demonstrating that the active site of enzymes also binds the transition state of the reactants.

18. Define enzyme saturation and inhibition. What is the probable molecular basis for these effects on enzymes?

19. What mechanisms regulate the active site? What is allosteric inhibition? What is end-product, or feedback, inhibition?

20. What effects do changes in temperature have on enzyme activity? Why?

21. What effects do changes in pH have on enzyme activity? Why?

22. What is an RNA-based catalyst? What types of reactions are catalyzed by these ribozymes?

Supplement 3-1

A Quantitative Treatment of Enzyme Saturation and Inhibition: The Michaelis–Menten Equation

An equation derived in 1913 by Leonor Michaelis and Maud L. Menten describes quantitatively the rate or velocity of an enzyme-catalyzed reaction, as plotted in Figure 3-9:

$$\text{velocity} = V = V_{max} \frac{[S]}{[S] + K_M} \quad (3\text{-}12)$$

in which V_{max} is the maximum velocity of the reaction when the enzyme is saturated (horizontal dashed line in Fig. 3-9), $[S]$ is the substrate concentration in moles, and K_M is a constant that reflects the affinity of an enzyme for its substrate (that is, the strength by which the enzyme binds to its substrate). Generally, the lower the value of K_M, called the *Michaelis constant*, the greater the affinity of the enzyme for its substrate. (Some representative values for K_M appear in Table 3-7.) The value of V_{max} is related to the turnover number for the enzyme being studied (the turnover number = $V_{max}/[S]$; Table 3-4 gives representative turnover numbers).

If the substrate concentration is adjusted so that it equals K_M, K_M can be substituted for $[S]$ in Equation 3-12:

$$V = V_{max} \frac{K_M}{K_M + K_M} = V_{max} \cdot \frac{1}{2} \quad (3\text{-}13)$$

Since this algebraic manipulation shows that $V = (1/2)V_{max}$ when $K_M = [S]$, the reverse statement is also true: $K_M = [S]$ when $V = (1/2)V_{max}$. Therefore, K_M *is equal to the substrate concentration when an enzyme-catalyzed reaction is running at half of its maximum velocity.* Figure 3-17 graphically depicts the relationships between V, V_{max}, $[S]$, and K_M.

In practice it is difficult to use experimental results to generate a saturation curve of the type shown in Figures 3-9 and 3-17 because many points must be plotted to draw the curve accurately (the resulting curve obtained is hyperbolic). It is also often difficult to de-

termine the saturation level, or V_{max}, directly from experiments. To solve these problems, a different approach, called the *Lineweaver–Burk plot*, is often used. For this plot reciprocals are taken of both sides of Equation 3-12:

$$\frac{1}{V} = \frac{K_M + [S]}{V_{max}[S]}$$

$$\frac{1}{V} = \frac{K_M}{V_{max}[S]} + \frac{[S]}{V_{max}[S]}$$

$$\frac{1}{V} = \frac{K_M}{V_{max}} \cdot \frac{1}{[S]} + \frac{1}{V_{max}} \quad (3\text{-}14)$$

Equation 3-14 has the form of a linear equation with slope equivalent to K_M/V_{max} and the intercept of the y-axis equivalent to $1/V_{max}$. This mathematical technique converts the hyperbola of Figure 3-17 into a straight line that can be plotted accurately from a relatively few experimental points (Fig. 3-18). Since the y-intercept is equivalent to $1/V_{max}$ and the x-intercept to $1/K_M$, the values for V_{max} and K_M can be determined quickly and accurately.

The Michaelis–Menten equation and its Lineweaver–Burk conversion accurately describe the behavior of many enzymes. The Lineweaver–Burk plot is especially effective as a means for determining whether inhibition is competitive or noncompetitive. Competitive inhibitors operate by combining with the active site of the enzyme. Because the combination is reversible and concentration-dependent, addition of

Table 3-7 The K_M Values of Some Enzymes	
Enzyme	K_M Value
Aminotransferase	0.0009
Carbonic anhydrase	0.008
Catalase	0.025
Chymotrypsin	0.005
Glutamic acid dehydrogenase	0.00012
Hexokinase	0.00015
Pyruvic acid carboxylase	0.004

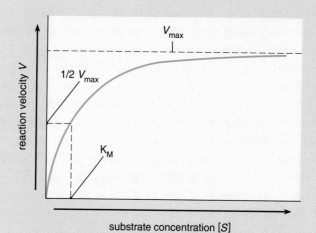

Figure 3-17 Enzyme saturation in terms of reaction velocity V, substrate concentration $[S]$, maximum velocity V_{max}, and the Michaelis constant K_M (see text).

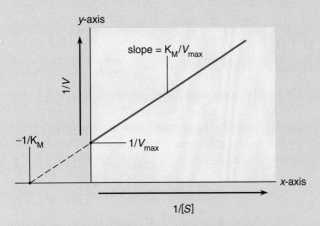

Figure 3-18 A plot of the reciprocals of the velocity ($1/V$) and of the substrate concentration ($1/[S]$). This plot, the Lineweaver–Burk plot, produces a straight line with the y-intercept equivalent to $1/V_{max}$ and the x-intercept equal to $-1/K_M$. The slope of the line is equivalent to K_M/V_{max} (see text).

more substrate reduces the amount of inhibitor bound to the enzyme. This, in turn, reduces the relative concentration of the inhibitor molecules. At very high substrate concentrations, practically all the enzyme molecules form enzyme-substrate complexes rather than enzyme-inhibitor complexes. As a result, with the addition of more substrate V_{max} can eventually reach its normal level, and the value obtained for $1/V_{max}$ in the Lineweaver–Burk plot (the y-intercept) remains unchanged (Fig. 3-19). At substrate concentrations below

the levels required to reach V_{max} in the presence of the inhibitor, the velocity of the reaction will be slower for a given substrate concentration because of interference with the active site by the inhibiting molecule. In other words, the affinity of the enzyme for the substrate (K_M) is altered. The reduction in velocity has the effect in the Lineweaver–Burk plot of increasing the slope of the line $1/V$ against $1/[S]$ and altering the x-intercept ($-1/K_M$). Thus, a Lineweaver–Burk plot of a reaction taking place in the presence of a competitive inhibitor has the same y-intercept ($1/V_{max}$) but an altered slope and x-intercept ($-1/K_M$).

In noncompetitive inhibition, since the inhibitor slows the enzyme by combining outside the active site, adding more substrate does not reduce the amount of inhibitor bound to the enzyme. Therefore, adding more substrate does not restore the reaction rate, and V_{max} never reaches the level attained in the absence of the inhibitor. Since the ability of the enzyme to combine with the substrate (enzyme affinity) remains the same, however, K_M has its usual value. Thus, a Lineweaver–Burk plot for a reaction slowed by noncompetitive inhibition will show the same x-intercept ($-1/K_M$) but an altered y-intercept ($1/V_{max}$) and slope (Fig. 3-20). In summary:

competitive inhibitors: increase in K_M, no change in V_{max}

noncompetitive inhibitors: reduction in V_{max}, no change in K_M

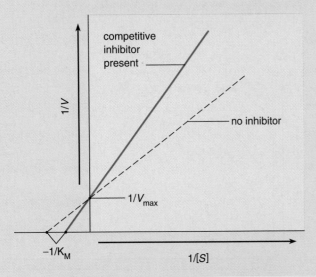

Figure 3-19 Alterations of the Lineweaver–Burk plot for an enzyme-catalyzed reaction as a result of competitive inhibition. The plot has the same y-intercept ($1/V_{max}$) as a reaction with no inhibition but an altered slope and x-intercept ($-1/K_M$).

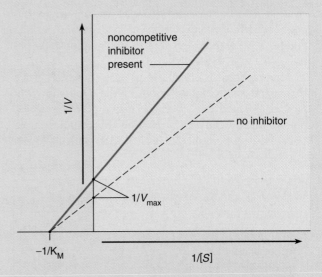

Figure 3-20 Alterations of the Lineweaver–Burk plot for an enzyme-catalyzed reaction as a result of noncompetitive inhibition. The plot has the same x-intercept ($-1/K_M$) as a reaction taking place without inhibition but an altered y-intercept ($1/V_{max}$) and slope.

4

MAJOR INVESTIGATIVE METHODS OF CELL AND MOLECULAR BIOLOGY

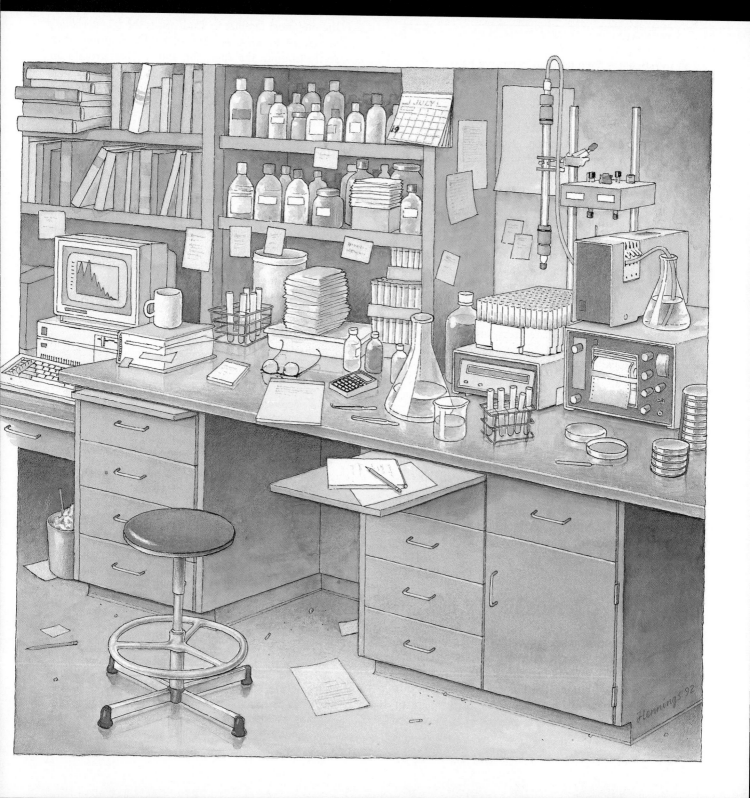

Investigators have used an extensive battery of experimental methods to develop the observations, hypotheses, and conclusions of cell and molecular biology—the subjects of this book. These techniques have been used to observe the structure and behavior of whole cells, cell organelles, and individual cellular molecules. In recent years technical approaches to the study of molecules, particularly nucleic acids and proteins, have become so powerful that molecular biology is now a distinct field of study. Methodology is presently so interwoven with results that it is impossible to understand cell and molecular biology, and the information developed in other areas of biology by the molecular approach, without a knowledge of the techniques used to study cells and molecules.

This chapter surveys these methods, beginning with light and electron microscope techniques employed to study cells and cell structures, and analytical approaches developed around these techniques. Methods for growing cells in quantity in tissue culture are then taken up, followed by the techniques used for breaking cells and obtaining purified fractions of cellular organelles and molecules. The chapter then discusses methods for studying individual cell molecules and concludes with an extended explanation of the molecular methods that have revolutionized biology. These include techniques for obtaining the DNA of specific genes; for increasing the number of copies by cloning and the polymerase chain reaction; for sequencing nucleotides in DNA and RNA molecules; and, through these sequences, for determining the amino acid sequences of proteins.

LIGHT AND ELECTRON MICROSCOPY

Light and electron microscopes have been highly important tools in many of the discoveries of cell and molecular biology. Until the early 1960s microscopical techniques were among the primary methods used to study cells. Although the importance of these instruments to research has been overtaken by molecular

methodology in recent years, many technical approaches still employ light or electron microscopy to analyze cell structures and functions. The light microscope, in particular, is also among the equipment that students in cell and molecular biology are most likely to use.

The Light Microscope and Its Applications

Light microscopes are used in several applications that take advantage of different light sources and patterns of image formation. The most common application, and the one most familiar to students, is the *brightfield light microscope*, in which a beam of light is transmitted directly through the specimen and the background appears bright. Other applications commonly used to meet specialized viewing or specimen requirements are the *darkfield, phase contrast, polarizing,* and *fluorescence microscopes*.

The Brightfield Light Microscope In a brightfield light microscope (Fig. 4-1) light rays from an illumination source such as the sun or an incandescent bulb are focused on the specimen by a condenser lens. The rays leaving the specimen are focused into a magnified image by two lenses placed at either end of a tube. (Supplement 4-1 outlines the principles governing image formation by the glass lenses used in light microscopes.) The lens nearest the specimen in the tube is the *objective*. The lens at the opposite end of the tube is the *ocular*.

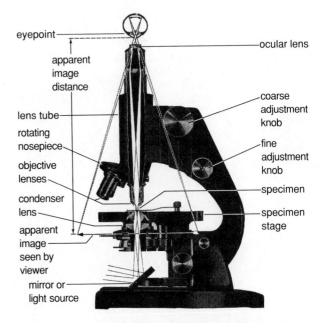

Figure 4-1 The brightfield light microscope (see text). Objective lenses of different magnifying power can be selected by rotating the nosepiece. Redrawn from an original courtesy of the American Optical Company.

To correct for inherent faults in the glass lenses used in light microscopes, the objective and ocular are constructed from a series of lenses placed close together, usually as many as eight to ten for the objective and two to three for the ocular. The individual lenses in the objective or ocular are mounted so close together that they act in effect as a single highly corrected lens.

The image is observed by looking directly into the ocular lens. Coarse and fine controls move the specimen stage or lens tube in order to place the specimen in the correct position for focusing by the objective lens. The position of the condenser can also be adjusted so that the light from this lens converges on the specimen and spreads into a cone that exactly fills the objective lens.

The extent to which a microscope can distinguish fine details in the specimen as separate, distinct image points is termed its *resolution*. This property depends on the interaction of several factors in a light microscope:

$$\text{resolution} = d = \frac{0.61\lambda}{n \sin \alpha} \qquad (4\text{-}1)$$

In this equation λ is the wavelength of the light illuminating the specimen, n is the refractive index[1] of the transmitting medium surrounding the specimen and filling the space between the condenser and objective lens, and α is the half-angle of the cone of light entering the objective lens (Fig. 4-2). The quantity 0.61 is a constant describing the degree to which image points can overlap yet still be recognized as separate points by an observer. Since resolution is a measure of the ability of a microscope to image fine details, the quantity d becomes smaller as resolution improves. Therefore, for best resolution 0.61λ should take on the smallest possible value and $n \sin \alpha$ the largest value.

In a good light microscope operated under the best conditions, all the quantities in Equation 4-1 take on limits at fixed values. The value for n can be pushed to its maximum by placing a drop of immersion oil (refractive index of approximately 1.5) in the space between the objective lens and the specimen. In some microscopes immersion oil also can be placed between the condenser lens and specimen to provide optimum

---

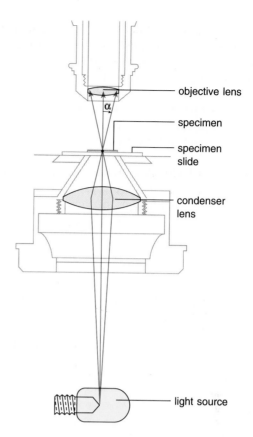

Figure 4-2 Light path from the condenser to the objective lens. For maximum resolution the half-angle (α) of the cone of light entering the objective lens should be as large as possible. The maximum half-angle obtainable is about 70°.

conditions for observation. The half-angle of the cone of light entering the objective lens in the best microscopes is about 70°, which gives a maximum value for $\sin \alpha$ of about 0.94. The quantity $n \sin \alpha$, called the *numerical aperture* of the objective lens, takes on a maximum value of about 1.4.

This leaves λ, the wavelength of the light used as an illumination source, as a variable in the equation. To keep d as small as possible, λ must take on a minimum value. The lower limit of this value is fixed by the shortest wavelength of visible light usable for illumination, which is light in the blue range at a wavelength of about 450 nm. Substituting this value and $n \sin \alpha = 1.4$ in Equation 4-1 shows that the best resolution possible with microscopes using visible light approaches 0.2 μm, or 200 nm. Using ultraviolet light as an illumination source ($\lambda = 250$ nm) pushes this value down to about 0.1 μm. At this level cell organelles such as nuclei, chloroplasts, and mitochondria are clearly resolved, but smaller details such as microtubules and ribosomes remain invisible.

Objects resolved at 0.2 μm by a high-quality light microscope must be magnified about a thousand times to be distinguished by the human eye. This is accomplished in most light microscopes by the combination of a 100 × oil-immersion objective lens with a 10 × ocular. Magnification beyond this level does not improve the resolution of small objects viewed in the light microscope. For this reason enlargement of the image beyond 1000 × is often termed "empty" magnification in light microscopy.

The Darkfield Light Microscope Darkfield illumination improves the visibility and apparent resolution of very fine specimen points in the light microscope. In this application an opaque disc is placed in the center of the condenser so that light can pass only around its edges. As a result, a hollow cone of light illuminates the specimen. The hollow illuminating cone is adjusted to an angle wide enough so that no light transmitted directly through the specimen can enter the objective lens (Fig. 4-3). Only light scattered to smaller angles by specimen points enters the objective lens, making these points appear bright against a dark background.

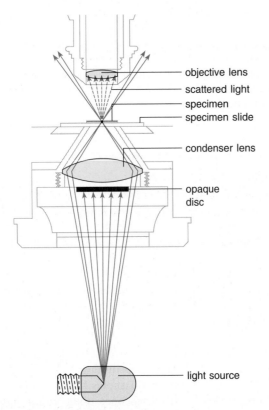

Figure 4-3 Light path from the condenser to the objective lens in darkfield illumination. An opaque disc blocks the central region of the condenser so that rays transmitted through the specimen without scattering cannot enter the objective lens. As a result, only scattered rays contribute to image formation. Specimen points that scatter light into the objective lens therefore appear bright against a dark background.

The method improves apparent resolution because objects too small to be resolved directly can still scatter and divert the paths of light waves. The criterion for visibility is simply the amount of scattering by the point. Darkfield microscopy allows objects as small as individual microtubules, with diameters of about 25 nm, to be visualized (as, for example, in Fig. 11-24).

The Phase Contrast Light Microscope Many of the resolvable structures in living cells, although differing in refractive index, are equally transparent and colorless and thus show little contrast in the ordinary brightfield light microscope. The phase contrast microscope, in which differences in the refractive index of the specimen appear as regions of differing brightness in the image, greatly improves the visibility of these structures.

The phase contrast microscope takes advantage of the fact that light rays both refract and slow as they pass through transparent objects of differing refractive index. On the average the light waves passing through the components of higher refractive index in cells are delayed by about one-fourth wavelength and are bent to a greater degree than the paths followed by waves passing through less refractive components. In an ordinary brightfield microscope the more greatly refracted rays follow a different path through the objective lens but are focused back into the same image as the less refracted rays. Since the rays refracted to greater and lesser degrees differ in phase by only one-fourth wavelength on the average, they do not interfere with each other significantly and cause few visible alterations in the brightness of image points.

Interference between the rays is increased in the phase contrast microscope by the introduction of a *phase plate* into the paths followed by the more greatly refracted rays. The phase plate is coated with a layer of transparent material of sufficient refractive index and thickness to delay the more greatly refracted rays by an additional one-fourth wavelength. Because the phase plate is uncoated in the central regions corresponding to the paths followed by the less refracted rays, these rays pass through the plate without additional delay.

The objective lens focuses all rays into the same image points. At image points corresponding to components of higher refractive index in the specimen, the crests of the more greatly refracted rays, since they now differ in phase from the unrefracted rays by one-half wavelength, coincide in the image with the troughs of the less refracted waves. The two waves therefore cancel each other, producing an interference that reduces the brightness of the image at these points. The effect over the entire image is to create differences in brightness corresponding to differences in the refractive index of the specimen. The phase contrast image makes many structures in living cells clearly visible without staining or alterations of any kind, and without significant effects on resolution (Fig. 4-4).

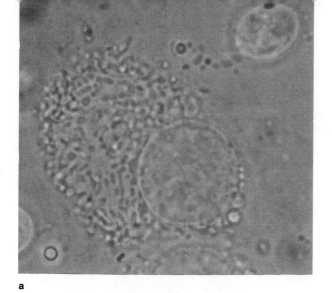
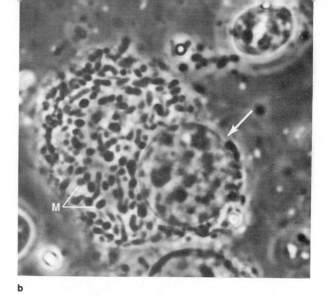

a   **b**

Figure 4-4 The effects of phase contrast on the light microscope image. **(a)** Brightfield image of a living cell from a frog liver. **(b)** The same living cell viewed with phase contrast optics. The nucleus is marked by an arrow; numerous bodies, including mitochondria (M), are visible in the cytoplasm. ×2,000. Courtesy of B. R. Zirkin.

The Polarizing Light Microscope The light sources used for brightfield microscopy contain individual light waves that vibrate in all planes. For some specimens, particularly those containing molecules or structures arranged in highly ordered patterns, valuable information can be obtained by viewing the specimen in light vibrating in only one plane. This is accomplished in the polarizing light microscope by the introduction of a filter in or below the condenser lens that blocks all waves except those vibrating in one plane. The light filtered in this way is said to be *polarized.*

In the polarizing light microscope a second polarizing filter is usually placed on the side of the specimen toward the objective lens. If this filter, called the *analyzer,* is rotated to the same orientation as the one in the condenser lens, so that the direction of filtering coincides, the light rays leaving the condenser also pass through the analyzer, and the field of view appears bright. If the analyzer is rotated 90° from this position, the light passing from the condenser without alteration in the specimen is blocked, and the background appears dark.

Crystals and some highly ordered biological structures can also alter the vibration planes of transmitted light. In effect, the light waves passing through such objects become polarized into two planes vibrating at right angles to each other. The light is also slowed by the passage, with the degree of retardation different in the two planes leaving the object. This has the effect of giving the object two differing refractive indices. Such objects are said to be *birefringent.*

Birefringent objects are examined in a polarizing microscope initially set with first and second filters placed at 90° to each other, so that the entire field of view is dark. The microscope stage is then rotated until one of the planes of vibration in the light leaving the object coincides with the polarizing direction of the analyzer. As an angle of coincidence is reached, the light leaving the object passes through the analyzer, and the object appears bright against the dark background. Because the light leaving a birefringent object is polarized in two planes 90° apart, there are four positions at which the microscope stage can be set to make the object appear bright.

The parallel alignment of microfilaments in striated muscle, and of microtubules in the spindle, for example, makes these structures birefringent and visible in the polarizing microscope (as in Fig. 24-14). This observation, made long before the electron microscope revealed the existence of either microfilaments or microtubules, indicated that both striated muscle and the spindle contain submicroscopic fibers arranged in highly ordered, parallel arrays.

The Fluorescence Microscope The fluorescence microscope is designed to detect the light emitted by fluorescent molecules tagged to a cell structure of interest, such as a microtubule or microfilament. One of the most-used applications of the method is *immunofluorescence,* in which fluorescent dye is attached to an antibody molecule that can recognize and bind specifically to the structure of interest.

Fluorescence results from the absorption and release of light energy by electrons in a molecule. On absorption of light, which must be at a particular wavelength for an electron occupying a given orbital, an electron rises from a lower to a higher energy level (from the ground state to an excited state; see p. 370). Excited electrons are highly unstable and tend to return quickly to the ground state, releasing the energy of the absorbed wavelength as light and heat. The energy released as light is termed *fluorescence.* Because some of the energy is also lost as heat, the wavelength of the light released as fluorescence contains less energy than the light originally absorbed and is therefore at a longer wavelength (for details, see p. 370).

In fluorescence microscopy a specimen is stained with a fluorescent dye and illuminated with light filtered to remove all wavelengths except the wavelength absorbed by the dye. In response to light absorption the dye fluoresces, releasing light at a different but characteristic wavelength. The fluorescent light leaving the specimen passes through another filter, the *barrier filter,* which eliminates all wavelengths except the one fluoresced by the dye. Since the illuminating wavelength is different from the wavelength of the fluorescence, little light passes through the barrier filter except for the fluoresced light. As a consequence the sites marked by the fluorescent dye glow strongly against a dark background (as, for example, in Fig. 1-20).

Ultraviolet light is frequently used as an illuminating source in fluorescence microscopy. This application uses a fluorescent dye that absorbs light in the ultraviolet range and fluoresces at a longer, visible wavelength. This fluorescence can be seen without a barrier filter because the illuminating ultraviolet light is invisible to the human eye.

Electron Microscopy

Electron microscopes use electrons as an illumination source to take advantage of the very short wavelengths attainable in electron beams, which permit a considerable improvement in resolution over the light microscope. The electron beams are focused by magnetic or electrostatic fields to produce an image. In the most common systems precisely controlled electric currents are passed through massive coils of wire to generate magnetic focusing fields. An iron *pole piece* (Fig. 4-5) inserted into the axis of a wire coil shapes the magnetic field into the three-dimensional configuration required for focusing electrons. Because electron lenses can be varied in focus by altering the current applied to their wire coils, electron microscopes are focused by changing lens current rather than by moving the lenses as in light microscopy.[2] For the same reason alterations in

[2] Technically, changes in lens current alter the *focal length* (the distance from the center of the lens to the point of focus) of an electron lens. Although electrons follow helical and highly complex paths through the field of a magnetic lens, the net effect is the same as the focusing of light rays by a convex glass lens. Electrons approaching the lens in a parallel beam are focused into a *focal point* just beyond the lens. By adjusting the current applied to the electron lenses, the position of the focal point, and thus the focal length, changes. As the focal length becomes shorter in response to increases in lens current, the magnification of the lens increases.

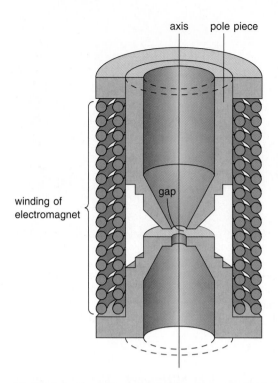

Figure 4-5 A magnetic lens. The magnetic field, generated by a current passed through the winding, is given a three-dimensional form by the metallic pole piece that focuses electrons. The gap in the pole piece is particularly important in shaping the magnetic field. Redrawn from R. B. Setlow and E. C. Pollard, *Molecular Biophysics*, courtesy of R. B. Setlow.

lens current rather than substitution of lenses are used to adjust the magnification of an electron microscope.

Because the illumination source is a beam of electrons, all the space traversed by electrons inside the electron microscope must be kept under a high vacuum. Otherwise, the electrons of the beam, which have relatively poor power to pass through matter, would be completely scattered and absorbed by gas molecules in the microscope. This places special requirements on the specimen, which must be dry and nonvolatile.

Electron microscopes are employed in several different applications. Of these, two—the *transmission electron microscope (TEM)* and the *scanning electron microscope (SEM)*—are most common.

The Transmission Electron Microscope (TEM) The TEM is so called because the electron beams forming the image pass through the specimen. In construction a TEM resembles an inverted light microscope (Figs. 4-6 and 4-7). At the top of the central column is the illumination source, the *electron gun,* consisting of a *filament* and an *anode.* An electric current heats the filament, a thin tungsten wire, to a high temperature, which causes electrons to be driven from its surface. The filament and its holder, which are well insulated from the rest of the column, are maintained at a high

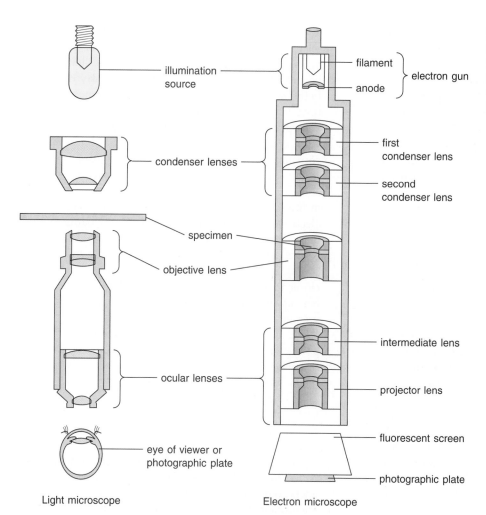

Figire 4-6 The arrangement of the illumination source, lenses, specimen, and viewing screen in a transmission electron microscope (right) resembles an inverted light microscope (left).

illumination source

filament
anode
electron gun

condenser lenses

first condenser lens

second condenser lens

specimen

objective lens

intermediate lens

projector lens

ocular lenses

fluorescent screen

eye of viewer or photographic plate

photographic plate

Light microscope

Electron microscope

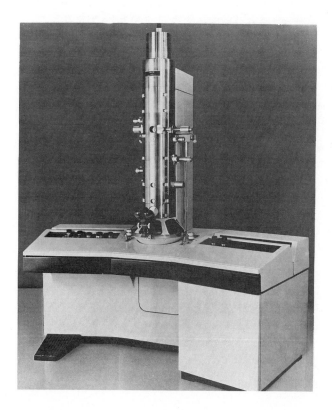

negative voltage, −50,000 to −100,000 volts (V) in most microscopes. The anode is grounded and is thus positive with respect to the filament. Consequently, electrons leaving the filament are strongly attracted to the anode. As they travel from the filament to the anode, the electrons accelerate to a velocity that depends on the voltage difference between the two locations. In electron microscopes operated at voltages

◀ **Figure 4-7** A high-performance transmission electron microscope. The central column, which is maintained under a vacuum, houses the electron lenses. To the right and left of the column are banks of controls used to regulate the lens current and thus the magnifying power of the lenses. The image, formed on a fluorescent screen at the base of the column, is viewed through the windows with a binocular microscope. Photographic plates can be exposed at a level just below the screen to make permanent records of the image. Airlocks are provided for the introduction of specimens and photographic plates. Courtesy of the Perkin-Elmer Corporation.

Light and Electron Microscopy   **113**

between $-50,000$ and $-100,000$ V, the velocity attained produces wavelengths in the range between 0.005 and 0.003 nm.

High-velocity electrons leave the gun through a hole in the center of the anode. Just below the gun (see Fig. 4-6), a series of two condenser lenses focuses a very small, intense spot of electrons on the specimen. The specimen can be moved, allowing different regions to be illuminated by the spot. It can also be held in a stable position, with movements no greater than an angstrom or so, to allow photographs to be made. Another set of lenses, usually including an objective, intermediate, and projector lens, focuses the electrons passing through the specimen. Each lens in the train forms successively magnified images, with total magnification varying from a few thousand to 300,000 times or more depending on the current applied to the lenses.

The final lens in the train, the projector, focuses the magnified image onto a fluorescent screen at the bottom of the column. This screen, similar to the screen of a television tube, is coated with crystals that respond to electron bombardment by emitting visible light. This process converts the electron image to a visual image. Exposure of a photographic plate to the electron beam at the level of the screen permanently records the image. Because photographic emulsions respond to electrons and light in essentially the same way, ordinary films and plates can be used for this purpose. Airlocks are provided so that specimens and photographic plates can be exchanged without disturbing the microscope vacuum.

Although the TEM resembles a brightfield light microscope in construction and operation, the interaction between specimen points and the illuminating beam in the two instruments is significantly different. In the brightfield light microscope development of contrast in the image depends primarily on differential absorption of light by structures within cells. In the TEM electron *scattering* rather than differences in absorbance produces contrast in the image. Scattering results from an interaction between specimen atoms and electrons of the illuminating beam. Atomic nuclei in the specimen, which are positively charged, scatter electrons by attracting them from their paths in the illuminating beam. The negatively charged electron clouds around atomic nuclei scatter electrons by repelling them. These effects increase as electrons pass closer to specimen atoms. Nuclei of high atomic number, as in atoms of heavy metals such as lead and uranium, cause the widest scattering.

If electrons are scattered widely enough by image points (Fig. 4-8a), they are eliminated from the electron beam leaving the specimen, some by collisions with the sides of the lenses and some by a narrow aperture (Fig. 4-8b) placed for this purpose below the objective lens. The aperture is small, 50 μm or less, so that only the unscattered electrons and a relative few of the scattered electrons pass through.

Elimination of the scattered electrons creates a "hole" in the beam corresponding to a scattering specimen point. After magnification the image of the specimen point, focused on the fluorescent screen, appears

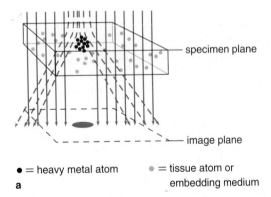

• = heavy metal atom • = tissue atom or embedding medium

a

Figure 4-8 Image formation in the TEM. **(a)** Scattering of electrons by an object point, represented by a cluster of heavy metal atoms in the specimen. As a result of scattering, fewer electrons fall on the corresponding region of the focused image, producing a "shadow" of the object point. **(b)** Elimination of scattered electrons (dashed lines) by an aperture below the objective lens. The aperture is so small that a large proportion of the scattered electrons, which otherwise would fall on the image plane as a general background fog, are screened out. **(b)** Redrawn from F. S. Sjostrand, *Electron Microscopy of Cells and Tissues*, 1967. Courtesy of F. S. Sjostrand and Academic Press, Inc.

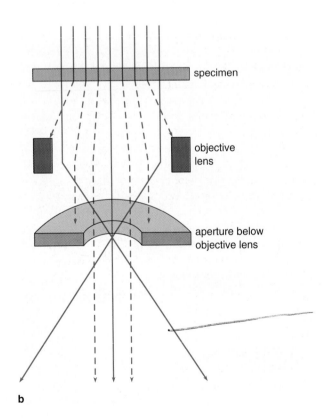

b

darker than surrounding regions because fewer electrons per unit time strike the screen in this area. The images of all other parts of the specimen containing scattering points also appear dark on the screen. These points collectively trace out the electron image in micrographs such as in Figure 1-1.

The improvement in resolution offered by electron microscopes can be calculated from the resolution equation, which differs slightly in its application:

$$\text{resolution} = d = \frac{0.61\lambda}{\alpha} \qquad (4\text{-}2)$$

Note that the equation uses α, rather than $n \sin \alpha$. Because the medium separating the image and the objective is a vacuum, the refractive index (n) has a value of 1 and is omitted from the equation. α is used rather than $\sin \alpha$ because the angle of the cone of light entering the objective lens is so small that α and $\sin \alpha$ have equivalent values. (The value for α, approximately 4×10^{-3} rad $= 0.23°$, is kept small to minimize the effects of inherent and presently uncorrectable defects in magnetic lenses.)

For a TEM operated at 50,000 V, the wavelength of the electron beam is 0.005 nm. Substituting $\alpha = 4 \times 10^{-3}$ rad and $\lambda = 0.005$ nm in Equation 4-2 gives an expected resolution of about 0.7 to 0.8 nm for a TEM operating at 50,000 V. Operation at 100,000 V, which shortens λ to 0.0037 nm, improves resolution to about 0.5 to 0.6 nm. Commercially available TEMs regularly achieve resolution in this range, which is about 100 times better than the best resolution of the light microscope.

At these levels the TEM can easily resolve structures such as ribosomes, microtubules, microfilaments, and large molecules such as proteins. Even images of individual heavy metal atoms have been produced under special operating conditions. Some gains in resolution beyond these levels have been obtained by increasing the accelerating voltage to a million volts or more. However, because scattering by specimen atoms is inversely proportional to the velocity of the beam electrons, the contrast of specimen details decreases as the accelerating voltage increases. At 1,000,000 V, the contrast of most biological objects is so poor that the added resolving power is frequently essentially useless. Specimen details are also often destroyed by heating and by chemical effects of the electron beam. As a consequence, high-voltage TEMs have produced little practical improvement in the resolution of details in biological specimens.

The apparent resolution of electron microscopes can be improved by darkfield operation, as in light microscopy. A darkfield effect is produced by placing a metal barrier above the specimen to produce a hollow beam of electrons, focused so that unscattered electrons fall outside the aperture of the objective lens (equivalent

in principle to Fig. 4-3). Alternatively, the condenser lens is tilted so that electrons focused on the specimen by the condenser and transmitted through the specimen without scattering fall outside the aperture of the objective lens. In either method only scattered electrons pass through the objective lens and contribute to the image. Because specimen points with dimensions below the resolving power of the TEM may still scatter electrons, the darkfield method allows such points to be visualized. Individual atoms and details of molecular structure have been made visible by this approach (Fig. 4-9).

The Scanning Electron Microscope (SEM) The SEM, widely used to examine the surfaces of cells or isolated cellular structures, differs extensively in its construction and operation from the TEM (Fig. 4-10). Only the illumination source and condenser lenses are similar. An electron gun produces an electron beam in the SEM, which is focused into an intense spot on the specimen surface by a magnetic lens system analogous to the condenser lens of a TEM. Rather than being stationary, however, the focused spot moves rapidly or *scans* back and forth over the specimen. The scanning movement is accomplished by beam deflectors, charged plates between the condenser lenses and the specimen.

The intense spot of electrons scanning the specimen excites molecules on its surface to high energy levels. The excited molecules release this energy in several forms, including high-energy electrons called *secondary electrons*. The image in an SEM is formed from the secondary electrons rather than the illuminating beam.

There are no other lenses in the SEM. The secondary electrons leaving a particular spot on the specimen surface are picked up by a *detector* at one side of the specimen (see Fig. 4-10). The detector consists of three elements: a small fluorescent screen, a photomultiplier, and a photoelectric cell. The screen emits

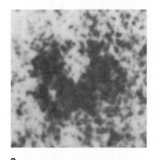 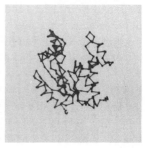

a   b

Figure 4-9 Darkfield electron microscopy. **(a)** A darkfield electron micrograph of a molecule of the enzyme myokinase. The contrast has been reversed, so that the molecule appears dark against a light background. ×5,250,000. **(b)** The structure of myokinase, as deduced from X-ray diffraction. Courtesy of F. P. Ottensmeyer; reproduced, with permission, from *Ann. Rev. Biophys. Bioeng.* 8: 129 (1979) © 1979 by Annual Reviews, Inc.

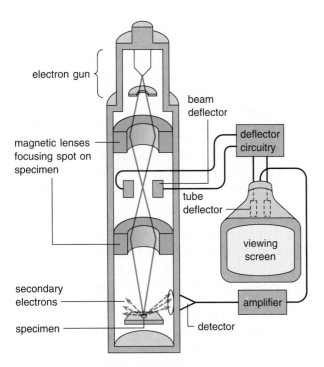

Figure 4-10 The scanning electron microscope (see text).

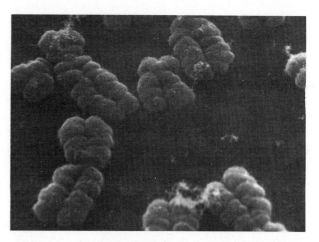

Figure 4-11 A scanning electron micrograph of metaphase chromosomes prepared by the G-banding technique (see p. 1043). Courtesy of C. J. Harrison, from *Cytogenet. Cell. Genet.* 35:21 (1983).

visible light when bombarded by secondary electrons leaving the specimen surface. The photomultiplier amplifies the emitted light and projects it into the photoelectric cell. The photoelectric cell, which works on the same principle as the light meter on a camera, emits a current proportional to the amount of light striking it.

The current leaving the photoelectric cell is used to create an image in a television tube. In the tube an electron gun produces a narrow beam of electrons that is focused into a spot at the front of the tube. Deflectors inside the tube, connected to the same electronic circuits scanning the beam in the microscope, move the spot back and forth across the screen at the same rate and in the same direction as the spot scanning the specimen. As a result, the spots on the screen and specimen are at corresponding points at any instant.

When the scanning spot in the microscope strikes a point on the specimen surface that emits large numbers of secondary electrons, the detector adjusts the scanning spot on the television screen at that point and instant to a high level of brightness. At specimen points emitting fewer secondary electrons, the detector current is reduced, and the spot scanning the screen at the same point and instant dims. Each point on the screen therefore corresponds in brightness to the number of electrons emitted by the same relative points on the specimen. The scanning is so rapid that an apparently instantaneous image of the specimen surface appears on the screen, with bright and dark areas on the screen

corresponding to ridges and valleys on the specimen surface. The result is an apparent three-dimensional reconstruction of the specimen surface (Fig. 4-11).

Resolution in the SEM depends primarily on the size of the spot scanning the specimen. As the spot becomes smaller, the number of secondary electrons emitted will be determined by smaller details on the specimen surface, allowing these to be recognized as separate points with different degrees of brightness on the viewing screen. A scanning spot 5 to 10 nm in diameter, typical of contemporary SEMs, resolves surface details of about the same dimensions.

The fact that the SEM uses electrons emitted from surfaces rather than transmitted electrons for image formation means that relatively thick specimens can be observed in the microscope. However, the lens system, specimen, and detector must be kept at high vacuum to avoid scattering and absorption of the secondary electrons emitted by the specimen surface. This operating restriction requires that the specimen be dry or that it release gas or water molecules too slowly to disturb the vacuum. This requirement makes it impractical to examine most living organisms in the SEM. However, a few organisms that are highly resistant to desiccation, such as tardigrades, have survived the vacuum and electron bombardment long enough to produce a usable image in the SEM.

The SEM produces excellent images of object surfaces ranging in size from whole cells to small insects. Since its resolution is about 20 times better than the light microscope, the surfaces of objects in this size range are imaged with significantly greater fidelity. The limited resolution of the SEM as compared to the TEM makes it less useful for observing structures smaller than whole cells, however.

PREPARATION OF SPECIMENS FOR MICROSCOPY

Living specimens are observed routinely in the light microscope, particularly with phase contrast microscopy. However, for many applications of light microscopy, and essentially all uses of the electron microscope, operating limitations of the instruments make it difficult or impossible to observe living material. These limitations are circumvented to a large degree by an extensive battery of preparation techniques that greatly improve the visibility and resolution of structures inside cells. Although some of these methods do allow living material to be observed, most are designed to kill and preserve cells in forms as close to the living state as possible.

Among the more important limitations of light and electron microscopes is the fact that both light and electrons have limited ability to penetrate biological structures. As a result, specimens must be very thin. This requirement is especially stringent for electron microscopy—specimens to be viewed in the TEM cannot be thicker than about 60 to 70 nm, placing their thickness well below the resolving power of the light microscope! Most biological specimens, as a result, must be cut into thin slices for examination, about 10 μm for light microscopy and 20 to 40 nm for electron microscopy. To maintain the structural integrity of such thin slices, biological material must be stabilized by treatment with *fixatives*, chemicals that crosslink or coagulate specimen details to hold them in place. For some applications specimens are quick-frozen as part of the fixation process. If the material is to be sectioned, it is embedded in a supportive medium, usually paraffin or methocrylate plastic for light microscopy or an epoxy plastic for electron microscopy. These techniques are complicated in electron microscopy by the fact that the specimen and its surroundings are suspended in a near-perfect vacuum inside the microscope. As a result, the fixed and embedded specimen must be in a dry, nonvolatile form. For most applications, these operating restrictions prevent observation of living material.

Other important limitations stem from the requirement for specimen thinness. With few exceptions biological structures in thin specimens produce relatively little contrast in the image. This limitation is especially severe in electron microscopy, because almost all atoms present in biological materials scatter electrons to approximately the same extent. This limitation is circumvented for light and electron microscopy by specimen *staining*—the addition of colored pigments for light microscopy or heavy metal ions for electron microscopy. The contrast of particulate electron microscope specimens is also increased by *shadowing*, in which a coat of heavy metal atoms is cast along one side of specimen details.

Many of the techniques used to increase specimen contrast, particularly in light microscopy, have been modified so that they stain or mark only certain structures or molecular types within specimens. Among the most important analytical methods are *autoradiography*, in which specimen details are marked by combination with radioactive atoms, and *immunochemistry*, in which antibodies are used to attach pigments or heavy atoms to specific structures or molecules in the specimen. These approaches are often combined with *enzymatic digestion* to remove molecules of specific types from the specimen. As a result of these applications, staining techniques can be used analytically to identify the molecules present in cell structures as well as to increase contrast.

Specimen Preparation for Light Microscopy

Fixation The chemical fixatives used in light microscopy are reactive substances, such as acetic acid, alcohol, formaldehyde, and glutaraldehyde. Acetic acid and alcohol work primarily by precipitating or coagulating cell structures. The action of these fixatives anchors the structures in place and frequently improves their visibility by making their outlines more pronounced. Formaldehyde and glutaraldehyde introduce covalent linkages that stabilize specimen molecules with little or no coagulation. The linkages preserve cell structures in very fine detail, often to molecular levels. The two aldehydes also frequently fix molecules such as proteins and nucleic acids in place without destroying their biological activities.

Staining and Sectioning Before or after chemical fixation light microscope specimens are usually stained with dyes that produce contrasting colors in cell components. Dyes can be chosen to color organelles such as nuclei or mitochondria differently or to color structures containing specific molecules such as RNA, DNA, or various proteins.

Some specimens are made thin enough to view in the light microscope simply by smearing them over a microscope slide or pressing them between two glass slides. Others are cut into thin slices. For this purpose fixed tissue blocks are usually dehydrated by exposure to alcohol or acetone and embedded in paraffin or plastic. The blocks are then sliced by a *microtome*, an instrument that uses a metal or glass knife to cut sections of uniform thickness. For light microscopy sections are routinely cut to thicknesses of 10 μm or less.

Analytical Techniques: Autoradiography Autoradiography involves tagging a specimen molecule with a radioactive isotope, usually ^{3}H, ^{14}C, ^{32}P, ^{35}S, ^{125}I, or ^{131}I (see p. 77). In the most common application of the technique labeling is accomplished by exposing living cells to a chemical substance containing one of the radioisotopes. Ideally, the substance should be a precursor molecule used by the cell to make only the molecules of interest. For example, most cells use thymidine only as a precursor for synthesis of DNA. By exposing cells to tritiated thymidine, that is, thymidine made radioactive by the inclusion of ^3H (tritium) atoms, the DNA can be specifically labeled by radioactivity. Similarly, RNA can be specifically labeled by exposing cells to tritiated uridine; proteins can be labeled by exposure to amino acids made radioactive by the inclusion of ^3H or ^{14}C atoms.

After exposure to the radioactive precursor, cells are washed in an unlabeled medium to remove any excess label and then fixed, embedded, sectioned, and placed on microscope slides. The slides are coated in a darkroom with a photographic emulsion that has been liquefied by heating. After coating, the slides are cooled and stored in the dark for a period of several days to weeks. During the storage period radioactive decay at labeled sites exposes crystals in the photographic emulsion lying over the sites. The emulsion is then developed by standard photographic techniques, and the sections are examined in the microscope. Developed silver grains mark radioactive areas in the section (for example, see Fig. 25-31, in which DNA molecules, rather than a section, have been labeled with radioactivity). The developed pattern of grains, called an *autoradiograph,* marks the sites containing the molecules of interest. (Fig. 4-12 summarizes the steps in preparing cells for light microscope autoradiography.)

Immunochemistry Antibodies are protein molecules made by higher animals in response to invasion by molecules from another individual or species, primarily proteins or protein fragments and some polysaccharides. The foreign substance inducing the antibody reaction is called the *antigen.* The antibodies, once made, are highly specific and usually react and bind only to the molecule used as an antigen. The technique is so

Figure 4-12 The techniques used to prepare cells for light microscope autoradiography.

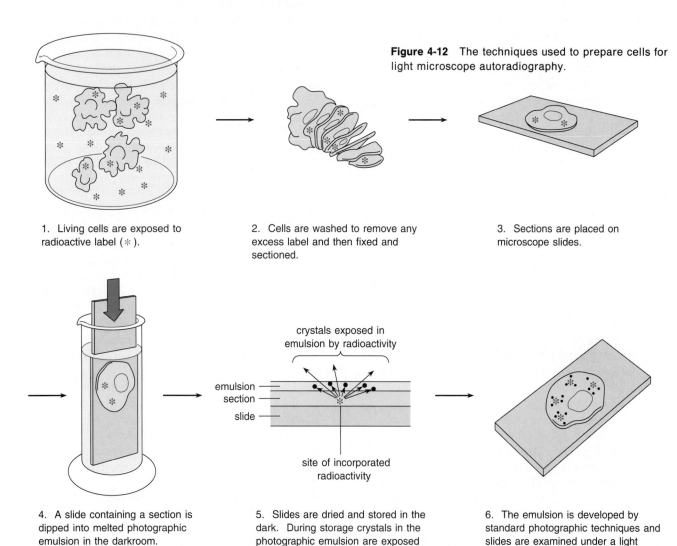

1. Living cells are exposed to radioactive label (∗).

2. Cells are washed to remove any excess label and then fixed and sectioned.

3. Sections are placed on microscope slides.

4. A slide containing a section is dipped into melted photographic emulsion in the darkroom.

5. Slides are dried and stored in the dark. During storage crystals in the photographic emulsion are exposed over radioactive sites in the section.

6. The emulsion is developed by standard photographic techniques and slides are examined under a light microscope. Grains exposed in the emulsion mark sites of radioactivity.

crystals exposed in emulsion by radioactivity

emulsion
section
slide

site of incorporated radioactivity

sensitive that in many cases proteins with sequences differing by only one or two amino acids can be separately detected and identified.

Antibodies are prepared by injecting a suitable animal, usually a rabbit or mouse, with the purified antigen of interest. Two or three injections of the antigen are given over a period of one to two months. After this time an antibody specific in its reaction to the antigen can be detected in the bloodstream of the injected animal. Extraction of a relatively small quantity of blood usually yields large quantities of the antibody, which is easily purified from the blood proteins.

For research using the light microscope antibodies are frequently marked by radioactivity or with fluorescent dyes. (Fig. 1-20 shows cells prepared for light microscopy by the fluorescent antibody technique.) Two major techniques are used in fluorescence marking. In *direct immunofluorescence* an antibody is directly linked chemically to a fluorescent group or molecule. The antibody is then reacted with cells, usually after treating the cells with an agent, such as glycerol or acetone, that disturbs membrane structure enough to permit penetration by the antibody.

In *indirect immunofluorescence* the fluorescent dye is not linked directly to the antibody developed against an antigen of interest. Instead, the antibody is used as an antigen to produce a second antibody in a source animal. Once developed, isolated, and purified, the second antibody is specific in its binding activity for the first antibody. The second antibody is then linked to the fluorescent marker. The first antibody, specific in its binding for the cellular molecules of interest, is reacted with tissue by the same techniques employed in direct immunofluorescence. The tissue is then reacted with the second antibody, which carries the fluorescent marker and is specific in its binding reaction to the first antibody. Cell structures containing the molecule reacting with the first antibody fluoresce strongly.

Indirect immunofluorescence provides two primary advantages over the direct method. The indirect technique eliminates the chance that the specificity of the first antibody will be altered by the chemical reactions that attach the fluorescent group. In addition, several of the fluorescent second antibody molecules may link to each of the first antibodies, yielding much brighter fluorescence for a given concentration of the antigen.

Enzymatic Digestion   Enzymes are used as analytical probes for the presence of specific molecules by testing the susceptibility of cellular structures to enzymatic breakdown. If a cell structure is destroyed or extensively altered after exposure to ribonuclease, for example, it can be assumed that RNA forms a part of the affected structure. Alternatively, the presence of enzymes as components of cell structures can be tested by adding an appropriate substrate for the enzyme, in combination with a chemical that produces a specific color or visible precipitate in the presence of a product of the reaction catalyzed by the enzyme.

For some preparations in which enzymes are to be analyzed, living tissues are quick-frozen without chemical fixation by exposing them to an extremely cold medium such as liquid nitrogen. The tissues are sectioned while still frozen and placed on slides. After thawing, the sections are analyzed for enzymatic activity as in chemically fixed tissues. The quick-freezing technique preserves essentially all cellular enzymes in native and active form, without chemical alterations resulting from fixation. However, the lack of chemical fixation and staining limits the stability and visibility of cell organelles and structures, so that it is often difficult to discern cellular details in the specimens.

Specimen Preparation for Electron Microscopy

Several preparative techniques satisfy the requirement that specimens for the TEM be thin, dry, and nonvolatile. Of these, the most widely employed include embedding and thin sectioning, positive and negative staining, shadowing, and freeze-fracture.

Embedding, Sectioning, and Staining   Producing thin sections for electron microscopy requires several steps. Tissues are chemically fixed by placing them in osmium tetroxide, formaldehyde, or glutaraldehyde, either singly or in combination. Osmium tetroxide acts as a stain as well as a fixative because osmium atoms, precipitated or chemically linked to cell structures, scatter electrons extensively and add significant contrast to the electron image. For most preparations tissues are fixed first in glutaraldehyde and then postfixed with osmium tetroxide.

Once fixed, the tissue is dehydrated by exposure to alcohol or acetone, embedded in plastic, and cut into thin sections by a microtome using a glass or diamond knife. After sectioning, the tissue is stained by immersing the sections in solutions of heavy metal salts. Osmium tetroxide, uranyl acetate, or lead citrate are used most often. There are only minor differences in the staining patterns produced by these and other heavy metal stains. As a result, relatively few analytical techniques based on staining differences exist as yet for electron microscopy.

Cell organelles or molecules are also examined in the electron microscope in isolated form rather than in thin sections of intact tissues. The isolated structures are often prepared for viewing by negative staining, a method in which a heavy metal stain is allowed to dry around the surfaces of an isolated cell particle or molecule. The stain, usually uranyl acetate or phosphotungstic acid, outlines surface details by depositing in depressions and crevices during the drying process.

The technique typically produces a "ghost" image, such as that shown in Figure 9-12, in which the specimen appears light against a dark background.

Shadowing Techniques   The contrast and surface details of isolated specimens are also increased by shadowing, in which a thin layer of a heavy metal is deposited on specimen surfaces. In this technique isolated cell organelles or molecules are first dried on a supporting film of plastic and then coated under vacuum by a heavy metal such as platinum, evaporated from a source located to one side (Fig. 4-13). Atoms of the metal, evaporated by electrically heating a small quantity of the metal to the boiling point, deposit on raised surfaces of the specimen facing the source. Because of the heavy-metal coat, the raised surfaces have high contrast in the electron microscope. Depressions in the specimen, located in the shadow of higher points, are not coated by heavy-metal atoms and appear transparent. When a negative image of the electron image is produced photographically, the effect is as if a strong light has been directed toward the specimen from one side, placing surface depressions in deep shadow (as, for example, in Fig. 11-1).

The Freeze-Fracture Technique   The freeze-fracture technique combines some characteristics of both sectioned and isolated preparations. In this method a small block of fixed or living tissue is rapidly frozen by immersing it in liquid nitrogen. The freezing, which is almost instantaneous, locks specimen molecules in place with little or no disturbance. Next, the frozen tissue block is transferred to a vacuum chamber and fractured by a knife edge. The fracture travels through the specimen, primarily following the hydrophobic interiors of membranes. The fractured surface is then shadowed with metal evaporated from one side, either immediately after fracturing or after a delay of some seconds or minutes. During the period between fracturing and shadowing, water escapes from the fractured surface, "etching" the specimen to produce greater surface relief and more clearly outlined details.

After shadowing, the entire specimen is coated with a thick layer of carbon evaporated from directly above. The carbon layer, which is transparent to electrons, forms a supportive backing for the shadowing metal deposited on the fractured surface. The specimen is then removed from the vacuum and treated with an acid to dissolve the tissue. The acid extraction leaves the shadowing metal, supported by the carbon film, as a *replica* of the fracture surface. The replica, viewed under the electron microscope, shows shadowed patterns that correspond to the ridges and depressions in the original fractured tissue (as in Fig. 7-21).

Analytical Techniques   Although the battery of analytical techniques available for electron microscopy is less extensive than the methods used in light microscopy, a number of similar approaches—including enzymes, autoradiography, and antibodies—have been used with success to identify molecules in tissue preparations. Enzymes are employed as chemical probes, as in light microscopy. In this technique tissue fixed in formalin or glutaraldehyde is exposed to hydrolysis by enzymes such as proteinases, ribonucleases, or deoxyribonucleases. After embedding and sectioning, cell structures are inspected for destruction by the enzymes. Other techniques, again as in light microscopy, test for the presence of enzymes in cell structures by adding an appropriate substrate in combination with a chemical reacting with a product of substrate breakdown to produce a visible precipitate. For example, adding ATP in combination with a lead salt such as lead citrate is frequently used as a test for the presence of an ATPase enzyme. In sites containing an ATPase, the ATP is hydrolyzed, releasing inorganic phosphate. The phos-

Figure 4-13   The preparation of a shadowed specimen for electron microscopy. The specimen, supported on a plastic film, is exposed to particles of a metal such as platinum evaporated from one side. The layer of deposited metal particles makes the surfaces of the specimen appear to be outlined as if a bright light has been directed from one side, with raised areas highlighted and recesses cast in deep shadow.

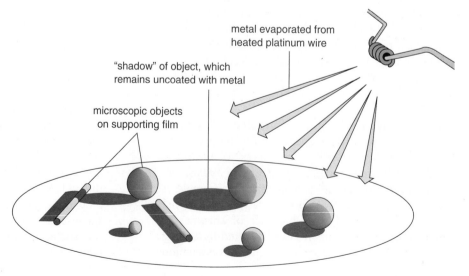

metal evaporated from heated platinum wire

"shadow" of object, which remains uncoated with metal

microscopic objects on supporting film

phate reacts with the lead salt, producing a dense, insoluble precipitate of lead phosphate that is directly visible in the electron microscope and marks the sites of enzymatic activity.

Autoradiography has been used with considerable success in electron microscopy. The preparative methods used are similar to those employed for light microscopy except that a thinner layer of emulsion is applied and the preparation is stored longer before development. (Fig. 25-31 shows a specimen prepared for electron microscope autoradiography.)

Antibodies are widely used as analytical probes for electron microscopy. To make antibodies visible under the electron microscope, they must be combined with an atom that scatters electrons, usually a metal atom. Gold particles or an iron-containing protein, *ferritin*, are used most often. In either case small, dense spots that are readily visible in the electron microscope mark the sites reacting with the antibody.

CELL CULTURE

A great part of the molecular revolution has resulted from the development of techniques for isolating cells and maintaining them in culture vessels under conditions in which they can grow and divide. Many unicellular microorganisms such as yeast, protozoa, algae, and bacteria can be cultured in large quantities relatively easily. The cells of higher eukaryotes, including both plants and animals, have more stringent requirements; growing many of these cells successfully seems to be a mixture of science, art, and luck. However, many animal and plant cell types have been grown in culture; these are indispensable as experimental subjects as well as sources of cellular organelles and molecules.

Culturing Microorganisms

The growth media and conditions required for culturing microorganisms such as yeast and bacteria are relatively simple. The much-cultured human intestinal bacterium

E. coli and the yeast *Saccharomyces cerevisiae* (brewer's yeast), for example, require only a carbon source such as glucose, a source of nitrogen, and inorganic salts. Under optimum conditions *E. coli* completes a cycle of cell growth and division every 20 minutes, allowing cultures to reach very large numbers of individuals in a relatively short time. The cells may be grown in liquid suspension or on the surface of a solid growth medium such as an agar gel (agar is a polysaccharide extracted from an alga). In either case sterile techniques must be used to ensure that foreign microorganisms do not contaminate the culture.

For many experimental purposes it is desirable to work with a genetically uniform culture of bacterial or yeast cells. For these purposes a culture may be started from a single individual. Such cultures are called *clones*, and the process that establishes the cultures is called *cloning*. The techniques used for cloning microorganisms are readily adapted to the selection of mutant strains that are resistant to antibiotics, or that require substances which normal or wild-type strains can make for themselves. (Figs. 4-14 and 4-15 show how these selections are made.) The many thousands of bacterial mutant strains selected by these and similar techniques are widely used in genetic, biochemical, and molecular studies.

Culturing Cells from Higher Eukaryotes

Higher eukaryotic cells are more difficult both to obtain and to culture than microorganisms. Cells from higher animals are usually isolated for culturing by treating dissected tissues with a protein-digesting enzyme such as trypsin, in combination with a chemical such as EDTA (ethylenediamine tetraacetate) that specifically binds and removes calcium ions from the medium. The proteinase digests away extracellular proteins holding cells together in masses; the removal of Ca^{2+} breaks the linkages binding animal cells to each other or to extracellular materials. (Many such linkages are dependent on the presence of Ca^{2+} for their formation and maintenance; see p. 255.) After these treatments,

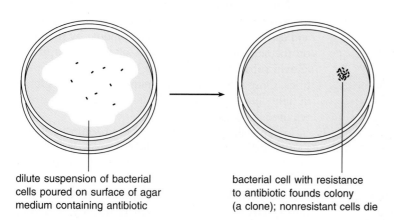

dilute suspension of bacterial cells poured on surface of agar medium containing antibiotic

bacterial cell with resistance to antibiotic founds colony (a clone); nonresistant cells die

Figure 4-14 Selection of a mutant bacterial strain resistant to antibiotics. A diluted sample of cells is poured over an agar medium containing the antibiotic; alternatively, the antibiotic may be mixed with the medium used to culture the bacteria. Any colonies growing on the agar surface may be assumed to be mutant strains with resistance to the antibiotic.

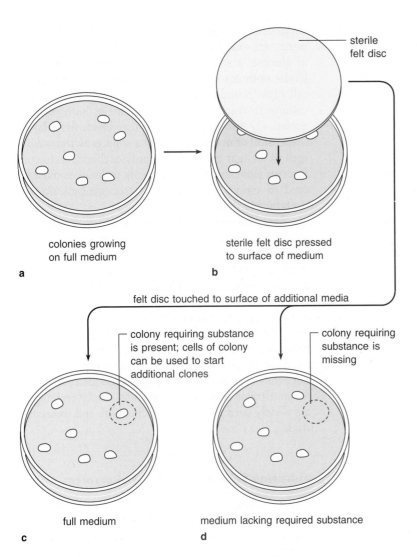

a

colonies growing
on full medium

b

sterile felt disc

sterile felt disc pressed
to surface of medium

felt disc touched to surface of additional media

colony requiring substance
is present; cells of colony
can be used to start
additional clones

colony requiring
substance is
missing

c

full medium

d

medium lacking required substance

Figure 4-15 Selection of a mutant strain requiring a substance that wild-type cells can make for themselves. **(a)** Cells are first grown on agar containing a full complement of nutrients, including amino acids and other substances that wild-type individuals can synthesize. **(b)** After colonies have grown on the agar surface, a replica of the colonies is made by pressing a piece of sterile felt against the agar. The felt is then pressed against the surface of two agar preparations, one with full nutrients **(c)** and one lacking only the substance of interest **(d)**. Any colonies absent in the preparation lacking the substance but present on the agar with the full complement can be assumed to contain mutant cells that have lost the capacity to make the substance. The colony growing on the full medium can be used to start clones of the mutant cells.

gentle shaking separates the cells. Embryonic tissues are often used as starter material for animal cell cultures because the extracellular materials and cell-cell linkages in embryos are relatively sparse, making the cells easier to separate.

The cells obtained by such methods are usually of mixed types. Several ingenious methods are used to sort out individual cell types for culture. Some use antibodies that specifically recognize and bind to surface molecules of a cell type of interest. The antibodies may be attached to a supporting surface, so that the cells of interest are linked to the surface and immobilized by the antibodies. Repeated washings then remove the unwanted cell types from the surface. In another technique antibodies developed against surface molecules of the cell type of interest are combined with a fluorescent dye, so that the desired cells glow brightly when irradiated with light at the correct wavelength. Separators automatically extract the fluorescent cells from a mixed culture at the rate of thousands per minute.

Once separated, animal cells are placed in growth media that are considerably more complex than those used for microorganisms. These media must contain the amino acids animal cells cannot synthesize for themselves (arginine, histidine, leucine, isoleucine, lysine, methionine, phenylalanine, threonine, tryptophan, and valine), plus cysteine, glutamine, and tyrosine, which are usually included for best growth. Also required are several vitamins (see p. 70), salts, and glucose. Blood serum, the fluid part of blood with the blood cells removed, must often be added to ensure growth. Many of the components supplied by serum have not been identified. Among them are probably proteins acting as growth factors—molecules that stimulate growth and division after recognition and binding by receptors on the surfaces of animal cells—and a variety of unknown substances required by the cells in trace quantities. Only a relatively few animal cells can be grown in fully defined media, that is, in media in which every required chemical is known.

Even with the addition of blood serum, cells removed from normal tissues do not grow indefinitely in culture. After 50 to 100 cycles of growth and division most normal cell lines become dormant and eventually die. Cells removed from tumors, however, often have the capacity for unlimited growth. Frequently, cells from normal tissues can be stimulated to grow indefinitely by exposing them to cancer-causing viruses or to carcinogens, chemicals that can induce some of the characteristics of cancer cells. The induction of the characteristics of tumor cells is called *transformation* (see p. 940). Of these characteristics the most significant for cell cultures is rapid growth and cell division. Cells with the capacity to grow indefinitely are said to be "immortalized."

Very few animal cells, whether normal or immortalized, can be cultured in free suspension. Instead, they need a supporting surface such as plastic beads, slides, or the inside surface of a culture vessel to survive and grow efficiently. Coating the plastic surfaces with extracellular molecules such as collagen greatly improves the adherence and growth of many animal cell types.

Cells from higher plants are cultured after their cell walls have been removed by exposure to a cellulase, an enzyme that digests away cell wall material. The naked plant cells, called *protoplasts,* can then be grown in chemically defined media. These techniques have been applied with greatest success to the cells of carrots and a few other higher plants. The cultures are limited in duration because it has as yet proved impossible to transform plant cells to an immortalized form.

Many differentiated eukaryotic cells retain specialized functions when grown in culture, allowing study of their activities and patterns of synthesis under controlled chemical and physical conditions. Because cultures can be cloned from single cells, sources of error resulting from the presence of mixed cell types can be avoided. Cell cultures are also invaluable for evaluating the toxicity of substances and for testing the ability of chemicals to cause or cure cancer. In addition, genetic engineering studies in which DNA containing genes of interest is introduced into the cells use cell cultures (see p. 783). Plant cell cultures provide an advantage for many of these studies because, with the addition of plant growth hormones, complete plants can be grown from treated and cultured cells. As a consequence, the results of experimental treatments can be observed in adult plants as well as protoplasts in culture.

Cell Fusion

Some cultured animal and fungal cells can be fused to produce composite cells containing the nucleus and cytoplasm of both cell types. Exposing the cells to certain enveloped viruses (see p. 23) or to polyethylene glycol induces the fusion. (The mechanisms of membrane fusion are detailed in Supplement 5-1.) Once fused, the composite cells can be maintained in culture as clones. The nuclei of fused fungal cells remain intact and separate in the composite cytoplasm, producing a *heterokaryon.* In fused animal cells the nuclei eventually fuse, so that both a composite cytoplasm and a single composite nucleus are formed.

Fused animal cells have proved invaluable in studies of factors regulating gene activity, particularly in dividing cells. A peculiarity of the composite cells produced by fusing mouse and human cells has allowed many human genes to be mapped to individual chromosomes. In the mouse-human clones chromosomes originating from the human cell used in the fusion are gradually lost during cycles of cell growth and division until only mouse chromosomes remain. Individual human chromosomes are retained in the clone, however, if they contain a gene that is present in an inactive or otherwise faulty mutant form in the mouse chromosomes. Thus, by using a mouse cell containing a mutant form of a gene of interest for the fusion, the equivalent human chromosome containing the normal form of the gene can be identified.

Cell fusion techniques have revolutionized the production of antibodies for scientific research and medical applications. In this application a white blood cell producing an antibody of interest is fused with a cancerous white blood cell (a *myeloma* cell) that has the capacity for unlimited growth, creating a composite cell known as a *hybridoma.* The hybridoma cell founds a cell line that grows and divides rapidly. Each cell in the clone produces the desired antibody in large quantity (for details, see p. 805). The antibodies produced by a single clone, which are all the same type and react with the same molecule or molecular group, are *monoclonal* antibodies. All the antibodies in a monoclonal preparation, for example, react with exactly the same site on a protein molecule. The uniform reactivity of monoclonal antibodies provides distinct advantages over the antibodies obtained by injecting an antigen into a test animal, which usually consist of a large family of antibody types with similar but not identical reactivity. The group of antibodies obtained by injecting a protein as antigen, for example, usually includes individual types that react with different sites on the protein.

Cultures of many animal cell types are available either commercially or from individual scientific laboratories. This availability allows investigators to skip the steps required to separate cells from tissues. Most of the available cultures have been developed from tumor cells or have been transformed into an immortalized form.

FRACTIONATION AND PURIFICATION OF CELL STRUCTURES AND MOLECULES

Once obtained in quantity from either cultures or tissues removed from an experimental organism, cells can be fractionated to yield cell organelles, structures such as ribosomes or microtubules, or individual molecular types. The methods available for this work are capable of producing highly purified cellular and molecular fractions.

Cells are disrupted by breaking the plasma membrane in a buffered solution at physiological pH. Among the several techniques used for breaking plasma membranes are sonication (exposure to high-frequency sound waves), grinding in fine glass beads or other abrasive materials, forcing cells through a narrow orifice, osmotic pressure, and exposure to detergents.

The suspension of cell organelles, structures, and molecules produced by the breakage is usually separated into fractions by spinning it in a centrifuge. *Centrifugation* can separate organelles as large as nuclei or mitochondria or molecules as small as proteins into separate groups. For molecular studies the fractions obtained by centrifugation are often further separated into pure samples by either of two methods—*gel electrophoresis* or *chromatography*.

Centrifugation

A centrifuge consists of a *rotor* driven at high speed by an electric motor (Fig. 4-16). The rotor holds tubes containing solutions or suspensions of the materials to be centrifuged. For centrifugation at higher speeds the centrifuge is enclosed in an armored chamber that can be cooled and held at a vacuum to reduce heat and friction caused by air resistance. The centrifugal forces generated by the spinning rotor in the most powerful instruments can reach 500,000 times the force of gravity.

Movement in response to centrifugal force is related to the mass, density, and shape of the structures or molecules being centrifuged, and to the mass, density, and viscosity of the surrounding solution. If a structure or molecule has greater mass or is denser than the surrounding solution, it moves downward in the centrifuge tube. If it has lesser mass or is less dense than the surrounding solution, it remains at the top of the tube, or if present at lower levels, it is displaced by molecules of the solution and moves upward in the tube. The velocity of movement is modified by the shape of the particle, with rodlike or elliptical forms moving more slowly than spherical particles. Increasing the viscosity of the suspending solution also slows the rate of movement of structures or molecules in the centrifuge tubes, particularly for nonspherical forms.

Because many discrete cell structures, such as the nucleus, mitochondria, chloroplasts, and ribosomes, differ in density and are reasonably resistant to mechanical disruption, they can be separated by centrifugation. Several centrifugations at successively higher speeds separate and purify the structures (Table 4-1). At lower speeds large, dense bodies such as nuclei are driven down and concentrated. Low-speed centrifugation can also be used to remove debris from the preparation. Centrifugation of the remaining solution at higher speeds drives down and concentrates structures of intermediate size and density, such as mitochondria and chloroplasts. Final centrifugation at very high speeds concentrates structures such as microtubules, microfilaments, ribosomes, and larger molecules such as nucleic acids and proteins.

Depending on conditions such as the ionic composition and pH of the surrounding medium, the organelles and smaller structures isolated by fractionation and centrifugation may retain some or all of their biological activity. Active preparations of this type, in which no whole cells are present, are *cell-free systems*. Biochemical studies, particularly those investigating RNA transcription and protein synthesis, frequently use these systems.

Molecules more dense than the suspending medium move downward or sediment at a constant rate when the speed of the centrifuge is held uniform. This constant[3] is usually expressed in Svedberg, or S, units, named in honor of one of the pioneers of the centrif-

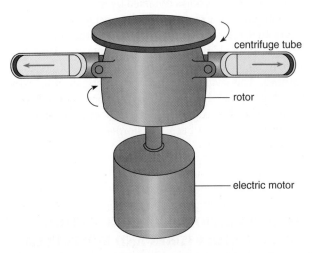

Figure 4-16 The arrangement of tubes and rotor in a centrifuge. In some centrifuges the tubes are held solidly at an angle within the head instead of hinged as this diagram shows.

---

[3] The constant is the *sedimentation constant, s,* equal to the speed at which molecules descend in a centrifuge in centimeters per second divided by the centrifugal acceleration $(2\pi\omega)^2 x$, in which ω is the rotation of the centrifuge in radians per second and x is the distance in centimeters of the molecule from the center of rotation:

$$\text{sedimentation constant} = s = \frac{\text{speed of sedimentation}}{(2\pi\omega)^2 x}$$

s values have units of seconds and are expressed in Svedberg, or S, units to avoid the use of small numbers:

$$S = 1 \times 10^{-13} \text{ sec}$$

Table 4-1   Fractionation of Cell Structures by Centrifugation

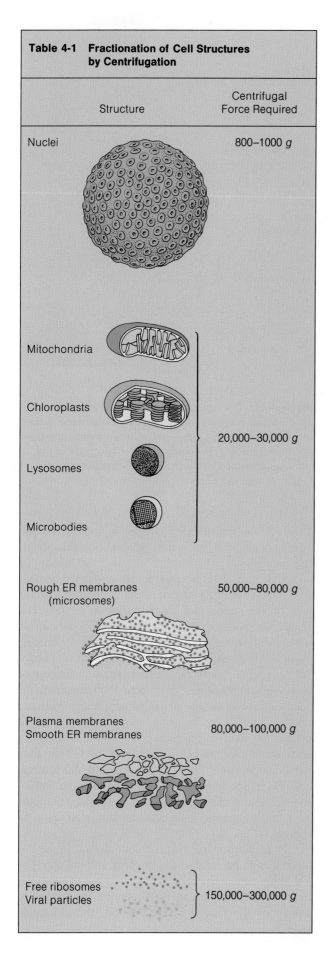

Structure	Centrifugal Force Required
Nuclei	800–1000 *g*
Mitochondria	
Chloroplasts	
Lysosomes	20,000–30,000 *g*
Microbodies	
Rough ER membranes (microsomes)	50,000–80,000 *g*
Plasma membranes Smooth ER membranes	80,000–100,000 *g*
Free ribosomes Viral particles	150,000–300,000 *g*

ugation technique. Larger molecules generally have higher S values. Often the S value is used to designate molecular size rather than molecular weight, particularly for nucleic acid molecules. Eukaryotic ribosomal RNAs, for example, are typically designated as 5S, 5.8S, 18S, and 28S rRNAs in order of increasing molecular size, rather than in molecular weights.

Density Gradient Centrifugation   For more precise separation of molecules, *density gradient centrifugation* is used. In the most widely applied version of this technique, a plastic centrifuge tube is filled with a solution of sucrose or cesium chloride (CsCl), mixed in gradually denser concentrations toward the bottom of the tube. The concentration gradient, made by laboratory equipment designed for this purpose, is adjusted so that its most dense region at the bottom of the tube is less dense than any of the molecules under study. A mixture of the molecules to be separated is carefully layered at the top of the tube.

During centrifugation the gradient separates the molecules into distinct bands that migrate downward at different rates. The bands remain sharply separated because the molecules at the leading edge of a band constantly encounter a denser solution than the molecules in the trailing portion and are slowed slightly in their progress down the tube. The molecules at the trailing edge of a band move slightly faster than the molecules at the leading edge because they encounter a less dense medium. The tendencies of the leading molecules to slow slightly and the trailing molecules to speed up narrows the band to the smallest width possible, and maintains this sharply defined arrangement as the band moves downward through the tube.

Centrifugation is stopped before any of the bands has moved to the bottom of the tube. After centrifugation a hole is punched in the bottom of the tube, and the bands are collected in separate fractions as they drain off.

Buoyant Density Centrifugation   For highly critical work molecules may be separated by *buoyant density centrifugation.* The technique usually employs a CsCl solution, adjusted initially to approximately the same density as the molecules under study. The sample is layered on top of the solution and the tube is spun at very high speed, sometimes for several days. As the tube spins, the CsCl molecules gradually pack more densely toward the bottom of the tube, producing a highly even gradient in density from top to bottom. Because the initial, uniformly dispersed CsCl solution approximately matched the density of the sample, the final CsCl gradient is less dense than the sample at the top of the tube and more dense at the bottom. As a result, the molecules in the sample descend until they reach the level at which their density and the density of the CsCl gradient are the same, where they form a sharp band. The method is so sensitive that molecules of the same type but containing different atomic isotopes can be separated into different bands. For example, a classic study of DNA replication in which DNA molecules containing the ^{14}N and ^{15}N isotopes of nitrogen were clearly and sharply separated employed this method (see p. 953).

Gel Electrophoresis

The centrifuges used for molecular separation become more expensive, required speeds become higher, and centrifuge times become longer as the size of the molecules under study becomes smaller. Even with centrifuges capable of the highest speeds, molecules with molecular weights smaller than about 10,000 daltons cannot be sedimented because of the opposing effects of diffusion. Gel electrophoresis circumvents these problems by using an electrical field instead of centrifugal force to move molecules through a viscous, gellike medium. The technique, using relatively inexpensive equipment, rapidly and precisely separates molecules from the size of large proteins down to very short polypeptide and nucleic acid chains or even single nucleotides.

In a gel electrophoresis apparatus (Fig. 4-17) the gel is enclosed in a glass tube or sandwiched as a slab between glass or plastic plates. For the separation of proteins, gels are usually cast from polyacrylamide plastic; for the separation of nucleic acids, gels are cast from agarose, a complex carbohydrate molecule extracted from seaweed. The tube or slab gel is suspended so that its ends are in contact with separate buffered salt solutions. The salts used are electrolytes, which conduct an electric current in water solutions. The conducting solutions at either end of the gel are connected to the positive and negative electrodes of an electrical power source so that current flows from one solution to the other through the gel. Any molecules added to the gel migrate toward the positive or negative electrode in accordance with their own electrical charge.

The gel forms an open molecular network with spaces just large enough to allow passage of the molecules under study along with the buffered salt solution used in the apparatus. The consistency of the gel effectively eliminates diffusion and convection, which tend constantly to disturb bands of molecules sedimenting in a centrifuge.

When the current is held constant, the rate at which molecules move through the gel depends primarily on three factors: *molecular size*, *shape*, and *charge density*. The rate of migration with respect to size reflects the ability of different molecules to thread through the gel network. Larger molecules move more slowly than smaller molecules because they meet more resistance in passing through the gel. The second factor, molecular

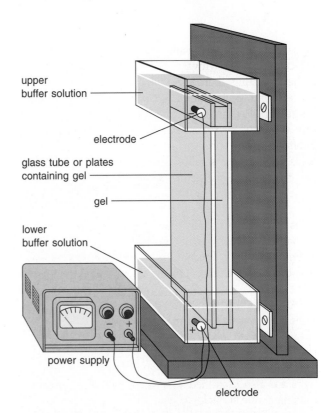

Figure 4-17 The apparatus used for gel electrophoresis. The gel, confined in a glass tube or between glass or plastic plates, is suspended between electrolyte solutions. The electrolyte solutions at the two ends of the gel are connected to the positive and negative electrodes of a power source.

shape, has a similar effect. Molecules with extended fibrillar or elliptical forms move more slowly than spherical ones.

The third factor, charge density, refers to the relative numbers of positive or negative charges per unit area on the surfaces of the migrating molecules. The higher the charge density, the more rapidly a molecule moves toward the end of the gel with opposite charge. Because the charge of many biological molecules varies with pH, the adjustment of pH is critical to the rate of migration. In solutions more acid than the *isoelectric point* of a molecule, the pH at which the molecule has no net positive or negative charge, primarily basic groups are ionized, promoting movement toward the negative end of the gel. In solutions more basic than the isoelectric point, primarily acid groups are ionized, and the molecule moves toward the positive end of the gel. If the pH is adjusted to the isoelectric point of a molecule, the molecule remains stationary in the gel.

In application of the technique a mixture of the molecules to be separated is carefully layered onto the gel at the top of the tube or slab. The buffer solutions at the ends of the tube are then connected to the power source so the charge at the bottom of the tube is opposite that of the molecules under study. The layered molecules move downward through the gel in response to the attracting charge. Because the relative rates at which the molecules move depend on size, shape, and charge density, the initial mixture of molecules gradually separates into a series of distinct bands moving at different rates through the gel. The fastest-moving bands contain the smallest, most compact molecules with the highest charge density. Because most or all of the molecules in the mixture form invisible bands, a dye molecule slightly smaller than any of the molecules under study usually is added to the initial mixture to enable the operator to keep track of the progress of the separation. The apparatus is turned off as the band containing the dye molecule reaches the bottom of the gel.

After separation into distinct bands the molecules in the gels may be examined in various ways. Gels containing proteins are frequently stained with an organic dye to reveal the positions of the bands. (Fig. 4-18 shows a tube gel prepared in this way.) Protein bands are also identified by radioactive labels. Bands containing nucleic acids are identified by radioactive labels or by reacting them with ethidium bromide, which fluoresces under ultraviolet light. For further analysis the molecules in individual bands can be isolated by slicing the gel into pieces or by making what is known as a *blot* (see below).

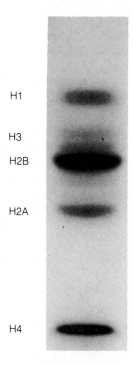

Figure 4-18 Proteins separated into distinct bands in a tube gel by electrophoresis. The proteins shown are the five histone proteins associated with DNA in eukaryotic chromatin. Courtesy of W. F. Reynolds.

Determining Relative Molecular Weight by Gel Electrophoresis Since the rate of migration through the gels reflects molecular size, the relative positions of migrating bands can provide an approximation of molecular weight. The approximation becomes more exact if the charge density and shape of the molecules under study are uniform. Nucleic acids satisfy this requirement nicely because of their uniform structure and even distribution of negatively charged phosphate groups.

Proteins, in contrast, vary widely in charge density and shape. Reacting proteins with the detergent sodium dodecyl sulfate (SDS) circumvents this problem. The negatively charged SDS molecules have hydrocarbon chains that interact with the hydrophobic portion of protein molecules. Because most proteins contain roughly the same proportion of hydrophobic amino acid residues, they bind equivalent quantities of SDS, in a ratio of SDS to protein of about 1:4 by weight. The bound SDS effectively swamps out the native charge of the proteins and gives them a more or less uniform negative charge density.

The SDS coat causes most proteins to lose their native secondary structures and unwind into a random coil. The unwinding is aided if proteins are treated with a reducing agent such as mercaptoethanol to break disulfide linkages. As a consequence of the unwinding and the uniform charge density imposed by the SDS coat, different proteins move toward the positive end of the gel primarily according to the length of their amino acid chains, that is, according to molecular weight.

In practice molecular weights are estimated by mixing unknowns with several readily identifiable molecules of known molecular weight. The knowns and unknowns move through the gel in such a way that the relative distance traveled plotted against the logarithm of the molecular weight produces a straight line. Placing the knowns and unknowns in position in the plot gives a reasonably accurate estimate of the molecular weights of the unknowns. Molecular weights determined in this way are designated M_r, indicating that the weight is an estimate made relative to a known molecular weight used as a standard.

Two-dimensional Gel Electrophoresis Running electrophoretic gels in one direction is often ineffective in separating large numbers of different proteins of similar size and shape. The proteins cannot travel far enough to separate into distinct bands before they reach the bottom of the gel. Longer gels provide a partial solution to the problem. However, there is a practical limit to gel length imposed by the higher currents required to drive molecules through long gels. At some point heat generated by the current destroys both the gel and the proteins.

Two-dimensional gel electrophoresis solves these problems. In this technique a protein mixture is first run in a tube gel to which a solution of *ampholines* has been added. Ampholines are chainlike molecules containing different numbers of positive and negative charges. Under the influence of the electric current, individual ampholines migrate according to charge in the gel. Those of more negative charge move toward the positive end of the tube; those of more positive charge move toward the negative end. The distribution produces an even charge gradient and, in effect, a smooth pH gradient from the top to the bottom of the gel.

Proteins are added to the top of the first gel in native form without exposure to SDS. As the proteins migrate through the gel in response to the current, they encounter a gradually changing pH. Eventually, each protein reaches the pH at which it has no net charge, that is, its isoelectric point. At this point the proteins stop moving and each becomes concentrated into a sharply defined stationary band. Concentration of proteins into stationary bands by the pH gradient in this

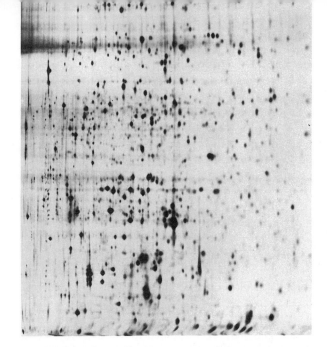

Figure 4-20 The proteins of a human (HeLa) cell, separated by two-dimensional gel electrophoresis. Courtesy of R. Duncan.

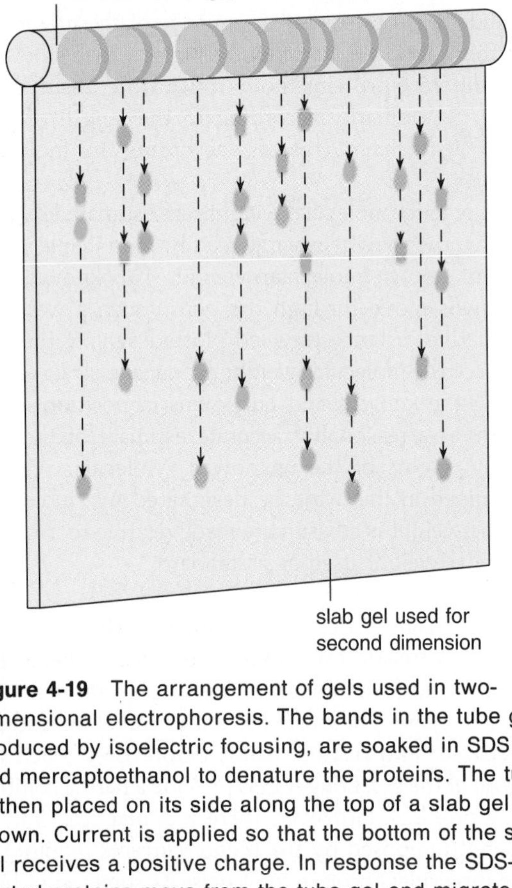

way is called *isoelectric focusing*. In starting mixtures that contain large numbers of different proteins, many bands are likely to contain more than one protein.

Isoelectric focusing produces the first dimension in the gel. The tube gel is removed from the apparatus and soaked in SDS and mercaptoethanol. This treatment denatures the proteins and gives them a negative charge coat that varies with the size of each molecule. The tube is then placed on its side along the top of a slab gel (Fig. 4-19), and current is applied so that the bottom of the slab gel is given a positive charge. In response the SDS-coated proteins move from the tube gel into the slab. As they move, the proteins from each band of the tube gel separate into distinct groups according to molecular size, with the smallest molecules toward the bottom. The migration produces a series of small bands or spots of proteins in the slab gel, distributed downward in rows corresponding to the former locations of bands in the tube gel. This distribution is the second dimension.

The second dimension effectively sorts out all the proteins of the sample because it is unlikely that different proteins with the same isoelectric point, which places them in the same band in the first dimension, will also have the same molecular weight. The method has been used successfully to separate all the proteins of a cell into separate bands or spots (as in the two-dimensional gel shown in Fig. 4-20).

Blotting Techniques Proteins or nucleic acids separated by gel electrophoresis are removed from tube or slab gels for further study by blotting. Blots are made by pressing gels against a piece of filter paper selected

tube gel containing proteins banded by isoelectric focusing (first dimension)

slab gel used for second dimension

Figure 4-19 The arrangement of gels used in two-dimensional electrophoresis. The bands in the tube gel, produced by isoelectric focusing, are soaked in SDS and mercaptoethanol to denature the proteins. The tube is then placed on its side along the top of a slab gel as shown. Current is applied so that the bottom of the slab gel receives a positive charge. In response the SDS-coated proteins move from the tube gel and migrate through the slab. As they move, the proteins originating from each band of the tube gel separate into distinct groups in the slab gel according to molecular size.

or treated so as to bind and immobilize the proteins or nucleic acids in the bands. For a gel containing unpaired DNA nucleotide chains separated into bands, a piece of filter paper soaked in salt solution is placed on one side of the gel (Fig. 4-21). A piece of nitrocellulose filter paper, on which the blot will form, is placed on the other side, backed by several sheets of dry filter paper. (Nitrocellulose is used because it can directly bind unpaired DNA nucleotide chains.) As the salt solution is drawn through the gel to the dry filter paper, it carries the molecules in the bands into the nitrocellulose paper, where the bands are deposited and tightly bound as blots. Individual blots containing DNA molecules of interest can then be identified by a technique known as hybridization (see below). A blot of DNA bands produced and identified in this way is called a *Southern blot*, after E. M. Southern, the investigator who originated the blotting technique for use with DNA.

Essentially the same technique is used to make blots from a gel containing RNA bands. An RNA blot produced in this way, in which individual RNA molecules are transferred to nitrocellulose paper and identified by hybridization, is termed a *northern blot*.

A more extensive modification is used to make blots of protein bands (Fig. 4-22). A gel containing protein bands is placed on a piece of nitrocellulose paper soaked in a buffer solution. (Nitrocellulose paper strongly binds proteins as well as unwound DNA chains.) The gel and nitrocellulose paper are sandwiched between soaked pieces of ordinary filter paper and two

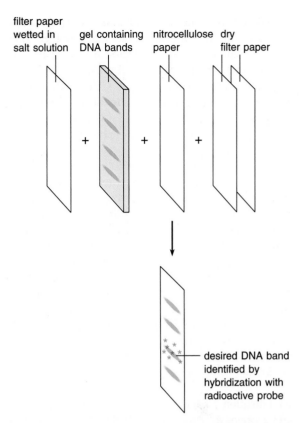

Figure 4-21 The Southern blotting technique for identifying DNA bands separated by gel electrophoresis. As the salt solution is wicked through the gel and onto the dry filter paper, the DNA bands are carried onto the nitrocellulose paper, where they are trapped as blots. Individual bands are then identified by hybridization with a radioactive probe consisting of a DNA or RNA molecule that is complementary to the sequences of interest.

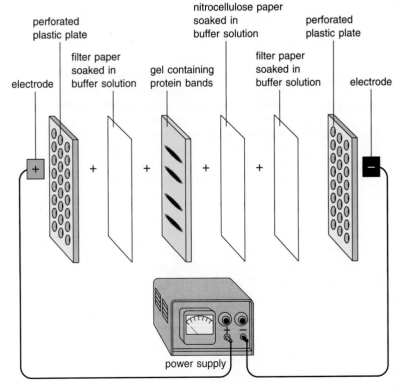

Figure 4-22 The western blotting technique for removing protein bands from an electrophoretic gel. A gel containing protein bands is placed on a piece of nitrocellulose paper soaked in a buffer solution. The gel and nitrocellulose paper are sandwiched between additional pieces of filter paper soaked in buffer and two perforated plastic plates. Current applied across the plates drives the proteins from the gel onto the nitrocellulose paper, producing a pattern of blots corresponding to the bands.

perforated plastic plates. When current is applied across the plates, the proteins are driven from the gel onto the nitrocellulose paper, producing a pattern of blots corresponding to the bands. A protein blot formed in this way is called a *western blot.* Once in the paper, individual proteins are usually identified by means of antibodies.

Chromatography

Chromatography is another method widely used to separate proteins according to size, shape, or charge. The various chromatographic methods involve passing a solution of proteins through a tube or column packed with beads or a matrix of material. Flow through the column depends on gravity or pressure; electrical currents are not used. The solution and packing material retard the passage of some proteins to a greater or lesser extent and allow others to pass freely.

Chromatography frequently allows proteins to be purified in their undenatured, native state. Three different applications of the technique commonly used are *gel filtration, ion exchange,* and *affinity chromatography.*

Gel Filtration Chromatography   Gel filtration separates proteins from mixtures according to size and shape (Fig. 4-23). For most applications the filtration columns are packed with porous plastic beads that have openings similar in dimensions to the molecules under study.

When a buffered solution containing proteins passes through the column, proteins larger than the pores move rapidly through the spaces between the beads and leave the column without further separation (Fig. 4-23a). Proteins with dimensions smaller than the pores enter the beads and are retarded in their passage through the column. In general the smaller the molecule, the more space available inside the beads and the greater the retardation. To move the differentially retarded proteins through the column, more buffer solution is added or pumped through the beads. As the retarded proteins pass from bead to bead downward through the column, they separate into bands according to size and shape, with the largest and most compact proteins reaching the bottom of the column first (Fig. 4-23b). The proteins are collected in separate fractions as they exit the bottom of the column.

The beads used for gel filtration are available with pore sizes that exclude proteins, or other substances being separated, with molecular weights ranging from as little as 500 to more than 1 million. Thus, if the proteins of interest have molecular weights below 50,000, for example, a filtering gel can be chosen that excludes all proteins above this value. The larger proteins do not enter the pores in the beads and quickly pass through the column.

Ion Exchange Chromatography   The ion exchange technique separates proteins according to charge rather

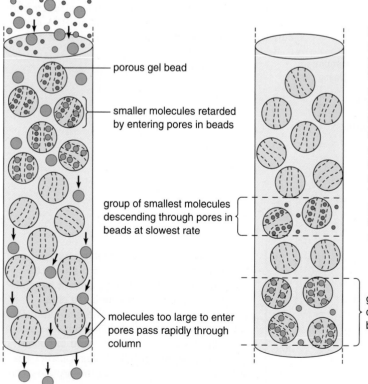

Figure 4-23   Gel filtration chromatography. **(a)** A mixture of proteins of different sizes is poured into a gel filtration column. Proteins too large to enter pores in the gel are excluded and pass rapidly through the column. Smaller molecules enter the pores in the gel and are retarded. **(b)** After traveling some distance through the column, the molecules able to fit into the pores are retarded differentially according to size, with the smallest molecules traveling most slowly. The differential retardation separates the proteins into distinct groups, or bands, descending through the column at different rates.

porous gel bead

smaller molecules retarded by entering pores in beads

group of smallest molecules descending through pores in beads at slowest rate

molecules too large to enter pores pass rapidly through column

group of larger molecules descending through pores in beads at faster rate

a

b

than size or molecular weight. The packing material used in this method consists of minute plastic beads that contain negatively or positively charged groups. For separation of positively charged proteins, beads are usually made from carboxymethyl cellulose (CM-cellulose), which is negatively charged; for separation of negatively charged proteins, beads made from diethylaminoethyl cellulose (DEAE-cellulose), which carries a positive charge, are commonly used.

A buffered solution containing a mixture of proteins to be separated is added to a column packed with charged beads. As the solution flows through the column, the proteins carrying an opposite charge attach to the beads (Fig. 4-24a). Those with no charge or the same charge as the beads pass through the column without retardation. The column is then washed with salt solutions of gradually increasing concentration. The salt ions displace the proteins by competing for the binding sites on the beads (Fig. 4-24b). The first proteins to be displaced by the competition are those with the fewest charged groups. These leave the column first; as the salt concentration of the wash is increased, additional proteins are displaced in order of charge. The proteins are collected in individual fractions as they pass in succession from the bottom of the column.

Affinity Chromatography Affinity chromatography is a highly effective method for separating proteins that can recognize and bind specific molecules. In this technique a molecule recognized and bound by the protein of interest is attached covalently to the beads in the column, which are otherwise inert (Fig. 4-25a). A mix-

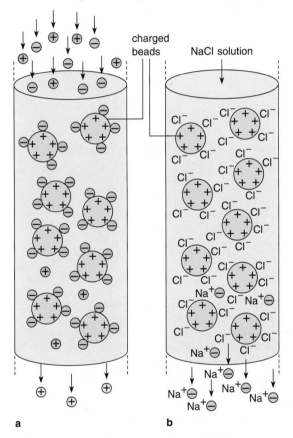

Figure 4-24 Ion exchange chromatography. The column is packed with beads opposite in charge to the proteins of interest. **(a)** A mixture of positively and negatively charged proteins is added to the column. The proteins of opposite charge stick to the beads; those with the same charge pass rapidly through the column. **(b)** Displacement of the proteins attached to the beads by an NaCl solution. The salt ions compete for charged groups on the surface of the beads, causing release of the proteins of interest. By gradually increasing the salt concentration, proteins of increasingly greater charge can be released in succession.

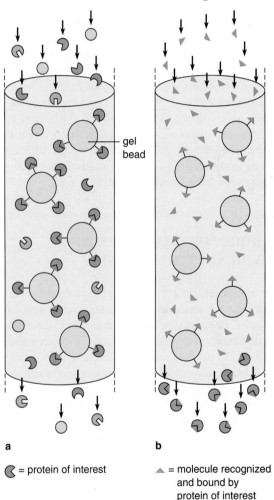

= protein of interest

= molecule recognized and bound by protein of interest

Figure 4-25 Affinity chromatography. A molecule recognized and bound by the protein of interest is covalently linked to the beads in the column. **(a)** A mixture of proteins is added to the column. Only the protein binding the molecule attached to the beads is trapped by the beads; other proteins pass through the column without hindrance. **(b)** The attached proteins are released by adding a solution of the molecules recognized and bound by the protein of interest. The added molecules compete for the binding site on the protein, releasing the protein from the beads.

ture of proteins is then passed through the column. Among the proteins in the mixture, only the protein recognizing and binding the substance attaches to the beads; the remainder pass through unhindered. After the unbound proteins are washed from the column, the protein of interest can be removed from the beads by adding a solution in which the substance bound by the protein is suspended in free, unattached form. As the solution passes through the column, the freely suspended molecules of the substance compete with the molecules attached to the beads for the binding site on the protein. The competition eventually releases most or all of the protein molecules from the beads for collection as a highly purified sample at the bottom of the column (Fig. 4-25b).

Affinity chromatography can rapidly separate a single protein from crude cell extracts containing hundreds or thousands of proteins. For example, if the receptor for the insulin hormone is the protein of interest, the hormone is attached covalently to the beads. The insulin receptor, if present in a preparation of cellular proteins, will attach specifically to the bound insulin when the mixture passes through the column; all other proteins exit the column without delay. Addition of a solution containing unbound insulin then removes the insulin receptor from the beads.

METHODS FOR CHARACTERIZING INDIVIDUAL CELL MOLECULES

Once cell molecules are isolated and purified by centrifugation, gel electrophoresis, or chromatography, techniques are available to determine their structure, their chemical and physical properties, and often their cellular functions. Functional studies have been particularly advanced by combinations of molecular and genetic techniques that permit segments of nucleic acid or protein molecules to be changed at will. For example, it is possible to analyze the functions of individual nucleotides in control regions of genes or amino acids in the active sites of enzymes.

X-Ray Diffraction

The most accurate method for working out molecular structures is X-ray diffraction, which uses X-rays to probe the positions of atoms and their structural configurations in molecules. For molecules that can form crystals, X-ray diffraction reveals molecular details as small as 1.5 A. This technique has led to some of the most far-reaching and significant discoveries of cell and molecular biology, including the discovery of DNA structure.

X-rays are a form of electromagnetic radiation of very short wavelength. The waves have so much

energy that they are able to pass through many substances, including solid objects of considerable thickness. Some X-rays travel through a substance without disturbance and emerge in the same direction as the entering beam. Others are deflected, scattered, or absorbed by atoms in the substance. The degree of disturbance of X-rays depends primarily on the number of electrons traveling in orbitals around atomic nuclei. The more electrons surrounding an atom, the greater the probability that an X-ray passing near the atom will be deflected from its path. Heavy atoms, particularly those of the heavy metals, are therefore most effective in deflecting X-rays.

X-rays passing through relatively unstructured materials in which individual atoms are arranged in nonrepeating patterns are deflected in random directions. It is possible to obtain a gross image of such materials by placing a photographic plate or viewing screen on

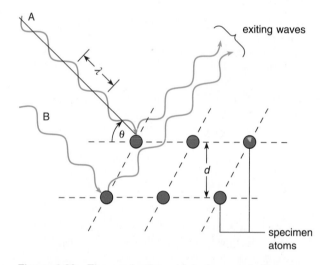

Figure 4-26 The production of interference in X-ray diffraction. For interference to occur, the X-rays in the entering beam must be at the same wavelength and parallel. In the diagram wave A in the entering beam is deflected by an atom in the top layer of the crystal lattice. Wave B is deflected by an atom in a corresponding location in the second layer to the same angle and direction as wave A. However, wave B must travel a longer distance to reach the atom in the second layer. Depending on the wavelength, angle of incidence θ, and the distance d between the layers of atoms in the crystal lattice, the deflected waves may leave the crystal in or out of phase. If d is equivalent to the wavelength, the deflected rays leave the crystal more or less in phase; the crests and troughs coincide and the waves reinforce, producing a reflected beam of higher intensity. If d is equivalent to one-half wavelength, the deflected waves leave the crystal more or less out of phase; the troughs of one wave coincide with the crests of the other, interfering to reduce the intensity of the exiting beam. Maximally interfering beams are shown exiting the crystal in the diagram.

the side from which the beam emerges. Such images are essentially shadows, with areas of greater or lesser density depending on the relative degree of absorption or deflection of X-rays by different regions. Metals or dense body tissues such as bone allow relatively few X-rays to pass without diffraction and cast a denser shadow; objects containing lighter atoms, such as plastics, paper, or soft body tissues, allow more X-rays to pass and cast a lighter shadow. Shadow images of this type are the basis for X-ray machines used in medicine and at airports to examine luggage.

X-rays passing through a substance with a highly ordered atomic structure, such as a crystal, are deflected in a different pattern. As the beam enters a crystal, the X-rays encounter atoms arranged in regularly repeated planes and positions. Each atom located in a specific position and plane in the repeating crystal lattice deflects X-rays to the same angle and direction (Fig. 4-26). As a result, the entering beam is split into a number of smaller beams that depart the crystal at different angles. The troughs and crests of the waves in a beam leaving the crystal at the same angle may *interfere*, that is, reinforce or cancel each other, depending primarily on the wavelength of the X-rays and the distance separating successive planes that deflect X-rays to the same angle (see Fig. 4-26). The degree of reinforcement or cancellation produces differences in the intensity of the smaller beams leaving the crystal. The combined effect of the deflection and interference, called *diffraction*, is most pronounced if the X-rays entering the crystal are all at the same wavelength and travel in parallel rather than diverging paths.

The number of emerging beams depends on the number of atoms in each repeating unit of the crystal. Simple crystals such as those of sodium chloride, with a repeating lattice unit containing only a few atoms, split the beam into a few smaller beams. Crystals with more complex lattices, such as those cast from proteins, diffract the entering beam into proportionally greater numbers of smaller beams, as many as hundreds or thousands in some cases. The number, positions, and intensities of the beams leaving a crystal are recorded on a photographic plate placed on the opposite side of the crystal from the X-ray source. The resulting distribution of spots is called an *X-ray diffraction pattern* (Fig. 4-27).

Because the angles and intensities of the emerging beams depend on the locations of atoms in the repeating crystal lattice, the diffraction pattern produced by the emerging beams can be used to reconstruct the positions of individual atoms in the crystal. The reconstruction is relatively simple if the repeating unit of the lattice contains only a few atoms that form a diffraction pattern with a small number of spots. Reconstruction becomes more difficult as the complexity of the repeating unit and the number of spots in the diffraction pattern increase. The diffraction patterns produced by

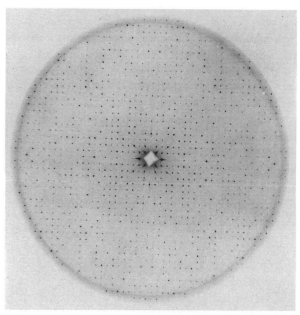

Figure 4-27 The X-ray diffraction pattern produced by a crystal of the protein myosin. Courtesy of H. M. Holden and I. Rayment, from *Prot. Struct. Funct. Genet.* 9: 135 (1991). Copyright © 1991 Wiley-Liss, a division of John Wiley and Sons, Inc.

highly complex lattices, such as those of proteins (as shown in Fig. 4-27), may take literally months of calculations by the most advanced computer systems to unravel.

In many cases the diffraction patterns produced by complex crystals are compatible with several different lattices. The possibilities can be narrowed by the addition of metal ions to introduce faults in the crystal lattice. The positions taken by the metal atoms, reflected in definite alterations in the pattern of spots, are frequently compatible with only one of the several possible lattices.

Many proteins and other biological molecules such as larger nucleic acid molecules are difficult or impossible to cast into crystals. X-ray diffraction can still provide information about the molecular structure of such substances if the molecules can be roughly ordered in some fashion. In the studies leading to the discovery of DNA structure, for example, DNA molecules were roughly aligned in parallel arrays by pulling a partially dried DNA sample into a fiber. The X-ray diffraction pattern produced by such substances blurs into partial or complete concentric rings rather than discrete spots. However, the spacing of the rings and their relative intensity still provide information about the dimensions of regularly repeating major patterns within the specimen. For example, the rings produced by X-ray diffraction of DNA indicated regular periodicities in the molecule at 3.4 A and 34 A. These periodicities were deduced to represent the distance between base pairs and the length of one full turn of the DNA double helix, respectively. Scale models, which allowed dimensional limits to be assigned to possibilities for the DNA double helix, contributed extensively to the deduction.

More recently, regular crystals have been cast from very short DNA molecules of known sequence. The clear X-ray diffraction patterns formed by these crystals confirmed the results of the earlier work and allowed many other details about DNA structure to be worked out. Small RNA molecules, such as tRNAs, have also been successfully cast into crystals and analyzed in detail by X-ray diffraction.

Obtaining DNA Sequences in Quantity

DNA, an abundant molecule in both prokaryotic and eukaryotic cells, can be extracted easily in bulk quantities for studies such as X-ray diffraction. However, obtaining specific DNA sequences, such as the DNA of a single gene, in quantities sufficient for molecular studies presents special problems because the cells of an organism may have as few as one or two copies of individual sequences. Several methods are available for selecting such sequences. Many of these techniques are based on *hybridization,* in which a sequence of interest is identified among other DNA sequences by pairing it with a complementary sequence used as a "probe." Once an individual sequence is isolated, either of two ingenious techniques, *cloning* or the *polymerase chain reaction,* may be used to increase the quantity of the sequence to levels allowing biochemical analysis.

Nucleic Acid Hybridization   Hybridization depends on the fact that DNA nucleotide chains unwind from the double helix at elevated temperatures and will rewind with complementary sequences if the DNA is cooled (see Information Box 4-1). In the application of this phenomenon to identification of individual DNA sequences, a DNA sample, if not already in single-chain form, is unwound by heating. Then a radioactive probe is added to the unwound sample under conditions that promote DNA rewinding. The probe consists of a single-chain RNA or DNA molecule that is complementary to part or all of the sequence of interest. If the sequence of interest is a gene, the probe may consist of part or all of the messenger RNA copied from the gene, a DNA copy of the mRNA made by the reverse transcriptase enzyme, which can use RNA as a template for making a DNA copy, or an artificially synthesized segment of complementary DNA. A probe may range from as few as 15 or so to many thousands of nucleotides in length.

Hybridization of the probe marks the sequence of interest with radioactivity, allowing it to be identified among the remaining DNA sequences in the sample. The hybridization technique is sensitive enough to detect and mark a sequence occurring only once within the entire DNA complement of a cell.

Hybridization with probes is often combined with gel electrophoresis of DNA samples in the Southern and northern blotting techniques. In Southern blotting a solution containing a radioactive probe is poured over single-chain DNA bound in blots to the nitrocellulose paper. The probe will hybridize only with a blot containing a complementary DNA chain. The paper is washed to remove unbound probe molecules, leaving the DNA molecule of interest as an individual blot marked with radioactivity. The probe used to mark individual RNA molecules in a northern blot is a radioactive single-chain DNA molecule complementary to the RNA molecule of interest.

DNA hybridization is also used to determine how closely DNA samples from different species are related in sequence. In this technique DNA molecules of the two species are mixed, heated to the melting point, and cooled. If the two species are closely related, their DNAs share many sequences in common. During cooling, the nucleotide chains will wind into many mixed double helices that contain a nucleotide chain from one species and a complementary chain from the other. DNA from organisms that are less closely related share fewer sequences and will form proportionately fewer mixed helices. The proportion of mixed double helices provides a measure of the extent to which the two DNA samples share the same sequence information and, therefore, a measure of the "relatedness" of the two organisms. DNA hybridization studies of this type have been of great value in several areas of research, particularly in the reconstruction of evolutionary lineages.

DNA hybridization can also provide a measure of the number of repeated sequences in the DNA of an organism. Unique, single-copy sequences have only one complementary mate when the DNA of a species is broken into short pieces and melted (Fig. 4-28*a*). As a result, random collisions leading to pairing of complementary single-copy sequences occur infrequently, and rewinding takes place relatively slowly when the preparation is cooled (Fig. 4-28*b*). Repeated sequences have several possible pairing partners (Fig. 4-28*c*). Consequently, chance pairing of these sequences takes place more frequently, and they rewind more rapidly (Fig. 4-28*d*). Comparing the rates at which single-copy and repeated sequences rewind allows the number of repeats to be estimated. This method has revealed that some sequences in eukaryotes are repeated from thousands to millions of times (for details, see p. 750).

DNA Cloning   In DNA cloning, developed in the laboratories of H. W. Boyer, P. Berg, and S. N. Cohen, a selected DNA sequence is introduced into a bacterium or virus, which is then cultured under conditions that promote maximum growth. During the growth the DNA sequence of interest is replicated along with the native DNA of the bacterium or virus. When growth reaches desired levels, the DNA sequence can be ex-

DNA Melting and Reannealing

The fact that hydrogen bonds are one of the primary forces holding the two nucleotide chains of a DNA double helix together makes the integrity of DNA structure sensitive to temperature. As the temperature rises, kinetic motion of the two nucleotide chains and collisions with molecules in the surrounding solution increasingly disturb the hydrogen bonds holding the double-helical structure together. If the temperature rises high enough, so many hydrogen bonds are disturbed that the two chains come apart. As the chains separate, DNA loses its highly ordered structure and is said to *melt*.

Because G-C base pairs are held together by three hydrogen bonds, and A-T base pairs by two, G-C base pairs are more resistant to separation. Therefore, the greater the proportion of G-C pairs in a DNA sample, the higher the temperature required to melt the DNA. The ef-

fect is approximately linear; each additional G-C base pair adds about 0.4°C to the temperature required to melt one-half of the DNA sample (T_m). Mammalian DNA, which contains on the average about 40% G-C and 60% A-T pairs, melts with a T_m of 87°C. Prokaryotic DNA, which typically has about 60% G-C base pairs, melts with a T_m of 95°C. These temperatures can be lowered significantly by adding an agent that reduces the strength of hydrogen bonding, such as formamide.

The process reverses if melted DNA is cooled slowly. During the cooling process random collisions between complementary chains provide opportunities for complementary base pairing and rewinding of the chains into intact double helices. The rewinding process is called *reannealing* or *renaturation*.

tracted in numbers equivalent to the number of bacterial cells or viruses produced. By providing selected DNA sequences in large quantities for sequence analysis, DNA cloning has been responsible for many of the successes of molecular biology.

The cloning technique depends on a group of enzymes, the *restriction endonucleases*, occurring in bacterial cells. These enzymes recognize short DNA sequences and cut the DNA at or near these sequences (Table 4-2). The most useful restriction endonucleases are

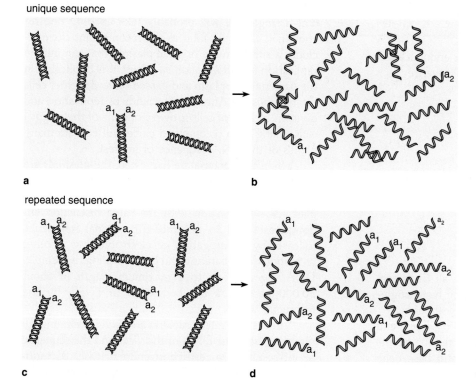

Figure 4-28 Hybridization of DNA fragments from unique or repeated sequences. The a_1 and a_2 nucleotide chains are complementary. **(a)** DNA containing unique, nonrepeated sequences. **(b)** After unwinding by heating, each nucleotide chain has only one pairing partner in the solution. On cooling, the random collisions of complementary sequences required for pairing are relatively rare, and rewinding proceeds slowly. **(c)** DNA fragments containing repeated sequences. **(d)** After unwinding by heating, each single chain has many possible pairing partners in the solution, numbering in proportion to the degree of sequence repetition. The probability of a collision involving complementary chains is greatly increased, and rewinding proceeds relatively rapidly.

Table 4-2 Bacterial Restriction Endonucleases

Enzyme	Source	Sequences Recognized and Sites Cleaved
EcoRI	E. coli	↓ G–A–A–T–T–C C–T–T–A–A–G ↑
HindIII	Haemophilus	↓ A–A–G–C–T–T T–T–C–G–A–A ↑
HpaII	Haemophilus	↓ C–C–G–G G–G–C–C ↑
HhaI	Haemophilus	↓ G–C–G–C C–G–C–G ↑
BamHI	Bacillus	↓ G–G–A–T–C–C C–C–T–A–G–G ↑

those that attack *inverted sequences* in such a way that the fragments produced by the digestion have complementary, single-strand ends. Inverted sequences contain a sequence element that repeats in reverse order on opposite nucleotide chains of a DNA molecule:

$$\text{---C–G–A–C–A–A–T} \ | \ \text{A–T–T–G–T–C–G---}$$
$$\longleftarrow$$

$$\text{---G–C–T–G–T–T–A} \ | \ \text{T–A–A–C–A–G–C---}$$
$$\longrightarrow$$

For example, the much-used restriction endonuclease *Eco*RI recognizes and cuts the inverted sequence

$$\downarrow$$
$$\text{---G–A–A–T–T–C---}$$
$$\text{---C–T–T–A–A–G---}$$
$$\uparrow$$

at the arrows, producing DNA fragments with the complementary single-chain ends

$$\text{---G} \qquad\qquad \text{A–A–T–T–C---}$$
$$\text{---C–T–T–A–A} \qquad\qquad \text{G---}$$

Since *Eco*RI attacks only this sequence, all fragments produced by the enzyme's cutting action have single-chain ends capable of complementary base pairing.

In most DNAs, the sequences cut by *Eco*RI and other restriction endonucleases appear at random points spaced, on the average, several thousand base pairs apart. The enzymes can therefore cut essentially any natural DNA molecule into a finite number of fragments. When the restriction endonuclease cuts inverted repeats in the pattern catalyzed by *Eco*RI, all the fragments produced are capable of pairing with any other fragment produced by the same enzyme.

The DNA fragments produced in this way are cloned in bacteria by taking advantage of *plasmids*, small, circular DNA molecules that exist in the cytoplasm of bacterial cells without connection to the major DNA circle of the bacterial nucleoid (see p. 567). Although unconnected to the nucleoid DNA, plasmids are replicated and passed on during bacterial cell division. To clone a sequence (Fig. 4-29), both the DNA sequence of interest and plasmids isolated from a bacterium are digested with a restriction endonuclease such as *Eco*RI, producing fragments with complementary ends (Fig. 4-29*a* and *b*). When the DNA is exposed to conditions that promote base pairing, DNA circles form that contain both the plasmid DNA and the DNA of interest (Fig. 4-29*c*). The single-chain breaks remaining in the composite circles are then covalently closed by *DNA ligase*, an enzyme with this activity (Fig. 4-29*d*). The circles including introduced DNA are termed *recombinant DNA* because the end product of the technique, in which DNA segments from different sources are joined covalently into a continuous DNA molecule, resembles the outcome of genetic recombination (see p. 1062).

The recombinant DNA circles are introduced into a bacterium such as *E. coli*. Exposing *E. coli* cells to an elevated temperature (42°C) for a few minutes in the presence of Ca^{2+} induces the living cells to take up the recombinant plasmids in intact form. The recombined plasmids are added at concentrations low enough so that a single cell will probably take up only one recombined circle. As long as the recombinant circle includes a site from the original plasmid that can initiate replication (called a *replication origin*), it is duplicated in the bacterial cytoplasm and passed on to daughter cells during division. After many rounds of replication and division, a clone of cells grown from one of the original cells taking up a recombined circle will contain many copies of the DNA sequence of interest.

When the bacteria are harvested, the plasmids are extracted and purified by techniques such as centrifugation. The sequence fragment is then cut from the plasmids by digestion with the same restriction endonuclease used to generate the original fragments, followed by centrifugation, gel electrophoresis, or other techniques for separating and concentrating the sample. DNA segments up to a maximum length of about 50,000 base pairs can be cloned in *E. coli* by the technique.

Viruses that infect bacteria are used for cloning by digesting both the DNA of the virus and the sequence of interest with a restriction endonuclease. Bacteriophage λ, which infects *E. coli*, is a common choice for

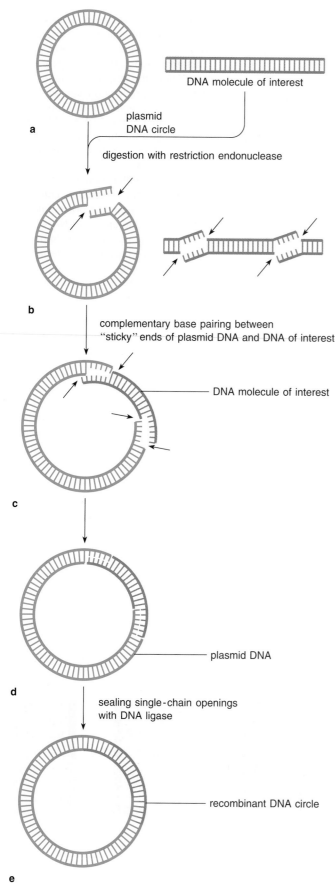

Figure 4-29 Production of recombinant DNA for cloning by restriction endonuclease digestion (see text).

this method. *Eco*RI digestion cuts a nonessential segment from the center of the λ-DNA that can be replaced with a fragment from a DNA sample digested with the same enzyme. Both the λ-DNA and the DNA of interest are digested (Fig. 4-30a). The end fragments broken from the λ-DNA and the fragments produced by digestion of the DNA sample of interest are mixed under conditions that promote rewinding of the complementary ends (Fig. 4-30b). The recombinant molecules, consisting of viral end pieces and a central region derived from the DNA sample, are closed into continuous molecules by DNA ligase. The recombinant DNA is coated with λ coat proteins to produce infective viral particles. Infecting an *E. coli* culture with the particles leads to the production of millions to billions of viral particles containing the DNA sequence of interest. After cloning, the DNA, now also multiplied into millions or billions of copies, is extracted from the virus particles and separated from the λ vector by digestion with *Eco*RI.

When the entire DNA complement of an organism is used in the bacteriophage λ cloning system, as shown in Figure 4-30a and b, the technique creates what is known as a *DNA library* containing cloned fragments of all the DNA sequences of the organism. Following extraction of the cloned DNA, individual sequences can be identified and isolated from the library by hybridization with DNA probes complementary to the sequences of interest.

Equivalent techniques using a virus infecting eukaryotic cells have also been developed. In this case DNA molecules of interest are recombined with the DNA of SV40, a virus that infects Green monkey cells. Green monkey cell cultures are then infected with the recombinant SV40 viral DNA. In the infected cells the inserted DNA sequences are replicated along with the viral DNA as the cells synthesize and assemble new viral particles.

The Polymerase Chain Reaction Producing multiple DNA copies by cloning recombinant molecules is a demanding procedure that requires elaborate biochemical techniques and considerable time. The polymerase chain reaction, developed in 1983 by K. B. Mullis and F. Faloona (see the Experimental Process box, p. 140), has greatly simplified the production of multiple DNA copies. The reaction takes advantage of the features of *DNA polymerase*, the enzyme replicating DNA.

In the replication reaction catalyzed by DNA polymerase, the two nucleotide chains of the molecule being duplicated unwind (Fig. 4-31a). Each chain serves as the template for synthesis of a complementary copy assembled by sequential base pairing (Fig. 4-31b and c). Wherever a thymine (T) is the base being copied in the template, a nucleotide containing adenine (A) is placed in the copy; wherever a cytosine (C) appears in the template, a nucleotide containing guanine (G) is placed in the copy. Similarly, an A in the template

The Unlikely Origin of the Polymerase Chain Reaction

Kary B. Mullis

KARY B. MULLIS earned a B.S. in Chemistry at Georgia Institute of Technology in 1966 and a doctorate in biochemistry from the University of California at Berkeley. After postdoctoral work at the University of Kansas in pediatric cardiology, and the University of California at San Francisco in pharmaceutical chemistry, he joined Cetus Corporation in Emeryville, California. He is currently living in La Jolla working as a private consultant to a number of companies. Dr. Mullis sees himself as a generalist with a chemical prejudice, and his publications include fiction and cosmology as well as biochemistry. He has received a number of awards for his invention of the polymerase chain reaction including a Gairdner International Award in 1991. His hobbies include photography, surfing, and his Institute For Further Study in rural Northern California.

Since its introduction in 1985, the polymerase chain reaction (PCR) has become a tool without which many molecular biologists would walk out of the lab in frustration. Despite occasional assertions that it has taken the challenge and subtlety out of their research, PCR is a surprisingly powerful tool that has made life more fun. It has rendered the enormous complexity of genetic material manageable and helped propel molecular biology into medicine, paleontology, forensic science, and a number of other disciplines. Together with such technologies as cloning and sequencing, and the indispensable aid of computers, PCR has strengthened our assurance that by sometime early in the next century entire genomes, including that of humans, will be known in their entirety, and if we so desire, they will be manipulated.

A great deal of technology develops through the efforts of a laboratory or a number of labs working toward a particular goal, beginning with an existing process and making progress by way of incremental modifications and improvements. Sometimes exciting breakthroughs occur, but from the outset the final goal is clear, and most of those involved have at least a general idea of how the goal will be accomplished. This kind of thing is called development, and sometimes it is a lot like work.

The polymerase chain reaction did not result from a long period of development. It was invented by accident late one night in May 1983 by the driver of a grey Honda Civic following California Highway 128 through the mountains between Cloverdale and the Anderson Valley. It happened at a particular moment, and I count myself among the luckiest of people that it was in my head that a number of disconnected notions, mostly true but some false, all of them quite ordinary, came together in this new way. Everything else I had been thinking about that night came to a halt, as did the car by the side of the road about 46 miles from the coast.

I had been thinking about the fact that one of my new ponds was leaking and also about an experiment I was planning, an experiment I never got around to doing. This experiment was what you might call development work. It wasn't something anyone else had done exactly, and it was interesting enough to compete with the pond crisis, but none of the parts were brand new things; they were just pieces of the repertoire of the DNA chemist and I was arranging them in my head into patterns that would fit my particular purpose.

I was trying to develop a method for looking at single-base-pair mutations in human DNA. A single base pair is only one of three billion in human DNA—what could be done about all those other irrelevant base pairs to focus on the base pair of interest? I was brand new to the kind of DNA that organisms have and particularly unencumbered by excessive experience with human DNA. Thus, I didn't really grasp the immense difference between a plasmid of maybe 5000 base pairs, to which you can hybridize an oligonucleotide fairly selectively, and a generous genome of 3.3 billion base pairs like people have. This was a Very Real Problem.

But somehow the importance of that problem had escaped me. I had innocently thought to myself, why can't you use a simple modification of Sanger's dideoxy sequencing method [see p. 143]; that is, leave out the deoxynucleosides altogether and thereby determine one base pair of a sequence at the point of a possible single-base-pair mutation? Why can't you arrange that an oligonucleotide primer will be labeled in one lane of a gel and not the other three if it adds a particular alpha-labeled dideoxynucleoside triphosphate (ddNTP) and not one of the other three? (This is because the 3'-end of the primer would hybridize only adjacent to a particular complementary base on the DNA strand of interest.) Today I would know better, and if PCR had not yet been invented, I probably wouldn't invent it.

Instead of seeing the real drawback—that the primer I would use for the extension reaction in the dideoxy sequencing approach would hybridize not only to the intended target but also to about ten thousand similar

sequences—I was going sharply to my left across the inside of a turn, cutting directly across the raised yellow dots, knowing that my driving was the best way around the turn and that if anyone was coming the other way I'd see him. I was thinking that if there was any question as to whether the primer had landed on the right spot it could be resolved by using a second primer directed to the same base pair as the first, but made to come in on the other strand. The two primers, after both had picked up a single labeled ddNTP, could be separated on a short gel if they were made to have different sizes. The way the primers were then arranged in my mind on the DNA template looked like a PCR reaction, but I hadn't noticed it yet. (Thinking just now about the way I write, I am aware of my ponderous sentences, which didn't do well in high school English. But I realize they hint at a mind with a propensity for holding many loosely connected facts at once. And this, I suppose, is a good trait for someone who would invent.)

An old piece of information floating around in my memory could have caused trouble for the method I was planning. The DNA samples I would need to look at would come from blood. From my work in a lab at UCSF I knew that John Maybaum had been designing assays to look for DNA nucleotides in blood, so I thought there must be some there, and they could be a source of problems for my experiment. Any nucleotides (dNTPs) present might be added one after the other to the primer, and then the chain-terminating ddNTP could have added to a string of one or more of these appended to the 3' end of a primer rather than to the primer itself. I didn't want them to add at any other location than the immediate 3'-end of the primer. Otherwise they would mark some other site in the sequence than the one directly adjacent to the end of the primer. What could I do with the possible contaminating dNTPs?

A very cheap enzyme called bacterial alkaline phosphatase (BAP) will take the phosphates off a dNTP down to the nucleoside, rendering it unsuitable as a substrate for DNA polymerase and harmless, therefore, to my reaction. So if I treated the samples with BAP before adding them I would be all right. Except that I needed some way to get rid of the dNTPs without later destroying the somewhat pricey ddNTPs, which are really expensive when alpha-labeled. So after using BAP on the dNTPs I would have to get rid of it. The party line on BAP was that, unlike most enzymes, it could not be permanently destroyed by boiling. It turns out that the party line is wrong, though. You *can* get rid of BAP by boiling; you just can't use the "standard" zinc-containing buffer. Without zinc the protein can't do its perfect refolding routine. But I didn't know that at the time—few people did. Good thing, too. Because it meant that I had to figure out some other way to get rid of the dNTPs, and that turned out to provide another element of PCR.

Now, the polymerase chain reaction was about to be discovered, as the Surprising Consequence of a Possible Solution to a Hypothetical Problem that might have arisen in a Proposed Experiment which itself contained an Implicit Assumption the Truth of which would have denied the Existence of the Very Real Problem the polymerase chain reaction so neatly solved. Got that?

It occurred to me that a much cleverer solution than heat denaturation was to use DNA polymerase to destroy any unwanted dNTPs that came in with the sample DNA. If I put the DNA sample and the primers together, denatured the DNA, and then let the reaction cool, the primers would bind to the DNA. Now if I added polymerase, and if there were traces of dNTPs in the reaction, they would be added to the primers until the dNTPs were depleted. Then, I thought, I can heat again to boiling to denature the primers, which may now have been extended from the DNA template. The primers would have been added in an immense excess over the template, so when I cooled the reaction new primers would land on the target sites rather than the ones that might have been extended. Now if I added the ddNTPs and the polymerase, the primers would be extended by one base, and in one of the four tubes each of the primers would be labeled, depending on which nucleoside it added.

And what about the primers that had possibly been extended in the dNTP depletion reaction? What if they had been extended enough to give rise to a sequence which, necessarily being the same as the sequence on the other target strand, would therefore be a site where the other primer would now hybridize? What if you added your own dNTPs in excess to cause this—intentionally left out the ddNTPs and repeated the process? What if you designed the primers to have more than just one base between their binding sites?

The answer was PCR. The sequence defined by the primer sites would be replicated during each cycle, and the molecules resulting from this would also be replicated each cycle. After ten cycles there would be a thousand copies of a double-stranded DNA molecule with a discrete size corresponding to the distance on the target between the ends of the primers. After twenty cycles there would be a million, and so on.

I spent the next three years demonstrating that the reaction did in fact work as predicted. During that time there was a patent and papers to write; eventually, there was a large group working on applications of the process at Cetus. Like anything else it got a little slow at times, and I headed up north to watch my pond leak. I was obliged by contract to assign the patent rights to Cetus. Nine years later, I read in the *Wall Street Journal* that Cetus had sold the patent, US 4,683,202, to Hofmann La Roche for about a third of a billion dollars.

Not bad for a night's work.

DNA Sequencing

The information obtained from DNA sequencing is one of the primary sources of the molecular revolution. The sequences provide insights into gene functions and the mechanisms by which genes are regulated. Sequence comparisons allow estimation of the degree to which genes of the same or different organisms are related, confirm evolutionary lineages determined by other, more traditional methods, and establish lineages that were previously unsuspected. In some cases, comparisons of normal and mutant gene sequences have revealed the molecular basis of hereditary disease. The results of gene sequencing have been so valuable that work is now in progress to determine the complete DNA sequence of humans and several other organisms.

Once obtained in large quantities by cloning or the polymerase chain reaction, a DNA sample can be sequenced by one of several methods. One of the most frequently employed, developed by A. Maxam and W. Gilbert, uses chemical reagents that break DNA molecules at specific bases. The DNA fragments are run on electrophoretic gels, and the sequence is read directly from the series of bands sorted out by the gels. Since DNA copies can be made from RNA molecules by the reverse transcriptase enzyme (see p. 585), the method also can be used indirectly to sequence RNA.

In the Maxam and Gilbert technique, a quantity of the DNA molecule to be sequenced is first digested by a restriction endonuclease. The digestion breaks the DNA into groups of fragments. Within each group all the fragments are of uniform length and contain the same subpart of the DNA sequence. The fragment groups are separated by centrifugation or gel electrophoresis.

The DNA lengths of each fragment group are labeled at one end (the 5' end; see p. 527) by enzymatic attachment of a radioactive phosphate group. A frag-

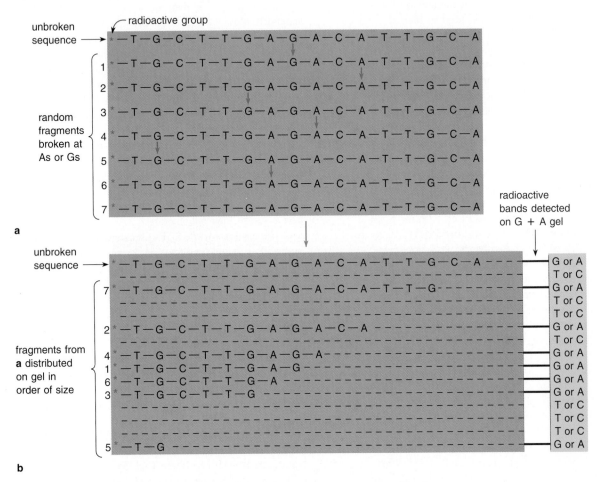

Figure 4-33 DNA sequencing by the Maxam and Gilbert technique (see text). To produce the G + A gel, a uniform class of DNA fragments **(a)** is broken randomly wherever guanine or adenine occurs in the sequence (arrows). The fragments are then sorted out according to length by running them on an electrophoretic gel **(b)**. The bands on the gel indicate points where a guanine or adenine occurs in the sequence; empty spaces show where T or C occurs in the sequence (see text). Fragments missing the beginning end of the sequence do not show up on the gel because they lack the radioactive marker and are not detected.

ment group is then treated by reagents that modify and break the DNA chain at sites where specific bases appear. One sample from a fragment group is treated with dimethyl sulfate, which adds methyl groups to guanines and adenines (both purines) in the DNA fragments. The dimethyl sulfate treatment is carried out under conditions that methylate some but not all of the guanine and adenine bases. The resulting positions of the methylated and unmethylated guanines and adenines are random. The DNA chain is unstable at sites containing the methyl groups and is easily broken at these sites by treatment with an alkali at 90°C. The alkali treatment produces a family of DNA pieces from a fragment group, with each DNA piece broken at a different guanine or adenine in the sequence (Fig. 4-33a).

The pieces from a DNA fragment group are then run on an electrophoretic gel, which separates the pieces into individual bands. A photographic emulsion is placed over the gels to detect the radioactive bands. Since the only bands detected will be those containing DNA segments labeled at their 5' end, the fragments are sorted out in order of the position of an adenine or guanine from the labeled end, with successively shorter lengths toward the bottom of the gel (Fig. 4-33b). Reading the gel from the bottom to the top gives the positions of guanines and adenines in the sequence (the G + A gel in Fig. 4-34).

A modification of the same technique produces segments broken only at adenines. Points in the DNA chain containing methylated adenines are less stable in an acid solution than points containing methylated guanines. Gentle treatment with dilute acid breaks the chain at only these adenines. Running the fragments sorts them out by length, with successively shorter lengths toward the bottom of the gel (the A gel in Fig. 4-34).

A similar series of treatments using the reagent hydrazine produces DNA pieces ending in T or C. Hydrazine removes the T and C bases (both pyrimidines) from the DNA fragments, producing weak points that break when treated with alkali. Hydrazine is used under conditions in which the reaction is incomplete, so that only some of the Ts or Cs are removed at random points. The resulting fragments, ending at either T or C, produce the T + C gel shown in Figure 4-34.

A final step distinguishes T from C. By using the hydrazine reagent in the presence of 1 molar salt, the Ts are protected and only Cs are removed randomly from the DNA fragment group. The fragments are then broken at the weak points by alkali, yielding a series of pieces ending with nucleotides that contain C. This preparation produces the C gel shown in Figure 4-34.

The DNA sequence of the fragment group is read directly from the bottom to top of the aligned gels by comparing the G + A, A, T + C, and C gels at each

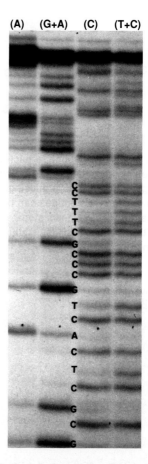

Figure 4-34 A, G + A, C, and T + C gels showing the sequence of a DNA fragment from a bacterial cell. The sequence is read directly from bottom to top. Courtesy of A. M. Maxam.

level. Combining the sequences of all the fragment groups from the original molecule allows the entire sequence of the DNA molecule, or of a cDNA copy of an RNA molecule, to be reconstructed.

An alternate sequencing technique developed by F. Sanger and his colleagues is similar except that it uses DNA replication to provide the consecutive sequence lengths for gel electrophoresis. In the replication, modified forms of the four DNA nucleotides are used in which a single —H is bound to the 3'-carbon of the deoxyribose sugar instead of an —OH (Fig. 4-35). A nucleotide in this form is known as a *dideoxynucleotide*. During DNA replication a new nucleotide is normally added to the 3'-OH group of the most recently added nucleotide in the copy. Because the dideoxynucleotides have no 3'-OH available for addition of the next base, DNA replication stops wherever one of these nucleotides is inserted instead of an unmodified nucleotide.

In practice groups of sequence fragments are obtained by digesting the DNA to be sequenced with a restriction endonuclease, as in the Maxam and Gilbert technique. Four different replications of a given frag-

lenses of an electron microscope focused? Why must specimens viewed in the TEM be thin and nonvolatile? Compare image formation in the TEM and brightfield light microscope. Why is α rather than $n \sin \alpha$ used in the resolution equation for electron microscopy?

8. Outline the construction of a scanning electron microscope. What is scanned in an SEM? What are secondary electrons? Outline the principles of image formation in the SEM. What determines resolution in the SEM? Why are specimens to be viewed in the SEM frequently coated with a thin film of metal?

9. What are fixatives? What kinds of fixatives are used in specimen preparation for light and electron microscopy? What is autoradiography? Outline the steps used in preparing tissue for light microscope autoradiography. Outline the difference between direct and indirect immunofluorescence. What is the primary advantage of indirect immunofluorescence?

10. Outline negative staining, shadowing, and freeze-fracture techniques used for preparing specimens for electron microscopy.

11. Compare the components of culture media for bacterial and animal cells. What is a clone? How is a clone prepared? What steps are followed in selecting a bacterial strain resistant to an antibiotic? What steps are followed in selecting a bacterial strain that requires a substance that bacteria can normally make for themselves?

12. How are animal cells obtained for culturing? How are cultures "immortalized"? Why is this necessary? What is cell fusion? How is it accomplished? In what ways have fused cells proved valuable in research? What is a hybridoma? A monoclonal antibody?

13. What is cell fractionation? How is centrifugation used in cell fractionation? What is a cell-free system? What factors affect the rate at which molecules move downward or upward in a spinning centrifuge tube? What is the sedimentation constant? How is it related to a Svedberg unit?

14. What is density gradient centrifugation? Why do groups of molecules form highly distinct bands in density gradients? What is buoyant density centrifugation? How is the density gradient established in this technique?

15. Outline the structure of a gel electrophoresis apparatus. What factors determine the rate at which molecules move through electrophoretic gels? What advantages does gel electrophoresis provide over centrifugation as a technique for isolating and purifying molecules from mixed samples?

16. How can relative molecular weights be determined by gel electrophoresis? Compare the methods used for determining relative molecular weights of protein and nucleic acid samples. Why are proteins treated with SDS and mercaptoethanol before electrophoresis?

17. Outline the steps followed in two-dimensional gel electrophoresis. What is isoelectric focusing?

18. What is the difference between a Southern, northern, and western blot? What is the difference between gel filtration, ion exchange, and affinity chromatography? Outline the principles underlying molecular separation by each of these chromatographic methods.

19. What features of molecular structure split an X-ray beam into separate subbeams? What determines the locations and intensity of spots in an X-ray diffraction pattern? What is the difference between diffraction and refraction?

20. What is nucleic acid hybridization? What is DNA melting? Why do DNA sequences rich in G-C pairs melt at higher temperatures than sequences rich in A-T pairs? Why does a heated DNA sample containing repeated sequences rewind more rapidly after cooling than a sample containing only unique, nonrepeated sequences?

21. What is a nucleic acid probe? How are probes used to identify individual spots in a Southern or northern blot?

22. Outline the steps followed in cloning a DNA sequence. What is a restriction endonuclease? Why is *Eco*RI especially useful for cloning DNA? What is a DNA library? How are DNA libraries constructed? What advantages does bacteriophage λ provide to making DNA libraries?

23. Outline the steps in a polymerase chain reaction. What is a DNA primer? How are primers supplied for a polymerase chain reaction?

24. Summarize the steps in DNA sequencing by the Maxam and Gilbert and the Sanger techniques. What is a dideoxynucleotide? Why does replication stop at a site at which a dideoxynucleotide is inserted?

25. How are protein sequences deduced from nucleic acid sequences? What is the significance of the AUG start codon and the terminator codons UAG, UUA, and UGA to this deduction?

Supplement 4-1
Lenses and Image Formation in the Light Microscope

The effects of glass lenses in focusing light rays and forming images in the light microscope depend on the behavior of light as it passes through transparent media of differing refractive index. As light rays pass through the boundary between two such media, the rays change in velocity and are bent (refracted) from their paths. The degree of refraction depends on two factors: the difference in the refractive indices of the transmitting media and the angle at which the light strikes the surface of the second medium (Fig. 4-39). If the light rays strike at an angle exactly perpendicular to the surface, no refraction occurs and the light is transmitted without a change in path (Fig. 4-39a). At any other angle some bending occurs (Fig. 4-39b). As the angle of incidence decreases from the perpendicular, the degree of bending increases. For a given medium, light passing through the boundary into the interior at a fixed angle is always refracted to the same degree. Therefore, if parallel rays strike a flat glass plate at an angle to the surface, as shown in Figure 4-39b, the refracted rays transmitted through the glass remain parallel and do not converge or diverge.

This is not the case if the glass surface transmitting the light is curved. In this case each point on the surface is presented to the approaching light beam at a different angle, and the angle of refraction at each point is different. If the glass surface is convex and part of a sphere, making it act as a lens, the light rays of a parallel beam strike the surface at a greater angle of incidence toward the periphery of the lens and are bent to greater angles. As a result, the rays converge toward a point called the *focal point* (Fig. 4-40). The distance between this point and the center of the lens is the *focal length* of the lens. For lenses with a smaller radius of curvature, the focal point is closer to the center of the lens. Lenses with the same curvature on their front and back surfaces have a focal point on either side of the lens, located at the same distance from the lens center.

Light rays traveling either direction through a lens follow the same paths. Therefore, if a light source is placed at a focal point on one side of a lens, the rays passing through the lens are refracted into a parallel beam as they leave the other side of the lens.

Light rays reflected or issuing from an object placed outside the focal point of a converging lens are focused into an *image* of the object in a plane that lies beyond the other side of the lens (Fig. 4-41). Consider an object placed at point A outside the focal point of the lens in Figure 4-41. Whether this object is a light source or reflects light rays emitted by another source, rays radiate outward in all directions from each point on the object surface. Rays originating from point B on the object surface and traveling parallel to the lens axis (ray BC) are refracted by the lens through the focal point F_2 on the opposite side of the lens. Rays from point B passing through the focal point F_1 on the same

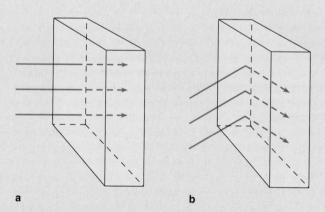

Figure 4-39 Refraction of light at the boundary between two media of different refractive index, as for example between air and a flat glass plate.
(a) Light striking the boundary at an angle perpendicular to the surface is unrefracted. **(b)** Light striking at any other angle is refracted as it passes through the boundary.

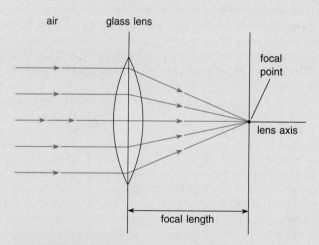

Figure 4-40 Convergence of parallel rays entering a convex lens to a point—the focal point—on the other side of the lens. The distance of this point from the center of the lens is the focal length of the lens. Rays originating at the focal point are bent to parallel paths on the opposite side of the lens.

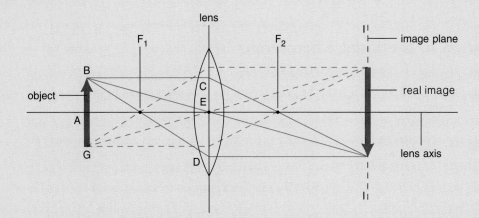

Figure 4-41 Formation of a real image by an object placed outside the focal point of a converging lens (see text).

side of the lens as the object (ray BD) are bent to a line parallel to the lens axis on the other side of the lens. Rays from point B passing through the axis of the lens (ray BE) enter and leave the lens at the same angle. The rays BC, BD, and BE converge at a location beyond the focal point F_2. This point lies on a plane defined as the *image plane* (I in Fig. 4-41). All other rays leaving a different point on the object, such as point G, also converge at a corresponding point in the image plane (dashed lines in Fig. 4-41). Because rays leaving all points in the object converge, or are *focused*, to corresponding points in the image plane, an image of the object is constructed in this plane. The image can be seen directly if a reflecting screen is placed in the image plane.

An image that can be focused on a screen in the image plane is defined as a *real image*. Real images are formed only by objects placed outside the focal point of a converging lens. A real image is typically inverted and larger in size than the original object; that is, it is *magnified*. The degree of magnification depends on the

focal length of the lens and the placement of the object with respect to the focal point. As the object moves closer to the lens (but remains outside the focal point), the image plane moves farther from the opposite side of the lens, and the image increases in size. The greatest magnification is obtained when objects are placed as close as possible to the focal point of a lens of very short focal length.

Objects placed inside the focal point of a converging lens do not form a real image. However, an observer can still see an image by looking directly into the opposite side of the lens. This image, called a *virtual image*, cannot be focused on a screen, and no points exist at any plane in space at which rays radiating from object points are brought to points of focus. Figure 4-42 shows why such images are visible to an observer looking into a lens. The object is placed at point A, inside the focal point of the lens. Ray BC represents a ray leaving point B and passing through the lens. This ray, if extended backward, would pass through the focal point. It therefore follows the same path as a ray orig-

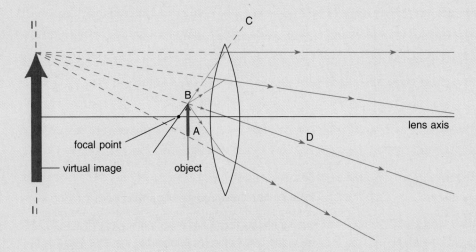

Figure 4-42 Formation of a virtual image by an object placed inside the focal point of a converging lens (see text).

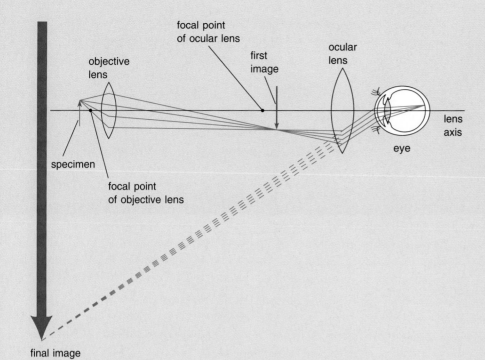

objective lens

focal point of ocular lens

first image

ocular lens

specimen

focal point of objective lens

lens axis

eye

final image

Figure 4-43 The arrangement of lenses and image formation in a compound light microscope. The specimen is placed just outside the focal point of the objective lens. As a result, the objective lens forms a real image of the specimen (indicated as the first image). The lenses are placed so that the real image formed by the objective lens falls in a plane inside the focal point of the ocular lens. As a result, the ocular lens forms a virtual image of the first image that can be seen by an observer looking into the ocular lens. (The virtual image seen by the observer is indicated as the final image.)

inating from the focal point and is bent parallel to the lens axis on the other side of the lens. Another ray leaving object point B and passing through the central axis of the lens (ray BD) is not bent from its path and leaves the lens at the same angle. The two rays BC and BD diverge on the opposite side of the lens from the object and are not brought to a point of focus. If lines are extended backward from rays BC and BD, however (dashed lines in Fig. 4-42), they converge at a point on the same side of the lens as the object at a location outside the focal point. All other rays leaving object point B also diverge on the side of the lens opposite the object point. If these lines are extended backward, they converge at the same point as rays BC and BD on the same side of the lens as the object. To an observer looking into the lens on the opposite side from the object, the rays from object point B would, therefore, appear to be originating from the point of convergence of the dashed lines. Construction of a similar diagram for all other object points would show the same pattern, with apparent points of origin falling in the same plane as the convergence point for rays BC and BD. This plane is the image plane for the virtual image (I in Fig. 4-42).

The total collection of points forming the virtual image can be directly seen by the eye because the rays apparently radiate outward from all points in the image plane. This is precisely equivalent to the rays that radiate outward from all points on a real physical object. These rays are focused into a real image on the retina by the converging lens of the eye, and the virtual image

is seen as a real object. The virtual image is magnified but not inverted.

The compound light microscope uses both real and virtual images (Fig. 4-43). The specimen is placed just outside the focal point of the objective lens, which has a very short focal length. Because of this placement, the objective lens focuses a real image on a plane on the opposite side of the lens from the specimen (indicated as the first image in Fig. 4-43). This real image of the specimen acts as the object for the ocular lens, which is placed so that the real image formed by the objective lens falls in a plane inside its focal point. As a result, the ocular lens forms a virtual image of the specimen (indicated as the final image in Fig. 4-43). This image is observed by looking into the ocular lens on the opposite side from the first image.

Light microscopes can also be constructed with a single lens. In this type of microscope the object is placed inside the focal point of the lens, and the image is seen as a virtual image by looking into the lens. Leeuwenhoek's microscopes (see p. 29), constructed in this way, consisted of a single, small sphere of glass mounted between two brass plates. The object was held on the point of a pin just inside the focal point of the lens. Although capable of magnifying objects between 200 and 300 times, the resolution of Leeuwenhoek's microscopes was relatively poor. It is still not understood how Leeuwenhoek was able to observe and correctly record the structure of objects that were technically much smaller than the resolving power of his instruments.

5

THE STRUCTURE OF
CELLULAR MEMBRANES

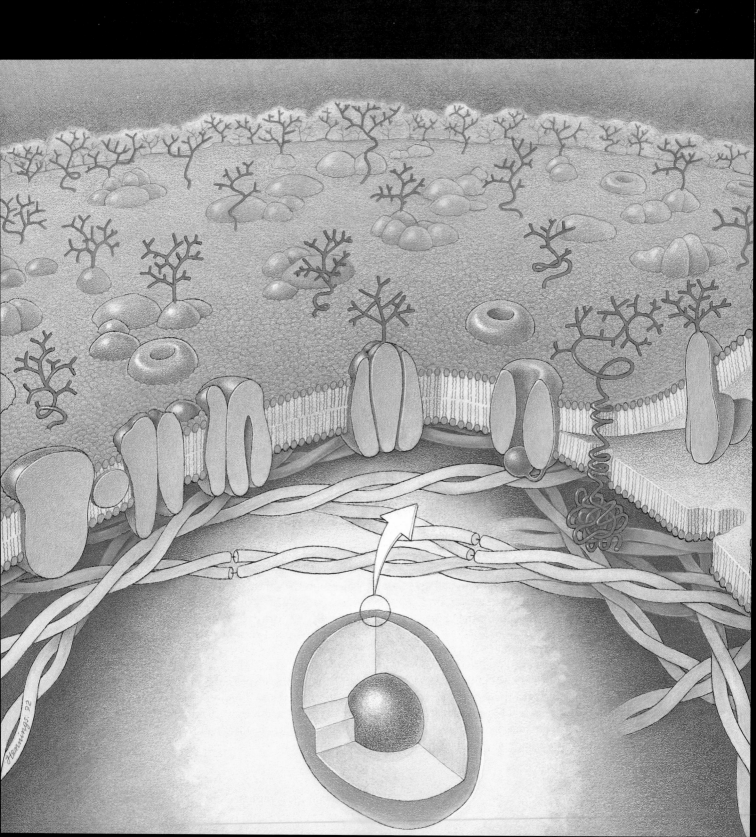

Membranes organize and maintain all cells as separate and distinct molecular environments. The plasma membrane, the outermost membrane of both prokaryotic and eukaryotic cells, keeps the cell contents from mixing freely with molecules outside the cell and serves as the primary zone of contact between the cell and the outside. In eukaryotic cells internal membranes also divide the cell interior into compartments with distinct biochemical environments and specialized functions. Regions within the narrow membrane interior provide hydrophobic environments in which certain reactions take place almost exclusively. Among the reactions of the membrane interior are some of the most vital interactions providing cellular energy, including major reactions of mitochondria and chloroplasts.

These fundamental roles depend on the basic structural plan of membranes. The contemporary view of this structure is based on the *fluid mosaic model*, a hypothesis first advanced in 1972 by S. J. Singer and G. L. Nicolson. According to the model (see Fig. 5-17), membrane structure is based on a double layer of lipid molecules, the *bilayer* (see Fig. 5-15). The lipids of the bilayer have both hydrophobic and hydrophilic qualities. Their hydrophobic portions, concentrated in the membrane interior, create a barrier that prevents free movement of polar molecules across the membrane. Their hydrophilic portions face the polar environments at the membrane surfaces. Under physiological conditions the bilayer is in a fluid state, that is, the lipid molecules forming the bilayer are free to exchange places and undergo motions, such as rotation, characteristic of fluids. Membrane proteins float individually in the lipid bilayer, suspended like "icebergs in the sea." The segregation of the membrane into separate regions, some occupied by proteins and others by lipids, is the "mosaic" part of the fluid mosaic model.

The protein molecules embedded in the lipid framework provide membranes with specialized functions. Some proteins associated with the plasma membrane transport selected polar molecules and ions between the cell interior and exterior, thereby maintaining the internal balance of molecules required for cellular life. Similar proteins carry out transport functions in the boundary membranes of internal organelles. Other proteins of the plasma membrane are markers, which identify the cell as part of the same individual or as foreign, or receptors, which can recognize and bind molecules originating as chemical signals from other cells. (Peptide hormones, for example, are among the signal molecules bound by receptors in the plasma membrane.) Some membrane receptors also recognize and bind chemical marker groups on other cells. These receptors allow cells to adhere in tissues and organs. Still other proteins give membranes the capacity to carry out important reactions of respiration and photosynthesis. Membrane proteins also take part in the sorting, distribution, and release of secretory products from cells. Each membrane type, including the plasma membranes of different cells and the membranes of cell organelles, has a characteristic group of proteins that is responsible for its specialized functions.

This chapter describes the structure of biological membranes, including the lipid bilayer and its association with proteins. The varied cellular roles of membranes are the subjects of later chapters in this book.

LIPID AND PROTEIN MOLECULES OF BIOLOGICAL MEMBRANES

All known biological membranes contain both lipid and protein molecules. The relative amounts of lipids and proteins vary from the extremes of myelin (a membrane found in nerve tissue), with about 80% lipids and 20% proteins by weight, to the inner membrane of mitochondria, which contains more than 75% proteins (Table 5-1). Some bacterial membranes, such as those of photosynthetic bacteria, contain about 75% proteins. Most

Table 5-1 Amounts of Lipid, Protein, and Carbohydrate in Different Biological Membranes

Membrane	Approximate Percent of Dry Mass		
	Protein	Lipid	Carbohydrate
Plasma membranes			
erythrocyte	49	43	8
nerve myelin	18	79	3
liver cell	54	36	10
Nuclear envelope	66	32	2
Endoplasmic reticulum	62	27	10
Golgi complex	64	26	10
Mitochondrion			
outer membrane	55	45	Trace
inner membrane	78	22	——
Chloroplast	70	30	——

Adapted from G. Guidotti; reproduced, with permission, from *Ann. Rev. Biochem.* 41:731 (1972). © 1972 by Annual Reviews, Inc., and R. Lotan and G. L. Nicolson, *Advanced Cell Biology* (Schwartz, L. M., and Azar, M. M., eds.), New York: Van Nostrand Reinhold,1981.

Table 5-2 Percent Lipid Composition of Some Animal, Plant, and Bacterial Membranes

	Myelin	Erythrocyte	Mitochondria	Microsomal Fractions* 1	Microsomal Fractions* 2	Escherichia coli	Bacillus megaterium	Chloroplasts
Cholesterol	25	25	5	6	——†	0	0	——
Phosphatidyl ethanolamine	14	20	28	17	18	100	45	0.6
Phosphatidyl serine	7	11	0	0	9	0	0	——
Phosphatidyl choline	11	23	48	64	48	0	0	4.1
Phosphatidyl inositol	0	2	8	11	6	0	0	1.4
Phosphatidyl glycerol	0	0	1	2	0	0	45	5.3
Cardiolipin	0	0	11	0	2	0	0	0.6
Sphingomyelin	6	18	0	0	9	0	0	——
Cerebroside	21	0	0	0	0	0	0	——
Cerebroside sulfate	4	0	0	0	0	0	0	——
Ceramide	1	0	0	0	0	0	0	——
Galactosyl diglyceride	——	——	——	——	——	——	——	14.7
Digalactosyl diglyceride	——	——	——	——	——	——	——	35.3
Sulfoquinovosyl diglyceride	——	——	——	——	——	——	——	4.9
Chlorophyll	——	——	——	——	——	——	——	23.5
Carotenoids	——	——	——	——	——	——	——	4.9

Adapted from E. D. Korn, Structure of biological membranes, *Science* 153:1491 (1966); chloroplast data from J. S. O'Brien, *J. Theoret. Biol.* 15:307 (1967).
* Primarily endoplasmic reticulum.
† Undetected or not analyzed.

plant and animal membranes fall midway between these extremes, with lipids making up from 30% to 50% of the total membrane molecules and proteins the remainder.

Carbohydrates also occur in membranes, as chemical groups in glycolipids and glycoproteins (see p. 68). Membrane carbohydrates are detected in greatest abundance in eukaryotic plasma membranes, where they occur almost exclusively on the outer membrane surface, facing the cell exterior. When present, carbohydrates make up from 1% to 10% of total membrane components by dry weight.

Membrane Lipids

The lipid part of membranes is highly varied (Table 5-2). Depending on the species and cell type, various plant and animal membranes may include *phosphoglycerides*, *sphingolipids*, and *sterols* as major classes. Phosphoglycerides and some sphingolipids contain phosphate groups and are collectively called *phospholipids* (see p. 55). Lipids containing a carbohydrate group, with or without a phosphate group, are the *glycolipids*.

Phosphoglycerides Phosphoglycerides, the most abundant and characteristic membrane lipids, are based on a glycerol framework. To that framework bind two fatty acyl chains and a phosphate group. The phosphate group acts as a linker binding choline, ethanolamine, glycerol, inositol, serine, or threonine to the lipid structure (see Figs. 5-1 and 2-13; Information Box 5-1 gives details of phosphoglyceride structure).

The most significant feature of phosphoglycerides from the standpoint of membrane structure is that the end of the molecule containing the hydrocarbon chains is nonpolar and hydrophobic, while the end containing the phosphate group is polar and hydrophilic (see Fig. 2-13a). The polar end of a phosphoglyceride extends approximately to the level of the bonds connecting the fatty acid residues to glycerol (represented by the dashed horizontal line in Fig. 5-1a; the circle in Fig. 5-1c represents the polar portion of the molecule). This end of a phospholipid molecule readily associates with water molecules if exposed to an aqueous medium. In contrast, nonpolar hydrocarbon chains (below the dashed line in Fig. 5-1a; represented by zigzag lines in Fig. 5-1c) prefer a nonpolar environment from which water is excluded. Substances with such dual polar-nonpolar properties are termed *amphiphilic* or *amphipathic*. Almost all membrane lipids and proteins possess this dual-solubility property, which is of fundamental importance to their interactions in membranes.

Although membrane phosphoglycerides are primarily structural—they are responsible for most of the bilayer structure of biological membranes—some have functional roles as well. Perhaps phosphatidyl inositol, a phosphoglyceride with the alcohol inositol (see Fig. 2-14) connected to the phosphate group, plays the most significant functional role. This phosphoglyceride takes part in receptor-response mechanisms linked to the plasma membrane. When an external signal molecule such as a peptide hormone or growth factor is bound by a receptor at the cell surface, a cascade of reactions initiated by the receptor triggers an internal cellular response. One of the major pathways sending signals

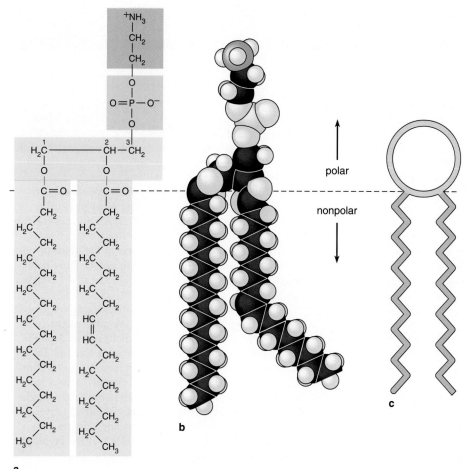

Figure 5-1 Phosphoglycerides. **(a)** Structure of a typical phosphoglyceride, phosphatidyl ethanolamine. The bonds linking the glycerol and fatty acyl chains appear in red. **(b)** Space-filling model of phosphatidyl ethanolamine. The "kink" in the hydrocarbon chain is caused by a double bond in the *cis* configuration. In many animal membrane phospholipids the fatty acyl chain at the 1-carbon of the glycerol backbone is saturated, and the chain at the 2-carbon has one or more double bonds in the *cis* configuration, as in this diagram. **(c)** The diagram used to indicate phospholipids in membrane diagrams. Portions of the molecule above the dashed line in this figure are polar; portions below the dashed line are nonpolar. (See Fig. 2-13.)

from the cell surface to the interior includes reactions in which phosphatidyl inositol is phosphorylated and split into smaller molecules that trigger an internal response (for details, see Ch. 7).

Why membranes contain several kinds of phosphoglycerides instead of a single type is not understood. The combinations appear to be important to membrane function, however, because the proportions of the various phosphoglycerides are maintained in a membrane of a given type. This indicates that there are biochemical mechanisms that detect changes in phosphoglyceride content and make compensatory adjustments if required. Maintenance of a particular group of phosphoglycerides may reflect the fact that the activity of membrane proteins such as enzymes and transport proteins appears to peak only when the proteins are associated with certain phosphoglyceride types.

Membrane Glycolipids Some membrane glycolipids are based on a glycerol framework, with fatty acyl chains attached to the 1- and 2-carbons, as in phosphoglycerides. However, the carbohydrate group of these glycolipids is linked directly to the remaining glycerol site, with no bridging phosphate group (Fig. 5-2). These glycolipids are the primary carbohydrate-

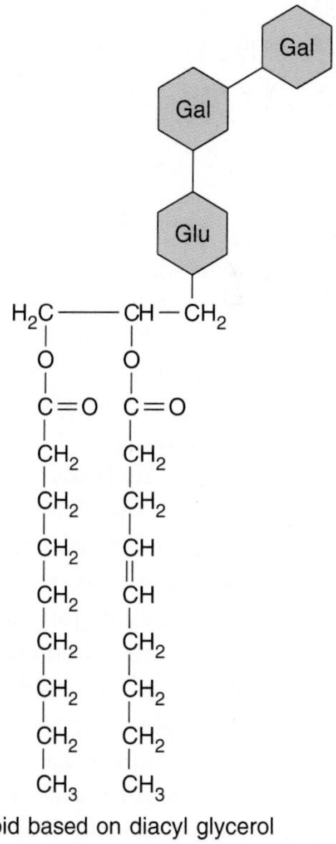

glycolipid based on diacyl glycerol

Figure 5-2 A glycolipid based on diacyl glycerol. The lipid consists of a glycerol residue with fatty acyl chains at the 1- and 2-carbons and a carbohydrate group at the 3-carbon. Glycolipids of this type are common in plants and bacteria but not in animals. Gal, galactose; Glu, glucose.

Phosphoglyceride Structure

Phosphoglycerides are built on a central chemical framework derived from glycerol, to which two fatty acids and a phosphate group link (see Fig. 5-1). The carbon of glycerol that links to the phosphate group is identified by convention as the 3-carbon. The phosphate group binds in turn to one of a group of alcohols, including choline, ethanolamine, glycerol, or inositol, or to the amino acids serine or threonine.

Choline, ethanolamine, and serine are the most common groups in membrane phosphoglycerides. Choline- and ethanolamine-containing phosphoglycerides are known as *phosphatidyl choline* and *phosphatidyl ethanolamine*, respectively, and the type containing serine as *phosphatidyl serine*. The phosphate group and the groups carried by some of the polar alcohols or amino acids linked to phosphoglycerides are charged at physiological pH. The net effect of these groups contributes to the characteristically negative surface charge of biological membranes.

The 1- and 2-carbons of the glycerol framework of phosphoglycerides are bound to fatty acyl chains, nonpolar hydrocarbon chains derived from saturated or unsaturated fatty acids (see Table 2-5). Fatty acyl chains containing 16, 18, or 20 carbons are most common in membrane phosphoglycerides. Unsaturated chains have "kinks" at positions occupied by a double bond. When carbon atoms sharing the double bond take on the *cis* configuration, which is by far the most common, the kink is pronounced—approximately 30°, as shown in Figure 2-10. These bends prevent the fatty acid chains of membrane lipids from packing tightly in the membrane interior, thereby altering the fluidity and other physical properties of the membrane. In animal membranes the majority of phosphoglycerides have a straight-chain, saturated fatty acyl chain at the 1-carbon of the glycerol backbone. An unsaturated chain with one or more double bonds in the *cis* configuration usually links at the 2-carbon (see Fig. 5-1b).

The particular fatty acyl chains occurring at the 1- and 2-carbon of a given phospholipid type vary extensively. For example, a phospholipid with choline at its 3-carbon can potentially link any of the fatty acids listed in Table 2-5 at its 1- and 2-carbons. For this reason a given phospholipid type such as phosphatidyl choline forms a family or class of closely related substances rather than a single molecular type. In animals the combinations of fatty acyl chains appearing in membrane phosphoglycerides vary with diet, particularly those fatty acids that the organism cannot make for itself.

linked lipids of plants and photosynthetic protists; they also occur as a major glycolipid type in bacteria. Although glycolipids based on glycerol also occur in animals, the primary animal glycolipids are based on sphingolipids.

The various sphingolipids are built on a sphingosine backbone with a fatty acyl chain attached (Fig. 5-3; Information Box 5-2 details sphingolipid structure). The *glycosphingolipids* are formed by the addition of carbohydrate units to sphingolipids (Fig. 5-4).

The nonpolar hydrocarbon chains at one end of a sphingolipid or glycosphingolipid give that end of the molecule a hydrophobic character, and the —OH or the various polar groups linked at the 3-carbon site make the other end hydrophilic. These lipids are therefore amphipathic substances that fit into membrane structure in positions analogous to the phosphoglycerides.

Glycosphingolipids occur in plasma membranes throughout the plant and animal kingdoms. They are especially abundant in the plasma membranes of animal nerve and brain cells. In plasma membranes, glyco- sphingolipids orient with their polar carbohydrate groups extending from the outer membrane surface. These carbohydrate groups may form highly stable, interlocked networks of hydrogen bonds extending over the membrane surface. Such networks are highly developed on the plasma membranes of cells exposed to physical or chemical stress, such as the cells lining the small intestine in mammals. These characteristics suggest that one of the functions of glycosphingolipids is membrane stabilization.

Several glycosphingolipids of the ganglioside group have been identified as sites recognized by antibodies in immune reactions or as antigens responsible for blood group interactions. Some cell surface markers of red blood cells responsible for the ABO blood groups of humans, for example, are glycosphingolipids (see Fig. 2-17). Some glycosphingolipids at the cell surface also act as binding sites for substances taken up by cells. A number of peptide hormones and some molecules that act as cell poisons, including the cholera and tetanus toxins, are among the molecules bound by glycosphingolipids. Several viruses and bacteria also

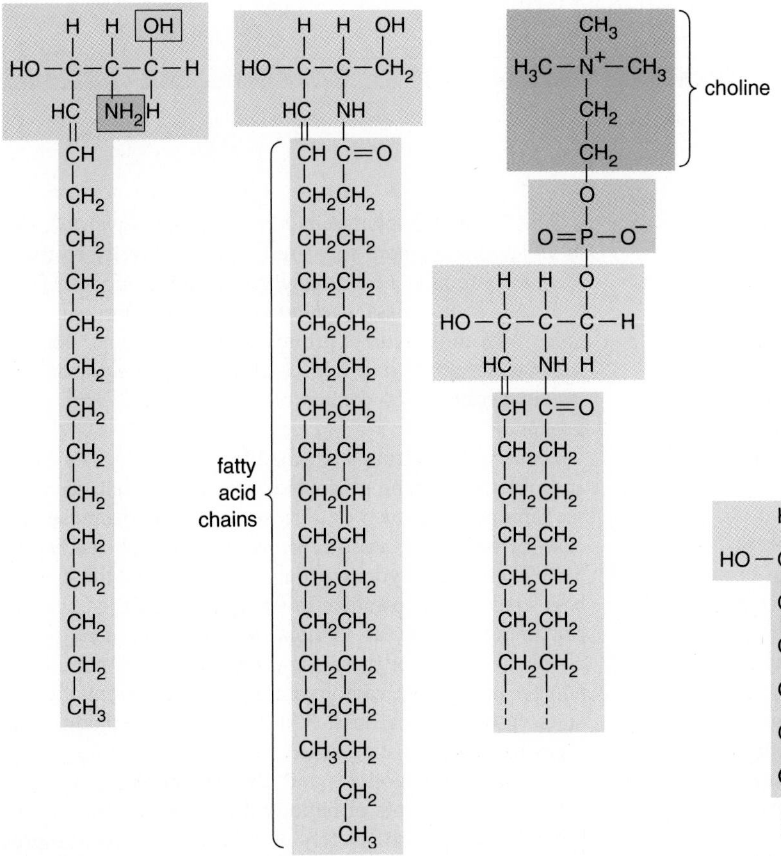

a sphingosine   **b** ceramide   **c** sphingomyelin

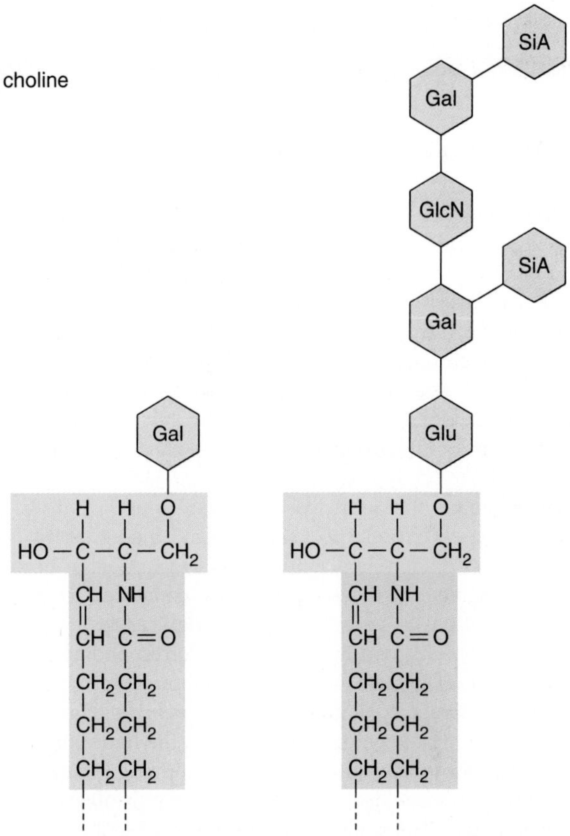

a cerebroside   **b** ganglioside

Figure 5-3 Sphingolipids. These lipids are built on a sphingosine backbone **(a)**, which contains an amino (—NH$_2$) and hydroxyl (—OH) group (boxed) as reactive sites. In membrane sphingolipids a fatty acyl chain attaches to the nitrogen of the amino group. The hydroxyl may be unbound, as in ceramides **(b)**, or linked by a phosphate group to one of the same polar alcohols occurring in phosphoglycerides. **(c)** Sphingomyelin, a phosphate-containing sphingolipid with choline as the polar alcohol. The hydroxyl may also link directly to a carbohydrate group to form a glycosphingolipid, as in Figure 5-4.

Figure 5-4 Glycosphingolipids, formed by the addition of carbohydrate units to a sphingolipid. The groups of neutral glycolipids formed by linkage of a single sugar unit such as a galactose to the sphingosine unit are called *cerebrosides* **(a)**; those with more complex straight or branched sugar chains containing sialic acid residues are termed *gangliosides* **(b)**. The presence of sialic acid residues gives gangliosides an acidic reaction at physiological pH. The names given to cerebrosides and gangliosides reflect their abundance in the plasma membranes of nerve and brain cells of animals. However, they are not restricted to this location and are found in membranes throughout the plant and animal kingdoms. Glu, glucose; Gal, galactose; GlcN, glucosamine; SiA, sialic acid.

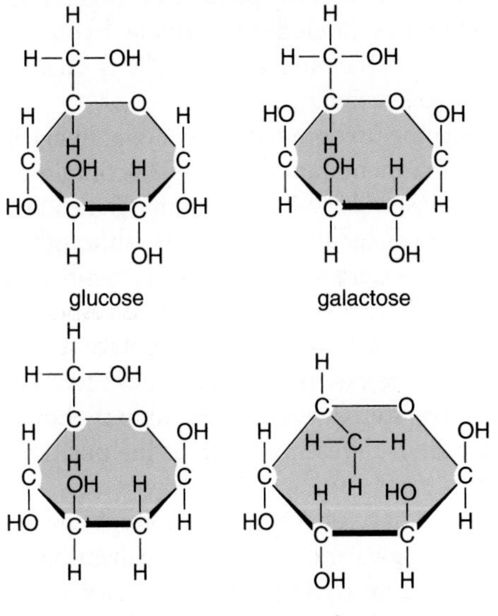

glucose   galactose

mannose   fucose

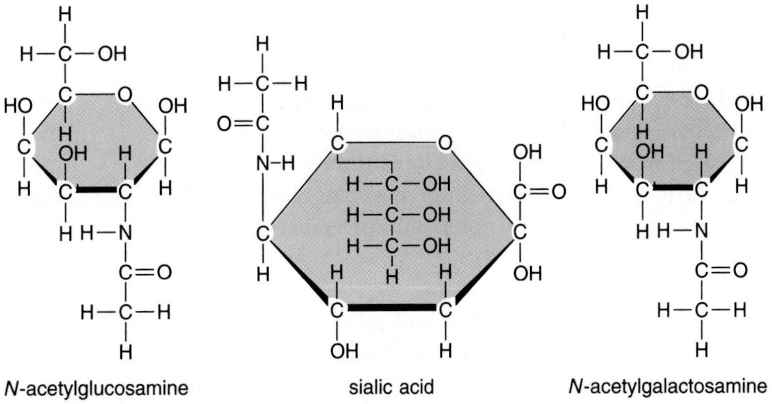

N-acetylglucosamine   sialic acid   *N*-acetylgalactosamine

Figure 5-5 Some of the sugars forming the carbohydrate groups of glycolipids and glycoproteins.

Sphingolipid Structure

Sphingolipids are based on a long-chain, nitrogen-containing alcohol called *sphingosine*, which consists of a fatty acid linked to the amino acid serine (Fig. 5-3a). Although distinct in structure from glycerol, sphingosine contains two reactive chemical groups in positions analogous to the —OH groups at the 2- and 3-carbons of glycerol. One of these reactive groups, in a position analogous to the 2-carbon of glycerol, is an amino (—NH$_2$) group; the second is an —OH group as in glycerol (blocked in green in Fig. 5-3a).

In membrane sphingolipids the amino group links to a fatty acyl chain (Fig. 5-3b). This chain and the nonpolar hydrocarbon chain of the sphingosine structure lie in positions analogous to the fatty acyl chains connected to the 1- and 2-carbons of glycerol in phosphoglycerides.

The hydroxyl group in the position corresponding to the 3-carbon of glycerol may remain unbound, forming a class of membrane sphingolipids called *ceramides* (Fig. 5-3b). In one more complex sphingolipid the hydroxyl site links via a phosphate group to choline, forming *sphingomyelin* (Fig. 5-3c), which—with phosphatidyl choline, phosphatidyl ethanolamine, and phosphatidyl serine—is one of the most common phospholipids in biological membranes. Ceramides and sphingomyelin are the only common membrane sphingolipids that do not contain carbohydrate groups.

The remaining sphingolipid types are *glycosphingolipids*, in which one or more sugar residues link directly to the 3-carbon site with no bridging phosphate group (see Fig. 5-4). Within this class, *cerebrosides* (Fig. 5-4a) are neutral lipids with carbohydrate groups containing from 1 to 20 sugar units; *gangliosides* (Fig. 5-4b) are negatively charged (acidic) because they contain one or more sialic acid groups.

Carbohydrate groups attached to glycosphingolipids and other membrane glycolipids are usually built up from six-carbon sugars such as glucose, galactose, mannose, fucose, or one of the various forms of sialic acid (see Fig. 5-5). These carbohydrate groups link singly to the lipid basal molecule or combine in straight or branched chains containing as many as 15 sugar residues. The many possible combinations of these sugar units make the potential number of different carbohydrate structures essentially infinite. In mammals at least 300 different glycosphingolipid types have been distinguished.

Although sphingolipids and glycosphingolipids are common constituents of biological membranes, their functions are still not completely understood. The *sphingo* prefix, derived from the word *sphinx*, reflects the mystery surrounding their functions when this group of lipids was first named.

---

use glycosphingolipids of animal plasma membranes as recognition and attachment sites. Breakdown products of sphingolipids have been implicated in metabolic regulation, including systems that regulate the activity of surface receptors, blood platelets, and growth factors.

Deficiencies in the metabolism of glycosphingolipids cause several important human disorders. One example is Tay-Sachs disease, in which normal breakdown and turnover of these molecules are deficient. The glycosphingolipids that accumulate as a result interfere with nerve and brain function, leading to paralysis and severe mental impairment. Glycosphingolipids of the plasma membrane may also change radically—old ones may disappear or new ones may appear—when normal cells are transformed into cancer cells. Programmed changes in glycosphingolipid types at cell surfaces are also noted during embryonic development in animals.

Sterols Sterols (see Fig. 2-16 and p. 57), based on a framework of four carbon rings, are found in both plant and animal membranes. The arrangement of the four rings makes sterols rigid, flattened structures; only the side chain extending from one end of the ring structure is flexible. In contrast to phospholipids and sphingolipids, sterols are only slightly polar. The polar segment of cholesterol is limited to a single hydroxyl group at one end of the molecule, which is characteristic of the sterols (see Fig. 2-16c).

Cholesterol is the predominant sterol of animal cell membranes. Plasma membranes in animals contain almost as much cholesterol as phospholipids. Cholesterol also occurs in internal cellular membranes, although in reduced quantities as compared to plasma membranes (see Table 5-2). Plant membranes contain small amounts of cholesterol and larger quantities of related sterols called *phytosterols*. The few protists and fungi studied, such as yeast, have sterols closely resembling cholesterol except for minor differences such as the positions of double bonds.

Among the primary effects of cholesterol and related sterols on biological membranes is the disturbance of the close packing of hydrocarbon chains of membrane phospholipids. As a result of this disturbance,

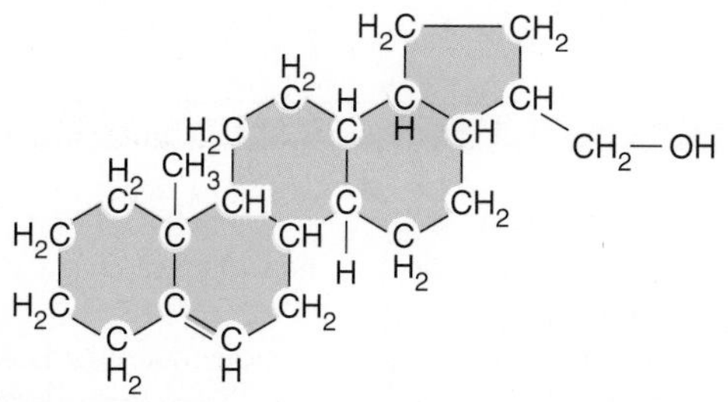

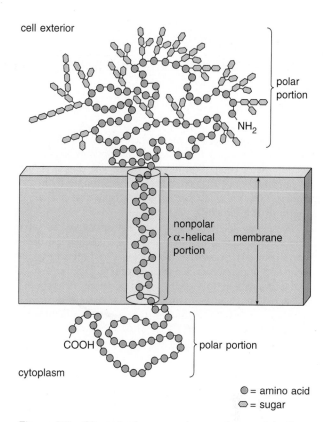

Figure 5-6 A hopanoid, a class of sterollike molecules substituting for sterols in prokaryotes.

membranes containing cholesterol remain fluid at reduced temperatures rather than "freezing" into nonfluid forms. Cholesterol also appears to increase both the flexibility and mechanical stability of membranes.

Sterols of any kind are rare or completely absent in the membranes of bacteria and cyanobacteria. However, in some prokaryotes a similar group of molecules, *hopanoids* (Fig. 5-6 and Information Box 5-3), substitutes for sterols in plasma membranes.

Membrane Proteins

The proteins of membranes occur in types and numbers that reflect the functional activity of the membrane. Most membranes contain from 10 to 50 different major protein types with molecular weights ranging from as little as 10,000 to more than 250,000. The relatively simple plasma membranes of erythrocytes (red blood cells), with about 11 different major types of membrane proteins, lie near one end of this distribution; the plasma membranes of HeLa cells, a cultured line of human cells, with more than 50 major protein types, lie near the other extreme. Many more proteins probably occur in amounts too small to be readily detected in these and other membrane types.

The Structure of Membrane Proteins Membrane proteins contain about the same proportion of hydrophobic amino acids as the soluble proteins of the internal cytoplasm—slightly less than 50%. Although the proportions are approximately the same, the distribution of hydrophobic and hydrophilic amino acids in membranes and soluble proteins is quite different. The hydrophobic amino acids of membrane proteins are frequently clustered in segments containing about 20 to 25 residues, separated by stretches of hydrophilic amino acids of varying length. In most soluble proteins, in contrast, hydrophobic residues tend to be scattered among hydrophilic amino acids without clustering. When twisted into an alpha helix, the hydrophobic blocks of membrane proteins are long enough to span the thickness of the membrane. Recent studies of membrane proteins have confirmed that the hydrophobic segments do indeed span membranes, forming anchors that hold the proteins in stable alignment in the lipid bilayer.

Figure 5-7 Glycophorin, a membrane glycoprotein that contains a single alpha helix of hydrophobic amino acids spanning the membrane. This diagram shows the distribution of both amino acids and sugar groups.

Glycophorin, for example (Fig. 5-7), a major plasma membrane protein of vertebrate erythrocytes, contains a single transmembrane block of hydrophobic amino acids. The remainder of the protein contains primarily hydrophilic amino acid residues that extend into the polar regions at either membrane surface. Another membrane protein, bacteriorhodopsin, which is concerned with light absorption in a few bacteria, contains no fewer than seven alpha-helical transmembrane segments (Fig. 5-8). These segments extend back and forth across the membrane, each connected to the next by a short stretch of hydrophilic amino acids that makes a loop at the membrane surface. Other membrane proteins have as many as 20 or more transmembrane segments.

Some proteins also have one or more beta-strand segments (see p. 63) that span the hydrophobic membrane interior. Porin, a protein forming a pore in the outer membrane of gram-negative bacteria (see p. 302), and the adenine nucleotide translocator, a transport protein of the mitochondrial inner membrane, probably contain membrane-spanning beta strands, for example. Fewer amino acids, about 10 to 11 as compared to the 20 or so in an alpha helix, are required to span the membrane in a beta strand. Single beta strands seem

Lipid and Protein Molecules of Biological Membranes **159**

Hopanoids

Hopanoids (see Fig. 5-6) are rigid molecules that are believed to play a role in prokaryotic membranes analogous to the role of cholesterol in eukaryotes. The ring structure of hopanoids, like that of cholesterol, is strongly hydrophobic; only the side chain extending from one end of the rings is hydrophilic.

These sterollike molecules were first discovered as substances produced as intermediates in the biochemical pathways of plants. Somewhat later, they were found to be abundant in petroleum deposits; according to some estimates as much as 15% of the world's reserves consists of hopanoids or their derivatives. If this estimate is accurate, fossil hopanoids might amount to as much as a hundred

billion to a trillion tons! This enormous quantity is equivalent to the total mass of organic carbon compounds in present-day organisms, and would make hopanoids the most abundant single class of organic compounds on the earth.

Because of their abundance in petroleum, hopanoids were suspected to be a membrane component of ancient prokaryotes contributing to the sediments forming fossil fuels. This conclusion prompted a search for hopanoids in the membranes of contemporary prokaryotes, where they have been found in more than 60 varieties of bacteria and cyanobacteria.

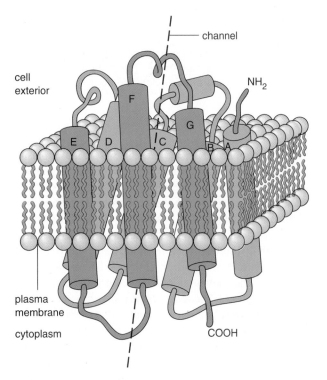

Figure 5-8 Bacteriorhodopsin, a protein absorbing light energy in plasma membranes of the halobacteria. The molecule contains seven alpha-helical transmembrane segments (shown as cylinders A to G) that span the membrane. Each alpha-helical segment is connected to the next by a loop of hydrophilic amino acids that extends into or through the polar membrane surfaces. The three-dimensional arrangement of the alpha helices was deduced from electron diffraction patterns by R. Henderson and coworkers.

unlikely to stand alone as membrane-spanning segments, however, because the edges of beta strands have hydrogen-bonding requirements that cannot be satisfied readily in a hydrophobic environment. Therefore, structures such as the beta barrel (Fig. 5-9), in which

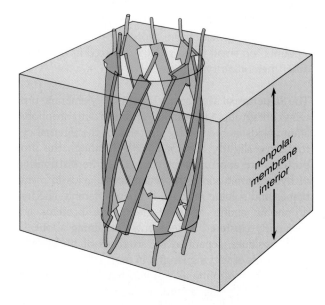

Figure 5-9 A bacterial protein consisting of beta strands. The strands form a cylindrical beta barrel that spans the hydrophobic membrane interior. A beta barrel is typically surrounded by segments of alpha helix, as in the triosephosphate isomerase molecule shown in Figure 2-25.

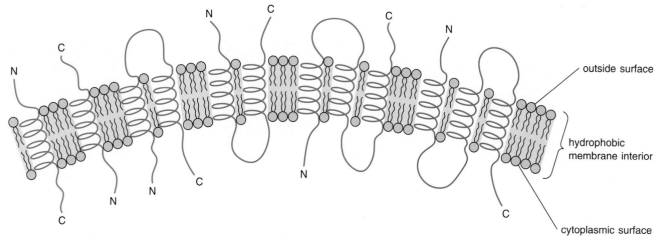

outside surface

hydrophobic
membrane interior

cytoplasmic surface

Figure 5-10 The arrangements of membrane-spanning hydrophobic segments and sur-face hydrophilic segments observed in different membrane proteins. N and C indicate the N- and C-terminal ends of the amino acid chains.

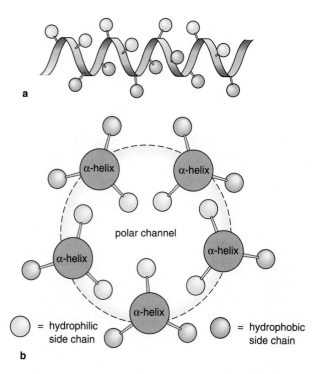

a

α-helix α-helix

polar channel

α-helix α-helix

α-helix

○ = hydrophilic side chain   ◉ = hydrophobic side chain

b

Figure 5-11 Amphipathic alpha helices in membrane-spanning segments. **(a)** An amphipathic alpha helix, in which hydrophobic amino acid side groups are located on one side of the helix and hydrophobic groups on the other. **(b)** A protein with four amphipathic alpha helices spanning the membrane, each with the arrangement of side groups shown in **(a)**. The helices are oriented with their hydrophilic side groups clustered inside the protein, forming a polar channel through the membrane. Arrangements of this type are believed to be typical of many proteins transporting polar substances across membranes.

beta strands are wound into a cylinder that allows interstrand hydrogen bonds to form, are considered more likely as membrane-spanning configurations. The beta barrel is further stabilized by a surrounding ring of alpha helices. (Fig. 5-10 summarizes the various patterns in which the hydrophobic and hydrophilic segments of membrane proteins are distributed in membranes.)

Some membrane proteins contain transmembrane segments in which one to three hydrophobic amino acids alternate in a repeating pattern with similar numbers of hydrophilic residues. When wound into an alpha helix, this arrangement places the hydrophobic residues on one side of the helix and the hydrophilic ones on the other (Fig. 5-11a). Such alpha helices, which have a polar and a nonpolar face, are amphipathic. These helices occur only in membrane proteins with multiple transmembrane segments. Evidently, in these proteins the segments align so that the hydrophobic sides of the amphipathic helices face the membrane interior, and the hydrophilic sides cluster around the central axis of the protein (Fig. 5-11b). This arrangement creates a polar channel that extends through the protein from one membrane surface to the other. Such arrangements are believed to be typical of transport proteins, which conduct charged and polar molecules across the membrane via the polar channel (see Ch. 6 for details).

The structure of membrane proteins thus establishes separate domains with hydrophobic amino acids facing the nonpolar membrane interior and hydrophilic amino acids facing the inside of the protein or the polar membrane surfaces. This segregation produces proteins with amphiphilic properties similar to membrane lipids, that is, with distinctly polar and nonpolar regions.

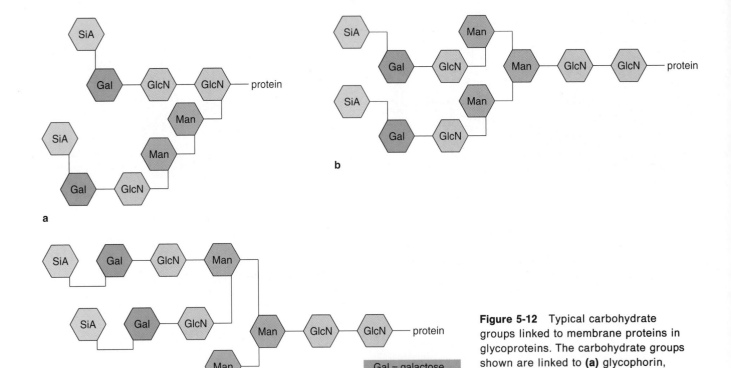

Figure 5-12 Typical carbohydrate groups linked to membrane proteins in glycoproteins. The carbohydrate groups shown are linked to **(a)** glycophorin, **(b)** serum transferrin, and **(c)** human fibroblast surface glycoprotein. Gal, galactose; Man, mannose; GlcN, glucosamine; SiA, sialic acid.

Gal = galactose
Man = mannose
GlcN = glucosamine
SiA = sialic acid

Glycoproteins Proteins with covalently attached carbohydrate groups occur in many types of cellular membranes. The carbohydrate groups of proteins, which include essentially the same monosaccharide units as those of glycolipids (see Fig. 5-5), link into straight or branched chains containing from 2 to 60 residues. (Fig. 5-12 shows several representative carbohydrate units of membrane glycoproteins.)

The carbohydrate groups of glycoproteins link to the protein segments of the molecules in two different patterns. One involves linkage to an amino (—NH₂) group on the side chain of an asparagine residue (Fig. 5-13a); carbohydrate groups with this pattern are *N*-

linked. The other group links to a side-chain —OH group on a serine or threonine residue (Fig. 5-13b) or, to some extent, to the side-chain —OH group of the modified amino acids hydroxyproline or hydroxylysine. The carbohydrates in this group are called *O-linked.* Both *N*- and *O*-linked carbohydrate groups may occur on the same protein. Glycophorin, for example, contains 1 *N*-linked and no less than 15 *O*-linked carbohydrate groups bound at different points in its amino acid sequence.

Glycoproteins are most abundant in plasma membranes; in fact, probably all plasma membrane proteins of eukaryotes are glycoproteins. The carbohydrate

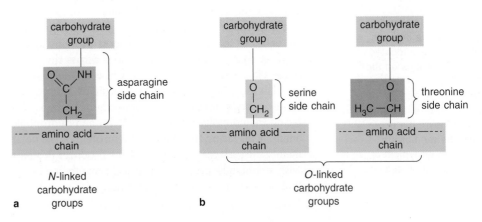

Figure 5-13 Linkage patterns of carbohydrate groups. **(a)** *N*-linked carbohydrate groups bind to a side-chain amino group in asparagine. **(b)** *O*-linked carbohydrate groups bind to an —OH group in the side chains of serine or threonine or, in some cases, to an —OH group in the modified amino acids hydroxyproline and hydroxylysine.

groups of these glycoproteins occur almost exclusively on the outer surface of the plasma membranes. Along with the carbohydrate groups of glycolipids, they give the cell surface what is often described as a "sugar coating." More technically, the layer of carbohydrates at the cell surface is called the *glycocalyx* (from the Greek *glykys* = sweet and *calyx* = cup or vessel).

In animals glycoproteins are predominant in the glycocalyx; in plants glycolipids make up most of the membrane coat. Relatively small amounts of glycoproteins occur in internal membranes such as those of the endoplasmic reticulum (ER), Golgi complex, and nuclear envelope. Although some glycoproteins are permanent components of these membrane systems, many are transient molecules being assembled for transport to the plasma membrane when complete. (The ER and the Golgi complex are the sites of glycoprotein and glycolipid synthesis in eukaryotic cells.)

The functions of the carbohydrate groups of membrane glycoproteins are uncertain. However, they are considered likely to add stability to membranes, provide resistance to attack by proteinases in the extracellular medium, and contribute to the forces folding and maintaining membrane proteins in their correct three-dimensional forms. The carbohydrate groups of glycoproteins probably also act as parts of the recognition sites of membrane receptors involved in binding extracellular signal molecules and cell-to-cell adhesion. Like those of glycolipids, the carbohydrate groups of membrane proteins are used as recognition and binding sites by infective bacteria and viruses. A segment of the carbohydrate group of lysosomal enzymes is known to act as a distribution signal directing the enzymes to this organelle (see p. 845).

PHOSPHOLIPID BILAYERS AND BIOLOGICAL MEMBRANES

In water, phospholipids such as many of the phosphoglycerides spontaneously assemble into bilayers (see Fig. 5-15). Within a bilayer individual phospholipid molecules pack with their long axes at right angles to the plane of the bilayer, in an orientation that satisfies their dual-solubility properties. The polar head groups face the surrounding aqueous medium, and the nonpolar hydrocarbon chains associate end to end in the membrane interior in a nonpolar region that excludes water.

Bilayers are stable when suspended in water and resist any disturbance that would expose their hydrophobic hydrocarbon chains to the surrounding water molecules. Membranes disperse almost instantaneously if exposed to a nonpolar environment or to detergents, which are amphipathic molecules that can form a hydrophilic coat around the hydrophobic portions of membrane lipids and proteins in water solutions (Fig. 5-14).

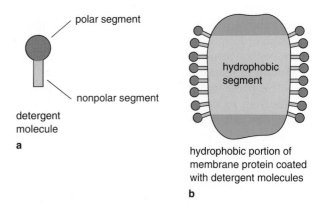

Figure 5-14 The activity of detergent molecules in suspending hydrophobic membrane proteins and lipids by forming a hydrophilic coat. **(a)** A detergent molecule, which has polar and nonpolar ends. **(b)** A coat of detergent molecules covering the hydrophobic surfaces of a membrane protein in a water solution. The detergent molecules become aligned so that their hydrophobic portions associate with the hydrophobic surfaces of the protein, and their hydrophilic surfaces face the surrounding polar medium.

Not all phospholipids or other lipid types form bilayers readily. Phosphatidyl ethanolamine and cholesterol, for example, do not form bilayers when placed individually in water under physiological conditions. However, these lipids are stabilized in bilayers by hydrophobic interactions with surrounding bilayer-forming phospholipids.

Physical Characteristics of Artificial and Natural Bilayers

The study of artificial phospholipid bilayers has provided a great deal of information about the physical properties of biological membranes. Such bilayers can readily be formed by painting a drop of phospholipid dissolved in a nonpolar solvent over a narrow opening in a Teflon plate. When held under a water surface, the droplet gradually thins out until an artificial bilayer only two molecules thick stretches across the opening.

Artificial bilayers present a more manageable experimental subject than natural membranes because they can be assembled from single lipids or a completely defined lipid mixture. Artificial systems have also been invaluable in the study of membrane proteins. In many instances it has been possible to study the functions of individual membrane proteins by extracting them from natural membranes and inserting them into artificial phospholipid bilayers.

Studies of the physical properties of artificial bilayers have revealed two fundamental characteristics that apply to both natural and artificial membranes: (1) bilayers undergo what is known as a *phase transition*, in which they "melt" or "freeze" above or below certain temperatures; and (2) at temperatures above the phase transition phospholipid bilayers are in a highly fluid state. Natural membranes must be held above the phase transition to remain fully functional.

The phase transition of a phospholipid bilayer results from an alteration in the mobility of hydrocarbon chains induced by changes in temperature. At low temperatures the hydrocarbon chains of bilayer phospholipids are tightly packed and thus restricted in movement (Fig. 5-15a). This tightly packed state is termed the *gel* or *crystalline* phase of a bilayer. As the temperature rises, a level is reached at which both the space separating adjacent phospholipid molecules and the movements of their hydrocarbon chains increase abruptly. At this temperature the phospholipid undergoes a phase transition in which it melts and becomes fluid (Fig. 5-15b). The increase in fluidity includes rotation of entire phospholipid molecules around their long axis, flexing of hydrocarbon chains, and the exchange of positions, so that individual molecules diffuse from place to place in the bilayer.

The temperature at which the phase transition occurs in a phospholipid bilayer depends primarily on the length and degree of saturation of the hydrocarbon chains. Generally, the longer the hydrocarbon chains, the higher the temperature at which the bilayer melts. This effect is evidently due to the increased opportunities for the generation of stabilizing van der Waals forces between longer hydrocarbon chains. The sharp bends introduced in hydrocarbon chains by double bonds in the *cis* configuration, which interfere with tight packing of the chains, reduce the temperature at which the phase transition takes place and allow the bilayer to remain fluid at lower temperatures.

The fluidity of both artificial and natural membranes has been amply demonstrated by several chemical and physical approaches (see pp. 170 and 172). In general these methods have confirmed that phospholipid molecules are highly mobile in natural and artificial phospholipid bilayers at temperatures above the phase transition. The rate of lateral diffusion of individual phospholipid molecules in bilayers in the fluid state is surprisingly high; measurements indicate a diffusion coefficient of about 10^{-8} cm^2/sec. At this rate, a phospholipid molecule can move the entire length of a small cell such as a bacterium in about one second!

The results of these studies apply to the *lateral* movement of phospholipid molecules *within the same bilayer half.* Movement of a molecule from one bilayer half to the other, called *flip-flop*, has different characteristics. Significantly, these characteristics differ widely between artificial bilayers and natural membranes. In artificial bilayers the half-time for flip-flop of phospholipid molecules from one bilayer half to another—the time required for half of the molecules present to flip-flop—is on the order of hours to days, depending on the temperature and phospholipid type. The observed rates are so low, in fact, that phospholipid molecules in artificial bilayers are essentially restricted to their own bilayer half. Some natural membranes, such as those of the ER, frequently show much faster flip-flop rates, with half-times often on the order of seconds. These rapid rates have been shown to be a result of the activity of proteins specialized for this function. When introduced into artificial bilayers, the proteins increase flip-flop to rates comparable to natural membranes (see below).

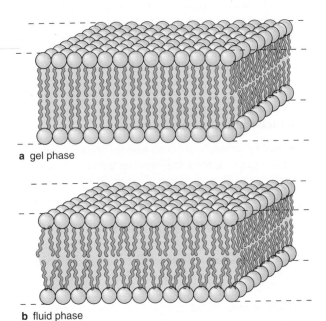

a gel phase

b fluid phase

Figure 5-15 The effects of temperature on phospholipid packing in lipid bilayers. **(a)** At temperatures below the phase transition, the fatty acyl chains are tightly packed with little freedom of movement; in this condition the bilayer is "frozen" into a semisolid gel. **(b)** At temperatures above the phase transition, the bilayer "melts." Chains have greater freedom of movement and are less tightly packed. Under these conditions the bilayer interior is fluid, and the phospholipid molecules are able to rotate in place or exchange places to diffuse through the bilayer.

Effects of Cholesterol on the Phase Transition

Adding cholesterol to artificial phospholipid bilayers has a marked effect on the phase transition. Added in small amounts, up to about 20% of the total bilayer lipid, cholesterol widens the temperature range over which the phase transition takes place. At levels above about 20%, it eliminates the gel phase entirely at most temperatures likely to be encountered by living organisms (bilayers can accommodate cholesterol up to a 1:1 molar ratio with phospholipid). At the same time cholesterol restricts rapid movement of the hydrocarbon chains of membrane phospholipids, which has the effect of reducing membrane fluidity at elevated temperatures. Cholesterol therefore has the dual effect of increasing membrane fluidity at low temperatures and reducing it at elevated temperatures. For these reasons cholesterol is said to be a "plasticizer" of natural membranes.

The plasticizing effects of cholesterol are believed to be due to the position it takes in bilayers and its interaction with surrounding bilayer lipids. X-ray diffraction indicates that cholesterol lines up between phospholipid molecules of a bilayer half, with the long axis of the cholesterol molecule paralleling the hydrocarbon chains of the phospholipids (Fig. 5-16a). The

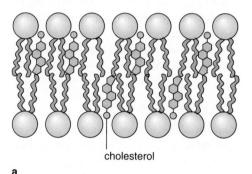

cholesterol

a

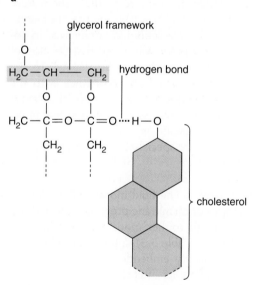

glycerol framework

hydrogen bond

cholesterol

b

Figure 5-16 The position taken by cholesterol in bilayers. **(a)** The hydrophilic —OH group, represented by the small circle at one end of the molecule, lies at the level of the ester bonds linking glycerol to the fatty acyl chains of the membrane phospholipids. The remainder of the cholesterol molecule, which is completely hydrophobic, lies parallel to the fatty acyl chains in the nonpolar membrane interior. **(b)** A stabilizing hydrogen bond between the hydrogen of the cholesterol —OH group and the oxygen of the ester linkage binding a fatty acyl chain.

cholesterol molecule orients in the bilayer with its single polar —OH group at the level of the bonds linking fatty acyl chains to glycerol. In this position, the —OH group can form a stabilizing hydrogen bond with the oxygen of an ester linkage between the glycerol backbone and a fatty acyl chain (Fig. 5-16b). The remainder of the rigid, hydrophobic cholesterol ring structure extends into the nonpolar membrane interior. In this position the cholesterol molecule interferes with tight packing of the hydrocarbon chains of nearby phospholipids and inhibits formation of the gel phase. At elevated temperatures at least three forces—van der Waals forces, hydrophobic interactions between the cholesterol ring structure and the hydrocarbon chains, and hydrogen bonding at the acyl bonds—stabilize phospholipid molecules enough to reduce fluidity.

The plasticizing effects of cholesterol are most significant for the plasma membranes of animal cells, which may contain cholesterol molecules in numbers approaching the maximum 1:1 ratio obtainable with phospholipids. In other biological membranes, such as the nuclear envelope, which has a cholesterol to phospholipid ratio of only 0.1 to 0.2, or in mitochondrial membranes, with a ratio as low as 0.02 to 0.05, the effects of cholesterol are much less significant.

Cholesterol has other effects in phospholipid bilayers. It forces the polar head groups of phospholipid molecules farther apart, increasing access to the bilayer interior by molecules at the membrane surface. Through this effect, cholesterol may aid the attachment of molecules to the membrane surface, such as proteins that bind by extending hydrophobic groups into the nonpolar membrane interior. Even though it increases the separation of phospholipid molecules in the bilayer, cholesterol helps to seal the nonpolar part of bilayers against the penetration of ions and small polar molecules such as glucose. This may result from the ability of cholesterol to fit into the spaces between hydrocarbon chains of membrane phospholipids, eliminating "faults" through which ions and small polar molecules and ions may pass (see p. 204).

Regulation of the Phase Transition in Living Organisms

Many organisms take advantage of the effects of chain length, degree of unsaturation, or cholesterol content to depress the phase transition and keep their membranes fluid as environmental temperatures fall. This adaptation has been best studied in bacteria, in which such adjustments are accomplished by increasing chain length or the number of double bonds in membrane phospholipids.

Most bacteria control the phase transition by adjusting the number of double bonds in fatty acyl chains. This is accomplished by controls regulating the cytoplasmic concentration and activity of an enzyme that introduces double bonds in fatty acyl chains. At elevated temperatures the gene encoding the enzyme is relatively inactive, and the enzyme is synthesized in very limited quantities. The relatively few enzymes present are down-regulated by a protein that inhibits their activity at elevated temperatures. As temperatures fall, the gene encoding the enzyme is activated, the mRNA is transcribed, and quantities of the enzyme are assembled on ribosomes in the cytoplasm. The protein down-regulating activity of the enzyme is also inactivated as temperatures fall. As a consequence, more fatty acids containing double bonds are made in the cytoplasm and introduced into membrane lipids. In the plasma membrane the higher proportion of unsaturated fatty acyl chains shifts the phase transition so that the membrane remains fluid as temperatures fall. The human intestinal bacterium E. coli is among the many bacteria using this mechanism to regulate the phase transition.

Adjustments in fatty acyl chain length are used by other bacteria, such as *Micrococcus*. In this bacterium the ratio of fatty acyl chains containing 18 carbons to those containing 16 carbons changes from 3:1 to 1:1 as temperatures fall. The chains are shortened by an increase in the activity of an enzyme that removes two carbons from the end of the 18-carbon chains.

The ability of organisms to regulate the phase transition to keep membranes fluid in cold environments is termed *homeoviscous adaptation.* This adaptation also has been detected in many eukaryotic organisms, including algae, higher plants, protozoa, and animals. In poikilothermic animals, in which body temperature fluctuates with environmental temperature, the proportions of both unsaturated fatty acyl chains and cholesterol are increased at lower temperatures as a means to regulate membrane fluidity.

Homeoviscous adaptation has been detected in hibernating mammals. In nonhibernating mammals the phase transition from fluid to gel takes place at about 15°C. As golden hamsters enter hibernation, their body temperature may fall from about 37°C to as low as 5°C. As body temperature falls, the proportion of double bonds in the membrane phospholipids increases, keeping the membranes fluid at the reduced temperatures.

The ability to keep membranes fluid at low temperatures is particularly significant to the nervous system of hibernating mammals. As a result of homeoviscous adaptation, the cells of the nervous system remain active in transmitting nerve impulses, and the animal, although sluggish, can maintain its basic body functions and respond to external stimuli.

Higher plants with resistance to chilling also contain membrane lipids with greater proportions of double bonds. Because resistance to chilling is a critical factor in the survival and growth of crop plants in colder regions, there is considerable interest in the possibility of adding cold resistance by genetic engineering. Some preliminary experiments have achieved limited success. For example, H. Wada and his coworkers increased the resistance of a cyanobacterium to low temperatures by introducing a gene for an enzyme increasing the number of double bonds in membrane lipids.

THE FLUID MOSAIC MODEL AND ITS SUPPORTING EVIDENCE

Singer and Nicolson's fluid mosaic model introduced new ideas about the distribution of lipids and proteins in membranes and thereby revolutionized scientific thinking about membrane structure. The model broke new ground with its convincing proposition that proteins are suspended directly in the membrane bilayer. The burst of research triggered by the hypothesis has fully supported its basic proposals about membrane structure and has provided additions that have clarified and greatly extended the model.

Features of the Fluid Mosaic Model

Membrane Bilayers The bilayer structure proposed by Singer and Nicolson for the lipid portion of biological membranes (Fig. 5-17) was in large part an affirmation of earlier models (see p. 182). The feature emphasized by Singer and Nicolson was that the bilayer is highly fluid under physiological conditions. In addition, their original model proposed that flip-flop of phospholipid molecules from one bilayer half to the other was highly restricted, in accordance with the observation that phospholipids in artificial bilayers rarely flip-flop. Later observations indicated that phospholipid flip-flop is greatly speeded in some natural membranes by proteins specialized for this activity (see below for details).

Membrane Proteins The fluid mosaic model defined two major classes of membrane proteins—*integral* and *peripheral*—according to their mode of association with membrane bilayers (Table 5-3). Integral proteins were proposed to be deeply embedded in the bilayer and held in place by nonpolar interactions with membrane lipids. Suspension in the bilayer depends on the dual-solubility properties of membrane proteins. At polar membrane surfaces, protein molecules fold to expose only hydrophilic amino acid side chains. Nonpolar side chains are exposed on the protein surfaces facing the hydrophobic membrane interior, and are held in this position by their association with the nonpolar hydrocarbon chains of membrane lipids. Integral proteins remain in stable suspension in the bilayer because, like membrane lipids, any change in orientation would expose their hydrophobic regions to the watery surroundings. Because of their intimate association with

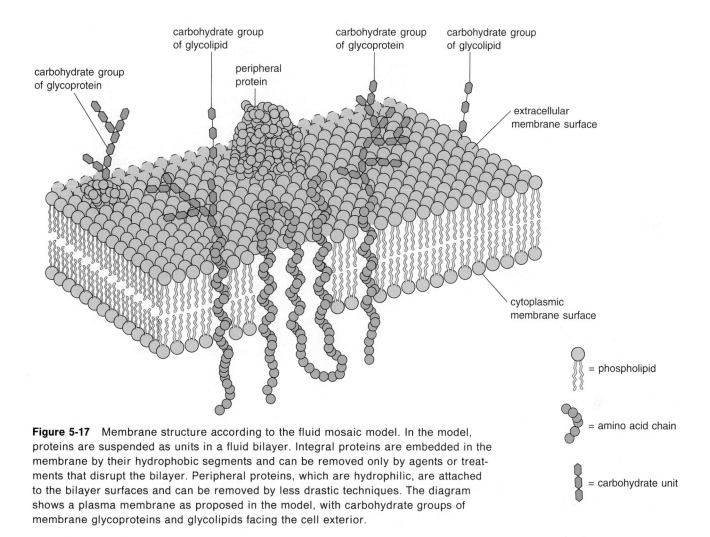

carbohydrate group of glycolipid

carbohydrate group of glycoprotein

peripheral protein

carbohydrate group of glycoprotein

carbohydrate group of glycolipid

carbohydrate group of glycoprotein

extracellular membrane surface

cytoplasmic membrane surface

= phospholipid

= amino acid chain

= carbohydrate unit

Figure 5-17 Membrane structure according to the fluid mosaic model. In the model, proteins are suspended as units in a fluid bilayer. Integral proteins are embedded in the membrane by their hydrophobic segments and can be removed only by agents or treatments that disrupt the bilayer. Peripheral proteins, which are hydrophilic, are attached to the bilayer surfaces and can be removed by less drastic techniques. The diagram shows a plasma membrane as proposed in the model, with carbohydrate groups of membrane glycoproteins and glycolipids facing the cell exterior.

the nonpolar membrane interior, integral membrane proteins can be removed from membranes only by disruptive agents, such as detergents or nonpolar solvents, that disperse the bilayer. Within these limitations integral proteins were proposed to be potentially free to displace phospholipid molecules and move laterally through the fluid bilayer.

Peripheral proteins are hydrophilic molecules that bind noncovalently to polar membrane surfaces. Be-

cause of their hydrophilic nature and polar associations, peripheral membrane proteins can be removed by relatively mild treatments that do not disrupt the bilayer, such as adjustments in the salt concentration or pH of the suspending medium. Singer proposed that peripheral proteins bind only to portions of integral membrane proteins exposed at the membrane surfaces. Others found that peripheral proteins may also bind to polar groups of membrane phospholipids.

Table 5-3 Characteristics of Integral and Peripheral Membrane Proteins		
Characteristic	Integral Protein	Peripheral Protein
Location in membrane	Buried in hydrophobic membrane interior	Bound to membrane surface
Requirements for release from membrane	Released only by agents that disrupt membrane bilayer, such as detergents	Released by treatments that leave bilayer intact, such as increases in salt concentration
Association with lipids when released	Usually associated with lipids	Not associated with lipids
Solubility	Usually insoluble in aqueous media	Usually soluble in aqueous media

The Fluid Mosaic Model and Its Supporting Evidence **167**

The Singer–Nicolson model originally proposed that integral proteins are free to move in the fluid bilayer. It has become apparent that the free movement may be restricted, however, by linkage of some integral proteins into aggregates or combinations between integral and peripheral proteins. In the latter case peripheral proteins may join in a cytoskeletal lattice or network at the membrane surface that holds many of the integral proteins in fixed position. Recent evidence has shown that such anchoring networks of peripheral proteins occur widely in cellular membranes, particularly in plasma membranes. In the plasma membranes studied so far, cytoskeletal networks immobilizing membrane proteins have been found exclusively on the inner membrane surface facing the cytoplasm. Other networks formed by extracellular materials form immobilizing linkages on the outer membrane surface. (Ch. 13 presents details of the cytoskeletal networks that immobilize integral membrane proteins; Ch. 8 describes extracellular structures and their linkages to integral membrane proteins.)

Membrane Carbohydrates In the model, hydrophilic carbohydrate groups of membrane glycoproteins and glycolipids were considered to extend into the polar environment at membrane surfaces (see Fig. 5-17). In plasma membranes, in accordance with observations that sugar groups are restricted to the outside surfaces of cells, these carbohydrate groups were thought to extend only from the exterior surface of membranes. Recent evidence supports this feature of the model, and indicates that carbohydrate groups are probably distributed so thickly over the exterior surface of plasma membranes that little of the phospholipid bilayer is exposed directly to the surrounding aqueous medium. In membranes forming internal vesicles that communicate directly or indirectly with the cell exterior, such as those of the ER or Golgi complex, carbohydrate groups extend from the membrane into the space enclosed by the vesicles (Fig. 5-18).

Interactions among carbohydrate groups at the membrane surface can also form networks that anchor integral membrane proteins and restrict their movement. Because carbohydrate groups are located on the outside of the plasma membrane, anchoring networks of this type are expected to occur primarily or exclusively on the outer surfaces of cells.

Membrane Asymmetry Another major feature of the fluid mosaic model proposed that the arrangement of phospholipids and both integral and peripheral proteins is *asymmetric* in biological membranes. For membrane

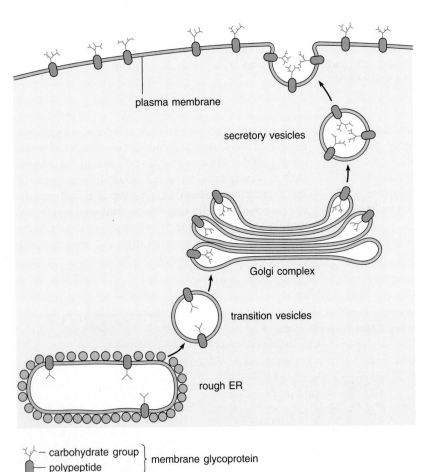

Figure 5-18 Orientation of carbohydrate groups of glycoproteins in internal vesicles of the cytoplasm including the rough ER, Golgi complex, and secretory vesicles. The central stem structures of the carbohydrate groups are added in the rough ER, and the branches are added in the Golgi complex.

plasma membrane

secretory vesicles

Golgi complex

transition vesicles

rough ER

— carbohydrate group
— polypeptide } membrane glycoprotein

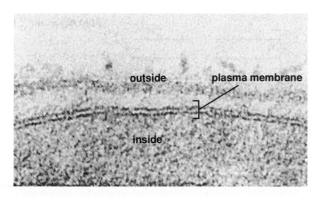

Figure 5-19 A plasma membrane at high magnification, showing the typical trilaminar or "railroad track" image seen in thin-sectioned membranes. × 240,000. Courtesy of R. B. Park.

lipids asymmetry means that the inner and outer bilayer halves contain different proportions of various lipid types. Lipid asymmetry is considered to be maintained by restrictions on flip-flop except for those phospholipids moved from one side to the other by membrane proteins specialized for this function. For integral membrane proteins asymmetry means that these molecules have specific orientations with respect to the two membrane surfaces and do not tumble or rotate to change their orientation. Such movements would be energetically unfavorable because rotation would expose the

hydrophobic portions of the proteins to polar membrane surfaces, and polar regions of the proteins, including any attached carbohydrate groups, to the nonpolar membrane interior. Peripheral proteins are also asymmetrically distributed. Because of their hydrophilic nature, these proteins are restricted to either membrane surface. (The Experimental Process essay by S. J. Singer on p. 174 describes his research showing that transport proteins do not rotate in the membrane.)

Evidence Supporting the Fluid Mosaic Model

Evidence That Bilayers Are Present in Membranes A variety of sources provide evidence that membrane lipids actually occur in bilayers. Electron microscopy shows that thin-sectioned natural membranes and artificial phospholipid bilayers have equivalent dimensions, with the thickness expected for a double layer of lipid molecules—about 7 to 8 nm. Moreover, artificial bilayers and many cross-sectioned membranes appear in electron micrographs as two dark surface lines separated by a less dense interzone (Fig. 5-19). This image is consistent with a bilayer structure if the dark lines represent the polar membrane surfaces, and the less dense interzone the nonpolar membrane interior.

Another line of supporting evidence comes from comparisons of artificial bilayers and natural membranes prepared for electron microscopy by the freeze-fracture

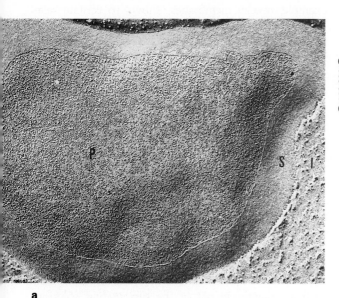

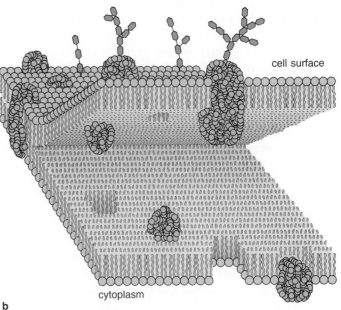

Figure 5-20 Membrane structures visible in freeze-fracture preparations. **(a)** Preparation of a human erythrocyte plasma membrane, with the bilayer interior exposed. The particles are integral proteins embedded within the membrane bilayer. P, membrane particles in bilayer; S, external membrane surface; I, ice crystals surrounding the specimen. × 38,000. Courtesy of T. W. Tillack, from *J. Cell Biol.* 45:649 (1970), by permission of the Rockefeller University Press. **(b)** The path followed by a fracture that splits a membrane into bilayer halves.

technique. In this method artificial bilayers or tissue samples are rapidly frozen by placing them in liquid nitrogen. The frozen specimen is then fractured or split by striking it with a sharp knife edge (see p. 120 for details). Because the nonpolar interior of artificial or natural bilayers produces weakly frozen "faults" in the preparations, which fracture more readily than the surrounding polar regions, a fracture tends to follow planes along membrane interiors, splitting the bilayers into inner and outer halves (Fig. 5-20). The image obtained is entirely consistent with the presence of bilayer structure in biological membranes.

X-rays are scattered by both artificial and natural membranes in patterns consistent with a double layer of lipid molecules in which the nonpolar ends of the lipids are associated in the membrane interior. The X-ray evidence indicates further that the lipid molecules are arranged in parallel fashion, with their long axes perpendicular to the bilayer surfaces.

Evidence That Membranes Are Fluid Evidence that membrane bilayers are fluid also comes from several sources. We have already noted that lipid molecules are able to diffuse rapidly through both natural membranes and artificial bilayers. Diffusion rates of the magnitudes observed are compatible only with fluid bilayers. A method using *photobleaching* provides a graphic demonstration of the fluidity of both natural and artificial membranes. In this technique fluorescent dyes are attached to membrane lipids or proteins. The dye markers are then bleached by a laser beam adjusted to illuminate a spot only a few micrometers in diameter. The spot, initially sharply defined on the membrane, gradually spreads and fades as the bleached molecules diffuse and exchange places with unbleached molecules from the surroundings. The time required for the spot to fade provides a measure of the fluidity of the membrane. The rate of diffusion indicated by the photobleaching method confirms that both natural and artificial membranes held above the phase transition are about as fluid as olive oil or light machine oil. Observations of the movements of membrane proteins (see below) also clearly support the conclusion that membrane bilayers are in a fluid state at temperatures above the phase transition.

While the bulk lipid of biological membranes is fluid, it is possible that lipid molecules in some regions may be restricted in mobility. At temperatures just above the phase transition, depending on the types of lipids present, bilayers may contain "islands" of gel-phase lipids distributed among more fluid regions. Some investigators have proposed that lipid movement may also be restricted by interactions with nearby membrane proteins. Such interactions might occur between polar groups on phospholipids and membrane proteins or between hydrophobic parts of membrane lipids and proteins. These interactions may produce a boundary

layer or "shell" of immobilized lipids surrounding a membrane protein. Whether such immobilized lipid shells actually exist in biological membranes remains controversial.

One curious aspect of membrane fluidity is its possible relationship to substances used as anesthetics. Many of these substances, which interrupt consciousness or the sensations of pain, touch, temperature, and other sensory inputs, are nonpolar. Their nonpolar nature makes them soluble in the hydrophobic membrane interior and increases the fluidity of lipid bilayers. Although the exact effects of anesthetics on nerve cells are uncertain, they may make membranes so fluid that the arrangement of receptors and other membrane proteins in nerve synapses (see p. 221) is disrupted. This disruption might prevent transmission of nerve impulses in the anesthetized region and interrupt the flow of sensory information to the brain.

Evidence That Proteins Are Suspended Within Membranes The best evidence that proteins are suspended in biological membranes comes from the electron microscopy of freeze-fracture preparations. When a fracture follows the interior of a membrane, splitting the bilayer into inner and outer halves, globular particles the size of protein molecules are clearly seen embedded in the bilayer (as in Fig. 5-20a).

The particles exposed in fractured membranes have been identified as proteins by various methods. For example, myelin membranes, known to contain relatively few proteins, show almost no globular particles in freeze-fracture preparations (Fig. 5-21). Other membranes known to contain proteins in greater quantity, such as the plasma membrane shown in Figure 5-20a, are crowded with particles when freeze-fractured.

Direct support that the particles are proteins comes from freeze-fracture studies of artificial bilayers carried out by D. W. Deamer, D. Branton, and others. Artificial bilayers created by suspending pure phospholipids in water show smooth interior surfaces without particulate units when freeze-fractured (Fig. 5-22a). When proteins are added to the artificial bilayers, globular particles closely resembling those in natural membranes appear in the freeze-fracture preparations (Fig. 5-22b).

Other experiments have shown parallels in the behavior of membrane proteins and the particles in freeze-fractured membranes. By various treatments, such as changes in pH of the surrounding solution, proteins in living membranes can be induced to pack into clumps. Freeze-fracture preparations of these membranes show that the membrane particles are also clumped together. As a group, these experiments establish that the particles in freeze-fractured membranes are proteins. Their distribution clearly supports the idea that proteins are suspended within the bilayers of biological membranes.

Figure 5-21 Freeze-fracture preparation of myelin membranes. The fracture has exposed successive interior and surface levels of the myelin membranes, which are wrapped in multiple layers around nerve axons (see Fig. 6-21). The exposed surfaces are smooth, with little or no evidence of particles, reflecting the fact that myelin membranes contain relatively few integral proteins. × 69,000. Courtesy of D. Branton, from *Exptl. Cell Res.* 45:703 (1967).

As the fluid mosaic model proposes, integral membrane proteins can be removed from membranes only by treatments, such as exposure to detergents, that disperse the membrane bilayer. Even when removed from membranes, integral proteins frequently retain a layer of lipid molecules covering their nonpolar surfaces. Many such lipid-coated integral membrane proteins remain active and can be identified and analyzed

biochemically in this form. If they are cleaned completely of their lipid coat and placed in a polar environment, however, most integral proteins are forced to refold so extensively that they are completely denatured.

Several lines of evidence have established that integral proteins span the lipid bilayer and protrude from both membrane surfaces. In some of the experiments a radioactive label or chemical group was attached to parts of proteins exposed at either membrane surface. Other experiments detected segments of proteins that extend from membrane surfaces by combining them with fluorescent antibodies (see p. 118) or digesting them with proteolytic enzymes. For example, E. Rechstein and A. Blostein added radioactive iodine to protein segments exposed at the surfaces of membranous vesicles (Fig. 5-23). Their experiment took advantage of the fact that membrane fragments can be induced to seal into right-side-out or inside-out forms by simple adjustments in the salt concentration and pH of the suspending solution. By this means, they could identify the parts of proteins that extend from the inside and outside surfaces of the membrane vesicles. Their experiments, and the results of similar tests using other techniques, confirm that integral proteins extend entirely through membranes and have segments exposed on both membrane surfaces. The same experimental approaches also provide a major part of the evidence that integral proteins take an asymmetric orientation in membranes (see below).

Proteins generally seem to extend entirely through membranes. Only one example has been found of a protein that may extend only partway through a membrane. This protein, cytochrome b_5, has a short, hydrophobic hairpin loop in its amino acid chain that apparently extends into the bilayer from one side but does not pass entirely through the membrane.

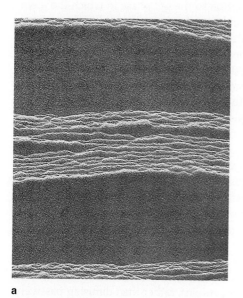

Figure 5-22 Artificial bilayers under the electron microscope. **(a)** Freeze-fracture preparation of layered artificial bilayers without added proteins; none of the layers show evidence of particulate substructure. **(b)** Artificial bilayers to which proteins have been added. Particles similar to those seen in freeze-fractured natural membranes are now visible. Courtesy of D. W. Deamer.

a

b

The Fluid Mosaic Model and Its Supporting Evidence **171**

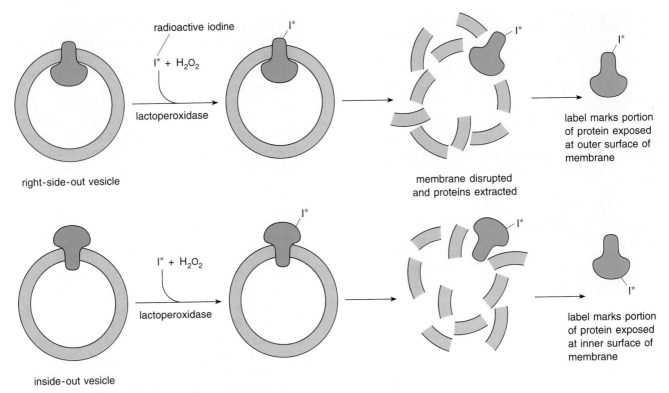

Figure 5-23 Rationale for Rechstein and Blostein's experiment identifying the parts of proteins that extend from the inside and outside surfaces of membranes. They used the enzyme lactoperoxidase to attach a radioactive iodine label to the exposed portions of the proteins. (Lactoperoxidase catalyzes the covalent linkage of halogens to proteins.)

Evidence That Membrane Proteins Can Move Laterally in the Bilayer The classic evidence that some integral proteins are free to move laterally in cellular membranes comes from an experiment carried out in 1970 by L. D. Frye and M. A. Edidin. These investigators worked with mouse and human cells, using a technique that fuses the plasma membranes of separate cells (see p. 123 and Supplement 5-1). The fusing technique involved exposure to the Sendai virus, which alters cells so that their plasma membranes flow together upon collision, joining the cells into a larger structure with a composite cytoplasm and plasma membrane.

Frye and Edidin used the cell fusion technique in combination with antibodies that bind specifically to plasma membrane proteins of either the mouse or human cell type. The antimouse antibodies were attached to molecules that fluoresce green under ultraviolet light, and the antihuman antibodies to molecules that fluoresce red (Fig. 5-24a; fluorescence microscopy is described on p. 111). The mouse and human cells were then fused, forming cells with a composite mouse–human plasma membrane (Fig. 5-24b). At first the fluorescent colors were segregated on the surfaces of the fused cells, with one-half of the membranes fluorescing red and one-half green. After 40 minutes at 37°C the fluorescent colors were completely intermixed on 90% of the cells, indicating that the mouse and human pro-

teins had become uniformly distributed throughout the composite membranes (Fig. 5-24c). Lowering the temperature gradually slowed intermixing; at 15°C the membrane proteins of the two species remained separate. From these results Frye and Edidin concluded that intermixing resulted from simple lateral diffusion of the mouse and human proteins through the fluid membrane bilayer. The interruption of diffusion at 15°C reflected transition to the gel phase in which the membrane lipids and proteins were immobilized.

Edidin later established that the diffusion coefficient for the intermixing proteins is about 2×10^{-10} cm²/sec, about 100 times slower than the rate observed for diffusion of membrane lipids. Other experiments investigating the mobility of membrane proteins produced equivalent results. For example, photobleaching experiments in which a fluorescent label is attached to membrane proteins reveal diffusion coefficients in the range of about 10^{-9} to 10^{-11} cm²/sec, indicating that proteins diffuse from 10 to 1000 times more slowly than membrane lipids. The slower diffusion rates for membrane proteins probably reflect several factors, including their larger size, interactions between charged and other groups on protein and membrane surfaces, and linkage of membrane proteins to the cytoskeleton.

Some additional experiments have shown that certain membrane proteins, rather than diffusing passively

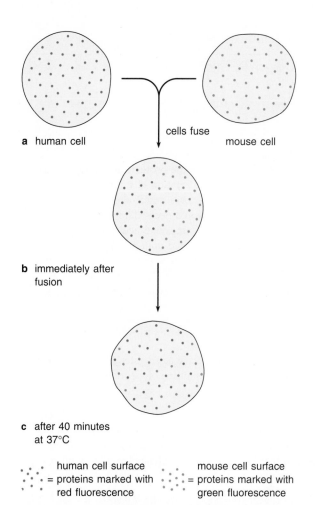

a human cell

cells fuse

mouse cell

b immediately after fusion

c after 40 minutes at 37°C

```
. . .  human cell surface
. . . = proteins marked with
. .    red fluorescence
```

```
. . .  mouse cell surface
. . . = proteins marked with
. . .  green fluorescence
```

Figure 5-24 Evidence for the mobility of membrane proteins in the combined plasma membranes of fused mouse-human cells. **(a)** The cells before fusion. **(b)** Immediately after fusion, membrane proteins from the mouse and human cells are segregated into distinct regions in the plasma membrane. **(c)** After 40 minutes at 37°C, the mouse and human membrane proteins are completely intermixed in most of the fused cells.

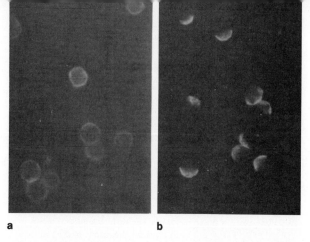

a **b**

Figure 5-25 Capping in lymphocytes (white blood cells). **(a)** Binding of fluorescent antibodies (light areas) to proteins distributed over the entire cell surface. **(b)** Formation of the cap, in which the marked proteins are actively swept to one end of the cell. Courtesy of G. M. Edelman.

in the membrane bilayer, are actively moved by systems that require energy. This active movement, called *capping*, also has been demonstrated with fluorescent antibodies that bind to specific proteins of the plasma membrane in cells such as lymphocytes. When antibodies are first added, the membrane proteins binding them are distributed randomly over the entire cell surface (Fig. 5-25a). Within minutes, however, the marked proteins are swept to one end of the cell, where they form a dense cap (Fig. 5-25b). The cap may remain in position or be taken into the cell interior by endocytosis (see p. 857). Inhibitors of reactions supplying ATP arrest the capping reaction, indicating that it is an active process that requires the expenditure of cellular energy. (For further details of the capping reaction, which depends on the activity of microfilaments, see Ch. 12.)

Some membrane proteins and glycoproteins also prove to be immobilized in biological membranes.

These immobile or only slightly mobile molecules are held in place by linkages to cytoskeletal networks or extracellular materials or by interactions between the carbohydrate groups of glycoproteins extending from the membrane surface. For example, *band 3 protein*, an integral protein of erythrocyte plasma membranes, is held essentially immobile through its linkage to cytoskeletal proteins that form a network underneath the plasma membrane (for details, see p. 509). Other examples are proteins forming parts of nerve synapses (see Ch. 6) or membrane junctions (see Ch. 7), which are also more or less rigidly fixed in the membrane.

Evidence That Membrane Lipids Are Asymmetric An extensive series of experiments has demonstrated that lipids are distributed asymmetrically in natural membranes. A group of enzymes that hydrolyze phospholipids, *phospholipases*, has been used extensively in this research. Using these enzymes L. L. M. van Deenen and his coworkers found that 76% of the total membrane phospholipids containing choline and 20% of those containing ethanolamine were hydrolyzed in intact erythrocytes (Fig. 5-26). These results indicate that choline- and ethanolamine-containing phospholipids occur in these proportions in the outside half of the erythrocyte bilayer. If so, the inner half of the bilayer contains 24% of the choline-containing and 80% of the ethanolamine-containing phospholipids. No serine-containing phospholipids were hydrolyzed by the enzymes as long as the erythrocytes were intact. However, if the plasma membranes were broken, exposing the inner membrane surface to the enzymes, all the phospholipids containing serine were hydrolyzed, indicating that this phospholipid type is confined to the inner bilayer half.

These results confirm that the distribution of plasma membrane phospholipids is asymmetric in erythrocytes. Equivalent results have been obtained for the plasma

The Experimental Process

How Do Ions Get Across Membranes?

S. J. Singer

S. JONATHAN SINGER is a University Professor in the Department of Biology at the University of California at San Diego. His early scientific training was in physical chemistry, and he obtained A.B. and A.M. degrees from Columbia University and a Ph.D. from the Polytechnic Institute of Brooklyn. After postdoctoral training with Dr. Linus Pauling at Caltech, Dr. Singer was on the faculty of chemistry at Yale University before joining the newly developing U.C. campus at San Diego in 1961. He is a member of the National Academy of Sciences and a former Research Professor of the American Cancer Society. In 1991, he received the E. B. Wilson Award of the American Society for Cell Biology for his pioneering work on the fluid mosaic model of membrane structure.

The transport of ions and small hydrophilic molecules (such as sugars and amino acids) across otherwise impermeable and hydrophobic membrane lipid bilayers is a critical phenomenon in cell biology. Such transport is involved in the regulation of cell metabolism and in all signal transmission processes, including transmission in the nervous system. Until the early 1970s, a prevalent view of the molecular mechanism of transport was that of the "rotating carrier" (Fig. A). The ion or hydrophilic molecule (the ligand) was supposed to be bound to a specific site on a transport protein molecule, which then rotated and diffused across the bilayer to release the ligand on the other side. However, the same thermodynamic arguments that we used in proposing the fluid mosaic model for the molecular organization of the proteins and lipids of membranes indicated that the rotation of a membrane protein molecule across a bilayer was not at all likely to occur, because such rotation would necessitate the transient insertion of some of the ionic amino acid residues of the protein molecule into the hydrophobic interior of the bilayer, an event that would be energetically quite unfavorable. We suggested instead[1] that transport proteins in membranes would have three main structural properties (see Fig. 6-3 on p. 194):

1. They would be specific aggregates of a small number (2 to 6) of identical or similar polypeptide chains traversing the bilayer.

2. Down the central axis through the subunit aggregate a narrow transmembrane channel would form that would be lined by ionic and hydrophilic amino acid residues of the protein subunits, and this channel would be filled with water molecules.

3. A specific binding site for the ligand to be transported would be present on one or more of the subunits within the channel, such that a *quaternary rearrangement* of the subunits, requiring only small amounts of energy, would transfer the ligand from one side of the membrane to the other (Fig. B).

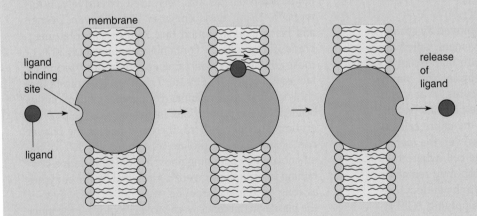

Figure A The rotating carrier mechanism, the prevalent hypothesis of membrane transport until the early 1970s. According to this hypothesis, a transport protein molecule rotates to carry a ligand across the bilayer, where it is released.

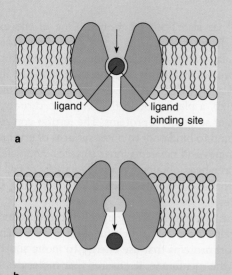

a

b

Figure B Active transport of a molecule through a membrane protein channel. A molecule interacts with the active site of a protein subunit **(a)**. This triggers an energy-yielding enzyme reaction that produces a shift in the subunit conformation **(b)**, "squeezing" the molecule through the membrane.

This was at a time when the structure of no transport protein was known. We decided to do an experiment that might rule out the rotating carrier model. The idea of the experiment was that if a large protein molecule, such as an antibody molecule, was attached to the exposed surface of a transport protein in an intact membrane, it should strongly inhibit any rotation of the transport protein across the membrane, and therefore greatly reduce the rate of ligand transport through the membrane. However, if transport occurred by way of a protein aggregate and a mechanism such as pictured in Figure B, little if any effect of the bound antibody on the transport rate might be expected.

We chose for our experimental system the transport of Ca^{2+} across the membranes of the sarcoplasmic reticulum (SR) of muscle cells. This transport is a critical part of the process that controls the Ca^{2+} concentration in the muscle cell cytoplasm and thereby regulates muscle contractility. The Ca^{2+} transport is carried out by a membrane protein, Ca^{2+}-ATPase, which is an enzyme that uses the energy of ATP hydrolysis to power Ca^{2+} import into the lumen of the SR. Vesicles of the SR, whose membranes consist of over 80% just the one protein, the Ca^{2+}-ATPase, can be isolated from the muscle. We could have produced an antibody to the Ca^{2+}-ATPase, but we reasoned that, with the methods available at the time, it

would be uncertain that an antibody that did not inactivate transport was really directed to the Ca^{2+}-ATPase. We therefore chose instead to modify the Ca^{2+}-ATPase in the intact SR vesicles to contain an exposed covalently bound 2,4-dinitrophenyl (DNP) group, for which highly purified specific antibodies (anti-DNP) were already available in our laboratory.

The experimental procedure[2] involved, first, the chemical modification of the intact SR vesicles using the enzyme liver transglutaminase and [³H]-DNP-cadaverine to catalyze the attachment of a DNP-cadaverine molecule by its amino group in amide linkage to some carboxyl group on the SR membrane protein. Radioactive counting of the attached [³H] allowed us to show that, depending upon the particular reaction conditions used, between 0.55 and 0.80 mol of DNP group was covalently bound per mol of Ca^{2+}-ATPase. Second, we showed that upon the addition of excess anti-DNP to the modified SR vesicles, $0.55 \pm .05$ mol of antibody was bound per mol of covalently-bound DNP group, probably indicating that each molecule of the antibody, bearing two anti-DNP binding sites, was bound to two DNP groups simultaneously. Third, we showed that such antibody binding to the DNP-modified SR vesicles had no significant effect either on the Ca^{2+}-ATPase enzyme activity or on the rate of radioactive [⁴⁵Ca^{2+}] transport into the vesicles.

These experiments, and similar ones carried out independently on other transport systems by J. Kyte,[3] and by A. Martonosi and F. Fortier,[4] indicated that "rotating carrier mechanisms for protein-mediated membrane transport should be laid to rest." Since 1976, when these experiments were performed, many transport proteins have been isolated and structurally characterized, and in many cases their amino acid sequences have been determined from the analysis of their isolated cDNAs. From these and other results, the general features of the model proposed in Figure B (see also Fig. 6-3) have been confirmed. (For an example, see ref. 5 concerning the acetylcholine receptor, its structure and mechanism of transport.)

References

[1] Singer, S. J. In *Structural and function of biological membranes*, Chapter 4. Ed. L. I. Rothfield. New York: Academic Press, pp. 145–222 (1971).

[2] Dutton, A. H.; Rees, E. D.; and Singer, S. J. *Proc. Natl. Acad. Sci. USA* 73:1532 (1976).

[3] Kyte, J. *J. Biol. Chem.* 249:3652 (1974).

[4] Martonosi, A., and Fortier, F. *Biochem. Biophys. Res. Commun.* 60:382 (1974).

[5] Unwin, N.; Toyoshima, C.; and Kubalek, E. *J. Cell. Biol.* 107: 1123 (1988).

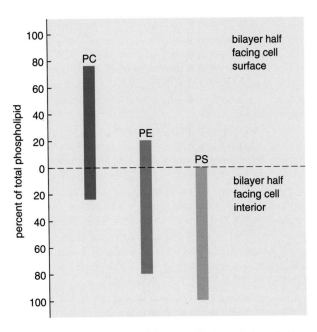

Figure 5-26 The asymmetric distribution of phosphatidyl choline (PC), phosphatidyl ethanolamine (PE), and phosphatidyl serine (PS) in erythrocyte membranes, revealed by the experiments of van Deenen and his coworkers.

membranes of various other cell types (Fig. 5-27a). Although most studies have concentrated on the plasma membrane, a few studies of organelle membranes such as those of mitochondria, the Golgi complex, and the endoplasmic reticulum indicate that phospholipid distribution in all cellular membranes is probably asymmetric (Fig. 5-27b).

The functional significance of lipid asymmetry is not completely understood. However, the asymmetric distribution of lipids between the two bilayer halves may maintain necessary differences in the charge, fluidity, or phase transition of the two membrane surfaces, or may be associated in some way with the asymmetric orientation of membrane proteins.

Evidence That Membrane Proteins Are Asymmetric
Experiments labeling the segments of proteins exposed at either membrane surface have provided ample evidence that proteins are asymmetrically suspended. Typically, these experiments have found that all membrane proteins of a given type have the same asymmetric orientation. A prime example is the glycophorin protein of erythrocytes, which is invariably arranged with its C-terminal end extending into the cytoplasm and its N-terminal end projecting from the outer membrane surface.

The fact that the carbohydrate groups of membrane glycoproteins, such as glycophorin, always extend asymmetrically from the outer surface of plasma mem-

branes has also been mentioned. One of the experiments demonstrating this orientation came from the Singer laboratory. Nicolson and Singer developed a glycoprotein marker for electron microscopy by linking ferritin, a large electron-dense molecule, to *concanavalin A (con A)*, a plant protein that binds to carbohydrate groups of some glycoproteins. The ferritin-con A label was found to bind only to outer surfaces of erythrocyte membranes and never to inner surfaces. Equivalent results have been obtained in a wide variety of cell types using this and other techniques for identifying carbohydrates.

The asymmetric arrangement of proteins and glycoproteins is central to their function in membranes. In transport proteins that use energy to move substances across the membrane, it ensures that molecules move preferentially in only one direction. The absorption of light in photosynthesis and its conversion to chemical energy in chloroplasts and the synthesis of ATP in both chloroplasts and mitochondria are equally dependent on membrane "sidedness" and the asymmetric arrangement of the proteins carrying out these activities. Moreover, the asymmetric orientation of glycoproteins on the outside surfaces of plasma membranes provides the cell with receptors and with recognition and adhesion groups directed toward its surroundings. The asymmetry of membrane proteins and glycoproteins is thus fundamental to the specialized functions of membranes in living cells.

There is ample evidence that some membrane proteins are restricted to different regions of cellular membranes. For example, in cells lining the mammalian intestine a membrane channel that admits sodium ions, and the enzyme alkaline phosphatase, occur only in the part of the plasma membrane facing the intestinal cavity. Other proteins of the intestinal cells, such as a surface recognition marker and a different sodium transporter, one that uses energy to pump sodium ions out of the cell, are restricted to the part of the plasma membrane facing the circulatory system. Similarly, in mammalian sperm cells several proteins or glycoproteins are restricted to subregions of the plasma membrane, such as that covering the head or tail (see p. 1095).

This unequal lateral distribution of membrane proteins may be maintained by one or more of several mechanisms. In the intestinal epithelial cells the unequal distribution is maintained by a type of surface junction (a *tight* or *sealing* junction; see p. 260) formed between adjacent cells. In the junction, plasma membranes of adjacent cells fuse so closely that membrane proteins cannot diffuse through the junction region. Because the junction extends in a belt entirely around the cell, it forms a complete barrier to diffusion of membrane proteins between the sides of the cell facing the intestinal cavity and the bloodstream. Destruction of the junction allows proteins of the two membrane regions

Figure 5-27 The asymmetric distribution of membrane lipids **(a)** in plasma membranes of different cell types and **(b)** in cell organelle membranes. PC, phosphatidyl choline; PE, phosphatidyl ethanolamine; PS, phosphatidyl serine; PI, phosphatidyl inositol; SP, sphingolipid; CL, cardiolipin, a lipid type characteristic of mitochondria (see Fig. 9-33). Compiled from data in M. D. Houslay and K. K. Stanley, *Dynamics of Biological Membranes,* New York: John Wiley and Sons, Inc., 1982; J. W. Pierre and L. Ernster, *Ann. Rev. Biochem.* 46:201 (1977); and G. Berga and R. P. Holmes, *Prog. Biophys. Molec. Biol.* 43:1195 (1984).

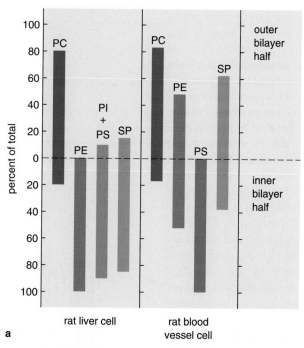

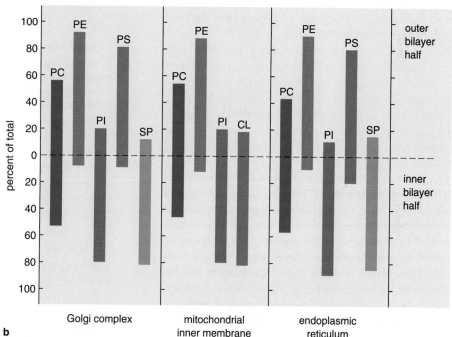

to intermix, as expected if the junction is responsible for establishing the different membrane domains.

Interactions such as hydrogen bonding between membrane proteins or linkages to the cytoskeleton may be responsible for maintaining the unequal distribution of proteins and glycoproteins in some membranes. In any case, however the unequal lateral distribution of proteins is maintained, its effect is to create subregions of the membrane that are specialized for different functions. For example, in cells lining the intestinal tract, the sodium channels facing the intestinal cavity specialize the plasma membrane in this region for free

entry of sodium ions from the intestinal contents. The location of active sodium pumps on the side facing the bloodstream specializes the plasma membrane in this region for removal of sodium ions from the cytoplasm. As a consequence of these specializations, sodium ions flow through the cells from the intestinal lumen to the bloodstream.

Evidence That Some Proteins Are Peripheral to Membranes Investigations of many membrane types, particularly plasma membranes, have identified an extensive group of proteins peripherally associated with membrane surfaces. On the cytoplasmic side, animal plasma membranes have been found to be associated with proteins of the cytoskeleton, including those of microfilaments, microtubules, intermediate filaments, and a variety of linker proteins tying these elements together (for details, see Ch. 13). Typically, these cytoskeletal elements are held by noncovalent linkages to the regions of integral membrane proteins projecting from the inner surface of the plasma membrane. On the outer surface of the plasma membrane, noncovalent linkages tie the membrane surface directly or indirectly to extracellular molecules, such as collagen, proteoglycans, and fibronectin (for details, see Ch. 8).

Similar noncovalent associations with peripheral proteins also have been observed in internal membrane systems. These linkages may involve interactions with either integral membrane proteins or the polar head groups of membrane lipids. One cytoskeletal element, a protein called *ankyrin*, for example, forms noncovalent linkages to an integral protein of the plasma membrane. Another protein, the so-called *basic protein* that adheres to the outer surfaces of plasma membranes forming myelin sheaths in nerve tissues, interacts with the head groups of membrane phospholipids.

Recently, one group of exceptional peripheral proteins has been discovered to associate with membranes by covalent linkage to an "anchor" that ties them to the nonpolar membrane interior. Only the N- or C-terminal ends of the proteins are covalently linked to the lipid membrane anchor; the remainder of the protein extends into the medium surrounding the membrane surface.

Two major types of anchors have been detected in these membrane-linked proteins. One is simply a fatty acid residue, usually derived from either palmitic or myristic acid (see Table 2-5), that links covalently to the N-terminal end of the protein chain. This anchor inserts in the membrane bilayer and, through nonpolar associations with the membrane interior, ties the protein to the membrane (Fig. 5-28a). Several eukaryotic proteins and some proteins inserted into the plasma membrane by infecting viruses, including the HIV retrovirus responsible for AIDS (see p. 820), have been discovered to be tied to membranes by this type of anchor. Covalent linkage of these proteins to the fatty acid chain classifies these complexes as *proteolipids*, a group of substances previously thought to be rare or nonexistent in eukaryotes.

The second anchor type, found to tie more than 40 different eukaryotic proteins to plasma membranes, is a short carbohydrate chain linking the C-terminal ends of the proteins to a membrane phospholipid, either phosphatidyl inositol or, more rarely, phosphatidyl ethanolamine. Ethanolamine may also be present as a polar group forming part of the anchor, as in the phosphatidyl inositol anchor shown in Figure 5-28b. The list of pro-

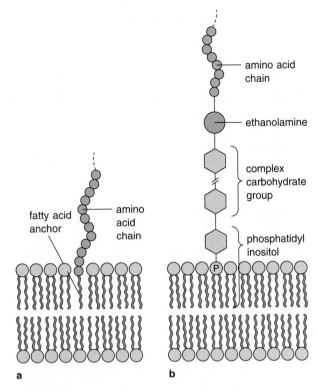

Figure 5-28 Anchors linking peripheral proteins to the hydrophobic interior of plasma membranes. **(a)** An anchor consisting of a single fatty acid residue. **(b)** A more complex anchor based on phosphatidyl inositol; the anchor includes a complex carbohydrate group and the polar alcohol ethanolamine.

teins linked to membranes by the phosphatidyl inositol anchor includes such well-characterized types as alkaline phosphatase and the acetylcholinesterase receptor of nerve and muscle cells (see Table 5-4).

The anchors linking these proteins to plasma membranes are readily degraded by cellular enzymes. The glycosyl–phosphatidyl inositol anchor, for example, can be hydrolyzed by phospholipase C, an enzyme commonly found in association with plasma membranes. The enzyme catalyzes breakage of the phosphate linkage binding inositol to the polar part of the anchor, releasing the protein from the membrane. The relative ease by which the anchored proteins are released may provide cells with a quick means to terminate a function provided by an anchored protein, such as an enzyme or surface marker, or remove a receptor once it has bound its target substance at the membrane surface.

Research carried out by a variety of methods clearly supports the major tenets of the fluid mosaic model. Lipids are arranged in membranes in bilayers that are fluid under physiological conditions. Integral membrane proteins are suspended in the lipid bilayer and can be removed only by treatments that disperse the bilayer.

Many of the lipid and protein molecules of membranes are linked to carbohydrate groups to form glycolipids or glycoproteins. Peripheral proteins, except for those with hydrophobic anchors, are linked to the membrane surfaces by noncovalent bonds that can be interrupted without bilayer disruption. Both lipid and protein components of membranes are arranged asymmetrically, making the two surfaces of a membrane different in molecular composition.

MEMBRANE BIOGENESIS

Membranes are dynamic structures that can rapidly assemble and disassemble. One investigator, C. E. Bracker, calculated that in a growing cell of a fungus, new membrane is added at the rate of 32 μm^2/min. In dividing cells that are roughly spherical in shape, the surface area of the plasma membrane increases by about 1.6 times during the growth phase between divisions.

Where and how is the new membrane material synthesized to accommodate such growth? Revelations about membranes arising from the fluid mosaic model and its supporting research have complicated attempts to answer this question, an old one in biology. In particular, the fact that membrane proteins have strongly hydrophobic and hydrophilic regions raises questions about where these proteins are assembled and how they get into membranes in the first place. How are the hydrophilic ends of proteins pushed through the hydrophobic membrane interior during their initial insertion? How do proteins move from their point of insertion in membranes to more distant locations in the cell? Essentially the same questions apply to membrane lipids. Where are they synthesized, and how are the polar parts of amphipathic lipids pushed from one side of a bilayer to the other as they are placed in membranes?

Lipid Biogenesis and Membrane Asymmetry

Using radioactively labeled precursors of lipids as markers, in 1970 D. J. Morré showed that in both plant and animal cells, labeled precursors are first incorporated into membrane lipids in the endoplasmic reticulum (ER). The label can be traced to the Golgi complex, and later to the plasma membrane. These observations indicate that membrane lipids are synthesized in the ER and are first inserted into membrane bilayers in this location. The enzymes required for membrane lipid synthesis have also been found in ER membranes isolated by cell fractionation (see p. 123).

Newly synthesized lipids enter ER membranes at the bilayer half facing the surrounding cytoplasm. They are then flip-flopped to the opposite bilayer half by enzymes, variously called *flippases, transfer proteins,* or *transporters,* that form an integral part of the membrane. These enzymes, which show differing specificities for glycerol-based phospholipids, sphingolipids, and cholesterol, greatly speed the flip-flop of lipid molecules between bilayers. (Cholesterol, which is only slightly amphipathic, can apparently also flip-flop readily without enzymatic catalysis.) Although lipid transfer by the flippases is readily detected, the mechanisms by which the enzymes move lipid molecules between bilayer halves remain unknown. Presumably, conformational changes in the flippases, or burial of polar lipid groups in the active site, cover the polar groups during their transfer across the nonpolar bilayer interior.

J. M. Backer and E. A. Dawidowicz successfully removed a flippase from ER membranes of the rat and inserted it into artificial phospholipid membranes. The ER flippase was capable of rapidly transferring phospholipids between bilayer halves in the artificial membranes, with no requirement for an energy input. Whether all flippases are similarly energy-independent remains unknown.

The newly synthesized and inserted lipids then move from the ER membranes through the Golgi complex to the plasma membrane. Movement from the ER to the Golgi complex occurs via membranous vesicles that pinch off from the ER and fuse with the Golgi; similarly, movement from the Golgi complex to the plasma membrane occurs via vesicles that pinch off from the Golgi complex, move through the intervening cytoplasm, and fuse with the plasma membrane. (Supplement 5-1 describes possible molecular mechanisms underlying membrane fusion; see Ch. 20 and Fig. 20-9 for further details of vesicle movement between the ER, Golgi complex, and plasma membrane.) Flow from the rough ER to the nuclear envelope is also possible through the direct connections between these membrane systems (see p. 826). The fluidity of the membrane bilayer provides the basis for the flow of lipids through these permanently or intermittently connected membrane systems.

Membrane flow along the ER → Golgi complex → plasma membrane or ER → nuclear envelope routes readily explains how lipids, newly synthesized in the ER, are transported between these membranes. However, it does not explain how the membranes of mitochondria and chloroplasts, which rarely show connections to other cellular membranes, receive their lipids. These organelles contain enzymes only for synthesis of a few minor membrane lipids; most of their membrane lipids apparently originate from sites of assembly in the ER.

An extensive series of experiments has shown that lipid flow between the ER and totally disconnected organelles, such as mitochondria, is promoted by soluble cytoplasmic proteins called *lipid transfer* or *exchange proteins.* For example, K. W. A. Wirtz and D. B. Zilversmit found that mitochondria take up label very rapidly when cells are exposed to labeled lipid precursors, confirming that lipids newly synthesized in the ER are rapidly transferred to mitochondria even though direct connections cannot be detected between the two membrane systems. Subsequently, Wirtz and Zilversmit isolated a group of lipid transfer proteins that can promote the exchange of lipid molecules between isolated ER and mitochondria and, within mitochondria, between the inner and outer mitochondrial membranes. Others have detected lipid transfer proteins that can stimulate movement of lipids between the ER and the plasma membrane or between the ER and chloroplast. In fact, proteins with this function have been detected in every cell type in which they have been investigated, including those of animals, plants, yeast, and bacteria. Many of the transfer proteins are identical in structure in these groups, indicating that their evolutionary origins are very ancient.

The specificities of lipid transfer proteins detected to date varies: some are specific in their activity for only a single phospholipid type or a group of closely related phospholipids; others can transfer essentially any lipid, including cholesterol, between different membrane systems.

The mechanism by which lipid transfer proteins move hydrophobic lipids through the aqueous cytoplasm is unknown. Presumably, they undergo a conformational change on binding a lipid that buries the hydrophobic portions of the lipid molecule in the protein interior, so that only hydrophilic portions, if any, face the surrounding medium. On contacting a target membrane, another conformational change releases the lipid from the protein interior and introduces it into the membrane bilayer.

Lipid transfer proteins can apparently distribute lipid molecules only to the bilayer half facing the soluble cytoplasm. Transfer to the opposite bilayer half is moderated by flippases in the target membranes. (Fig. 5-29 summarizes the pathways of lipid movement.)

Origins of Membrane Proteins

Experiments tracing the incorporation of labeled amino acids have also implicated the ER in the synthesis of integral membrane proteins. Experiments by W. W. Franke and his colleagues showed that, within minutes of exposure to radioactive amino acids, assembly of amino acids into labeled proteins could be detected in the rough ER. Label subsequently appeared in the Golgi complex and, after about 20 min, in the plasma membrane. Equivalent experiments by Morré with plant tissues showed the same pattern: label was incorporated

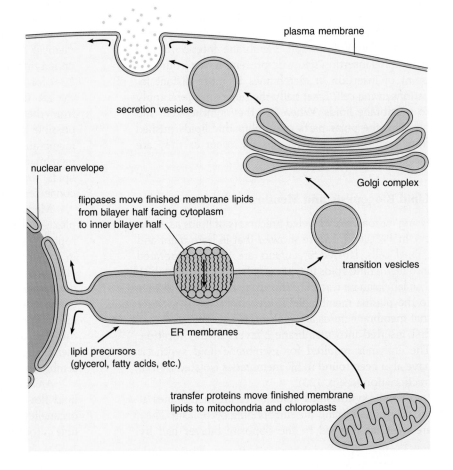

Figure 5-29 The pathway followed by newly synthesized membrane lipids. Lipids are assembled from soluble precursors in the ER on the bilayer half facing the surrounding cytoplasm. Flippases catalyze flip-flop of newly synthesized membrane lipids to the opposite bilayer half. From the ER the lipids move to the nuclear envelope by direct membranous connections, or to the Golgi complex and plasma membrane via membranous vesicles that travel between these structures. Mitochondria, chloroplasts, and other cytoplasmic organelles not connected to the ER → Golgi → plasma membrane traffic system receive membrane lipids by means of lipid transfer proteins, which are capable of moving lipids from the ER to these structures through the aqueous cytoplasmic medium.

plasma membrane

secretion vesicles

nuclear envelope

Golgi complex

flippases move finished membrane lipids from bilayer half facing cytoplasm to inner bilayer half

transition vesicles

ER membranes

lipid precursors (glycerol, fatty acids, etc.)

transfer proteins move finished membrane lipids to mitochondria and chloroplasts

first in the rough ER and moved to the Golgi complex and plasma membranes within 20 to 60 min.

Exposing cells to precursors of carbohydrate groups, such as labeled glucose, shows that the addition of complex sugars to proteins in glycoprotein formation occurs in both the ER and the Golgi complex, with greatest concentration of activity in the Golgi complex. Analysis of enzymatic activity in isolated membrane fractions also shows that the greatest concentration of enzymes attaching carbohydrate groups to proteins occurs in the Golgi complex (for details, see Ch. 20).

These results indicate that proteins, like membrane lipids, are first incorporated into cellular membranes in the ER. The new membrane proteins then flow through the Golgi complex to reach the plasma membrane by the same routes as membrane lipids. Presumably, flow through the rough ER membranes to the nuclear envelope also occurs.

Proteins are inserted in ER membranes through the *signal mechanism.* a process first proposed and demonstrated by G. Blobel and his associates. The proteins destined to be inserted into ER membranes contain a *signal,* a sequence of amino acids that includes a strongly hydrophobic segment. The signal is usually located at the N-terminal, or "front," end of a newly synthesized protein. The signal causes the protein to attach to the ER membranes as it is synthesized on ribosomes. Once attached, the signal, by virtue of its hydrophobic segment, penetrates into the membrane. The remaining segments of the protein extend through the membrane as they are assembled. Evidently, some or all of the energy required for this penetration is supplied by the protein itself as it folds in response to its arrangement of hydrophilic and hydrophobic amino acids. The final folding arrangement of the protein fixes its asymmetric location in the membrane.

The initial portions of carbohydrate groups are added to the ends of newly synthesized membrane proteins that project into the interior of the ER sacs (see Fig. 5-18). These carbohydrate groups are completed as the proteins move through the Golgi complex. As in the ER, the finished carbohydrate units face the inside of vesicles pinching off from the Golgi. Fusion of the vesicles with the plasma membrane places the carbohydrates on the exterior surface of the cell. (For details of the reactions synthesizing the carbohydrate groups of membrane proteins, see Supplement 20-2 and Figs. 20-41 and 20-42.)

Proteins of mitochondrial and chloroplast membranes, and those of other cytoplasmic organelles not part of the ER—Golgi complex—nuclear envelope group, originate from ribosomes freely suspended in the soluble cytoplasm without association with membranes. After their synthesis is complete these proteins enter the organelles by a process analogous to the signal mechanism. In this case the signal recognizes and binds specifically to the organelles rather than to the ER,

causing insertion of the newly synthesized proteins into the organelle membranes. (Ch. 20 describes in more detail the signal mechanism and the processes sorting proteins to the various membrane systems of the cell.)

HISTORICAL DEVELOPMENTS LEADING TO THE FLUID MOSAIC MODEL

The importance of membranes to cell structure and function was recognized early in the development of cell biology. From the late 1800s onward investigations into the nature of cell boundaries gradually revealed the molecules that are present in membranes, and provided indications of the manner in which these molecules are put together to form the thin surface coats of cells and internal cellular organelles. The fluid mosaic model includes several elements that are direct descendants of this early research, some of it dating back to the turn of the century.

The Discovery of Membrane Lipids and Bilayers

The first significant experiments indicating the presence of lipids at the cell boundary were carried out by E. Overton in the late 1890s. These experiments revealed that cells absorb small, nonpolar lipid molecules much more rapidly than most polar molecules. On this basis Overton proposed that molecules enter cells by dissolving or penetrating a layer of lipid molecules covering cell surfaces.

E. Gorter and F. Grendel derived the first clues about the physical arrangement of lipid molecules in the surface layer in the 1920s. For their experiments Gorter and Grendel used mammalian erythrocytes, which contain no nucleus or internal organelles and are of uniform size and shape. Because no nuclei or internal organelles are present, all the lipid in the preparations could be assumed to originate from the plasma membranes. Bursting erythrocytes by exposing them to distilled water releases the cell contents, which consist of little more than a solution of hemoglobin; the empty plasma membranes are called erythrocyte "ghosts." By measuring the ghosts under a light microscope, Gorter and Grendel were able to estimate the total area of a single erythrocyte plasma membrane. They then extracted lipids from a preparation of erythrocyte ghosts, and spread the lipids in a monolayer one molecule in thickness on a water surface. (Hydrophobic substances readily form monolayers when dropped on a clean water surface.) Comparing the surface area of the lipid monolayer with the surface area of the ghosts from which the lipid was extracted indicated that enough lipid is present to make a layer two molecules in thickness around each cell. Using this information, Gorter and Grendel made the first proposal that the lipid molecules of cell membranes occur in a bilayer.

Membrane Proteins and the Danielli–Davson Model

J. Danielli and H. Davson first implicated proteins as parts of membranes through research carried out in the 1930s and 1940s. These investigators measured the surface tension of triglycerides in water and compared the values obtained with the surface tension of living cells. Triglycerides are totally hydrophobic, oily substances that round up into spherical droplets when placed in water because a spherical shape exposes the smallest possible surface area to the surrounding water molecules. The resistance of the droplets to flattening, which increases the surface area per unit volume, is reflected in their relatively high surface tension. Danielli and Davson reasoned that if cells are actually covered with a layer of lipids, as the earlier work indicated, then the surface tension of cells ought to be similar to that of oil droplets. But their results turned out to be quite different: the surface tension of cells was consistently found to be much lower than that of oil droplets. Danielli and Davson discovered, however, that they could mimic the surface properties of living cells by adding proteins to the oil droplets. The proteins formed a hydrophilic film over the surface of the droplets and reduced their surface tension to levels similar to living cells.

From this information Danielli and Davson proposed a model for membrane structure that was to shape biological thinking for many years. They agreed that the lipid molecules of membranes are arranged in a bilayer. To account for the low surface tension of living cells, they proposed that the bilayer is coated on both the internal and external surfaces by a layer of protein that reduces its surface tension. To accommodate a complete layer of protein on both surfaces within the very thin dimensions actually observed for biological membranes, Danielli and Davson proposed that the proteins extend over the lipid bilayer as completely unfolded and extended peptide chains.

The Danielli–Davson hypothesis was later modified to include pores of small dimensions extending through the bilayer. The surface coat of extended polypeptide chains was considered to extend into the pores, giving the pores a polar quality. This modification accounted for the unexpectedly high rate at which water penetrates through membranes (see p. 204). By 1954 the Danielli–Davson model had reached the form shown in Figure 5-30.

The Danielli–Davson model provided the conceptual framework for essentially all the thought and experimentation in membrane structure until the 1960s. During these years a variety of experiments confirmed that phospholipids and proteins are important constituents of membranes. Particularly important was the work of J. D. Robertson, who drew attention to the almost uniform appearance of sectioned biological

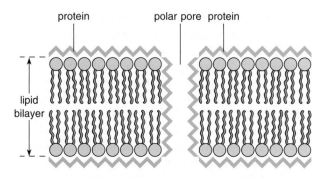

Figure 5-30 The Danielli–Davson model for membrane structure. According to the model, the bilayer is coated by protein molecules that extend into a layer one amino acid thick. At intervals polar "pores" coated by extended proteins perforate the membrane.

membranes as a "railroad track"—two dark lines separated by a lighter interzone—in electron micrographs such as Figure 5-19. He proposed that the dark lines in the electron image represent the protein layers on the inner and outer membrane surfaces in the Danielli–Davson model, and that the less dense interzone is the lipid bilayer. As part of his *unit membrane* concept advanced in 1959, Robertson hypothesized that all cellular membranes are built up from a structural unit of this type.

Robertson also noted that plasma membranes are covered on their external surfaces with a layer of carbohydrates. He took this unequal distribution of carbohydrate groups into consideration in his modification of the Danielli–Davson model by showing a thicker protein layer on the outside membrane surface (Fig. 5-31). By doing so, Robertson introduced the idea that cellular membranes may be asymmetric.

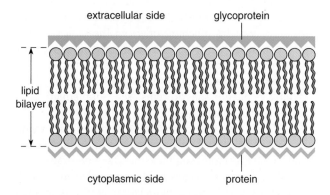

Figure 5-31 Robertson's modification of the Danielli–Davson model. The modification includes a layer of glycoproteins on one side, making the membrane asymmetric.

Final Steps Toward the Fluid Mosaic Model

Inadequacies in the Robertson–Danielli–Davson models became increasingly apparent during the 1960s. In 1966 J. Lenard and Singer showed that as much as 30% of the amino acid chains of membrane proteins is twisted into an alpha helix. This made it unlikely that membrane proteins could be spread over the membrane surface in completely extended form as Danielli and Davson had proposed. Singer noted that the total content of alpha helix in membrane proteins, in fact, is typical of polypeptides with a spherical rather than a flattened shape. However, spherical proteins seemed incompatible with the Danielli–Davson model because placing proteins in this form on the two sides of a bilayer would build a structure several times thicker than the 7- to 8-nm thickness actually observed for biological membranes.

Further problems for the Danielli–Davson model came from Singer's observation that unfolding proteins into fully extended form on membrane surfaces would inevitably expose hydrophobic amino acid side chains to the aqueous medium. He pointed out that this condition is thermodynamically unlikely, since large amounts of energy must be expended to maintain polar or nonpolar groups in a position that exposes them to the opposite environment. Instead, proteins tend to fold into minimum energy states, with hydrophilic and hydrophobic groups in separate regions facing environments of like polarity. Singer calculated that the energy required to maintain proteins in the fully extended form would make membranes highly unstable and very unlikely to remain intact.

Observations of the behavior of phospholipids in water solved some of these problems. Investigators found that phospholipids, in contrast to the triglycerides used by Danielli and Davson in their experiments, can take up a bilayer arrangement in which all the exposed surfaces are hydrophilic. This arrangement reduces their surface tension with no requirement for a surface protein coat. In fact, pure phospholipid bilayers can spontaneously take on forms with a very high ratio of surface area to lipid volume, such as large vesicles and extended sheets that often stack into multiple layers. Thus, as often happens in scientific research, Danielli and Davson arrived at a correct conclusion—that proteins are important in membrane structure—for what turned out to be the wrong reasons.

The freeze-fracture technique, developed in the 1960s, provided some of the first clues to the actual arrangement of protein molecules in membranes. Freeze-fractured membrane preparations revealed that large numbers of particles, with the dimensions expected for protein molecules, are suspended in the membrane interior. Later work, as noted, established that the particles are actually proteins. These findings suggested that proteins are embedded in membranes as globular units rather than extended on membrane surfaces.

In 1972 Singer and Nicolson combined all the available evidence and arguments into a new hypothesis for membrane structure, the fluid mosaic model. In simplest terms the Singer–Nicolson model retained the phospholipid bilayer advanced in earlier models as the basic structure underlying biological membranes, and proposed that the bilayer is fluid. Proteins were considered to be suspended in the fluid bilayer as discrete, individual units. Research carried out since its proposal has fully supported the fluid mosaic model, and it is now accepted as an accurate depiction of the fundamental structure of biological membranes.

For Further Information

Some biological membranes commonly undergo fusion, a process in which the bilayers of two separate membranes join to form a single, continuous membrane. The fusion, in effect, dumps the lipids and proteins of one membrane into another, allowing the membrane components to mix freely. During secretion, for example, cytoplasmic vesicles containing substances to be released to the cell exterior contact the plasma membrane on its cytoplasmic side (see Fig. 20-9). On contact the vesicle and plasma membrane bilayers join and flow together, adding the lipids and proteins of the vesicle membrane to the plasma membrane and releasing the vesicle contents to the cell exterior. Membrane fusion occurs regularly when vesicles pass enclosed material from the ER to the Golgi complex, when vesicles enclosing materials from outside the cell join with lysosomes, and when sperm and egg plasma membranes fuse to place the sperm and egg nuclei in a common

cytoplasm during fertilization. Membrane fusion also is a part of pathological processes such as cell infection by enveloped viruses (see below).

For fusion to occur, the phospholipid bilayers of the joining membranes must come into direct contact. At least two barriers must be overcome for direct contact to take place. One is a film of water molecules covering polar groups of phospholipids at the membrane surface. The water molecules in this *hydration layer* are more highly ordered than surrounding water molecules and resist disruption. The second barrier is electrostatic repulsion by phospholipid head groups, most of which carry a negative charge under physiological conditions. Together, the hydration layer and charge repulsion normally keep the surfaces of membranes that collide randomly from fusing.

Membrane fusion has been extensively studied in connection with enveloped viruses, which are sur-

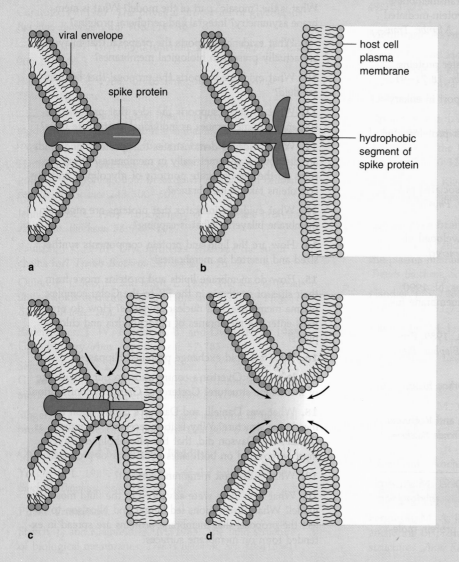

Figure 5-32 Fusion of the surface membrane of an enveloped virus with the plasma membrane of the host cell. **(a)** The spike protein in the viral envelope that recognizes and binds sites on the host cell plasma membrane. **(b)** On contact with the host cell membrane, the spike protein undergoes a conformational change; the change exposes a hydrophobic segment that becomes buried in the hydrophobic interior of the host membrane. **(c)** Lipid molecules of the outer bilayer halves of the two membranes flow and meet over the surface of the hydrophobic spike segment. **(d)** Separation and flow of the inner bilayer halves complete membrane fusion.

rounded by a surface membrane coat (see p. 23). These viruses fuse with cellular membranes during infection by one of two patterns. A few, including the measles, herpes, Sendai, and HIV viruses, attach and fuse directly with the plasma membrane (Fig. 5-32). The attachment takes place by means of *spike glycoproteins* that extend 10 to 15 nm from the viral membrane coat (Fig. 5-32a). The spike glycoproteins, which resemble segments of molecules normally reacting with the cell surface, are bound by the cell's receptors. On binding to the cell surface, the viral proteins undergo a conformational change that exposes a hydrophobic segment (Fig. 5-32b). This segment probably inserts into the hydrophobic interior of the plasma membrane and, in some as yet unknown way, promotes fusion of the host cell and viral membrane bilayers. Possibly, the hydrophobic segment of the viral spike protein provides a surface along which lipids of the outer bilayer half can flow in order to fuse (Fig. 5-32c). Once the outer bilayer halves are fused, the inner halves flow over them to form a completely fused membrane (Fig. 5-32d). Proteins able to promote membrane fusion by this or a related mechanism are known as *fusogens*.

The majority of enveloped viruses follow a different pathway in which fusion occurs inside cells, with membranes of endocytotic vesicles. The initial steps are the same—the viruses attach to surface receptors on the host cell by means of their spike proteins. Binding does not immediately stimulate conversion of the spikes to fusogens, however, because conversion of the spikes requires an acid environment. Instead, attachment to the receptors stimulates endocytosis, in which the attached viruses are taken into the cell in vesicles that invaginate from the plasma membrane (Fig. 5-33a and b; for details of the virus–receptor interaction, see p. 859). Once inside the cell, the vesicle contents typically become acidic through the action of membrane pumps that move H^+ into the vesicle. The acidic environment triggers a conformational change in the spike proteins that exposes their hydrophobic segments and converts them into fusogens. Fusion of the viral envelope and vesicle membranes then takes place (Fig. 5-33c and d), releasing the uncoated viral particle directly into the cytoplasm.

The fusion of cellular membranes in activities such as exocytosis, although less well studied than the viral mechanisms, also appears to be promoted by fusion proteins. Among these are cytoplasmic proteins collectively called *annexins*. These proteins are activated by local increases in Ca^{2+} concentration that typically occur just before membrane fusion takes place. After activation the annexins bind to the vesicle membranes, possibly through the activity of hydrophobic segments that are exposed by the activation. After binding, the steps in membrane fusion may resemble those taking place in the fusion of enveloped viruses with cellular membranes.

Ca^{2+} may also exert an effect directly on membranes, because the ion alone can promote fusion between artificial phospholipid films, with no proteins in evidence. The Ca^{2+} may neutralize the negative charges at the membrane surfaces or, as some investigators have suggested, may aggregate membrane phospholipids into "islands" that create faults in the

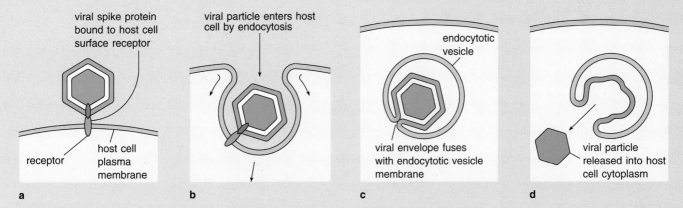

Figure 5-33 Fusion of the surface membrane of an enveloped virus with the endocytotic membrane of the host cell. **(a)** The viral particle attaches to a receptor at the host cell surface. Binding the receptor stimulates endocytosis **(b)** of the host cell plasma membrane. The invaginated pocket containing the virus subsequently pinches off as an endocytotic vesicle that sinks into the host cell cytoplasm. Within the endocytotic vesicle the change to an acid pH triggers a conformational change in the spike protein that converts it to a fusogen. **(c)** The viral envelope and the endocytotic vesicle membrane fuse, possibly by a mechanism resembling that shown in Figure 5-32. Fusion releases the uncoated viral particle into the host cell cytoplasm **(d)**.

bilayer. The faults may open regions that lack water and electrostatic barriers and promote membrane approach and contact.

Membrane fusion between vesicles entering and leaving the Golgi complex has also been investigated in some detail. Research with these membranes has revealed an extensive series of reactions during membrane fusion, including the activity of several proteins that may combine to act as fusogens (see p. 833 for details).

Suggestions for Further Reading: See p. 185.

THE FUNCTIONS OF MEMBRANES IN IONIC AND MOLECULAR TRANSPORT

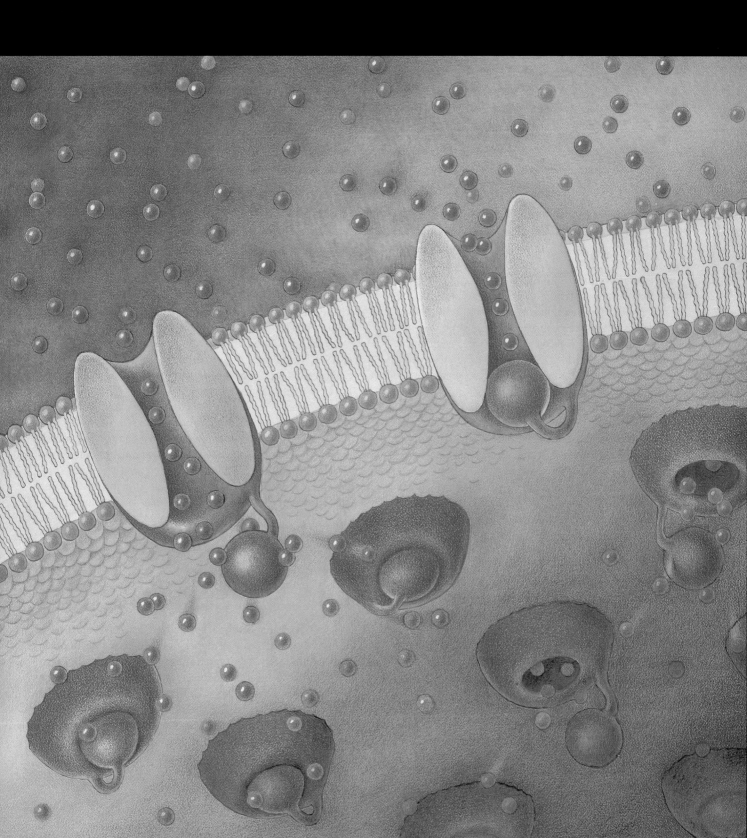

- *Passive transport* ▪ *Diffusion* ▪ *Osmosis*
- *Facilitated diffusion* ▪ *Ion channels and gating*
- *Direct active transport* ▪ *P-type and V-type*
active transport pumps ▪ *Indirect active*
transport ▪ *Membrane potentials* ▪ *Nerve*
conduction ▪ *Generation and propagation of*
the nerve impulse ▪ *Synaptic transmission*
- *Bacterial transport*

Cellular life depends on the organization of molecules inside cells. Any severe disturbance in the concentrations of substances inside cells or in the kinds of molecules present will impair function or lead to cell death. The internal concentrations of molecules and ions are maintained in their correct numbers and kinds by membranes, which regulate the passage of all substances moving in and out of cells and among their membrane-bound interior compartments.

As part of this maintenance, substances move constantly in both directions across cellular membranes. Metabolites, including all necessary fuel substances and raw materials, enter the cell from the outside, and waste materials and cell secretions exit in the opposite direction. Ions flow constantly in both directions and between the different compartments in the cell interior.

Two primary mechanisms underlie ionic and molecular transport. One, *passive transport*, depends simply on differences in the concentration of substances inside and outside cells. If molecules are more concentrated outside, the direction of movement is from outside to inside. If concentration is higher inside, movement is in the opposite direction. The difference between inside and outside concentrations, which drives the movement of the molecules in either direction, is called a *concentration gradient*. Passive transport of ions is influenced by charge differences as well as concentration differences on either side of a membrane. Because passive transport depends on concentration gradients, it requires no expenditure of cellular energy.

In the second mechanism, *active transport*, substances move *against* their concentration gradients. Unlike passive transport, active transport requires cells to expend energy; it stops if the reactions supplying energy for cellular activities are experimentally inhibited. Both molecules and ions are transported actively.

The rates at which molecules and ions are transported reflect the properties of both the lipid and protein parts of membranes. Transport through the lipid component of membranes, which is strictly passive, is influenced by the hydrophobic, nonpolar character of the lipid component. In general nonpolar molecules diffuse passively through the lipids of membranes much more readily than polar or charged substances. The only exception to this pattern is water, which passes rapidly through the lipid component even though it is strongly polar.

Transport by membrane proteins primarily involves movement of ions and polar molecules. Some of this transport is passive, driven by concentration gradients. Other protein-mediated movement of polar substances is active and requires the expenditure of cellular energy. Transport by membrane proteins has the additional characteristic of *specificity:* in general membrane proteins transport only certain types or classes of ions and polar molecules.

For most cells the great majority of transported substances are specifically selected hydrophilic molecules and ions that move through membranes via transport proteins. Relatively few nonpolar molecules pass through the lipid part of membranes; the nonpolar component acts primarily as a seal that prevents passage of hydrophilic substances for which no selective protein carriers exist in a particular membrane type.

Because ions carry a charge, the passive and active movement of these substances can produce significant electrical effects on either side of a membrane. These electrical effects provide the basis for many important cellular functions, including the activities of nerve and brain cells in communication. The currents generated by ion movements across membranes can be surprisingly high. In the electric organs of fish such as the electric eel, the currents are strong enough to stun or kill animals as large as humans.

Not all movement of substances between the cell's interior and exterior depends on transport directly across cellular membranes. Some substances, particularly large molecules or molecular complexes, are moved instead in vesicles that form from or fuse with the plasma membrane. Inward movement by this mechanism is termed *endocytosis,* and outward movement is *exocytosis* (see Fig. 1-15). (Ch. 20 discusses in detail endocytosis and exocytosis, and the internal routes followed by materials taken or released by these pathways.)

PASSIVE TRANSPORT

Diffusion as the Basis for Passive Transport

Passive transport is a specialized form of *diffusion*, a physical process that depends on the constant motion of molecules at temperatures above absolute zero ($-273°C$; see p. 86). Molecules confined within a space travel in straight lines until they collide with other molecules or with the boundaries of the space. After colliding, the molecules rebound and move in a direction dependent on the angles of the collisions. Each molecule has a definite amount of energy, which depends on the velocity of its movement. This energy of movement, called *kinetic energy*, can be expressed in average terms for the entire collection, even though individual molecules may possess different amounts of energy.

Kinetic motion may result in diffusion—that is, a net movement of molecules from one region to another—if two spaces in communication have different initial *concentrations* of molecules (concentration = number of molecules per unit volume). Consider two spaces of equal volume that are initially separated by a barrier that molecules cannot pass. The absolute temperature of the spaces is the same, but one of them contains more molecules. The space with the higher concentration of molecules contains a greater amount of kinetic energy because it includes more moving particles with mass and energy.

If the barrier between the two compartments is removed, the movement and collisions on the more concentrated side propel the molecules to the other side. Molecules also move from the less concentrated side into the more concentrated side, but over any interval of time there are more collisions and movement from the more concentrated side. As a result, there is a net movement of molecules from the side of greater concentration to the side of lesser concentration. This net movement or diffusion in response to concentration differences continues until the molecules are evenly distributed throughout the available space. Molecules continue to move from one space to the other even after the concentration is the same on both sides. However, no net increase on either side occurs after the distribution becomes uniform.

The net movement occurs because the unequal distribution of molecules before the barrier is removed represents a more ordered state than the uniform distribution in the final state. After diffusion takes place, the molecules are evenly distributed throughout the available space and have reached a condition of minimum energy and maximum entropy or disorder. As the molecules diffuse to the evenly distributed state, they release energy to their surroundings. This energy is *free energy* (see p. 83) and can accomplish work. The energy released as substances run down concentration gradients is actually used to accomplish work in cells—in fact, in most cells the primary source of energy for ATP production is derived from concentration gradients (for details, see Ch. 9).

The free energy released by the movement of a mole of molecules down a concentration gradient can be calculated from a modification of the equation used to determine free energy changes occurring in chemical reactions:

$$\text{free energy} = \Delta G = -RT \ln C_1/C_2 \qquad (6\text{-}1)$$

in which R is the gas constant, T is the absolute temperature, and C_1 and C_2 are the high and low concentrations of the gradient. At a temperature of 25°C (absolute temperature = 298 Kelvin) and concentrations of 0.1 M and 0.0001 M,

$$\Delta G = -[1.98 \text{ cal/(degree} \cdot \text{mol)}](298) \ln (0.1/0.0001)$$
$$= -(590)(\ln 1000)$$
$$= -(590)(6.908)$$
$$= -4075 \text{ cal/mol}$$

The free energy released as the mole of molecules travels down the gradient, -4075 cal/mol, exceeds the energy available from hydrolysis of a mole of ATP at 25°C, which would be about -3000 cal/mol. The concentration difference used for this calculation, which amounts to 1000 times, would not be unusual for biological systems.

The movement of charged particles such as ions in response to concentration gradients is modified by electrical attractions and repulsions if spaces in communication initially differ in the number of positive and negative charges. In response to charge differences, net movement tends to proceed spontaneously toward a condition in which all parts of the space contain the same number of positive and negative charges, and the space is thus electrically neutral. The free energy made available by such systems is called *electrical potential*, or *voltage*, and can also do work. The work of the nervous system, for example, is powered by free energy released by gradients of electrical charge.

The Effects of Semipermeable Membranes on Diffusion

An artificial or natural membrane placed between two regions containing molecules at different concentrations has no effect on the final outcome of diffusion if all the molecules or ions can pass through the membrane with equal ease. However, the net movement may be altered, sometimes in unexpected ways, if some molecules or ions pass through the membrane less readily or are excluded entirely. Membranes having this effect, which include some artificial and all biological membranes, are said to be *semipermeable*.

Semipermeable Membranes and Osmosis The presence of a molecule that cannot diffuse across a semipermeable membrane affects the movement of water molecules in an unexpected way, producing the phenomenon of *osmosis*. Consider two spaces of equal volume separated by a semipermeable membrane in which one side contains pure water and the other a solution of protein molecules in water (Fig. 6-1). The pores in the semipermeable membrane are large enough to admit water molecules but too small to allow the protein

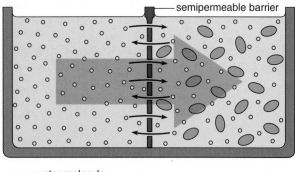

o = water molecule
⬭ = protein molecule

Figure 6-1 Osmotic flow of water in a system in which a semipermeable barrier separates two compartments of equal volume, with pure water on the left and a solution of protein molecules in water on the right. Although the water molecules can move freely through the barrier in either direction, the protein molecules cannot pass. On the right some of the available space is taken up by protein molecules. As a result, there are fewer water molecules per unit of volume in the right compartment than in the left, and therefore a concentration gradient for water exists between the two compartments. In response a net movement of water molecules occurs from left to right (large arrow).

molecules to pass. It is obvious that a concentration gradient exists for the protein molecules; however, no net movement of the protein can occur because of the semipermeable barrier. It is less obvious that there is also a concentration gradient for water molecules. This gradient exists because some of the available space on one side is taken up by the protein molecules. Therefore, there are fewer water molecules per unit volume on the side containing the protein. In response, a net movement of water occurs from the side containing only water molecules to the side containing water and protein molecules. Osmosis is the net movement of water molecules in response to a gradient of this type. Because osmosis occurs in response to a concentration gradient, it releases free energy and can accomplish work.

Osmosis is often demonstrated with an apparatus that consists of an inverted thistle tube closed at its lower end by a sheet of cellophane (Fig. 6-2). Inside the tube is a sucrose solution in water; the tube is suspended in a beaker of distilled water. The cellophane film acts as a semipermeable barrier—its pores are large enough to allow water molecules but not sucrose to pass in either direction. A concentration gradient exists for water molecules in this system because some of the available space for water is taken up by the sucrose molecules in the tube. After a short time the solution in the tube rises as water passes across the cellophane

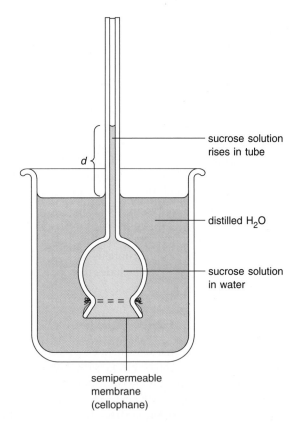

Figure 6-2 An apparatus demonstrating osmotic flow and pressure. The beaker contains pure water, and the inverted thistle tube contains a solution of sucrose in water. The cellophane film covering the bottom of the thistle tube allows water molecules, but not sucrose molecules, to pass in either direction. The level of solution in the tube will rise the distance *d* because of the net movement of water molecules from the beaker into the thistle tube. Osmotic flow continues until the weight of the water column *d* develops sufficient pressure to counterbalance the osmotic movement of water molecules.

"membrane" from the beaker in response to its concentration gradient. The level of the solution continues to rise until the pressure created by the weight of the raised solution exactly balances the tendency of water molecules to move from the beaker into the tube. At this point the system is in balance, and although water molecules still move in both directions across the cellophane membrane, no further net movement of water occurs. The work accomplished by osmosis in the apparatus is equivalent to the height reached by the solution rising in the tube. **The pressure required to counterbalance exactly the tendency of water molecules to move into the tube is the *osmotic pressure* of the solution in the tube.**

The amount of osmotic pressure in a system such as the one demonstrated by the thistle tube apparatus can be calculated from a simple relationship. For a system in which a semipermeable membrane separates pure water from a dilute solution of a solute that cannot pass across the membrane, the pressure developed in the solution in atmospheres is given by the *van't Hoff equation*:

$$\text{osmotic pressure} = \pi = RTC$$

in which R is the gas constant, T the absolute temperature, and C the concentration of the solute in moles (or the total concentration of all nonpermeable solutes on the other side). Conversely, the osmotic pressure developed by a solution can be used to determine the concentration of its solutes in moles or even to approximate the molecular weight of a solute.

Osmosis in Living Cells   Cells act as osmotic devices similar to the apparatus shown in Figure 6-2 because they contain solutions of proteins and other molecules that are retained inside by a membrane impermeable to them but freely permeable to water. The resulting osmotic movement of water into cells produces a force that operates constantly in living cells. This force may be used as an energy source for some of the activities of life, or it may be a disturbance that must be counteracted for survival.

The root cells of most land plants, for example, contain proteins in solution but are surrounded by almost pure water. As a result, water flows into the root cells by osmosis. The pressure developed in response contributes part of the force required to raise water into the stems and leaves of the plant. In stems and leaves the tendency of water to move into cells develops pressure that pushes the cells tightly against their walls. The pressure supports the softer tissues of stems and leaves against the force of gravity. Bacterial and cyanobacterial cells are also kept tightly pressed against their walls by osmotic pressure.

To keep from bursting, cells without rigid walls, such as those of protozoa and other small freshwater organisms, must expend considerable energy to excrete the water constantly entering by osmosis. Organisms living in surroundings that contain highly concentrated salt solutions have the opposite problem and must constantly expend energy to replace water lost by osmosis. Within the bodies of many-celled animals, ions, proteins, and other molecules occur in significant concentrations in the extracellular fluids as well as inside cells, so that the concentration of water inside and outside cells is more closely balanced. However, the concentration difference is great enough that animal cells must still expend considerable energy to counteract the effects of the inward movement of water by osmosis. (In animals osmotic pressure is reduced by the active transport of Na^+ out of cells; see below.)

Semipermeable Membranes and Movement of Charged Particles   The presence of a semipermeable barrier can also have unexpected effects on the movement of charged particles if not all the particles can move across the barrier with equal ease. Consider again two spaces separated by a semipermeable barrier. Water fills the two spaces, and a collection of negatively charged proteins is held on one side by the barrier. If a salt such as NaCl is added to either side, the Na^+ and Cl^- ions will tend to diffuse until their concentrations are the same on both sides of the barrier. However, reaching a uniform distribution of ions would produce more total negative charges on the side containing both Cl^- and proteins. As a result, the buildup of negative charge on the side containing the proteins opposes the tendency of the system to run toward an even distribution of particles. When the two sources of net movement are in opposition, as they are in this example, the final ion concentration is a balance between the tendency of ions and molecules to diffuse toward an even distribution and electrical neutrality. Consequently, neither electrical neutrality nor uniform distribution of particles is achieved in such systems—both a charge difference and a concentration gradient remain for the ions in the final state.

Biological Membranes and Passive Transport: Facilitated Diffusion

Biological membranes have complex effects on passive transport by diffusion because they are structured as a mosaic of regions with distinct hydrophobic and hydrophilic properties. As a consequence, polarity significantly affects the ability of molecules to pass through biological membranes by passive diffusion. The ability of membrane proteins to specifically admit some polar molecules and exclude others also greatly modifies the responses of biological membranes to concentration gradients.

The effects of biological membranes on diffusion were first studied at the turn of the century by Ernst Overton, who measured the rate at which different substances penetrate into plant and animal cells. Among the substances under study, Overton observed that molecules soluble in lipid solvents penetrated into cells more rapidly than water-soluble molecules, up to a limit determined by molecular size. This behavior contrasted with the passage of molecules across artificial barriers such as cellophane, in which the relative degree of water solubility had no noticeable effect. Overton's work provided the first clue that cells are surrounded by a surface layer that is lipidlike in nature (see p. 181).

Later work confirmed that many substances penetrate across biological membranes according to lipid solubility. However, certain polar molecules were found to enter many cells much more rapidly than expected according to their solubility in lipids.[1] Several important metabolites, such as glucose and amino acids, were among the polar molecules penetrating much more rapidly than expected. Ions such as Na^+ were also found to penetrate rapidly across some natural membranes in spite of their hydrophilic nature. This passive transport of substances at rates higher than predicted from their lipid solubility is termed *facilitated diffusion* (see Fig. 6-4).

Proteins and Facilitated Diffusion Proteins were directly implicated in facilitated diffusion by comparisons of natural membranes with artificial membranes consisting of pure phospholipid films (see p. 163). When pure phospholipid films were used as membranes, all molecules except water penetrated strictly according to lipid solubility and molecular size. (The exceptional behavior of water is discussed further below.) Ions were essentially excluded from passage, and only a very few polar substances occurring naturally in cells, such as ethyl alcohol, urea, and glycerol, were soluble enough in the phospholipid films to be able to penetrate to any extent. Addition of membrane proteins, however, frequently allowed many other polar and charged substances to penetrate at rates comparable to those of natural membranes.

Cells control the particular group of polar and charged molecules and ions passing through their membranes by regulating the types of transport proteins that are synthesized in the ER, modified in the Golgi complex, and placed in cellular membranes (see pp. 831 and 836). As a result, each cell type has its own spectrum of plasma membrane proteins passing a characteristic group of hydrophilic substances by facilitated diffusion (or by active transport; see below). The membranes surrounding interior compartments, such as those of mitochondria and chloroplasts, also have specific protein channels that control the inward and outward flow of polar and charged substances.

The transport proteins of natural membranes are integral membrane proteins that completely span the lipid bilayer (Fig. 6-3). The hydrophilic channels in these proteins are set up by an arrangement of amino acid chains that opens the channels and lines their sides with

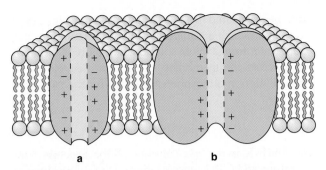

Figure 6-3 Formation of a polar or charged channel through the interior of a single membrane protein **(a)** or by alignment of several membrane proteins **(b)**.

polar groups. The channels may be formed by a polar opening through a single protein (Fig. 6-3*a*) or by several proteins that combine to form a channel between them (Fig. 6-3*b*).

The Characteristics of Facilitated Diffusion Facilitated diffusion has several important characteristics (see Table 6-1). Although membrane proteins may greatly enhance the diffusion of polar molecules, the energy required for facilitated diffusion is provided only by a favorable concentration gradient, and the process stops if the gradient falls to zero. Because facilitated diffusion depends entirely on favorable concentration gradients, it is passive and requires no direct expenditure of cellular energy.

Comparisons of the rate of facilitated diffusion and

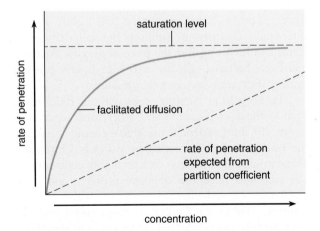

Figure 6-4 Facilitated diffusion. At low concentrations substances penetrate through natural membranes by facilitated diffusion (solid line) at rates much higher than expected from the partition coefficient (dashed line). At increasingly higher concentrations penetration gradually drops off until it reaches a maximum rate that does not respond to further increases in concentration. At this rate the facilitated diffusion mechanism is saturated.

---

[1] Lipid solubility can be evaluated by determining the lipid–water *partition coefficient* for a substance:

$$\text{partition coefficient} = \frac{\text{amount dissolving in a lipid}}{\text{amount dissolving in water}}$$

The substance under investigation is shaken in a mixture of water and oil or a lipid solvent such as benzene. The mixture is allowed to stand until the lipid and water separate into distinct layers. The amount of the substance dissolved in each layer is then determined, and the ratio is used to calculate the partition coefficient.

the concentrations of the transported molecules reveal another fundamental characteristic of the mechanism (Fig. 6-4). At successively higher concentrations the degree of enhancement drops off until at some point the mechanism becomes *saturated:* further increases in concentration cause no further rise in the rate of penetration. This saturation is in sharp contrast to the behavior of nonpolar molecules, which penetrate across the phospholipid portion of membranes by simple diffusion according to lipid solubility. For these molecules the rate of penetration across membranes is directly proportional to concentration, with no marked dropoff at high concentrations.

The saturation of facilitated diffusion at elevated concentrations closely resembles the behavior of enzymes in catalyzing biochemical reactions (see p. 95 and Fig. 3-9). In enzyme-catalyzed reactions the enzyme gradually becomes saturated as substrate concentration increases, and the rate of the reaction levels off. The similarity in behavior between the two systems indicates that facilitated diffusion is carried out by membrane proteins with properties similar to those of enzymes.

Another similarity between facilitated diffusion transporters and enzymes is the property of specificity: each molecule transported by facilitated diffusion is carried by a separate protein specific only for that substance or a group of closely related substances. For example, the protein facilitating the transport of glucose will also transport the closely related sugars mannose, galactose, xylose, and arabinose. However, it will carry only the naturally occurring D-isomers and not the L-

isomers of these sugars (see Ch. 2). The specificity of facilitated diffusion further resembles the specificity of enzymes in that substances with structures closely related to the normally transported molecule can inhibit the rate of transport of that molecule.

Like enzymes, the transport proteins carrying out facilitated diffusion have been demonstrated to combine briefly with their transported substances. The facilitated diffusion transporters combine with molecules or ions on the side of the membrane of higher concentration and release them on the side of lower concentration. The transporters have also been demonstrated to undergo conformational changes in restricted regions of the protein. These changes alternate the protein between two conformations in which the binding site for the transported molecule faces either the inside or outside surface of the membrane. In some systems the binding sites have been shown to have significantly different affinities for the transported substance in the two conformations.

How Facilitated Diffusion Transporters Are Believed to Operate   The molecular basis for facilitated diffusion is not known. However, the characteristics of facilitated diffusion transporters (summarized in Table 6-1) and the general properties of integral membrane proteins have provided the basis for a hypothesis, the *alternating conformation model,* proposed by S. J. Singer and others for the mechanism by which transport proteins carry out facilitated diffusion (Fig. 6-5). According to the model, a transport protein shifts between two conformations that direct the binding site of the protein al-

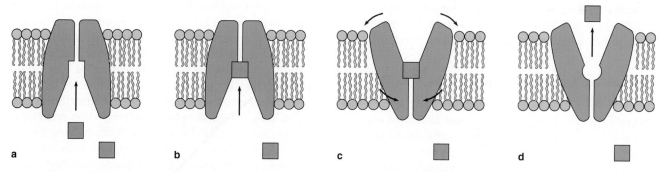

Figure 6-5   Facilitated diffusion, according to the alternating conformation model. **(a)** At the beginning of the cycle the protein is folded so that the site binding the transported substance faces the side of higher concentration. The binding site is in a high-affinity state in which it binds strongly to molecules of the transported substance that collide randomly with the transport protein **(b)**. Binding causes the protein to shift to the alternate folding conformation in which the binding site faces the side of lower concentration **(c)**. The change in conformation also alters the binding site to a low-affinity state in which it binds the transported molecule relatively weakly. The weak binding facilitates release of the transported substance to the medium **(d)**. On releasing the transported molecule, the protein returns to the conformation in **(a)**. It is now ready to begin another cycle of facilitated diffusion.

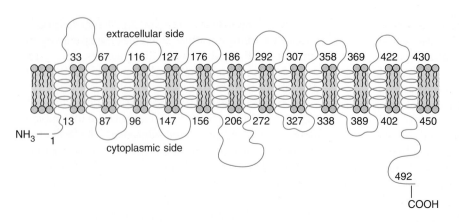

Figure 6-6 The distribution of transmembrane segments and surface loops in the glucose transporter of erythrocyte and other plasma membranes. The numbers above and below the membrane indicate the amino acid positions at which the transmembrane alpha helices enter and leave the membrane. The diagram is two dimensional; the three-dimensional pattern into which the helices fold to form the channel across the plasma membrane remains unknown. From data in M. Mueckler et al., *Science* 229:941 (1985).

ternately toward the inside and outside of the membrane. In the initial step of the process, the protein is folded so that the site binding the transported molecule is exposed toward the membrane surface facing the region of higher concentration (Fig. 6-5a). In this position the binding site is in a *high-affinity state*, in which it binds strongly to the transported molecule. Collisions resulting from kinetic movements cause the transported substance to bind to the transporter (Fig. 6-5b). Binding causes a change in the folding of the transporter protein, so that it shifts to the alternate conformation in which the binding site is exposed to the membrane surface facing the region of lower concentration (Fig. 6-5c). The conformational change also alters the active site to a *low-affinity state*, in which it binds the transported molecule relatively weakly. As a result, the transported substance is readily released to the surrounding medium (Fig. 6-5d). The release returns the protein to its original folding conformation, with the binding site facing the opposite membrane surface again (as in Fig. 6-5a), ready to initiate another cycle of facilitated diffusion. (See also S. J. Singer's essay on p. 174 describing his experiment showing that membrane proteins do not rotate or flip during transport.)

Transporter Types Operating in Facilitated Diffusion

Several types of transporters carry out facilitated diffusion in biological membranes. One type, found primarily in animal cells, is specialized for the facilitated transport of relatively small organic molecules such as glucose and other hexose sugars (glucose is also actively transported; see below). A second animal transport protein, called an *anion carrier*, facilitates the diffusion of negatively charged inorganic groups such as carbonate or phosphate. A third type forms channels admitting ions such as Na^+, K^+, and Ca^{2+}. Most of these ion channels, which occur in both plant and animal cells, are *gated*; that is, they are controllable and exist in open or closed states. Gated ion channels are particularly important in animals because they provide the basis

for such fundamental processes as conduction of impulses along and between nerve cells.

All facilitated diffusion transporters share several structural characteristics. All are integral membrane proteins containing several to many alpha-helical segments that zigzag back and forth across the membrane. Most, particularly those active in plasma membranes, are glycoproteins with carbohydrate groups directed toward the cell exterior.

A Carrier of Organic Molecules: The Glucose Transporter The plasma membranes of most vertebrate cell types contain proteins transporting glucose and other six-carbon sugars, including mannose, galactose, xylose, arabinose, fucose, and rhamnose, by facilitated diffusion. Mammals have at least five different forms of the glucose transporter, encoded in separate genes.

M. Mueckler and his coworkers and others isolated and sequenced the genes coding for the glucose transporters in human liver and rat brain cells. The amino acid sequence of the protein, deduced from the gene sequence, reveals that this transport protein probably has no less than 12 hydrophobic segments long enough to span the membrane when wound into alpha helices. This structure indicates that the protein zigzags back and forth across the plasma membrane 12 times, with the membrane-spanning segments connected by loops extending into the cytoplasm or the extracellular space (Fig. 6-6). The shift between alternate conformational states may involve the ninth and tenth transmembrane segments in Figure 6-6; these segments undergo shifts in position that may open the glucose-binding site to one side or the other of the membrane. Transmembrane segments 3, 5, 7, 8, and 11 form amphipathic alpha helices; these segments may combine to form a polar pore extending through the glucose transporter.

The amino acid sequence of the vertebrate glucose transporter is related to the sequences of several proteins that transport sugars and other molecules in bacteria. A human glucose transporter (470 amino acid residues) and the proteins transporting arabinose and xylose in *E. coli* (about 580 residues), for example, have identical amino acid residues at 75 positions, and conservatively substituted residues at another 121 positions. (Conservative substitution means that one amino acid has been replaced by another of similar size and chemical properties.) In the glucose transporter of the yeast *Saccharomyces cerevisiae* (884 residues), the amino acids at positions 86 to 581 are about 30% identical to those of both the bacterial and mammalian glucose transporters. These observations suggest that bacterial, yeast, and mammalian types are members of an ancient family of transport proteins descended from a single ancestral type. Interestingly, the bacterial proteins related to mammalian and yeast glucose transporters are active transport proteins, indicating that the capacity for active transport was lost as this family of transport proteins evolved in early eukaryotes.

Facilitated diffusion of glucose is possible for most animal cells because glucose is present in higher concentrations in the circulation than inside cells. In mammals, glucose is about seven times more concentrated outside than inside cells. The concentration difference is maintained partly by a biochemical "trick" in which glucose is quickly phosphorylated to glucose 6-phosphate after it enters the cytoplasm. Because glucose 6-phosphate cannot pass backward through the glucose transporter (phosphorylated molecules generally are unable to pass through biological membranes), the phosphorylation effectively removes the glucose entering cells from the concentration inside. Cells also use the phosphorylation trick to effectively lower the inside concentrations of other molecules moved inward by facilitated diffusion.

Some facilitated glucose transporters, such as those in muscle and fat cells, are regulated by the hormone *insulin.* When secreted into the circulation in response to elevated blood glucose levels, the hormone stimulates muscle and fat cells to insert more glucose transporters in the plasma membrane. The glucose transporters of most other body cells in mammals, such as those of the brain and liver, are unaffected by insulin; they are present in plasma membranes in the same numbers whether the hormone is present in the bloodstream or not. In one common form of diabetes, mutations alter the sequences of transmembrane segments of the glucose transporter, inhibiting or blocking its ability to transport glucose molecules into cells. Glucose accumulates in the bloodstream as a result.

Certain vertebrate cells, such as those of the small intestine and kidney, contain proteins capable of actively transporting glucose against its concentration gradient (see below). These active transporters move glucose from the kidney filtrate back into the bloodstream or from the intestinal contents into the circulation. The active glucose transporters in these cells appear to be unrelated in sequence to the transport protein moving glucose by facilitated diffusion.

The transporter carrying glucose by facilitated diffusion has been extracted, purified, and added to artificial phospholipid bilayers (see p. 163). When the transporter is inserted, artificial bilayers are able to carry out the facilitated diffusion of glucose by essentially the same patterns as natural membranes.

Types of Ion Channels Transport proteins forming channels for the facilitated diffusion of ions are widely distributed among eukaryotes. Although almost all these channels conduct positively charged ions such as Na^+, K^+, or Ca^{2+}, a channel conducting Cl^- has been detected and is believed to be widely spread. The channels are not completely specific for single ions. Instead, they are selective—the Ca^{2+} channel, for example, also admits Na^+ but passes it about 1000 times more slowly than Ca^{2+}.

The mechanism by which ion channels admit one ion freely and restrict the passage of others is not understood. Part of the selectivity possibly depends on binding sites that recognize and bind only a single ion or group of related ions. If this is the case, the binding sites would have to accept and release ions at very high rates because ion channels conduct ions at rates approaching a million particles per second when fully open.

The shape of the polar channel extending through the ion transporters may also contribute to the selectivity. The ion channels investigated in detail have proved to be shaped like a double funnel, with a wider opening at either membrane surface leading to a narrow, constricted region at the channel midpoint (Fig. 6-7).

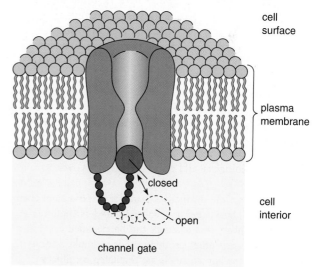

Figure 6-7 Structure of a gated ion channel. The narrow midpoint in the channel may operate as a selective filter that admits only ions of the type transported by the channel. Conformational changes triggered by voltage changes, ligand binding, or mechanical stress open the gate to admit ion flow. In most channels the gate is on the cytoplasmic side of the membrane as shown.

The Experimental Process

Identification of the Amino Acids Forming the Conduction Pathway in Potassium Channels

Hali A. Hartmann and Arthur M. Brown

HALI A. HARTMANN received her Ph.D. in biomedical sciences from the University of Hawaii in 1986. Her post-doctoral training has included electrophysiology of calcium channels and the nicotinic receptor-channel complex in other labs, and the molecular biology of potassium channels in Dr. Arthur M. Brown's lab. She is presently Research Assistant Professor in the Department of Molecular Physiology and Biophysics, Baylor College of Medicine. ARTHUR M. BROWN received his M.D. from the University of Manitoba in 1956 and his Ph.D. from the University of London in 1964. He is presently Chairman of the Department of Molecular Physiology and Biophysics at Baylor College of Medicine. This work on the chimeric K⁺ channel was done in collaboration with John A. Drewe, Rolf H. Joho, Glenn E. Kirsch, and Maurizio Taglialatela at Baylor College of Medicine.[1]

Ion channels are proteins that selectively conduct a single ion species through external membranes at the surface of cells or internal membranes. This is an amazing capability, when one considers the minute differences in size between the primary conducted ions—potassium, sodium, and calcium ions. The radius of a potassium ion is only 1.33A compared to 0.93A for a sodium ion, yet the protein differentiates between them. A second amazing capability is that these proteins conduct ions at extremely fast rates of approximately 10^6 ions/sec. These properties of ion selectivity and fast transport, which characterize all ion channels, have puzzled cell and molecular biologists, electrophysiologists, and other scientists for over 50 years.

Although the distinctive character of potassium channels, both ligand-gated and voltage-gated, is the ion conduction pathway, the channels can be classified according to some of their other properties. For example, various types of potassium channels are regulated by different intracellular macromolecules such as ATP and G proteins. Also, K⁺ channels have varying sensitivities to certain macromolecules which block K⁺ ions or decrease the rate at which they flow through their pores. Some K⁺ channels have an external site for blockage by these agents

while others have an internal site. This suggests structural differences in the proteins. The kinetics of the processes by which the channel opens and closes also differ between K⁺ channels. Finally, voltage-gated K⁺ channels can be activated at different voltages during the action potential. The DNA encoding some of these different types of K⁺ channels had already been isolated and sequenced, largely from the *shaker* gene of *D. melanogaster* and from rat brain tissue.

In Mackinnon and Miller's labs in Boston, the K⁺ channel cloned from the *shaker* gene of *Drosophila* was found to be sensitive to a scorpion venom toxin, charybdotoxin.[2] After the K⁺ channels were exposed to the charybdotoxin, K⁺ ions were still conducted through the channel, but at a significantly reduced rate. From these results, it was concluded that the toxin must be binding to sites in close proximity to the part of the protein forming the actual ion conduction pathway or pore. These researchers were then able to detect which amino acids had a role in binding the toxin using site-directed mutagenesis [see p. 91]. These experiments gave one of the first clues about the location of the channel pore.

Tetraethylammonium (TEA) was known to block some K⁺ channels in a manner similar to charybdotoxin. However, certain K⁺ channels were sensitive to TEA from the extracellular side, while others were more sensitive from the intracellular side. One type of K⁺ channel in our labs at Baylor, NGK2, had been cloned from a hybrid cell line of mouse and rat neuroblastomaglioma cells. It had high sensitivity to extracellular TEA, but low sensitivity to intracellular TEA. A second channel cloned in our labs from rat brain, DRK1, was more sensitive to intracellular TEA and less sensitive to extracellular TEA. Another difference between these two proteins was the rate of ion conductance through a single channel. NGK2 conducted K⁺ ions at 2.5 times the rate of DRK1, 25 pS (picosiemens) versus 10 pS, respectively.

These distinct channel properties presented an ideal experimental design to locate the pore of voltage-gated K⁺ channels. By swapping sections of the DNA in DRK1 with corresponding sections from NGK2, the hybrid, or chimeric, channel should have the pore properties of the donor channel. This chimeric channel could only work if the amino acids maintained a similar tertiary and quaternary structure to form a functional ion channel protein. If DRK1 could be converted into an NGK2-like channel, i.e., with a 25 pS K$^+$ conductivity rate, high sensitivity to external TEA, and low sensitivity to intracellular TEA, then we would know which parts of the DNA composed the actual pore segment. Thus, the chimeric channel would consist largely of DRK1's DNA, with only small transplanted sections of NGK2's DNA.

Fortunately, since the primary sequence of K$^+$ channel proteins had already been known for several years, computer modelers had already developed potential structures for the proteins. By aligning the primary sequences of all the known K$^+$ channels to date, B. L. Tempel and colleagues,[3] H. R. Guy and P. Seetharamulu,[4] and R. E. Greenblatt and coworkers[5] had actually hypothesized that a specific region between two of the hydrophobic transmembrane segments formed the ion-conducting pore. Now we had the tools to test this prediction and hypotheses about other areas of the cDNA that may have important functional roles in the protein.

One of the more straightforward ways to exchange DNA from one cDNA to another is through the cassette method. Using bacterial restriction endonucleases [see p. 135], specific fragments of DNA can be cut out of the cDNA molecule. Unfortunately, there may not be restriction sites flanking both sides of the cDNA segment one might want to cut out and swap with another cassette. In fact, we encountered this problem in our experimental design. We already had restriction maps of the sites where the endonucleases would cut DRK1 cDNA and the plasmid vector pBluescript, which houses DRK1. It was evident that we would have to engineer DRK1's so that it could be recognized and cut by the restriction enzyme but would still code for the same amino acid as the original codon. These so-called silent restriction sites were engineered into the cDNA of DRK1 by site-directed mutagenesis. The DNA was sequenced to confirm that the mutations were successful. It was also digested with the respective restriction enzymes, and the identification of the correct size fragment also confirmed that the new sites had been correctly added.

A second, more laborious part of the experiments concerned the preparation of NGK2's DNA to be inserted into the corresponding area of DRK1. To be certain, although there were some very similar areas in their primary sequences, these two cDNAs did not have similar restriction maps. The regions of NGK2 DNA that needed to be transferred into DRK1 were not flanked by the same restriction sites as those we had engineered into DRK1 cDNA; therefore, the segment of NGK2's DNA could not be ligated into DRK1. To circumvent this technical problem, we used the polymerase chain reaction [PCR; see p. 137] to engineer NGK2's DNA so it could be ligated into DRK1 at the same restriction sites without a frameshift occurring in the codon.

The forward and reverse primers for the PCR were designed with the codons for the restriction sites followed by approximately 8 to 10 bases of the NGK2 sequence.

Directly before the DNA sequence where the restriction enzymes would recognize their active site, about 6 to 8 extra bases of NGK2's sequence were included in the primers. These bases were included so there would be extra DNA on the other side of the recognition site for stable attachment of the restriction enzyme during digestion of the PCR fragments. The same compatible ends on the PCR fragment and the cut vector were absolutely essential for a correct ligation. Hopefully, with all the primers correctly designed and the optimum conditions calculated for each set of primers, the PCR would work perfectly the first time for all the different segments of NGK2's DNA we wanted to transplant into DRK1.

Two of the reactions succeeded the first time. Since we knew what the PCR fragments should be, aliquots of the reactions could be run on agarose gels. We could crudely assess whether a successful reaction had occurred by running alongside these aliquots a suitable DNA ladder with bands in the ranges of the expected PCR fragments. Although we did not know at the time, the band representing the fragment of NGK2's DNA that would transfer its pore phenotype to DRK1 was correctly made by the PCR the very first time!

Before ligation of the PCR fragments into the cut vector of DRK1, the fragments also had to be cut with the restriction enzymes to provide the same sites to ligate into DRK1. A ligation would be the final step to confirm that Taq polymerase had correctly amplified the primers and NGK2's DNA in the PCR. The reactions that did amplify the fragment of the correct length successfully ligated into their respective vectors. The new chimeric cDNAs were amplified in E. coli XL1-blue cells and the amplified DNA was purified. The last and definitive test to assure that the different chimeras were exactly correct was a good autoradiogram of the sequence of the ligated segment of NGK2's DNA. The sequencing autoradiogram waved the go-ahead flag and we prepared in vitro cRNA of the different chimeras of DRK1 and NGK2 to inject into oocytes.

It was a great accomplishment for biologists, physiologists, and computer protein-structure modelers that one of the NGK2's segments transplanted into DRK1 had changed the phenotype of DRK1. The electrophysiologists measured not only NGK2's conductance rate of 25 pS in one of the chimeric channels in the oocyte membrane, but also found that its sensitivity to external TEA had decreased ten-fold, essentially similar to that of NGK2. The transferred segment changed something on the inside of the transmembrane protein as well. DRK1 now was sensitive to internal TEA, as is NGK2.

However, the chance that what we were measuring was due to a mutation or change in the cDNA of DRK1 other than what we had engineered was a possibility. The DNA of the chimeric channels had only been sequenced at the ligation sites of the PCR fragments; and a change in the primary sequence, remote from the areas of ligation, could certainly affect protein structure and function. There were two ways we could establish that the transplanted cassette was responsible for the new phenotype of DRK1. One way would be to sequence the entire cDNA of the chimeric channel. This approach is labor-intensive due to the length of the cDNA, and we had already expended a great deal of time and effort on lengthy preparation techniques. The second method was to cut out NGK2's segment and replace it with the original DRK1 cassette in the same construct of DNA. The phenotype should revert back and demonstrate the functions characteristic of wild-type DRK1. This revertant experiment was a more clear and decisive approach, and it proved to be a convincing control. The pore properties of conduction and sensitivity to blockade by both external and internal TEA were reverted back to those of DRK1 by replacing the section of NGK2's DNA that was inserted into DRK1 with the original cassette of DRK1 DNA. The segment exchanged between DRK1 and NGK2 was only 21 amino acids, yet pore properties of both the intracellular and extracellular membrane were affected.

What did we really learn from all these results? Our chimeric K^+ channel and the complementary work of other labs defined a segment of DNA whose corresponding amino acid sequence formed the pore properties of ion conduction and sensitivity to macromolecules blocking that conduction in these K^+ channels. The fact that this highly conserved segment in all K^+ channels was predicted earlier from inspection of the primary sequence and computer modeling is a remarkable scientific advance.

In testing any hypothesis, one hopes that the results will open the doors for many additional questions to be answered. We still need to know how these amino acids in the pore region interact with each other to allow only K^+ to pass through, excluding other ions like Na^+ or Ca^{2+} ions. Which of these amino acids interact with the different therapeutic agents that block K^+ conduction, or change the kinetics of the opening and closing of the pore? Most importantly, what is the exact three-dimensional structure of the pore region? What we have discovered to date should facilitate answering these significant questions.

References

[1] Hartmann, H. A.; Kirsch, G. E.; Drewe, J. A.; Taglialatela, M.; Joho, R. H.; and Brown, A. M. *Science* 251:942–45 (1991).

[2] MacKinnon, R., and Miller, C. *Science* 245:1382 (1989).

[3] Tempel, B. L.; Papazian, D. M.; Schwarz, T. L.; Jan, Y. N.; and Jan, L. Y. *Science* 237:770–75 (1987).

[4] Guy, H. R., and Seetharamulu, P. *PNAS USA.* 83:508–12 (1986).

[5] Greenblatt, R. E.; Blatt, Y.; and Montal, M. *FEBBS LETT.* 193: 125–34 (1985).

Presumably, the narrow funnel neck operates as a "selective filter" that allows one ion type to pass readily and excludes others to a greater or lesser extent. One hypothesis maintains that the neck is shaped to strip hydrating water molecules from the surface of the penetrating ion. When separated from its hydrating water layer, the ion fits exactly in the channel and traverses the narrow region rapidly. Larger ions, presumably, are too wide in diameter to fit the narrow channel neck whether hydrated or not; smaller ions fail to lose their hydrating layer and remain trapped in the funnel-shaped openings.

Most ion channels are gated by conformational changes. Gating is rapid; ion flow can change from rates of millions of particles per second to essentially zero in a matter of milliseconds. The gating can also be modulated—that is, regulated upward or downward—through adjustment of the gates to positions between fully open and fully closed states. Modulation is typically a short-term, rapid change. Ion channels are also subject to long-term controls exerted indirectly by extracellular signals such as hormones. The hormones interact with receptors at the cell surface and trigger a series of internal chemical alterations that adjust the ability of ion channels to open or close. Many long-term changes are produced by the addition or removal of one or more phosphate groups from the channels (for details of the control pathways, see Ch. 7).

Three types of gated ion channels have been detected in animal cells. One, the *voltage-gated channel*, opens or closes in response to changes in membrane voltage or electrical potential.[2] In most cases the voltage changes result from alterations in the distribution of ions or charged molecules on either side of the membrane. The second type, the *ligand-gated channel* (from *ligare* = to bind or tie), opens or closes in response to binding specific control molecules. Many hormones and molecules released by nerve cells as *neurotransmitters*

[2] Voltage is a measure of the tendency of charged particles to move from one region to another in response to a gradient of electrical potential. Thus, voltage and electrical potential are equivalent terms. Electrical terms like voltage can sometimes be understood most easily when compared to the flow of water in a pipe. In these terms, voltage is equivalent to the pressure of water in a pipe. Electrical current, defined as a movement of charged particles from one region to another in response to a voltage difference between the two regions, is equivalent to the flow of water through a pipe in response to pressure.

(see below) act as control molecules that open or close ligand-gated channels. The third channel type is the *mechanosensitive*, or *stretch-gated, channel*, in which mechanical stresses on the membrane cause conformational changes that open the channel gates.

Gated Na⁺ Channels Voltage-gated Na^+ channels occur characteristically in cells that can respond to stimuli by undergoing rapid changes in plasma membrane voltage. Cells with this property are termed *excitable*. The Na^+ channels, opened by a relatively small voltage disturbance at some point on the plasma membrane, admit very large inflows of Na^+. These inflows greatly magnify the voltage change and extend it over the cell surface. (Na^+ is characteristically high in concentration outside cells and low inside.) The inward ion flow produced when voltage-gated Na^+ channels open is responsible for much of the voltage change involved in conduction of nerve impulses (for details, see p. 214). The importance of Na^+ channels in excitable cells is underscored by the fact that many neurotoxins in naturally occurring poisons and venoms target the Na^+ channels in nerve and muscle. When bound, the neurotoxins lock the channels in an open or closed state. In either case the neurotoxins so seriously disturb the functions of the nervous and muscular systems that their effects are usually fatal.

Voltage-gated Na^+ channels occur in multiple, closely related forms in animal cells. Mammalian brain cells, for example, contain at least three different but closely related forms of the voltage-gated Na^+ channel, encoded in different but related genes. Many other cell types have distinct Na^+ channels with slight but significant differences in their gating properties.

Many epithelia, such as those of kidney tubules, colon, lungs, trachea, and sweat ducts, have ligand-gated Na^+ channels. Entry of Na^+ through these channels is regulated by hormones such as aldosterone and vasopressin and by cytoplasmic ions such as Ca^{2+}. It is not certain whether these substances act directly as ligands binding to the Na^+ channels of this type or whether they work indirectly by activating other substances that gate the channels. Stretch-gated Na^+ channels have been detected in muscle cells in mammals.

Gated K⁺ Channels Gated K^+ channels occur in essentially all animal cells. More than 30 types of K^+ channels have been identified, some gated by voltage changes and some by ligands. In nerve cells, for example, voltage-gated K^+ channels, along with Na^+ channels, are responsible for the generation and conduction of an impulse along nerve cells; a ligand-gated K^+ channel contributes to communication between nerve cells in some systems. K^+ channels have also been detected in fungi and in stomatal cells of higher plants. In one plant, *Chara*, a voltage-gated K^+ channel is associated with an excitable response.

Gated Ca²⁺ Channels Ca^{2+} channels also occur ubiquitously in eukaryotic cells. Voltage-gated Ca^{2+} channels contribute to generation of impulses in some nerve cells. In some cell types voltage-gated Ca^{2+} channels provide important links between nerve impulses and the regulation of major cell functions. In muscle cells, for example, a voltage change induced by the arrival of a nerve impulse at the cell surface opens voltage-gated Ca^{2+} channels, releasing Ca^{2+} into the muscle cell cytoplasm. The resulting increase in Ca^{2+} concentration is the immediate trigger for muscle contraction (for details, see p. 477). In secretory cells, such as those of the pituitary gland, Ca^{2+} released into the cytoplasm through the same mechanism leads to fusion of vesicles with the plasma membrane, thereby releasing the secretory product to the cell exterior. A similar secretory mechanism triggered by the opening of voltage-gated Ca^{2+} channels takes part in the transmission of impulses across nerve synapses (see p. 221). Voltage-gated Ca^{2+} channels are also suspected to occur in higher plants.

Ca^{2+} channels also occur in both ligand-gated and stretch-gated form. Ligand-gated Ca^{2+} channels are opened directly by binding molecules on the cell surface such as glutamate. Other ligand-gated Ca^{2+} channels occur on internal membranes, such as those of the ER. Some of these channels are also activated by electrical changes in the membrane, so that they operate as combined ligand- and voltage-gated transporters. Ligand-gated Ca^{2+} channels located in both the plasma membrane and internal cell membranes are important in many cellular regulatory pathways, particularly those triggered by the activation of receptors at the cell surface (for details, see Ch. 7).

Stretch-gated Ca^{2+} channels have been detected in fungi and animals and are suspected to occur in higher plants. In yeast the channels may be part of pathways regulating responses to osmotic pressure and growth by budding. In animals stretch-gated Ca^{2+} channels have been detected in cells of the epithelia lining blood vessels, where their activity may be linked to the regulation of blood pressure.

Like Na^+ and K^+ channels, Ca^{2+} channels of all types occur in multiple forms in different cells and tissues. Striated muscle cells, for example, have two primary voltage-gated Ca^{2+} channels, one associated with the plasma membrane and one with the ER.

A Gated Cl⁻ Channel The first known voltage-gated Cl^- channel was recently discovered in the electric organ of the ray *Torpedo* by T. J. Jentsch and his colleagues. The newly discovered Cl^- channel is an unusual membrane protein that shows no sequence relationships to the voltage-gated Na^+, K^+, and Ca^{2+} channels, or indeed to any other known protein. It may be the first known member of a separate family of voltage-gated Cl^- channels with wide distribution among higher animals.

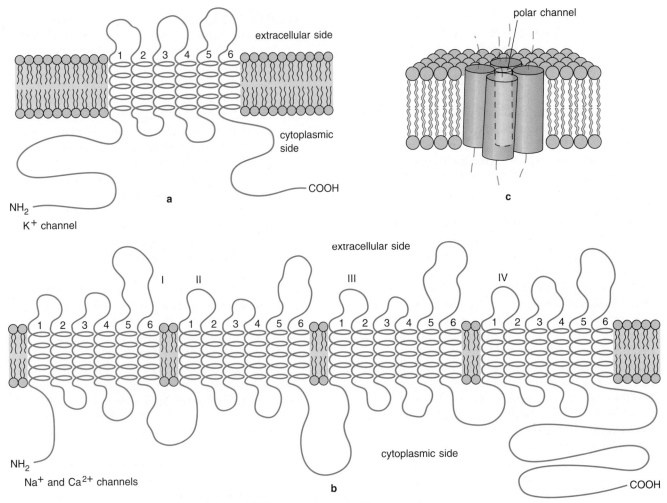

Figure 6-8 Voltage-gated channel structure. **(a)** The arrangement of transmembrane segments in the single domain of the K$^+$ channel. **(b)** The arrangement of transmembrane segments in the four domains of Na$^+$ and Ca^{2+} channels. **(c)** The probable arrangement of the four domains of Na$^+$ and Ca^{2+} channels in the membrane, which forms a central ion-conducting channel. Four of the K$^+$ polypeptides probably associate in the same arrangement to form a polar channel.

Another recently discovered chloride channel is implicated in the hereditary disease *cystic fibrosis*. In this disease, regulation of the channel is faulty; the altered operation of the channel in epithelia leads to unusually high levels of sodium and chloride in sweat and, through unknown processes, to accumulation of mucus in the respiratory tract and failure of exocrine secretion in glands such as the pancreas. Blockage of airways leads to chronic lung infections that, with other effects of the Cl$^-$ transport deficiency, are fatal by about age 20 for half the persons born with the disease (see also pp. 230 and 308).

Gated Channel Structure and Regulation The peptides of both ligand- and voltage-gated channels from several species have been fully sequenced. An analysis of amino acid sequences indicates that most ligand-gated channels are related proteins that contain a cluster of three alpha-helical segments that span the membrane and form a hydrophilic channel.

The voltage-gated Na$^+$, K$^+$, and Ca^{2+} channels also appear to be members of a related group. The simplest of the three, the K$^+$ channel, is a relatively small polypeptide with a single domain. The domain contains six segments of hydrophobic amino acids capable of winding into membrane-spanning alpha helices (Fig. 6-8*a*). The more complex channel-forming polypeptide in the Na$^+$ and Ca^{2+} transporters (Fig. 6-8*b*) contains four domains, each with six transmembrane segments arranged in patterns similar to the K$^+$ channel.

The manner in which the alpha-helical transmembrane segments align to form gated channels is still uncertain. Probably, amphipathic helices (see p. 161 and

Fig. 5-11) among the transmembrane segments associate so that their polar surfaces form an inner channel through the core of the protein. Calculations by M. Montal, M. S. Montal, and their coworkers indicated that four amphipathic helices could be aligned in a minimum-energy structure creating a central polar channel. To test this hypothesis, the Montal group made a synthetic channel protein by assembling four identical amphipathic alpha helices on a small protein backbone. The synthetic channel, when inserted into an artificial phospholipid bilayer, exhibited several characteristics of natural channels. These included the ability to conduct ions across the bilayer, selectivity for cations, and rapid transitions between open and closed states.

Several lines of evidence implicate a region between transmembrane helices 5 and 6 as the pore-forming part of the K^+ channel. Among the most significant experiments was the work of H. A. Hartmann and A. M. Brown and their coworkers, in which mutations induced in this region were shown to be most effective in altering ion flow through K^+ channels (see the Experimental Process essay by Hartmann and Brown on p. 198).

The voltage-gated Na^+, K^+, and Ca^{2+} channels share a conserved sequence forming the fourth transmembrane alpha helix that is believed to act as a voltage positively charged amino acid (arginine or lysine) occurs at every third or fourth position. According to a hypothesis developed by several investigators working with the channels, the S4 sequence is neutralized and stabilized by association with negatively charged residues located elsewhere in the polypeptides. Voltage changes in the membrane are considered to destabilize the neutralizing association and propel the S4 sequence toward the outside of the membrane. The movement, according to the hypothesis, induces the conformational change that opens the channel to ion flow. A recent experiment by D. M. Papazian and her colleagues, in which site-directed mutations (see p. 91) in the S4 sequence of a K^+ channel reduced the channel's sensitivity to voltage changes, directly supports this hypothesis.

Several experiments have provided clues to the possible structure and operation of the channel gate. In 1977 C. M. Armstrong and F. Bezanilla found that the ability of voltage-gated channels to close could be blocked by a proteinase introduced on the cytoplasmic side of the membrane. From this finding they proposed that the gate consists of a segment of the protein that dangles from the cytoplasmic side of the channel, in a position that makes it susceptible to removal by the proteinase. According to their hypothesis, the dangling amino acid segment folds into a ball suspended by an extended chain of amino acids, that fits into the opening of the ion channel on its cytoplasmic side. Gating is accomplished by conformational changes that swing the ball to or from the opening (see Fig. 6-7).

Recent experiments by R. W. Aldrich and his coworkers lend support to the ball-and-chain gating hypothesis. Using site-directed mutagenesis to change individual amino acids in the N-terminal end of a voltage-gated Na^+ channel, the Aldrich group found that alterations in the last 20 amino acids could destroy channel gating. This group may form the ball at the end of the chain. Deletions of amino acids in a segment preceding the terminal group, instead of destroying gating, speeded the rate at which the channel closed. Presumably, this segment is the chain; shortening the chain would reduce the time required for the ball to swing into the channel mouth. In another series of experiments, Aldrich and his colleagues investigated a naturally occurring *Drosophila* mutation (*shaker*) in which a voltage-gated K^+ channel is defective because of deletions in the channel protein at its N-terminal end. To test whether an artificial ball could close the mutant channels, they synthesized a short peptide with the same amino acid sequence as the presumed ball of a normal Na^+ channel. Addition of a solution of the artificial peptide enabled the mutant channels to close, in keeping with the ball-and-chain hypothesis.

The Na^+, K^+, and Ca^{2+} voltage-gated channels are subject to long-term regulation by phosphorylation. The addition of phosphate groups makes K^+ and Ca^{2+} channels more sensitive to voltages opening these channels, so that they are more readily activated to ion transport. Phosphorylation has the opposite effect on the Na^+ channel, making it less readily activated. Removal of the phosphate groups reverses these effects. Some accessory polypeptides associated with the Na^+ and Ca^{2+} channels are known to contain sites at which phosphate groups are added or removed. These polypeptides may therefore function in the long-term regulation of the major polypeptide forming the gated ion channel. Some forms of the K^+ channel are also regulated by the addition or removal of Ca^{2+}. In this case addition of Ca^{2+} fully opens the channel.

Methods for Studying Ion Channels Ion-conducting channels have been studied extensively by inserting them into artificial phospholipid films after isolation and purification. For many channel types isolation has been aided by the use of naturally occurring toxins that recognize and combine specifically with the channel. The toxins serve as a marker that allows the channel to be distinguished from other proteins released when cell membranes are disrupted.

Voltage-gated channels, in particular, have been studied by a technique in which a plasma membrane is sealed across the end of a glass micropipette (Fig. 6-9). For unknown reasons the membrane forms an almost perfect seal around the mouth of the micropipette, so that ions can move in or out of the pipette and surrounding solutions only through channels in the

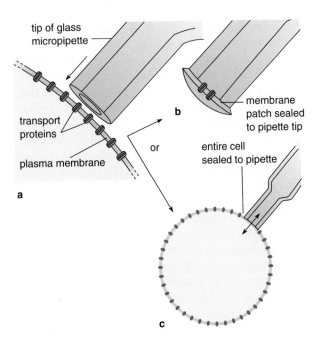

tip of glass
micropipette

transport
proteins

plasma membrane

a

b

or

membrane
patch sealed
to pipette tip

entire cell
sealed to pipette

c

Figure 6-9 Attachment of membrane segments to the tip of a glass micropipette. **(a)** The pipette is touched to a cell surface to seal the membrane to the pipette. Variations in the technique can produce a small membrane patch attached to the pipette tip **(b)** or an entire cell **(c)**. After attachment the effects of solutions introduced on either side of the attached membrane can be followed; electrodes can also be introduced on either side of the membrane. The technique can be used to study both natural membranes and artificial phospholipid films. The pipette can be made small enough (5 μm or less) to attach a membrane patch containing a single transport protein. This is the only method available to study the activity of a single protein molecule.

membrane. The technique can be varied so that the membrane sealed to the pipette includes only a small patch (Fig. 6-9*b*) or the entire plasma membrane (Fig. 6-9*c*). Segments of artificial phospholipid films with or without added membrane proteins can also be picked up and sealed to the end of the pipette. The solutions inside and outside the pipette can be varied at will, and electrodes introduced on either side. The technique using small patches sealed to the mouth of the pipette, which can be less than 5 *μm* in diameter, allows ion flows through single channel types to be recorded and varied experimentally. For example, the experiments by Aldrich and his colleagues with artificial peptide gates were carried out with membrane patches containing the mutant *shaker* K$^+$ channel.

The micropipette method, known as the *patch clamp technique*, was developed in the mid-1970s by the German scientists E. Neher and B. Sakmann. Neher and Sakmann received the Nobel Prize in 1991 for their work, which revolutionized membrane studies in cell physiology.

The Passive Transport of Water Through Lipid Bilayers

Water is an exceptional substance because it penetrates through both natural membranes and artificial phospholipid bilayers at rates about 100 to 1000 times greater than expected for a strongly polar substance. The rate of penetration of water through natural membranes is reduced but still remains unexpectedly high if membrane proteins are removed or denatured. These observations indicate that although there is undoubtedly facilitated diffusion of water by membrane proteins, relatively large amounts can penetrate directly through nonpolar regions of lipid bilayers. The basis for this exceptional behavior is unknown.

Two hypotheses, neither clearly supported by experimental evidence, have been advanced to explain the unusual ability of water to penetrate lipid bilayers. One hypothesis proposes that lipid bilayers have stable openings or "pores" just large enough to admit water molecules but no other polar substances. This idea has fallen from favor in recent years because the measured electrical resistance of both artificial bilayers and natural membranes is much higher than expected for a membrane perforated by open, water-filled pores.

The second hypothesis proposes that water molecules move through temporary "faults" in the bilayer created by flexing and bending movements of fluid hydrocarbon chains of phospholipids. At any instant these faults might transiently open narrow spaces in one bilayer half or the other, allowing water molecules to slip from the surface into the bilayer halves or from one bilayer half to the other. At no time, however, would these movements be likely to open a continuous channel through the bilayer. This mechanism is more compatible with the high electrical resistance noted for both natural and artificial bilayers, but there is as yet no evidence directly supporting it.

Increases in the proportion of cholesterol in natural and artificial membranes reduce the passive transport of water. The basis for this effect is also unknown. However, it is considered possible that by restricting movement of hydrocarbon chains in the bilayer (see Fig. 5-16*a* and p. 165) cholesterol reduces the number of transient spaces that admit water molecules to the nonpolar bilayer interior. It has also been suggested that cholesterol molecules may help to "seal" membranes against water penetration by fitting into crevices in the surfaces of integral membrane proteins.

Passive transport, involving both simple and facilitated diffusion, accounts for the movement of a wide variety of substances through cellular membranes. By this mechanism cells are able to absorb many of the hydrophobic and hydrophilic molecules required for their biological reactions and to release waste materials or secreted products to the outside. Among the most important substances entering cells by facilitated diffusion are ions and fuel molecules such as sugars (both ions and sugars also enter by active transport). Passive transport, because it is driven by concentration gradients, takes place without an expenditure of cellular energy. It greatly enhances the transport of key substances that would otherwise penetrate cells too slowly to support cellular life.

ACTIVE TRANSPORT

Many substances are actively transported into or out of cells, that is, moved across membranes against their gradients of chemical or electrical potential. Among the substances transported actively are Na^+, H^+, Ca^{2+}, and other ions; glucose and other six-carbon sugars; and amino acids. In some cases the "pumps" moving these substances are powerful enough to continue pushing their transported molecules against concentration differences as great as 1 million times.

The energy required for active transport can be calculated as essentially the reverse of diffusion along concentration gradients. If a mole of a substance releases 4075 cal in traveling down a concentration difference initially amounting to 1000 times, as shown in Equation 6-1, pushing a mole to a concentration difference this great would require at least this much energy.

Characteristics of Active Transport

The most fundamental characteristic of active transport is its requirement for an input of cellular energy. For most active transport the required energy is derived directly or indirectly from the hydrolysis of ATP, supplied in turn by the oxidation of cellular fuel substances. As a consequence of its ultimate dependence on oxidative metabolism, active transport is sensitive to metabolic poisons and, in contrast to simple or facilitated diffusion, stops if the cellular reactions supplying ATP are inhibited.

Active transport resembles facilitated diffusion in other features. The process depends on membrane proteins and stops if these proteins are denatured or removed. Like the proteins carrying out facilitated diffusion, active transport proteins exhibit specificity: only a certain molecule or closely related group of molecules is moved across membranes by a given active transport system. In summary, active transport

1. goes against gradients of chemical or electrical potential;

2. requires metabolic energy and is sensitive to metabolic poisons;

3. depends on the presence and activity of membrane proteins; and

4. is specific for certain substances or closely related groups of substances.

During active transport the carrier proteins combine briefly with the transported substance, as in facilitated diffusion and enzymatic catalysis. (Table 6-1 compares the characteristics of simple diffusion, facilitated diffusion, and active transport.)

Table 6-1 Characteristics of Transport Mechanisms

Characteristic	Simple Diffusion	Facilitated Diffusion	Active Transport
Membrane component responsible for transport	Lipids	Proteins	Proteins
Binding of transported substance	No	Yes	Yes
Energy source	Concentration gradients	Concentration gradients	ATP hydrolysis or concentration gradients
Direction of transport	With gradient of transported substance	With gradient of transported substance	Against gradient of transported substance
Specificity for molecules or molecular classes	Nonspecific	Specific	Specific
Saturation at high concentrations of transported molecules	No	Yes	Yes

Active transport proteins exhibit many of the same structural characteristics as facilitated diffusion transporters. They have active sites that bind the transported molecules. They also alternate between two major conformational changes in which the active sites are directed to either side of the membrane during the pumping cycle. Depending on the side faced, the active sites bind the transported molecules with high or low affinity.

Active transport takes place in eukaryotes by two primary mechanisms—one directly dependent on ATP hydrolysis and one indirectly dependent. In directly dependent mechanisms, ATP hydrolysis and active movement of the transported substance are carried out by the same transporter protein, and the energy released is used directly by the transporter to push substances across the membrane. All the known pumps working in this way transport positively charged ions across cellular membranes. Indirectly dependent mechanisms, in contrast, do not directly break down ATP as their energy source for active transport. Instead, indirect transporters use the concentration gradient of an ion, built up by other, directly ATP-dependent transporters, as their energy source for active transport of a different substance. Indirect pumps move a variety of organic substances as well as either positively or negatively charged ions across membranes.

P-Type Direct Active Transport Pumps

The best-known direct active transport pumps temporarily bind a phosphate group removed from ATP during the pumping cycle. Because they bind phosphates, these carriers are known as *P-type pumps*. The three best-known P-type pumps of eukaryotes move positively charged ions against their concentration gradients. One, the *H^+-ATPase pump* (ATPase = adenosine triphosphatase, an enzyme that breaks down ATP) actively moves H^+ out of cells. The second, the *Ca^{2+}-ATPase pump*, hydrolyzes ATP to move Ca^{2+} against its concentration gradient out of cells or into cytoplasmic vesicles such as those of the ER. The third, the *Na^+/K^+-ATPase pump*, moves Na^+ outward and K^+ inward across the plasma membrane by the same pumping action.

P-type pumps share several features. All are transmembrane proteins in which one polypeptide carries out all functions of the pump including ATP breakdown, phosphate binding, and movement of the pumped ion across the membrane. ATP hydrolysis by the pumps requires the presence of Mg^{2+}; in each of the transporters the phosphate group is added during the pumping cycle to an aspartic acid residue located in a similar position in a loop of the amino acid chain that extends into the cytoplasm.

Representatives of each pump have been sequenced through gene cloning techniques; all share sequence similarities, indicating that they probably evolved from a single ancestral type. Because a member of this family of active transport proteins also exists in bacteria (the bacterial K^+-ATPase pump; see Supplement 6-1), the ancestral gene must have very ancient origins, possibly in the first living cells. Eukaryotic P-type pumps consist of a single polypeptide containing eight to ten transmembrane segments that zigzag back and forth across the membrane (Fig. 6-10a); the slightly smaller bacterial version has six. Although the pumps consist of a single polypeptide, they may operate in the form of twounit dimers that coordinate their pumping cycles; the Na^+/K^+-ATPase pump contains an auxiliary polypeptide subunit (Fig. 6-10b).

Figure 6-11 illustrates how P-type pumps are believed to operate, using the H^+-ATPase transporter as an example. The transporter is initially folded so that its H^+-binding site faces the inside of the cell (H^+ is pumped from inside to outside in most systems). The pump first binds and hydrolyzes ATP; as the ATP is hydrolyzed, its terminal phosphate is bound covalently to the transporter (Fig. 6-11a). Much of the energy released by the ATP breakdown is retained in the complex formed between the transporter protein and the phosphate, which can be regarded as a "high energy" complex. Attaching the phosphate also converts the H^+-binding site to the high-affinity state in which it readily binds H^+ colliding with the transporter on the cytoplasmic side of the membrane (Fig. 6-11b). In response to binding H^+, the transporter protein undergoes a conformational change that exposes the H^+-binding site to the opposite side of the membrane (Fig. 6-11c). At the same time the conformational change alters the H^+-binding site to a low-affinity state in which its binding is so weak that H^+ is released to the outside even when the concentration of the ion is relatively high (Fig. 6-11d). The phosphate group is also released at the inside of the membrane at this time. Release of the H^+ and phosphate induces a final conformational change that returns the protein to the initial conformation shown in Figure 6-11a. The pumping cycle is now ready to begin again.

One of the most fundamental and far-reaching discoveries made through the study of P-type pumps is that *they can be made to synthesize ATP by forcing them to run in reverse*. For example, the Na^+/K^+-ATPase pump normally moves Na^+ out of animal cells. I. M. Glynn and his coworkers showed that if the external Na^+ concentration is raised to extremely high levels, Na^+ is pushed backward through the pump by the gradient, forcing the pump to run in reverse. Under these conditions the pump protein *adds phosphate groups to ADP molecules to form ATP*. Equivalent observations have been made by others studying H^+ and Ca^{2+} active transport pumps. These observations are signif-

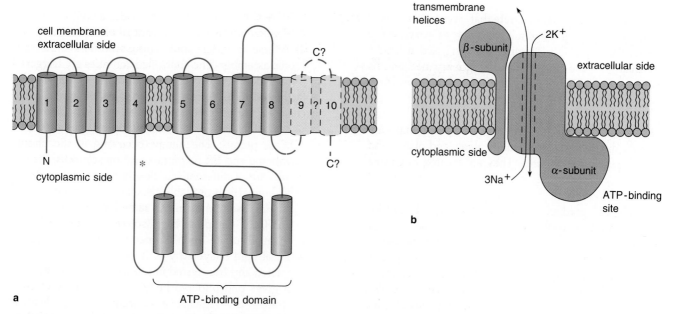

Figure 6-10 P-type active transport pumps. **(a)** The possible arrangement of transmembrane helices and domains in P-type pumps. At least eight alpha helices zigzag back and forth across the membrane; an additional two helices are possible. Although the N-terminal end is known to extend into the cytoplasm, it is still uncertain whether the C-terminal end is directed toward the cytoplasm or the cell exterior. A segment of the protein containing five alpha helices forms a large cytoplasmic domain that contains the ATP-binding site. The location of the aspartic acid residue to which the phosphate group is reversibly bound is marked by an asterisk. A phosphatase activity that can remove the phosphate group is associated with the cytoplasmic loop between transmembrane helices 2 and 3. **(b)** Structure of the Na^+/K^+-ATPase pump. The additional β-subunit may regulate the pumping activity of the α-subunit, which is similar in amino acid sequence and structure to other P-type pumps.

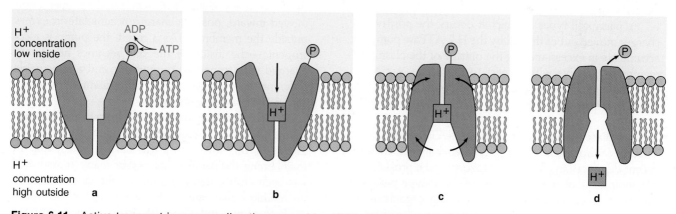

Figure 6-11 Active transport in a pump directly powered by ATP hydrolysis, using the H^+-ATPase pump as an example. The transporter is initially folded so that its H^+-binding site faces the inside of the cell. **(a)** In the first step of the cycle, the pump binds and hydrolyzes ATP. The terminal phosphate of the ATP is transferred and attached covalently to the transporter, forming a "high energy" complex. **(b)** Attachment of the phosphate converts the H^+-binding site to the high-affinity state in which it readily binds H^+ colliding with the transporter on the cytoplasmic side of the membrane. **(c)** The protein then undergoes a conformational change that exposes the H^+-binding site to the opposite side of the membrane. **(d)** The conformational change alters the H^+-binding site to a low-affinity state in which the H^+ is released to the outside even when the outside concentration of the ion is relatively high. The phosphate group is released simultaneously at the inside of the membrane. Release of the H^+ and phosphate induces a final conformational change that returns the protein to the initial conformation shown in **(a)** so that the pumping cycle is ready to begin again.

icant because they demonstrate that *a concentration gradient can be used by membrane proteins as an energy source for synthesizing ATP*. This principle, discussed at length in Chapters 9 and 10, underlies the primary mechanisms synthesizing ATP in all eukaryotic and most prokaryotic organisms.

H^+-ATPase Pumps   P-type H^+-ATPase pumps are widely distributed among fungi, algae, and plants, and also occur in prokaryotes. They are located in plasma membranes, where they push H^+ from the cytoplasm to the cell exterior.

The H^+ gradient created, high in the cell exterior and low in the cytoplasm, provides the primary energy source for indirect active transport in fungi and plants (see below). In yeast cells the H^+-ATPase pump may comprise as much as 5% to 10% of the total membrane protein. The H^+-ATPase pump of plants and fungi also regulates cytoplasmic pH by moving H^+ out of cells. (Surplus H^+ is constantly produced by many metabolic reactions in cells, and must be eliminated to maintain cytoplasmic pH at neutral values.) P-type H^+ pumps are more limited in distribution in animals. However, a P-type H^+ pump is responsible for the movement of H^+ from cells lining the stomach into the gastric juice. The stomach contents may become as acidic as pH 0.8 as a result of this movement. The pH of the gastric juice, as compared to that of the cells lining the stomach (about 7.2), indicates a difference in H^+ concentration of about 1 million times. The gastric H^+-ATPase responsible for this difference, which has been analyzed in some detail, is a 114,000-dalton protein that may have as many as nine transmembrane alpha-helical segments.

In many cell types in which it occurs, the positive charges moved out of the cell by the H^+-ATPase pump accumulate in excess, making the outside of the plasma membrane positive with respect to the inside. Active transport pumps producing electrical potential in this way are termed *electrogenic*.

Ca^{2+}-ATPase Pumps   Although Ca^{2+}-ATPase pumps were once thought to occur only in animal cells, their recent discovery in plants, a protozoan, and a prokaryote indicates that representatives of this P-type pump are much more widespread among living organisms. The action of the Ca^{2+}-ATPase pumps of eukaryotic cells pushes Ca^{2+} from the cytoplasm to the cell exterior and into the vesicles of the ER. As a result, the Ca^{2+} concentration is high outside cells and in ER vesicles and very low in the cytoplasmic solution. In muscle cells, for example, Ca^{2+} levels are as low as 10^{-6} to 10^{-7} M, values typical of most cells; in the ER Ca^{2+} concentrations are as high as 10^{-2} M. The Ca^{2+}-ATPase pump of muscle cell smooth ER (called the *sarcoplasmic reticulum* in muscle cells; see p. 472) makes up 80% of the total ER membrane protein.

The gradient resulting from the activity of Ca^{2+}-ATPase pumps is used as a control of cellular activities as diverse as muscle contraction, secretion, and microtubule assembly. **The controlled activities are triggered by a sudden release of Ca^{2+} into the cytoplasm, by facilitated diffusion through voltage- or ligand-gated channels in the plasma membrane and ER.**

Most animal cells have multiple types of Ca^{2+}-ATPase pumps. The pumps occurring in the plasma membrane and ER membranes of muscle cells, for example, are of different but closely related types. Most other animal cells probably also have distinct but related Ca^{2+}-ATPase pumps in the same locations.

A Ca^{2+} pump has been isolated and purified from both erythrocyte and ER membranes and added to artificial lipid bilayers by E. Racker and others. When extracted and transferred in this way, the purified protein gives phospholipid bilayers the capacity to concentrate Ca^{2+} against a concentration gradient if supplied with ATP. In such systems the Ca^{2+}-ATPase pump can also be made to run in reverse by abnormally high Ca^{2+} gradients. Under these conditions the pump synthesizes ATP from ADP and phosphate.

Na^+/K^+-ATPase Pumps   Almost all animal cells actively pump Na^+ outward and K^+ inward across the plasma membrane. These ion movements are maintained in animal cells by the Na^+/K^+-ATPase pump, which moves Na^+ and K^+ simultaneously in opposite directions. (Fig. 6-12 shows a model explaining how the pump may operate.) Three Na^+ are pumped out and two K^+ are pumped into the cell by each cycle of the pump. Because the Na^+/K^+-ATPase pump moves three positively charged ions outward for every two moved inward, positive charges accumulate in excess outside the membrane. As a result, the pump is electrogenic—the inside of the cell becomes negatively charged with respect to the outside, and a measurable difference in voltage, or electrical potential, develops across the plasma membrane.

The constant removal of Na^+ from the cell interior by the Na^+/K^+-ATPase pump counteracts the osmotic effects of this ion. The importance of the pump in maintaining the osmotic balance of many animal cells has been clearly demonstrated by the use of *ouabain*, an inhibitor that specifically binds to the Na^+/K^+-ATPase pump and stops its action. When exposed to ouabain, many types of animal cells swell and burst because of the inward osmotic movement of water in response to the increased cytoplasmic Na^+ concentration.

The Na^+/K^+-ATPase pump has two other important functions in animal cells. The voltage set up across the plasma membrane by the pump provides the basis for the conduction of electrical impulses by nerve and muscle cells. In addition, the high outside/low inside Na^+ gradient supplies the energy source for

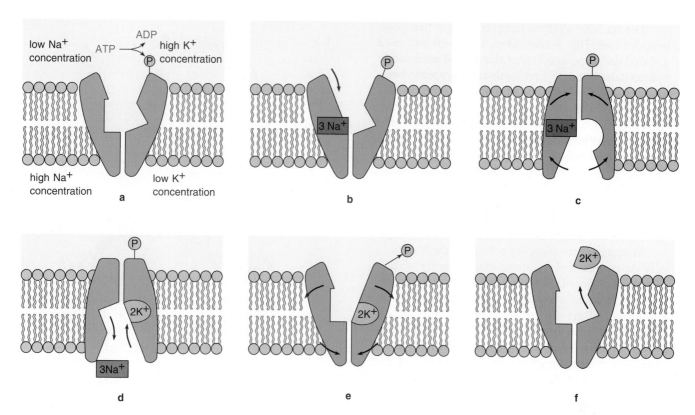

Figure 6-12 A possible mechanism for operation of the Na^+/K^+-ATPase pump. In the first step of the cycle **(a)** the pump binds and hydrolyzes ATP; the phosphate group is transferred from ATP to the transporter protein in the process. The phosphorylated pump may be considered as a high-energy complex. **(b)** Binding the phosphate converts the Na^+-binding site to its high-affinity state, in which it avidly binds Na^+ on the cytoplasmic side of the membrane. **(c)** Binding Na^+ triggers a conformational change that exposes the Na^+-binding site to the medium outside the cell. **(d)** As the conformational change becomes complete, the Na^+-binding site is altered to a low-affinity state, releasing Na^+ to the outside medium. At the same time the K^+-binding site is activated so that it now binds K^+ at the outside surface with high affinity. **(e)** K^+ binding induces a conformational change that exposes the K^+-binding site to the cell interior and releases the phosphate group. At the same time the K^+-binding site is converted to a low-affinity state in which K^+ is released **(f)**. Completion of this step returns the transporter to the initial state, in which it is ready to repeat the cycle. The transporter operates in such a way that three Na^+ are moved outside and two K^+ inside for each turn of the cycle.

indirect active transport of sugars and amino acids (see p. 211). In terms of indirect active transport, the Na^+ gradient set up by the Na^+/K^+-ATPase pump is the functional equivalent of the H^+ gradient established by the H^+-ATPase in fungi and plants.

Despite its importance in animal cells, the Na^+/K^+-ATPase pump evidently has no equivalent in plants, fungi, or bacteria. One line of evidence supporting this conclusion is that the specific Na^+/K^+-ATPase pump inhibitor, ouabain, has no effect on Na^+ transport in these organisms. In plants, fungi, and bacteria, Na^+ concentrations are controlled instead by the combined activity of the H^+-ATPase pump and an indirect active transporter that uses the energy stored in the H^+ gradient to drive Na^+ outward.

Some investigators have proposed that the Na^+/K^+-ATPase pump is an animal adaptation that made cellular life possible without cell walls. Even with the combined activity of the H^+-ATPase pump and indirect Na^+ transport, plant and bacterial cells swell and burst in most natural environments if their cell walls are ruptured or removed. The Na^+/K^+-ATPase pump, evolved as part of the adaptations leading to the appearance of animal cells, allows internal Na^+ concentrations to be maintained at levels low enough to prevent cells without walls from bursting because of osmotic pressure. Some idea of the importance of the Na^+/K^+-ATPase pump can be gained from the fact that most animal cells use about 30% of their entire energy budget to operate this pump alone.

The Na$^+$/K$^+$-ATPase pump consists of two polypeptides (see Fig. 6-10b). One, the α-subunit, is a 112,000-dalton polypeptide of about 1020 amino acids that contains the ATP hydrolysis and phosphate-binding sites, three Na$^+$-binding sites, and two sites binding K$^+$. The α-subunit, which carries out the pumping cycle, has eight to ten alpha-helical segments that cross the membrane. At least three genes encode closely related forms of the α-subunit in humans. The second polypeptide, the β-subunit, is a 50,000-dalton glycoprotein of slightly more than 300 amino acids. It probably has accessory or possibly regulatory functions that are not directly part of the pumping cycle. About 30% of the total molecular weight of the β-subunit is carbohydrate.

In mammals the pumping activity of the Na$^+$/K$^+$-ATPase is regulated by an extensive variety of hormones and growth factors, including vasopressin, insulin, glucagon, catecholamines and glucocorticoids, and epidermal growth factor. Some of the pathways triggered as these hormones and factors bind to receptors at the cell surface culminate in the addition of regulatory phosphate groups to the α-subunit of the pump. (Ch. 7 presents details of some of these pathways.)

V-Type Direct Active Transport Pumps

A recently discovered family of active transport pumps that directly uses ATP without binding a phosphate group, the *V-type* (V = vesicle) *pumps*, has been detected in the membranes of internal structures such as endocytotic vesicles, the Golgi complex, lysosomes, and secretory vesicles in all eukaryotic cells examined to date. In plants and lower eukaryotes such as yeasts, pumps of this type also occur in the boundary membrane of the large central vacuole. In all these locations, the V-type pumps work to push H$^+$ from the cytoplasm into the space enclosed by the organelle or vacuole. As a consequence, these pumps lower the pH of the vesicle or vacuole contents and contribute toward maintenance of neutral pH in the cytoplasm. A V-type pump with structure closely related to eukaryotic V-types was recently discovered in a bacterial group, *archaebacteria*, indicating very ancient evolutionary origins for the pumps in this group. Sequence analyses by E. J. Bowman and B. J. Bowman and others showed that the amino acid sequences in two of the polypeptide subunits of V-type pumps (the A and B subunits; see below) are highly conserved in fungi, plants, archaebacteria, and human kidneys.

V-type pumps are large, complex structures with two segments, one buried in the membrane and one extending into the cytoplasm on the outer side of the vesicle membrane (Fig. 6-13). The segment buried in the membrane, composed of two major polypeptides, one of them in six copies, forms a channel conducting

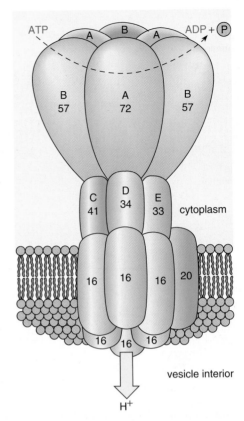

Figure 6-13 Structure of a V-type active transport pump. A membrane channel conducting H$^+$ from the cytoplasm into an organelle or vacuole is formed from six copies of a 16,000-dalton polypeptide subunit in association with one copy of a 20,000-dalton polypeptide subunit. The large cytoplasmic segment consists of at least five different subunits, with two of them, the A and B subunits, present in three copies each. The catalytic activity associated with ATP binding and hydrolysis is believed to be associated with the A subunit. The V-type pumps show clear structural similarities to F$_o$F$_1$ ATPase, which uses an H$^+$ gradient as an energy source for ATP synthesis in bacteria, cyanobacteria, mitochondria, and chloroplasts (compare Fig. 9-31).

H$^+$ through the membrane. The large cytoplasmic segment consists of at least five different subunits, with two of them, the A and B subunits, present in three copies each. Although the functions of the cytoplasmic subunits are uncertain, most of the catalytic activity associated with ATP binding and hydrolysis appears to be associated with the A subunit.

In structure V-type pumps are remarkably similar to the F$_o$F$_1$ ATPases of bacteria, cyanobacteria, mitochondria, and chloroplasts, which use an H$^+$ gradient as the energy source for synthesizing ATP from ADP and phosphate (compare Figs. 6-13 and 9-31). The F$_o$F$_1$ ATPases in these locations have membrane and cytoplasmic segments with functions and polypeptide subunits equivalent to those of V-type pumps. These similarities indicate that V-type active transport pumps, which break down ATP as an energy source to pump H^+, and F_oF_1 ATPase enzymes, which use an H^+ gradient as an energy source for ATP synthesis, evolved from a single ancestral type.

Indirect Active Transport Pumps

In indirect active transport a concentration gradient maintained by the active transport of an ion serves as the energy source driving the active transport of another substance. Because molecules carried in or out of cells by indirect active transport always move across membranes in conjunction with an ion supplying the driving force, this pathway of active transport is also termed *cotransport*.

Indirect active transport takes place in either of two patterns, *symport* and *antiport* (Fig. 6-14). In symport, the cotransported substance moves in the same direction as the driving ion. Among the most important metabolites moved actively into cells by symport are sugars and amino acids. Ions are also transported inward by this mechanism in many cell types. In antiport, the cotransported substance moves in the direction opposite to the driving ion. This pattern is generally restricted to ions.

The driving force for indirect active transport in animal cells is the high outside/low inside Na^+ gradient set up by the Na^+/K^+-ATPase pump. The equivalent energy source for cotransport in plants and fungi is the high outside/low inside H^+ gradient set up by the H^+-ATPase pump. Bacteria also carry out H^+-driven cotransport, but in bacteria the high outside/low inside H^+ gradient is established through oxidative electron transport instead of an active transporter (see Supplement 6-1).

Figure 6-15 shows the mechanism by which an indirect pump of animal cells may use an Na^+ gradient as the energy source for the active transport (symport) of glucose. The mechanism assumes that Na^+ exists at a high outside/low inside gradient established by the Na^+/K^+-ATPase pump. In the first step in the pumping cycle, the transporter protein is folded so that sites tailored to fit Na^+ and glucose are directed toward the outside surface of the membrane (Fig. 6-15a). The glucose-binding site is considered to be in a high-affinity state. Binding Na^+ and glucose (Fig. 6-15b) triggers a conformational change in the transporter that closes access to the outside and directs the Na^+- and glucose-binding sites toward the inside membrane surface (Fig. 6-15c). Energy for the conformational change may be derived from the movement of Na^+ down its concentration gradient. In this conformation the binding site for glucose is altered to a low-affinity state in which glucose and Na^+ are released simultaneously to the inside of the membrane (Fig. 6-15d). In response the transporter returns to the original conformation (Fig. 6-15a), ready to enter another cycle. In the transport mechanism Na^+ moves passively in the direction of its concentration gradient, and glucose molecules are moved actively against their gradient. Antiport works by a similar mechanism except that binding and release of the driving ion and transported substances occur on opposite sides of the membrane.

Various metabolites and ions are carried across membranes by indirect active transport. In eukaryotes practically all the organic substances actively transported into cells are moved by cotransport. In vertebrates the primary sites of the indirect active transport of glucose and other six-carbon sugars are in the intestine and kidney. Carriers cotransporting all the amino acids, either singly or in related groups, have been detected in animals, plants, fungi, and bacteria. In an-

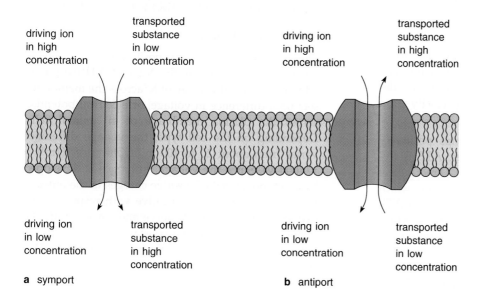

driving ion in high concentration

transported substance in low concentration

driving ion in high concentration

transported substance in high concentration

driving ion in low concentration

transported substance in high concentration

driving ion in low concentration

transported substance in low concentration

a symport

b antiport

Figure 6-14 Indirect active transport by symport and antiport, in which a favorable concentration gradient of an ion is used as the energy source for active transport of another substance. **(a)** In symport the transported substance moves in the same direction as the driving ion. Sugars, amino acids, and some ions are important metabolites moved actively into cells by symport. **(b)** In antiport the transported substance moves in the direction opposite to the driving ion. This pattern of cotransport is generally restricted to ions.

Coordination of Passive and Active Transport in the Maintenance of Cellular Ion Concentrations

The combined activity of passive and active transport proteins maintains the concentrations of ions inside cells at the levels required for cellular functions. Na^+ is constantly pushed outward in animal cells by the Na^+/K^+-ATPase pump; the resulting high outside/low inside Na^+ gradient effectively regulates intracellular osmotic pressure and supplies the Na^+ gradient driving indirect active transport. Activity of this pump is also the primary source of the voltage difference established across the plasma membranes of animal cells. Na^+ enters constantly through leakage across the plasma membrane, through the activity of the Na^+-driven cotransport pumps, and in excitable cells through voltage- and ligand-gated Na^+ channels. In animal cells the inward traffic of Na^+ is easily counteracted by the Na^+/K^+-ATPase active transport pump.

K^+ enters animal cells actively in two ways: by the activity of the Na^+/K^+-ATPase pump and by indirect active transport. As a result, K^+ is maintained at a high inside/low outside gradient. Outward movement along this concentration gradient occurs passively through slow leakage across the plasma membrane and through the activity of voltage- and ligand-gated K^+ channels, which occur in both excitable and nonexcitable cells. As in the case of Na^+, constant activity of the Na^+/K^+-ATPase pump quickly counteracts any internal loss of K^+.

H^+ is pushed out of the cytoplasm in animal cells by the activity of the indirect active transporter that exchanges Na^+ outside for H^+ inside. This transporter counteracts the constant production of surplus H^+ by metabolic reactions, effectively regulating the internal pH of the cytoplasm. In plants and fungi H^+ is constantly removed from the cytoplasm by the H^+-ATPase active transport pump. The activity of this pump regulates cytoplasmic pH, and supplies the high outside/low inside H^+ gradient that drives H^+-linked indirect active transport in plants.

In all eukaryotes H^+ is also removed from the cytoplasmic solution by V-type active transport pumps, which push H^+ from the cytoplasm into vesicles such as the ER, Golgi complex, secretory vesicles, and lysosomes, and into the large central vacuoles of plants and fungi. The activity of the V-type pumps simultaneously lowers the pH of the internal contents of these organelles and vacuoles and raises the pH of the surrounding cytoplasm. The increased acidity of the vesicles and vacuoles is critical to their functions in storage, in recycling surface receptors trapped in the vesicles during endocytosis, and in activation of enzymes held within the vesicles (for details, see Ch. 20).

Ca^{2+} is constantly removed from the cytoplasmic solution of all eukaryotic cells by the activity of Ca^{2+}-ATPase pumps in the plasma membrane and in vesicles of

Experiments Establishing the Role of Na^+ and K^+ Flows in the Action Potential   The fact that the interior of the nerve axon becomes positive for a brief instant during the action potential led an early investigator of nerve function, A. L. Hodgkin, to conclude that a positive ion, probably Na^+, rushes inward to produce the potential change. This conclusion was supported by experiments carried out in the 1940s by Hodgkin and Bernard Katz, who found that if axons were placed in a medium without Na^+, the action potential did not appear.

The actual ionic events taking place during generation of the action potential in squid axons were worked out by Hodgkin and A. L. Huxley in 1952. In their work the flow of current that produces the action potential was of major interest, since current is a measure of the flow of charges through a conductor or across a membrane (see footnote 2, p. 200). The charges flowing in this case are the ions moving to produce the action potential. (Information Box 6-3 describes the *voltage clamp* technique used by Hodgkin and Huxley.)

Figure 6-18 shows the flow of charges Hodgkin and Huxley detected during generation of an action potential. Immediately after a stimulus generating an action potential in a nerve axon, a rapid inward flow or current of positive charges occurs across the membrane (Fig. 6-18a). The inward flow, lasting for less than 1 msec, is followed by a change to a slower outward flow of positive charges.

The contributions of Na^+ and K^+ to these current flows were sorted out by further experimentation. By replacing Na^+ with choline in the solution outside the axon, any effects due to Na^+ could be eliminated. (Choline is a positively charged organic molecule that does not penetrate the membrane.) Under these conditions the initial inward flow of positive charges did not occur, and only the outward flow was recorded (Fig. 6-18b). Therefore, the initial inward flow of current

the ER. Mitochondria can also take up and store large quantities of Ca^{2+} by the formation of inorganic calcium deposits in the mitochondrial matrix. The resultant Ca^{2+} gradient, high in the cell exterior and in the ER and mitochondria, and low in the soluble cytoplasm, controls a wide range of cellular mechanisms in both plants and animals. At the low Ca^{2+} concentrations maintained by the pumps removing the ion from the cytoplasm, the controls of these mechanisms are inactive. Sudden influxes of Ca^{2+}, triggered by a variety of systems, set the control mechanisms in operation. Many of these control mechanisms are triggered by receptor molecules embedded in the plasma membrane. (Ch. 7 details the major regulatory mechanisms that trigger Ca^{2+} release.) Entry of Ca^{2+} through voltage- or ligand-gated channels also contributes to the generation of electrical impulses in nerve cells.

Mg^{2+}, an essential cofactor for many enzymes, is also important in the regulation of membrane transporters such as the Ca^{2+} active transport pumps. It appears that Mg^{2+} leaks steadily and slowly into most cell types from relatively high concentrations in the extracellular medium. Free cytoplasmic concentrations of the ion are kept at relatively low levels through binding to cellular molecules such as ATP and RNA and by an antiport system driven by the high outside/low inside Na^+ gradient. As Na^+ ions move inward along their gradient, the antiport pumps Mg^{2+} outward.

Negatively charged ions (anions) may enter cells by either passive or active transport. In animal cells facilitated diffusion transporters permit passive entry of anions such as Cl^-, phosphate, and sulfate. The same anions can also be actively pumped into animal cells by Na^+-driven cotransport. Another group of transporters exchanges anions across the plasma membranes of animal cells. One important representative of this group is the Cl^-/HCO_3^- anion exchanger of erythrocytes and other cells, which exchanges extracellular chloride for intracellular bicarbonate. The combined activities of the passive and active transporters delivering anions to the cell interior complement the various carriers moving positively charged ions in and out of the cytoplasm so that most cells have a significant surplus of anions in the cytoplasm. This unequal distribution of positive and negative charges produces the voltage difference characteristic of living cells, in which the cell interior is negative relative to the exterior.

The work of active transport pumps contributing to the maintenance of cellular ion concentrations may account for a major part of the total energy expenditure, in some cases more than 60%. In mammals as much as 20% of the heat maintaining body temperature is generated by the transformation of chemical energy to heat as active transporters move ions across the plasma membrane. Of the various ion pumps working in animals, the greatest direct consumer of cellular energy is the Na^+/K^+-ATPase pump. The high consumption of energy by this transporter reflects its importance in controlling osmotic balance and setting up the Na^+ gradient used for cotransport.

could be attributed to sudden movement of Na^+ across the membrane from outside to inside. The outward current flow recorded in Figure 6-18b began immediately after the stimulus and built up relatively slowly, reaching a maximum in about 3 msec. This outward flow was identified with movement of K^+ from inside to outside by the use of labeled potassium (^{42}K). During buildup of the outward current flow in Figure 6-18b, radioactive ^{42}K accumulated outside the membrane at the same rate as the outward flow of charges. This accumulation directly linked the outward flow to K^+ movement.

From these observations Hodgkin and Huxley concluded that the action potential results from a sudden, explosive inflow of Na^+, lasting for less than 1 msec. This flow changes the potential inside the axon from negative to positive. At the same time a slower outflow of K^+ gradually builds up; within two msec after the inward Na^+ flow reaches its peak, the outward movement of K^+ restores membrane potential to the resting value.

Early evidence for the existence of separate channels conducting Na^+ and K^+ was provided by the use of two chemical agents. *Tetrodotoxin*, a deadly poison extracted from the Japanese puffer fish, binds to the membrane and eliminates Na^+ inflow without affecting current changes due to K^+. A second agent, *tetraethylammonium ion*, blocks K^+ flow without affecting Na^+. The two agents bound to different proteins, confirming that separate channels exist for the two ions. More recently, the flow blocked by these agents has been shown to result from their interaction with the voltage-gated Na^+ and K^+ channels of nerve cell membranes.

How Voltage-Gated Na^+ and K^+ Channels Produce the Action Potential   The ion flows generating an action potential depend on the characteristics of the response of the voltage-gated Na^+ and K^+ channels to changes

Quantitative Estimations of Membrane Potential: The Nernst Equation

The magnitude of the electrical potential or voltage difference across the plasma membrane can be calculated from the inside and outside concentrations of the ions responsible for the membrane potential. The potential difference resulting from any single ion can be calculated from the *Nernst equation:*

$$\text{potential difference} = E = \frac{RT}{z\mathscr{F}} \ln \frac{C_o}{C_i} \qquad (6\text{-}2)$$

in which R is the gas constant, T is the absolute temperature, z is the charge carried by the ion, $\mathscr{F}$ is the Faraday constant (23,062 cal/mol · V, or 96,000 coulombs/mol · V), and C_o and C_i are the outside and inside concentrations of the ion in moles.

For K^+ ions, for example, which in squid axons have outside and inside concentrations of 20 and 400 millimoles (mM), substitution in the Nernst equation gives

$$\text{potential difference due to } K^+ = E_K = \frac{RT}{z\mathscr{F}} \ln \frac{20}{400}$$

Since $\log_{10} = 2.3 \ln$, and $RT/\mathscr{F}$ at $18°C = 25$ and $z = 1$

$$E_K = 2.3 \times 25 \log_{10} \frac{1}{20} = -75 \text{ mV}$$

The result, -75 mV, gives a reasonably good approximation of the -60 mV actually measured for the membrane potential of resting squid axons.

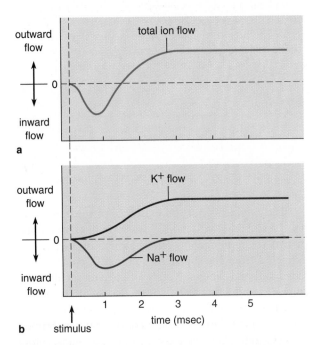

Figure 6-18 Ion flows across the plasma membrane, recorded as changes in current as an action potential develops in a squid nerve axon. **(a)** Total changes noted during an action potential. An inward flow of positive charges is followed by an outward flow of positive charges. **(b)** Current flow when Na^+ is eliminated from the solution surrounding an axon (purple curve). The first rapid inward flow is absent, indicating that the initial inward flow noted in **(a)** is due to Na^+ (red curve).

in the voltage difference across the plasma membrane. In resting neurons or muscle cells both the Na^+ and K^+ channels in the plasma membrane are closed. The channels open if the voltage difference across the plasma membrane is reduced sufficiently. The reduction in voltage, which acts as the stimulus, can be produced experimentally either by making the cytoplasm inside the plasma membrane less negative or by making the outside less positive. This is done by applying an electrical current across a second pair of electrodes placed inside and outside nerve axons. Relatively small experimental reductions in the resting potential, on the order of 10 to 15 mV, are effective in opening the Na^+ and K^+ channels.

The very rapid rise of the membrane voltage and the somewhat slower return to the resting value during an action potential depend on differences in the rate at which the Na^+ and K^+ channels open and close. The Na^+ channels open almost instantly, allowing Na^+ to flow rapidly from outside to inside in response to the high outside/low inside concentration gradient, while the slower K^+ channels are just beginning to open. The sudden inward flow of positive charges as the Na^+ channels open causes the inside of the membrane to become momentarily positive, producing the spike in Figure 6-17. Almost as quickly, while the K^+ channels are still opening, the Na^+ channels snap shut again.

By this time the K^+ channels have opened enough

The Voltage Clamp

Measurement of current flow during an action potential is complicated because membranes act as *capacitors*. Capacitance is a measure of the ability of a nonconducting substance to take on an electrical charge if a potential difference is applied across it. If the voltage across a capacitor changes after the charge is set up, some of the charge is released as an electrical current.

The nonpolar interior of the plasma membrane is a nonconductor that acts as a capacitor charged by the resting potential. During development of the action potential, the changes in voltage across the membrane cause some release of current by the membrane capacitor. This current contributes to the electrical changes noted at the recording electrodes. When electrical changes are recorded directly from electrodes implanted in nerve cells, it is impossible to separate the contributions of ion movements from current due to the discharge of the membrane capacitor.

In the Hodgkin–Huxley experiments, the effects of membrane capacitance were eliminated by a *voltage clamp*. In this technique the electrodes detecting voltage changes across the plasma membrane (electrodes 1 and 2 in the figure in this box) are connected to a feedback device. The feedback device detects voltage changes and, through another electrode (electrode 3 in the figure), instantly applies exactly the amount of current required to maintain the membrane voltage at a constant value. The voltage difference across the membrane is thus "clamped" at a steady value by the feedback device. Because the voltage is held constant, the clamp prevents discharge of membrane capacitance. As a consequence, all electrical effects noted can be assumed to result from ionic movements.

In the clamped situation the voltage changes characteristic of the action potential do not develop. A stimulus can be applied, however, and the resulting ion flows noted by recording the current required to compensate for

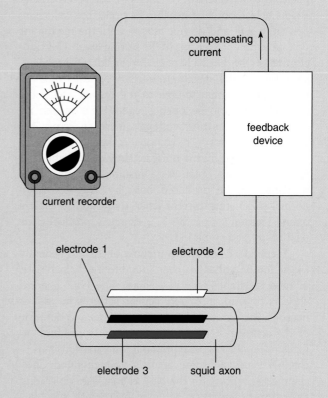

them. (The compensating current is equivalent and opposite to the ion flows taking place across the nerve cell membrane during generation of the action potential.) The voltage clamp made the measurements of positive ion flow possible, allowing Hodgkin and Huxley to quantify the electrical changes responsible for the action potential. The technique is still used routinely in experiments analyzing ion flows during nerve conduction.

to release the rapid flow of K^+ from inside the membrane to the outside in response to the high inside/low outside concentration gradient of this ion. The outward flow of positive charges gradually increases as the K^+ channels open more fully; within about 2 msec enough K^+ has moved outside to compensate for the Na^+ ions moving inward, and the inside of the membrane changes to negative again. The flow of K^+ to the outside becomes so rapid as the K^+ channels open fully that the inside of the membrane dips briefly below the resting value (see Fig. 6-17). As the inside of the membrane becomes fully negative, the K^+ chan-

nels close and the membrane equilibrates at the resting potential.

During their measurements of the action potential, Hodgkin and Huxley noted that for a very brief interval after the action potential reaches its peak, a second action potential cannot be generated by the same region of the axon no matter how large the stimulating current. This brief unreactive time span is called the *absolute refractory period*. Following this period, which lasts only 1 msec or less, a second action potential can be generated, but only by a larger stimulating current than usual. This second period, requiring larger-than-normal stimuli, lasts 1 to 2 msec, and is called the *relative refractory period*. The refractory periods are critical to the mechanism by which action potentials move along nerve axons (see below).

The two refractory periods are considered to be due to the characteristics of the Na^+ and K^+ channels. As the action potential reaches its peak, the Na^+ channels snap shut and do not reopen until the membrane voltage difference returns to the resting level. This tight closure of the Na^+ channels is considered to be responsible for the absolute refractory period. Once the membrane potential returns to the resting value, the Na^+ channels can be opened again by a second reduction in the resting voltage difference across the membrane. However, reductions in the voltage difference are resisted by the outward flow of K^+ while the K^+ channels are open. As a result, a second action potential can be produced only by a greater-than-normal stimulating current when the K^+ channels are open. This effect of the K^+ channels produces the relative refractory period.

Dependence on an inward flow of Na^+ to generate the action potential has been found to be characteristic of most excitable cells in animals. The Na^+ channel appears to work in closely similar patterns in various neurons; the differences in firing patterns and duration of the action potential in different cell types and organisms seem to depend primarily on variations in K^+-channel types. In some animals voltage-gated Ca^{2+} channels substitute for the Na^+ channels in generating an action potential or act in concert with the Na^+ channels. For example, the embryonic muscle cells of tunicates show evidence of both Na^+ and Ca^{2+} inflows in the production of an action potential. In adult tunicates generation of the action potential seems to depend only on Ca^{2+} inflow in muscle cells.

Movement of Impulses Along an Axon: Propagation of the Action Potential

Once an action potential develops, it passes along the surface of a nerve or muscle cell as a wave of electrical change traveling away from the stimulation point. According to the Hodgkin–Huxley model, movement or *propagation* of the action potential results from the electrical effects that the segment initially generating an action potential has on adjacent segments of an axon (Fig. 6-19). Within the segment initiating an action potential (the blue region in Fig. 6-19), the outside of the membrane becomes negative and the inside positive. The negative region at the membrane surface in the area generating the action potential attracts positive ions from the adjacent resting regions, which are positive. Similarly, the positive ions inside the membrane in the region generating the action potential flow under the membrane toward the adjacent, negatively charged resting regions of the axon (arrows, Fig. 6-19). The net

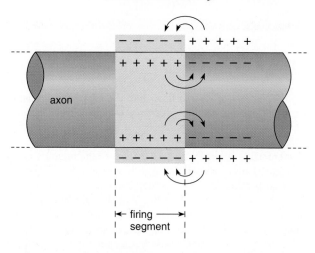

direction of propagation

axon

firing segment

Figure 6-19 Propagation of an action potential by the flow of charges between a firing segment (shaded area) and an adjacent unfired region of an axon. Within the firing segment the flow of ions makes the outside of the membrane negative and the inside positive. The negative region at the membrane surface in the firing segment attracts positive ions from the adjacent resting regions, which are positive. Similarly, the positive ions inside the membrane in the region generating the action potential flow under the membrane toward the adjacent, negatively charged resting regions of the axon. The net effect of these flows reduces the potential difference across the membrane in the adjacent regions, pushing these regions over the threshold for production of an action potential. Each segment induces the next to fire in this way, causing the action potential to move directionally along the axon.

effect of these flows of positive charges is to make the adjacent, resting segments of the axon membrane less positive on the outside and more positive on the inside or, in other words, to reduce the potential difference across the membranes in the adjacent regions.

The reduction in potential produced by these ion flows is easily large enough to open the Na^+ and K^+ channels in the adjacent membrane segments, and the action potential develops in these regions. In this way each segment of the axon stimulates the next segment to "fire," and the action potential propagates rapidly along the axon as a nerve impulse.

The impulse does not reverse at any point because of the refractory periods. For a brief period of 2 to 3 msec, when K^+ outflow restores the resting potential, Na^+ channels are held shut and regions that have just fired are unable to open and generate a second action potential. By the time capability is restored, the firing region is too distant for its electrical disturbance to open Na^+ gates in the recently fired region.

An action potential normally develops through the activity of ligand-gated channels in the dendrites of a neuron. The action potential is then propagated over the neuron through the action of voltage-gated channels and moves along the axon in a one-way direction toward the axon terminals. However, if an axon is artificially stimulated at any point between its ends, two action potentials are generated that separate and move in opposite directions. Because neurons are normally stimulated only by ligand-gated channels in the dendrites, bidirectional conduction from a point along an axon does not occur.

The individual action potentials traveling along a neuron are produced by equivalent ion flows, involve the same degree of voltage change, and occupy the same period of time from initiation to completion of the electrical disturbance. Thus, changes in the intensity of neuron activity are not reflected in modulation of the size or extent of action potentials. Instead, changes in intensity are produced by alterations in the number of action potentials generated per unit time, from zero up to a limit determined by the refractory period.

Some evidence for the generation and propagation of action potentials has also been obtained in higher plants able to respond to stimuli. In the sensitive plant *Mimosa*, for example, which closes its leaves in response to touch, action potentials are generated and propagated by voltage-sensitive K^+ and Cl^- channels.

Saltatory Conduction in Myelinated Neurons

In most invertebrate and some vertebrate neurons, axons work in the pattern shown in Figure 6-19. Impulses are conducted by a smooth and continuous flow of the action potential over the axon membrane surface, much like a burning fuse. The rate of conduction is proportional to the diameter of the axon, so increases in conduction rate are achieved by increases in axon diameter. The giant axons of squids and some other invertebrates represent the maximum development of this mechanism for increasing conduction rate.

Conduction rate is increased in most vertebrate neurons by a different but related mechanism. The axons of these cells are surrounded by a layer of *myelin* membranes, which act as an electrical insulator (Figs.

6-20 and 6-21). The covering is complete except for regularly spaced interruptions called *nodes* (Fig. 6-20 and Fig. 6-21b and c), which expose the axon membrane directly to the surrounding extracellular fluids.

Na^+ channels are restricted almost entirely to the segments of plasma membrane exposed at the nodes, where their density reaches levels as high as 2000 to 12,000 per μm^2. Channel density in the myelinated regions between the nodes is much lower, less than 25 per μm^2; even the relatively few Na^+ channels present in the myelinated regions are evidently prevented from opening by the insulating myelin sheath. The sheath also minimizes leakage of ions through the membrane in opposition to the active transport pumps. These characteristics essentially eliminate current leakage and capacitance changes in the membrane and give the axon in the myelinated regions the properties of an electrical cable, with the ability to conduct a current without significant losses from one node to the next. In unmyelinated axons, by contrast, current leakage and capacitance greatly reduce the ability of the axon to conduct an electrical signal.

Because of the myelin sheath and concentration of Na^+ channels in the nodes, propagation of a nerve impulse proceeds from node to node in a jumping, or *saltatory*, fashion. Application of a stimulus at a node at one end of the axon develops an action potential at this node; the voltage change is conducted electrically and almost instantaneously through the myelinated portion of the axon to the next node. Arrival of the electrical change stimulates the next node to fire an action potential, producing a current that fires the next node, and so on. The result is very rapid travel of the impulse along the axon as the action potential skips from one node to the next. The rate, at up to 100 meters per second (m/sec), is much faster than steady conduction through unmyelinated axons, which depends primarily on ion movements and travels at velocities of only 20 to 50 m/sec.

Reversal of travel is prevented by the same mechanism operating in an unmyelinated neuron. After a node fires, it takes 2 to 3 msec to recover its resting potential. By that time the impulse has skipped too far

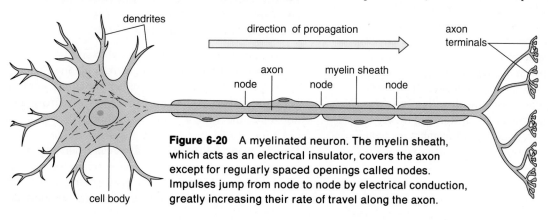

Figure 6-20 A myelinated neuron. The myelin sheath, which acts as an electrical insulator, covers the axon except for regularly spaced openings called nodes. Impulses jump from node to node by electrical conduction, greatly increasing their rate of travel along the axon.

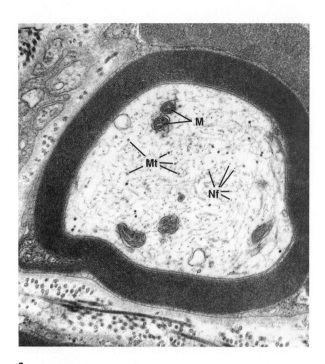

a

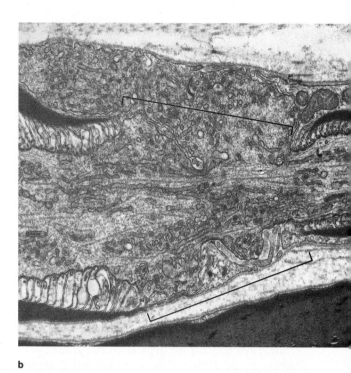

b

Figure 6-21 Myelinated axons and nodes. **(a)** A myelinated axon of the newt in cross section. The myelin sheath appears as successive layers of membranes surrounding the axon. M, mitochondria; Mt, microtubules; Nf, neurofilaments. × 27,000. Courtesy of T. L. Lentz, from *J. Cell. Biol.* 52:719 (1972). **(b)** A myelinated axon of the mouse in longitudinal section. The sheath is interrupted at the node (brackets). × 16,400. Courtesy of S. Tsukita and H. Ishikawa, from *J. Cell Biol.* 84:513 (1970), by permission of the Rockefeller University Press. **(c)** Diagram of a node from the same view as **(b)**. Na⁺ channels are concentrated in the segment of the axon exposed at the node.

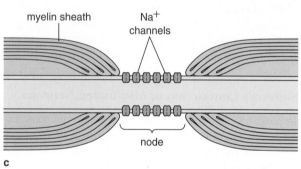

c

along the axon for its electrical current to stimulate a previously fired and recovered node.

Under the electron microscope myelin appears as a layer of densely packed membranes surrounding an axon (see Fig. 6-21). By tracing the formation of myelinated neurons during embryonic growth, in 1954 B. B. Geren showed that myelin membranes develop from the plasma membranes of accessory cells called *Schwann cells.* These membranes spiral in flat sheets around the developing axon (Fig. 6-22). As the spiraling progresses, the cytoplasm between successive layers of the Schwann cell membranes is eliminated. The membranes fuse into a multilayered sheath with as many as 300 membrane layers completely surrounding the axon. The phospholipid bilayers of myelin membranes have good insulating properties because there are essentially no proteins forming channels to permit the movement of ions. Most nerve cell processes of mammals and other vertebrates are myelinated.

Myelinization and the branching of dendrites and axons greatly increase the total surface area of membranes in the nervous system. In the human brain the total membrane area is estimated at between 1000 and 10,000 square meters!

Ligand-Gated Membrane Channels and the Conduction of Nerve Impulses Between Neurons and Other Cells

An axon may connect at its terminal to another neuron, to the surface of a muscle cell, or to a gland cell. Two types of connections occur, either by direct contact between plasma membranes or across a narrow gap known as the *synapse.*

Direct Contact Between Excitable Cells Direct contact between neurons and other cells is relatively rare. In this type of connection, an impulse arriving at the end of an axon is transmitted as an electrical stimulus di-

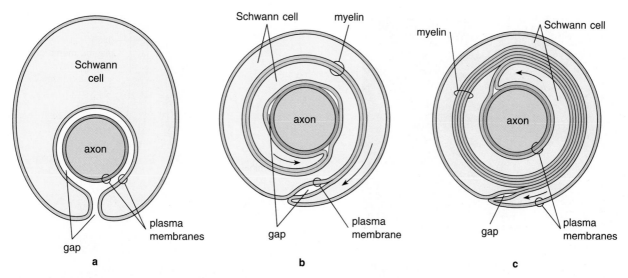

Figure 6-22 Development of the myelin sheath. **(a)** The earliest stage of development, in which a single axon is embedded in a Schwann cell. **(b)** The Schwann cell membrane has begun to form an overlapping spiral around the axon. Note that the intercellular gap between plasma membranes is eliminated. **(c)** A later stage, at which the spiral is more extensive. The membranes of the Schwann cell become so closely packed that both intercellular and cytoplasmic spaces are eliminated. Redrawn from an original courtesy of J. D. Robertson.

rectly to the recipient cell, which fires its own action potential as a result. Some nerve connections in dental pulp and between neurons and muscle cells in fishes are "wired" directly together in this way. The direct connection allows an almost instantaneous reaction by the receiving cell to an arriving nerve impulse. (Ch. 7 describes the surface junctions responsible for direct electrical connections of this type, termed *gap* or *communicating junctions*.)

Connections by a Synapse   Most connections between neurons and other cells occur via a synapse. In a synapse the membranes of the two cells making the connection, although closely aligned, remain separated by a regular space measuring from 26 to 40 nm, the *synaptic cleft* (Fig. 6-23). As a result of this separation, and the electrical characteristics of the substance filling the gap, an action potential arriving at the axon terminal of one cell fails to leap the intervening distance and stops at this point. The impulse is conducted across the gap by the release of neurotransmitters and their action on ligand-gated channels in the cell on the receiving side of the synapse.

The neurotransmitters, which include acetylcholine, various amino acids, and other substances (Table 6-2), are stored in *synaptic vesicles* within the cytoplasm of the axon terminal (arrows in Fig. 6-23*a*). Arrival of an impulse causes the vesicles to fuse with the plasma membrane of the axon terminal and release their neurotransmitter molecules into the synaptic cleft. The

neurotransmitter molecules diffuse through the synaptic cleft to make contact with the plasma membrane of the receiving cell at the other side of the synapse. There they are bound by either ligand-gated membrane channels or receptors that act in a similar manner. The ligand-gated channels or receptors open in response to binding the neurotransmitter and initiate an action potential in the receiving cell, which may be another neuron or a muscle cell.

The movement and fusion of synaptic vesicles with the plasma membrane of the sending cell are regulated by a voltage-gated Ca^{2+} channel that opens when an action potential arrives at the synapse. The role of Ca^{2+} in the axon terminal has been nicely demonstrated with *aequorin*, a protein that emits light in response to

Table 6-2	Some Neurotransmitters and Their Mode of Removal from Synaptic Clefts
Neurotransmitter	Primary Mode of Removal
Acetylcholine	Hydrolyzed by acetylcholinesterase
Dopamine	Resorption
Norepinephrine	Resorption; chemical modification
Serotonin	Resorption
Histamine	Chemical modification or breakdown
Glutamine, aspartic acid, glycine	Resorption; chemical modification

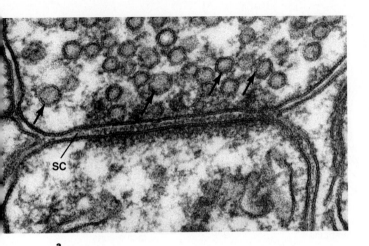

a

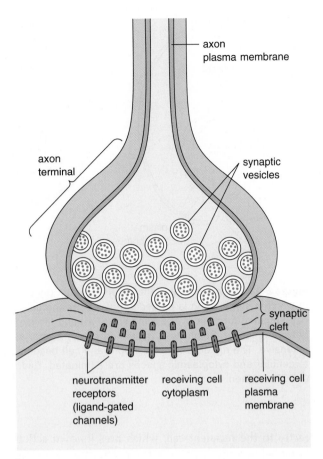

b

Figure 6-23 The axon terminal and the synapse. **(a)** A thin-sectioned synapse as seen in the electron microscope. Large numbers of synaptic vesicles (arrows) are visible inside the axon terminal. SC, synaptic cleft. × 103,000. Courtesy of C. Sotelo, from *International Cell Biology*, ed. B. R. Brinkley and K. R. Porter, 1977, by permission of the Rockefeller University Press. **(b)** Major structures of a synapse.

Ca^{2+} binding. As an action potential arrives at the terminal, aequorin injected into the cytoplasm at the terminal emits light strongly, indicating a sudden Ca^{2+} inflow. R. R. Llinás and his colleagues demonstrated direct linkage between Ca^{2+} inflow and neurotransmitter release. They found that removal of Ca^{2+} from the extracellular medium, so that no calcium inflow accompanies arrival of an impulse at the axon terminal, completely inhibits transmitter release. The same investigators showed that the total amount of neurotransmitter released is directly proportional to the amount of Ca^{2+} flowing into the axon terminal cytoplasm from the cell exterior.

Ca^{2+} influx is followed in microseconds by fusion of synaptic vesicles with the plasma membrane and release of the vesicle contents into the synaptic cleft. Although the molecular processes linking Ca^{2+} inflow to vesicle release are not completely understood, the ion is known to activate *synapsin*, a protein of the annexin family, all of which promote membrane fusion (see p. 856). Additional steps leading to fusion of synaptic vesicles with the plasma membrane involve the addition of phosphate groups to proteins in the vesicle membranes. (The release of synaptic vesicles is a specialized form of exocytosis; for further details, see Ch. 20.)

The ligand-gated channels or receptors at the receiving side of the synapse contain a binding site directed toward the synaptic cleft that is tailored to fit the neurotransmitter molecule (Fig. 6-24a). Binding the neurotransmitter induces a conformational change in the channel or receptor that opens the gate (Fig. 6-24b), and positively charged ions, usually Na^+ or K^+, flow inward or outward along their concentration gradients. If the electrical disturbance created by the ions flowing through the channels is large enough, voltage-gated Na^+ and K^+ channels open in membrane regions adjacent to the synapse, and a new action potential is triggered in the recipient cell. Once generated, the action potential is propagated over the plasma membrane of the receiving cell by voltage-gated Na^+ and K^+ channels. Inhibitory neurotransmitters have the opposite effect and make the ligand-gated channel resistant to opening. Although the mechanism seems cumbersome, the entire process of synaptic transmission proceeds in a matter of milliseconds, and is almost as rapid as propagation of a nerve impulse along an axon.

Recovery of the synapse after action potentials cease arriving at the axon terminal involves several steps. First, the voltage-gated Ca^{2+} channels in the terminal snap shut, and any excess Ca^{2+} in the axon

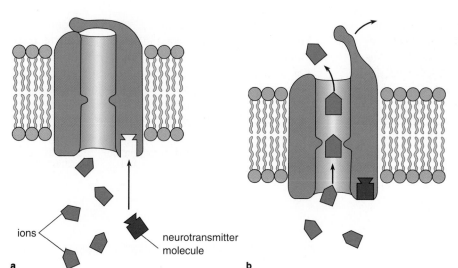

Figure 6-24 Action of a ligand-gated channel controlling the diffusion of ions in response to binding a neurotransmitter molecule. **(a)** Channel closed. **(b)** Binding the neurotransmitter induces a conformational change that opens the channel gate, allowing the ions to flow inward along their concentration gradient.

ions

neurotransmitter molecule

a   b

cytoplasm is quickly removed by Ca^{2+}-ATPase pumps. The drop in Ca^{2+} concentration inactivates synapsin. This inhibits vesicle fusion with the plasma membrane and release of neurotransmitter molecules into the synaptic cleft. Any neurotransmitter molecules remaining in the cleft are quickly removed by enzymatic breakdown, diffusion away from the cleft region, or resorption into the axon terminal, which proceeds by means of specialized membrane proteins that bind and transport the neurotransmitters from the synapse back into the axon cytoplasm. These mechanisms remove the neurotransmitter from the cleft within several microseconds.

The drop in concentration promotes release of the neurotransmitter molecules bound to the ligand-gated channels or receptors in the receiving membrane (the binding is reversible), and the channels close in response. These changes stop the generation of action potentials in the receiving membrane and restore the synapse to its resting condition.

Synaptic Transmission Based on Acetylcholine
Synaptic transmission has been most completely studied in axons releasing acetylcholine as a neurotransmitter (Fig. 6-25a). Synapses of this type are typical of many motor neurons and neuromuscular junctions in vertebrates. Each synaptic vesicle in the axon terminals serving these synapses contains some 6000 to 10,000 molecules of acetylcholine. An arriving action potential

releases the contents of about 300 synaptic vesicles; only 2 msec or so are required for the released acetylcholine to diffuse across the synaptic cleft.

The acetylcholine receptor is a complex protein containing four different polypeptides, each with several transmembrane segments, encoded in four separate but related genes. One of the polypeptides occurs in two copies; the remaining three occur in single copies in the acetylcholine receptor. The two identical polypeptides each contain an acetylcholine-binding site and form a channel through the membrane. Binding acetylcholine opens the receptor channel to the flow of positively charged ions, primarily Na^+, which initiates the action potential on the receiving side of the synapse. The channel opens for only about 1 msec. At peak conductance, however, the channel can admit Na^+ at the rate of 30,000 ions/msec.

Acetylcholine released into the synaptic cleft is rapidly cleared by *acetylcholinesterase*. This enzyme, which is constantly active in the cleft, hydrolyzes the neurotransmitter into choline and acetate (Fig. 6-25b).

The acetylcholine receptor is the target of several venoms and toxins. *Curare* and *bungarotoxin*, a component of snake venom, compete directly with acetylcholine for the binding site on the receptor and, once bound, lock the receptor in an open or closed condition. Other agents, such as the drug *procaine*, combine noncompetitively (see p. 96) by binding with other sites on the receptor to block its action.

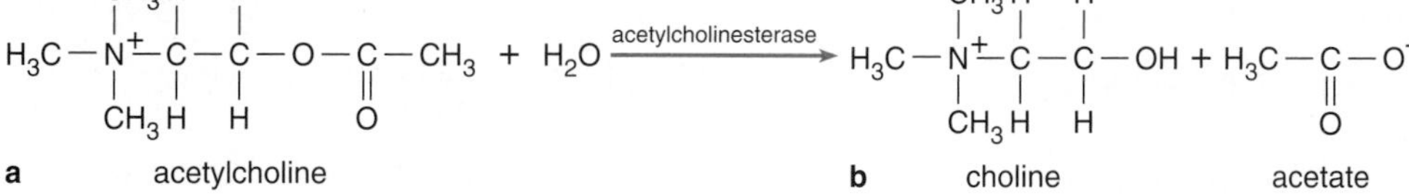

$$H_3C-\overset{\overset{\displaystyle CH_3}{|}}{\underset{\underset{\displaystyle CH_3}{|}}{N^+}}-\overset{\overset{\displaystyle H}{|}}{\underset{\underset{\displaystyle H}{|}}{C}}-\overset{\overset{\displaystyle H}{|}}{\underset{\underset{\displaystyle H}{|}}{C}}-O-\overset{\overset{\displaystyle}{}}{\underset{\underset{\displaystyle O}{\|}}{C}}-CH_3 + H_2O \xrightarrow{\text{acetylcholinesterase}} H_3C-\overset{\overset{\displaystyle CH_3}{|}}{\underset{\underset{\displaystyle CH_3}{|}}{N^+}}-\overset{\overset{\displaystyle H}{|}}{\underset{\underset{\displaystyle H}{|}}{C}}-\overset{\overset{\displaystyle H}{|}}{\underset{\underset{\displaystyle H}{|}}{C}}-OH + H_3C-\overset{\overset{\displaystyle}{}}{\underset{\underset{\displaystyle O}{\|}}{C}}-O^-$$

a   acetylcholine   b   choline   acetate

Figure 6-25 Acetylcholine and the reaction deactivating it by enzymatic breakdown. **(a)** The acetylcholine molecule. **(b)** The reaction breaking down acetylcholine, catalyzed by the enzyme acetylcholinesterase.

Transmission across synapses depends on the combined activity of voltage- and ligand-gated channels or receptors on either side of a synapse. The flow of ions through the gated channels occurs by facilitated diffusion along concentration gradients; the Na^+, K^+, and Ca^{2+} gradients responsible for the flow, in turn, are set up by the activity of Na^+/K^+-ATPase and Ca^{2+}-ATPase active transport pumps. In nerve cells gated ion flow produces the action potential and, through propagation of the action potential, generates nerve impulses. The nerve impulses, passed from cell to cell via synapses or, more rarely, by direct connections between membranes, are the primary route of communication between excitable cells. All the functions of animal nervous and muscular systems, including those of your eyes and brain in reading and comprehending the words on this page, depend on these nerve impulses; in turn, the impulses result from the activities of passive and active membrane transporters, embedded as integral proteins in the plasma membranes of nerve and muscle cells.

For Further Information

ATP synthesis in mitochondria and chloroplasts, *Chs. 9 and 10*
Changes in membrane potential during fertilization, *Ch. 26*
Endocytosis and exocytosis, *Ch. 20*
F_oF_1 ATPase in chloroplasts and cyanobacteria, *Ch. 10*
F_oF_1 ATPase in mitochondria and bacteria, *Ch. 9*
Membrane structure, *Ch. 5*
Muscle contraction, *Ch. 12*
Transport in mitochondria and chloroplasts, *Chs. 9 and 10*
Transport through the nuclear envelope, *Ch. 14*

Suggestions for Further Reading

General Books and Articles

Benga, G. 1988. Water transport in red blood cell membranes. *Prog. Biophys. Molec. Biol.* 51:193–245.

Finkelstein, A. 1987. *Water movement through lipid bilayers, pores, and plasma membranes.* New York: Wiley.

Frace, A. M., and Gargus, J. J. 1991. Molecular biology of membrane transport proteins. *Curr. Top. Membr. Transport* 39:3–36.

Gennis, R. B. 1989. *Biomembranes: molecular structure and function.* New York: Springer-Verlag.

Morell, P., and Norton, W. T. 1980. Myelin. *Sci. Amer.* 242:88–118 (May).

Okazaki, Y., and Tazawa, M. 1990. Calcium ion and turgor regulation in plant cells. *J. Membr. Biol.* 114:189–94.

Pasternak, C. A. 1989. Membrane transport and disease. *Molec. Cellular Biochem.* 91:3–11.

Quinton, P. M. 1990. Cystic fibrosis: a disease in electrolyte transport. *FASEB J.* 4:2709–17.

Stein, W. D. 1990. *Channels, carriers, and pumps: an introduction to membrane transport.* New York: Academic.

Sussman, M. R., and Harper, J. F. 1989. Molecular biology of the plasma membrane in higher plants. *Plant Cell* 1:953–60.

Wiggins, P. M. 1990. Role of water in some biological processes. *Microbiol. Rev.* 54:432–49.

Glucose Transporter (Facilitated Diffusion)

Baly, D. L., and Horuk, R. 1988. The biology and biochemistry of the glucose transporter. *Biochim. Biophys. Acta* 947:571–90.

Gould, G. W., and Bell, G. I. 1990. Facilitative glucose transporters: an expanding family. *Trends Biochem. Sci.* 15:18–23.

Pessin, J. E., and Bell, G. I. 1992. Mammalian facilitative glucose transporter family. *Ann. Rev. Physiol.* 54:911–30.

Widdas, W. F. 1988. Old and new concepts of the membrane transport for glucose in cells. *Biochim. Biophys. Acta* 947:386–404.

Ion Channels

Alper, S. L. 1991. The band 3-related anion exchanger (AE) gene family. *Ann. Rev. Physiol.* 53:549–64.

Bean, B. P. 1989. Classes of calcium channels in vertebrate cells. *Ann. Rev. Physiol.* 51:367–84.

Begenisch, T. 1987. Molecular properties of ion permeation through sodium channels. *Ann. Rev. Biophys. Biophys. Chem.* 16:247–63.

Blott, M. R. 1991. Ion channel gating in plants: physiological implications and integration for stomatal function. *J. Membr. Biol.* 124:95–112.

Catterall, W. A. 1986. Molecular properties of voltage-sensitive sodium channels. *Ann. Rev. Biochem.* 55:953–85.

———. 1988. Structure and function of voltage-sensitive ion channels. *Science* 242:50–61.

———. 1991. Functional subunit structure of voltage-gated calcium channels. *Science* 253:1499–1500.

Eisenberg, R. S. 1990. Channels as enzymes. *J. Membr. Biol.* 115:1–12.

Flatman, P. J. 1991. Mechanisms of magnesium transport. *Ann. Rev. Physiol.* 53:259–71.

Hosey, M. M., and Lazdunski, M. 1988. Ca^{2+} channels: molecular pharmacology, structure, and regulation. *J. Membr. Biol.* 104:81–105.

Jan, L. Y., and Jan, Y. N. 1989. Voltage-sensitive ion channels. *Cell* 56:13–25.

———. 1992. Structural elements involved in specific K^+ channel function. *Ann. Rev. Physiol.* 54:537–55.

Krueger, B. K. 1989. Toward an understanding of structure and function of ion channels. *FASEB J.* 3:1906–14.

Latorre, R. 1991. Metabolic control of K^+ channels: an overview. *J. Bioenerget. Biomembr.* 23:493–97.

Latorre, R.; Oberhauser, A.; Labarca, P.; and Alvarez, O. 1989. Varieties of calcium-activated potassium channels. *Ann. Rev. Physiol.* 51:386–99.

Mackinnon, R. 1991. Using matagenesis to study potassium channel mechanisms. *J. Bioenerget. Biomembr.* 23:647–63.

McClesky, E. W., and Schroeder, J. E. 1991. Functional properties of voltage-dependent calcium channels. *Curr. Top. Membr. Transport* 39:296–326.

Miller, C. 1991. 1990: Annus mirabilis of potassium channels. *Science* 252:1092–96.

Montal, M. 1990. Molecular anatomy and molecular design of channel proteins. *FASEB J.* 4:2623–35.

Morris, C. E. 1990. Mechanosensitive ion channels. *J. Membr. Biol.* 113:93–107.

Schroeder, J. I., and Hedrich, R. 1989. Involvement of ion channels and active transport in osmoregulation. *Trends Biochem. Sci.* 14:187–92.

Schroeder, J. I., and Thuleau, P. 1991. Ca^{2+} channels in higher plant cells. *Plant Cell* 3:555–59.

Smith, P. R., and Benos, D. J. 1991. Epithelial Na^+ channels. *Ann. Rev. Physiol.* 53:509–30

Stühmer, W. 1991. Structure-function studies of voltage-gated ion channels. *Ann. Rev. Biophys. Biophys. Chem.* 20:65–78.

Trimmer, J. S., and Agnew, W. S. 1989. Molecular diversity of voltage-sensitive Na channels. *Ann. Rev. Physiol.* 51:401–18.

Tsien, R. W., and Tsien, R. Y. 1990. Ca^+ channels, stores, and oscillations. *Ann. Rev. Cell Biol.* 6:715–60.

Tyerman, S. D. 1992. Anion channels in plants. *Ann. Rev. Plant Physiol. Plant Molec. Biol.* 43:351–73.

P-Type and V-Type Active Transport Pumps

Briskin, D. P. 1990. The plasma membrane H^+-ATPase of higher plant cells: biochemical and transport functions. *Biochim. Biophys. Acta* 1019:95–109.

Carafoli, E., and Chiesi, M. 1992. Calcium pumps in the plasma and intracellular membranes. *Curr. Top. Cellular Regulat.* 32:209–24.

Dhalla, N. S., and Zhao, D. 1988. Cell membrane Ca^{2+}/Mg^{2+} ATPase. *Prog. Biophys. Molec. Biol.* 52:1–37.

Horisberger, J.-D.; Lemas, V.; Kraehenbul, J.-P.; and Rossier, B. C. 1991. Structure-function relationships of the Na,K-ATPase. *Ann. Rev. Physiol.* 53:565–84.

Jencks, W. P. 1989. How does a Ca^{2+} pump work? *J. Biolog. Chem.* 264:18855–58.

Lingrel, J. B; Orlowski, J.; Shull, M. M.; and Price, E. M. 1990. Molecular genetics of Na,K-ATPase. *Prog. Nucleic Acid Res. Molec. Biol.* 38:37–89.

Nakamoto, R. K., and Slayman, C. W. 1989. Molecular properties of the fungal plasma membrane [H^+]-ATPase. *J. Bioenerget. Biomembr.* 21:621–32.

Nelson, N. 1989. Structure, molecular genetics, and evolution of vacuolar H^+-ATPases. *J. Bioenerget. Biomembr.* 21:553–71.

———, and Taiz, L. 1989. The evolution of H^+-ATPases. *Trends Biochem. Sci.* 14:113–16.

Rabon, E. C., and Reuben, M. A. 1990. The mechanism and structure of the gastric H,K-ATPase. *Ann. Rev. Physiol.* 52:321–44.

Rossier, B. C.; Geering, K.; and Krahenbuhl, J. P. 1987. Regulation of the sodium pump: how and why. *Trends Biochem. Sci.* 10:483–87.

Serrano, R. 1988. Structure and function of proton translocating ATPase in plasma membranes of plants and fungi. *Biochim. Biophys. Acta* 947:1–28.

Stone, D. K.; Crider, B. P.; Sudhof, T. C.; and Xie, X.-S. 1989. Vacuolar proton pumps. *J. Bioenerget. Biomembr.* 21:605–20.

Strehler, E. E. 1991. Recent advances in the molecular characterization of plasma membrane Ca^{2+} pumps. *J. Membr. Biol.* 123:93–103.

Indirect Active Transport Pumps (Cotransport)

Baldwin, S. A., and Henderson, P. J. F. 1989. Homologies between sugar transporters from eukaryotes and prokaryotes. *Ann. Rev. Physiol.* 51:459–71.

Kimmich, G. A. 1990. Membrane potentials and the mechanism of intestinal Na^+-dependent sugar transport. *J. Membr. Biol.* 114:1–27.

Lienhard, G. E.; Slot, J. W.; James, D. E.; and Mueckler, M. M. 1992. How cells absorb glucose. *Sci. Amer.* 266:86–91 (January).

Saier, M. H.; Daniels, G. A.; Boerner, P.; and Lin, J. 1988. Neutral amino acid transport systems in animal cells: potential targets of oncogene action and regulators of cell growth. *J. Membr. Biol.* 104:1–20.

White, M. F. 1985. The transport of cationic amino acids across the plasma membrane of mammalian cells. *Biochim. Biophys. Acta* 822:356–74.

Generation and Propagation of the Action Potential in Nerve Cells

Catterall, W. A. 1984. The molecular basis of neuronal excitability. *Science* 223:653–61.

Llinás, R. R. 1988. The intrinsic electrophysiological properties of mammalian neurons: insights into central nervous system function. *Science* 242:1654–63.

Synaptic Transmission and Neurotransmitters

Bloom, F. E. 1988. Neurotransmitters: past, present, and future directions. *FASEB J.* 2:32–41.

DeCamilli, P., and Jahn, R. 1990. Pathways to regulated exocytosis in neurons. *Ann. Rev. Physiol.* 52:625–45.

Pollard, H. B.; Burns, A. L.; and Rojas, E. 1990. Synexin (annexin VII): a cytosolic Ca^{2+}-binding protein which promotes membrane fusion and forms Ca^{2+} channels in artificial bilayers and natural membranes. *J. Membr. Biol.* 117:101–12.

Bacterial Transport

Ames, G. F.-L. 1988. Structure and mechanism of bacterial periplasmic transport systems. *J. Bioenerget. Biomembr.* 20:1–18.

Benz, R. 1988. Structure and function of porins from gram negative bacteria. *Ann. Rev. Microbiol.* 42:359–93.

Henderson, P. J. F. 1990. Proton-linked sugar transport systems in bacteria. *J. Bioenerget. Biomembr.* 22:571–92.

Higgins, C. F.; Hyde, S. C.; Mimmack, M. M.; Gileadi, U.; Gill, D. R.; and Gallagher, M. P. 1990. Binding protein-dependent transport systems. *J. Bioenerget. Biomembr.* 22:571–92.

Maloney, P. C. 1990. Anion exchange reactions in bacteria. *J. Bioenerget. Biomembr.* 22:509–23.

Nakai, T. 1986. Outer-membrane permeability of bacteria. *CRC Crit. Rev. Biochem.* 13:1–62.

Nikaido, H. 1992. Porins and specific channels of bacterial outer membranes. *Molec. Microbiol.* 6:435–42.

Shuman, H. A. 1987. The genetics of active transport in bacteria. *Ann. Rev. Genet.* 21:156–77.

Silver, S.; Nucifora, G.; Chu, L.; and Misra, T. K. 1989. Bacterial resistance ATPases: primary pumps for exporting toxic cations and anions. *Trends Biochem. Sci.* 14:76–80.

Review Questions

1. What are the differences between passive and active transport?

2. What is a concentration gradient?

3. What is diffusion? What causes diffusion?

4. What is a semipermeable membrane? What is the relationship of osmosis to semipermeable membranes?

5. What conditions are necessary for osmosis to occur? What provides the energy for osmosis?

6. What is osmotic pressure? How is osmotic pressure related to cellular life? Compare a cell with the osmosis apparatus shown in Figure 6-2.

7. What is passive transport? What is the relationship between passive transport and lipid solubility?

8. What is facilitated diffusion? What is the relationship between membrane proteins and facilitated diffusion? How does facilitated diffusion resemble the activity of enzymes?

9. Outline the mechanism by which facilitated diffusion is believed to work.

10. Give examples of membrane transport proteins carrying substances by facilitated diffusion.

11. What are ion channels? What is gating? What are voltage-gated channels? Ligand-gated channels? What functions do gated channels carry out in cells?

12. What is unusual about the transport of water through cellular membranes? What might account for the unusual behavior of water?

13. What is the relationship between passive transport and the fluid mosaic structure of cellular membranes?

14. Compare facilitated diffusion and active transport. What supplies the energy for active transport? What is the difference between direct and indirect active transport?

15. How are direct active transport pumps believed to work? What are P-type active transport pumps? V-type? Give examples of both types and outline their importance in living cells. Compare the structure of P-type and V-type pumps and the F_oF_1 ATPase. In what organisms do these structures occur?

16. How are indirect active transport pumps believed to work? What kinds of substances are moved across membranes by indirect active transport?

17. Discuss the balance of Na^+, H^+, K^+, and Ca^{2+} in cells and how this balance is maintained.

18. What happens if a direct active transport protein is forced to run in reverse? What significance does this have for cellular energy metabolism?

19. How do facilitated diffusion and active transport depend on the ability of proteins to take on different folding conformations?

20. How can transport produce a voltage difference across membranes? What is voltage? Electrical potential? Current? Capacitance? What do these terms mean in relationship to membrane transport?

21. Outline the structure of a nerve cell.

22. What is the resting potential? The action potential? What ion movements are responsible for generating the action potential? What is the role of gated membrane

channels in generation of the action potential? What experimental evidence supports the conclusion that these ion movements actually produce the action potential?

23. What is propagation of the action potential? What is responsible for this propagation? What is myelin? A node?

Outline the difference between continuous and saltatory conduction. How does saltatory conduction work?

24. What are refractory periods? What is the relationship of refractory periods to propagation of the action potential in continuous and saltatory conduction?

Supplement 6-1
Transport in Bacteria

Bacteria possess complete transport systems that set up ion gradients across the plasma membrane and move metabolites such as sugars and amino acids into the cell. As in eukaryotes bacterial transport is accomplished by a combination of passive and active transport. Passive transport of polar substances depends on facilitated diffusion catalyzed by membrane proteins, as it does in eukaryotes; however, bacteria have relatively few facilitated diffusion transporters. Further, although passive transporters exist for ions and metabolites, such as sugars and amino acids, these substances rarely occur in bacterial environments in concentrations high enough to drive diffusion into the cell. As a consequence, most ions and metabolites move across bacterial plasma membranes by active transport.

Direct active transport in bacteria is driven by mechanisms more diverse than those found in eukaryotes. Some active transport is carried out by pumps using ATP as a direct energy source. Other pumps appear to be driven by phosphate-bond energy, in forms such as polyphosphates (inorganic phosphates linked into chains) or by a combination of ATP hydrolysis and a driving H^+ gradient, that is, by a combination of direct and indirect active transport. In a few photosynthetic bacteria active transport is directly propelled by light. In these cells a carrier absorbs light and uses the absorbed energy to push H^+ out of the cell.

Indirect active transport in bacteria is driven primarily by an H^+ gradient. However, in contrast to fungi and plants, which also use an H^+ gradient to drive cotransport, the high outside/low inside H^+ gradient of bacteria is established primarily through the activity of membrane proteins taking part in oxidative reactions. These proteins use the energy of electrons removed from substances undergoing oxidation to drive H^+ from the cytoplasm to the cell exterior. In this sense the electron-driven systems represent yet another class of active transport pumps operating in bacterial plasma membranes. Equivalent systems use the energy of electrons removed in oxidations to establish an H^+ gradient in mitochondria and chloro-

plasts. In these organelles the H^+ gradient, in turn, is used to drive ATP synthesis through activity of the F_oF_1 ATPase (for details, see Chs. 9 and 10).

A few bacterial cotransport systems use a high outside/low inside Na^+ gradient as their energy source. These Na^+-dependent transporters are actually indirectly dependent on the H^+ gradient set up by electron transport. This is because the Na^+ gradient itself is created by an indirect transport pump that uses the H^+ gradient to move Na^+ outward across the plasma membrane.

Bacterial transport has been studied most extensively in *E. coli*. Transport in this bacterium is complicated by the fact that, as in other gram-negative bacteria, the cytoplasm is surrounded by two boundary membranes (see p. 302 and Fig. 8-31*b*). The inner membrane, which lies just under the cell wall, is the plasma membrane, and is structurally and functionally equivalent to the plasma membranes of other cells. The outer membrane lies just outside a layer of cell wall material. It effectively seals the outer cell surface against the entry of most molecules in the extracellular medium except for those that enter passively through channel-forming proteins called *porins* (Fig. 6-26). Entry through the porin channels of *E. coli* is limited to ions and relatively small organic substances with molecular weights below about 600 to 650. Once through the outer membrane, substances cross the intervening cell wall region, called the *periplasmic space*, and enter the cytoplasm through specific passive and active transporters embedded in the plasma membrane. Thus there are two sets of transport proteins in gram-negative bacteria: the porins of the outer membrane, which admit substances by passive diffusion, and the passive and active transporters of the plasma membrane. Gram-positive bacteria, which lack the outer membrane, have only plasma membrane–associated transport systems.

Porins and Transport Through the Outer Membrane of Gram-Negative Bacteria

The channels formed by porins allow hydrophilic molecules below a certain size to pass through the outer membrane of gram-negative bacteria. Some porins are

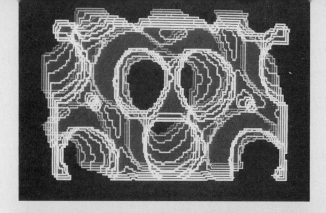

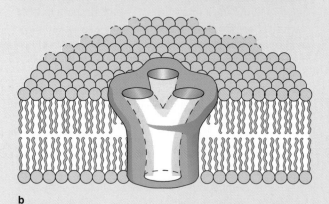

a

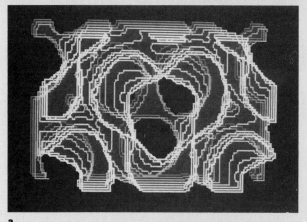

b

Figure 6-26 A nonselective bacterial porin with three openings facing the outside surface, which merge to form a single channel on the cytoplasmic surface of the membrane. **(a)** A reconstruction from high-resolution analysis of an electron microscope image. Courtesy of A. Engel, reprinted by permission from *Nature* 317: 643 (1985); copyright © 1985 Macmillan Magazines Ltd. **(b)** A porin molecule in position in the outer membrane.

enzymatic proteins. The outer membrane also protects intestinal bacteria such as *E. coli* from the detergent effects of bile secreted into the digestive tract. (For further details of the structure and function of the outer membrane of gram-negative bacteria, see Ch. 8.)

E. coli normally has only a single, nonselective porin type in its outer membrane. Each cell may have as many as 100,000 or more copies of the porin molecule, however, making the porin one of the most abundant cellular proteins. Stressful environmental conditions such as high salinity or limitations in the supply of inorganic phosphates may induce other nonselective porin types to appear in *E. coli*.

The selective porins, which preferentially admit substances such as lactose- or maltose-containing carbohydrates, ribose, or short peptides, occur in bacterial outer membranes only in association with active transport pumps in the underlying plasma membrane. The pumps push the substances admitted by the porins across the plasma membrane. The selective porins are encoded by a separate gene family with no sequence similarities to the nonselective porins. At least 20 different types of selective porins have been identified in *E. coli*, each one associated with a specific active transport pump that moves the molecules admitted by the porin through the plasma membrane.

Porins with characteristics similar to the sieving porins of gram-negative bacteria also occur in the outer membranes of mitochondria and chloroplasts. These porins admit molecules of larger dimensions than bacterial porins—up to about 6000 daltons for mitochondrial porins and as much as 10,000 to 13,000 daltons for chloroplast porins. At these dimensions the mitochondrial and chloroplast porins have channels that are large enough to allow free passage of all the metabolites entering and leaving the organelles but still small enough to exclude larger molecules such as enzymes. The existence of porins in the outer membranes of mitochondria and chloroplasts is one of the many structural and functional similarities between these eukaryotic organelles and prokaryotes (for details, see Chs. 21 and 27).

Transport Through the Bacterial Plasma Membrane

Two major systems of transport proteins exist in bacterial plasma membranes. One system, found only in gram-negative bacteria, occurs in conjunction with the selective class of outer membrane porins. These active transport pumps, built up from multiple polypeptide subunits, transport only organic substances admitted by selective porins. The second system, characteristic of all bacteria, transports both ions and organic substances. In general, the second class, which includes both passive and active transporters, is formed from

selective and admit only certain types or classes of molecules; others act simply as molecular sieves that admit essentially any substance small enough to pass through their channels. In either event the size ranges admitted are small enough to exclude potentially damaging molecules of larger size such as antibiotics and single polypeptides unrelated in structure to the multiple-subunit carriers of the first class.

Active Transport Occurring in Conjunction with Selective Porins

The complex active transport systems working in conjunction with selective porins in gram-negative bacteria contain several components (Fig. 6-27). One component is the selective porin itself. The second component is a soluble *periplasmic protein*, so called because it is located in the periplasmic space between the outer and plasma membranes. Each transport system has its own periplasmic protein, which combines with the substances admitted by the porin associated with the system and greatly increases the efficiency of its transport. The final component of the system is the active transport pump embedded in the plasma membrane. Most of these pumps, termed *periplasmic protein-related transporters*, consist of four polypeptide subunits. Two of the subunits are transmembrane proteins with five or six transmembrane segments, and two are hydrophilic subunits located on the cytoplasmic side of the plasma membrane. The two cytoplasmic segments have binding sites for ATP.

One of the best-characterized periplasmic protein-dependent transporters is the maltose-transporting system of *E. coli*. The porin of the maltose system, called *maltoporin*, is highly selective for maltose and small polysaccharides with up to six or seven maltose units. The maltoporin is also called the *lamB* protein because it is encoded in a gene given this name in *E. coli*. The periplasmic protein for the maltose system, encoded in the *malE* gene, is a 370-amino acid polypeptide with one binding site that specifically recognizes maltose sugars. Like other periplasmic proteins, the malE protein is a kidney-shaped molecule with two globular domains connected by a narrow neck or hinge, which is the maltose binding site (see Fig. 6-27). Studies of conformational changes occurring when malE binds maltose indicate that the molecule bends at the hinge as the substrate attaches, clamping the sugar between the two lobes. The periplasmic protein with its bound sugar molecule diffuses across the periplasmic space and in some as yet unknown manner greatly speeds uptake of the sugar by the active transporter in the plasma membrane. Presumably, combination with the sugar activates a binding site on the periplasmic protein that specifically recognizes the active transporter. Binding between the periplasmic protein and the active transporter releases the sugar in a position that greatly favors its entry into the channel of the active transporter. Experimental elimination of the periplasmic protein

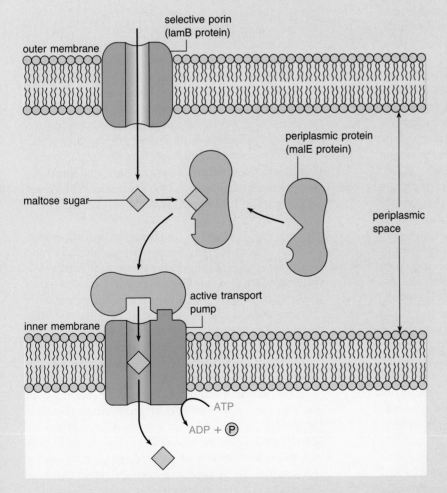

Figure 6-27 The periplasmic active transport system of *E. coli* and other gram-negative bacteria, including a selective porin in the outer membrane, the periplasmic protein, and the active transport complex of the inner membrane. The system transporting maltose sugars in *E. coli* is shown.

selective porin
(lamB protein)

outer membrane

maltose sugar

periplasmic protein
(malE protein)

periplasmic space

active transport pump

inner membrane

ATP

ADP + P

markedly slows active transport of maltose sugars, confirming its importance to the transport system.

The active transport protein, like other active transport pumps, undergoes conformational changes that alternately expose the maltose-binding site on the outside and inside surfaces of the plasma membrane as the pumping cycle turns (as in Fig. 6-11). Energy for the pumping cycle is evidently derived from a combination of the H^+ gradient and hydrolysis of ATP or other phosphates. If either of these energy sources is eliminated in experimental systems, pumping stops. Other porin-associated active transport systems work similarly to the maltose system in *E. coli* and other gram-negative bacteria.

Several active transport pumps recently discovered in humans, yeast, *Drosophila*, a plant chloroplast, and the protozoan responsible for malaria evidently have structural relationships to the bacterial pumps of the porin-associated group. Each of these pumps, although structured from a single protein, has four domains with structures corresponding to the periplasmic protein-dependent pumps of bacteria. The correspondence extends to strongly conserved amino acid sequences in bacterial and eukaryotic types and to the ability of the two cytoplasmic domains to bind ATP. One of these eukaryotic active transport pumps, the *MDR* pump (*MDR* = *MultiDrug Resistance*) appears in increased numbers in the plasma membranes of human cells exposed to a wide variety of toxic substances. Unfortunately, the toxic substances removed from the cytoplasm by the pump include drugs used to treat cancer. The MDR pump frequently nullifies the effect of anticancer drugs by transporting them rapidly out of tumor cells. A *CQR pump* (*CQR* = *ChloroQuinone Resistance*) recently discovered in the malaria parasite has a similar function, allowing the parasite to pump out chloroquinone, a drug used to treat malaria.

Another human transporter related to the same family of active transport pumps is encoded in the *CF* gene, which in mutant form is responsible for defects in chloride transport associated with cystic fibrosis (see p. 202). The gene, discovered in the laboratories of L.-C. Tsui, F. S. Collins, and J. R. Riordan, encodes a membrane protein called *CFTR* (for *Cystic Fibrosis Transmembrane conductance Regulator*). The name appears to be a misnomer, because, rather than being a regulator, CFTR appears to be a transport protein that directly conducts Cl^- across plasma membranes. Furthermore, although related transporters are active pumps, CFTR apparently operates as a passive gated ion channel. Mutations in *CF* are common—about 1 in 25 persons of Northern European extraction are unaffected carriers, and 1 in 2500 are homozygous recessives born with the disease.

Other Bacterial Transporters The remaining bacterial transport systems occur in the plasma membranes of both gram-negative and gram-positive bacteria. The active transporters of this group are single polypeptides that may use phosphate bond energy, H^+ or Na^+ gradients, or a combination of both gradients and phosphate hydrolysis to drive substances across the plasma membrane. The various pumps may transport ions or organic substances. Although ion transport may be direct or indirect, most transporters of organic molecules use the H^+ ion gradient established by electron transport as their energy source.

Ion Transport Active transport systems have been detected in *E. coli* and other bacteria that push K^+ and Na^+ across the plasma membrane. K^+ is concentrated inside bacterial cells by a transporter with clear structural and functional affinities to the P-type pumps of eukaryotes. This K^+-ATPase directly uses ATP as its energy source, and attaches a phosphate group derived from ATP during the pumping cycle, as do the P-type pumps of eukaryotes. The inward movement of K^+ driven by the bacterial K^+-ATPase provides potassium necessary for metabolic reactions and maintains internal osmotic pressure at positive values. Depending on conditions, Na^+ may be pushed inward or outward. Active inward movement is driven by an Na^+-ATPase pump that directly uses ATP as its energy source; active outward movement takes place by an antiporter using the H^+ gradient as its energy source.

Bacteria also have a series of ATP-driven active transport pumps that specifically transport toxic ions out of the cell. Pumps in this group, collectively known as *resistance ATPases*, push out such toxic ions as arsenate, arsenite, and antimony ions. The resistance ATPases, which provide part of the environmental defenses of both gram-positive and gram-negative bacterial cells, form a family of structurally related pumps that has no counterpart in eukaryotes and shows no sequence relationships to other classes of ATP-driven active transport pumps in bacteria.

Transport of Organic Substances Inward movement of organic molecules, primarily sugars and amino acids, may take place either passively or actively in bacteria. As noted, passive transporters are limited in occurrence because favorable concentration gradients are rare for organic substances in bacterial environments. However, passive transporters facilitating the inward diffusion of organic molecules can be detected in the plasma membrane when their substrates exist in plentiful supply in the medium.

The active transporters of organic molecules in bacteria are cotransporters; although some are Na^+ dependent, most use the H^+ gradient established by electron transport as their energy source. In *E. coli*, cotransporters capable of concentrating all 20 amino acids in the cytoplasm have been detected. Some of these pumps are specific for single amino acids; others

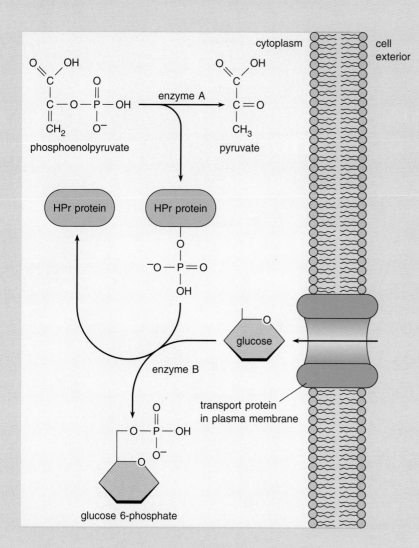

Figure 6-28 A typical group translocation reaction converting glucose to glucose 6-phosphate on its entry into the cytoplasm, shown in simplified form. A phosphate group is transferred from phosphoenolpyruvate, an intermediate compound in cellular oxidative reactions, to a carrier protein, HPr, by the activity of enzyme A. The phosphate group is then transferred to a glucose molecule entering the cytoplasm through a transport protein located in the plasma membrane. The reaction, catalyzed by enzyme B, produces glucose 6-phosphate and the unphosphorylated form of the HPr protein. Phosphorylation of glucose to glucose 6-phosphate effectively removes the sugar from the glucose concentration gradient.

carry groups of amino acids with related chemical structures.

H^+-driven transporters for sugars such as lactose, arabinose, galactose, and xylose also occur in *E. coli.* All these sugar transporters have similar amino acid sequences capable of forming 12 to 14 alpha-helical transmembrane segments. As noted, the sugar transporters of *E. coli* also share sequence homologies with the protein moving glucose into erythrocytes and other mammalian cells by facilitated diffusion (see p. 196).

In many bacterial transport systems sugars are phosphorylated as they enter the cytoplasm. As in eukaryotic systems (see p. 197) this phosphorylation effectively removes the sugar from the concentration gradient because the phosphorylated types cannot pass through the membrane. In bacteria the systems attaching a phosphate group to sugars, called *group translocation reactions*, derive the phosphate group from an intermediate in oxidative reactions, phosphoenolpyruvate (see p. 316) instead of ATP (Fig. 6-28).

Light-Driven Active Transport of H^+ in Photo-Synthetic Bacteria The plasma membranes of one

bacterial group, the purple photosynthetic bacteria (see p. 408), possess an active transport pump that uses light energy to push H^+ from the cytoplasm to the cell exterior. The protein, *bacteriorhodopsin* (see Fig. 5-8), contains a light-absorbing organic group identical to *rhodopsin*, the characteristic visual pigment of animal cells. Absorption of light by rhodopsin induces a conformational change in the protein that results in expulsion of H^+ across the plasma membrane. The bacteriorhodopsin system supplements the H^+ gradient established by oxidative electron transport in the purple bacteria, especially under conditions in which the oxygen necessary for electron transport is in limited supply. Bacteriorhodopsin has been successfully isolated and placed in operating condition in artificial phospholipid films. These isolated bacteriorhodopsin systems have been critical to experiments demonstrating that an H^+ gradient, set up normally by electron transport in bacteria, mitochondria, and chloroplasts, can be used as a direct energy source for the synthesis of ATP (for details, see Ch. 9).

Suggestions for Further Reading: See p. 226.

7

THE CELL SURFACE AND
INTRACELLULAR COMMUNICATION

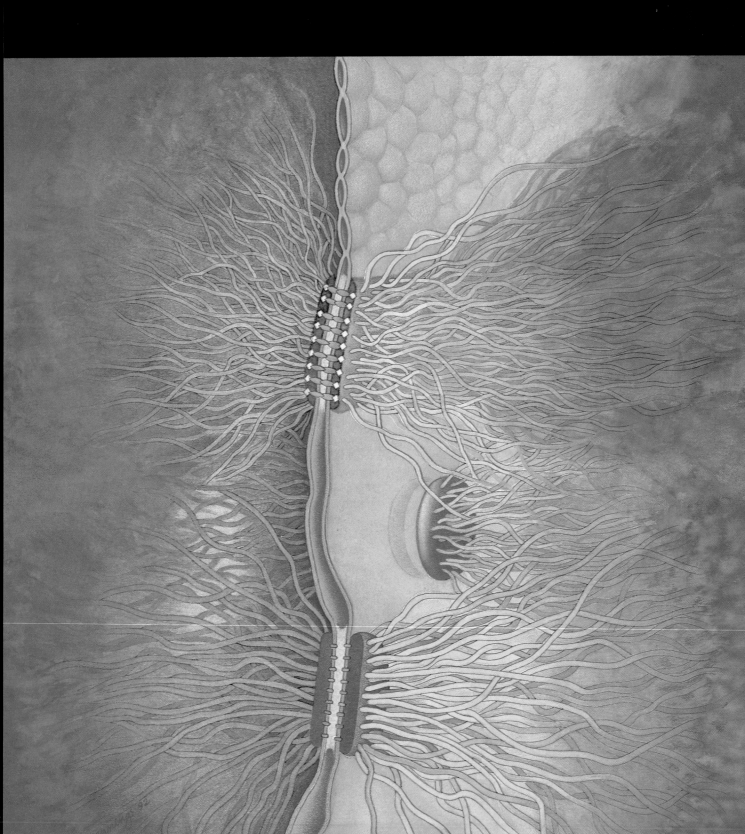

• Glycolipids and glycoproteins • Recognition markers • Lectins • Surface receptors and cellular responses • G proteins • Second messengers • Calmodulin • Protein kinases • Cell-cell adhesions • Desmosomes • Adherens junctions • Septate junctions • Tight junctions • Gap junctions • Recognition and response mechanisms in bacteria

C ells make contact with the outside world through the cell surface, which includes the plasma membrane, its lipid and protein molecules, and parts of these molecules that extend from the membrane into the extracellular medium. Through the activities of this contact layer, cells recognize other cells as part of the same individual or as foreign, send and receive chemical and physical signals, and adhere to other cells or to extracellular materials. The cell surface also provides an outer line of defense against attack by other cells or viruses.

The cell surface is specialized in various ways to carry out these tasks. Cell-to-cell recognition is based on membrane glycolipids and glycoproteins. The carbohydrate portions of these molecules extend like antennae from the plasma membrane, forming a "sugar coating" or *glycocalyx* (see p. 163) at the cell surface. In combination with membrane proteins the carbohydrate groups in effect provide a cell with a name and address.

Other membrane glycoproteins function in cell reception. These receptors, on binding an extracellular chemical signal such as a peptide hormone, trigger complex internal responses that range from adjustments in transport and the rate of oxidation and other metabolic reactions to secretion, cell growth, and division. Surface receptors and internal response mechanisms are the primary elements providing the cell coordination necessary for multicellular existence in animals.

Other glycoproteins are specialized for cell-cell adhesions. Some adhesions depend on the interactions of individual glycoprotein molecules between cells or between cells and extracellular materials. Other adhesions depend on highly specialized arrangements of membrane proteins, glycoproteins, and lipids into *cell junctions* of various kinds. The junctions, each with a distinctive structure, anchor cells to each other, seal cell boundaries to the flow of ions and molecules, and form channels for direct transport and communication between cells.

The cell surface of eukaryotes and its role in recognition, reception, and adhesion, including the various types of cell junctions, are the subjects of this chapter. Supplement 7-1 discusses bacterial recognition, reception, and adhesion.

MOLECULES OF THE CELL SURFACE

The glycolipid and glycoprotein molecules active in cellular recognition, reception, and adhesion consist of a basal phospholipid or protein structure embedded within the membrane and one or more carbohydrate chains that extend from the cell surface. (For details of glycolipid and glycoprotein structure, see Chs. 2 and 5.) The proteins bearing the carbohydrate chains are typically transmembrane proteins with segments that protrude from both the exterior and cytoplasmic surfaces of the plasma membrane. The transmembrane segments are structured primarily from one or more alpha-helical segments that connect the cytoplasmic and extracellular extensions.

The carbohydrate groups of membrane glycolipids and glycoproteins are built up from seven common monosaccharide types (see Fig. 5-5). Four of the monosaccharide units are individual six-carbon sugars: glucose, galactose, mannose, and fucose. The remaining three types are classes made up from members with closely related structures. Two of the classes are formed by *amino sugars*, glucosamine and galactosamine, which consist of a glucose or galactose unit in which one or more of the —OH groups have been replaced by an amino (—NH_2) group. Individual amino sugars differ in the positions of the amino groups and side groups at other points on the molecules. The third sugar class is formed by *sialic acids*, a family based on a seven-carbon framework to which additional short side chains are added. In addition, all sialic acids have a —COOH group attached to the 1-carbon (see Fig. 5-5). The —COOH group is responsible for the acidic properties of this sugar class. Substitutions and additions to the side groups attached to the carbons at other points form more than 20 different molecules in the sialic acid family.

The seven types of sugar units combine in straight- or branched-chain patterns to form the carbohydrate groups of membrane glycolipids and proteins. Individual sugar units may join in $1 \rightarrow 1$, $1 \rightarrow 2$, $1 \rightarrow 3$, $1 \rightarrow 4$, $1 \rightarrow 6$, $2 \rightarrow 3$, and $2 \rightarrow 6$ linkages in either the α- or β-form. (In a $1 \rightarrow 2$ linkage, for example, the 1-carbon of one sugar is linked to the 2-carbon of the next sugar in line; see p. 52). The number of possible linkages allows even a small number of sugar molecules to form an almost endless variety of straight and branched structures. For example, sugar units of only four different types can link together to form more than 35,000 distinct tetrasaccharides! The possibilities for variation in polysaccharide structure, in fact, are much more diverse than even those of proteins. Different membrane glycolipids and glycoproteins may bear single or multiple carbohydrate segments, ranging in length from a single monosaccharide to chains of 70 or more. On proteins the sugar groups are termed *N-*

linked or *O*-linked, depending on the amino acid side groups to which they attach (see p. 162).

Although the functions of the complex carbohydrate groups in membrane glycoproteins and glycolipids are incompletely understood, they probably provide part of the specific structure allowing these molecules to act in recognition, reception, and adhesion. In addition, hydrogen bonds set up between carbohydrate structures stabilize the membrane surface.

Added to the carbohydrate variability are differences in the amino acid sequences in the protein portion of cell surface glycoproteins. These two sources of variability provide so many possible combinations that the potential number of different membrane glycoproteins is essentially infinite. Thus, the carbohydrate structures of membrane glycolipids and both the carbohydrate and protein segments of membrane glycoproteins easily provide the variability required for the diverse functions of the cell surface.

MEMBRANE GLYCOLIPIDS AND GLYCOPROTEINS IN CELL RECOGNITION

The cell surfaces of many organisms, particularly in the animal kingdom, contain glycolipid and glycoprotein molecules that serve as sites of recognition. Some of these surface molecules are markers that identify cells as belonging to a single individual or to a particular tissue type. Other surface glycolipids and glycoproteins have the capacity to recognize other cells and to identify them as part of the same individual or as foreign. In vertebrates recognition of surface molecules as foreign triggers an immune response that normally destroys the invading cell. Rejection of organ transplants in mammals, for example, results from recognition of foreign surface molecules by cells of the immune system, followed by an immune response that kills the transplanted cells. The reactions of humans to blood transfusions also depend on recognition of cell surface molecules.

Major Histocompatibility Complex (MHC) Molecules

The molecules primarily responsible for recognition of mammalian cells as part of self or foreign are a group of surface glycoproteins termed *MHC molecules* (Fig. 7-1). The *MHC*, or *major histocompatibility complex*,[1]

[1] In humans the surface molecules responsible for recognition of cells as self or foreign are also called *HLA markers* (*HLA = Human Leucocyte Antigens*) because they were first identified in leucocytes. Because surface molecules are now known to be present on all human cells, they are more commonly named for the MHC genes encoding them. Similar surface molecules have been found in all vertebrates studied to date.

refers to a group of genes encoding these molecules. The molecules of this group occur in two different types—*class I* and *class II MHC proteins* (see Fig. 7-1). In mammals class I MHC proteins occur on the surfaces of all body cells except those of the immune system. Cells responsible for the immune response—primarily leucocytes of various types—have class II MHC proteins on their surfaces. Both protein types contain constant and variable sequence regions. The constant regions are closely similar or identical between different individuals of a species. The variable regions differ significantly in amino acid sequence, so that no two individuals (except identical twins) are even remotely likely to have the same MHC proteins. The differences arise from random combinations of a large number of *alleles* (see p. 1053) of the MHC genes.

The primary role of class I MHC proteins is "presentation" of foreign molecules (antigens) to cells of the immune system containing class II MHC proteins. In this presentation the antigen, usually a polypeptide, is taken into a cell bearing a class I MHC molecule and broken into fragments. The fragments then appear on the cell surface in combination with the class I MHC molecules. Recognition of the class I MHC-antigen combination stimulates a T cell, a specialized class II MHC-bearing leucocyte of the immune system, to initiate an immune response against the antigen. If the cell bearing the class I MHC molecule is from the same individual, the immune response is mounted only against the antigen, unless the antigen is derived from a virus infecting the cell. If the cell bearing the class I MHC-antigen combination is from a different individual, as it might be as a result of an organ or tissue transplant, both the antigen and MHC protein are recognized as foreign, and an immune response is mounted against the cell as well as the antigen. MHC molecules are also called *transplantation antigens* because of their involvement in transplant rejection. The combination of a class I MHC molecule with an antigen derived from an infecting virus may also lead to destruction of the cell bearing the MHC protein, even if it is from the same individual.

Class I and II MHC molecules and antibodies share significant structural relationships. Antibody molecules have both constant and variable regions, as MHC molecules do. The constant and variable regions of MHC molecules and antibodies are divided into similar domains, and the constant regions of both show clear sequence similarities (see Figs. 7-1 and 19-12), indicating that all evolved from common ancestral genes. Several other cell surface glycoproteins, including some of the molecules involved in cell-cell adhesions, also have domains related to the constant and variable domains of antibodies. The large group of cell surface glycoproteins related to antibodies, which is constantly expanding as new discoveries are made, is known as the *immunoglobulin superfamily*.

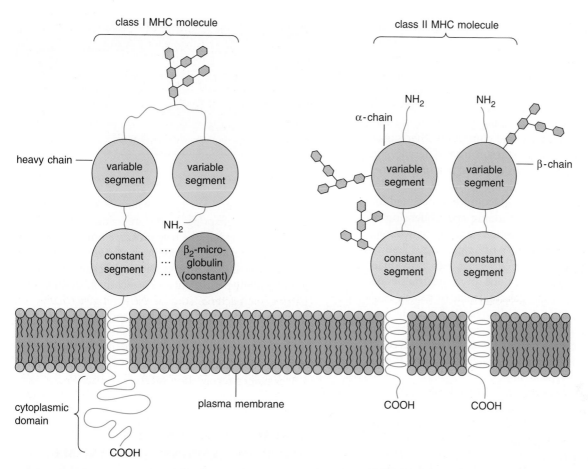

Figure 7-1 The class I and class II MHC molecules of mammalian cells. Both glycoproteins consist of two polypeptide chains. The extracellular portions of each molecule consist of two constant and two variable domains. The constant domains are built up from amino acid sequences that are similar in all members of the immunoglobulin superfamily. The variations in amino acid sequence responsible for the differences in MHC molecules between different individuals are concentrated in the variable domains. Class I proteins consist of a heavy chain forming three of the four major domains of the molecule, with an additional small polypeptide, *β₂-microglobulin*, that completes the structure. Class II markers are built up from two polypeptides of approximately equal size, each of which contains one constant and one variable domain.

A cell surface molecule working in coordination with MHC proteins is used by HIV, the AIDS virus, as a binding site for attachment to cells of the immune system. A coat protein of the virus, identified as *gp120*, binds to a cell surface molecule of *helper T cells* of the immune system. This attachment is the first step by the virus in the pathway of infection of immune system cells. The molecule involved in the binding, a surface glycoprotein known as *CD4*, interacts with class II MHC molecules during recognition of foreign proteins in its normal role in lymphocytes. (For further details of the MHC, the immunoglobulin superfamily, HIV, AIDS, and their relationships to the immune system, see Ch. 19.)

Blood Groups and Other Surface Markers

The glycolipid and glycoprotein markers responsible for blood groups in humans and other vertebrates provide another important example of surface groups involved in recognition. The glycophorin glycoprotein of erythrocytes (see Fig. 5-7) carries the markers responsible for the MN blood groups of humans; evidently both the protein and carbohydrate portions of glycophorin contribute to the MN markers.

The ABO blood groups, in contrast, depend on small differences in the carbohydrate structure attached to a glycolipid in erythrocyte plasma membranes. Persons with type A blood have the amino sugar *N*-

acetylgalactosamine at one position in the carbohydrate complex of this glycolipid. (Fig. 2-17c shows this form of the carbohydrate structure.) Persons with type B blood have a galactose unit at the same position. Individuals with type AB blood have glycolipid molecules of both types on their red blood cells. In persons with type O blood, the corresponding position in the carbohydrate group is empty and contains neither sugar.

The ABO blood groups result from differences in glycosyl transferase enzymes encoded in a single gene. Glycosyl transferases, which are active in the Golgi complex, add sugar units to glycoproteins destined for the cell surface (see p. 873). Cloning and sequencing of the ABO gene by F. Yamamoto and his coworkers revealed that a small number of single-base substitutions in the gene of type A, B, or AB individuals produces alternate forms of the glycosyl transferase that place either N-acetylgalactosamine or galactose at the critical position in the carbohydrate group. In persons with type O blood, deletion of a single base in the ABO gene causes production of a faulty glycosyl transferase that is incapable of catalyzing transfer of a sugar unit to the critical site.

The differences in the ABO blood types, although small, can have tragic consequences if blood types are mixed in the wrong combinations during transfusions. In general, individuals do not react against blood containing cells with the same surface markers as their own red blood cells or against O type, which contains neither the A nor B marker type (Table 7-1). Type A persons may receive blood from other type A individuals, for example, or from those with type O. Type B blood, however, cannot be safely introduced into type A individuals.

Similar surface recognition groups are involved in sperm-egg interactions in animals and in recognition of mating types between reproductive cells in algae and fungi. Surface recognition groups in higher plants are involved in graft acceptance or rejection and in the recognition of symbiotic bacteria such as *Rhizobium*, which stimulates the formation of root nodules in peas, clover, and other legumes.

A number of pathogenic protozoa, bacteria, and viruses have developed variable surface markers that enable them to avoid an immune response. The protozoan responsible for African sleeping sickness, *Trypanosoma bruceii*, has a coat of glycoproteins called *VSG markers* (VSG = Variable Surface Glycoprotein) that completely covers the surface of the organism. The VSG markers are so thickly packed in the coat that no other molecules of the surface are exposed to the immune system. At regular intervals, offspring of an infecting population of the protozoan appear with changes in the amino acid sequence of the VSG markers. These altered individuals survive the immune response directed against the original VSG type. The process continues, so that as an immune response develops

| Table 7-1   Human ABO Blood Types ||||
Blood Type	Antigens	Antibodies	Blood Types Accepted
A	A	Anti-B	A or O
B	B	Anti-A	B or O
AB	A and B	None	A, B, AB, or O
O	None	Anti-A, anti-B	O

against a given VSG type, a new VSG appears that allows some offspring in the infecting population to survive. As a result, the trypanosome population keeps one step ahead of the immune system. The system works so well that African sleeping sickness infections are essentially permanent and presently incurable. Some pathogenic bacteria, such as *Neisseria gonorrhoeae*, the pathogen producing gonorrhea, and the HIV-1 virus, use a similar system of constantly altering surface markers to escape detection and destruction by the immune system. (For further details, see Ch. 19; Supplement 7-1 describes recognition markers of bacteria.)

SURFACE RECEPTORS AND RECEPTOR-RESPONSE MECHANISMS

Receptors are membrane glycoproteins tailored to recognize and bind molecules colliding with the cell surface. The molecules recognized and bound, called the *ligands* of the receptors, may be suspended in solution in the extracellular medium, form parts of the extracellular matrix, or be linked to the surface of another cell. These receptors occur in several major types, each with distinct functions.

One large and highly important group of receptors binds peptide hormones, growth factors, and certain neurotransmitters (known as *first messengers*; see below) as ligands (Table 7-2). In most systems receptors binding one of these ligands trigger a biochemical response inside the cell, such as an increase or decrease in the rate of transport, secretion, oxidative metabolism, initiation of cell division, or cell movement. Usually, only a small proportion of the available surface receptors need to bind a ligand for production of a full cellular response.

A second major receptor group binds molecules suspended in solution or linked to cell surfaces that are to be taken into cells by endocytosis. In this process, called *receptor-mediated endocytosis*, specific molecules are recognized and bound by the surface receptors. Once tied to the cell surface by linkage to their receptors, the substances are taken into the cell by pockets that form in the plasma membrane and pinch off as cytoplasmic vesicles. Peptide hormones

Table 7-2 Some Hormones, Growth Factors, and Neurotransmitters Acting as First Messengers for Receptor-Response Pathways

First Messenger	Origin	Targets and Effects
Insulin	Pancreatic β-cells	Stimulates glucose uptake by body cells, carbohydrate metabolism, lipid synthesis in fat cells, and protein synthesis and cell division.
Platelet-derived growth factor (PDGF)	Vascular endothelial cells	Stimulates growth and cell division of fibroblasts and vascular smooth muscle cells.
Epidermal growth factor (EGF)	Unknown	Stimulates growth and division of cells in skin.
Glucagon	Pancreatic α-cells	Stimulates glucose release from glycogen stores in liver cells.
Angiotensin	Enzymatic breakdown of precursor in blood plasma	Increases blood pressure by constricting blood vessels.
Adrenaline (epinephrine)	Adrenal medulla	Increases heart rate and blood pressure; stimulates glucose release from glycogen stores in liver and fatty acid release from adipose cells.
Norepinephrine	Axon terminals of neurons	Operates as neurotransmitter in central and peripheral nervous systems.
Acetylcholine	Axon terminals of neurons	Operates as neurotransmitter from motor nerves to muscles and between neurons in central nervous system.
Vasopressin	Posterior pituitary	Increases blood pressure and water resorption in kidney tubules.

and growth factors are also usually taken into cells by endocytosis along with their receptors once their binding has stimulated a cellular response.

A third major receptor group binds cell surface molecules of other cells or molecules of the extracellular matrix and thereby contribute to the linkages holding cells in place in the tissues and organs of animals. These receptors differ from those binding peptide hormones and neurotransmitters in that receptor binding, in most cases, does not induce a complex biochemical response by the cell. (Details of the linkage of receptors to the extracellular matrix and receptor-mediated endocytosis are given in Chs. 8 and 20, respectively.)

All the surface receptors accomplishing these varied activities have been identified as glycoproteins. Initially, this identification was carried out by noting the effects of enzymes on the activity of surface receptors. Of the enzymes studied, those attacking either carbohydrates or proteins altered the recognition and binding response most efficiently. Two carbohydrate-digesting enzymes, *galactosidase*, which breaks off galactose units from carbohydrates, and *neuraminidase*, which attacks sialic acid units, were particularly effective in destroying the recognition and binding of specific molecules at the cell surface.

Many individual receptors have been isolated and identified. In some cases ligands have been marked with a radioactive label or another molecular group that can be identified biochemically. For this work the association between the receptor and ligand is usually stabilized by techniques that introduce a covalent bond. After the ligand is bound, cell membranes are broken down by detergents, sonication, or other means (see p. 124). The labeled receptors can be separated from other cellular molecules by a technique such as gel electrophoresis and identified. Receptors are also routinely isolated by affinity chromatography (see p. 131), in which the ligand, or a molecule closely resembling the ligand in structure, is attached to the beads in a chromatography column (Fig. 7-2). Running through the column a preparation of proteins released by membrane disruption leads to combination of the receptor and ligand, which traps the receptor in the column. Other proteins pass through the column without hindrance.

Receptors have also been identified through searches of gene libraries (see p. 137) for sequence groups that occur commonly in different receptor types. By cloning and sequencing receptor genes identified in this way the amino acid sequence of many receptor types

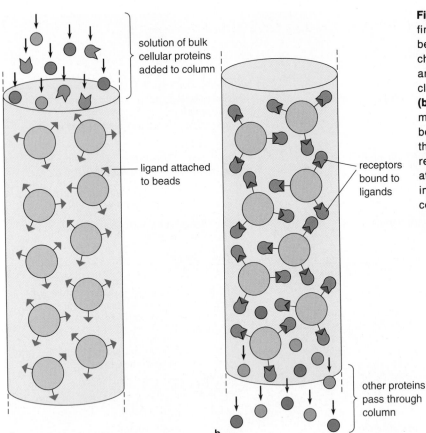

Figure 7-2 Separation of receptors by affinity chromatography. **(a)** A ligand that can be bound by the receptor of interest is chemically linked to beads in the column, and a solution of cellular proteins that includes the receptor is added to the column. **(b)** The receptor binds the ligand, which immobilizes the receptor molecules on the beads in the column; other proteins pass through the column without hindrance. The receptor can be removed from the beads after other proteins are eliminated by pouring a solution of the ligand in excess concentration through the column.

solution of bulk cellular proteins added to column

ligand attached to beads

receptors bound to ligands

other proteins pass through column

a

b

has been established. The sequences, in turn, have provided insights into the structure and function of the receptor molecules. The genetic studies have also revealed previously unknown relationships between different receptors, showing that many are members of large families with evolutionary lineages stemming from single ancestral genes. These highly productive experiments have made receptor research one of the most exciting areas in cell and molecular biology.

Receptors Binding Peptide Hormones, Growth Factors, and Neurotransmitters

The surface receptors binding peptide hormones, growth factors, and neurotransmitters form part of a complex signaling system that coordinates the activities of animal cells. The segments of these receptors extending from the outer membrane surface contain the site that recognizes and binds the hormone or neurotransmitter. In response to binding, the receptor undergoes a conformational change, which activates an enzymatic site at the end of the protein extending into the cytoplasm. The cytoplasmic site catalyzes a reaction that serves as the first step in a series of reactions leading to the cellular response.

The cell surface may contain from hundreds to thousands of individual receptor molecules. Receptors for the different peptide hormones, for example, may

number from 500 to 100,000 or more per cell, with 10,000 to 20,000 receptors per ligand type being typical. Different cell types contain distinct combinations of receptors, allowing them to react individually to only certain molecules among the spectrum of hormones and neurotransmitters circulating in the bloodstream or extracellular fluids. Nearly 80 different receptor types have been detected on lymphocytes, for example. The combination of surface receptors on particular cell types is not static but changes as the cells develop. Changes in receptors are also frequently noted as part of the alterations transforming a normal cell into a cancer cell.

In no case is the total series of reactions leading from receptor-ligand binding to the cellular response completely understood. However, receptor-response mechanisms have several characteristics that give important clues to their operation. First, the peptide hormone, growth factor, or neurotransmitter does not have to penetrate into the cell to induce a response; binding to the receptor at the cell surface is fully sufficient. Second, no response is produced if the ligand is directly injected into the cytoplasm. Third, the receptor remains in position in the plasma membrane while the cellular response is initiated; it does not enter the cell with its bound ligand to initiate the response. These characteristics indicate that the ligand has no function other than binding to the receptor at the cell surface.

Another significant feature of receptor-response mechanisms is that under certain conditions a full cellular response may be triggered by the receptors without participation of the ligand. This has been demonstrated by the use of *lectins*, a group of glycoproteins that binds with varying degrees of specificity to surface carbohydrate groups. Frequently, lectins binding to a surface receptor trigger the same internal response as the ligand. The same type of response has also been produced by antibodies developed against receptors. This characteristic confirms that the mechanism initiating the cellular response is built into the receptor; the only function of the ligand is to act as an external signal that sets the mechanism into motion.

Other important characteristics of the receptor-response pathways reflect operation of the internal response mechanisms triggered by the receptors. Although a great variety of ligands may be bound by many different receptor types, the response mechanisms follow similar pathways and share several common features.

One feature shared by most pathways is the primary means by which cellular responses are produced and controlled. This is through activation of *protein kinases*, enzymes that attach phosphate groups derived from ATP to specific target proteins. Addition of phosphate groups inhibits or stimulates the activity of target proteins in carrying out a cellular response. Among the target proteins controlled by these receptor-mediated phosphorylations are (1) enzymes carrying out critical steps in metabolic pathways, (2) transport proteins such as ion channels, (3) ribosomal proteins, (4) proteins regulating gene activity, and (5) the receptors themselves. Some protein kinases catalyzing the phosphorylations are associated with membranes. Others are suspended in solution in the cytoplasm or nucleus. The particular cellular response elicited by a pathway depends on the types of protein kinases present in a cell, controlled ultimately by the systems regulating genetic activity in the cell nucleus. All the protein kinases associated with the pathways have catalytic domains that share sequence homologies, indicating that all stem from a common ancestral protein.

In a relatively few systems the protein kinase activity carrying out the phosphorylations forms an integral part of the receptor segment extending into the cytoplasm under the plasma membrane. More commonly, the protein kinases inducing cellular responses are separate enzymes working in the final steps of a reaction sequence activated by the receptor.

In all response pathways the phosphate groups added by protein kinases can be removed by the activity of a varied group of enzymes known collectively as *protein phosphatases*. Although relatively little is known about the pathways controlling these enzymes, their activity is highly significant to cell functions because they inhibit or reverse the effects of the receptor-response pathways triggering the activity of protein kinases.

Receptors with Integral Protein Kinase Activity

Although relatively few receptors have integral protein kinase sites, cellular responses controlled by these receptors are among the most fundamental processes of animal cells. One receptor of this group recognizes and binds the peptide hormone *insulin* (see Table 7-2). This hormone controls glucose uptake and the rate of many metabolic reactions in its target cells. Other receptors of this group bind *epidermal growth factor (EGF)* and *platelet-derived growth factor (PDGF)*, both of which regulate growth and division of specific cell types in higher animals.

All receptors in this group are related proteins with similarities in amino acid sequence and structure (Fig. 7-3). The EGF and PDGF receptors (Fig. 7-3a) consist of a single transmembrane polypeptide. The extracellular portion of the polypeptide forms a domain containing the binding site for the ligand. On the cytoplasmic side of the plasma membrane the polypeptide folds into a large domain containing the protein kinase site. The extracellular and cytoplasmic domains are connected by a single alpha-helical segment that spans the membrane.

The more complex insulin receptor (Fig. 7-3b) is built up from two polypeptides, each of which occurs in two copies. Two copies of the α-*chain* polypeptide are entirely extracellular and form the insulin-binding site. Two copies of the β-*chain* polypeptide extend through the membrane; on the cytoplasmic side of the membrane; these polypeptides form the protein kinase site. The four polypeptides are held in a single, stable structure by disulfide linkages on the extracellular surface of the membrane. Although at first glance the insulin receptor seems to be fundamentally different in structure from the EGF and PDGF receptors, the polypeptides of the two receptor types are closely related. The α-chain of the insulin receptor closely resembles the extracellular segment of the EGF and PDGF receptors. The β-chain is closely related to the transmembrane segment and cytoplasmic domain of the EGF and PDGF receptors. The two polypeptides of the insulin receptor are encoded in a single gene, as is the polypeptide of the EGF and PDGF receptors. Processing reactions that take place after the gene is transcribed split the precursor of the insulin receptor into α- and β-chains.

Figure 7-3c and d shows how receptors with integral protein kinase activity are believed to operate. The protein kinase site on the cytoplasmic side of the membrane is inactive when the extracellular site is unbound (Fig. 7-3c). Binding a hormone or growth factor at the cell surface induces a conformational change,

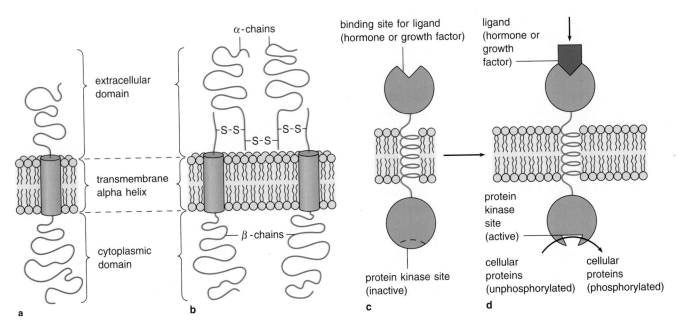

Figure 7-3 Structure and function of receptors with integral protein kinase activity. **(a)** Structure of the EGF or PDGF receptor, which both contain a single polypeptide. The extracellular domain includes the binding site for the ligand. The cytoplasmic domain of the polypeptide contains the protein kinase site. A single membrane-spanning alpha-helical segment connects the extracellular and cytoplasmic domains. **(b)** Structure of the insulin receptor, built up from two polypeptides, each occurring in two copies. The combination is stabilized on the extracellular side of the membrane by disulfide (—S—S—) linkages. The two α-chain polypeptides, which are entirely extracellular, form the insulin-binding site. The two β-chain polypeptides form transmembrane segments and, on the cytoplasmic side of the membrane, the protein kinase site. **(c and d)** Operation of the receptors. When the extracellular site is unbound, the protein kinase site on the cytoplasmic side of the membrane is inactive **(c)**. Binding a hormone or growth factor at the cell surface induces a conformational change that is transmitted through the transmembrane segment and activates the protein kinase site **(d)**, which adds phosphate groups to tyrosine residues in the target proteins. The proteins phosphorylated are usually enzymes that control key steps in metabolic pathways or reaction sequences related to gene activation. Phosphorylation may either stimulate or inhibit activity of the target proteins.

transmitted in some unknown way through the alpha-helical segment spanning the membrane, that activates the protein kinase site (Fig. 7-3d). The helical segment may slide or rotate in response to the conformational change in the receptor domain; this movement may adjust the conformation of the cytoplasmic domain to switch the protein kinase site from the inactive to active state. The proteins phosphorylated by the protein kinase site, which adds phosphate groups to tyrosine residues in the target proteins, are usually enzymes controlling key steps in metabolic pathways or reaction sequences related to gene activation. Addition of the phosphate groups may either stimulate or inhibit activity of the target proteins.

The receptors in this group phosphorylate a wide variety of target proteins, so many that it has proved difficult to isolate and identify them individually. However, recent work indicates that several of these proteins form important links in other receptor-response path-ways. Among these proteins are phospholipase C and the G proteins (see below); protein kinases that add phosphate groups to serine and threonine residues in their target molecules; and proteins that regulate gene activity. The phosphate groups added to these proteins can be removed by various protein phosphatases. The regulatory pathways controlling the activity of these enzymes remain largely unknown.

Relatives of the EGF, PDGF, and insulin receptors have been discovered in *Drosophila,* indicating that this receptor family has roots extending back through at least 800 million years of evolution. One indication of the fundamental importance of the receptors in this family to normal cell activity is that many *oncogenes* (see Ch. 22) promoting cancer are altered forms of genes encoding either EGF or PDGF receptors. The faulty receptors encoded in the oncogenes have protein kinase sites that are constantly active, whether the growth factor is bound to the receptor or not. For example,

Figure 7-4 The pathway used by receptor-response systems in which the kinase is separate from the receptor. Binding the first messenger, usually a peptide hormone, growth factor, or neurotransmitter, activates the receptor. By a series of steps involving G proteins, the receptor then activates the effector. The effector is an enzyme, either adenylate cyclase or phospholipase C, that converts precursor substances into second messengers. The second messengers, in turn, activate one or more protein kinases that add phosphate groups to enzymes or other molecules. The phosphorylations, by promoting or inhibiting enzymatic and other activities, produce a cellular response. Ca^{2+}, released into the cytoplasm by $InsP_3$, acts as a supplementary second messenger. cAMP, cyclic AMP; $InsP_3$, inositol triphosphate; DAG, diacylglycerol.

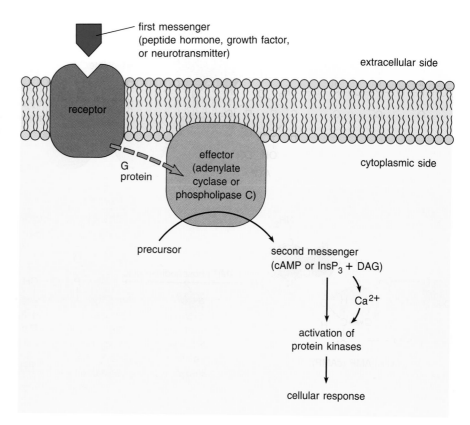

the oncogene responsible for *erythroblastosis*, a form of cancer in chickens, encodes a faulty EGF receptor that lacks most of the external segment binding the EGF hormone. The remaining transmembrane and cytoplasmic segments assume a conformation that continuously activates the protein kinase site, leading to rapid, uncontrolled cell division.

Hereditary defects in the insulin receptor are responsible for some forms of diabetes, a disease in which glucose cannot be taken up in sufficient quantity by body cells and accumulates in the bloodstream. Most commonly, relatively minor defects in the receptor impair but do not completely interfere with insulin binding or kinase activity. In rarer instances a mutation eliminates most of the protein kinase domain from the receptor, so that the enzymatic segment is inactive even though the receptor binds insulin. In other forms of diabetes insufficient quantities of insulin are released to the body circulation because production of the hormone by the pancreas is faulty.

Receptors with Separate Protein Kinase Activity

Receptors activating protein kinases that are separate proteins, unconnected to the receptor structure, are much more common than integral receptors. On binding their ligands, these receptors activate protein ki-

nases via a complex pathway with several to many steps, operating in a sequence known as a *cascade*. Although many different receptors and protein kinases act in this pattern, almost all response cascades follow one of two primary pathways. Within the two pathways several steps use similar elements or operate in the same manner (Fig. 7-4).

The receptors of the pathways are all large proteins built up from a single polypeptide with seven transmembrane segments. The cytoplasmic domain of the receptors contains an enzymatic site that is activated by conformational changes induced when a ligand acting as a first messenger binds to the extracellular domain. Rather than acting directly as a protein kinase, the active site catalyzes the first step in a reaction cascade leading to activation of one or more protein kinases.

The reaction cascades generate one or more *second messengers*, substances that serve as the primary functional links between the receptors and protein kinase activation. As part of these cascades the receptors directly activate *G proteins*, a group of molecules that stimulate enzymes catalyzing production of second messengers. Operation of both pathways also involves *amplification*, a process in which the number of molecules activated or generated is magnified at successive steps. The protein kinases activated by the pathways add phosphate groups to serine or threonine residues in their target proteins.

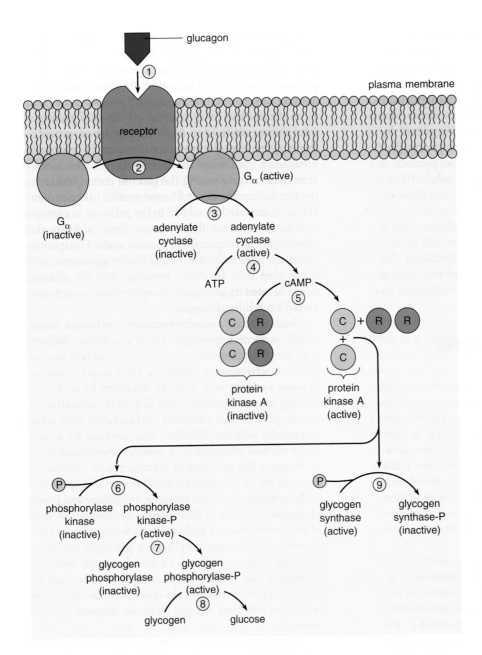

Figure 7-8 The cAMP-based pathway triggered in liver cells by the peptide hormone glucagon. Glucagon is secreted by the pancreas when blood sugar falls to low levels. Binding of the hormone by a receptor on a liver cell (step 1) activates the receptor, which triggers the sequence of reactions that synthesizes cAMP (steps 2 to 4; see also Fig. 7-7). The water-soluble cAMP activates a series of protein kinases by diffusing through the cytoplasm. The protein kinases activated by cAMP, the protein kinase A family, consist of two regulatory (R) and two catalytic (C) subunits in the inactive form. The cAMP combines with R subunits (step 5), inducing a conformational change that releases and activates the C subunits. In active form the protein kinase A phosphorylates two enzymes, phosphorylase kinase and glycogen synthase. Phosphorylase kinase is activated by the phosphate groups added by the protein kinase A (step 6). The activated phosphorylase kinase adds a phosphate group to glycogen phosphorylase (step 7), an enzyme that catalyzes the breakdown of glycogen into glucose units (step 8). Glycogen synthase is inactivated by the phosphorylation catalyzed by protein kinase A (step 9), stopping the reaction catalyzed by glycogen synthase, which is the assembly of glycogen from glucose units.

sponse pathways. One, *PP2B*, is activated by Ca^{2+} released in the $InsP_3$/DAG pathway; the second, *PP1*, is activated by both cAMP and Ca^{2+}. Thus, cAMP may activate enzymes removing phosphate groups as well as the kinases adding them; which activity predominates depends on the relative concentrations and types of enzymes present in a cell carrying out a cAMP-based response pathway.

Both cAMP and the effector of the cAMP pathway, adenylate cyclase, occur in higher plants. However, the protein kinases through which cAMP exerts its effects in animal cells have not been discovered in plants; neither has any link been detected between cAMP concentrations and the known physiological reactions of plants. Therefore, although some elements of the cAMP-based pathway exist in plants, the degree to

which this system works in them is presently unknown.

A recently discovered group of receptors acts much like those using cAMP as second messenger, except that their cytoplasmic extension contains an enzymatic site with *guanylate cyclase* activity. On binding a ligand, the guanylate cylase site of these receptors converts cytoplasmic GTP to *cyclic GMP* (*cGMP*), a substance identical to cAMP except that guanine substitutes for adenine in its structure. Among other effects, cGMP acts in opposition to cAMP-based control pathways by activating a cGMP-dependent phosphodiesterase that breaks down cAMP (see below). Cell-surface receptors are not the only source of cGMP; this cyclic nucleotide is also generated by soluble guanylate cyclase enzymes suspended in the cytoplasm without connection to receptors.

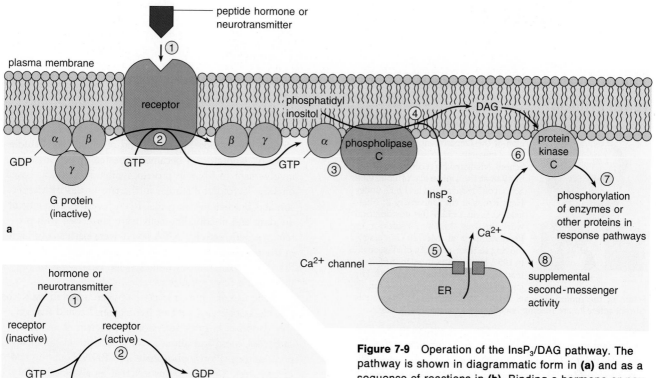

Figure 7-9 Operation of the InsP₃/DAG pathway. The pathway is shown in diagrammatic form in **(a)** and as a sequence of reactions in **(b)**. Binding a hormone or neurotransmitter activates the receptor (step 1), which in turn activates G proteins (step 2) and phospholipase C (step 3) and produces the second messengers InsP₃ and DAG (step 4). The InsP₃ messenger, which is water soluble, diffuses through the cytoplasm to the ER, where it acts as the ligand for a ligand-gated Ca²⁺ channel. InsP₃ binding opens the channel and releases Ca²⁺ from the ER into the cytoplasm (step 5). The released Ca²⁺ and the DAG produced in step 4 together activate the protein kinase C enzymes of the pathway (step 6). The released Ca²⁺ also directly or indirectly regulates a variety of cellular mechanisms by working as a supplemental second messenger for the InsP₃/DAG pathway (step 8). The activated protein kinases add phosphate groups to serine and threonine residues in the target proteins controlled by the InsP₃/DAG pathway (step 7).

transport, movements such as contraction of smooth muscle, and glucose metabolism. The list of more than 40 known peptide hormones and neurotransmitters acting as first messengers for the InsP₃/DAG pathway includes *vasopressin, angiotensin,* and *norepinephrine* (see Table 7-2 and the Experimental Process essay by L. E. Hokin on p. 248).

Figure 7-9 illustrates operation of the InsP₃/DAG pathway. Binding of a hormone or neurotransmitter activates the receptor (step 1 in Fig. 7-9), in turn leading to sequential activation of G proteins and phospholipase C, and production of the InsP₃ and DAG second messengers (steps 2 to 4).

The water-soluble InsP₃ messenger is released from the membrane and diffuses through the cytoplasm to

The InsP₃/DAG Pathway Apparently, the InsP₃/DAG pathway is universally distributed among eukaryotic organisms, including both vertebrate and invertebrate animals, fungi, and plants. The system, first detected by M. N. Hokin and L. E. Hokin and R. Michell, controls responses as varied as secretion of hormones and neurotransmitters by gland and nerve cells, cell division, early events in fertilization, sugar and ion

The Road to the Phosphoinositide-Generated Second Messengers

Lowell E. Hokin

LOWELL E. HOKIN is Professor and Chairman of the Department of Pharmacology, University of Wisconsin Medical School, Madison. He received his M.D. in 1948 at the University of Louisville School of Medicine and his Ph.D. under Hans Krebs at the University of Sheffield, U.K., in 1952. After postdoctoral work, he was Assistant Professor of Pharmacology at McGill University, moving to the University of Wisconsin in 1957. He was appointed to the Chairmanship of Pharmacology in 1968. His current emphasis is the mechanism of action at the molecular level of the antimanic drug lithium. His current interests are reading, swimming, and downhill skiing.

Neurotransmitters, growth factors, and hormones generally activate intracellular processes by combining with cell surface receptors. This interaction in turn activates membrane enzymes, usually via a GTP binding protein, to produce "second messengers" that diffuse to an intracellular target. In recent years, it has become clear that on binding of a variety of agonists [or ligands] to their receptors, inositol lipids in cell membranes, called phosphoinositides, are cleaved to inositol phosphates and diacylglycerol or DAG, which is the lipid backbone of phosphoinositides. There are three main phosphoinositides in the cell membrane: phosphatidylinositol (PI), which comprises about 95% of membrane inositol lipid, phosphatidylinositol 4-phosphate (PIP), and phosphatidylinositol 4,5-bisphosphate (PIP_2). Activation of an appropriate receptor stimulates cleavage of PIP_2 to form DAG and inositol 1,4,5-trisphosphate or $Ins(1,4,5)P_3$. $Ins(1,4,5)P_3$ diffuses to a Ca^{2+} storage site (endoplasmic reticulum), binds to its membrane receptor, and releases Ca^{2+} into the cytosol. Ca^{2+} then activates physiological responses by Ca^{2+}/calmodulin protein kinases. DAG activates protein kinase C, which phosphorylates a different set of proteins, leading to a variety of biological responses.

This is the story of the discovery of the phosphoinositide effect in the early 1950s, a "discovery before its time." Twenty years elapsed after our initial demonstration of agonist-stimulated PI turnover before Bob Michell was to rekindle interest in this effect—an effect that has answered (and posed!) so many questions about signal transduction mechanisms.

The story really begins in 1949, when I arrived at Hans Krebs' laboratory in Sheffield to study for a Ph.D in biochemistry. Krebs' policy was to have a student formulate his own problem, pursue it independently with a minimum of supervision, and publish by himself. With some background in gastroenterology, which I had received in the laboratory of Warren S. Rehm while attending medical school, I looked into whether I could use pancreatic tissue as a model to study the synthesis and secretion of proteins, using amylase as a measure of these two processes.

Pigeon pancreas slices proved to be an excellent system for this purpose. Near the end of my doctoral studies in Krebs' laboratory, I became interested in the possible involvement of RNA in protein synthesis. This was based on some cytological studies in the late forties by Caspersson and Brachet showing that RNA levels in a variety of dividing and nondividing cells were correlated with protein synthetic activity. RNA levels were particularly high in the pancreas. (My interest in this problem antedated the Watson-Crick DNA double helix, which was to be published a few years later.)

The use of ^{32}P as a tracer had been adopted by Krebs and his associates a year or two earlier. I found that on stimulation of enzyme secretion in pancreas slices with carbachol, there was about a 100% increase in the incorporation of [^{32}P]orthophosphate into RNA. Near the end of this work, I began to suspect that an alkaline hydrolytic product of phospholipids was contaminating the then crude RNA fraction, and that this might be responsible for the stimulation of ^{32}P incorporation into RNA.

This idea came to me shortly after I had completed my requirements for the Ph.D. and was about to set sail with Mabel Hokin for Halifax, Nova Scotia. I did manage to do one experiment, but I did not have time to go through the necessary procedures to rid the lipid fraction of contaminating [^{32}P]orthophosphate and to count the radioactivity. I literally took with me on the boat a rack of about a dozen rather large test tubes containing the ethanol-ether extracts. (This would be prohibited with the strict regulations governing radiation safety now in place.) When we arrived at J. H. Quastel's laboratory at McGill, we counted the purified total lipid fractions and found an enormous increase in specific radioactivity in the lipids from pancreas slices that had been stimulated with carbachol. The 1953 *Journal of Biological Chemistry* paper arising from this work is often incorrectly quoted as demonstrating the stimulated turnover of PI. We did not actually show this until 1955, when techniques for measuring radioactivity in individual phospholipids from small samples of tissue became available.

The discovery of the PI effect launched Mabel Hokin and me on a 15-year quest for its explanation, as well as a pursuit of its biochemical mechanism. Mabel Hokin continued this search after I had left the field around 1965. I returned in the early 1980s. We made several key observations in the fifties and early sixties concerning the PI effect, including: (1) the effect occurred with many agonists and in many cell types; (2) the effect for the most part was due to a turnover and not a net synthesis of PI and phosphatidic acid (PA), with DAG being partly or completely conserved; and (3) omission of Ca^{2+} blocked secretion but did not inhibit PI turnover to any great extent. This was part of the underpinnings of Michell's hypothesis relating PI turnover to Ca^{2+} gating.

With the aid of kinetic studies on the turnover of PI and PA in the avian salt gland on cholinergic stimulation (followed by atropine clamping), we were able to present the first version of the PI cycle. The main points of this cycle were that on cholinergic stimulation of salt gland slices phospholipase C catalyzed the breakdown of PI to DAG and inositol-1-phosphate (IP). Diacylglycerol kinase then formed PA. On quenching with atropine, the resting steady-state level of PI was restored at the expense of PA by the sequential actions of PA-cytidyl transferase and PI synthase.

Mabel Hokin made some key observations in the 1970s further supporting the PI cycle, including the findings that on stimulation in pancreas, there was a loss in the mass of PI and a rise in the mass of PA. Also, on stimulation, the fatty acid composition of PA approached more closely that of PI. These data gave additional strong support to the view that PA was derived from PI via DAG during stimulated turnover of PI and PA. She further found an increase in mass of DAG in pancreas on agonist stimulation. At about the same time, Michell showed independently a fall in mass in PI in the parotid gland on stimulation. Mabel Hokin also confirmed the salt gland kinetic data in the pancreas. Her recalculation of some old measurements of PI in the salt gland (expressed as percentages of controls rather than as absolute values) showed a significant loss in mass in PI in this tissue as well. The original PI cycle has, of course, been modified to incorporate the polyphosphoinositides. There does appear, however, to be direct breakdown of PI as well, which appears to be an additional source of second messenger DAG.

In the early 1960s we and others studied the turnover of PIP and PIP$_2$, including our demonstration of a fall in steady-state level of ^{32}P-labeled polyphosphoinositides on stimulation of salt gland slices with acetylcholine. Alas, we did not carry this any further! We also showed for the first time the presence of the kinases for PI and PIP (in the erythrocyte membrane), as well as the rapid exchange of ^{32}P into the monoesterified phosphate groups of the polyphosphoinositides in erythrocyte membranes on incubation with [^{32}P]ATP. In the late 1960s, Durell demonstrated the formation of IP and inositol 1,4-bisphosphate (IP$_2$) in synaptic vesicles isolated from neurons stimulated by acetylcholine. He suggested that the primary response might be the cleavage of phosphodiester bonds in the polyphosphoinositides, releasing inositol phosphates. Abdel-Latif pursued this line in the late 1970s and showed that cholinergic or adrenergic stimulation of iris smooth muscle caused a rapid breakdown of PIP$_2$, which was accompanied by an increase in IP, IP$_2$, and inositol trisphosphate.

The discovery of the PI effect in the early 1950s was indeed a "discovery before its time." Very little was known about membrane structure. The classic experiments from the laboratories of Douglas and Katz pointing to the importance of Ca^{2+} in stimulus-response coupling were not carried out until the early 1960s. We did not know for certain the structure of the "phosphoinositide" that showed increased turnover in pancreas and brain until around 1958, when we showed it was PI. It was not until the early 1960s that the identification and the full struc-

tural elucidation of PIP and PIP$_2$ were made in the laboratories of Ballou and Dawson. The existence of the intracellular "trigger pool" of Ca^{2+}, which rapidly releases part of its stores of Ca^{2+} on agonist stimulation, was not known until the work of Schulz and others in the late 1970s. The technique of cell permeabilization to allow entry of phosphorylated compounds into the cell was not developed until the late 1970s and early 1980s. This, of course, permitted Streb, Irvine, Berridge, and Schulz (1983) to demonstrate the release of nonmitochondrial intracellular stores of Ca^{2+} by Ins(1,4,5)P$_3$.

Apparently, there was some derision of the PI effect in the early years. At a meeting a few years ago, John Fain said that when he arrived at Berridge's laboratory in the late 1970s, the common attitude toward the PI response (presumably not that of Berridge) was that it was "Hokin's hokum and Michell's folly."

We are currently interested in the effects of the antimanic drug lithium on Ins(1,4,5)P$_3$ levels in brain. We have found that in species ranging from the mouse to the monkey, lithium elevates levels of Ins(1,4,5)P$_3$ in cerebral cortex slices. The Ins(1,4,5)P$_3$ levels can be measured by prelabeling the slices with [^{3}H]inositol. After incubation with lithium, the slices are extracted with perchloric acid, the inositol phosphates are separated by high-performance liquid chromatography, and their radioactivity is determined. Another method is an Ins(1,4,5)P$_3$ receptor binding assay. Extracts of brain cortex slices incubated without [^{3}H]inositol are mixed with standard [^{3}H]Ins(1,4,5)P$_3$, the Ins(1,4,5)P$_3$ receptor, and the unknown sample. The Ins(1,4,5)P$_3$ in the unknown sample will displace [^{3}H]Ins(1,4,5)P$_3$ from the receptor, and the level of Ins(1,4,5)P$_3$ in the sample can be calculated.

Naturally, I am elated and I know I speak for Mabel Hokin as well, when I contemplate that what we started 40 years ago has led to the discovery of two phosphoinositide-derived second messengers—Ins(1,3,5)P$_3$ and DAG (by Nishizuka). Many additional inositol phosphate compounds formed on stimulation have been found. It appears that there may be more phosphoinositide-derived second messengers. For example, Ins(1,3,4,5)P$_3$ also appears to be involved in Ca^{2+} movements. Provocative studies on the involvement of the phosphoinositide system in fertilization, growth, and oncogene action are emerging. Functions for phosphoinositides other than as second messenger generators are beginning to surface. For example, PI is an important anchor for many proteins on the cell surface.

References

Berridge, M. J., and Irvine, R. F. *Nature* 341:197–205 (1989).

Rana, R. S., and Hokin, L. E. *Physiol. Rev.* 70:115–64 (1990).

Farago, A., and Nishizuka, Y. *FEBS Lett.* 268:350–54 (1990).

Bansal, V. S., and Majerus, P. W. *Annu. Rev. Cell. Biol.* 6:41–67 (1990).

Lee, C. H.; Dixon, J. F.; Reichman, M.; Moummi, C.; Los, G.; and Hokin, L. E. *Biochem. J.* 282:377–85 (1992).

Early indications that surface glycoproteins are involved in cell adhesion were obtained from experiments with proteinases and enzymes attacking carbohydrates. Enzymes of both types inhibit cell adhesion and aggregation and break up existing cell aggregates. More recent experiments have isolated and identified many of the glycoproteins involved in cell adhesion.

Characteristics of Cell Adhesion

Cell attachments are normally permanent once embryonic development is complete, but they may change in highly programmed patterns during development. Embryonic cells regularly make initial attachments that are broken and remade as individual cells or tissues change position in the developing embryo. Some cells, such as erythrocytes, lose their attachments when mature and become freely suspended in a fluid medium. A return to the unattached state characteristic of embryonic cells often occurs in cancer, when malignant cells break their attachments in a tumor and move to new locations in the body.

The capacity for cell adhesion extends throughout the animal kingdom, from sponges to the most advanced vertebrate and invertebrate species. Sponges, in fact, were among the organisms used extensively to demonstrate cell adhesion in early experiments by A. A. Moscona and T. Humphreys. The bodies of many sponges are so loosely organized that they can easily be dispersed into single cells. Under certain conditions the dispersed cells of some sponges are able to sort out and reassemble again into intact, functional individuals, even when cells from two different species are mixed together.

Sorting and adhesion by cell type have also been demonstrated in higher animals. For example, brain cells of mammalian embryos can reassemble after dispersion and reconstruct typical brain cell associations to a remarkable degree. Similar experiments have been carried out with kidney, heart, and other tissues of animal embryos. Cells dispersed from these tissues and mixed in random combinations can frequently sort out and form clusters showing internal organization typical of their tissue type. The classical experiments of Moscona and J. Holtfreter demonstrated that even whole embryos of many vertebrates, such as frogs and salamanders, have the capacity to reaggregate if dispersed into single cells at early stages of development. In the reaggregation the cells sort out into separate endoderm, mesoderm, and ectoderm layers, often in the correct orientation.

Types of Cell Adhesion Molecules

The molecules responsible for cell-cell adhesions have been most extensively studied in mammals and other vertebrates. Among the best-characterized molecules are the *CAMs* (*CAM = Cell Adhesion Molecules*) dis-

covered by G. M. Edelman and others. Mature neural cells contain two CAMs, *N-CAM*, the neural adhesion molecule, and *Ng-CAM*, the neuron-glial adhesion molecule. (Glial cells form a supportive tissue surrounding neurons.) A third CAM, *L-CAM*, occurs on liver cells. A fourth, *I-CAM*, binds a ligand known as *LFA-1* and occurs on many cell types. Another type, *LEC-CAM*, is found on leucocytes and other cells of the circulatory system. Each of the CAMs is a glycoprotein built up from an extracellular carbohydrate group linked covalently to one or more polypeptides with transmembrane segments.

N-CAM molecules on the surface of a neuron bind directly to N-CAMs on other neurons. The N-CAM interaction is one of the associations setting up and maintaining nerve tissues in organized arrays. The Ng-CAMs on neurons bind to Ng-CAM receptors on glial cells. The Ng-CAM–receptor linkages effectively fix neurons and glial cells in place in nerve and brain tissues. In adult tissues Ng-CAMs are distributed only on axons and dendrites and not on cell bodies of neurons (see Fig. 6-16), so that the neuron-glial interactions are restricted to these regions. The association between axons and Schwann cells (a type of glial cell; see Fig. 6-22) forming myelin sheaths is a primary example of an adhesion maintained by bonds between Ng-CAMs and Ng-CAM receptors.

L-CAMs contribute to the binding interactions holding cells in organized masses in liver tissues. Whether L-CAMs react directly with other L-CAMs on adjacent cells or are bound by a separate receptor type remains unknown. However, M. Takeichi and his coworkers showed that a related molecule, *E-cadherin*, binds directly to E-cadherins on adjacent cells. L-CAMs, which belong to the same *cadherin* family of cell surface molecules (see below), are likely to form similar direct linkages.

Molecules forming cell-cell adhesions are not limited to the CAMs discovered by Edelman and others. Neurons of the brain, for example, develop other adhesion molecules as well, among them *L1* and *J1* (Table 7-3). These glycoproteins contribute to the cell-cell adhesions holding neurons in place—L1 is a neuron-neuron adhesion molecule, and J1 is a neuron-glial cell adhesion molecule. Multiple connections of this type, involving several to many different adhesion molecules, are typical of most body cells held in solid tissues. The initial cell-cell adhesions set up during embryonic development by CAMs and other surface glycoproteins are reinforced in many tissues by adhesive junctions of various kinds (see below).

Many cell adhesion molecules are members of related families of glycoproteins. N-CAM, Ng-CAM, I-CAM, and L1, for example, are all members of the immunoglobulin superfamily (Fig. 7-13). Each has constant domains with clear sequence and structural relationships to the constant domains of antibody proteins.

Table 7-3   Some Cell Adhesion Molecules

Type	Molecular Weight	Occurrence
N-CAM	140,000–250,000	Neurons
Ng-CAM	135,000	Neurons
I-CAM	90,000	Many cell types
L1	200,000	Neurons
J1	160,000	Neurons
N-cadherins	135,000	Nerve and muscle cells
L-CAM (an E-cadherin)	124,000	Liver, other epithelial cells
Ovomorulin (an E-cadherin)	120,000	Early embryos, adult epithelial cells
Cell-CAM 120/180 (an E-cadherin)	120,000	Embryos, adult epithelial cells
Arc-1 (an E-cadherin)	130,000	Epithelial cells
P-cadherins	130,000	Placental and epithelial cells
R-cognin	50,000	Embryonic neural retinal cells

Compiled primarily from data presented in B. Obrink, *Exptl. Cell Res.* 163:1 (1986).

L-CAM is a member of a different family of cell-cell adhesion molecules, the *cadherins* (see Table 7-3). In contrast to the receptors of the immunoglobulin superfamily, which have no Ca^{2+} requirement, cadherins form cell-cell adhesions only when Ca^{2+} is pres-ent. Other members of the cadherin family include *E-cadherins* of epithelial cells, *N-cadherins* of nerve and muscle cells, and *P-cadherins* of placental and epithelial cells.

Developmental Interactions of Cell Adhesion Molecules

The pattern of developmental appearance and linkage of cell adhesion molecules illustrates the importance of these surface glycoproteins in development, and the dynamic nature of their interactions as development proceeds. N-CAM and L-CAM appear as early as the blastoderm stage in chick embryos. At this stage L-CAM occurs in cells in all three primary germ layers; N-CAM is generally distributed among ectoderm and mesoderm. Later in development N-CAMs become concentrated in cells forming parts or derivatives of the nervous system. L-CAMs become restricted to liver cells and other structures of mesodermal origin. Ng-CAMs appear much later in development as glial cells differentiate and take up their positions in the embryo.

The dynamic nature of the adhesions formed by CAMs is evident in the movements of neural crest cells, a cell type that migrates from the developing central nervous system (Fig. 7-14). These cells display N-CAMs and adhere to their neighbors in the nerve tube early in development. As mesodermal structures begin to differentiate, N-CAMs disappear from their surfaces, and the neural crest cells break loose from their attachments in the nerve tube. They migrate to other locations in the embryo, where they develop into a wide variety of structures. The N-CAMs and connections to neurons reappear as the neural crest cells reach their final locations.

In some instances, particularly in invertebrate animals, cell-cell adhesions are set up by a specialized extracellular molecule that serves as a bridge linking cells together. Connections to the extracellular linker

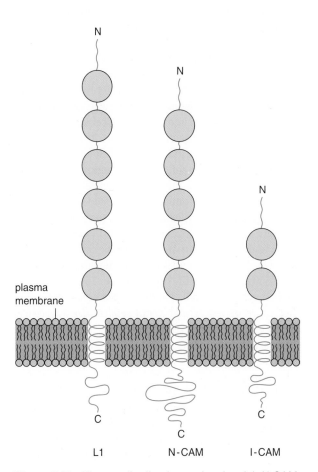

Figure 7-13   Three cell adhesion molecules, L1, N-CAM, and I-CAM, which are members of the immunoglobulin (Ig) superfamily. The domains related to immunoglobulin molecules are shown as circles.

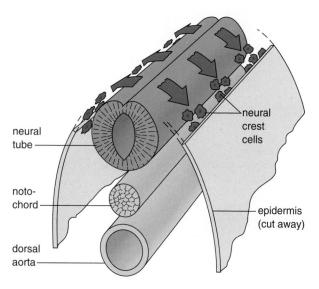

Figure 7-14 Neural crest cells and their origins in the dorsal margins of the developing neural tube. From this location they migrate to various final destinations, where they give rise to such diverse structures as cranial nerves and root ganglia, glial and Schwann cells of the central nervous system, spinal sensory ganglia, sympathetic and parasympathetic ganglia, facial bones and cartilage, skin pigment cells, and the adrenal medulla.

are made by surface receptors that can recognize and bind it. This pattern of linkage occurs in sponges, for example. The extracellular linker is a glycoprotein that depends on Ca^{2+} for its linking activity. In addition to receptors establishing cell-cell adhesions, extensive families of surface receptors bind and attach cells to molecules of the extracellular matrix. (Ch. 8 describes these receptors.)

Some pathogens use cell adhesion molecules as sites for attachment to cell surfaces. The rhinovirus causing the common cold, for example, has a coat protein that is recognized and bound by I-CAM. The poliovirus also attaches to a cell adhesion molecule. Although unidentified, the cell adhesion molecule binding the poliovirus has constant domains that make it a member of the immunoglobulin superfamily. Cell attachment is the first step in the cycle of infection of these viruses.

An experiment by S. D. Marlin and his colleagues shows promise of leading to the long-sought cure for the common cold. These investigators found that I-CAM molecules, made soluble by removal of a segment anchoring them to the plasma membrane, effectively tie up the recognition sites on free rhinovirus particles. The particles blocked in this way are unable to attach and infect host cells. Whether the soluble I-CAMs, which could be produced in commercial quantities in genetically engineered bacteria (see p. 787), can be applied to prevent cold infections without serious side effects remains to be seen.

CELL JUNCTIONS

In many animal tissues the cell-cell adhesions set up during embryonic development are quickly followed by formation of intercellular junctions. In contrast to the interactions responsible for initial cell-cell adhesion, which occur at the level of individual molecules, cell junctions are relatively large, complex structures that are built up by the cooperative activity of many different molecules.

Junctions occur in three major types, each with different functions. *Adhesive junctions* hold cells together, acting as intercellular "buttons" or "zippers" that maintain cells in their fixed positions in tissues. *Tight junctions* close the space between cells to diffusion, forming a sort of dam that prevents flow of molecules and ions through the extracellular space. *Gap* or *communicating junctions* set up open channels that allow ions and small molecules to flow directly from one cell to another. Junctions of all three types occur widely in the animal kingdom.

Adhesive Junctions

There are three main forms of adhesive junctions. Two of them, *desmosomes* and *adherens* junctions, occur widely in the animal kingdom. The third type, the *septate* junction, is limited to invertebrates.

Desmosomes First described in 1954 by K. R. Porter, desmosomes (from *desmos* = bond; also called *macula adherens*) are usually circular or elliptical structures (Fig. 7-15). In a region containing a desmosome a thick layer of dense material, the *plaque*, lies just under the plasma membrane on either side of the junction. *Intermediate filaments*, which are common cytoskeletal structures in animal cells, extend from plaques for some distance into the underlying cytoplasm. Called *tonofilaments* when they are connected to desmosomes, intermediate filaments evidently serve as cytoplasmic anchors for the junction. Depending on the cell, they may be of the cytokeratin or vimentin type (see Ch. 13). The plasma membranes of the cells sharing a desmosome are parallel, more or less straight, and separated by a regular distance in the region of the junction. The extracellular space separating them is filled with faint filaments and granular material collectively called the *core* of the desmosome.

Half desmosomes, or *hemidesmosomes* (Fig. 7-16), occur where cells are anchored to extracellular materials such as the *basal lamina* secreted by some epithelial cells (see p. 277). Hemidesmosomes also appear along the plasma membrane in cells grown in tissue culture, where they anchor cells to the glass or plastic substrate on which they are grown.

The protein complement of desmosomes has been isolated from rat, bovine, and other animal tissues by

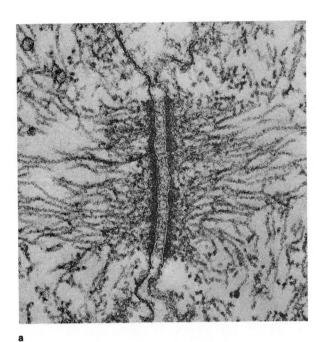

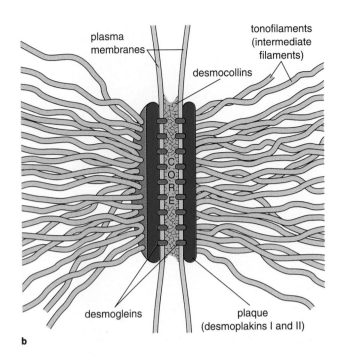

a

b

Figure 7-15 Desmosomes. **(a)** A desmosome binding adjacent cells in the skin of a salamander. ×22,000. Courtesy of D. E. Kelly, from *J. Cell Biol.* 28:51 (1966), by permission of the Rockefeller University Press. **(b)** Desmosome structure. The plaques contain at least two major proteins, desmoplakins I and II. The desmogleins serve as a transmembrane bridge between the plaques and the extracellular material. The extracellular space (the core region) between the plaques contains glycoproteins known as desmocollins.

W. W. Franke, M. S. Steinberg, and others. Among the several major proteins consistently found in desmosomes in different animals and tissues, two of the largest, *desmoplakins I* and *II,* are structurally related molecules associated with plaques. Another group of glycoproteins, the *desmogleins,* which Franke and his coworkers

recently found to be members of the cadherin family, serves as a transmembrane bridge between plaques and extracellular material. Carbohydrate groups are attached to the extracellular portions of desmogleins. Three different glycoproteins occur in the extracellular space of desmosomes. Presumably, these glycoproteins, called *desmocollins,* which are also members of the cadherin family, bind the opposing plasma membranes in the region of a desmosome. Desmoplakins I and II and desmoglein cannot be detected in hemidesmosomes, indicating that hemidesmosomes are not simply half desmosomes, as the name and electron micrographs such as Figure 7-16 suggest, but are molecularly distinct structures.

If living cells held together by desmosomes are digested briefly by dilute solutions of a proteinase such as trypsin, only the intercellular filamentous material is attacked, allowing the cells to separate without apparent damage. This method is routinely used to disperse cells linked together by desmosomes, such as tissue culture cells or the cells of early embryos. Reducing extracellular Ca^{2+} concentration to very low levels also causes desmosomes to separate. A part of the sensitivity of desmosomes to Ca^{2+} concentration may depend on desmogleins, which require Ca^{2+} for their binding activity.

Desmosomes are present between many cell types in multicellular animals. They are especially abundant in tissues normally subjected to shear or lateral stress,

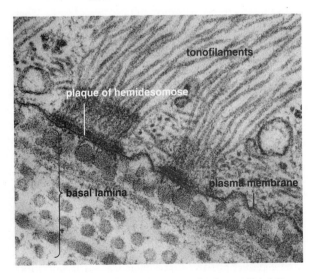

Figure 7-16 Hemidesmosomes anchoring cells of the skin to the basal lamina in a salamander. ×68,000. Courtesy of D. E. Kelly, from *J. Cell Biol.* 28:51 (1966), by permission of the Rockefeller University Press.

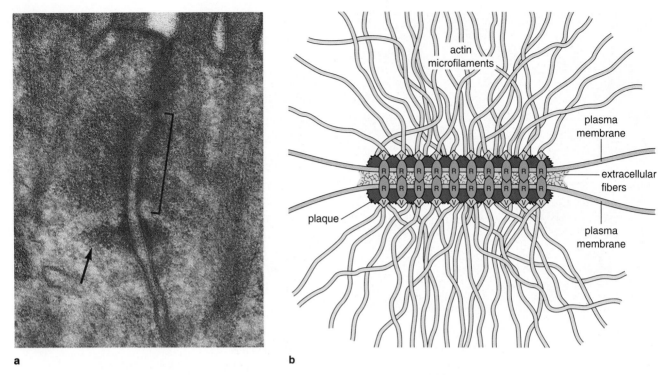

a   b

Figure 7-17 Adherens junctions. **(a)** An adherens junction (bracket) between intestinal cells. A portion of a desmosome (arrow) is also visible just below the adherens junction. ×99,000. Courtesy of T. S. LeCount. **(b)** Possible structural elements of an adherens junction. R, an integral membrane glycoprotein of the cadherin family, which acts as a receptor linking adjacent plasma membranes in the region of the junction; V, a complex of molecules including catenins, which binds the junction to microfilaments of the cytoskeleton.

such as skin cells and the epithelia lining the internal surfaces of body cavities. Desmosomes appear early in embryonic development and contribute to the maintenance of cell position from the earliest stages. They also form readily between cells maintained in cultures.

Desmosomes frequently disappear when cells are transformed into cancerous types. This change undoubtedly contributes to the tendency of cancer cells to *metastasize*, that is, to break loose from tumors and move to new locations in the body (see p. 943). Another human disability, *pemphigus*, in which blisters develop in the skin over much of the body, also appears to result from deficiencies in desmosomes. In this case the deficiencies are caused by autoantibodies developed against desmosome components. (Autoantibodies are antibodies that react with an individual's own proteins rather than foreign substances; see p. 815.) J. C. R. Jones and his coworkers found that autoantibodies against two polypeptides in hemidesmosomal plaques, produced in some patients with pemphigus, do not cross react with desmosomes. This finding provides another line of evidence indicating that hemidesmosomes are molecularly distinct.

Adherens Junctions Adherens junctions, also called *zonula adherens*, bind cells in certain animal tissues such

as heart muscle and the thin layers covering organs and lining body cavities (Figs. 7-17 and 7-18). As in desmosomes the region just under the plasma membrane in an adherens junction is thickened into a plaque. However, the plaques of adherens junctions appear as loosely structured mats, not solidly dense layers. Fibers also radiate from the plaques of adherens junctions into the underlying cytoplasm. These anchoring fibers, which are noticeably thinner than the tonofilaments of desmosomes, have been identified as *actin microfilaments*. Faintly visible filaments fill the extracellular space in the region of an adherens junction as in desmosomes.

Adherens junctions occur in forms ranging from buttonlike structures to extended, beltlike configurations that completely encircle the cell. The presence of microfilaments in adherens junctions suggests that these junctions, in addition to their adhesive function, may form part of actin-based motile systems in some cells. The extracellular connections holding plasma membranes together in the region of an adherens junction, like those of desmosomes, are calcium dependent; extreme reductions in extracellular Ca^{2+} concentrations release the connections.

Adherens junctions also occur in a "hemi" form consisting of an apparent half junction. These structures occur on the plasma membranes of cells linked to ex-

Figure 7-18 Adherens junctions (brackets) linking cardiac muscle cells in heart tissue. Courtesy of N. J. Severs and Plenum Publishing Corp., from *Adv. Myocardial.* 5:223 (1985).

tracellular structures such as the basal lamina, and also at the margins of cells bound to the surfaces of culture vessels.

Adherens junctions contain high concentrations of a transmembrane glycoprotein of the cadherin family, and other proteins including α–, β–, and γ– *catenin.* The extracellular portion of the cadherin probably sets up cell–cell linkages in the region of the junction; the cytoplasmic segment of the cadherin organizes the catenins in a complex that forms links with tactin microfilaments. (α– catenin is related to vinculin, a microfilament–binding protein; see Table 12-1.)

Adherens junctions in cell layers lining body cavities such as the digestive tract often occur along with desmosomes and tight junctions. Adherens junctions also link cardiac muscle cells in heart tissue (Fig. 7-18). In these cells the actin microfilaments of the junctions are continuous with the internal microfilaments of heart muscle cells, which are responsible for the cell contractions producing the heartbeat (see p. 478). These connections directly implicate adherens junctions in the motile systems of these cells.

Septate Junctions First described in *Hydra* by R. L. Wood in 1959, septate junctions (from *septum* = wall) are so called because in cross section the membranes forming the junction are connected by regularly spaced crossbars or *septa* in the extracellular region (Fig. 7-19). The space between the membranes is held at a very regular 15 to 17 nm by the dense bars, which alternate with lighter spaces throughout the length of the junction. All septate junctions have distinctive crossbars in the extracellular region, but the detailed morphology

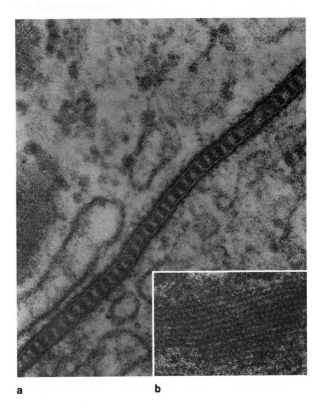

a   b

Figure 7-19 Septate junctions between epithelial cells of a mollusk. **(a)** A septate junction in cross section, showing the typical regularly spaced, dense bars alternating with less dense spaces in the extracellular region. × 160,000. **(b)** A septate junction in a section parallel to the membrane surfaces. The dense bars form a pleated or zigzag pattern. × 80,000. Courtesy of N. B. Gilula, from *Cell Communication* (Cox, R. P., ed.). New York: John Wiley and Sons, Inc., 1974.

Hannun, Y. A., and Bell, R. M. 1989. Functions of sphingolipids and sphingolipid breakdown products in cellular regulation. *Science* 243:500–507.

Merril, A. H., and Stevens, V. L. 1989. Modulation of protein kinase C and diverse cell functions by sphingosine: a pharmacologically interesting compound linking sphingolipids and signal transduction. *Biochim. Biophys. Acta* 1010:131–39.

Mooibroek, M. J., and Wang, J. H. 1988. Integration of signal-transduction processes. *Biochem. Cell. Biol.* 66:557–64.

Putney, J. W. 1990. The integration of receptor-regulated intracellular Ca^{2+} release and Ca^{2+} entry across the plasma membrane. *Curr. Top. Cellular Regulat.* 31:111–27.

Rasmussen, H., and Rasmussen, J. E. 1990. Calcium as intracellular messenger: from simplicity to complexity. *Curr. Top. Cellular Regulat.* 31:1–109.

Sibley, D. R.; Benovic, J. L.; Caron, M. G.; and Lefkowitz, R. J. 1987. Regulation of transmembrane signaling by receptor phosphorylation. *Cell* 48:913–22.

Tsien, R. W., and Tsien, R. Y. 1990. Ca^{2+} channels, stores, and oscillations. *Ann. Rev. Cell Biol.* 6:715–60.

Cell Adhesion Molecules

Albelda, S. M., and Buck, C. A. 1990. Integrins and other cell adhesion molecules. *FASEB J.* 4:2868–80.

Boyer, B., and Thiery, J.-P. 1989. Epithelial cell adhesion molecules. *J. Membr. Biol.* 112:97–108.

Edelman, G. M., and Crossin, K. L. 1991. Cell adhesion molecules: implications for a molecular histology. *Ann. Rev. Biochem.* 60:155–90.

Faissner, A. 1989. Cell-cell adhesion in the nervous system: structural groups emerge. *Bioess.* 10:79–81.

Hynes, R. O. 1992. Integrins: versatility, modulation, and signaling in cell adhesion. *Cell* 69:11–25.

McClay, D. R., and Ettensohn, C. A. 1987. Cell adhesion in morphogenesis. *Ann. Rev. Cell Biol.* 3:319–45.

Takeichi, M. 1990. Cadherins: a molecular family important in selective cell-cell adhesion. *Ann. Rev. Biochem.* 59:237–52.

Cell Junctions, General

Larsen, W. J., and Risinger, M. A. 1985. The dynamic histories of intercellular membrane junctions. *Mod. Cell Biol.* 4:151–216.

Desmosomes and Hemidesmosomes

Garrod, D. R. 1986. Desmosomes, cell adhesion molecules and the adhesive properties of cells in tissues. *J. Cell Sci.* Suppl. 4:221–37.

Schwarz, M. A.; Owaribe, K.; Kartenbeck, J.; and Franke, W. W. 1990. Desmosomes and hemidesmosomes: constitutive molecular components. *Ann. Rev. Cell Biol.* 6:461–91.

Adherens Junctions

Geiger, B.; Avnur, Z.; Volberg, T.; and Volk, T. 1985. Molecular domains of adherens junctions. In *The cell in contact*, ed. G. M. Edelman and J.-P. Thiery, pp. 461–89. New York: Wiley.

Tight Junctions

Handler, J. S. 1989. Overview of epithelia polarity. *Ann. Rev. Physiol.* 51:729–40.

Larsen, W. J., and Risinger, M. A. 1985. The dynamic life histories of intercellular membrane junctions. *Mod. Cell Biol.* 4:151–216.

Madara, J. L. 1988. Tight junction dynamics: is paracellular transport regulated? *Cell* 53:497–98.

Gap Junctions

Boyer, E. C.; Paul, D. L.; and Goodenough, D. A. 1990. Connexin family of gap junction proteins. *J. Membr. Biol.* 116:187–94.

Hertzberg, E. L., and Johnson, R. G., eds. 1988. *Gap junctions. Mod. cell biol.*, vol. 7. New York: Alan R. Liss.

Kistler, J., and Bullivant, S. 1988. The gap junction proteins: vive la difference. *Bioess.* 9:167–68.

Pitts, J. D., and Finbow, M. E. 1986. The gap junction. *J. Cell Sci.* Suppl. 4:239–66.

Warner, A. 1988. The gap junction. *J. Cell Sci.* 89:1–7.

Cell Recognition and Adhesion in Bacteria

Ceri, H., and Westra, Y. 1988. Host binding proteins and bacterial adhesion: ecology and binding model. *Biochem. Cell. Biol.* 66:541–48.

Karlsson, K.-A. 1989. Animal glycosphingolipids as membrane attachment sites for bacteria. *Ann. Rev. Biochem.* 58:309–50.

Sharon, N. 1987. Bacterial lectins, cell-cell recognition, and infectious disease. *FEBS Lett.* 217:145–57.

Receptor-Response Mechanisms in Bacteria

Albright, L. M.; Huala, E.; and Ausubel, F. M. 1989. *Ann. Rev. Genet.* 23:311–36.

Bourret, R. B.; Hess, J. F.; Borkovich, K. A.; Pakula, A. A.; and Simon, K.-M. I. 1989. Protein phosphorylation in chemotaxis and two-component regulatory systems of bacteria. *J. Biolog. Chem.* 264:7085–88.

Eisenbach, M. 1990. Functions of the flagellar modes of rotation in bacterial motility and chemotaxis. *Molec. Microbiol.* 4:161–67.

Gross, R.; Aricó, B.; and Rappnoli, R. 1989. Families of bacterial signal-transducing proteins. *Molec. Microbiol.* 3:1661–67.

Hess, J. F.; Bourret, R. B.; and Simon, M. I. 1988. Histidine phosphorylation and phosphoryl group transfer in bacterial chemotaxis. *Nature* 336:139–43.

Parkinson, J. S. 1988. Protein phosphorylation in bacterial chemotaxis. *Cell* 53:1–2.

Stock, J. B.; Stock, A. M.; and Mottonen, J. M. 1990. Signal transduction in bacteria. *Nature* 344:395–400.

Review Questions

1. Outline the structure of membrane glycolipids and glycoproteins. How are these molecules embedded in the fluid mosaic structure of the plasma membrane?

2. How is the capacity for cell-to-cell recognition related to membrane surface glycolipids and glycoproteins?

3. What are MHC markers? What are the major classes of these markers? How are these markers involved in the rejection of organ transplants? In the immune response to viral infections? What is the relationship between MHC markers and antibodies?

4. Outline the molecular basis for the ABO blood groups of humans. What blood types can be mixed successfully in transfusions? Why?

5. What is a cell surface receptor? What evidence indicates that receptors transmit a signal to the cytoplasm? What types of molecules are bound by receptors?

6. What are protein kinases? What is the relationship of protein kinases to surface receptors? What two major types of receptors exist with reference to protein kinase activity?

7. What are first messengers? Second messengers? G proteins? Effectors? How do these elements act in receptor-response mechanisms?

8. Outline the role of G proteins in receptor-response pathways.

9. What two types of effectors operate in receptor-response pathways? What second messengers are products of the enzymatic activity of these effectors? What activates the effectors?

10. What does amplification mean with reference to receptor-response pathways?

11. Outline the reaction steps taking place in a pathway using cAMP as second messenger. What types of first messengers stimulate operation of cAMP-based pathways?

12. Outline the reaction steps taking place in a pathway using $InsP_3$ and DAG as second messengers. What types of first messengers stimulate $InsP_3$/DAG-based pathways? What is the role of Ca^{2+} in these pathways?

13. Compare the activities of the protein kinases activated in cAMP- and $InsP_3$/DAG-based pathways.

14. What is the relationship between the receptor-response pathways and cancer? Between substances acting as toxins or poisons?

15. What is the importance of cell adhesion in animals? What kinds of molecules are responsible for cell adhesion? Outline the role of cell adhesion molecules in development. How is cell adhesion related to surface reception?

16. Outline the major types of junctions formed between animal cells. What are the major functions of each junction type?

17. What is a desmosome? An adherens junction? In what ways are these two junction types similar, and in what ways are they different? What is a septate junction? What evidence indicates that septate junctions probably have an adhesive function?

18. Where do desmosomes and adherens junctions occur in animals? What is the relationship between adhesive junctions and cancer?

19. What is a tight junction? What evidence supports the idea that tight junctions close intercellular spaces to the flow of molecules?

20. Outline the possible structure of a tight junction. What is the relationship between tight junctions and the distribution of membrane lipids and proteins?

21. Where do tight junctions occur in animals?

22. Describe the structure of a gap junction. What evidence indicates that gap junctions provide open channels for the flow of molecules and ions between cells?

23. How are gap junctions regulated? What keeps damaged cells from causing leakage from other cells in tissues interconnected by gap junctions?

24. Where do gap junctions occur in animals? List some major body functions that depend on cell communication via gap junctions. What evidence indicates that gap junctions are important in embryonic development?

Supplement 7-1

Recognition, Reception, and Cell Adhesion in Bacteria

Bacteria also have systems for recognition, reception, and cell adhesion. Cell wall molecules active in recognition and adhesion take part in infective mechanisms, in which bacterial cells recognize and adhere to cells of a host eukaryote, and in sexual conjugation, in which bacterial cells of opposite mating types recognize and adhere. Reception depends on receptor molecules that bind external molecules acting as signals, as in eukaryotic systems. Cellular responses triggered by binding the signal adjust functions as diverse as transport, nitrogen metabolism, sporulation, and *chemotaxis*, a motile response in which cells swim toward or away from attractive or repellent substances in the medium.

Bacterial Recognition and Adhesion

Cell recognition and adhesion in many bacteria depend on *pili* (singular = *pilus*; pili are also called *fimbriae*), specialized protein fibers about 5 to 7 nm in diameter and 100 to 200 nm in length (Fig. 7-30). Pili, composed of linear chains of protein subunits called *pilins*, extend entirely through the cell wall into the surrounding medium. Associated with pili are other proteins that act as lectins (see p. 239). Bacterial pili and their associated proteins are capable of binding carbohydrate groups on the surfaces of both prokaryotic and eukaryotic cells; they also serve as recognition groups bound by receptors on eukaryotic cells and the coat proteins of bacterial viruses.

Individual cells may occur without pili or possess from 1 to 300 of the same or different types. In some bacteria the presence or absence of pili determines the "sex" of conjugating cells. During conjugation, in which a cytoplasmic bridge forms between mating pairs, bacterial cells with pili act as donor cells that transfer DNA to cells without pili. Because cells with pili are the donors, they are considered as "male" or + cells in the exchange. Receiving cells, considered as "female" or − in the exchange, possess a surface receptor that can recognize and bind the pilin protein.

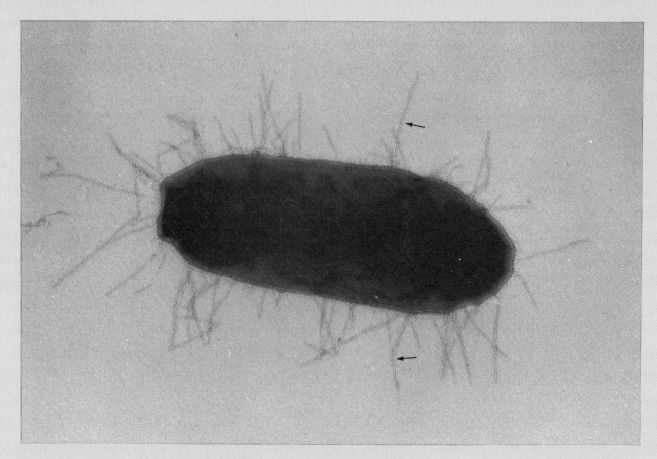

Figure 7-30   Pili (arrows) extending from the surface of an *E. coli* cell prepared for electron microscopy by negative staining. Courtesy of A. Gbarah and N. Sharon.

In *Neisseria gonorrhoeae*, the bacterium that causes gonorrhea, surface pili are responsible for a recognition and binding reaction during initial stages of infection. The binding makes the bacterial cells adhere strongly to the mucosal lining of the urinary tract, so strongly that they are not readily dislodged by urine flow. Like the surface glycoproteins of the trypanosomes causing sleeping sickness (see p. 816), the proteins of the gonococcal pili have variable regions that constantly undergo changes in amino acid sequence because of genetic changes in the infecting cells. The changes in the variable regions, which are the protein segments recognized by the immune system, enable the infecting bacterial population to keep one step ahead of the immune system and escape destruction. For this reason gonorrheal infections are highly persistent unless treated by antibiotics.

Pili occur in gram-positive and gram-negative bacteria. Both bacterial types also have cell wall molecules that serve as recognition sites. Gram-positive bacteria form wall components known as *teichoic acids*, consisting of linear polymers built up from repeating sugar units (see p. 305). The teichoic acids, which extend from the surface of the cell wall, are used as recognition and binding sites by bacterial viruses infecting these cells. The teichoic acids also act as antigens stimulating antibody formation in response to infections in humans and other mammals. In gram-negative bacteria the *O antigens*, which are subunits of lipopolysaccharides, act similarly in recognition and surface binding. The lipopolysaccharides form a major component of the outer membrane of gram-negative bacteria. (See p. 305; Ch. 8 presents further information on pili, teichoic acids, O antigens, and wall structure in gram-positive and gram-negative bacteria.)

Bacterial Receptors Inducing Internal Responses

More than 20 different surface receptors triggering internal responses have been detected in gram-negative bacteria such as *E. coli* and *Salmonella*, in which receptor mechanisms have been best studied. Each receptor mechanism consists of a pair of proteins: a *sensor* and a *regulator* (Fig. 7-31). The sensor in most systems is a transmembrane protein that combines with a molecule acting as a sensory signal. In a few systems the sensor is cytoplasmic, and is activated indirectly by additional membrane proteins that combine directly with a signal molecule. The second member of the pair, the regulator, is cytoplasmic in all known systems. Other cytoplasmic components are also present in some systems.

Although much of the sensor protein varies in amino acid sequence between different sensor-regulator pairs, these proteins share a homologous domain of about 200 amino acids at their C-terminal ends. In the

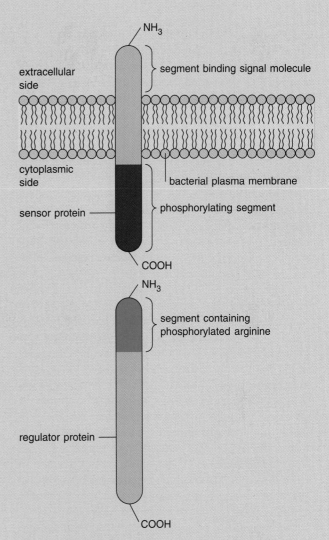

Figure 7-31 Sensor-regulator pairs in bacteria. The sensor proteins of different pairs share a homologous domain at their C-terminal ends of about 200 amino acids (dark purple). In most bacterial systems the sensors are transmembrane proteins; the N-terminal domain, which varies in sequence in the sensors of different pairs, contains the site binding signal molecules at the outer surface of the plasma membrane. In the regulators the homologous domain forms an element of approximately 120 amino acids at the N-terminal end of these proteins (dark green). The homologous domain contains an arginine residue that is phosphorylated by an activated sensor protein; phosphorylation activates the regulator, which triggers an internal response. In most sensor-regulator pairs the regulator controls transcription of one or more genes.

sensors suspended in the plasma membrane, the variable element, which combines with molecules acting as sensory signals, extends from the outside membrane surface. The homologous domain extends into the cytoplasm on the other side of the membrane. The regulators also have homologous and variable sequence elements. In the regulators the homologous domain includes a sequence element of approximately 120 amino acids at the N-terminal end.

Binding a molecule that acts as a sensory signal induces a conformational change in the sensor, which is transmitted through the protein's structure to its C-terminal domain. The conformational change activates a self-phosphorylation in the C-terminal domain, in which a phosphate group is transferred from ATP to a histidine residue in the protein. The sensor then transfers the phosphate group to an aspartic acid residue in the N-terminal domain of the regulator. This phosphorylation activates or modifies the activity of the regulator, which triggers an internal response. For many bacterial systems the activated regulator promotes transcription of one or more genes by combining directly with control regions in the DNA.

The sensor-regulator pair responding to changes in solute concentration of the surrounding medium in *E. coli* is one of the better understood bacterial systems (Fig. 7-32). When the medium has concentrations of ions and molecules that produce low osmotic pressure inside the cell, a sensor protein located in the plasma membrane, EnvZ, is activated at relatively low levels. In this condition the transfer of phosphate groups to the regulator protein, OmpR, is limited, and the regulator combines with elements in the DNA that activate transcription of *ompF*, a gene encoding the porin OmpF. The porins, which are channellike proteins inserted in the outer membrane of gram-negative bacteria, control the upper limit of molecular sizes able to penetrate through the outer membrane to reach the plasma membrane (see p. 227). When the concentrations of ions and molecules in the medium produce high osmotic pressure inside the cell, EnvZ is highly activated and heavily phosphorylates OmpR. In the highly phosphorylated form OmpR combines with sites in the DNA regulating a gene encoding a different porin, OmpC. The presence of either OmpF or OmpC as the predominant outer membrane protein makes compensating adjustments in the osmotic pressure of the cell.

Because each response pathway has its own set of sensors and receptors, the bacterial systems differ fundamentally from the receptor mechanisms of eukaryotes, in which a wide variety of cellular responses is controlled through a small number of second messengers common to many pathways. There are, in fact, no known bacterial equivalents to the second messengers of eukaryotes; although cAMP occurs in bacteria, it functions in pathways regulating genetic activity that are not a direct part of the sensor-regulator systems (for details, see p. 730).

Suggestions for Further Reading: See p. 226.

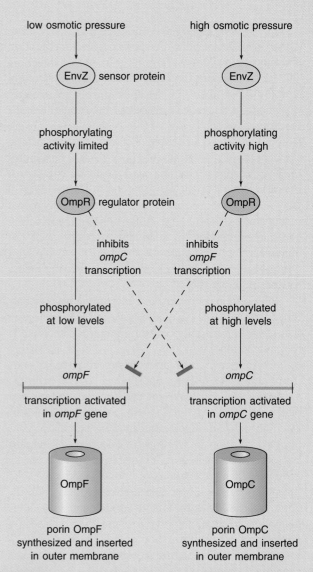

Figure 7-32 The sensor-regulator pair responding to changes in osmotic pressure in *E. coli*. When osmotic pressure is reduced, the sensor protein EnvZ, located in the plasma membrane, is activated only at relatively low levels. In this condition the transfer of phosphate groups to the regulator protein, OmpR, is limited. In response the regulator takes on a conformation that promotes its combination with elements in the DNA that activate transcription of *ompF*, a gene encoding the outer membrane porin OmpF. When osmotic pressure is high, EnvZ is highly activated and heavily phosphorylates OmpR. In this form OmpR combines with sites in the DNA regulating *ompC*, the gene encoding the porin OmpC. The presence of either OmpF or OmpC as the predominant outer membrane protein alters the substances reaching the plasma membrane and results in compensating adjustments in the osmotic pressure of the cell. Lightly phosphorylated OmpR also inhibits *OmpC* transcription, and heavily phosphorylated OmpR inhibits *OmpF* transcription as shown.

THE EXTRACELLULAR MATRIX
OF EUKARYOTIC CELLS

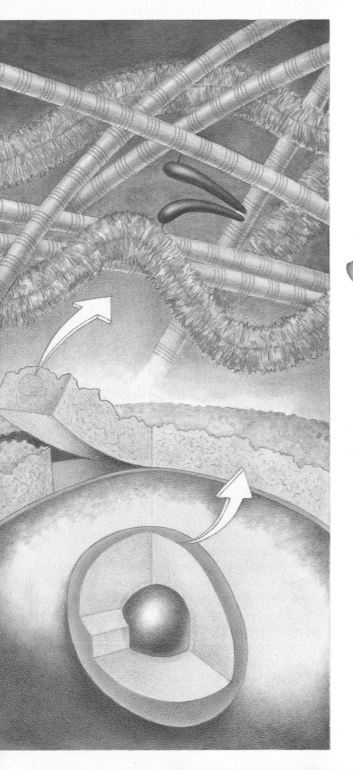

M any types of animal cells, and most cells of other organisms, secrete materials into the regions outside the plasma membrane. These secreted materials form extracellular structures, collectively called the *extracellular matrix*, that carry out a variety of functions. The extracellular matrix secreted by animal cells forms structures as diverse as the tough, elastic framework of tendons, cartilage, and bone; the clear, crystallike cornea of the eye; and supportive elements in epithelia and other tissues and organs. In plants and prokaryotes extracellular materials form the walls surrounding and separating the cells of these organisms. (The cell walls of plants are described in the body of this chapter; the structures and functions of cell walls in prokaryotes are outlined in Supplement 8-1.)

The primary function of the extracellular matrix is support. However, it has a variety of additional functions in eukaryotes and prokaryotes. The patterns and types of extracellular materials laid down in animals may regulate cell division, adhesion, cell motility and migration, and differentiation during embryonic development. Extracellular material also figures importantly in the reactions to wounding and disease. During the development of cancer, the extracellular matrix becomes altered in structure, and cells frequently lose the connections anchoring them to this material. Some functions of extracellular materials are unexpected. For example, a meshlike extracellular layer acts as a filter that excludes larger molecules such as proteins from passing through cell layers in the walls of capillaries and kidney glomeruli.

In plants, fungi, algae, and prokaryotes extracellular materials of the walls may carry sites that recognize and bind molecules from the surrounding medium. Plant cell walls also function in cell-to-cell communication. The external walls in all these forms provide protection from mechanical damage and attack by infective organisms and viruses.

In spite of their diverse functions, the extracellular structures of eukaryotes include common elements. Many contain long, semicrystalline *fibers*, which provide resistance to stretching and other tensile forces. The fibers are embedded in a second element common to many extracellular structures, a more or less elastic *network* of branched molecules. The network holds the

fibers in place and resists compression by trapping and retarding the flow of water molecules. Variations in the kinds of fiber and network molecules, the degree of crosslinking, the amount of trapped water, and the types of added substances such as the calcium phosphate crystals of bone or the lignin of plant cell walls produce extracellular structures ranging in consistency from soft, watery gels to materials almost as dense and hard as rock.

ANIMAL EXTRACELLULAR STRUCTURES

The *collagens*, a large family of glycoprotein molecules, are the primary fibers of animal extracellular structures. The main components of the surrounding network are the *proteoglycans*, a group of glycoproteins of unusually high carbohydrate content.

The Primary Fiber: Collagens

The collagens form semicrystalline fibers that hold cells in place, provide tensile strength and elasticity to the extracellular matrices supporting body cells, and serve functions related to cell motility and development. Collagens occur in all animal phyla from sponges to chordates. In vertebrates the collagens of tendons, cartilage, and bone are the single most abundant protein of the body, making up about half of the total body proteins by weight. Outside vertebrates collagen fibers are particularly abundant in the extracellular material of sponges and the thick surface cuticle of nematode and annelid worms.

Collagen Structure Collagens make up a unique group of predominantly insoluble glycoproteins characterized by a high content of glycine and two modified amino acids, hydroxylysine and hydroxyproline (see Fig. 2-20). In much of the molecule, glycine is repeated at every third position in the amino acid sequence in the pattern (gly-X-Y)$_n$, in which X and Y are other amino acids, most commonly proline or lysine in either unmodified or modified form. Almost all the carbohydrate portion of collagens is built up from two sugars, glucose and galactose (see Fig. 5-5). These sugars link in short two-unit chains attached to the protein component of the molecules. About 10% of the total weight of collagens is carbohydrate.

Individual collagen molecules are linear structures built up from three polypeptide chains. The individual chains, the α-*chains*, wind in a shallow left-handed helix (Fig. 8-1*a*). The regularly placed proline residues are important in stabilizing this helix. Three of the α-chains twist together into a relatively rigid right-handed triple helix that forms about 95% of the central portion of the molecule (Fig. 8-1*b* and *c*). This helical structure is

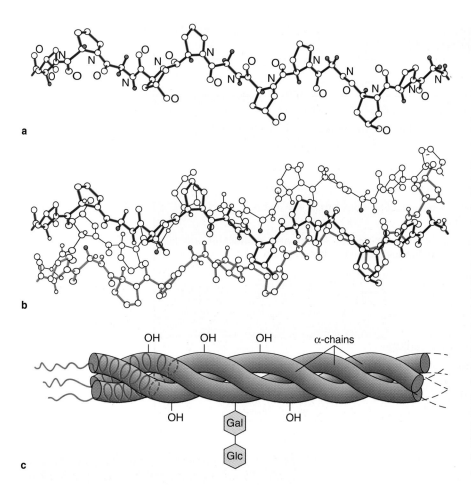

Figure 8-1 Collagen structure. **(a)** An α-chain, a single polypeptide subunit of a collagen molecule, considered as a (glycine-proline-hydroxyproline)$_n$ sequence. The chain winds into a shallow left-handed helix, the *polyproline helix*, determined by the large numbers of proline and hydroxyproline residues. The two hydrogens associated with the glycine residues are shown in green; other hydrogens capable of forming stabilizing hydrogen bonds are shaded. **(b)** A collagen molecule, a triple helix of three α-chains wound together in a right-handed twist. The spacing of glycine residues (in green) places them in the axis of the triple helix, where their small size allows close packing of the three α-chains. **(c)** A collagen molecule in diagrammatic form. The carbohydrate portion of collagens consists almost completely of chains of short two-sugar galactose-glucose (Gal-Glc) units attached to the protein component of the molecule. The hydroxyl (—OH) groups on hydroxyproline and hydroxylysine residues form hydrogen bonds to join collagen molecules into collagen fibers. **(a** and **b)** redrawn from originals courtesy of R. D. B. Fraser, from *J. Molec. Biol.* 129:463 (1979).

possible because the regularly spaced glycine residues are small enough to fit along the central axis of the helix (shown in green in Fig. 8-1*b*).

The triple helix is responsible for the fibrous nature of collagen. Depending on the collagen type, the triple helix may be almost continuous, or it may contain numerous segments that give way to less-ordered conformations. The less-structured segments act as hinges that give the central region greater flexibility.

At the tips of the central triple helix, the amino acid chains fold into globular segments that cap the ends of the molecule. Although the globular end caps in some collagen types include limited segments of triple helix, most of the globular region is folded into other conformations. The entire structure is held together by hydrogen bonds between the α-chains in both the triple helix and globular caps. The —OH groups added to the modified forms of proline and lysine are particularly important as hydrogen-bonding sites that stabilize the triple helix.

As many as 25 different kinds of α-chains, all closely related but of distinct amino acid sequence, occur in vertebrates. These α-chains twist by threes in different combinations to form at least 14 distinct types of collagen molecules, designated types I to XIV (Table 8-1). The molecules associate through both noncovalent

and covalent bonding to produce collagen fibers of various structures and dimensions. Most of these fibers are assembled primarily from one collagen type, with smaller amounts of other types in fibers of some cellular locations.

Type I, II, and III collagen molecules make up the main fibers of animal extracellular structures. These molecules, in which the central triple helix is almost uninterrupted and contains about 1000 amino acids, are the most fibrous and least flexible of the collagen types. In the remaining collagens the central triple helix has more interruptions and is more flexible. All these more flexible collagen types, with the exception of type IV, are minor constituents of supportive structures.

Type I collagen molecules, the primary constituents of bone, skin, and tendons, form about 90% of the body's collagen. In bone collagen fibers are constructed from essentially pure type I molecules; in skin and tendons collagen fibers contain type I in combination with lesser amounts of type III molecules. Fibers containing type I and smaller amounts of type V occur in the cornea. Type II collagen makes up the major fibers of cartilage.

Collagen fibers are arranged in rigid plates in bones, in parallel bundles in tendons and in a dense meshwork in cartilage. In the cornea collagen occurs in clear, crys-

Synthesis of Cellulose and Network Components

Biochemical research has shown that cellulose molecules and microfibrils are assembled at the plasma membrane. Electron microscopy of freeze-fracture preparations by R. M. Brown and others filled in some of the details of the process. In freeze-fracture preparations the positions of newly synthesized cellulose microfibrils can be identified in groovelike depressions in the plasma membrane at the sites of microfibril assembly (Fig. 8-26). Fractures splitting the membrane into bilayer halves show protein-sized particles embedded within the membrane at the ends of the grooves. The particles may be arranged in linear rows (as in Fig. 8-26) or in circular patterns called *rosettes* (Fig. 8-27). The linear rows and rosettes are believed to be enzyme complexes that synthesize cellulose microfibrils from glucose (Fig. 8-28).

The cellulose microfibrils assembled by the enzyme complexes are believed to be more or less stationary as they grow in length in the wall. As a result, the enzyme complexes are considered to be pushed through the fluid membrane bilayer by the growing end of the cellulose microfibril.

Each rosette of enzymes, consisting of a circle of six protein subunits arranged in a hexagon, is thought to cast a microfibril about 5 nm in diameter. Larger microfibrils are assembled by groups of rosettes that may number from as few as 2 or 3 to as many as 175 or more. (The group shown in Fig. 8-27 contains more than 100 rosettes.) The 5-nm-diameter microfibrils assembled by individual rosettes fuse as they form, producing the much larger cellulose microfibrils that occur in most cell walls. When the cellulose microfibrils are fully formed, they are released from the plasma membrane grooves and are bound into place in the growing

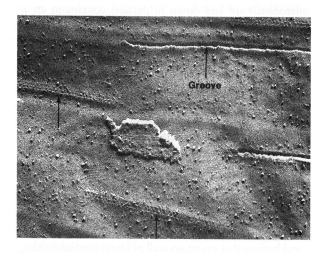

Figure 8-26 Freeze-fracture preparation of a plasma membrane in regions synthesizing cellulose microfibrils. New microfibrils are believed to form in the grooves. The rows of particles (arrows) are considered to be the enzyme complexes synthesizing cellulose. From the alga *Oocystis*; ×64,000. Courtesy of R. M. Brown, Jr.

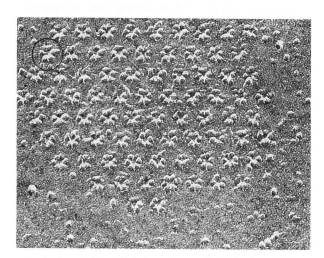

Figure 8-27 Group of rosettes (one rosette is circled) arranged in a regular hexagonal pattern in the membrane of the green alga *Microsterias*. Each rosette is believed to synthesize a cellulose microfibril 5 nm in diameter; the 5-nm-diameter microfibrils secreted by the entire group fuse to form a larger, composite microfibril. ×220,000. Courtesy of T. H. Giddings, reproduced from *J. Cell Biol.* 84:327 (1980), by copyright permission of the Rockefeller University Press.

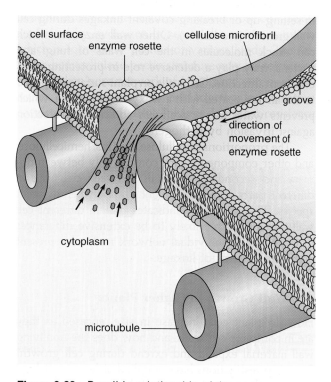

Figure 8-28 Possible relationship of the enzyme rosette and plasma membrane groove in generation of cellulose microfibrils. The rosette may be guided by parallel tracks of microtubules as shown.

wall by the hemicelluloses and pectins of the surrounding network.

Synthesis of the network molecules holding cellulose fibers in place in cell walls has been followed by experiments using labeled glucose. The label is first incorporated into precursors of pectins and hemicelluloses in the Golgi complex. From this location the molecules, still in the form of unfinished intermediates, move to the plasma membrane in vesicles derived from the Golgi complex. As they reach the cell border, the vesicles fuse with the plasma membrane and release their contents to the cell exterior. There, enzymes located in the cell wall complete synthesis of pectins and hemicelluloses and set up covalent bonds linking them into the wall. Few of the steps in these reactions are known; nor is it known how the enzymes catalyzing them are regulated, or how the enzymes take up their locations in their growing cell wall.

Microtubules in Cell Wall Synthesis In 1963 M. C. Ledbetter and K. R. Porter discovered a layer of microtubules in the cytoplasm just beneath growing cell walls. Subsequent investigations confirmed the presence of microtubules as a characteristic feature of cell wall formation that is related in some way to the direction of cellulose microfibril synthesis. Where microfibrils are laid down in a parallel array, the microtubules just inside the plasma membrane can be seen to extend in the same direction (Fig. 8-29).

Interference with these microtubules usually, but not always, causes disarray of the cellulose microfibrils in the walls. This was first noted by P. B. Green, who studied the effects of colchicine on cells of the alga *Nitella*. (Colchicine is a drug that interferes with microtubule assembly; see p. 427.) Growing *Nitella* cells normally deposit microfibrils in parallel arrays at an angle to the long axis of the cell. When colchicine is added during cell wall growth, the microfibrils are laid down at random. With some exceptions other experiments with microtubule inhibitors have had the same result: interference with microtubules destroys cellulose microfibril orientation but does not inhibit cellulose synthesis.

How microtubules determine the orientation of cellulose fibrils remains unknown. The most likely possibility is that microtubules bind to integral membrane proteins on either side of the enzyme rosettes, establishing boundaries or "fences" (as shown in Fig. 8-28). The enzyme complexes synthesizing microfibrils are presumably confined to linear pathways between the microtubule fences.

Plasmodesmata: Communicating Junctions of Plant Cells

Both primary and secondary plant cell walls retain minute openings, called *plasmodesmata* (singular = *plas-*

a

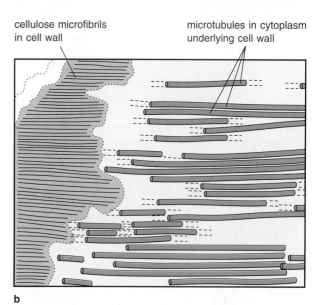

cellulose microfibrils in cell wall

microtubules in cytoplasm underlying cell wall

b

Figure 8-29 The parallel orientation of microtubules and cellulose microfibrils in plant cell walls. **(a)** An electron micrograph of sectioned plant cell. **(b)** A diagram of the structures in the micrograph. Toward the left side of the micrograph and diagram the plane of section passes through the cell wall, revealing the cellulose microfibrils. Toward the right the plane of section passes through the cytoplasm underlying the wall and shows microtubules oriented in the same direction as the microfibrils. × 57,000. Micrograph courtesy of E. H. Newcomb.

modesma), through which the cytoplasm of adjacent cells apparently remains in open communication (Fig. 8-30). A plasmodesma is usually roughly cylindrical in shape, with the cylinder narrowed in diameter at both ends. The plasma membranes of adjacent cells completely line the cylinder walls, so that there is no break in the membranes between cells. A narrow, tubelike extension of the ER runs through the central axis of the channel formed by a plasmodesma. Surrounding

The Experimental Process

Cell/Cell Communication in Plants: An Open and Shut Case

Peter Goodwin

PETER GOODWIN is Reader in Horticulture in the School of Crop Sciences, University of Sydney, NSW. He obtained his Bachelor and Master of Science in Agriculture Degrees at Sydney University, and then went on to the University of Nottingham for his Ph.D. He was employed at the Scottish Horticultural Research Institute and at the Research School of Biological Sciences, Australian National University, before moving to his present position. As well as interests in the developmental biology of plants, he has been involved in the selection and breeding of native Australian plants.

The work described in this paper sprang from what, in the late 1970s, was our almost total ignorance of the actual properties of the cell-to-cell pathway via plasmodesmata in plants. The state of knowledge in this now distant era was reviewed by various authors in Gunning and Robards (1976).[1] In general, while a large conjectural framework had been built up over the previous century, the actual characteristics of the system were unknown. Cells of higher plants, had been shown to be connected by a relatively low electrical resistance pathway—for example by Spanswick in 1972[2]—but this technique was too insensitive to give any detailed understanding of the physiology of plasmodesmata.

Meanwhile, progress in the understanding of cell-to-cell communication in animal cells, although begun much later than the plant studies, had proceeded much more rapidly. A key advance was the development of a graded series of fluorescent probes to examine the size and permeability of the intercellular pores, the gap junctions. Application of this technique to plant cells showed that the molecular exclusion limit is similar in both plants and animals.[3]

If the cytoplasm of all plant cells is linked by plasmodesmata, which allow molecules of the order of 3 nm (700 daltons) to move from cell to cell, then most small metabolite molecules should be able to move from cell to cell, so complicating the normal processes of tissue differentiation. In addition, when some cells interconnected by plasmodesmata in a tissue are damaged—a very common occurrence—then the same small metabolites should leak from the uninjured to the damaged cells, making the tissue vulnerable to metabolite leakage through damage to only a fraction of its cells. The obvious suggestion is that perhaps the plasmodesmata can shut when they are exposed to the "outside" or to a damaged cell. In addition, the same valving perhaps operates during differentiation.

At this moment an able student, Michael Erwee, arrived and was given the challenge of discovering whether plasmodesmata can be caused to shut, if this is reversible, and what is the key signal which causes closure and opening. The obvious candidate was the cytoplasmic calcium ion concentration, which is far higher "outside" than "inside" the cytoplasm, and which was already known to be involved in many processes in the cellular biology of both animals and plants. Of special relevance was the demonstration by Rose, Simpson, and Loewenstein in 1977[4] that calcium ions regulate gap junction permeability. In addition, a number of treatments likely to increase cytoplasmic calcium ion concentrations were known to decrease electrical coupling.

The approach used combined trials with various inhibitors, which would be expected to increase cytoplasmic calcium ion concentrations, and also the direct injection of various ions. The work employed leaves of the simple water plant *Egeria densa*. The leaves have only two layers of cells, so that all cells can be easily seen under the microscope. In addition, one can avoid a major possible artifact in fluorescent probe movement, where the probe moves from cell to cell in the cell wall. In *Egeria* the cells are all exposed to water, so that probes show very limited movement in the cell wall—they diffuse out into the solution. In addition, because of the large, clearly defined cells, it is easy to tell if the probe is in the cell wall, cytoplasm, vacuole, or nucleus.

the central tubule is a sleevelike extension of the cytoplasm. The structure of a plasmodesma thus apparently sets up continuities between the plasma membranes, soluble cytoplasm, ER membranes, and ER cisternae of adjacent plant cells.

The experiments linking plasmodesmata to direct communication between plant cells closely resemble those linking gap junctions to direct cytoplasmic communication in animal cells. R. M. Spanswick and J. W. F. Costerton placed electrodes in adjacent plant cells and found that electrical resistance between the cells is about 50 times lower than that expected if the cells were separated by continuous plasma membranes. Experiments by R. L. Overall and B. E. S. Gunning

The next possible source of artifact was that the probes might simply leak across the plasmalemma (plasma membrane). The tissues were soaked in solutions of the probes and washed. No detectable dye entry occurred within 1 hour. The next step was to inject the probes. Those of molecular weight up to 665 daltons moved rapidly from cell to cell. At the same time it was possible to check for another possible source of artifacts—that the probes damage the cells and then leak to successive cells. As judged by a continuation of cytoplasmic streaming, the cells were not damaged by the probes. The largest probe dye able to move, fluorescein glutamyl glutamic acid ($F(GLU)_2$), was chosen to assess the effects of various ions and inhibitors, since its movement would be sensitive to even a small restriction in the pores. The problem was how to inject both the divalent ions and the probes. In fact, a simple solution presented itself: to inject the divalent ions and the probes from the same electrode, with the divalent cations injected using a positive current, and the probes (which are anions) injected using a negative current. This overcame the possibility that the different molecules would be injected into different parts of the cell, or that the use of two electrodes would cause major damage to the cell.

It was found that Ca^{2+}, Mg^{2+}, and Sr^{2+} injected a short time (0–1 min) before the probe inhibited its movement. Perhaps any cation would have the same effect? Not K^+ or Na^+. Was only the largest mobile probe affected? No, the movement of all probes tested was inhibited, suggesting that the plasmodesmata shut, rather than showing a small fall in the size of molecule able to pass. The inhibition of movement by the divalent ions was an all-or-none effect. Either the dye moved as well as in the control (in about a third of cases) or it did not move at all. Presumably, in a third of cases the cation either did not shut the plasmodesmata, or shut them too slowly. The inhibiting effect of the cations was greater if there was a 1 minute delay between injection of cation and probe, suggesting that shutting is time dependent, i.e., being greater after 1 minute.

The next question was, is this a reversible process? Since the cell is very competent to reduce divalent ion levels, one might expect the plasmodesmata to eventually reopen. This was found: if there was a delay of 5–30 minutes between divalent cation and dye injection, then the dye moved from cell to cell, showing that the calcium-1 or magnesium-mediated closing of the plasmodesmata is a reversible process. It also showed that the injection of the divalent ions had not caused some type of permanent damage to the cell or the plasmodesmata.

There is always uncertainty in the use of inhibitors, as they invariably have multiple effects. However, a number of chemicals likely to increase the concentration of cytoplasmic calcium ions, notably the mitochondrial uncoupler carbonyl cyanide p-trifluoromethoxy-phenyl hydrazone, the inhibitor of mitochondrial Ca^{2+} uptake trifluralin, and the ionophore A23187, all inhibited probe movement.

The significance of the work was that it showed for the first time that plasmodesmata are highly dynamic structures, able to respond rapidly to changes in the cellular environment. The work, of course, raised as many questions as it answered. How do the divalent ions work? Do other gating molecules exist? Is the observed response the normal system? Does it occur in other species?

The whole field of cell-to-cell communication has expanded since these early studies. For a recent review see Robards and Lucas (1990).[5] Of special interest has been the demonstration of changes in cell/cell communication with development,[6] and the viral regulation of plasmodesmata.[7] Unexpectedly, in view of the work on *Egeria*, it has been shown that plasmodesmata can stay open in isolated bundle sheath cell strands,[8] a system offering considerable potential for the direct study of plasmodesmatal function.

References

[1] Gunning, B. E. S., and Robards, A. W. *Intercellular communications in plants: studies on plasmodesmata.* Berlin: Springer (1976).

[2] Spanswick, R. M. *Planta* 102:215–27 (1972).

[3] Goodwin, P. B. *Planta* 157:124 (1983); Tucker, E. B. *Protoplasma* 113:193 (1982).

[4] Rose, B.; Simpson, I.; and Loewenstein, W. R. *Nature* 267:625 (1977).

[5] Robards, A. W., and Lucas, W. J. *Annu. Rev. Plant Phys. Plant Mol. Biol.* 41:369 (1990).

[6] Erwee, M. G., and Goodwin, P. B. *Planta* 163:9 (1985).

[7] Wolf, S.; Deom, C. M.; Beachy, R. N.; and Lucas, W. J. *Science* 246:377 (1989).

[8] Burnell, J. N. *J. Exp. Bot.* 39:1575 (1988); Weiner, H.; Burnell, J. N.; Woodrow, I. E.; Heldt, H. W.; and Hatch, M. D. *Plant Physiol.* 88:815 (1988).

showed that the flow of current, reflecting ion movements between adjacent cells, is directly proportional to the number of plasmodesmata linking the cells. Fluorescent dyes with molecular dimensions too large to pass directly across membranes move readily from one cell to its neighbors if injected; P. B. Goodwin and E. B. Tucker showed that the upper limit for free movement is about 700 to 800 daltons, a figure coincidentally similar to that obtained for gap junctions in animal cells. Goodwin and his colleagues also found that injected Ca^{2+} drastically reduced the flow of substances between adjacent plant cells, indicating that movement through plasmodesmata is regulated (see the Experimental Process essay by Goodwin on p. 296). Further

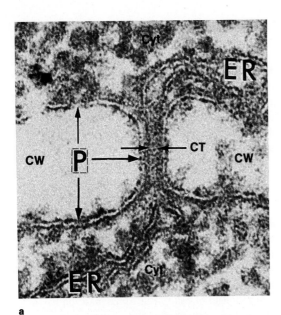

a

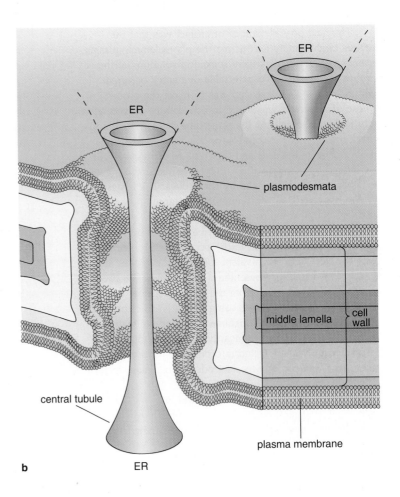

b

Figure 8-30 Plasmodesmata. **(a)** A plasmodesma in the wall separating two onion root cells. ×240,000. Courtesy of B. E. S. Gunning and Springer-Verlag, from *Protoplasma* 111:134 (1982). **(b)** Plasmodesma structure. A central tubule, evidently derived from cisternae of the ER on either side of the cell wall, lies in the central channel of a plasmodesma. Channel diameter at the narrowest point is about 25 to 30 nm. CW, cell wall; P, plasma membrane; ER, endoplasmic reticulum; CT, central tubule of plasmodesma; Cyt, cytoplasm.

similarities between plasmodesmata and gap junctions are suggested by the work of A. Yahalom and his coworkers, who found that antibodies against the connexins of animal gap junctions (see p. 262) cross-react with the proteins of maize plasmodesmata.

Infecting plant viruses evidently travel from cell to cell through the plasmodesmata. In some as yet unknown way, the viruses enlarge the plasmodesmatal openings enough to allow either the genetic material or entire viral particles complete with coat proteins to pass. In some viruses, such as the tobacco mosaic virus, passage depends on a 30,000-dalton protein encoded in the viral genome and synthesized in the host cell cytoplasm. Using genetic engineering techniques, S. Wolf and his colleagues showed that insertion of the viral gene for the 30,000-dalton protein into the DNA of plant cells increased the size of molecules passed by plasmodesmata from the normal 700 to 800 daltons to about 9400 daltons.

Plasmodesmata develop during deposition of the primary cell wall between dividing cells at sites where strands of the ER are trapped (for details, see p. 1035). As later growth converts primary to secondary cell walls, plasmodesmata persist in their original numbers, which in mature walls may vary from 1 to 140 per square micrometer. At these levels a mature plant cell may have as many as 100 to 100,000 plasmodesmata connecting it to its neighbors.

Plasmodesmata are found in all plant groups from bryophytes to angiosperms. Equivalent structures have also been observed in some fungi and multicellular algae. In the higher plants only a few exceptional cell types lack plasmodesmata, among them the generative and vegetative cells of pollen grains (see p. 1115) and the mature guard cells surrounding stomata, in which plasmodesmata become plugged during maturation.

Plant cell walls protect and bind cells together in the body structure of plants, give permanent shape to plant cells, resist the internal forces developed by osmosis, and protect cells from infection. The primary fiber of plant cell walls is cellulose; pectins, hemicellulose, and extensins are the primary network molecules. Although plant cells are completely surrounded by walls, they are able to expand and grow through processes that allow cell wall components, particularly cellulose microfibrils, to slide to accommodate the expansion. New cellulose microfibrils are laid down by enzyme complexes embedded in the plasma membrane, guided by microtubules on the cytoplasmic side of the membrane. Plasmodesmata persist in cell walls as sites where the cytoplasm of adjacent cells is continuous. As such, they provide a function in plants equivalent to the communicating or gap junctions of animal cells. The extracellular matrices of animals, plants, and other eukaryotes are unusual among biological structures in that they are assembled, in all their complexity, primarily by enzymatic systems located outside the cell. (The extracellular materials of prokaryotes are described in Supplement 8-1.)

For Further Information

Carbohydrate group synthesis in the ER and Golgi complex, *Ch. 20*
Cell division, *Ch. 24*
Cell wall formation during division, *Ch. 24*
Cytoskeleton, *Ch. 13*
Gap or communicating junctions, *Ch. 7*
Glycoprotein structure, *Chs. 2 and 5*
Intermediate filaments, *Ch. 13*
Membrane structure, *Ch. 5*
Microfilaments, *Ch. 12*
Microtubules, *Ch. 11*
Receptors and the cell surface, *Ch. 7*
Transport in eukaryotes and prokaryotes, *Ch. 6*

Suggestions for Further Reading

General Books and Articles

Edelman, G. M., and Thiery, J.-P., eds. 1985. *The cell in contact.* New York: Wiley.

Fessler, J. H., and Fessler, L. I. 1989. *Drosophila* extracellular matrix. *Ann. Rev. Cell Biol.* 5:309–39.

Ginsburg, V., and Robbins, P., eds. 1981. *Biology of carbohydrates*, vols. 1 and 2. New York: Wiley-Interscience.

Weber, K., and Osborn, M. 1985. The molecules of the cell matrix. *Sci. Amer.* 253:110–20 (October).

Collagens: Structure and Synthesis

Burgeson, R. E. 1988. New collagens, new concepts. *Ann. Rev. Cell Biol.* 4:551–77.

Byers, P. H. 1990. Brittle bones-fragile molecules: disorders of collagen gene structure and expression. *Trends Genet.* 6:293–96.

Kucharz, E. J. 1991. *The collagens: biochemistry and pathophysiology.* New York: Springer-Verlag.

Kuivaniemi, H.; Tromp, G.; and Prockop, D. 1991. Mutations in collagen genes: causes of rare and some common diseases in humans. *FASEB J.* 5:2052–60.

Martin, G. R.; Timpl, R.; Muller, P. K.; and Kuhn, K. 1985. The genetically distinct collagens. *Trends Biochem. Sci.* 10:285–87.

Prockop, D. 1990. Mutations that alter the primary structure of type I collagen. *J. Biolog. Chem.* 265:15439–52.

Rest, M. van der, and Garrone, R. 1991. Collagen family of proteins. *FASEB J.* 5:2814–23.

Sykes, B. 1985. The molecular genetics of collagen. *Bioess.* 3:112–17.

Vuorio, E., and de Crombrugghe, B. 1990. The family of collagen genes. *Ann. Rev. Biochem.* 59:837–72.

Proteoglycans

Fransson, L.-A. 1987. Structure and function of cell-associated proteoglycans. *Trends Biochem. Sci.* 12:406–11.

Goetnick, P. F. 1991. Proteoglycans in development. *Curr. Top. Devel. Biol.* 25:111–31.

Hassell, J. R.; Kimura, J. H.; and Hascall, V. C. 1986. Proteoglycan core protein families. *Ann. Rev. Biochem.* 55:539–67.

Kjellen, L., and Lindahl, U. 1991. Proteoglycans: structures and interactions. *Ann. Rev. Biochem.* 60:443–75.

Kuettner, K. E., and Kimura, J. H. 1985. Proteoglycans: an overview. *J. Cellular Biochem.* 27:328–36.

Ruoslahti, E. 1988. Structure and biology of proteoglycans. *Ann. Rev. Cell Biol.* 4:229–55.

Scott, J. E. 1992. Supramolecular organization of extracellular matrix of glycosaminoglycans, in vitro and in the tissues. *FASEB J.* 6:2639–45.

Bone, Cartilage, and Basal Laminae

Caplan, A. I. 1984. Cartilage. *Sci. Amer.* 251:84–94 (October).

Inoue, S. 1989. Ultrastructure of basement membranes. *Internat. Rev. Cytol.* 117:57–98.

Weiner, S. 1986. Organization of extracellularly mineralized tissues: a comparative study of biological crystal growth. *CRC Crit. Rev. Biochem.* 20:365–408.

Yurcheno, P. D., and Schittny, J. C. 1990. Molecular architecture of basement membranes. *FASEB J.* 4:1577–90.

**Fibronectin, Laminin, and Other Molecules
Linking Cells to the Extracellular Matrix**

Albelda, S. M., and Buck, C. A. 1990. Integrins and other cell adhesion molecules. *FASEB J.* 4:2868–80.

Beck, K.; Hunter, I.; and Engel, J. 1990. Structure and function of a laminin: anatomy of a multidomain glycoprotein. *FASEB J.* 4:148–60.

Hynes, R. O. 1986. Fibronectins. *Sci. Amer.* 254:42–51 (June).

———. 1992. Integrins: versatility, modulation, and signalling in cell adhesion. *Cell* 69:11–25.

Mosher, D. R., ed. 1988. *Fibronectin.* New York: Academic Press.

Extracellular Matrix Receptors

Akiyama, S. K.; Nagata, K.; and Yamada, K. M. 1990. Cell surface receptors for extracellular matrix components. *Biochim. Biophys. Acta* 1031:91–110.

Buck, C. A., and Horwitz, A. F. 1987. Cell surface receptors for extracellular matrix molecules. *Ann. Rev. Cell Biol.* 3:179–205.

Hynes, R. O. 1987. Integrins: a family of cell surface receptors. *Cell* 48:549–54.

Liotta, L. A.; Wewer, U. M.; Rao, C. N.; and Bryant, G. 1985. Laminin receptor. In *The cell in contact,* ed. G. M. Edelman and J.-P. Thiery, pp. 333–44. New York: Wiley.

Mecham, R. P. 1991. Receptors for laminin on mammalian cells. *FASEB J.* 5:2538–46.

Extracellular Matrix Assembly

McDonald, J. A. 1988. Extracellular matrix assembly. *Ann. Rev. Cell Biol.* 4:183–207.

The Extracellular Matrix and Development

Ekblom, P.; Vestveber, D.; and Kemler, R. 1986. Cell-matrix interactions and cell adhesion during development. *Ann. Rev. Cell Biol.* 2:28–47.

Sanes, J. R. 1989. Extracellular matrix molecules that influence neural development. *Ann. Rev. Neurosci.* 12:521–46.

General Reviews of Plant Cell Wall Structure

Brown, R. M., ed. 1982. *Cellulose and other natural polymer systems.* New York: Plenum.

Knox, P. 1990. Emerging patterns of organization at the plant cell surface. *J. Cell Sci.* 96:557–61.

McNeil, M.; Darvill, A. G.; Fry, S. C.; and Albersheim, P. 1984. Structure and function of the primary cell walls of plants. *Ann. Rev. Biochem.* 53:625–63.

Pectins and Hemicelluloses

Darvill, A.; McNeil, M.; Albersheim, P.; and Delmer, D. P. 1980. The primary cell walls of flowering plants. In *The biochemistry of plants,* ed. P. K. Stumpf and E. E. Conn, vol. 1, pp. 91–162. New York: Academic Press.

Hayashi, T. 1989. Zyloglucans in the primary cell wall. *Ann. Rev. Plant Phys. Plant Molec. Biol.* 40:139–68.

Extensin and Other Cell Wall Glycoproteins

Cassab, G. I., and Varner, J. E. 1988. Cell wall proteins. *Ann. Rev. Plant Physiol.* 329:321–53.

Cooper, J. B.; Chen, J. A.; van Holst, G.-J.; and Varner, J. E. 1987. Hydroxyproline-rich glycoproteins of plant cell walls. *Trends Biochem. Sci.* 12:24–27.

Lignins

Lewis, N. G., and Yamamoto, E. 1990. Lignin: occurrence, biogenesis, and biodegradation. *Ann. Rev. Plant Physiol. Plant Molec. Biol.* 41:455–96.

Cell Wall Structure

Fry, S. C. 1986. Cross-linking of matrix polymers in the growing cell walls of angiosperms. *Ann. Rev. Plant Physiol.* 37:165–86.

McNeil, M.; Darvill, A. G.; Fry, S. C.; and Albersheim, P. 1984. Structure and function of the primary cell walls of plants. *Ann. Rev. Biochem.* 53:625–63.

Varner, J. E., and Lin, L.-S. 1989. Plant cell wall architecture. *Cell* 56:231–39.

Cell Wall Growth and Cellulose Microfibril Synthesis

Delmer, D. P. 1987. Cellulose biosynthesis. *Ann. Rev. Plant Physiol.* 38:259–90.

Taiz, L. 1984. Plant cell expansion: regulation of cell wall mechanical properties. *Ann. Rev. Plant Physiol.* 35:585–657.

Plasmodesmata

Gunning, B. E. S., and Overall, R. L. 1983. Plasmodesmata and cell-to-cell transport in plants. *Biosci.* 33:260–65.

Robards, A. W., and Lucas, W. J. 1990. Plasmodesmata. *Ann. Rev. Plant Physiol. Plant Molec. Biol.* 41:369–419.

Plant Defense Mechanisms

Bohlmann, H., and Apel, K. 1991. Thionins. *Ann. Rev. Plant Physiol. Plant Molec. Biol.* 42:227–40.

Bohlmann, H.; Clausen, S.; Behnke, H. G.; Hiller, C.; Reimann-Philipp, U.; Schrader, G.; Barkholt, V.; and Ape, K. 1988. Leaf-specific thionins of barley—a novel class of cell wall proteins toxic to plant-pathogenic fungi and possibly involved in the defence mechanism of plants. *EMBO J.* 7:1559–65.

Bowles, D. J. 1990. Defense-related proteins in higher plants. *Ann. Rev. Biochem.* 59:873–907.

Dixon, R. A., and Lamb, C. J. 1990. Molecular communication in interactions between plants and microbial pathogens. *Ann. Rev. Plant Physiol. Plant Molec. Biol.* 41:339–67.

Lamb, C. J.; Lawton, M. A.; Dron, M.; and Dixon, R. A. 1989. Signals and transduction mechanisms for activation of plant defenses against microbial attack. *Cell* 56:215–24.

Ryan, C. A., and Farmer, E. E. 1991. Oligosaccharide signals in plants: a current assessment. *Ann. Rev. Plant Physiol. Plant Molec. Biol.* 42:65–74.

Prokaryotic Extracellular Structures: General Information

Beveridge, T. J., and Graham, L. L. 1991. Surface layers of bacteria. *Microbiol. Rev.* 55:684–705.

Ferris, F. G., and Beveridge, T. J. 1985. Functions of bacterial cell surface structures. *Biosci.* 35:172–77.

Rogers, H. J.; Perkins, H. R.; and Ward, J. B. 1980. *Microbial cell walls and membranes.* London: Chapman and Hall.

Gram Stain

Scherrer, R. 1984. Gram's staining reaction, gram types and cell walls of bacteria. *Trends Biochem. Sci.* 9:242–45.

Gram-Negative and Gram-Positive Wall Structure

Ferris, F. G., and Beveridge, T. J. 1985. Functions of bacterial cell surface structures. *Biosci.* 35:172–77.

Hayashi, S., and Wu, H. C. 1990. Lipoproteins in bacteria. *J. Bioenerget. Biomembr.* 22:451–70.

Outer Membrane Structure in Gram-negative Bacteria

Baker, K.; Mackman, N.; and Holland, I. B. 1987. Genetics and biochemistry of the assembly of proteins into the outer membrane of *E. coli. Prog. Biophys. Molec. Biol.* 49:89–115.

Lugtenberg, B., and Van Alphen, L. 1983. Molecular architecture and functioning of the outer membrane of *Escherichia coli* and other gram-negative bacteria. *Biochim. Biophys. Acta* 737:51–115.

Wall Growth in Bacteria

Koch, A. L. 1985. Bacterial wall growth and division or life without actin. *Trends Biochem. Sci.* 10:11–15.

————. 1988. Biophysics of bacterial walls viewed as stress-bearing fabric. *Microbiol. Rev.* 52:338–53.

Capsules and Other Surface Polysaccharides

Boulnois, G. J., and Jann, K. 1989. Bacterial polysaccharide capsule synthesis, export, and evolution of structural diversity. *Molec. Microbiol.* 3:1819–23.

Sutherland, I. W. 1988. Bacterial surface polysaccharides: structure and function. *Internat. Rev. Cytol.* 113:188–231.

Cell Surface Recognition in Bacteria

Sharon, N. 1987. Bacterial lectins, cell-cell recognition, and infectious disease. *FEBS Lett.* 217:145–57.

Review Questions

1. What functions do fiber and network molecules have in the extracellular matrix of animal cells? Of plant cells?

2. What molecules provide the fiber and network of animal extracellular structures?

3. Outline the structures of collagens. List each collagen type and its overall functions in the extracellular matrix.

4. How do collagen polypeptides combine to form collagen molecules? How do collagen molecules combine to form collagen fibers? Where and how are collagen molecules and fibers synthesized?

5. What are proteoglycans? GAGs? How do proteins and GAGs combine in proteoglycan structure? How do proteoglycans and hyaluronic acid combine into molecular superstructures?

6. In what structures do proteoglycans occur in the animal extracellular matrix? What functions do proteoglycans carry out in these locations?

7. Outline the structures of fibronectin and laminin. What is the relationship between these molecules and the extracellular matrix?

8. How do animal cells form linkages to the extracellular matrix? How is the cytoskeleton involved in these linkages?

9. What changes occur in the extracellular matrix as cells become transformed to cancerous forms?

10. What are basal laminae? What functions does this extracellular structure have in animal cells?

11. What are the fiber and network molecules of plant cell walls?

12. What is the difference between primary and secondary cell walls?

13. Outline the structure of cellulose. How are cellulose molecules believed to combine to form cellulose microfibrils? What other kinds of polysaccharides form fiber molecules in plant cell walls?

14. What are pectins? Hemicelluloses? Extensins? Lignins? How do these molecules function in cell wall structure?

15. What kinds of enzymes occur in plant cell walls?

16. How do plant cell walls accommodate cell growth and expansion?

17. How are new cellulose microfibrils laid down in growing cell walls?

18. What evidence indicates that microtubules are involved in the deposition of cellulose microfibrils? How might microtubules determine the direction in which cellulose microfibrils are laid down in cell walls?

19. What are plasmodesmata? In what ways are plasmodesmata similar to the gap junctions of animal cells?

20. What evidence indicates that plasmodesmata provide direct cytoplasmic connections between plant cells?

Supplement 8-1
Cell Walls in Prokaryotes

With the exception of mycoplasmas, all prokaryotic cells have cell walls. Like the walls of plant cells, prokaryotic cell walls are assembled from complex polysaccharides linked with other substances into a "supermolecule" that completely surrounds the cell. The walls of prokaryotes support and give shape to cells and prevent them from bursting as a result of osmotic pressure. In bacteria osmotic pressure may easily reach levels of 3 to 5 atmospheres—the equivalent of the pressure in a bicycle tire. In addition to these supportive roles, prokaryotic walls protect cells from infection by viruses and provide sites for cell recognition and adhesion. The cell wall also serves as a partial permeability barrier and houses some molecules involved in molecular transport in gram-negative bacteria. Because cell walls are so important to the viability of prokaryotes, chemical interference with the walls or wall synthesis is an effective and much-used way to control the growth of disease-causing bacteria (see p. 305).

Cell Wall Structure in Gram-Positive and Gram-Negative Bacteria

Almost all bacteria fall into two major structural groups that are distinguished by a technique called the *gram stain,* devised in the 1800s by Hans Christian Gram, a Danish bacteriologist. Although the chemical basis for the gram stain is still uncertain, the technique and its results are straightforward. After fixation, usually by heating, cells are stained with crystal violet and iodine and then exposed to a decolorizing organic solvent such as alcohol or acetone. Gram-positive cells, resistant to decolorization, retain a deep blue color. Gram-negative cells are rapidly decolorized. The color is retained inside gram-positive cells and not in the wall; in some way the structure of the wall prevents decolorization of the protoplasm in these cells. The difference may be simply that gram-positive walls are much thicker and retard the removal of the stain more effectively than the very thin walls of gram-negative bacteria. This interpretation is supported by the observation that some bacteria with gram-positive wall structure (see below) but very thin walls, such as *Clostridium,* frequently stain as gram negative rather than gram positive.

Electron microscopy and biochemical analysis reveal clear, distinct differences in wall morphology between gram-positive and gram-negative bacteria. Walls in gram-positive bacteria (Fig. 8-31*a*) consist of a single, relatively thick layer of apparently homogeneous polysaccharide wall material that is not visibly differentiated into sublayers. Gram-negative bacteria have walls containing two distinct layers (Fig. 8-31*b*). The most unusual of these layers is an extra boundary membrane,

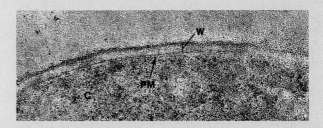

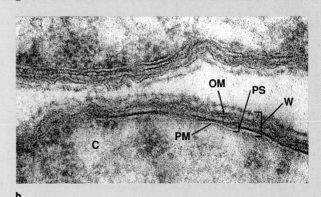

Figure 8-31 Wall structure in bacteria. **(a)** The single-layered wall characteristic of gram-positive bacteria. × 103,000. **(b)** The multilayered wall of gram-negative bacteria. × 112,000. W, wall; PM, plasma membrane; OM, outer membrane; PS, periplasmic space; C, cytoplasm. Courtesy of J. W. Costerton, reproduced, with permission, from *Ann. Rev. Microbiol.* 33:459 (1979). Copyright 1979 by Annual Reviews, Inc.

the *outer membrane,* lying outside the plasma membrane. The region between the plasma and outer membranes, the *periplasmic space,* is filled with a thin layer of polysaccharide wall material.

One or more structures may extend outward from cell walls of both gram-positive and gram-negative bacteria. Many bacteria have an external *capsule* or *slime layer* surrounding the cell wall (Fig. 8-32*a*; see also Fig. 1-3). The capsule, which may be from tens of nanometers to several micrometers in thickness, shows great structural and chemical diversity. The wall may also bear motile flagella of the bacterial type, consisting of long, filamentous fibers of protein molecules (see Fig. 1-8 and pp. 447–50), and shorter, nonmotile rods of protein called *pili* (see Fig. 7-30). The pili, which stick out like rigid hairs from the surface, are involved in cell recognition and adhesion (for details, see p. 270).

Other external layers are present in some bacteria. In some, the cell wall or capsule is surrounded by a *sheath,* a structure that is similar to a capsule but more rigid and ordered. Some motile bacteria move by sliding through the sheath. A number of bacteria, particularly among pathogenic types, have a highly ordered *S layer* immediately surrounding the cell wall (Fig. 8-32*b*). The protein or glycoprotein molecules of the S layer are tightly packed in a semicrystalline array that completely covers the surface except for regularly spaced, porelike

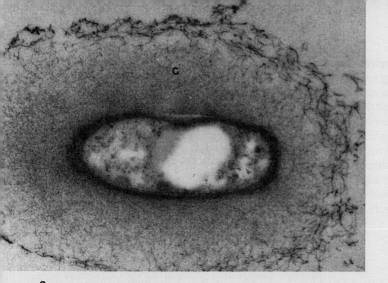

c

a

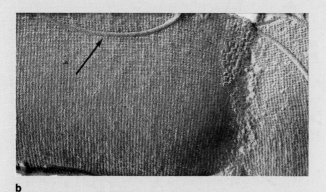

b

Figure 8-32 Layers located outside the cell wall in bacteria. **(a)** The capsule (C) surrounding the cell wall in the soil bacterium *Rhizobium.* × 30,000. Courtesy of W. J. Brill, from *J. Bact.* 137:1362 (1979). **(b)** The S layer covering the surface of the cell wall in the bacterium *Clostridium.* Part of a bacterial flagellum (arrow) is also visible, × 86,000. Courtesy of U. B. Sleytr and A. M. Glauert, © Academic Press, Inc., from *J. Ultrastr. Res.* 50:103 (1975).

openings. Among the S layers that have been analyzed, the openings block access to the cell by molecules larger than about 50,000 to 80,000 daltons.

Molecular Constituents of Prokaryotic Cell Walls

Chemical analysis of bacterial walls reveals a variety of substances, some with no counterparts in other living organisms. Gram-positive and gram-negative bacterial walls share a unique molecular group, the *peptidoglycans,* which provide rigidity and strength to bacterial walls. Both gram-positive and gram-negative bacterial cell walls also contain an extensive battery of enzymes concerned with wall synthesis, transport, and the breakdown of organic molecules absorbed by the cell.

Other molecular groups are found in only one of the two bacterial types. Almost all gram-positive bacteria contain *teichoic acids* as a wall constituent. These molecules, among other functions, may serve as recognition and binding sites in gram-positive cell walls. Gram-negative bacteria lack teichoic acids but contain several molecular types not found in gram-positive bacteria. *Lipopolysaccharides,* which occur as a part of the outer membrane, have no counterparts in any other living organism. *Lipoproteins,* which are also limited to gram-negative bacteria, link the outer membrane to the underlying peptidoglycan layer. Gram-negative outer membranes contain, in addition, *porins,* channels that allow small, water-soluble molecules such as sugars and amino acids to pass. (Table 8-4 summarizes the major characteristics of gram-positive and gram-negative bacterial walls.)

Peptidoglycans Peptidoglycans, also called *mureins, mucopeptides,* or *glycopeptides,* are a class of biological macromolecules found only in prokaryotic cell walls. All are built up from a backbone of repeating two-sugar units with attached short peptide chains. Adjacent backbones are crosslinked by a single peptide bond or a short sequence of amino acids (Figs. 8-33 and 8-34). The polysaccharide backbones and peptide crosslinks make peptidoglycans the equivalent of both the fiber and network molecules of eukaryotic extracellular structures. With few exceptions the same fundamental peptidoglycan structure occurs in all gram-negative and gram-positive bacteria and also in cyanobacteria.

Table 8-4   Major Cell Wall Characteristics of Bacteria and Cyanobacteria

Characteristic	Gram-positive Bacteria	Gram-negative Bacteria	Cyanobacteria
Peptidoglycans	+	+	+
Teichoic acids	+	−	−
Lipopolysaccharides	−	+	+
Lipoproteins	−	+	+
Porins	−	+	+
Capsule	+	+	+
Gram stain	+	−	+
Wall layers	Single	Complex	Complex

Abe   = abequose
Rha   = rhamnose
Man   = mannose
Gal   = galactose
GlcNAc = N-acetyl glucosamine
Glc   = glucose
Hep   = heptulose
EtA   = ethanolamine
KDO   = 2-keto-3-deoxyoctonate
GlcN   = glucosamine

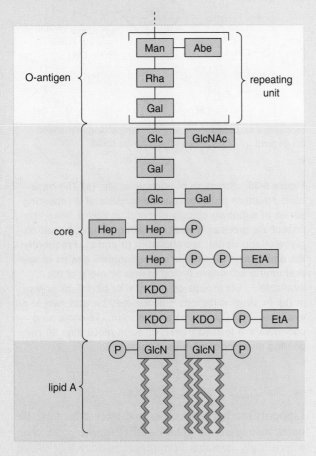

Figure 8-36 The structure of lipopolysaccharides. The basal unit, lipid A, contains a short backbone constructed from a pair of glucosamine subunits linked end to end. Connected to the backbone are several nonpolar fatty acid chains, frequently lauric or palmitic acid (see Table 2-5). Next to lipid A is the core subunit, a branched polysaccharide chain containing the common six-carbon sugars glucose, galactose, and glucosamine, and several seven- and eight-carbon sugars found nowhere else among living organisms. (The *Hep* and *KDO* units shown in the figure are examples of these unusual sugars.) Connected to the core and extending outward from the main body of the molecule is a long carbohydrate chain forming the O-antigen. Within the O-antigen are repeating units of three to six sugars, all six-carbon hexoses, that may link end to end into a chain with as many as 40 or more repeats. The sugars and the form of the repeating units in the O-antigen are highly variable among different gram-negative bacteria and even among different genetic lines of the same species.

pattern of the repeats are highly variable among different gram-negative bacteria, and even among different genetic lines of the same species. The chain, called the O-antigen because it provides the primary group recognized by viruses and antibodies in gram-negative cell walls, is equivalent to the teichoic acids in gram-positive bacteria.

Because of its nonpolar and polar ends, a lipopolysaccharide molecule has dual-solubility properties (see p. 154). As such, it fits readily into the membrane, with the nonpolar fatty acid chains buried in the nonpolar bilayer interior and the polar core and O-antigen subunits extending outward from the membrane surface. Analysis of outer membranes shows that lipopolysaccharides and their hydrophilic tails are confined to the outermost bilayer half of the outer membrane; the O-antigen segments may extend outward as far as 30 nm from the surface of the wall (see Fig. 8-37).

The layer of lipopolysaccharides in the outer membrane is much more resistant than phospholipids to disruption by detergents and lipid solvents. The highly resistant outer membrane is believed to account for the ability of *E. coli* to tolerate the strongly detergent effect of bile salts released into the human digestive tract.

The extreme variability of the O-antigen segment of outer membrane lipopolysaccharides can present severe problems to development of an immune response. *E. coli*, for example, is normally a harmless gram-negative inhabitant of the human large intestine. Occasionally, *E. coli* becomes an infective, pathogenic bacterium in other regions of the body, particularly in the urogenital tract, the bloodstream, or the fluid within the brain and spinal cord. In these locations *E. coli* infections are often very persistent because of the variability in the O-antigen. The O-antigen mutates so rapidly from generation to generation that, in effect, the bacterium keeps ahead of the body's antibody defenses by continually developing new types that are not recognized by previously developed antibodies. As a consequence, infections of pathogenic *E. coli* are resistant to the immune system and usually persist unless controlled by antibiotics.

Lipoproteins Lipoproteins of the outer membrane in gram-negative bacteria consist of a protein unit to which is attached one or more fatty acid chains. The nonpolar fatty acid chains anchor lipoproteins in the lipid bilayer of the outer membrane, in this case in the inner bilayer half facing the underlying peptidoglycan sheath (see Fig. 8-37). In *E. coli* about one-third of the lipoprotein molecules form covalent links with peptidoglycans, firmly anchoring the outer membrane to the underlying peptidoglycan sheath. There are as many as 70,000 lipoprotein molecules in the outer membrane of an *E. coli* cell, making it one of the most abundant proteins in the organism.

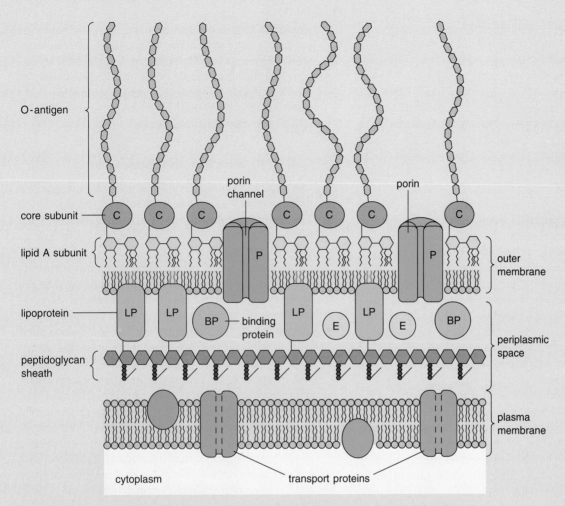

O-antigen

porin
channel

porin

core subunit — C C C C C C C

lipid A subunit

P

P

outer
membrane

lipoprotein

LP LP

BP — binding
protein

LP LP

E E

BP

periplasmic
space

peptidoglycan
sheath

plasma
membrane

cytoplasm

transport proteins

Figure 8-37 A summary of the structural units of gram-negative bacterial cell walls. C, core subunit of lipopolysaccharide; P, porin; LP, lipoprotein; BP, binding protein; E, hydrolytic enzyme. The binding proteins, located in the periplasmic space, specifically bind and promote the transport of substances into the cell. The substances bound, primarily sugars and amino acids, are transferred from the binding proteins to active transport systems of the plasma membrane (for details, see p. 229).

Porins Porins are channel-forming proteins that extend entirely through the outer membrane, providing an opening that allows small polar molecules to penetrate from the medium into the wall and eventually to the plasma membrane (see Fig. 6-26). *E. coli* outer membranes contain about 10,000 copies of different porin molecules, also placing this class among the most abundant proteins of the bacterium. Each porin type forms a pore of slightly different size and characteristics. Some act as simple molecular sieves or filters, admitting all molecules below a certain size by passive diffusion along concentration gradients. Other porins admit only certain classes of molecules (for details, see p. 227).

The cell walls of gram-positive bacteria consist primarily of a peptidoglycan sheath with teichoic acid molecules extending from the sheath into the region outside the wall. The gram-negative wall has two distinct layers—the peptidoglycan sheath and, outside it, the complex outer membrane. The outer membrane, with its porins, lipopolysaccharides, and lipoproteins, sets up an effective barrier to essentially all substances except ions and small molecules admitted by porins. The barrier protects the cell from chemical attack by larger molecules such as enzymes and antibiotics. The outer membrane is also less susceptible than the underlying plasma membrane to the disruptive effects of agents such as bile and detergents.

Phospholipids are also present in gram-negative outer membranes, in quantities amounting to about half of the lipopolysaccharide content by weight. They are confined with the lipoproteins almost entirely to the inner bilayer half of the outer membrane; the outer bilayer half contains essentially only lipopolysaccharides. Most of the protein complement of the outer membrane is supplied by porins; enzymatic activity is limited to a few proteins able to catalyze the breakdown of lipids and proteins, and an enzyme associated with capsule assembly. Most enzymes required for synthesis of molecules in the outer membrane are concentrated in the plasma membrane. (Fig. 8-37 summarizes the structure of the gram-negative outer membrane.)

The Capsule

The polysaccharide material forming the capsule, which occurs in both gram-positive and gram-negative bacteria, varies extensively in chemical structure between different species. However, within a given species the capsule is of uniform structure and usually contains a single polysaccharide type assembled primarily from glucose, galactose, fucose, mannose, or glucuronic acid subunits. Generally, two or more of these sugars link in various combinations to form a tetrasaccharide unit. The unit repeats in end-to-end linkages to form the capsule polysaccharide.

Capsule polysaccharides absorb many times their weight in water molecules, forming a highly hydrated layer of slimelike material coating the surface of the wall. The capsule provides protection against attack by bacterial viruses. In many bacterial species the capsule's adhesive properties allow cells to attach to each other in colonies or to a substrate.

The capsule figures importantly in bacterial diseases of both plants and animals. The thick layer of capsular material impedes the diffusion and attachment of antibodies to the bacterial cell surface. The capsule also interferes with the ability of phagocytes such as white blood cells to attach and engulf bacterial cells by endocytosis (see p. 857). The capsule is so effective in these protective roles that the difference between the virulent and noninfective forms of many disease-causing bacteria simply reflects the presence or absence of the capsule. For example, *Pneumococcus* is nonvirulent and can easily be eliminated by natural defense mechanisms if it is injected into mice or other mammals as an unencapsulated mutant. The normal encapsulated form, however, is a highly pathogenic bacterium that avoids destruction by the immune system and causes severe pneumonia in humans and other mammals.

The capsular material in some cases contributes directly to the pathogenic effects of bacterial infections. Some species of *Pseudomonas* bacteria infecting plants, for example, produce capsular material that greatly increases the viscosity of fluids in the vascular system of the plants. The highly viscous fluid blocks the vascular system to such an extent that water cannot flow from roots to stems and leaves and causes severe wilting. A *Pseudomonas* species that infects humans suffering from cystic fibrosis causes an analogous condition. The bacteria accumulate in the air passages of the lungs, where their capsular material makes the mucus so viscous that the lungs cannot be cleared. Bacterial capsules are also a major component of dental plaque, the slimy material that collects on the surfaces of teeth and figures as a major cause of tooth decay.

Cell Wall Structure in Cyanobacteria

The walls of cyanobacteria (blue green algae) resemble those of gram-negative bacteria in most characteristics. Each of the complex layers of gram-negative bacterial walls has a counterpart in cyanobacteria, including an outer membrane (Fig. 8-38).

Although some chemical differences are noted, particularly in the lipid A subunit, the lipopolysaccharides of the cyanobacterial outer membrane generally resemble those of gram-negative bacteria. The periplasmic space between the plasma and outer membranes is filled with a thin layer of peptidoglycans as in the gram-negative bacteria. Curiously, in spite of their chemical and morphological similarities to gram-negative bacteria, cyanobacteria are gram positive. The chemical basis for this unexpected reaction to the gram stain is uncertain. However, it may result from reduced permeability of the peptidoglycan layer in these organisms. A polysaccharide capsule surrounds cyanobacterial walls as in bacteria. (Table 8-4 compares the wall characteristics of bacteria and cyanobacteria.)

Suggestions for Further Reading: See p. 301.

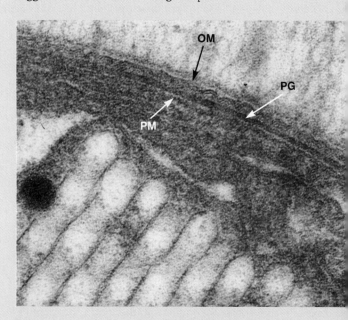

Figure 8-38 Cell wall of a cyanobacterium. The layers correspond to the multilayered wall of gram-negative bacteria. PM, plasma membrane; OM, outer membrane; PG, peptidoglycan layer; C, capsule. × 120,000. Courtesy of N. J. Lang.

ENERGY FOR CELL ACTIVITIES: CELLULAR OXIDATIONS AND THE MITOCHONDRION

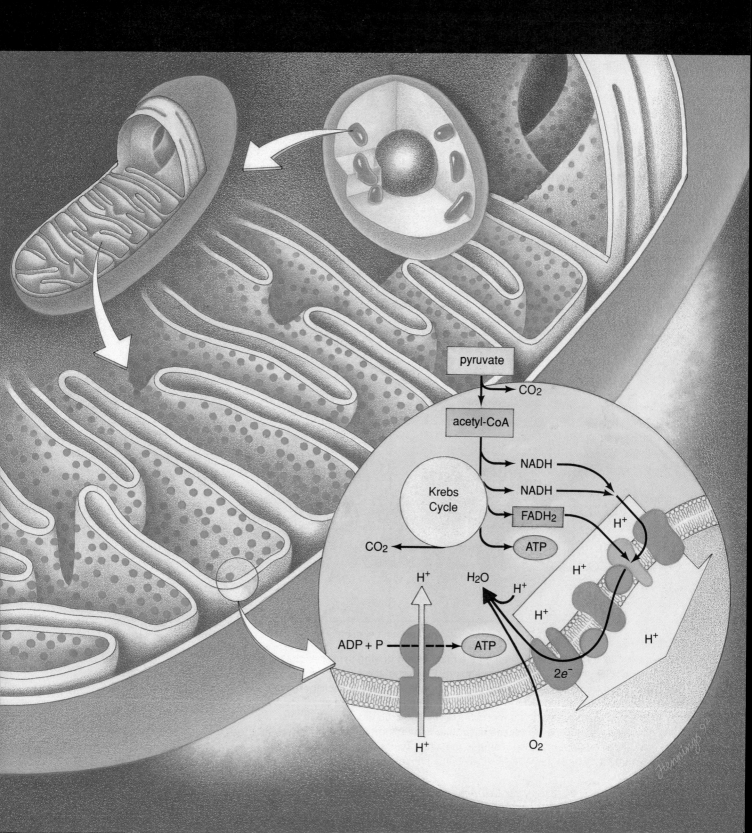

The activities of life require a continuous input of energy. At the cellular level the required energy is derived ultimately from the oxidation of complex organic fuel molecules such as carbohydrates and fats. Green plants also use light as an energy source. Energy from either of these sources is converted into the chemical energy of ATP (adenosine triphosphate; see p. 83), the "dollar" of the cellular energy economy.

Reactions that release energy, such as those oxidizing organic molecules, drive the synthesis of ATP from ADP and inorganic phosphate. Reactions that require energy obtain it by splitting ATP into ADP or AMP and inorganic phosphate. The reactions requiring energy constitute the chemical and physical work of the cell: the synthesis of molecules needed for cellular growth; the maintenance of active transport, cell movements, and responses; and cell division.

ATP thus cycles between reactions that release energy and those that require energy. This chapter discusses the major reaction systems that synthesize ATP by harnessing the energy released from the oxidation of organic molecules. These ATP-generating mechanisms proceed by one or both of two mechanisms in living organisms. One, *oxidative phosphorylation,* is common to eukaryotic cells and many prokaryotes. This oxygen-dependent mechanism is highly efficient in converting the chemical energy of fuel substances into energy bound into ATP. The second mechanism, *substrate-level phosphorylation,* is common to all organisms. Although less efficient, substrate-level phosphorylation serves as the primary source of ATP for cells living temporarily or permanently without oxygen.

PATHWAYS PRODUCING ATP: AN OVERVIEW

Oxidative phosphorylation can be conveniently divided into three overall parts common to all living organisms possessing this pathway (Fig. 9-1):

1. a series of reactions providing a *source of electrons at elevated energy levels;*

2. an *electron transport system,* embedded in a membrane, that uses the energy of the electrons provided in (1) to build an H^+ gradient across the membrane housing the system; and

3. an *ATP-synthesizing enzyme,* also embedded in the membrane, that uses the H^+ gradient produced in (2) as an energy source to phosphorylate ADP to ATP.

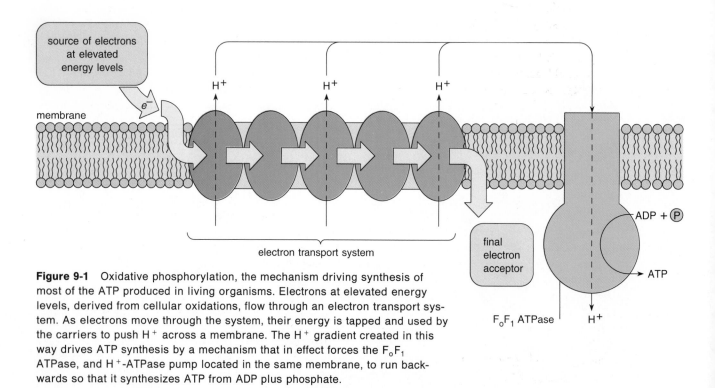

Figure 9-1 Oxidative phosphorylation, the mechanism driving synthesis of most of the ATP produced in living organisms. Electrons at elevated energy levels, derived from cellular oxidations, flow through an electron transport system. As electrons move through the system, their energy is tapped and used by the carriers to push H^+ across a membrane. The H^+ gradient created in this way drives ATP synthesis by a mechanism that in effect forces the F_oF_1 ATPase, and H^+-ATPase pump located in the same membrane, to run backwards so that it synthesizes ATP from ADP plus phosphate.

The Source of Electrons at Elevated Energy Levels

Cellular oxidations, in which electrons are removed from organic molecules used as fuel substances, are the immediate source of electrons at elevated energy levels. (Oxidation and reduction are explained in Information Box 9-1.) Although any of the four major classes of biological molecules—carbohydrates, lipids, proteins, or nucleic acids—may be oxidized as electron sources, the primary fuel substances used in most cells are carbohydrates and lipids. After electrons are removed from these substances, they are passed on to the electron transport system setting up the H^+ gradient. The oxidative reactions take place in the soluble background cytoplasm of both prokaryotic and eukaryotic cells and within the mitochondria of eukaryotes.

The Electron Transport System

The system transporting the electrons removed in cellular oxidations consists of a series of carrier molecules embedded in a membrane (see Fig. 9-1). Each electron carrier in the series accepts and releases electrons at lower energy levels than the carrier preceding it in the series. The energy lost by the electrons as they pass from one carrier to the next is released as free energy, that is, energy that can do work. Some of the released energy is used by the carriers to move H^+ from one side of the membrane to the other against its concentration gradient. This transport establishes and maintains the H^+ gradient used to drive ATP synthesis.

Most of the molecules transporting electrons are integral membrane proteins complexed with a nonprotein organic group. The organic group, which in many carriers includes a metal ion, is the part of the carrier that picks up and releases electrons to accomplish electron transport. The activity that uses energy released by electrons to push H^+ across the membrane is concentrated in the protein portion of the carrier molecules.

After traveling through an electron transport system, electrons are passed to a final electron acceptor, which may be either an organic or inorganic substance. In an electron transport system associated with oxidative phosphorylation, the final electron acceptor is oxygen. The oxygen combines with hydrogen as it accepts electrons, producing water in the final reaction of electron transport.

Although electron flow through the carriers of the electron transport system is known to set up an H^+ gradient across the membrane containing the system, some uncertainty remains about the details of the mechanism. The best indications are that the electron transport carriers work as active transport pumps that use energy released by electrons to push H^+ across the membrane. Electron transport systems of this type, capable of producing an H^+ gradient in response to electron flow, occur in the plasma membrane and its derivatives in prokaryotes and in the internal membranes of mitochondria and chloroplasts.

ATP Synthesis

The molecules directly using the H^+ gradient built up by electron transport can be considered H^+-ATPase pumps (see p. 208) that are forced to operate in reverse. If operating in the forward direction, the pumps would use energy released by ATP hydrolysis to drive H^+ across a membrane. However, in cellular systems producing ATP, the pumps are driven in reverse by the magnitude of the H^+ gradient produced by electron transport. When running in reverse, the pumps, instead of breaking down ATP, use the energy of the H^+ gradient to drive the synthesis of ATP from ADP and phosphate. This ability to reverse is not unique to the H^+-ATPase pumps operating in ATP synthesis; many active transport pumps directly using ATP as an energy source can be forced to run in reverse and synthesize ATP (see p. 206).

In oxidative phosphorylation both the electron transport system and the H^+-ATPase pumps are embedded in the same membrane (see Fig. 9-1). The H^+-ATPase pumps synthesizing ATP in response to very high H^+ gradients are called F_oF_1 ATPases. These pumps occur in the plasma membrane and its derivatives in prokaryotes and in the innermost membranes of mitochondria and chloroplasts in eukaryotes.

The overall mechanism linking an H^+ gradient to ATP synthesis was first proposed in 1961 by Peter Mitchell, in a model he called the *chemiosmotic hypothesis.* Since then, the experimental tests of Mitchell's model have all supported its basic tenets. Mitchell received the Nobel Prize in chemistry in 1975 for the chemiosmotic hypothesis and his experimental work providing evidence in support of the model.

Substrate-level phosphorylation differs fundamentally from oxidative phosphorylation. In ATP synthesis at the substrate level, phosphate groups are transferred from an organic molecule directly to ADP. Removal of the phosphate group from the organic donor is typically associated with a large release of free energy; some of this energy is captured in attachment of the phosphate group to ADP to form ATP. Substrate-level phosphorylation is the primary source of ATP for organisms that do not possess an oxidative electron transport system, or it becomes the primary source if oxygen is unavailable as a final acceptor for cells with the electron transport pathway.

OXIDATIVE REACTIONS SUPPLYING ELECTRONS FOR ELECTRON TRANSPORT

The primary substances used as electron sources—carbohydrates and fats—are split into their component subunits in preliminary steps that precede their

Information Box 9-1

Oxidation and Reduction

Many substances, both organic and inorganic, can gain or lose electrons. Removal of electrons is called *oxidation;* acceptance of electrons is termed *reduction.* A substance from which electrons are removed is *oxidized;* an accepting substance is *reduced.* In general, the energy content of a given substance is greater in its reduced state than in its oxidized state. The greater energy content of the reduced state represents the energy associated with the added electrons. When any substance is oxidized, another molecule is usually reduced by combining with the removed electrons; that is, each oxidation is usually accompanied by a simultaneous reduction. Frequently, one or two hydrogen ions (protons) as well as electrons are removed from a molecule during an oxidation. The molecules acting as acceptors for the removed electrons may also combine with one or both of these hydrogens.

The amount of energy associated with electrons removed in an oxidation depends on the orbitals they occupied in the oxidized molecule. The energy of the removed electrons can be measured and expressed as a relative potential or voltage by comparison with an arbitrary standard (see the scale diagram in this box). The standard used is the characteristic energy of electrons removed from hydrogen in the reaction $H_2 \rightarrow 2H^+ + 2e^-$. The potential arbitrarily assigned to these electrons is 0.00 V. All other potentials, called *redox* or *reduction-oxidation potentials,* are measured and assigned a value with re-

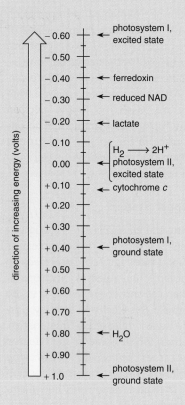

oxidation. Complex carbohydrates such as starches are split into individual monosaccharides, and fats are broken into glycerol and fatty acids. Proteins to be oxidized as an electron source are hydrolyzed into individual amino acids.

These fuels are oxidized in a series of reactions that may include one or both of two major stages, the first occurring outside mitochondria and the second inside (Fig. 9-2). In the first stage, *glycolysis* (from *glykys* = sweet and *lysis* = breakdown), fuels are partially oxidized and converted to three-carbon segments. These short carbon chains are completely oxidized to carbon dioxide and water in the second stage inside mitochondria, which includes two reaction sequences, *pyruvate oxidation* and the *citric acid cycle.* A relatively small amount of ATP is produced by substrate-level phosphorylation during both stages. However, the major synthesis of ATP is driven as the electrons removed during these oxidative stages traverse an electron transport system.

Monosaccharides and many other carbohydrate units follow a main line of oxidation through both

stages. Glycerol enters the glycolytic pathway in the first stage, and fatty acids enter in the second stage. Amino acids are converted by *deamination* (removal of the amino, or —NH$_2$, group) into molecules that, depending on the product, may enter the pathway in either stage.

The First Major Stage: Glycolysis

The reactions of glycolysis yield electrons and the three-carbon substance *pyruvic acid* or *pyruvate.*[1] This three-carbon molecule is the primary fuel substance for the second stage of oxidative reactions taking place

---

[1] Pyruvic *acid* refers to the undissociated form of the acid and pyruv*ate* to the dissociated form:

$$CH_3CH_2COOH \rightleftharpoons CH_3CH_2COO^- + H^+$$

pyruvic acid   pyruvate

Because organic acids typically dissociate under physiological conditions, the dissociated *-ate* form is used in the text and diagrams in this book.

Oxidation	Redox Potentials (in volts)
Acetaldehyde → acetate + $2H^+$ + $2e^-$	−0.58
Isocitrate → α-ketoglutarate + CO_2 + $2H^+$ + $2e^-$	−0.38
β-Hydroxybutyrate → acetoacetate + $2H^+$ + $2e^-$	−0.346
NADH + H^+ → NAD^+ + $2H^+$ + $2e^-$	−0.320
NADPH + H^+ → $NADP^+$ + $2H^+$ + $2e^-$	−0.324
Ethanol → acetaldehyde + $2H^+$ + $2e^-$	−0.197
Lactate → pyruvate + $2H^+$ + $2e^-$	−0.185
Malate → oxaloacetate + $2H^+$ + $2e^-$	−0.166
Succinate → fumarate + $2H^+$ + $2e^-$	−0.031
Coenzyme Q_{red} → coenzyme Q_{ox} + $2H^+$ + $2e^-$	+0.10
2 cytochrome $b_{K(red)}$ → 2 cytochrome $b_{K(ox)}$ + $2e^-$	+0.030
2 cytochrome c_{red} → 2 cytochrome c_{ox} + $2e^-$	+0.254
2 cytochrome $a_{3(red)}$ → 2 cytochrome $a_{3(ox)}$ + $2e^-$	+0.385
H_2O → $\frac{1}{2} O_2$ + $2H^+$ + $2e^-$	+0.816

SOURCE: From A. L. Lehninger, *Biochemistry*, 2nd ed. Worth Publishers, New York, 1975.

spect to this standard. (The table in this box lists the redox potentials of the electrons removed from some important intermediates in cellular oxidations.)

Although all electrons carry a negative charge, their relative voltage may be greater or less than the 0.00 standard. The voltage of electrons with energy greater

than the 0.00 V standard is given a minus value; the voltage of electrons with energy lower than the standard is written as a positive value. Therefore, an electron with a potential of −0.2 V has greater energy than one with a potential of +0.2 V; the difference in voltage between these electrons totals 0.4 V.

Electrons at elevated energy levels, as defined in this chapter, are those that have relatively high voltage or electrical potential. Electrons at elevated energy levels travel at higher velocities and at shorter wavelengths than electrons at lower energy levels. Electrons at the most elevated energy levels normally found in living systems have an electrical potential of about −0.6 V; the lowest usable energies are at about +0.8 V.

In general electrons are transferred from one substance to another located lower on the scale, that is, the donor is higher on the scale than the acceptor. As electrons are transferred from a donor to an acceptor, the amount of free energy released is equivalent to the distance between the donor and acceptor on the redox scale (or the difference in their redox potentials). In the series of electron carriers that are alternately oxidized and reduced as electrons flow through electron transport systems of mitochondria, chloroplasts, and prokaryotes, much of the free energy released during redox reactions is used to build an H^+ gradient and, ultimately, to synthesize ATP.

The movement of electrons in oxidation and reduction reactions is also frequently described in terms of the *affinity* of molecules for electrons. Substances high on the scale are said to have relatively little affinity for electrons and to give them up easily. Substances lower on the scale have greater affinity for electrons and act readily as acceptors. In this interpretation the redox potentials in the table are regarded as a measure of the difference in electron affinity between the members of each pair.

inside mitochondria. Glycolysis also produces a small quantity of ATP by substrate-level phosphorylation. Although the amount of ATP directly synthesized in glycolysis is relatively small, it can be vitally important to survival when oxygen supplies are low.

Overview of Glycolysis Glycolysis has two major parts (Fig. 9-3). In the initial part (the uphill portion in Fig. 9-3) six-carbon derivatives of glucose are raised to an energy level high enough for entrance into the second part of the glycolytic sequence. The energy for the first part is derived from the hydrolysis of ATP.

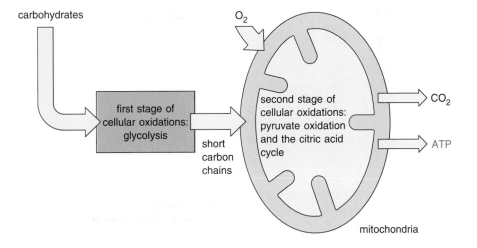

Figure 9-2 The two major stages of cellular oxidations. In the first stage, glycolysis, fuel molecules such as glucose are oxidized and broken into three-carbon organic segments. In the second stage, which includes pyruvate oxidation and the citric acid cycle, the three-carbon segments are oxidized completely to CO_2. Small amounts of ATP are synthesized directly by substrate-level phosphorylation in glycolysis and pyruvate oxidation.

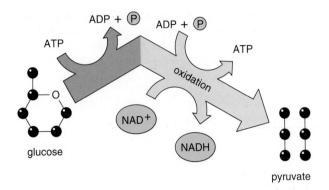

Figure 9-3 The overall reactions of glycolysis, which take place in two major parts. In the first, uphill part six-carbon molecules derived from glucose are raised to higher energy levels at the expense of ATP. In the second part the products of the first part are oxidized and split into three-carbon pyruvate molecules. The second part yields NADH and a net gain in ATP.

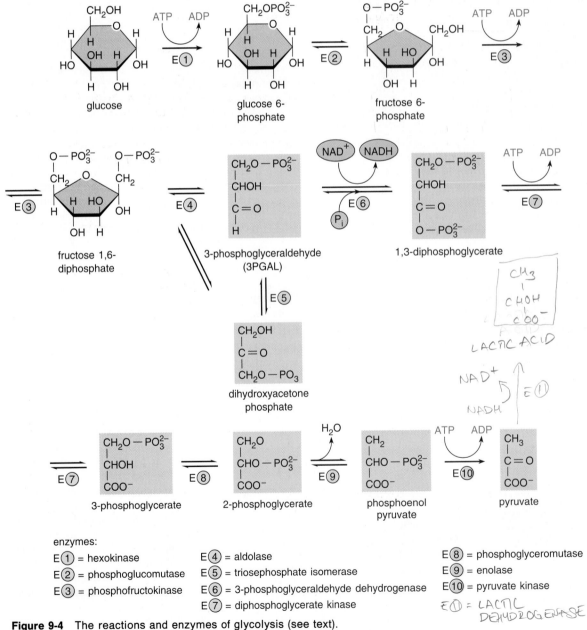

enzymes:

E① = hexokinase   E④ = aldolase   E⑧ = phosphoglyceromutase
E② = phosphoglucomutase E⑤ = triosephosphate isomerase   E⑨ = enolase
E③ = phosphofructokinase E⑥ = 3-phosphoglyceraldehyde dehydrogenase   E⑩ = pyruvate kinase
                         E⑦ = diphosphoglycerate kinase

Figure 9-4 The reactions and enzymes of glycolysis (see text).

the first part is derived from the hydrolysis of ATP. In the second part of the glycolytic sequence (the downhill portion in Fig. 9-3), electrons are removed from the derivatives of glucose, and the energy expended in the first part is recovered with a net gain in ATP by substrate-level phosphorylation. In the process the glucose derivatives are split into two molecules of pyruvate. Each step in the glycolytic pathway is catalyzed by a specific enzyme.

The various enzymes, reactants, and products of glycolysis are suspended in the cytoplasmic solution outside mitochondria. Although they may be loosely organized in the cytoplasm by attachment to the cytoskeleton, they interact primarily through random collisions.

Reactions of the Glycolytic Sequence The first reactions of glycolysis (reactions 1 through 3 in Fig. 9-4) hydrolyze two molecules of ATP to convert glucose into a highly reactive six-carbon, two-phosphate sugar. These initial reactions go uphill in terms of the energy content of the products and proceed only because they are coupled to the breakdown of ATP.

The first reaction of glycolysis transfers a phosphate group from ATP to glucose, producing *glucose 6-phosphate*. The activity of the enzyme catalyzing this reaction, *hexokinase*, illustrates one of the many feedback controls (see p. 96) regulating the rate of oxidation in cells. If glucose 6-phosphate accumulates because the remainder of the glycolytic sequence is running slowly, hexokinase is inhibited, blocking further entry of glucose into the pathway.

The glucose 6-phosphate produced in the first reaction is rearranged (reaction 2) and then phosphorylated (reaction 3) at the expense of a second molecule of ATP, producing *fructose 1,6-diphosphate*. The characteristics of this reaction illustrate another mechanism regulating glycolysis, this one linked to the ATP supply. The enzyme catalyzing the third reaction, *phosphofructokinase*, is allosterically inhibited (see p. 98) by high concentrations of ATP. If ATP is present in elevated concentrations in the cytoplasm, inhibition of phosphofructokinase slows or stops the subsequent reactions of glycolysis. Because the glycolytic sequence produces a three-carbon fuel substance for the oxidations taking place in mitochondria, slowing glycolysis can retard the oxidative reactions driving ATP synthesis inside mitochondria. A surplus of other products or intermediates of oxidative metabolism, such as NADH or citrate (see below), also inhibits the enzyme. If energy-requiring activities take place in the cell, resulting in extensive conversion of ATP to ADP, the reduction in ATP concentration relieves the inhibition of phosphofructokinase. In addition, ADP acts as an allosteric activator of the enzyme. The rate of glycolysis and ATP production therefore increases proportionately as ADP concentration rises. Regulation of phosphofruc-

tokinase is probably the most sensitive and significant control of glycolysis because it is directly keyed to the relative concentrations of ATP and ADP and thus to the rate at which cells use energy for their activities. (The conformational changes occurring in phosphofructokinase in response to combination with its allosteric inhibitors or activators, which have been extensively investigated, are described in Ch. 3.)

The ATP used in the initial steps of glycolysis is recovered with a net gain in the remaining reactions of the pathway (reactions 4 through 10). In reaction 4, fructose 1,6-diphosphate is broken into two different three-carbon sugars, each with one phosphate group. Only one of these molecules, *3-phosphoglyceraldehyde* (*3PGAL*), directly enters the next step in glycolysis. However, as this three-carbon sugar is used, a rearranging enzyme converts the second three-carbon sugar, *dihydroxyacetone phosphate*, into 3PGAL (reaction 5). As a result, both products of reaction 4 are used in the remainder of the sequence.

At the next step (reaction 6) two electrons and two hydrogens are removed from 3PGAL. Some of the released energy is trapped, as part of the reaction, by the addition of an inorganic phosphate group from the medium (not from ATP) to form a three-carbon, two-phosphate product, *1,3-diphosphoglycerate*. The electrons removed in reaction 6 have a relatively high potential and are accepted by a carrier molecule, *nicotinamide adenine dinucleotide*[2] (*NAD*; Fig. 9-5). NAD is one of a group of carriers based on nucleotides that shuttle electrons between major reaction systems. Two hydrogens are also removed from 3PGAL in reaction 6. One of the hydrogens binds to NAD^+ in the position shown in Figure 9-5 to form NADH. The second is released to the medium as H^+. The reduced NAD, or NADH, formed at this step can be considered a high-energy substance by virtue of the electrons it gains in reaction 6. When oxygen is abundant, the electrons carried by NADH eventually enter the electron transport system in mitochondria, where much of their energy is captured in the conversion of ADP to ATP.

The 1,3-diphosphoglycerate product of reaction 6 may also be regarded as a high-energy substance because removal of the two phosphates from its derivatives later in the sequence releases large increments of free energy. Much of the released energy is captured in the substrate-level conversion of ADP to ATP.

The reaction removing the first of the phosphates (reaction 7) provides one of the best-studied examples of substrate-level phosphorylation. If 1 mole (see footnote 1, p. 40) of 1,3-diphosphoglycerate were hydrolyzed directly to the three-carbon, one-phosphate product of reaction 7 (*3-phosphoglycerate*) and inorganic

[2] The oxidized form of this electron carrier is symbolized as NAD^+, the reduced form as NADH. A similar convention is used for several other nucleotide-based electron carriers.

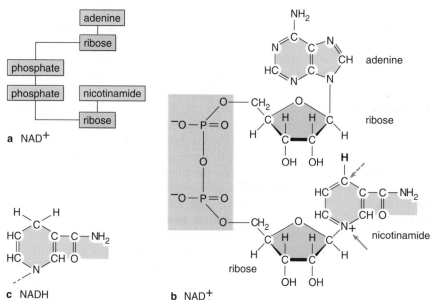

Figure 9-5 NAD (nicotinamide adenine dinucleotide), an electron carrier reduced in glycolysis and other cellular reactions. The molecule consists of two nucleotides linked end to end. (**a** and **b**) The oxidized form of the carrier, designated NAD $^+$. One electron is added at each of the two positions marked by an arrow as the carrier is reduced; a hydrogen is also added at the position marked by the dashed arrow. (**c**) The reduced form of the nicotinamide portion of the carrier. In reduced form the carrier is designated NADH.

phosphate, the reaction would release about 10,000 to 15,000 cal under standard conditions. In glycolysis the breakdown of the two-phosphate sugar is linked to ATP synthesis by the substrate-level transfer of the removed phosphate to ADP. The combined reaction still releases energy but at the reduced level of about 4500 cal/mol:

1,3-diphosphoglycerate + ADP $\longrightarrow$

3-phosphoglycerate + ATP   (9-1)

The difference in free energy (see p. 83) released by the two reactions is energy captured in the substrate-level formation of ATP.

The 3-phosphoglycerate then enters the final series of reactions, which produces another ATP and pyruvate. Reaction 8 is a rearrangement that shifts the phosphate group from the 3- to the 2-carbon, producing *2-phosphoglycerate.* Reaction 9 is an internal oxidation in which electrons are removed from one part of 2-phosphoglycerate and delivered to another part at lower energy levels; most of the energy lost by the electrons is retained in the product of the reaction, *phosphoenolpyruvate.* The shift greatly increases the energy made available when its phosphate group is removed in the next reaction. The final reaction of the pathway (reaction 10) removes the phosphate group from phosphoenolpyruvate and yields pyruvate, the end product of glycolysis. The phosphate group is transferred directly to ADP to form ATP in the second substrate-level phosphorylation of the pathway. ATP is an allosteric inhibitor of the enzyme catalyzing the final reaction, *pyruvate kinase,* providing another control point at which elevated ATP concentrations stop or

slow the glycolytic pathway. Phosphoenolpyruvate is another allosteric inhibitor of phosphofructokinase, the enzyme catalyzing the critical third reaction of the glycolytic pathway.

The substrate-level phosphorylations of the second part of glycolysis provide a net gain in ATP because each glucose molecule entering the pathway ultimately produces two molecules of 3PGAL. As these two 3PGALs are oxidized to two molecules of pyruvate, a total of four ATP molecules are produced. Two of these ATPs "pay back" the two ATP molecules required to convert glucose to fructose 1,6-diphosphate in the initial steps of glycolysis (reactions 1 through 3). The remaining two ATP molecules represent a net gain for each molecule of glucose oxidized to two molecules of pyruvate. The second segment of glycolysis (reactions 4 through 10) also produces two molecules of NADH, which represent a significant reservoir of energy that can also be captured in the conversion of ADP to ATP. The total reactants and products of glycolysis are:

glucose + 2ADP + 2P$_i$ + 2NAD$^+$ $\longrightarrow$

2 pyruvate + 2NADH + 2ATP   (9-2)

where P$_i$ indicates inorganic phosphate.

Important Variations of the Glycolytic Pathway A wide variety of substances can be used as inputs for glycolysis. Glycogen in animals and starch in plants are both long-chain polysaccharides with chains of repeating glucose links (see Fig. 2-9). Glycogen, stored in the liver and striated muscle cells of humans and other vertebrates, is hydrolyzed to yield glucose 6-phosphate, which enters the glycolytic pathway as an input into reaction 2. The phosphate added to the 6-carbon of the glucose units removed from glycogen is derived from inorganic supplies rather than ATP. (Glycogen breakdown is regulated by a cAMP-based second-messenger pathway; see p. 245 for details.) Plant starch is hydrolyzed into individual glucose molecules, which enter as inputs into the first reaction in the glycolytic pathway. The enzymes necessary for complete hydrolysis of plant starch, *amylase* and *maltase*, occur as digestive enzymes in humans and other animals.

The 3PGAL molecule also serves as an important entry point for glycolsis. Besides taking part in the glycolytic sequence, it is formed as the first net product of photosynthesis (see pp. 355–358 for details). In plants the 3PGAL produced in photosynthesis may serve as a cellular fuel by entering glycolysis directly in reaction 6. The glycerol released by fat breakdown is also converted indirectly into 3PGAL, and enters cellular oxidations in reaction 6 of glycolysis.

The most significant variations in glycolysis from the standpoint of cellular energy economy occur at the end of the pathway. They enable glycolysis to operate as the primary source of ATP when oxygen supplies are too low for ATP production in mitochondria.

One of the most important of these variations involves conversion of the end product of glycolysis, pyruvate, into lactate:

$$\text{pyruvate} + \text{NADH} \longrightarrow \text{lactate} + \text{NAD}^+ \quad (9\text{-}3)$$

In the reaction pyruvate acts as an acceptor for the electrons carried by NADH. The NAD^+ is then free to recycle to the oxidative reaction of glycolysis (reaction 6) to accept further electrons from 3PGAL. Because NAD^+ is constantly regenerated by lactate formation, glycolysis can continue to run with the net production of ATP. Lactate forms in the muscle cells of animals, including humans, if intensive, sustained physical activity is carried out before increases in breathing and heart rate are able to meet the demand for oxygen in the muscle tissue. The lactate accumulating under these conditions serves as a temporary storage site for electrons that can later be passed on to the mitochondrial electron transport system when the oxygen content of the muscle cells returns to normal levels. This occurs by a sequence that essentially reverses Reaction 9-3 and regenerates pyruvate and NADH.

Another important variation with the same outcome occurs in microorganisms such as yeasts. In this modification pyruvate accepts electrons from NADH and is converted by additional reactions into *ethyl alcohol* (two carbons) and CO_2:

$$\text{pyruvate} + \text{NADH} \longrightarrow$$
$$\text{ethyl alcohol} + CO_2 + \text{NAD}^+ \quad (9\text{-}4)$$

This variation has been of central importance to human activities since the earliest days of recorded history. It provides alcohol for brewing and CO_2 for baking. The Sumerians used yeast to make beer as early as 7000 B.C.; the Egyptians recorded the use of yeast in baking as early as 4000 B.C.

When electrons removed from 3PGAL in glycolysis are delivered to an organic molecule by NADH, as they are in lactate or ethyl alcohol formation, the reaction system is termed a *fermentation*. Various fermentations, producing a wide range of products, generate ATP in different fungi and bacteria. Some of these species, called *strict anaerobes,* produce ATP only by fermentations and cannot use oxygen at any time as a final electron acceptor. Others can switch between fermentations and full oxidation pathways that depend on the oxygen supply. Cells in this category are called *facultative anaerobes.* Some species, termed *strict aerobes,* are unable to live solely by fermentation at any time.

Cells in higher eukaryotes also vary in their oxygen requirements. Many, including the muscle cells of vertebrates described above, are facultative and can switch between fermentation and complete oxidation depending on the availability of oxygen. Others, such as the brain cells of humans and other vertebrates, are strict aerobes and cannot survive unless oxygen is available.

Reversing Glycolysis: Gluconeogenesis The reactions of glycolysis can be reversed in effect, so that glucose or glucose 6-phosphate can be synthesized from pyruvate. Other noncarbohydrate substances such as fats and amino acids can also be converted into glucose after conversion into pyruvate or intermediates of the glycolytic pathway. Synthesis of glucose from noncarbohydrate precursors by this pathway is called *gluconeogenesis.*

Not all the reactions of glycolysis can be directly reversed. Three of the steps, reactions 1, 3, and 10 (which are shown in color in Fig. 9-4), have equilibria so far in the direction of products (see p. 82) that they are essentially irreversible when catalyzed by glycolytic enzymes. Gluconeogenesis uses different enzymes to circumvent these steps (Fig. 9-6). For example, the irreversibility of glycolytic reaction 1, in which hexokinase catalyzes the conversion of glucose to glucose 6-phosphate, is circumvented by an alternate enzyme, *glucose 6-phosphatase,* which can catalyze the reverse reaction. The reverse pathway synthesizing glucose from pyruvate or other substances is an energy-requiring mechanism that proceeds only through the expenditure of ATP and GTP and conversion of other phosphate groups from organic to inorganic form.

Although requiring an energy input to proceed, gluconeogenesis is a vital source of glucose when an organism's glucose supplies are limited. For example, in humans and other vertebrates, the reserves of glycogen stored in the liver are depleted by only a few

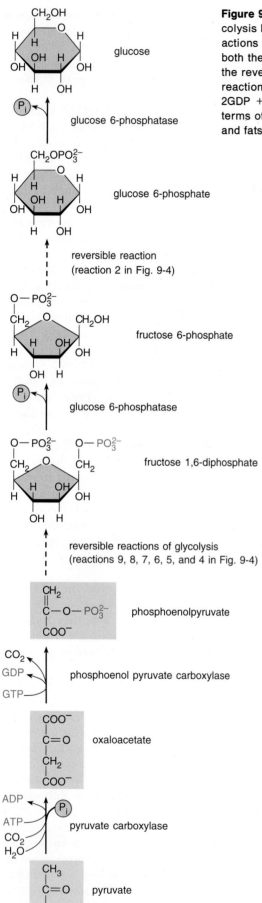

Figure 9-6 Gluconeogenesis. The reactions of this series effectively reverse glycolysis by substituting other pathways for the irreversible glycolytic reactions (reactions 1, 3, and 10 in Fig. 9-4). To yield one molecule of fructose 1,6-diphosphate, both the reactions producing phosphoenolpyruvate from pyruvate in this figure and the reversed reactions 9 through 4 of Figure 9-4 must be doubled. Doubling these reactions gives a total expenditure of 4ATP and 2GTP converted to 4ADP + 2GDP + 6P$_i$ for the entire sequence from pyruvate to glucose. Although costly in terms of energy, gluconeogenesis provides a means for conversion of amino acids and fats into glucose and other carbohydrates.

hours of fasting. Under these conditions liver cells compensate for the glycogen depletion by gluconeogenesis in which breakdown products of fats and proteins are converted into glucose. The maintenance of blood glucose by gluconeogenesis is vital to survival because brain cells can continue to function only if glucose remains available at normal concentrations in the blood.

The balance between glycolysis and gluconeogenesis in liver cells is closely regulated by hormones in mammals and other higher vertebrates. When dietary carbohydrate intake is inadequate, several hormones stimulate gluconeogenesis in liver cells and inhibit glycolysis and glycogen formation. Some, such as glucagon, work through cAMP-based receptor-response pathways (see p. 245). Increases in cAMP, induced when glucagon binds to its receptor, stimulate protein kinases that add phosphate groups to the enzyme (*pyruvate kinase*), catalyzing reaction 10 in the glycolytic pathway. This phosphorylation inhibits the enzyme; phosphorylations stimulated by cAMP also greatly reduce the quantities of fructose 1,6-diphosphate produced in reaction 3 of glycolysis. Both these effects slow glycolysis. At the same time inhibition of glycolytic reaction 3 increases the concentrations of fructose 6-phosphate and glucose 6-phosphate, the primary inputs for the steps reversing reactions 1 and 2 of glycolysis. As a result, gluconeogenesis is stimulated. Other hormones, such as vasopressin and angiotensin, which work through InsP$_3$/DAG-based pathways (see p. 247), have the same inhibitory effects on pyruvate kinase. In this case Ca^{2+}/calmodulin-dependent protein kinases, activated in the InsP$_3$/DAG-based pathways, add phosphate groups to pyruvate kinase. These controls operate primarily in liver cells when dietary carbohydrate intake is inadequate; other cells, such as those in the brain, do not have receptors for the hormones controlling liver function and continue to run glycolysis in the forward direction.

The net effect of the inhibition of glycolysis and glycogen formation, and stimulation of gluconeogenesis by these hormones, is the release of glucose from the liver into the bloodstream. When carbohydrates are ingested in adequate quantities, glycolysis predominates and gluconeogenesis is inhibited. Part of the inhibition of gluconeogenesis depends on the effects of the hormone insulin, which operates by "crosstalk" between receptor-response pathways (see p. 252) to drastically reduce the concentrations of cAMP in liver

The Pentose Phosphate Pathway

Not all the glucose metabolized in cellular oxidations is oxidized by the glycolytic pathway. Certain inhibitors, such as fluoride or iodoacetate, block the glycolytic pathway but do not completely halt glucose oxidation. This finding led to the discovery of several alternate pathways for glucose metabolism. The most important is the *pentose phosphate pathway*, in which glucose is oxidized and converted into a five-carbon sugar, pentose. NADP$^+$ (see Fig. 9-16) serves as electron acceptor for the pathway.

The metabolism of glucose via the pentose phosphate pathway begins with glucose 6-phosphate, which is oxidized to *6-phosphogluconate* in a two-step process (see reactions 1 and 2 of the figure in this box). The oxidation takes place in the first step; NADP$^+$ acts as the acceptor for the electrons. After the elements of a molecule of water have been added, the product, 6-phosphogluconate, enters the second oxidation of the pathway (reaction 3 of the figure). NADP$^+$ is also the electron acceptor for this oxidation; as part of the reaction, the carbon chain is shortened to yield the five-carbon sugar *ribulose 5-phosphate*. The carbon removed is released as CO_2. Ribulose 5-phosphate is rearranged to *ribose 5-phosphate* in the final step of the pathway (reaction 4 in the figure).

The two major products of this pathway, ribose 5-phosphate and NADPH, are both of great significance in cellular metabolism. The pentose can be used as the starting point for the synthesis of a variety of three-, four-, five-, six-, and seven-carbon sugars in both plants and animals. Five-carbon sugars form part of the structure of nucleotides, including those of ATP, FAD, FMN, and coenzyme A, and of the nucleic acids DNA and RNA. A derivative of ribose 5-phosphate combines with atmospheric CO_2 as the first step in photosynthesis. Ribose 5-phosphate may also enter a complex series of reactions with its own three-, seven-, and four-carbon derivatives to regenerate glucose 6-phosphate, which may enter the pentose phosphate pathway for another series of oxidations. In this way all the carbons of an original glucose molecule may eventually be oxidized to CO_2 by the pentose phosphate pathway, with further generation of NADPH.

Because NADPH is important in synthetic reactions in which a reduction is required, the pentose phosphate pathway may be considered as an alternative to glycolysis that generates reducing power in the form of NADPH. The NADH produced in glycolysis, in contrast, drives ATP synthesis by delivering electrons removed from glucose to the mitochondrial electron transport system.

Whether glucose is oxidized via glycolysis or the pentose phosphate pathway depends on the enzyme most active in conversion of glucose 6-phosphate (the product of the first reaction in the glycolytic pathway). If NADPH is present in high concentrations in the cytoplasm, the first enzyme of the pentose phosphate pathway (*glucose 6-phosphate dehydrogenase*) is inhibited, and glucose 6-phosphate is converted to fructose 6-phosphate by phosphoglucomutase, as it normally would be in the glycolytic pathway. If NADPH is present in low concentrations and the levels of NADP$^+$ are high, glucose 6-phosphate dehydrogenase is stimulated and glucose 6-phosphate is oxidized, shunting it into the pentose phosphate pathway. The effects of the relative concentrations of NADP$^+$ and NADPH on the enzyme tightly couple the pentose phosphate pathway to the cell's needs for reducing power in the form of NADPH. The pathway operates only when NADPH is required by synthetic reactions somewhere else in the cell. Because the pentose phosphate pathway diverts glucose 6-phosphate from glycolysis when cells need reducing power, it is sometimes called the pentose phosphate *shunt*.

cells. As a consequence, glycolysis and glycogen formation predominate, and gluconeogenesis is inhibited in liver cells.

Besides being reversed, glycolysis can be bypassed by a reaction sequence that oxidizes glucose 6-phosphate into a five-carbon sugar, *ribose 5-phosphate*. This sequence, called the *pentose phosphate pathway*, is an important source of five-carbon sugars for the synthesis of nucleotides and other substances. (Details of the pathway are given in Information Box 9-2.)

Unlike all other enzymes of glycolysis, gluconeogenesis, or glycogen metabolism, which are suspended in solution in the cytoplasm, glucose 6-phosphatase is a transmembrane protein embedded in the ER. Its active site is directed toward the internal ER compartment so that glucose 6-phosphate must be transported across ER membranes from the cytoplasm for dephosphorylation to proceed. After the reaction is complete, the products, glucose and an inorganic phosphate group, must be transported in the opposite direction, from the ER compartment into the cytoplasm. Several transport carriers specialized for these movements are associated with glucose 6-phosphatase in ER membranes of liver and kidney cells.

A series of mutations in genes encoding glucose 6-phosphatase or the transport proteins are responsible for several inherited human maladies collectively called *glycogen storage diseases.* Mutations producing deficient forms of glucose 6-phosphatase, for example, cause massive enlargement of the liver and kidneys because of glycogen accumulation. Children with the abnormality are typically short in stature, with a greatly protruding abdomen and multiple skin lesions. Adults with the disease are subject to gout, arthritis, and renal failure. Although the disease is presently incurable, the symptoms can be alleviated by a diet rich in carbohydrates.

Discovery of the Glycolytic Pathway The research leading to the discovery of glycolysis began in the late 1800s. In Germany Eduard and Hans Buchner discovered that alcoholic fermentation could be carried out by nonliving extracts of yeast cells. Through research spurred by this discovery, the first enzymes were discovered and described. Intensive research in alcoholic fermentation continued well into the 1930s. At that time Gustave Embden assembled the known reactions into a provisional reaction sequence that explained the events of glycolysis. Other investigators, most notably Otto Meyerhoff, Carl Cori, and Otto Warburg, fitted the final pieces in the puzzle. By 1940 the reactions of glycolysis were known essentially as they are today. Glycolysis is frequently called the *Embden–Meyerhoff pathway* in honor of the two scientists considered to have made the most significant contributions to solving the glycolytic puzzle.

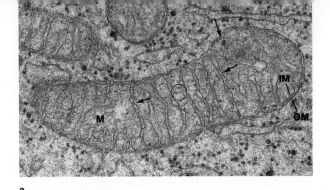

a

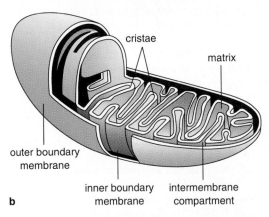

b

Figure 9-7 Mitochondrial structure. **(a)** A mitochondrion from a chick intestinal cell. The outer boundary membrane (OM) is smooth and covers the entire mitochondrion; the inner boundary membrane (IM) folds into cristae (arrows) that extend into the mitochondrial interior. The matrix (M), the innermost mitochondrial compartment, contains proteins and other molecules in solution, including the enzymes and intermediates of pyruvate oxidation and the citric acid cycle. Mitochondrial ribosomes (circle) and DNA are also present in the matrix. × 45,000. Courtesy of J. Mais. **(b)** The membranes and compartments of mitochondria.

The Second Major Stage: Pyruvate Oxidation and the Citric Acid Cycle

In the second major stage of cellular oxidations, the pyruvate produced by glycolysis is oxidized into carbon dioxide inside mitochondria. The mitochondrial oxidations occur in two overall parts. In the first, *pyruvate oxidation*, pyruvate is oxidized into two-carbon *acetyl* (CH_3—C—) units and CO_2. In the second part,

the *citric acid cycle*, the acetyl units produced by pyruvate oxidation are completely oxidized to CO_2. Because oxygen is normally the final acceptor for electrons removed in pyruvate oxidation and the citric acid cycle, the oxidative activities of mitochondria are frequently termed *respiration*. The use of this term is distinct from the more common physiological reference to breathing. However, the primary use of the oxygen we inhale in physiological respiration is actually as the final electron acceptor for mitochondrial oxidation.

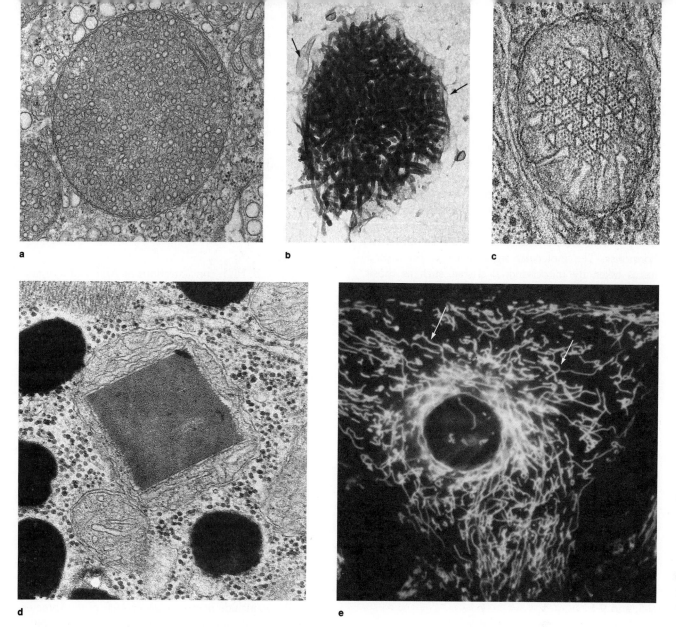

Figure 9-8 A gallery of mitochondrial types. **(a)** A large, spherical mitochondrion from the adrenal cortex of the rat, with cristae in the form of tubes or vesicles. In vertebrates tubular cristae are typical of cells synthesizing steroid hormones. ×32,000. Courtesy of D. S. Friend. **(b)** An isolated, partially spread mitochondrion from the protozoan *Euplotes*. This mitochondrion, which was released from a cell disrupted on a water-air interface, contains the tubular cristae typical of many ciliate protozoa. (Tubular cristae in protozoan mitochondria are not related to steroid synthesis.) The outer membrane (arrows) has been lost except at a few locations. ×28,000. Photograph by the author. **(c)** Mitochondrion from the sphincter muscle of the crayfish vas deferens, in which cristae have a triangular form. ×86,500. Courtesy of M. A. Cahill, from *J. Cell Biol*. 74:326 (1977), by permission of the Rockefeller University Press. **(d)** Mitochondrion from the frog oocyte containing a crystalline yolk body. ×47,000. Courtesy of P. B. Armstrong. **(e)** Threadlike mitochondria (arrows) in a living cell of the gerbil, stained with a fluorescent dye (rhodamine 3B). The dye, a small, positively charged molecule, accumulates in the mitochondria. ×460. Courtesy of Lan Bo Chen, reproduced from *J. Cell Biol*. 88:526 (1981), by copyright permission of the Rockefeller University Press.

Basics of Mitochondrial Structure   In most cells mitochondria appear as spherical or elongated bodies about 0.5 μm in diameter and 1 to 2 μm in length, about the size of a bacterium. They may be much larger in some cells—up to several micrometers in diameter and as long as 10 μm.

Two separate membrane systems, the *outer* and *inner boundary membranes*, define mitochondrial structure (Fig. 9-7; see also Fig. 1-13). The outer boundary membrane forms a single, continuous layer around the mitochondrion. The inner membrane is thrown into folds or tubular extensions, the *cristae* (singular = *crista*), which project into the interior of the mitochondrion and greatly increase the surface area of this membrane system. Cristae take many forms in the mitochondria of different tissues and species. Among the most common is the arrangement shown in Figure 9-7, in which cristae consist of flattened, saclike folds extending across the interior of the mitochondrion, more or less at right angles to the boundary membranes. Tubular cristae (Fig. 9-8a and b) also occur in many plants, in protozoa, and in some cells of higher animals. In higher animals including mammals, tubular cristae appear regularly in the mitochondria of cells synthesizing steroid hormones. The molecular significance of the varied forms taken by mitochondrial cristae, such as those shown in Figure 9-8, remains unknown.

The outer and inner boundary membranes separate the mitochondrial interior into two distinct regions (see Fig. 9-7b): the *intermembrane compartment*, located between the inner and outer membranes, and the *matrix*, the innermost compartment enclosed by the inner boundary membrane.

Part 1 of Mitochondrial Respiration   In pyruvate oxidation a molecule of pyruvate is converted into an acetyl unit in a cyclic series of reactions that removes two electrons, two hydrogens, and one carbon in the form of CO_2. (Fig. 9-9 shows the overall cycle; details of the cycle are given in Fig. 9-10.) The two electrons and one of the hydrogens removed are accepted by NAD^+, which is converted to NADH. The acetyl unit is transferred to a molecule called *coenzyme A* (Fig. 9-11), which transfers acetyl units between major reaction systems. (Coenzyme A, like ATP and NAD, is based on a nucleotide structure; compare Figs. 2-31, 9-5, and 9-11.)

pyruvate + coenzyme A + NAD^+ $\longrightarrow$

$\qquad$ acetyl-coenzyme A + NADH + CO_2 (9-5)

The CO_2 and the remaining hydrogen are released to enter the surrounding medium. Because each molecule of glucose entering glycolysis produces two molecules of pyruvate, all the reactants and products in Reaction 9-5 are multiplied by a factor of 2 if pyruvate oxidation is considered as a continuation of glycolysis.

Pyruvate oxidation is catalyzed by the *pyruvate dehydrogenase complex*, a molecular assembly containing three enzymes. (See Fig. 9-10; details of the cycle are described in the caption to the figure.) Activity of the complex, whose structural and functional relationships were unraveled by L. J. Reed, P. J. Randle, and their coworkers and others, is inhibited by elevated concentration of substances critically involved in mitochondrial reactions. Elevated concentrations of acetyl-coenzyme A and high levels of NADH inhibit the complex; these substances accumulate when the rate of oxidation in the citric acid cycle or electron transport

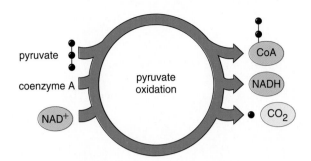

Figure 9-9   Overall reactants and products of pyruvate oxidation (see text).

is reduced. High concentrations of ADP and inorganic phosphate, which accumulate when cells increase their energy consumption, stimulate activity of the complex, as do elevated concentrations of pyruvate. Some of this regulation is exerted through reversible phosphorylation of enzyme E1 of the complex by a protein kinase: addition of phosphate groups inhibits E1 and slows activity of the pyruvate-dehydrogenase complex; removal of phosphate groups has the opposite effect. For example, direct combination of ADP or pyruvate with the protein kinase inhibits its activity, leading to reduced phosphorylation of E1 and increased activity of the pyruvate dehydrogenase complex. NADH and acetyl-coenzyme A inhibit later steps in the cycle. The combination of regulatory mechanisms provides a sensitive control that closely matches the rate of pyruvate oxidation to the energy requirements of the cell.

The pyruvate dehydrogenase complex exists in mitochondria in multiple assemblies containing as many as 100 or more polypeptide chains. The assemblies, which may have a total molecular weight of many millions, include many copies of each of the three enzymes catalyzing pyruvate oxidation. Bacteria able to carry out pyruvate oxidation possess similar assemblies. (Fig. 9-12 shows the structure of pyruvate dehydrogenase assemblies isolated from *E. coli*.)

Both organic products of pyruvate oxidation are important elements in cellular energy metabolism. The acetyl units carried by coenzyme A serve as the immediate fuel for the remaining oxidative reactions of mitochondria. The electrons carried by NADH represent potential energy that is eventually tapped to drive ATP synthesis.

Part 2 of Mitochondrial Respiration: The Citric Acid Cycle   The acetyl units carried by coenzyme A are oxidized to CO_2 in the *citric acid cycle* (also called the *Krebs cycle* in honor of Hans Krebs, who was the first to piece together the reactions of the series). In the overall reactions of the cycle (Fig. 9-13) there is a continuous input of ADP, NAD^+, another oxidized electron carrier—*flavin adenine dinucleotide* (*FAD*; Fig. 9-14), and two-carbon acetyl units carried to the cycle by coenzyme A. The overall products are CO_2, ATP,

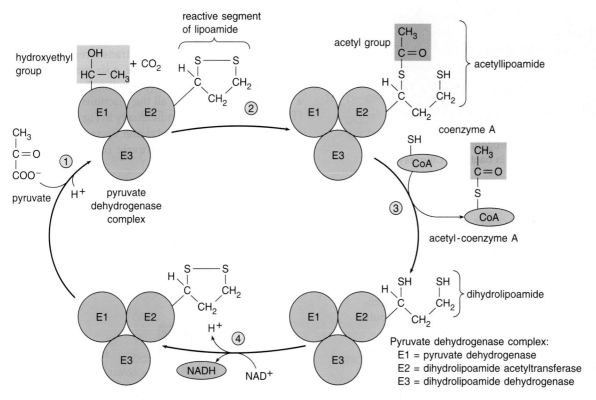

Figure 9-10 A simplified outline of pyruvate oxidation. In reaction 1 of the cycle, the pyruvate dehydrogenase complex, which consists of the three enzymes E1, E2, and E3, binds a molecule of pyruvate. After binding, E1 (pyruvate dehydrogenase) catalyzes removal of one carbon from pyruvate; the carbon is released as CO_2. The two-carbon segment produced remains attached to the prosthetic group (thiamin pyrophosphate, or TPP) of E1 as a hydroxyethyl group. In reaction 2, E1 oxidizes the hydroxyethyl group, forming an acetyl group. As part of this reaction, the acetyl group is transferred to lipoamide, the prosthetic group of the second enzyme of the complex, E2 (dihydrolipoamide acetyltransferase). The transfer converts lipoamide to *acetyllipoamide*. In reaction 3, E2 catalyzes transfer of the acetyl group to coenzyme A, forming acetyl-coenzyme A. The transfer converts the prosthetic group of E2 to dihydrolipoamide, a fully reduced form that may be considered to retain both electrons removed from pyruvate during its oxidation. These electrons are transferred to NAD^+ in reaction 4 of the cycle, which is catalyzed by E3 (dihydrolipoamide dehydrogenase). The electron transfer converts dihydrolipoamide back to lipoamide, and the pyruvate dehydrogenase complex is ready to repeat the cycle.

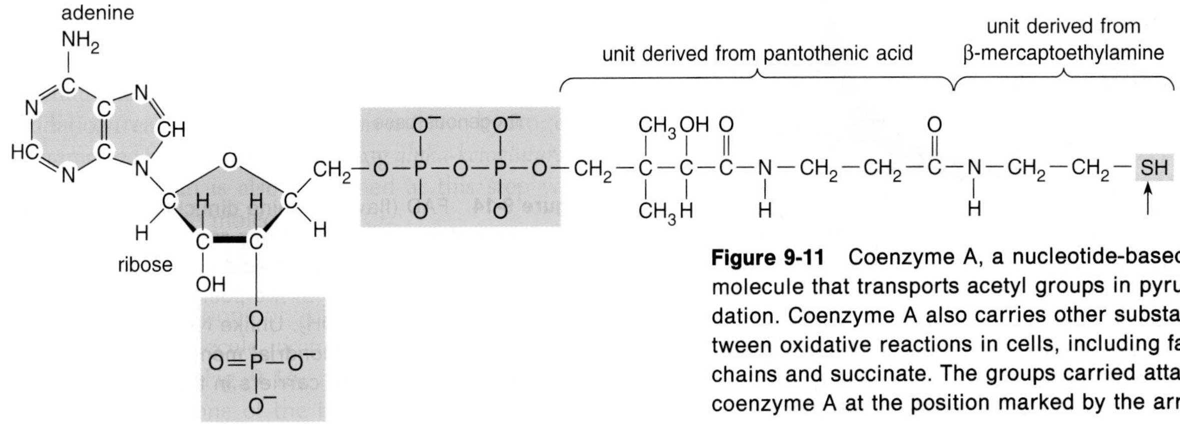

Figure 9-11 Coenzyme A, a nucleotide-based carrier molecule that transports acetyl groups in pyruvate oxidation. Coenzyme A also carries other substances between oxidative reactions in cells, including fatty acid chains and succinate. The groups carried attach to coenzyme A at the position marked by the arrow.

Placing Electron Carriers in Sequence

Part of the research placing the carriers of the mitochondrial electron transport system in sequence is based on measurements of the voltage, or potential, of electrons removed from purified prosthetic groups. By arranging the carriers in order of decreasing voltage, their sequence in electron transport can be approximated. This method introduces some error because the potentials measured under standard conditions, when the carriers are in isolated form, differ to an unknown extent from the potentials in the natural situation in the inner boundary membrane.

A more direct and accurate method for sequencing the carriers is based on the fact that most carrier molecules absorb light at characteristic wavelengths. The wavelengths absorbed change as the carriers cycle between oxidized and reduced states. The graph in this box shows the light absorption by cytochrome *c* in the oxidized state, in which the spectrum has a single major peak, and in the reduced state, in which the spectrum has three peaks. The primary advantage of the method is that the absorption patterns of the carriers can be detected, and their level of oxidation or reduction determined, in intact mitochondria.

The characteristic light absorption patterns are used in different ways to place the carriers in sequence. One of the most effective was used by the pioneer of this method, B. Chance, and others. It involves the use of inhibitors known to block the activity of specific electron carriers (see the diagram in this box). Carriers located after the inhibited carrier in the sequence gradually become oxidized by removal of electrons, and show the characteristic light absorption patterns of their oxidized forms. Carriers located before the blocked carrier remain in

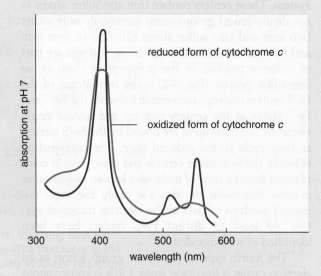

reduced form and show absorption patterns characteristic for this state.

For example, the drug *antimycin A* blocks the release of electrons by cytochrome *b*. In mitochondria exposed to antimycin A, NAD^+ and the flavoproteins remain in the reduced state, and cytochromes *a* and *c* become oxidized. This indicates that NAD and the flavoproteins fall before cytochrome *b*, while cytochromes *a* and *c* fall after it, in the sequence. Using inhibitors for different electron carriers, and noting which are held in the reduced or oxidized state in mitochondria exposed to the inhibitors, allowed many of the carriers to be placed in sequence in the mitochondrial system.

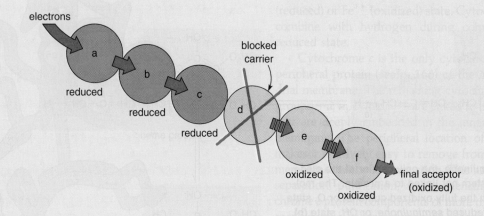

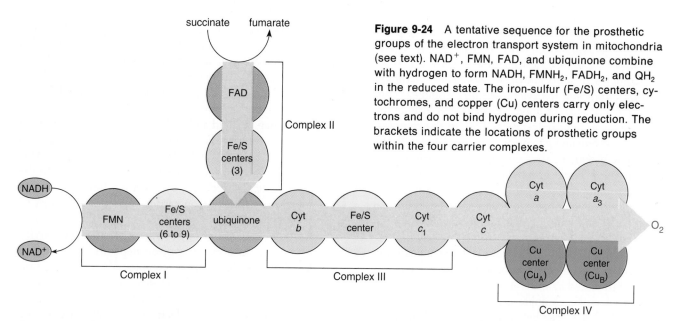

Figure 9-24 A tentative sequence for the prosthetic groups of the electron transport system in mitochondria (see text). NAD^+, FMN, FAD, and ubiquinone combine with hydrogen to form NADH, $FMNH_2$, $FADH_2$, and QH_2 in the reduced state. The iron-sulfur (Fe/S) centers, cytochromes, and copper (Cu) centers carry only electrons and do not bind hydrogen during reduction. The brackets indicate the locations of prosthetic groups within the four carrier complexes.

these approaches has established the sequence shown in Figure 9-24. Note that electrons have two routes of entry into the chain, one at FMN and one at FAD, which join at ubiquinone. Electrons flow spontaneously along the chain, losing energy as they move from one carrier to the next.

Organization of Prosthetic Groups with Proteins in Electron Transport

Organization of the prosthetic groups into carrier complexes was revealed by isolation of proteins from the inner membranes of mitochondria. This research, pioneered in the laboratories of Y. Hatefi, E. Racker, Y. Kagawa, and D. E. Green, demonstrated that most of the carriers shown in Figure 9-24, rather than existing individually in mitochondrial membranes, are combined with proteins into the four major complexes shown in Figure 9-18. Each complex has been purified and placed in active form in artificial phospholipid membranes.

The four complexes include:

1. *complex I,* conducting electrons from NADH to the ubiquinone pool;

2. *complex II,* carrying electrons from the oxidation of succinate in the citric acid cycle to ubiquinone;

3. *complex III,* conducting electrons from ubiquinone to cytochrome *c*;

4. *complex IV,* conducting electrons from cytochrome *c* to oxygen.

Work with labels, or reagents attacking or digesting parts of the complexes exposed at inner mitochondrial membrane surfaces, has established that each complex

extends entirely through the inner membrane (see p. 171 for details of these techniques).

Complex I (also known as *NADH-ubiquinone oxidoreductase* and *NADH dehydrogerase*), contains at least 16 and as many as 26 polypeptides. Among them is an enzymatic activity that catalyzes transfer of electrons from NADH to FMN, a prosthetic group bound to the complex. Also present in complex I are as many as nine iron-sulfur centers. These centers participate in the oxidation of $FMNH_2$, in the movement of electrons from complex I to the ubiquinone pool, and possibly also in the initial reduction of FMN. Some of the polypeptides of complex I form parts of the Fe/S centers; others may act as structural subunits or possibly as part of the mechanism moving H^+ across the membrane.

The oxidation of reduced NAD occurs on the segment of complex I that faces the mitochondrial matrix. Purified complex I particles, when inserted into artificial phospholipid membranes, are capable of moving H^+ across the membrane when supplied with NADH and a suitable electron acceptor.

Complex II (also known as *succinate-ubiquinone oxidoreductase* and *succinate dehydrogerase*) contains two polypeptides with succinate dehydrogenase activity and two to three additional peptides. Because the complex is located within the inner boundary membrane, succinate must collide with the membrane surface facing the matrix for oxidation in the citric acid cycle. The acceptor for the electrons removed in the oxidation, FAD, is bound as a prosthetic group to the complex. Three iron-sulfur centers are also present. Electrons are believed to pass from FAD to the iron-sulfur centers within the complex, and from there to the ubiquinone pool.

Complex II fails to pump H^+ when it is purified, inserted into artificial phospholipid films, and supplied with succinate and a suitable electron acceptor. Evidently the complex functions solely as an entry point for the electrons removed from succinate, which contain insufficient energy to reduce NAD^+.

Complex III (also known as *ubiquinone-cytochrome c oxidoreductase* and the *cytochrome b/c₁ complex*) contains as many as eight polypeptides. These include the major cytochrome c reductase activity of the complex. Some of the remaining peptides are associated with the prosthetic groups of the complex, which include two cytochrome b heme groups, one cytochrome c_1, and one iron-sulfur center. There are also indications that one or two ubiquinone molecules may be bound to complex III and participate in the movement of electrons within the complex.

Complex III picks up electrons from the ubiquinone pool and delivers them to cytochrome c. Research with complex III placed in artificial phospholipid membranes confirms that the complex can move H^+ across the membrane as it cycles between reduced and oxidized states. The complex, which plays a central role in the electron transport systems of mitochondria, chloroplasts, and most aerobic bacteria, has been highly conserved in evolution.

The final complex of the electron transport system, complex IV (also known as *cytochrome c-O_2 oxidoreductase* or *cytochrome oxidase*), conducts electrons from cytochrome c to oxygen. The complex contains as many as 13 polypeptides, including a cytochrome c oxidase activity. Other polypeptides are linked to the four prosthetic groups of complex IV, which include cytochrome a, cytochrome a_3, and two copper centers identified as Cu_A and Cu_B. Recent evidence indicates that the copper centers are closely coordinated with the heme groups of the cytochromes in the transfer of electrons from cytochrome c to oxygen. In this transfer, electrons from cytochrome c are evidently initially accepted by a complex between the cytochrome a heme group and Cu_A. The electrons then move to the major oxidation-reduction center of complex IV, consisting of the cytochrome a_3 heme group in association with Cu_B. This center passes the electrons to oxygen, along with H^+ taken up from the matrix, to form H_2O. Like complexes I and III, complex IV extends entirely through the inner mitochondrial membrane.

When isolated and added to artificial phospholipid bilayers, complex IV moves H^+ across the bilayers as it cycles between reduced and oxidized states. This complex therefore provides the third site adding to the H^+ gradient produced by electron transport. In its function as the electron carrier delivering electrons to O_2 in mitochondria, complex IV accounts for 90% of the oxygen consumed by living organisms. The activity of cyanide as a deadly poison depends on its ability to block the function of complex IV in transferring electrons to oxygen. (Table 9-1 summarizes the components of the four electron transport complexes.)

The carriers of the electron transport system were originally conceived as being fixed in the inner boundary membrane, held more or less rigidly in the order of their interactions. More recent evidence indicates that the four major complexes, ubiquinone, and cytochrome c are free to move and interact by random collisions. Some of this evidence comes from studies by C. R. Hackenbrook and his coworkers, who added lipids to isolated mitochondrial membranes to increase the lipid/protein ratio. The effects of the added lipids were followed by both electron microscopy and measurements of the rate of electron transfer. As the phospholipid content was increased from 30% to 700% above normal levels, distances between membrane particles visible in the electron microscope increased proportionately (Fig. 9-25). At the same time the rate of electron transfer from reduced NAD or succinate to oxygen, rather than stopping, decreased in the same proportion. These results directly support the idea that the components of the electron transport are freely

Table 9-1 Complexes of the Mitochondrial Electron Transport System

Complex	Total Molecular Weight	Polypeptide Subunits	Prosthetic Groups	H^+ Transport
I	700,000–850,000	16–26	1 FMN 6 to 9 Fe/S centers	Yes
II	97,000	4–5	1 FAD 3 Fe/S centers	No
III	280,000–300,000	8	2 cytochrome b 1 cytochrome c, 1 Fe/S center	Yes
IV	200,000	10–13	1 cytochrome a 1 cytochrome a_3 2 Cu centers	Yes

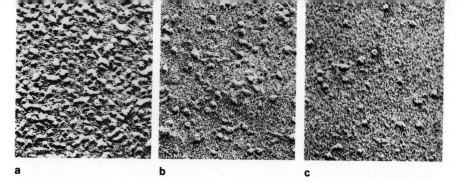

a   b   c

Figure 9-25 Freeze-fracture preparations of inner mitochondrial membranes with **(a)** no added lipids, **(b)** 80% increase in lipids, and **(c)** 700% increase in lipids. The spacing between the integral membrane particles visible in the micrographs increases proportionately with the amount of lipid added to the membranes, indicating that the electron transport complexes are not arranged in rigidly fixed groups. ×200,000. Courtesy of C. R. Hackenbrook, from *Trends Biochem. Sci.* 6:151 (1981).

suspended in the membrane and interact by random collisions. Other evidence from Hackenbrook's laboratory supporting this interpretation comes from the fact that antibodies made against the different complexes can aggregate the complexes separately from the mixture of particles in the inner mitochondrial membrane.

In the random collisions that accomplish electron transport, the molecules of the ubiquinone pool, moving more rapidly because of their relatively small size, collide with the large complexes and act as shuttles transferring electrons from complexes I and II to complex III. Similarly, the cytochrome *c* molecules, also moving more rapidly because of their relatively small size, shuttle electrons from complex III to complex IV. (Cytochrome *c* is bound loosely to the surface of the inner mitochondrial membrane facing the intermembrane compartment.)

H^+ Pumping by the Electron Transport System

The mitochondrial electron transport system pumps H^+ from the matrix into the intermembrane compartment as electrons flow through the carriers from NAD to oxygen (Fig. 9-26a). As noted, three of the carrier complexes—I, III, and IV—when isolated, purified, and inserted into artificial phospholipid membranes, can pump H^+ if supplied with a suitable source of electrons and an electron acceptor. Transport of H^+ by the electron transport system is also clearly demonstrated, as P. C. Hinkle and L. L. Horstman have shown, by work with right-side-out and inside-out vesicles (see p. 171) formed from isolated inner mitochondrial membranes. The sidedness of the vesicles can be easily identified by noting whether the F_oF_1 ATPase extends from the inside or outside surfaces of the vesicle membranes. If the vesicles are right side out, H^+ is pushed to the outside by the transport system (Fig. 9-26b). If inside out, H^+ is pumped into the vesicles (Fig. 9-26c).

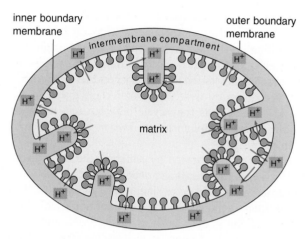

intact mitochondrion

a

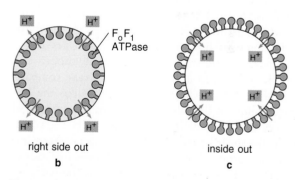

right side out   inside out

b   c

Figure 9-26 The direction in which H^+ is pumped by the electron transport carrier complexes in intact mitochondria **(a)** and vesicles derived from inner mitochondrial membranes **(b and c)**. The "sidedness" of the vesicles can be determined from the orientation of the F_oF_1 ATPase. In right-side-out vesicles **(b)** the F_oF_1 ATPase extends from the membrane toward the vesicle interior, as it does in intact mitochondria. In inside-out vesicles **(c)** the F_oF_1 ATPase extends from the outside surfaces of the vesicle membranes. The arrows show the direction of H^+ pumping.

Measurements of the number of H^+ moved across membranes by isolated complexes and by the intact system indicate that an average of at least three H^+ are pushed across the inner boundary membrane as a pair of electrons moves through complex I, III, or IV. The entire system, therefore, adds at least nine H^+ to the gradient for each pair of electrons traveling from NADH to oxygen. Electron-pairs moving from $FADH_2$ to oxygen, since they bypass complex I, account for a total of six H^+ added to the concentration gradient.

Two more H^+ are removed from the mitochondrial matrix for each pair of electrons delivered to oxygen as part of the reaction producing water:

$$2H^+ + \tfrac{1}{2}O_2 \longrightarrow H_2O \qquad (9\text{-}9)$$

The removal of these two H^+ contributes to the H^+ gradient created by mitochondrial electron transport.

Although the available evidence clearly supports Mitchell's proposal that electron transport creates an H^+ gradient across the inner mitochondrial membrane, the mechanism by which the major complexes add to the H^+ gradient as they cycle between reduced and oxidized states is still uncertain. Two mechanisms have been proposed.

In the first, advanced by Mitchell, the movement of H^+ across the membrane depends on combinations of carriers into *redox loops* (Fig. 9-27). Activity of the redox loops depends on the fact that some electron carriers of the system, such as FAD, FMN, and ubiquinone, carry hydrogen as well as electrons in the reduced state, and others, such as cytochromes and iron-sulfur proteins, carry only electrons. Each time electrons pass from a nonhydrogen to a hydrogen carrier in the pathway, H^+ is picked up from the mitochondrial matrix (Fig. 9-27a). As electrons pass from a dual hydrogen-electron carrier to a nonhydrogen carrier, H^+ is released into the intermembrane compartment (Fig. 9-27b). The net movement builds the H^+ concentration in the intermembrane compartment and reduces it in the matrix, producing an H^+ gradient across the inner boundary membrane as electrons traverse the carriers. Each dual hydrogen-electron carrier is paired with a nonhydrogen carrier in the pathway to form a redox loop that transports H^+ across the membrane. It follows from the hypothesis that each complex able to transport H^+ should contain a redox loop.

The alternate mechanism, known as the *conformational hypothesis*, maintains that the major carrier complexes capable of transporting H^+ act as active transport pumps powered by alternate cycles of oxidation and reduction (Fig. 9-28; compare with Fig. 6-11). According to this idea, proposed by P. D. Boyer and others, reduction of a carrier complex induces a conformational change in one or more peptides of the

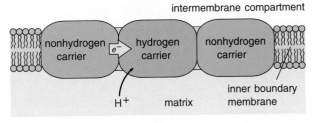

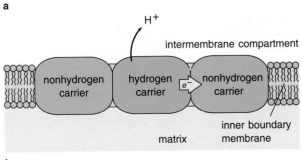

Figure 9-27 The redox loop model for the movement of H^+ from the matrix to the intermembrane compartment, as proposed by Mitchell. Each loop is formed by a dual hydrogen-electron carrier followed by a nonhydrogen carrier, that is, one that transports only electrons. **(a)** As electrons pass from a nonhydrogen to a hydrogen carrier, H^+ is removed from the mitochondrial matrix. **(b)** As electrons pass from a hydrogen carrier to a nonhydrogen carrier, H^+ is expelled into the intermembrane compartment.

complex. The change activates a site that binds H^+ with high affinity on the matrix side of the membrane (Fig. 9-28a and b). Subsequent oxidation of the carrier changes its conformation in such a way that the binding site now faces the opposite side of the membrane (Fig. 9-28c). At the same time the binding site is altered so that its affinity for H^+ is reduced. As a consequence, H^+ is released to the intermembrane compartment (Fig. 9-28d).

Of the two hypotheses the weight of recent experimental evidence clearly favors the conformational hypothesis. One difficulty for Mitchell's hypothesis is the lack of sufficient dual H^+-electron carriers to form redox loops in all parts of the transport system. Of the approximately 20 known carriers, only three—FMN, FAD, and ubiquinone—are dual hydrogen-electron carriers. Of these, only two—FMN and ubiquinone—are in segments of the transport system that actually pump H^+ in response to electron transport. No known hydrogen carriers exist in complex IV, which is capable of pumping H^+ when placed in artificial membranes.

Pumping by conformational changes, on the other hand, has many parallels in active transport. In these systems the movement of ions across membranes has

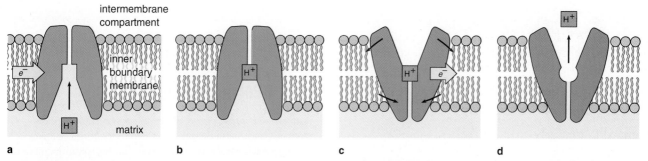

Figure 9-28 The conformational model for H$^+$ pumping by mitochondrial electron transport complexes. Reduction of a carrier complex induces a conformational change in one or more peptides of the complex. The conformational change activates a site that binds H$^+$ with high affinity on the matrix side of the membrane (**a** and **b**). Subsequent oxidation of the carrier changes its conformation in such a way that the binding site now faces the opposite side of the membrane (**c**). In this position the binding site is altered so that its affinity for H$^+$ is reduced, and the H$^+$ is released to the intermembrane compartment (**d**).

been demonstrated to be accompanied by conformational changes in the pump proteins. Among the ions moved by such active transport pumps is H$^+$, pumped by an H$^+$-ATPase that occurs in the plasma membranes of many eukaryotes (see p. 208). It seems likely that conformational changes are also responsible for the movement of H$^+$ across the inner mitochondrial membrane in the electron transport system.

The H$^+$ gradient set up by electron transport provides the energy required for the synthesis of ATP. Measurements of the quantities of ATP synthesized in response to electron flow through the electron transport system, made by A. L. Lehninger and others, established that the H$^+$ gradient produced by travel of one electron pair through the entire pathway from NAD to oxygen is sufficient to drive synthesis of as much as three ATP. The shorter pathway followed by electrons removed from succinate, from the FAD of complex II to oxygen, results in the synthesis of as many as two ATP per electron pair.

Discovery of Mitochondrial Electron Transport

Research leading to discovery of the mitochondrial electron transport system began in the 1920s with the identification by Warburg of an iron-containing enzyme that catalyzes the reduction of oxygen to water in aerobic cells. In 1925 D. Keilen identified the enzyme as a cytochrome. Keilen discovered that other cytochromes also exist in cells, and proposed they act in chains to transfer electrons in stepwise fashion from donor molecules to oxygen. His concept was extended by the research of Green and his coworkers, who found that ubiquinone, iron-sulfur proteins, and copper centers also act as carriers in the chain. The work of Britton

Chance was critical in the assignment of the carriers to a tentative sequence; Green's laboratory and the laboratory of Y. Hatefi were instrumental in the isolation of carriers and the discovery of the major carrier complexes. Research continues today in a worldwide effort to work out the number and types of polypeptides and prosthetic groups in the carriers, their structural arrangements, the pathways followed by electrons within and between the carrier complexes, and the mechanisms by which the carriers transport H$^+$ in response to electron flow.

THE MITOCHONDRIAL F$_o$F$_1$ ATPASE AND ATP SYNTHESIS

The H$^+$ gradient developed through electron transport acts as the direct energy source for the F$_o$F$_1$ ATPase synthesizing ATP from ADP and phosphate. Like the carrier complexes of the electron transport system, F$_o$F$_1$ ATPase consists of a large number of polypeptides, as many as 20 in eukaryotes. These polypeptides are organized into a complex that completely spans the inner mitochondrial membrane.

Structure of F$_o$F$_1$ ATPase

The name given to F$_o$F$_1$ ATPase is based on the fact that the enzyme complex can easily be separated into two subparts, the F$_o$ and F$_1$ *subunits*. (The o in F$_o$ is the small letter o, not zero; see below.) The F$_1$ subunit consists of a group of polypeptides that can be removed from inner mitochondrial membranes in stable, soluble form by mild treatments, such as adjustments in salt concentration of the isolating medium. The F$_o$ subunit,

Divide and Conquer: Deciphering the Functions of a Polypeptide of the Chloroplast ATP Synthase

Richard E. McCarty

RICHARD E. McCARTY received an A.B. degree in Biology and a Ph.D. degree in Biochemistry from The Johns Hopkins University in Baltimore, Maryland. After postdoctoral training at the Public Health Research Institute of the City of New York, he joined the faculty of Cornell University in Ithaca, New York, in 1966. He was Chairman of the Section of Biochemistry, Molecular, and Cell Biology from 1981 to 1985 and Director of the Biotechnology Program from 1988 to 1990. In 1990, Dr. McCarty returned to Johns Hopkins as Professor and Chairman of the Department of Biology. His research is in the area of photosynthesis, with an emphasis on chloroplast membrane proteins.

ATP is required for photosynthesis. The green energy-converting membranes of chloroplasts, known as thylakoid membranes, carry out the synthesis of ATP from ADP and inorganic phosphate. This energy-requiring reaction is driven by the energy-liberating flow of protons down their electrochemical gradient. The electrochemical proton gradient is generated across the thylakoid membrane by light-dependent electron flow in much the same way that electron transport powers proton gradient formation in mitochondria.

The enzyme that couples the synthesis of ATP to proton movement across thylakoid membranes is known as the ATP synthase and also as H^+-ATPase. An ATPase is an enzyme that catalyzes the hydrolysis of ATP, and ATP hydrolysis is the reverse of ATP synthesis. Under special conditions, the ATP synthase can act as an ATPase.

The ATP synthase is remarkable in its structure and complexity. The enzyme is made up of nine different polypeptides ranging in molecular weight from 56,000 to 8,000 and a total of about 20 polypeptide chains. The synthase is thus a rather large protein complex with a molecular weight of approximately 550,000. The enzyme consists of two parts, CF_o and CF_1. CF_o is a collection of rather hydrophobic membrane proteins that anchor CF_1 to the thylakoid membrane. CF_o also functions in the rapid transport of protons across the membrane. CF_1 (molecular weight, 400,000) bears the catalytic sites of the ATP syn-thase and is comprised of a total of nine polypeptide chains and five different kinds of polypeptides. Unlike CF_o, which can be removed from the thylakoid membrane only by destroying the membrane with detergents, CF_1 may be readily removed from membranes and purified. Once removed from the membrane CF_1 is soluble in aqueous solution.

The polypeptide composition of CF_1 is highly unusual. There are three copies each of the larger α and β subunits, but only one copy of the three smaller CF_1 subunits (γ, δ, and ϵ). This composition means that the enzyme is structurally asymmetric. The β subunits, which likely bear the catalytic centers, for example, cannot be in equivalent environments.

An intriguing feature of the ATP synthase is its complexity. What are the functions of the individual polypeptide subunits? Which polypeptides are required for catalytic activity? Which may be involved in proton transport? Which are required for CF_1-CF_o interactions?

A classical biochemical approach to the study of complex systems is to dissect the system into its components and put it all back together again. This approach is sometimes called "resolution and reconstruction analysis," but I prefer "divide and conquer." The idea is quite simple. Find ways to remove polypeptides selectively from the ATP synthase (or parts thereof) and test the depleted enzyme for function. If the depleted ATP synthase is defective, readdition of the polypeptide removed from the enzyme should restore normal function.

The divide-and-conquer approach will be illustrated by the selective removal of the smallest CF_1 subunit, ϵ (molecular weight, 14,700). CF_1 may be isolated and purified in relatively large amounts (200–400 mg) from only 2 kg of a convenient, inexpensive source—spinach leaves purchased at a local market. The chloroplast thylakoid membrane is by far the most prevalent membrane in green leaves and, thus, is very easy to prepare. The leaves are ground in a blender with the buffered sucrose solution and, after filtration to remove debris, the homogenate is centrifuged. Thylakoid membranes are large and relatively dense. Thus, they sediment much more readily than all of the other membranes in the homogenate. To remove contaminating soluble proteins, the thylakoid membranes are extensively washed by repeated resuspension in fresh buffer and centrifugation to pellet the membranes.

Dilution of the washed membranes into a medium of very low salt concentration that also contains a reagent that binds Mg^{2+} causes CF_1 to pop off the thylakoid membranes. The membranes are removed and the protein solution, which is at least 50% CF_1, is concentrated. Further purification is achieved by ion exchange chromatography.

A fascinating feature of the chloroplast ATP synthase is that it is virtually inactive in the dark, but very active in the light. In the light the enzyme makes ATP. In the dark, the enzyme should, in principle, catalyze the reverse reaction, ATP hydrolysis. It does not. Elaborate mechanisms have evolved that prevent ATP hydrolysis from occurring. If the ATP synthase were active in the dark in ATP hydrolysis, it would deplete the ATP pools in both the chloroplast and cytoplasm of leaf cells and strongly inhibit metabolism. In the light, the enzyme operates as an ATP synthase not as an ATPase because the electrochemical proton gradient drives the reaction in the synthetic direction.

We were interested in finding out more about the ways in which the ATP synthase is switched on in the light and off in the dark. That is, how is the enzyme interconverted between inactive and active forms? One question of interest was, which of the polypeptides of the ATP synthase are involved in this regulation of activity?

We were fortunate in that CF_1 in solution has properties in common with membrane-bound CF_1. CF_1 has low ATPase activity. Several treatments, including the incubation of CF_1 with certain detergents or organic solvents, increase the ATPase activity markedly. Could this activation be the result of either the denaturation or release of an inhibitory subunit?

We decided to test this idea. CF_1 was bound to a column of cellulose to which positively charged diethylaminoethyl groups were attached. CF_1 is negatively charged and sticks tightly to the column at low salt concentrations. The column was then eluted with a buffered solution containing 20% (by volume) ethanol and 30% (by volume) glycerol. This concentration of ethanol was chosen since work in other laboratories had established that 20% ethanol gave the maximum stimulation of the ATPase activity of CF_1.

We were delighted to find[1] that the ethanol-glycerol buffer caused the elution of the ϵ subunit of CF_1. If ethanol were omitted from the elution buffer, no ϵ was eluted. CF_1 now depleted in ϵ ($CF_1(-\epsilon)$) could then be washed from the column by elution with a buffer of high ionic strength.

The $CF_1(-\epsilon)$ had very high ATPase activity. Also, the addition of small amounts of purified ϵ to the $CF_1(-\epsilon)$ strongly inhibited the ATPase. This experiment shows rather conclusively that the ϵ subunit is an inhibitor of the enzyme. Having divided the enzyme, we attempted to conquer the system. Our ability to remove ϵ selectively allowed us to determine the binding stoichiometry of ϵ (one ϵ per CF_1) and to gain information about the position of ϵ within the CF_1 structure.

It was also now possible to show that the ϵ subunit is absolutely required for ATP synthesis. Thylakoid membranes from which CF_1 had been removed cannot catalyze ATP synthesis. Adding CF_1 to the depleted membranes restores ATP formation. $CF_1(-\epsilon)$ binds to the membranes (really to CF_0) equally as well as CF_1, but does not restore ATP synthesis.

It was a surprise that an inhibitor of the ATP synthase is required for ATP synthesis. Logic dictates that an inhibitor cannot block only one direction of a reversible reaction. This, however, appeared to be the case. The ϵ subunit seemed to inhibit ATP hydrolysis, but not ATP synthesis. We are still uncertain as to the exact mechanism of this unusual regulation. However, polyclonal antibodies prepared against the ϵ polypeptide have given some clues. The incubation of thylakoid membranes with an anti-ϵ serum in the dark had no effect on either ATP synthesis or hydrolysis. If, however, the membranes were incubated in the light with the antiserum, ATP synthesis was inhibited. The antibodies actually removed part of the ϵ subunit from the CF_1 in the membrane.[2]

This result suggests that illumination of thylakoid membranes causes changes in the structure of CF_1 that expose the ϵ subunit to the antibodies. The effect of light is indirect. Very likely, the enzyme senses the electrochemical proton gradient generated by light-dependent electron flow and alters its structure. In particular, the interactions between the inhibitory ϵ subunit and other CF_1 polypeptides are changed. It seems plausible that these changes overcome the inhibitory effects of ϵ. CF_1 is simply not the same enzyme in the dark as it is in the light.

The divide-and-conquer approach has given new insights into the functions of the ϵ subunit of CF_1. In the future we will attempt to use recombinant DNA methods to engineer mutant forms of ϵ. In this way we will be able to exploit our reconstitution systems to define further the roles of the ϵ subunit in the regulation of ATP synthesis in chloroplasts.

References

[1] Richter, M. L.; Patrie, W. P.; and McCarty, R. E. *J. Biol. Chem.* 259:7371 (1984).

[2] Richter, M. L., and McCarty, R. E. *J. Biol. Chem.* 262:15037 (1987).

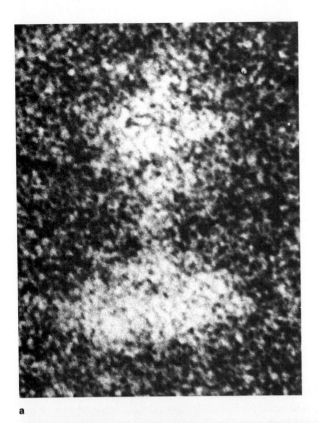

a

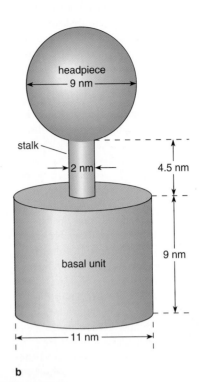

headpiece
9 nm

stalk

2 nm

4.5 nm

basal unit

9 nm

11 nm

b

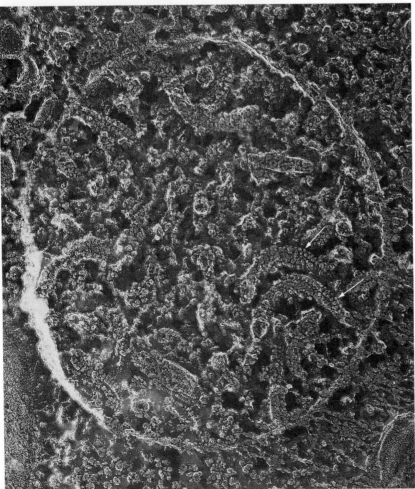

c

Figure 9-29 The F_oF_1 ATPase complex. **(a)** An F_oF_1 ATPase complex isolated from a mitochondrion and prepared for electron microscopy by negative staining. ×4,000,000. Courtesy of R. L. Pedersen, The Johns Hopkins University, from *J. Biological Chem.* 254:11170 (1979). **(b)** Overall structure and dimensions of the F_oF_1 ATPase. **(c)** F_oF_1 ATPase units (arrows) in the inner mitochondrial membranes of *Paramecium*. ×104,000. Courtesy of R. D. Allen, C. C. Schroeder, and A. K. Fok; reproduced from *J. Cell Biol.* 108:2233 (1988) by copyright permission of the Rockefeller University Press.

in contrast, is built up from polypeptides that can be removed from the inner membranes only by harsh treatments, such as the use of detergents, that destroy the membrane. The polypeptides of the F_o subunit are therefore suspended in the nonpolar membrane interior.

The two subunits can be identified with major substructures of F_oF_1 ATPase visible in the electron microscope as a lollipop-shaped particle with a clearly defined *headpiece* connected to a *basal unit* by a *stalk* (Fig. 9-29). When F_1 subunits are removed from inner mitochondrial membranes, the headpiece and stalk disappear from preparations made for electron microscopy. Therefore, the headpiece and stalk contain the F_1 subunit, and the segment remaining in the membrane contains the F_o subunit.

The active site of ATP synthesis was identified with the F_1 subunit primarily in the laboratory of Racker. Racker's group first prepared inner mitochondrial membranes, employing conditions that produced sealed inside-out vesicles (Fig. 9-30a). These preparations, which clearly showed headpieces and stalks extending from the vesicles when prepared for electron microscopy, were fully capable of ATP synthesis. Treatment of the isolated membranes with reagents that removed the headpieces and stalks from the membranes (Fig. 9-30b) yielded vesicles that could no longer make ATP. The removed headpieces and stalks, when separately

purified, were found to catalyze the reversible reaction synthesizing ATP:

$$ADP + P_i \rightleftharpoons ATP + H_2O \qquad (9\text{-}10)$$

Returning these particles to the inner mitochondrial membranes (Fig. 9-30c) restored their ability to synthesize ATP. Racker called the purified headpiece-stalk particles *coupling factor 1*, later shortened to the F_1 subunit of the ATPase complex. Analysis of *E. coli* F_1 subunits (Tables 9-2 and 9-3), which are the best characterized, indicates that they are constructed from five different polypeptides, designated as the α-, β-, γ-, δ-, and ϵ-*polypeptides* of the headpiece and stalk. Two of these, the α- and β-polypeptides, occur in three copies each in the headpiece; the remaining polypeptides occur in single copies in the stalk (Fig. 9-31). The minimum combination of subunits able to work fully as an ATPase appears to be the $\alpha_3\beta_3\gamma$-combination in *E. coli*. (Tentative functions for the *E. coli* polypeptides are listed in Table 9-3.)

Mitochondrial F_1 subunits are assembled from similar polypeptides, with closest sequence relationships to the *E. coli* headpiece noted for the α-, β-, and γ-polypeptides (see Table 9-2). Less similarity to the bacterial F_1 is noted for three additional polypeptides—OSCP, δ, and ϵ—occurring as part of the mitochondrial

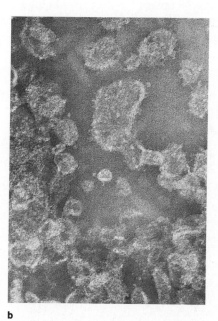
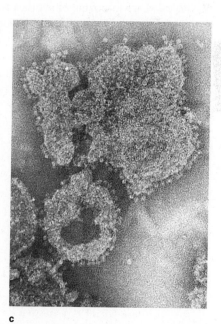

a   b   c

Figure 9-30 The Racker experiment equating the headpieces of F_oF_1 ATPase complexes with ATPase activity. **(a)** Inner mitochondrial membranes prepared as inside-out vesicles retaining headpieces and stalks of F_oF_1 ATPase complexes (arrows). This preparation was fully capable of ATP synthesis. **(b)** Removal of the headpieces and stalks from the membranes yielded vesicles that could no longer make ATP. The removed headpieces and stalks, when separately purified, were found to catalyze ATP hydrolysis. **(c)** Returning the headpieces and stalks to the membranes restored their ability to synthesize ATP. × 130,000. Courtesy of E. Racker.

Table 9-2 Polypeptide Subunits of the *E. coli* and Mitochondrial F_oF_1 ATPases

E. coli F_oF_1 ATPase	Mitochondrial F_oF_1 ATPase
F_1 subunit	
α	α
β	β
γ	γ
δ	OSCP
ϵ	δ
	ϵ
F_o subunit	
a	a (subunit 6)[1]
b	b (subunit 4)
c	c (subunit 9)

Additional polypeptides detected in the mitochondrial F_o subunit:
subunit 8, ATPase inhibitor, 9000d protein, 15,000d protein, factor B, F6, subunit d

[1] Terms in parentheses are additional names for the same polypeptide.

headpiece. The Experimental Process essay by R. E. McCarty on page 338 describes a major approach used to assign functions to subunits of the chloroplast ATPase, which is closely similar to its mitochondrial and bacterial counterparts.

F_o channels inserted into membranes introduce an H^+ leak and are incapable of synthesizing ATP unless capped by the stalk proteins and headpiece polypeptides. The F_o subunit is therefore considered to supply

an H^+-conducting channel for the F_oF_1 ATPase. Conductance of H^+ through the channel is relatively slow—about 40 to 80 ions per second in the bacterial F_o subunit and about 400 ions per second in the mammalian F_o. Conductance at these rates indicates that the channel probably binds H^+ as part of the mechanism moving the ion and works as a carrier rather than a simple, open pore.

The *E. coli* F_o unit has three polypeptides, designated a, b, and c. Equivalents to the a, b, and c polypeptides occur in mitochondrial F_o units (see Table 9-2), along with as many as seven additional polypeptide subunits. Of the F_o subunits, research with *E. coli* indicates that the a, b, and c polypeptides are most likely to contribute to channel formation. Six to 12 of the c polypeptides, arranged in a circle, are believed to form the channel (see Fig. 9-31). The a and b polypeptides may stabilize the channel; the b polypeptide, which occurs in two copies in the bacterial F_o subunit, is also considered likely to bind the channel to the stalk portion of the F_1 unit. The other peptides of the mitochondrial F_o subunit may act as factors regulating the channel.

The F_o subunit gets its name from *oligomycin*, a drug that acts as an inhibitor of the ATPase complex. Racker's group found that of the two ATPase subunits, oligomycin binds to the portion firmly attached to the membrane. This subunit was subsequently termed the *oligomycin-sensitive factor* or F_o.

In the *E. coli* F_oF_1 ATPase the polypeptide subunits occur in the combination $\alpha_3\beta_3\gamma\delta\epsilon ab_2c_{6-10}$, with ten copies of the c subunit considered as the most likely num-

Table 9-3 Structure and Function of the *E. coli* F_oF_1 ATPase Polypeptides

Polypeptide or Subunit	Amino Acid Residues	Molecular Weight	Function
F_1 subunit	3517	381,759	
α	513	55,259	ATP/ADP binding site; promotes ATP synthesis by β-subunit
β	459	50,117	Contains site that catalyzes ATP synthesis
γ	286	31,414	May stabilize F_o-F_1 assembly; necessary for full catalytic activity of β-subunit; may regulate flow of H^+ through proton channel
δ	177	19,303	Required for F_o-F_1 assembly; primary component of stalk
ϵ	138	14,914	Required for F_o-F_1 assembly; inhibits ATP hydrolysis by β-subunit; binds δ-subunit
F_o subunit	1373[1]	147,223[1]	
a	271	30,275	Stabilizes H^+ channel
b	156	17,244	Stabilizes H^+ channel; binds δ-subunit
c	79	8,246	Forms H^+ channel

Complete F_oF_1 ATPase: 4890 amino acids; molecular weight = 528,982

SOURCE: Residue counts and molecular weights from data presented in M. Futai and H. Kanazawa, *Microbiol. Rev.* 47:285 (1983).
[1] Assumes ten copies of the c polypeptide in the F_o subunit.

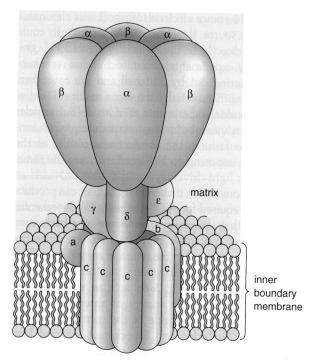

Figure 9-31 A possible arrangement of the polypeptide subunits of F_oF_1 ATPase. α, β, γ, δ, and ϵ are subunits of the headpiece and stalk (F_1); a, b, and c are subunits of the basal unit (F_o). The subunits shown are those of the *E. coli* F_oF_1 ATPase; mitochondrial F_oF_1 ATPase complexes have three additional subunits in the basal unit. The β-subunits contain the sites converting ADP to ATP; a circle of 6 to 10 c subunits probably forms the channel conducting H^+ through the inner mitochondrial membrane.

The Mechanism of ATP Synthesis by F_oF_1 ATPase

Chapter 6 describes the active transport pumps that directly hydrolyze ATP as an energy source to move ions and other substances across membranes against their concentration gradients (see p. 206). These active transport pumps, which have enzymatic activity as ATPases, remove a phosphate group from ATP when working in the forward direction:

$$ATP + H_2O \longrightarrow ADP + P_i \qquad (9\text{-}11)$$

Many of the ATP-dependent active transport pumps, especially those transporting ions, can be reversed if the concentration of the ion is raised to very high levels on the side of the membrane toward which it is normally pumped. Under these conditions the ion moves backward through the membrane channel, and the active transport pump *adds phosphate groups to ADP to form ATP.*

Probably this is also what happens in the F_oF_1 ATPase of mitochondrial membranes. The very high H^+ gradient set up by the electron transport system "overpowers" the F_oF_1 ATPase, which, if working in the forward direction, would hydrolyze ATP to pump H^+ in the same direction as the electron transport system. When pushed in the reverse direction by the high H^+ gradient, the F_oF_1 ATPase instead synthesizes ATP from ADP and phosphate:

$$ADP + P_i \longrightarrow ATP + H_2O \qquad (9\text{-}12)$$

The free energy required to run this reaction to the right—uphill—is supplied by the H^+ gradient. The likelihood that pumping reversal underlies ATP synthesis by F_oF_1 ATPase is strengthened by the fact that bacterial F_oF_1 ATPase works regularly in both directions. The V-type ATPase pumps, which are structurally similar to F_oF_1 ATPase (see p. 210), evidently work entirely in the forward direction as H^+-ATPase pumps.

The details of the molecular mechanism by which F_oF_1 ATPase converts the energy of an H^+ gradient into ATP synthesis remain unknown. It is considered likely that movement of H^+ through the channel of the F_o subunit induces a conformational change that is transmitted through the stalk, activating catalysis of ATP synthesis by the headpiece. This view is supported by extensive conformational changes that can actually be detected in F_oF_1 ATPase as ATP synthesis proceeds.

Each of the three β-polypeptides evidently contains an active site that both binds ADP and phosphate and synthesizes ATP. These sites are believed to work in a cooperative fashion so that, at any instant, each site is at a different stage of the reactions binding ADP and

ber. In 1983 J. E. Walker, M. Futai, and their colleagues successfully isolated the gene complex coding for the entire group of polypeptides forming F_oF_1 ATPase in *E. coli*. Sequencing the genes allowed deduction of the amino acid sequence and molecular weight of each peptide of the *E. coli* complex (see Table 9-3). Assuming that ten copies of the c polypeptides occur in the F_o subunit, the complete F_oF_1 ATPase contains 4890 amino acid residues and has a total molecular weight of 528,982, placing it among the largest and most complex enzymes known. S. D. Dunn and L. A. Heppell and E. Schneider and K. Altendorf successfully reassembled the *E. coli* F_oF_1 ATPase in functional form from its polypeptide subunits.

The F_oF_1 ATPase is a member of a family of H^+-ATPase pumps that includes the V-type pumps located in vesicle membranes in the cytoplasm of eukaryotic cells (see p. 208). Each pump in this family includes headpieces, stalks, and membrane segments of similar structure, assembled from related proteins.

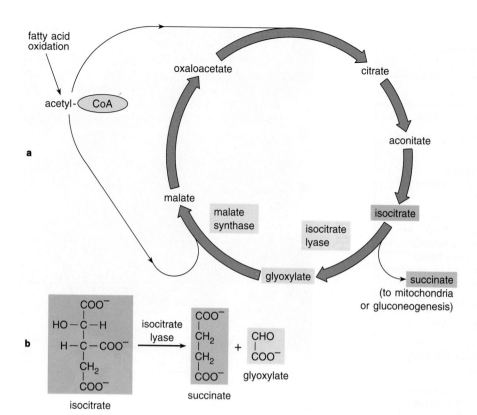

Figure 9-38 The glyoxylate cycle. **(a)** In the cycle acetyl-coenzyme A units produced by fatty acid oxidation combine with oxaloacetate to form citrate, as in the citric acid cycle. Isocitrate is then produced by interconversion between citrate and aconitate, also as in the citric acid cycle. From this point on the glyoxylate cycle diverges from the citric acid cycle through the action of two enzymes that are unique to microbodies. One, *isocitrate lyase*, splits isocitrate into succinate and glyoxylate **(b)**. Glyoxylate then interacts with a second molecule of acetyl-coenzyme A to produce malate. This reaction is catalyzed by the second enzyme unique to glyoxisomes, *malate synthase*. Malate is oxidized to oxaloacetate by the same enzyme oxidizing this substance in the citric acid cycle, malate dehydrogenase, and the cycle is ready to begin again.

of blood fat in persons with faulty lipid metabolism. One of the primary effects associated with many of the drugs is a marked increase in the number of microbodies in liver cells. Normal persons living on a high-fat diet also show a similar increase in liver cell microbodies. These observations led to a search for enzymes of lipid metabolism in microbodies, which soon turned up all the enzymes and intermediates of fatty acid oxidation. The microbodies of some species also contain the enzymes necessary for the initial breakdown of fats.

The fatty acid oxidation pathway in microbodies is almost identical to the mitochondrial pathway described in Supplement 9-1 (see Figs. 9-40 and 9-41 and p. 360). In microbodies the pathway increases the cell's capacity to oxidize fats. It also apparently increases the efficiency with which long-chain fatty acids can be oxidized. Fatty acid chains longer than 10 to 12 carbons are oxidized relatively slowly by mitochondria. In microbodies, in contrast, fatty acid chains as long as 22 carbons are rapidly oxidized. The long chains are clipped two carbons at a time by the peroxisomal pathway until they reach the shorter 10- to 12-carbon length that can be oxidized by mitochondria; in many species the shortened fatty acid chain is then transferred from microbodies to mitochondria for further oxidation.

After discovery of the fatty acid oxidation pathway in liver cell microbodies, the same enzymatic capabilities were identified in the microbodies of protozoa and plants. The pathway is particularly important in oily seeds such as the peanut and castor bean, where fatty

acid oxidation provides an important link in pathways converting stored fats into glucose and other carbohydrates via the glyoxylate cycle. In humans defects in microbody formation and function underlie a group of diseases characterized by faulty metabolism of long-chain fatty acids.

Microbodies and the Glyoxylate Cycle: Glyoxisomes

The glyoxylate cycle, which occurs typically in the microbodies of plant seeds, represents a partial circuit around the citric acid cycle in which some reactions are included and others bypassed. (Fig. 9-38 outlines the cycle.) Each turn of the cycle converts two molecules of acetyl-coenzyme A into one molecule of succinate. The succinate may diffuse from microbodies to mitochondria, where it can enter the citric acid cycle. Alternatively, it may serve as an indirect input for reactions that synthesize glucose by gluconeogenesis (see p. 317), thus leading to the generation of a sugar from the fat originally entering oxidation in the microbody. The gluconeogenic pathway is the probable fate of most of the succinate produced in the microbodies of oily plant seeds, where acetyl-coenzyme A yielded by fatty acid oxidation is the primary input for the cycle. Through this pathway the stored fat of the seeds contributes to products formed from glucose, such as the cellulose of new cell walls. Microbodies containing the glyoxylate cycle are frequently termed *glyoxisomes.*

Glycolysis and Gluconeogenesis

Burchell, A. 1990. Molecular pathology of glucose-6-phosphatase. *FASEB J.* 4:2978–88.

Granner, D., and Pilkis, S. 1990. The genes of hepatic glucose metabolism. *J. Biolog. Chem.* 265:10173–76.

Hers, G. A., and Hue, L. 1983. Gluconeogenesis and related aspects of glycolysis. *Ann. Rev. Biochem.* 52:617–53.

Masters, C. J.; Reid, S.; and Don, M. 1987. Glycolysis—new concepts in an old pathway. *Molec. Cellular Biochem.* 76:3–14.

Pilkis, S. J., and Granner, D. K. 1992. Molecular Physiology of the regulation of hepatic gluconeogenesis and glycolysis. *Ann. Rev. Physiol.* 54:885–909.

Mitochondrial Electron Transport System

Babcock, G. T. and Wikstrom, M. 1992. Oxygen activation and the conservation of energy in cell respiration. *Nature* 356:301–9.

Beinert, H. 1990. Recent developments in the field of Fe-S proteins. *FASEB J.* 4:2483–91.

Brink, J.; Boekema, E. J.; and van Bruggen, E. F. J. 1988. Electron microscopy and image analysis of the complexes I and V of the mitochondrial respiratory chain. *Electron Micr. Rev.* 1:175–99.

Cooper, C. E.; Nicholls, P.; and Freedman, J. A. 1991. Cytochrome *c* oxidase: Structure, function, and membrane topology of the polypeptide subunits. *Biochem. Cell Biol.* 69:586–607.

Lenas, G. 1988. Role of mobility of redox components in the inner mitochondrial membrane. *J. Membr. Biol.* 104:193–209.

Ohnishi, T. 1987. Structure of the succinate-ubiquinone oxidoreductase (complex IV). *Curr. Top. Bioenerget.* 15:37–65.

Patel, M. S., and Roche, T. E. 1990. Molecular biology and biochemistry of pyruvate dehydrogenase complexes. *FASEB J.* 4:3224–33.

Ragan, C. I. 1987. Structure of NADH-ubiquinone reductase (complex I). *Curr. Top. Bioenerget.* 15:1–36.

Weiss, H. 1987. Structure of mitochondrial ubiquinol-cytochrome c reductase (complex II). *Curr. Top. Bioenerget.* 15:67–90.

Wilson, M. T., and Bickar, D. 1991. Cytochrome oxidase as a proton pump. *J. Bioenerget. Biomembr.* 23:755–71.

Mechanism of H$^+$ Pumping

Capaldi, R. A. 1990. Structure and function of cytochrome *c* oxidase. *Ann. Rev. Biochem.* 59:569–96.

Rich, P. R. 1986. A perspective on Q-cycles. *J. Biomembr. Bioenerget.* 18:145–56.

F$_o$F$_1$ ATPase

Brink, J.; Boekema, E. J.; and van Bruggen, E. F. J. 1988. Electron microscopy and image analysis of the complexes I and V of the mitochondrial respiratory chain. *Electron Micr. Rev.* 1:175–99.

Futai, M.; Noumi, T.; and Maeda, M. 1989. ATP synthase (H$^+$-ATPase): results by combined biochemical and molecular biology approaches. *Ann. Rev. Biochem.* 58:111–36.

Glaser, E., and Norling, B. 1991. Chloroplast and plant mitochondrial ATP syntheses. *Curr. Top. Bioenerget.* 16:223–63.

Hashimoto, T.; Yoshida, Y.; and Tagawa, K. 1990. Regulator proteins of F$_o$F$_1$ ATPase: role of ATPase inhibitor. *J. Bioenerget. Biomembr.* 22:27–38.

Nagley, P. 1988. Eukaryotic membrane genetics: the F$_o$ sector of mitochondrial ATP synthase. *Trends Genet.* 4:46–52.

Ysern, X.; Amzel, L. M.; and Pedersen, P. L. 1988. ATP synthesis: structure of the F$_1$-moiety and its relationship to function and mechanism. *J. Bioenerget. Biomembr.* 20:423–68.

Mechanism of ATP Synthesis by F$_o$F$_1$ ATPase

Boyer, P. D. 1988. A perspective of the binding change mechanism for ATP synthesis. *FASEB J.* 3:2164–78.

Slater, E. C. 1987. The mechanism of the conservation of energy of biological oxidations. *Eur. J. Biochem.* 166:489–504.

Mitochondrial Transport

Benz, R. 1987. Porin from bacteria and mitochondrial outer membranes. *CRC Crit. Rev. Biochem.* 19:145–90.

Kinnally, K. W.; Antonenko, Y. N.; and Zorov, D. B. 1992. Modulation of inner mitochondrial membrane channel activity. *J. Bioenerget. Biomembr.* 24:99–110.

Mannella, C. A., and Tedeschi, H. 1992. The emerging picture of mitochondrial membrane channels. *J. Bioenerget. Biomembr.* 24:3–5.

Human Diseases Due to Mitochondrial Deficiencies

Scholte, H. R. 1988. The biochemical basis of mitochondrial diseases. *J. Bioenerget. Biomembr.* 20:161–91.

Wallace, D. C. 1992. Diseases of the mitochondrial DNA. *Ann. Rev. Biochem.* 61:1175–1212.

Microbodies

Borst, P. 1989. Peroxisome biogenesis revisited. *Biochim. Biophys. Acta* 1008:1–13.

van den Bosch, H.; Schutgens, R. B. H.; Wanders, R. J. A.; and Tager, J. N. 1992. Biochemistry of peroxisomes. *Ann. Rev. Biochem.* 61:157–97.

deDuve, C. 1983. Microbodies in the living cell. *Sci. Amer.* 248:74–84 (May).

Moser, H. W.; Bergin, A.; and Cornblath, D. 1991. Peroxisomal disorders. *Biochem. Cell Biol.* 69:463–74.

Osumi, T., and Hashimoto, T. 1984. The inducible fatty acid oxidation system in mammalian peroxisomes. *Trends Biochem. Sci.* 9:317–19.

Tolbert, N. E. 1981. Metabolic pathways in peroxisomes and glyoxisomes. *Ann. Rev. Biochem.* 50:133–57.

Thermogenesis in Brown Fat Mitochondria and Plants

Cannon, B., and Nedergaard, J. 1985. The biochemistry of an inefficient tissue: brown adipose tissue. *Essays Biochem.* 20:110–64.

Himms-Hagen, J. 1990. Brown adipose tissue thermogenesis: interdisciplinary studies. *FASEB J.* 4:2890–98.

Klingenberg, M. 1990. Mechanism and evolution of the uncoupling protein of brown adipose tissue. *Trends Biochem. Sci.* 15:108–12.

Riquier, D.; Castiella, L.; and Bonilland, F. 1991. Molecular studies of the uncoupling protein. *FASEB J.* 5:2237–42.

Oxidative Reactions in Prokaryotes

Anraku, Y. 1988. Bacterial electron transport chains. *Ann. Rev. Biochem.* 57:101–32.

———, and Gennis, R. B. 1987. The aerobic respiratory chain of *E. coli. Trends Biochem. Sci.* 12:262–66.

Curtis, S. E. 1988. Structure, organization, and expression of cyanobacterial ATP synthase genes. *Photosynth. Res.* 18:223–44.

Futai, M.; Noumi, T.; and Maeda, M. 1988. Molecular genetics of F_1-ATPase from *Escherichia coli. J. Bioenerget. Biomembr.* 20:31–58.

———; ———; and ———. 1989. ATP synthase (H^+-ATPase): results by combined biochemical and molecular biology approaches. *Ann. Rev. Biochem.* 58:111–36.

Gabellini, N. 1988. Organization and structure of the genes for the cytochrome b/c_1 complex in purple photosynthetic bacteria. A phylogenetic study describing the homology of the b/c_1 subunits between prokaryotes, mitochondria, and chloroplasts.

Scherer, S.; Almon, H.; and Boger, P. 1988. Interaction of photosynthesis, respiration, and nitrogen fixation in cyanobacteria. *Photosynth. Res.* 15:95–114.

Senior, A. E. 1990. The proton-translocating ATPase of *E. coli. Ann. Rev. Biophys. Biophys. Chem.* 19:7–41.

Skulachev, V. P. 1989. The Na^+ cycle: a novel type of bacterial energetics. *J. Bioenerget. Biomembr.* 21:635–47.

Review Questions

1. Define oxidation and reduction. What are redox potentials?

2. What are the sources of electrons in cells? How is the energy of these electrons tapped off in electron transport systems? How is this energy used to drive the synthesis of ATP?

3. What is substrate-level phosphorylation? Oxidative phosphorylation?

4. Outline the sequence of events in glycolysis. What are the net inputs and outputs of the glycolytic sequence? List some of the mechanisms regulating glycolysis. Where does glycolysis take place?

5. What is fermentation? What advantages does the ability to carry out fermentations have for cells? Can cells live by fermentations alone? Do human cells carry out fermentations?

6. What substances can act as inputs to glycolysis? What is gluconeogenesis?

7. What are aerobes? Anaerobes? Facultative anaerobes? Strict anaerobes?

8. Outline the chemical inputs and outputs of pyruvate oxidation.

9. Outline the chemical inputs and outputs of the citric acid cycle. List some of the mechanisms regulating the citric acid cycle. Where do the citric acid cycle and pyruvate oxidation take place?

10. Compare the structure and function of ATP, NAD, NADP, FAD, and coenzyme A.

11. What is a cytochrome? A heme group? The prosthetic group of a flavoprotein? An iron-sulfur center? A quinone? How are these prosthetic groups placed in a tentative sequence in electron transport systems?

12. Outline the structure of complexes I, II, III, and IV of the mitochondrial electron transport system. Where are these complexes located in mitochondria?

13. How are electron transport complexes believed to set up an H^+ gradient in response to electron flow? Compare the redox-loop and conformational hypotheses as explanations for H^+ pumping by the electron transport system.

14. Outline the structure of F_oF_1 ATPase of mitochondria. Where is F_oF_1 ATPase located in mitochondria? How is F_oF_1 ATPase believed to use an H^+ gradient as its energy source for the synthesis of ATP?

15. How many molecules of ATP are synthesized through the complete oxidation of a molecule of glucose to CO_2 and H_2O? (Consider that the H^+ gradient set up by electron transport is used at maximum efficiency for the synthesis of ATP.)

16. Consider that the maximum efficiency of glucose oxidation to CO_2 and H_2O is 39%. What proportion of this efficiency is contributed by glycolysis? By pyruvate oxidation? By the citric acid cycle? (Assume that all electrons accepted by NAD or FAD are delivered to oxygen.)

17. Why do electrons carried by NAD potentially result in the synthesis of greater amounts of ATP than those carried by FAD?

18. Outline the relationship between ADP/ATP concentrations and the rate of mitochondrial oxidations.

19. Outline the major evidence supporting the chemiosmotic hypothesis.

20. Where and how are proteins oxidized?

21. How does the ATP made inside mitochondria get outside? How do the electrons carried by NAD outside mitochondria get inside? What is the energy source for active transport of substances in or out of mitochondria? What kinds of molecules are transported through mitochondrial membranes? What are mitochondrial porins?

22. Outline the structure and function of the membranes and compartments of mitochondria.

23. What are microbodies? Peroxisomes? Glyoxisomes? What enzymatic reactions are common to most microbodies? What important links do microbodies supply in fatty acid oxidation? In gluconeogenesis? What is the glyoxylate cycle? The glycolate pathway? How are both pathways related to photosynthesis?

Supplement 9-1
Fatty Acid Oxidation

Fats (triglycerides), which consist of three fatty acid chains linked to a glycerol backbone (see Fig. 2-12), are important energy sources in all living organisms. In the pathways oxidizing these substances, fats are first hydrolyzed in the cytoplasm outside mitochondria into separate glycerol and fatty acid molecules. Glycerol is converted indirectly into 3PGAL, one of the intermediates of the glycolytic pathway (see p. 315).

The long fatty acid chains produced by fat breakdown enter mitochondria via the carnitine shuttle (see Fig. 9-35). As part of the operation of this shuttle, fatty acid chains are linked to coenzyme A and delivered to the mitochondrial interior as fatty acyl-coenzyme A complexes. Attachment of the fatty acid chain to coenzyme A is driven by the breakdown of ATP to AMP:

$$\text{fatty acid} + \text{ATP} + \text{coenzyme A} \longrightarrow$$
$$\text{fatty acyl-coenzyme A} + \text{AMP} + 2P_i \quad (9\text{-}19)$$

Once in the mitochondrial matrix, fatty acyl-coenzyme A serves as the input for the reactions of the fatty acid oxidation pathway.

The pathway shortens the fatty acid chain two carbons at a time, yielding units of acetyl-coenzyme A as products (Fig. 9-40). If the fatty acid is saturated (with no double bonds between its carbons) and contains an even number of carbon atoms, the reaction series resembles the segment of the citric acid cycle between succinate and oxaloacetate; that is, an oxidation, addition of the elements of a molecule of water, and a second oxidation occur in sequence.

In reaction 1 of the pathway, two electrons and two hydrogens are removed from a fatty acyl-coenzyme A unit. FAD is the electron acceptor for the reaction, which introduces a double bond between the

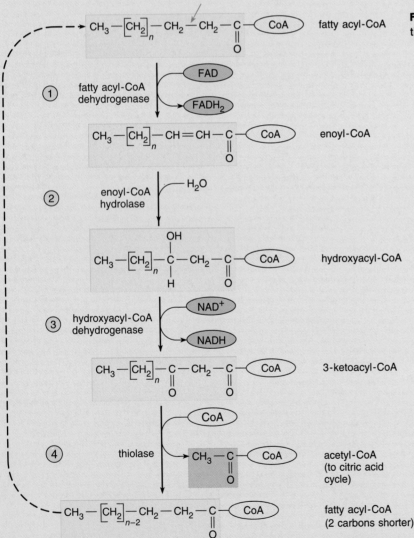

Figure 9-40 The pathway of fatty acid oxidation inside mitochondria (see text).

two carbon atoms involved (red arrow). In reaction 2 the product of the first step is hydrated by the addition of the elements of an H_2O molecule, eliminating the double bond and preparing the group for the second oxidation. The product is oxidized in reaction 3, with another electron pair and two hydrogens being removed. NAD^+ is the electron acceptor.

Much of the energy released in the second oxidation is conserved in the structure of the product, *3-ketoacyl-coenzyme A*. Most of this energy is transferred to a second molecule of coenzyme A in reaction 4. In this step the last two carbons of the chain are split off and attached to coenzyme A, forming acetyl-coenzyme A. The other product of the reaction is the remaining fatty acid chain attached to the second molecule of coenzyme A. This product is identical to the fatty acyl-coenzyme A complex entering the first reaction of the series except that it is now two carbons shorter.

The shortened fatty acyl-coenzyme A complex then repeats the four steps, which split off another two carbons and produce another molecule of acetyl-coenzyme A. The pathway continues to operate until the entire fatty acid chain is split into two-carbon acetyl units attached to coenzyme A. The acetyl-coenzyme A produced by the pathway directly enters the citric acid cycle in the same manner as acetyl-coenzyme A units originating from pyruvic acid oxidation. As net products one cycle of the overall reaction sequence yields:

fatty acyl-coenzyme A + coenzyme A +

$\qquad$ FAD + NAD^+ $\longrightarrow$

fatty acyl-coenzyme A (2 carbons shorter) +

acetyl-coenzyme A + $FADH_2$ + NADH $\quad$ (9-20)

Unsaturated fatty acids with even numbers of carbon atoms are oxidized by essentially the same sequence. However, wherever the fatty acyl groups contain a double bond, the sequence skips the first step and proceeds directly to the second, missing the initial oxidation. As a result, unsaturated fatty acids reduce one FAD less for each double bond in their carbon chains.

Odd-numbered fatty acid chains are oxidized into two-carbon units by the same sequences as even-numbered fatty acids until the last cycle of the pathway. At this point the final turn of the cycle splits the remaining five-carbon fatty acid into one molecule of acetyl-coenzyme A and a three-carbon segment (a *proprionyl* group), also attached to coenzyme A:

fatty acyl-coenzyme A $\longrightarrow$
$\quad$ (five carbons)

$\qquad$ acetyl-coenzyme A + proprionyl-coenzyme A
$\qquad$ (two carbons) $\qquad\qquad$ (three carbons)

$$(9-21)$$

The acetyl-coenzyme A enters the citric acid cycle as usual. The proprionyl-coenzyme A is converted to succinyl-coenzyme A by a reaction involving addition of CO_2 (a carboxylation):

proprionyl-coenzyme A + CO_2 $\longrightarrow$

$\qquad\qquad$ succinyl-coenzyme A $\quad$ (9-22)

The succinyl-coenzyme A then enters the citric acid cycle directly at reaction 5 in Figure 9-15, as a part of the complex side cycle that yields one ATP by substrate-level phosphorylation.

The total ATP produced by complete oxidation of a saturated, even-numbered fatty acid chain to CO_2 and H_2O gives an idea of the energy available from these substances. Each two-carbon acetyl unit yields three NADH and one $FADH_2$ (from Reaction 9-7). Adding to this the single NAD^+ and FAD reduced in Reaction 9-20 gives a total of four NAD^+ and two FAD reduced for the oxidation of each two-carbon segment.

One ATP molecule is produced directly in the citric acid cycle for each acetyl unit oxidized. The electrons carried in each of the four NADH drive synthesis of three ATP as they travel through the mitochondrial electron transport system. Electrons carried by each of the two $FADH_2$ drive the synthesis of two ATP. Therefore, each two-carbon segment removed from a saturated, even-numbered fatty acid will result in the synthesis of $(4 \times 3) + (2 \times 2) + 1 = 17$ ATP. Since the average fatty acid is about 18 carbons long, the potential energy available from fats is comparatively high—about twice the energy yield of carbohydrates by weight. This explains why fats are such an excellent source of energy in the diet and why they are used so extensively as a low-weight form of stored energy in animals. (Storage of the equivalent amount of energy as carbohydrates rather than fats would add more than 100 pounds to the weight of an average man or woman!) The pathway is also a valuable source of "metabolic" water for desert insects and mammals that essentially never drink water—one net H_2O molecule is produced for each acetyl-coenzyme A unit made through fatty acid oxidation.

Breakdown of fats in vertebrate animals is triggered by hormones through a cyclic AMP-based second-messenger pathway (see p. 245). Binding of the hormones, including epinephrine, norepinephrine, or glucagon, to the surface of a fat storage cell activates the effector converting ATP to cAMP in the cytoplasm. Cyclic AMP then activates a protein kinase, which in its activated form adds a phosphate group to the lipase breaking down fats. Phosphorylation activates the lipase, which breaks the triglyceride into glycerol plus fatty acids and triggers the entire fatty acid oxidation pathway. The pathway is turned off by disappearance of the activating hormones from the bloodstream or by insulin, which reduces cAMP concentration and thus inhibits the breakdown and oxidation of fats.

Fatty acids are oxidized inside microbodies by the same sequence shown in Figure 9-40 except for the first reaction. In microbodies the oxidase catalyzing this step transfers electrons directly to oxygen instead of FAD, yielding H_2O_2 as a product (Fig. 9-41). Since the electrons removed at this step are delivered directly to oxygen, all the free energy released at this step is lost as heat. The NAD^+ or FAD accepting electrons at later steps in the peroxisome pathway may eventually deliver their electrons to the electron transport system of mitochondria by means of various shuttles.

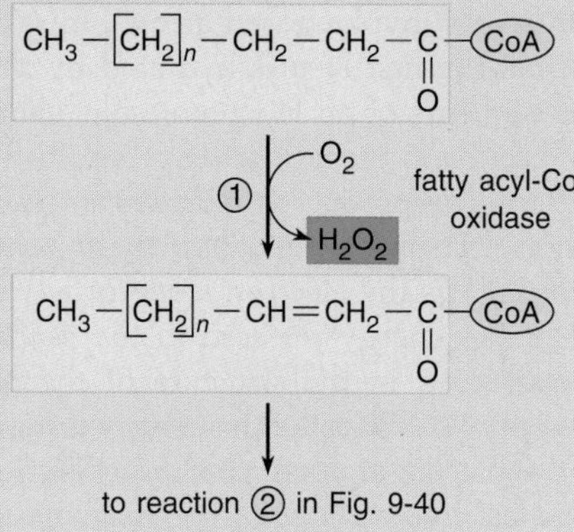

Figure 9-41 The variation in the first reaction of fatty acid oxidation occurring in microbodies (see text).

Supplement 9-2
Oxidative Metabolism in Prokaryotes

Most bacteria possess all the enzymes necessary to carry out the oxidative reactions of glycolysis, pyruvate oxidation, and the citric acid cycle. Glycolysis takes place in the cytoplasmic solution as it does in eukaryotic cells. The reactions of glycolysis follow essentially the same routes in prokaryotes and eukaryotes, except that the final products of fermentations in bacteria are more varied. Pyruvate oxidation and the citric acid cycle also occur by the same pathways in bacteria and eukaryotes. The location of the latter reactions in the cytoplasm is quite different, however. Since bacteria have no mitochondria, the enzymes and intermediates of the citric acid cycle are distributed throughout the cytoplasm without being enclosed in separate membrane-bound compartments. The pyruvate-dehydrogenase complex and the carriers of the electron transport system are associated with the plasma membrane in bacteria. (Fig. 9-12 shows the pyruvate-dehydrogenase complex of bacteria.)

There are few organic molecules that cannot be oxidized by at least some bacterial species. Some bacteria oxidize inorganic substances such as iron (as Fe^{2+}), H_2S, or hydrogen (as H_2) as electron sources. In many of these oxidative pathways, the electrons removed from the oxidized substances are passed directly to the electron transport system with no intermediate steps in the reaction pathway. The wide variety of oxidative reactions in bacteria is another reflection of the biochemical versatility of these minute but highly successful organisms.

Electron Transport in Bacteria

The electron transport pathways of bacteria are as diverse as their oxidative pathways. Although the electron transport systems of many bacteria include the same prosthetic groups present in mitochondria, including FMN, FAD, iron-sulfur groups, quinones, and cytochromes, different bacteria vary considerably in the types within these groups. For example, although most bacteria use cytochromes as electron carriers, unique types such as cytochromes o, d, and a_2 may be present in their electron transport systems.

The bacterial electron carriers are organized with proteins into carrier complexes as in mitochondria (Fig. 9-42). Some bacteria contain a complex similar to mitochondrial complex I, with FMN and Fe/S carriers. Some of the complex I-like bacterial carriers, however, may contain FAD instead of FMN, or neither of these prosthetic groups. Carriers may also be present with similarities to complex II (with FAD and Fe/S carriers), complex III (with carriers similar to cytochrome b_1), and complex IV (with cytochromes o, a, a_2, a_3, or d in various combinations, along with Cu centers). Typically, these carriers are simpler in polypeptide composition than their mitochondrial equivalents—a bacterial complex equivalent to mitochondrial complex IV, for example, may contain as few as two polypeptides. As with mitochondrial carrier complexes, most bacterial electron transport complexes are capable of pumping H^+ as they cycle between oxidized and reduced states.

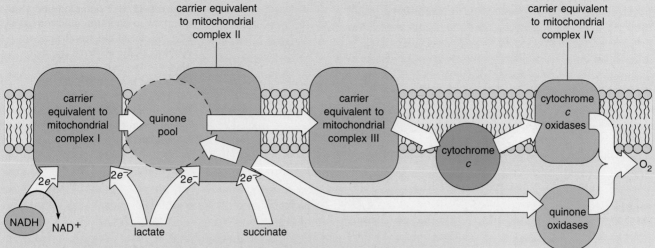

Figure 9-42 A generalized system showing pathways of electron transport in bacteria. Carriers equivalent to mitochondrial complexes I and II are not always present in bacterial systems. Carrier I, when present, contains Fe/S centers; however, FAD or FMN may be present or absent in the carrier complex. Complex II, when present, closely resembles mitochondrial complex II. A quinone pool and a carrier equivalent to mitochondrial complex III are usually present in bacterial systems. Like its mitochondrial equivalent, the bacterial complex III usually contains cytochromes b and c_1 and an Fe/S center. As equivalents of mitochondrial complex IV, bacterial systems frequently contain oxidases that deliver electrons from cytochrome c or directly from the quinone pool to oxygen.

Like their mitochondrial counterparts, bacterial carrier complexes are tightly bound to membranes as integral membrane proteins. Unlike mitochondrial carriers, however, bacterial complexes are located in the plasma membrane. In this location the carrier complexes move H^+ from the cytoplasm to the outside of the plasma membrane as they cycle between oxidized and reduced states (Fig. 9-43).

In some bacteria, equivalents of one or more of the mitochondrial electron transport complexes may be missing; a few bacterial electron transport systems contain only a single carrier complex. In species with a more extensive electron transport system, the pathways of electron flow are frequently more branched than the system in eukaryotic mitochondria. That is, they have supplemental electron inputs and exits in addition to or in place of the eukaryotic routes.

The electron transport pathway is also capable of change in some bacterial species. *E. coli*, for example, synthesizes and inserts cytochrome d at the end of its pathway in place of cytochrome o when oxygen is in short supply.

Bacteria also differ in the substance used as the final acceptor for electrons carried by their transport systems. Aerobic bacteria by definition use oxygen as the final electron acceptor. However, some anaerobic species use inorganic substances such as sulfur, sulfate, or nitrate, or organic molecules such as fumarate in this role instead. A few anaerobic bacteria entirely lack a functional electron transport system.

Bacterial F_oF_1 ATPase

The most striking similarity between bacterial and eukaryotic oxidative mechanisms is the form taken by F_oF_1 ATPase. The polypeptides of the headpiece, stalk, and basal unit of bacterial F_oF_1 ATPase have similar counterparts in the mitochondrial F_oF_1 ATPase (see Table 9-2). Sequencing studies by Walker, Futai, and their colleagues (see p. 134), and research carried out with isolated and purified polypeptides, confirmed that the bacterial ATPase complex shares many of the same major subunits as its mitochondrial counterpart.

In mitochondria an inhibitor protein linked to the ATPase complex prevents the complex from operating as an H^+-ATPase active transport pump, that is, as a pump pushing H^+ to the outside of the plasma membrane at the expense of ATP. In many bacteria, however, no equivalent of an inhibitor protein is present and the enzyme is fully reversible. In these bacteria the enzyme operates to synthesize ATP when electron transport maintains a sufficiently strong H^+ gradient, and reverses, using ATP energy to push H^+ across the plasma membrane when electron transport and the H^+ gradient fall off. The reversible F_oF_1 ATPase in these bacteria allows them to maintain the H^+ gradient whether electron transport proceeds or not. This sustains the H^+ gradient necessary to power H^+-linked active transport and flagellar motility, both of which depend directly on the H^+ gradient in bacteria. (The role of H^+ in powering flagellar movement is described

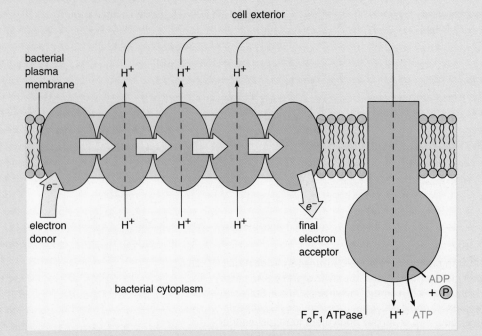

cell exterior

bacterial plasma membrane

H⁺   H⁺   H⁺

e⁻

electron donor

H⁺   H⁺   H⁺

e⁻

final electron acceptor

bacterial cytoplasm

F_oF_1 ATPase

H⁺   ATP

ADP + Ⓟ

Figure 9-43 H^+ pumping and electron flow in bacterial electron transport. Bacterial electron transport systems pump H^+ from the cytoplasm to the cell exterior as electrons flow through the carrier complexes. The headpiece and stalk of bacterial F_oF_1 ATPase, as shown, extend from the membrane into the cytoplasm.

in Supplement 11-1.) In bacteria lacking a functional electron transport system, the H^+ gradient necessary to power transport and flagellar motion is produced entirely by the F_oF_1 ATPase complex at the expense of ATP. In this case, the required ATP is supplied primarily by anaerobic fermentations.

In a number of bacteria, including *Vibrio, Bacillus, Salmonella, Clostridium,* and *Streptococcus,* Na^+ supplements or substitutes for H^+ as an ion pumped across the plasma membrane by electron transport carriers. Some of these bacteria, for example, contain a complex equivalent to mitochondrial complex I that pumps Na^+ across the plasma membrane as electrons are transferred from NADH to a quinone. A gradient of this ion is also maintained by an Na^+-ATPase pump in the plasma membrane. Apparently, the F_oF_1 ATPase of these bacteria can use the Na^+ gradient as an energy source as well as an H^+ gradient.

Oxidations and ATP Synthesis in Cyanobacteria

Relatively little is known about oxidative metabolism in cyanobacteria. When placed in the dark, many cyanobacteria are unable to oxidize fuel substances such as glucose. Those that can do so only very slowly, and are limited to a very few molecules that can be used as fuels. These include little more than glucose and fructose and, in a few species, ribose or glycerol.

The limited oxidative reactions taking place in cyanobacteria evidently proceed primarily or exclusively by a reaction series in which glucose is oxidized by the pentose phosphate pathway (see Information Box 9-2). Electrons removed during oxidations in the path-

way are accepted by $NADP^+$, which in cyanobacteria may deliver them in turn to the electron transport system. Through this means the oxidations carried out in the pentose phosphate pathway may lead to ATP synthesis. Because enzymes of the glycolytic pathway can be detected in some cyanobacteria, these organisms are probably also capable of glycolysis. Several key enzymes of the citric acid cycle are either rare or undetectable in cyanobacteria, indicating that the complete cycle does not occur in these prokaryotes.

The limited oxidative capabilities of cyanobacteria are probably a reflection of the fact that these prokaryotes depend primarily on photosynthesis for their ATP requirements. All cyanobacteria contain a membrane-linked electron transport system as part of their photosynthetic apparatus, including essentially the same carriers as the system operating in photosynthesis in higher plants (see p. 412).

Some cyanobacteria can slowly take up oxygen in the dark, indicating that these organisms are capable of respiratory electron transport. Several carriers capable of respiratory electron transport have been detected in these organisms, including a carrier equivalent to complex IV of mitochondria. The respiratory carriers are considered to be combined with other carriers that also form parts of the photosynthetic electron transport system of cyanobacteria. The combination of carriers sets up pathways that probably conduct electrons from both NADH and NADPH to oxygen. (Fig. 9-44 shows a respiratory pathway proposed for cyanobacteria by P. Boger and his colleagues.)

The electron transport systems of cyanobacteria may be located in the plasma membrane, as in bacteria, or in membranous vesicles suspended in the cytoplasm

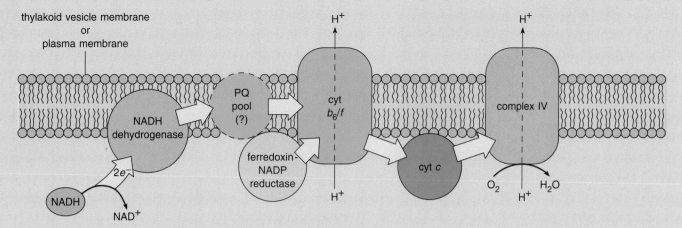

Figure 9-44 A pathway for respiratory electron transport in cyanobacteria, as proposed by Boger and colleagues. The carriers acting only in the respiratory pathway, NADH dehydrogenase and complex IV, are shown in red. NADH dehydrogenase contains FAD as a prosthetic group, and may function similarly to complex II in the mitochondrial electron transport system. The remaining carriers also function as part of the photosynthetic electron transport system. PQ is a quinone with properties and function similar to ubiquinone; the cytochrome b_6/f complex acts similarly to complex III in the mitochondrial electron transport system. (For positions and functions of these carriers in the photosynthetic electron transport systems of cyanobacteria and higher plants, see Fig. 10-10.) Adapted from S. Scherer et al., *Photosynth. Res.* 15:95 (1988).

(see pp. 362 and 414). In either location flow of electrons through the carrier complexes is assumed to establish an H^+ gradient. The H^+ gradient in turn drives ATP synthesis by a membrane-bound F_oF_1 ATPase, as in other organisms. Recently, the genes encoding F_oF_1 ATPase in two cyanobacteria, *Anabaena* and *Synechococcus*, were isolated and sequenced by A. L. Cozens, J. E. Walker, S. E. Curtis, and others. The amino acid sequences deduced for the polypeptides of the cyanobacterial F_oF_1 ATPase indicate that the ATP-synthesizing enzyme of cyanobacteria has a structure closely similar to those of bacteria, mitochondria, and chloroplasts. Of the various F_oF_1 ATPases the cyanobacterial enzyme is most closely related in amino acid sequences to the F_oF_1 ATPase of chloroplasts. This similarity provides another of the many functional and structural similarities between cyanobacteria and chloroplasts, and strengthens the conclusion that chloroplasts are evolutionary descendants of cyanobacteria (see Ch. 27 for details).

Suggestions for Further Reading: See p. 358.

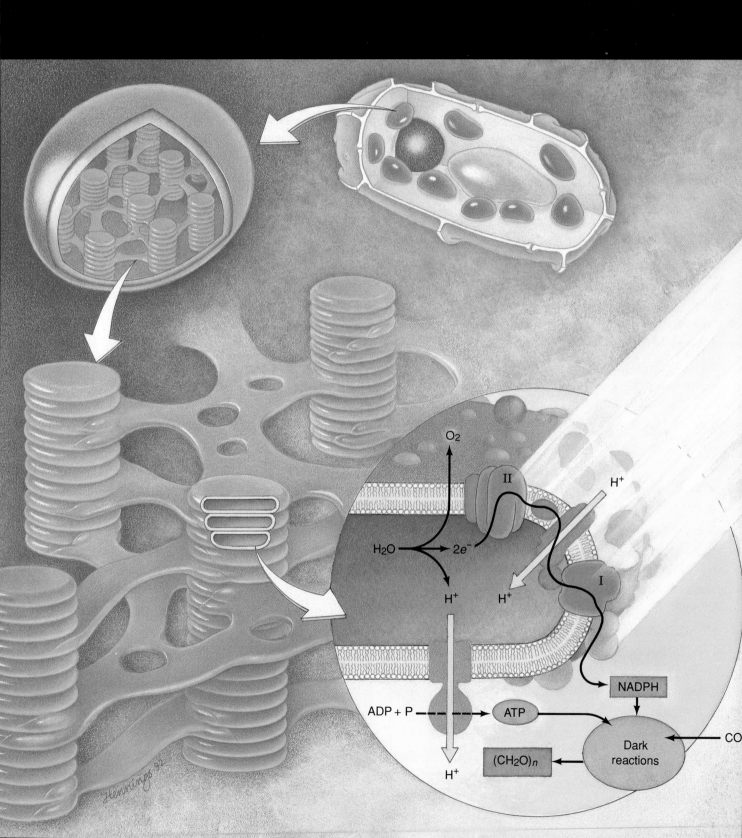

O_2

II

H^+

H_2O — $2e^-$

H^+

H^+

I

H^+

NADPH

ADP + P - - - → ATP

CO

$(CH_2O)n$

Dark reactions

H^+

Hennings 92

P hotosynthesis, carried out by green plants, eukaryotic algae, cyanobacteria, and photosynthetic bacteria, constantly replenishes the organic molecules oxidized by all organisms as a source of cellular energy. The reactions of photosynthesis use the energy of sunlight to convert inorganic substances into organic molecules. In most photosynthetic organisms organic molecules are assembled from raw materials no more complex than water, carbon dioxide, and a supply of inorganic minerals.

Photosynthetic organisms typically synthesize much greater quantities of organic substances than they require for their own activities. Much of the remainder is used as a fuel source by animals and other organisms that live by eating plants. These plant-eating forms are consumed in turn by other organisms, and so on down the line until the last of the organic matter assembled by photosynthesis is completely oxidized to carbon dioxide and water. Because the reactions capturing light energy provide the first step in this extended pathway of energy flow, photosynthesis is the vital link between the energy of sunlight and the vast majority of living organisms. Without the activity of photosynthetic prokaryotes and eukaryotes capturing light and converting it to chemical energy, most of the earth's creatures, including the human population, would soon cease to exist.

This chapter describes the capture and utilization of light energy in photosynthesis, emphasizing the photosynthetic pathways in eukaryotic plants. Supplement 10-1 outlines photosynthesis in prokaryotes.

PATHWAYS OF PHOTOSYNTHESIS: AN OVERVIEW

Photosynthesis proceeds by two major steps in which electrons play a primary role. In the first step the energy of sunlight is absorbed and used to push electrons to an elevated energy level.[1] In cyanobacteria, eukaryotic algae, and higher plants, water is the source of electrons

[1]For a definition of electrons at an "elevated energy level," see Information Box 9-1 on p. 312.

for this process. After being elevated in energy level, the electrons are used in the second step of photosynthesis as a source of energy for ATP synthesis, and also for synthetic reactions in which reductions are required. (See Information Box 9-1 for a definition of oxidation and reduction.) In cyanobacteria, algae, and plants the primary substance reduced by the addition of electrons in photosynthesis is carbon dioxide. The reduction, which also adds hydrogens, converts carbon dioxide into organic substances. Oxygen, derived from the H_2O used in step 1, is released to the environment as an important by-product of photosynthesis. In eukaryotes both major steps of photosynthesis take place inside chloroplasts.

The First Major Step: The Light Reactions

The reactions of the first major step of photosynthesis follow the same fundamental pathway as the energy-trapping mechanisms of mitochondria. These mechanisms, outlined in Chapter 9, include: (1) provision of electrons at elevated energy levels; (2) an electron transport system, embedded in a membrane, that taps off energy from the electrons and uses it to build an H^+ gradient across the membrane; and (3) an ATPase that uses the H^+ gradient as an energy source for ATP synthesis. In the operation of these mechanisms in chloroplasts, only part of the energy of the electrons is tapped off and used to drive ATP synthesis. The remainder is used as reducing power for the second overall step in photosynthesis, the reduction of CO_2. Because the mechanisms of the first major step are dependent on light, and stop if the source of light is interrupted, they are called the *light* or *light-dependent reactions* of photosynthesis.

The Source of Electrons in Water The electrons used in eukaryotic photosynthesis are removed from water in the reaction:

$$H_2O \longrightarrow 2H^+ + \tfrac{1}{2}O_2 + 2e^- \qquad (10\text{-}1)$$

The electrons removed from water initially exist at energy levels too low to enter the electron transport system directly. These electrons are elevated in energy level by transfer to a molecule capable of absorbing light. In all photosynthetic organisms, prokaryotes as well as eukaryotes, the molecules pushing electrons to higher energy levels by absorbing light are green pigments of the *chlorophyll* family (see Fig. 10-4).

Electrons pushed to an elevated energy level by light absorption in chlorophyll are highly unstable. They are converted to stable form by immediate transfer from chlorophyll to a molecule called a *primary acceptor*. Once the electrons are stabilized in the primary acceptor, the energy of light has been captured as chemical energy.

The provision of electrons for eukaryotic and cyanobacterial photosynthesis thus takes place in three unique steps: (1) removal of electrons at low energy levels from an inorganic donor, water; (2) elevation of the electrons to higher energy levels by the absorption of light energy, and (3) transfer of the electrons to stable form in a primary acceptor, which converts light to chemical energy. The molecules carrying out these reactions are embedded in the internal membranes of chloroplasts.

Electron Transport and ATP Synthesis in the Light Reactions Like its mitochondrial counterpart the electron transport system of photosynthesis consists of a series of carriers that accept and release electrons at successively lower energy levels. At the end of the pathway, electrons are delivered to a final acceptor. As electrons move through the carrier sequence, some of the released energy is used by the carrier molecules to push H^+ across the membrane housing the electron transport system. The resulting H^+ gradient provides the energy source driving the synthesis of ATP.

Photosynthetic electron transport differs from the mitochondrial system in that the electrons delivered to the final acceptor at the end of the pathway still exist at relatively high energy levels. Further, the final electron acceptor in photosynthetic electron transport in chloroplasts is an organic substance, *NADP* (*nicotinamide adenine dinucleotide phosphate;* see Fig. 9-16). By contrast, the final electron acceptor in mitochondrial electron transport is an inorganic substance, O_2. The NADPH produced at the end of photosynthetic electron transport carries electrons to the reactions converting CO_2 to organic molecules in the second major step of photosynthesis.

Utilization of the H^+ gradient for ATP synthesis proceeds in chloroplasts by exactly the same mechanism as in mitochondria (see Fig. 9-1). The chloroplast ATPase is an active transport pump that, if working in the forward direction, would break down ATP as an energy source to push H^+ across the membrane housing the electron transport system. When working in the forward direction, the chloroplast ATPase would push H^+ in the same direction as the electron transport system. The H^+ gradient set up by electron transport is so high, however, that the ATPase is forced to run in reverse; in this reversal the enzyme synthesizes ATP from ADP and phosphate. The chloroplast ATPase, called the CF_oCF_1 *ATPase,* contains molecular subunits similar to those of the mitochondrial F_oF_1 ATPase, arranged in the same headpiece-stalk-basal unit structure that extends from the photosynthetic membranes as a lollipop-shaped particle (Table 10-1; see also Figs. 9-29 and 10-15).

The light reactions of photosynthesis thus use energy derived from light to produce two substances, ATP and NADPH, that serve as reactants for the second

Table 10-1 Equivalence Between Chloroplast CF_oCF_1 ATPase, Bacterial F_oF_1 ATPase, and Mitochondrial F_oF_1 ATPase Polypeptide Subunits

Chloroplast CF_oCF_1ATPase Subunits	Bacterial F_oF_1 ATPase Subunits	Mitochondrial F_oF_1ATPase Subunits
Headpiece and stalk		
α	α	α
β	β	β
γ	γ	γ
δ	δ	OSCP
ϵ	ϵ	δ
		ϵ
Basal unit		
subunit II		
subunit IV	a	a (subunit 6)
subunit I	b	b (subunit 4)
subunit III	c	c (subunit 9)

major step. The oxygen released when water is split in the initial step of the light reactions (Reaction 10-1) is also an important product of photosynthesis.

The Second Major Step: The Dark Reactions

In the second major step of photosynthesis, CO_2 is reduced and converted into complex organic substances. Since the reactions of the second major step depend only on a supply of ATP, NADPH, and CO_2, and do not directly require light, they are called the *dark* or *light-independent reactions* of photosynthesis. In the dark reactions electrons carried by NADPH are used directly to reduce CO_2 into carbohydrates and other organic molecules. The ATP produced in the light reactions also supplies energy for the dark reactions.

Figure 10-1 summarizes the interrelationships of the light and dark reactions in photosynthesis. Note from the figure that the ATP and NADPH produced by the light reactions, along with CO_2, are the reactants of the dark reactions. The ADP, inorganic phosphate, and $NADP^+$ produced by the dark reactions, along with H_2O, are the reactants for the light reactions. The light and dark reactions thus form a cycle in which the net inputs are H_2O and CO_2, and the net outputs are organic molecules and O_2:

$$H_2O + CO_2 \longrightarrow \text{organic molecules} + O_2 \quad (10\text{-}2)$$

For sugars the organic molecules can be considered to be carbohydrate units (CH_2O), so that this reaction can be written in balanced form as:

$$H_2O + CO_2 \longrightarrow (CH_2O) + O_2 \quad (10\text{-}3)$$

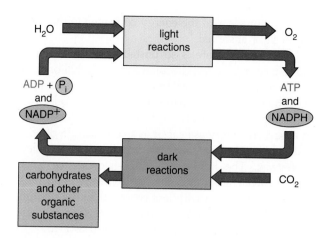

Figure 10-1 Relationships between the light and dark reactions of photosynthesis. The two reaction pathways are linked by ATP and NADP (see text).

Location of the Light and Dark Reactions in Eukaryotes: The Chloroplast

Chloroplasts are lens-shaped, membranous bodies in eukaryotic algae and higher plants that range from about 3 to 10 μm in length and 1 to 5 μm in diameter (Fig. 10-2). Chloroplasts are surrounded by two continuous membranes, the *outer* and *inner boundary membranes.* The two boundary membranes are separated by an *intermembrane compartment.* The outer boundary membrane is smooth; frequently, the inner boundary membrane folds or extends into tube- or vesiclelike forms (as in Fig. 10-2). The two boundary membranes enclose an inner compartment, the *stroma,* analogous in location to the mitochondrial matrix. Within the stroma is a third membrane system consisting of small, flattened sacs, the *thylakoids.* Chloroplasts thus have three separate internal compartments delineated by membranes: the intermembrane compartment between the boundary membranes, the stroma, and the compartments enclosed within the thylakoids.

In green algae and higher plants thylakoids in most chloroplasts are arranged in stacks called *grana* (singular = *granum;* see Figs. 10-2 and 10-3). Chloroplasts may contain from a few to 40 to 60 grana, each formed from stacks of 2 or 3 to as many as 100 individual thylakoids. The thylakoids of grana stacks are interconnected by flattened, tubular membranes called *stromal lamellae* (circled in Fig. 10-2). Stromal lamellae probably connect the thylakoid compartments into a single, continuous compartment within the stroma (diagramed in Fig. 10-3; further details of chloroplast structure are given later in this chapter).

Usually, chloroplasts are considerably larger than mitochondria in the same plant cell and occur in smaller numbers. The number may vary from a single chloroplast, as in *Micromonas* and several other green algae, to several hundred in the cells of most higher plants.

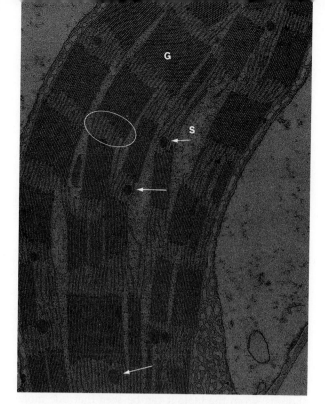

Figure 10-2 A chloroplast from maize. The chloroplast is surrounded by a smooth outer boundary membrane and a much-folded and convoluted inner boundary membrane. Thylakoids stacked into grana (G) are clearly visible inside the chloroplast. Connections called stromal lamellae (circled) run between the thylakoids of adjacent grana. Thylakoid membranes are suspended in the stroma (S), an inner solution of proteins and other molecules. Numerous osmiophilic granules (arrows) are suspended in the stroma. × 24,500. Courtesy of L. K. Shumway, Washington State University.

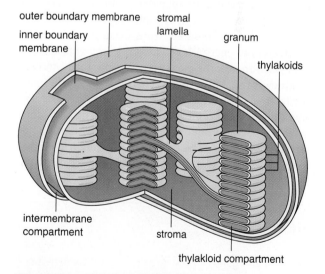

Figure 10-3 The membranes and compartments of chloroplasts (see text). Thylakoids are stacked into grana; stromal lamellae interconnect the individual thylakoids, making the compartment enclosed within the thylakoids a continuous space in much or all of the chloroplast.

Pathways of Photosynthesis: An Overview **369**

In higher plant cells active in photosynthesis, chloroplasts may take up as much as 25% of the total cell volume.

THE LIGHT REACTIONS

Light and Light Absorption

Visible light is a form of radiant energy with wavelengths ranging from about 400 nm, seen as blue light, to 680 nm, seen as red. Although radiated in apparently continuous beams, the energy of light actually flows in discrete units called *photons*. The photons of a beam of light contain an amount of energy that is inversely proportional to the wavelength of the light. The shorter the wavelength, the greater the energy of a photon.

The energy content of photons at various wavelengths is given in Table 10-2 in cal/Einstein. The Einstein relates light energy to a gram molecular weight and is equivalent to a "mole" of light, 6.023×10^{23} photons. From the table it can be seen that one mole of chlorophyll, containing 6.023×10^{23} molecules and absorbing one Einstein of red light at 650 nm (equivalent to absorption by one electron per chlorophyll molecule), absorbs 43,480 cal of energy.

Molecules such as chlorophyll appear colored or pigmented because they absorb light at certain wavelengths and transmit light at other wavelengths. The color of a pigment is produced by the transmitted light. Chlorophyll, for example, absorbs blue and red light and transmits intermediate wavelengths that are seen in combination as green.

Light is absorbed in pigments by electrons occupying certain orbitals in a pigment molecule. In darkness or when exposed to light at wavelengths not absorbed by the molecule, these electrons occupy orbitals at a relatively low energy level known as the *ground state*. If an electron absorbs the energy of a photon, it moves to a new orbital at a higher energy level. In the new orbital the electron is said to be in an *excited state*. Typically, excited orbitals are farther than ground-state orbitals from the atomic nuclei associated with the light-sensitive electrons. The difference in energy level between the ground-state and excited-state orbitals is exactly equivalent to the energy contained in the photon of light absorbed.

The excited state is so unstable that an electron can remain in an excited orbital for only a billionth of a second or less. Return from an excited state may occur by one of several pathways. The excited electron may simply drop back to its ground state, releasing all its absorbed energy as heat or as a combination of heat and light. Light energy released in this way is called *fluorescence*. The light released in fluorescence is at longer wavelengths than the light absorbed and thus represents a photon with less energy than the photon absorbed. The difference in energy between the absorbed and fluoresced photon is released as heat.

Two other possible fates for an excited electron have greater significance for photosynthesis. If a molecule containing an excited electron is situated close enough to another molecule, the excited orbital may overlap an orbital in the second molecule. The excited electron may then transfer to an orbital in the second molecule, particularly if the orbital in the second molecule represents a ground rather than excited state and is thus stable. If an excited electron is transferred to a stable orbital in a second molecule, the energy of the electron is effectively trapped as chemical energy. Any difference in energy level between the excited orbital in the light-absorbing molecule and the stable ground-state orbital in the acceptor is released as heat.

In photosynthesis light is converted to chemical energy through transfer of electrons, excited by light absorption in chlorophyll, to stable orbitals in primary acceptor molecules. The transfer occurs within a few trillionths of a second after the chlorophyll is excited by the absorption of light energy.

The third possible fate for excitation energy is transmission to a nearby molecule by a mechanism called *inductive resonance*. On absorbing light a pigment molecule sets up an electromagnetic field because of the rapid vibration of its excited electrons. Electrons in equivalent orbitals of an adjacent pigment molecule, lying within the vibrational field created by the excited molecule, are induced to vibrate or resonate at the same frequency. In so doing, the electrons in the second molecule absorb the excitation energy of the first. As the excitation energy is transferred to the second molecule, electrons in this molecule are raised to an excited state, and the electrons in the first molecule return to the ground state. The second molecule may transfer its excitation energy to a third pigment molecule by any of the pathways: by inductive resonance, by passing its electron to a stable orbital in an acceptor molecule, or by releasing the energy as a combination of light and heat as its excited electron returns to the ground state. Transfer from one pigment to another by inductive resonance, which occurs in billionths of a second with relatively little energy loss, is important in the light reactions as a mechanism transferring light energy

Table 10-2	Energy Content of Light at Various Wavelengths	
Wavelength (nm)	Color	Cal/Einstein
395	Violet	71,800
490	Blue	57,880
590	Yellow	48,060
650	Red	43,480
750	Far red	37,800

Figure 10-4 The structure of chlorophylls *a* and *b*. Chlorophyll *b*, which occurs along with chlorophyll *a* in green algae and higher plants, acts as an accessory pigment, passing absorbed light energy to chlorophyll *a* for conversion to chemical energy. The tetrapyrrole ring of the chlorophylls is similar to the ring structure of cytochromes and hemoglobins except for the additional ring structure shown in blue. The distribution of light-absorbing electrons in chlorophyll is shown in red. The nonpolar character of the chlorophylls is due primarily to the phytol side chain, which is strongly hydrophobic.

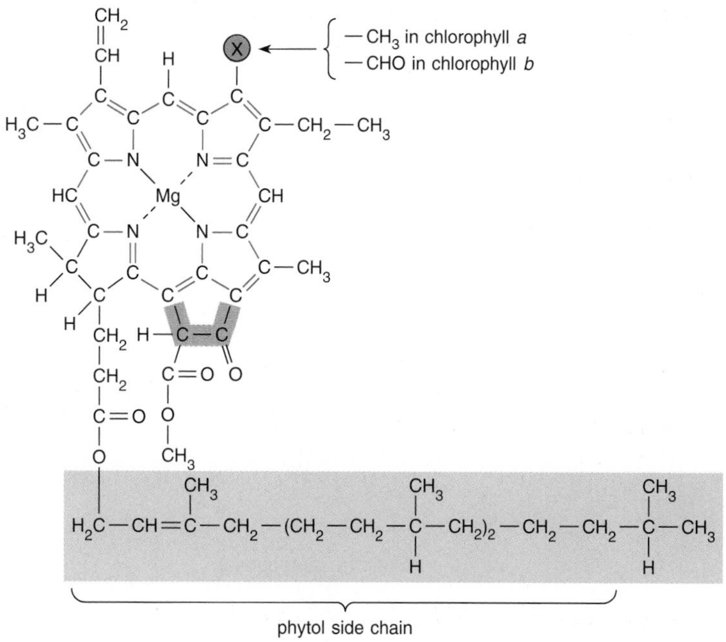

absorbed in other pigment molecules to the chlorophylls passing electrons to stable orbitals in acceptor molecules. The pigment molecules passing absorbed energy to these chlorophylls are termed *accessory pigments*.

The Structure of Chlorophyll and Other Light-Absorbing Molecules

Although chlorophylls are the molecules directly involved in light absorption and the transfer of electrons to primary acceptors in photosynthesis, other pigments, called *carotenoids*, also absorb light energy and pass it on to the chlorophylls. Both chlorophylls and carotenoids are lipid molecules bound to thylakoid membranes and stromal lamellae in chloroplasts.

Chlorophylls The chlorophylls are a family of closely related molecules based on a tetrapyrrole ring (Fig. 10-4). The ring structure of chlorophylls is similar to those of cytochromes and hemoglobin except for the presence of an extra subunit (in blue in Fig. 10-4). A magnesium atom is bound at the center of the chlorophyll ring. Attached to the ring is a long, hydrophobic side chain that gives the two major chlorophylls found in higher plants, *chlorophyll a* and *chlorophyll b*, their lipidlike solubility characteristics. Chlorophylls *a* and *b* differ only in the side groups attached to one carbon of the tetrapyrrole ring. Of the two types chlorophyll *a* plays the central role in the conversion of light to chemical energy in all photosynthetic eukaryotes and also in cyanobacteria; chlorophyll *b* is one of several pigments that pass the energy of absorbed light to chlorophyll *a*. The closely related chlorophyll *c* occurs as an accessory pigment with chlorophyll *a* in brown algae, diatoms, and dinoflagellates.

Chlorophylls contain many electrons capable of moving to excited orbitals by absorbing light. These electrons are distributed in a network of alternating single and double bonds extending around the tetrapyrrole ring (shown in red in Fig. 10-4). Electrons in the network each absorb light at different wavelengths. The combination of wavelengths broadens the distribution absorbed into an extended curve rather than a single, sharp peak (Fig. 10-5). The broad curve is called an *absorption spectrum*. Each chlorophyll type has a distinct absorption spectrum.

Absorption spectra are modified by association of chlorophylls with other molecules in chloroplast membranes, particularly with proteins. For example, purified chlorophyll *a* absorbs red light most strongly at 665 nm when dissolved in acetone. In chloroplast membranes

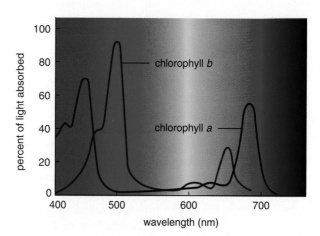

Figure 10-5 The absorption spectra of chlorophylls *a* and *b* at different wavelengths. The positions of the peaks change to some extent depending on the combination of chlorophylls with proteins.

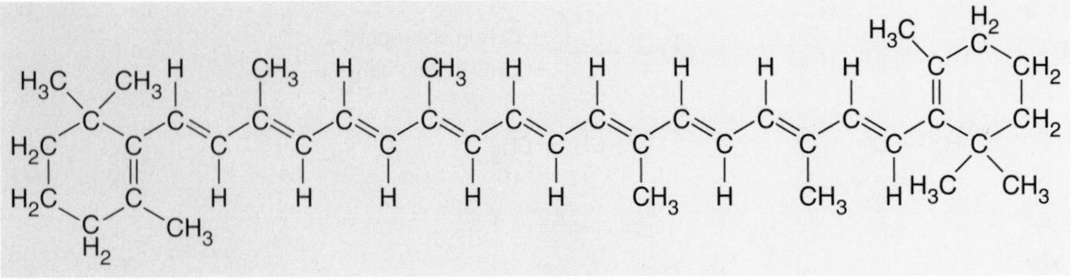

Figure 10-6 Two examples of carotenoid pigments, which absorb light at blue wavelengths and pass the absorbed light energy to the chlorophylls. The carotenoids consist of **(a)** the carotenes, which are pure hydrocarbons, and **(b)** the xanthophylls (also called carotenols), which contain oxygen atoms in their terminal structures.

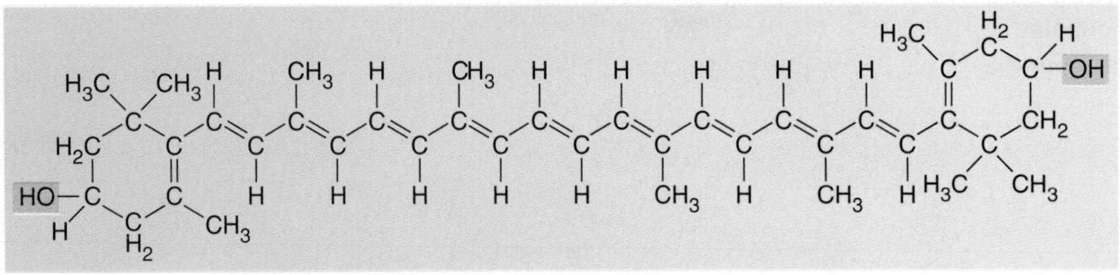

a β-carotene

b lutein (a xanthophyll)

in which the pigments are closely associated with other molecules, individual chlorophyll *a* molecules may absorb light strongly at other wavelengths such as 660, 670, 680, and 700 nm. Although their absorption spectra are altered, the chemical structure of the chlorophylls is not changed by the associations. Some of the associations responsible for modifying light absorption, particularly those producing absorption peaks in chlorophyll *a* at 680 and 700 nm, figure importantly in the light reactions of photosynthesis.

Although chlorophylls absorb light in the blue end of the spectrum as well as in the red, the energy level of an excited electron passed from chlorophyll *a* to a primary electron acceptor is equivalent to that of photons in the red wavelengths between 680 and 700 nm. The difference in energy between any shorter light wavelengths absorbed and the energy of excited electrons released by chlorophyll is lost as heat. This characteristic makes all wavelengths absorbed by chlorophylls equally effective in photosynthesis, even though photons at the shorter wavelengths contain more energy.

Carotenoids The carotenoids (Fig. 10-6) are a separate family of light-absorbing lipids built upon a single, long carbon chain containing 40 carbon atoms. Various substitutions in side groups attached to the 40-carbon backbone give rise to different carotenoid pigments. Two types of carotenoids occur in all green plants. The *carotenes* are pure hydrocarbons that contain no oxygen atoms; chief in abundance among these pigments in higher plants is β-*carotene* (Fig. 10-6a). The second major carotenoid type, the *xanthophylls* (also called *carotenols*), are closely similar except for the presence of oxygen atoms in their terminal structures (Fig. 10-6b).

Carotenoids of both types have multiple light-absorbing electrons associated with the alternating single and double bonds of the backbone chain. These

electrons absorb at blue-green wavelengths from 400 to 550 nm (Fig. 10-7) and transmit other wavelengths in combinations that appear yellow, orange, red, or brown.

The carotenoids, in their role as accessory pigments, absorb light at wavelengths weakly absorbed by chlorophylls in the blue end of the spectrum. The light energy absorbed by carotenoids, and by chlorophyll *b*, is eventually transferred by inductive resonance

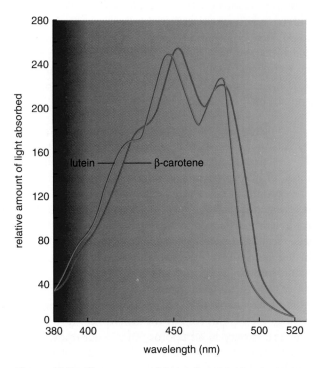

Figure 10-7 The amount of light absorbed by the carotenoids β-carotene and lutein (a xanthophyll) at different wavelengths. Note that the horizontal scale is expanded as compared to Figure 10-5, and includes only the wavelengths in the range from 380 to 520 nm.

to the chlorophyll *a* molecules involved in transforming light into chemical energy. The net effect of the entire combination of chlorophylls and carotenoids is to broaden the spectrum of wavelengths used efficiently as energy sources for photosynthesis.

The Organization of Photosynthetic Pigments in Chloroplasts

Carotenoid and chlorophyll molecules are organized with proteins and other molecules into large complexes that play central roles in the light reactions. Two of these complexes, called *photosystems I* and *II*, act directly in the absorption of light and the conversion of light to chemical energy in eukaryotic plants. The two photosystems have been isolated and purified from chloroplast membranes; work with the isolated photosystems has revealed that each contains from about 50 to 100 chlorophyll *a* molecules and a smaller number of carotenes. These pigment molecules are combined with polypeptides—as many as 11 in photosystem I, and 20 in photosystem II—that carry out structural, enzymatic, regulatory, and other functions in the photosystems. One or two chlorophyll *a* molecules within the photosystems form a *reaction center* in which the central event of the light reactions—the transfer of excited electrons to stable orbitals in a primary acceptor—takes place. The photosystems also contain a sequence of carriers involved in the transport of electrons away from the primary acceptor. Photosystem II, in addition, contains the mechanism splitting water as an electron source (Reaction 10-1).

The internal membranes of chloroplasts also contain pigment-protein assemblies called *light-harvesting complexes (LHCs)*, which act as accessory light-gathering "antennas." The pigments of the accessory antennas are linked with proteins in complexes without reaction centers or primary acceptors. Because LHCs have no reaction centers or primary acceptors, no conversion of light to chemical energy takes place within them. Instead, the energy of absorbed light is passed on to the two photosystems. Most eukaryotes have two types of LHCs: *LHC-I* is associated with photosystem I and *LHC-II* with photosystem II.

Photosystem I Photosystem I (see Fig. 10-12) contains a light-absorbing group of about 130 chlorophyll *a* molecules and 16 *β*-carotenes combined with 10 to 11 polypeptides into a structure known as the *core antenna.* Individual pigment molecules absorbing light energy in the core antenna pass the absorbed energy to the reaction center, which in photosystem I consists of a specialized form of chlorophyll *a* called *P700* (*P* = pigment). P700 is given this name because its light absorption spectrum changes sharply at a wavelength of 700 nm as electrons are passed to the primary acceptor of the photosystem. The photosystem I reaction

center is believed to contain a pair of P700 molecules associated with a pair of large 80,000-dalton polypeptides. The genes encoding the two large polypeptides, which form part of the chloroplast DNA in higher plants, have been sequenced in maize and several other plants. Amino acid sequences deduced from the gene sequences show that the two polypeptides, which are about 45% identical in amino acid sequence, contain 751 and 735 residues and calculated molecular weights of 83,217 and 82,634. The sequences indicate that both polypeptides contain 11 alpha-helical segments that span the membrane.

Photosystem I also contains a series of built-in electron carriers, including several iron-sulfur (Fe/S) centers. These carriers conduct electrons away from the reaction center within the system (see below).

Photosystem II The core antenna of photosystem II (see Fig. 10-12) contains about 50 chlorophyll *a* molecules and a much smaller number of *β*-carotenes. The reaction center of this photosystem contains a specialized form of chlorophyll *a*, *P680*, which undergoes a conspicuous change in light absorption at 680 nm as electrons are passed to the primary acceptor. A pair of P680 chlorophylls is probably located at the reaction center of photosystem II, along with two other chlorophyll *a* molecules acting in an accessory role. A pair of chlorophyll molecules lacking Mg^{2+} (*pheophytins*; see below) participate in the transfer of electrons from the reaction center to the primary acceptor. These pigments are bound to two related proteins, D1 and D2, to form the reaction center. A short series of carriers based on quinones (see below) is also associated with the D1 and D2 polypeptides. Another major polypeptide of photosystem II is the protein segment of a cytochrome, *cytochrome* b_{559}. The function of this cytochrome in the photosystem remains unknown.

The water-splitting mechanism is also associated with photosystem II. This system includes three soluble polypeptides and four manganese atoms that form a *manganese center* linked to the surface of the photosystem II complex. By a process that is still incompletely known, these elements promote the oxidation of water as the first step in the light reactions. The total protein complement of photosystem II, including the water-splitting complex, may have as many as 20 different polypeptides.

Absorption and Migration of Light Energy Within the Photosystems After a photon of light is absorbed by a pigment molecule associated with either photosystem, the absorbed energy is passed to the chlorophyll *a* molecules at the reaction center. Although the mechanism transferring the energy is not completely understood, the energy of an absorbed photon is considered to "walk" by inductive resonance from one pigment molecule to another in a core antenna until it reaches

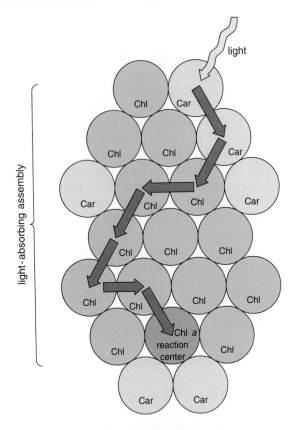

Figure 10-8 The "random walk" of excitation energy by inductive resonance from an absorbing pigment molecule in a core antenna to the reaction center. An entire walk typically occurs within a billionth of a second or less, without significant loss of the absorbed energy as heat. Chl = chlorophyll, Car = carotene.

the reaction center (Fig. 10-8). The entire walk typically occurs within a billionth of a second or less, without significant loss of the absorbed energy as heat.

Once absorbed light energy reaches the chlorophyll *a* molecules at the reaction center, it is apparently trapped in this location and does not migrate back to other pigment molecules of the assembly. One factor in the trapping is the rapid transfer of electrons from the chlorophylls to a primary acceptor. Another may depend on slightly lower energy levels associated with the excited orbitals in P680 or P700. After dropping into these orbitals, the excited electrons may not contain enough energy to move back from P680 or P700 to other pigment molecules in the photosystems.

Light-Harvesting Complexes (LHCs) The two LHC antennas contain both chlorophylls *a* and *b*. LHC-I antennas contain about 80 to 120 chlorophyll molecules in a ratio of about 3 to 4 chlorophyll *a* molecules to each chlorophyll *b*. The chlorophyll *a* and *b* molecules are combined with three or four polypeptides to form the LHC-I complex. LHC-II antennas, according to a recent analysis by W. Kühlbrandt and D. N. Wang, contain three closely similar or identical polypeptides, with each binding 8 chlorophyll *a* and 7 chlorophyll *b*

molecules. About 50% of the chlorophyll *a* content and 75% of the chlorophyll *b* of higher plant chloroplasts is concentrated in LHC-II antennas. LHC-II antennas also contain a few carotenoid molecules, including both β-carotene and xanthophylls. Although the proportion of xanthophylls in LHC-II is relatively small, the total collection of LHC-II antennas contains almost all the xanthophylls of higher plant chloroplasts. The protein complement of LHC-II makes up about one-third of the total protein of photosynthetic membranes in chloroplasts.

LHC antennas are large particles that completely span the photosynthetic membranes. The LHC-I antenna forms a more or less permanent complex with photosystem I. The LHC-II antenna, however, dissociates from photosystem II complexes under certain conditions, so that the accessory antenna and photosystem may occur as either a tight complex or separately in chloroplast membranes. Separation of LHC-II from photosystem II, which greatly reduces the light-harvesting efficiency of photosystem II, provides an important mechanism balancing the light reactions (see below; the structure of LHC-photosystem complexes is shown in Fig. 10-12). The energy of photons absorbed within the LHCs is presumably passed to the photosystems by the same inductive resonance walk taking place in the core antennas of the photosystems.

The Electron Transport System Linking the Photosystems

The two photosystems are linked in chloroplasts by an electron transport system that conducts electrons from one photosystem to the next, and delivers electrons to NADP$^+$ at the end of the pathway. Most of the molecules oxidized and reduced in the photosynthetic system, like the electron carriers of mitochondria (see p. 328), consist of nonprotein prosthetic groups organized with proteins into large complexes. The prosthetic groups of the chloroplast system include the same types active in mitochondrial electron transport— flavoproteins, cytochromes, Fe/S centers, copper (Cu) centers, and quinones (see pp. 327–331 for a description of these electron carriers). One unusual electron carrier of the chloroplast system consists of a single amino acid residue within a protein forming part of photosystem II.

The same battery of techniques used to place mitochondrial carriers in sequence (see Information Box 9-3) has been used to assign tentative positions to electron carriers of the photosynthetic electron transport system. Among the inhibitors used to block steps in the pathway are various herbicides that target the photosystems or electron transport carriers. These techniques have revealed that the photosynthetic carriers are arranged in a Z *pathway* (Fig. 10-9), first advanced as a possibility by R. Hill and F. Bendall. The first leg of the Z is the pathway of electron flow from water

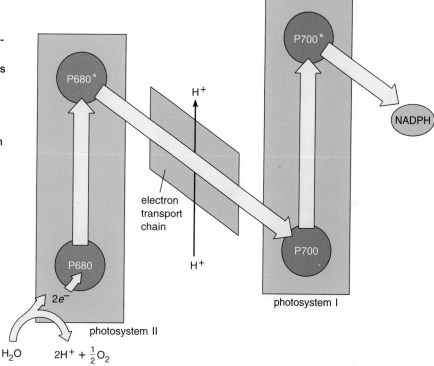

Figure 10-9 The overall pattern of electron flow in the Z pathway. The first leg of the Z is the flow of electrons from water through photosystem II. The diagonal is formed by the flow of electrons through a series of carriers that connects photosystems II and I; some of the energy released by the electrons in this passage is used to pump H^+ across the membrane housing the carriers. The electrons next pass through photosystem I and are finally transferred to a short transport chain leading to $NADP^+$. The asterisks indicate the excited forms of P680 and P700.

through photosystem II.[2] The electrons then flow through a long series of carriers that connect photosystems II and I; this series forms the diagonal of the Z. As they pass through this series, electrons lose energy; some of the released energy is used to pump H^+ across the membrane housing the carriers. The electrons then pass through photosystem I and are transferred to a short transport chain leading to the final acceptor of the chloroplast system, $NADP^+$. The tentative positions of individual prosthetic groups acting as carriers within the Z pathway are shown in Figure 10-10.

Individual Carriers and Their Activities in the Z Pathway Electrons entering the Z pathway are removed from water at low energy levels by the water-splitting complex in photosystem II. The electrons removed from water are delivered to the reaction center of photosystem II by an electron carrier identified as Z. There are indications that Z is the side chain of a tyrosine residue within the D1 polypeptide of photosystem II. This side chain picks up electrons from the water-splitting reaction and releases them to a P680 chlorophyll at the reaction center. P680 accepts single electrons and raises them to excited orbitals through the absorption of light energy.

[2] Photosystems I and II are numbered in the order in which they were discovered. However, in the overall pattern of electron flow in the light reactions, photosystem II occurs before photosystem I in the pathway. Thus, the identifiers given to the two photosystems are reversed in the sense of electron flow. Given scientific tradition, the pathway of electron flow in photosynthesis is far more likely to reverse than the nomenclature.

After excitation electrons are released to the primary acceptor of photosystem II, identified as a *plastoquinone*, a quinone type typical of chloroplasts (Fig. 10-11). The plastoquinone forming the primary acceptor of photosystem II is identified as Q_A. Delivery of electrons from P680 to Q_A occurs via pheophytin, the molecule identical to chlorophyll a except for the absence of Mg^{2+} in the center of the tetrapyrrole ring. Pheophytin evidently acts as an intermediary in the transfer by housing electrons in unstable orbitals for a fraction of a second. From Q_A electrons flow to a second plastoquinone, Q_B, which is the final electron carrier in the photosystem II complex. The entire pathway of electron flow within the photosystem II complex thus includes:

$$H_2O \longrightarrow 2e^- \longrightarrow Z \longrightarrow P680 \longrightarrow$$
$$pheophytin \longrightarrow Q_A \longrightarrow Q_B$$

The H^+ removed from water by photosystem II is released to add to the H^+ gradient produced by electron transport in chloroplasts (see below).

The transfer of electrons from P680 to Q_B in photosystem II is believed to follow a three-dimensional molecular pathway closely similar to the system operating in a purple photosynthetic bacterium, *Rhodopseudomonas*. The pathway was worked out by X-ray diffraction of crystals cast from the bacterial photosystem (see Supplement 10-3 and Fig. 10-38).

Electrons are transferred from the Q_B carrier of photosystem II to a pool of plastoquinone molecules that forms the first carrier of the transport system link-

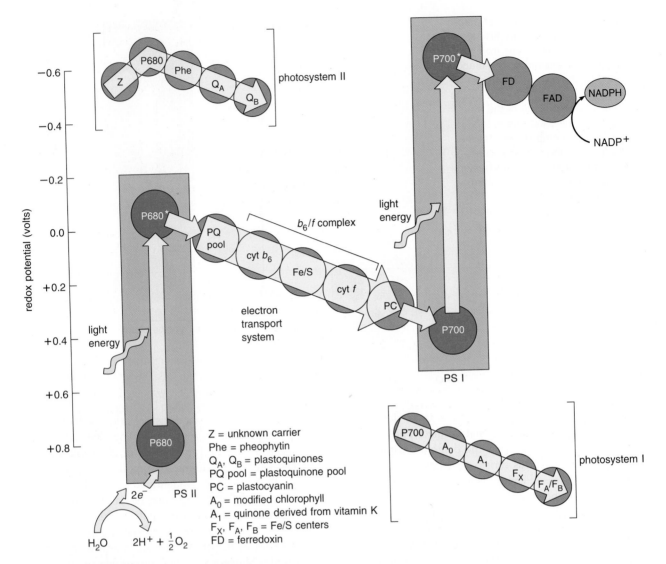

Z = unknown carrier
Phe = pheophytin
Q_A, Q_B = plastoquinones
PQ pool = plastoquinone pool
PC = plastocyanin
A_0 = modified chlorophyll
A_1 = quinone derived from vitamin K
F_X, F_A, F_B = Fe/S centers
FD = ferredoxin

Figure 10-10 The prosthetic groups acting as electron carriers and their tentative sequence in the Z pathway. The asterisks indicate the excited forms of P680 and P700.

ing the two photosystems (see Fig. 10-12). The pool, in which individual plastoquinone molecules are un-associated with proteins and free to diffuse within the membrane interior, is analogous to the ubiquinone pool of the mitochondrial electron transport system. Electron transfer to the pool probably occurs simply by de-tachment and entry into the pool of the reduced Q_B molecule from photosystem II. The reduced plasto-quinone diffusing into the pool is replaced in photo-system II by an oxidized plastoquinone from the pool.

From the plastoquinone pool electrons flow through three carriers—cytochrome b_6, an Fe/S pro-tein, and cytochrome f. In structure and function the complex containing these carriers, known as the b_6/f complex of the chloroplast system, is closely similar to complex III of the mitochondrial system. (Cytochrome f of the b_6/f complex, although given the f identifier,

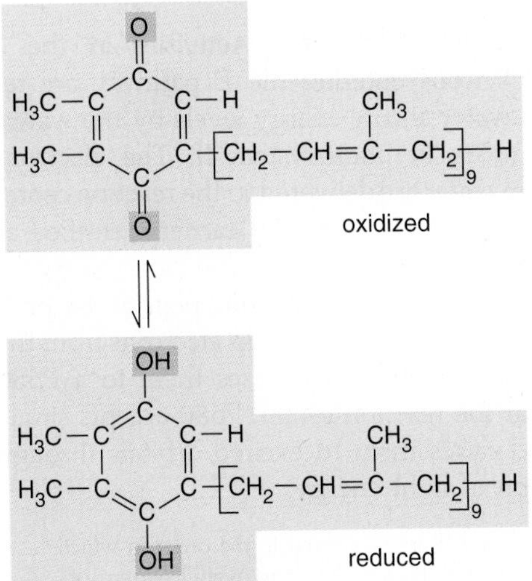

Figure 10-11 Plastoquinone, a carrier that transports electrons within the photosystem II complex and in the electron transport system linking the photosystems in the Z pathway. A related but different quinone derived from vitamin K_1 carries electrons within photosystem I.

is a c-type cytochrome closely related to cytochrome c_1 of mitochondrial complex III.) The b_6/f complex contains several major polypeptides. Two of these are the protein components of the two cytochromes; another is the polypeptide component of the Fe/S protein. Other polypeptides may take part in H^+ pumping by the b_6/f complex.

From the b_6/f complex electrons pass to the final carrier linking the photosystems, *plastocyanin*. This carrier, a protein containing a copper atom that varies between the Cu^+ and Cu^{2+} states during alternate cycles of oxidation and reduction, shuttles electrons from the b_6/f complex to photosystem I. Plastocyanin thus occupies a position in the chloroplast electron transport system equivalent to cytochrome c of the mitochondrial electron transport system. (In many cyanobacteria and some eukaryotic algae cytochrome c substitutes for plastocyanin at this point in the Z pathway.) The pathway from the plastoquinone pool to plastocyanin is essentially the same as the mitochondrial pathway from ubiquinone to cytochrome c (compare Figs. 10-10 and 9-24).

From plastocyanin electrons flow to the P700 chlorophyll molecule at the reaction center of photosystem I. After excitation by light absorption electrons are transferred from P700 to the primary acceptor of this photosystem, a modified form of chlorophyll known as A_0. From A_0 electrons flow to A_1, a quinone derived from vitamin K. The electrons then flow through a

chain of three Fe/S centers, designated F_X, F_A, and F_B. The sequence of electron flow within the three Fe/S centers remains uncertain but may proceed from F_X to either of the other two. The pathway of electron flow within photosystem I is therefore:

$$P700 \longrightarrow A_0 \longrightarrow A_1 \longrightarrow F_X \longrightarrow (F_A/F_B)$$

Electrons are next transferred to the first carrier outside photosystem I, *ferredoxin*, an Fe/S protein that acts as a separate, highly mobile electron carrier in the chloroplast system. From ferredoxin electrons flow along the short final chain from FAD to NADP to complete the Z pathway. FAD (flavin adenine dinucleotide; see Fig. 9-14) is the prosthetic group of an enzyme called *ferredoxin-NADP oxidoreductase*. This oxidoreductase, although a separate protein of the chloroplast electron transport system, evidently remains closely associated with photosystem I.

Electrons receive two boosts in energy as they move through the Z pathway, one in photosystem II and one in photosystem I (see Fig. 10-10). The two consecutive boosts raise the electrons to energy levels high enough to reduce NADP. Along the way some energy is tapped from the electrons and used to produce an H^+ gradient. The gradient is established primarily through the activity of the b_6/f complex, which, like its structural counterpart in mitochondria (mitochondrial complex III; see p. 334), acts as an electron-driven

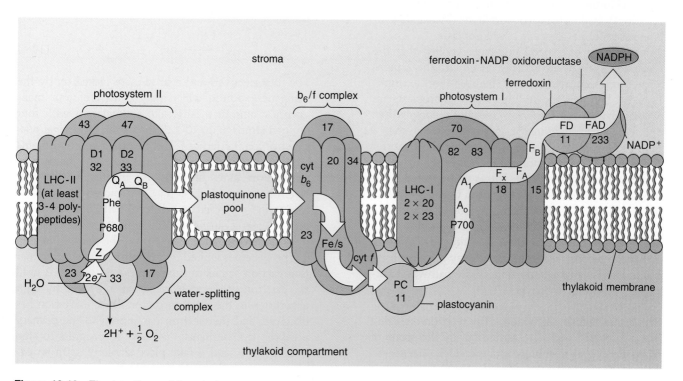

Figure 10-12 The locations of the photosystems and electron transport carriers in thylakoid membranes. The numbers give the approximate molecular weight of polypeptides in thousands.

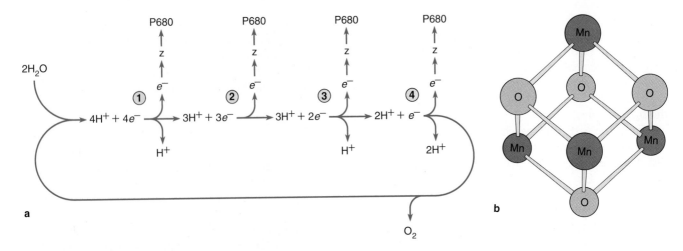

Figure 10-13 The reaction pathway splitting water to generate electrons for the Z pathway. **(a)** The series of stepwise reactions removing electrons from water and delivering them one at a time to the P680 reaction center of photosystem II. **(b)** A possible arrangement of manganese and oxygen atoms in the manganese center.

pump actively transporting H^+. In chloroplasts H^+ is pumped from the stroma into thylakoid compartments.

The pattern of electron flow in the Z pathway from H_2O to NADP is frequently called *noncyclic photosynthesis* because electrons are removed from water and travel in a one-way direction to NADP. Figure 10-12 summarizes the organization of the individual electron carriers within the photosystems and carrier complexes of the chloroplast system. All the components of the Z pathway are located on or within the thylakoid membranes or stromal lamellae of chloroplasts.

A variety of experiments have verified that the two photosystems are actually linked into the Z pathway proposed by Hill and Bendall. For example, L. N. M. Duysens showed that when chloroplasts are irradiated with red light at wavelengths longer than 680 nm (expected primarily to excite photosystem I but not II), cytochrome f becomes oxidized. At shorter wavelengths expected primarily to excite photosystem II, cytochrome c_1 is reduced. This is the expected result if cytochrome f occurs in an electron transport system linking the two photosystems, as proposed by the Z scheme. Equivalent experiments have supported other segments of the Z pathway. Recently, E. Lam and R. Malkin reconstructed the Z pathway in active form by adding isolated components of the pathway to artificial phospholipid bilayers.

The Water-Splitting Reaction The reaction oxidizing water into $2H^+$ and $\frac{1}{2}O_2$, carried out by the group of three peripheral membrane polypeptides attached to the surface of photosystem II and the manganese center, is believed to involve two molecules of water and to proceed in a four-step pathway first proposed by B. Kok and his coworkers (Fig. 10-13). One of the major

lines of evidence used to develop the model is the fact that if photosystem II is illuminated by light in very brief flashes, four sequential flashes are required to complete the water-splitting reaction, and O_2 is released at only one of the flashes. Movement from one step to the next in the pathway is intimately linked to the excitation of electrons in photosystem II.

On this basis Kok proposed that at the beginning of the pathway (Fig. 10-13a), two molecules of water are split:

$$2H_2O \longrightarrow 4H^+ + 4e^- + O_2 \qquad (10\text{-}4)$$

The four electrons and four H^+ are picked up by the manganese center, which may consist of four manganese atoms in a latticelike combination with four oxygen atoms. (Fig. 10-13b shows one of several possibilities for the lattice.) The electrons and hydrogens are then removed one at a time from the manganese center in a series of steps. As the first of the four electrons is removed (step 1), it is picked up by the Z carrier and delivered to a chlorophyll P680 at the reaction center. Transfer to P680 reoxidizes Z, and readies it for acceptance of the electron to be released in step 2. The system is limited by excitation, however, so that the reaction sequence cannot progress to step 2 until the electron delivered from step 1 is excited by light absorption in the reaction center. Once excitation has occurred, and the electron has passed to the primary acceptor, the mechanism is ready to progress to step 2. Steps 2, 3, and 4 take place similarly, with an excitation completing each step. Hydrogens are removed from the manganese center at some but not all of the steps; current indications are that one H^+ is released at steps 1 and 3, and the remaining two H^+ at step 4,

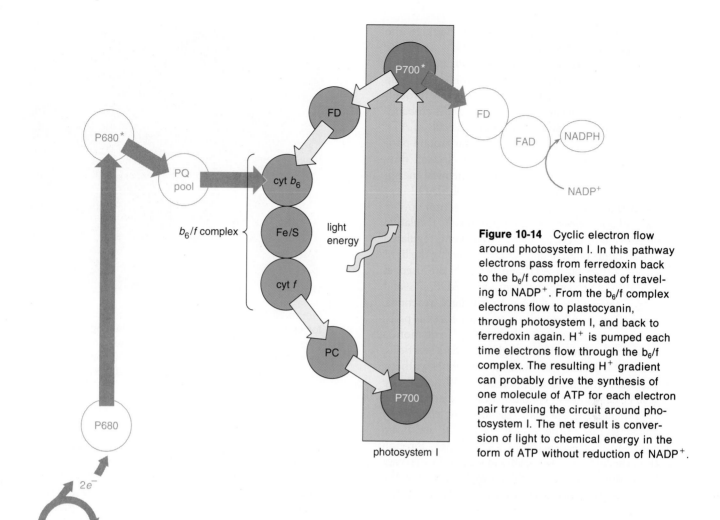

Figure 10-14 Cyclic electron flow around photosystem I. In this pathway electrons pass from ferredoxin back to the b_6/f complex instead of traveling to $NADP^+$. From the b_6/f complex electrons flow to plastocyanin, through photosystem I, and back to ferredoxin again. H^+ is pumped each time electrons flow through the b_6/f complex. The resulting H^+ gradient can probably drive the synthesis of one molecule of ATP for each electron pair traveling the circuit around photosystem I. The net result is conversion of light to chemical energy in the form of ATP without reduction of $NADP^+$.

as shown in Figure 10-13a. The two oxygen atoms removed in the preliminary reaction of the pathway are released as O_2 at the last step.

Many features of the pathway are still not understood. Both Ca^{2+} and Cl^- are required for the reaction to proceed, at levels of about two to three Ca^{2+} and four to five Cl^- per manganese center. The role of these ions is unknown; either may play structural or regulatory roles, or may contribute to catalysis of the reaction. The precise structure of the manganese center and its linkage to photosystem II are also unknown. It is considered likely that the binding site is located in a pocket on the surface of either of the D1 or D2 polypeptides or in the interface between the two polypeptides. Valence changes of the manganese ions of the center between $2+$, $3+$, and $4+$ states undoubtedly contribute to the uptake and release of electrons. However, the combinations and sequence of these changes remain unknown. The function of the three soluble polypeptides is also uncertain. Curiously, the water-splitting reaction can proceed without the polypeptides if Ca^{2+} and Cl^- concentrations are raised to abnormally high levels. This may indicate that the polypeptides simply promote Ca^{2+} and Cl^- binding, so that the reaction can proceed at physiological concentrations of these ions.

The water-splitting reaction carried out by photosystem II, developed in cyanobacteria some 2 to 3 billion years ago, is one of the most significant and fundamentally important of all biological interactions. The reaction enabled photosynthetic organisms to use one of the most abundant substances on earth—water—as an electron source. The reaction is also responsible for the oxygen present in the atmosphere of our planet. The appearance of oxygen in the atmosphere paved the way for the evolution of aerobic organisms, in which oxygen serves as the final acceptor for electrons removed in cellular oxidations.

Cyclic Electron Flow Within the Z Pathway

Electrons may flow cyclically within a segment of the Z pathway by operation of a "shunt" that forms a closed circuit around photosystem I (Fig. 10-14). In this pathway electrons pass from ferredoxin back to the b_6/f complex instead of traveling to $NADP^+$. From the b_6/f complex electrons flow to plastocyanin, through photosystem I, and back to ferredoxin. This closed cycle pumps additional H^+ each time electrons flow through the b_6/f complex. The resulting H^+ gradient is probably sufficient to drive the synthesis of one molecule of ATP for each electron pair traveling the circuit around photosystem I. The net result, as long as electron flow remains cyclic, is conversion of light to chemical energy in the form of ATP without reduction of $NADP^+$.

A balance of cyclic and noncyclic electron flow in the Z pathway allows requirements for ATP relative to NADPH to be precisely met. The balance is regulated by the amounts of ADP and $NADP^+$ entering the light reactions. When both ATP and NADPH are used in quantity by the dark reactions, and ADP, P_i, and $NADP^+$ are readily available to the light reactions, electrons flow predominantly along the noncyclic pathway. Under conditions in which NADP oxidation is low but ATP consumption is high in the dark reactions, much of the NADP surrounding the photosynthetic membranes is in reduced form but ADP and P_i are available. Under these conditions cyclic electron flow is predominant. If the dark reactions are completely inhibited, so that neither $NADP^+$ nor ADP and P_i are available to the light reactions, electron flow through both cyclic and noncyclic pathways halts. In this case light energy absorbed by chlorophylls in any of the light-harvesting assemblies is released as fluorescence and heat.

H^+ Pumping by the Z Pathway

Although the b_6/f complex is the primary H^+ pump of the Z pathway, the H^+ gradient is increased at two other sites as well. One is the water-splitting reaction (Reaction 10-1) of photosystem II, which takes place on the side of thylakoid membranes facing the thylakoid compartment. The reaction releases two H^+ on this side for each electron pair entering the Z pathway. The second reaction takes place as electrons flow from ferredoxin to NADP via the ferredoxin-NADP oxidoreductase carrier. As $NADP^+$ is reduced to NADPH, it removes one H^+ from the stroma. Removal of this H^+ adds to the gradient created by electron transport.

As many as four H^+ per electron pair are pumped by the b_6/f complex as it cycles between oxidized and reduced forms. Recently, E. C. Hart and his colleagues inserted isolated and purified b_6/f complexes into artificial lipid vesicles. When supplied with reduced plastoquinone as an electron donor and oxidized plastocyanin as an acceptor, the reconstituted system pumped H^+ from the medium into the vesicle interiors as the b_6/f complex cycled between oxidized and reduced states.

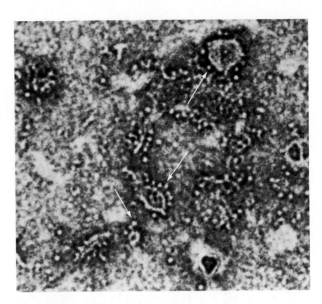

Figure 10-15 Isolated thylakoid membranes prepared for electron microscopy by negative staining. The headpieces and stalks of the CF_oCF_1 ATPase can be seen extending from the membrane surfaces (arrows). Courtesy of E. Racker, from *J. Biolog. Chem.* 248:8281 (1973).

Using the H^+ Gradient to Synthesize ATP

The H^+ gradient established by the Z pathway is used as the energy source for ATP synthesis by the chloroplast CF_oCF_1 ATPase. The CF_1 subunit of the chloroplast ATPase contains five subunits equivalent to the α-, β-, γ-, δ-, and ϵ-polypeptides of the bacterial F_oF_1 ATPase, and similar to those of the mitochondrial enzyme (see Table 10-1). These polypeptides are arranged into a spherical headpiece built up from three α- and three β-polypeptides, carried on a stalk formed from one each of the γ-, δ-, and ϵ-polypeptides (see Fig. 9-31). The stalk anchors the CF_1 subunit to the CF_o basal unit buried in the membrane. The CF_1 subunits can be seen as "lollipop" structures extending from the outer surfaces of photosynthetic membranes prepared for electron microscopy by negative staining (Fig. 10-15).

Less is known about the structure of the CF_o basal unit of the chloroplast ATPase. At least four polypeptides have been identified in isolated CF_o preparations; three are considered to be equivalent to the a, b, and c polypeptides of the bacterial and mitochondrial F_o subunits (see Table 10-1).

As in the mitochondrial F_oF_1 ATPase, the active site of the enzymatic mechanism converting ADP to ATP is concentrated in the β-polypeptides of the CF_1 head unit (see p. 337). The CF_o subunit provides the membrane channel conducting H^+ through the CF_oCF_1 ATPase. Conformational changes resulting from the H^+ flow are transmitted through the stalk to the headpiece; in some unknown manner the conformational changes drive the synthesis of ATP from ADP and inorganic phosphate (see p. 343).

The chloroplast CF_oCF_1 ATPase is essentially irreversible, even more so than the mitochondrial F_oF_1 ATPase. The complete irreversibility of the chloroplast CF_oCF_1 ATPase may be connected with the fact that no H^+ gradient established by electron transport can be set up across thylakoid membranes under conditions of darkness. Because the H^+ gradient is nonexistent in darkness, no opposing force acts as a brake to prevent the CF_oCF_1 ATPase from acting as an active transport pump. (In mitochondria electron transport is essentially continuous so that an H^+ gradient always exists across the membrane housing the F_oF_1 ATPase.) An easily reversible CF_oCF_1 ATPase, unopposed by an H^+ gradient, might run backward in the dark, hydrolyzing all available ATP to push H^+ across the photosynthetic membranes. The irreversibility of the chloroplast CF_oCF_1 ATPase evidently depends on the ϵ-subunit (see the Experimental Essay by R.E. McCarty on p. 338).

Organization of Light-Reaction Components in Thylakoid Membranes

The molecules of the Z pathway are organized into four multimolecular complexes and three individual electron carriers. The four multimolecular complexes include photosystem I, photosystem II, the b_6/f complex, and the ferredoxin-NADP oxidoreductase. Each photosystem is tightly complexed with its LHC antenna. (Photosystem II may separate from its LHC-II antenna under certain conditions; see below.) Photosystem I is also evidently associated with ferredoxin-NADP oxidoreductase. The three individual carriers of the Z pathway—plastoquinone, plastocyanin, and ferredoxin—exist as single molecules in or on the photosynthetic membranes.

The locations of these Z-pathway components within thylakoid membranes have been analyzed by the same techniques used for other cellular membranes (see p. 171). These techniques include exposure of inside or outside membrane surfaces to radioactive or other marker groups, digestive enzymes, antibodies developed against the Z-pathway components, or artificial electron donors or acceptors. For example, artificial electron donors that can substitute for water work only on inside-out and not right-side-out vesicles produced from thylakoids, demonstrating that the water-splitting segment of photosystem II faces the thylakoid compartments.

The distribution of Z-pathway components in thylakoid membranes, as determined by localization studies, is shown in Figure 10-12. The two photosystems and the b_6/f complex are integral membrane units that extend entirely through thylakoid membranes. The plastoquinone pool is suspended in solution in the hydrophobic membrane interior. The remaining components are located on thylakoid membrane surfaces. Plastocyanin is located on the membrane surface facing the thylakoid compartments, and ferredoxin and ferredoxin-NADP oxidoreductase on the side facing the stroma. The LHC-I and LHC-II antennas are membrane-spanning complexes associated closely with their respective photosystems. The three polypeptides forming the water-splitting complex, as noted, are located on the part of photosystem II extending into the thylakoid compartment. NADP reduction occurs in the stroma surrounding thylakoid membranes. Most components of the Z pathway have the same orientation and distribution in thylakoids and stromal lamellae.

Although there are enough photosystem complexes and electron transport elements to form several hundred Z pathways in each thylakoid, the light reactions do not seem to be organized into rigid Z assemblies. With certain exceptions the components are evidently free to move through or over the fluid thylakoid membrane and interact by random collisions.

The exceptions to free movement of the Z-pathway components were discovered by B. Andersson, J. M. Anderson, and others by an application of the freeze-fracture technique for electron microscopy (see p. 120), used in combination with several biochemical and genetic approaches. Among the most useful have been genetic techniques using mutants that lack one or more of the Z-pathway components. These techniques allowed the particles visible in freeze-fracture preparations to be identified by correlating the absence of a particle of characteristic size and shape with a mutant lacking a particular Z-pathway component.

The most striking restriction to free mobility revealed by these techniques involves photosystems I and II. Photosystem I is apparently confined almost entirely to stromal lamellae and regions where outer thylakoid surfaces directly face the stroma. Most photosystem II complexes lie in regions of thylakoid membranes that are fused to other thylakoid membranes within stacks (Fig. 10-16; see also Fig. 10-2). The remaining components of the Z pathway appear to be distributed more or less uniformly throughout thylakoid membranes and stromal lamellae. The segregation of photosystem II complexes to stacked regions of thylakoids is evidently an evolutionary adaptation that balances the activities of the two photosystems in light absorption and electron flow (see Information Box 10-1).

Segregation of photosystems I and II in distinct regions of thylakoid membranes means that noncyclic electron transport from photosystem II to I in the Z pathway must proceed via mobile electron carriers that can shuttle between the stacked and unstacked regions of thylakoids. The most likely candidates for this shuttle are plastoquinone and plastocyanin, both relatively small molecules that can diffuse rapidly through or along thylakoid membranes. In their role as shuttles

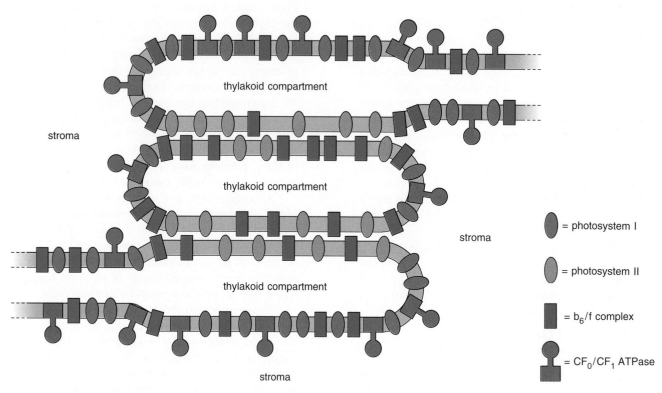

stroma

thylakoid compartment

stroma

thylakoid compartment

thylakoid compartment

stroma

stroma

= photosystem I

= photosystem II

= b_6/f complex

= CF_0/CF_1 ATPase

Figure 10-16 Distribution of Z-pathway components in photosynthetic membranes. Photosystem II complexes are largely confined to membrane segments that are fused in thylakoid stacks; photosystem I complexes and the CF_0CF_1 ATPase are located in unfused thylakoid regions that face the stroma, and in stromal lamellae. The remaining components are distributed approximately equally throughout thylakoid membranes and stromal lamellae.

reduced plastoquinone molecules would transport electrons from photosystem II in stacked thylakoid regions to b_6/f complexes. Mobile plastocyanin carriers would make the necessary connections between the b_6/f complexes and photosystem I in unstacked regions. The shuttle activity of plastoquinone and plastocyanin molecules probably proceeds simply through random motions and collisions, aided by the relatively small size of the two shuttle molecules and the highly fluid nature of the thylakoid bilayer.

Photosystem II complexes are evidently held in the fused regions of stacked thylakoid membranes by interactions between LHC-II antennas of adjacent membranes (Fig. 10-17). Among the evidence supporting this conclusion are observations of mutants lacking the LHC-II complex. In these mutants thylakoids do not fuse into grana stacks.

In freeze-fracture preparations the CF_0CF_1 ATPase can be identified in stromal lamellae and in unfused segments of thylakoid membranes facing the stroma (see Fig. 10-16). Extension of the ATPase headpieces into the stroma and the direction of H^+ pumping in

chloroplasts make thylakoid compartments the functional equivalents of the intermembrane compartment in mitochondria (Fig. 10-18).

The Light Reactions and the Chemiosmotic Mechanism

In 1954 D. I. Arnon and his colleagues first demonstrated the connection between electron transport and ATP synthesis in chloroplasts. These investigators supplied isolated spinach chloroplasts with ADP and inorganic phosphate, and detected ATP synthesis in the light. Since the quantity of ATP synthesized was correlated with the amount of electron flow in response to light, Arnon and his coworkers proposed that ATP synthesis in the isolated chloroplasts was somehow coupled to electron transport. This light-driven ATP synthesis was termed *photophosphorylation*.

The nature of the connection between electron transport and ATP synthesis was not understood until Mitchell advanced his chemiosmotic hypothesis in 1961

Figure 10-17 Relationships between thylakoid stacking and LHC-II antennas. **(a)** Binding attractions between LHC-II antennas hold thylakoids in stacks and confine photosystem II (PS II) complexes to fused regions of thylakoid membranes. **(b)** Conformational changes in LHC-II antennas alter their binding affinities, releasing both the attractions that hold thylakoids in stacks and the linkage of the antennas to the photosystem complexes. As a consequence, thylakoids unstack and LHC-II antennas and photosystem II complexes separate and are free to diffuse individually through the thylakoid membranes. Detachment of LHC-II antennas from the photosystem complexes greatly reduces the efficiency and rate of conversion of light to chemical energy in photosystem II.

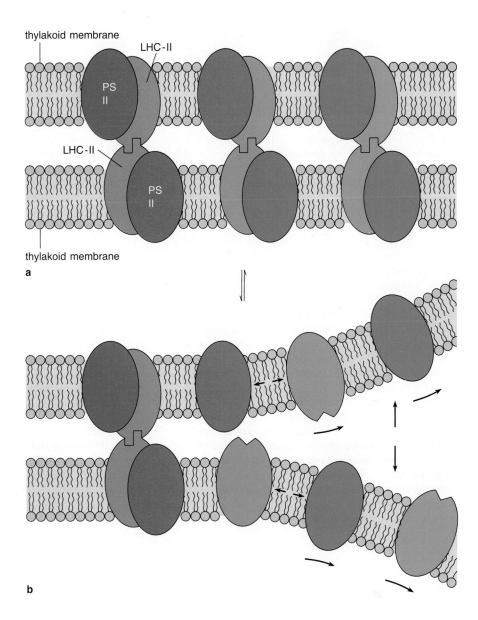

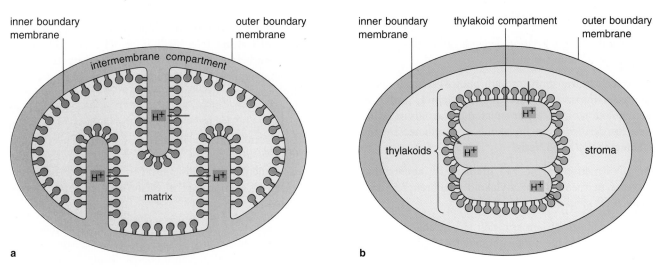

Figure 10-18 The direction of H^+ pumping by the electron transport systems of mitochondria **(a)** and chloroplasts **(b)**. H^+ is transported into the intermembrane compartment in mitochondria and into the thylakoid compartments in chloroplasts.

Balancing the Photosystems

For the Z pathway to operate most efficiently, the activity of the two photosystems must be finely balanced. If photosystem I lags behind photosystem II in light absorption and energy conversion, carriers in the electron transport pathway linking the two photosystems will accumulate in reduced form and the efficiency of the light reactions will drop. If photosystem II lags, the linking carriers will accumulate in oxidized form and electrons will be unavailable to photosystem I. The balancing mechanism, related to physical segregation of the two photosystems in stacked grana, was recently discovered by a group of investigators including P. Horton, J. F. Allen, J. Bennet, K. E. Steinback, and C. J. Arntzen. The ingredients of the balancing mechanism include the plastoquinone pool, a protein kinase (an enzyme that adds phosphate groups to a protein) specific in its activity for the LHC-II antenna, a phosphatase (an enzyme that removes phosphate groups), and LHC-II antennas.

The phosphatase is continually active in chloroplasts. Activity of the protein kinase depends on the redox state of the plastoquinone pool. When the plastoquinones are relatively oxidized, the protein kinase is inhibited. Accumulation of reduced plastoquinones activates the kinase. In activated form the kinase adds phosphate groups to portions of the LHC-II complex exposed at the thylakoid membrane surface.

In unphosphorylated form LHC-II antennas remain associated with photosystem II complexes and function normally in their two roles: passing excitation energy to photosystem II and binding thylakoids into stacks. Phosphorylation of the LHC-II antenna inhibits its binding to both the photosystem II complexes and the adjacent thylakoids. As a consequence, the LHC-II antennas separate from the photosystem II complexes and the grana stacks come apart.

The entire mechanism provides a feedback control regulating the activity of photosystem II (see the figure in this box). If photosystem II absorbs a greater proportion of the available light and becomes more active than photosystem I, the proportion of reduced plastoquinone molecules increases in the plastoquinone pool. This condition stimulates the protein kinase, leading to phosphorylation of LHC-II antennas. In phosphorylated form the LHC-II antennas dissociate from the photosystem II complexes. As a result, the efficiency of photosystem II in absorbing light energy is reduced, and the balance between the photosystems in the Z pathway is restored. At the same time the degree of thylakoid stacking falls because the binding of adjacent thylakoid membranes by LHC-II antennas is inhibited.

If photosystem II drops below photosystem I in activity, the control mechanism adjusts the system in the opposite direction. Under these conditions the plastoquinone pool becomes relatively oxidized, and activity of the protein kinase is inhibited. LHC-II antennas are then dephosphorylated by the continuing activity of the phosphatase.

(see pp. 311 and 344). The applicability of the chemiosmotic hypothesis to chloroplasts is supported by several major lines of evidence:

1. Photophosphorylation depends on the presence of closed membranous vesicles.

2. Electron transport through photosystems II and I, or cyclic transport around photosystem I, results in accumulation of H^+ inside isolated thylakoids or inside vesicles derived from thylakoids, as predicted by the chemiosmotic hypothesis.

3. Destruction of the H^+ gradient by agents that increase permeability of thylakoid membranes to H^+ halts ATP synthesis.

4. Imposition of an artificial H^+ gradient across closed thylakoid membranes stimulates ATP synthesis.

The supporting evidence mentioned in (4), stemming from experiments conducted by A. T. Jagendorf and E. Uribe in 1966, is among the strongest in favor of the chemiosmotic hypothesis. Jagendorf and Uribe lowered the internal pH of isolated chloroplasts by placing them in a solution containing acids (at pH 4) that could penetrate the chloroplast membranes. Subsequent transfer to a medium at pH 8 created an H^+ gradient between the thylakoid compartments and the stroma, with H^+ concentration much higher in the thylakoid compartments. Creation of the gradient produced an immediate burst of ATP synthesis that continued until the H^+ concentration difference between the stroma and thylakoids fell to zero. To eliminate light as a source of energy for ATP synthesis, the experiment was conducted in the dark. (For other experiments supporting the chemiosmotic hypothesis, see p. 344.)

The light reactions absorb light energy and convert it into chemical energy in the form of electrons held in stable orbitals at elevated energy levels. The electrons are derived from water in a reaction that releases oxygen as an important by-product. As electrons pass along the electron transport system toward their final

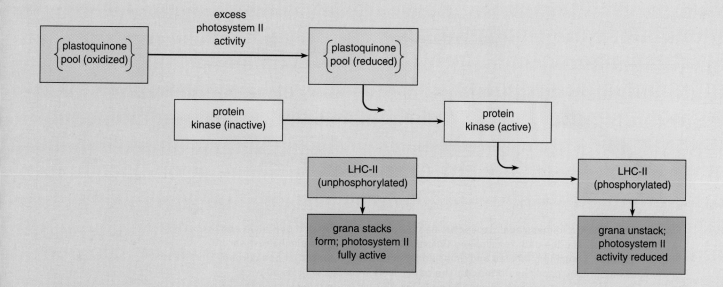

In dephosphorylated form the LHC-II antennas associate with photosystem II complexes, thereby increasing the activity of photosystem II and restoring the balance of the Z pathway. At the same time the proportion of thylakoids in stacked form increases.

The feedback mechanism balancing the two photosystems explains an older observation concerning the relative distribution of thylakoids between the stacked and unstacked form in higher plants. In shade-grown plants about 60% of the thylakoids in chloroplasts are fused into stacks. In full sunlight the proportion of stacked membranes falls to about 40%. This stacking difference depends on the distribution of wavelength intensity char-acteristic of full sunlight and shaded light, and on the mechanism balancing photosystem activity in the light reactions. In full sunlight all the wavelengths in the visible spectrum are approximately equal in intensity, and the two photosystems operate at equivalent levels of efficiency. In shade the far red wavelengths are relatively more intense, a condition that reduces the efficiency of photosystem II. Under this condition LHC-II complexes are extensively dephosphorylated, which activates the binding affinity of LHC-II antennas for photosystem II complexes and adjacent thylakoid membranes. Both photosystem II activity and the proportion of thylakoids stacked into grana then increase in the shade-grown plants.

acceptor, $NADP^+$, some of their energy is tapped off and used to establish an H^+ gradient across thylakoid membranes. The H^+ gradient serves as the immediate energy source for ATP synthesis by the chloroplast CF_o/CF_1 ATPase. The primary products of the light reactions, ATP and NADPH, serve as the sources of energy and reducing power for the dark reactions of photosynthesis.

THE DARK REACTIONS

In the dark or light-independent reactions chemical energy produced in the light reactions is used to convert carbon dioxide into carbohydrates and a variety of other organic products. Although commonly called the dark reactions because they do not depend directly on light, the reactions of this series actually take place primarily in the daytime, when ATP and NADPH are available from the light reactions.

Tracing the Pathways of the Dark Reactions

The first significant progress in unraveling the dark reactions was made in the 1940s, when radioactive compounds became available to biochemists. One substance in particular, CO_2 labeled with the radioactive carbon isotope ^{14}C, was critical to this research.

Melvin Calvin, Andrew A. Benson, and their colleagues began using radioactive CO_2 in 1945 to trace the pathways of dark reactions. In their experiments, which required ten years of intensive effort, Calvin and his colleagues allowed photosynthesis to proceed in a green alga, *Chlorella*, in the presence of radioactive CO_2. At various times after exposure to labeled CO_2, cells were removed and placed in hot alcohol, which killed the cells and stopped the dark reactions instantly. Extracts of carbohydrates and other substances were made and separated by two-dimensional paper chromatography (Fig. 10-19). The substances in the radioactive spots separated by the chromatography were then identified.

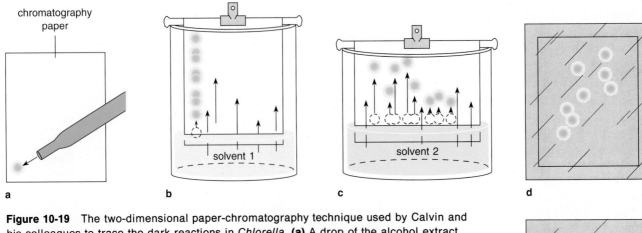

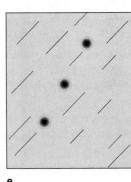

Figure 10-19 The two-dimensional paper-chromatography technique used by Calvin and his colleagues to trace the dark reactions in *Chlorella*. **(a)** A drop of the alcohol extract of *Chlorella* cells, containing labeled and unlabeled compounds, is placed at one corner of the chromatography paper and dried. **(b)** The paper is then placed with its edge touching the solvent used to produce separation in the first direction (a water solution of butyl alcohol and propionic acid was used). The compounds in the extract separate and are carried upward by the migrating solvent at rates that vary according to molecular weight and solubility. **(c)** The paper is dried, turned 90°, and placed in the second solvent (a water solution of phenol was used). The compounds migrate upward at rates that are different from their mobility in the first solvent. This step separates compounds that migrated similarly in the first solvent. The final positions allow some of the compounds to be identified by comparison with previously established standards. **(d)** The dried paper is covered with a sheet of photographic film. Any radioactive compounds expose spots on the film. **(e)** The film is developed and compared with the paper chromatograph to identify radioactive compounds.

If the carbohydrate extracts were made within a few seconds after exposure of *Chlorella* cells to labeled CO_2, most of the radioactivity could be identified with the three-carbon substance *3-phosphoglycerate (3PGA)*. The rapidity with which the labeled CO_2 was incorporated into 3PGA indicated that it is one of the earliest products of photosynthesis. If extracts were made after longer periods, radioactive label showed up in more complex substances, including a variety of six-carbon sugars, sucrose, and starch.

In other experiments Calvin and his colleagues reduced the amount of CO_2 available to the *Chlorella* cells so that photosynthesis was delayed even though light was adequate. Under these conditions a five-carbon sugar, *ribulose 1,5-bisphosphate (RuBP)*, accumulated in the chloroplasts. This result suggested that RuBP is the first substance to react with CO_2 in the dark reactions, and accumulates if CO_2 is in short supply. By similar methods most of the remaining compounds intermediate between CO_2 and six-carbon sugars were identified.

Using this information Calvin, with his colleagues Benson and J. A. Bassham, pieced together the dark reactions of photosynthesis. The cycle is now called the *Calvin–Benson* or *C_3 cycle* because the first product of the cycle, 3PGA, is a three-carbon molecule. Calvin was awarded the Nobel Prize in 1961 for his brilliant work deducing the reactions of the cycle.

The C_3 Cycle

The reactions worked out by Calvin and his colleagues resemble the citric acid cycle in that the intermediate compounds of the sequence are continuously regenerated as the cycle turns. The cycle uses CO_2, ATP, and NADPH as net reactants and releases ADP, $NADP^+$, and a three-carbon carbohydrate derived from 3PGA, *3-phosphoglyceraldehyde (3PGAL)*, as net products (Fig. 10-20).

The individual reactions of the C_3 cycle are shown in Figure 10-21. In reaction 1, CO_2 combines directly with RuBP. This reaction, catalyzed by the enzyme *ribulose 1,5-bisphosphate carboxylase (RuBP carboxylase)*, produces two molecules of 3PGA, the substance detected by the labeling experiments as the first product of the dark reactions. One of the 3PGAs contains newly incorporated CO_2 in the position marked by an asterisk in Figure 10-21. The reaction requires no input of energy because the two three-carbon products exist at a much lower energy level than RuBP, which can be considered as a high-energy substance. The first reaction of the C_3 cycle accomplishes carbon *fixation*: the conversion of the carbon of an inorganic molecule, CO_2, into the carbon of an organic substance.

The fixation reaction actually proceeds in two steps. Initially, CO_2 combines with RuBP to produce a six-carbon intermediate that remains bound to the

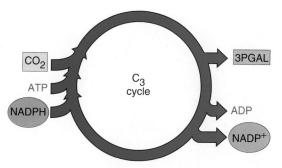

Figure 10-20 Overall reactants and products of the C_3 cycle. The cycle uses CO_2, ATP, and NADPH as net reactants and releases ADP, $NADP^+$, and 3-phosphogly-ceraldehyde (3PGAL) as net products.

active site of RuBP carboxylase. This unstable intermediate is immediately split, by the addition of a molecule of water, into the two molecules of 3PGA. This hydrolysis accounts for the H_2O molecule shown as a reactant in reaction 1.

The next two reactions of the cycle, which proceed at the expense of ATP and NADPH, yield the net carbohydrate product of the cycle, 3PGAL. In the first

of these reactions (reaction 2 in Fig. 10-21), a phosphate group is added to each of the two 3PGA molecules produced by reaction 1. These phosphate groups are derived from ATP, which is converted to ADP in the process. The products of the second reaction each accept an H^+ and two electrons from NADPH in the next reaction (reaction 3). One phosphate is removed from each of the reactants at the same time, yielding two molecules of 3PGAL for each molecule of CO_2 added to RuBP in reaction 1. The reaction sequence to this point uses two ATP and two NADPH:

$$CO_2 + RuBP + 2ATP + 2NADPH \longrightarrow$$
$$2(3PGAL) + 2NADP^+ + 2ADP + 2P_i \quad (10\text{-}5)$$

Some of the 3PGAL produced at this step is required to regenerate the RuBP used in the first step in the cycle, and some is released as a surplus to the cycle.

Regeneration of RuBP occurs through a series of reactions (reaction series 4, not diagrammed in Fig. 10-21) that yields as an initial product the five-carbon, one-phosphate sugar *ribulose 5-phosphate.* Another phosphate is removed from ATP and added to this product in reaction 5 to produce RuBP. This reaction

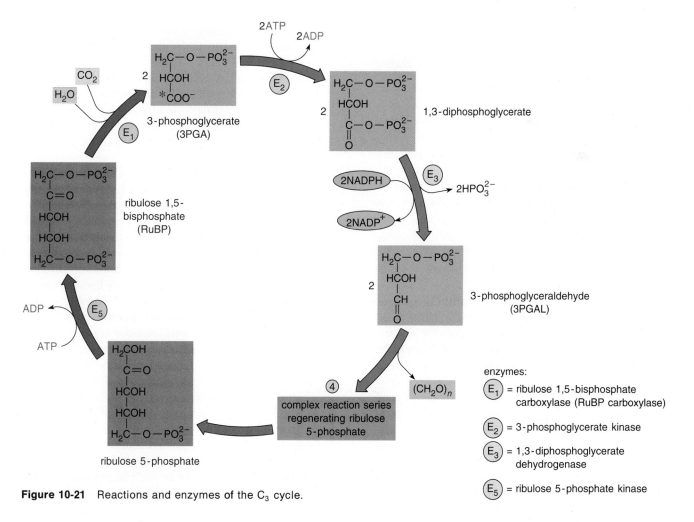

Figure 10-21 Reactions and enzymes of the C_3 cycle.

replaces the RuBP used in the first step in the cycle, and the entire series is ready to begin again.

Three turns of the C_3 cycle are required to yield one molecule of 3PGAL as a surplus that can be used for synthesis of more complex carbohydrates. In three turns of the cycle through reaction 3 in Figure 10-21, six molecules of 3PGAL are formed (for a total of 18 carbons). Five of these molecules (totaling 15 carbons) enter the complex series that regenerates the three RuBP molecules used in the three turns. The remaining molecule of 3PGAL is surplus and can enter reaction pathways yielding glucose, sucrose, starch, and a variety of other complex organic substances. The C_3 cycle can be considered as a "breeder" reaction because it makes a greater quantity of one of its own intermediates (3PGAL) than it uses.

A single turn of the cycle thus takes up one molecule of CO_2 and yields one surplus "unit" of carbohydrate (CH_2O). Three turns are required to form one surplus molecule of 3PGAL; six turns are required to make enough (CH_2O) units to synthesize a six-carbon sugar such as glucose.

For each turn of the cycle two ATP and two NADPH are used in reactions 2 and 3, and one ATP is used in reaction 5, for a total of three ATP and two NADPH required for each turn. Although one of the phosphates derived from ATP is attached to 3PGAL, this phosphate is eventually released as 3PGAL is converted into other substances. As net reactants and products one complete turn of the cycle therefore includes:

$$CO_2 + 3ATP + 2NADPH \longrightarrow$$
$$(CH_2O) + 2NADP^+ + 3ADP + 3P_i \quad (10\text{-}6)$$

The cycle is about 80% efficient in conserving the energy of the three ATP and two NADPH in the (CH_2O) carbohydrate product. Synthesis of a molecule of glucose, which requires six turns of the C_3 cycle, involves:

$$6CO_2 + 18ATP + 12NADH \longrightarrow$$
$$C_6H_{12}O_6 + 12NADP^+ + 18ADP + 18P_i \quad (10\text{-}7)$$

The Key Enzyme of the C_3 Cycle: RuBP Carboxylase and Its Regulation RuBP carboxylase (also called *RuBisCo* and *RuBPCase*), which catalyzes the first reaction of the C_3 cycle, is unique to photosynthetic organisms including photosynthetic bacteria, cyanobacteria, algae, and plants. There are so many copies of the enzyme in chloroplasts that RuBP carboxylase may make up 50% or more of the total protein of plant leaves. As such, it is probably the world's most abundant protein. It is estimated to exist in an amount of some 40 million tons, equivalent to about 20 pounds for every human on the earth! In this quantity the enzyme fixes about 100 billion tons of CO_2 into car-

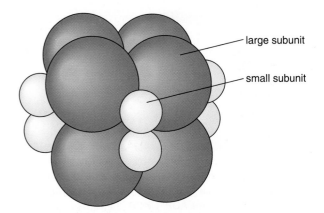

Figure 10-22 The possible three-dimensional structure of RuBP carboxylase, which contains eight copies each of its large and small subunits. The large subunit is encoded in chloroplast DNA and synthesized on ribosomes inside the organelle; the small subunit is encoded in the nucleus and synthesized on cytoplasmic ribosomes outside the chloroplast.

bohydrates annually. Its abundance probably compensates for the fact that the enzyme catalyzes CO_2 fixation relatively slowly, at rates of only about three molecules of CO_2 per enzyme per second.

RuBP carboxylase is assembled from two polypeptides. One, the *large subunit*, is a 56,000-dalton polypeptide encoded in DNA molecules included in the chloroplast (see below). The second, the *small subunit*, is a 14,000-dalton polypeptide encoded in the DNA of the plant cell nucleus. An active RuBP carboxylase enzyme is an octamer assembled from eight each of the large and small subunits, forming a complex with a total molecular weight of more than 550,000 (Fig. 10-22).

The position of RuBP carboxylase as the initial enzyme of the C_3 cycle makes it a key regulatory site. One mechanism regulates the C_3 cycle by controlling the amount of RuBP carboxylase available to the dark reactions. The adjustments are carried out by regulation of the genes encoding the two polypeptide subunits of RuBP carboxylase. This entails gene regulation in both the cell nucleus and chloroplasts because one of the subunits is encoded in nuclear DNA and the other in chloroplast DNA. The mechanisms coordinating gene regulation in the two locations are unknown.

A second major pathway regulating RuBP carboxylase involves activators and inhibitors that adjust activity of the enzyme. At least four different controls balance RuBP carboxylase activity to match the rate of the light reactions. One is the pH of the stroma. In darkness, when no electrons flow through the Z pathway, the stroma and thylakoid compartments are both at the same pH, about 7. At this pH RuBP carboxylase is essentially inactive. As daylight begins and the light reactions start, H^+ is pumped from the stroma into thylakoid compartments by the photosynthetic electron transport system. As a consequence, pH in the thylakoid compartments gradually falls to about 4, and the stroma, by virtue of the H^+ pumped from this region into the thylakoid compartments, rises to about pH 8. At this pH RuBP carboxylase reaches its pH optimum (see p. 98) and is potentially able to catalyze the first reaction of the C_3 cycle at its maximum rate. Any reduction in stroma pH because of slower H^+ pumping by the light reactions causes a proportionate reduction in activity of RuBP carboxylase as it departs from its pH optimum.

The second control involves Mg^{2+}, which is also required for full activity of RuBP carboxylase. In darkness Mg^{2+} is present in relatively low and equal concentrations in the stroma and thylakoid compartments. Under these conditions RuBP carboxylase is nearly inactive. As light intensity increases and the light reactions pump H^+ into thylakoid compartments, Mg^{2+} flows from the thylakoid compartments into the stroma to maintain electrical neutrality across the thylakoid membranes (thylakoid membranes are freely permeable to Mg^{2+}). The resultant increase in Mg^{2+} concentration in the stroma activates the RuBP carboxylase, once again in proportion to the rate of the light reactions.

The third control regulating RuBP carboxylase depends on the concentration of NADPH in the stroma. NADPH is an allosteric regulator (see p. 98) for RuBP carboxylase. Unless NADPH is available to the allosteric site on the enzyme, RuBP carboxylase remains inactive. This regulatory control also keys RuBP carboxylase activity to the light reactions. Unless the light reactions operate, too little NADPH is present in the stroma to combine with the allosteric sites, and the C_3 pathway remains inactive.

A fourth control depends on the relative amount of ATP as compared to ADP or phosphate. As ATP concentrations rise, and ADP or phosphate concentrations fall, due to increases in the rate of the light reactions, the activity of RuBP carboxylase is stimulated. High concentrations of ADP or phosphate relative to ATP have the opposite effect, and inhibit the enzyme. This control keys activity of RuBP carboxylase to both the light reactions and the C_3 cycle.

Thus, no fewer than four controls—stroma pH, the concentrations of Mg^{2+} and NADPH, and the relative amounts of ATP, ADP, and inorganic phosphate—are superimposed to adjust RuBP carboxylase activity and the rate of the C_3 cycle to match the rate of the light reactions. Similar controls involving availability of ATP and NADPH also regulate the enzymes catalyzing reactions 2 and 3 (see Fig. 10-21) of the C_3 cycle. Other controls closely linking the C_3 cycle to the light reactions affect several enzymes of the pathway regenerating RuBP.

Controls linking the activity of the C_3 cycle to the light reactions ensure that initial steps of the dark reactions operate only in the daytime, when ATP and NADPH are available for CO_2 fixation. Thus, the idea that the dark reactions are light independent is incorrect in this regard—although all of the dark reactions can take place in darkness, stringent requirements for activation of key enzymes of the C_3 pathway limit the dark reactions to the daytime, when the light reactions are operating. Why such elaborate safeguards arose in evolution to prevent operation of the C_3 pathway in periods of darkness is difficult to answer. Probably, shutdown of the C_3 cycle when the light reactions are inoperative ensures that no substances that can be oxidized as fuels enter the C_3 pathway at night. This restriction funnels fuel substances into the oxidative reactions of glycolysis and respiration at night, when plants are forced to burn fuels to obtain ATP energy. Thus in darkness plants in effect become the biochemical equivalents of animals.

RuBP carboxylase can work as an oxygenase as well, in a reaction that converts RuBP into 3-phosphoglycerate and phosphoglycolate:

$$RuBP + O_2 \longrightarrow$$
$$\text{3-phosphoglycerate} + \text{phosphoglycolate} \quad (10\text{-}8)$$

The phosphoglycolate is subsequently dephosphorylated to glycolate, which is oxidized inside microbodies (see p. 351).

The ability of RuBP carboxylase to work as an oxygenase is an evolutionary defect that greatly reduces the efficiency of photosynthesis because glycolate cannot enter subsequent steps in the dark reactions. Glycolate oxidation, among other effects, may also lead to conversion of the CO_2 fixed in the dark reactions back to inorganic form in a process known as *photorespiration* (see Information Box 10-2).

Formation of Complex Carbohydrates and Other Substances from 3PGAL Phosphoglyceraldehyde is the starting point for synthesis of a variety of complex carbohydrates and polysaccharides. It may either remain inside chloroplasts for conversion into starch and other complex substances, or it may enter the surrounding cytoplasm to act as the input for various synthetic pathways. Alternatively, 3PGAL may be oxidized in the second half of glycolysis, producing pyruvate that subsequently enters mitochondria for complete oxidation to CO_2 and H_2O (for details, see p. 315).

Glucose and other sugars are formed from 3PGAL through a series of reactions that, in effect, reverses part of glycolysis (Fig. 10-23). Glucose, although traditionally regarded as the end product of photosynthesis, is actually formed in limited quantities in many plants. Instead, glucose and other six-carbon sugars are converted into sucrose, starch, cellulose, and a wide variety of other organic molecules. Sucrose, formed from a combination of glucose and fructose units (see Fig. 10-23), is the main form in which the products of

The Oxygenase Activity of RuBP Carboxylase and Photorespiration: An Evolutionary Fly in the Photosynthetic Ointment

The normal progress of photosynthesis is impeded in many plant species because RuBP carboxylase can also act as an oxygenase, catalyzing a reaction that combines O_2 instead of CO_2 with RuBP in the first reaction of the C_3 cycle (see figure in this box). The oxygenase activity of RuBP carboxylase increases as O_2 concentration rises inside leaves and as CO_2 concentration falls. The phosphoglycolate product of the oxygenase activity, after conversion to glycolate by removal of a phosphate group, enters an oxidative pathway that can eventually yield CO_2 (see p. 355). Since the oxygenase activity of RuBP carboxylase uses O_2 and the remainder of the pathway evolves CO_2, the entire pathway is termed *photorespiration.* The pathway returns CO_2 from the fixed, organic form to the inorganic form and thus has the effect of reversing the dark reactions. Depending on the temperature and the concentrations of O_2 and CO_2 in the atmosphere, as much as 15% to 50% of the CO_2 fixed in photosynthesis may be lost through photorespiration.

The oxygenase activity of RuBP carboxylase may reflect the fact that the enzyme evolved during primeval times, in an atmosphere rich in CO_2 and depleted in oxygen. Selection against the oxygenase activity may therefore be a relatively recent evolutionary development in the history of the enzyme, one that has perhaps not yet had time, in the evolutionary sense, to have much effect. It is also possible that the combined carboxylase-oxygenase activity is the only pattern by which the enzyme can work, and that greater efficiency cannot be attained.

It is unclear why nature evolved the costly photorespiration pathway converting the product of oxygenation of RuBP to CO_2. Although there is much controversy surrounding this point, the oxidative pathway apparently serves to rid cells of glycolate, which is toxic at high concentrations. That is, rather than evolving a means to eliminate the oxygenase activity of RuBP, plants with high photorespiration rates have instead developed a pathway to eliminate the glycolate product of the faulty oxygenase reactions. This adaptation allows plants to survive in the contemporary high O_2/low CO_2 atmosphere in spite of the oxygenase activity of RuBP carboxylase.

Plants with high photorespiration rates, especially at elevated temperatures, include many important crops—rice, barley, wheat, soybeans, tomatoes, potatoes, and tobacco, to name a few. When grown experimentally in hothouses at CO_2 concentrations high enough to eliminate photorespiration, some of these crops increase in growth by as much as five times by dry weight. This demonstrates the extent to which photorespiration reduces the efficiency of photosynthesis in these plants.

The deleterious effects of photorespiration on crop yields have prompted efforts to find commercially workable methods to eliminate the problem. With the exception of growing plants in hothouses with elevated CO_2 concentrations, which has proved commercially successful with tomatoes, none of the various approaches have produced results. Efforts to reduce the oxygenase activity of RuBP carboxylase by genetic engineering have also proved fruitless; similarly, attempts to eliminate photorespiration by inhibiting enzymes of the pathway oxidizing glycolate to CO_2 have had lethal rather than beneficial results. In fact, some inhibitors of photorespiration actually show more promise as herbicides than fertilizers. This outcome emphasizes the importance of the pathway eliminating the glycolate product of faulty RuBP carboxylase activity in plants with active photorespiration.

It is possible that the problem may take care of itself. If CO_2 concentration in the atmosphere doubles in the next 50 years, as some ecologists predict, photorespiration in most plants will automatically be reduced, with possible beneficial effects on many crops. The effects of increased CO_2 concentrations on plant growth may also provide a natural buffer system regulating atmospheric CO_2 and O_2 concentrations. Presumably, as plant growth increases in response to increased atmospheric CO_2 concentrations, more CO_2 will be removed from the atmosphere and fixed into organic matter. At the same time, greater quantities of O_2 will enter the atmosphere as a result of increased photosynthesis. The combined effects of increased CO_2 fixation and O_2 release will tend to restore the atmospheric concentrations of these gases to normal levels.

Figure 10-23 Reaction pathways leading to glucose, sucrose, starch, and other carbohydrates from 3PGAL, the net product of the C_3 cycle. Sucrose is the primary form in which the products of photosynthesis circulate from cell to cell in plants. Carbohydrates are stored as either sucrose or starch or a combination of the two in most higher plants. UTP, uridine triphosphate; in some plants ATP is used instead of UTP.

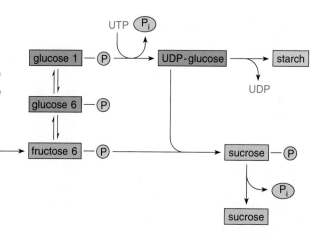

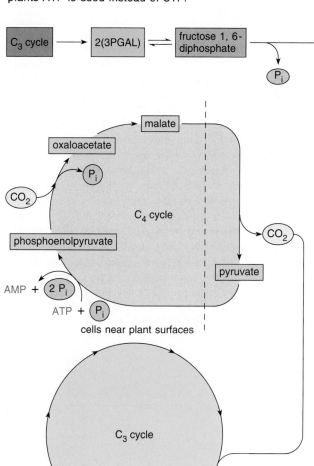

cells near plant surfaces

cells in deeper layers

Figure 10-24 The overall pathway followed by the C_4 cycle, a mechanism that circumvents photorespiration in some plants. In the cycle CO_2 combines with a three-carbon molecule, phosphoenolpyruvate, eventually producing malate, a four-carbon acid. The step fixing CO_2 is catalyzed by phosphoenolpyruvate carboxylase, a characteristic enzyme of the C_4 cycle. The malate product diffuses to other cells of the plant, where it is oxidized to pyruvate, a three-carbon molecule, with release of CO_2. The CO_2 release raises the concentration of this substance enough to prevent significant activity of RuBP carboxylase as an oxygenase. Pyruvate then diffuses back to the first set of cells to enter another turn of the C_4 cycle. The phosphoenolpyruvate used in the initial carboxylation is regenerated from pyruvate at the expense of one molecule of ATP converted to AMP.

photosynthesis circulate from cell to cell within higher plants. Nutrients are stored in most higher plants as sucrose or starch or a combination of the two in proportions that depend on the plant species. In the sugar beet, for example, sucrose is the major storage product; in barley starch is the primary storage form. Many of these substances can be synthesized inside chloroplasts as well as in the surrounding cytoplasm.

In addition to starch and other carbohydrates, chloroplasts can also synthesize amino acids, fatty acids, and lipids. Lipids synthesized inside chloroplasts include chlorophylls and carotenoids, which are assembled from simple precursors originating from the C_3 cycle. Protein synthesis, occurring on ribosomes suspended in the chloroplast stroma (see below), can also be detected inside chloroplasts, along with linkage of nucleotides into DNA and RNA.

The balance among carbohydrates, fats, and amino acids synthesized inside chloroplasts varies widely among different species of plants. In some nearly all the CO_2 absorbed is incorporated into carbohydrates. In others, such as the alga *Chlorella,* synthesis of fats and amino acids greatly exceeds carbohydrate synthesis, which may account for 5% or less of the CO_2 absorbed. In any event the chloroplasts of most species contain the enzymes and intermediates required to synthesize many substances in addition to carbohydrates. Chloroplasts, in fact, have synthetic capacity practically equivalent to entire cells.

The C_4 Cycle

During the 1960s several groups working independently, including M. D. Hatch and C. R. Slack, and H. P. Kortschak and his colleagues, discovered that certain plants carry out a series of reactions, the C_4 *cycle,* which circumvents problems caused by the oxygenase activity of RuBP carboxylase (Figs. 10-24 and 10-29; Hatch's Experimental Process essay on p. 392 describes experiments leading to discovery of the C_4 cycle).

The C_4 cycle uses a series of enzymes that initially combines CO_2 with a three-carbon molecule, *phos-*

The C₄ Pathway: A Surprise Option for Photosynthetic Carbon Assimilation

Marshall D. Hatch

MARSHALL D. HATCH completed his undergraduate studies at Sydney University in 1954 and was awarded a Ph.D. from the same University in 1959. After two years post doctoral studies at the University of California, Davis, he joined the laboratory of the Colonial Sugar Refining Company in Brisbane, Australia, to study aspects of the biochemistry and physiology of sugarcane. In 1970 he moved to the Division of Plant Industry, CSIRO, in Canberra, where he is currently a Chief Research Scientist. Since 1965 he has worked almost exclusively on aspects of the mechanism and function of C₄ photosynthesis. He was elected to the Australian Academy of Science in 1975, the Royal Society in 1980, and as a Foreign Associate of the National Academy of Science of the United States in 1990. He was awarded the Rank Prize in 1981 and the International Prize for Biology in 1991.

By 1960, Melvin Calvin and colleagues Benson, Bassham, and many others had worked out the essential features of the path of CO_2 assimilation now known as the Calvin cycle, or C₃ pathway. The critical first step in this process is the assimilation of CO_2 by a reaction with the 5-carbon sugar ribulose 1,5-bisphosphate giving two molecules of the 3-carbon compound 3-phosphoglyceric acid. This compound is then transformed through many steps to give sucrose, starch, and various other metabolites.

The general rule in biochemistry is for uniformity rather than diversity so that biochemical mechanisms for particular processes are usually the same both within and between the different classes of living organisms. Therefore, there was no reason to expect that a photosynthetic process differing from the Calvin cycle might occur in plants. However, it is interesting to note, in retrospect, that there was in fact a considerable amount of physiological and anatomical evidence available at the time which pointed to that possibility.

In the early 1960s Dr. Roger Slack and I were working in Brisbane, Australia, in the research laboratory of the Colonial Sugar Refining Company on various aspects of sugarcane metabolism. We were aware, through private communication, of some unusual results of colleagues working on sugarcane in Hawaii. In these studies, initiated in the late 1950s, Hugo Kortschak and collaborators looked at the radioactive products formed when sugarcane leaves were allowed to assimilate radioactive carbon dioxide ($^{14}CO_2$) in the light. What they noted was that after short periods in $^{14}CO_2$ more radioactivity appeared in the 4-carbon (C₄) acids malate and aspartate than in the C₃ acid 3-phosphoglyceric acid. Of course the latter compound should be the first product labeled with radioactivity if the normal Calvin cycle was operating alone.

There were various possible explanations of this result, which Dr. Slack and I had often discussed. However, to cut a long story short, when Kortschak and colleagues finally published their results in 1965[1] Dr. Slack and I set about trying to repeat these observations and resolve which of these options might be correct. To complete the background history, there is one other interesting twist to this story. Three or four years after starting our study on this process we discovered a report published in 1960 in an obscure Proceedings of a Russian agricultural institute. This report, by Russian scientist Yuri Karpilov, clearly described the predominant radioactive labeling of malate and aspartate in maize (corn) leaves exposed to $^{14}CO_2$ in the light. This observation was not followed up by Karpilov until several years after our first publication, however.

Our initial experiments were designed to trace the exact fate of carbon assimilated during photosynthesis. Following the general procedure developed in Calvin's studies, we exposed sugarcane leaves to $^{14}CO_2$ under natural conditions in the light and then stopped photosyn-

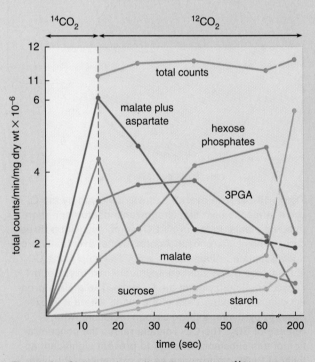

Figure A Changes in radioactive carbon (^{14}C) content in photosynthetic intermediates and products in sugarcane leaves following a 15-second pulse in $^{14}CO_2$ and then a "chase" in normal ($^{12}CO_2$) air. Data from Hatch and Slack (1966).[2]

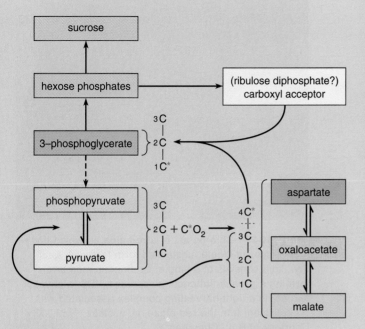

Figure B Proposal for the pathway of photosynthetic CO_2 fixation in sugarcane leaves as it was presented in our initial paper on C_4 photosynthesis.[2]

thesis after various periods up to 90 seconds by boiling or freezing leaves in killing mixtures containing ethanol. The leaf contents were then extracted and the compounds containing radioactivity were identified by paper chromatography. Later, these compounds were purified and degraded to find out which particular carbons were labeled with radioactivity. These studies confirmed the original observation of Kortschak and colleagues that in shorter times most of the radioactive carbon incorporated from $^{14}CO_2$ appears in malate and aspartate. In addition, we showed that the first product formed is actually an unstable dicarboxylic acid, oxaloacetic acid, and that malate and aspartate were formed from oxaloacetate.

We went on to show that this radioactivity appears exclusively in the C-4 carboxyl of these C_4 acids. So called pulse-chase experiments were also performed where the fate of ^{14}C incorporated in a short pulse in $^{14}CO_2$ is followed in a subsequent "chase" period after leaves are transferred back to air containing normal nonradioactive CO_2. The results of these pulse-chase experiments appearing in our first publication on this subject[2] are shown in Fig. A. It can be inferred from this data that the C-4 carboxyl of C_4 acids is transferred to the C-1 carboxyl of 3-phosphoglycerate, which is then metabolized to hexose phosphates and finally sucrose and starch. The pattern of radioactive labeling of 3-phosphoglycerate and sugar molecules was consistent with incorporation via a reaction similar to the one catalysed by the Calvin cycle carboxylase, ribulose 1,5-bisphosphate carboxylase. Several control experiments were conducted to make sure that the results we got were due to photosynthesis and were not an artifact of the way leaves were killed or extracted.

Our conclusions were summarized in a scheme published in the original paper[2] and reproduced here (Fig. B). This proposes a novel path of carbon assimilation differing substantially from the Calvin cycle. The simplest interpretation of the data was that CO_2 is initially assimilated by a carboxylation reaction involving a 3-carbon acceptor, possibly pyruvate or phosphoenolpyruvate, giving rise to oxaloacetate and then other C_4 dicarboxylic acids as first products. The carbon assimilated into the C-4 carboxyl is apparently transferred to appear in 3-phosphoglycerate and then into sugar phosphates, sucrose, and starch. The pattern of slower secondary incorporation of radioactive carbon into the other carbons (C-1, C-2, and C-3) of C_4 acids was consistent with it originating by exchange between 3-phosphoglycerate and phosphoenolpyruvate. We named this process the C_4 dicarboxylic acid pathway of photosynthesis, since abbreviated to the C_4 pathway. The Calvin cycle is now commonly referred to as the C_3 pathway on the grounds that its first product is a 3-carbon compound.

In the following three or four years our studies elaborated on these findings and showed that the pathway operated in a number of other higher plant species. We identified several of the key enzymes involved. For instance, we showed that phosphoenolpyruvate carboxylase was the enzyme responsible for the primary carboxylation reaction giving oxaloacetate. Amongst the others were two previously undiscovered enzymes, pyruvate P_i dikinase (which converts pyruvate to phosphoenolpyruvate) and NADP-malate dehydrogenase (which reduces oxaloacetate to malate). It was also shown that the carboxylation reaction leading to C_4 acid formation occurs in mesophyll cells while the decarboxylation reaction leading to transfer of carbon to 3-phosphoglycerate occurs in adjacent bundle sheath cells. We reviewed these early studies in 1970.[3]

In the intervening time the full complexity of this process has been revealed. As outlined in a recent review,[4] there are three distinct variants of C_4 photosynthesis and these options are species-specific. We now know that the function of the reactions unique to C_4 photosynthesis is to concentrate CO_2 in bundle sheath cells.[4] The primary purpose is to suppress the oxygenase reaction catalysed by Rubisco, and hence to eliminate the process known as photorespiration.[4] This endows C_4 plants with advantages in terms of photosynthetic capacity and water use efficiency. C_4 plants are at a particular advantage in drier and especially hotter locations. The recognition of the C_4 process and its physiological significance has been a powerful influence in understanding the factors affecting plant dry matter production and growth.

References

[1] Kortschak, H. P.; Hartt, C. E.; and Burr, G. O. *Plant Physiol.* 40:209 (1965).

[2] Hatch, M. D., and Slack, C. R. *Biochem. J.* 101:103 (1966).

[3] Hatch, M. D., and Slack, C. R. *Annu. Rev. Plant Physiol.* 21:141 (1970).

[4] Hatch, M. D. *Biochim. Biophys. Acta* 895:81 (1987).

phoenolpyruvate, eventually producing *malate*, a four-carbon acid. (The C_4 cycle gets its name from this four-carbon product.) An enzyme critical to the C_4 pathway, *phosphoenolpyruvate carboxylase*, catalyzes the step fixing CO_2. The C_4 cycle occurs instead of the C_3 cycle in cells in which oxygen is abundant enough to stimulate the oxygenase activity of RuBP carboxylase. The malate product of the C_4 cycle then diffuses to a deeper cell layer, where oxygen concentrations are lower. Here malate is oxidized to pyruvate, a three-carbon molecule, with release of CO_2. The CO_2 enters the C_3 cycle in these cells, in which the relatively low oxygen concentration defeats the oxygenase activity of RuBP carboxylase. The pyruvate diffuses back to the first cell layer to enter another turn of the C_4 cycle. As part of the reentry, pyruvate is converted to phosphoenolpyruvate at the expense of one molecule of ATP converted to AMP. The C_4 cycle occurs in many plants, including important crop grasses such as corn and sugar cane. (The C_4 cycle and CAM metabolism, a variation of the C_4 pathway, are discussed in more detail in Supplement 10-1.)

SOME DETAILS OF CHLOROPLAST STRUCTURE AND OCCURRENCE

Although fusion of thylakoids into grana is typical of most green algal and higher plant chloroplasts, elongated, single thylakoids appear along with grana in a few cell types of these groups, most notably in the bundle sheath cells of plants with the C_4 photosynthetic pathway (see Supplement 10-1). In nongreen algae, thylakoids are found exclusively in the single, unstacked form. (Fig. 10-25 shows the elongated, unstacked thylakoids of a red algal chloroplast.) Despite these variations the basic structure of thylakoids is uniform. Each thylakoid consists of a single, continuous membrane enclosing an inner compartment that is completely separated from the surrounding stroma by the thylakoid membrane.

The chloroplast stroma may contain a variety of inclusions. Most conspicuous of these are *starch granules* (Fig. 10-26), found in many chloroplasts after a period of active photosynthesis in the light. The stroma also frequently contains dense, spherical particles called *osmiophilic granules* or *droplets* (visible in Fig. 10-2). These dense granules are probably aggregates of lipids suspended in the stroma.

The chloroplast stroma, like the mitochondrial matrix, also contains DNA, ribosomes, and all the factors required for DNA replication, RNA transcription, and protein synthesis. Of the visible elements of this system, the DNA appears at scattered locations in the stroma as faintly visible threads (see Fig. 21-3). Chloroplast

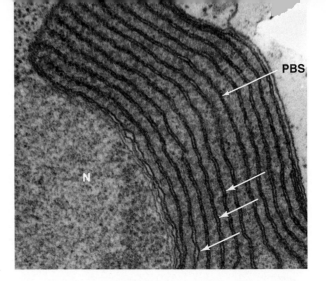

Figure 10-25 Chloroplast of a red alga, in which thylakoids occur in single, unstacked form (arrows). Each thylakoid consists of a single, continuous membrane forming a greatly flattened, closed sac. PBS, a phycobilisome, is a light-harvesting complex associated with photosystem II in the red algae. N, nucleus. ×62,000. Courtesy of R. L. Chapman.

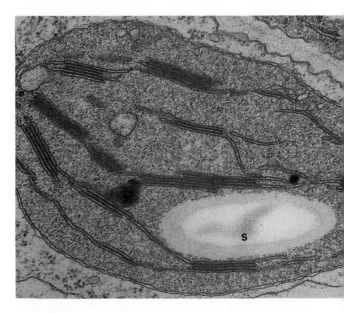

Figure 10-26 A starch granule (S) in a tobacco chloroplast. Courtesy of D. Stetler.

ribosomes, conspicuously smaller than ribosomes in the surrounding cytoplasm, are distributed throughout the stroma (see Fig. 21-17; the functions of chloroplast DNA and ribosomes in replication, transcription, and protein synthesis are discussed at length in Chapters 21 and 23.)

Localizing the light and dark reactions in specific parts of chloroplasts was first accomplished by R. B. Park and N. G. Pon, who used a centrifuge to separate broken chloroplasts into thylakoid and stroma fractions. Thylakoids, which are released intact when boundary membranes are broken, were found to contain the light-absorbing pigments and the system synthesizing ATP. The stroma, which remained in suspension as a colorless solution of proteins, was found to contain the major enzymes of CO_2 fixation. These results clearly demonstrated that the light reactions are concentrated in thylakoids and the dark reactions in the stroma.

Further research with isolated chloroplast fractions confirmed that all subparts of the light reactions are associated with thylakoid membranes and their stromal lamellae connections. These reactions include the mechanism splitting water, the pigments capturing and converting light to chemical energy, all the molecules and complexes transporting electrons and reducing $NADP^+$, and the CF_oCF_1 ATPase.

Proteins occur in outer boundary membranes in approximately half the amount of lipids by weight. The proteins of this system include as many as 75 different polypeptides; many of these are involved in the transport of substances between the chloroplast stroma and the surrounding cytoplasm (see below). Enzymatic proteins associated with synthesis of fatty acids and chloroplast lipids, including carotenoids and quinones, are also concentrated in chloroplast boundary membranes.

Thylakoid membranes contain proteins in a much higher ratio, approaching a 2:1 ratio of proteins to lipids by weight. Analysis of thylakoids reveals more than 50 distinct polypeptides; most of these can be identified as enzymes and other factors associated with the light reactions.

Some of the proteins inside chloroplasts are encoded in DNA inside the organelles and synthesized on ribosomes in the stroma. Others are encoded in the cell nucleus and synthesized on ribosomes in the cytoplasm outside chloroplasts. The proteins synthesized outside get into chloroplasts by a mechanism that is unrelated to the transport systems moving metabolites. Like their counterparts in mitochondria, chloroplast proteins made in the cytoplasm contain a *signal* consisting of an amino acid sequence that is recognized and bound by receptors in chloroplast boundary membranes. Depending on the information coded in the signal, the proteins are routed to final destinations in the chloroplast membranes or compartments. (For details of the mechanism routing proteins to chloroplasts, see Ch. 20; the genetic system of chloroplasts and the proteins encoded and synthesized inside the organelle are described in Ch. 21.)

Glycolipids (see p. 155) form as much as three-quarters of the total chloroplast membrane lipid content by weight. Typically, the carbohydrate groups of chloroplast glycolipids are relatively simple structures containing only one or two sugar units, usually galactose (Fig. 10-27). Some glycolipids, about 10 to 15% of the total lipid fraction, are *sulfolipids*, which include sulfur atoms in their carbohydrate groups (see Fig. 10-27*d*).

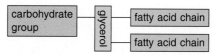

a

monogalactosyl diacylglycerol
(a glycolipid)

b

digalactosyl diacylglycerol
(a glycolipid)

c

sulfoquinovosyl diacylglycerol
(a sulfolipid)

d

Figure 10-27 Glycolipids of chloroplast membranes. **(a)** Structural plan of a glycolipid. **(b)** A glycolipid with a carbohydrate unit containing a single sugar; **(c)** a glycolipid with a carbohydrate unit containing two sugar units. Galactose units are the most common sugars of chloroplast membrane glycolipids. **(d)** A sulfolipid, a glycolipid containing a sulfur group.

Table 10-3 Chloroplast Membrane Lipids

Lipid Type	Percent of Total Chloroplast Lipid
Glycolipids	
monogalactosyl diacylglycerol (MGDG)	40–48
digalactosyl diacylglycerol (DGDG)	20–25
sulfolipids	10–15
Phospholipids*	10–15
Pigments (chlorophylls and carotenoids)	20–25
Quinones	6

* Phospholipids include phosphatidyl choline, phosphatidyl glycerol, diphosphatidyl glycerol, phosphatidyl inositol, and phosphatidyl ethanolamine.

Glycolipids are typically distributed asymmetrically between the inner and outer bilayer halves in thylakoid membranes. Glycolipids with carbohydrate groups containing a single sugar unit are more concentrated in the outer half of the thylakoid bilayer; glycolipids with two sugar units in their carbohydrate groups are more concentrated in the inner bilayer half. Sulfolipids are confined almost entirely to the inner bilayer half.

Phospholipids, at about 10% to 15% of the total lipid fraction, form a much smaller proportion of chloroplast lipid content than in mitochondria. Chloroplast quinones, including plastoquinones (see Fig. 10-11), make up about 6% of the total chloroplast lipids. The remaining lipids of chloroplasts are primarily chlorophylls and carotenoids, which comprise 20% to 25% of the total. Small amounts of a sterol, stigmast-7-enol, may also be present. The characteristic structural lipid of mitochondria, cardiolipin, is absent from chloroplast membranes. (Table 10-3 summarizes the lipid content of chloroplast membranes.)

The boundary and thylakoid membranes contain similar types of phospholipids and glycolipids. However, lipid pigments active in light absorption in photosynthesis are associated primarily with thylakoid membranes and stromal lamellae. Chlorophylls are confined entirely to these membranes; although carotenoid pigments can be detected among the lipids of all chloroplast membranes, most carotenoids occur in thylakoids and stromal lamellae. The function of carotenoids in chloroplast boundary membranes remains unknown. Their presence may simply reflect the fact that they are synthesized in this location.

The entire spectrum of chloroplast structural lipids, including both glycolipids and phospholipids, is unusual among cellular membrane lipids because of the high degree of unsaturation of their nonpolar fatty acid chains. The many "kinks" introduced by double bonds in these unsaturated fatty acid chains (see p. 54) greatly increase the fluidity of chloroplast membranes. This characteristic, along with the absence of sterols, makes chloroplast membranes, particularly thylakoid membranes, among the most fluid of any biological membranes known. The nonpolar interior of these membranes is about as fluid as water, a characteristic that greatly increases the frequency of random collisions between the proteins and protein complexes responsible for the light reactions.

Chloroplasts are members of a family of related organelles known collectively as *plastids*. The various plastid types are related developmentally as well as structurally, so that one plastid form may change into another depending on the developmental stage and the growth conditions of the plant. (The major plastid types and their development are described in Supplement 10-2.)

CHLOROPLAST TRANSPORT

All substances entering and exiting chloroplasts must cross the double barrier presented by the outer and inner boundary membranes. The permeability and transport functions of the two chloroplast boundary membranes resemble those of mitochondria (see p. 347). In both organelles the outer boundary membrane acts essentially as a nonspecific molecular sieve that admits all molecules below a certain size limit. The inner boundary membrane of both organelles contains transport systems that select specific molecules for transport into or from the organelle interior.

Porins similar to those in the outer boundary membrane of mitochondria (see p. 349) are responsible for the ability of the outer boundary membrane of chloroplasts to act as a molecular sieve. The limit for transport through the porin channels of chloroplasts falls at molecular weights of about 10,000 to 13,000, somewhat higher than the 6000-dalton limit of mitochondrial porins.

The inner boundary membrane is impermeable except to water, gases such as CO_2 and O_2, and a relatively small number of substances admitted by specific transport proteins embedded in the membrane. All the carrier systems of the inner boundary membrane appear to be driven by favorable concentration gradients rather than active transport. This is in distinct contrast to mitochondria, in which much of the transport is active, driven directly or indirectly by the H^+ gradient established by electron transport (see p. 350).

Inorganic phosphate is required in quantity by chloroplasts because the products of photosynthesis are released to the surrounding cytoplasm primarily in the form of phosphorylated molecules such as 3PGAL. This transport is accomplished by a *phosphate exchange carrier*, a protein that exchanges inorganic phosphate from the

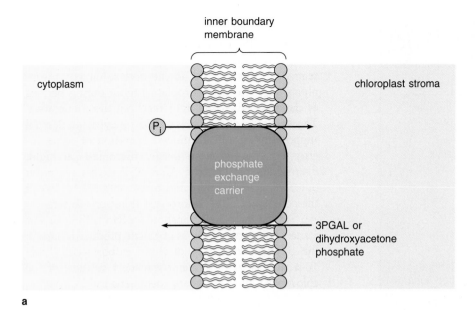

a

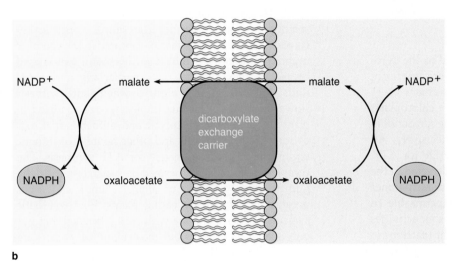

b

Figure 10-28 Transport carriers of the chloroplast inner boundary membrane. **(a)** The phosphate exchange carrier, which transports inorganic phosphate ions (P_i) into chloroplasts in exchange for 3PGAL transported outward into the cytoplasm. **(b)** The dicarboxylate exchange carrier, a shuttle that exchanges acids containing two carboxyl groups on a one-for-one basis between the chloroplast interior and the cytoplasm. The shuttle can, in effect, move high-energy electrons from NADPH inside the chloroplast to $NADP^+$ in the cytoplasm.

outside for a three-carbon phosphorylated sugar made inside on a strict one-for-one basis (Fig. 10-28a). The exchange is so strict that unless phosphate ions enter chloroplasts, the products of photosynthesis cannot readily be exported and both the light and dark reactions are strongly inhibited. The phosphate exchange carrier, at 12% of the total, is one of the most abundant proteins of the chloroplast boundary membranes.

Small amounts of the products of photosynthesis are also exported to the cytoplasm in the form of glucose by a *glucose carrier* in the inner boundary membrane. Activity of the glucose carrier is relatively limited in comparison to the phosphate exchange carrier, which accounts for most of the organic carbon exported by chloroplasts to the cytoplasm.

Another major transport protein of the chloroplast inner boundary membrane, the *dicarboxylate exchange carrier*, exchanges acids containing two carboxyl groups—malate, succinate, oxaloacetate, fumarate, glu-

tamate, and aspartate—on a one-for-one basis between the stroma and cytoplasm. The primary activity of this transport protein is apparently to participate in *shuttles* of various kinds. One of these shuttles transfers electrons between NADP inside and outside the chloroplast by exchanging malate in the cytoplasm for oxaloacetate in the stroma, in combination with side reactions that oxidize and reduce NADP (Fig. 10-28b). The malate/oxaloacetate shuttle compensates for the fact that, like its mitochondrial counterpart, the inner boundary membrane of chloroplasts is completely impermeable to NADP and NAD in either reduced or oxidized form.

Other shuttles based on the dicarboxylate exchange carrier participate in the movement of NH_4^+ or, in C_4 plants, of bound CO_2 into chloroplasts. In the latter shuttle CO_2 bound into malate enters chloroplasts in exchange for another dicarboxylate; once inside, malate is oxidized to pyruvate, delivering both CO_2 and electrons that reduce $NADP^+$ in the chloroplast in-

Table 10-4 Transport Systems of the Chloroplast Inner Boundary Membrane

Carrier	Function
ADP/ATP exchange carrier	Exchanges ATP for ADP between the cytoplasm and the chloroplast stroma.
Dicarboxylate exchange carrier	Exchanges dicarboxylic acids between the cytoplasm and the chloroplast stroma.
Glucose carrier	Transports glucose from the stroma to the cytoplasm.
Glycolate carrier	Transports glycolate from the stroma to the cytoplasm.
Phosphate exchange carrier	Exchanges cytoplasmic inorganic phosphate for three-carbon sugars (3PGAL, dihydroxyacetone phosphate) of the stroma.

terior. Pyruvate is subsequently exported to the cytoplasm to complete the C_4 cycle by another carrier protein specialized for this purpose (see Supplement 10-1).

Whether the ATP synthesized inside chloroplasts can be directly used for cell activities in the surrounding cytoplasm has not been determined conclusively. Although a transport protein exchanging ADP for ATP, the *ADP/ATP exchange carrier*, has been detected in the chloroplasts of a few species, chloroplast membranes in most plants appear to be highly impermeable to either ADP or ATP. If so, ATP for cytoplasmic activities cannot be obtained directly from the light reactions in most plants. However, since 3PGAL can enter the cytoplasm for oxidation, reactions inside chloroplasts may lead indirectly to the synthesis of large quantities of cytoplasmic ATP through glycolysis, pyruvate oxidation, and the citric acid cycle.

The ADP/ATP exchange carrier, when present, seems actually to work in the reverse direction, supplying ATP to the chloroplast interior during periods of darkness when the light reactions are inoperative. This exchange of chloroplast ADP for cytoplasmic ATP supposedly permits continued operation of some energy-requiring reactions inside chloroplasts at night. Even when present, the ADP/ATP exchange carrier operates at levels estimated to be only 5% or less of the activity of other chloroplast transport carriers.

A final important transport protein is the *glycolate carrier*. This carrier exports glycolate produced through the oxygenase activity of RuBP carboxylase. (Table 10-4 summarizes the major transport systems of the chloroplast inner boundary membrane.)

The reactions of photosynthesis enable eukaryotic plants to use the energy of sunlight to drive the assembly of organic molecules from simple inorganic precursors. Some chemical energy captured in the complex organic molecules assembled inside chloroplasts is used by plants themselves as fuels, particularly during periods of reduced light, at night, or when photosynthesis slows or stops. These fuel molecules are oxidized in plants by the same glycolytic and mitochondrial oxidations used by animals and other nonphotosynthetic organisms to provide energy for ATP synthesis. The complex molecules synthesized by plants also form the primary energy source of animals, fungi, and other organisms that live by eating plants. Photosynthetic systems of plants operate with such efficiency that not only plants themselves but most living organisms are supported by the light energy captured and converted into chemical energy inside chloroplasts. (Photosynthesis in prokaryotes is outlined in Supplement 10-3.)

For Further Information

DNA, ribosomes, and protein synthesis in chloroplasts, *Ch. 21*

Electron transport and ATP synthesis in mitochondria, *Ch. 9*

Evolutionary origins of chloroplasts, *Ch. 27*

Mechanisms delivering proteins to chloroplasts, *Ch. 20*

Membrane structure, *Ch. 5*

Microbodies (peroxisomes) and photorespiration, *Ch. 9*

Mitchell's chemiosmotic hypothesis, *Ch. 9*

Porins in mitochondria, *Ch. 9*
  in bacteria, *Ch. 6*

Transport, active and passive, *Ch. 6*

Suggestions for Further Reading

General Books and Articles

Arnon, D. I. 1984. The discovery of photosynthetic phosphorylation. *Trends Biochem. Sci.* 9:258–62.

Bogorad, L. 1981. Chloroplasts. *J. Cell Biol.* 91:256s–70s.

Ginsburg, V., and Robbins, P., eds. 1981. *Biology of carbohydrates.* New York: Wiley.

Govindjee, ed. 1982. *Photosynthesis.* Vols. I and II. New York: Academic Press.

Hatch, M. D., and Boardman, N. K., eds. 1987. *The biochemistry of plants.* Vol. 10. *Photosynthesis.* New York: Academic Press.

Hoober, J. K. 1984. *Chloroplasts*. New York: Plenum.

Huber, R. 1989. A structural basis of light energy and electron transfer in biology. *Biosci. Rep.* 9:635–73.

Lawler, D. W. 1986. *Photosynthesis*. New York: Wiley.

Youvan, D. C., and Marrs, B. L. 1987. Molecular mechanisms of photosynthesis. *Sci. Amer.* 256:42–48 (June).

Chlorophylls and Carotenoids

Bendich, A., and Olson, J. A. 1989. Biological actions of carotenoids. *FASEB J.* 3:1927–32.

Siefermann-Harms, D. 1985. Carotenoids in photosynthesis I. Location in photosynthetic membranes and light-harvesting function. *Biochim. Biophys. Acta* 811:325–55.

Photosystems

Andersson, B., and Styring, S. 1991. Photosystem I: molecular organization, function, and acclimation. *Curr. Top. Bioenerget.* 16:2–81.

Andréaesson, L.-E., and Vänngard, T. 1988. Electron transport in photosystems I and II. *Ann. Rev. Plant Physiol. Plant Molec. Biol.* 39:379–411.

Barber, J. 1987. Photosynthetic reaction centers: a common link. *Trends Biochem. Sci.* 12:321–26.

Ghanotakis, F., and Yocum, C. F. 1990. Photosystem II and the oxygen-evolving complex. *Ann. Rev. Plant Physiol. Plant Molec. Biol.* 41:255–76.

Glazer, A. N., and Melis, A. 1987. Photochemical reaction centers: structure, organization, and function. *Ann. Rev. Plant Physiol.* 38:11–45.

Golbeck, J. H., and Bryant, D. A. 1991. Photosystem I. *Curr. Top. Bioenerget.* 16:83–177.

Mathis, P. 1990. Compared structure of plant and bacterial photosystem reaction centers. Evolutionary implications. *Biochim. Biophys. Acta* 1018:163–67.

Mattoo, A. K.; Marder, J. B.; and Edelman, M. 1989. Dynamics of the photosystem II reaction center. *Cell* 56:241–46.

Rochaix, J.-D., and Erickson, J. 1988. Function and assembly of photosystem II: genetic and molecular analysis. *Trends Biochem. Sci.* 13:56–59.

Rutherford, A. W. 1989. Photosystem II, the water-splitting enzyme. *Trends Biochem. Sci.* 14:227–32.

Light-Harvesting Complexes (LHCs)

Chitnis, P. R., and Thornber, J. P. 1988. The major light-harvesting complex of photosystem II: aspects of its molecular and cell biology. *Photosynth. Res.* 16:41–63.

Kühlbrandt, W., and Wang, D. N. 1991. Three-dimensional structure of plant light-harvesting complex determined by electron crystallography. *Nature* 350:130–34.

Water-Splitting Reaction

Beck, W. F., and de Paula, J. C. 1989. Mechanism of photosynthetic water oxidation. *Ann. Rev. Biophys. Biophys. Chem.* 18:25–46.

Brudvig, G. W. 1987. The tetranuclear manganese complex of photosystem II. *J. Bioenerget. Biomembr.* 19:91–104.

Dekker, J. P., and van Gorkom, H. J. 1987. Electron transfer in the water-oxidizing complex of photosystem II. *J. Bioenerget. Biomembr.* 19:125–42.

Ghanotakis, F., and Yocum, C. F. 1990. Photosystem II and the oxygen-evolving complex. *Ann. Rev. Plant Physiol. Plant Molec. Biol.* 41:255–76.

Govindjee and Coleman, W. J. 1990. How plants make O_2. *Sci. Amer.* 262:50–58 (February).

Hohmann, P. H. 1987. The relations between the chloride, calcium, and polypeptide requirements of photosynthetic water oxidation. *J. Bioenerget. Biomembr.* 19:105–23.

Rutherford, A. W. 1989. Photosystem II, the water-splitting enzyme. *Trends Biochem. Sci.* 14:227–32.

Yocum, C. F. 1991. Calcium activation of photosynthetic water oxidation. *Biochem. Biophys. Acta* 1059:1–15.

Electron Transport in Chloroplasts: the Z Pathway

Arnou, D. T. 1991. Photosynthetic electron transport: emergence of a concept, 1949–59. *Photosynth. Res.* 29:117–31.

Cramer, W. A.; Furbacher, P. N.; Szczepaniek, A.; and Tae, G.-S. 1991. Electron transport between photosystem II and photosytem I. *Curr. Top. Bioenerget.* 16:180–222.

Haehnel, W. 1984. Photosynthetic electron transport in higher plants. *Ann. Rev. Plant Physiol.* 35:659–93.

Karplus, P. A.; Daniels, M. J.; and Herriott, J. R. 1991. Atomic structure of ferredoxin-NADP$^+$ reductase: prototype for a structurally novel flavoenzyme family. *Science* 251:60–66.

Knaff, D. B., and Hirasawa, M. 1991. Ferredoxin-dependent chloroplast enzymes. *Biochim. Biophys. Acta* 1056:93–125.

O'Keefe, D. P. 1988. Structure and function of the chloroplast cytochrome b-f complex. *Photosynth. Res.* 17:189–216.

Pschorn, R.; Ruhle, W.; and Wild, A. 1988. Structure and function of ferredoxin-NADP$^+$ oxidoreductase. *Photosynth. Res.* 17:217–29.

Wiley, D. L., and Gray, J. C. 1988. Synthesis and assembly of the cytochrome b-f complex in higher plants. *Photosynth. Res.* 17:125–44.

Organization of Thylakoid Membranes

Anderson, J. M. 1986. Photoregulation of the composition, function, and structure of thylakoid membranes. *Ann. Rev. Plant Physiol.* 37:93–136.

Melis, A. 1991. Dynamics of photosynthetic membrane structure and function. *Biochim. Biophys. Acta* 1058:87–106.

Murphy, D. J. 1986. The molecular organization of the photosynthetic membranes of higher plants. *Biochim. Biophys. Acta* 864:33–94.

Supplement 10-1
The C₄ Cycle and CAM Metabolism

Two groups of plants, the C_4 and CAM (for *Crassulacean Acid Metabolism*) plants, have evolved different but related mechanisms to compensate for the potentially damaging ability of RuBP carboxylase to act as an oxygenase. In C_4 plants, initial CO_2 uptake and final CO_2 fixation take place in separate locations in photosynthetic tissues. Oxygen is limited in the region containing RuBP carboxylase, so that the oxygenase activity of the enzyme is suppressed. In CAM plants, which also employ the C_4 pathway, initial CO_2 uptake is confined largely to the hours of darkness, and final CO_2 fixation to the daytime to place RuBP carboxylase in a low-oxygen environment.

The C₄ Cycle

The C_4 cycle was discovered when Kortschak, Hatch, Slack, and their coworkers looked for the earliest labeled intermediates produced in corn, sugar cane, and other tropical grasses after the plants were exposed to labeled CO_2. Surprisingly, the earliest label appeared in a pool of malate and other four-carbon acids instead of 3PGA as in the C_3 cycle. Intermediates of the C_3 cycle were also found to be labeled a few seconds after the appearance of label in the four-carbon acids.

Hatch and Slack proposed that the appearance of label in the pool of four-carbon acids and in other intermediates was part of a side cycle linked to the main C_3 cycle (Fig. 10-29). As a preparatory step in the C_4 cycle (reaction 1 in Fig. 10-29), pyruvate is phosphorylated at the expense of ATP, which is hydrolyzed to AMP. The product of this reaction, *phosphoenolpyruvate (PEP)*, then reacts with CO_2 (reaction 2) to produce oxaloacetate, another four-carbon molecule. Because carbon is converted from inorganic to organic form in this reaction, it provides a second route to carbon fixation in plants with the C_4 pathway. The enzyme catalyzing the fixation, *PEP carboxylase*, unlike RuBP carboxylase, does not use oxygen as an alternate substrate and is therefore unaffected by elevated oxygen concentrations.

Oxaloacetate produced in reaction 2 follows a pathway that varies among different plants using the C_4 cycle. In the most frequently observed route oxaloacetate is reduced to malate (reaction 3). The electrons required for this reduction come from NADPH.

Malate produced by reaction 3 acts as a carrier delivering CO_2 to the C_3 cycle. In the vicinity of the C_3 cycle, malate is oxidized to pyruvate in a reaction that splits off a carbon as CO_2 (reaction 4). Electrons removed from malate in the reaction are picked up by $NADP^+$. The NADPH produced in reaction 4 replaces the NADPH used in reaction 3. The CO_2 released by reaction 4 enters the C_3 cycle by the regular route—by combination with RuBP in a reaction catalyzed by RuBP carboxylase—and the C_3 cycle turns as usual. The other product of reaction 4 of the C_4 cycle, pyruvate, replaces the pyruvate used in reaction 1, and the C_4 cycle is ready to turn again. In terms of net reactants and products, the C_4 cycle uses two high-energy phosphate groups as it splits one molecule of ATP into AMP and $2P_i$.

The C_4 pathway initially appeared to be a "futile" cycle that resulted only in the net breakdown of ATP. However, it gradually became clear that the C_4 pathway is an evolutionary adaptation that effectively circumvents the oxygenase activity of RuBP carboxylase and eliminates or reduces photorespiration (see Information Box 10-1).

The effects of the C_4 pathway became obvious when the various parts of the C_4 pathway were found in *mesophyll* cells that occur in regions of photosynthetic tissues that are more directly exposed to oxygen (Fig. 10-30). In these cells CO_2 enters the C_4 cycle, catalyzed

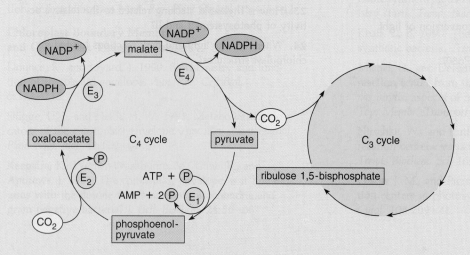

Figure 10-29 The C_4 cycle and its integration with the C_3 cycle (see text).

enzymes:

E_1 = pyruvate kinase

E_2 = phosphoenolpyruvate kinase

E_3 = NADP-malate dehydrogenase

E_4 = malate dehydrogenase

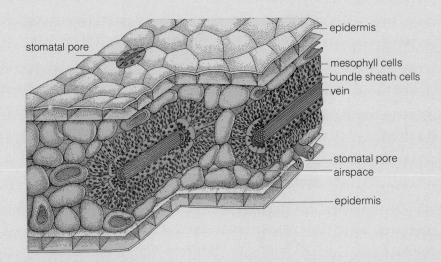

Figure 10-30 The locations of C_4 and C_3 reactions in leaves of C_4 plants. Carbon fixation in the C_4 cycle occurs in mesophyll cells, in which O_2 concentrations are relatively high because these cells are exposed to atmospheric oxygen entering the stomata. The C_3 cycle takes place in bundle sheath cells, in which CO_2 released by the C_4 cycle raises this substance to concentrations high enough to defeat the oxygenase activity of RuBP carboxylase. From "Photosynthesis" by O. Bjorkman and J. Barry. Copyright © 1973 by Scientific American, Inc. All rights reserved.

by enzymes that are insensitive to oxygen. No RuBP carboxylase is present in mesophyll cells, and consequently, neither the C_3 cycle nor photorespiration takes place in the relatively O_2-rich environment of this cell layer.

The C_4 cycle proceeds as far as malate in mesophyll cells. Malate is then exported to *bundle sheath cells,* which are completely surrounded by mesophyll cells and thus sequestered from oxygen (see Fig. 10-30). In bundle sheath cells the oxidation of malate releases CO_2 that diffuses into chloroplasts. Release of CO_2 into bundle sheath cells, coupled with the limited diffusion of oxygen into this layer, produces an environment for the RuBP carboxylase enzyme that is relatively CO_2 rich. (CO_2 concentration in bundle sheath cells of C_4 plants is estimated to reach levels seven times greater than atmospheric CO_2.) Under these conditions the oxygenase activity of RuBP carboxylase is inhibited, and little or no fixed carbon is lost in photorespiration. The pyruvate product of malate decarboxylation diffuses back to the mesophyll cell layer to enter another turn of the C_4 cycle. Movement of malate and other components of the C_4 pathway between cells is believed to proceed through plasmodesmata (see p. 295).

The ATP lost in the C_4 cycle is evidently the price paid to circumvent the oxygenase activity of RuBP carboxylase. Because two phosphates are removed from ATP for each turn of the cycle, 12 phosphates must be removed for the six turns required to make a molecule of glucose. Adding these to the 18 ATP used in the C_3 cycle (from Reaction 10-7) gives a total of 30 phosphate groups removed from ATP molecules for the synthesis of a molecule of glucose by the combined activities of the C_4 and C_3 cycles. The C_4 cycle thus significantly increases the amount of ATP required to synthesize sugars and other organic molecules in the dark reactions.

In spite of the price paid in ATP, the C_4 pathway provides an overall increase in photosynthetic efficiency at elevated temperatures, at which plants possessing

only the C_3 pathway are penalized by greatly increased oxygenase activity of RuBP carboxylase. (The oxygenase activity of the enzyme increases under these conditions because the solubility of oxygen in plant tissues is higher at elevated temperatures.) The crossover point seems to lie at about 30°C. Above this temperature C_4 plants become significantly more efficient than C_3-limited plants; below 30°C the ATP penalty paid by C_4 plants evidently makes C_3-limited species more efficient in spite of photorespiration.

Chloroplasts in the bundle sheath cells of C_4 plants are distinguished by a reduction in thylakoid stacking; in some species almost all the thylakoids in these cells occur in single, unstacked form. The reduction in thylakoid stacking is correlated with a reduced proportion of photosystem II and a preponderance of photosystem I complexes (see Information Box 10-1). The prevalence of photosystem I complexes in bundle sheath cells suggests that the light reactions proceed primarily by the cyclic pathway, eliminating O_2 production in these cells and perhaps replacing, to some extent, the ATP used in driving the C_4 pathway.

A side effect of the C_4 cycle reduces water loss in C_4 plants growing in hotter climates. The greater efficiency in CO_2 utilization reduces the size of the *stomata,* minute openings in leaves that allow the exchange of gases and water vapor between the plant and the surrounding atmosphere. Smaller stomata reduce the amount of water lost by the plant. As a result, C_4 plants are about twice as economical in water use as C_3-limited plants—C_4 plants lose only about 250 to 300 grams of water for every gram of carbon fixed into organic molecules, as compared to 400 to 500 grams for C_3 plants.

The 30°C crossover point and reduced water consumption place C_4 plants at an advantage in the tropics and in temperate regions with elevated summer temperatures, as in the central United States. Several important tropical and temperate crop plants benefit from the C_4 adaptation, including sugar cane, corn, sorghum,

The C_4 Cycle and CAM Metabolism **403**

and some pasture grasses. The reduced efficiency of plants with the C_4 pathway in cooler regions makes C_3-limited plants dominant in the northern United States and in Canada.

A wide variety of weeds—among them bermuda grass, crab grass, barnyard grass, water hyacinth, foxtail, Johnson grass, and morning glory—have evolved the C_4 pathway, making them formidable pests in warmer zones. In fact, the ten worst weed pests of the world are all C_4 plants.

The CAM Pathway

The *CAM* plants, named for the Crassulaceae family in which the CAM pathway was first discovered, use essentially the same C_4 cycle shown in Figure 10-29 to shuttle CO_2 from the atmosphere to the C_3 cycle. However, rather than placing the C_3 and C_4 cycles in different compartments in the plant, CAM plants run the two cycles at different times of the day.

CAM plants are typically succulents that live in regions that are hot and dry during the day and cool at night. They have fleshy leaves with a low surface to volume ratio and relatively few stomata. At night stomata open in the leaves, and CO_2 from the atmosphere enters the C_4 pathway. Subsequent C_4 steps proceed as far as malate, which accumulates throughout the night and is stored in large cell vacuoles until the leaves are saturated with the acid. In some CAM plants malate is converted to citrate and isocitrate for storage.

Daylight initiates the second phase of CAM me-tabolism. As the light reactions begin, stomata close and malate diffuses from cell vacuoles into the cyto-plasm. The remainder of the C_4 pathway now proceeds, with oxidation of malate and liberation of CO_2 in high concentrations. Because stomata are closed, no oxygen enters the leaves from the atmosphere. The low O_2 concentration and massive release of CO_2 by malate decarboxylation inhibit oxygenase activity of the RuBP carboxylase enzyme, and photosynthesis proceeds at maximum efficiency with no loss of organic carbon from photorespiration. Closure of stomata during the daytime and reduced leaf surface area also make CAM plants extremely resistant to water loss; CAM plants lose only 50 to 100 grams of water for every gram of carbon fixed into organic molecules. As a result, CAM species, which include the pineapple, many species of cactus such as prickly pear, and other succulents such as ice plant, can tolerate extreme heat and dryness.

The C_4 pathway and its CAM variation are strat-egies that compensate effectively for the oxygenase activity of RuBP carboxylase. These adaptations are spread among at least 16 different plant families, all angiosperms appearing more recently in evolution. Some of these families are unrelated enough to suggest that the C_4 pathway may have arisen independently several times in the evolution of higher plants. Evo-lution may have thus repeatedly followed pathways increasing CO_2 concentration and reducing O_2 in the vicinity of RuBP carboxylase rather than directly elim-inating the oxygenase activity of the enzyme.

Suggestions for Further Reading: See p. 400.

Supplement 10-2
Plastids and Plastid Development

Plant cells may contain a variety of plastid types that are closely related to chloroplasts, including *amyloplasts, elaioplasts, proteinoplasts, chromoplasts,* and *etioplasts*. All these plastid types have outer boundary membranes that are similar in molecular structure. However, the stroma is specialized and distinctly different in each plastid type.

In amyloplasts, which are colorless, the stroma is filled with starch grains (Fig. 10-31). These plastids store starch in roots and other tissues in plants. One highly unusual plastid in the tropical plant *Cecropia* (the trum-pet tree) stores starch in the form of glycogen that is chemically indistinguishable from the glycogen of an-imals. The tissues containing the glycogen-storing plas-tids are used as food by biting ants that live in a symbiotic relationship with *Cecropia*. Elaioplasts and proteinoplasts, which contain stores of oils and proteins, respectively, are also colorless plastids with a storage function. The stroma of these plastids appears struc-tureless except for the deposits of stored materials.

The stroma of chromoplasts, responsible for the red, yellow, or orange color of fruits, flower petals, and roots such as the carrot and sweet potato, contains deposits of carotenoid pigments, usually in the form of small granules. As much as 50% or more of the total weight of chromoplasts may consist of carotenoids.

All plastid types develop from *proplastids* (Fig. 10-32), small organelles in which a relatively undiffer-entiated stroma is surrounded by inner and outer boundary membranes. Few or no thylakoids are visible in the stroma of proplastids. DNA deposits and ribo-somes are usually visible, in some cases with starch

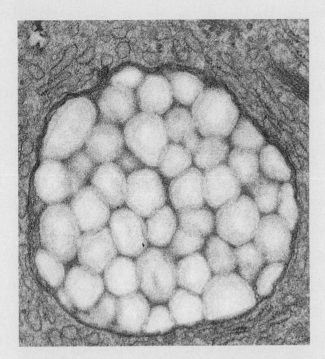

Figure 10-31 An amyloplast in the cytoplasm of a *Magnolia* pollen grain. The stroma of this storage plastid is completely packed with starch granules.

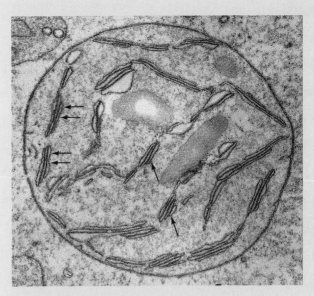

Figure 10-33 An early stage in the development of a proplastid into a chloroplast in the leaf meristem of *Oenothera* (evening primrose). Initial folding of the inner membranes into grana stacks has occurred (arrows). ×32,000. Courtesy of W. Menke.

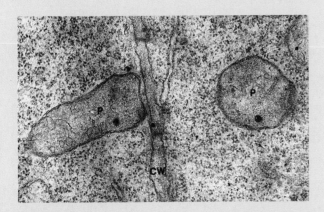

Figure 10-32 Proplastids (P) in the cytoplasm of adjacent tobacco cells. CW, cell wall. ×26,000. Courtesy of D. von Wettstein.

grains or osmiophilic granules. About 10 to 20 proplastids occur in meristematic cells of a growing stem. Somewhat larger numbers, up to 40, are found in root meristems. Proplastids also occur in pollen and egg cells of plants (see p. 1113). In stems and leaves grown in the light, proplastids in cell lines produced by division of meristems quickly develop into chloroplasts.

Transformation of proplastids into chloroplasts occurs first by tubelike growths of the inner boundary membrane that gradually extend into the stroma and flatten. At various sites the flat, thylakoidlike vesicles become doubly and triply folded into multiple layers (Fig. 10-33), eventually forming typical grana. Synthesis

of chlorophyll and accessory pigments takes place during growth of the thylakoid membranes.

In plants grown in the dark proplastids in leaves and stems develop into etioplasts instead of chloroplasts. An etioplast (Fig. 10-34) contains a regular lattice of internal membranes and a fibrous deposit of proteins. The lattice, called a *prolamellar body*, is formed from a regular network of connected, membranous tubules. The large protein deposit, the *stromacenter*, consists almost entirely of crystallized RuBP carboxylase molecules. DNA deposits and ribosomes may be scattered in the stroma around the prolamellar bodies.

Under the light microscope etioplasts, and the leaves and stems containing them, appear yellow. The yellow color is due to carotenoid pigments, which may be present in etioplasts in quantities amounting to as much as one-third of the carotenoid content of mature chloroplasts. Also present is *protochlorophyllide*, a colorless molecule identical to chlorophyll *a* except for the absence of the long phytol tail and two hydrogens in the tetrapyrrole ring. These pigments are concentrated in the membranes of the prolamellar body, which also contain most of the protein components of photosystems I and II and the electron carriers linking them. Although most of the elements necessary for photosynthesis are present, the photosystems and the electron transport system in etioplasts are inactive.

When etiolated plants are placed in light, etioplasts rapidly develop into chloroplasts, and the leaves containing them turn green. The wavelengths of light that

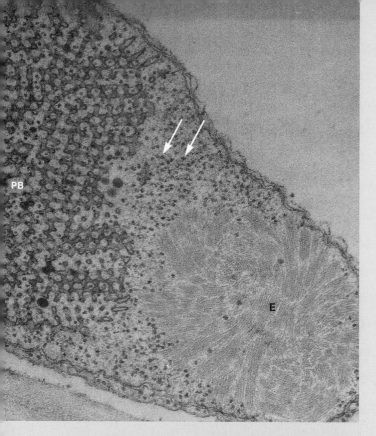

Figure 10-34 An etioplast from an *Avena* (oat) leaf. A highly ordered prolamellar body (PB) is present in the stroma; also visible is the stromacenter, a semicrystalline deposit containing RuBP carboxylase (E). Numerous ribosomes (arrows) are also visible in the stroma. × 55,000. Courtesy of B. E. S. Gunning, from *Protoplasma* 60:111 (1965).

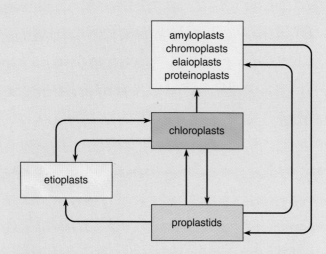

Figure 10-35 Developmental interrelationships of the various plastid types. Proplastids may develop into etioplasts, chloroplasts, or storage plastids. Chloroplasts may revert to etioplasts or develop into one of the storage types. Under certain conditions, such as the development of embryonic plants from mature plant tissue, chloroplasts or other mature plastid types may revert to proplastids as cells dedifferentiate and return to embryonic form.

induce the transformation of etioplasts are those absorbed by protochlorophyllide, which, under the influence of light, is rapidly converted to *chlorophyllide a* by the addition of two electrons and two hydrogens. Chlorophyllide *a* is subsequently converted to chlorophyll *a* by addition of a phytol tail. Complete conversion of protochlorophyllide into chlorophyll *a* may be completed in as little as one minute after exposure to light. Appearance of chlorophyll *b* is much slower; this pigment, which is synthesized by modification of chlorophyll *a*, cannot be detected until several hours after transfer of plants to light. As protochlorophyllide is converted to chlorophyll *a*, the tubules of the prolamellar body gradually develop into thylakoids.

These changes are correlated with the appearance of full biochemical activity. Within 30 minutes after etioplasts begin to green, the activity of the photosystems is detectable. Biochemical activity of photosystem II is correlated with development of thylakoid stacks. Many of these changes, which involve gene activation and protein synthesis, are controlled by two additional light-sensitive substances, *phytochrome* and *blue light receptors*. Phytochrome consists of a linear tetrapyrrole associated with a protein; absorption of

light at red wavelengths converts this substance into an activated form that directly or indirectly triggers the transcription of genes encoding enzymes and other components of the light and dark reactions. Absorption of blue wavelengths by the blue light receptors has similar effects. The mechanisms by which these light-sensitive substances regulate gene activity remain unknown. However, there are indications that Ca^{2+} is released into the cytoplasm following activation of phytochrome; this Ca^{2+} may activate enzymes taking part in a reaction cascade linking phytochrome activation to cellular responses. Regulatory proteins controlling gene transcription, which are directly or indirectly activated by phytochrome, have also been identified in a few plants.

All the various plastid types are potentially interconvertible, depending on growth conditions in the plant (Fig. 10-35). Proplastids may develop into etioplasts or chloroplasts, or directly into amyloplasts, chromoplasts, or other storage plastids. Similarly, chloroplasts may revert to etioplasts or develop into one of the storage types. Under certain conditions, such as the development of embryonic plants from mature plant tissue, chloroplasts or other mature plastid types such as chromoplasts or amyloplasts may revert to proplastids as cells dedifferentiate.

Suggestions for Further Reading: See p. 400.

Photosynthesis in Prokaryotes

The photosynthetic prokaryotes, *green* and *purple bacteria* and cyanobacteria, differ fundamentally in the pathways of their photosynthetic reactions. Photosynthetic bacteria have comparatively primitive systems limited essentially to the activities carried out by photosystem I in eukaryotic plants. Lacking a water-splitting activity equivalent to photosystem II, photosynthetic bacteria cannot use H_2O as an electron donor and do not evolve oxygen in photosynthesis. Nevertheless, in terms of their internal pathways of electron transport, the photosystems of these bacteria may resemble either photosystem I or II of eukaryotic plants.

Cyanobacteria, although typically prokaryotic in cellular organization, have two photosystems equivalent to eukaryotic photosystems I and II and carry out photosynthesis by essentially the same mechanisms as eukaryotic plants. Cyanobacteria can use H_2O as an electron donor and evolve oxygen in photosynthesis.

All photosynthetic prokaryotes fix CO_2 through the activity of RuBP carboxylase enzymes working in a C_3 cycle as in eukaryotic algae and higher plants. Many of these prokaryotes also possess the enzymes of the C_4 pathway, indicating that these organisms may circumvent photorespiration by evolutionary strategies similar to those of eukaryotes. In addition to the C_3 and C_4 pathways, some photosynthetic bacteria are able to fix CO_2 by reactions that essentially reverse the CO_2-generating reactions of pyruvate oxidation or the citric acid cycle (for details of these reactions, see p. 320).

The Photosynthetic Bacteria

Purple and green photosynthetic bacteria comprise less than 50 known species. Purple bacteria occur in two recognized subgroups, *purple sulfur bacteria*, which use sulfur-containing compounds such as H_2S as electron donors for noncyclic photosynthesis, and *purple nonsulfur bacteria*, which use complex sulfur-free organic substances such as malate and succinate as electron donors. Different species among green bacteria may use either inorganic sulfur-containing compounds or nonsulfur organic molecules as electron donors for noncyclic photosynthesis. However, the different green bacterial types have not as yet been formally classified as sulfur and nonsulfur subgroups.

The pigments, light-harvesting antennas, photosystems, electron carriers, and F_oF_1 ATPase carrying out the light reactions in photosynthetic bacteria are embedded in the plasma membrane or in tube- or saclike invaginations of the plasma membrane that extend into the cytoplasm (Fig. 10-36; see also Fig. 1-4).

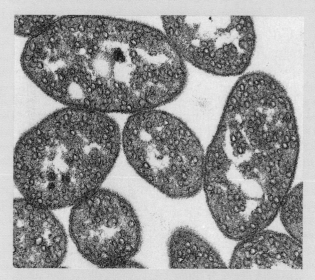

Figure 10-36 Saclike photosynthetic vesicles in the cytoplasm of the purple bacterium *Rhodopseudomonas.* × 32,500. From *Biochemistry of Chloroplasts*, vol. 1, T. W. Goodwin, ed., 1966. Courtesy of W. Menke and Academic Press, Inc.

Bacterial Photosynthetic Pigments Both purple and green bacteria use a distinct family of chlorophylls, the *bacteriochlorophylls* (Fig. 10-37), as primary photosynthetic pigments. Bacteriochlorophylls, like chlorophylls of eukaryotic plants, are built on a tetrapyrrole ring containing a central magnesium atom, and differ only in minor substitutions in side groups attached to

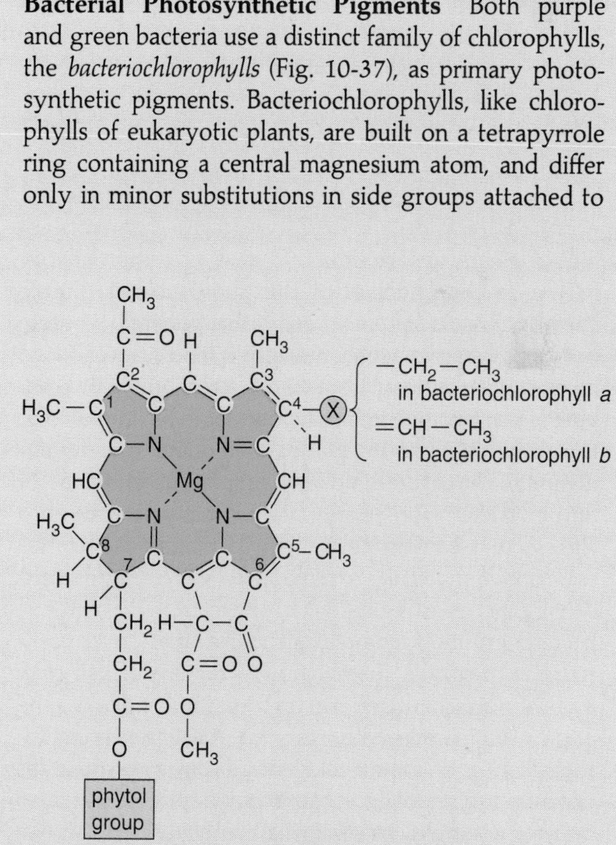

Figure 10-37 Bacteriochlorophylls *a* and *b*. An additional —H occurs at the 4-carbon in bacteriochlorophyll *a*. One form of bacteriochlorophyll *a* is linked to a different hydrophobic, long-chain alcohol instead of the phytol chain. Other side-group substitutions produce bacteriochlorophylls *c, d,* and *e*.

the ring. Purple bacteria contain bacteriochlorophylls *a* and *b*. Bacteriochlorophylls *c*, *d*, and *e* predominate in the green bacteria in combination with small quantities of bacteriochlorophyll *a*, which occur in reaction centers.

Bacteriochlorophyll pigments absorb light most strongly in the near ultraviolet and far red regions of the spectrum and transmit most of the visible wavelengths. As a result, they contribute little color to either bacterial group. The distinctive colors of green and purple bacteria come primarily from different carotenoids occurring as accessory pigments in association with bacteriochlorophylls.

Bacteriochlorophylls and carotenoids of photosynthetic bacteria are organized with proteins into photosystems and light-harvesting complexes as in eukaryotic plants. Within the photosystems specialized bacteriochlorophylls form a reaction center that raises electrons to excited energy levels and passes them on to primary acceptors. The primary acceptors are followed by short sequences of electron carriers that, in different purple and green bacteria, may resemble the systems of either eukaryotic photosystem I or II. Because the single photosystem of bacteria cannot be reduced by the very low-energy electrons released from water, electron donors with higher reducing power must be used. These donors, as we have noted, include H_2S and other sulfur-containing compounds and complex organic molecules such as malate and succinate in different green and purple bacterial groups.

The light-harvesting complexes of both purple and green bacteria, like the LHC-I and LHC-II antennas of higher plants, absorb light and pass excitation energy to the reaction centers of the photosystems. Purple bacteria possess simple antennas that generally contain only two bacteriochlorophyll *a* or *b* molecules and two carotenoids (either carotenes or xanthophylls), combined with two or three polypeptides in a complex that is tightly bound to the plasma membrane or its invaginations. Though individually simple, the antennas of purple bacteria are organized in many species into "lakes" that contain as many as a thousand bacteriochlorophyll molecules serving 20 to 40 reactioncenters. The light-harvesting antennas of green photosynthetic bacteria are more complex structures that are concentrated into saclike, membranous structures called *chlorosomes*, attached to the cytoplasmic side of the plasma membrane. As many as 1500 bacteriochlorophyll *c*, *d*, or *e* molecules may be in a chlorosome, organized with one or more polypeptides that also occur in multiple copies in the complex.

Electron Transport in Bacterial Photosynthesis

Electrons flow both cyclically and noncyclically in the electron transport systems associated with photosynthesis in purple and green bacteria. Although the carriers active in bacterial electron transport include most

of the major carriers of eukaryotic photosynthesis—quinones, cytochromes, and Fe/S proteins—there is great variety in the types and combinations of these carriers. Many different *b*- and *c*-type cytochromes, most of them distinct from their eukaryotic counterparts, carry electrons in the bacterial systems; even quinones occur in four or more types. Complicating the situation is the fact that some carriers transport electrons in both photosynthetic and respiratory pathways.

Cyclic flow around the single photosystem builds up an H^+ gradient that is subsequently used as an energy source for ATP synthesis. Noncyclic flow leads from an electron donor directly or indirectly to NAD^+ rather than $NADP^+$ as in cyanobacteria and eukaryotic plants. After reduction NADH may donate its electrons to synthetic reactions requiring a reduction, such as those of the dark reactions fixing CO_2, or to an electron transport chain producing an H^+ gradient for ATP synthesis.

Light Reactions in Purple Bacteria   The single photosystem of purple bacteria is built around three membrane-spanning polypeptides known as the *light* (*L*), *medium* (*M*), and *heavy* (*H*) *polypeptides*. These proteins organize a reaction center containing either bacteriochlorophyll *a* or *b* and a short series of electron carriers closely resembling those of photosystem II of eukary-

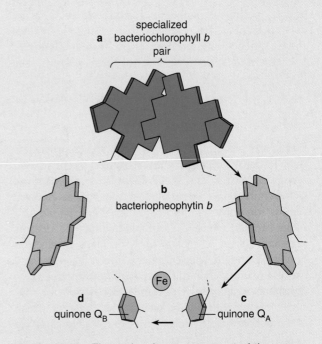

Figure 10-38   The molecular arrangement of the reaction center and internal electron carriers of the *Rhodopseudomonas viridis* photosystem. After excitation in the reaction center, which consists of a pair of bacteriochlorophyll *b* molecules **(a)**, electrons flow to bacteriopheophytin *b* **(b)**, and through two quinones, Q_A and Q_B, in sequence **(c** and **d)**.

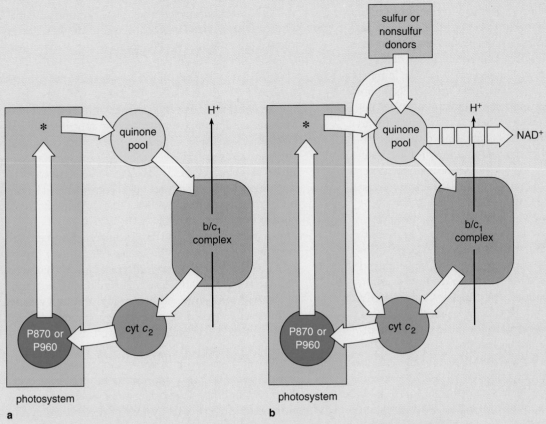

Figure 10-39 Patterns of photosynthetic electron transport in purple photosynthetic bacteria. **(a)** Cyclic electron transport; **(b)** noncyclic electron transport. Electron donors for noncyclic photosynthesis may be sulfur-containing compounds such as H_2S or nonsulfur organic substances such as succinate. The asterisk indicates the excited form of the photosystem.

otic plants. In most purple bacteria the bacteriochlorophyll molecules at the reaction center undergo a conspicuous change in absorption at a wavelength of 870 or 960 nm, depending on the species, as they undergo cycles of oxidation and reduction in connection with the excitation of electrons. The reaction-center bacteriochlorophylls of these bacteria are identified accordingly as *P870* or *P960*.

Recently, H. Michel, J. Deisenhofer, and R. Huber crystallized the photosystem of one purple bacterium, *Rhodopseudomonas viridis*, and analyzed the crystals by X-ray diffraction. Their analysis allowed the arrangement of the reaction center and electron carriers of this bacterium to be visualized at the atomic level (Fig. 10-38; Michel, Deisenhofer, and Huber received the Nobel Prize in 1988 for their analysis). The reaction center consists of a pair of specialized bacteriochlorophyll *b* molecules. After excitation in the reaction center (Fig. 10-38a), electrons flow to *bacteriopheophytin b* (Fig. 10-38b), which resembles bacteriochlorophyll *b* without the central magnesium atom. From bacteriopheophytin *b* electrons flow through two quinones, Q_A and Q_B, which are associated with an iron atom (Fig. 10-38c and d). At this point electrons pass from the photosystem to carriers of the electron transport system.

Thus the electron pathway within the *R. viridis* photosystem is equivalent to the P680 → pheophytin → Q_A → Q_B pathway of eukaryotic photosystem II (and cyanobacterial photosystem II; see below). The similarities between the photosystem of *R. viridis* and other purple bacteria and eukaryotic photosystem II are reinforced by the fact that the L and M polypeptides of the *R. viridis* system are similar in amino acid sequence to the D1 and D2 polypeptides forming the reaction center of photosystem II (see Fig. 10-12). The general similarities between the *R. viridis* photosystem and photosystem II of cyanobacteria and eukaryotic plants suggest that the more advanced photosystem evolved from its bacterial counterpart primarily by development of the three polypeptides carrying out the water-splitting reaction (see p. 378).

Electrons may flow cyclically or noncyclically around the single photosystem of purple sulfur bacteria. In cyclic electron transport (Fig. 10-39a), electrons released from the photosystem enter a quinone pool. Although variations are noted in different purple bacteria, in many species electrons are later transferred from the quinone pool to a b/c_1 complex (see pp. 334 and 376). Like its chloroplast and mitochondrial counterparts, the bacterial b/c_1 complex contains a *b*-type and a c-type cytochrome linked with an iron-sulfur protein and a group of polypeptides. Electron flow through the bacterial b/c_1 complex pumps H^+ across the bacterial membrane, building an H^+ gradient linked to electron transport as in eukaryotic systems. In most purple bacteria electrons flow from the b/c_1 complex to another c-type cytochrome, cytochrome c_2, a peripheral membrane protein that takes a position in bacterial cyclic electron flow equivalent to that of plastocyanin in eukaryotic plants. From cytochrome c_2 electrons return at lower energy levels to the reaction center of the single photosystem. After another energy boost through light absorption, they may repeat the cyclic pathway.

In noncyclic flow in purple bacteria (Fig. 9-39b), electrons derived from various sulfur or nonsulfur donors, depending on their energy level, may be passed by a carrier, usually a cytochrome, to the photosystem and then to the quinone pool. Or, they may directly enter the quinone pool. In either case electrons in the quinone pool initially contain too little energy to directly reduce NAD^+. Some electrons in the pool, however, evidently receive an additional energy boost from the membrane potential built up by cyclic electron transport. Transport of H^+ is electrogenic (see p. 208) and creates a voltage difference across the membrane as well as an H^+ gradient. The electrons boosted by the membrane voltage attain energy levels high enough to reduce NAD^+. Although the mechanism by which the membrane potential boosts electrons in purple bacteria is not understood, there is no doubt that this process occurs. Experiments have shown that light-dependent NAD^+ reduction in purple bacteria relies on the membrane potential built up by electron flow through the transport system; uncoupling agents, which make the membrane "leaky" to H^+ and destroy the H^+ gradient, completely block NAD^+ reduction.

Noncyclic flow results in the one-way transfer of electrons from donor substances to NAD^+. The NADH produced by the reduction, like the NADPH produced in eukaryotic systems, provides a source of electrons for reductions elsewhere in the cell, as in the dark reactions fixing CO_2 into carbohydrates. Alternatively, electrons carried by NADH can enter electron transport linked to the synthesis of ATP.

The same F_oF_1 ATPase active in oxidative phosphorylation (see p. 363) uses the H^+ gradient built up by photosynthetic electron transport as an energy source for ATP synthesis in purple bacteria. The bacterial F_oF_1 ATPase is similar in structure to the chloroplast CF_oCF_1 ATPase and mitochondrial F_oF_1 ATPase (see Table 10-1).

ATP produced by the bacterial F_oF_1 ATPase, along with NADH formed by noncyclic photosynthesis, provide the energy and reducing power for CO_2 fixation in the dark reactions. The molecules and complexes of the light reactions, including light-harvesting antennas, photosystems, and electron transport carriers, are associated with saclike invaginations of the plasma membrane in purple bacteria (as in Fig. 10-36).

Light Reactions in Green Bacteria   The photosynthetic systems of green bacteria, although less well known than those of purple bacteria, appear to fall into two fairly well-defined groups with respect to photosystem structure and electron transport systems. One group is anaerobic and possesses a photosystem resembling photosystem II of eukaryotic plants. The second group is aerobic, with a photosystem similar to eukaryotic photosystem I.

Within the photosystem of anaerobic green bacteria, bacteriochlorophyll a molecules forming the reaction center change in absorbance at a wavelength of 870 nm and are identified as $P870$. Following excitation in the reaction center, electrons flow through a series of internal carriers including bacteriopheophytin and iron-associated quinones as in purple bacteria or eukaryotic photosystem II.

Electron transport around the photosystem of anaerobic green bacteria appears to be primarily cyclic (Fig. 10-40). Only a few different cytochromes occur in the electron transport system of these bacteria. The connection between this electron flow and ATP synthesis or reduction of NAD^+ remains to be established. It is doubtful that electrons excited by the photosystem have enough energy to reduce NAD directly. However,

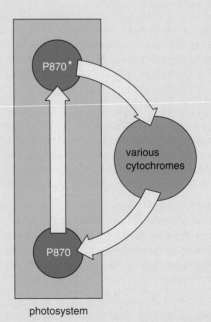

Figure 10-40 Photosynthetic electron flow in anaerobic green bacteria, which progresses primarily or exclusively by a cyclic pathway (see text).

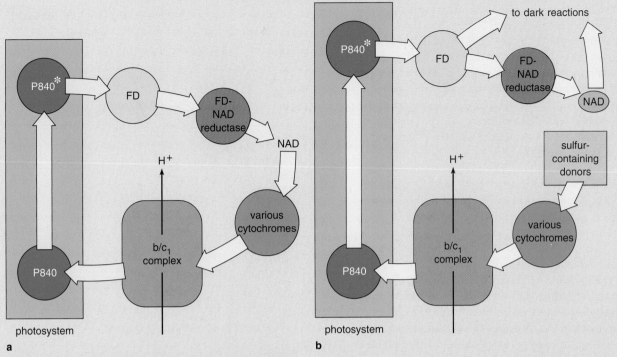

Figure 10-41 Cyclic (a) and noncyclic (b) electron flow in aerobic green bacteria (see text). FD, ferredoxin; FD-NAD reductase, ferredoxin-NAD$^+$ oxidoreductase.

NADH may still be formed by a mechanism similar to that proposed for purple bacteria. Potential electron donors for noncyclic electron flow, if this pattern indeed occurs in anaerobic green bacteria, are limited to organic molecules that release electrons at relatively high energy levels.

The photosystem of aerobic green bacteria contains specialized bacteriochlorophyll *a* molecules absorbing light at 840 nm. These molecules, identified as *P840*, pass electrons to a primary acceptor and a chain of Fe/S centers rather than quinones. The electron transport system conducting excited electrons from the photosystem varies greatly among different aerobic green bacteria. The more complex electron transport systems of these bacteria may include ferredoxin, a b/c$_1$ complex, and the ferredoxin-NAD oxidoreductase complex.

One pattern in which these carriers are arranged in aerobic green bacteria may accomplish either cyclic or noncyclic electron flow (Fig. 10-41). In cyclic transport (Fig. 10-41a), electrons flow to ferredoxin after excitation and movement through the internal carriers of the photosystem. The direct reduction of ferredoxin as the first carrier reflects the fact that electrons are excited to significantly higher energy levels by the photosystem in these bacteria. Electrons then flow to NAD$^+$ via the ferredoxin-NAD oxidoreductase complex, which may contain FAD as an internal carrier as in the equivalent complex of higher plants. From NADH electrons are transferred via one or more cy-

tochromes to the b/c$_1$ complex, and then back to P840 molecules at the reaction center of the photosystem. Transfer from the b/c$_1$ complex to the photosystem may be direct or may occur via additional cytochromes; the cytochromes on this side of the b/c$_1$ complex, like those delivering electrons to the complex from NADH, are generally types that do not occur in eukaryotic systems. Because the b/c$_1$ complex is present in the loop, H$^+$ is pumped across the membrane housing the system each time electrons cycle around the photosystem.

Noncyclic electron flow in aerobic green bacteria (Fig. 10-41b) uses electrons removed from inorganic sulfur compounds, including many of the substances used as donors by purple sulfur bacteria. This flow occurs through cytochromes that vary widely in different species. In the pattern shown in Figure 10-41b, electrons pass from the donors through one or more of these cytochromes to reach the b/c$_1$ complex. From this point electrons enter the photosystem and, after excitation, are delivered at high-energy levels to ferredoxin. The electrons may remain with ferredoxin, which serves directly as an electron donor for dark reactions in green bacteria (see below). Alternatively, electrons may be delivered from ferredoxin to NAD$^+$ by the same pathway used in noncyclic photosynthesis. The NADH produced may provide electrons for the dark reactions or may enter the respiratory electron transport system leading to oxygen as the final electron acceptor. Alternatively, the electrons may reenter the photosynthetic pathway from NADH and travel cyclically through one or more loops around the photosystem.

As in purple bacteria the same F_oF_1 ATPase active in oxidative phosphorylation uses the H^+ gradient established by photosynthetic electron transport as the energy source for ATP synthesis. All components of the light reactions are associated with the plasma membrane in green bacteria.

Bacterial Dark Reactions The C_3 and C_4 pathways fixing CO_2 in most photosynthetic bacteria, both purple and green, proceed by the same mechanisms as in eukaryotes. Green bacteria can also fix CO_2 into organic molecules by reactions that reverse oxidative steps in pyruvate oxidation or the citric acid cycle (Fig. 10-42). These reactions are catalyzed by unique enzymes that use ferredoxin, reduced in noncyclic photosynthesis, as a donor of high-energy electrons for CO_2 fixation.

For example, reaction 4 of the citric acid cycle (see Fig. 9-15), which oxidizes isocitrate to α-ketoglutarate, can be reversed in green bacteria by electrons supplied by reduced ferredoxin:

$$\text{α-ketoglutarate} + CO_2 + \text{reduced ferredoxin} \longrightarrow$$
$$\text{isocitrate} + \text{oxidized ferredoxin} \quad (10\text{-}9)$$

Figure 10-42 indicates other points at which green bacteria can fix CO_2 by reversing oxidative reactions. These unusual reactions, using electrons carried by ferredoxin, apparently provide the major pathway for CO_2 fixation in green bacteria.

Photosynthesis in Halobacteria One exceptional bacterial group, the *halobacteria*, possesses an incomplete and primitive photosynthetic mechanism that differs radically from those of purple and green bacteria. The halobacteria, which typically live in extreme environments, usually at levels of salinity and temperature too high to be tolerated by other forms, have no light-harvesting antennas, no photosystems, and no light-driven electron transport system. Photosynthetic membranes of *Halobacterium* contain a light-absorbing molecule known as *bacteriorhodopsin* (see Fig. 5-8), consisting of a polypeptide chain with the light-absorbing unit. The light-absorbing unit of bacteriorhodopsin is *retinal*, a molecule almost identical to the visual pigment of animals (Fig. 10-43a).

Bacteriorhodopsin responds to light by pumping H^+ across the membrane containing the complex. The pumping mechanism is incompletely understood. During a pumping cycle the retinal unit alternately picks up and releases a hydrogen. This pickup and release may be combined with conformational changes in the protein component that expose the retinal unit to the cytoplasm during H^+ binding, and to the cell exterior

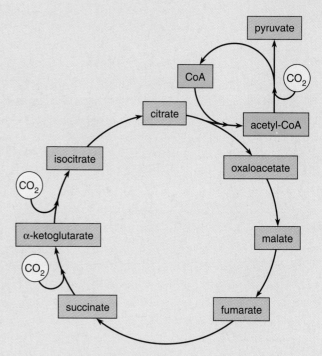

Figure 10-42 Points at which green bacteria can fix CO_2 by reversing reactions of pyruvate oxidation and the citric acid cycle (compare with Fig. 9-15).

during H^+ release. The H^+ gradient established by light-induced pumping through the bacteriorhodopsin molecule drives ATP synthesis by a membrane-bound F_oF_1 ATPase as in other bacteria.

Studies conducted with isolated bacteriorhodopsin by E. Racker and W. Stoeckenius supplied some of the most compelling evidence supporting Mitchell's chemiosmotic hypothesis. Racker and Stoeckenius added purified bacteriorhodopsin to artificial phospholipid vesicles; while illuminated, the reconstituted vesicles pumped H^+ and established a measurable H^+ gradient. Addition of F_oF_1 ATPase to the reconstituted membranes allowed the system to synthesize ATP in response to the gradient (Fig. 10-43b), all in accordance with expectations of the Mitchell hypothesis.

The artificial vesicles used by Racker and Stoeckenius in their experiments were assembled from phospholipids extracted from a plant, and the F_oF_1 ATPase was purified from beef heart mitochondria. The experiments thus created a functional light-driven, ATP-synthesizing system with molecular components isolated from a bacterium (bacteriorhodopsin), a plant (the membrane phospholipids), and an animal (the F_oF_1 ATPase)!

Photosynthesis in Cyanobacteria Cyanobacteria carry out the light and dark reactions of photosynthesis by essentially the same pathways as eukaryotic algae and higher plants. Photosystems I and II resembling their eukaryotic counterparts, including chlorophyll *a*

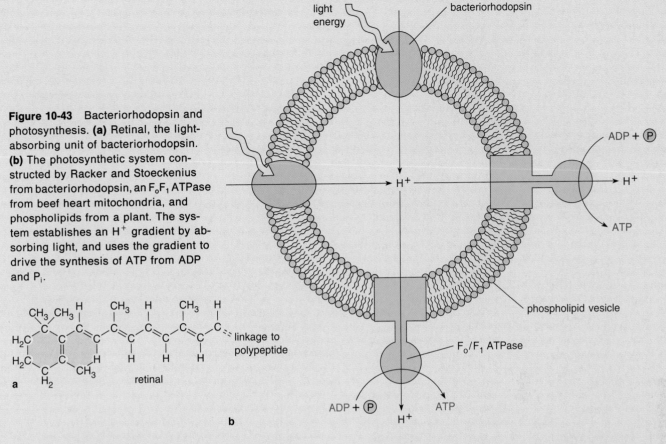

Figure 10-43 Bacteriorhodopsin and photosynthesis. **(a)** Retinal, the light-absorbing unit of bacteriorhodopsin. **(b)** The photosynthetic system constructed by Racker and Stoeckenius from bacteriorhodopsin, an F_oF_1 ATPase from beef heart mitochondria, and phospholipids from a plant. The system establishes an H^+ gradient by absorbing light, and uses the gradient to drive the synthesis of ATP from ADP and P_i.

rather than bacteriochlorophylls, are present in these organisms. The cyanobacterial photosystem II includes an enzymatic assembly allowing water to be split as the source of electrons for photosynthesis. As a consequence, cyanobacteria release oxygen as a by-product of photosynthesis.

Cyanobacterial photosystems are connected into a Z pathway of the eukaryotic type. With the exception of a few cyanobacteria that substitute cytochrome c_2 for plastocyanin, the Z pathway in these prokaryotes includes the same electron carriers and complexes as eukaryotic systems. The dark reactions in cyanobacteria also proceed by the same pathways used by the eukaryotic plants.

Genes encoding the F_oF_1 ATPase of cyanobacteria were recently isolated and sequenced by A. L. Cozens and J. E. Walker, S. E. Curtis, and others (see p. 134). The sequences revealed that the cyanobacterial F_oF_1 ATPase is more closely related to the CF_oCF_1 ATPase of chloroplasts than to the F_oF_1 ATPase of bacteria. The cyanobacterial F_oF_1 ATPase also contains the additional F_o subunit (subunit II, see Table 10-1) found in chloroplast CF_oCF_1 ATPase but not in the bacterial enzyme.

Molecules carrying out the light reactions in cyanobacteria are bound to saclike membranes suspended in the cytoplasm (Fig. 10-44; see also Fig. 1-5). Thus the location of photosynthetic membranes in cyanobacteria is similar to that of purple bacteria, even though

the reactions of photosynthesis proceed according to the eukaryotic pattern.

Another difference between cyanobacterial and most eukaryotic systems is in the form taken by the light-harvesting antennas. Chlorophyll b and the light-harvesting complexes typical of eukaryotes are absent in cyanobacteria. Taking their place are protein-pigment combinations known as *phycobilisomes*, which can be seen as spherical or ovoid particles attached to the cytoplasmic surfaces of photosynthetic membranes (arrows, Fig. 10-44). The primary pigments of phycobilisomes are *phycobilins* (Fig. 10-45). Phycobilisomes serve as light-harvesting antennas for photosystem II in cyanobacteria; two photosystem II complexes are associated with each phycobilisome. No accessory light-absorbing structures of any kind appear to be associated with photosystem I in cyanobacteria.

Phycobilins are hydrophilic, water-soluble pigments that contain a light-absorbing structure resembling the tetrapyrrole ring of hemoglobin. The pyrroles of the phycobilins occur in linear form, however. The light-absorbing structure of phycobilins resembles two linear-chain products of hemoglobin breakdown that appear in animal bile, *biliverdin* and *bilirubin* (hence, the phyco*bilin* name applied to these photosynthetic pigments). In phycobilins the linear-chain pyrrole forming the light-absorbing structure is covalently linked to two related proteins known as the α- and β-polypeptides. Either polypeptide may carry one or more phycobilin

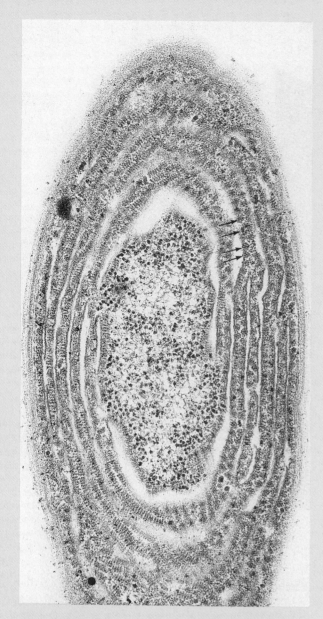

Figure 10-44 Photosynthetic vesicles in the cytoplasm of the cyanobacterium *Synechococcus*. Numerous phycobilisomes (arrows), the light-harvesting antennas of cyanobacteria, are attached to the vesicle membranes. ×51,000. Courtesy of M. R. Edwards, New York State Department of Health, from *J. Cell Biol.* 50:896 (1971), by permission of the Rockefeller University Press.

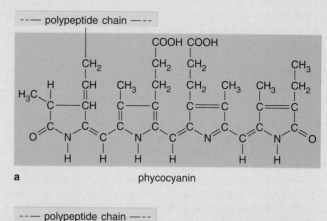

a phycocyanin

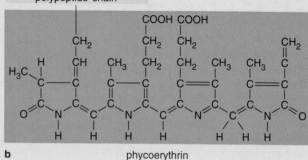

b phycoerythrin

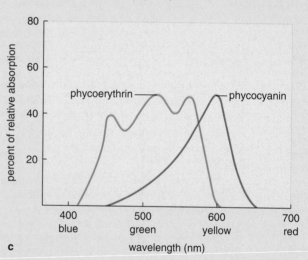

c

Figure 10-45 The light-absorbing structures of the phycobilins phycocyanin **(a)** and phycoerythrin **(b)**. The pyrrole groups of the phycobilins are in linear, rather than ring, form. **(c)** The absorption spectra of phycocyanin and phycoerythrin. Both phycocyanin and phycoerythrin occur in cyanobacteria; red algae may have either or both of these pigments.

units. The covalent linkage of phycobilins to the polypeptides is unusual, because most other known pigment-protein complexes, such as the associations between chlorophylls and proteins, are noncovalent.

The primary phycobilin occurring as an accessory pigment in cyanobacteria is *phycocyanin* (see Fig. 10-45a). Other related phycobilins, including *phycoerythrin* (see Fig. 10-45b), *phycoerythrocyanin,* and *allo-*

phycocyanin, are organized into a light-harvesting assembly that consists of a core structure with radiating arms (Fig. 10-46). Light energy absorbed by the pigments in the arms is conducted to the core structure, and from there to the photosystem II complex. About 400 phycobilin molecules are associated with each phycobilisome.

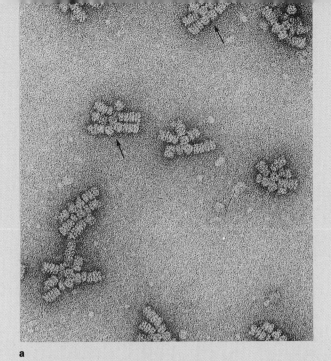

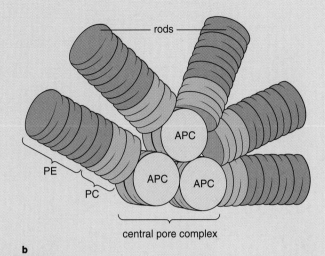

rods

APC

APC

APC

PE

PC

central pore complex

a

b

Figure 10-46 Phycobilisomes. **(a)** Phycobilisomes isolated from the cyanobacterium *Synechocystis*. The arrows indicate the central core complexes from which the rods extend. ×225,000. Courtesy of A. N. Glazer; reproduced from *J. Cell Biol.* 85:558 (1980), by copyright permission of the Rockefeller University Press. **(b)** A model for phycobilisome structure showing pigment locations. PE, phycoerythrin; PC, phycocyanin; APC, allophycocyanin.

Among eukaryotes only one group, the red algae, uses phycobilisomes instead of LHC antennas. In red algae phycobilisomes occur inside chloroplasts on the surfaces of thylakoid membranes facing the stroma (see Fig. 10-25).

The dark reactions of cyanobacteria proceed primarily by the C_3 cycle. Although the PEP carboxylase enzyme necessary for the C_4 cycle (see Supplement 10-1) can also be detected in cyanobacteria, it is not certain whether these prokaryotes possess the C_4 pathway. The requirement for the C_4 pathway may be reduced by the fact that cyanobacteria are able to concentrate CO_2 inside their cells by active transport, through a carrier that probably uses ATP as its energy source.

Cyanobacteria were the first organisms to release oxygen as a by-product of photosynthesis. As such they were critical to the first appearance of oxygen in the earth's atmosphere and to the evolutionary development of aerobic organisms. The relatively limited development of oxidative metabolism in cyanobacteria (see p. 364) may be related to the fact that these prokaryotes evolved in an atmosphere that initially did not contain oxygen. Evidence from various sources indicates that cyanobacteria are the evolutionary progenitors of chloroplasts in eukaryotic algae and green plants. The presence of phycobilisomes and phycobilins suggests that red algae may be the eukaryotic algae most closely related to the ancient cyanobacteria believed to have given rise to chloroplasts. (See Ch. 27 for details of the probable evolutionary origins of chloroplasts in cyanobacteria.)

Suggestions for Further Reading: See pp. 400 and 401.

MICROTUBULES AND
MICROTUBULE-BASED CELL MOTILITY

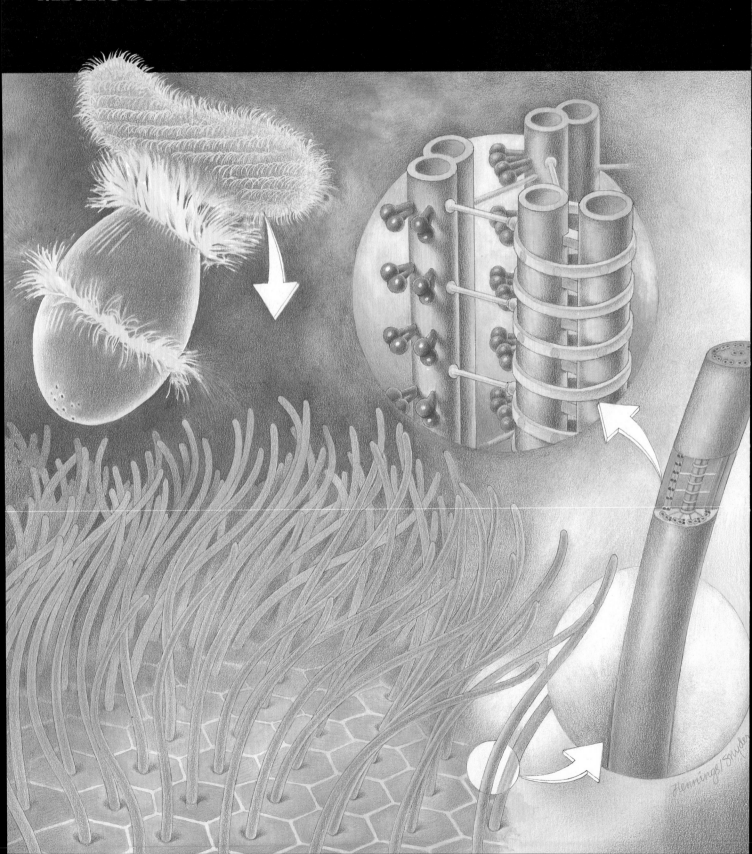

- *Microtubule structure and biochemistry*
- *Dynamic instability* ▪ *Microtubule-associated proteins (MAPs)* ▪ *Microtubule polarity*
- *Microtubule-organizing centers (MTOCs)*
- *Dynein and microtubule sliding* ▪ *Flagella and cilia* ▪ *Kinesin, cytoplasmic dynein, and other microtubule crossbridges* ▪ *Centrioles, basal bodies, and the generation of cilia and flagella*
- *Bacterial flagella*

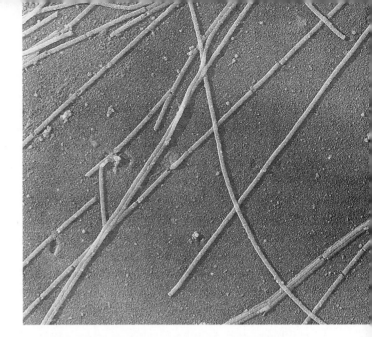

Figure 11-1 Microtubules isolated from erythrocytes of the salamander *Taricha* and prepared for electron microscopy by shadowing. ×42,000. Photograph by the author.

All cells are capable of movement. In the simplest case cellular movements take the form of slow and limited rearrangements of internal structures. More elaborate movements include extended translocations of organelles in the cytoplasm, overall changes in cell shape, contraction of muscle cells, and the rapid motions of cell appendages such as flagella and cilia. In many-celled organisms individual cellular movements are coordinated to produce the movements of body parts or the entire individual.

Two different structures, *microtubules* and *microfilaments*, are responsible for most of these movements. Microtubules are unbranched cylinders about 25 nm in diameter with an open central channel (Figs. 11-1 and 11-2; see also Fig. 1-17). Microfilaments (see Fig.12-1) are extremely fine, unbranched fibers of much smaller diameter, about 5 to 7 nm—not much thicker than the wall of a microtubule. The two motile elements are built from different proteins—microtubules from *tubulin* and microfilaments from *actin*.

Microtubules and microfilaments act both separately and in coordination to produce cellular movements. The swimming movements of flagella and cilia, for example, are based on microtubules. Microfilaments are responsible for muscle contraction in animals and cytoplasmic streaming in both animals and plants. The two motile elements coordinate their activities in animal cell division, in which microtubules divide and distribute chromosomes and microfilaments divide the cytoplasm.

The molecular mechanisms generating motile forces are similar in microtubules and microfilaments; both produce movement by an active sliding mechanism. Microtubules slide over microtubules, or microfilaments over microfilaments; and cell organelles slide over the surface of either motile element. These motions are produced by active, ATP-driven crossbridges that "walk" over the surfaces of microtubules or microfilaments by making attachments, swiveling actively and forcefully over a short distance, and then releasing (see Figs. 11-18 and 12-12). Distinct proteins form crossbridges for the two motile elements. Several different proteins form microtubule crossbridges; best known of these are *dynein* and *kinesin*. Microfilament crossbridges consist of a single protein type, *myosin*.

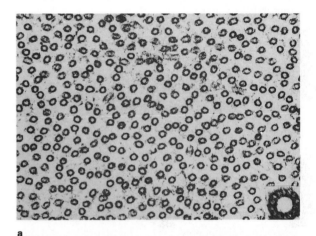

a

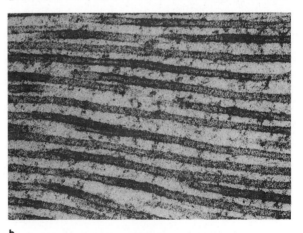

b

Figure 11-2 Microtubules (a) in cross section and (b) in longitudinal section. The microtubules in this figure were polymerized from tubulin heterodimers in the test tube. ×42,000; inset to a, ×380,000. Courtesy of W. L. Dentler, Jr.

Like enzymatic catalysis and membrane transport, the crossbridging mechanisms depend on the ability of proteins to undergo conformational changes in response to chemical interactions. In the microtubule and microfilament systems, the conformational changes, which produce the oarlike, swiveling motions, occur in response to ATP binding and hydrolysis. The effect of the conformational changes is to transform the chemical energy of ATP into the mechanical energy of movement.

The dynein crossbridges of microtubules may extend between microtubules or may move other structures such as mitochondria or vesicles along microtubules. Kinesin crossbridges typically extend between cytoplasmic structures such as vesicles and microtubules. Myosin crossbridges may extend between microfilaments or may connect a variety of cell structures such as mitochondria or chloroplasts to microfilaments. Apparently none of the crossbridges directly link microtubules and microfilaments. Thus, although the two motile elements may coordinate their activities in processes such as cell division, they remain structurally separate.

Both microtubules and microfilaments also produce motion by controlled growth or disassembly. The resultant changes in length push or pull attached structures from place to place within the cell or impose stretching or compressive forces. Both structures also act as cytoskeletal elements providing support rather than motility to cell structures.

Three chapters describe the structure and activities of microtubules and microfilaments in nondividing eukaryotic cells. The structure, function, and biochemistry of microtubules and the motile structures containing them, including flagella and cilia, are described in this chapter. The motile systems of bacterial cells are discussed in Supplement 11-1. The structure and biochemistry of microfilaments and their functions in providing movement are covered in Chapter 12. Chapter 12 also provides examples of the relatively few eukaryotic motile systems based on elements other than microtubules or microfilaments. The structural roles of microtubules and microfilaments in the cytoskeleton are described in Chapter 13. Chapters 24 and 25 outline the activities of microtubules and microfilaments in cell division.

MICROTUBULE STRUCTURE AND BIOCHEMISTRY

The structure of microtubules is similar in all eukaryotic cells, including those of animals, plants, fungi, and protists. The similarities extend to the molecular level: microtubules of all eukaryotes are built up from related tubulins, which assemble and disassemble by essentially the same mechanisms.

Microtubule Structure

Highly magnified cross sections show that the wall of a microtubule consists of a circle of spherical subunits, each about 4 to 5 nm in diameter (Fig. 11-3). The circle in most cells contains 13 subunits. However, circles of 11, 12, 14, 15, or 16 subunits have been observed in some systems. Longitudinal views show that the subunits line up in parallel, lengthwise rows, called *protofilaments*, to form the walls (see Figs. 11-4 and 11-6*b*).

Each spherical wall particle is a unit of tubulin, the microtubule protein, which occurs in microtubules in

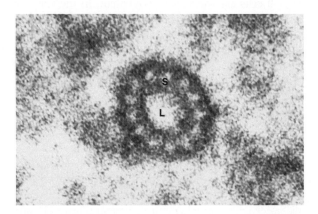

Figure 11-3 A single microtubule from a nerve axon of the goldfish in a highly magnified cross section. The microtubule wall is a circle of 13 subunits (S), with each subunit a molecule of either α- or β-tubulin. This microtubule was fixed in glutaraldehyde containing tannic acid. Tannic acid deposits around and inside the microtubules and outlines the walls and wall subunits in a sharp, negative image. L, microtubule lumen, or central channel. × 830,000. Courtesy of P. R. Brown.

Figure 11-4 Negatively stained microtubules isolated from sperm cells of the plant *Pteridium*. The arrangement of the wall subunits into longitudinal rows, called protofilaments, is clearly visible in the region enclosed by brackets. × 170,000. Courtesy of R. Barton and Academic Press, Inc., from *J. Ultrastr. Res.* 20:6 (1967).

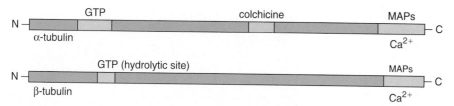

Figure 11-5 The α- and β-tubulin polypeptides, diagramed with their amino acid chains in unfolded and extended form. Segments of the amino acid chains forming binding sites are shown in yellow. Both the α- and β-tubulin polypeptides have binding sites for GTP; as a result, the tubulin heterodimer, which consists of one α- and one β-polypeptide, binds two molecules of GTP. The site on α-tubulin binds GTP stably; the site on the β-polypeptide is responsible for the GTP hydrolysis associated with microtubule assembly. Colchicine, a microtubule "poison" that interferes with microtubule assembly, binds to a site on the α-polypeptide. Both polypeptides bind Ca^{2+} and microtubule-associated proteins (MAPs) at their C-terminal ends; these proteins modify microtubule assembly, in most cases stabilizing microtubules in assembled form. Binding sites for other elements, such as the active crossbridges responsible for microtubule-based motion and bracing elements inside flagella, are also formed by regions of the amino acid chains. The locations of these sites remain uncertain.

two related forms known as α- and *β-tubulin* (Fig. 11-5). These proteins link into a dumbbell-shaped *heterodimer*, about 4 nm wide and 8 nm long, consisting of one α- and one β-tubulin unit (Fig. 11-6). Each visible wall subunit is half of a dumbbell. The heterodimer dumbbells line up end to end in longitudinal rows to form the protofilaments of microtubule walls.

In some cellular structures microtubules occur in a double or triple form consisting of one complete microtubule with one or two partial microtubules fused to it in a row at one side. In cross section the additional microtubules appear as C-shaped structures that butt at their tips into the adjacent microtubule (Fig. 11-7). Microtubules fused in double rows occur in flagella and cilia (see Fig. 11-15) and in triple rows in *centrioles*, or *basal bodies*, structures that give rise to flagella and anchor them in the cytoplasm (see Fig. 11-28).

The Tubulins

The α- and β-tubulins both have molecular weights near 50,000. A major α-tubulin of pig brain, for example, has a molecular weight calculated at 50,086 for its series of 450 amino acids; a primary β-tubulin from the same source contains 455 amino acids with a calculated molecular weight of 49,861. In most species the two tubulin types share from 40% to 50% of their sequences in common, indicating that the genes encoding them must have evolved from a single ancestral gene. Because all known eukaryotes have the two tubulin types, the split of this ancient gene occurred very early in the evolutionary history of eukaryotes. A third tubulin type, *γ-tubulin,* was recently discovered by C. E. Oakley and B. R. Oakley to be encoded in the *mipA* gene of the fungus *Aspergillus nidulans.* The new tubulin was

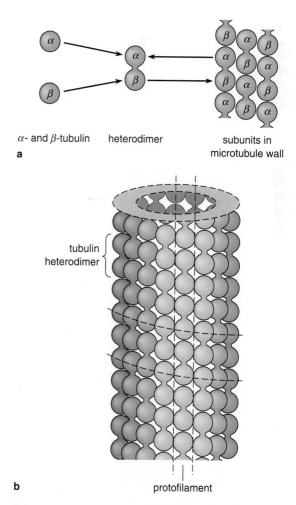

Figure 11-6 Tubulin, heterodimers, and protofilaments. **(a)** The relationship among α- and β-tubulin, the tubulin heterodimer, and the visible wall subunits of microtubules. **(b)** The arrangement of heterodimers in microtubule walls.

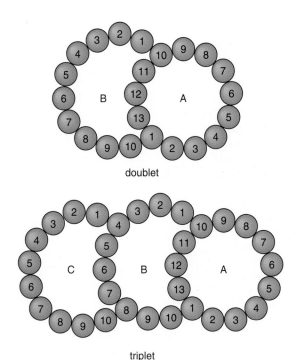

doublet

triplet

Figure 11-7 The arrangement of protofilaments in microtubule doublets and triplets. Doublets occur in flagella; triplets occur in centrioles (compare with Figs. 11-15 and 11-28). Doublet and triplet microtubules are labeled as A, B, and C subtubules as shown, beginning with the complete microtubule of the set as the A subtubule.

Table 11-1 Tubulin Isotypes

Organism	Number of α-tubulin Isotypes	Number of β-tubulin Isotypes
Fungi		
Aspergillus	2	2
Physarum	4	3
Saccharomyces	1	2
Chlamydomonas (alga)	2	2
Drosophila	4	4
Chicken	5–6	7
Mouse	6	5
Human	5–6 (?)	6

subsequently detected in humans, *Drosphila*, the amphibian *Xenopus*, maize, and the yeast *Aspergillus*, and is probably generally distributed among eukaryotes. Rather than assembling into microtubules, γ-tubulin appears to form part of an organizing region, the *spindle pole body* in *Aspergillus* and probably other lower eukaryotes, and the *cell center* in higher eukaryotes, which give rise to new microtubules containing α-tubulin and β-tubulin at various stages in the cell cycle (see below and p. 1007).

Most species have several variants or *isotypes* of both the α- and β-tubulins (Table 11-1). The sequence differences determining the isotypes are typically clustered in limited segments of the proteins, while other regions are constant in sequence, or nearly so, between different isotypes. For example, in both α- and β-tubulins the sequence near the C-terminal end is highly variable; in β-tubulins the sequence from amino acids 401 to 425 is identical in all eukaryotes.

Isolation of tubulin genes in species as diverse as slime molds, algae, *Drosophila*, sea urchins, and vertebrates such as the chicken, mouse, and human has shown that the tubulin isotypes are products of different but related genes. The green alga *Chlamydomonas*, for example, has two α- and two β-tubulin genes; *Drosophila* has four genes coding for each tubulin; vertebrates have about six isotypes of α- and β-tubulin. Genes encoding

the isotypes are activated in controlled patterns as embryonic development proceeds, as if the particular tubulin variants have significance for microtubule function at different developmental stages and in different cell structures and types. If the particular tubulin isotypes used in microtubule assembly have a functional significance, it remains unknown, however. In most cell types no regular differences in isotype composition are noted between microtubules with different functions, such as those of the spindle or flagella.

Both tubulin polypeptides occur in chemically modified forms. The α-tubulins are modified by the addition of acetyl groups or by a unique mechanism in which the final amino acid at the C-terminal end of the protein, always initially a tyrosine, is removed or added back. The β-tubulins are regularly modified by the addition of phosphate groups. As with tubulin isotypes, the significance of chemical modifications is uncertain. Fully assembled microtubules contain tubulin polypeptides in both modified and unmodified form. However, correlations are noted between the degree of modification and microtubule stability in some cell types. In these cells unassembled α-tubulin or α-tubulin in unstable microtubules retains the terminal tyrosine; α-tubulin subunits in stable microtubules assembled into long-term structures are more frequently detyrosinated. Stable microtubules also frequently contain the largest proportions of acetylated α-tubulins. In other cell types, however, stable microtubules remain relatively unmodified.

Because microtubules can exist in stable form without chemical modification, these alterations, although possibly contributing to stability, do not appear to be absolutely required. It is even possible that modification may simply reflect the fact that, in cells possessing the enzymes, stable microtubules are the only subset present in cells long enough to be chemically altered by detyrosination or acetylation.

Polymerization of Tubulins into Microtubules

In living cells, microtubules assemble from a pool of heterodimers in the cytoplasm. The equilibrium between assembled microtubules and the pool of subunits is finely balanced, so that microtubule assembly and disassembly take place continually at all stages of the cell cycle.

The Biochemistry of Microtubule Assembly Investigations into microtubule assembly began with the study of spindles in cell division, even before microtubules were identified with spindles. In the early 1950s S. Inoué noted that certain treatments, such as exposure to temperatures near 0°C, to elevated pressure, or to the microtubule poison *colchicine* (see below), cause the spindle in dividing cells to disassemble and disappear. The spindle reassembles quickly after removal of the disturbing treatment, sometimes within seconds or minutes. The reassembly can take place in cells in which protein synthesis is inhibited and, in any event, takes place much too rapidly to be attributed to synthesis of new polypeptides. From these observations Inoué proposed that the spindle fibers, only later discovered to be microtubules, exist in equilibrium with a pool of unassembled subunits in the cytoplasm. The treatments disassembling the spindle shift the equilibrium in the direction of the unassembled pool; discontinuation of the treatment shifts the equilibrium in the direction of assembled microtubules.

The subunits in the cytoplasmic pool were later established to be tubulin heterodimers, allowing the equilibrium to be written as:

$$\text{tubulin heterodimers} \rightleftharpoons \text{microtubules}$$

A number of chemical and physical factors affect the heterodimer-microtubule equilibrium. The nucleoside triphosphate GTP must be available and Ca^{2+} must be kept at very low concentrations for the reaction to progress significantly in the direction of microtubule assembly. The assembly reaction proceeds most rapidly at a pH of about 6.8. Elevated temperatures in physiological ranges promote microtubule assembly; reduced temperatures push the equilibrium in the opposite direction. Pressure also has an effect on equilibrium: as pressure rises, microtubules disassemble at progressively faster rates.

The experiments of R. C. Weisenberg and others demonstrated that tubulin heterodimers can assemble into microtubules in the test tube as well as in living cells. Microtubules, in fact, were among the first cellular structures to be assembled in a cell-free system. (Fig. 11-2 shows microtubules assembled in the test tube.) When tubulin heterodimers are placed in the test tube under conditions that promote microtubule as-

sembly, polymerization proceeds slowly at first until small aggregates of the subunits form in the solution. These aggregates serve as *nucleation sites* from which microtubule assembly continues. Once the aggregates form, assembly occurs more or less rapidly, depending on conditions such as the concentration of heterodimers, temperature, pH, and so on. The nucleation process can be speeded greatly by "seeding" the reaction mixture with short lengths of previously assembled microtubules, which supply nucleation sites.

The Role of GTP in Microtubule Assembly Tubulin heterodimers bind GTP before they assemble into microtubules. One GTP is bound by the α-tubulin and one by the β-tubulin subunit of each heterodimer. At some time during or after microtubule assembly, the GTP bound to the β-polypeptide is hydrolyzed to GDP plus phosphate. This observation, made early in the investigation of microtubule assembly, led to an initial hypothesis that microtubule formation requires energy and that each heterodimer added to a growing microtubule binds at the expense of one molecule of GTP hydrolyzed to GDP.

Researchers later discovered that microtubules can also assemble, although more slowly, when heterodimers are combined with nonhydrolyzable GTP analogs—molecules that resemble GTP in structure but lack phosphate groups that can be split off by hydrolysis. These findings indicated that GTP hydrolysis is not absolutely and directly required for microtubule assembly. Instead, GTP binding probably induces conformational changes in the heterodimer that push the equilibrium significantly in the direction of polymerization. Moreover, GTP does not break down until some time after heterodimers assemble into microtubules, a finding consistent with the idea that GTP hydrolysis is not directly required for the polymerization reaction.

Microtubule Polarity, Dynamic Instability, and Treadmilling

Detailed investigations of the characteristics of microtubule assembly by L. Bergen, G. C. Borisy, M. W. Kirschner, and others revealed that microtubules are polar structures with respect to their rates of assembly and disassembly. At most tubulin concentrations one end of a microtubule assembles or disassembles about two to four times faster than the other. The end with the faster rate of assembly or disassembly is the *plus end* of a microtubule; the end with the slower rate is the *minus end.* This polarity has been found to be significant for cellular motility as well as microtubule assembly because the direction in which microtubules generate motile forces depends directly on their polarity.

that disrupt microtubules in other locations. In contrast, spindle microtubules are easily disassembled by these and other treatments. The importance of MAPs in producing these differences in stability is underscored by the fact that tubulins from flagella and spindles generally contain similar mixtures of tubulin variants, and, when assembled into microtubules in the test tube, are equally sensitive to disruption by colchicine, temperature, and pressure if no MAPs from either source are present. Because MAPs are proteins that are synthesized and added to microtubules in stable structures, they afford relatively long-term regulation of microtubule assembly, in contrast to the short-term regulation provided by alterations in Ca^{2+} concentration.

Brain-cell microtubules provide an interesting example of the relationship of MAPs to microtubule stability. The majority of microtubules isolated from mammalian brain tissue are stable only at temperatures above $15°C$ and rapidly disassemble below this temperature. However, a small fraction of mammalian brain microtubules resists disassembly even at temperatures approaching $0°C$. The cold stability of this fraction has been shown to depend on the presence of a MAP not present in the cold-sensitive brain microtubules. MAPs with similar properties also account for the general cold stability of brain, spindle, and cytoplasmic microtubules in animals such as arctic fishes, which normally live with body temperatures near $0°C$. The MAPs with this function are sometimes called STOPs (STOPS = Stable Tubule Only Proteins).

In many cellular systems the ability of MAPs to stabilize microtubules is modified by the addition of phosphate groups. The phosphorylations, carried out by protein kinases specific for certain MAPs, generally reduce the ability of MAPs to stabilize microtubules and push the equilibrium in the direction of disassembly. In effect, the phosphorylations add a system that fine-tunes the regulation of microtubule assembly and stability by MAPs. The ability of the MAP mentioned above to stabilize mammalian brain microtubules at cold temperatures, for example, is reduced through phosphorylation by a specific protein kinase.

MAPs and Ca^{2+} interact extensively in the regulation of microtubule assembly. MAPs may increase or decrease the sensitivity of microtubules to Ca^{2+} concentration or may make microtubules sensitive to Ca^{2+} only when the ion is combined with calmodulin, forming an active Ca^{2+}/calmodulin complex (see Information Box 7-2). In most or all of the systems in which Ca^{2+} exerts its effects through calmodulin, the Ca^{2+}/calmodulin complex regulates microtubule assembly by modifying the activity of a MAP.

The brain-cell microtubules sensitive to reduced temperatures, for example, disassemble rapidly if exposed to Ca^{2+}. The effect of Ca^{2+} is evidently exerted directly on the tubulins or heterodimers of the cold-sensitive microtubules. The cold-stable fraction of brain microtubules, in contrast, is insensitive to elevated Ca^{2+} concentrations as long as the MAP that provides cold stability is present. Calmodulin by itself has no effect on the cold-stable brain microtubules. However, addition of both calmodulin and Ca^{2+}, permitting formation of the active Ca^{2+}/calmodulin complex, causes immediate disassembly of the cold-stable microtubule fraction. Combination with the Ca^{2+}/calmodulin complex evidently causes a conformational change in the MAP that reduces its ability to stabilize microtubules.

Some of the protein kinases adding phosphate groups to MAPs are cAMP-dependent. The phosphate groups added by these kinases may be removed by phosphatases dependent for their activity on the presence of Ca^{2+} or calmodulin. A MAP-2 of brain-tissue microtubules, for example, is modified by phosphorylation via cAMP-dependent protein kinases at as many as 13 different sites. The activity of cAMP-dependent protein kinases and Ca^{2+}/calmodulin-dependent phosphatases links microtubule regulation by the MAPs to the major cAMP and $InsP_3$/DAG regulatory pathways controlled by cell surface receptors (see Ch. 7).

Not all MAPs are regulatory in function. Some act as linkers between microtubules or between microtubules and other cell structures. Some of these MAPs, such as those linking microtubules, microfilaments, and intermediate filaments, are important contributors to the cytoskeletal network. The linking ability of many MAPs, including the tau group and MAP-2, depends on their structure as two-domain proteins. One domain recognizes and binds sites on the heterodimers of a microtubule; the second domain recognizes and links to other microtubules or cell structures. Another factor recently detected by R. D. Vale, which may fall into the MAP group, apparently contributes to disassembly of microtubule-based structures by severing microtubules. Although proteins severing microfilaments are commonly observed, Vale's report is the first known instance of a microtubule severer. The factor may be important in the controlled disassembly of such microtubule-based structures as the spindle and flagella.

Cellular Centers of Microtubule Polymerization: MTOCs

Microtubules may occur at essentially any location in the nucleus or cytoplasm. In many locations where they assemble, particularly when the numbers generated are small, no visible structures are associated with microtubule assembly. The microtubules polymerize from apparently structureless background material in the cytoplasm or nucleus. At locations where microtubules are generated in quantity, sites of microtubule generation are marked by deposits of more or less structured masses of dense material. These specialized sites of

Figure 11-14 The microtubule poisons colchicine, vinblastine, vincristine, and taxol. In vinblastine X = CH₃; in vincristine X = CHO.

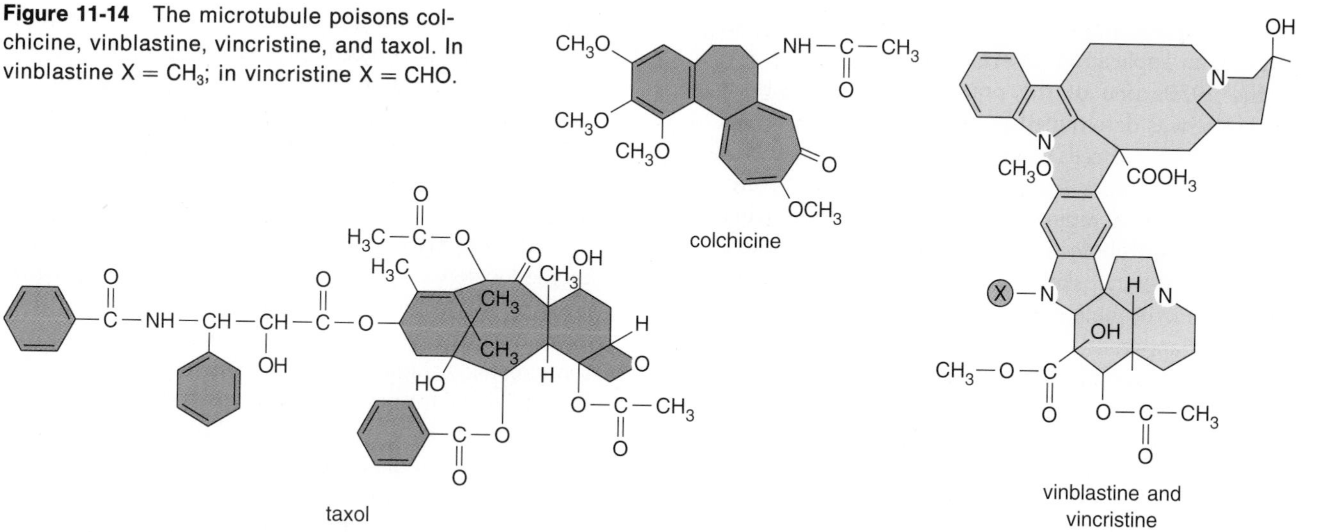

colchicine

taxol

vinblastine and vincristine

microtubule organization, termed *microtubule-organizing centers (MTOCs)*, are found in four different major types:

1. *Centrioles* (also termed *basal bodies*; see Fig. 11-28 and 11-29), which generate the microtubules of flagella.

2. *Cell centers* (also termed *centrosomes* or *pericentriolar material*), collections of dense material located near the nucleus, which organize cytoplasmic microtubules in animal cells (see Fig. 13-16). The same centers also organize spindle microtubules during animal cell division, where they are known as *asters* (see Fig. 24-11).

3. *Kinetochores*, disc- or platelike structures on the surfaces of chromosomes, which attach and may also give rise to spindle microtubules (see Fig. 24-6).

4. *Spindle pole bodies*, dense, plaquelike layers of material, which give rise to the spindle in many fungi (see Fig. 24-15).

Each of the MTOC types has been successfully isolated from cells. When supplied with tubulin under polymerizing conditions, all are capable of generating microtubules in the test tube. In all these structures, microtubules grow from the MTOCs with their minus ends anchored in the dense matter of the MTOC, and their plus ends directed outward.[1]

Because the plus end, at which further subunits are added, grows away from the MTOC, MTOCs are likely to act as nucleation centers, providing the "seeds" from which microtubule assembly proceeds. Once assembly is under way, subunits can add to the newly formed plus end, and the microtubule grows rapidly outward from the MTOC.

[1] Inside living cells spindle microtubules connect to kinetochores at their plus ends, so that their minus ends extend outward, away from the kinetochore. This arrangement is believed to result from the activity of kinetochores in attaching to previously assembled microtubules rather than acting as MTOCs (see p. 1004 for details).

MTOCs may provide nucleation centers by adjusting local conditions, such as Ca^{2+} concentration, to greatly favor heterodimer polymerization. Alternatively, they may work by providing organizing sites for MAPs that promote the initiation of microtubule assembly. Some support for the latter mechanism is provided by the fact that antibodies developed against some MAPs react positively with MTOCs. The recently discovered γ-tubulin, which has been found to be localized in microtubule-organizing regions of the cell center, may also promote microtubule assembly, possibly by acting as a template.

Tests of the molecular components of MTOCs reveal that proteins and, surprisingly, RNA are present. Destruction of either component by enzymes such as trypsin or ribonuclease greatly inhibits or destroys the ability of MTOCs to generate microtubules, indicating that both proteins and RNA are active in MTOC function. Although the proteins might function as MAPs, the possible role of RNA in MTOCs remains unknown.

Microtubule Poisons

The importance of microtubules to cellular activities makes them a sensitive target for biological defenses, among them several produced by plants (Fig. 11-14). Some of these microtubule "poisons," which interfere with the motile or cytoskeletal roles of microtubules when ingested by animals, have proved to be highly useful in medicine and research. One group of poisons, which includes *colchicine*, inhibits microtubule assembly and promotes disassembly of existing microtubules. Another poison, *taxol*, has opposite effects and pushes the equilibrium far in the direction of assembled microtubules.

Colchicine, an alkaloid extracted from plants in the genus *Colchicum* (including autumn crocus and meadow saffron), has been known since ancient times as an animal poison. (Alkaloids are cyclic organic compounds that react as bases.) During the 1700s mild doses of colchicine were found to be effective in the treatment of gout, an application for which it is still used today. In the 1930s one of the primary cellular effects of colchicine was determined to be the complete arrest of mitosis in animal cells. Further research revealed that colchicine combines specifically with tubulin heterodimers and causes rapid disassembly of the more labile types of microtubules in animal cells, including those of the spindle, cytoskeleton, and nerve cells. Existing flagellar microtubules, in contrast, are unaffected by colchicine, apparently because of the presence of MAPs that confer extra stability on heterodimer assembly. In general microtubules in plants and many types of fungi, algae, and protozoa are more resistant to the alkaloid than those in animals.

The molecular basis for the disruption of microtubules by colchicine is not completely understood. The poison combines with unpolymerized heterodimers in the ratio of one colchicine molecule per heterodimer. The colchicine-heterodimer complexes are capable of binding to the plus ends of existing microtubules; when bound, the complexes block the addition of further subunits to the plus ends. Presumably, the capped microtubules then disassemble from their minus ends.

Colchicine combines with tubulin specifically enough to make it an invaluable tubulin or microtubule "marker." The first effective methods for isolating and purifying tubulins depended on the use of colchicine made radioactive by the attachment of tritium (^{3}H) atoms. Once marked by the radioactive colchicine, tubulin could be extracted and purified from cells simply by collecting the radioactive fraction separated by centrifugation.

The destruction of many cellular microtubule types by colchicine makes the poison a highly useful probe for microtubule-based motility. Since it interacts with few or no other cellular molecules, inhibition of a particular cellular motion by colchicine provides a good first indication that the motion is based on microtubule structure and function. The method is not foolproof since the microtubules of some species are generally insensitive to the poison, and some types of microtubule-based motion—flagellar motion, for example—are unaffected by colchicine.

The general insensitivity of microtubules to colchicine among plants, algae, and fungi appears to result from a greatly reduced binding affinity between the poison and tubulin heterodimers in these groups. In plants, for example, colchicine binds so weakly to tubulin that inhibition of microtubule assembly and function requires colchicine concentrations as much as 1000 times higher than in animal cells. The reduced binding affinity for colchicine explains why plants such as *Colchicum* can survive with relatively high concentrations of colchicine in their tissues.

The reduced affinity of plant heterodimers for colchicine apparently depends on limited substitutions in

the amino acid sequence of the α-tubulins, which may alter the folding conformation of the site binding colchicine sufficiently to inhibit linkage by the poison. This conclusion is supported by colchicine-resistant animal mutants, in which resistance also depends on one or more substitutions in the amino acid sequence of α-tubulin.

Another plant alkaloid, *nocodazole*, acts as a microtubule poison by a mechanism similar to colchicine. Two others, *vinblastine* and *vincristine*, extracted from *Vinca rosea* (the periwinkle), link heterodimers into large, semicrystalline aggregates. The linkage removes the heterodimers from the cytoplasmic pool and pushes the heterodimer-microtubule equilibrium in the direction of microtubule disassembly. Several of these alkaloids have also proved useful as laboratory probes for microtubule structure and function or as medicines.

Taxol, an alkaloid extracted from *Taxus brevifolia* (the western yew) is of special interest because its effects on microtubules are opposite to those of colchicine and the other alkaloids. Taxol stabilizes microtubules by pushing the equilibrium far in the direction of the polymer, in a mechanism similar to MAPs. It is so effective in this role that most of the available tubulin heterodimers assemble into microtubules. The stabilizing effects of taxol have made it valuable as a means for

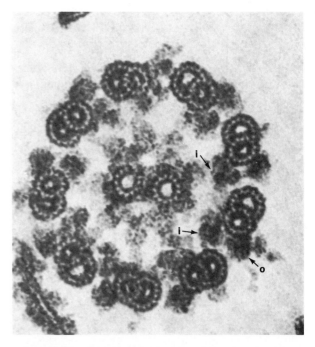

Figure 11-15 Dynein arms (arrows) in a sea urchin (*Lytechinus*) flagellum in cross section. Tannic acid has been added to the fixing solution to outline the microtubule subunits in a negative image. i, inner dynein arms; o, outer dynein arms. ×590,000. Courtesy of K. Fujiwara, from *J. Cell Biol.* 59:267 (1963), by permission of the Rockefeller University Press.

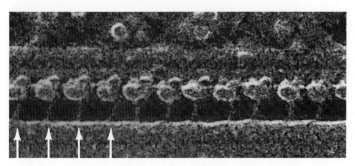
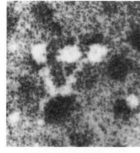
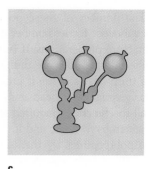

a   b   c

Figure 11-16   Dynein structure. **(a)** Dynein arms (arrows) in a flagellum of the protozoan *Tetrahymena* prepared by the freeze-fracture technique. ×235,000. Courtesy of U. W. Goodenough; reproduced from *J. Cell Biol.* 95:798 (1982), by copyright permission of the Rockefeller University Press. **(b)** A single, isolated dynein arm as seen in the scanning transmission electron microscope. ×425,000. Courtesy of K. A. Johnson; reproduced from *J. Cell Biol.* 96:669 (1983), by copyright permission of the Rockefeller University Press. **(c)** Tracing of the dynein arms in **(b)**. Although some variations are noted in different species, the polypeptides usually include two or three heavy components (with molecular weights in excess of 400,000), which correspond to the globular head units seen in electron microscope preparations. Also present are variable numbers of polypeptides of intermediate (55,000 to 125,000) and low (about 20,000) molecular weight. **(d)** Tentative positions of the various polypeptides in the dynein complex. Molecular weights are given in thousands.

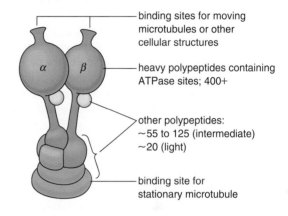

binding sites for moving microtubules or other cellular structures

heavy polypeptides containing ATPase sites; 400+

other polypeptides: ~55 to 125 (intermediate) ~20 (light)

binding site for stationary microtubule

d

isolating intact microtubules from a variety of cell types, including many plants, in which microtubules are so sparsely distributed and unstable that isolation and purification of microtubule proteins had previously been impossible.

In medicine colchicine, taxol, and other microtubule poisons have proved to be useful treatments reducing the growth of both benign and malignant tumors. They are effective in this role primarily because, by interfering with the spindle, they slow or stop the uncontrolled cell division characteristic of tumors.

THE MOTILE FUNCTIONS OF MICROTUBULES

One of the two fundamentally different mechanisms by which microtubules produce motion depends on dynein, kinesin, and other microtubule crossbridges capable of converting chemical to mechanical energy. The second mechanism is based on microtubule growth and disassembly: as microtubules lengthen or shorten due to controlled adjustments in the heterodimer-microtubule equilibrium, cellular structures attached to the microtubules are pushed or pulled through the cytoplasm.

Dynein and Microtubule Sliding

Of the various motors that act on microtubules by forming active crossbridges, dynein is the best known. The structure of this crossbridge, and its activity in converting ATP energy into the mechanical energy of movement, are probably representative of kinesin and other microtubule motors. The crossbridging mechanism of dynein also has direct parallels to the activity of myosin in microfilament-based motile systems.

Dynein Structure   Dynein has been studied most extensively in flagella, where it can be seen in sections as short hooked or bobbed *arms* extending from one set of double microtubules to the next (Fig. 11-15). Freeze-fracture preparations (see p. 120) reveal some details of dynein substructure. In these preparations a dynein arm appears as two or three globular heads connected to a basal unit by short stalks, much like a bouquet of flowers (Fig. 11-16).

Analysis of dynein isolated from flagella shows that an arm contains as many as 12 polypeptides with combined molecular weight ranging from 1.2 to 1.9 million (see Fig. 11-16*d*). ATPase activity, which depends on the presence of Mg^{2+}, can usually be detected in the *heavy chain* polypeptides of the dynein arms, which have molecular weights in excess of 400,000.

The number of these polypeptides, two to three, is equivalent to the number of head units seen in freeze-fracture preparations. If two heavy chains are present, the polypeptides are designated α- and β-*polypeptides*; a third heavy chain, if present, is designated as the γ-*polypeptide.* The head units evidently contain most or all the functions required for active crossbridging. Research by W. Sale and L. A. Fox showed that isolated β-polypeptides by themselves are able to induce microtubule movements. Other polypeptides are two or more *intermediate chains,* with molecular weights between 55,000 and 125,000, and four to eight *light chains,* with molecular weights of about 20,000.

The two dynein arms attached to flagellar microtubules at any level are known as the *inner* and *outer arms* (see Fig. 11-15). Separate analysis of the outer and inner arms has been made possible by the use of mutants lacking one of the two arm types or by chemical techniques allowing them to be removed separately. These methods reveal that the two arms have different groups of polypeptides and that the inner arms occur in at least three different forms, each with distinct polypeptide types. A recently characterized nonflagellar dynein, called *cytoplasmic dynein,* also contains a distinct group of polypeptides. The dyneins therefore make up a family of energy-transducing molecules in which the members have related but different properties. One or more dyneins are probably distributed universally among eukaryotic cells.

Dynein Crossbridges and Microtubule Sliding

Much remains to be established about the molecular basis for the crossbridging mechanism. However, observations of the biochemistry of ATP breakdown by the arms, and the effects of this breakdown on crossbridge binding and structure, have allowed researchers to fill in some details of the crossbridging cycle.

Several of these observations come from experiments with flagella or cilia in sea urchin sperm, the protozoan *Tetrahymena,* and the green alga *Chlamydomonas,* all favorite subjects for dynein studies. In all these cells flagella or cilia become locked into rigid bends or waves if ATP supplies are rapidly depleted. Under the electron microscope the dynein arms in this *rigor state,* as it is called, extend at approximately right angles to the microtubules and are tightly attached at both ends to the surfaces of two adjacent microtubules. (Fig. 11-16a shows dynein arms locked in rigor.) Addition of fresh ATP to flagella in rigor restores their flexibility and often restarts flagellar beating. As part of the restoration, ATP is hydrolyzed. Examination of ATP-restored flagella shows that the arms have released their attachment at one end, and some arms now extend from the microtubule surface at an angle approximating 45° (Fig. 11-17).

These and other observations establish several important characteristics of the crossbridging mechanism:

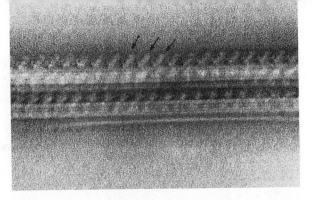

Figure 11-17 Dynein arms (arrows) extending at a 45° angle on microtubule doublets isolated from *Tetrahymena* and prepared for electron microscopy by negative staining. The image in this micrograph has been reinforced by two successive printings of the original negative. For the second printing, which was made directly over the first on the same piece of photographic paper, the paper was moved in the direction of the microtubules through a distance equivalent to the separation between dynein arms. × 165,000. Courtesy of F. D. Warner.

1. Dynein arms undergo a conformational change that swivels their angle of attachment to the microtubule surface between approximately 90° and 45°.

2. ATP binding, but not hydrolysis, is required to release dynein arms attached in the 90° position.

3. ATP is hydrolyzed at some point after release of the dynein arms; hydrolysis precedes or accompanies shift of the arms to the 45° position.

4. Release of the products of ATP hydrolysis (ADP plus phosphate) does not occur until the dynein arms reattach to the adjacent microtubule after (3) above.

The hypothetical crossbridging cycle shown in Figure 11-18, advanced by P. Satir and others, combines these observations and shows how dynein arms may operate to produce a sliding force between microtubules. The crossbridging cycle shown in the figure is greatly simplified; the actual mechanism, which is not completely understood, is undoubtedly more complex. The cycle may be considered to begin (Fig. 11-18a) at the close of a previous cycle, with a dynein arm tightly linked to an adjacent microtubule at the 90° position. Binding an ATP molecule from the medium causes a conformational change at the tip of the dynein arm, altering the binding site so that it no longer fits the adjacent microtubule. Under these conditions the arm releases (Fig. 11-18b) and then hydrolyzes its bound ATP. The products of hydrolysis remain bound to the arm (Fig. 11-18c). ATP hydrolysis causes a conformational change in the arm, shifting it to the 45° position and reactivating the microtubule-binding site at its tip (Fig. 11-18d). In this condition the arm is like a compressed molecular spring storing much of the energy released by ATP hydrolysis. Reattachment to the ad-

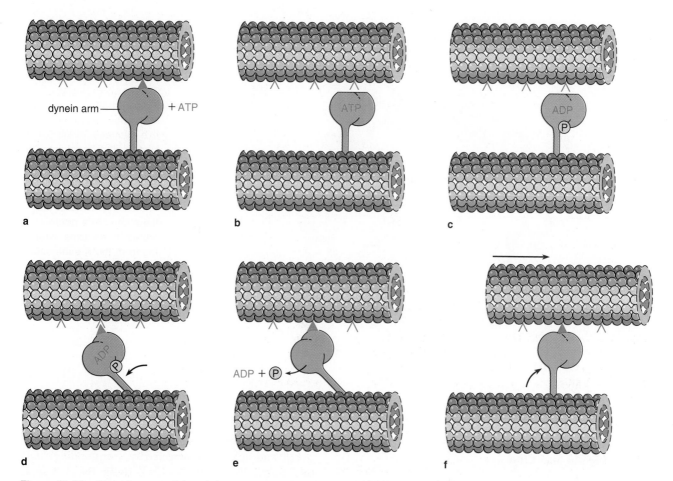

Figure 11-18 Steps in a provisional dynein crossbridging cycle, shown in simplified form. **(a)** The cycle begins with a dynein arm linked to an adjacent microtubule at the 90° position, the rigor position. **(b)** Binding an ATP molecule from the medium causes a conformational change in the binding site at the tip of the dynein arm, so that it no longer fits the adjacent microtubule. **(c)** The arm releases and hydrolyzes its bound ATP, but the products of hydrolysis remain bound to the arm. **(d)** Hydrolysis causes a conformational change, shifting the arm to the 45° position and reactivating the microtubule-binding site at its tip. In this state the arm stores much of the energy released by ATP hydrolysis. **(e)** Reattachment to the adjacent microtubule releases the products of ATP hydrolysis and triggers release of the arm from its 45° position. **(f)** The arm swivels forcefully through an arc of 45°, sliding the attached microtubule along by an equivalent distance. The arm is now ready to bind a second ATP and repeat the cycle. Dynein crossbridges also move cell structures such as mitochondria along microtubules instead of sliding one microtubule over another. In this type of motility the structure would be substituted for the bottom microtubule in each part of this figure. The crossbridge base would be fixed to the structure, and the tip would move the structure along the microtubule by attaching, swiveling, and releasing from the microtubule surface.

jacent microtubule releases the products of ATP hydrolysis (Fig. 11-18e) and triggers release of the arm from its 45° position. In response the arm swivels forcefully through a 45° arc, sliding the attached microtubule along by an equivalent distance (Fig. 11-18f). The arm is now ready to bind a second ATP molecule and repeat the cycle. Each attach-swivel-release cycle slides the adjacent microtubules past each other by an increment of about 16 nm, approximately the length of two heterodimers. The combined action of hundreds

or thousands of arms between the doublets produces a smooth and continuous sliding motion. All the different dyneins probably cycle by a similar mechanism.

In motile systems in which dynein or other crossbridges move cytoplasmic structures such as mitochondria along microtubules, the base of the crossbridge is attached to the structure. The opposite end of the crossbridge walks the structure along a microtubule by attaching, swiveling, and releasing from the microtubule surface.

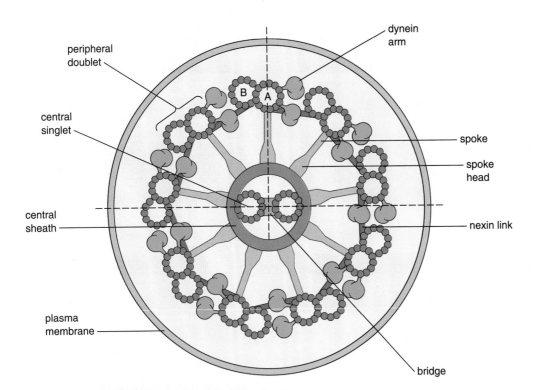

Figure 11-19
Microtubules and connective elements in the 9 + 2 system of a flagellum. The direction pointed by the dynein arms allows the point of view with respect to the tip or base of the flagellum to be determined. If the arms extending from the A subtubule point clockwise, the flagellum is being viewed from base to tip.

Labels on figure: dynein arm; peripheral doublet; central singlet; central sheath; plasma membrane; B; A; spoke; spoke head; nexin link; bridge

Motile Systems Based on Microtubule Sliding: Flagella and Cilia

Much of the available information about dynein structure and function comes from the study of flagella and cilia. These studies give the clearest support to the conclusion that dynein crossbridges produce microtubule sliding.

The 9 + 2 System of Microtubules in Flagella and Cilia

The arrangement of microtubules in flagella and cilia, called the *9 + 2 system* or *flagellar axoneme*, is remarkably uniform in different species. With few exceptions, the system (see Figs. 11-15 and 11-19) consists of a circle of nine fused, double microtubules, the *peripheral doublets*, arranged around a center containing two single microtubules, the *central singlets*. The circle of peripheral doublets has an outer diameter of about 200 nm, placing the entire 9 + 2 system just within the resolving power of the light microscope.

Each of the two central singlets is a complete microtubule. The peripheral doublets contain one complete microtubule with 13 protofilaments, to which is joined a second, incomplete C-shaped microtubule built up from 10 to 11 protofilaments (see Fig. 11-7). In the 9 + 2 system the complete microtubule of each peripheral doublet is the *A subtubule*; the incomplete C-shaped microtubule is the *B subtubule* (see Fig. 11-19).

The peripheral doublets and central singlets are held together in the 9 + 2 system by a complex system of connective elements. The central singlets, which lie about 9 nm apart, are connected by a surrounding *sheath* and a *bridge* that runs directly between them. A *spoke* attaches the A subtubule of each doublet to the central sheath. The spokes broaden into a *head* where they meet the sheath.

Another connective element, the *nexin link* (from the Latin *nexere* = to bind), runs between the peripheral doublets, usually attaching the A subtubule of one doublet to the B subtubule of the next. The A subtubules also link to a pair of dynein arms. The dynein arms often appear hooked, with the hooks bent inward toward the center of the 9 + 2 system. In some flagella linking elements also extend outward from each doublet to contact the overlying plasma membrane.

Longitudinal sections of flagella reveal that the arms, spokes, and other connective elements repeat at regular intervals (Fig. 11-20). The distances between elements are all multiples of the basic 8-nm length of a tubulin heterodimer, suggesting that a heterodimer dumbbell has a variety of binding sites that may link to dynein arms, nexin links, spokes, or central sheath elements and bridges (see caption to Fig. 11-20 for details). Depending on the position of a heterodimer in the walls of flagellar microtubules, one or more or none of these elements may attach.

In most cell types both the peripheral doublets and central singlets are capped at the outermost tip of the flagellum (Fig. 11-21). The capping structures, which vary widely from species to species, probably anchor the tip of the 9 + 2 system to the plasma membrane

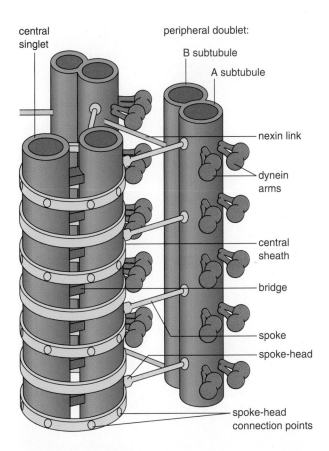

central
singlet

peripheral doublet:

B subtubule

A subtubule

nexin link

dynein
arms

central
sheath

bridge

spoke

spoke-head

spoke-head
connection points

Figure 11-20 The periodic arrangement of arms and linking elements in the 9 + 2 system. Depending on the species, the spokes are spaced evenly in groups of two or three along the A subtubules of the doublets. Entire spoke groups, containing two or three spokes per repeating unit, repeat at a 96-nm interval. The sheath elements contacted by the spoke heads are spaced more closely at intervals of 16 nm. Dynein arms occur in pairs at intervals of 24 nm along the A subtubule of the peripheral doublets; nexin links are more widely spaced at intervals of 96 nm.

a

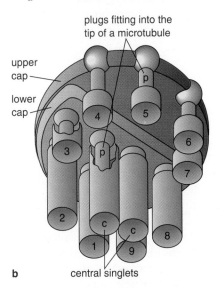

plugs fitting into the
tip of a microtubule

upper
cap

lower
cap

b   central singlets

Figure 11-21 Capping structures in the 9 + 2 system. **(a)** Cap structure at the tip of a frog palate cilium. U, upper cap; L, lower cap; and P, plug. ×150,000. **(b)** Reconstruction by E. L. LeCluyse and W. L. Dentler of the cap in frog palate cilia. 1–9, A subtubules of the peripheral doublets (the B subtubules of the doublets do not extend to the flagellar tip). Courtesy of W. L. Dentler, from *J. Ultrastr. Res.* 86:75 (1984).

surrounding the flagellum. The caps may also block flagellar microtubules from further growth at their tips. During flagellar movement the capping structures slip from side to side to accommodate microtubule sliding.

As might be expected from these structural complexities, a wide range of polypeptides, more than 200 different types, can be detected in flagella. An analysis of *Chlamydomonas* mutants lacking spokes indicates that these structures alone are built from as many as 17 different polypeptides.

The 9 + 2 system appears to be the same in cilia and flagella, which differ primarily in their number per cell and their beat pattern. When only one or a few are present, they are called *flagella;* when present in large numbers, they are called *cilia* (Fig. 11-22). Some cells may possess hundreds to thousands of cilia. Cilia are relatively short, ranging from 5 to 25 μm in length. Flagella are usually two to three times longer. In insects

sperm flagella may be several to many millimeters in length. In some insects, such as *Drosophila*, the sperm flagellum is longer than the entire adult male animal! Flagella produce forces that propel swimming cells through a liquid medium. Cilia have the same function in single-celled organisms such as ciliate protozoa and in many early animal embryos. Cilia also move liquid media over the surfaces of cells that remain stationary within the body of an animal.

Sperm flagella beat in a series of waves that travel continuously in a roughly flat plane from the base to the tip (Fig. 11-23a and b). In a few protozoan species, flagellar waves follow a helical path rather than lying

Figure 11-22 Cilia covering epithelial cells in the frog palate, as seen in the scanning electron microscope. ×3,200. Courtesy of W. L. Dentler, from *J. Ultrastr. Res.* 86:75 (1984).

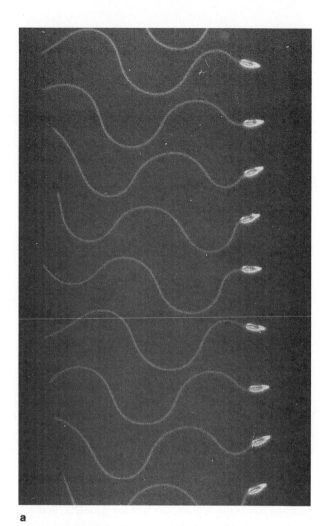

a

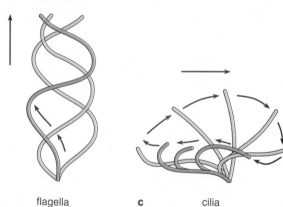

b   flagella   **c**   cilia

Figure 11-23 Flagellar and ciliary beating patterns. **(a)** The flagellar beating pattern of a sperm cell, photographed on a moving film under a strobe light flashing at regular intervals. The pattern consists of a series of smooth waves vibrating in a flat plane. The waves start at the base of the flagellum and move toward the tip. Courtesy of C. J. Brokaw. **(b)** The flagellar beat in diagrammatic form. **(c)** The ciliary beating pattern—the oarlike power stroke is followed by a recovery stroke (dark green) in which a single bend travels from the base to the tip. Flagella of some organisms, such as the alga *Chlamydomonas*, can switch between the two beating patterns.

in a flat plane. Cilia, in contrast, typically bend stiffly from the base in an oarlike "power" stroke, with the body of the cilium extended and almost straight (Fig. 11-23*c*). The bend then travels from the base to the tip to accomplish the recovery stroke. The beating in either case may range between 10 and 50 or more cycles per second.

Both beat patterns can be observed in the flagella of some species. In the single-celled alga *Chlamydomonas*, for example, the two flagella beat in the ciliary pattern to propel the cell forward and switch to the flagellar pattern to move it in reverse.

Almost all eukaryotic organisms possess at least some cell types that develop flagella or cilia at various stages of the life cycle. Among the major plant and animal groups, only flowering plants have no flagellated cells of any type. Many protists have flagella or cilia; the sperm cells of most animals and the male gametes of some fungi, and of plants from algae to primitive gymnosperms, swim by means of flagella. Cells with flagella or cilia also occur regularly in the body tissues of most animals. In humans ciliated cells line the respiratory tract, the ventricles of the brain, ducts in the male and female reproductive systems, and epithelia in other parts of the body. In these locations cilia move liquid media over the surfaces of the epithelial cells. Cilia also occur in highly modified form in the rods and cones of the retina and hair cells of the organ of Corti in the inner ear; these structures have a sensory rather than a motile function.

Microtubule Sliding and the Flagellar Beat As long ago as 1959, B. A. Afzelius suggested that the beating movements of flagella are produced by active microtubule sliding, powered by the arms extending from

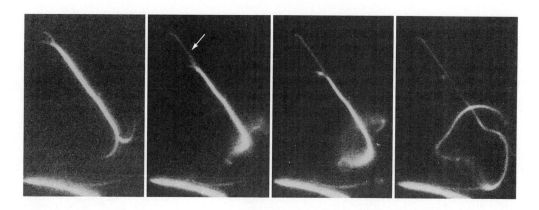

Figure 11-24 Successive light micrographs showing microtubule doublets (arrow) sliding actively from an isolated 9 + 2 system. ATP was added in the first frame. × 1,350. Courtesy of I. R. Gibbons, from *Proc. Nat. Acad. Sci.* 68:3092 (1971).

the doublets. A series of major experiments has since provided clear and elegant support for an active sliding mechanism in flagella.

The ATPase activity of dynein was first discovered in flagella. During the 1970s I. R. Gibbons noticed that when flagella were stripped of their plasma membranes by exposure to detergents, the arms could be removed or added back by adjustments in salt concentration. When the arms were removed, the 9 + 2 systems lost their ATPase activity. The lost ATPase activity could be detected among the proteins released from the demembranated flagella. Adding the arms back to the 9 + 2 systems restored their ATPase activity. Gibbons called the flagellar ATPase *dynein,* meaning "force protein."

Several experiments later demonstrated that microtubules slide actively and forcefully in the production of flagellar movements. The most graphic was carried out by K. E. Summers and Gibbons, who isolated flagella and removed their membranes by treating them with a detergent. The isolated, demembranated flagella were then treated briefly with a protein-digesting enzyme, which left the dynein arms intact but partly digested the spokes and nexin links. Addition of ATP to these digested preparations caused the peripheral microtubule doublets, no longer held by their links, to slide actively and forcefully out of the 9 + 2 system (Fig. 11-24). The sliding involves movement of one doublet along another, not the movement of A subtubules over B subtubules.

From their work Summer and Gibbons proposed, in agreement with other investigators, that dynein arms produce the force sliding doublets in the 9 + 2 system. Spokes and nexin links, according to their model, normally act as elastic connectors that hold the doublets together and prevent them from sliding out of the 9 + 2 system during movement. The restriction imposed by the spokes and links forces the entire flagellum to bend to accommodate the internal displacement of the doublets due to sliding. Thus, according to the Summer and Gibbons model, dynein arms produce the force for sliding, and spokes and nexin links convert the doublet sliding into flagellar bends.

More recent experiments generally support the conclusions of these experiments. Among the most definitive are experiments by R. B. Vallee, B. M. Paschal, and their colleagues, who isolated dynein arms from sea urchin flagella and attached them to the surface of a glass slide. Microtubules were then added to the slide in a solution containing ATP and other factors required for active dynein crossbridging. The dynein arms cycled actively between the glass and the microtubules, causing the microtubules to slide smoothly over the surfaces of the slides.

Other confirming evidence comes from the study of mutants, which gives insights into the possible functions of other flagellar structures. For example, in *Chlamydomonas* mutants lacking spokes and central singlets, flagella are paralyzed and incapable of movement even though dynein arms are present. Additional "bypass" mutations are able to cancel the effects of the missing spokes and central singlets, producing flagella that can move in altered waveforms and only in the flagellar pattern. These results indicate that spokes and central singlets may contribute to the shape of flagellar bends but do not generate the force for bending. The results also suggest that spokes or central singlets may contribute "switches" turning the flagellar beat on or off; when missing, the beat is locked in the off position unless a second mutation releases the doublets and dynein arms from the control.

Chlamydomonas mutants have also revealed differences in structure and function of outer and inner dynein arms. In mutants lacking outer arms, flagellar beats take the normal form but are reduced in frequency. Flagella lacking inner arms beat at normal frequencies but have an altered waveform. Analysis of mutants lacking inner or outer arms has revealed that inner arms are more complex and contain more polypeptide subunits than outer arms. Differences have even been noted among inner arms. These findings confirm that flagellar dyneins constitute a family of related motors rather than a single type.

Factors Regulating Flagellar or Ciliary Beating
Many flagellated cells, particularly among protozoa and

algae, are able to start and stop flagellar beating or switch between different beating patterns. *Paramecium*, for example, can quickly stop and reverse its ciliary beating upon encountering an obstacle; *Chlamydomonas* can switch almost instantly between a ciliary and flagellar beating pattern. Similarly, sperm cells of most animals are quiescent until their release, when a "switch" activates movement of the flagellum.

Some parts of the control mechanism have been pieced together. J. R. Tash and A. R. Means showed that the signal initiating movement in sperm and other flagellated cells depends on phosphorylation of flagellar polypeptides. The phosphate groups are added by a cAMP-dependent protein kinase, probably activated by a membrane receptor. Flagellar beating can be turned off again by a Ca^{2+}/calmodulin-activated phosphorylase that removes the phosphate groups. The involvement of phosphorylation and dephosphorylation in the regulation of movement may be extensive because as many as 80 of the some 200 polypeptides of *Chlamydomonas* flagella can be detected in phosphorylated form. These polypeptides include subunits of dynein arms, peripheral doublet microtubules, spokes, and central singlets. For example, in paralyzed *Chlamydomonas* mutants lacking spokes and central singlets, several inner-arm polypeptides are phosphorylated; in the bypass mutants in which the flagella beat without spokes and central singlets, inner-arm polypeptides are dephosphorylated.

Adjustments in Ca^{2+} concentration are also linked to the mechanism switching between flagellar and ciliary beat patterns. In both *Paramecium* and *Chlamydomonas*, exposure of cells to low Ca^{2+} concentrations induces the ciliary pattern typical of normal swimming; elevated Ca^{2+} concentrations trigger the flagellar pattern typical of avoidance. In general these are the results obtained with other cells: low Ca^{2+} concentrations regulate generation of the usual, forward-swimming beat pattern of the organism, and elevated Ca^{2+} induces the beating pattern characteristic of avoidance maneuvers. The molecular steps directly linking alterations in Ca^{2+} concentration to beat switching remain unknown.

A different switching mechanism is probably responsible for the back-and-forth beating movements of the flagellar shaft (Fig. 11-25). To bend the flagellum in one direction, doublets on one side of the 9 + 2 system must be activated, and those on the opposite side must be deactivated (Fig. 11-25a). Once the bend in this direction is complete, and the flagellum is ready to bend in the opposite direction, the pattern of activation must be reversed (Fig. 11-25b). This pattern of opposite activation and deactivation is necessary because dynein arms can exert their pulling force only in one direction; that is, they cannot push on one side of a flagellum and pull on the other. Although the basis for the back-and-forth switching is unknown, it probably is transmitted mechanically through the various

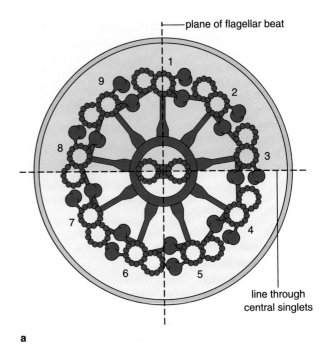

a

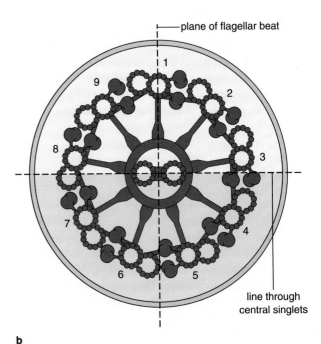

b

Figure 11-25 The switching mechanism creating flagellar bends. **(a)** To bend in one direction, the doublets on one side of the 9 + 2 system must be active (darker color) and those on the opposite side inactive. **(b)** For the return stroke, in which the flagellum bends in the opposite direction, the active and inactive sides of the 9 + 2 system reverse. The central singlets and the spokes connected to them apparently determine the doublets switched as a group, which are divided along a line running parallel to the plane of the singlets (horizontal dashed line).

linkages holding the 9 + 2 system together rather than chemically through changing concentrations of cAMP or Ca^{2+}. As one side reaches the limits of its stretch, for example, a trigger, perhaps involving phosphorylation or dephosphorylation, might stop contraction on one side and initiate it on the other. One set of observations indicating a mechanical linkage is the preservation of normal back-and-forth beating in demembranated flagella, in which Ca^{2+} and cAMP are lost by leakage into the suspending medium.

Observations of the direction of flagellar beating in relation to the orientation of the 9 + 2 system indicate that the plane of the flagellar beat lies perpendicular to a line drawn through the central singlets (the horizontal dashed line in Fig. 11-25), as if the position of these elements determines the sides of the flagellum switched on or off during back-and-forth beating. This impression is strengthened by observations of the central singlets in protozoa with flagella that beat in a helical waveform, in which the central singlets rotate around the flagellar axis. Evidently, the rotation moves the plane of beating around a circle centered on the flagellar axis. Sperm flagella with a 9 + 0 pattern of microtubules, in which the central singlets are missing (see below), also beat in nonplanar waves.

Deviations from the 9 + 2 Pattern in Sperm Flagella Although the 9 + 2 arrangement of microtubules is found universally in eukaryotic flagella, there are exceptions with special significance for flagellar motility. Most exceptions occur in the flagella of animal sperm cells. One unusual arrangement is the 9 + 0 pattern, in which the central singlets are missing, found, for example, in the sperm tails of a mayfly and an annelid worm. Some other species have sperm cells with a 9 + 1 pattern, in which the single central tubule is modified into a complex cylinder with multiple, concentric walls. A few other variations, some that differ considerably from the 9 + 2 pattern, are found in the sperm cells of insect species. Generally, these modifications are limited to sperm cells; cilia in body tissues of species with unusual sperm flagella have the usual 9 + 2 arrangement.

All sperm cells with variations of the standard 9 + 2 system are motile. However, the flagellar beat is slower and less powerful and differs in form to a greater or lesser extent from that of sperm with the standard pattern. The fact that sperm cells with unusual microtubule arrangements are still motile indicates that the 9 + 2 arrangement, rather than being essential for flagellar movement, is probably the most efficient of several possible arrangements.

Elements in addition to the standard 9 + 2 complex appear in the flagella of some species. The sperm cells of many animals contain nine longitudinal *accessory fibers* spaced immediately outside each doublet of the axial complex. Accessory fibers are most prominent in mammalian sperm, but are also found in insects and some birds, molluscs, and snakes.

In mammals accessory fibers are dense structures of considerable mass (Fig. 11-26). The nine accessories are of unequal size and shape, giving the mammalian sperm tail an asymmetrical appearance in cross section. In some insect sperm extra microtubules, arranged in a regular pattern outside the 9 + 2 system, appear as accessory elements (Fig. 11-27).

No trace of tubulin, actin, dynein, myosin, or other proteins with ATPase activity can be detected in the accessory fibers of mammalian sperm. Instead, the fibers contain polypeptides similar in composition to *keratin*, a cytoskeletal protein of intermediate filaments (see Ch. 13). As a consequence, accessory fibers are believed to act as cytoskeletal, rather than motile, elements; they may function to stiffen and provide elasticity to mammalian sperm flagella.

Centrioles, Basal Bodies, and the Generation of Cilia and Flagella The microtubules of cilia and flagella arise directly from centrioles. These structures, which occur in the cytoplasm of all eukaryotic cells capable of giving rise to flagella or cilia, contain a system of microtubules clearly related to the 9 + 2 system (Fig. 11-28). Nine sets of triplet microtubules make up the outer circle of the centriole. Each triplet contains one complete microtubule, the A subtubule, to which are attached a row of two incomplete microtubules, the B and C subtubules (see Figs. 11-7 and 11-28). Centriole triplets are thus analogous in structure, location, and arrangement to doublets of the flagellar 9 + 2 system of microtubules. The triplets are embedded in deposits of dense material that vary in pattern from one species to another. There are no central microtubules equivalent to the central singlets of flagella. Centrioles appear almost structureless in the center except near one end, where the interior contains a cartwheel pattern. The cartwheel has a central hub with nine radiating spokes that connect to the A subtubule of each peripheral triplet (see Fig. 11-28).

During generation of a flagellum the A and B subtubules of each triplet lengthen at the end of the centriole opposite the cartwheel, giving rise to the A and B subtubules of the 9 + 2 system (Fig. 11-29). The central singlets of the 9 + 2 system arise at the border between the centriole and the growing axial complex without any connections to centriole microtubules. In most flagella no structures arise from the C subtubules. The exceptions are in insects, in which the accessory microtubules (as in Fig. 11-27) arise from the C subtubules.

The elongating 9 + 2 system is covered by an extension of the plasma membrane. When development is complete, the centriole giving rise to a flagellum, which remains attached to the microtubules at the base of the flagellum, is called the *basal body* (Fig. 11-30).

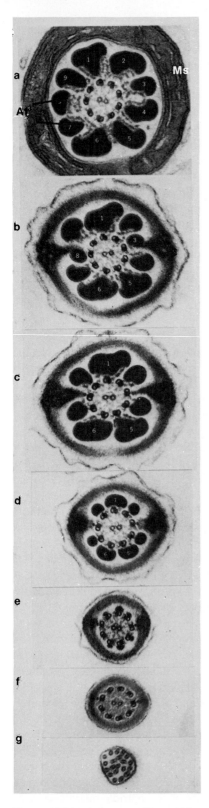

Figure 11-26 Accessory fibers (Af, numbered from 1 to 9) in the flagellum of a guinea pig sperm. Ms, mitochondrial sheath surrounding the sperm flagellum at its base. (**a** to **g**) Successive sections made from the base to the tip of the sperm tail. ×50,000. Courtesy of D. W. Fawcett.

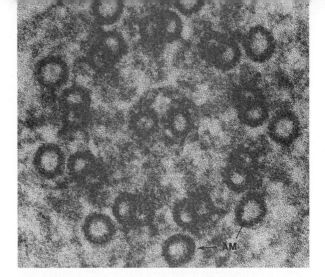

Figure 11-27 Accessory microtubules (AM) in the flagellum of a sperm cell of the field cricket *Acheta assimilis*. ×270,000. Courtesy of J. S. Kaye, from *J. Cell Biol.* 45:416 (1970), by permission of the Rockefeller University Press.

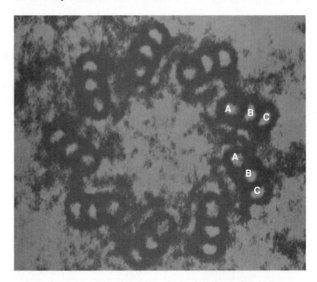

Figure 11-28 A centriole from a mouse cell in cross section. The cartwheel is faintly visible in the centriole lumen. The A, B, and C subtubules are marked for two of the triplets. ×360,000. From *The Nucleus*, ed. A. J. Dalton and P. Haguenau, 1968. Courtesy of E. de Harven and Academic Press, Inc.

Centrioles apparently have little or no motile function in cilia or flagella, because flagella broken from cell surfaces without their basal bodies can beat in a normal wave pattern when supplied with ATP. In a few organisms, such as the alga *Chlorogonium*, basal bodies regularly detach from flagella during mitosis and return when cell division is complete. Temporary removal of the basal bodies has no apparent effects on the flagellar beat pattern in this organism. Centrioles persisting as basal bodies in mature flagella and cilia may therefore serve simply as cytoplasmic anchors. Basal bodies may

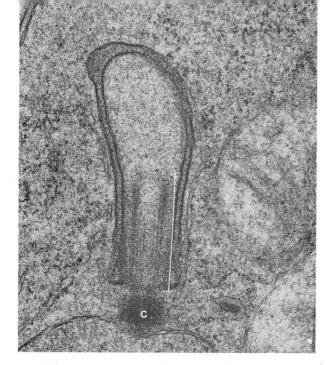

Figure 11-29 A centriole giving rise to a flagellum in the mosquito *Culex*. The developing 9 + 2 system (bracket) is encased in an elongated, cuplike vesicle that will eventually fuse with the plasma membrane to form the flagellar membrane. In many cell types that develop flagella, the growing 9 + 2 system simply pushes the plasma membrane into a fingerlike extension rather than becoming encased in a vesicle as in this example. C, centriole giving rise to the 9 + 2 system. ×49,000. Courtesy of D. M. Phillips, from *J. Cell Biol.* 44:243 (1970), by permission of the Rockefeller University Press.

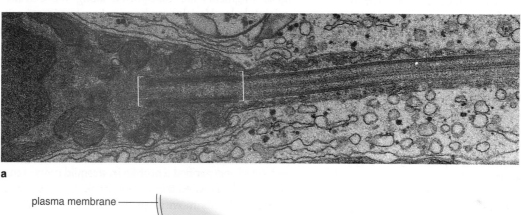

a

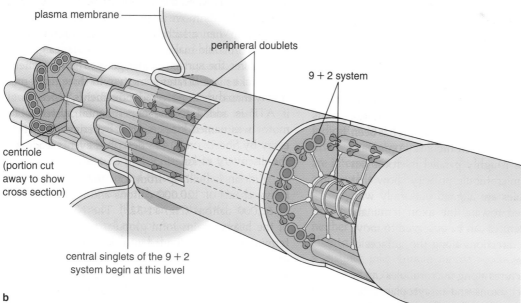

plasma membrane

peripheral doublets

9 + 2 system

centriole (portion cut away to show cross section)

central singlets of the 9 + 2 system begin at this level

b

Figure 11-30 Basal bodies and flagella. **(a)** A centriole persisting as the basal body (bracket) in the flagellum of a frog sperm. ×19,500. Courtesy of B. R. Zirkin. **(b)** Relationship of the microtubules of the basal body to the 9 + 2 system of microtubules in a flagellum.

The Experimental Process

Identification of Microtubule Motors Involved in Axonal Transport

Michael P. Sheetz

MICHAEL P. SHEETZ received his B.A. from Albion College in 1968 and his Ph.D. from the California Institute of Technology in 1972. He was a research fellow for two years at the University of California, San Diego. Before joining Duke University Medical Center as Chairman of the Department of Cell Biology, Dr. Sheetz had been a professor at Washington University Medical School in St. Louis since 1985. Before that, he was in the Department of Physiology at the University of Connecticut Health Center at Farmington. Dr. Sheetz' major research contributions have been in the fields of cell motility and membrane structure.

By 1983, the question of how neurons maintain long axonal processes had evolved into the question of how membranous vesicles are transported back and forth between the synapse and the cell. In the 1970s studies of polypeptide assembly in neurons found no detectable protein synthesis in axons or synapses. Instead, newly synthesized proteins are carried from the cell body to the synapse in two major transport pools at two quite different rates. The slower pool (with a rate of 0.01–0.05 μm/sec) contains components of the cytoskeleton (actin, tubulin, and neurofilament proteins). The fast pool (with a rate of 1–5 μm/sec) is composed of vesicular proteins, which presumably move as small axonal vesicles. Over the next two years, two motor activities, one of which was later identified with the protein kinesin,[1] were identified in axons. These findings together with other morphological analyses of axons provided convincing evidence that membranous organelles are transported by ATP-dependent motors along microtubules.

As is often the case, it was the combination of such factors—and some serendipity—which resulted in the identification of the microtubule motor kinesin. Before 1985 no microtubule motors had been identified except for axonemal dynein, which is restricted to cilia and flagella. Although regions of axons that are rich in the vesicles transported by fast axonal transport are also rich in microtubules, other transport filaments such as actin filaments could have been aligned along microtubules. The identification of the microtubule motor proteins was made possible by the combination of a number of factors including a new technique in light microscopy called video-enhanced differential interference contrast (DIC) microscopy, the conceptual advance of *in vitro* motility systems, and the unique advantages of squid axoplasm—the cytoplasm extruded from squid giant axons [see p. 213].

The technology of video-enhanced DIC microscopy came from developments in the video and defense industries which allowed subtraction of video images in real time and also contrast enhancement. Images from the video camera were contrast enhanced well beyond what the eye could do, and this brought out noise in the microscope optics. Since much of the noise was the same in both out-of-focus and in-focus images, subtracting an out-of-focus from an in-focus image would give a clean image.[2] Single microtubules (0.025 μm in diameter) or vesicles of 0.050 μm in diameter could be detected routinely, which enabled the first visualization of fast axonal transport.[2] Video-enhanced DIC microscopy revealed small vesicles that moved rapidly along transport filaments at 1–2 μm/sec.

At the same time, an *in vitro* assay for myosin motility was developed using myosin-coated latex spheres. The behavior of the spheres was similar to that of the small vesicles in axoplasm. Namely, the myosin-coated spheres would show Brownian movement until they attached to actin cables from the alga *Nitella*. After binding to the actin, the myosin beads would move at a rate of about 2 μm/sec, much like the axonal transport vesicles.[3] This suggested that the transport vesicles were powered by myosin or by another motor that moves on the transport filaments. This also suggested that the motor protein interacts with the vesicle surface, moving the vesicles in the direction defined by the transport filament. The major difference between the actin/myosin-based motility and axoplasmic transport was that although actin-based movement is unidirectional, the transport filaments support bidirectional movements.

A minor digression is warranted here to describe the limitations of light microscopy, including video-enhanced DIC microscopy. From basic physical principles, the limit of resolution of light microscopy is about 0.2 μm [see p. 109]; thus, objects separated by less than that distance appear as a single object. This means that video-enhanced DIC microscopy can detect a single microtubule—which is 0.025 μm in cross section—by inflating its diameter (utilizing the phenomenon of light diffraction), but it cannot resolve individual filaments within bundles of actin microfilaments or microtubules. Thus, it was necessary to analyze the same transport filament in both the light and electron microscopes to confirm that a single microtubule can support bidirectional transport (Fig. A).

Efforts to isolate the axoplasmic transport motor(s) were greatly aided by the serendipitous discovery that microtubules would move on glass coated with axo-

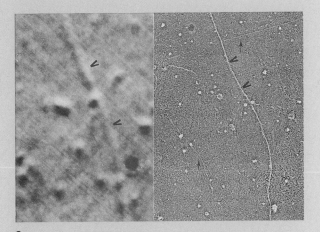

a

b

Figure A **(a)** Corresponding video-enhanced DIC image (left) and electron micrograph (right) of a single microtubule (open arrowheads). **(b)** A series of video-enhanced DIC images showing bidirectional movement of two vesicles (triangles) on a microtubule.

plasmic supernatant freed of vesicles. Because of the *in vitro* myosin motility studies, this suggested that axoplasmic supernatants contained a microtubule motor which would bind to glass in an active form. An assay was developed following the movement of added microtubules over axoplasm fractions bound to a glass slide.[4] This led to the development of a myosin motility assay analyzing the movement of actin filaments on myosins bound to a glass surface. By adding various fractions derived from axoplasm and several brain sources, kinesin was identified as a motor driving fast axonal transport.

The direction of movement driven by the kinesin motor remained to be identified. This movement was determined by using microtubules polymerized from a centrosome. Centrosomes seed the polymerization of microtubules with their plus or fast-growing ends outward. In this system, purified kinesin moved toward the plus end of the microtubules. In the axon it was found that the microtubules are arrayed in parallel with their plus ends pointed toward the periphery. Thus, kinesin must produce fast transport in the anterograde direction, that is, toward the axon tips. At the same time a factor was defined which was separable from kinesin and would move objects in the retrograde direction, away from the axon tips.

Using the same assay of microtubule movement on glass it was subsequently found that MAP1C was the retrograde motor and that it shared many properties with dynein (hence the name *cytoplasmic dynein*, to distinguish it from axonemal dynein).[5] Many motors have been added to the list and it seems that a large number of different cellular processes involve the directed movement of objects on microtubules.

References

[1] Vale, R. D.; Reese, T. S.; and Sheetz, M. P. Identification of a novel force-generating protein, kinesin, involved in microtubule-based motility. *Cell* 42:39–50 (1985).

[2] Allen, R. D.; Metuzals, J.; Tasaki, I.; Brady, S. T.; and Gilbert, S. P. Fast axonal transport in squid giant axon. *Science* 218: 1127–28 (1982).

[3] Sheetz, M. P., and Spudich, J. A. Movement of myosin-coated fluorescent beads on actin cables *in vitro. Nature* 303(5912):31–35 (1983).

[4] Kron, S. J.; and Spudich, J. A. Flourescent actin filaments move on myosin fixed to a glass surface. *Proc. Natl. Acad. Sci. USA* 83:6272–76 (1986).

[5] Paschal, B. M.; Shpetner, H. S.; and Vallee, R. B. MAP1C is a microtubule-activated ATPase which translocates microtubules in vitro and has dynein-like properties *J. Cell Biol.* 105:1273–128 (1987).

rections along microtubules. Additional microtubule motors will undoubtedly be discovered in the future.

Motile Systems Based on Microtubule Growth

Observations and experiments with the spindle provide some of the best evidence that microtubules can generate force by growth, that is, by controlled assembly and disassembly as well as by sliding. Some of the most significant evidence for this conclusion comes from an examination of spindle structure and function in primitive organisms such as the fungus *Phycomyces*, made by W. W. Franke and P. Reau and others. In *Phycomyces*, as in many other fungi, unbroken microtubules extend from one end of the spindle to the other with no overlap. During division the spindle elongates to many times its initial length and moves the dividing chromosomes a distance equivalent to the degree of elongation. Since no overlap is visible at any time between any of the microtubules in the spindle, this elongation may occur by microtubule growth. Microtubule growth may also contribute to the movement of chromosomes by the more complex spindles of eukaryotes (see Ch. 24 for details).

Other Microtubule-Powered Mechanisms

Microtubules have been implicated in a number of additional cellular movements. These include movements of the nucleus in the cytoplasm (including movements of sperm and egg nuclei within the egg cytoplasm during fertilization), mitochondrial movements, and the motions of pigment granules in skin pigment cells. A few cases of cytoplasmic streaming in rhizoids, the green alga *Caulerpa*, and in the filamentous growths (hyphae) of fungi have also been linked to microtubules. In none of these movements has it been established whether microtubules act to produce force by dynein, kinesin, or other crossbridging elements, by growth, or by other, as yet undiscovered, mechanisms.

The evidence supporting microtubules as the force-generating structures in these systems is largely circumstantial. In all of them microtubules are arranged in parallel arrays extending in the direction of move-

ment. Furthermore, exposure to colchicine or other agents disrupting microtubules inhibits the motion. However, in none of these examples has it yet been possible to determine conclusively whether microtubules generate the movements or merely provide cytoskeletal tracks guiding motions actually generated by other elements such as microfilaments.

Microtubules assemble by end-to-end polymerization of dumbbell-shaped heterodimers consisting of one α- and one β-tubulin polypeptide. The polymerization reaction, which proceeds as a finely balanced equilibrium between a heterodimer pool and assembled microtubules, is readily tipped in either direction by a large number of factors. Inside cells the primary factors regulating microtubule assembly appear to be alterations in Ca^{2+} concentration and combination of heterodimers with MAPs. Alterations in Ca^{2+} concentration provide short-term controls that may lead toward either assembly or disassembly of microtubules; combination with MAPs generally increases microtubule stability. Adjustment of the balance by these factors allows cells to regulate microtubule assembly and disassembly precisely, so that microtubules appear and disappear at controlled times and locations in the nucleus and cytoplasm.

Microtubules produce cellular motion by several mechanisms. The type of motion best documented by research is produced by crossbridging proteins such as dynein and kinesin. These proteins generate motile forces by hydrolyzing ATP and undergoing conformational changes that convert chemical energy released by ATP hydrolysis into mechanical energy. Other possible motile mechanisms involving microtubules are controlled growth and treadmilling.

Microtubules are characteristic eukaryotic structures that have no direct parallels among prokaryotes. The universal occurrence of microtubules in eukaryotes indicates that they arose as part of the structural and functional adaptations leading to the first appearance of eukaryotic cells. (Bacterial flagella, based on proteins and a motile mechanism that are completely different from those of eukaryotic flagella, are described in Supplement 11-1.)

For Further Information

Suggestions for Further Reading

General Books and Articles

Allen, R. D. 1987. The microtubule as an intracellular engine. *Sci. Amer.* 256:42−49 (February).

Dustin, P. 1980. Microtubules. *Sci. Amer.* 243:66−76 (August).

————. 1984. *Microtubules.* 2nd ed. New York: Springer-Verlag.

Schliwa, M. 1986. *The cytoskeleton. Cell biology monographs.* Vol. 14. New York: Springer-Verlag.

Warner, F. D.; Satir, P.; and Gibbons, I. R., eds. 1989. *Cell movement. Vol. I: The dynein ATPases.* New York: Alan R. Liss.

Warner, F. D., and McIntosh, R., eds. 1989. *Cell movement. Vol. II: Kinesin, dynein, and microtubule dynamics.* New York: Alan R. Liss.

The cytoplasmic matrix and the integration of cellular function. J. Cell Biol. Vol. 99, no. 1, pt. 2 (1984).

The Tubulin Polypeptides

Allende, J. E. 1988. GTP-mediated macromolecular interactions: the common features of different systems. *FASEB J.* 2:2356−67.

Burns, R. G. 1991. α-, β-, and γ-tubulins: sequence comparisons and structual constraints. *Cell Motil. Cytoskel.* 20: 181−89.

Cleveland, D. W. 1987. The multitubulin hypothesis revisited: what have we learned? *J. Cell Biol.* 104:381−83.

Fosket, D. E., and Morejohn, L. C. 1992. Structural and functional organization of tubulin. *Ann. Rev. Plant Physiol. Plant Molec. Biol.* 43:201−40.

Joshi, H. C., and Cleveland, D. W. 1990. Diversity among tubulin subgroups: toward what functional end? *Cell Motil. Cytoskel.* 16: 159−63.

MacRae, T. H. 1987. Nonneural microtubule proteins: structure and function. *Bioess.* 6: 128−32.

Stearns, T.; Evans, L.; and Kirschner, M. 1991. γ-tubulin is a highly conserved component of the centrosome. *Cell* 65:825−36.

Sullivan, K. F. 1988. Structure and utilization of tubulin is otypes. *Ann. Rev. Cell Biol.* 4:687−716.

Microtubule Assembly

Bayley, P. M. 1990. What makes microtubules dynamic? *J. Cell Sci.* 95:329−34.

Bulinski, J. C., and Gundersen, G. G. 1991. Stabilization and post-translational modification of microtubules during cellular morphogenesis. *Bioess.* 13:285−91.

Cassimeris, L. U.; Walker, R. A.; Pryer, N. K.; and Salmon, E. D. 1987. Dynamic instability of microtubules. *Bioess.* 7:149−54.

Correia, J. J., and Williams, R. C. 1983. Mechanism of assembly and disassembly of microtubules. *Ann. Rev. Biophys. Bioeng.* 12:211−35.

Frieden, C. 1985. Actin and tubulin polymerization: the use of kinetic methods to determine mechanism. *Ann. Rev. Biophys. Biophys. Chem.* 14:189−210.

Gelford, V. J., and Bershadski, A. D. 1991. Microtubule dynamics: mechanism, regulation, and function. *Ann. Rev. Cell Biol.* 7:93−116.

Kirschner, M., and Mitchison, T. 1986. Beyond self-assembly: from microtubules to morphogenesis. *Cell* 45:329−42.

Schultze, E.; Asai, D. J.; Bulinski, J. C.; and Kirschner, M. 1987. Posttranslational modification and microtubule stability. *J. Cell Biol.* 105:2167−77.

MAPs

Cleveland, D. W. 1990. Microtubule MAPping. *Cell* 60:701−2.

Lee, G. 1990. Tau proteins: an update on structure and function. *Cell Motil. Cytoskel.* 15:199−203.

Olmstead, J. B. 1986. Microtubule-associated proteins. *Ann. Rev. Cell Biol.* 2:421−57.

Vallee, R. B., and Bloom, G. S. 1984. High molecular weight microtubule-associated proteins (MAPs). *Modern Cell Biol.* 3:21−75.

Wiche, G.; Oberkanins, C.; and Himmler, A. 1991. Molecular structure and function of microtubule-associated proteins. *Internat. Rev. Cytol.* 124:217−73

MTOCs

Brinkley, B. R. 1985. Microtubule organizing centers. *Ann. Rev. Cell Biol.* 1:145−72.

Vallee, R. B. 1990. Molecular characteristics of high molecular weight microtubule-associated proteins: some answers, many questions. *Cell Motil. Cytoskel.* 15:204−209.

Dynein, Kinesin, and Other Microtubule Motors

Amos, L. A., and Amos, W. B. 1987. Cytoplasmic transport in axons. *J. Cell Sci.* 87:1−2.

Endow, S. A. 1991. The emerging kinesin family of microtubule motor proteins. *Trends Biochem. Sci.* 16:211−25.

Gibbons, I. R. 1988. Dynein ATPases as microtubule motors. *J. Biolog. Chem.* 263:15837−40.

McIntosh, J. R., and Porter, M. E. 1989. Enzymes for microtubule-dependent motility. *J. Biolog. Chem.* 264:6001−04.

Piperno, G. 1990. Functional diversity of dyneins. *Cell Motil. Cytoskel.* 17:147−49.

Porter, M. E., and Johnson, K. A. 1989. Dynein structure and function. *Ann. Rev. Cell Biol.* 5:119−51.

Schroer, T. A.; Dabora, S. L.; Steuer, E.; and Schroer, T. A. 1990. Control of organelle movements and endoplasmic reticulum extension powered by kinesin and cytoplasmic dynein. *Curr. Top. Membr. Transport* 36:117−28.

Schroer, T. A., and Sheetz, M. P. 1991. Functions of microtubule-based motors. *Ann. Rev. Physiol.* 53:629−52.

Schroer, T. A.; Steuer, E. R.; and Sheetz, M. P. 1989. Cytoplasmic dynein is a minus-end-directed motor for membranous organelles. *Cell* 56:937–46.

Sheetz, M. P. 1987. What are the functions of kinesin? *Bioess.* 7:165–68.

Stebbings, H. 1990. How is microtubule-based organelle translocation regulated? *J. Cell Sci.* 95:5–7.

Stewart, R. J., and Goldstein, L. S. B. 1991. Molecular genetic analyses of *Drosophila* kinesin. *Curr. Top. Membr. Transport* 38: 1–11.

Vale, R. D. 1987. Intracellular transport using microtubule-based motors. *Ann. Rev. Cell Biol.* 3:347–78.

Vallee, R. B., and Shpetner, H. S. 1990. Motor proteins of cytoplasmic microtubules. *Ann. Rev. Biochem.* 59:909–32.

Wiche, G.; Oberkanins, C.; and Himmler, A. 1991. Molecular structure and function of microtubule-associated proteins. *Internat. Rev. Cytol.* 124:217–73.

9 + 2 System or Flagellar Axoneme

Brokaw, C. J. 1990. Flagellar oscillation: new vibes from beads. *J. Cell Sci.* 95:527–30.

Luck, D. J. L. 1984. Genetic and biochemical dissection of the eukaryotic flagellum. *J. Cell Biol.* 98:789–94.

Murray, J. M. 1991. Structure of flagellar microtubules. *Internat. Rev. Cytol.* 125:47–93.

Satir. P., and Sleigh, M. A. 1990. The physiology of cilia and mucociliary interactions. *Ann. Rev. Physiol.* 52:137–55.

Tash, J. S. 1989. Protein phosphorylation: the second messenger signal transducer of flagellar motility. *Cell Motil. Cytoskel.* 14:332–39.

Bacterial Flagella

Armitage, J. P. 1992. Behavioral responses in bacteria. *Ann. Rev. Physiol.* 54:683–714.

Eisenbach, M. 1990. Functions of the flagellar modes of rotation in bacterial motility and chemotaxis. *Molec. Microbiol.* 4:161–67.

Imae, Y., and Atsumi, T. 1989. Na^+-driven bacterial flagellar motors. *J. Bioenerget. Biomembr.* 21:705–16.

Macnab, R. M., and Aizawa, S.-I. 1984. Bacterial mobility and the bacterial flagellar motor. *Ann. Rev. Biophys. Bioeng.* 13:51–83.

Macnab, R. M., and Parkinson, J. S. 1991. Genetic analysis of the bacterial flagellum. *Trends Genet.* 7:196–200.

Namba, K.; Yamashita, I.; and Vonderviszt, F. 1989. Structure of the core and central channel of bacterial flagella. *Nature* 342:648–54.

Review Questions

1. Describe the structure of microtubules. What are protofilaments?

2. What are α- and β-tubulins? What relationships do these proteins have to the subunits visible in microtubule walls? What are tubulin variants or isotypes? What relationship do variants have to microtubule structure?

3. What are heterodimers?

4. What are MAPs? What effects do MAPs have on microtubule assembly?

5. Outline the effects of Ca^{2+} on microtubule assembly. How do Ca^{2+}, calmodulin, and MAPs interact in microtubule assembly?

6. How does GTP enter into microtubule assembly? In what ways are tubulin heterodimers and the G proteins active in cell surface reception (see p. 243) similar?

7. What is treadmilling? Microtubule polarity? What is the relationship of polarity to treadmilling? To cellular motility?

8. What are MTOCs? What major types of MTOCs occur in cells? How might MTOCs work in the initiation of microtubule assembly?

9. Outline the 9 + 2 system of microtubules and connecting elements in flagella.

10. What experiments demonstrated that the dynein arms attached to the microtubule doublets of the 9 + 2 system have ATPase activity?

11. Outline the structure of dynein. How is the dynein crossbridging cycle that produces microtubule sliding believed to work? Where and how does ATP enter the system? What property of proteins is responsible for transforming the chemical energy of ATP into the mechanical energy of movement? In what direction do dynein crossbridges "walk" with respect to microtubule polarity?

12. What experiments demonstrated that microtubules actually slide in flagella?

13. Compare the structure and function of cilia and flagella.

14. What mechanisms may regulate flagellar beating? What is the relationship of Ca^{2+} concentration to flagellar regulation? Of phosphorylation?

15. What deviations from the 9 + 2 system are observed in sperm flagella? What is the significance of these deviations to flagella motion?

16. What is fast axonal transport? Slow axonal transport? What substances or structures are moved by slow and fast axonal transport? What is the relationship of microtubules to fast axonal transport?

17. What is kinesin? What evidence links kinesin to fast axonal transport? Compare the activity of kinesin and dynein in microtubule-based motility. How are the activities of kinesin, cytoplasmic dynein, and microtubules coordinated in fast axonal transport?

18. How is microtubule growth related to cell motility? How might microtubule growth be involved in spindle function?

19. Outline the structure of centrioles. How are centrioles related to the 9 + 2 system of flagella? What are the relationships among centrioles, flagella, and basal bodies?

Supplement 11-1

Bacterial Flagella

Many bacterial species move through a liquid medium by means of long, fibrous appendages that extend from the cell surface (see Figs. 11-33 and 1-8). Although the bacterial swimming appendages are called flagella, they have no structural or functional relationships to the flagella of eukaryotic cells. In fact, bacterial flagella generate motion by a mechanism that is completely without parallels in eukaryotes.

Structure

Bacterial flagella have three parts, the *filament*, *hook*, and *basal body* (Fig. 11-34). The filament is a long, slender fiber of protein about 20 nm in diameter and from 10 or 20 to 100 μm in length. The diameter of a filament is thus smaller than a single microtubule in a eukaryotic flagellum.

Flagellar filaments consist of a tubular polymer of a single protein called *flagellin*, which occurs in a distinct type in each species. The flagellin of *Salmonella*, for example, is a protein with 489 amino acids and a molecular weight of about 51,000. Although the molecular weights are similar to those of the tubulins, the bacterial flagellins are unrelated to the tubulins or to actin.

Flagellin proteins assemble in 11 rows called *protofilaments* to form a flagellar filament (Fig. 11-35). The protofilament rows follow helical paths around a central lumen. The protofilaments of a flagellar filament are able to undergo side-by-side slippage, allowing them to unwind and rewind between left-handed and right-handed helices without disturbing the tubular structure of the filament. The alternate helical patterns allow swimming bacterial cells to change direction (see be-

a

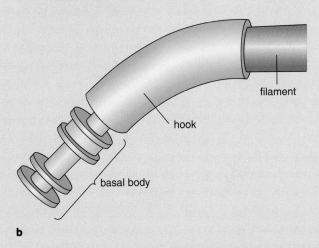

b

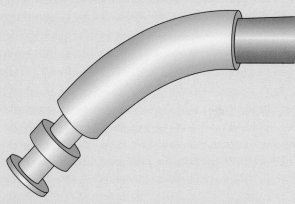

c

Figure 11-34 Structure of the flagellum in bacteria. **(a)** The basal body and hook of a flagellum isolated from *E. coli*, a gram-negative bacterium. × 46,000. Courtesy of M. L. DePamphilis and Julius Adler. Diagrams of the basal body and hook of a flagellum in **(b)** gram-negative and **(c)** gram-positive bacteria.

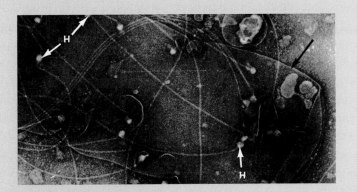

Figure 11-33 Flagella isolated with a portion of the cell wall (arrow) from the bacterium *Proteus mirabilis*. The basal structures of the flagella (H) appear as spherical particles in this preparation. × 50,000. Courtesy of W. Van Iterson and North-Holland Publishing Company, Amsterdam.

low). In either helical arrangement the entire filament follows a spiral path resembling a corkscrew.

The hook contains a polymer of a single polypeptide type that is different from the filament protein. Like the lattice formed by the filament monomers, the

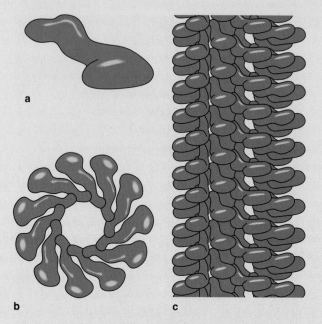

Figure 11-35 Arrangement of flagellin polypeptides in the filament of a *Salmonella* flagellum. **(a)** A single flagellin polypeptide. **(b)** A filament in cross section, showing the central lumen. The lumen is wide enough to allow passage of single flagellin polypeptides, which flow through the filament to reach the tip during filament growth. **(c)** The helical packing of flagellin polypeptides in the filament. Redrawn from an original courtesy of K. Namba, I. Yamashita, and F. Vonderviszt. Reprinted by permission from *Nature* 342:648 (1989); copyright © 1985 Macmillan Magazines Ltd.

hook polymer can alternate between several forms that change its overall shape. The hook structure is believed to form a sort of "universal joint" that connects the filament to the basal body.

Both the hook and filament extend from the surface of the bacterial cell wall. The remaining structure of the bacterial flagellum, the basal body (see Figs. 11-34 and 11-36), is completely embedded in the wall and plasma membrane of the bacterial cell. The basal body is a complex structure consisting of a central *shaft* bearing a series of *rings* that fit into the plasma membrane and cell wall.

The rings of the basal body are arranged differently in gram-positive and gram-negative bacteria (see p. 302). In gram-negative bacteria two pairs of rings link the basal body into the various layers of the cell border (see Figs. 11-34a and b and 11-36). The outermost ring, the *L ring*, fits into the outer membrane of the cell wall; the other ring of this pair, the *P ring*, fits into the peptidoglycan sheath underlying the outer membrane (bacterial cell wall structure is outlined in Supplement 8-1). The innermost ring of the other pair, the *M ring*, fits into the plasma membrane; the other ring of this pair, the *S ring*, lies on the outside surface of the plasma membrane. The entire basal body contains as many as

20 or more different polypeptides in gram-negative bacteria.

Less differentiation is visible in the basal body of gram-positive bacteria (see Fig. 11-34c). However, the innermost ring is considered to have two functional parts, equivalent to the S and M rings of gram-negative basal bodies. The single outer ring, which fits into the relatively thick peptidoglycan sheath of gram-positive bacteria, is probably equivalent to the P ring of gram-negative cells.

The Flagellar Motor

The bacterial basal body forms a biological motor that rotates a flagellum in its socket in the wall, much like the propeller of a boat. The S and M rings are believed to interact in some way to produce the rotational force, and the remaining rings to form a bearing through which the shaft rotates in the cell wall.

Evidence that bacterial flagella rotate like propellers rather than beating in waves like eukaryotic flagella was developed in several experiments by M. R. Silverman and M. I. Simon. They used ingenious techniques to circumvent the fact that, because bacterial flagella lie at the limits of visibility in the light microscope, they cannot be directly seen to rotate. In one experiment Silverman and Simon tethered bacterial cells to a glass slide by coating the slide with antibodies against the flagellar protein. Reaction with the antibodies fixed the flagellar filaments to the slide and prevented them from rotating. Under these conditions the bacterial cells, which can be fully resolved in the light microscope, could be seen to rotate instead. In a second experiment the two investigators attached polystyrene spheres, large enough to be visible in the light microscope, to the surface of an untethered bacterial flagellum by coating the spheres with antiflagellin antibody. Examination of living cells showed that the spheres revolved around the flagellar axis, exactly as expected if the flagella move by rotating in their wall sockets.

Several experiments have demonstrated that rotation of the flagellar motor in most bacteria is driven by an H^+ gradient across the plasma membrane. Initially, S. H. Larsen and his colleagues showed that agents which make the plasma membrane leaky to H^+, thereby destroying the H^+ gradient, completely inhibit flagellar motility even though ATP reserves remain high and available. Later, H. C. Berg and his colleagues found that an externally imposed H^+ gradient can restart flagellar motility in cells that have exhausted their energy reserves and stopped moving.

The flagellar motor has some unusual characteristics that limit possibilities for the unknown mechanism using the H^+ gradient to drive rotation. The speed of rotation and the twisting force remain constant over a wide temperature range. Over this range the efficiency of

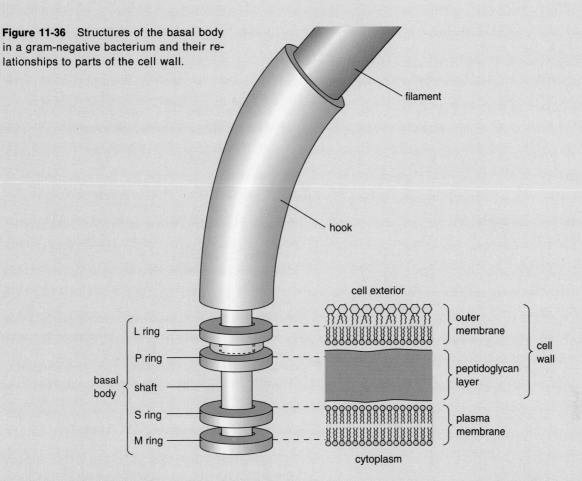

Figure 11-36 Structures of the basal body in a gram-negative bacterium and their relationships to parts of the cell wall.

energy conversion from the H^+ gradient to the mechanical energy of flagellar rotation is very high, approaching 100%. Moreover, the motor works equally well, and with the same temperature and efficiency relationships, in either the clockwise or counterclockwise direction. These observations collectively rule out possibilities such as enzymatically catalyzed formation and breakage of covalent or hydrogen bonds, or conformational changes in proteins, as the basis for motor function. These mechanisms are all temperature dependent and would be expected to increase significantly in rate with increases in temperature, rather than remaining constant as the flagellar motor does.

In view of these restrictions, S. Khan and H. C. Berg proposed an interesting model for the molecular mechanism propelling the bacterial motor. The model suggests an interaction between the M ring and another protein, the *mot* protein, recently found in genetic experiments to be necessary for flagellar function. (Bacterial mutants lacking the *mot* protein have paralyzed flagella.) In bacteria oxidative electron transport pushes H^+ to the cell exterior, producing an H^+ gradient that is high outside and low inside (see p. 362). According to the Khan and Berg model (Fig. 11-37), the *mot* protein forms a channel in the membrane that conducts hydrogen ions (protons) from outside the plasma membrane to the cell interior. The channel has two segments,

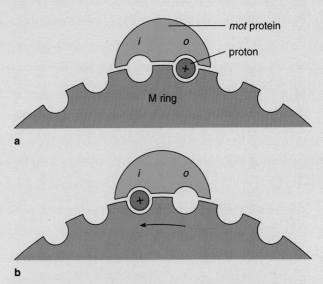

Figure 11-37 A simplified version of the Khan and Berg model for the prokaryotic flagellar motor. In **(a)** a proton has just arrived at the point where the *o* channel meets the M ring. In **(b)** rotation of the M ring carries the proton from the *o* to the *i* channel, from which it can enter the cell. Another proton-binding site—these sites are spaced evenly around the edge of the M ring—now faces the *o* channel, and the process is ready to repeat. Reversal of the motor could be accomplished by a flip in position of the *mot* protein, so that the positions of the *o* and *i* channels are reversed.

one (o in Fig. 11-37) that conducts protons from outside the cell to the M ring and another (i in Fig. 11-37) that conducts protons from the M ring to the cell interior. For a proton to pass from outside to inside the cell along the concentration gradient, it must shift in the region of the M ring from the o to the i channel in the *mot* protein. The shift occurs by movement of the M ring, which must rotate for the proton to pass from the o to the i channel.

Flagellar Rotation and Bacterial Swimming Behavior

The Silverman and Simon experiments answered a long-standing question about bacterial swimming behavior. Bacterial cells typically swim smoothly in a more or less straight direction for one to several seconds and then tumble or "twiddle" randomly for a much shorter period of about 0.1 second. Smooth swimming and tumbling alternate in a regular fashion, so that the cells typically progress in short directional bursts interspersed with momentary tumbling motions. The tethering experiments revealed that the periodic shift between smooth swimming and tumbling reflects a change in direction of flagellar rotation. During straight running the motor turns counterclockwise. To initiate the tumbling motion, the motor switches almost instantaneously to clockwise rotation. Tumbling ceases as the motor switches back to counterclockwise rotation.

Molecular changes underlying the switch between smooth swimming and tumbling depend on the helical winding of protofilaments in the flagellar filaments. During the counterclockwise rotation that produces smooth swimming, the protofilaments of a flagellar filament are twisted into a left-handed helix. When the motor reverses direction, the left-handed helix is forced to unwind and rewind relatively slowly into a right-handed helix. Of the two helices the left-handed one is more stable; if undisturbed or rotated in a counterclockwise direction, flagellar filaments quickly snap into a helix in this direction.

The unwinding and rewinding of the protofilament helix begins at the base of the flagellar filament and proceeds toward the tip. At the point where the helical change is taking place, the smooth corkscrew traced by the flagellar filament assumes a random kink or bend. Bends in all the flagella destroy directional force and cause the cell to tumble randomly. If the clockwise rotation persists long enough, the entire flagellar filament rewinds into a right-handed helix, producing an unkinked corkscrew that allows a smooth, although somewhat wobbly, swimming pattern to resume. Usu-

ally, however, the motor rotation returns to the counterclockwise direction before the entire filament has time to wind to the right.

The return to smooth swimming takes place through a momentary pause in rotation. During the pause the untwisting force is removed from the flagellar filaments, and they snap back into the stable left-handed helix. As they snap back, counterclockwise rotation begins, and swimming returns to the smooth mode.

The shift between smooth swimming and tumbling allows bacterial cells to swim toward favorable chemical stimuli and avoid noxious ones. When swimming cells encounter increasing concentrations of substances, such as glucose or galactose, that are registered by cell receptors as a favorable chemical environment, the tumbling mode is suppressed. As a consequence, the cells continue to swim toward the region of greatest concentration of the favorable stimulus. If a substance is encountered that registers as unfavorable, such as acetic acid, the straight running mode is suppressed and tumbling occurs much more frequently. The greater frequency of tumbling, caused by more frequent motor switching, changes the swimming direction more often and increases the probability that the swimming cell will escape the unfavorable stimulus. The receptors that detect favorable or unfavorable chemicals in the environment, and a series of proteins linking them to the flagellar motors, form one of several *sensor-regulator pathways* in bacteria (see Supplement 7-1).

Y. Imae and his colleagues found that in the bacterium *Bacillus* an Na^+ gradient substitutes for the H^+ gradient driving the bacterial flagella. In *Vibrio* an Na^+ gradient drives a single flagellum at one end of the cell, and an H^+ gradient drives flagella located laterally. Whether driven by H^+ or Na^+, the motor and flagellar system of these bacteria seems to operate by the same mechanism as the more common H^+-driven systems.

One eukaryote, a protozoan that lives in the gut of termites, has developed a curious symbiotic relationship in which the flagellar motors of bacteria are used as propellers to push the protozoan cell through the liquid medium in the termite gut. In the relationship, investigated by S. L. Tamm, bacterial cells are carried by the hundreds in pits lined up in regularly spaced rows covering the surface of the as yet unnamed protozoan. The bacterial flagella extend from the cell surface, forming tufts that rotate to propel the protozoan through its liquid environment. Presumably, the symbiotic relationship creates an environment so favorable for the bacteria that tumbling is completely suppressed, and swimming always proceeds in the straight-running mode.

Suggestions for Further Reading: See p. 446.

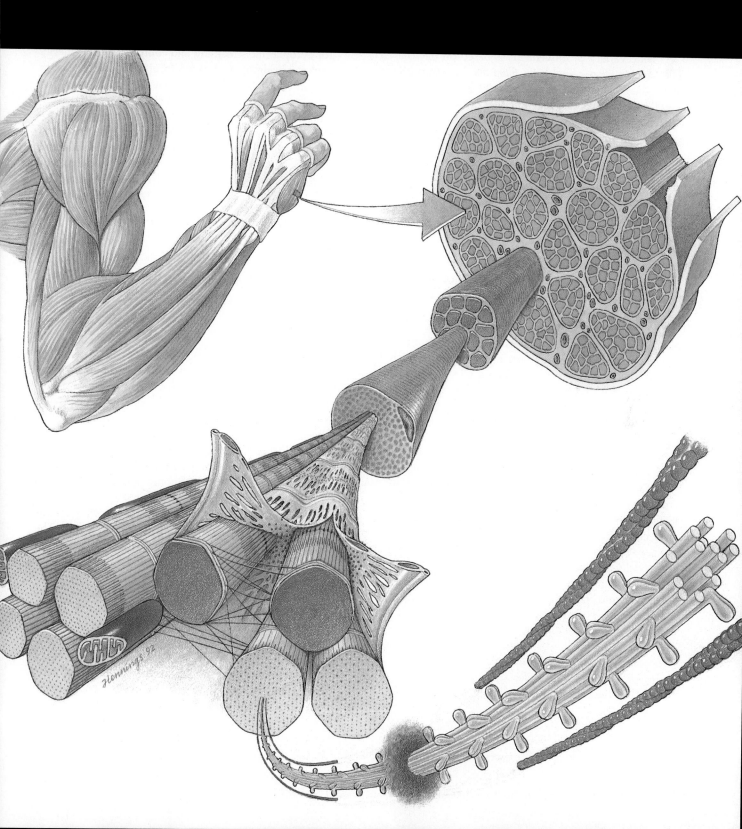

- *Microfilament structure and biochemistry*
- *Actins* ▪ *Microfilament polarity* ▪ *Actin-binding proteins* ▪ *Myosins and microfilament sliding*
- *Regulation of microfilament-based motility*
- *Striated muscle* ▪ *Smooth muscle* ▪ *Cytoplasmic streaming* ▪ *Ameboid motion* ▪ *Movements based on microfilament growth* ▪ *Movements without microfilaments or microtubules*

A wide variety of movements is based on microfilaments, also called *actin filaments*, in eukaryotic cells. Microfilament-based movements, like those generated by microtubules, depend either on active sliding or on pushing movements developed by the directional assembly of microfilaments. Besides generating motility, microfilaments, along with microtubules and intermediate filaments, form supportive structural networks in the cytoplasm.

Microfilament sliding, which accounts for most microfilament-generated motility, is powered by the action of crossbridges that cycle in an attach-pull-release cycle to move cellular structures along the microfilaments. The movements that result from structures sliding along microfilaments take two different major forms in eukaryotic cells, *cytoplasmic streaming* and *contraction*.

In cytoplasmic streaming segments of the cytoplasm flow directionally from one region to another. The elements moved by cytoplasmic streaming, which is a relatively slow process, include a wide variety of molecules, particles, and organelles within cells. More organized forms of cytoplasmic streaming may push cell borders into extensions or, in *ameboid motion*, may participate in movement of entire cells.

In contraction structures sliding over microfilaments forcibly constrict or shorten cell segments or whole cells. The most highly organized contractile microfilament systems underlie muscular movements in animals. These movements include the slow and powerful contractions of smooth muscle cells and, in the ultimate state of microfilament organization, the rapid and equally powerful contractions of the striated cells of skeletal and cardiac muscle. Less highly organized contractile microfilament systems produce the cytoplasmic movements that narrow or constrict cells during embryonic development and form the cleavage furrow that separates the cytoplasm of animal cells into two parts during cell division. Contractile networks arranged in layers just under the plasma membrane account for the capping movements that sweep receptors to one end of the cell surface and take part in ameboid motion.

In movements by directional assembly, subunits polymerize into new microfilaments or add to the ends of existing microfilaments. As microfilaments lengthen, cell structures attached to the microfilament tips are pushed in the direction of growth. This mechanism powers rapid and spectacular extension of a filament from the head of some invertebrate spermatozoa during fertilization, for example. Directional microfilament assembly may also be responsible for part of ameboid movement.

Microfilaments responsible for these varied cellular movements are assembled from subunits consisting of the protein *actin*. Another protein, *myosin*, forms the crossbridges that produce the sliding forces responsible for cytoplasmic streaming and contraction. The energy for crossbridging is derived from ATP hydrolysis, as it is in microtubule-based motility.

This chapter discusses the structure and biochemistry of microfilaments and their functions in cell motility. The roles of microfilaments in the cytoskeleton are described in the following chapter. Supplement 12-1 surveys several evolutionary adaptations that produce motion in eukaryotic cells without the involvement of either microfilaments or microtubules.

STRUCTURE AND BIOCHEMISTRY OF MICROFILAMENTS

Microfilament Structure

Microfilaments appear in sections as thin, dense lines about 7 to 9 nm in diameter in the electron microscope (Fig. 12-1; see also Fig. 1-18). When isolated and prepared for electron microscopy by negative staining (see p. 119), microfilaments can be seen to consist of two linear chains of roughly spherical subunits wound into a double helix (Fig. 12-2a and b). The pitch of the double helix is very long with respect to its diameter, making a complete turn only once in every 71 nm. The double helix makes a half turn every 35.5 nm, so the two chains appear to cross over each other at this interval when the helix is viewed from the side. The subunits in the twisted chains are individual actin molecules, each about 5 to 6 nm long and 4 nm in diameter (Fig. 12-2c).

Microfilaments vary considerably in length. They may contain from as few as 10 to 20 to more than a thousand actin subunits. In many locations in the cell microfilaments also include another protein, *tropomyosin*, as a regular constituent (see Fig. 12-13). Tropomyosin serves as a fibrous reinforcement along actin chains and in some systems participates in regulation of microfilament sliding (see below).

Microfilaments are often difficult to identify directly because they resemble other fiber types visible in the cytoplasm, such as intermediate filaments. This

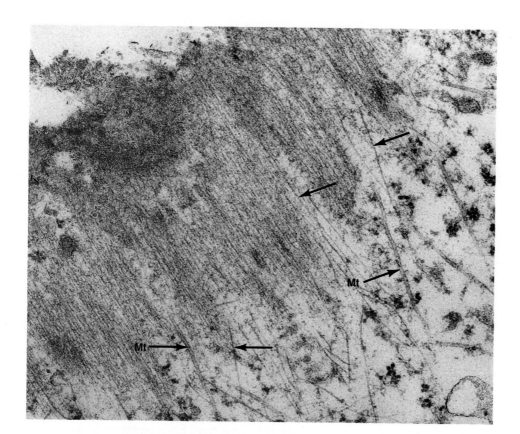

Figure 12-1 Microfilaments (arrows) in human capillary endothelial cells. Several microtubules (Mt) are also visible. ×50,000. Courtesy of K. G. Bensch, from *J. Ultrastr. Res.* 82:76 (1983).

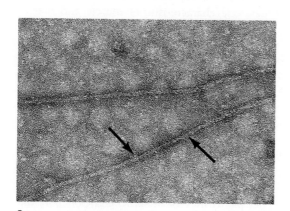

a

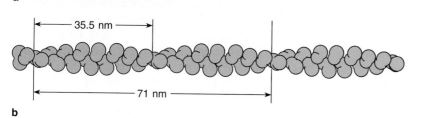

b

Figure 12-2 Microfilament structure. **(a)** Negatively stained microfilaments isolated from a protozoan, *Acanthamoeba*. Individual actin subunits are visible at some locations (arrows). Courtesy of T. D. Pollard. **(b)** The double helix of actin subunits in a microfilament. Each actin subunit is a bilobed or kidney-shaped particle approximating the form shown in **(c)**. Recent evidence from X-ray and optical diffraction studies suggests that each actin subunit is aligned with its long axis almost perpendicular to the long axis of a microfilament, turned so that the two lobes face the microfilament surface.

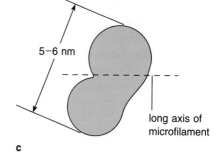

c

problem is solved by a technique known as *myosin decoration,* developed in 1963 by H. E. Huxley. In the technique microfilaments are reacted with a segment of the myosin molecule. Attachment of the myosin segments produces an unmistakable "arrowhead" pattern (Fig. 12-3*a*) that is characteristic only of microfilaments. The arrowheads reflect binding of the myosin segments, one for each actin subunit, in a double spiral that follows the actin double helix around a microfilament (Fig. 12-3*b*).

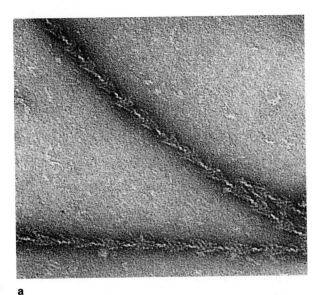

a

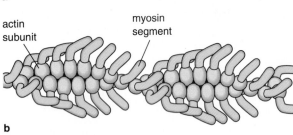

actin subunit

myosin segment

b

Figure 12-3 Use of the myosin decoration technique to identify microfilaments in the electron microscope. **(a)** The "arrowhead" pattern produced when actin microfilaments are reacted with the HMM or S1 segments of myosin molecules (see p. 462). Courtesy of J. A. Spudich, with permission from *J. Mol. Biol.* 72:619 (1972). Copyright by Academic Press, Inc. (London) Ltd. **(b)** The arrangement of myosin segments that produces the arrowhead pattern on an actin microfilament.

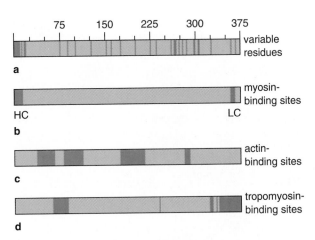

Figure 12-4 Structure of actin molecules. **(a)** Sites in the amino acid sequence at which variations are noted between different actins. **(b)** Sites at which actin molecules bind myosin. HC, heavy chain; LC, light chain. **(c)** Sites at which actin subunits bind other actins in microfilament assembly. **(d)** Sites binding tropomyosin. Redrawn from an original courtesy of B. D. Hambly and the Company of Biologists Ltd., from *Bioess.* 4:124 (1986).

Actins

Actins, which occur in all eukaryotes, comprise one of the most abundant groups of cellular proteins. X-ray and optical diffraction studies (see Ch. 4) show that actin folds into a three-dimensional structure consisting of two lobes of unequal size separated by a cleft or depression on one side (see Fig. 12-2c). Distributed over the structure are separate binding sites for myosin, other actin molecules, Mg^{2+}, and either ATP or ADP (Fig. 12-4). Actin also has the unique property of greatly stimulating the activity of myosin as an ATPase (the name *actin* is derived from this property).

Most actins contain 374, 375, or 376 amino acids and have total molecular weights near 42,000. One of the actins of skeletal muscle cells in mammals, for example, contains 375 amino acids with a molecular weight calculated from its amino acid sequence of 41,872. Differences in length between various actins occur entirely at the N-terminal end of the protein chain.

Although some unicellular eukaryotes possess only a single actin type, most fungi, plants, and animals have two or more distinct actins that vary to some extent in amino acid sequence. Most invertebrates have two actin types, one characteristic of muscle and one of nonmuscle cells. The same distribution is found in the most primitive chordates. The muscle actins of sharks and bony fishes are differentiated into cardiac and smooth types. Reptiles, birds, and mammals have no less than six different actins. In reptiles, birds, and mammals two different actins, both called α-actins, are characteristic of striated muscle cells; one of these, α_{sk}, occurs in skeletal muscle, and the other, α_c, in cardiac muscle. Another α-actin, α_{sm}, occurs with a different actin type, γ_{sm}, in smooth muscle cells. Two more types, β and γ_{nm}, make up the actins of nonmuscle cells.

Higher plants may have six to eight or more different actin types. The functional significance of the high actin diversity in higher plants is unknown. It has been suggested that distinct actins may be associated with different microfilament-based movements—such as cytoplasmic streaming, pollen tube growth, and organelle movement—and with different cytoskeletal systems such as those of dividing and nondividing cells.

All these actin types are closely related in sequence. Among all eukaryotes the various actins differ in sequence by only about 15%; those of animals and some protozoa vary by only 5%. The single actin of the protozoan *Amoeba proteus*, for example, is as similar to the six actins of higher vertebrates as the vertebrate actins are to each other.

Most amino acid substitutions responsible for different actin types are confined to limited *variable* regions in the sequence. The remaining sequences are *constant* or nearly so among actins of the same or different species (see Fig. 12-4*a*). Where substitutions occur in actin sequences, the majority are conservative; that is, they involve replacement of one amino acid by another with the same or similar chemical properties. The entire picture of sequence conservation indicates that most segments of actin molecules are critical to function of the protein and that gene mutations substituting amino acids in these segments are likely to be lethal. The close sequence relationships of actins suggest that all arose from a single ancestral type in very early evolutionary lines leading to eukaryotes.

Assembly of Actin Molecules into Microfilaments

Actin molecules exist inside cells both in an unassembled pool and in assembled microfilaments. In unassembled form actin molecules are frequently called *G actin* (from G = globular). Actin assembled into microfilaments is frequently termed *F actin* (from F = filamentous). Individual actin molecules may exchange between the unassembled pool and the assembled form, setting up an equilibrium similar to the tubulin-microtubule equilibrium (see p. 421):

$$\text{actin molecules} \rightleftharpoons \text{microfilaments} \quad (12\text{-}1)$$

To assemble into microfilaments, G-actin molecules must combine with Mg^{2+} and a molecule of ATP. The ATP is hydrolyzed as assembly proceeds; both the ADP and inorganic phosphate produced remain bound to the actin subunits linked into microfilaments. The ATP breakdown associated with microfilament assembly led researchers to conclude at first that it is an energy-requiring process. However, later work showed that molecules which resemble ATP in structure but cannot be hydrolyzed as an energy source can support the assembly reaction as readily as ATP. Even G-actin molecules linked to ADP rather than ATP can successfully assemble into microfilaments if their concentration is increased by about 20 to 50 times. These observations indicate that free energy released by ATP hydrolysis is not directly required for polymerization of microfilaments from actin. Instead, ATP binding, like the analogous binding of GTP during microtubule polymerization (see p. 421), probably induces a conformational change that facilitates assembly of actin microfilaments and pushes the equilibrium in the direction of polymerization.

The most recent observations of the role of ATP in microfilament assembly reveal that ATP breakdown lags significantly behind polymerization. As a result, the tips of polymerizing microfilaments contain actin subunits with ATP still attached. Just behind the tips hydrolysis occurs so that all or almost all the subunits

at points farther along the microfilaments are linked to ADP and phosphate instead of ATP. The ATP-linked subunits at the tips may help to stabilize the microfilaments, at least in the test tube, because ATP-actin subunits are about ten times slower to disassemble than ADP-actin subunits. The phosphate bound to subunits in which ATP has been hydrolyzed also contributes to microfilament stability; if this phosphate is released, microfilaments break down rapidly.

Under cellular conditions the actin-microfilament equilibrium is expected to tilt significantly in the direction of assembled microfilaments. The proportion of unassembled actin molecules in many cell types, however, is much larger than expected from the characteristics of the equilibrium reaction. This maintenance of unassembled subunits depends on a wide variety of *actin-binding proteins* that link either to G actin or to microfilaments and generally push the equilibrium toward the left (see below). Actin-binding proteins thus usually have activities opposite to those of microtubule-associated proteins (MAPs), which in general adjust the tubulin-microtubule equilibrium to the right and stabilize microtubules in the assembled form (see p. 425).

Microfilament Polarity and Treadmilling

Similarities between microfilament and microtubule assembly extend to the relative rates at which the two ends of a microfilament add or release actin subunits. As with microtubules the two ends of a microfilament have significantly different rates of assembly and disassembly. As a result, microfilaments have *polarity* and possess plus and minus ends with respect to rates of assembly. Depending on conditions, the plus end of microfilaments may assemble or disassemble from 10 to 40 times more rapidly than the minus end.

The relationship of polarity to microfilament assembly is essentially the same as in microtubules (Fig. 12-5 and p. 422). At all actin concentrations subunits constantly add and release at both ends of a microfilament. The concentration at which the equilibrium reaction is perfectly balanced at a microfilament end, so that neither net assembly nor disassembly proceeds, is termed the *critical concentration* for that end. At low actin concentrations (to the left side of the graph in Fig. 12-5) the rate of subunit release exceeds the rate of addition at both ends, and microfilaments become shorter in average length. As actin concentration reaches the critical level for the plus end (C_c^+ in Fig. 12-5), subunits add and release at equal rates from this end, while release still exceeds addition at the minus end. As actin concentration rises past this level, subunits add more rapidly than they release from the plus end, so that this end grows in length. However, release still exceeds addition at the minus end. At some point as actin concentration continues to increase, net addition

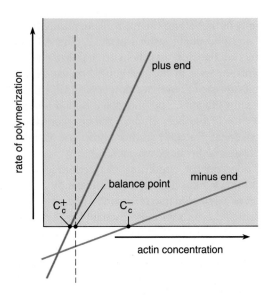

Figure 12-5 The relationship between actin concentration and the rates of polymerization at the plus and minus ends of microfilaments (see text).

at the plus end exactly equals net subunit loss at the minus end. At this balance point (see Fig. 12-5) microfilaments stay the same average length. Under these conditions actin molecules may *treadmill* through microfilaments by adding at the plus end, traveling through the microfilament as more subunits are added

at the plus end, and finally releasing at the minus end (Fig. 12-6). Once actin concentration increases to the critical level for the minus end (C_c^- in Fig. 12-5), net subunit loss ceases at this end and treadmilling stops. At concentrations above C_c^-, the rate of actin addition exceeds the rate of release at both ends, and microfilaments grow in average length in the solution.

Treadmilling, originally discovered by T. D. Pollard, A. Wegner, and their colleagues, has as yet been demonstrated only in the test tube and is not considered likely to occur or generate movements in living cells. Among the observations that make treadmilling seem unlikely as a basis for cellular movements is the fact that the direction in which particles or structures move over the surfaces of microfilaments, from the minus to the plus end, is opposite to the direction of treadmilling, which proceeds from the plus to the minus end. Treadmilling may be blocked inside cells by actin-binding proteins that cap and prevent assembly or disassembly at either or both ends of microfilaments (see below).

Microfilament polarity can be visualized by myosin decoration. If microfilaments previously decorated by myosin are placed in an actin solution at levels above the critical concentrations for the plus and minus ends, additional actin subunits, undecorated by myosin, clearly add more rapidly to one end than the other (Fig. 12-7). Myosin arrowheads on the decorated segment point toward the more slowly polymerizing mi-

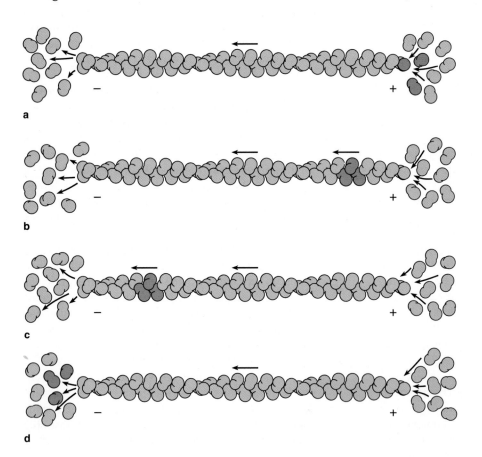

Figure 12-6 Movement of a group of actin subunits through a microfilament by treadmilling, in which microfilament subunits continuously add at the plus end and release at the minus end. A given group of subunits (in darker red) adds to the plus end in **(a)**; as further subunits add at the plus end, the added group is pushed toward the minus end **(b)**. The added group continues to treadmill through the microfilament **(c)** until its subunits release at the minus end **(d)**.

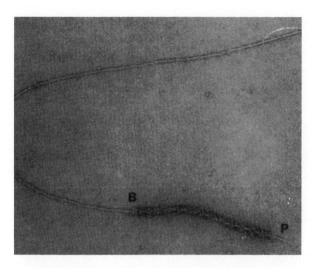

Figure 12-7 Microfilament polarity and myosin decoration. A microfilament marked by myosin arrowheads was placed in conditions favoring addition of actin subunits at both ends. The newly added actin appears smooth; the length added to the plus (**B**) end is much longer than that added to the minus (**P**) end. The arrowheads point to the minus end. (**P** and **B** refer to the *P*ointed and *B*arbed ends of the myosin arrowheads.) Courtesy of T. D. Pollard.

nus end; the plus end lies at the end marked by the tails of the arrowheads. Myosin decoration thus provides a method for directly determining the polarity of microfilaments.

Factors Modifying
Microfilament Polymerization

Actin-Binding Proteins   More than 60 known proteins bind to actin and adjust the actin-microfilament equilibrium in the direction of unassembled subunits. (Table 12-1 lists several of these proteins.) Some of them are widely distributed among eukaryotes; others occur only in a single cell type of one species. Most eukaryotic cells are estimated to contain at least ten actin-binding proteins at any time.

Although the various actin-binding proteins are numerous and diversified in structure, they adjust the actin-microfilament equilibrium by one or more of a relatively few mechanisms. In one mechanism a protein links to unassembled actin molecules and blocks their polymerization into microfilaments. The binding effectively removes actin molecules from the pool of subunits available for assembly, tipping the balance toward microfilament disassembly. Primary among the proteins acting in this way is *profilin* (see Table 12-1), which is widely distributed among eukaryotic cells.

The remaining mechanisms—including *capping*, *severing*, *crosslinking*, and *bundling*—affect assembled

microfilaments rather than actin subunits. Capping proteins bind to actin molecules at the end of existing microfilaments, most of them at the plus end, and prevent further assembly and microfilament growth. Many of these proteins cap very short microfilaments containing only a few subunits, effectively removing actin monomers from the pool available for polymerization. Several capping proteins, including *fragmin* (see Table 12-1), are so closely related to actin in size and structure that they cross react with anti-actin antibodies. These actinlike cappers may have appeared in evolution through mutations of duplicated actin genes.

Severing proteins break existing microfilaments into shorter lengths by binding to individual actin subunits. The binding, which may take place at any point between the plus and minus ends of a microfilament, alters the actin subunits so that they separate from their neighbors and introduce microfilament breaks. Severing proteins regulate microfilament length in the cytoplasm and also disassemble microfilament networks. Many proteins acting by this mechanism remain attached to the severed fragments and prevent their reassembly. They therefore function as capping as well as severing proteins (see Table 12-1).

Crosslinking proteins, which are typically elongated, fibrous molecules, form "crosswelds" where existing microfilaments cross over each other. The crosslinks, formed by proteins such as *fodrin* and *gelactin*, establish microfilament networks that make up parts of the cytoskeleton. In many cells these networks become extensive enough to convert the cytoplasmic regions containing them into a semisolid gel. A few crosslinking proteins also link microfilaments to the plasma membrane. Some MAPs linked to microtubules, such as MAP-2 and several tau proteins (see p. 425), also bind microfilaments, adding interlinks that tie microtubules and microfilaments into cytoskeletal networks.

Bundling proteins such as *villin* link microfilaments in parallel, side-by-side fashion. The microfilament bundles created by these modifiers form cytoskeletal structures such as the internal cores supporting fingerlike extensions of the cell surface called *microvilli* (see Fig. 13-13 and p. 512).

The activities of actin-binding proteins are readily reversible, allowing cells to push the actin-microfilament equilibrium quickly in either direction or to assemble or disassemble microfilament networks rapidly in local regions of the cytoplasm. Many actin-binding proteins are inactive at the low Ca^{2+} concentrations typical of the cytoplasm and become active only when Ca^{2+} levels are increased by release of this ion into the cytoplasm (see p. 208; actin-binding proteins controlled by Ca^{2+} are indicated in Table 12-1). Some proteins vary in type of activity depending on Ca^{2+} concentration. Villin, for example, acts as a microfilament bundler at low Ca^{2+} concentrations and severs

Table 12-1 Some Actin-binding Proteins

Protein	Molecular Weight	Calcium Sensitivity	Caps	Severs	Crosslinks	Bundles	Attaches to Plasma Membrane	Sources
Fragmin	42,000	+	+	+				Blood plasma
β-actinin	34,000–37,000	−	+					Muscle cells
Gelsolin	90,000–95,000	+	+	+				Macrophages, platelets, blood plasma, brain
Villin	95,000	+	+	+		+		Intestinal brush border
Brevin	90,000	+	+					Blood plasma
Severin	40,000	+	+	+				*Dictyostelium*
Filamin	250,000–270,000	−			+			Smooth muscle, macrophages
Spectrin	225,000–260,000	−	+		+		+	Many mammalian cell types
Fodrin	235,000–240,000	−			+			Widely distributed
α-actinin	100,000–105,000	+			+		+	Platelets, fibroblasts, muscle cells
Gelactin	23,000–38,000	−			+			Protozoans (soil amebae)
Fascin	58,000	−			+	+		Sea urchin oocytes
Vinculin	130,000	−			+	+	+	Muscle cells, fibroblasts
Talin	215,000	−			+	+	+	Smooth muscle
Fimbrin	68,000	−			+			Intestinal brush border
MAP-2	280,000	−			+ (To microtubules)			Brain
Tau	55,000–62,000	−			+ (To microtubules)			Brain
Profilin	12,000–15,000	−	Binds actin subunits, prevents assembly into microfilaments					Widely distributed

microfilaments at high Ca^{2+} levels. The effects of Ca^{2+} may be direct, through direct binding of the ion to the protein, or indirect, through binding of the ion to the Ca^{2+}-dependent control protein calmodulin (see Information Box 7-1). In some systems the Ca^{2+}/calmodulin complex activates a protein kinase, which adds phosphate groups to an actin-binding protein to adjust its activity.

The direct or indirect calcium sensitivity of actin-binding proteins adds the assembly of microfilaments and microfilament networks to the long list of cellular processes regulated by relatively small and practically instantaneous changes in Ca^{2+} concentration. The myosin crossbridging cycle, which produces microfilament sliding, is also regulated by rapid adjustments in Ca^{2+} concentration in muscle and many other cell types (see below). The effects of Ca^{2+} also link microfilament regulation to the $InsP_3$/DAG pathway triggered by cell surface receptors, in which Ca^{2+} is released as a second messenger (see p. 247). The effects of Ca^{2+} and the actin-binding proteins provide living cells with a delicately balanced control of the time, place, and form of microfilament organization in the cytoplasm.

Microfilament Poisons   The actin-microfilament equilibrium, like the tubulin-microtubule equilibrium, is the target of biological poisons. Compared to the lengthy list of antimicrotubule poisons (see p. 427), however, antimicrofilament poisons are relatively few in number. Only two types, *phalloidin* and the *cytochalasins* (Fig. 12-8), both produced by fungi, are presently known to be made by eukaryotes. Both poisons have been highly useful in microfilament research. A single type of microfilament poison, the *botulinum* toxins, is known to be made by bacteria of the *Clostridium* genus.

The cytochalasin family (from *cyto* = cell and *chalasis* = relaxation) is produced by *Helminthosporium* and other molds. Cytochalasins, like many microtubule poisons, are alkaloids—cyclic organic molecules that react as bases. They inhibit microfilament assembly and promote the breakdown of existing microfilaments. Twenty different cytochalasins, each with minor substitutions in chemical groups, have been isolated from fungal molds. Two of these, *cytochalasins B* and *D*, and a synthetic chemical derivative, *dihydrocytochalasin B*, are the most potent and widely used in microfilament research. (Fig. 12-8a shows the chemical structure of cytochalasin B.)

Cytochalasins inhibit microfilament polymerization by capping the plus end. The capping is so effective that one molecule of cytochalasin can completely block further growth of a microfilament. Cytochalasins may also sever microfilaments into shorter lengths and disrupt microfilament networks, converting gelled regions of the cytoplasm into a more liquid form.

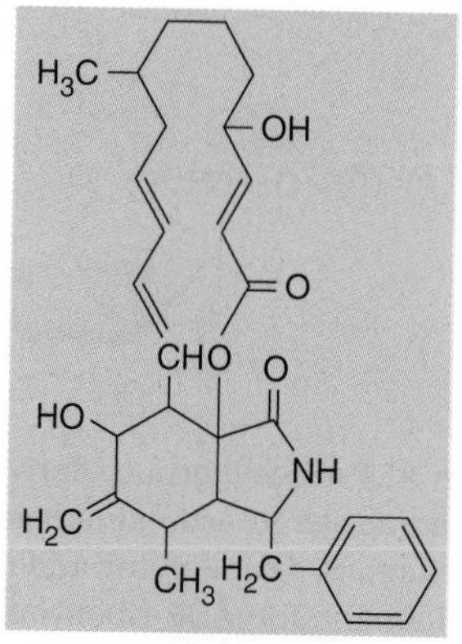

cytochalasin B

a

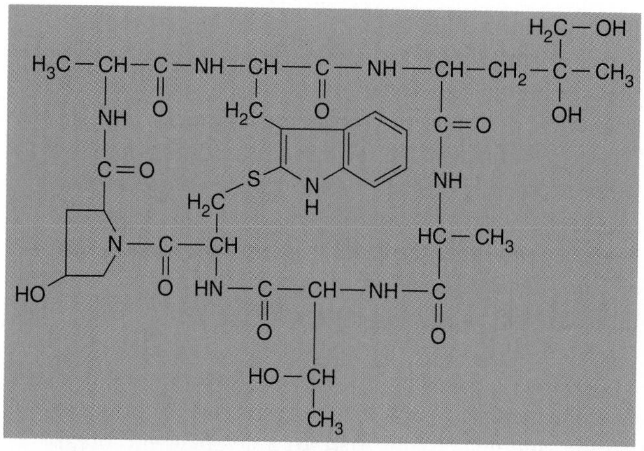

phalloidin

b

Figure 12-8   Microfilament poisons synthesized by fungi. **(a)** Cytochalasin B, which pushes the actin-microfilament equilibrium in the direction of monomers. The cytochalasins may also introduce breaks in microfilaments. **(b)** Phalloidin, which pushes the equilibrium in the direction of assembled microfilaments.

Phalloidin is produced by mushrooms, among them *Amanita phalloides*, the "death cap" mushroom. This microfilament poison is a cyclic peptide that binds to unassembled G-actin subunits and greatly increases their affinity for both the plus and minus ends of microfilaments. As a result, phalloidin has effects opposite to cytochalasins and pushes the actin-microfilament equilibrium far in the direction of assembly. The resulting microfilaments are highly stable, so much so that cells pretreated with phalloidin are insensitive to cytochalasins.

Both phalloidin and at least one cytochalasin (cytochalasin D) are strongly specific and combine only with actin or microfilaments. In doing so, both poisons are highly effective in arresting microfilament-based motility. These properties have made them indispensable as laboratory probes for microfilament-based motility inside living cells. If a motile activity is arrested

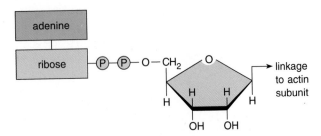

Figure 12-9 An ADP-ribosyl group, derived from NAD (see Fig. 9-5) and added to actin subunits in a reaction catalyzed by botulinum toxins. After addition to the plus end of a microfilament, the ADP-ribosylated subunit acts as a cap that blocks further polymerization.

or significantly inhibited by phalloidin or cytochalasin, it can reasonably be assumed as a first approximation that the motility is based on microfilaments. Because phalloidin promotes microfilament assembly and stabilizes microfilaments in the assembled form, it has also made it possible to identify and isolate microfilaments in groups such as plants in which identification and analysis were previously difficult or impossible.

Botulinum toxins are polypeptides that catalyze attachment of an *ADP-ribosyl* group, derived from NAD (see p. 542) to actin subunits (Fig. 12-9). The ADP-ribosylated actin subunits bind to the plus ends of microfilaments as caps that prevent further polymerization and eventually lead to microfilament breakdown. The botulinum toxins, which also have other lethal effects (among them blockage of acetylcholine secretion; see p. 223), are among the most poisonous substances known. They are responsible for about 15 deaths each year in the United States, usually through ingestion of poorly preserved home-canned foods in which *Clostridium* bacteria have become established.

Parallels in the Biochemistry of Microfilaments and Microtubules

There are many parallels in the biochemistry of microfilaments and microtubules. Both structures are assembled from a protein subunit, microtubules from tubulin and microfilaments from actin; in both cases the assembled and disassembled forms of the protein are balanced in an equilibrium that can be tipped in either direction. The balance is regulated and controlled inside living cells by a battery of regulatory proteins—MAPs in microtubules, and the long list of actin-binding proteins that cap, sever, crosslink, and bundle microfilaments. Many regulatory proteins in both systems are directly or indirectly activated by Ca^{2+}, providing a link between microtubule and microfilament activity and the fundamental cellular regulatory mechanisms based on calcium ions. The two motile elements are also the targets of natural poisons, most notably colchicine and cytochalasins, which disassemble microtubule and

microfilament polymers, respectively, and taxol and phalloidin, which push their respective microtubule and microfilament targets in the direction of stable polymers. As the next section shows, many parallels also exist between the crossbridging mechanisms that generate microfilament- and microtubule-based motility.

MYOSIN AND MICROFILAMENT-BASED MOTILITY

The power for movements based on microfilament sliding is supplied by myosin crossbridges, which turn through an attach-pull-release cycle to produce the sliding. The activity of myosin crossbridges has a parallel in microtubule-based motility, in which dynein, kinesin, or other crossbridging proteins provide "motors" that slide structures along microtubules.

Myosin Structure

The myosins are a family of molecules that occur in two primary forms, *myosin I* and *myosin II*. Myosin I is involved in several types of membrane-associated movement in nonmuscle cells, probably including ameboid motion and the movement of membrane-bound organelles in cytoplasmic streaming. In some cells myosin I serves as a linker molecule tying microfilaments into cytoskeletal structures. Myosin II is responsible for the microfilament sliding that takes place in muscle cells as well as several nonmuscle cell movements (including cytoplasmic division and the capping of substances bound to the plasma membrane), and possibly ameboid motion in coordination with myosin I in some organisms (see below).

Both myosin types are built up from one or two large polypeptide subunits, the *myosin heavy chains*, and one or more smaller peptide subunits, the *myosin light chains*. The heavy chains typically fold into a globular *head* unit. The light chains are associated with the head units. Myosin I consists of a single heavy chain polypeptide in association with one or more light chains; myosin II is a double-headed molecule containing two heavy chains, each in combination with two light chains.

The head segments of both myosin types contain similar amino acid sequences, with one site that binds actin and another that binds and hydrolyzes ATP (Fig. 12-10). The hydrolytic activity of myosins is unique among cellular ATPases because it is activated by actin and strongly inhibited by Mg^{2+}. In both myosin types the head units are responsible for the crossbridging action that produces microfilament sliding.

The primary differences between the heavy chain polypeptides making up myosins I and II appear to be in the protein domains that extend from the heads (see Fig. 12-10). All myosin I heavy chains so far investigated have a relatively short 180- to 250-amino-acid

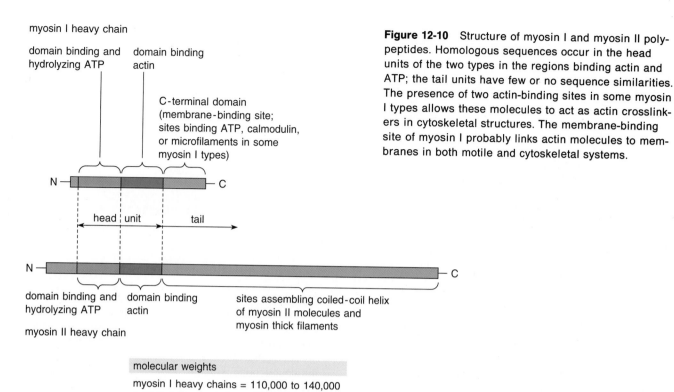

myosin I heavy chain

domain binding and hydrolyzing ATP

domain binding actin

C-terminal domain (membrane-binding site; sites binding ATP, calmodulin, or microfilaments in some myosin I types)

N —————————— C

head ┆ unit   tail

N ————————————————————————— C

domain binding and hydrolyzing ATP

domain binding actin

sites assembling coiled-coil helix of myosin II molecules and myosin thick filaments

myosin II heavy chain

molecular weights

myosin I heavy chains = 110,000 to 140,000
myosin II heavy chains = 175,000 to 240,000

Figure 12-10 Structure of myosin I and myosin II polypeptides. Homologous sequences occur in the head units of the two types in the regions binding actin and ATP; the tail units have few or no sequence similarities. The presence of two actin-binding sites in some myosin I types allows these molecules to act as actin crosslinkers in cytoskeletal structures. The membrane-binding site of myosin I probably links actin molecules to membranes in both motile and cytoskeletal systems.

sequence extending from the head unit at the C-terminal end. This small domain contains a membrane-binding site and, in some myosin I types, an additional site capable of binding and hydrolyzing ATP. A few myosin I types also have calmodulin-binding sites or an additional microfilament-binding site. The C-terminal extension has little or no alpha-helical structure, so that it does not constitute a fibrous tail.

In contrast, the heavy chain polypeptide of myosin II has an extended fibrous tail. The tail contains binding sites that assemble these heavy chains by twos into the double structures typical of this myosin type and into superstructures containing several to many myosin molecules.

Relatively little is known about the structure and occurrence of myosin I. This myosin, originally discovered by Pollard and E. D. Korn in *Acanthamoeba*, has since been detected in such diverse species as *Dictyostelium, Drosophila,* and vertebrates. Molecules suspected to be myosin I types have also been identified in algae and higher plants. In vertebrates it occurs in the microvilli of intestinal cells (see p. 512), where it may serve a primarily cytoskeletal role by linking microfilament bundles to the plasma membrane. At least some species, including *Acanthamoeba* and *Dictyostelium,* have several distinct genes encoding myosin I molecules.

Myosin II, which is the most common and best-known form of myosin in higher animals, is a double-headed structure with a long, fiberlike tail (Fig. 12-11) and a total molecular weight of about 460,000. The bulk of the structure is in the two identical heavy chain polypeptides, each with a globular head and a long tail. In different myosin II molecules the heavy chains vary from 175,000 to 240,000 daltons, with about 90,000 daltons of this mass concentrated in the head unit. The light chains are relatively small subunits forming parts of the heads in the assembled molecule.

The entire amino acid sequence of a myosin II heavy chain contains nearly 2000 amino acids. About 850 of these fold into the head, which is about 6 nm long and 4 nm in diameter. The remainder of the amino acid chain forms the tail, which, at about 150 nm, comprises most of the length of the polypeptide. The tail of a myosin heavy chain winds almost entirely into an alpha helix. Only a short 23 amino acid segment at the tip of the tail is globular.

Two heavy chains link together to form the bulk of a myosin II molecule. In the tail region the alpha-helical segments of the two chains line up side by side and twist around each other into a double helix or *coiled coil* (see Information Box 2-4).

One of the two light chains in each head unit of a myosin II molecule, called the *essential light chain,* can be removed only by relatively drastic treatments, such as exposure to a strong alkali, that denature myosin and destroy its function. The second light chain, the *regulatory light chain,* regulates the crossbridging activity of myosin in some systems, as in smooth muscle (see below). A regulatory light chain is easily removed from the remainder of the myosin molecule by mild treatment, such as adjustments in salt concentration.

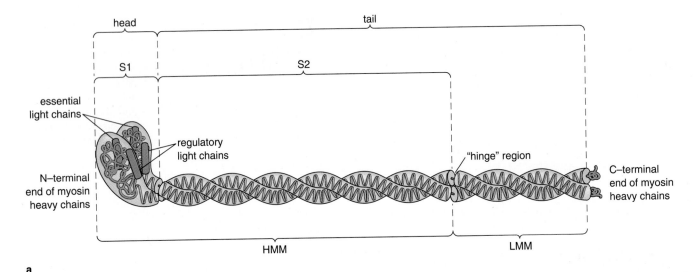

head | tail

S1 | S2

essential light chains

regulatory light chains

"hinge" region

N–terminal end of myosin heavy chains

C–terminal end of myosin heavy chains

HMM | LMM

a

b

Figure 12-11 Myosin II structure. **(a)** Two myosin heavy chains wind into a coiled-coil structure to form the tail and into a less regular conformation to form the double head structure. One essential and one regulatory myosin light chain are also present in each subunit of the double head structure. Little is known about the exact location and conformation of the light chains, except that the N-terminal ends of the regulatory light chains are located near the head-tail junction. HMM (heavy meromyosin), LMM (light meromyosin), S1, and S2 are fragments produced by protease digestion of myosin molecules. **(b)** Myosin molecules isolated from chick skeletal muscle and prepared for electron microscopy by freeze-drying and shadowing. The tail and double head structure (arrows) are visible. Courtesy of T. Wallimann, from *Eur. J. Cell Biol.* 30:177 (1983).

The light chains attach to the head units in positions near the junction between the heads and the tail, with the regulatory light chains nearest the point of juncture.

Myosin II molecules can be split into defined subparts by gentle exposure to protein-digesting enzymes such as trypsin. Trypsin digestion splits the molecule at a site about one-third of the way along the double tail from the C-terminal end (see Fig. 12-11*a*). The fragment containing the heads, called *heavy meromyosin (HMM)*, retains its ATPase activity and is soluble at pH ranges characteristic of living cells. The remaining tail segment, called *light meromyosin (LMM)*, is insoluble at physiological pH. The fact that the molecule splits into two parts at a particular site suggests that the alpha helices of the tail may be unwound to some extent in this region, producing a flexible site. Some investigators have proposed that this site serves as a "hinge" that may facilitate the crossbridging cycle between actin and myosin (see below).

More extensive digestion with trypsin or other proteinases splits the HMM fragment at the point where the heads attach to the tail, suggesting that another flexible hinge may exist in this region. The break produces two fragments, *S1* and *S2*: S1 fragments are the individual head units, and S2 fragments are the portion of the tail between the heads and the breakpoint between HMM and LMM. Isolation and purification of S1 fragments revealed that the ATPase activity of myosin is associated with these fragments and therefore the head structure, and that each head also contains a site capable of binding actin. Since there are two head units, each myosin II molecule has two ATPase and two actin-binding sites. The ability of S1 or HMM segments to bind actin forms the basis for the myosin decoration technique (see Fig. 12-3).

Myosin II molecules from different species and from different cell types, such as striated, smooth, and nonmuscle cells within the same species, differ in properties such as amino acid sequence of both heavy and light chains, activity as an ATPase, and solubility. In higher vertebrates at least 11 different myosin II heavy chain variants occur in specific cell types, as in fast-contracting and more slowly contracting skeletal muscle, in the atrium and ventricles of the heart, in smooth muscle, and in nonmuscle cells. Myosin light chains also occur in distinct types in most of the same tissues.

Differences are further noted in the heavy and light chains of embryos and adults. The functional significance of these differences in myosin II types remains unknown. Smaller numbers of myosin II genes occur in lower eukaryotes; the slime mold *Dictyostelium*, for example, has only a single myosin II gene.

Like actins and many other families of cellular proteins, the heavy and light chain polypeptides of myosin IIs contain some sequence segments that vary widely among different members of the myosin family, and other segments that are highly conserved from one myosin to the next. The most highly conserved segments of myosin heavy chains are in the head units, including the domain that binds and breaks down ATP. Sequences in the head units, as noted, are also similar or identical between myosin I and II. About 36% of the amino acid sequence in the head unit of the myosin I molecule of a protozoan, *Acanthamoeba*, for example, is identical with the myosin II of rat striated muscle cells.

Myosin II Superstructures: Thick Filaments

In many cellular systems both muscle and nonmuscle myosin II molecules assemble into superstructures called *thick filaments*. The assembly proceeds as an equilibrium reaction, in a manner similar to the assembly of microfilaments. In some cell types, however, the myosin equilibrium includes an extra step in which myosin II molecules assume a folded form incapable of assembly:

$$\begin{array}{ccc} \text{folded} & \text{extended} & \\ \text{myosin II} & \text{myosin II} & \text{thick} \\ \text{(assembly} \rightleftharpoons & \text{(can} \rightleftharpoons & \text{filaments} \quad (12\text{-}2) \\ \text{blocked)} & \text{assemble)} & \end{array}$$

In the folded form the tail bends over and covers the head groups. The folded form appears to occur only in smooth muscle and in nonmuscle cells. Striated muscle cells include only the portion of the reaction between extended myosin II and assembled thick filaments.

Conversion between folded myosin II, extended myosin II, and thick filaments is regulated by addition and removal of phosphate groups. Conversion between folded and unfolded myosin II is regulated by phosphorylation of the light chains. Addition of phosphate groups to the light chains pushes the folding-unfolding equilibrium to the right, toward the extended form. Removal of phosphates folds myosin II and pushes the equilibrium to the left. The light chain phosphorylation responsible for this equilibrium adjustment is regulated indirectly by Ca^{2+} concentration, through a protein kinase dependent on Ca^{2+}/calmodulin for its activity.

The conversion between extended myosin II and thick filaments is regulated by phosphorylation of the heavy chains and works in the opposite sense to the folding-unfolding portion of the equilibrium. Addition of phosphate groups inhibits assembly of extended myosin II molecules into thick filaments and pushes the equilibrium to the left; removal promotes assembly. In at least some organisms the protein kinase adding phosphate groups to myosin II heavy chains is inhibited by Ca^{2+}, so that the overall effect of increased Ca^{2+} concentration is to push both steps of the equilibrium in the direction of assembled thick filaments.

The folding-unfolding mechanism of the equilibrium is important in regulating microfilament activity in the cells in which it occurs. Conversion of myosin II to the folded form effectively removes it from interaction with microfilaments and thereby inhibits motility due to microfilament sliding.

The Myosin Crossbridge Cycle

Much of the evidence that myosin crossbridges move through an attach-pull-release cycle to slide structures along microfilaments comes from X-ray diffraction and electron microscope studies of striated muscle by H. E. Huxley and J. Hanson, and A. F. Huxley and R. Niedergerke. In 1954 these investigators proposed that microfilament sliding is responsible for muscle contraction. In 1969 H. E. Huxley suggested the basic principle of a myosin crossbridging cycle as the power source for the sliding. (Equivalent models for dynein-microtubule crossbridging cycles were based on hypotheses developed through the research on striated muscle.) The basic model advanced by H. E. Huxley was fleshed out with biochemical details by R. W. Lymn, E. W. Taylor, and others, who based their conclusions on the interactions of actin and myosin both inside muscle cells and in the test tube.

X-ray diffraction, electron microscopy, and biochemical observations established several important characteristics of the myosin-actin interaction, which proved critical for the crossbridging model:

1. Myosin crossbridges in striated muscle (see Figs. 12-19 and 12-20) extend toward microfilaments at angles between 45° and 90°.

2. Myosin unattached to actin can bind and hydrolyze ATP. However, the products of the reaction—ADP and inorganic phosphate—cannot be released, nor the reaction fully completed, until myosin binds to actin. Although ATP is hydrolyzed, most of the energy from hydrolysis is retained in the myosin molecule until the products are released.

3. Binding between myosin and actin promotes release of the products of ATP breakdown and makes available a large increment of free energy.

4. Once the products of ATP hydrolysis are released, myosin crossbridges do not release readily from actin until bound to fresh ATP.

Do the Two Heads of the Myosin Molecule Function Independently or Cooperatively?

Yoshie Harada and Toshio Yanagida

YOSHIE HARADA is a research associate of the Department of Biophysical Engineering, Osaka University, Osaka, Japan. She graduated from the Faculty of Science of Ibaraki University in 1982 and received her D. Eng. in biophysics from the Faculty of Engineering Science of Osaka University in 1988. TOSHIO YANAGIDA is a professor in the Department of Biophysical Engineering at Osaka University. He graduated from the Faculty of Engineering Science of Osaka University in 1969 and received his D. Eng. in biophysics there in 1975. He was the recipient of the Osaka Science Award in 1989 and the Tsukahara Award in 1990.

The myosin molecule consists of two heads, each of which contains an enzymatic active site and an actin-binding site. The fundamental problem of whether the two heads function independently or cooperatively in the generation of sliding force has been studied by methods using suspensions of purified proteins,[1] actinomyosin threads,[2] precipitated molecules,[3] and chemical modification of muscle fibers.[4] No clear conclusion was reached about the action of myosin heads from these experiments.

We approached this question using a new *in vitro* motility assay. Several years ago, we demonstrated that single actin filaments labeled with a complex of fluorescent dye and phalloidin, which stabilizes the filament structure of actin and is strongly fluorescent, can be resolved and clearly observed continuously under a fluorescent microscope.[5] This allowed us to follow the sliding movement of single actin filaments interacting with myosin bound to a substrate *in vitro*.[6] This *in vitro* motility assay is very simple and reproducible, and it has since been widely used for studies on the mechanism of movement in muscle and nonmuscle cells. In the further application of this technique, we developed a method to hold and manipulate a single actin filament by glass microneedles. The needles can also be used to measure sliding forces generated by single actin filaments.[7,8]

Using these techniques, we measured the sliding velocity of the actin filament on one-headed myosin molecules[9] and the force generated as the actin filament interacts with myosin subfragment-1 *in vitro*.[7] In these experiments, actin and myosin were obtained from rabbit skeletal muscle and purified by conventional methods.[10] Actin filaments at a concentration equivalent to 2.5 μM in actin monomers were labeled with phalloidin-tetramethylrhodamine by incubating them overnight at 4°C in a solution containing 5 μM fluorescent phalloidin, 100 mM KCl, and 10 mM HEPES buffer at pH7.0.[5] The fluorescently labeled actin filaments were observed under an inverted microscope equipped with epifluorescence optics and illuminated by a mercury arc lamp. The fluorescent images obtained were videotaped with a high-sensitivity camera.[5] Using this equipment, the actin filaments could be clearly and continuously observed on a TV monitor. In order to minimize photo-bleaching and denaturation of proteins by the strong illumination, oxygen was removed from the assay solution by adding to it glucose-glucoseoxidase, catalase, and a reducing agent.[10]

Single-headed myosin was obtained by digestion of molecules with papain and purified.[2] The purity was checked; contamination by double-headed myosin molecules was found to be only 1%. Short myosin filaments, formed from the purified preparation of single-headed myosin dissolved in a high ionic solution, were coated onto a glass microscope slide. After washing off unbound myosin filaments, fluorescently labeled actin filaments at a concentration equivalent to about 10 nM in monomers were added to the slide in the presence of ATP. Smooth and fast movement over the glass slide was observed at a velocity of 6 μm/s at 23°C, almost the same as that produced by double-headed myosin (7 μm/s). Although the preparation contained less than 2% molar ratio of double-headed myosin, we were able to confirm that the contaminating double-headed myosin was not responsible for the movement. This was done by observing the movement of actin filaments along hybrid myosin filaments that contained single- and double-headed myosin mixed in various molar ratios.

Another problem was the possibility that two adjacent single myosin heads might interact cooperatively to act as two-headed myosin molecules rather than acting individually and singly, because the density of heads on the thick filaments was very high. This possibility was excluded by observing the rate at which actin filaments moved over myosin molecules from which most of the heads had been removed. This was done in order to reduce the spatial concentration of single heads. The actin filaments moved just as fast along myosin filaments from which >90% of the heads had been thinned as before the thinning. Furthermore, the movement of thin filaments containing the tropomyosin-troponin complex over the single-headed myosin was found to be regulated by calcium ions, although the velocity curve was slightly different from that obtained when double-headed myosin was used. These results demonstrated collectively that cooperative interaction between two myosin heads is not essential to the development of the sliding movement.[9]

These tests did not tell us whether single-headed myosin can produce significant sliding force. For this purpose, we used a new technique for holding and manipulating single actin filaments. One end of the actin filament was caught and held by a very fine glass microneedle whose surface had been previously coated by N-ethylmaleimide-treated myosin to increase its affinity for actin.[7] The other end of the actin filament was brought into contact with the myosin-coated surface of a glass slide. In the presence of ATP, the actin filament moved and bent the needle. The force due to interaction between actin and myosin was determined by measuring the degree of bending of the needle. The stiffness of needles used was 5 to 20 pN/μm.

The method for producing myosin coats on glass slides had to be altered for this experiment. For previous observations of the movement of actin filaments at zero load, we had coated the surface with myosin molecules assembled into short filaments. But this preparation was not suitable for the force measurements because the density of myosin heads on the surface was not homogeneous and consequently the force varied. Therefore, we coated the glass surface, which had been treated with silicone, with myosin in monomeric form. Electron microscope observation showed that the surface was homogeneously coated with myosin heads.[10] It was found that monomeric myosin applied in this way could move actin filaments as fast as myosin filaments.[11]

Using the monomer-coated slides, the force generated by a single actin filament 1 μm long interacting with the myosin-coated surface was about 30 piconewtons (Harada et al. unpublished data). The number of myosin heads in the vicinity of an actin filament of this length was estimated to be about 100 from the density of myosin heads on the surface. Since the orientation of myosin heads was random and the heads bound to the surface would be able to interact with only one side of the actin filament, the number of heads that could participate in the force generation was probably one fourth of the 100 heads, i.e., about 25 heads per μm of actin filament. Thus, the force per head was estimated to be roughly 1 pN, which is comparable to the force exerted by double-headed myosin molecules in muscle.

To minimize damage to the myosin heads during the initial isolation and purification techniques, we digested myosin molecules with proteinase for only a brief period. This reduced the chance that the proteinase might introduce breaks within the heads. This was important to our results because myosin heads with breaks from digestion can move the actin filament as fast as normal ones, but can generate little force.

In conclusion, the results show that single-headed myosin subfragment-1 is sufficient to produce the movement of actin filaments at zero load.[11] The force measurements demonstrated that S-1 myosin fragments bound to the silicone-treated glass surface produced a force as large as intact, double-headed myosin. Therefore, cooperative interaction between the two heads of intact myosin is essential for inducing neither the sliding movement of actin filaments nor the force. But it is not yet known why myosin has double-headed structure.

References

[1] Tonomura, Y. *Muscle proteins, muscle contraction and cation transport.* Tokyo: University of Tokyo Press (1972).

[2] Cooke, R., and Franks, K. E. *J. Mol. Biol* 120:36 (1978).

[3] Marggossian, S. S., and Lowey, S. *J. Mol. Biol.* 74:312 (1971).

[4] Chaen, S.; Shimada, M; and Sugi, H. *J. Biol. Chem.* 261:13632 (1986).

[5] Yanagida, T.; Nakase, M.; Nishiyama, K.; and Oosaswa, F. *Nature* 307:58 (1984).

[6] Kron, S. J., and Spudich, J. A. *Proc. Nat'l Acad. Sci.* 83:6272 (1986).

[7] Kishino, A., and Yanagida, T. *Nature* 334:74 (1988).

[8] Ishijima, A.; Doi, T.; Sakurada, K.; and Yanagida, T. *Nature* 352:301 (1991).

[9] Harada, Y.; Noguchi, A.; Kishino, A.; and Yanagida, T. *Nature* 326:805 (1987).

[10] Harada, Y.; Sakurada, K.; Aoki, T.; Thomas, D. D.; and Yanagida, T. *J. Mol. Biol.* 216:49 (1990).

[11] Toyoshima, Y., et al. *Nature* 328:536 (1987).

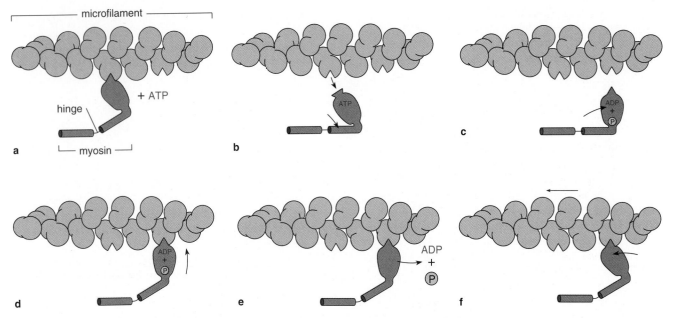

Figure 12-12 A simplified version of the crossbridging cycle, which hydrolyzes ATP to power microfilament sliding. The crossbridges are formed by the head units of myosin molecules. **(a)** As the cycle begins, the myosin crossbridge is assumed to be bent at a 45° angle and tightly linked to an actin subunit. The products of ATP breakdown have been released, and the ATP-binding site on the myosin crossbridge is active. **(b)** Binding an ATP molecule changes the conformation of the actin-binding site so that it no longer fits the actin subunit. The crossbridge releases from actin, and its bound ATP is hydrolyzed. The products of the hydrolysis, ADP + P_i, remain bound to the crossbridge. **(c)** ATP breakdown causes another conformational change in the crossbridge, shifting it to the 90° position and reactivating the actin-binding site at its tip. Attachment to an actin subunit in the microfilament **(d)** releases the products of ATP breakdown **(e)**. **(f)** The release triggers a third conformational change that swivels the crossbridge forcefully from the 90° to the 45° position. The swiveling movement pulls the attached microfilament an equivalent distance. The crossbridge is now ready to bind another ATP molecule and repeat the cycle.

The last of these observations was developed through experiments with muscles in rigor. In muscles deprived of ATP, myosin crossbridges become locked to actin microfilaments. (This is the cause of *rigor mortis*, the stiffness of the limbs noted in animals after death.) Adding fresh ATP releases the crossbridges and restores flexibility to the muscles. Whether the crossbridges in the rigor state are locked at 90° or 45°, or at various angles between these extremes, is now a matter of considerable debate. The slightly curved shape of crossbridges in muscle (see Fig. 12-20) adds to the difficulties in measuring their angles. For the purposes of this discussion we will assume that the locked, rigor position is 45°.

Figure 12-12, based on the Huxley–Lymn–Taylor model, combines these observations and shows in a simplified way how myosin crossbridges may use ATP energy to produce active microfilament sliding. The hypothetical cycle begins at the close of a previous cycle, with the myosin crossbridge bent at a 45° angle and tightly linked to an actin subunit (Fig. 12-12a). The products of ATP breakdown have been released, and

the ATP-binding site on the myosin crossbridge is active. Binding an ATP molecule (Fig. 12-12b) changes the conformation of the actin-binding site on the crossbridge, greatly reducing its affinity for the actin subunit, and the crossbridge releases. As the crossbridge separates from actin, its bound ATP is hydrolyzed, but the products of the hydrolysis—ADP + P_i—remain bound to the crossbridge. ATP breakdown causes another conformational change in the crossbridge, shifting it to the 90° position and reactivating the actin-binding site at its tip (Fig. 12-12c). In this condition the crossbridge may be considered as a compressed molecular spring storing much of the energy released by ATP breakdown. Attachment to an actin subunit in the microfilament (Fig. 12-12d) releases the products of ATP breakdown (Fig. 12-12e) and triggers release of the crossbridge from its 90° position. The crossbridge then swivels forcefully to the 45° position, pulling the attached microfilament an equivalent distance (Fig. 12-12f). The crossbridge is now ready to bind another ATP molecule and repeat the cycle. Movement of a crossbridge through the entire attach-pull-release sequence hydrolyzes one molecule of ATP and pulls the adjacent microfilament through a distance of about 10 to 20 nm.

Note in Figure 12-12e that as the myosin crossbridge makes its attachment to the adjacent microfilament during the crossbridge cycle, the head unit and part of the tail are shown as bending outward from a hinge point located partway along the tail. Whether a hinge actually exists at this point, or whether the entire myosin tail bends through a smooth arc to accommodate movement of the head toward actin, remains a subject of intense debate. Other areas of current uncertainty and debate center around the parts of the crossbridging cycle in which actin and myosin actually attach and release, how extensive the movements of the crossbridges are (the arc through which the head moves may be considerably less than 45°), the segments of the head undergoing conformational changes, and whether one or both heads of a myosin molecule are involved in a single cycle.

Whatever the uncertainties about the crossbridging cycle, the ability of myosin to slide structures along microfilaments was confirmed recently in an especially graphic fashion by the experiments of M. P. Sheetz and J. A. Spudich. These investigators obtained microfilament bundles by breaking open cells of the alga *Nitella.* The microfilament bundles in this alga form bands around the cell in a layer near the cell border. The bundles are responsible for cytoplasmic streaming, which circulates chloroplasts through the cytoplasm. Microscopic plastic beads coated with the HMM fragments of rabbit skeletal myosin II molecules were added to the microfilament bundles, which were laid out on a microscope slide and washed free of cytoplasm. When supplied with ATP, the beads "walked" actively along the microfilament bundles at speeds comparable to the rate at which microfilaments slide in skeletal muscle. (The Experimental Process essay by Sheetz on p. 442 discusses his experiments with myosin-coated beads, and describes equivalent experiments with microtubules and the microtubule motor, kinesin.)

Spudich and Sheetz and another group, T. Yanagida and T. Harada and their coworkers, repeated the same approach using only S1 head units (see Harada's essay on p. 464). The head units, by themselves, were able to slide beads along microfilaments. This finding indicates that the crossbridging cycle—including ATP binding and breakdown, binding and release between actin and myosin, and the conformational changes accomplishing movement—takes place in the head units. The finding also makes it seem unlikely that portions of the myosin tail, such as the hinge region, are necessary for generation of crossbridging power. Because single head units can produce sliding motion, the results indicate further that the crossbridging cycle does not necessarily require a coordinated "walking" interaction between both heads of a myosin molecule. Observa-

tions with S1-coated beads do not eliminate the possibility that cooperation between head units may occur in myosin molecules in living cells, however. Such cooperation might increase the efficiency of the crossbridging cycle.

Estimates of the force generated by a single myosin crossbridge have been made in striated muscle cells, in which the measured force can be related to a specific number of crossbridges arranged in a regular pattern (see below). Such estimates indicate that a single crossbridge generates a force equivalent to one piconewton. Yanagida and Harada obtained the same result (see Harada's essay), indicating that myosin transforms chemical energy into mechanical energy with an efficiency of about 10%.

Regulation of the Crossbridging Cycle

It is obvious from the crossbridging mechanism outlined in Figure 12-12 that if the cycle operated without controls, only two states would be expected: active microfilament sliding when ATP supplies are adequate and rigor when ATP supplies are low. However, we know from personal experience with the contractile mechanisms of our voluntary muscles that the mechanism is under delicate and precise control and can be readily modulated or switched on and off. What is the biochemical basis for this control?

A series of investigations extending over many decades has revealed that two major pathways in living systems regulate the myosin crossbridging cycle. *Actin-linked regulation* involves proteins that bind to actin and form regulatory units on the surfaces of microfilaments. *Myosin-linked regulation* is based on regulatory myosin light chains. Both systems are ultimately controlled by changes in cytoplasmic Ca^{2+} concentrations.

Actin-linked Regulation in Striated Muscle Regulation linked to actin is best understood in striated muscle, including both skeletal and cardiac muscle. This form of actin-linked regulation depends on the activity of two proteins, *tropomyosin* and *troponin*, which bind to microfilaments in striated muscle. The two proteins are believed to control the crossbridging cycle by undergoing Ca^{2+}-dependent conformational changes. The changes alternately block and expose the sites on actin bound by myosin during the crossbridging cycle.

Tropomyosin is a family of closely related, fibrous molecules built up from two alpha-helical chains wound into a coiled-coil structure similar to the myosin tail (Fig. 12-13a). Although its name suggests that tropomyosin is a form of myosin, the two molecular types are unrelated except for sharing the overall sequence structure underlying the coiled-coil pigtail, in which amino acids occur in repeating seven-unit groups (see Information Box 2-4). Distinct forms of tropomyosin occur in striated muscle, smooth muscle, and nonmuscle cells. The differences in amino acid sequence charac-

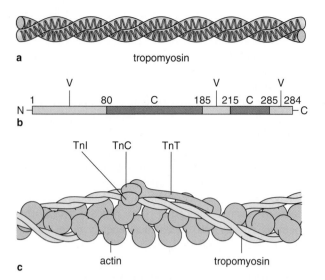

a tropomyosin

b

c

TnI TnC TnT

actin   tropomyosin

Figure 12-13 Tropomyosin and its association with tropomyosin. **(a)** The coiled-coil structure of tropomyosin. **(b)** Sequence structure of striated and smooth forms of tropomyosin, showing the locations in which the amino acid sequence is conserved (C) and variable (V) among different tropomyosin types. Nonmuscle tropomyosins are shorter in length but have conserved segments at equivalent distances from the C-terminal end. **(c)** The arrangement of tropomyosin and the TnC, TnI, and TnT subunits of troponin on striated muscle microfilaments.

terizing the various tropomyosin types are confined to several variable segments within the polypeptide chains. Other regions are practically identical in sequence (Fig. 12-13*b*).

The two polypeptide chains of a tropomyosin molecule, depending on the cell type, may be identical or different. In striated muscle the two chains are different. Each tropomyosin molecule of striated muscle contains one α- and one β-chain, both with 284 amino acids in distinct but similar sequences.

A tropomyosin molecule is slightly more than 40 nm in length, just long enough to stretch over a row of seven actin subunits in a microfilament. Tropomyosin molecules extend end to end along the microfilament in two chains, one on either side near the "grooves" in the actin double helix (Fig. 12-13*c*).

Troponin is built up from three polypeptide subunits called *troponin C (TnC), troponin T (TnT)*, and *troponin I (TnI*; see Fig. 12-13*c*). TnC and TnI are globular polypeptides; TnT is an elongated, fibrous molecule about one-third the length of tropomyosin. TnC, which closely resembles calmodulin in structure and function, provides the direct link between Ca^{2+} concentration and control of microfilament sliding by the actin-linked mechanism. (Calmodulin, in fact, can be substituted for TnC with little effect on troponin function.) Troponin occurs in equal numbers with tropomyosin molecules in striated muscle microfilaments.

Elegant work in the S. Ebashi and H. E. Huxley laboratories revealed how troponin and tropomyosin probably interact to regulate the myosin crossbridge cycle in muscle cells. The first clues to the role of these proteins in actin-linked regulation were discovered by Ebashi and his coworkers, who noted that Ca^{2+} is required for the interaction of actin and myosin isolated from striated muscle cells. At the very low Ca^{2+} concentrations typical of resting muscle cells, the actin and myosin of muscle do not interact even if ATP is present. At higher Ca^{2+} concentrations the crossbridge cycle proceeds if ATP is available.

Ebashi and his coworkers later found that Ca^{2+} exerts its regulatory effects in striated muscle through troponin and tropomyosin. Muscle actin and myosin stripped of troponin and tropomyosin, they discovered, react together without dependence on Ca^{2+} concentration. Adding troponin and tropomyosin proteins again establishes Ca^{2+} sensitivity. On this basis Ebashi and his colleagues proposed that Ca^{2+} regulates muscle contraction through troponin and tropomyosin. When combined with Ca^{2+}, troponin and tropomyosin turn muscle contraction on; when free of Ca^{2+}, the control proteins switch the system off.

The probable molecular basis for the activities of troponin and tropomyosin in regulating the crossbridging cycle was worked out by Huxley and his colleagues. X-ray diffraction indicated that tropomyosin molecules lie in or near the surface grooves in the F-actin double helix (see Fig. 12-13*c*). During the initiation of microfilament sliding in muscle, when the mechanism is switched from "off" to "on," the Huxley group noted that the X-ray patterns ascribed to tropomyosin change slightly. The change indicated that tropomyosin moves slightly closer to the grooves, being displaced through an arc of 10° to 15° around the actin chain.

From these observations Huxley, Ebashi, M. Greaser, and others proposed that when the Ca^{2+} concentration is at the resting level, troponin is tightly bound to tropomyosin. The binding holds the tropomyosin molecule slightly outside the microfilament grooves, in a position in which it covers the actin sites bound by myosin crossbridges (Fig. 12-14*a*). As Ca^{2+} concentration rises, troponin C binds Ca^{2+}, causing a conformational change that reduces its affinity for tropomyosin. Release of tropomyosin allows it to move into the groove. This shift exposes the myosin-binding sites on the actin chains (Fig. 12-14*b*), permitting attachment of actin and myosin and initiating the crossbridging cycle. (Fig. 12-14*c* and *d* outlines the interactions between the TnC, TnI, and TnT subunits of troponin that are believed to be responsible for the binding and release of tropomyosin.)

Actin-linked regulation through the Ca^{2+}-sensitive interactions between troponin and tropomyosin controls microfilament sliding in all types of striated muscle, including both voluntary and cardiac muscle. (Further

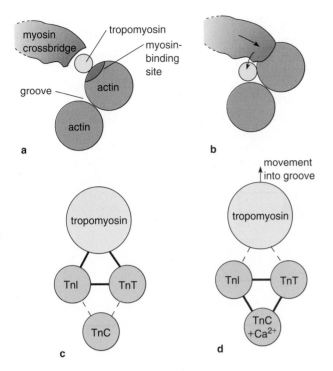

Figure 12-14 The actin-linked control mechanism involving blockage by tropomyosin of the crossbridge-binding sites on microfilaments, as seen in a cross section of a microfilament. **(a)** In the blocking position, tropomyosin covers binding sites for myosin crossbridges. **(b)** Movement of tropomyosin toward the microfilament groove exposes the myosin-binding sites and triggers the crossbridging cycle (see text). **(c** and **d)** Binding attractions between TnC, TnI, and TnT in the troponin complex in the absence **(c)** and presence **(d)** of Ca^{2+}. Dashed lines indicate weak bonding; solid lines indicate strong bonding. TnC has four Ca^{2+}-binding sites ($C = Ca^{2+}$ binding) and reacts with both TnT and TnI but not directly with tropomyosin. During Ca^{2+} binding TnC undergoes conformational changes that greatly increase its binding affinity for TnT and TnI. TnT (T = tropomyosin binding) and TnI (I = inhibition) both bind to tropomyosin. In the absence of Ca^{2+}, TnC is only loosely bound to TnT and TnI, leaving TnT and TnI strongly linked to tropomyosin. This linkage holds tropomyosin in a position in which it blocks the myosin-binding sites on microfilaments. At higher Ca^{2+} concentrations TnC binds Ca^{2+} and undergoes a conformational change that greatly increases its binding affinity for TnT and TnI. Binding of TnT and TnI to tropomyosin is thereby weakened, and tropomyosin can move more deeply into the microfilament groove, in the non-blocking position.

details of the mechanism regulating striated muscle contraction are presented later in this chapter.) Because troponin cannot be detected in smooth muscle or nonmuscle cells, actin-linked regulation involving troponin may be characteristic only of striated muscle. Other actin-linked regulatory pathways, however, may operate in smooth muscle and nonmuscle cells (see below).

Myosin-linked Regulation Control of the crossbridging cycle by elements linked to myosin involves the same regulatory principles as actin-linked regulation. Under the influence of Ca^{2+} a series of steps leads to conformational changes in a protein, in this case the regulatory myosin light chain, that either inhibit or activate the ability of myosin to bind actin and proceed through the crossbridging cycle. This control pathway was discovered primarily through research on smooth muscle cells by A. Szent-Gyorgyi and his coworkers.

Myosin-linked control systems operate through a pathway in which elevated Ca^{2+} concentrations induce phosphorylation of regulatory light chains. Interaction between myosin and actin in smooth muscle cells does not proceed unless phosphate groups are linked to myosin light chains at several positions. If the phosphate groups are removed, the myosin head units are completely inactive in the crossbridging cycle. Removal of regulatory light chains eliminates the control from the crossbridging cycle, so that crossbridge cycling and microfilament sliding proceed as long as ATP is available.

Phosphorylation of the light chains in myosin-linked control systems depends on the activity of an enzyme, *myosin light chain kinase (MLC kinase)*. MLC kinase includes a calmodulin subunit that provides Ca^{2+} sensitivity to the system. When Ca^{2+} is present at the low concentrations typical of resting muscle cells, calmodulin takes on a conformation that inhibits the activity of MLC kinase, and no phosphorylation of myosin light chains takes place. As Ca^{2+} concentration rises above the resting level, calmodulin binds Ca^{2+} and undergoes a conformational change that activates the kinase. Under these conditions phosphate groups are added to the light chains, and the crossbridging cycle is turned on.

Switching the mechanism off depends on *myosin light chain phosphatase (MLC phosphatase)*, an enzyme that removes phosphate groups from myosin light chains. MLC phosphatase is continually active. As the Ca^{2+} concentration falls to the resting level, Ca^{2+} is released from calmodulin and the phosphorylating activity of MLC kinase is inhibited. Under these conditions phosphate groups are quickly removed from regulatory light chains by the constant MLC phosphatase activity, and the crossbridging cycle stops. (Fig. 12-15 summarizes myosin-linked regulation.)

In many smooth muscle types myosin regulation via light chains is supplemented by an actin-linked mechanism based on another Ca^{2+}/calmodulin-activated protein, *caldesmon*. Caldesmon also occurs as an actin-linked regulatory protein in many nonmuscle cell types. Caldesmon links directly to microfilaments of smooth muscle cells, which also contain tropomyosin. When Ca^{2+} is present at resting levels, caldesmon takes a form in which it binds actin and blocks the sites bound by myosin. Elevated Ca^{2+} concentrations activate cal-

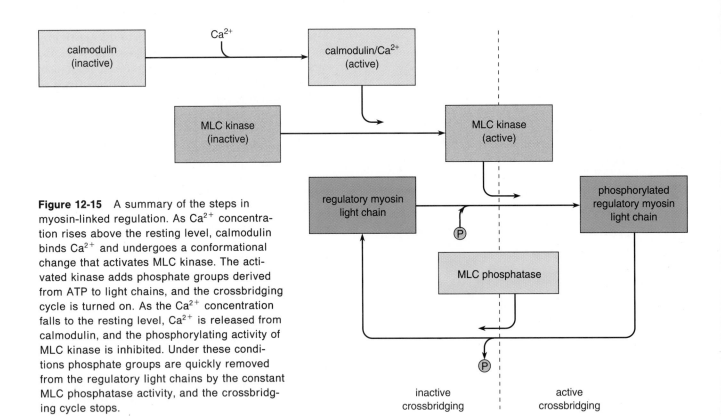

Figure 12-15 A summary of the steps in myosin-linked regulation. As Ca²⁺ concentration rises above the resting level, calmodulin binds Ca²⁺ and undergoes a conformational change that activates MLC kinase. The activated kinase adds phosphate groups derived from ATP to light chains, and the crossbridging cycle is turned on. As the Ca²⁺ concentration falls to the resting level, Ca²⁺ is released from calmodulin, and the phosphorylating activity of MLC kinase is inhibited. Under these conditions phosphate groups are quickly removed from the regulatory light chains by the constant MLC phosphatase activity, and the crossbridging cycle stops.

modulin, which in turn binds to caldesmon, inducing a conformational change that greatly reduces the affinity of caldesmon for actin. Caldesmon then releases from actin; the release exposes myosin-binding sites on microfilaments. Caldesmon may directly cover and uncover myosin-binding sites, or it may work indirectly by binding and releasing tropomyosin, in a mechanism similar to the troponin-based pathway of striated muscle cells. Two types of caldesmon have been detected: a heavy form weighing about 150,000 daltons in smooth muscle cells and a lighter form of about 80,000 daltons in nonmuscle cells.

In some smooth muscle cells Ca²⁺ release is regulated by hormones, through surface receptors that trigger the InsP₃/DAG pathway and cause release of Ca²⁺ as a second messenger (see p. 247). Controls of this type are responsible for the establishment and maintenance of extended, long-term contractions in these smooth muscle types. Controls working in the opposite direction are imposed in some smooth muscle cells through cAMP-based regulatory pathways activated by hormone binding at the cell surface (Fig. 12-16). Attachment of hormones to these receptors triggers the synthesis of cAMP, which activates a kinase that adds phosphate groups to MLC kinase. Phosphorylated MLC kinase is inactive, phosphorylation of myosin light chains does not take place even if Ca²⁺ concentration is elevated, and the crossbridging cycle stops. MLC kinase inhibition is reversed by a phosphatase that is also present in smooth muscle cells. The cAMP-based inhibitory pathway, which is triggered by

β-adrenergic hormones in higher vertebrates, produces long-term relaxation of smooth muscle cells.

Although myosin-linked regulation was discovered primarily through research on smooth muscle cells, this regulatory pathway is apparently also widely distributed among nonmuscle cells in eukaryotes. In many nonmuscle systems the myosin-linked pathway is supplemented by regulation of actin by caldesmon, as it is in smooth muscle cells.

A few significant variations have been observed in pathways of myosin-linked regulation. In clams and other bivalve molluscs, the powerful *adductor* muscles that close the shells may be striated or smooth. In these molluscs, the adductor muscles are controlled through the myosin-linked pathway even if they are striated. In some molluscs and other invertebrates myosin-linked control is imposed through direct combination of Ca²⁺ with regulatory light chains, without phosphorylation of light chains by an MLC kinase.

All these regulatory mechanisms involve myosin II molecules in various locations. Relatively little is known about systems regulating myosin I activity. In the soil protozoan *Acanthamoeba*, Korn and his coworkers found that the myosin I type of this organism is activated by phosphorylation of the heavy chain head unit, in a reaction catalyzed by a heavy chain kinase. At least some forms of myosin I are also believed to be regulated by Ca²⁺ because calmodulin forms a light chain associated with the head unit in these myosins. However, changes in Ca²⁺ concentration have not as yet been linked to myosin I activity in these types.

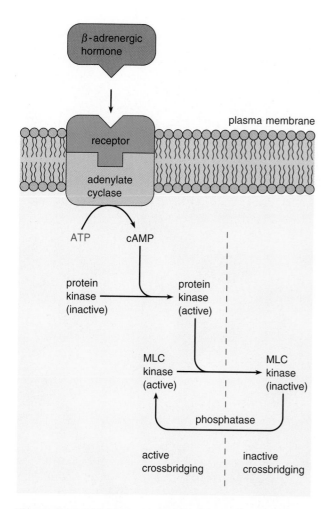

Figure 12-16 Inhibition of smooth muscle contraction by hormones triggering cAMP-based regulatory pathways (see p. 245 for details of the pathways). Attachment of a hormone to a surface receptor triggers the synthesis of cAMP, which activates a kinase that adds phophate groups to MLC kinase. In the phosphorylated form MLC kinase is inactive, and myosin light chains are not phosphorylated even at elevated Ca^{2+} concentrations. As a result, the crossbridging cyle is inhibited. MLC kinase inhibition is reversed by a phosphatase present in smooth muscle cells. The inhibitory pathway, triggered by β-adrenergic hormones in higher vertebrates, produces long-term relaxation of smooth muscle cells.

MICROFILAMENT-BASED MOTILE SYSTEMS

Striated Muscle Cells

Striated Muscle Structure   Striated muscle received its name because it shows a pattern of regularly spaced crossbands under the light microscope (Fig. 12-17a). A striated muscle, such as the biceps muscle of the upper arm, consists of a large number of greatly elongated, cylindrical or ribbonlike cells called *muscle fibers* (Fig.

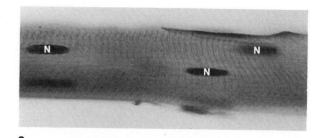

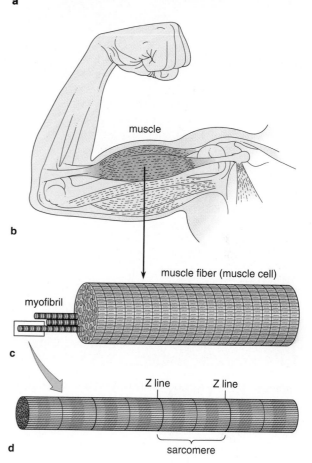

Figure 12-17 Striated muscle structure. **(a)** Striated chicken pectoralis muscle under the light microscope. Muscle fibers run horizontally in the micrograph. The cross striations from which striated muscle takes its name run vertically. ×920 **(b, c,** and **d)**. Relationships among muscles, muscle fibers, myofibrils, and sarcomeres. Micrograph courtesy of N. N. Malouf. Copyright by Academic Press, Inc., from *Exp. Cell Res.* 122:233 (1979). Diagrams redrawn from an original courtesy of D. W. Fawcett, from D. W. Fawcett and W. Bloom, *A Textbook of Histology,* 10th ed. Philadelphia: W. B. Saunders, 1975.

12-17*b* and *c*). Individual muscle fibers may be as long as 40 millimeters (mm). Muscle fibers are themselves striated; disruption shows that their appearance results from a striated or crossbanded structure in long, fibrous strands called *myofibrils,* which are arranged parallel to the long axis of the cells (Fig. 11-17*d*). Surrounding the

Figure 12-18 Myofibrils of fully contracted frog semitendinosus muscle in the electron microscope. ×34,000. Courtesy of J. G. Tidball.

myofibrils are cytoplasmic segments containing multiple nuclei (visible in Fig. 12-17a) and large numbers of mitochondria. The cytoplasm surrounding myofibrils also contains a complex system of membranous tubules called the *sarcoplasmic reticulum (SR)*, related to the smooth endoplasmic reticulum of other cell types. The SR forms part of the system controlling muscle contraction (see below). The multiple nuclei in muscle fibers, which may number as many as 100 per cell, reflect the fact that in embryos these elongated cells develop through the end-to-end fusion of many individual embryonic cells whose nuclei persist individually in mature fibers.

The repeating units responsible for the striated appearance of myofibrils and muscle cells are clearly visible in the electron microscope (Figs. 12-18 and 12-19a). Each unit, about 2.0 to 2.5 μm long in resting muscle, is called a *sarcomere*. The boundaries of a sar-

comere are marked by dark, slender transverse bands called *Z lines*. (Information Box 12-1 explains the specialized terms and letters used in muscle studies.) The broader light and dark bands between the Z lines are formed by overlapping *thin* and *thick filaments* that run lengthwise within the sarcomeres.

Thin filaments are single actin microfilaments in combination with troponin and tropomyosin. Myosin decoration shows that thin filaments connect by their plus ends to either side of a Z line and extend their minus ends toward the center of a sarcomere. In the midregion of a sarcomere, thin filaments overlap with thick filaments, which are assemblies of myosin molecules. In vertebrates each thick filament contains some 300 to 400 myosin molecules.

Myosin head units in the thick filaments form crossbridges with the thin filaments in the zones of overlap. The myosin crossbridges, which are only faintly visible

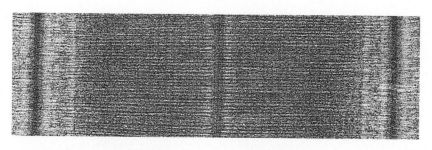

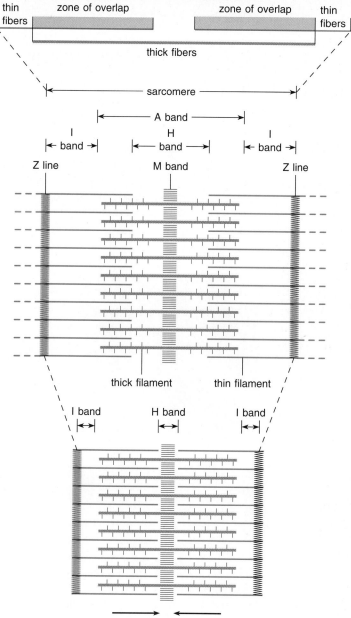

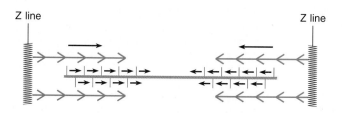

a

thin fibers | zone of overlap | zone of overlap | thin fibers

thick fibers

sarcomere

b

A band

I band | H band | I band

Z line | M band | Z line

thick filament | thin filament

c

I band | H band | I band

d

Z line | Z line

Figure 12-19 Microfilament sliding in sarcomere contraction. **(a)** A single striated muscle sarcomere from frog semitendinosus muscle. ×44,000. Courtesy of J. G. Tidball. **(b)** Sarcomere structure in the relaxed state. **(c)** Sarcomere under contraction; the thin filaments have slid inward along the thick filaments, reducing the width of the I and H bands by an equal distance and shortening the entire sarcomere. **(d)** Opposite polarity (arrows) of thin filaments and myosin crossbridges. Because of their opposite polarity, the attach-pull-release cycle of myosin head units pulls the Z lines together and actively shortens the sarcomeres.

A Primer of Muscle Cell Terms

Terms Used for Cell Structures

muscle cell = muscle fiber
plasma membrane = sarcolemma
smooth endoplasmic reticulum = sarcoplasmic reticulum
mitochondrion = sarcosome
segment of myofibril from Z line to Z line = sarcomere

Derivation of Letters Used for Sarcomere Bands

I band, from *isotropic*, meaning that the band has no directional filtering effect on light waves
A band, from *anisotropic*, meaning that the band filters out light waves vibrating in some directions
Z line, from the German *Zwischenscheibe* (from *zwischen* = between and *scheibe* = disc)
H band, from the German *helles*, meaning clear or light in color

in cross sections, are clearly seen in muscle tissue that has been prepared by the freeze-fracture technique (Fig. 12-20). A striated muscle sarcomere is estimated to contain about 2×10^{12} myosin crossbridges.

Cross sections of sarcomeres show that thick and thin filaments overlap in a highly regular pattern. In vertebrates this pattern forms a *double hexagon*, with thick and thin filaments in a ratio of 1:2 (Fig. 12-21*a*). Other patterns, both highly ordered and irregular, occur in invertebrates, giving a variety of thick-to-thin filament ratios and spacings in different invertebrate cell types and species. (Fig. 12-21*b* to *d* shows some of these patterns.)

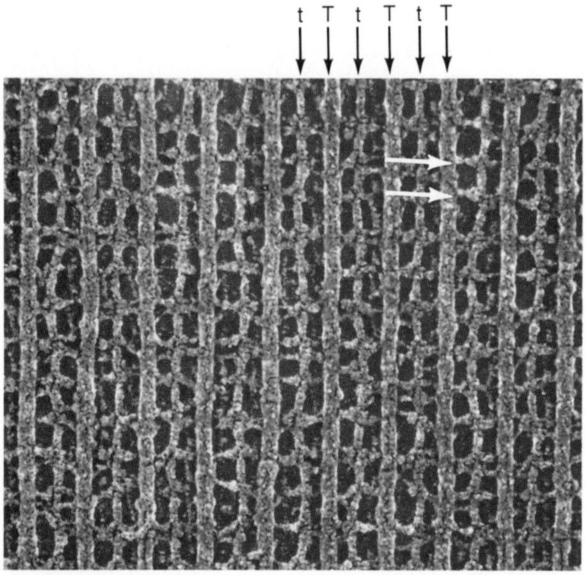

In relaxed muscle thin filaments extend only about halfway along the thick filaments. This arrangement leaves open spaces that appear as a broad, somewhat lighter band, the *H band*, running transversely across the center of each sarcomere (see Fig. 12-19*b*). A narrow,

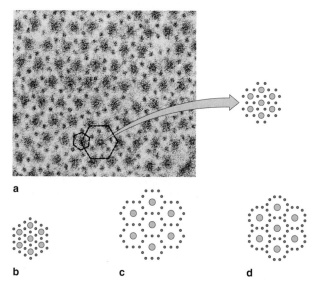

Figure 12-21 The arrangement of thick and thin filaments in sarcomeres. **(a)** A vertebrate sarcomere in cross section, made in the region of overlap between thick and thin filaments. The filaments form a double hexagon pattern (in outline). Each thick filament is surrounded by six thin filaments arranged hexagonally, and the thick filaments make up a larger hexagonal pattern. × 115,000. **(b)** Insect flight muscle (thick:thin = 1:3). **(c** and **d)** Various arthropod muscles (c, thick:thin = 1:5; d, thick:thin = 1:6). Micrograph courtesy of H. E. Huxley, with permission from *J. Mol. Biol.* 37:507 (1968). Copyright by Academic Press, Inc. (London) Ltd. Diagrams courtesy of F. A. Pepe; reproduced from *J. Cell Biol.* 37:445 (1968), by copyright permission of the Rockefeller University Press.

Figure 12-20 Freeze-fracture preparation of insect flight muscle from the giant water bug *Lethocerus*. The thick (T) and thin (t) filaments run vertically in the micrograph; myosin crossbridges (arrows) run horizontally. × 135,000. Courtesy of J. E. Heuser, from *J. Mol. Biol.* 169:97 (1983).

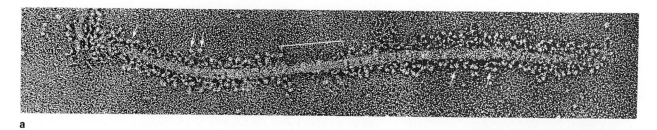

a

central bare zone

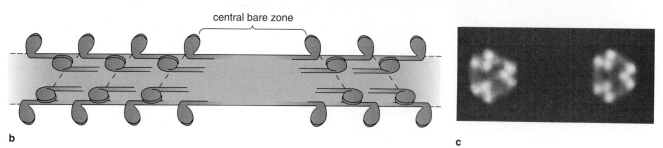

b

c

Figure 12-22 Myosin thick filaments in striated muscle. **(a)** An isolated thick filament prepared for electron microscopy by negative staining. Myosin head units (arrows) are distributed all along the thick filament except for the central bare zone (bracket). × 116,000. Courtesy of A. Elliott, from *J. Mol. Biol.* 131:133 (1979). **(b)** The antiparallel or end-to-end arrangement of myosin molecules believed to occur in thick filaments. **(c)** A composite, computer-averaged cross-sectional image of a thick filament from vertebrate striated muscle. The image shows the average density distribution of many superimposed, thick-filament cross sections. The density is reversed, so that regions of heaviest density appear lightest in the image. The image shows 9 subunits, indicating that the tails of 9 myosin molecules associate to form a thick filament. Courtesy of F. A. Pepe.

dense transverse band, the *M band*, runs through the middle of the H band at the center of the sarcomere. As muscle contracts, the width of the H band and the space between the ends of thick filaments and Z lines (the I bands in Fig. 12-19*b* and *c*) both narrow by an equivalent distance. This observation, and the fact that neither thick nor thin filaments shorten during sarcomere contraction, formed the basis for the first proposals that sliding, rather than contraction, underlies microfilament-based movements.

The identity of thick filaments with myosin and thin filaments with actin was established in 1953 by H. E. Huxley and J. Hanson, who took advantage of the fact that actin and myosin are released from striated muscle by different salt concentrations. At salt concentrations that release myosin, thick filaments disappeared or became fainter in muscle sections. At salt concentrations that release actin, thin filaments lightened or disappeared. Later investigations using labeled antibodies confirmed that thick filaments contain myosin and thin filaments are actin microfilaments.

Thick Filament Structure When isolated from striated muscle and prepared for electron microscopy by negative staining, crossbridges can be seen to extend from all regions of thick filaments except for a "bare zone" in the middle (Fig. 12-22*a*). This appearance is

believed to reflect an arrangement in which individual myosin molecules are held parallel to the long axis of the thick filament, running in opposite directions from the two ends. Heads are directed toward the filament ends, and tails extend toward the midregion (Fig. 12-22*b*). Only tails are present near the center, arranged in an overlapping pattern that produces the bare zone. Cross sections suggest that at any point vertebrate thick filaments are built up from a bundle of 9 parallel myosin molecules (Fig. 12-22c).

Vertebrate thick filaments also contain several other components of unknown function, identified as *C, H, I,* and *X proteins.* Another protein, *paramyosin,* occurs along with myosin in the thick filaments of invertebrate muscle cells. Paramyosin, a coiled coil resembling the tail units of myosin in structure and dimensions, is believed to be a "filler" molecule that forms the central core of thick filaments in invertebrates. (Invertebrate thick filaments typically have a greater diameter than those of vertebrates.) The paramyosin core in these thick filaments is believed to be covered by a myosin coat one molecule in thickness.

Z Lines and M Band Structure The actin and myosin filaments of striated muscle are held in register by elements in the Z lines and M bands, which actually appear as disclike structures rather than lines or bands

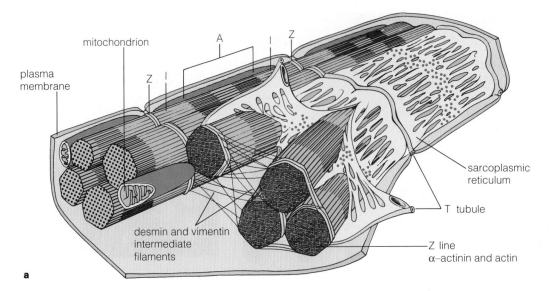

plasma
membrane

mitochondrion

A   Z

Z   I

I

sarcoplasmic
reticulum

T tubule

desmin and vimentin
intermediate
filaments

Z line
α–actinin and actin

a

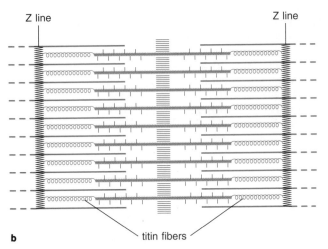

Z line   Z line

b

titin fibers

Figure 12-23 Muscle fiber structure. **(a)** The arrangement of sarcomeres and myofibrils in muscle fibers. The Z bands are held in place and reinforced by desmin and vimentin intermediate filaments. Redrawn from an original courtesy of E. Lazarides. Reprinted by permission from *Nature* 283:249; copyright © 1980 Macmillan Magazines Ltd. **(b)** Titin fibers connecting thick filaments of a sarcomere to the Z line. Titin fibers probably keep the thick and thin filaments in register and prevent the sarcomeres from separating if they are experimentally stretched to the point where thick and thin filaments no longer overlap. Titin fibers may also be responsible for maintaining the slight tension noted in relaxed muscle.

when viewed in three dimensions (Fig. 12-23*a*). Z lines contain several proteins, including α-actinin, *filamin, synemin,* and *Z protein,* which probably crosslink microfilaments and hold them in register in this region. Z protein and α-actinin, in particular, are believed to be important in this linkage. Two intermediate filament proteins, *desmin* and *vimentin* (see p. 503), also associate with Z lines, probably as structural elements that reinforce Z lines and anchor them to the cytoskeleton. Some intermediate filaments anchored to Z lines run lengthwise and reinforce muscle cells in this direction; others run transversely and connect the Z lines of adjacent myofibrils (see Fig. 12-23*a*).

M bands hold thick filaments in their hexagonal array in the central region of sarcomeres. Several proteins have been identified in M bands, including *M protein, B protein,* and *skelemin.* These proteins reinforce M bands and link directly to thick filaments, holding them in register in the center of the sarcomere. Several other polypeptides, possibly also concerned with anchoring thick filaments in position in the sarcomere, can be detected in the M band.

Longitudinal Reinforcements in Sarcomeres   If myofibrils are experimentally stretched far enough to slide thin and thick fibers completely apart, the sarcomeres are still held together by elastic fibers that extend from the tips of thick filaments to Z lines (see Fig. 12-23*b*). These fibers, consisting of a protein known as *titin* or *connectin,* evidently work as "rubber bands" that reinforce sarcomeres longitudinally and resist the overextension of striated muscle. Titin fibers may also help maintain thick filaments in register between thin filaments and produce the slight tension that can be detected in resting muscles. Titin, with a molecular mass of about 3 million, is the largest single polypeptide known. Another protein, *nebulin,* may reinforce thin filaments in the longitudinal direction.

Striated Muscle Motility and Its Regulation   As a consequence of the arrangement of thick and thin filaments in sarcomeres, the myosin head units slide thin filaments on either side of the sarcomere toward the M band. The sliding draws Z lines closer together and actively shortens or contracts the entire sarcomere. The con-

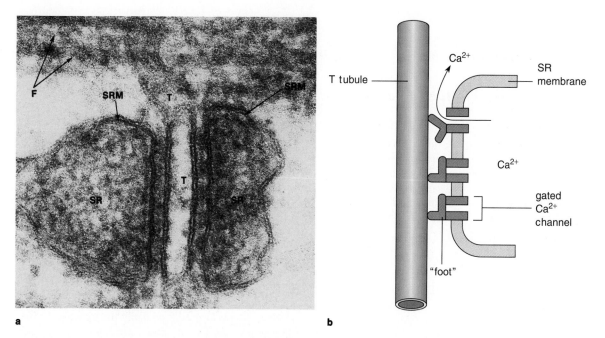

a   b

Figure 12-24 The relationship of T tubules and SR membranes in a mammalian muscle. **(a)** A T tubule and its associated SR membranes in human skeletal muscle. T, T tubule; SRM, SR membranes; F, thick and thin filaments of a myofibril. Courtesy of M. J. Cullen. Reprinted by permission from *Nature* 330:693; copyright © 1987 Macmillan Magazines Ltd. **(b)** A T tubule and an associated SR membrane in diagrammatic form. Each gated Ca^{2+} channel has a "foot" connecting it to the SR; the gate of the channel at the top is open.

traction of all sarcomeres in a muscle fiber forcibly shortens the entire muscle and moves structures to which the muscle is attached.

The control of striated muscle contraction by motor nerves is linked to the troponin-tropomyosin system by a complex mechanism that regulates Ca^{2+} concentration in muscle cells. The SR membranes of muscle cells contain a Ca^{2+}-ATPase pump (see p. 208) that, in resting muscle, continuously removes Ca^{2+} from the cytoplasm surrounding myofibrils and pushes it into SR vesicles. Ca^{2+}-ATPase pumps located in the plasma membrane also remove Ca^{2+} from the cytoplasm. This continuous pumping activity keeps the Ca^{2+} concentration of the muscle cell cytoplasm at the resting level, too low for muscles to contract.

SR membranes are connected to the muscle cell surface by long, tubular invaginations of the plasma membrane called *T tubules* (or *transverse tubules*; see Fig. 12-23). The T tubules closely adjoin but do not fuse directly with SR membranes (Fig. 12-24a).

At some point a motor neuron makes contact with the plasma membrane of a muscle cell via a synapse (see p. 221). An arriving impulse causes a voltage change in the muscle cell plasma membrane that is propagated over the entire muscle cell surface and downward into the T tubules. When the impulse arrives at the T tubule–SR junctions, gated Ca^{2+} channels in SR membranes snap open in response to the arriving impulse, and the SR membranes suddenly become "leaky" to their stored

Ca^{2+} (see Fig. 12-24b and p. 201). At the elevated concentrations produced by this sudden release, troponin binds Ca^{2+}, causing tropomyosin to move away from the position in which it blocks myosin-binding sites. This turns on the crossbridging cycle, and the muscle cell contracts.

As nerve impulses cease to arrive at the muscle cell surface, the gated Ca^{2+} channels close. Ca^{2+} ceases to flow into the muscle cell cytoplasm, and Ca^{2+}-ATPase pumps quickly remove the remaining Ca^{2+}. As a consequence, Ca^{2+} concentrations drop to the resting level, and the troponin-tropomyosin system loses its bound Ca^{2+}. Under these conditions tropomyosin is pulled into the blocking position, the crossbridging cycle is switched off, and contraction stops.

Among the several uncertainties in our knowledge of this regulatory mechanism are the manner in which T tubules are connected to the SR and the nature of the signal that opens the gated Ca^{2+} channels in SR membranes. Recent observations by T. Wagenknect and his associates indicate that the connection between T tubules and SR membranes is made by a footlike extension of Ca^{2+} channels in SR membranes (see Fig. 12-24). The extension bridges the gap between the T tubule and SR membranes, which appears to be 10 to 20 nm wide in most striated muscle cells. Movement of an electrical impulse down the T tubules may induce a conformational change in the footlike extensions that directly opens the Ca^{2+} channels.

Storage and Supply of ATP in Striated Muscle Cells

The amount of ATP in solution in striated muscle cells is sufficient for only a few seconds of contractile activity at most. Some phosphate-bond energy, equivalent to about six times the quantity of ATP in solution in muscle cells, is stored in a substance known as *creatine phosphate*. The phosphate group of this substance can be transferred directly to ADP to form ATP, in a reaction catalyzed by *creatine kinase*:

$$O^- \!-\! \overset{\overset{\textstyle O}{\|}}{\underset{\underset{\textstyle O}{\|}}{P}} \!-\! \underset{\underset{\textstyle H}{|}}{N} \!-\! \overset{\overset{\textstyle NH}{\|}}{C} \!-\! \underset{\underset{\textstyle CH_3}{|}}{N} \!-\! CH_2 \!-\! COO^- + ADP \longrightarrow$$

creatine phosphate

$$H_2N \!-\! \overset{\overset{\textstyle NH}{\|}}{C} \!-\! N \!-\! CH_2 \!-\! COO^- + ATP$$
$$\underset{\textstyle CH_3}{|}$$

creatine

This reaction replenishes some of the ATP hydrolyzed in muscle contraction. In animals carrying out a short burst of intensive muscular activity, as in a runner competing in a 100-meter dash, most of the ATP hydrolyzed in muscle tissue is obtained from creatine phosphate. In some invertebrates arginine phosphate serves as the reservoir of phosphate-bond energy instead.

As the store of creatine phosphate becomes exhausted, additional ATP is produced from ADP in a reaction catalyzed by adenylate kinase:

$$2ADP \longrightarrow ATP + AMP$$

ATP in much greater quantities is supplied by oxidation of glucose, obtained from the circulation or stored in muscle tissue as glycogen (see p. 521). Although essentially all muscle cells can carry out glucose oxidation by both glycolysis and the oxidative reactions of mitochondria, the balance of the two pathways differs depending on the muscle type. The cells of "red" muscle, which owes its color to large quantities of *myoglobin*, a hemoglobinlike protein that binds and stores oxygen, contain many mitochondria and produce ATP primarily through aerobic oxidation of glucose. The oxidation, which proceeds through glycolysis and mitochondrial oxidations, uses oxygen released from myoglobin as the final acceptor for electron transport (see p. 327). Red muscle cells also oxidize fatty acids as a major energy source.

"White" muscle, typified by the breast muscles of birds and the muscles of frog legs, contains only small quantities of myoglobin and relatively few mitochondria. These cells produce ATP primarily by anaerobic breakdown of glucose. In this breakdown, a fermentation (see p. 317), electrons removed during oxidative reactions are delivered to pyruvic acid, which is converted to lactic acid. The lactic acid is oxidized back to pyruvic acid during periods of rest. The amount of ATP produced by the anaerobic reactions of white muscle is limited compared to the large amounts generated by the fully oxidative capabilities of red muscle.

Unwieldy and complex as it sounds, the mechanism controlling nerve impulses, T tubules, and the SR provides almost instantaneous regulation of skeletal muscle activity. The mechanism provides humans and other animals with the ability to carry out voluntary movements, which are characteristically rapid and powerful yet delicately and precisely controlled. (Information Box 12-2 outlines the major biochemical mechanisms storing and supplying ATP in striated muscle cells.)

Cardiac Muscle Cardiac muscle cells also contain striated myofibrils consisting of sarcomeres linked end to end. The thick and thin filaments of cardiac sarcomeres are arranged in the same pattern as those of skeletal muscle sarcomeres, and they contract by the same mechanism. SR membranes and T tubules are also present. There are several significant differences, however, in other features of the structure and function of cardiac muscle cells.

Cardiac muscle cells are much shorter than skeletal muscle cells and contain only a single nucleus, reflecting the fact that the embryonic cells forming cardiac muscle do not fuse end to end. Within the cytoplasm the mitochondria surrounding myofibrils are generally larger and more numerous than in skeletal muscle. We have already noted that cardiac muscle contains a distinct actin, α_c; both myosin light and heavy chains also take distinct forms in the sarcomeres of the heart.

Cardiac muscle cells are branched, so that each cell may connect to several neighboring cells. At the connecting points cardiac muscle cells are held together by a specialized junction called an *intercalated disc* (Fig. 12-25). The microfilaments of sarcomeres nearest the junction anchor directly in the intercalated disc, tying

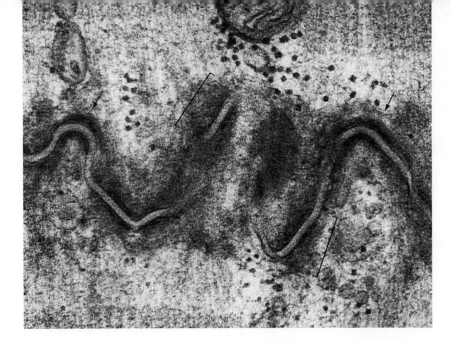

Figure 12-25 An intercalated disc connecting muscle cells in cat cardiac tissue. Thin filaments of the sarcomeres nearest the disc are anchored in junctions resembling the adherens junctions of other tissues (brackets); desmosomes are also present (arrows). × 72,000. Courtesy of D. W. Fawcett, from *An Atlas of Fine Structure: The Cell*, W. B. Saunders Co., 1966.

the connection to the contractile apparatus. The regions of intercalated discs that anchor microfilaments thus resemble the adherens junctions of nonmuscle cell types, in which microfilaments also make connections to plasma membranes (see p. 258). Also within intercalated discs are desmosomes in which intermediate filaments anchor to the structure (see Fig. 12-25) and gap junctions permitting the direct flow of ions between cells. Intercalated discs and the branched nature of the cell-cell connections link cardiac cells into a tightly integrated network that is mechanically reinforced in all directions instead of primarily lengthwise as in skeletal muscle. The branches also cause heart muscle to contract in all directions to produce a squeezing or pumping action rather than the lengthwise contraction characteristic of skeletal muscle.

No motor neurons connect directly to cardiac cells, and there are no synapses between neurons and muscle cells within the heart. The stimulus for contraction originates within the heart itself, in specialized groups of cardiac muscle cells called *pacemakers*, which generate impulses for contraction in a regular, periodic rhythm. An impulse triggered in a pacemaker travels rapidly throughout the heart by passing directly from the plasma membrane of one cardiac cell to the next. As in skeletal muscle cells, impulses passing over the plasma membranes are conducted into cardiac muscle cells by T tubules. Arrival of the impulse at the SR triggers Ca^{2+} release as in skeletal muscles, and contraction is switched on. As the signal originating from the pacemakers ceases, the flow of Ca^{2+} into the muscle cell cytoplasm stops, and contraction is switched off. The wave of contraction triggered by the pacemakers travels through the heart from top to bottom, first pumping blood from the auricles into the ventricles, and then from the ventricles into the arteries leaving the heart.

The sarcomeres of skeletal and cardiac muscle are the most complex and organized assemblies of microfilaments known. Like the equivalent arrangement of microtubules in the 9 + 2 system of flagella, sarcomeres evidently provide the most efficient assemblage of microfilaments for the generation of motility. Measurements of the power generated by striated muscle, in fact, indicate an output about ten times higher than flagella and place this microfilament-based system on a par with gasoline engines in power output per weight.

Mutations Affecting Striated Muscle The many protein types occurring in striated muscle cells, and the fact that regions of these proteins must be perfectly conserved in amino acid sequence to function with full efficiency, make striated muscle particularly susceptible to the effects of mutations. Many mutations in humans interfere with striated muscle function and are responsible for serious and debilitating diseases. Duchenne muscular dystrophy, for example, is a hereditary disease produced by one or more mutations in a gene carried on the X chromosome. The disease causes progressive muscle degeneration and eventual death. Recent work in several laboratories revealed that the affected gene codes for *dystrophin*, a protein associated with the cytoplasmic side of the plasma membrane in striated muscle cells. Because it is related in sequence to α-actinin and *spectrin*, another linker anchoring actin microfilaments to the plasma membrane, dystrophin is believed to anchor microfilaments into a cytoskeletal network that reinforces the plasma membranes of striated muscle cells against the stresses of muscle cell contraction. Absence of the protein makes the plasma membrane fragile, susceptible to tearing and separation. The frequent openings in the plasma membrane apparently lead to progressive muscle-fiber destruction and loss of muscle function.

a

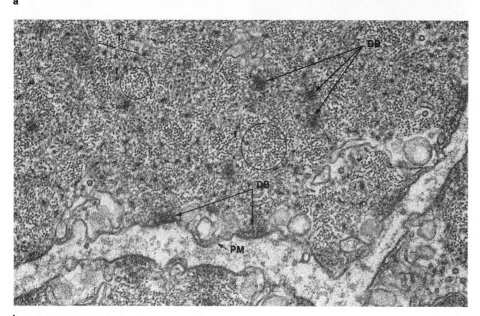

b

Figure 12-26 The arrangement of thick and thin fibers in smooth muscle cells. **(a)** Smooth muscle cell of rabbit vascular tissue. Myosin thick filaments (arrows) are surrounded by actin thin filaments in a pattern that is less highly ordered than the filament pattern of striated muscle. × 40,000. Courtesy of A. V. Somlyo, with permission from *J. Mol. Biol.* 98:17 (1975). Copyright by Academic Press, Inc. (London) Ltd. **(b)** Smooth muscle cell of the digestive system of the tapeworm *Taenia coli* in cross section. T, thick filaments; bundles of thin filaments are circled; DB, dense bodies; PM, plasma membrane. Courtesy of P. Cooke.

Smooth Muscle Cells

Smooth muscle occurs in the walls of the gut, air passages, blood vessels, and the urogenital tract of vertebrate animals. The cells of smooth muscle, so called because they show no regular striations in either the light or electron microscope, are slender, elongated, and pointed at their tips (Fig. 12-26). Like cardiac cells each smooth muscle cell has a single nucleus. However, no T tubules extend inward from the plasma membrane in smooth muscle cells and, although smooth ER membranes are present, they appear primarily as small, spherical vesicles scattered just under the plasma membrane and in groups in the cell interior.

Thin and Thick Filaments in Smooth Muscle Cells

Actin thin filaments (see Fig. 12-26) are distributed more or less uniformly in the cytoplasm of smooth muscle cells, aligned roughly with the long axis of the cell. These microfilaments contain distinct forms of actin and tropomyosin in the same proportions as in skeletal and cardiac muscle cells. Troponin is absent from the thin filaments of smooth muscle cells. Instead, caldesmon, which is present in one-quarter the amount of

tropomyosin, seems to play an equivalent role in regulation of smooth muscle cell contraction.

The myosin thick filaments of smooth muscle cells are somewhat flattened or ribbonlike. They lie parallel to the long axis of the smooth muscle cell, distributed among the thin filaments. No central bare zone is visible in isolated smooth muscle thick filaments. Instead myosin head groups are distributed along the whole length of the fibers. The bipolar structure necessary for muscle contraction is evidently produced by an arrangement of myosin molecules in opposing directions on opposite sides of a thick filament (Fig. 12-27).

Smooth muscle thick filaments are generally larger than their striated muscle counterparts in both diameter and length. These differences reflect the distinct characteristics of smooth muscle myosin II molecules, which differ from striated muscle myosins in the amino acid sequences of both heavy and light chains, solubility, and activity as an actin-activated ATPase. The proportion of thick filaments is much lower in smooth than in striated muscle cells—smooth muscle contains about 1 thick filament for every 12 to 18 thin filaments, as compared to 1:2 and similar thick:thin ratios in striated muscle.

a

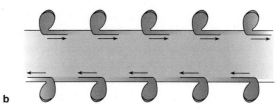

b

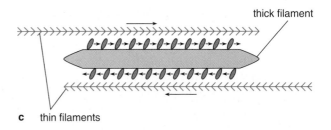

thick filament

c   thin filaments

Figure 12-27 Myosin thick filaments of smooth muscle. **(a)** A smooth muscle thick filament isolated from the guinea pig, prepared for electron microscopy by negative staining. ×150,000. Courtesy of J. V. Small. **(b)** Possible arrangement of myosin molecules that produces bipolar thick filaments with no central bare zone. **(c)** The sliding action between thick and thin filaments of smooth muscle.

Smooth muscle thin filaments, which typically occur in bundles, anchor in *dense bodies* within the cytoplasm (see Fig. 11-26b) that may be the functional equivalents of Z lines in striated muscle. Like Z lines, dense bodies contain α-actinin. The microfilaments anchor with their plus ends linked to the dense bodies, in an orientation equivalent to the attachment of thin filaments to Z lines in striated muscle. The microfilaments between any two dense bodies, with their associated thick filaments, therefore resemble a relatively unorganized sarcomere. Smooth muscle microfilaments also anchor at scattered points to specialized dense bodies that are linked to the plasma membrane. These membrane-attached dense bodies contain *vinculin* and *talin,* two proteins involved in crosslinks between microfilaments and the plasma membrane, as well as α-actinin.

The dense bodies of smooth muscle cells are also anchored to each other and to the plasma membrane by intermediate filaments of the desmin and vimentin types. These linkages tie the entire assemblage of thick and thin fibers into an elastic, interlinked assembly. This cytoskeletal network, besides providing mechanical reinforcement to the cytoplasm, may be partly responsible for the maintenance of long-term tension in smooth muscle cells once contraction has occurred (see below).

During contraction of smooth muscle cells, thin filaments slide over thick filaments, drawing the dense bodies closer together and, through the connections of dense bodies to each other and to the plasma membrane, forcibly shortening the entire cell. The irregular arrangement of thin and thick filaments allows smooth muscle cells to contract and relax over much greater lengths than skeletal and cardiac muscle cells. This ability to contract and stretch over wide limits allows organs containing smooth muscle in their walls, such as the stomach, intestine, urinary bladder, and uterus,

to make major adjustments in size to accommodate, move, or expel their contents.

Control of Smooth Muscle Contraction   There are no close connections between neurons and smooth muscle cells equivalent to the nerve-muscle synapses of striated muscle cells. Instead many smooth muscle cells are controlled by neurotransmitters (see p. 221) released from neurons that may be located at some distance from the cells. Hormones secreted into the body circulation, such as norepinephrine, histamine, angiotensin, vasopressin, and prostaglandin, also induce contraction of many smooth muscle cells. The response of smooth muscle to these substances depends on the presence of receptors in the muscle cell plasma membrane that are tailored to recognize and bind the controlling molecule.

Once bound to their activating substance, the receptors indirectly trigger the opening of gated Ca^{2+} channels in the ER or plasma membrane. In many smooth muscle cells receptors triggering Ca^{2+} release do so through the $InsP_3$/DAG second-messenger pathway. Opening the membrane channels allows Ca^{2+} to flow into the cytoplasm from the extracellular medium or stores inside the ER cisternae. In other smooth muscle cells receptor binding opens membrane channels admitting ions such as K^+, leading to a change in the potential across the membrane (see p. 213). The potential change opens voltage-gated Ca^{2+} channels in the plasma membrane, allowing Ca^{2+} to flow from the extracellular medium into the muscle cell cytoplasm. Smooth muscle cells in some organs are interconnected by gap junctions (see p. 262) that allow ions to flow directly from one cell to the next. In this way an ion flow altering plasma membrane potential and stimulating internal Ca^{2+} release may be transmitted directly between cells, spreading a wave of contraction through the smooth muscle of an entire organ.

Ca^{2+} released into the smooth muscle cytoplasm

by these stimuli triggers contraction primarily through the myosin-linked pathway. Ca^{2+} activates calmodulin, which in turn activates MLC kinase, causing phosphorylation of regulatory light chains and triggering the attach-pull-release cycle of myosin. This regulatory pathway is supplemented in many vertebrate smooth muscle cells by the actin-linked pathway involving caldesmon. Calmodulin, activated by combination with Ca^{2+}, induces a conformational change in caldesmon that releases it from actin and directly or indirectly uncovers the myosin-binding sites of thin filaments.

Relaxation of some smooth muscle types is induced by hormones bound by receptors at cell surfaces (see Fig. 12-16). In these cells receptors activate cAMP-based pathways, leading to stimulation of cAMP-dependent protein kinases that add phosphate groups to MLC kinase. The phosphorylations reduce the activity of MLC kinase, thereby reducing phosphorylation of myosin light chains and inhibiting smooth muscle contraction.

Smooth muscle typically contracts much more slowly than striated muscle. Many smooth muscle types can also remain in a contracted state long after the signal for contraction has stopped and cytoplasmic Ca^{2+} concentrations have returned to resting levels. The prolonged contraction in these cells continues with almost no consumption of ATP, indicating that a crosslinker of some kind "latches" microfilaments into a fixed position after contraction has taken place. Recent research by H. Rasmussen and his colleagues suggests that the "latch" may result from an alteration in the cytoskeleton surrounding thin and thick filaments. Once contraction is fully under way, protein kinases activated by the $InsP_3$/DAG second-messenger pathway add phosphate groups to proteins of the cytoskeleton. The phosphorylation crosslinks and locks the elastic network into a tight network that may maintain the tension of the contracted cells without further expenditure of ATP. Relaxation of the cells occurs through the activity of phosphatases that remove phosphate groups from cytoskeletal proteins.

Microfilament-Based Motility in Nonmuscle Cells

The distinctive biochemical properties of actin and myosin, as determined from studies of muscle cells, armed investigators with a means to detect the two proteins and microfilament-based motility in nonmuscle cells. The myosin decoration technique was used to detect actin microfilaments in protists, in the nonmuscle cells of a wide variety of animals, and in algae, fungi, and higher plants. Myosin has proved to be more difficult to identify than actin in nonmuscle cells because it occurs in much smaller quantities and exhibits greater variability in structure. However, myosin has also been isolated and identified among the cellular proteins of all eukaryotes. These discoveries led to the concept that all eukaryotic cells contain motile systems based on actin and myosin.

The amino acid sequences of nonmuscle actins are distinct but still closely related to those of striated and smooth muscle actins. Because of their large size, nonmuscle myosins have proved more difficult to sequence. However, the limited sequence data available indicate that nonmuscle type II myosins are distinct but related in sequence to muscle cell myosin IIs. Both type I and type II nonmuscle myosins differ from myosin IIs of striated muscle, however, in the location of sites at which regulatory phosphate groups are added. The heavy chains of type II myosins in nonmuscle cells also lack the flexible hinge region of muscle heavy chains. All nonmuscle myosins share with muscle myosin the ability to bind actins from any source and to hydrolyze ATP. In addition, all nonmuscle myosin IIs can be induced to assemble in the test tube into myosin thick filaments resembling those of striated muscle. However, thick filaments are often difficult to detect in nonmuscle cells, and the extent of myosin assembly into thick filaments in these cells is still uncertain.

Actin and myosin molecules detected in nonmuscle cells are responsible for a variety of movements, including cytoplasmic streaming, ameboid motion, capping, the movements of cell layers during embryonic development, and the furrowing movements that divide the cytoplasm during cell division. In all these systems, myosin crossbridges are believed to generate the force for movement, in a process equivalent to the sliding filament mechanism of striated muscle cells.

Cytoplasmic Streaming The active flowing movements of cytoplasm known as cytoplasmic streaming occur in all eukaryotic cells. The most spectacular movements of this type take place in plant cells, where they have been the subject of experimentation since their first observation by B. Corti in 1774. In cells such as the algae *Nitella* and *Chara*, cytoplasmic streaming may proceed at speeds of 100 μm per second, among the fastest rates observed for this type of cellular movement.

In algal cells active streaming takes place in a layer of liquid cytoplasm just under a thick, gelled cortical layer. Microfilaments are conspicuous at the interface between the cortical gel and the underlying motile cytoplasm. In many algae these microfilaments occur in bundles running in the direction of cytoplasmic streaming (Fig. 12-28).

B. A. Palevitz and P. K. Hepler used the myosin decoration technique to study cytoplasmic streaming in *Nitella*. In this alga chloroplasts move along microfilament bundles underlying the cortical gel. The technique confirmed that the bundles consist of microfilaments and established that the polarity of these microfilaments is always uniform and opposite to the

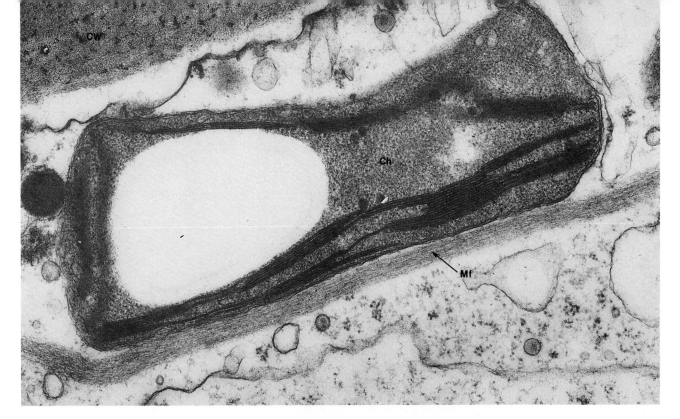

Figure 12-28 A microfilament bundle (Mf) underlying a chloroplast (Ch) at the surface of a *Chara* cell. CW, cell wall. ×34,000. Courtesy of J. D. Pickett-Heaps.

direction of cytoplasmic streaming. From these observations Palevitz and Hepler suggested that active streaming in *Nitella* is generated by myosin molecules attached to chloroplasts, which "walk" by their cross-bridging cycles over microfilaments anchored in the cortical gel. This interpretation is supported by the experiments of Sheetz and Spudich (see p. 467), in which myosin-coated beads were observed to move along the surfaces of microfilament bundles exposed by stripping the cortex from *Nitella* cells.

T. Kato and Y. Tonomura found that the myosin responsible for cytoplasmic streaming in *Nitella* shares several characteristics with animal myosin II, including similar molecular weight and the ability to form thick filaments in the test tube. In *Chara* F. Grolig and his coworkers showed that antibodies developed against mouse myosin heavy chains react with two different myosin heavy chains. One of the heavy chains had a molecular weight of 200,000 and the other 110,000, indicating that both myosin I and II are present in *Chara*.

Microfilament-based cytoplasmic streaming has also been demonstrated in higher plants, particularly in pollen tubes and in cambium, vascular, and root tip cells. J. Heslop-Harrison and Y. Heslop-Harrison showed recently that antibodies developed against bovine myosin react strongly with the surfaces of plant organelles transported by cytoplasmic streaming in pollen tubes, indicating that myosin "walks" the organelles along tracks supplied by microfilaments. When microfilaments in the moving regions are decorated by

HMM, the direction pointed by the arrowheads is always opposite to the direction of cytoplasmic streaming, as in algae.

Little is known about the mechanisms regulating cytoplasmic streaming in algae or higher plants. However, excitation of plant cell plasma membranes, by generation of an action potential (see p. 213), stops cytoplasmic streaming. Several experiments indicate that this control is exerted through release of Ca^{2+} into the cytoplasm in response to the action potential. One experiment by R. E. Williamson and C. C. Ashley used *aequorin,* a protein that emits light when it interacts with Ca^{2+}. The aequorin, injected into the cytoplasm of a plant cell, glowed brightly when an action potential was generated over the cell's plasma membrane. In another experiment T. Hayama and M. Tazawa were able to arrest cytoplasmic streaming by directly injecting Ca^{2+} into the cytoplasm of plant cells.

Cytoplasmic streaming can also be observed in protozoa, fungi, and animals. In these organisms the streaming movements are halted by drugs such as cytochalasins and other treatments that interfere with microfilament structure or function. Although cytoplasmic streaming is generated by microfilaments in these organisms, the organization and location of microfilaments and myosin molecules producing the movement remain unknown. In a few algae, such as the marine green alga *Caulerpa,* cytoplasmic streaming appears to be generated by bundles of microtubules rather than microfilaments.

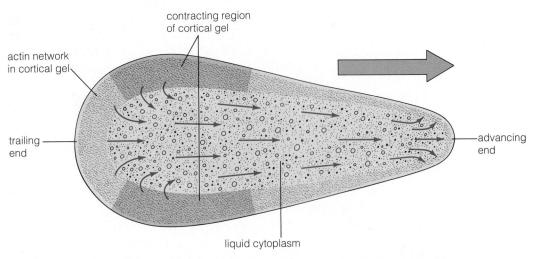

contracting region
of cortical gel

actin network
in cortical gel

trailing
end

advancing
end

liquid cytoplasm

Figure 12-29 A model for the microfilament interactions producing ameboid motion. At the trailing end the gelled cortical tube, consisting of a crosslinked network of microfilaments, continually disassembles by actin depolymerization or breakage of microfilament crosslinks. At the advancing end the tube continuously extends through microfilament assembly or formation of microfilament crosslinks. The liquid inner cytoplasm, containing the actin or microfilaments removed from the trailing end of the cortical gel, is forced passively through the tube by contraction of the cortical gel. Contraction of the gel, produced by myosin-microfilament interaction, occurs in the midregion and near the trailing end of the cell. Myosin II molecules may be responsible for contracting the cortical gel; myosin I may power microfilament sliding, advancing the leading edge of a pseudopod.

Ameboid Motion   Ameboid motion, which is most often associated with protozoans such as *Amoeba proteus*, also occurs in fungi and all animals. In animals ameboid motion of individual cells takes place regularly during embryonic development. Certain cells, such as the white blood cells of vertebrates, retain that capacity in adults. Other cell types, although nonmotile in their normal locations in adult tissues, may migrate by ameboid motion under certain conditions. Fibroblasts, for example, may break their normal attachments and migrate by ameboid motion into damaged tissue areas as part of the reconstruction taking place in response to wounding. Many other normally nonmotile body cells may become motile and travel by ameboid motion if they become transformed into cancer cells or if they are grown in tissue culture outside the body.

Ameboid motion seems simple when observed in the light microscope. At some point along the cell margin, a cytoplasmic lobe called a *pseudopod* forms and swells outward. The cytoplasm of the cell flows into the lobe, eventually carrying cytoplasmic organelles and the nucleus with it. Close examination of an ameboid cell shows that the cytoplasm just under the cell surface, except at the tip of an advancing lobe, is semisolid or gelled, forming a sort of cortical cytoplasmic tube through which the more liquid internal cytoplasm flows to reach the advancing pseudopod tip.

The rate at which the liquid cytoplasm flows through the cortical tube is generally much faster than the rates noted for cytoplasmic streaming.

Actin and myosin were linked to ameboid motion by a series of important discoveries in protozoan and slime mold amebas. In 1960 R. D. Allen and his colleagues showed that ameba cytoplasm retains the ability to move by cytoplasmic streaming when extracted from cells. Somewhat later, T. D. Pollard and others showed that extracted ameba cytoplasm contains microfilaments that form the typical arrowheads when reacted with HMM. Actin isolated from ameboid cells was found to be almost identical to striated muscle actin in structure and chemical properties. Myosin was also identified in the isolated ameba cytoplasm, typically in smaller amounts with respect to actin than in striated muscle. As far as can be determined, troponin is not present in ameboid motile systems. The movements generated through interaction of actin and myosin in ameboid cells require ATP, and they start, stop, and undergo alterations in rate in response to changes in Ca^{2+} concentration.

Thin sections of ameboid cells reveal that microfilaments form a multidirectional network just under the plasma membrane. This layer corresponds to the cortical layer of gelled or semisolid cytoplasm in these cells. The layer is thinnest at the tip of an advancing pseudopod and becomes gradually thicker toward the tail end of the cell. Studies using fluorescent antibodies reveal that myosin is also present in the cortical layer of gelled cytoplasm. Recent work by Korn and his colleagues with fluorescent antibodies against myosin I or II demonstrated that myosin II occurs at highest concentrations in midregions and near the tail of ameboid *Dictyostelium* cells; myosin I is concentrated in the cortical region near the leading edges of an advancing pseudopod. Exposure of the cells to cytochalasin disperses the cortical actin gel and slows or stops ameboid motion.

These observations clearly indicate that actin and myosin are responsible for generating the force for ameboid motion. Just how the actin-myosin system produces motion remains a matter of controversy, however. D. L. Taylor and his colleagues proposed that contraction of the cortical gel in midregions and near the rear of an advancing ameba forces the liquid internal cytoplasm passively through the tube toward the tip, much like squeezing a toothpaste tube (Fig. 12-29). The fact that myosin II concentrations are highest in midregions and at the rear of the cortical tube supports this proposal and indicates that this myosin may be responsible for squeezing the tube. Other supporting evidence comes from the flow rate of the internal cytoplasm, which is usually faster than that of cytoplasmic streaming, and from the observation that extracted cortical gels are capable of contraction when supplied with ATP.

In recent research Spudich, W. F. Loomis, and their colleagues altered the gene encoding myosin II in *Dictyostelium* into nonfunctional form. Although pseudopods still appeared in the myosin-II deficient amebas, they formed randomly around the cell periphery and did not produce clearly directional movement. This suggests that there may be a division of labor among the two myosin types, at least in *Dictyostelium*—myosin I may be responsible for initial pseudopod formation and for advancing the leading edge of a pseudopod, while myosin II is responsible for contracting the cell cortex and squeezing cytoplasm into the advancing lobe from the rear of a migrating cell.

There is good evidence that ameboid motion is controlled by local changes in Ca^{2+} concentrations. One experiment nicely demonstrating the involvement of Ca^{2+}, by Taylor and his coworkers, used aequorin (see p. 483). Bright luminescence from injected aequorin was noted in the tail regions of advancing amebas and also in pulses at other points of apparent contractile activity in the cells, indicating that Ca^{2+} was released in these regions.

Other experiments using antibodies against calmodulin and MLC kinase showed that both these elements of Ca^{2+}-based regulation by the myosin-linked pathway are present in actively moving regions of ameboid cells. Depending on the cell type, the kinase adds phosphate groups to myosin light chains, heavy chains, or both polypeptide types to regulate motility. In general phosphorylation of myosin light chains initiates or speeds movement, as it does in muscle cells. Heavy chain phosphorylation, depending on the cell type, may either speed or slow movement. The presence of caldesmon in some nonmuscle cells indicates that actin-linked regulation of movement by this protein may also operate in these cells.

During ameboid motion the cortical gel probably remains stationary as the underlying liquid cytoplasm moves toward the tip of the pseudopod. To accommodate forward movement of the cell, the gel is thought to disassemble continually in the thick region at the rear end of the cell. If this is the case, the actin subunits and short microfilament lengths released by the disassembly would be carried with the liquid cytoplasm toward the advancing tip. At the tip actin and microfilaments would then reassemble into a new network in the thin cortical layer. Formation and breakdown of the cortical gel at its front and rear ends could be regulated by calcium ions through their effects on Ca^{2+}-sensitive microfilament capping, crosslinking, and severing proteins. Many of these proteins, including gelsolin, fragmin, filamin, and gelactin (see Table 12-1), can be detected in ameboid cells of different types. Calcium ions triggering microfilament sliding and possibly regulating transitions between the gel and liquid states of the cytoplasm are thought to be stored and released by the ER. In general elevated Ca^{2+} concentrations would induce disassembly of the gel, and resting concentrations would favor cortical gel assembly.

Other Microfilament-Based Motile Systems Microfilaments have been established as the basis for several additional cellular movements. In all these motile mechanisms, including capping, division of the cytoplasm in animal cells (furrowing), and the contraction of cell layers during embryogenesis, microfilaments that can be identified by myosin decoration are attached to or associated with the elements being moved. Myosin can also be detected in the moving regions. Treatment with cytochalasins or antibodies developed against actin or myosin generally stop the motion. In many cases elements of the myosin-linked, but not the actin-linked, pathway for regulation of microfilament sliding have been detected in association with microfilament systems generating the motion.

Capping (see p. 173) involves movements that sweep substances bound to the cell surface into a localized mass, the "cap," at one end of the cell. Capped substances are later taken into the cytoplasm by endocytosis. Microfilaments responsible for capping are attached to the cytoplasmic portions of receptor molecules embedded in the plasma membrane. Sliding movements generated by these microfilaments sweep receptors, and substances bound to them on the outside surface of the plasma membrane, into the cap (see Fig. 5-25). Myosin and MLC kinase can also be detected in association with microfilaments in the capping region. Spudich and his coworkers found that *Dictyostelium* mutants lacking myosin II lose the ability to cap substances bound to the plasma membrane, indicating that this myosin type is responsible for the microfilament sliding that underlies capping.

The furrow that divides the cytoplasm during animal cell division is produced by a ringlike assemblage of microfilaments and myosin molecules lying just under the plasma membrane (see Fig. 24-34). The ring contracts by microfilament sliding, gradually cutting the cytoplasm in two. (Further details of the furrowing mechanism and its role in cell division are presented in Ch. 24.) Furrowing stops in *Dictyostelium* mutants lacking myosin II, which indicates that this myosin type is responsible for the progressive microfilament sliding that divides the cytoplasm. Similar microfilament rings extending around the margins of embryonic cells are responsible for contractions that bend or fold tissue layers, such as the fold that produces the neural tube early in embryonic development.

Many other cell movements have been linked to microfilaments. These include changes in shape that take place during the activation of blood platelets in response to wounding in invertebrates, contraction of blood clots, and movements of tube feet in echinoderms such as starfish. The great variety of these and other microfilament-based motile systems indicates that the interaction of myosin crossbridges with microfilaments has proven to be highly versatile as a cytoplasmic motor during the evolution of eukaryotes.

Movements Generated by Microfilament Growth

In systems in which microfilaments generate motion by directional growth rather than sliding, microfilaments extend forcefully by adding actin subunits. The prime example of motion produced by microfilament growth—extension of a process by sperm cells during fertilization—occurs in echinoderms and some other invertebrates. The process, called the *acrosomal filament*, firmly attaches the sperm cell to the surface of the egg. As the filament shoots outward, it remains covered with an extension of the plasma membrane so that it is at all times an internal cytoplasmic structure.

In sperm cells of *Thyone* (a sea cucumber), studied by L. G. Tilney and his colleagues, extension of the filament is spectacularly rapid, reaching a length of 90 μm within ten seconds. In unreacted *Thyone* sperm large quantities of G actin are stored in a cup-shaped deposit at the anterior end of the sperm head (Fig.

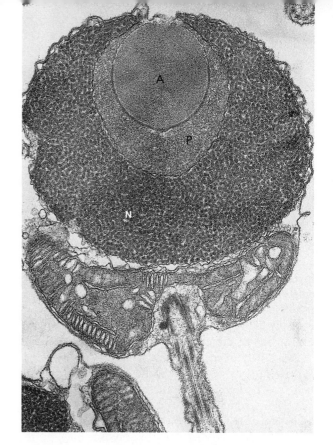

Figure 12-30 A *Thyone* sperm cell, showing the acrosome (A), the cup-shaped region containing unpolymerized G actin (P), and the nucleus (N). ×54,000. Courtesy of L. G. Tilney, from *J. Cell Biol.* 69:51 (1976), by permission of the Rockefeller University Press.

12-30). As the sperm cell contacts the surface of a *Thyone* egg, the material stored in the cup assembles explosively to form a parallel bundle of microfilaments (Fig. 12-31).

Tilney and his coworkers found that the G-actin molecules in unreacted sperm cells are linked to four different polypeptides that prevent their conversion into microfilaments. One of the polypeptides, in molecular weight and biochemical properties, closely resembles profilin (see Table 12-1), an actin-binding protein that stabilizes actin in unassembled form. According to Tilney, initiation of the acrosomal reaction in *Thyone* involves sudden release of the actin in the cup from its combination with the inhibiting proteins. No trace of myosin can be found in either the cup or the polymerized acrosomal filament; the abrupt extension of the filament appears to depend on actin polymerization alone.

Myosin decoration shows that microfilaments in the acrosomal filament are all oriented with their minus ends anchored in the sperm head and their plus ends directed toward the tip of the filament. This means that the rapidly advancing filament assembles from the tip rather than the base within the sperm cell. As a result, G-actin monomers must diffuse outward along the extending filament from their location in the storage cup

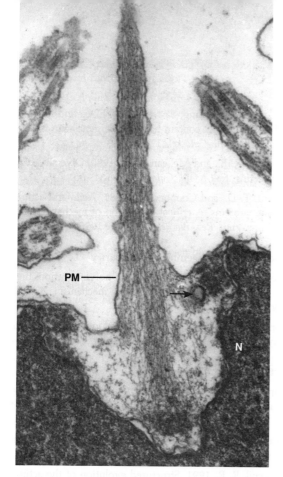

Figure 12-31 A *Thyone* sperm cell after extension of the acrosomal filament (arrow). Parallel microfilaments are visible within the filament. N, nucleus; PM, plasma membrane. ×64,000. Courtesy of L. G. Tilney, from *J. Cell Biol.* 77:536 (1978), by permission of the Rockefeller University Press.

to reach the steadily advancing site of polymerization at the tip. An analysis by Tilney confirms that simple diffusion can account for movement of actin molecules from the cup to the advancing tip within the time occupied by filament extension. In fact, the time required for diffusion of actin to the tip is probably the rate-limiting step in the reaction.

Tilney's group also studied filament extension in another invertebrate species in which the movement is due to a change in physical arrangement of existing microfilaments rather than polymerization. In this species, *Limulus* (the horseshoe crab), a long filament containing fully polymerized microfilaments coils around the nucleus in unreacted sperm (Fig. 12-32a). On contact with the egg, the filament unwinds explosively from the coil, extending to a length of 60 μm in about four seconds (Fig. 12-32b). The filament turns axially in a screwlike motion as it extends. Analysis of this species showed that conversion of the filament from the coiled to the extended state depends simply on a change in the twist of microfilaments within the ropelike bundle forming the filament. An actin-binding protein, which Tilney and his coworkers aptly named *scruin*, is believed

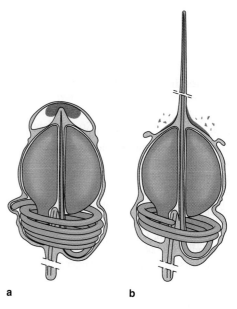

a   b

Figure 12-32 Extension of the filament in *Limulus*, in which force is produced by a change in the twist of actin microfilament bundles. (a) In an unreacted sperm the filament is coiled at the base of the nucleus. (b) On discharge the coil unwinds forcibly, thrusting the filament outward from the anterior end of the sperm cell. Courtesy of L. G. Tilney; reproduced from *J. Cell Biol.* 93:934 (1982), by copyright permission of the Rockefeller University Press.

to be involved in the change of twist that extends the filament.

Microfilaments produce a variety of movements in both muscle and nonmuscle cells. The motile function of microfilaments in almost all these systems is compatible with the sliding filament model derived from studies of striated muscle, in which motion is powered by myosin crossbridges walking along microfilaments by an oarlike attach-pull-release cycle. Energy for the movement is provided by ATP hydrolysis during the crossbridging cycle. In a few systems actin microfilaments produce motion by assembling rapidly and directionally or by undergoing structural changes.

The activities of microfilaments in generating motility clearly parallel the functions of microtubules in eukaryotic cells (as outlined in Chapter 11). The existence of these parallel systems raises one of the fundamental questions of cell biology. Why have two separate but equivalent mechanisms, based on entirely different proteins, evolved and persisted in all eukaryotic cells to accomplish the same apparent ends of motility and cell support? The answer must lie in the distinct characteristics of the polymerization and sliding reactions of the two systems, which make one or the other indispensable for the production of a given motion or supportive element at a certain time and place in the cell. (Supplement 12-1 describes motile systems that operate independently of microtubules and microfilaments.)

1. What types of cellular movements are based on microfilaments?

2. Outline the structure of microfilaments. How are actin subunits arranged in microfilaments?

3. Outline the structure of actin molecules. What binding sites occur on actin?

4. What is the difference between G actin and F actin? What conditions are necessary for G actin to assemble into microfilaments?

5. What is microfilament polarity? How is polarity related to the assembly of microfilaments? What is microfilament treadmilling?

6. What kinds of proteins modify the assembly of microfilaments? What are capping, severing, crosslinking, and bundling proteins? What functions do these modifiers have inside cells?

7. What effects do cytochalasins and phalloidin have on microfilaments? In what ways have these substances been useful in microfilament research?

8. In what ways are the structure and biochemistry of microfilaments and microtubules similar? Different?

9. Outline the structure of a myosin molecule. What are myosin heavy chains? Light chains?

10. What is the basis for the myosin decoration technique? What is the relationship of myosin decoration to microfilament polarity?

11. What are thick filaments? What factors regulate the assembly of myosin into thick filaments?

12. Outline the crossbridging cycle that powers microfilament sliding. What current uncertainties exist concerning the crossbridging cycle?

13. What are troponin and tropomyosin? How do these molecules interact with actin and myosin in the regulation of the crossbridging cycle? What is the relationship of Ca^{2+} to regulation of crossbridging by these proteins?

14. Outline the mechanism of myosin-linked regulation of the crossbridging cycle. What is the relationship of Ca^{2+} and phosphorylation to myosin-linked regulation?

15. How are actin microfilaments arranged in striated muscle cells? How is myosin arranged in striated muscle? What is the structural relationship between actin and myosin in striated muscle?

16. Define myofibril, muscle fiber, sarcomere, sarcoplasmic reticulum, T tubule, Z line, and M band. What molecular components occur in each of these structures?

17. How were thick and thin fibers of striated muscle identified as actin and myosin? What happens to thick and thin fibers when muscle contracts?

18. What cytoskeletal elements reinforce the structure of striated muscle fibers?

19. Outline the role of the SR and T tubules in the control of voluntary muscle contraction.

20. What are the similarities and differences between striated and cardiac forms of striated muscle? Compare the mechanisms regulating contraction of the two striated muscle types.

21. Outline the structure of smooth muscle cells. How are microfilaments and myosin thick filaments arranged in smooth muscle? What cytoskeletal elements reinforce the structure of smooth muscle cells? What role may this cytoskeleton have in smooth muscle contraction?

22. What mechanisms control the contraction of smooth muscle? What is caldesmon? Compare the regulation of smooth and striated muscle.

23. Outline the similarities and differences between the nonmuscle and muscle forms of actin and myosin.

24. What is cytoplasmic streaming? How are microfilaments and myosin believed to interact in the generation of cytoplasmic streaming?

25. What is ameboid motion? How might the cortical gel, the plasma membrane, and the liquid inner cytoplasm interact in the production of ameboid motion? What mechanisms regulate ameboid motion? In what kinds of cells does ameboid motion take place?

26. What is capping? Furrowing? What roles do microfilaments play in these movements?

27. What movements are generated by microfilament growth?

Motility Without Microtubules or Microfilaments

Not all cellular movements depend on the activities of microtubules and microfilaments. Other motile mechanisms have appeared as evolutionary adaptations, including contractions produced by conformational changes in Ca^{2+}-sensitive proteins and changes in cell shape caused by alterations in osmotic pressure. Although generally not as conspicuous as microtubule- or microfilament-based systems, motile mechanisms based on Ca^{2+}-sensitive contractile proteins are highly efficient and specialized in some groups, especially among protozoa. In plants changes in cellular osmotic pressure are widely employed as a motile device.

Motile Systems Based on Ca^{2+}-Sensitive Contractile Proteins

Movements produced by Ca^{2+}-sensitive proteins depend on conformational changes induced by Ca^{2+} binding and release. As Ca^{2+} binds to these molecules, segments of the protein undergo extensive conformational changes, leading to an active contraction of the protein in one or more directions. Reverse alterations are produced as Ca^{2+} is released. Conformational changes occurring as Ca^{2+} is bound or released are not unusual; many proteins, including calmodulin, respond to calcium binding or release by altering their folding patterns.

Motile systems driven by conformational changes in Ca^{2+}-sensitive proteins take several forms among protozoa. One form, the *spasmoneme,* involves a long, slender fiber built up from Ca^{2+}-sensitive proteins that can contract almost instantly into a tight coil. Another form, the *myoneme,* consists of a thick bundle of contractile proteins that shortens more slowly but still at rates comparable to striated muscle. Proteins with similar Ca^{2+}-dependent contractile properties may also occur in *flagellar rootlets,* which anchor the basal bodies of flagella in many eukaryotic species.

Spasmoneme Contractions in Vorticellids Vorticellids are protozoans with a bell-shaped cell body attached to a substrate by a long, slender stalk (Fig. 12-33). These protozoans respond to mechanical and chemical disturbances by contracting the stalk almost instantaneously into a tight coil. The contraction rate exceeds that of the fastest striated muscle by about 100 times. The movement has excited the curiosity of microscopists since its first description by Leeuwenhoek three centuries ago.

Ultrastructural studies by W. B. Amos showed that a fibrous bundle running the length of the stalk, built up from parallel masses of subunits about 3 to 6 nm in diameter, contracts to convert the stalk from ex-

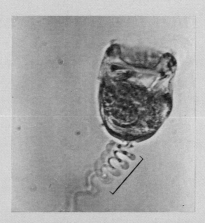

Figure 12-33 A vorticellid with its stalk contracted into a tight coil (bracket). ×300.

tended to helical form. Interspersed among the fibers of the stalk are smooth ER tubules. Amos found that the proteins isolated from the stalk do not include actin, myosin, or tubulin. Instead, two major Ca^{2+}-binding proteins, collectively called *spasmin,* make up 80% of the mass of isolated stalk protein. Among the remaining proteins is a Ca^{2+}-ATPase active transport pump.

Amos and his colleagues discovered that changes in Ca^{2+} concentration within ranges typical of the cytoplasm could induce repeated cycles of contraction and relaxation in spasmonemes in the total absence of ATP. From this information, and the fact that membranous vesicles in the stalk had previously been shown to be capable of Ca^{2+} storage and release, Amos proposed that contraction of the fibrous elements within the stalk is directly powered by changes in Ca^{2+} concentration. To initiate contraction, Ca^{2+} is released from the vesicles by a mechanism analogous to Ca^{2+} release by SR in striated muscle. The Ca^{2+} binds to contractile proteins of the stalk, initiating conformational changes that directly shorten the fiber, possibly by altering each protein molecule in the fiber from an extended to a helical form. Relaxation of the stalk is induced by the Ca^{2+}-ATPase pump, which returns Ca^{2+} to the vesicles. The resulting drop in Ca^{2+} concentration causes release of the Ca^{2+} bound to the contractile fiber, which responds by returning to the extended form.

Although stalk contraction is evidently directly powered by changes in Ca^{2+} concentration, the Ca^{2+} gradients necessary for the process are created by Ca^{2+}-ATPase pumps located in the vesicle membranes in the stalk. Therefore, the energy for the contractile process ultimately originates from ATP hydrolysis. In keeping with this conclusion, ATPase inhibitors were found by Amos and his coworkers to inhibit stalk contraction.

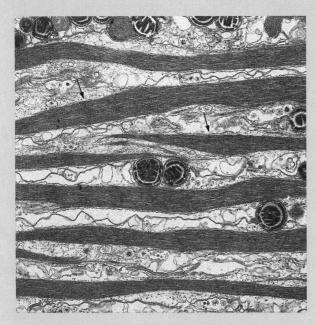

Figure 12-34 Myoneme bundles (arrows) responsible for body wall contraction in *Stentor*. × 13,000. Courtesy of B. Huang, from *J. Cell Biol.* 57:704 (1973), by permission of the Rockefeller University Press.

The inhibitors had no effect, however, if the stalks were exposed experimentally to repeated changes in Ca²⁺ concentration. Similar spasmoneme-based motile systems have been observed in a few other protozoans, including several dinoflagellates and *Actinocoryne*, a heliozoan that, like vorticellids, attaches to a substrate by a contractile stalk.

Myonemes in *Stentor* and Other Protozoans *Stentor* is an elongated, trumpet-shaped protozoan that, when disturbed, contracts into a compact, spherical ball. B. Huang and D. R. Pitelka discovered that the force for contraction in *Stentor* is generated by myonemes distributed throughout the cell just under the plasma membrane (Fig. 12-34). The fibers within myonemes measure 4 nm in diameter when *Stentor* is in the extended, relaxed state. On contraction the myonemes shorten; within the myoneme bundles fiber diameter increases to 10 to 12 nm. Cross sections of myoneme fibers in the contracted state show dense rings surrounding an open center, suggesting that during contraction the fibers wind into a tightly coiled helix.

Huang and Pitelka discovered a connection between Ca²⁺ concentration and myoneme contractions by experiments with killed *Stentor* cells. Elevating Ca²⁺ concentration caused slow contraction of myonemes in the killed *Stentor* cells. When the Ca²⁺ concentration in the bathing medium was lowered, the myonemes reextended. Similar experiments, with the same results, were later carried out with isolated myonemes by other investigators.

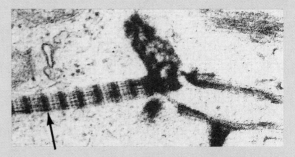

Figure 12-35 A ciliary rootlet (arrow) in a frog epithelial cell. The basal body and flagellar shaft are not visible in this section. × 75,000. Courtesy of W. L. Dentler from *J. Ultrastr. Res.* 86:75 (1984).

Myoneme-based contractile systems have been observed in other protozoans, including *Spirostomum* and dinoflagellates such as *Noctiluca*. Contraction of myonemes in all these systems, although much slower than spasmoneme contraction, is comparable in rate to that of striated muscle in animals. In at least one of these protozoans, *Spirostomum*, reextension of the cell after myoneme contraction is powered by microtubules.

Flagellar and Ciliary Rootlets The basal bodies of flagella and cilia in many cell types are anchored in the underlying cytoplasm by fibrous, tapered flagellar or ciliary rootlets (Fig. 12-35). The rootlets, which are widely distributed among eukaryotes, are marked by regular cross striations reminiscent of the striations of collagen fibers (see p. 276). The periodicity and pattern of the cross striations in flagellar rootlets and collagen are completely different, however. Depending on the cell type and species, flagellar rootlets vary in length from as few as 100 to 200 nm to more than 100 μm.

The rootlets, through their connections to basal bodies, are believed to maintain flagella or cilia in a position that orients the plane of flagellar beating. In normal cells of the green alga *Chlamydomonas*, for example, a short flagellar rootlet extends between the basal bodies of two flagella and apparently orients them at an angle to each other and to the cell surface (Fig. 12-36). In mutants lacking the rootlet, the basal bodies take irregular orientations and the flagella, while beating in normal wave patterns, beat in random directions. The mutants, as a consequence, swim randomly with no evidence of directional control.

There is a good possibility that flagellar rootlets are motile as well as cytoskeletal, and that movements of the structures are produced not by microtubules or microfilaments but by conformational changes triggered by alterations in Ca²⁺ concentration. J. L. Salisbury and his colleagues discovered that rootlets of the alga *Tetraselmis* change extensively in length depending on the presence of Ca²⁺ in the medium (Fig. 12-37). The rootlets are fully extended when cells are

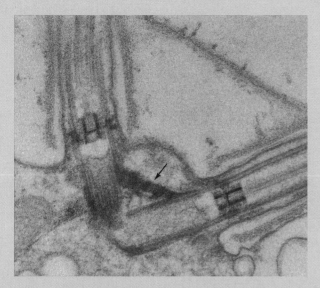

Figure 12-36 The rootlet (arrow) maintaining the two basal bodies in a V-shaped orientation in a normal *Chlamydomonas* cell. Courtesy of J. W. Jarvik.

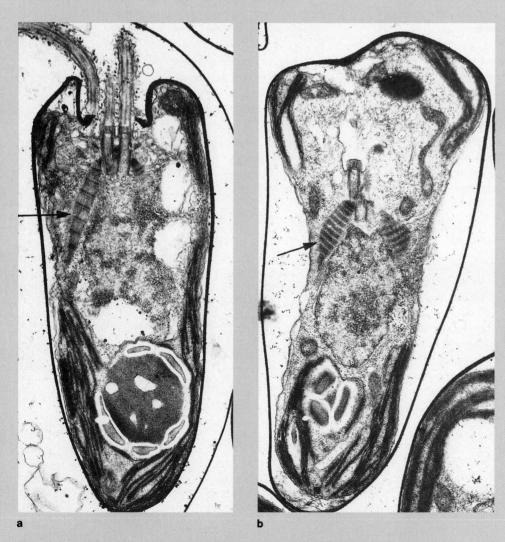

a

b

Figure 12-37 Flagellar rootlets (arrows) of the green alga *Tetraselmis* in extended **(a)** and contracted **(b)** form. The cell shown in **(a)** was fixed in a calcium-free medium; **(b)** was exposed to Ca^{2+} during fixation. ×15,000. Courtesy of J. L. Salisbury; reproduced from *J. Cell Biol.* 99:962 (1984) by copyright permission of the Rockefeller University Press.

fixed in calcium-free solutions (Fig. 12-37a). When exposed to Ca^{2+} during the fixation processes, the rootlets contract to less than half of their extended length (Fig. 12-37b). At the same time the two flagella of *Tetraselmis* undergo radical changes in position. Isolation and analysis of *Tetraselmis* flagellar rootlets yields a group of calcium-sensitive polypeptides with molecular weights near 20,000. None of the polypeptides has ATPase activity, and none has any apparent similarities to microtubule or microfilament proteins. Addition and removal of Ca^{2+} converts the purified rootlet proteins reversibly between fibrous (low Ca^{2+}) and globular (high Ca^{2+}) forms, indicating that the rootlet polypeptides can undergo Ca^{2+}-dependent conformational changes between extended and contracted states. Similar results have been obtained with flagellar rootlets in other algae and protozoa.

If the motility of flagellar rootlets is actually produced by Ca^{2+}-dependent proteins, the mechanism links the rootlets to the proteins of spasmonemes and myonemes. Significantly, Salisbury discovered that antibodies raised against flagellar rootlets also react with proteins in the cell centers of mammalian cells. Antibodies developed against spasmin, the contractile protein of spasmonemes, also cross react with flagellar rootlets and proteins forming part of the cell center. On this basis Salisbury suggested that a "heart" of spasmin may beat at the center of every animal cell! Thus, it may well be that Ca^{2+}-sensitive contractile proteins of the spasmin type, although reaching their highest evolutionary specialization and development among protozoa, are widely distributed as a minor, but still vital, motile system among animal cells.

Movements Generated by Osmosis in Plants

Observations that higher plants move their leaves on a daily cycle have been recorded for more than 2000 years. During the day the leaves of many plants change position to remain facing the sun as it moves across the sky. These movements have been shown to depend on changes in the internal osmotic pressure of cells located in a bulblike organ, the *pulvinus*, located at the base of the leaf. Cells located toward the side of the pulvinus opposite to the direction of movement swell, and cells on the side toward the direction of movement shrink to produce the motion. The alterations in osmotic pressure responsible for these changes result from massive movements of ions such as K^+ and Cl^- across the plasma membranes of pulvinus cells.

The signal for initiation of the ion movements in pulvini may involve elements of the $InsP_3$/DAG regulatory pathway (see p. 247). M. J. Morse and his co-workers discovered that in the plant *Samanea* the ion concentration of pulvini cells changes in response to conversion of phosphatidylinositol to IP_3, as in the IP_3/DAG system. The receptor in the *Samanea* system is not a membrane protein but a light-absorbing pigment in the cytoplasm, probably a phytochrome (see p. 406). The pigment converts from a red- to a far red-absorbing form on exposure to light; in some way the conversion leads to release of $InsP_3$ as a second messenger triggering changes in ionic balance within the pulvini.

In a few plants, such as the sensitive mimosa and Venus's-flytrap, leaf movements due to changes in cellular pressure take place in response to touch or stimuli such as heat or electrical shock. Movement of the touch-sensitive leaves is rapid, compared to the 24-hour cycle of light-induced leaf movement. In Venus's-flytrap leaf closure may be complete within 0.5 seconds of the stimulus. In mimosa the response is propagated from the point of stimulus at a rate of about 2 centimeters (cm) per second, so that leaves fold progressively along a stem or branch. Electrical measurements show that movements of K^+ and other ions are responsible for propagation of the response, in a change that resembles propagation of an action potential in neurons.

Evolution has explored a variety of pathways for the generation of motion in eukaryotes in addition to microtubule- and microfilament-based systems. Unique motile systems also occur in prokaryotes, including the concentration gradient-driven "motors" powering the rotation of bacterial flagella (see p. 447). Although some eukaryotic systems, such as Ca^{2+}-sensitive contractile proteins and the osmotic pressure changes of plant cells, are widely distributed in nature, microtubules and microfilaments remain the only motile systems employed universally in eukaryotic cells.

Suggestions for Further Reading: See p. 489.

MICROTUBULES, MICROFILAMENTS, AND INTERMEDIATE FILAMENTS IN THE CYTOSKELETON

- *Intermediate filaments, structure, and biochemistry* ▪ *Cytoskeletal roles of microtubules, microfilaments, and intermediate filaments* ▪ *Cytoskeleton of the plasma membrane* ▪ *Microvilli and brush borders* ▪ *Cytoskeleton of the cell cortex* ▪ *Inner cytoplasm* ▪ *Dynamic changes in the cytoskeleton* ▪ *The cytoskeleton and disease*

The cytoplasmic ground substance—the material surrounding the major cell organelles in the cytoplasm—was originally thought to be an unorganized solution of proteins. In the most extreme view cells were regarded simply as unstructured "bags of enzymes." The major cellular organelles were considered to be freely suspended within the cytoplasmic solution, with no fibrous elements or other supportive structures stabilizing their positions.

Recent research has revealed that rather than being an unstructured, freely mobile suspension or solution, the cytoplasmic ground substance is supported and stabilized by a complex network of cytoskeletal elements. One or more of at least three structures—microtubules, microfilaments, and intermediate filaments—make up the major scaffolding of the cytoplasmic network. These elements are crosslinked by a host of additional proteins and combined with other as yet unidentified fibers into an integrated system that orders and supports the membranes, organelles, and ground substance of the cytoplasm and, in many systems, helps to fix the cell in its environment.

Cytoskeletal systems extend from the plasma membrane to the nuclear envelope; a specialized matrix is suspected to support the nuclear envelope and the interior of the nucleus as well. The cytoskeleton also anchors the cell to extracellular structures via linkages that extend through the plasma membrane. A major fraction of total cellular protein, as much as one-third in some cells, is cytoskeletal material. This material, rather than being fixed in position in cells, has been found to be a dynamic entity that changes in makeup and structure as cells develop and differentiate, move, and alternate between cycles of growth and division. The importance of the cytoskeleton to cell function is underlined by the drastic effects of some of the known mutations in cytoskeletal elements, not only on individual cells but on the entire organism. Several human diseases, in fact, are caused directly by cytoskeletal faults.

Chapters 11 and 12 surveyed the structure and biochemistry of two of the three major elements forming cytoskeletal supports—microtubules and microfilaments. This chapter begins with a discussion of the remaining major element, the large and complex family of fibers classified as intermediate filaments. The interactions of intermediate filaments with microtubules and microfilaments are presented next, followed by examples of the cytoskeletal systems associated with major regions of the cytoplasm. Finally, dynamic changes in the cytoskeleton related to motility, development, and cell division are considered, along with cytoskeletal alterations causing or accompanying disease. The primary emphasis of this chapter is on the cytoplasm; the systems known or suspected to support the nuclear envelope and nuclear interior are described in Chapter 14.

STRUCTURE AND BIOCHEMISTRY OF INTERMEDIATE FILAMENTS

Intermediate filaments (Figs. 13-1 and 13-2) are so called because their diameters, which average about 10 nm, range between those of microtubules (about 25 nm) and microfilaments (about 5 to 7 nm). Although all are assembled from polypeptides belonging to one family, intermediate filaments are much more varied in structure than either microtubules or microfilaments. Five major classes of intermediate filaments have been discovered in the cytoskeletal structures of various animal cells, including the *keratin, vimentin, desmin, neurofilament,* and *glial filament* classes (Table 13-1). These filament classes are assembled from related but distinct types of proteins: *cytokeratins, vimentin, desmin, neurofilament,* and *glial fibrillary acid* proteins, respectively. Other filament

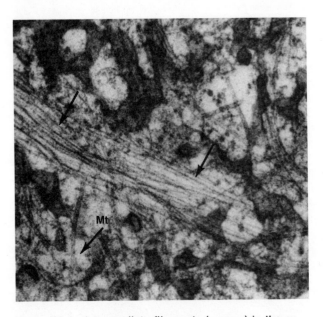

Figure 13-1 Intermediate filaments (arrows) in the cytoplasm of a cultured rat kangaroo cell. Microtubules (Mt) are also visible in the section. ×71,000. Courtesy of K. McDonald.

Table 13-1   Intermediate Filament Classes and Their Polypeptides

Class	Polypeptides	Number	Type	Molecular Weights	Occurrence
Keratin	Cytokeratins I and II	30	I and II	40,000–70,000	Epithelial cells
Vimentin	Vimentin	1	III	57,000	Cells of mesodermal origin
Desmin	Desmin	1	III	55,000	Skeletal, cardiac, and smooth muscle cells
Glial	Glial fibrillary acid protein	1	III	51,000	Glial cells
Neurofilament	Neurofilament	3	IV	70,000, 160,000, and 210,000	Neurons
Lamin	Lamins A, B, and C (in mammals)	3	V	72,000 (A), 66,800 (B), and 65,000 (C)	Possibly in all eukaryotic cells

types assembled from related polypeptides have been detected in cells, including a major class, the *lamins*, associated with the nuclear envelope. (Ch. 14 details lamins and their arrangement in the nuclear envelope.)

Intermediate filaments of all five major cytoplasmic classes occur throughout vertebrate animals. Intermediate filaments have also been observed in nervous, epithelial, and muscle tissue in cells of individuals representing most of the invertebrate phyla, in slime mold cells, and in protozoa. No cytoskeletal elements related to intermediate filaments have as yet been definitely identified in plants. However, C. W. Lloyd and his coworkers have found fiber bundles in carrot and other plant cells that cross react with monoclonal antibodies

developed against animal intermediate proteins. The intermediate filament proteins of lamins may also be generally distributed among eukaryotes, including protists, fungi, and plants as well as animals.

Intermediate filaments are most often identified in intact cells via the fluorescent antibody technique, in which antibodies made against intermediate filament proteins are linked chemically to a fluorescent marker (Fig. 13-3; see p. 119 for details of the technique). Preparations of this type have revealed the unexpectedly wide distribution of intermediate filaments in animal cells and have allowed the various classes of intermediate filaments to be identified with specific cell types.

Intermediate Filament Structure

Sectioned intermediate filaments appear in the electron microscope as solid, cylindrical fibers with diameters ranging between extremes of about 7 and 12 nm, and averaging about 10 nm. They may extend for long

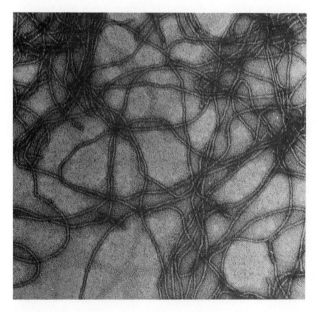

Figure 13-2 Intermediate filaments of the vimentin class isolated from Chinese hamster ovary cells and prepared for electron microscopy by negative staining. × 115,000. Courtesy of P. M. Steinert.

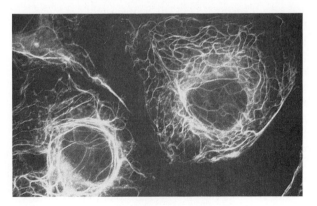

Figure 13-3 Keratin intermediate filaments stained by the fluorescent antibody technique in a cultured cell. Courtesy of B. Geiger.

The Experimental Process

Of Mice and Men: Genetic Skin Diseases Arising from Defects in Keratin Filaments

Elaine V. Fuchs

ELAINE FUCHS received her Ph.D. in Biochemistry from Princeton University in 1977, working under the supervision of Dr. Charles Gilvarg. From Princeton, Dr. Fuchs went to the Massachusetts Institute of Technology, where she was a Damon Runyon–Walter Winchell postdoctoral fellow, studying with Dr. Howard Green. In 1980, she became an Assistant Professor in Biochemistry at the University of Chicago; she was promoted to Associate Professor of Molecular Genetics and Cell Biology in 1985, and to Professor in 1989. She was also appointed to the Howard Hughes Medical Institute in 1988, where she is currently an investigator. She has received a number of academic honors and awards, including a Searle Scholar Award, NIH Career Development Award, Presidential Young Investigator Award, and R. R. Bensely Award from the American Association of Anatomists.

The epidermis provides the protective interface between various chemical and physical traumas of the environment and the rest of the bodily organs. It manifests its protective function by building an extensive cytoskeletal network of 10 nm intermediate filaments (IFs) composed of keratin proteins. Keratins (molecular mass 40–70 kilodaltons) are a family of proteins that can be subdivided into two distinct sequence groups, type I and type II, based on sequence homology. Type I and type II keratins are coexpressed as specific pairs, which first form obligatory heterodimers, then assemble into tetramers, and finally associate into higher ordered structures, leading to IFs with a 1:1 ratio of the two proteins.[1,2] In the early 1980s, Werner Franke's laboratory discovered that during epidermal development, embryonic basal cells are the first to express detectable levels of the type I keratin K14 (50 kd) and type II keratin K5 (58 kd). Subsequently, we found that the synthesis of this keratin pair increases greatly later in development, when cells begin to differentiate into epidermis. As basal epidermal cells differentiate and move outward toward the skin surface, they reduce K5/K14 expression and switch on a new pair of keratins, K1 (67 kd) and K10 (56.5 kd).[3] In the fully differentiated squamous cell, these keratins constitute approximately 85% of total cellular proteins. Thus, keratins are to an epidermal cell what globins are to a red blood cell.

Since embarking on keratin research nearly 15 years ago, a major question that we have asked is whether there might be defects in keratin genes giving rise to genetic skin diseases, as there are defects in globin genes giving rise to genetic blood diseases such as sickle cell anemia and thalassemias. One way to address this question is to genetically alter keratin sequences and evaluate the consequences of these defects on protein function. At the time we embarked on these studies, the function of keratin filaments was largely unknown. However, it was well established that these proteins could self-assemble into IFs, forming a distinct cytoskeletal network in keratinocytes. Therefore, after isolating and characterizing the genes encoding the human keratins K5 and K14, we began to engineer defects in the coding sequence of the K14 gene. In order to monitor the expression of the mutant keratin in cells that expressed the normal K14, we replaced sequences encoding the antigenic carboxy-terminal 5 residues of the K14 protein with sequences encoding the antigenic portion of another protein, the neuropeptide substance P. This enabled us to use an antibody to K14 to identify the wild-type K14 protein and an antibody to substance P to identify the mutant K14 protein.

We first introduced the tagged wild-type and mutant keratin genes into cultured epidermal cells. Using a technique known as gene transfection, a calcium phosphate-DNA precipitate was layered onto the surface of the keratinocytes. Approximately 10% of the cells engulfed this precipitate, enabling the mutant keratin genes to enter these cells. When these genes were expressed, the wild-type and mutant keratin proteins recognized the native keratin K5 in epidermal cells, formed heterodimers and then heterotetramers, and finally integrated into the keratin filament network.[4] The small substance-P tag did not interfere with the ability of the keratin to form keratin networks, as judged by double immunofluorescence staining with antibodies against substance P and K14 (Figure A, frames (A) and (B), respectively; note the two transfected cells in the center). In addition, some mutations that we engineered had no effect, and the keratin filament network containing the mutant protein appeared normal as judged by this same technique. However, some mutant keratins caused gross distortions in the keratin filament network, producing withdrawal from the plasma membrane and disruption of the network (frames (C) and (D); note the one transfected cell in the center).

Did these mutant proteins interfere with IF formation, or did they distort some intracellular interaction(s) involving keratins? To evaluate this, we conducted *in vitro* filament assembly studies using purified human K5 and

Figure A Expression of keratin 14 mutants in human keratinocytes using gene transfection. Tagged wild-type and mutant K14 genes were introduced into cultured human keratinocytes using gene transfection. The expression of these transgenes was detected with a primary rabbit antibody against the tag and a fluorescein (green fluorescence) secondary antibody against rabbit IgG (A and C). The expression of the native keratin network was detected with a primary guinea pig antibody against K5 and a Texas Red (red fluorescence) secondary antibody against guinea pig IgG (B and D). Cells in frames A and B were transfected with a P-tagged wild-type K14 gene, whose product integrated into, but did not disturb, the keratin network. Cells in frames C and D were transfected with a dominant negative mutant K14 gene, whose product integrated into and distorted the keratin network. Bar represents 25 μm. (Courtesy of Anthony Letai and Dr. Kathryn Albers; see also reference 4.)

mutant K14 obtained from bacterial clones. A bacterial promoter was used to express the human K14 and K5 cDNAs in *E. coli*, which do not make keratins or other IF proteins. In these bacteria, the keratins accumulated in large cytoplasmic aggregates, called inclusion bodies.[2] These bodies could be isolated and solubilized in a 6.5 *M* urea buffer. In the presence of both K5 and K14 (or mu-

tant K14), stable tetramers formed, which could be isolated by anion exchange chromatography. When dialyzed to remove the urea, these tetramers then assembled into IFs. In contrast to wild-type K14/K5 tetramers, which formed long, uniform IFs (Figure B, frame A), some mutant K14/K5 tetramers assembled into filaments that were aggregated and sometimes frayed at their ends (frame B),

Figure B *In vitro* assembly of human K14 and K5 proteins from genetically engineered bacteria. Bacteria were genetically engineered to produce wild-type K14 and K5 human proteins, and mutant K14 proteins. Purified K5 and K14 proteins were solubilized in a 6.5 *M* urea buffer, and dialyzed against assembly buffer. (A) Filaments assembled from wild-type K5 and K14; (B) filaments assembled from wild-type K5 and a K14 mutant that caused filament aggregation; (C) filaments assembled from wild-type K5 and a K14 mutant that completely disrupted IF formation; (D) filaments assembled from wild-type K5 and the same K14 mutant shown in C, but this time in the presence of wild-type K14, at a 99% wild-type to 1% mutant K14 ratio. Note that this was a very potent dominant negative mutant: even 1% of the mutant protein was enough to interfere with filament elongation, resulting in shorter filaments than wild-type. Bar represents 100 nm. (Courtesy of Dr. Pierre Coulombe; see also reference 4.)

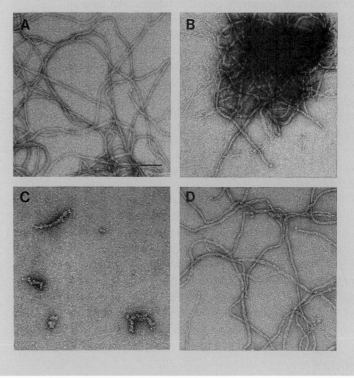

and some could not assemble into 10 nm structures at all (frame C).[4] Those mutants that were unable to assemble with K5 into IFs could nevertheless interfere with the assembly of wild-type K5 and K14: when as little as 1% of the mutant K14 protein shown in frame C was mixed with 99% wild-type K14 and 100% wild-type K5, the filaments formed were short and somewhat irregular (frame D). This is called a *dominant negative mutant*: the mutant recognized the wild-type proteins and interfered with their function. Most dominant negative mutants of K14 were deletions or point mutations within the ends of a 310-amino-acid-residue domain predicted to be largely α-helical and involved in the intertwining of K14 and K5 polypeptides into a coiled-coiled heterodimer.

At this point in our studies, we wondered how our dominant negative mutant keratins might behave in the epidermis of skin: what happens when the native keratin filament network of a basal epidermal cell is disrupted? To answer this question, we used the human K14 promoter to drive the expression of a dominant negative K14 mutant in the basal epidermal cells of the skin of transgenic mice.[5] For this technique, we introduced the mutant keratin gene, called a *transgene*, into fertilized single-cell mouse embryos. The mouse embryos were then transferred to the oviduct of a foster mother mouse, who was mated the night before with a vasectomized male. Thus, the foster mother had the right hormones to nurture and carry the embryos to birth. When the mice were born, some of them contained the transgene in every cell of their body. These mice were from embryos that integrated at least one copy of the transgene into their own chromosomal DNA before their first cell division.

Our transgenic animals exhibited morphological and biochemical symptoms typical of humans who have an autosomal dominant skin disease known as *epidermolysis bullosa simplex (EBS)*.[5,6] Upon incidental trauma such as suckling, the mouse skin blistered. Microscopic examination of blistered skin sections revealed extensive cell breakdown in the basal epidermal layer where the mutant keratin transgene was expressed. The basal cells also showed disorganized keratin filament networks, often with shorter filaments. In severe cases, clumps of keratin protein were abundant in the cytoplasm of these cells.[5] Interestingly, the cells always lysed in a specific zone of the columnar basal epidermal cell, beneath the nucleus and above the plasma membrane. The pattern of cell breakage suggested to us that the function of the keratin filament networks in these basal epidermal cells is to impart mechanical integrity, without which the columnar cells become fragile and then lyse upon incidental trauma. In addition, these studies provided us with an important clue to the possible genetic basis for a sometimes life threatening and psychologically devastating human skin disease that affects about 1 person in 50,000.

At this point, our studies turned to the human skin disease EBS.[7] In conjunction with dermatologists, we obtained skin samples from EBS patients and cultured their epidermal cells. We examined the keratin networks in cultured EBS cells and discovered changes similar to those we had seen when genetically engineered mutant keratins were expressed by gene transfection in cultured human keratinocytes. In addition, the IFs assembled from the keratins isolated from EBS epidermal cells were shorter than normal. When the K5 and K14 mRNAs and genes from two different EBS patients were isolated and characterized, we found two different point mutations in the K14 genes, both leading to an amino acid change at arginine 125. In one case, the mutation resulted in a cysteine 125 mutation, and in the other, a histidine 125 mutation.[7] Interestingly, when 51 different IF sequences sharing 25–99% amino acid identity were examined, the residue corresponding to this position was either arginine or lysine in 50 sequences. Moreover, this residue was within one of two small segments of the K14 protein that we had shown through our mutagenesis studies to be highly critical for IF assembly.

We do not yet know the extent to which EBS diseases have as their basis defects in keratin genes. EBS may have multiple causes, and defects in posttranslational modification of keratins or defects in as yet unidentified proteins that associate with, degrade, or sever keratin filaments in basal epidermal cells may also play a role in the disease. Finally, we would predict that some EBS cases should have defects in K5 as well as K14 genes. As more cases are examined, the etiology of EBS should become clearer.

In conclusion, our studies have shown that by starting with a protein and using a molecular genetic approach, we have been able to gain valuable insights into our understanding of the function of the protein, as well as developing an animal model for the study of a human skin disease. This approach should be broadly applicable to a number of fundamental questions in cell biology, where proteins have been extensively studied, but aspects of their function and relations to human diseases are unknown.

References

[1] Aebi, U.; Haner, M.; Troncoso, J.; Eichner, R.; and Engel, A. *Protoplasma* 145:73–81 (1988).

[2] Coulombe, P. A., and Fuchs, E. *J. Cell Biol.* 111:153–69 (1990).

[3] Fuchs, E. *J. Cell Biol.*, 111:2807–14 (1990).

[4] Coulombe, P. A.; Chan, Y.-M.; Albers, K.; and Fuchs, E. *J. Cell Biol.* 111:3049–64 (1990).

[5] Vassar, R.; Coulombe, P. A.; Degenstein, L.; Albers, K.; and Fuchs, E. *Cell* 64:365–80 (1991).

[6] Haneke, E., and Anton-Lamprecht, I. *J. Invest. Dermatol.* 78: 219–23 (1982).

[7] Coulombe, P. A.; Hutton, M. E.; Letai, A.; Hebert, A.; Paller, A. S.; and Fuchs, E. *Cell* 66:1301–11 (1991).

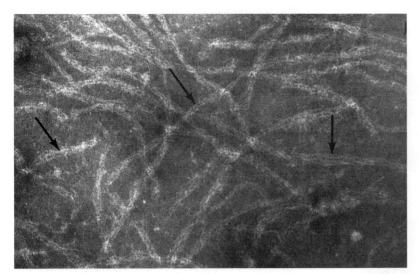

Figure 13-4 Human keratin intermediate filaments exposed to phosphate ions. The filaments have unwound slightly, showing subdivision into twisted subfibers called protofilaments (arrows). × 180,000. Courtesy of U. Aebi; reproduced from *J. Cell Biol.* 97:1131 (1983), by copyright permission of the Rockefeller University Press.

distances in the cytoplasm, lying singly or linked into bundles or networks. Microtubules and microfilaments usually run in more or less straight courses in the cytoplasm, but intermediate filaments are often bent into wavelike forms, suggesting that they are more elastic and more flexible than microtubules or microfilaments. Crosslinks can frequently be seen to extend between intermediate filaments and other cell structures, such as organelles, the plasma membrane, microtubules, microfilaments, or groups of ribosomes.

When isolated under mildly disruptive conditions, such as exposure to phosphate ions or other charged molecules at low concentrations, intermediate filaments unwind somewhat, revealing a twisted, ropelike structure (Fig. 13-4). This appearance indicates that intermediate filaments contain several to many subfibers wound into a helical structure, an impression borne out by molecular studies of intermediate filament assembly (see below).

Proteins of all major intermediate filament types have been at least partially sequenced, either directly or through the sequences of genes encoding them. The sequences confirm that protein subunits making up intermediate filaments are members of a single family. The sequences also reveal several structural features that are shared by all proteins of the family (Fig. 13-5). All contain a central region more than 300 amino acids in length that can wind into four relatively rigid, rodlike alpha-helical segments, separated by three less structured regions (Fig. 13-5a). The four alpha-helical segments contain a repeating a-b-c-d-e-f-g pattern of amino acid residues, in which the first and fourth residues are hydrophobic. This repeating sequence pattern is typical of alpha-helical chains that can wind by twos into a coiled-coil pigtail like those of tropomyosin and the rodlike myosin tail (Fig. 13-5b; the coiled-coil structure is discussed in detail in Information Box 2-4). The

fibrous nature of the central rodlike portion of intermediate filament proteins, which is about 40 nm long, sets them apart from the protein subunits of microtubules and microfilaments, which are globular rather than fibrous proteins.

At either end of the central region the amino acid chain folds into less ordered globular segments. The end segments, which contain from 50 to 200 amino acid residues in different intermediate filament proteins, are highly diverse in sequence and probably account for many of the distinctive properties of individual intermediate filament types. The globular segments, particularly those at the C-terminal end, are also primarily responsible for variations in molecular weight among the different intermediate filament proteins, which range from about 40,000 to as much as 210,000.

The central rod segment, with its characteristic pattern of alpha-helical central segments, and the globular end caps are hallmarks of this protein family. These features distinguish intermediate filament proteins from all other cellular proteins and provide the criteria necessary to identify a protein as an intermediate filament type. The lamins associated with the nuclear envelope, for example, were not known to be related to cytoplasmic intermediate filaments until sequencing studies and research with antibodies, carried out by M. W. Kirschner, F. D. McKeon, G. Blobel, M. Osborn, and K. Weber, revealed that they contain the internal arrangement of amino acids typical of intermediate filament proteins. Undoubtedly, more proteins belonging to the intermediate filament family remain to be discovered.

The amino acid sequences of different intermediate filaments fall into a number of related subgroups (see Table 13-1). Within subgroups as many as 90% to 95% of the amino acid sequences of the members are similar. Different subgroups, however, are related in only about

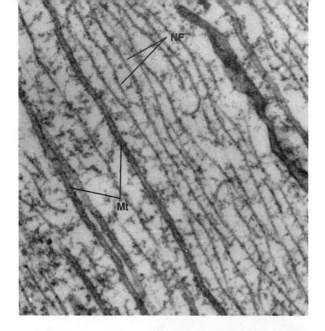

Figure 13-7 Neurofilaments (NF) and microtubules (Mt) in an axon of a human neuron. ×96,000. Courtesy of P. Dustin and J. Flament-Durand, from *Microtubules*, 2nd ed., New York: Springer-Verlag, 1984.

70,000, 160,000, and 210,000, respectively. Among mammalian and avian neurofilament proteins, only the light type can assemble by itself into intermediate filaments in the test tube (see below). Medium and heavy proteins are incapable of filament formation when separately isolated and purified. Although evidently not necessary for filament assembly, the medium and heavy proteins add in some way to the structure or interactions of neurofilaments when assembled with the light proteins. There are indications that the heavy proteins form a crosslinker extending from the surface of neurofilaments. This crosslinker probably ties adjacent neurofilaments into bundles and may maintain the regular 45-nm separation frequently visible between parallel neurofilaments in axons. Thus evolutionary changes may have modified the heavy neurofilament protein of birds and mammals into a role resembling an intermediate filament–associated protein (see below). There are as yet no hints about the possible roles of the medium protein in neurofilaments. There are indications that two additional proteins, *α-internexin* and *nestin*, may be part of the type IV subgroup. These proteins, which were recently shown to contain typical intermediate filament sequence structures, may contribute to neurofilament structure, particularly in embryonic neurons.

The neurofilament proteins of reptiles, and presumably also those of lower vertebrates, contain only two proteins corresponding to the light and medium types of birds and mammals. Some invertebrates, such as the squid, also produce two neurofilament protein types. Other invertebrates have only one.

Not all neurons, and not all regions within neurons, contain high concentrations of neurofilaments. In mam-

mals neurofilaments occur in greatest numbers in axons of the peripheral nervous system, particularly in those of larger diameter. Nerve cell bodies and dendrites have reduced numbers of neurofilaments or none at all. Many central nervous system cells, such as the Purkinje cells of the cerebellum, contain only very limited numbers of neurofilaments in any cellular region. In spite of their prevalence in some types of neurons, the function of neurofilaments remains uncertain. Presumably, the neurofilaments of axons contribute to the elasticity and tensile strength of these long cell extensions and prevent breakage or separation as they stretch to accommodate the movements of animals.

Intermediate Filament Assembly

In contrast to microtubules and microfilaments, cytoplasmic intermediate filament types are almost completely stable at the salt concentrations and pH ranges typical of cells. In general intermediate filament proteins assemble spontaneously into filaments in the test tube under cellular conditions of salt concentration and pH, with no requirements for particular ions, modifying proteins, or energy in the form of ATP or other nucleotides. In order to induce intermediate filaments to disassemble in the test tube, they must be exposed to salt at very low concentrations (much lower than the concentrations typical for living cells) or to strongly denaturing agents, such as urea, at very high concentrations.

Research by several groups, including those of E. Fuchs, Z. E. Zehner, N. Geisler, and S. A. Lewis and N. J. Cowan, indicates that during assembly into intermediate filaments individual protein subunits first associate by twos into the coiled-coil pigtails shown in Figure 13-5b. In these dimers the two polypeptides are in parallel alignment, with their N-terminal ends directed toward the same end of the structure. To form the next level of structure, two coiled-coil pigtails probably line up side by side to form a *tetramer* containing four intermediate filament proteins (Fig. 13-5c). Within the tetramer the two coiled-coil pigtails are slightly staggered and lined up in antiparallel fashion, with the C-terminal ends of one pigtail next to the N-terminal ends of the other. This antiparallel arrangement would seem to rule out the possibility that the tetramers might form intermediate filaments with polarity, that is, with plus and minus ends of different chemical properties.

The tetramers line up end to end during or after their assembly to form a cablelike structure called a *protofilament* (Fig. 13-5d). Measurements of protofilament mass indicate that they probably contain two rows of tetramers. Thus a protofilament, which measures 4.5 nm in diameter, may be an *octamer* of intermediate filament polypeptides. Examination of slightly unraveled intermediate filaments indicates that four protofilaments intertwine to form the complete structure

(Fig. 13-5e). If protofilaments are indeed octamers, a fully assembled intermediate filament would contain 32 individual polypeptides in cross section.

Although the extreme conditions required for disassembly in the test tube might indicate that intermediate filaments are permanent, static elements once assembled inside cells, they actually appear to be dynamic structures that can readily be taken apart and reassembled (see below). The mechanisms used to disassemble intermediate filaments in cells remain uncertain, but they may include phosphorylation of intermediate filament proteins. Phosphorylation at sites in the N-terminal segments has been observed to occur prior to the disassembly of vimentin and desmin filaments. The lamins, which regularly disassemble from the nuclear envelope when this membrane system breaks down during cell division (see p. 1017), become phosphorylated just before disassembly and lose their phosphate groups as they reassemble at the close of division (for details, see pp. 1017 and 1031).

Experiments have also indicated that protein subunits can exchange between a soluble pool and fully assembled intermediate filaments in the test tube. K. J. Angiledes and his coworkers, for example, detected an apparent exchange of subunits marked by a fluorescent dye between neurofilaments and a protein pool in the test tube. Other work by J. Ngai, T. R. Coleman, and E. Lazarides used living mouse cells into which a chicken vimentin gene had been introduced by genetic engineering techniques (see p. 782). The chicken vimentin subunits synthesized in the cells, which were traced by their reaction with specific antibodies, were found to insert at random points along the length of existing intermediate filaments rather than adding to the ends or forming completely new filaments. A small pool of unassembled subunits was also detected in the cells used in this study. The pattern of insertion and the subunit pool suggests that a subunit–intermediate filament equilibrium may exist in living cells, which may account for the ability of these structures to assemble and disassemble inside cells. If this is the case, the equilibrium is expected to be tilted far in the direction of assembled filaments.

Intermediate Filament–Associated Proteins (IFAPs)

A gradually lengthening list of proteins has been detected in association with intermediate filaments of the different classes. Generally, these proteins, which have been designated as *intermediate filament–associated proteins* (*IFAPs*; Table 13-2), are so closely linked to intermediate filaments that they are retained with the filaments as they are purified and cleaned of cellular debris and contaminants. Antibodies developed against isolated IFAPs also react strongly with intermediate filaments inside cells, verifying a close association between IFAPs and intermediate filaments.

The work of assigning functions to IFAPs identified by these and other methods has only just begun. However, the functions provisionally identified for some known IFAPs indicate that they set up crosslinks between intermediate filaments or cap the ends of intermediate filaments to arrest further assembly. The crosslinkers, which set up end-to-end or end-to-side linkages, produce intermediate filament networks in the cytoplasm. One crosslinking IFAP, *filaggrin* (see Table 13-2), acts as a *bundler* by tying keratin-based intermediate filaments into parallel arrays. Some IFAPs also act as *interlinkers*, tying intermediate filaments to other elements of the cytoskeleton (see also below).

Intermediate filament crosslinkers evidently act only by assembling networks and fibers from existing intermediate filaments in the cytoplasm; none of the IFAPs discovered to date has the capacity to modify the reactions assembling intermediate filaments from their individual proteins. Thus the functions of MAPs and actin-binding proteins in adjusting equilibrium reactions that assemble microtubules and microfilaments appear to have no parallels among the IFAPs.

Intermediate Filament Regulation

Although intermediate filaments are highly stable and resistant to chemical or physical disturbances in the test tube, cells are capable of quickly modifying or disassembling their intermediate filament systems under cer-

Table 13-2 Some Intermediate Filament–Associated Proteins (IFAPs)

IFAP	Molecular Weight	Intermediate Filament Associations	Cell Types	Tentative Function
Epinemin	44,500	Vimentin	Muscle and other mesodermally derived cells	Unknown
Filaggrin	~30,000	Keratin	Epidermal cells	Bundling
Paranemin	280,000	Desmin, vimentin	Embryonic muscle cells (myocytes)	Crosslinking
Plectin	300,000	Vimentin	Wide variety of cells of mesodermal origin	Crosslinking
Synemin	230,000	Desmin, vimentin	Muscle cells, some erythrocytes	Crosslinking

Structure and Biochemistry of Intermediate Filaments **505**

tain conditions. During mitotic cell division cytoplasmic intermediate filament networks of many cell types are extensively rearranged and reduced or, in some cell types, disappear completely. In damaged mammalian nerve axons neurofilaments break down and then reorganize as a new axon is formed. Reorganization appears to take place by assembly from a cytoplasmic pool of existing subunits, without synthesis of new intermediate filament proteins.

Other modifications occur as cells progress through development and differentiation. Both vimentin and desmin intermediate filaments coexist in some embryonic muscle cells in association with the IFAPs *paranemin* and *synemin*. As these muscle cells mature into adult form, vimentin and paranemin disappear. Vimentin is also found in some embryonic nerve cells but not in fully differentiated adult neurons. These modifications point to the developmental regulation of both the type and composition of intermediate filaments as cells progress from embryonic to fully differentiated form.

Because addition of phosphate groups has been correlated with disassembly of vimentin and desmin filaments in the cytoplasm, and with lamin disassembly in the nuclear envelope, at least some of the dynamic changes noted in intermediate filament structure and arrangement probably depend on reversible addition of phosphate groups. Other patterns of disassembly may depend on hydrolysis by proteinases, possibly regulated by adjustments in Ca^{2+} concentration in the cytoplasm. Changes in the distribution of existing intermediate filaments or the degree of their linkage into networks are probably moderated through alterations in the activities of the crosslinking and bundling IFAPs.

CYTOSKELETAL ROLES OF INTERMEDIATE FILAMENTS, MICROTUBULES, AND MICROFILAMENTS

In animal cells intermediate filaments occur in association with microtubules and microfilaments in cytoskeletal systems that extend throughout the cytoplasm (Figs. 13-8 and 13-9). These systems both support the major cytoplasmic regions and maintain the position of the nucleus in the cytoplasm. In their supportive roles each major cytoskeletal element contributes characteristics that depend on its particular structural and biochemical properties, as altered by MAPs, actin-binding proteins, and IFAPs.

Functions of Microtubules, Microfilaments, and Intermediate Filaments

Microtubules, arranged singly, in networks, and in parallel bundles, evidently provide rigidity to cytoplasmic regions. When arranged in parallel bundles, as they are in the spindle and nerve axons, microtubules also establish *polarity* by setting up definite ends to which elements being transported or separated may be delivered. They probably also provide elasticity to cellular projections and possibly contribute to the tensile strength of cellular parts and regions subjected to a stretching force. The cytoskeletal activities of microtubules are modified by a variety of MAPs (see p. 425), which regulate assembly and disassembly and establish crosslinks between microtubules and other elements in the cytoplasm. The activity of many MAPs is controlled by Ca^{2+}, working either directly or through calmodulin, and by cAMP-activated enzymes. These mechanisms tie microtubule polymerization and crosslinking in the cytoskeleton to the major Ca^{2+} and cAMP-linked systems regulating cellular activities (see p. 245).

Microfilaments also occur singly, in networks, and in parallel bundles in the cytoskeleton. The degree of crosslinking of microfilament networks, regulated by a seemingly endless variety of actin-binding proteins (see Table 12-1), controls the viscosity of the cytoplasmic background substance and gives the cytoplasm consistencies ranging from liquid to a rigid gel. In parallel bundles microfilaments provide tensile strength to cytoplasmic regions. Other microfilament bundles support extensions of the plasma membrane such as microvilli. Many of the actin-binding proteins are regulated by

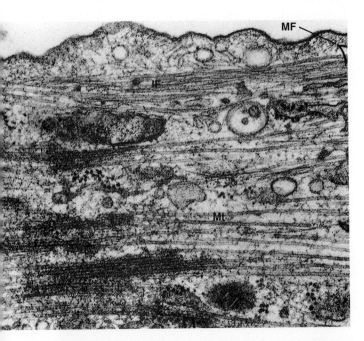

Figure 13-8 Microtubules (Mt), microfilaments (MF), and intermediate filaments (IF) in the region of the cleavage furrow of a dividing mammalian cell. ×52,000. Courtesy of K. McDonald, from *J. Ultrastr. Res.* 86:107 (1984).

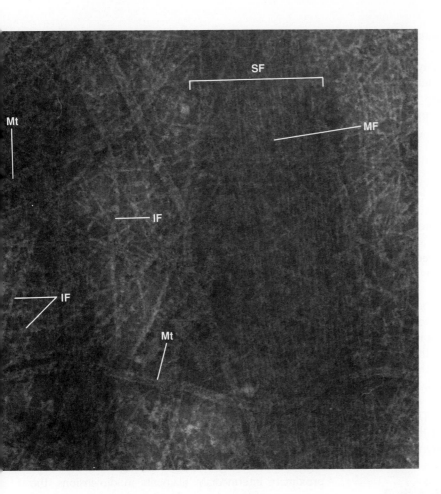

Figure 13-9 Microtubules (Mt), intermediate filaments (IF), and microfilaments (MF) in a fibroblast prepared for electron microscopy by a technique that produces a negatively stained image. The microfilaments are bundled into a stress fiber (SF). ×106,000. Courtesy of J. V. Small; reprinted with permission from *Electron Micr. Rev.* 1:155, copyright 1988, Pergamon Press Ltd.

Ca^{2+} and cAMP, also placing microfilament polymerization, crosslinking, and bundling under control of the major cellular regulatory systems.

Though intermediate filament networks and bundles are also almost universally assumed to act as supportive structures in the cytoplasm, there are as yet relatively few experiments confirming this conclusion (see Fuchs's essay on p. 498). Most of the supporting evidence is circumstantial—the filaments are found in locations and in arrangements expected for cytoskeletal supports, often in association with microtubules and microfilaments. One factor underlying the uncertainty is that, although anti-intermediate filament antibodies are available, there are no intermediate filament poisons equivalent to colchicine, cytochalasins, and other drugs that allow investigators to turn microtubule or microfilament function or assembly on and off at will. Further, in many cell types precipitation or coagulation of intermediate filament networks by exposure to antibodies has no detectable effects on cell shape or behavior, or on activities such as division or cell-to-cell interactions. In spite of these difficulties, there is little doubt among investigators that intermediate filaments are cytoskeletal supports; from their observed physical characteristics intermediate filaments are expected to provide cytoplasmic regions with elasticity and resistance to breakage during stretching.

Interlinkers and Integration of the Cytoskeleton

The three major cytoskeletal elements—microtubules, microfilaments, and intermediate filaments—are tied into a coordinated support system by a variety of interlinking proteins. Some of the interlinkers form ties between two cytoskeletal elements, such as microtubule-microfilament links, or microtubule–intermediate filament links, and some can interlink all three. The tau group of MAPs, for example, forms microtubule-microfilament interlinks; MAP-2 and another important interlinker, *ankyrin* (see below), have binding affinities for all three cytoskeletal elements. Some interlinkers, such as ankyrin and MAP-1, can form connections with membrane proteins as well, tying cytoskeletal elements to the plasma membrane and the boundary membranes of organelles in the cytoplasm. The list of known interlinking proteins is certain to be incomplete; more proteins with dual or triple binding affinities will undoubtedly be added as intensive research continues in this area. The known interlinkers are characteristically activated or inhibited by regulatory elements forming parts of the $InsP_3$/DAG-based and cAMP-based pathways linked to receptors at the cell surface. (See Ch. 7 for details of these pathways.)

Interlinking proteins integrate the cytoskeleton to such an extent that none of the cytoskeletal elements is likely to be completely independent as a supportive element in the cytoplasm. As a consequence, modification or disturbance of one element is probably transmitted by interlinkers to all parts of the system. The cytoskeleton is thus a complex network, tied by the interlinking proteins into an integrated supportive structure.

Other Cytoskeletal Elements

There are indications that other as yet unidentified cytoskeletal elements may occur in the cytoplasm. These unidentified elements appear in different cell types as fibers ranging in diameter from 2 to 3 nm to as much as 10 nm. Frequently, these unidentified fibers occur in networks in close association with the known elements of the cytoskeleton.

The most extensive of these additional cytoskeletal systems is visible in cells that have been exposed to certain detergents. The treatment removes lipids and 75% or more of cellular proteins, including those of microtubules and microfilaments, and leaves behind intermediate filaments embedded in a network of unidentified fibrous material extending into the nuclear interior as well as the cytoplasm (Fig. 13-10). The network produced by the technique in either of the two regions is called the *cytoplasmic* or *nuclear matrix*.

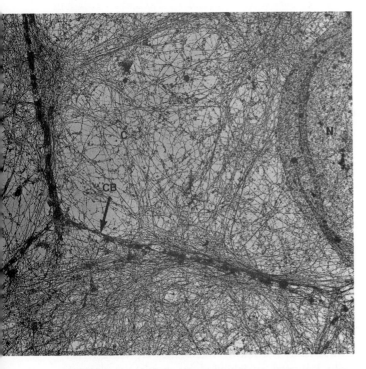

Figure 13-10 The cytoskeletal matrix in cells prepared by extraction with a nonionic detergent. N, nucleus; C, cytoplasm; CB, cell border. ×6,000. Courtesy of S. Penman; reproduced from *J. Cell Biol.* 98:1982 (1984), by copyright permission of the Rockefeller University Press.

There has been considerable debate about the reality of both the cytoplasmic and nuclear matrices. A major reason for the controversy is that the matrices are not visible in cells prepared for electron microscopy by the standard fixation, embedding, and thin-sectioning techniques or in freeze-fracture preparations (see p. 119). Further, no cytoskeletal proteins can be identified in the detergent-produced matrices other than microtubule, microfilament, and intermediate filament polypeptides. Thus any proteins that might form the unidentified fibers in the cytoplasmic or nuclear matrices cannot be detected. For these reasons many investigators feel that much of the unidentified fibrous material in cytoplasmic and nuclear matrices may be produced by precipitation of other, noncytoskeletal proteins by the preparation techniques and, therefore, may not actually exist inside living cells. (Further details of the nuclear matrix, including the controversy surrounding its reality and its proposed functions, are presented in Ch. 14.)

Another group of unidentified cytoplasmic fibers occurs in quantity in protozoa, particularly in those with complex arrangements of cilia at the cell surface. These fibers extend between the bases of individual cilia and anchor underlying cytoplasmic structures to the plasma membrane. In yeast cells a prominent band of unidentified fibers runs around the cell periphery in regions undergoing budding. Although these fibers approximate intermediate filaments in dimensions, the sequences of the proteins forming them have none of the characteristics of the intermediate filament group.

MAJOR DOMAINS AND SUBDIVISIONS OF THE CYTOSKELETON

The cytoskeleton is a fully integrated system that extends throughout the cytoplasm. However, it may be considered for convenience to have three major subdivisions: (1) the plasma membrane and cell surface, (2) the cortical region just beneath the plasma membrane, and (3) the inner cytoplasm between the cortex and the nucleus. By far the greatest body of information concerning these subdivisions has been developed through research with animal cells, particularly those of birds and mammals.

The cytoskeleton has been studied by an extensive battery of methods and techniques. The use of antibodies developed against cytoskeletal elements has been particularly valuable in light and electron microscope studies. The light microscope techniques in which antibodies are tagged with fluorescent dyes (see p. 119) have been used effectively in living as well as chemically fixed cells. These techniques allow the positions of intermediate filaments, microtubules, and microfilaments to be separately traced in cytoskeletal structures.

The major cytoskeletal subregions have also been isolated from cells and analyzed biochemically. Here again, antibodies have proved invaluable as a means for detecting individual cytoskeletal proteins. Other important methods involve the use of protein cross-linkers (see p. 547) to determine the "nearest neighbors" of cytoskeletal elements inside cells. These methods are frequently combined with microtubule and microfilament poisons such as colchicine, taxol, and cytochalasins, and anti-intermediate filament antibodies to inhibit or destroy individual elements of the cytoskeleton.

Cytoskeletal Systems Associated with the Plasma Membrane

Systems associated with the plasma membrane are the most complex and varied of the cytoskeletal structures of animal cells. Depending on the cell type, these systems may carry out many functions in addition to simply reinforcing the plasma membrane. Some provide support and connections for motile mechanisms acting on or through the plasma membrane, such as ameboid movement and capping, or stabilize surface receptors and contacts between cells and the extracellular matrix. Other membrane-associated cytoskeletal systems anchor cell junctions and maintain the structure of surface projections such as microvilli (see below).

Complex plasma membrane–associated cytoskeletons have not been detected in plants. Only microtubules regularly occur in association with the cytoplasmic side of the plasma membrane in plant cells. Several lines of evidence (discussed in Ch. 8) indicate that the primary function of these microtubules is to stabilize the molecular system depositing cellulose microfibrils in the cell wall rather than to reinforce the plasma membrane. The apparent absence of elaborate plasma membrane–associated cytoskeletal systems in plants probably reflects the fact that cell walls support and reinforce the cell perimeter in these organisms.

Membrane Support and Reinforcement in Mammalian Erythrocytes Much information about cytoskeletal systems associated with the plasma membrane comes from research with mammalian erythrocytes (red blood cells). These cells consist of little more than a plasma membrane enclosing ribosomes and a solution of hemoglobin; no nucleus or membranous cytoplasmic organelles are present. When placed in distilled water, mammalian erythrocytes burst, spilling out the hemoglobin and leaving cell "ghosts" consisting of essentially pure plasma membranes with their underlying cytoskeletons. The absence of nuclei and membranous cytoplasmic organelles eliminates these structures as possible sources of contaminants in the preparations and greatly simplifies the task of identifying the plas-

ma membrane-associated cytoskeletal proteins and working out their functions.

The cytoskeleton of mammalian erythrocyte ghosts is limited to short actin microfilaments and *spectrin*, a fibrous actin-binding protein that makes up most of the cytoskeletal network. Actin and spectrin are linked to each other and to the membrane by proteins called ankyrin and *band 4.1 protein* (Fig. 13-11a). The band 4.1 protein is named for its relative position when the proteins associated with erythrocyte membranes are isolated and run on electrophoretic gels (see p. 126). Myosin and tropomyosin are also present as part of the microfilaments. Another protein of the membrane cytoskeleton, *band 4.9 protein*, may link actin microfilaments into bundles.

Each spectrin molecule is a four-part structure built up from two copies each of two closely related polypeptides, α- and β-spectrin. The polypeptides contain as many as 36 repeats of a 106 amino acid sequence. The repeats, although all precisely 106 amino acids in length, vary to a greater or lesser extent in sequence. The two spectrin polypeptides twist into an α-β pigtail; two pigtails line up end to end to form a complete spectrin tetramer (Fig. 13-11b). A number of other proteins, including the actin-binding protein α-actinin, also contain the 106-amino acid repeat characteristic of spectrin.

An entire spectrin molecule is a highly elastic fiber 100 nm long with a total molecular weight approaching 1 million. Each molecule has binding sites for actin, ankyrin, band 4.1 protein, and other spectrin molecules. The spectrin-binding sites, which occur at the tips of the molecules, allow spectrin molecules to link together into an extended, meshlike network (Fig. 13-11c). The cytoskeletal network of a single erythrocyte contains as many as 100,000 spectrin molecules.

The short microfilaments of the erythrocyte membrane cytoskeleton contain only 10 to 20 actin subunits held in association with tropomyosin. The microfilaments link directly to spectrin fibers through an association that is stabilized by band 4.1 protein. (Band 4.1 protein increases the affinity of actin for spectrin by as much as 1 million times.) The band 4.9 protein, as noted, may crosslink some of the actin microfilaments into bundles. The myosin molecules in the membrane cytoskeleton may also crosslink microfilaments.

The spectrin-actin network is tied to the plasma membrane by ankyrin and the band 4.1 protein. Ankyrin sets up links between spectrin and a major transport molecule of the erythrocyte plasma membrane, the *anion transporter* (also known as *band 3 protein*; see p. 196). This link may be stabilized by another protein known as *band 4.2*. Ankyrin has an attached fatty acid, palmitate, which may allow this protein to anchor directly in the plasma membrane as well as to link the anion transporter. Band 4.1 protein also forms crosslinks between spectrin and another integral protein of the

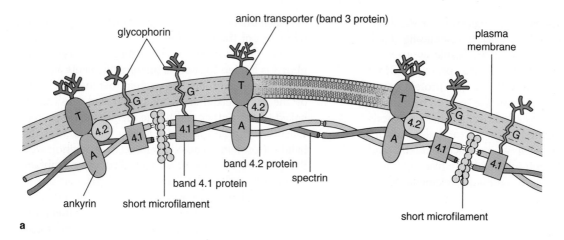

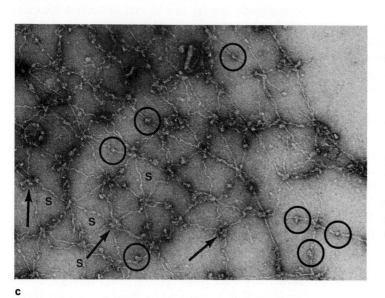

Figure 13-11 The cytoskeletal network underlying erythrocyte plasma membranes. **(a)** Interaction of microfilaments, spectrin, ankyrin (A), band 4.1 protein (4.1), band 4.2 protein (4.2), the anion transporter (T), and glycophorin (G) in the cytoskeletal system reinforcing the erythrocyte plasma membrane. **(b)** Structure of spectrin, which consists of four fibrous polypeptides wound into a loose double pigtail. Each pigtail consists of one α- and one β-spectrin polypeptide. **(c)** A cytoskeletal network isolated from a human erythrocyte membrane. The spectrin molecules (s) form the network itself; actin and band 4.1 proteins occur at the junctions of the network (arrows). Ankyrin molecules are visible on the spectrin fibers at the midpoints between the junctions (circles). ×70,000. Micrograph courtesy of J. Palek and S.-C. Liu; reproduced from *J. Cell Biol.* 104:527 (1987), by copyright permission of the Rockefeller University Press.

plasma membrane, *glycophorin* (see p. 159 and Fig. 5-7). The anion transporter and glycophorin are immobilized in the membrane by their linkages to the underlying cytoskeleton. The entire network, including spectrin, ankyrin, actin, and the band 4.1, 4.2, and 4.9 proteins, makes up as much as 40% of the total protein content of mammalian erythrocyte ghosts.

The degree of crosslinking of the membrane cytoskeleton is regulated by phosphorylation of major components, including ankyrin and band 4.1 and 4.9 proteins. Addition of phosphate groups reduces both the binding affinities of these components and the extent and rigidity of the cytoskeleton.

A mammalian erythrocyte survives about four months in the circulatory system, during which it makes about 500,000 circuits around the body. The membrane cytoskeleton probably maintains the typically flattened, discoid shape of these cells and reinforces the plasma membrane against the stresses of passage through narrow blood vessels and turbulent regions of the circulatory system, such as the heart valves.

Several common hereditary deficiencies in humans have been related to mutations in the proteins forming the plasma membrane–associated cytoskeleton of erythrocytes. Most of these mutations cause various forms of *hereditary hemolytic anemia*, in which red blood cells are fragile because of their weakened cytoskeletal supports. The mutant cells tend to rupture during passage through narrow vessels and turbulent regions in the circulatory system. The mutations responsible for the inherited fragility, including *hereditary pyropoikilocytosis, elliptocytosis,* and *spherocytosis,* involve alterations that reduce the binding affinities of spectrin, ankyrin, band proteins 4.1, 4.2, and 4.9, glycophorin, or the anion transporter in the membrane cytoskeleton.

Plasma Membrane Cytoskeletons in Other Cells Spectrin, ankyrin, and band 4.1 and 4.9 proteins were once thought to be limited to red blood cells. However, research with antibodies developed against these components has revealed that the plasma membranes of many animal cell types are supported by an underlying network containing these or related proteins. These discoveries led to the concept that a cytoskeleton similar to that of mammalian erythrocytes may be associated with the plasma membranes of all animal cells. In many of these cells, and in the nucleated erythrocytes of vertebrates other than mammals, microtubules and intermediate filaments also occur in the cytoskeletal network underlying the plasma membrane. At least some of the same plasma membrane—associated elements have also been detected in protozoa and fungi.

The cytoskeletal framework underlying the plasma membranes of brain cells, for example, contains *fodrin,* a major structural element that was originally thought to be distinct from spectrin. However, fodrin proved to cross react with antispectrin antibodies. Later sequencing studies revealed that fodrin is built up from polypeptides with the same 106 amino acid repeat structure as spectrin. Other plasma membrane—associated proteins, such as α-actinin and the *260/240 protein* of intestinal brush border cells (see below), were also found to contain the typical 106-amino acid repeat structure and to be related to spectrin. A molecule equivalent to erythrocyte spectrin has even been detected in *Acanthamoeba,* a protozoan living in the soil. The different spectrins share an α-polypeptide subunit that is closely similar in all cases and have variant β-polypeptides that are responsible for the distinct structure and properties of individual spectrin types. Spectrins thus form a multimember family of proteins that functions as plasma membrane reinforcements in a variety of eukaryotic cells.

The plasma membrane—associated cytoskeletons of nonerythrocyte cells function along the same lines as those of erythrocytes—they reinforce the plasma membrane and maintain the positions of integral membrane proteins. Some integral membrane proteins of these cells link at their outer ends to elements of the extracellular matrix. The fibronectin receptor, for example, which is a member of the integrin family of surface receptors (see p. 285), links to fibronectin in the extra-

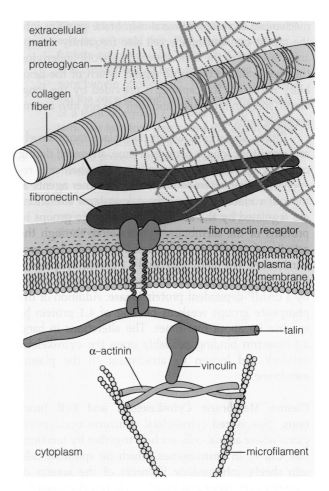

Figure 13-12 Linkage of the cytoskeleton to the extracellular matrix via membrane receptors. On the external side of the plasma membrane, the receptor binds fibronectin, which links in turn to major elements of the extracellular matrix including collagen and proteoglycans. On the cytoplasmic side of the membrane, the receptor is believed to bind talin, which forms linkages through vinculin and α-actinin to microfilaments of the cytoskeleton. Microfilaments may also bind directly to the cytoplasmic end of the fibronectin receptor.

cellular matrix (Fig. 13-12). This linkage ties the cell to collagen fibers and other elements of the extracellular matrix. On the cytoplasmic side of the plasma membrane the fibronectin receptor makes linkages via the actin-binding proteins *talin, vinculin,* and α-actinin to microfilaments in the membrane cytoskeleton. These transmembrane attachments between the cytoskeleton and the extracellular matrix further stabilize the plasma membrane and anchor cells in position in tissues and organs.

Several observations made it obvious that the cytoskeletal networks supporting the plasma membrane are flexible structures capable of short-term adjustments and dynamic change. During capping (see p. 173) the

The Cytoskeleton of the Cell Cortex

In many cell types the cytoplasm immediately beneath the plasma membrane is markedly "stiffer" than more internal regions. The gel-like nature of the cortical region depends in most cells on a network of microfilaments held together by crosslinking proteins such as filamin. These cortical networks are most highly developed in protozoan and animal cells that move by ameboid motion and in some animal oocytes. However, at least some cells in higher plants also possess a cortical microfilament network that gives this region a gel-like consistency.

In ameboid cells the cortical network forms a "tube" through which the more liquid internal cytoplasm flows during movement (see Fig. 12-29). Actin-binding proteins such as gelsolin, depending on the local calcium concentration, constantly crosslink microfilaments to extend the cortical tube at the advancing end of an ameboid cell and sever microfilaments to disassemble and liquefy the tube at the trailing end. The cortical gel of ameboid cells probably has a motile as well as cytoskeletal function—contraction of the cortical tube, through an interaction between actin and myosin, is believed to supply the force moving the liquid internal cytoplasm (see p. 484).

Both intermediate filaments and microfilaments frequently appear as parts of the cortical networks of animal cells. Although the intermediate filaments may reinforce the microfilament network of the cortex, they have not as yet been identified as contributing to any particular function such as ameboid motion.

The cortical region of cells that swim by means of flagella is frequently reinforced by bundles or networks of microtubules. The microtubules provide a cytoskeletal scaffold that anchors the basal bodies and rootlets (see p. 492) of the flagella. Cortical microtubule reinforcements of this type occur in protozoa and in the swimming gametes of many animals, fungi, and plants. The most complex of these cortical systems occur in ciliate and flagellate protozoans, which have complicated arrays of microtubules that run in various directions and interconnect basal bodies and flagellar rootlets (an example is shown in Fig. 13-15). The cortical microtubules of these protozoans, along with complex networks of unidentified fibers of various kinds, evidently reinforce the cell against the stresses of flagellar beating and maintain the basal bodies and shafts of flagella in a fixed latticelike pattern at the surface of the organism.

Cytoskeletal Structures of the Inner Cytoplasm

Interior Cytoskeletal Structures of Animal Cells In animal cells the cytoplasm between the cortex and the nucleus is supported by cytoskeletal systems that include all three cytoskeletal elements. These systems

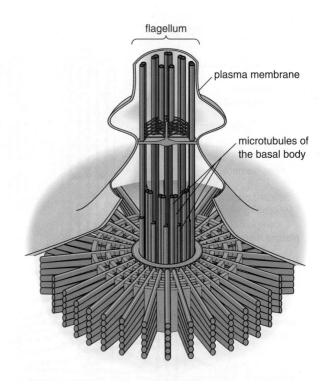

Figure 13-15 The system of microtubules and other, unidentified cytoskeletal structures supporting a basal body and flagellum in the flagellate protozoan *Codosiga botrytis*. The system is representative of the complex cortical arrays of cytoskeletal elements occurring in ciliate and flagellate protozoa. Redrawn from an original courtesy of D. J. Hibberd and The Company of Biologists Ltd., from *J. Cell Sci.* 17:191 (1975).

evidently hold the nucleus and many of the cytoplasmic organelles and structures in position and reinforce the greatest cell volume of the major cytoskeletal systems.

Microtubules and intermediate filaments are the most prominent elements in the interior cytoskeleton of animal cells. Frequently, both elements lie in more or less parallel bundles radiating from one or a few centers near the nucleus (see Figs. 13-3 and 13-16). The microtubules and intermediate filaments radiating from these centers surround the nucleus and extend through the cytoplasm to the cell cortex. For microtubules the center of radiation is the *cell center* or *centrosome*, a region that acts as a microtubule-organizing center (MTOC) from which cytoskeletal microtubules originate and extend (see pp. 426 and 1012). The microtubules radiating from the cell center are generally arranged with their minus ends nearest the cell center and their plus ends toward the periphery.

The centers of intermediate filament organization are frequently located near cell centers so that microtubules and intermediate filaments appear to radiate from almost the same points. The radiating arrangement of intermediate filaments in these cells seems to be dependent on the microtubule system. Dispersal of microtubules by antimicrotubule agents such as

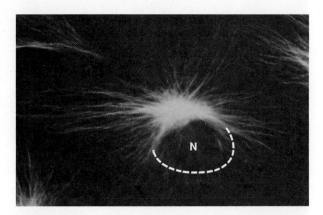

Figure 13-16 Microtubules, marked by fluorescent antibodies, radiating from the cell center of a cultured African green monkey cell. The dashed line marks the boundary of the nucleus (N). × 1,800. Courtesy of J. Izant, from *Modern Cell Biology*, vol. 2 (J.R. McIntosh, ed.). Copyright © 1983, Wiley-Liss, a division of John Wiley and Sons, Inc.

colchicine, low temperature, or high pressure causes the intermediate filament network to collapse. Destruction of the intermediate filament networks, in contrast, has no detectable effect on the pattern of radiating microtubules.

Microtubules radiating from the cell center apparently also fix the positions of several major cytoplasmic organelles, including segments of the ER, the Golgi complex, secretory vesicles, lysosomes, and at least some mitochondria. Usually, the Golgi complex is held in position between the cell center and the plasma membrane; the ER extends throughout much of the cytoplasm. Disruption of microtubules causes Golgi membranes to disperse and ER membranes to cluster near the nucleus, as expected if microtubules radiating from the cell center are responsible for maintaining the normal positions of the Golgi complex and ER. Injection

of anti-intermediate filament antibodies, which causes aggregation and collapse of the intermediate filament network, usually has no effect on the positions of cytoplasmic organelles. This indicates that even though intermediate filaments are located in many of the same regions as microtubules in the deeper cytoplasm crosslinks occur only between organelles and microtubules.

Ameboid motility in cultured animal cells seems to be related to the position of the cell center and its radiating microtubules. In these cells formation of advancing cytoplasmic lobes occurs at the side of the cell containing the cell center. Ameboid motion, as a consequence, takes place in the direction "pointed" by the cell center. As expected from this relationship, ameboid movements of cultured cells become random in direction if microtubules radiating from the cell center are destroyed. Microtubules organized by the cell center are therefore considered to provide a cytoskeletal foundation that anchors and provides polarity and direction to the ameboid motion of cultured cells.

Specialized Interior Cytoskeletal Structures of Muscle Cells Highly specialized networks of intermediate filaments occur in the inner cytoplasm of both striated and smooth muscle cells. In striated muscle cells intermediate filaments of the desmin class parallel myofibrils and connect the Z lines of adjacent myofibrils (as in Fig. 13-6). Other desmin filaments make connections between the Z lines of adjacent myofibrils and between Z lines and the plasma membrane. The linkages set up by intermediate filaments evidently hold sarcomeres in register and transmit contractile forces to the plasma membrane. Microtubules also reinforce the inner cytoplasm of striated muscle cells, particularly in regions containing T tubules and vesicles of the sarcoplasmic reticulum. Microtubules surround the nucleus and mitochondria in these cells, running in parallel bundles that extend in the direction of contraction (Fig. 13-17).

Figure 13-17 Microtubules (arrows) surrounding mitochondria (M) and extending parallel to the direction of contraction in striated muscle. × 58,000. Courtesy of M. A. Goldstein and Plenum Publishing Co., from *Cell and Muscle Motility*, vol. 2, 1982.

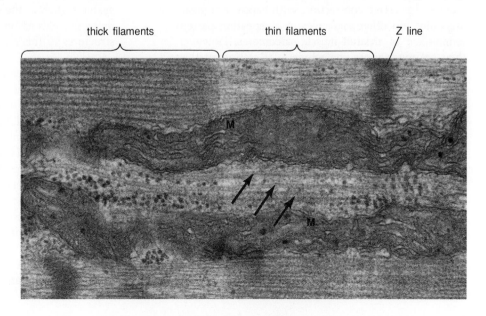

thick filaments thin filaments Z line

In smooth muscle cells intermediate filaments of the desmin class extend between dense bodies, which act as a structural and functional parallel to the Z lines of striated muscle cells (see p. 472). Intermediate filaments also extend from dense bodies to connections at the plasma membrane. The entire intermediate filament network, like the equivalent system of striated muscle cells, maintains the organization of the contractile system and transmits contractile forces through the cell and to the plasma membrane.

Moreover, an intermediate filament network may provide the "latch" that maintains tension in some smooth muscle cells after contraction. In these cells the contracted tension is maintained without ATP expenditure and without further cycling of myosin cross-bridges. One current hypothesis for the latch maintains that crosslinks are set up in the intermediate filament network once the cell is fully contracted, locking the network in an arrangement that holds the cell in the contracted state (see p. 482 for details).

Microfilaments in the Interior Cytoskeleton: Stress Fibers Microfilament networks, although present in deeper regions of the cytoplasm, are usually less concentrated than in the cortex. One microfilament system, however, the *stress fiber* (see Figs. 13-9 and 13-18), is prominent in deeper layers of a few animal cell types, including epithelial cells and fibroblasts. Stress fibers are thick bundles of microfilaments that extend in straight paths in the cytoplasm, usually making connections to the plasma membrane at points where the cell contacts and attaches to an external surface. Besides microfilaments, stress fibers contain myosin, α-actinin, tropomyosin, and, at their points of contact with the plasma membrane, vinculin and talin. Caldesmon, a protein regulating the motile interactions of actin and myosin (see p. 469), is concentrated along the margins of the stress fibers. Stress fibers contain all the elements required for active contraction, with bands of myosin, α-actinin, and other molecules in a repeating pattern reminiscent of striated muscle sarcomeres. However, they are much more characteristic of stationary than motile cells. In fact, stress fibers are dismantled when cells containing them break loose from their extracellular connections and become motile. Microfilaments reorganize into stress fibers as these cells attach, flatten, and cease moving. For these reasons many investigators consider it likely that stress fibers are primarily cytoskeletal structures that reinforce the characteristically flattened or extended shape of the cells containing them. Any tension developed in stress fibers by myosin cross-bridging probably contributes to the flattening. Stress fibers also disappear when the cells containing them enter cell division. At that time the cells lose their flattened shape, round up, and break most of their contacts with the substrate.

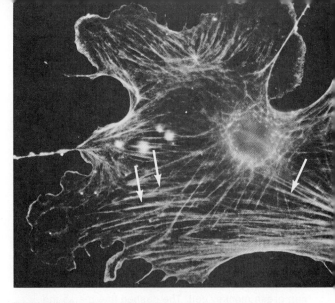

Figure 13-18 Stress fibers (arrows) in a cultured gerbil fibroblast cell. × 1,000. Courtesy of J. J.-C. Lin and Plenum Publishing Co., from *Cell and Muscle Motility*, vol. 2, 1982.

The sites where stress fibers contact the plasma membrane, called *focal contacts*, are specialized sites of interaction with the extracellular matrix. Cells grown on glass or plastic surfaces without added fibronectin or another extracellular linker, *vitronectin*, do not form focal contacts. If fibronectin or vitronectin is added as a film over the glass or plastic surfaces, focal contacts are set up by fibronectin or vitronectin receptors at the cell surface. On the cytoplasmic side of the plasma membrane, these receptors connect to the microfilaments of stress fibers through crosslinks set up by vinculin and talin (as in Fig. 13-12). Focal contacts are disassembled when stress fibers break down during cell movement or division.

The cytoskeletal elements of the inner cytoplasm of animal cells may not be limited to the three major cytoskeletal elements. It is possible that other fiber systems, such as the cytoplasmic matrix detected after detergent extraction of cells, may occur as supportive structures of this region.

Interior Cytoskeletons of Protozoa, Fungi, and Plants Cytoskeletal systems also support deeper regions of the cytoplasm in protozoan, fungal, and plant cells. Microtubules and microfilaments are the prominent supportive elements in the interior cytoskeletal systems of most cells in these groups. In at least some protozoa and fungi, however, limited networks of intermediate filaments occur along with microtubules and microfilaments.

In plants cytoskeletal elements of the inner cytoplasm are apparently limited to microtubules and microfilaments. Microtubules occur singly in widely scattered arrays in the inner cytoplasm. Microfilaments are present in a sparse network that extends throughout the cytoplasm and, in some cell types, surrounds the nucleus in a cagelike array. Microtubule poisons have no noticeable effect on the locations taken by the Golgi complex or ER in plant cells. W. Hensel and others found, however, that exposing cells to the microfilament poison cytochalasin disturbed the distribution of ER vesicles. Thus, microfilaments instead of microtubules may maintain the position of the ER in plant cells.

Some microfilaments in the deeper cytoplasm probably function in cytoplasmic streaming and are thus motile rather than cytoskeletal. The large central vacuole of plant cells also provides significant support for the deeper layers of the cytoplasm. If the vacuole is ruptured or otherwise disturbed, the organization of cytoplasmic organelles collapses.

There is also some possibility that intermediate filaments or at least proteins related to intermediate filaments occur in the cytoplasm of plant cells. C. W. Lloyd and his coworkers found that an antibody with specificity for an amino acid segment common to all five types of animal intermediate filament proteins also reacted with plant cytoplasm. Although the antibody, traced in its reactions by a fluorescent marker, reacted in regions that also contained microtubules, there was no detectable reaction with microtubule proteins themselves. The proteins interacting with the antibody in the plant cell cytoplasm were therefore apparently distributed in the same regions as microtubules, in a relationship similar to the codistribution of microtubules and microfilaments in animal cells.

As yet unidentified cytoskeletal elements may possibly support deeper regions of the cytoplasm in plant cells. The Lloyd group, for example, discovered a network of fibers about 7 nm in diameter in carrot cells prepared by techniques similar to those used to reveal the cytoplasmic matrix in animal cells, including extraction with detergents (Fig. 13-19). The fibers, although similar in diameter to microfilaments, did not decorate with myosin and were insensitive to microfilament poisons.

DYNAMIC CHANGES IN THE CYTOSKELETON

Cytoskeletal Alterations Related to Motility

Cells in tissue culture undergo a variety of cytoskeletal alterations related to motility. During initiation of movement the cell center, with its aggregates of radiating microtubules, migrates to the side of the nucleus facing the direction of movement. The disassembly of stress fibers and focal contacts that takes place as fibroblasts break connections to a substrate and become mobile and the re-formation of these structures as the cells cease movement and reattach have already been

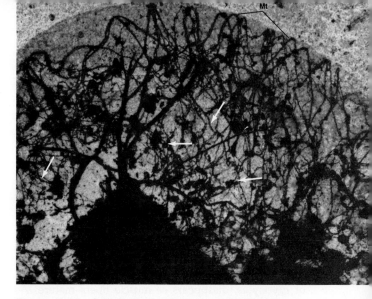

Figure 13-19 Cytoplasmic matrix of unidentified 5-nm to 7-nm fibers (arrows) in a carrot cell extracted with a nonionic detergent. Cortical microtubules (Mt) are also visible. ×3,500. Courtesy of C. W. Lloyd and The Company of Biologists, Ltd., from *J. Cell Sci.* 56:319 (1982).

noted. In some cell types intermediate filaments also change in distribution from an extended network into a dense cap over one side of the nucleus as cells leave fixed positions and begin to move.

Contact between cultured cells is often responsible for the arrest of movement and return of the cytoskeleton to the immobile state. In some cell types cell contact brings about changes in the synthesis of cytoskeletal elements as well as their arrangement. For example, A. Ben-Ze'ev showed that bovine epithelial cells in the motile, unattached state synthesize large quantities of vimentin and limited amounts of cytokeratins. When contact is made with other cells, which arrests the movement, cytokeratins are synthesized in large quantities, and vimentin synthesis is greatly reduced. These changes demonstrate that signals from the cell surface can influence the types, distribution, and arrangement of cytoskeletal elements.

Cytoskeletal Alterations in Developing Cells

The cytoskeleton of many cell types changes extensively during development. Both the arrangement of existing cytoskeletal elements and their rate of synthesis may be altered. Striated muscle, one of the cell types in which cytoskeletal development has been studied most completely, illustrates both types of alteration.

Striated muscle fibers develop through fusion of individual embryonic cells called *myocytes*. Before fusion myocytes have a cell center from which microtubules radiate throughout the cytoplasm. Intermediate filaments also radiate from one or more centers near the nucleus. The intermediate filaments of myocytes are primarily of the vimentin class, but some desmin filaments are also present. As striated muscle develops, myocytes fuse end to end and develop into muscle fibers. During these developmental changes vimentin synthesis ceases, and intermediate filament assembly becomes concentrated almost entirely on the desmin type. The resulting desmin intermediate filaments become associated with Z-line margins and regions of the plasma membrane overlying Z lines. In some striated muscle cells synemin becomes concentrated in Z lines as the predominant IFAP when desmin fibers take up the distribution characteristic of the fully differentiated cell. Another IFAP, paranemin, which is characteristic of myocytes, disappears as muscle fibers mature. As these changes occur, microtubules become aligned in roughly parallel bundles extending primarily in the long axis of the muscle fibers instead of radiating from cell centers.

Changes in the cytoskeleton have been followed in other cell types, as in the development of macrophages, ameboid white blood cells that engulf foreign particles in the circulation. D. J. Kwiatkowski found that as these cells differentiate, an mRNA encoding gelsolin increases in amount by 5 to 10 times, and gelsolin protein level rises by 40 to 50 times. Filamin mRNA increases by 2 to 4 times, and filamin protein levels by 3 to 7 times.

Cytoskeletal Rearrangements During Cell Division

The cytoskeleton also changes in highly ordered patterns as cells cycle through interphase growth and cell division. Of the three major cytoskeletal elements mi-

crotubules undergo the most dramatic alterations. During late interphase in animal cells, the cell center (Fig. 13-20a) splits into two parts, which move apart and take up positions on opposite sides of the nucleus. At the same time, almost the entire complement of cytoskeletal microtubules breaks down and reorganizes into the spindle, which stretches between the separated cell centers (Fig. 13-20b). After division of the replicated chromosomes by the spindle, cytoplasmic division cuts through the spindle midpoint and segregates the cell centers into separate daughter cells (Fig. 13-20c). As division is completed, the spindle microtubules disappear, and the cytoskeletal arrangement of microtubules grows from the cell center, which takes up its characteristic position near the nucleus.

Microfilaments are also rearranged in dividing animal cells. In many cell types interior microfilament networks break down, and microfilaments become concentrated in the spindle. Their role in spindle function, if any, is unknown. Experimental interference with microfilaments generally has no effect on division of the chromosomes (for details, see p. 1027). In animal cells a band of microfilaments organizes around the cell periphery and contracts to divide the cytoplasm. Once division is complete, the microfilament networks typical of nondividing cells reappear. The loss of stress fibers and focal contacts in fibroblasts undergoing division has been noted.

The degree of change in intermediate filament organization and distribution during division varies greatly in different animal cell types. In some cells no rearrangements occur and the existing intermediate filament network is simply separated into two parts as the cytoplasm divides. In other cells the intermediate network collapses into a layer over the nuclear envelope

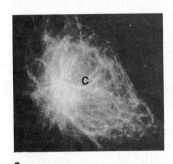 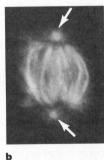 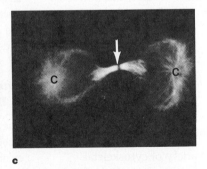

a   b   c

Figure 13-20 Reorganization of microtubules in dividing mammalian cells. **(a)** During the interphase preceding division, microtubules radiate from a single cell center (C). **(b)** Just before division begins, the cell center divides into two parts, which are separated by the developing spindle. The arrows in **(b)** point toward the spindle poles. **(c)** At the close of division, the spindle breaks down, and the cytoplasm divides to place the two cell centers (C) in separate daughter cells. The arrow in **(c)** marks the plane of cytoplasmic division. **(a** and **c)** Courtesy of S. H. Blose, from *Cell Motil.* 1:417 (1981). Copyright © 1981 Wiley-Liss, a division of John Wiley and Sons, Inc. **(b)** Courtesy of B. R. Brinkley; from *J. Cell Biol.* 113:1091 (1991), by copyright permission of the Rockefeller University Press.

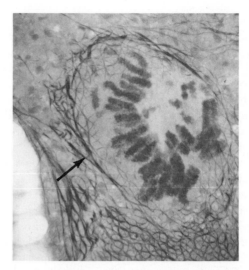

Figure 13-21 An intermediate filament "cage" (arrow) surrounding the spindle during mitosis in a mouse cell. × 1,600. Courtesy of J. C. R. Jones; reproduced from *J. Cell Biol.* 100:93 (1985), by copyright permission of the Rockefeller University Press.

as cytoskeletal microtubules disassemble and reorganize into the spindle. In some of these cells, the intermediate filaments surrounding the nucleus remain in the same position during cell division, forming a "cage" around the spindle (Fig. 13-21). In a few cell types intermediate filaments disassemble and disappear almost completely during cell division. Generally, intermediate filaments become more highly phosphorylated as cells make the transition from interphase to cell division.

Major changes in the organization of microtubules and microfilaments also accompany growth and divi-

sion in higher plant cells. Research by R. W. Seagull, M. Osborn, K. Weber, B. E. S. Gunning, and others demonstrated that during interphase most microtubules in plant cells are concentrated in a layer under the plasma membrane (Fig. 13-22*a*). These are the microtubules stabilizing enzyme complexes that synthesize cellulose microfibrils. As mitotic cell division nears, many of the microtubules break their linkages to the plasma membrane and assemble into a thick belt surrounding the cell (Fig. 13-22*b*). Groups of microtubules run from the belt to the nucleus. This peripheral belt of microtubules, the *preprophase band*, marks the level at which a new cell wall will form to divide the cytoplasm after nuclear division is complete. As cell division begins, the preprophase band disassembles and microtubules collect around the margins of the nucleus. These microtubules grow, increase in number, and eventually organize into the spindle as the nuclear envelope breaks down early in mitosis (Fig. 13-22*c*). After division is complete, the spindle disassembles.

Spindle disassembly proceeds rapidly until all that remains is a residual layer of short microtubules at the former spindle midpoint. In the layer the short microtubules are oriented in the direction of the former spindle axis. The microtubule layer extends laterally until it contacts the cell walls (Fig. 13-22*d*). Cytoplasmic division and formation of the new wall separating the cytoplasm into two parts proceed in this residual layer of microtubules. After cytoplasmic division is complete, the microtubule layer disassembles and the cell's complement of microtubules reorganizes into the cortical arrangement typical of interphase.

Investigations by Lloyd, A. M. Lambert, and others showed that in at least some plant cells the distribution

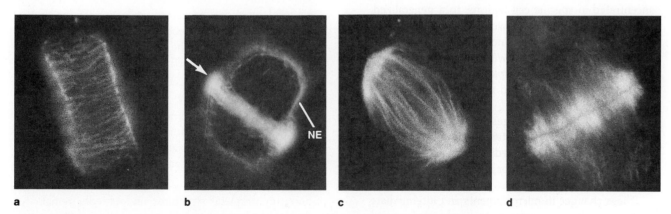

a   b   c   d

Figure 13-22 Reorganization of microtubules during cell division in onion root tip cells. **(a)** The predominantly cortical distribution of microtubules at interphase. × 1,200. **(b)** Rearrangement of microtubules into the preprophase band (arrow). Microtubules also faintly outline the nuclear envelope (NE). × 1,500. **(c)** Concentration of microtubules into the spindle during cell division. × 1,500. **(d)** Organization of microtubules into the phragmoplast. × 1,600. Courtesy of S. M. Wick; reproduced from *J. Cell Biol.* 89:685 (1981), by copyright permission of the Rockefeller University Press.

of microfilaments parallels the microtubule rearrangements during the cell cycle. Microfilaments are distributed in a network in the cell cortex and in sparse collections in the cytoplasmic interior of nondividing cells. As the preprophase band forms preceding division, many of the networks break down, and microfilaments become concentrated in the preprophase band along with microtubules. Later, as the spindle forms, many of the microfilaments in the preprophase band become distributed in the spindle. As in animal spindles the function of microfilaments in higher plant spindles, if any, remains unknown. Not all the microfilaments disappear from the region of the preprophase band. Some persist in the region of the band and possibly contribute later to the location of the new wall separating the cytoplasm at the close of division (see p. 1034 for details).

The total spectrum of changes taking place in mitosis confirms the dynamic nature of the cytoskeleton and the fact that cells can disassemble and reassemble or rearrange all their cytoskeletal elements. The regular cycle of cytoskeletal alterations accompanying cell division indicates that the rearrangements are highly programmed through a master schedule, controlled ultimately through changes in gene activity in the cell nucleus. (The characteristics of the master schedule are discussed further in Ch. 22.)

Cytoskeletal Abnormalities and Disease

A number of diseases are associated with disturbances of the cytoskeleton, particularly in cells undergoing transformation from the normal to the cancerous state. Typically, microfilaments in transformed cells change from their organized arrangement, in which they are concentrated in specific cell regions, to a generalized, more evenly distributed network. The proportion of actin in disassembled form increases. Stress fibers, if present, generally break down and disappear. Linkages between actin and the plasma membrane, made via proteins such as talin and vinculin, frequently are broken.

Other cytoskeletal changes associated with transformation to the cancerous state involve intermediate filaments. Frequently, intermediate filaments revert to arrangements typical of embryonic rather than fully differentiated adult cell types.

These changes in microfilaments and intermediate filaments are often accompanied by breakage of contact with neighboring cells; as part of these surface alterations, cell junctions such as desmosomes and tight junctions disappear. Breakage of connections to other cells removes the normal contact inhibition of motility and cell division. As a result, the transformed cells divide more rapidly and when not dividing move constantly among neighboring cells. These characteristics—break-

age of normal cell contacts, increased division rate, and invasive movements—are the hallmarks of cancer cells. They are also responsible for much of the disruptive effect of cancer cells on bodily functions. As yet, it is unknown whether the changes observed in the cytoskeleton of transformed cells are among the causes of cancer or merely part of a spectrum of effects produced by more fundamental changes in genes, cell structure, and biochemistry.

Intermediate filaments have proved to be a useful indicator of the tissue of origin of cancer cells, which is often important in determining the type of treatment to be employed. Because many body cells have distinct types of intermediate filaments, the tissue of origin for tumor cells can frequently be determined by identifying their intermediate filament types. For example, the presence of desmin intermediate filaments indicates that the tumor is a sarcoma originating from muscle tissue. Vimentin intermediate filaments are typical of lymphomas and nonmuscle sarcomas. Cytokeratins indicate that the tumor is a carcinoma originating from epithelial tissues. The particular cytokeratins present often allow carcinoma subtypes to be identified—squamous cell carcinomas can be distinguished from adenocarcinomas, for example, by the combination of cytokeratin types in the tumor cells. The determinations are routinely carried out by means of antibodies developed against each intermediate filament protein type. (Ch. 22 presents further details of the development, characteristics, and treatment of cancer.)

The cytoskeletal system of each cell type provides the foundation for cellular motility, determines cell shape, and anchors the nucleus and major cytoplasmic organelles in the cytoplasm. Cytoskeletal frameworks also reinforce the plasma membrane and fix the positions of junctions, receptors, and connections to the extracellular matrix. In addition, elements of the cytoskeleton may organize the enzyme systems responsible for major cytoplasmic activities such as glycolysis and protein synthesis. Even cytoplasmic water may be ordered to some extent by the cytoskeleton.

The cytoskeletal systems accomplishing these functions are dynamic, not static, formations that change in composition as cells develop and differentiate, enter mobile and immobile stages, and pass through cycles of growth and division.

Some cytoskeletal structures undoubtedly remain to be identified and chemically characterized. Many questions also persist about the cellular controls regulating the assembly and crosslinking of individual cytoskeletal elements and the unknown cellular factors that integrate the cytoskeleton into a functional whole. And, last but not least, much is yet to be discovered concerning the role of the cytoskeleton in diseases such as cancer.

For Further Information

Ameboid motion, *Ch. 12*
Calmodulin, Ca²⁺ and cellular regulation, *Information Box 7-2 and Ch. 7*
cAMP and cellular regulation, *Ch. 7*
Cancer, *Ch. 22*
Cell cycle, *Ch. 22*
Cell junctions, *Ch. 7*
Cell surface and surface receptors, *Ch. 7*
Cell walls, *Ch. 8*
Coated vesicles, *Ch. 20*
Endocytosis and exocytosis, *Ch. 20*
Extracellular matrix, *Ch. 8*
Microfilament structure and biochemistry, *Ch. 12*
Microtubule-organizing centers (MTOCs), *Ch. 11*
Microtubule structure and biochemistry, *Ch. 11*
Mitotic cell division, *Ch. 24*
Muscle cells, *Ch. 12*
Nuclear envelope, *Ch. 14*
Spindle, *Ch. 24*

Suggestions for Further Reading

General Books and Articles

Bershadsky, A. D., and Vasilev, J. M. 1988. *Cytoskeleton.* New York: Academic Press.

Dustin, P. 1984. *Microtubules.* 2nd ed. New York: Springer-Verlag.

Fulton, A. B. 1984. *The cytoskeleton.* New York: Chapman and Hall.

Schliwa, M. 1986. *The cytoskeleton. Cell biology monographs.* Vol. 13. New York: Springer-Verlag.

Weber, K., and Osborn, M. 1985. The molecules of the cell matrix. *Sci. Amer.* 253:110–20 (October).

Intermediate Filaments and IFAPs

Albers, K., and Fuchs, E. 1992. The molecular biology of intermediate filament proteins. *Internat. Rev. Cytol.* 134:243–79.

Carmo-Fonseca, M. C., and David-Ferreira, J. F. 1990. Interactions of intermediate filaments with cell structures. *Electron Micr. Rev.* 3:115–41.

Columbe, P. A., and Fuchs, E. 1990. Elucidating the early stages of keratin filament assembly. *J. Cell Biol.* 111:153–69.

Fliegner, K. H., and Liem, R. K. H. 1991. Cellular and molecular biology of neuronal intermediate filaments. *Internat. Rev. Cytol.* 131:109–67.

Fuchs, E. 1988. Keratins as biochemical markers of epithelial differentiation. *Trends Genet.* 4:277–81.

Fuchs, E.; Tyner, A. L.; Guidice, G. J.; Marchuk, D.; Ray-Chaudhury, A.; and Rosenberg, M. 1987. The human keratin genes and their differential expression. *Curr. Top. Devel. Biol.* 22:5–34.

Klymkowski, M. W.; Bachant, J. B.; and Domingo, A. 1989. Functions of intermediate filaments. *Cell Motil. Cytoskel.* 14:309–31.

Markl, J. 1991. Cytokeratins in mesenchymal cells: impact on functional concepts of the diversity of intermediate filament proteins. *J. Cell Sci.* 98:261–64.

Quinlan, R. A., and Stewart, M. 1991. Molecular interactions in intermediate filaments. *Bioess.* 13:597–600.

Shaw, G. 1986. Neurofilaments: abundant but mysterious neuronal structures. *Bioess.* 4:161–70.

Skalli, O., and Goldman, R. D. 1991. Recent insights into the assembly, dynamics, and function of intermediate filament networks. *Cell Motil. Cytoskel.* 19:67–79.

Steinert, P. M., and Liem, R. K. H. 1990. Intermediate filament dynamics. *Cell* 60:521–23.

Steinert, P. M., and Roop, D. R. 1988. Molecular and cellular biology of intermediate filaments. *Ann. Rev. Biochem.* 57:593–625.

Wiche, G. 1989. Plectin: general overview and appraisal of its potential role as a subunit of the cytomatrix. *Crit. Rev. Biochem. Mol. Biol.* 24:41–67.

Organization of Cellular Water

Clegg, J. S. 1984. Intracellular water and the cytomatrix: some methods of study and current views. *J. Cell Biol.* 99:167s–71s.

Cytoskeletal Systems Associated with the Plasma Membrane

Bennett, V.; Otto, E.; Davis, J.; Davis, L.; and Kordeli, E. 1991. Ankyrins: a family of proteins that link diverse membrane proteins to the spectrin skeleton. *Curr. Top. Membr. Transport* 38:65–77.

Bretscher, A. 1991. Microfilament structure and function in the cortical cytoskeleton. *Ann. Rev. Cell Biol.* 7:337–74.

Chasis, J. A., and Shohet, S. B. 1987. Red cell biochemical anatomy and membrane properties. *Ann. Rev. Physiol.* 49:237–48.

Davies, K. A., and Lux, S. E. 1989. Hereditary disorders of the red cell membrane skeleton. *Trends Genet.* 5:222–27.

Fox, J. E. B., and Boyles, J. K. 1988. The membrane skeleton: a distinct structure that regulates the function of cells. *Bioess.* 8:14–17.

Lazarides, E., and Woods, C. 1989. Biogenesis of the red blood cell membrane skeleton and the control of erythroid morphogenesis. *Ann. Rev. Cell Biol.* 5:427–52.

Marchesi, V. T. 1985. Stabilizing infrastructure of cell membranes. *Ann. Rev. Cell Biol.* 1:531–61.

Moon, R. T., and McMahon, A. P. 1987. Composition and expression of spectrin-based membrane skeletons in non-erythroid cells. *Bioess.* 7:159–64.

Niggli, V., and Berger, M. M. 1987. Interaction of the cytoskeleton with the plasma membrane. *J. Membrane Biol.* 100:97–121.

Microvilli and the Brush Border

Bretscher, A. 1991. Microfilament structure and function in the cortical cytoskeleton. *Ann. Rev. Cell Biol.* 7:337–74.

Friedrich, E.; Pringault, E.; Arpin, M.; and Louvard, D. 1990. From the structure to the function of villin, an actin-binding protein of the brush border. *Bioess.* 12:403–8.

Proulx, P. 1991. Structure-function relationships in intestinal brush-border membranes. *Biochim. Biophys. Acta* 1071:255–71.

Cytoskeleton of the Cortex and Deeper Cytoplasm

Burridge, K., and Fath, K. 1989. Focal contacts: transmembrane links between the extracellular matrix and the cytoskeleton. *Bioess.* 10:104–8.

Grain, J. 1987. The cytoskeleton in protists: nature, structure, and functions. *Internat. Rev. Cytol.* 104:153–249.

Kelley, R. B. 1990. Microtubules, membrane traffic, and cell organization. *Cell* 61:5–7.

Kreis, T. E. 1990. Role of microtubules in the organization of the Golgi apparatus. *Cell Motil. Cytoskel.* 15:67–70.

Lloyd, C. W. 1987. The plant cytoskeleton: the impact of fluorescence microscopy. *Ann. Rev. Plant Physiol.* 38:119–39.

———. 1988. Actin in plants. *J. Cell Sci.* 90:185–88.

Roberts, T. M. 1987. Fine (2–5 nm) filaments: new types of cytoskeletal structures. *Cell Motil. Cytoskel.* 8:130–42.

Ross, J. H. E.; Hutchings, A.; Butcher, G. W.; Lane, E. B.; and Lloyd, C. W. 1991. The intermediate-filament related system of higher plant cells shares an epitope with cytokeratin 8. *J. Cell Sci.* 99:91–98.

Stearns, T. 1990. The yeast microtubule cytoskeleton: genetic approaches to structure and function. *Cell Motil. Cytoskel.* 15:1–6.

Terasaki, M. 1990. Recent progress on structural interactions of the endoplasmic reticulum. *Cell Motil. Cytoskel.* 15:71–75.

Tiwari, S. C.; Wick, S. M.; Williamson, R. E.; and Gunning, B. E. S. 1984. Cytoskeleton and integration of cellular function in cells of higher plants. *J. Cell. Biol.* 99:63s–69s.

Vorobjev, T. A., and Nadezhdina, E. S. 1987. The centrosome and its role in the organization of microtubules. *Internat. Rev. Cytol.* 106:227–93.

Williamson, R. E. 1991. Orientation of cortical microtubules in interphase plant cells. *Internat. Rev. Cytol.* 129:135–206.

Dynamic Changes in the Cytoskeleton

Bourguignon, L. Y. W., and Bourguignon, G. J. 1984. Capping and the cytoskeleton. *Internat. Rev. Cytol.* 87:195–224.

Kakiuchi, S., and Sobue, K. 1983. Control of the cytoskeleton by calmodulin and calmodulin-binding proteins. *Trends Biochem. Sci.* 8:59–62.

The Cytoskeleton and Disease

Davies, K. A., and Lux, S. E. 1989. Hereditary disorders of the red cell membrane skeleton. *Trends Genet.* 5:222–27.

Kellie, S. 1988. Cellular transformation, tyrosine kinase oncogenes, and the cellular adhesion plaque. *Bioess.* 8:25–30.

Marchesi, S. L. 1991. Mutant cytoskeletal proteins in hemolytic disease. *Curr. Top. Membr. Transport* 38:155–74.

Review Questions

1. Compare the dimensions and structure of intermediate filaments, microtubules, and microfilaments.

2. Outline the classes of intermediate filaments and their constituent proteins. In what cell types does each class of intermediate filament occur?

3. Compare the reactions assembling intermediate filaments, microtubules, and microfilaments. What factors might regulate intermediate filament disassembly inside cells? What information indicates that intermediate filaments can be disassembled inside cells?

4. Outline the structure of an intermediate filament protein. In what ways are the different intermediate filament proteins similar, and in what ways do they differ? How are intermediate filament proteins believed to assemble into intermediate filaments?

5. What are IFAPs? What functions do IFAPs carry out in the cytoskeleton?

6. Outline the roles and structures of microtubules, microfilaments, and intermediate filaments in the cytoskeleton. How are cAMP and Ca^{2+} concentrations related to the cytoskeleton?

7. What are cytoskeletal interlinkers? What roles do these molecules play in the cytoskeleton?

8. What structures may form parts of the cytoskeleton in addition to microtubules, microfilaments, and intermediate filaments?

9. What are the major subdivisions of the cytoskeleton? What methods have been employed to study the arrangement of the cytoskeleton in these subdivisions?

10. Outline the cytoskeletal system supporting the plasma membrane in red blood cells. Why have red blood cells been used to study the cytoskeleton associated with the plasma membrane? Compare the cytoskeleton of red blood cells with that of other cell types.

11. What is the relationship of the cytoskeleton to cell junctions?

12. What are microvilli? Brush borders? What is the terminal web? How are microtubules and intermediate filaments organized in these structures? What are the possible functions of microvilli and brush borders? How does the cytoskeleton contribute to these functions?

13. Describe the cytoskeleton of the cell cortex. How does the cytoskeleton in this region contribute to ameboid motion? To flagellar motion?

14. What is the cell center? What is the relationship of the cell center to the distribution of microtubules and intermediate filaments in the cytoskeleton? To cell motility?

15. What are stress fibers? Focal contacts? What molecular types occur in these structures? What evidence indicates that stress fibers are cytoskeletal rather than motile structures?

16. Outline the arrangement of cytoskeletal elements in smooth and striated muscle cells. What are the functions of cytoskeletal elements in muscle cells?

17. Compare the cytoskeletons supporting the inner cytoplasm in plant and animal cells.

18. What changes occur in the cytoskeleton in relation to motility? To cell development? To cell division? What is the relationship of these changes to the development of cancer?

19. Outline the changes taking place in the microtubule-based cytoskeleton of plant cells in relation to the division cycle.

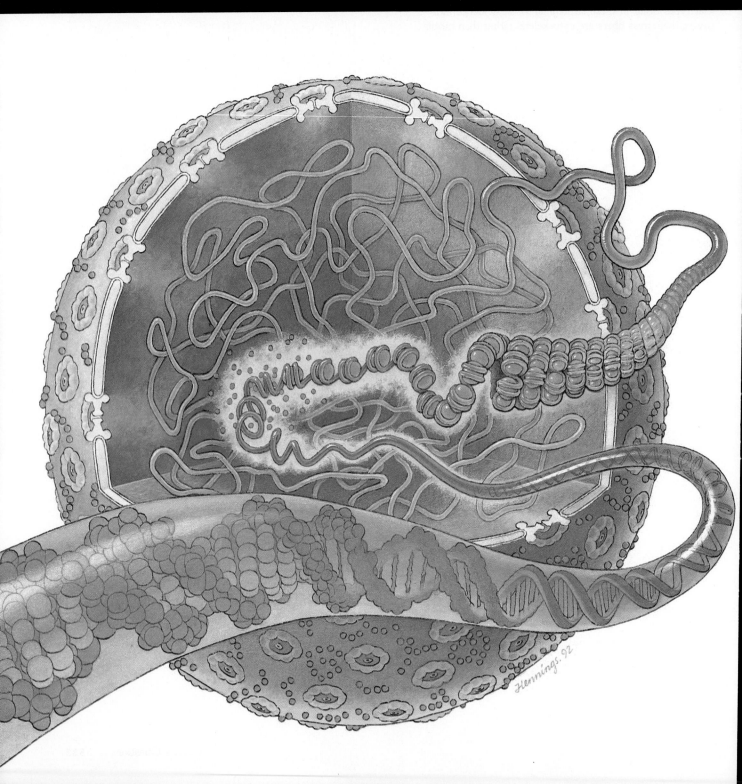

The discovery of DNA structure in 1953 by J. D. Watson and F. H. C. Crick set off the molecular revolution in biology. The discovery not only elucidated the structure of DNA but also indicated how information is encoded in DNA, copied and utilized in cells, and replicated and passed on in cell division. Subsequent findings, accelerated by the development of methods for sequencing DNA and RNA, revealed the structure of genes, the RNA molecules copied from them, and the amino acid sequences of the proteins encoded in the genes. The excitement continues today with current emphasis on the mechanisms regulating gene activity, the total sequence structure and organization of the cell nucleus, the reactions processing RNA copies into finished form, and the detailed structure and activity of the mechanisms assembling and distributing proteins in the cell.

The major discoveries of this molecular revolution and the emphasis of current research are the topics of Chapters 14 through 18. This chapter concentrates on the structure of DNA, its associated proteins, and the organization of DNA and proteins into chromatin in the cell nucleus. The nuclear envelope, the membrane system that separates the nucleoplasm from the cytoplasm, is also described. The equivalent structures of prokaryotes are outlined in Supplement 14-1. Chapter 15 describes the genes coding for mRNA and the accessory RNAs, their transcription, and the chemical processing of the different RNAs from initial to finished form. Chapters 16 and 17 outline protein synthesis and the mechanisms regulating transcription and translation. These chapters concentrate on the synthetic activities of cells during *interphase*, the period of growth between cycles of cell division. The mechanism of DNA replication and the molecular and structural events accomplishing cell division are the subjects of later chapters.

DNA STRUCTURE

The Watson–Crick Model

The hypothesis proposed by Watson and Crick for DNA structure relied on several lines of evidence. The most important came from X-ray diffraction (see p. 132) of DNA in the laboratory of M. H. F. Wilkins. Wilkins's evidence, with important contributions from another investigator, R. Franklin, indicated that the nucleotide chains forming the backbone of the DNA molecule twist into a regular spiral, or helix. Other evidence obtained by E. Chargaff from chemical analysis of DNA indicated that the bases adenine and thymine occur in equal quantities in this nucleic acid; the quantities of guanine and cytosine are also equal. Investigations before 1900 had revealed the structure of the individual DNA nucleotides (see Ch. 2) and the pattern by which the nucleotides link into chains.

From this information Watson and Crick proposed the fundamental structure of DNA, in which two nucleotide chains wind into a double-helical structure (Fig. 14-1). The "backbones" of the two nucleotide chains, which consist of alternating sugar and phosphate groups, are separated from each other by a regular distance, approximately 1.1 nm across the center of the helix. This space is exactly filled by base pairs consisting of one pyrimidine (adenine or guanine) and one purine (thymine or cytosine; see Figs. 14-2 and 2-35). The model proposed that each pair lies in a flat plane roughly perpendicular to the long axis of the molecule. (The planes are viewed from their edges in Fig. 14-1*b*). Each complete turn of the double helix, which was proposed to twist in a right-handed or clockwise direction, includes ten base pairs.

The packing of atoms in the DNA double helix produces two conspicuous grooves that spiral around the surface of the molecule. The two grooves are of equivalent depth, but one, the *major* or *wide groove*, is significantly wider than the other, the *minor* or *narrow groove* (see Fig. 14-1).

The structures of adenine and thymine allow the two bases to fit together like the pieces of a jigsaw puzzle. Guanine and cytosine also fit together perfectly within the space available in the center of the helix. The fit allows three stabilizing hydrogen bonds (see p. 42) to form between guanine and cytosine and two to form between adenine and thymine. Adenine-cytosine and guanine-thymine pairs do not fit stably within the available space and cannot form stabilizing hydrogen bonds when the bases are in their most common configuration. These factors, which effectively restrict the purine-pyrimidine base pairs to adenine-thymine and guanine-cytosine, explain why Chargaff found these bases to be present in equal quantities in DNA.

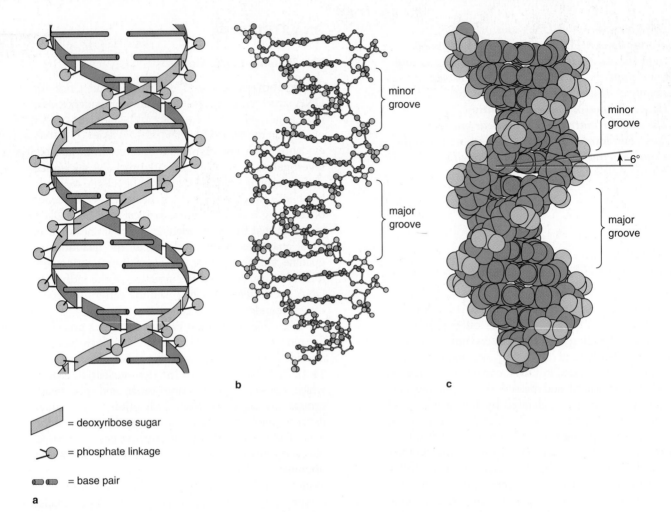

= deoxyribose sugar

= phosphate linkage

= base pair

a

Figure 14-1 The DNA double helix in the B conformation. **(a)** Arrangement of sugars, phosphate groups, and base pairs in the DNA double helix. **(b)** Positions of atoms and bonds in the DNA double helix. Atoms are shown as circles and bonds as straight lines. The bases, which lie in flat planes, are seen edge-on in this view. **(c)** Space-filling diagram of the DNA double helix. The spaces occupied by individual atoms are shown as spheres. The "tilt," the angle from the horizontal made by the planes of the base pairs, is −6°, with the minus sign indicating that the direction of tilt is counterclockwise.
(b and **c)** Redrawn from original art courtesy of W. Saenger, from *Principles of Nucleic Acid Structure.* New York: Springer-Verlag, 1983.

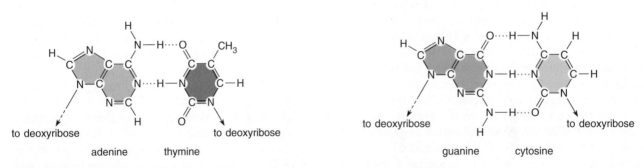

Figure 14-2 The A-T (adenine-thymine) and G-C (guanine-cytosine) base pairs of DNA. Adenine and guanine are purines; thymine and cytosine are pyrimidines. Hydrogen bonds between the purine-pyrimidine base pairs are shown by dotted lines. A-T base pairs form two stabilizing hydrogen bonds, and G-C pairs form three.

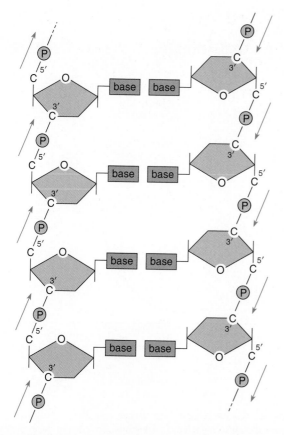

Figure 14-3 The antiparallel arrangement of the two nucleotide chains in DNA. The arrows point in the 5' → 3' direction.

A second fundamental deduction by Watson and Crick is that the two nucleotide chains of the DNA double helix are *antiparallel;* that is, they run in opposite directions. This conclusion is most easily understood if the double helix is unwound and laid out flat, as in Figure 14-3. If the phosphate linkages holding together

the chain on the left are traced from the bottom to the top of the figure, they extend from the 5'-carbon of the sugar below to the 3'-carbon of the sugar above each linkage. On the other chain the 5' → 3' linkages run from the top to the bottom of the figure. This feature of DNA structure is significant for both DNA duplication and RNA transcription because it requires that a new RNA or DNA chain being copied *must run in the opposite direction from the template.*

Several stabilizing forces hold the DNA double helix together. One is hydrogen bonding between base pairs in the interior of the molecule. Although individually relatively weak, the collective effect of the many hydrogen bonds along a length of double helix sets up a strong and stabilizing attraction between the two nucleotide chains. (The effects of temperature on hydrogen bonding in the DNA double helix and the application of this effect to DNA studies are outlined in Information Box 14-1.) A second stabilizing force is provided by hydrophobic associations (see p. 41) between the base pairs in the interior of the molecule. In this region the bases, which are primarily nonpolar, pack tightly enough to exclude water and form a stable, nonpolar environment. Van der Waals forces (see p. 43) established between the closely packed base pairs also help stabilize the DNA double helix.

The limitation to adenine-thymine and cytosine-guanine pairs suggested to Watson and Crick a possible basis for the duplication of DNA, known to occur as part of cell reproduction. Because of the pairing restriction, a given sequence in one chain is compatible with only a single sequence in the opposite chain (Fig. 14-4a). In the language of molecular biology, the sequence of the two chains is said to be *complementary.* To account for DNA replication, Watson and Crick proposed that the two chains of a replicating molecule separate (Fig. 14-4b), and each serves as a *template* for

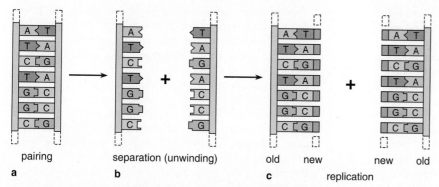

pairing a   separation (unwinding) b   old new c   new old

replication

Figure 14-4 Complementarity and DNA replication. **(a)** The base-pairing rules determine that the sequence A-T-C-T-G-G-C in one chain restricts the opposite chain to the complementary sequence T-A-G-A-C-C-G. Complementarity provides the basis for replication. Each of the old chains unwinds **(b)** and serves as a template for the synthesis of a new complementary chain, according to the base-pairing rules **(c)**. The result is two molecules, each consisting of one old and one new chain, that are exact duplicates of the original molecule.

DNA Melting and Renaturation

The fact that the two nucleotide chains of a DNA double helix are held together partly by hydrogen bonds makes DNA structure highly sensitive to temperature. As the temperature rises, kinetic motions of the two nucleotide chains and collisions with molecules in the surrounding solution increasingly disturb the hydrogen bonds holding the double-helical structure together until the two chains come apart. As the chains separate, DNA loses its highly ordered structure and is said to *melt*.

Melting is monitored by measuring the amount of ultraviolet light absorbed by a DNA solution as its temperature rises. The DNA bases absorb ultraviolet light at a wavelength of 260 nm. The absorbence is relatively weak when nucleotide chains are wound into a double helix; as the helix melts, the bases are exposed so that they absorb light about 40% more strongly. With gradually increasing temperatures the ultraviolet absorption increases as shown by the S-shaped or sigmoid pattern on the graph in this box. The midpoint of the S-shaped curve is defined as the melting temperature or T_m of a given DNA sample.

Because G-C base pairs are held together by three hydrogen bonds, and A-T base pairs by two, G-C base pairs show greater resistance to the effects of temperature. Therefore, the more G-C as opposed to A-T base pairs in a DNA sample, the higher the T_m observed as the sample is melted. The effect is approximately linear; each additional G-C base pair adds about 0.4°C to the T_m. Mammalian DNA, which averages about 40% G-C and 60% A-T base pairs, melts with a T_m at 87°C; prokaryotic DNA, which typically has about 60% G-C base pairs, has a T_m of 95°C. These temperatures can be lowered significantly by adding agents that reduce the strength of hydrogen bonds, such as formamide or urea.

Gradual reduction in the temperature of a melted DNA sample allows the separated DNA chains to rewind into intact double helices. The rewinding process, called *renaturation* or *reannealing*, is facilitated if the DNA sample is broken into short lengths of a hundred nucleotides or so. The short lengths increase the number of random collisions between nucleotide chains in the cooling solution and thereby increase the probability that complementary chains will "find" each other and rewind in the solution.

DNA renaturation is used often as a method for determining how closely DNA samples from different sources are related in sequence. Because they share many sequences in common, DNA nucleotide chains from two closely related organisms can wind into *hybrid* double helices if their DNAs are melted and then cooled (see also p. 135). In hybrid double helices one nucleotide chain orig-

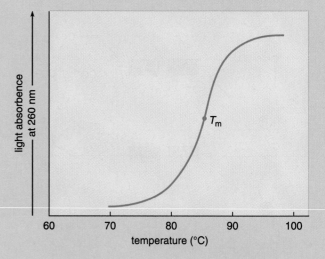

inates from one of the two species and the opposite chain from the other species. DNA from less closely related organisms will share fewer sequences and will form proportionately fewer hybrid helices when mixed and cooled after melting.

The accuracy and sensitivity of DNA hybridization are greatly improved if two DNA nucleotide chains from the same species are prevented from winding into "native" double helices, in which both nucleotide chains originate from the same organism. To accomplish this, DNA from one species is melted and fixed in single-chain form by pouring onto nitrocellulose filter paper. (Nitrocellulose strongly binds single nucleotide chains.) This immobilizes the nucleotide chains of the DNA sample and prevents them from rewinding into native double helices. The filter paper is then washed free of any unbound DNA and treated chemically so that any remaining binding sites for single nucleotide chains are covered. The second sample is then melted and added. Because no other binding sites on the nitrocellulose paper are available, the only DNA chains absorbed by the filter will be those that can form hybrid double helices with nucleotide chains fixed to the paper. The amount of the second sample attaching to the paper provides a measure of the extent to which the two DNA samples share the same sequence information and, therefore, a measure of the "relatedness" of the two organisms. DNA hybridization studies of this type have been of great value in several areas of research, particularly in evolution.

making a complementary copy. The result is two new double helices with a sequence of base pairs identical to that of the original double helix (Fig. 14-4c). Much evidence, discussed in detail in Chapter 23, confirms that replication actually occurs in this pattern. Complementarity also underlies transcription of RNA copies from DNA (see Fig. 15-2).

The detailed X-ray diffraction and other work carried out since Watson and Crick announced their model have supported the fundamental double-helical structure they proposed for DNA. The two investigators, with Wilkins, received the Nobel Prize in 1962 for the model, which has proved to be one of the most significant and far-reaching discoveries in the history of biology.

Refinements to the Model: B-DNA and Other DNA Conformations

The X-ray data used by Watson and Crick were obtained from partially hydrated DNA molecules that had been oriented in parallel fashion by drawing them into a fiber. Such preparations, which are at best only semicrystalline, produce somewhat blurred X-ray diffraction patterns. (The clearest patterns are produced by the highly ordered atoms of crystals.) In the years since Watson and Crick's original discoveries, methods have been developed for casting true crystals of short DNA molecules of fixed sequence. X-ray diffraction of these crystals has produced a greatly refined picture of DNA structure and has allowed the precise positions of individual atoms to be established.

This research also revealed that the structure worked out by Watson and Crick, now known as *B-DNA*, is only one of several possible helical forms of DNA (Table 14-1). Of the additional conformations, termed *A-, C-, D-, E-,* and *Z-DNA,* only the A and Z conformations are known to be common in nature. The Watson–Crick B conformation is the form taken by most DNA in both eukaryotic and prokaryotic cells. The A conformation is the primary form taken by double-helical RNA molecules and by *hybrid* double

helices consisting of one RNA and one DNA chain wound together. The Z conformation, in which DNA winds to the left instead of the right, is believed to occur in a small proportion of cellular DNA along with B-DNA. The various DNA conformations, rather than being rigidly fixed structures, are flexible helices that vary to a limited degree in essentially all their dimensions. The primary source of variations is the particular sequence of nucleotides within a region of a DNA molecule; depending on the sequence, there are local variations in such characteristics as groove width, base packing, pitch, and numbers of base pairs per turn of the helix. These local variations are critical to recognition of individual sequences by regulatory and other proteins combining with DNA (see p. 542).

The Right-Handed DNA Conformations The dimensions and structure now established for B-DNA closely resemble those proposed by Watson and Crick. In solutions approximating cellular conditions of salt concentration and pH, the number of base pairs per turn in the B conformation averages about 10.5, very close to the 10 base pairs per turn originally proposed in the Watson–Crick model. Within the molecule the bases tilt slightly from the perpendicular, at an angle of $-6°$ (see Fig. 14-1c; the minus sign indicates that the tilt is counterclockwise). All DNA molecules of any source, natural or artificial, and of any sequence, can take on the B conformation.

In DNA taking the A conformation (Fig. 14-5), the tilt of individual base pairs shifts to $+20°$. Changes in the structure of the helix deepen the minor groove and make the major groove more shallow than in the B form. DNA of any sequence and type can also wind into the A conformation. Double-helical regions in RNA molecules are limited to the A conformation because of the presence of an —OH rather than an —H at the 2'-carbon of the ribose sugar (see Fig. 2-31); this —OH interferes with atomic packing enough to prevent RNA from winding into the B conformation. RNA double helices form regularly in regions of RNA molecules that fold back upon themselves in "hairpin" for-

Table 14-1 DNA Conformations

Conformation	Base Pairs per Turn	Vertical Rise per Base Pair (A)	Angle Turned per Base Pair	Tilt of Base Pairs	Groove Width Major (A)	Groove Width Minor (A)	Groove Depth Major (A)	Groove Depth Minor (A)	Approximate Diameter (A)
A	11	2.56	32.7°	+20°	10.9	3.7	2.8	13.5	23
B	10.1–10.6	3.38	36.0°	−6°	5.7	11.7	7.5	8.5	20
C	9.3	3.32	38.6°	−8°	4.8	10.5	7.9	7.5	19
D	8	3.04	45.0°	−16°	1.3	8.9	6.7	5.8	
E	7.5	3.25	48.0°						
Z	12	3.70	−30.0°	−7°	8.8	2.0	3.7	13.8	18

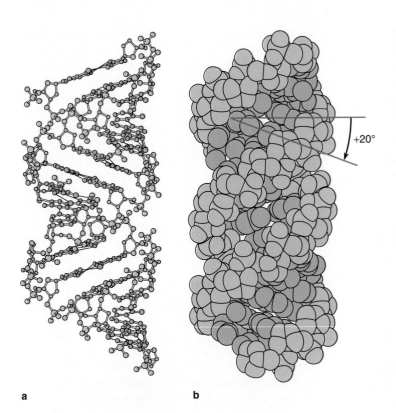

a   **b**

Figure 14-5 DNA in the A conformation. In **(a)** atoms are shown as small circles and bonds as straight lines. **(b)** Space-filling model of A-DNA. RNA double helices or hybrid double helices consisting of one DNA and one RNA nucleotide chain also take on the A conformation. The tilt of the base planes from the horizontal in A-DNA is +20°. Redrawn from original art courtesy of W. Saenger, from *Principles of Nucleic Acid Structure.* New York: Springer-Verlag, 1983.

+20°

mations (see Information Box 14-2 and p. 574). Hybrid double helices produced by the winding together of one DNA and one RNA nucleotide chain take on the A conformation because this is the only helical structure common to the two nucleic acids. Hybrid helices occur as temporary intermediates during transcription, when an RNA copy is made from a DNA chain.

Of the remaining right-handed conformations, C-DNA seems only to be a laboratory curiosity, formed under extreme conditions that are not believed to occur inside living cells. The D and E conformations can be taken on only by DNAs that contain certain sequences, such as regularly alternating A-T base pairs. D-DNA has been detected in nature only in bacteriophage T2, a virus infecting bacterial cells.

The Left-Handed Z Conformation Z-DNA, once also thought to be a laboratory curiosity, is now considered to be a likely alternate conformation for a small but significant portion of the DNA in cell nuclei. The Z conformation was discovered in 1979 by A. Rich and his coworkers, through research originally intended to refine details of B-DNA structure. For their research Rich cast crystals using short DNA molecules with an alternating G-C sequence, in the form

$$-C-G-C-G-C-G-$$
$$\vdots \; \vdots \; \vdots \; \vdots \; \vdots \; \vdots$$
$$-G-C-G-C-G-C-$$

To their surprise the X-ray diffraction patterns produced by the crystals indicated that the DNA was twisted into a left-handed double helix. (The events leading to Rich's discovery of Z-DNA are related in his essay on p. 532.)

Detailed analysis established that the left-handed helix of the DNA in the crystals is more slender than B-DNA and contains more base pairs per turn (Fig. 14-6 and Table 14-1). The helix was designated as Z-DNA because a line connecting the phosphate groups follows a zigzag path rather than the relatively smooth line observed in B-DNA. Within Z-DNA the base pairs are flipped through 180° and are thus upside down with respect to their position in B-DNA. The edges of the flipped base pairs fill in most of the surface region analogous to the major groove of B-DNA, so that Z-DNA has a shallow major groove and a relatively deep minor groove.

Z-DNA is less stable than B-DNA, primarily because the Z conformation brings the negatively charged phosphate groups closer together than in B-DNA. Repulsion between the phosphate groups favors a shift to B-DNA. Certain cellular conditions, however, increase stability of the Z conformation. One is DNA sequence; Z-DNA is favored by sequences that contain alternating purines and pyrimidines, such as the alternating G-C sequence used by Rich and his coworkers for their studies. Consecutive sequences of A-T base pairs also favor the Z conformation. A second factor increasing the stability of Z-DNA is the addition of a

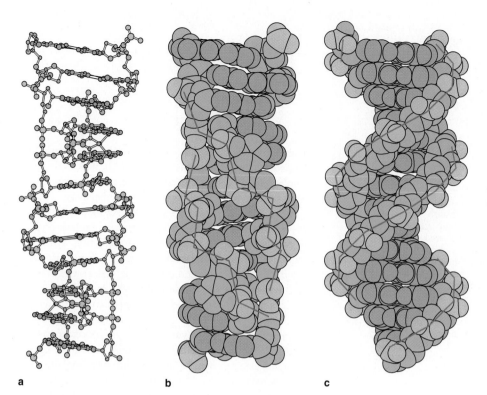

a   b   c

Figure 14-6 Z-DNA and B-DNA compared. **(a** and **b)** DNA wound into the left-handed Z conformation. Note the zigzag pathway (red lines) followed by the phosphate groups in Z-DNA in **(b)**. **(c)** The smooth curve followed by the phosphate groups in B-DNA (red lines). Redrawn from original art courtesy of W. Saenger, from *Principles of Nucleic Acid Structure.* New York: Springer-Verlag, 1983.

methyl (—CH₃) group to cytosines in the DNA sequence (Fig. 14-7). This chemical modification, common in nature, adds a strongly hydrophobic group that is exposed to water molecules in B-DNA (Fig. 14-8) but protected from water in a hydrophobic pocket in Z-DNA. If enough cytosines in a series of G-C base pairs are methylated, the Z form becomes the lower energy state, and DNA may rewind spontaneously from the B to Z conformation and remain stable in the Z form under cellular conditions.

A third cellular condition favoring Z-DNA formation is *negative supercoiling* (see below). Negative supercoiling can be regarded as a force that attempts to unwind a B-DNA molecule, much as if one end of a B-DNA helix is held in a fixed position and the

Figure 14-8 The position of methyl groups (red) in the major groove of the B-DNA double helix. In this position the hydrophobic methyl groups are exposed to the surrounding watery medium. Redrawn from original art courtesy of W. Saenger, from *Principles of Nucleic Acid Structure.* New York: Springer-Verlag, 1983.

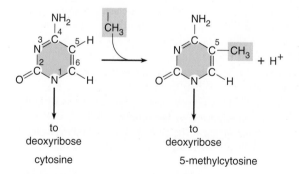

Figure 14-7 The addition of a methyl group to cytosine to form the modified base 5-methylcytosine. The modification is common in natural DNAs.

The Discovery of Left-Handed Z-DNA

Alexander Rich

ALEXANDER RICH has an A.B. and M.D. from Harvard University. After post-doctoral work at Caltech he went to NIH from 1954 to 1958 as section head in Physical Chemistry. In 1955–56, he worked at the Cavendish Laboratory in Cambridge, England. At NIH and in England, he discovered the three-dimensional structure of collagen. During that period, he and his NIH coworkers discovered the formation of triple-stranded nucleic acid complexes. In 1958 he moved to the Biology Department at M.I.T. His work included discovery of polysomes in protein synthesis, discovery of DNA in chloroplasts, and determination of the three-dimensional structure of yeast phenylalanine transfer RNA. He was a member of the Viking mission team that searched for active biology on the surface of Mars.

One of the most important ways we have of learning about the three-dimensional structure of biological molecules, especially macromolecules, is through single-crystal x-ray diffraction analysis. In the early years of research on the nucleic acids, most of the x-ray diffraction work was carried out on DNA fibers. These studies are often useful for obtaining the general conformation of a polymer molecule but they are unable to produce a detailed picture. The first single-crystal x-ray diffraction analysis of the double helix came in 1973 when my colleagues and I solved the structure of two dinucleoside phosphates, ApU and GpC.[1] Both of these crystallized as double helical fragments held together by Watson–Crick hydrogen bonds. Furthermore, the crystals diffracted to 0.8 A resolution, that is, atomic resolution. Every atom could be seen in the lattice, including water molecules and ions. If we could obtain this amount of detail from even larger nucleic acid fragments, we would learn a great deal more. By 1974 we were able to determine the three-dimensional structure of yeast phenylalanine transfer RNA.[2] However, this was carried out at about 2.5 A resolution, which gave a reasonable view of the molecule, but the details still eluded us.

Shortly thereafter I met Jacques van Boom, a Dutch organic chemist who was highly skilled in synthesizing oligonucleotides. We resolved to collaborate in solving some nucleic acid structures. He would do the synthesis and we would try to crystallize and solve them. Little was known about the properties of small oligonucleotides, so we decided to concentrate on those containing guanine-cytosine base pairs, which are somewhat more stable than

adenine-thymine base pairs. The first material he made for us was the alternating sequence d(CGCG), which is self-complementary. It crystallized, and we decided then to go a step further, to the self-complementary hexamer d(CGCGCG). My postdoctoral fellow Andrew Wang discovered that this would crystallize readily. Remarkably, the crystals diffracted x-rays to 0.9 A resolution, again atomic resolution. This excited us very much because we thought that we could find out for the first time in some detail what the DNA double helix really looks like.

In x-ray diffraction studies, the intensity of the diffracted beams is registered on photographic film or with radiation counters. However, we lose the phases of the beams relative to each other. To reconstruct a three-dimensional electron density map and actually "see" the atoms, we have to solve the phases. When you believe you know the structure of the molecule, you can use a technique called a "translation-rotation search" to solve the diffraction pattern. Using a computer program, we took various models—A-DNA, B-DNA, and other variants—and searched through the diffraction data looking for a good fit. None of these models fit the diffraction data we obtained for the hexamer, so we decided to do the more laborious but more certain method of obtaining heavy atom derivatives. These would modify the diffraction pattern enough so that we could determine the relative phases of the diffracted beams and thus solve the structure. We found that a number of heavy metal ions could be diffused into the crystal lattice, and their position was determined by measuring the difference between their diffraction pattern and the diffraction pattern of the native molecule. Given enough of these, we eventually were able to determine the relative phases of the diffracting rays and from this reconstruct a three-dimensional electron density net.

In an x-ray diffraction experiment, solution of the structure consists of obtaining a three-dimensional electron density map. The electrons are clustered around the atoms and from this we can discern the structure of the molecule. In those days, the electron density was drawn on transparent plastic sheets in layers, which were then piled atop one another to view the entire three-dimensional distribution. When this was done with the first crystal of d(CGCGCG), we observed a very unusual molecule. We had expected to see ordinary B-DNA, but in fact it became apparent that the chains coiled about each other in a left-handed rather than a right-handed array.

Nonetheless, the familiar guanine-cytosine base pairs were clearly visible in the electron density map, so the map couldn't be entirely wrong. I asked Andrew Wang whether he was certain that he had piled up the plastic sheets in the correct order. If, for example, they were piled in inverted order, the helix would have an inverted sense. However, after careful checking, he concluded that they had indeed been correctly assembled and that the molecule was, to our great surprise, a left-handed double helix![3]

It took us some time to understand the relationship of this left-handed helix with its zigzag backbone to right-handed B-DNA with its smooth backbone [see Fig. 14-6]. It quickly became apparent that all the guanines in the structure had rotated about the glycosyl bond into the less common *syn* conformation, while the cytosines had the normal *anti* conformation. [*Syn* and *anti* refer to two energetically favored positions of the base with respect to the sugar in a nucleotide. In the more common *anti* conformation, the bulk of the base is positioned away from the sugar; in the *syn* conformation, the base rotates to a position closer to the sugar.] Further inspection revealed that in order to go from right-handed B-DNA to left-handed Z-DNA, the base pairs actually have to turn upside down. This act of turning upside down converts the sense of the helix from right to left and also is responsible for the change in the guanine residues to the *syn* conformation.

We began to read more widely about this sequence, and I rediscovered a paper of Pohl and Jovin,[4] written some seven years previously, in which they studied the circular dichroism of the polymer poly (dG-dC). They clearly showed that in a high salt solution, 5 M NaCl, the circular dichroism spectrum nearly inverted itself. Although the theory of circular dichroism is not well enough developed to come to a firm conclusion, it seemed likely to us that the crystal of left-handed Z-DNA (named for the zigzag backbone) might be the high salt form of poly (dG-dC).

Proving that required a different kind of analysis. For this we chose laser Raman spectroscopy.[5] The Raman spectrum is related to the manner in which the atoms in a molecule vibrate. These vibrations are more or less independent of whether the molecule is constrained in a lattice as in a crystal or freely tumbling in solution. We were able to show that the DNA in the crystal had a Raman spectrum identical to that found in 5 M NaCl and quite distinct from that of poly (dG-dC) in a 0.1 M NaCl solution. Thus it was clear that left-handed Z-DNA was the form of poly (dG-dC) stabilized in 5 M NaCl.

Although the first crystalline work was done with sequences with alternating cytosine and guanine residues, later work incorporated adenine-thymine base pairs. These can convert to the Z form, although not as readily as CG base pairs. The sequences that prefer Z-DNA formation are those that have alternating purines and pyrimidines

with CG sequences being most important for this process. Z-DNA can form even in the absence of regularly alternating purines and pyrimidines, but somewhat more energy is needed for it to form. Therefore, Z-DNA may be regarded as a higher energy state of the DNA double helix. The B-DNA conformation is the ground state, and energy has to be applied to the molecule to stabilize its conversion into the Z form.

The discovery of Z-DNA was startling to many research workers. It had been known for quite a while that DNA could undergo conformational changes. However, the magnitude of the conformational change from right-handed to left-handed DNA is much greater than other changes, and it stimulated a great deal of chemical and molecular biological investigation into its properties.

Several studies have been carried out that describe the formation of Z-DNA *in vivo*. The method for detecting the change often involves changes in the chemical reactivity of DNA when it goes from the B to the Z conformation. These have proven especially useful for studies of DNA in prokaryotic organisms.[6] In eukaryotic organisms, monoclonal antibodies have been used to study Z-DNA formation in physiologically active mammalian nuclei.[7] Nuclei have been prepared that are permeable to macromolecules but still maintain the ability to carry out DNA replication and transcription. The amount of Z-DNA formed in these nuclei is determined by the torsional strain of the DNA. Processes such as transcription that generate negative supercoiling increase the amount of Z-DNA in the nuclei, while the continued activity of the relaxing enzyme, topoisomerase I, results in a decrease of Z-DNA formation. Selective regions of individual genes have been identified that flip from the B to the Z form during transcriptional activity. However, further work will be needed to fully define the biological role of these conformational changes within cells.

References

[1] Day, R. O.; Seeman, N. C.; Rosenberg, J. M.; and Rich, A. *Proc. Nat. Acad. Sci. USA* 70:849−53 (1973); Rosenberg, J. M.; Seeman, N. C.; Kim, J. J. P.; Suddath, F. L.; Nicholas, H. B.; and Rich, A. *Nature* 243:150−54 (1973).

[2] Kim, S. H.; Suddath, F. L.; Quigley, G. J.; McPherson, A.; Kim, J. J.; Sussman, J. L.; Wang, A. H.-J.; Seeman, N. C.; and Rich, A. *Science* 185:435−39 (1974).

[3] Wang, A. H.-J.; Quigley, G. J.; Kolpak, F. J.; Crawford, J. L.; van Boom, J. H.; van der Marel, G.; and Rich, A. *Nature* 282:680−86 (1979).

[4] Pohl, F. M., and Jovin, T. M. *J. Mol. Biol.* 67:375−96 (1972).

[5] Thamann, T. J.; Lord, R. C.; Wang, A. H.-J.; and Rich, A. *Nuc. Acids Res.* 9:5443−57 (1981).

[6] Rahmouni, A. R., and Wells, R. D. *Science* 246:358−63 (1989).

[7] Wittig, B.; Dorbic, T.; and Rich, A. *J. Cell. Biol.* 108:755−64 (1989).

Inverted Sequences (Palindromes)

Inverted sequences are symmetrically repeated DNA sequences that are the same in both chains of a double helix when read in the 5′ → 3′ direction:

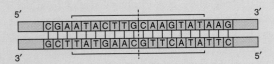

In this example the inverted repeat is enclosed by brackets. The vertical dotted line marks the center of symmetry or *dyad axis* of the inverted sequence. An inverted sequence is usually not contiguous on either side of the dyad axis. Instead, a noninverted sequence (boxed in the diagram below) usually separates the two halves of the inverted sequence:

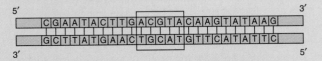

Inverted sequences, which occur commonly in natural DNA, can wind into a regular DNA helix in the B conformation. Because inverted sequences are symmetrical on both sides of the dyad axis, they can also wind into an X-shaped, or cruciform, structure. The noninverted sequence separating the inverted sequences extends as unpaired loops at the top and bottom of a cruciform structure:

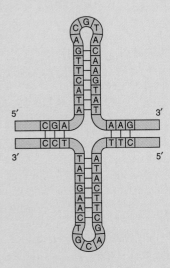

Each of the paired segments winds into a double helix, with the loop extending from the tip:

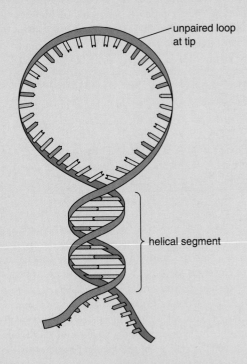

An RNA molecule copied from an inverted sequence may fold back on itself to form a double-helical segment in the region of the sequence. Because the RNA copy exists only as a single chain, the structure formed is a *hairpin* instead of a cruciform. For example, the RNA sequence copied from the top half of the DNA palindrome shown above:

can fold into the hairpin structure:

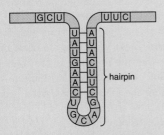

Hairpins copied from inverted sequences in DNA are a common and important feature of RNA structure, particularly in rRNA and tRNA molecules (as, for example, in the rRNA shown in Fig. 15-26).

opposite end is twisted in the left-handed direction against the right-handed turns of the B conformation. Negative supercoiling adds a "strain" to B-DNA that can be relieved by segments that rewind from the B to the Z conformation. The DNA molecules of most cells are negatively supercoiled, a condition that increases the possibility that Z-DNA may occur in nature. A final important cellular factor stabilizing Z-DNA is a group of recently discovered *Z-DNA–binding proteins*. These proteins, discovered in the nuclei of *Drosophila*, human, wheat, bacterial, and other cells, can combine specifically with Z-DNA and "fix" the double helix in stable left-handed form.

It has proved difficult to establish that Z-DNA actually exists inside cells, and researchers are still not absolutely certain that this conformation occurs in nature. The observation that proteins exist in cells that can specifically recognize, bind, and stabilize Z-DNA supports the possibility that DNA does indeed wind into left-handed helices inside living cells. Z-DNA has been demonstrated indirectly in *E. coli* by R. D. Wells and his colleagues through the activity of one enzyme that adds methyl groups and another that introduces breaks in B-DNA. Neither enzyme is effective in catalyzing these reactions in Z-DNA. Some of the DNA in *E. coli* is protected from both enzymes, suggesting that it is Z-DNA.

Another line of evidence comes from the existence of anti-Z-DNA antibodies in persons with certain autoimmune diseases; production of these antibodies would not seem likely unless the Z form actually exists in the body. Anti-Z-DNA antibodies can also be produced by injecting Z-DNA into a test animal. These antibodies react positively with eukaryotic cells, indicating that Z-DNA is present. When linked to a fluorescent dye, the anti-Z antibodies produce a generally distributed fluorescence in the nucleus and in regions of chromosomes, such as polytene chromosomes (see p. 736). The antibody work is not conclusive, however, because some of the preparative methods used in these experiments are suspected to induce the formation of Z-DNA artificially. For example, acid fixation (see p. 117) is often used to prepare cells used in antibody studies; exposure to acid conditions evidently causes a significant fraction of cellular DNA to wind from the B to Z conformation. In spite of these difficulties, most investigators believe that Z-DNA is actually present in limited quantities inside both prokaryotic and eukaryotic cells.

Several functions have been proposed for Z-DNA in cell nuclei. One possibility is that Z-DNA segments supply a reservoir of left-handed turns that are used to unwind right-handed turns in B-DNA. This would allow long DNA molecules to unwind locally, greatly facilitating processes such as RNA transcription. A second proposal is that Z-DNA segments, through their ability to promote DNA unwinding, form part of the

control mechanisms regulating RNA transcription. Addition or removal of methyl groups or Z-DNA–binding proteins, according to this proposal, would regulate the shift between the Z and B forms. (The possible influences of Z-DNA on transcriptional regulation are discussed further in Ch. 15.) A third possibility is that stretches of Z-DNA promote the unwinding of the double helix necessary for genetic recombination, in which two helices unwind, break, and exchange segments (see Ch. 25 for details of the recombination mechanism).

The A, B, C, D, E, and Z conformations by no means exhaust the possibilities for DNA conformations inside and outside living cells. There is little doubt that other conformations form in DNA, induced by the attachment of binding proteins or local differences in DNA sequence.

One artificial DNA conformation that promises to have important applications in research and medicine is the *triple* helix. Rich and his colleagues were the first to show that RNA nucleotide chains can wind into a stable triple helix; DNA chains were also later shown to have this capability. Experiments now in progress are testing the possibility that induced triple helix formation may be used to turn off specific genes. The idea is to construct a nucleotide chain complementary to the control region of a gene of interest. Introducing the chain into cells and inducing triple helix formation would presumably block the control region to the enzymes and factors initiating transcription, making the targeted gene inactive. Genes implicated in tumor formation, for example, might be targeted and turned off by this approach.

Cruciform DNA

DNA may also take on conformations in which additional folding patterns are superimposed on the existing conformations. One of these is *cruciform DNA*, in which a DNA segment folds back on itself locally to form a cross-shaped segment (Fig. 14-9). To form a cruciform segment, the DNA within the folded-back regions must contain a sequence structure known as an *inverted repeat* or *palindrome* (Information Box 14-2). Although the inverted repeats necessary to form cruciform structures might at first glance seem unlikely, they are quite common in DNA and the RNA molecules copied from DNA.

Cruciform DNA, like Z-DNA, has proved difficult to detect with certainty in cells. One experiment, similar to the enzyme experiment used to detect Z-DNA in *E. coli*, supports the existence of cruciforms in this bacterium. This experiment used an enzyme capable of recognizing and cutting the short single-stranded DNA segments at the center of the cruciform pattern (arrows in Fig. 14-9). The enzyme induced cuts in the DNA of

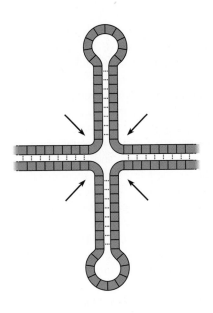

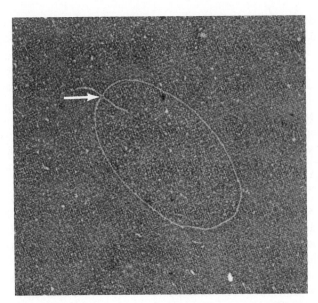

a   b

Figure 14-9 DNA cruciforms. **(a)** A cruciform structure created in DNA by fold-back pairing around an inverted repeat sequence. (Inverted repeats are described in Information Box 14-2.) The arrows show points at which the center of the cruciform may be attacked by enzymes hydrolyzing single-stranded DNA. **(b)** A cruciform (arrow) visible in a circular DNA molecule. Courtesy of M. Gellert.

living *E. coli* cells, suggesting that cruciforms actually exist in this organism. The cruciform structures, if they indeed exist, have been proposed as sites recognized or induced by some regulatory proteins as they bind to DNA.

Chemical Modification of DNA

Both DNA and RNA are modified by the addition of chemical groups to purine and pyrimidine bases or by chemical conversion of the bases to other forms. Although alterations are frequent and highly varied in RNA molecules, creating a long list of modified bases that may appear in this nucleic acid (see p. 581), DNA modifications are limited to a relatively few types.

Addition of a methyl group to cytosine, creating the modified base *5-methylcytosine* (see Fig. 14-7), is the most common chemical modification of DNA. In some organisms, particularly in prokaryotes, a similar reaction adds a methyl group to the 6-carbon of adenine to form the modified base *6-methyladenine*. Other modifications are so rare that they are restricted to one or a few known species or viruses.

The degree to which cytosines are methylated varies widely in different groups of organisms. In bacteria less than 1% of the cytosines in DNA are methylated; the DNA molecules in protozoa, fungi, and most invertebrates similarly contain few or no methylated cytosines. Vertebrates show more extensive methylation of cytosines, ranging from about 2% to 8% of the total cytosine content. The most extensive modification is noted in higher plants, in which one-third or more of the cytosines may be methylated.

DNA methylation is of interest because a clear correlation has been established in vertebrates between the methylation of cytosines and the level of mRNA transcription. The DNA of many mRNA-encoding genes is methylated at certain sites when the genes are inactive in transcription and demethylated at these sites when active. In many cell lines the patterns of methylation in these genes are reproduced and passed on during cell division. For these reasons it is considered possible that methylation and demethylation of cytosines may be part of transcriptional regulation in vertebrates. One mechanism by which methylation may regulate DNA, as noted, is by favoring the conversion from B- to Z-DNA. (The possible relationships between methylation and the regulation of mRNA transcription are discussed further in Ch. 17.)

Methylation may have various additional effects on DNA. Methylation modifies the binding of proteins to DNA; in some cases it inhibits binding and in others stimulates it. For example, in bacteria methylation inhibits the binding of a group of enzymes, *restriction endonucleases*, that introduce breaks in the sugar-phosphate backbone of unmethylated DNA at certain sequences (see p. 135 for details). Methylation in bacteria, in fact, is restricted to the sequence elements recognized as break points by the enzymes. The protection allows unmethylated foreign DNAs, such as those of bacterial viruses, to be attacked by the restriction endonucleases without danger to the bacterial DNA. Binding of another protein, the *lac repressor*, is stimulated by methylation. (The *lac* repressor regulates a gene group coding for enzymes that metabolize lactose sugars; see p. 727.) Methylation has also been linked to genetic recombination and to the repair of damaged or mismatched DNA sequences in prokaryotes.

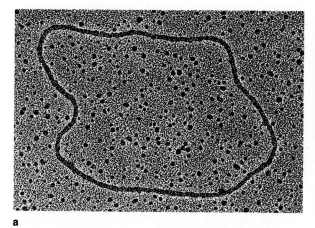

a

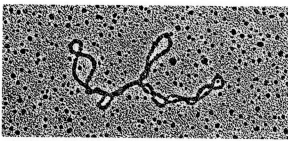

b

Figure 14-10 A circular DNA molecule in relaxed **(a)** and supercoiled **(b)** forms. Each twist in the DNA circle is a supercoil. Courtesy of L. W. Coggins.

DNA Supercoiling

In B-DNA the number of base pairs per turn and the length occupied by one full turn can vary over only relatively narrow limits. Turns cannot be compressed or expanded in B-DNA as in a rubber band; more or fewer turns per unit length cannot easily be packed into the helix. Instead, if a twisting force is applied in either the positive or negative direction, DNA flips into supercoils to compensate for the strain induced by the twisting (Fig. 14-10). One supercoil forms for each full turn imposed on the DNA. The loops resemble the "knots" that appear when a coiled telephone cord is twisted. As noted, all natural B-DNAs are exposed to forces that tend to twist them in the direction opposite to the right-handed turns of the double helix, forming negative supercoils. (Fig. 14-10*b* shows a negatively supercoiled DNA molecule under the electron microscope.)

The origins of negative supercoils differ in prokaryotes and eukaryotes. In prokaryotes supercoils are induced by enzymes called *topoisomerases*.[1] When working in the direction that introduces supercoils, these enzymes open a break in one or both nucleotide chains of a B-DNA double helix, rotate or twist the chains, and close the breaks so that both nucleotide chains are continuous again. (Fig. 14-11 shows the operation of a topoisomerase I enzyme, which breaks one of the two nucleotide chains; topoisomerase II enzymes break both chains.) Energy for the supercoiling is supplied by ATP, which is hydrolyzed at a rate of at least one molecule of ATP for each supercoil wound into DNA. Topoisomerases can also work in reverse to remove supercoils from DNA. Removal of supercoils, called "relaxation" of the DNA, is powered by release of the supercoils and does not require an energy input.

In prokaryotes the total amount of supercoiling, which approximates one negative supercoil for every

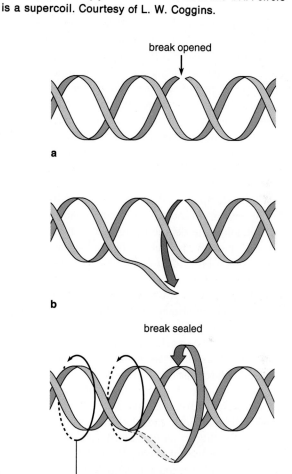

break opened

a

b

break sealed

counterclockwise rotation to compensate for rotation of free end in **(b)**, or to introduce negative
c supercoil

Figure 14-11 Activity of topoisomerase I in the introduction of supercoils in DNA. **(a)** The enzyme opens a single break in one of the two chains of a DNA double helix by opening a phosphate linkage. **(b)** A free end is rotated through one full turn and covalently rejoined **(c)**. If the ends of the DNA molecule are not free to turn, one supercoil will be introduced for each complete rotation of the free end. In the example shown, the supercoil will be negative because the direction of rotation of the free end in **(b)** is opposite to the right-handed turns of the DNA double helix.

---

[1] Topoisomerases are named for *topology*, the mathematical study of the possible ways in which objects of constant geometry, such as DNA, can be bent or twisted without disturbing their fundamental structure.

200 DNA base pairs, is maintained by an automated mechanism that balances the activity of two different topoisomerases. One topoisomerase uses ATP energy to induce negative turns, and the other releases the turns. The activity of the two topoisomerases is regulated in part by the degree of supercoiling. If negative supercoiling is greater than approximately one supercoil per 200 base pairs, the topoisomerase inducing supercoils is inhibited, and the one releasing supercoils is stimulated. The opposite condition, when supercoils fall below the level of one per 200 base pairs, stimulates the topoisomerase inducing supercoils and inhibits the one releasing them.

The supercoil balance is also controlled by gene regulation in prokaryotes. When negative supercoiling falls below the optimum level, transcription of the gene encoding the topoisomerase adding supercoils is stimulated. An increase in negative supercoils above the optimum level has the opposite effect: transcription of the gene encoding the topoisomerase removing supercoils is stimulated (for details, see p. 733).

Although topoisomerases also occur in eukaryotes, they probably are not the primary source of negative supercoils in these organisms. In eukaryotes topoisomerases evidently participate in the reactions unwinding DNA during replication and genetic recombination rather than maintaining the degree of negative supercoiling (see pp. 967 and 1072 for details). Instead, eukaryotic DNA appears to be held in negative supercoils by combination with histone proteins. These proteins combine with DNA to form a beadlike unit called a *nucleosome*, the fundamental structural unit of chromatin (see below and Fig. 14-17). The DNA turns in a left-handed direction around the nucleosomes, much like a rope around a pulley. In effect the winding imposes one negative supercoil for each nucleosome. There are enough nucleosomes to produce about the same number of negative supercoils per unit length of DNA as in prokaryotes—about one for each 200 base pairs.

Negative supercoiling in natural DNAs produces several effects. The negative supercoils favor local unwinding of the DNA helix, exposing single, unpaired nucleotide chains for processes such as transcription, replication, and genetic recombination. Negative supercoiling also favors the conversion of B- to Z-DNA and modifies the binding affinity of proteins for DNA, in some cases enhancing protein-DNA binding and in other cases inhibiting it. The degree of supercoiling may be a factor in gene regulation, as it is in the topoisomerase gene in prokaryotes.

PROTEINS ASSOCIATED WITH DNA IN EUKARYOTIC NUCLEI

The nuclear DNA of eukaryotic cells occurs in close association with two classes of proteins, *histones* and *nonhistones*. The three elements—DNA, histones, and nonhistones—collectively form the *chromatin* of eukaryotic nuclei. DNA and histone proteins occur in approximately equal quantities by weight in chromatin; the more variable nonhistone proteins are found in quantities ranging from as little as 20% to approximately the same as the DNA by weight.

Though not as well characterized as DNA, the histones have been identified, isolated, and completely sequenced in many eukaryotes, and their general pattern of interaction with DNA has been established. Little is known about the three-dimensional structure of individual histone proteins in intact chromatin, however. The histones evidently act in chromatin primarily as structural elements maintaining DNA in specific three-dimensional coiling and folding conformations.

Many nonhistone proteins have also been identified and sequenced. The proteins in this group have proved to be primarily functional molecules that catalyze or regulate the activity of DNA in transcription and replication. Current research has just begun to reveal the structure of regulatory nonhistones, how they interact with DNA, and how they control the activity of genes in transcription.

Histones

The histones are distinguished from most other cellular proteins by their abundance, relatively small size, and strongly basic charge. These distinctive features contributed to their early discovery as a nuclear constituent—the histones were first isolated by Albrecht Kossel in 1884, not long after the discovery of DNA. (DNA was discovered by F. Miescher in 1871.)

The histones are easily isolated by exposing chromatin to acids or salts at elevated concentrations. Removal in this way demonstrates that the histones are held in chromatin by electrostatic attractions rather than covalent bonds. Five major histone types, *H1*, *H2A*, *H2B*, *H3*, and *H4*, are released in sequence from chromatin as the salt concentration is gradually raised over cellular levels.

The strongly basic charge of the histones results from the presence of lysine and arginine residues, which make up a large part of the amino acid complement of these proteins (Table 14-2). All the histones except H1 have molecular weights between about 11,000 and 15,000, placing them among the smaller proteins of cells. H1 varies from molecular weights of 17,000 to 19,000 in lower eukaryotes and 21,000 to 28,000 in higher eukaryotes. Even so, H1 is still a relatively small protein.

Isolation of the histones from all eukaryotes, with very few exceptions, releases proteins equivalent to the same H1, H2A, H2B, H3, and H4 types. The DNA of many mammalian viruses, such as the SV40 (*simian virus*

Table 14-2   Histones

Histone Type	Molecular Weight	Number of Amino Acids	Approximate Content of Basic Amino Acids
H1	17,000–28,000	200–265	27% lysine, 2% arginine
H2A	13,900	129–155	11% lysine, 9% arginine
H2B	13,800	121–148	16% lysine, 6% arginine
H3	15,300	135	10% lysine, 15% arginine
H4	11,300	102	11% lysine, 4% arginine

40) and *polyoma viruses,* is also associated with the same five histones, in this case derived from their host cells. The few known exceptions occur among unicellular algae and fungi. Algae, although possessing histones identifiable as H1, H3, and H4, have only a single histone with properties similar to both H2A and H2B. Yeasts such as *Saccharomyces cerevisiae* either have no H1 or have an H1 equivalent so different in structure that it is not recognizable as a histone.

Histone Sequences and Structure   The histones of many eukaryotes, including representatives of all the major kingdoms, have been completely sequenced. The sequencing studies reveal fundamental structural plans for each histone type that are more or less closely shared in all species (Fig. 14-12). In all types, basic and nonpolar amino acids are clustered in separate segments of the sequence. The nonpolar segment of each histone forms a single globular region when the histones are suspended in solution. The basic segments extend as unfolded "arms" from one or both sides of the globular region. Although the interactions of the charged and nonpolar segments are still largely unknown, it appears that both regions coordinate in binding histone molecules to each other and to DNA.

H2A, H2B, H3, and H4 are called the *core histones* because they form the beadlike core structure around which DNA wraps to form nucleosomes. H1 is sometimes called the *linker histone* because it is believed to combine near a segment of DNA that extends as a link from one nucleosome to the next (see Fig. 14-17).

Comparisons among histone sequences of different species show that the histones, particularly those of higher eukaryotes, are among the most highly con-

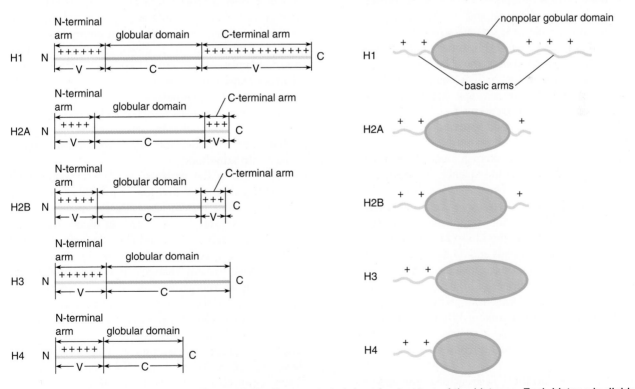

Figure 14-12 Sequence and domain structure of the histones. Each histone is divided into extended, basic domains and a globular, nonpolar domain. The basic domains (V) are more variable in sequence between different species than the nonpolar domains (C), which are much more highly conserved.

served proteins known. The degree of conservation runs roughly in reverse of the numbering system. H3 and H4 are the most highly similar in overall sequence among eukaryotes; H2A and H2B are intermediate in conservation; H1 is least conserved. Many of the substitutions in core histones that appear in different species are conservative; that is, one amino acid is replaced by another of similar size and chemical properties. Further, most of the sequence differences in core histones are restricted to the charged arms; few or no sequence substitutions occur in the nonpolar globular domains. Thus, the core histones are segregated into *variable* and *constant segments* (see Fig. 14-12) that correspond to the arm and globular domains.

Similarities in core histones are most marked among higher eukaryotes. For example, only 2 conservative substitutions out of a total of 102 amino acids occur in H4 histones of species as evolutionarily diverse as the calf and garden pea. The H3 histones of different plant and animal species, although somewhat more variable than H4, are still almost identical. The H3 histones of the calf and garden pea, for example, differ only in 4 conservative substitutions among 135 amino acids. H2A and H2B, though more variable among higher eukaryotic species, still show extensive sequence conservation, especially in the globular, nonpolar region of the proteins.

More extensive sequence differences are noted in the core histones of invertebrates and lower eukaryotes, suggesting that the structural and functional requirements for core histones are somewhat less restrictive among these organisms. The H3 histone of yeast shows 15 sequence differences from mammalian H3; yeast H4 is similarly altered in sequence. The apparent replacement of H2A and H2B by a single histone in some unicellular algae has been mentioned.

The H1 histones of different species, although still clearly related in sequence, show much more variation than core histones. Among mammals alone H1 histones show about a 20% variation in sequence. As vertebrate evolutionary relationships become more distant, the sequence diversity in H1 histones becomes more marked. Trout H1 histones, for example, have only 194 amino acids as compared to the 213 to 215 of mammals, with 140 identical amino acids and 85 differences; only 14 of the 85 differences, which include deletions, are conservative. The H1 histones of higher plants vary in about 75% of their sequences from animal H1s. As in core histones, the charged arms of H1s from different sources vary most extensively in sequence. The central globular segment is more highly conserved.

Histone Sequence Variants   Each histone type except H4 commonly exists in several variants that differ slightly in sequence within individual species. The number of variants per histone type, like general sequence conservation, runs roughly in reverse of the numbering order. Only a few species, such as wheat, have been observed to have H4 variants. H3, H2A, and H2B variants are more common. Most vertebrate species have two or three versions of these histones that differ at one or two positions in their amino acid sequences. H1 histones are by far the most variable; most cell types have four to six H1 variants, some of which differ extensively in sequence. Among the core histones the variants are usually identified by the histone type followed by a decimal point and a number. For example, the three H2B types commonly observed in vertebrate cells are identified as *H2B.1*, *H2B.2*, and *H2B.3*.

The H1 variants are named by the same convention or are identified with a following letter as *H1a*, *H1b*, *H1c*, and so on. Two additional, highly distinctive types commonly detected in vertebrates are given special designations as *H1⁰* and *H5*.

The existence of sequence variants in cells of the same species indicates that most histones are encoded in more than one gene. This indication is confirmed by DNA sequencing studies, which reveal that the histones of all eukaryotic species are encoded in multiple gene copies usually numbering in the tens to hundreds, with many duplicates of the genes coding for each variant.

The histone variants have several interesting characteristics. In vertebrate species, particularly among mammals, the same variants are highly conserved and repeated over a wide range of species. The variants change in relative amounts and appear and disappear at fixed times in coordination with embryonic development. During sea urchin development, for example, two H1 types—H1.1 and H1.2—are present at early cleavage stages; as the developing embryo reaches 400 cells in size, H1.1 synthesis stops. H1.2 synthesis continues through the blastula stage and ceases at gastrulation. Two more variants—H1.3 and H1.4—appear as the embryo reaches 200 to 300 cells; these variants persist into adulthood.

Changes in the variants are also correlated with cycles of growth and division of individual cell lines. Slowly dividing mammalian cells, for example, have H2A.2, H2A.3, H2B.1, and H3.3 in higher quantities than H2A.1, H2B.2, H3.1, and H3.2. In embryonic and rapidly dividing cells these levels are reversed. Some of the most striking changes in histone variants in relationship to cell growth and division are noted in H1. Among vertebrates most H1 variants are replaced by H5 in the nuclei of mature erythrocytes (red blood cells) except in mammals, where the nucleus is lost during erythrocyte development. This change is believed to reflect, in some way, the total repression of transcription characteristic of fully mature erythrocytes. H1⁰ increases in relative concentration in some mammalian cell lines that stop replicating their DNA and dividing. Characteristic histone variants also appear in many species during meiosis, particularly during meiotic prophase, when chromosome pairing and recombination take place (for details, see Ch. 25).

The controlled changes appearing in coordination with development and cell division indicate that the variants are not simply random, nonfunctional differences in the histone proteins. Rather, they are probably required for adjustments in chromatin structure that are critical to the normal progress of cells through development and cycles of growth and division. Although this is likely to be the case, the structural changes brought about by controlled changes in the histone variants and their significance remain to be established.

As part of the biochemical and structural alterations that produce motile male gametes in both plants and animals, histones may be partly or completely replaced by a group of much smaller and even more basic proteins, the *protamines*. The details of this replacement, which is apparently related to the extreme compaction of DNA that takes place during the development of motile male gametes, are discussed in Chapter 26.

Histone Modification The five histones are extensively modified by the addition of chemical groups to individual amino acid residues. Simple modifications include the addition of acetyl, methyl, or phosphate groups to amino acids in the charged arms of histone chains. Complex modifications involve the addition of relatively large chemical structures—one, an entire protein called *ubiquitin*, and the other, a complex nucleotide-based unit, the *ADP-ribosyl group*. (Details of these modifications are presented in Information Box 14-3.)

The various histone modifications are correlated with many important cellular mechanisms, including genetic regulation, DNA replication, chromatin assembly and compaction, and cell division (see Information Box 14-3). In no case, however, has it been possible to determine whether the modifications are part of the control mechanisms regulating these activities or whether they merely appear as effects of other mechanisms of cell regulation.

Nonhistone Proteins

Nonhistones are defined broadly as the proteins, excluding histones, that are associated with DNA in chromatin. Most proteins in this highly varied group are negatively charged or neutral at physiological pH. A few are basic, positively charged proteins, but none are as basic as the histones. Among this group are proteins with the most important function of any in the cell: they regulate the activity of individual genes. As such, they provide the central controls of cell activity and differentiation.

Numbers and Types of Nonhistone Proteins The total numbers and kinds of nonhistone proteins have proved

difficult to establish. Because most nonhistones, unlike histones, have chemical properties resembling other cellular proteins, it is often difficult to determine which of the proteins or polypeptides in bulk preparations are actually nonhistone chromosomal proteins and which are contaminants from other cellular sources. Distinguishing between intact proteins and polypeptide fragments also presents difficulties because many nonhistones break easily into subparts that register as apparently distinct nonhistone proteins. Moreover, many nonhistone proteins, particularly those involved in transcriptional regulation, occur in the nucleus in numbers too small to be readily detected.

These difficulties have produced wide variations in the numbers and kinds of polypeptides identified as nonhistone proteins. Depending on isolation methods, the proteins identified as nonhistones may vary from as few as 20 to as many as several hundred different kinds, ranging in molecular weight from 7000 to 200,000 or more. Many of these proteins appear in forms that, like the histones, are modified by acetylation, methylation, phosphorylation, ubiquitination, or ADP ribosylation. It seems likely that the actual number of nonhistone proteins ranges toward the higher end of the distributions obtained in bulk preparations, with most cells probably possessing at least hundreds of different nonhistone types.

The proteins appearing as nonhistones fall into several clearly defined groups:

1. Proteins regulating the activity of genes in RNA transcription.

2. Enzymes and factors functioning in transcription, replication, recombination, DNA repair, and the modification of DNA and the chromosomal proteins.

3. Proteins associated with the RNA products of transcription.

4. Proteins contributing to the maintenance of chromatin structure or to the conversion of chromatin between extended and tightly folded or condensed states.

The best known and characterized of these groups are the enzymes active in transcription or replication and other enzymatic activities associated with chromatin. Many of the proteins temporarily associated with the RNA products of transcription have also been identified. Intensive research into nonhistones regulating RNA transcription is rapidly adding to the list of proteins in this group. Least known are nonhistones contributing to the maintenance of chromatin structure.

Isolation and identification of regulatory nonhistones have been greatly accelerated by techniques in which the DNA of specific genes is used as a "trap" to catch individual protein types. In one of the most

Histone Modifications

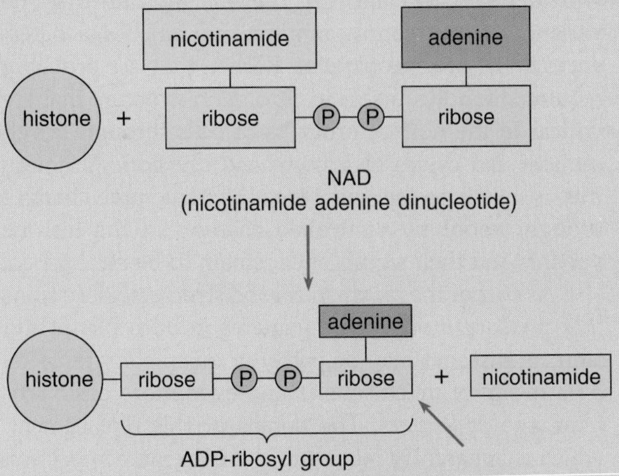

Histones are among the most highly modified cellular proteins. Chemical structures added covalently to these proteins include acetyl, methyl, phosphate, and ADP-ribosyl groups, and an entire protein, ubiquitin. Correlations noted between the various modifications and important cellular mechanisms such as genetic regulation and cell division suggest that the alterations are highly significant to cellular functions.

Acetyl, methyl, or phosphate groups are added to positively charged amino acid residues in the extended arms of the histones. The modifications eliminate one positive charge for each addition and presumably reduce the strength of the electrostatic attraction of histones to DNA.

Addition of acetyl groups to lysine residues in core histones increases in coordination with several cellular functions, including nucleosome assembly and conversion of genes from inactive to active form. Histones are temporarily acetylated before being added to DNA during nucleosome assembly and then deacetylated when assembly is complete. This temporary acetylation may reduce the electrostatic attraction between histones and DNA enough to permit precise control of nucleosome assembly. Histones associated with many but not all genes become more highly acetylated as the genes change from inactive to active form. In this case acetylation may loosen chromatin structure, through reduction of the DNA-histone attraction, enough to allow access to the DNA by enzymes catalyzing RNA transcription. Extensive histone acetylation also precedes their replacement by protamines during sperm cell development. Histone acetylation, in this case, probably facilitates their removal from chromatin and replacement by the protamines (see Supplement 26-1).

Addition of methyl groups to lysines and histidines in H1, H2B, H3, and H4 is correlated with DNA replication

NAD
(nicotinamide adenine dinucleotide)

ADP-ribosyl group

in some cells. Histone phosphorylation is correlated with progress through the cycle of growth and division in many cell types. In these cells phosphate groups are added, particularly to H1, just before and during cell division. The added phosphates are removed as cells finish dividing and enter the next phase of growth (for details, see Ch. 22 and Fig. 22-8). Increases in H1 phosphorylation are thought to promote the packing of chromatin into assemblies such as the clumps of heterochromatin seen in interphase nuclei (see Fig. 14-23a) or the rodlike chromosomes appearing during cell division (see Fig. 24-2).

Ubiquitin, a small protein containing 76 amino acid residues, is added covalently to H2A and, to a lesser extent, to H2B. Ubiquitin is so called because it occurs in all eukaryotes in almost the same form. The protein attaches by a peptide linkage to an amino group forming part of the side chain of a lysine residue in H2A or H2B. The addition, since it links one protein to the side of another, produces a branched protein. The modification is extensive; as much as 5% to 15% of the H2A molecules in chromatin may be converted to the ubiquitinated form. In this

effective of these techniques, the DNA of a specific gene, produced in many copies by cloning or the polymerase chain reaction, is attached to a piece of filter paper or to agarose beads in a column (see pp. 132 and 586 for details). A crude preparation of nonhistone proteins is then poured over the paper or through the column. From the many proteins in the preparation, those that can recognize control sequences in the DNA, often as few as one or two different regulatory types, are immobilized and trapped on the paper or in the column by binding to the DNA.

Interactions of Regulatory Nonhistone Proteins with DNA
Much current interest centers around the patterns by which regulatory nonhistone proteins interact

with DNA. There is no doubt that regulatory nonhistones can recognize and bind individual DNA sequences even in DNA helices that are fully wound.

Regulatory proteins evidently bind to specific DNA sequences in fully wound DNA primarily by "reading" the edges of base pairs exposed in the major DNA groove. Reading involves the formation of hydrogen bonds between the edges of the base pairs and side groups of amino acids in the regulatory protein. (Fig. 14-13 shows how several of these base-pairing relationships can form.) The hydrogen-bonding patterns are distinct enough to allow A-T, T-A, G-C, and C-G pairs to be distinguished in a fully wound DNA double helix. Regulatory proteins may also recognize local distortions in groove width and other helix diform the H2A-ubiquitin complex is known as *protein A24* or *uH2A*. In spite of the relatively large size of the added protein group, ubiquitination has little or no detectable effect on chromatin structure.

For cellular proteins other than histones, ubiquitin addition serves as a marker for identification and destruction by the scavenging systems of the cell (for details, see p. 720). This does not seem to be the case for histones because the rate of histone turnover—the rate at which old histone molecules are destroyed and replaced by new ones—is very slow. It is much slower, in fact, than the rate at which ubiquitin groups are added and removed from H2A and H2B. Ubiquitination of H2A and H2B decreases in some transcribed genes and disappears entirely from the chromatin during cell division. For these reasons addition of the protein group is thought by some investigators to be related in some unknown way to gene regulation or the chromosome compaction that occurs during cell division.

ADP ribosylation occurs by the transfer of subparts of NAD (nicotinamide adenine dinucleotide; see p. 316) to the histones. In the transfer the nicotinamide group is split from NAD, and the remaining structure, a ribose sugar linked to ADP, is attached to the histone at a glutamic acid, glutamine, or lysine residue (see the figure in this box). Although H1, H2A, and H2B are the primary histones modified by ADP ribosylation, H3 may also be altered in this way. In some cells H2A may be modified simultaneously by ubiquitination, ADP ribosylation, methylation, acetylation, and phosphorylation!

Researchers have observed both negative and positive correlations between the degree of ADP ribosylation and gene transcription, DNA repair, cell development and maturation, and cell division. In all cases their results are contradictory; some noted an increase in the degree of ADP ribosylation as a part of these activities, some noted a decrease, and some saw no change at all. The upshot is that although ADP ribosylation may be related to some or all of these functions the precise effects of this extensive, and, in some respects, bizarre modification are as yet unknown.

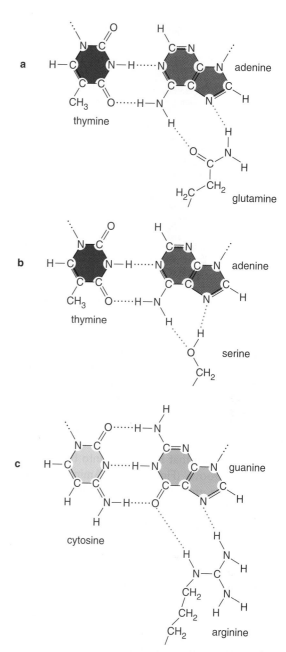

Figure 14-13 Formation of hydrogen bonds **(a)** between the edges of an A-T base pair exposed in the major groove of DNA and a glutamine residue in a protein; **(b)** between the edges of an A-T base pair and a serine residue; and **(c)** between the edges of a G-C base pair and an arginine residue. **(a** and **c)** from N. C. Seeman et al., *Proc. Nat. Acad. Sci.* 73:804 (1976); **(b)** from D. H. Ohlendorf et al., *Nature* 298:718 (1982).

mensions that result from differences in base sequence. For example, sequences containing G-C base pairs slightly open the minor DNA surface groove, and sequences containing A-T pairs slightly narrow the groove. Initial combination of the regulatory protein with the DNA may induce further distortions that open the grooves or the entire helix more extensively, allowing more precise pairing of the protein with individual sequences. Many regulatory proteins, for example, produce a distinct kink or bend in the DNA helix as they bind.

An examination of the interactions between regulatory proteins and the sequences they recognize, using X-ray diffraction and other biophysical techniques, has revealed structural features common to several regulatory subgroups. These structural features, or "motifs" as they are called, include the *helix-turn-helix*, *zinc finger*, and *leucine zipper* (Fig. 14-14).

The helix-turn-helix motif (Fig. 14-14), discovered by T. A. Steitz, B. W. Matthews, and their coworkers, involves two alpha-helical segments in the regulatory protein, linked together by a tight bend in the amino acid chain. One of the helices, called the *recognition helix,*

DNA Endonucleases and Nucleosome Periodicity

The endonuclease digestion technique used by Hewish and Burgoyne to detect periodicity in chromatin has become one of the fundamental methods of chromatin research. The technique uses a group of enzymes that break sugar-phosphate bonds in both nucleotide chains of the DNA double helix. The breaks cause the DNA double helix to separate completely at the points attacked by the enzymes.

Three different DNA endonucleases are commonly applied in this research: *DNase I, DNase II,* and *micrococcal nuclease* (also called *Staphylococcal nuclease*). The breaks induced by these enzymes, rather than being in exact register in the two nucleotide chains of the double helix, are staggered by either two bases, for DNase I and micrococcal nuclease, or four bases, for DNase II. Although the breaks are staggered, too few bases remain paired in the region between breaks to resist unwinding, and the DNA separates completely at the sites of enzymatic attack.

In application of the technique chromatin in intact form, with histones still attached, is exposed to one of the DNA endonucleases. After digestion has proceeded for the allotted time, the enzyme is removed from the solution or inhibited, and the histones are separated from the DNA by exposure to high salt. Histone removal liberates the broken DNA pieces, which are run on an electrophoretic gel (see p. 126) to separate them by length. Typically, the DNA pieces from such preparations separate into a series of regularly spaced bands called a "ladder," with the shortest DNA in the band at the bottom of the gel (see the photo in this box). The regular spacing of bands in the ladder indicates that the DNA length of each heavier band is a regular multiple of the first band. If the first band contains DNA pieces 200 base pairs in length, for example, the next higher band contains pieces 400 base pairs in length, and so on.

Micrograph courtesy of S. Spiker, reproduced with permission from *Ann. Rev. Plant. Physiol* 36:235, © 1985 by Annual Reviews, Inc.

somes. More prolonged digestion slowly hydrolyzes the DNA associated with H1 in the nucleosome, releasing H1 and leaving a resistant piece about 140 + base pairs long. The 140 + base-pair segment resisting more prolonged digestion is the DNA wound around the core particle, protected from enzymatic attack by its tight association with the histone octamer. The core particle, consisting of two pairs of each of the core histones, and the location of a single H1 outside the core particle explained the crosslinking data and the 1:2:2:2:2 ratio observed for the histones in chromatin.

After Kornberg published his model, subsequent detailed work with DNA endonucleases revealed that once the initial digestion producing the rough 200 base-pair lengths is accomplished, the DNA is cut gradually to a relatively constant length of about 165 base pairs. This is believed to represent the two full turns of DNA wrapped around the core particle, held in place by an H1 molecule still tied to the nucleosome (see Fig. 14-17). Digestion past this point, which proceeds more slowly, removes about 10 base pairs from each end of the two DNA turns, releasing H1 and shortening the DNA to the highly constant 146 ± 1 base-pair length, which strongly resists further digestion.

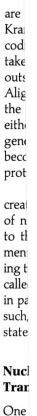

are
Kra
cod
take
outs
Alig
the
eith
gene
becc
prot

crea
of n
to th
men
ing t
calle
in pa
such,
state

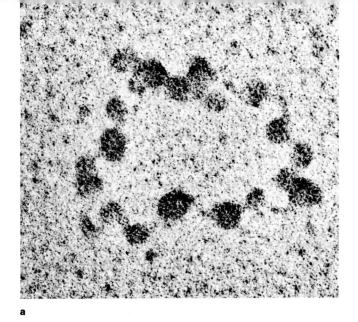

a

b

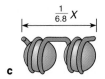

c

Figure 14-19 The SV40 minichromosome and DNA packing in nucleosomes. **(a)** The circle of nucleosomes in the SV40 minichromosome. ×1,300,000. Courtesy of J. D. Griffith, from *The Molecular Biology of the Mammalian Genetic Apparatus*, ed. P. O. P. Ts'o, Elsevier/North Holland, 1977. The difference between **(b)** DNA at its fully extended length *(x)*, and **(c)** DNA wrapped around nucleosomes (1/6.8x).

Nucl
Tran

One
struc
trans
some
to tal

E
doub
line o
DNA
from
and r
to dig
respo
may
struct
somes
points
tones
pletel
DNA.
H1, m
becon
of chr
moval
slide a
a rope
Tl
bined
modifi
are ty
active
static

Several experiments immediately confirmed the basic tenets of Kornberg's model and verified predictions developed from his hypothesis. One particularly illuminating experiment by J. D. Griffith employed the SV40 virus, a mammalian virus in which the DNA circle of the virus is complexed with histones (Fig. 14-19a). This experiment tested a prediction of the model regarding what is known as the *packing ratio* of DNA in nucleosomes. Kornberg predicted that wrapping DNA around nucleosomes compacts its length to about 1/6.8 of its fully extended length, or to a packing ratio of 6.8:1 (Fig. 14-19b). To test this prediction, Griffith first measured the circumference of the viral DNA circle with nucleosomes present and then measured the DNA circle in the fully extended state, with histones removed and nucleosomes absent. The two circumferences obtained, 0.2 μm with nucleosomes present and 1.48 μm for the extended state, indicated a DNA packing ratio in nucleosomes of 7:1. This is very close to the 6.8:1 packing ratio predicted by Kornberg.

Other experiments, including X-ray diffraction and detailed electron microscopy, have confirmed that the DNA is wound around the outside of the core particle in a two-turn coil with a diameter of 8.6 nm and extends between nucleosomes as a linker. X-ray diffraction studies by A. Klug, P. J. G. Butler, J. T. Finch, and their colleagues, carried out with crystals cast from core particles retaining their tightly bound DNA, indicated locations of individual histones within the core octamer (see Fig. 14-17b). Their work also showed that the core particle is a flattened, wedge-shaped structure that probably introduces sharp kinks or bends at points in the DNA wrapped around it.

Nucleosome Assembly

Nucleosomes with apparently typical properties can be assembled in the test tube from only the five histones and DNA; no other proteins are required. Essentially any B-DNA molecule, including DNA molecules of viruses and bacteria that are not complexed with histones in their natural locations, can wind with histone proteins into nucleosomes. The ability of nucleosomes to assemble from DNA of essentially any natural or artificial source indicates that DNA sequence is of little or no significance in the assembly process. RNA does not form nucleosomes in either the single- or double-helical form.

In the test tube, H3 and H4 alone can induce DNA to coil into nucleosomelike structures that can be completed by addition of H2A and H2B. DNA will also wrap around fully assembled core particles, including those that have been stabilized by reaction with protein crosslinkers.

Purified histones and DNA do not form nucleosomes readily in the test tube in solutions adjusted to salt concentrations typical of living cells. The test-tube assembly reactions work well only at elevated salt concentrations approaching those required to dissociate the histones from DNA. Evidently, at physiological salt concentrations electrostatic attractions between DNA and histones are so strong that weaker associations necessary for core particle assembly, such as hydrogen bonds and hydrophobic attractions, are "swamped out" and prevented from forming. Under these conditions DNA and histones form only a random, unstructured precipitate. Raising the salt concentration probably surrounds the charged groups of DNA and histones with

Bacterial DNA circles may be organized into looplike domains similar to those suspected to exist in eukaryotic chromatin. Evidence for the domains is taken from the fact that the supercoils of native bacterial circles do not entirely unwind if a break is introduced experimentally at some point. Instead, only a relatively small portion of the DNA, containing about 100,000 base pairs on the average, unwinds its supercoils when a single break is introduced. From the number of breaks required at different sites to unwind all supercoils, R. Sinden and D. Pettijohn estimated that the *E. coli* DNA circle contains 43 ± 10 looplike domains.

Bacterial Nucleoid Proteins

A number of different proteins can be detected in association with bacterial DNA. These include enzymes of transcription and replication, and regulatory proteins equivalent in function to some of the regulatory nonhistone proteins of eukaryotes.

A number of proteins roughly equivalent to eukaryotic histones can also be detected in bacterial nucleoids. These possible structural proteins are small, basic molecules with chemical properties similar to histones. Several different types have been isolated from *E. coli* (Table 14-3). Although these proteins bind DNA less strongly than histones, two of them, *HU* and *IHF*, wind bacterial DNA into compact, nucleosomelike structures in the test tube. Another *E. coli* protein, *H*, contains segments that cross react with antibodies against eukaryotic H2A. The *E. coli* H does not appear to interact directly with prokaryotic DNA, however.

HU contains basic and hydrophobic amino acids in about the same proportions as eukaryotic histones. However, no clustering into distinct basic and hydrophobic domains is discernible as in the histones, and there are no similarities in amino acid sequence between HU and histones at any points. Whatever its structural similarities and differences to eukaryotic chromosomal proteins, HU appears to be generally distributed and highly conserved among different prokaryotes, including both bacteria and cyanobacteria. More distantly

related forms of HU have also been detected in mitochondria and chloroplasts. Estimates of the total quantity of HU and other histonelike proteins in bacterial cells vary from enough to bind a minimum of 20% to as much as 100% of bacterial DNA.

Although HU and IHF can wind bacterial DNA into nucleosomelike structures in the test tube, there is no evidence directly indicating that these proteins have the same function inside living bacterial cells. Experiments with DNA endonucleases, for example, fail to reveal any regular periodicities in bacterial DNA similar to those reflecting nucleosome structure in eukaryotes. Despite the lack of direct evidence, HU, in particular, is widely believed to be a structural protein compacting DNA in bacteria and cyanobacteria. One or more of the histonelike proteins may segregate bacterial DNA circles into looplike domains.

Although the precise roles of the bacterial histonelike proteins are largely unknown, there is little doubt that these or other elements increase the packing of bacterial DNA. Repulsion by the negatively charged phosphate groups spaced along the double-helical backbone limits the minimum volume into which naked, uncomplexed DNA can be packed. The volume inside an entire *E. coli* cell is about 1000 times smaller than the volume expected to be occupied by an uncomplexed *E. coli* DNA circle, indicating that the histonelike proteins or other molecules with similar properties neutralize the phosphate groups and allow packing to at least this level.

One interesting bacterial group, *Thermoplasma*, contains a histonelike protein, HT_a, that may be linked to the ability of these organisms to survive in hot springs with temperatures approaching 60°C. HT_a, a small, strongly basic protein containing 89 to 90 amino acids, evidently stabilizes *Thermoplasma* DNA from heat denaturation at elevated temperatures. HT_a contains one segment with slight but statistically significant sequence similarities to eukaryotic histones. Another bacterial genus living in hot springs, *Sulfolobus*, was recently found also to contain histonelike proteins capable of stabilizing DNA at elevated temperatures.

Table 14-3	Histonelike Proteins of *Escherichia coli*	
Designation	Molecular Weight	Possible Function
H	28,000	May inhibit DNA replication; cross reacts with antibodies against eukaryotic histone H2A
HU	9,000	DNA folding; stimulation of transcription
IHF	10,000–11,000	Participates in DNA packing; stimulates recombination
FIS	11,200	May participate in DNA packing
H1 (or H-NS)	15,000	May participate in DNA packing
HLPI	17,000	Unknown

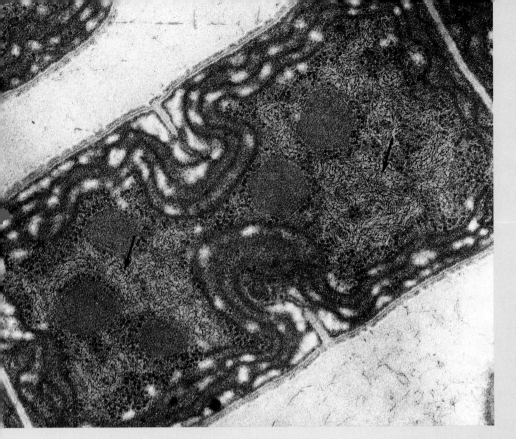

Figure 14-35 Nuclear material (arrows) in a dividing cell of the cyanobacterium *Phormidium.* ×62,000. Courtesy of M. R. Edwards.

Nucleoid Structure in Cyanobacteria

Comparatively little is known about the structural organization of DNA in cyanobacteria. Different cyanobacterial cells contain DNA quantities ranging from about the same to as much as twice the amount in *E. coli.* Fibers visible in the nucleoids of cyanobacteria (see Fig. 14-35) are of slightly larger diameter than those of bacterial nucleoids, about 5 to 7 nm, suggesting that cyanobacterial DNA may have greater supercoiling or a more elaborate superstructure of some kind.

Although cyanobacterial DNA usually appears in linear form when isolated, T. M. Roberts and K. E. Koths detected a small number of DNA circles among the DNA isolated from the cyanobacterium *Agmenellum.* This finding suggests that the DNA of these prokaryotes may actually occur in the same circular form as in bacteria. The linear molecules usually isolated may reflect breakage of longer cyanobacterial DNA circles during isolation.

Cyanobacteria contain at least two histonelike proteins. One, already noted, is closely related to the HU protein of bacteria. The second, a 16,000-dalton protein, may also interact with cyanobacterial DNA to form compact structures.

Dinoflagellates: Eukaryotes with Prokaryotic DNA Organization

One group of organisms has a nuclear system intermediate in structural complexity between prokaryotes and eukaryotes. These are the *dinoflagellates* (Figs. 14-36 and 14-37), single-celled organisms that are variously classified as protists, algae, or in a separate division of their own. They are primarily marine creatures that live by photosynthesis.

The cytoplasmic structures of dinoflagellates are typically eukaryotic. The cytoplasm contains mito-

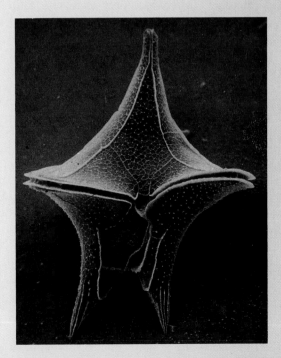

Figure 14-36 The dinoflagellate *Peridinium.* ×1,400. Courtesy of M. Ricard.

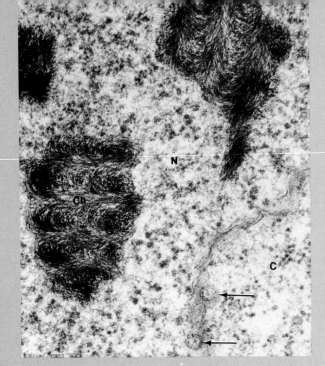

Figure 14-37 Chromosoids (Ch), which resemble bacterial nucleoids in appearance, in the nucleus of the dinoflagellate *Gyrodinium*. Other structures visible in the nucleus are typically eukaryotic, such as the nucleoplasm (N) surrounding the chromosomes and a nuclear envelope complete with pore complexes (arrows). C, cytoplasm. ×47,000. Courtesy of D. F. Kubal and H. Ris, from *J. Cell Biol.* 40:508 (1969), by copyright permission of the Rockefeller University Press.

chondria, chloroplasts, and other internal membrane systems characteristic of eukaryotes, and the cells are propelled by typically eukaryotic flagella containing the 9 + 2 system of microtubules (see p. 432). Dinoflagellate nuclei also have certain eukaryotic features. They are separated from the cytoplasm by a nuclear envelope complete with pore complexes and have a background substance equivalent to the nuclear matrix of eukaryotes.

The organization of dinoflagellate DNA, however, is unique among eukaryotes. Suspended within the nucleus in different dinoflagellate species are from 4 to 6 to more than 200 small, rod-shaped bodies, the *chromosoids*, that under the light microscope resemble the fully condensed chromosomes of dividing eukaryotic cells. Until the advent of electron microscopy, in fact, chromosoids of dinoflagellates were thought to be typical eukaryotic chromosomes that are always fully condensed.

Electron microscopy revealed the surprising fact that dinoflagellate chromosoids closely resemble bacterial nucleoids. Within the chromosoids are tightly packed masses of 3-nm to 8-nm fibers, folded into transverse bands. The bands give chromosoids a curly appearance in thin sections that resembles the ordered folds visible inside some bacterial nucleoids (compare Figs. 14-34 and 14-37).

No proteins equivalent to the histones of other eukaryotes can be detected in dinoflagellates. Instead, P. J. Rizzo and coworkers found that dinoflagellate DNA is complexed with one or two histonelike proteins that differ in each species examined. These proteins occur in a 0.1:1 ratio with dinoflagellate DNA by weight, much smaller than the 1:1 histone to DNA ratio typical of other eukaryotes. No nucleosomelike bodies are produced by the interaction between dinoflagellate histonelike proteins and DNA.

Although the structural organization of dinoflagellate DNA appears to be prokaryotic, the DNA sequence organization in these organisms is evidently typically eukaryotic. J. R. Allen and others discovered that a large proportion of dinoflagellate DNA consists of repeated sequences that apparently have no coding function (for details, see p. 750). Sequences of this kind are not found in any significant proportion in the DNA of bacteria or cyanobacteria. In dinoflagellates, however, Allen found that repetitive sequences make up 55% to 60% of the DNA, a figure comparable to the amount of repeated DNA in other eukaryotes. Dinoflagellate DNA, though more or less typically prokaryotic in structure, is therefore eukaryotic in this respect. The mechanisms of nuclear division in some dinoflagellates also share prokaryotic and eukaryotic characteristics (see p. 1028 for details).

The part-prokaryotic–part-eukaryotic nature of dinoflagellates makes these unusual organisms the subject of much evolutionary interest and speculation. Do these organisms represent a step in the main line of evolution of eukaryotes from prokaryotes or are they an evolutionary offshoot that has persisted until today? Although no definite answer can be given, most evolutionists believe that present-day dinoflagellates are living fossils, survivors of a side branch of the evolutionary tree that diverged just before the appearance of fully eukaryotic cells.

Suggestions for Further Reading: See p. 565.

Annulate Lamellae

The cytoplasm of some cell types contains masses of membranes with many structural similarities to the nuclear envelope. These membranes, called *annulate lamellae*, consist of stacks of layered membranes that resemble segments of the nuclear envelope, complete with closely spaced pore complexes (Fig. 14-38). The pore complexes of annulate lamellae, as far as can be determined, are identical in dimensions and structure to pore complexes of the nuclear envelope. Frequently, the pores of annulate lamellae stack in vertical register, with the annular material of adjacent pores in contact or apparently continuous (as in Fig. 14-38*a*).

Annulate lamellae sometimes lie next to the nucleus, with the nuclear envelope forming the innermost layer of the lamellar stack. More frequently, lamellae occur at a distance from the nucleus, without visible association with the nuclear envelope. Connections with membranes of the ER are often seen in these locations, however (Fig. 14-39). Annulate lamellae also appear in single layers inside the nucleus.

Masses of annulate lamellae are sometimes large enough to be recognizable under the light microscope as dense granules within the cytoplasm. Cytochemical tests reveal that these granules contain RNA in high concentrations. ATPase activity is also associated with the pore complexes of annulate lamellae, as in the pores of the nuclear envelope. Curiously, substances can also be seen to traverse the pore complexes of annulate lamellae. For example, in Feldherr's experiments with transport of nucleoplasmin molecules (p. 559), gold-tagged particles were also detected in the pore complexes of annulate lamellae in the cell type, an oocyte, used for this study.

Annulate lamellae may develop from the nuclear envelope or the ER. The lamellar membranes first appear as flattened, closed sacs without pore complexes that extend from the nuclear envelope or ER. Pore complexes then differentiate rapidly in the sacs. Tests for the presence of RNA become positive as pore complexes appear. Although the pore complexes and membranes of annulate lamellae appear to be identical, or nearly so, to those of the nuclear envelope, experiments by T.-Y. Chen and E. M. Merisko showed that antibodies to lamins fail to react with annulate lamellae. The skeletal support presumably provided by lamins therefore appears to be absent from annulate lamellae.

Annulate lamellae are observed most frequently in animal oocytes, where they appear early in oocyte development and persist through fertilization into the first few division cycles of the embryo. They also appear in the cytoplasm of developing sperm cells in some animals; however, they disappear as these cells reach maturity. Annulate lamellae have also been detected as transient structures in a wide variety of body cells in animals, including epithelial, secretory, muscle, uterine, placental, and nerve cells, and in cultured plant cells,

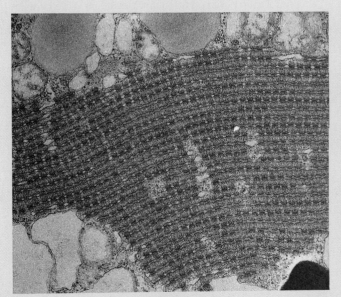

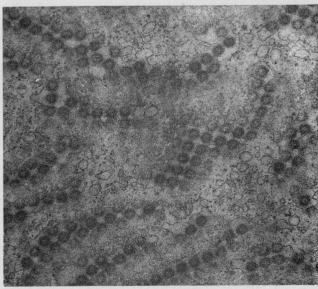

a

b

Figure 14-38 Annulate lamellae. **(a)** A stack of annulate lamellae in the cytoplasm of a frog oocyte in cross section. ×47,000. **(b)** Annulate lamellae of a frog oocyte sectioned so that the pore complexes of successive layers are seen face on. ×22,000. Courtesy of R. G. Kessel and Academic Press, Inc., from *J. Ultrastr. Res.* Suppl. 10, 1968.

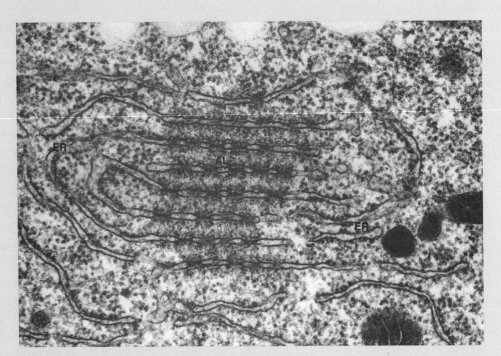

Figure 14-39 Annulate lamellae (AL) in direct connection with rough ER membranes (ER) in the cytoplasm of a *Xenopus* oocyte. ×38,000. Courtesy of J. P. Stafstrom, reproduced from *J. Cell Biol.* 98:699 (1984), by copyright permission of the Rockefeller University Press.

developing pollen cells, and some algae and protozoa. Significantly, many types of animal tumor cells develop annulate lamellae. They may also be induced to appear in many cell types by exposure to such diverse stimuli as reduced temperatures, certain hormones or antibiotics, virus infections, antitubulin drugs such as colchicine, and cAMP. The variety of cell types in which they appear has led to the opinion that essentially any cell type may, under certain conditions, develop annulate lamellae.

Many hypotheses have been proposed for the functions of annulate lamellae, including transport or storage of ribonucleoprotein complexes originating in the nucleus. However, annulate lamellae do not appear to be a vehicle for transferring ribonucleoprotein from the nucleus to the cytoplasm because, as noted, RNA does not appear until after pore complexes differentiate, usually at some distance from the nucleus.

Storage seems a more likely possibility. According to this idea, ribonucleoprotein complexes, such as particles containing mRNA in combination with proteins, are stored in the annulate lamellae of oocytes and early embryonic cells by binding to the pores of these structures after they enter the cytoplasm. As the lamellae disperse later in embryonic development, the mRNA

is released to the cytoplasm in an active form. Their presence in tumor cells, according to this line of reasoning, is related to the fact that tumor cells typically undergo partial reversion to an embryonic state.

Another idea proposes that annulate lamellae, as an addition or alternative to the storage function, may serve as reservoirs of membranes and pore complexes needed for rapid nuclear envelope assembly. This hypothesis is supported by the observation of an inverse relationship between the quantities of annulate lamellae and nuclear envelope membranes in some species. In the forameniferan *Hastigerina*, for example, a large nucleus containing many copies of the chromosome set divides in a few hours to form about 300,000 gamete nuclei. The division increases the total surface area of nuclear envelopes by 60 to 70 times. Before division begins, the large nucleus is surrounded by many layers of annulate lamellae. These lamellae disappear as the gamete nuclei form. A similar pattern of apparent conversion of annulate lamellae into nuclear envelopes occurs in many insects, including *Drosophila*, in which thousands of nuclei are produced by rapid mitotic divisions within a few hours of fertilization.

Suggestions for Further Reading: See p. 565.

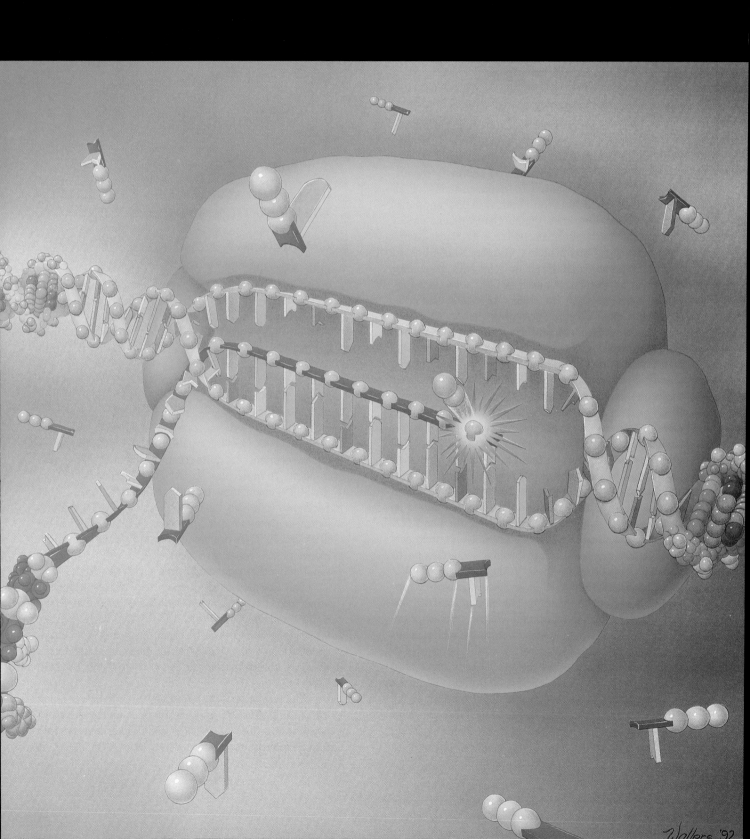

T ranscription, the mechanism by which RNA copies of genes encoded in the DNA are made, produces three major RNA types that interact in protein synthesis—messenger (mRNA), ribosomal (rRNA), and transfer RNA (tRNA). Messenger RNAs carry the encoded information required to make proteins of all types. Ribosomal RNAs form both structural and functional parts of ribosomes, the multienzyme complexes that use the information encoded in mRNAs to assemble amino acids into proteins. Transfer RNAs link to amino acids and match the amino acids with the coding sequences in mRNAs. As such, tRNAs serve as a molecular dictionary that translates the nucleic acid code into the amino acid sequences of proteins.

The production of finished RNA molecules proceeds by a series of steps. Each major RNA type is transcribed initially in a preliminary form known as a *precursor*. The precursor, which is typically a larger molecule than the finished RNA product, is then *processed* by several biochemical steps. One step involves clipping and splicing reactions that remove surplus nucleotide sequences from the precursor. Additional processing steps may add nucleotides to one or both ends of the precursor and chemically modify individual bases within the precursor. Proteins that form structural and functional parts of the finished product are added as processing takes place. The finished product, an RNA molecule linked to proteins in a ribonucleoprotein particle, is transported through the nuclear envelope to the sites of protein synthesis in the cytoplasm.

The production of RNA molecules is closely regulated at each of these steps. The most important regulatory controls are imposed during transcription. These controls, which in effect determine which genes are active in the cell nucleus, lie at the heart of embryonic development and the specialized functions of differentiated cells. Regulation is also imposed during RNA processing, in which individual steps may be accelerated, delayed, or varied to alter the sequence of the RNA product. The regulatory mechanisms acting during RNA processing, called *posttranscriptional* con-

trols, fine-tune the results of the primary regulation mechanisms acting at the transcriptional level. Malfunctions of transcriptional and posttranscriptional controls are responsible for differentiation and developmental pathways that go awry. In some cases the misdirected pathways are responsible for hereditary diseases and cancer in humans and other organisms.

This chapter describes general features of RNA structure and the overall pathways of transcription. Also discussed are the individual RNA types, the genes encoding them, and the processing reactions converting the initial products of transcription to the final, mature mRNAs, rRNAs, and tRNAs that interact in protein synthesis in eukaryotes. The regulation of RNA transcription is discussed in Chapter 17. Prokaryotic mRNA, tRNA, and rRNA genes and their transcription are described in Supplement 15-1.

GENERAL FEATURES OF RNA STRUCTURE, TRANSCRIPTION, AND PROCESSING

RNA Structure

Both eukaryotic and prokaryotic RNAs exist almost entirely in the form of single nucleotide chains. However, many regions of RNA molecules typically wind into a variety of double-helical structures by folding back on themselves. Most common of these structures is a *hairpin* or *stem-loop* configuration in which a foldback double helix is capped at its tip by an unpaired loop. (Fig. 15-1 shows some of the primary types of fold-back structures.) Hairpins and other double-helical structures form primarily in regions containing a symmetrically reversed sequence known as an *inverted repeat* or *palindrome* (see Information Box 14-2). These sequence elements are quite common and probably appear in all RNA molecules. In tRNAs as much as 90% of the RNA chain may consist of inverted repeats that wind into fold-back helices. Base pairs in RNA double-helical regions are not limited to A-U and G-C pairs; flexibility in RNA double helices and spatial rearrangements of the bases also allow G-U, G-A, A-C, A-A, G-G, and other nonstandard base pairs to form.

The double-helical segments of RNA molecules wind into a conformation equivalent to A-DNA (see p. 529 and Fig. 14-5). RNA-DNA hybrid sequences, which form as temporary structures during RNA transcription, also take the A conformation because this is the only conformation shared by the two nucleic acid types.

The sequence of bases in an RNA molecule is called its *primary structure*, and the number and position of hairpins its *secondary structure*. RNA molecules also fold into higher-level conformations. However, little is

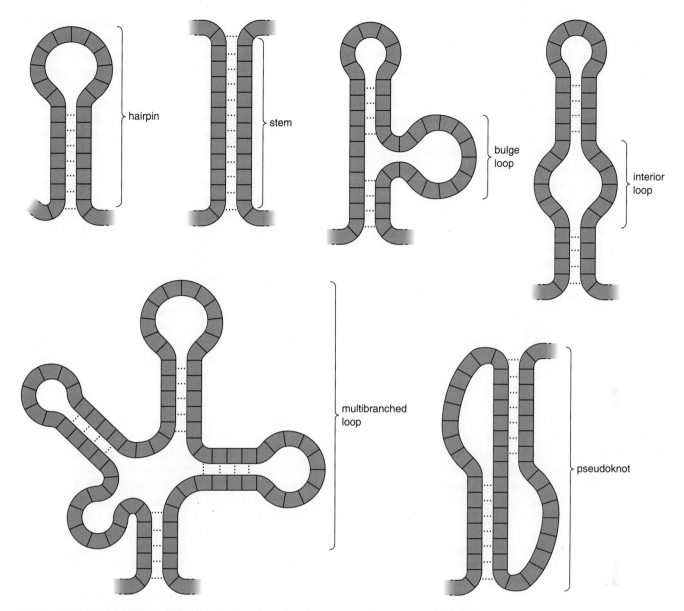

Figure 15-1 The most common secondary structures formed in RNA molecules by fold-back helices. In the paired regions the RNA chain winds into a double helix in the A conformation. Flexibility in the RNA double helices and spatial rearrangements of the bases allows G-U, G-A, A-C, A-A, G-G, and other nonstandard base pairs to form in addition to the standard A-U and G-C pairs.

known about the three-dimensional arrangement of RNA into these *tertiary structures* except for a very few of the smallest RNA types such as tRNAs (see Fig. 15-41).

Depending on the RNA type, secondary and tertiary structures may be more important to RNA functions than primary sequence. In many regions of ribosomal RNAs, for example, sequences can be altered extensively without detriment to function as long as the complex pattern of hairpins and other secondary structures is preserved. The secondary and tertiary regions function in recognition and pairing between RNA

and DNA molecules and between RNA molecules and proteins. These interactions are important at all levels of RNA function, from transcription, processing, and RNA transport through the mechanisms of protein synthesis.

RNA Transcription

RNA transcription is a polymerization reaction in which individual nucleotides link sequentially into a chain. The reaction, catalyzed by enzymes called *RNA polymerases*, requires a DNA *template* (the DNA molecule from which the RNA copy is made), Mg^{2+} or Mn^{2+}, and the four RNA nucleoside triphosphates *adenosine triphosphate* (*ATP*), *guanosine triphosphate* (*GTP*), *cytidine triphosphate* (*CTP*), and *uridine triphosphate* (*UTP*). RNA nucleotides contain ribose rather than deoxyribose as a five-carbon sugar; the ATP used in RNA transcription is identical to the ATP synthesized in glycolysis and mitochondrial oxidations.

A broad outline of transcription is shown in Figure 15-2. The mechanism is initiated when an RNA polymerase binds to the DNA template and recognizes the first base to be copied, which in Figure 15-2a is shown as guanine. According to base-pairing rules, the presence of guanine at this site causes the RNA polymerase to bind CTP from among the pool of four nucleoside triphosphates constantly colliding with the enzyme (Fig. 15-2b).

In response to binding CTP, the enzyme undergoes a conformational change causing it to "read" the next base exposed on the DNA template, which in Figure 15-2 is shown as adenine. The presence of adenine at this site induces the enzyme to bind UTP from the nucleotides in the surrounding pool. Once bound to the enzyme, the UTP is held opposite the adenine on the DNA template, in a position favoring formation of the first linkage of the new RNA chain (Fig.15-2c). RNA polymerase then catalyzes the linking reaction, in which the last two phosphates of the nucleotide most recently bound to the enzyme (UTP in Fig. 15-2c) are split off. The remaining phosphate binds to the 3'-carbon of the first nucleotide (Fig. 15-2d). The reaction creates a *phosphodiester linkage* between the 3'-carbon of the first sugar and the 5'-carbon of the second. Removing the terminal phosphates from UTP releases a considerable increment of free energy and so greatly favors formation of the linkage that the reaction is essentially irreversible.

In response to the formation of the first phosphodiester linkage, the enzyme undergoes another conformational change, causing it to move to the next base exposed on the DNA template, shown as thymine in Figure 15-2. The enzyme then binds an ATP from the medium and catalyzes formation of the second phosphodiester linkage. This linkage, as before, is formed at the expense of two phosphate groups split off from the nucleotide most recently bound by the RNA polymerase. The polymerization continues, binding nucleotides one at a time into the growing chain until the enzyme reaches the end of the DNA segment being copied. At this point the newly synthesized RNA molecule and the enzyme are released from the DNA template, terminating transcription.

Figure 15-2 shows that the first nucleotide in the RNA chain retains all three of its phosphates linked to the 5'-carbon of the sugar. These phosphates mark the beginning or 5' *end* of a newly synthesized RNA molecule. The opposite, 3' *end* is marked by an —OH

group retained at the 3'-carbon of the last nucleotide bound into the RNA chain. Because a 5'-carbon marks the beginning and a 3'-OH marks the end of a newly synthesized RNA chain, transcription is said to proceed in the 5' → 3' *direction.*

The mechanism of RNA transcription proceeds in three distinct phases—*initiation, elongation,* and *termination.* Initiation includes attachment of RNA polymerase to the DNA template and binding by the enzyme of the first nucleotide to be placed in the RNA molecule. Elongation involves the reactions in which RNA nucleotides are added sequentially according to the DNA template. Termination is the cessation of RNA polymerization and the release of the enzyme and completed RNA transcript from the DNA template. All three phases are carried out by RNA polymerase in response to sequence information in the DNA.

Accessory proteins called *factors* are also important in RNA transcription. These proteins are not directly part of RNA polymerase, but they are required for the most efficient progress of the reactions. In a sense the factors act as catalysts that speed individual steps of transcription.

Initiation Random collisions between RNA polymerase molecules and DNA lead to initiation if they occur in a specific region of the DNA called the *promoter.* This region contains sequence elements that bind RNA polymerase and indicate the first base to be copied into an RNA transcript. The promoter also includes sequences involved in the regulation of transcription.

Initiation factors are necessary for successful initiation to take place. Without these accessory proteins RNA polymerases may bind loosely to DNA sequences of essentially any type without initiating RNA transcription. The presence of the initiation factors allows RNA polymerase to bind tightly to the recognition sequences in the promoter. Evidently, eukaryotic initiation factors bind first to the promoter, forming an active complex that fits part of the RNA polymerase. The enzyme then binds tightly to the initiation factors and the promoter.

The DNA of the promoter is still wound into a double helix during attachment of the initiation factors and the RNA polymerase. Complete assembly of the initiation complex destabilizes the DNA, causing it to unwind in the promoter region. The destabilization may involve an initial configuration in which the DNA is bent in a left-handed, or negative, turn around the initiation complex. The strain imposed by the reverse bend may force the helix to unwind.

The enzyme is now ready to add the first RNA base. Of the two DNA nucleotide chains fully exposed by the unwinding, only one, called the *sense chain*, acts as a template for transcription. The location of the promoter sequences indicates which chain is the sense

Figure 15-2 The overall mechanism of RNA transcription. **(a)** An RNA polymerase binds to the DNA template and recognizes the first base to be copied, guanine. **(b)** Base pairing and a conformational change in the enzyme cause RNA polymerase to bind CTP from the pool of nucleoside triphosphates in the surrounding medium. The enzyme then pairs a nucleoside triphosphate with the next base exposed on the DNA template, adenine. **(c)** The adenine in the template is paired with a UTP from the nucleotide pool. The UTP is held opposite the adenine of the template, in a position favoring formation of the first linkage of the new RNA chain. **(d)** The enzyme catalyzes the linking reaction, in which the last two phosphates of the nucleotide most recently bound to the enzyme, UTP in **(c)**, are split off. The remaining phosphate is linked to the 3′-carbon of the first nucleotide, creating a phosphodiester linkage between the 3′-carbon of the first sugar and the 5′-carbon of the second. The enzyme then moves to the next base exposed on the DNA template, thymine; an ATP is bound from the medium, and a second phosphodiester linkage forms. The process continues, binding nucleotides successively into the growing chain until the enzyme reaches the end of the DNA segment being copied.

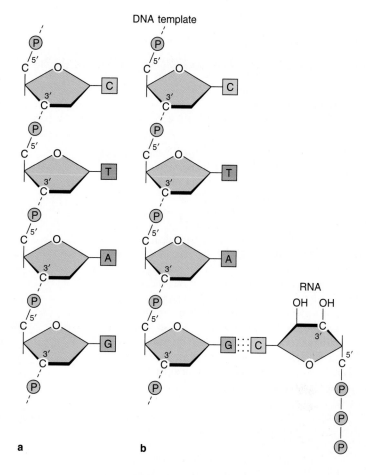

Information Box 15-1

Footprinting

Footprinting is widely used as a method to reveal sites on DNA bound by RNA polymerases and DNA-binding proteins, including transcription factors and regulatory proteins. In the technique a DNA segment containing the region bound by a protein is produced in many copies by cloning or the polymerase chain reaction (see pp. 134 and 137). One sample of the DNA, unbound by protein, is treated with a DNA endonuclease or an alkali adjusted in concentration so that each DNA segment in the sample is broken at only one or a few sites. The breakage produces a number of fragments that are sorted out according to length by running them through an electrophoretic gel (see part *a* of the figure in this box). For the most definitive results the DNA is radioactively labeled at its 5' end, and autoradiography is used to detect the bands on the gel. This ensures that the only fragments seen on the gel are those that include the labeled end and are, therefore, sorted out by length from the beginning of the DNA sequence.

A second end-labeled sample of the DNA is then combined with the protein of interest and treated with the DNA endonuclease or alkali. The same collection of fragments is produced by the treatment except in regions where the DNA is protected from attack by combination with the protein (see part *b* of the figure in this box). When run on the gel, the fragments sort out by length, but missing from the gel are fragments corresponding to the region protected by combination with the protein. The missing fragments correspond to a region in which no bands appear in the gel. This blank region is the "footprint" of the protein. The position of the missing bands and the region bound by the protein can be fixed by comparing the gels made from the two samples. The footprint can be precisely located in a DNA sequence by comparing its sequence with sequencing gels made from the same DNA sample (see p. 134).

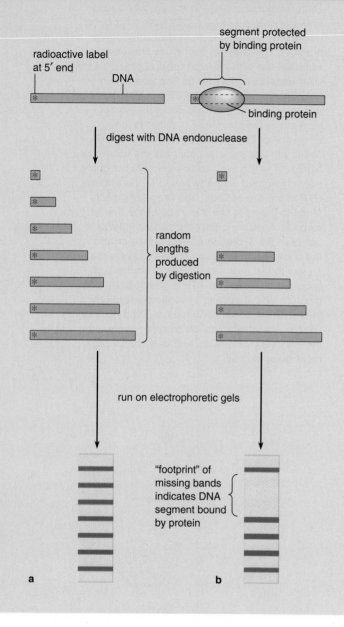

chain by fixing the initiation complex to this chain. The sequence of the opposite chain, the *missense chain*, is complementary to the template but does not contain encoded information. "Sense" is not limited to one of the two DNA nucleotide chains; different genes may have their sense chains on either chain. For a given gene, however, the nucleotide chain being copied remains the same; the RNA polymerase does not switch back and forth between the two nucleotide chains within the boundaries of a gene.

Binding between RNA polymerase and the promoter is tight enough to protect the DNA in the region

of contact from attack by enzymes or chemicals such as strong bases. This characteristic has been used to determine the length of the promoter and as a means to isolate promoters for analysis through a technique called *DNA footprinting* (see Information Box 15-1). Basically, the technique involves exposing DNA to chemical or enzymatic attack after binding a molecule such as RNA polymerase. After the attack the only DNA left intact in the sample is the segment attached to the binding molecule. The intact segment, called the *footprint* of the binding protein, is then removed and identified by comparison with DNA molecules including the entire gene. Analysis of the footprint made during initiation of an mRNA-encoding gene, for example, reveals that RNA polymerase initially binds tightly to a DNA segment of about 75 base pairs, extending from about 55 nucleotides in advance of ("upstream") to about 20 nucleotides past ("downstream") the first nucleotide to be copied. About 17 DNA base pairs unwind within the 75 base pairs tightly bound to the enzyme.

The final step in initiation takes place as RNA polymerase binds the first RNA nucleotide to be placed in the RNA transcript. Much interest centers on the almost perfect fidelity with which the first and subsequent complementary bases are inserted into the growing RNA chain. Some of the accuracy depends on the formation of correct base pairs between the template DNA and the incoming RNA bases. However, research with *base analogs*, molecules that are similar to the nucleoside triphosphates, indicates that correct base pairing in transcription rests on conformational changes in the enzyme as well as base pairing. Some analogs closely resemble ATP, UTP, CTP, or GTP in structure but cannot form the usual hydrogen bonds with DNA bases. Nevertheless, these analogs are correctly inserted in place of the regular RNA bases by the RNA polymerase. Evidently, conformational changes occur in the active site in response to the template DNA base encountered by the enzyme. The changes alter the active site so that only the "correct" complementary RNA nucleoside triphosphate can fit and bind.

Elongation and Termination Once the first base is added, the elongation phase begins, and RNA nucleotides are added sequentially until the enzyme reaches the end of the template. During each addition the enzyme (1) binds a nucleoside triphosphate; (2) splits off two phosphate groups from the nucleoside triphosphate; (3) forms the phosphodiester linkage; and (4) moves or *translocates* to the next base to be copied in the DNA template. During elongation some of the initiation factors are released, and one or more *elongation factors* may add to the enzyme complex to speed steps in the sequential addition of bases.

During elongation the DNA template unwinds continuously just in front of the RNA polymerase complex and rewinds just behind it (Fig. 15-3). How the local DNA unwinding is accomplished remains unknown. The RNA polymerases may, as some investigators have suggested, include "unwindase" and "rewindase" capabilities as part of their active sites. Alternatively, DNA topoisomerases (see p. 537) may form part of the enzyme-factor complex carrying out RNA transcription. These topoisomerases would unwind the DNA by introducing breaks in one or both nucleotide chains and allowing the free ends to rotate around each other. Some evidence for this possibility is taken from the fact that DNA topoisomerases become

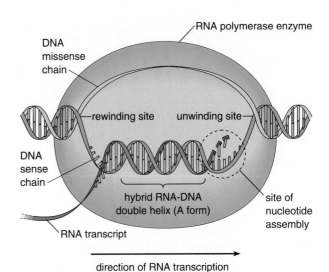

Figure 15-3 DNA unwinding and rewinding at the site of transcription. The transcript, as shown, forms a hybrid RNA-DNA double helix with the DNA sense chain as part of the transcription mechanism.

concentrated in regions of chromosomes that are active in RNA transcription. In addition, S. J. Brill and his colleagues showed that transcription is halted in yeast mutants in which the topoisomerases thought to be involved in transcription are inactive.

As the new RNA molecule is transcribed, it winds temporarily with the sense chain of the DNA into a short hybrid RNA-DNA helix of about 10−20 base pairs (Fig. 15-3). Beyond this length the growing RNA chain unwinds from the DNA. The template DNA chain rewinds with the missense chain into an intact DNA molecule as the RNA polymerase passes.

On the average about 40 nucleotides are added per second to a growing RNA chain during elongation. The rate of elongation varies depending on local differences in secondary structure of the DNA template or RNA copy or on the sequences being transcribed. At some points, as in regions in which an inverted sequence is copied into the RNA transcript, the enzyme complex may slow its progress or even pause temporarily.

In prokaryotes RNA transcription terminates in response to DNA sequences that signal the end of the region to be copied into RNA (see Supplement 15-1 for details). Sequences signaling termination also appear in some eukaryotic genes, such as those encoding rRNAs. In eukaryotic mRNA genes termination may be coupled to processing reactions rather than occurring in response to specific signal sequences in the DNA (see below). *Termination factors* contribute to destabilization of the transcription complex and release of the transcript in prokaryotes and possibly also in eukaryotes.

Table 15-1	Eukaryotic RNA Polymerases	
RNA Polymerase Type	RNAs Transcribed	Ionic Requirement
I	28S, 18S, and 5.8S rRNAs	Mg^{2+}
II	mRNAs, most snRNAs	Mg^{2+}
III	tRNAs, 5S rRNAs, one snRNA, and scRNAs	Mn^{2+}

The RNA Polymerases The RNA polymerases, with molecular weights ranging from 500,000 to 700,000, are among the largest and most complex proteins of living organisms. These enzymes are similar in overall structure in both prokaryotes and eukaryotes; each polymerase consists of two large polypeptide subunits and about six to ten smaller ones. Typically, the two largest polypeptides and a few small polypeptides form a *core enzyme* that is capable of transcribing essentially any DNA sequence, natural or artificial, into an RNA copy. The remaining small polypeptides confer specificity and restrict the transcribing capabilities of the enzyme to natural DNAs containing a promoter, and in eukaryotes to certain gene types.

Only one RNA polymerase occurs in bacteria. The single bacterial enzyme transcribes all the RNA classes (see Supplement 15-1 for details). Eukaryotes, in contrast, contain three types, designated *RNA polymerase I, II,* and *III* (Table 15-1).

Eukaryotic RNA polymerase I, which transcribes all rRNAs except one type, 5S rRNA, is active only in the nucleolus. This enzyme recognizes promoters that lie primarily upstream, in the 5'-flanking region of rRNA genes. RNA polymerase I accounts for the largest fraction, about 50% to 70%, of the mature RNAs entering the cytoplasm in most eukaryotic cells.

RNA polymerase II transcribes mRNA genes and also most genes encoding small RNA molecules (*small nuclear RNAs*, or *snRNAs*; see p. 626) that participate in RNA processing. This enzyme also initiates transcription from promoters that lie primarily upstream, in 5'-flanking regions of its target genes. RNA polymerase II transcribes significantly less, about 10% to 40%, of mature cellular RNAs.

RNA polymerase III transcribes 5S rRNAs, all tRNAs, and the remaining snRNAs and other RNAs not copied by RNA polymerase II. The various RNAs transcribed by this enzyme are all relatively small molecules ranging in length from 80 to 90 to a maximum of about 550 nucleotides. In contrast to the promoters recognized by RNA polymerases I and II, the promoters of most genes copied by RNA polymerase III lie inside the gene, within the sequence copied into RNA. Se-

quences within these genes thus serve double duty as both promoter and coding template. Although most genes copied by RNA polymerase III have internal promoters, many have additional 5'-flanking sequences that modify the rate of transcription. A few genes transcribed by RNA polymerase III have upstream rather than internal promoters. The rRNA, tRNA, snRNA, and other small RNA molecules transcribed by RNA polymerase III make up about 5% to 10% of mature cellular RNAs.

The three eukaryotic RNA polymerases are common to all eukaryotic organisms, including animals, fungi, protozoa, and plants. Although some differences in total polypeptide number are detected in the RNA polymerases of the different major eukaryotic groups, the same division of labor between RNA polymerases I, II, and III is noted in all cases.

The structure of the RNA polymerases reflects the complexity of their activities in RNA transcription. The enzymes have sites that recognize promoters, react with initiation, elongation, and termination factors, recognize DNA bases for correct pairing, bind and hydrolyze RNA nucleotides, form phosphodiester linkages, terminate transcription, and perhaps unwind and rewind DNA. The proteins housing these activities are large— eukaryotic RNA polymerases measure about 10 by 15 nm, making them larger than individual nucleosomes. RNA polymerase enzymes are large enough, in fact, to be easily recognized as individual particles in electron micrographs of transcribing chromatin (see Fig. 15-49).

For example, RNA polymerase II has at least eight to ten subunits ranging in molecular weight from 10,000 to 240,000. The largest subunit, L', which ranges from about 220,000 to 240,000 daltons in different species, is highly conserved among all eukaryotes. L' also shares sequences with the L' subunits of eukaryotic RNA polymerases I and III and with the largest polypeptide subunit (β'; see p. 639) of the bacterial RNA polymerase. Among the structures of L' is a zinc-finger motif (see p. 543) that may be involved in recognizing and binding the DNA template. The second-largest subunit, L, which ranges from about 140,000 to 150,000 daltons, shares sequences with the L subunits of RNA polymerases I and III and with the second-largest subunit (β) of the bacterial RNA polymerase. Three of the smaller polypeptide subunits, 27,000, 23,000, and 14,500 daltons, are common to all three eukaryotic polymerases; two additional ones occur only in RNA polymerases I and III. Each polymerase also has one or more polypeptide subunits unique to the type. The L' polypeptide of RNA polymerase II is heavily phosphorylated; at least some of the phosphorylations are carried out by protein kinases stimulated by cAMP. RNA polymerase II may thus be regulated in its activity by control systems stimulated by cAMP-based pathways (see p. 245).

Accessory factors are associated with all three eukaryotic polymerases. The factors have been best studied in the initiation phase of transcription, in which separate factors allow each polymerase type to recognize its promoters in DNA. For example, at least five different initiation factors are necessary for attachment of RNA polymerase II to the promoters of mRNA genes. The factors occur in two fundamental types. One type is *general* and acts in conjunction with one of the three polymerases in most or all of the gene types transcribed by the enzyme. The other type is *gene specific* and indicates a particular gene or group of genes to be copied from among the types transcribed by a given polymerase.

RNA Processing

All major eukaryotic RNA types are modified by processing reactions—clipping and splicing, addition of nucleotides, and chemical modification of bases. These reactions begin while transcription is still in progress and continue after assembly of the RNA precursor is complete. Therefore, RNA precursors normally do not actually exist inside cells in fully defined or complete form; they are in a constant state of change as transcription and processing proceed.

Clipping is catalyzed by *RNA exonucleases* and *endonucleases* and, in some cases, by *ribozymes*—RNA molecules with the capacity to act as enzymes (see p. 101). RNA exonucleases remove nucleotides one at a time from the 3' end of the precursors. RNA endonucleases make cuts at points within an RNA molecule. Ribozymes also cut RNA precursors at points within the chains. The points at which cuts are made by endonucleases or ribozymes are precisely fixed, so that the piece removed by a clipping reaction contains not one base too few or too many. The positions at which cuts are made are determined by either sequence information or secondary structures, or a combination of both within the RNA precursors. Once cuts are made to remove an internal segment, the free ends are rejoined into a continuous RNA molecule.

Addition of nucleotides, catalyzed by many specific enzymes, may occur at either the 5' or 3' ends of RNA precursors. The added nucleotides do not correspond to any sequences in the DNA template because they are linked to the RNA precursors after transcription is complete.

Chemical modification of individual bases is catalyzed by a battery of enzymes. Most frequent of the modifications is the addition of methyl groups to the bases or to the ribose sugars of RNA. Interestingly, addition of a methyl group at the 5-position of uracil converts the uracil into thymine. Thus, thymine, usually considered characteristic only of DNA, may appear as a modified base in RNA molecules. Other frequent modifications are removal of amino groups and addition

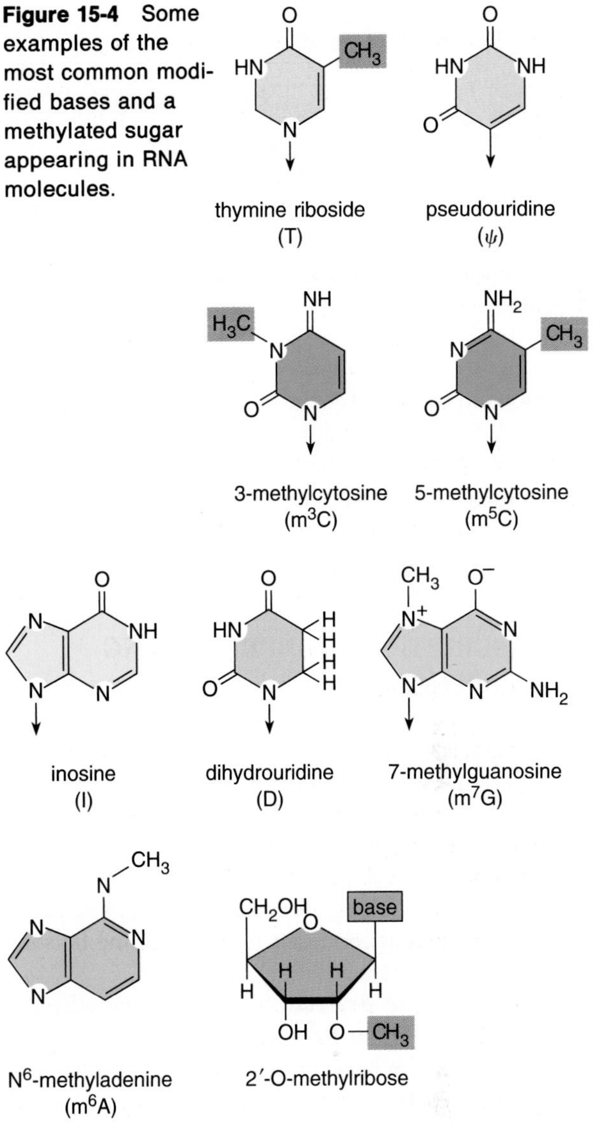

Figure 15-4 Some examples of the most common modified bases and a methylated sugar appearing in RNA molecules.

thymine riboside (T)

pseudouridine (ψ)

3-methylcytosine (m³C)

5-methylcytosine (m⁵C)

inosine (I)

dihydrouridine (D)

7-methylguanosine (m⁷G)

N⁶-methyladenine (m⁶A)

2'-O-methylribose

of a hydroxymethyl (—CH₃OH) group. These and other less frequent chemical changes produce a wide variety of modified bases in finished RNA molecules. (Fig. 15-4 shows some of the most common modified bases.)

Base modifications are most frequent in tRNA and rRNA molecules. However, mRNAs also contain a few modified bases, particularly at the 5' and 3' ends. Base modifications have a significant effect on the three-dimensional conformations taken by RNA molecules. Undoubtedly, they are also important as recognition signals in reactions between RNAs and other molecules, particularly proteins.

Proteins associate with the various RNA types as processing takes place. These proteins participate in the processing reactions, fold the RNAs into their finished three-dimensional format, and contribute to the activity of the RNAs in processes such as protein synthesis. The proteins associated with an RNA type vary as the processing reactions proceed and the RNA takes up its

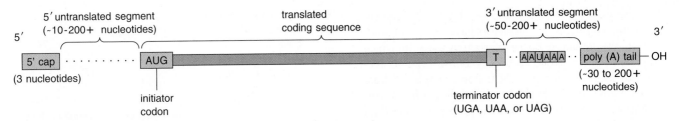

Figure 15-5 Structural features of a eukaryotic mRNA molecule. The poly(A) tail is missing from some eukaryotic mRNAs; the 5' cap is absent in some mRNAs of viruses infecting eukaryotic cells.

functions in the cell. Of the proteins added to RNAs, by far the greatest numbers are added to rRNAs—between 70 and 80 proteins are added during processing of eukaryotic rRNAs to finished ribosomal subunits.

TRANSCRIPTION AND PROCESSING OF MESSENGER RNA

Messenger RNA transcription is the first step in the sequence of events leading to protein synthesis and, ultimately, to cell differentiation and the determination of cell structure and function. The precursors from which finished mRNA molecules are processed, called *pre-mRNAs* or *hnRNAs* (*hn* = *heterogeneous nuclear*), may be as much as ten times longer than the finished product.

Studies of the structure of pre-mRNAs, mRNAs, and the genes coding for them have produced many unexpected results. Most unexpected was the discovery that most mRNA genes are laid out in interrupted form, with intervening segments that are copied into pre-mRNAs but removed during processing. These intervening segments may be removed in different combinations, producing mRNAs copied from the same gene but containing different sequences and coding for related but different proteins. Also unexpected was the finding that finished mRNAs contain long spacer sequences at their leading and trailing ends. These sequences are copied from the gene but do not contain codes for amino acids and are not translated into proteins. Other segments added to the extreme 5' and 3' ends of finished mRNAs during processing do not appear in complementary form in the DNA coding for pre-mRNAs.

mRNA Structure

All eukaryotic mRNAs contain a continuous sequence of nucleotides coding for synthesis of a protein or polypeptide. The coding sequence lies internally in a eukaryotic mRNA molecule, separated from the ends by stretches of nucleotides that are copied from the template but are not translated into amino acid sequences. The *5' untranslated segment* or *region* lying in advance of the coding sequence is enclosed by a structure at the 5' end of an mRNA known as the *5' cap*. The *3' untranslated segment* or *region* following the coding sequence is terminated in most, but not all, mRNAs by a series of adenines at the 3' end known as the *poly(A) tail.* Starting from the 5' end, a typical eukaryotic mRNA thus consists of a 5' cap → 5' untranslated segment → coding sequence → 3' untranslated segment → poly(A) tail (Fig. 15-5).

The 5' Cap   The cap structure at the 5' end of eukaryotic mRNAs consists of three nucleotides (Fig. 15-6). The first is a guanine nucleotide added during processing, after transcription of the 5' end. This guanine is reversed so that its 3'-OH faces the beginning end of the mRNA. As a result, both ends of a finished eukaryotic mRNA terminate with a 3'-OH group. The 5'-carbon of the added guanine is joined to the 5'-carbon of the second nucleotide in the 5' cap by a chain of three phosphate groups that forms a 5' → 5' linkage at this point rather than the usual 3' → 5' linkage. The other two nucleotides in the cap were transcribed from the DNA template and were originally the first and second nucleotides of the unprocessed pre-mRNA. They remain in their original orientation, joined to each other and to the succeeding nucleotides of the 5' untranslated segment by the usual 3' → 5' linkages containing a single phosphate group. The three nucleotides of the cap are also modified by the addition of methyl groups at various positions (see caption to Fig. 15-6 for details).

In different mRNAs, the two nucleotides following the initial guanine may carry any of the four A, U, G, or C bases. These variations in the nucleotides at the second and third positions in the cap and differences in the points at which methyl groups are added create a family of cap structures. In most cases a given mRNA type has only one of the several 5' cap structures.

All eukaryotic mRNAS have a 5' cap. The 5' caps also occur on most, but not all, mRNAs of the viruses infecting eukaryotic cells. Among the viral mRNAs lacking a 5' cap are polio and picornavirus mRNAs. (Prokaryotic mRNAs are typically uncapped; see p. 643.)

Figure 15-6 The 5' cap of a eukaryotic mRNA molecule. Methyl groups are shaded. The initial guanine, which is added to the 5' end of the transcript during processing, is reversed so that its 3'-OH faces the beginning end of the mRNA. The base of this guanine nucleotide is methylated at its 7-position. The second and often the third nucleotides of the cap are also methylated, but the methyl group is usually attached to the sugar rather than the base. The base of the second nucleotide is also methylated in some mRNAs.

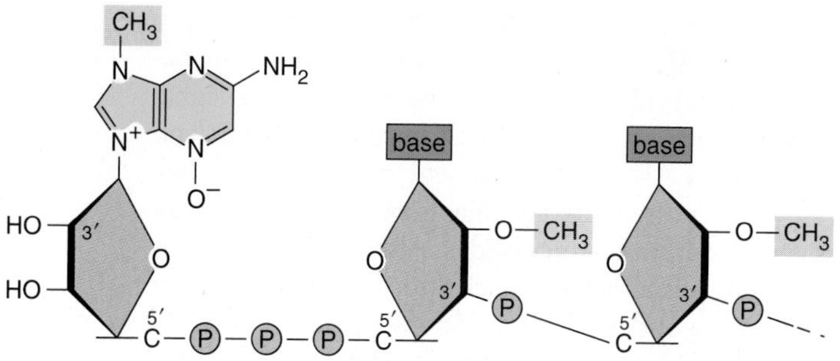

The 5' cap evidently facilitates the initiation phase of eukaryotic protein synthesis (for details, see p. 654). Removal of the cap slows the rate at which protein synthesis is initiated by about 85% to 90%. Addition of 5' caps to mRNAs that normally lack them, such as bacterial mRNAs, increases their rate of utilization in protein synthesis by eukaryotic ribosomes by an equivalent factor.

The 5' Untranslated Segment The 5' untranslated segment directly following the 5' cap varies in length from as few as 10 to more than 200 nucleotides in different eukaryotic mRNAs. A variety of inverted sequences in this segment provide ample opportunities for secondary structures to form. Relatively few sequence features are conserved in the 5' untranslated segment of different mRNAs, even among the mRNAs of closely related proteins such as variants of the globin, tubulin, or actin families. This lack of conservation suggests that much of the 5' untranslated segment may function in a manner similar to the leader of a movie film, by threading through the ribosome and holding the beginning of the coding segment in a position allowing protein synthesis to start.

One sequence that is conserved among different mRNAs lies just upstream of the AUG initiator codon for protein synthesis. Functions have been worked out for sequences of this type by genetic techniques in which individual bases are substituted or deleted. The effects of this approach, called *substitution-deletion analysis*, are then noted in living cells or in the test tube. Substitution-deletion analysis by M. Kozak and A. Shatkin showed that a purine, usually A, must be present at the position three nucleotides upstream from the AUG for the initiator codon to be recognized by a ribosome. Other nucleotides, although not absolutely required, occur frequently enough in the region just upstream of the AUG to allow construction of the consensus sequence . . . GCCGCCPuCCAUGG . . . for this region (the initiator codon is underlined). This sequence evidently acts as an indicator pointing out the initiator codon. In some mRNAs additional AUGs lie upstream of the initiator codon in the 5' untranslated segment. These lack the indicator sequence, however, and are ignored as starting sites by ribosomes.

In some mRNAs the 5' untranslated segment contains sequence elements that promote entry of the mRNA into protein synthesis. The 5' untranslated segment of the AIDS virus mRNA, for example, contains a segment that increases translation of this mRNA by about a thousand times if proteins of the AIDS virus are present. Other mRNAs are increased in translation rate by an equivalent factor if supplied with the 5' untranslated segment of the AIDS mRNA and the viral proteins.

The Coding Sequence The coding sequence begins with the AUG initiator codon and is read consecutively, three nucleotides at a time, until one of three *terminator codons*, UGA, UAA, or UAG, appears. Substitutions in the initiator codon are extremely rare in eukaryotes. However, a few have been observed; for example, CUG is the initiator codon for synthesis in c-*myc*, a gene that in a faulty form or location is implicated in the transformation of some normal cells to cancerous types (see p. 934). The terminator codon, which does not code for an amino acid, acts as a signal telling the ribosome to stop protein synthesis.

The coding sequence bounded by the initiator and terminator codons may include from less than a hundred to thousands of nucleotides, depending on the size of the protein or polypeptide encoded in the message. The coding sequences of most eukaryotic mRNAs are about 300 to 1500 nucleotides in length and code for proteins containing about 100 to 500 amino acids. Among the shortest coding sequences are those of mRNAs coding for protamines (see p. 1119), which, in some cases, include less than 100 nucleotides. At the other extreme are coding sequences with 14,000 nucleotides, such as that of the silk fibroin protein of silkworms.

The 3' Untranslated Segment The 3' untranslated segment directly following the coding sequence shows few similarities in sequence or length among different eukaryotic mRNAs. However, a variety of reverse-repeat sequences provide opportunities for secondary structures to form as in the 5' untranslated segment. In addition, the 3' untranslated segment of almost all mRNAs contains the short sequence *AAUAAA*, located from 10 to 20 nucleotides before the poly(A) tail (see

Fig. 15-5). The AAUAAA sequence, which is highly conserved (AUUAAA, the only common variation, occurs in about 12% of mRNAs), is a processing signal fixing the length of the 3' untranslated segment and indicating the site for attachment of the poly(A) tail.

The 3' untranslated segments of some mRNAs are as short as 50 to 60 nucleotides. Although the limit for most others is about 200 nucleotides, a few are as long as 600 nucleotides or more. Some variation is noted in the length of this segment even in mRNAs coding for the same protein. These length variations have no apparent effect on the function of the mRNAs in protein synthesis.

The 3' untranslated segment may be the mRNA equivalent of the trailer of a movie film—it may thread through the ribosome to hold an mRNA in place on a ribosome during the termination of protein synthesis. In addition, sequences within the 3' untranslated segment appear to regulate mRNA stability. Some mRNAs persist for a few hours or less in the cytoplasm; others are retained for as long as a hundred hours or more (see Table 17-2). In many cases the stability characteristics of a given mRNA can be transferred to other mRNAs by transferring its 3' untranslated segment.

The poly(A) Tail The unbroken stretch of adenine nucleotides forming the poly(A) tail varies in length from about 30 to more than 200 nucleotides. Most mRNAs have a poly(A) tail. However, some normally tailed mRNAs can also be detected in the cytoplasm in tailless form. These alternate forms are called *poly(A)*$^+$ (tailed) and *poly(A)*$^-$ (tailless). A relatively few mRNAs, including those for protamines and most histones, occur regularly without poly(A) tails. The mRNAs for histones synthesized during DNA replication are primarily tailless; the relatively few histones synthesized at all times in the cell cycle have poly(A) tails.

The poly(A) tail is complexed with a *poly(A) binding protein* that evidently winds the tail into nucleosomelike structures. Without the binding proteins poly(A) tails are digested by T2 ribonuclease into random lengths. When the binding protein is present, the enzyme breaks the tail into regular, 27-base pair lengths, as if this is the unit length wrapped around each protein molecule. The binding protein, which contains 577 amino acids in yeast and 633 in humans, is highly conserved in the two species. Mutations eliminating the protein in yeast are lethal, indicating that its functions in binding the poly(A) tail are vital. The tail is much more resistant to enzymatic breakdown when associated with the binding protein, which may account for the lethality of mutations eliminating the protein.

Although the poly(A) tail is one of the most striking and characteristic structures of mRNA molecules, its function remains uncertain. Provision of mRNA stability is considered a likely function, because mRNAs

that normally possess tails, if injected into cells with their tails removed, are quickly degraded. The same mRNAs in tailed form may persist for long periods of time. The tail may thus protect some mRNAs from breakdown by ribonuclease enzymes in the cytoplasm. Normally tailless mRNAs, such as those for histones, seem to be as stable in the cytoplasm as those with poly(A) tails. Other sequence structures, such as specific fold-back structures at the 3' end, may protect these mRNAs from degradation.

The tail of a poly(A)$^+$ mRNA gradually shortens as the messenger ages in the cytoplasm. The gradual shortening occurs whether or not protein synthesis takes place, indicating that there is no direct connection between reduction of the tail and the number of times or frequency with which mRNAs are translated by ribosomes. These observations have given rise to the hypothesis that the poly(A) tail may fix the ultimate life expectancy of mRNAs by acting like a slow-burning fuse. The poly(A) tail, according to this hypothesis, is shortened slowly and steadily by exonucleases until the tail becomes too short to wrap around one molecule of the poly(A) binding protein. At this length none of the protective beadlike structures can form, and the mRNA is rapidly degraded.

In support of this hypothesis are the results of experiments by U. Nadel and his coworkers, who injected rabbit globin mRNAs with tails of various lengths into oocytes. The mRNAs with tails 30 or more adenines long were relatively stable in the oocytes; those with shorter tails were degraded about ten times more rapidly.

Presence or absence of the poly(A) tail has varying effects on protein synthesis, depending on the cell type and developmental stage. In most cells both poly(A)$^-$ and poly(A)$^+$ mRNAs are able to direct protein synthesis; however, those without poly(A) tails are translated less frequently. In some cell types, such as fertilized eggs of the clam *Spisula*, some poly(A)$^-$ mRNAs, such as those encoding ribonucleotide reductase and cyclin A, remain untranslated until poly(A) tails are added. Addition of the tail has opposite effects in a few cases; for example, in some mRNAs encoding ribosomal proteins in *Spisula* embryos. Addition of the poly(A) tail in these embryos is correlated with inhibition of translation.

Proteins Associated with mRNAs Messenger RNAs are associated with proteins throughout transcription, processing, and protein synthesis. The proteins change as mRNAs leave the nucleus and enter the cytoplasm. Those initially associated with mRNAs are involved primarily with the processing reactions converting pre-mRNAs into finished form. Later proteins maintain mRNA structure, promote the interactions of mRNAs with ribosomes in protein synthesis, and possibly protect mRNAs from enzymatic breakdown. The total number of associated proteins is significantly reduced as mRNAs become mature and link with ribosomes.

Some proteins associated with mRNAs are common to most or all mRNA types, and some vary according to the cell type. Presumably, those that vary with the cell type regulate the rate at which mRNAs join with ribosomes in protein synthesis or control the rate at which mRNAs are degraded. Among the proteins common to all mRNAs are at least four associated with the 5' cap, which take part in the initial reactions linking mRNAs to ribosomes, and the binding protein associated with poly(A) tails.

Isolation of mRNAs

Messenger RNAs and their functions were proposed as entities in the 1950s, long before their actual identification and purification became possible. These molecules proved elusive until the 1970s, when techniques were developed that allowed mRNAs to be isolated and purified for study and biochemical analysis. In eukaryotes the first techniques to be successful took advantage of highly specialized cells, such as maturing muscle and red blood cells in animals and maturing seeds in plants, that synthesize only a few proteins in quantity. The relatively few different mRNAs in these cells greatly simplified the task of identifying single messengers against the general background of cellular RNAs. A number of eukaryotic messengers were successfully isolated in pure form by this approach, including the mRNAs for myosin and hemoglobin from muscle and blood cells (a hemoglobin mRNA from a blood cell was the first eukaryotic messenger to be purified); ovalbumin mRNA, from the cells lining the oviducts of birds; immunoglobulin mRNA, from mouse myeloma tumor cells; lens crystallin mRNA, from cells producing the lens protein of developing vertebrate eyes; seed storage proteins from legumes and cereals such as peas, beans, and maize; and a variety of mRNAs transcribed from viruses infecting eukaryotic cells. Virus-infected cells were and are particularly useful, because during infection by many virus types, cells stop making most of their own proteins and concentrate on a few viral proteins. It was also possible to isolate and purify the mRNAs encoding histones and protamines because the mRNAs for these very small proteins form a class of low molecular weight molecules that is relatively easy to distinguish from other mRNAs.

More recent methods take advantage of the poly(A) tail, which binds readily to poly(U) or poly(T) molecules by complementary base pairing. Poly(U) or poly (T) molecules attached to filter paper or to the plastic beads of a separatory column (see p. 130) readily bind mRNA molecules, in effect catching them by the tail and allowing their quick separation from a mixture of cellular RNAs (Fig. 15-7). Specific mRNAs can be isolated from these samples by a similar approach using complementary DNA segments as the "catching" agent bound to the filter paper.

Single mRNAs are also isolated by the use of antibodies against the protein encoded in the mRNA. When added to cell extracts, such antibodies will often react with incompletely finished proteins emerging from ribosomes. Isolation of the ribosomes tagged by the antibody also isolates the mRNA encoding the protein.

Once isolated and purified, mRNA molecules are sequenced in the form of a complementary DNA copy called a *cDNA*. The cDNA is made initially by a reaction catalyzed by *reverse transcriptase*, an enzyme that uses an RNA molecule as template for making a DNA copy. The cDNA copy is then increased to quantities large enough for sequencing by cloning or the polymerase chain reaction (see pp. 134 and 137).

Pre-mRNAs

Pre-mRNAs or hnRNAs, the precursors of finished mRNAs, average from five to ten times longer than their corresponding mature products. These long precursors were discovered to exist in the nucleus some time before they were established as the forerunners of mRNAs. Originally, these molecules were identified simply as an abundant nuclear RNA type that is extremely heterogeneous in size and is apparently degraded almost entirely in the nucleus (hence the term *heterogeneous nuclear RNA,* or *hnRNA*).

Early indications that hnRNA is actually pre-mRNA came from the fact that 5' caps and poly(A) tails can be detected in the long nuclear precursors. The suggestions from this information were strengthened by *pulse-chase* experiments conducted by J. E. Darnell and others. Cells were exposed briefly to tritiated uridine, a radioactive substance incorporated in quantity only into newly synthesized RNA. After a few seconds of exposure, the label was "chased" by placing the cells in a medium free of the radioactive uridine. In this way a group of hnRNA molecules, labeled during the brief exposure to tritiated uridine, could be traced by their radioactivity as they were processed. These experiments revealed that, although radioactivity was liberated in soluble form as much of the hnRNA was degraded in the cell nucleus, significant quantities appeared in mature mRNA molecules entering the cytoplasm. The appearance of radioactivity first in pre-mRNA and later in mRNA molecules in these experiments clearly indicated a precursor-product relationship between the two RNA classes.

More definitive evidence came from experiments using *RNA-DNA hybridization* (see p. 528). Either pre-mRNA or mRNA, if mixed with denatured (single-stranded) whole-cell DNA from the same cell type, will wind together or hybridize with the same DNA segments. This means that both pre-mRNA and mRNA

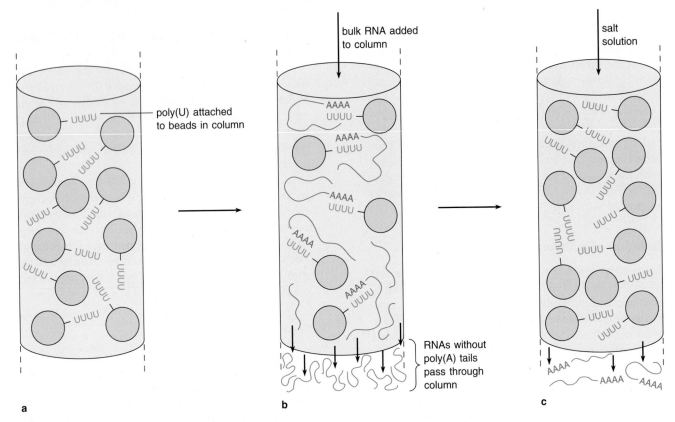

Figure 15-7 Use of affinity chromatography to separate mRNAs from a sample containing many types of cellular RNAs. **(a)** The chromatography column is packed with plastic beads to which segments of poly(U) RNA are attached. **(b)** The bulk RNA sample is added; mRNA molecules are trapped on the beads by base pairing between their poly(A) tails and the poly(U) segments on the beads; other RNA types pass through the column. **(c)** The mRNA is displaced from the beads and released from the column as a purified sample by a salt solution.

In figure: bulk RNA added to column — salt solution — poly(U) attached to beads in column — RNAs without poly(A) tails pass through column — a — b — c

contain sequences that are complementary to the same gene and are evidently copied from the same DNA. If mixed together before hybridization, pre-mRNA and mRNA *compete* for the same sequences in DNA during hybridization, providing further evidence that both are copied from the same DNA template and that one is the precursor of the other.

These experiments are frequently carried out by attaching the DNA of a particular mRNA gene, increased in quantity by cloning or the polymerase chain reaction (see pp. 134 and 137), to nitrocellulose filter paper. A solution containing a radioactive form of the mRNA encoded in the gene, in either pre-mRNA or mature form, is poured over the filter paper under hybridizing conditions. (The mRNA is made radioactive by growing cells in the presence of tritiated uridine.) After allowing time for hybridization, the preparation is washed to remove unhybridized RNA and treated with a ribonuclease. The enzyme digests away any single-chain RNA segments that are not hybridized with the DNA. The level of radioactivity of the filter paper gives a direct measure of the amount of mRNA binding to the DNA. The method shows that both an mRNA and its pre-mRNA bind to the gene in equiv-

alent quantities when added separately. Competition for the same DNA sequence is detected by determining the amount of either radioactive mRNA or pre-mRNA binding to a measured quantity of the DNA of a gene attached to the paper. If mRNA is used, for example, an equivalent quantity of radioactive mRNA is then mixed with nonradioactive pre-mRNA, and both are added to another sample of the DNA bound to nitrocellulose paper. If the amount of radioactivity retained on the paper after washing and ribonuclease treatment is significantly reduced (ideally, by about one-half), the mRNA and pre-mRNA can be said to compete for the same DNA sequence.

Hybridization experiments of another type produced some of the first results indicating that mRNA genes are interrupted by intervening sequences. In one of the most elegant of these experiments, P. Leder and his colleagues hybridized mouse β-globin pre-mRNA and mature mRNA to the DNA of the β-globin gene and examined the hybrids in the electron microscope. The β-globin pre-mRNA molecules hybridized with and covered essentially the entire β-globin gene. The much smaller β-globin mRNA molecules, as expected, hybridized with segments of the β-globin DNA totaling

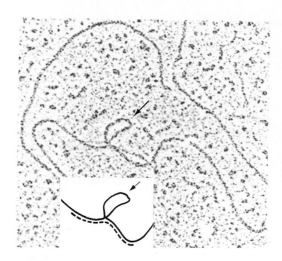

Figure 15-8 The DNA of a mouse β-globin gene (solid line in inset) hybridized with a mature mRNA molecule (dashed line in inset) encoded in the gene. A large, unpaired loop of DNA (arrow) extends outward from the middle of the hybridized region. The loop is an intron, a segment of the DNA that is copied into the pre-mRNA transcript but spliced out in the processing reactions producing the finished mRNA. Micrograph courtesy of S. M. Tilghman.

only about one-tenth of the length covered by the β-globin pre-mRNA.

Of major interest in this experiment was the fact that a large loop of unhybridized DNA extended from the middle of the DNA length hybridized to the mature mRNA (Fig. 15-8). The loop showed that a segment of the DNA that hybridized with the β-globin pre-mRNA had no counterpart in the processed mRNA. Thus, a sequence within the transcribed region of the β-globin DNA was copied into the pre-mRNA but was later removed during processing. The β-globin gene therefore occurs in the DNA in interrupted form; the segments retained in mature mRNAs, called *exons*, are separated from each other in the gene by intervening sequences, called *introns*, which are transcribed into pre-mRNAs but do not appear in finished mRNAs.

Direct sequencing of genes, pre-mRNAs, and mature mRNAs provided the final and most definitive evidence that pre-mRNAs are the precursors of mRNAs. The sequencing studies also revealed that most, but not all, eukaryotic protein-encoding genes are interrupted by introns. Depending on the gene, from 1 to more than 50 introns may be present. At the upper end of this distribution is the α2 collagen gene of chickens, which contains 51 introns (Fig. 15-9); the gene encoding dystrophin, a protein that in mutant form is responsible for muscular dystrophy, is interrupted by no less than 65. The entire dystrophin gene, with introns and exons, spans a distance of more than 2 million base pairs in the human genome!

Pre-mRNA molecules become associated with proteins as soon as they are transcribed from the DNA template. As many as ten different polypeptides, ranging from about 30,000 to 40,000 daltons, have been detected in pre-mRNA–protein complexes. Some of the polypeptides occur in multiple copies in the complexes. Antibodies developed against the proteins associated with pre-mRNAs of one vertebrate group also react with those of other vertebrate groups, indicating that these proteins have been highly conserved in vertebrate evolution. Some cross reactivity has even been noted between antibodies against the pre-mRNA proteins of *Drosophila* and human cells.

The proteins associated with pre-mRNAs are believed to hold pre-mRNA molecules in specific three-dimensional forms and to take part in the processing reactions. The proteins change as pre-mRNA molecules are processed. Most, if not all, of the pre-mRNA–associated proteins are replaced by those characteristic of cytoplasmic mRNA as processing is completed.

Also found in association with pre-mRNA particles is a group of snRNAs known as the *U snRNAs* because uridine occurs in high proportions among their bases. Like the pre-mRNA–associated proteins, these snRNAs play both structural and functional roles during processing reactions (see below).

Structure of mRNA Genes

The genes coding for mRNA molecules, like the pre-mRNA molecules copied from them, are much longer than their corresponding mRNAs (Fig. 15-10). Most of the extra length of these genes reflects the presence of introns. Additional, nontranscribed sequences that are important for gene function lie on either side of the segment actually copied into pre-mRNA.

The beginning, or 5′ end, of an mRNA gene (so called because it corresponds to the 5′ end of the pre-mRNA molecule copied from it) contains the promoter as its primary sequence element. The promoter provides regulatory, recognition, and binding sites for the RNA polymerase II enzyme and indicates the first DNA nucleotide of the gene to be transcribed into an RNA copy. The regulatory sequences immediately upstream of the transcribed portion of the gene are known as *5′-flanking sequences.*

The first nucleotide to be transcribed into a pre-mRNA, called the *startpoint,* is numbered as +1. All nucleotides of the gene are numbered from the startpoint. Those upstream of the startpoint are numbered −1, −2, −3, and so on, starting with the nucleotide immediately to the left of the startpoint. Those to the right, or downstream, of the startpoint are numbered +2, +3, +4, and so on.

The gene segment transcribed into a pre-mRNA contains the sequences corresponding to the 5′ untranslated segment, coding segment, and 3′ untranslated

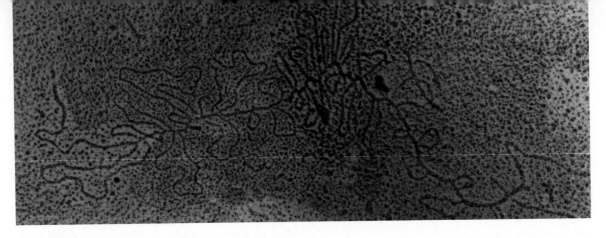

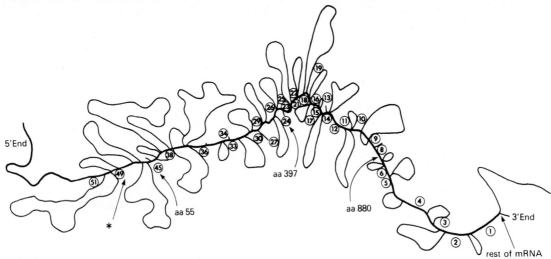

Figure 15-9 Loops produced by hybridization of the DNA of a chick α2 collagen gene with a mature mRNA molecule encoded in the gene. Each loop marks the position of an intron that is transcribed into a pre-mRNA molecule and removed during processing. Courtesy of I. H. Pastan, from *Cold Spring Harbor Symp. Quant. Biol.* 45:777 (1981).

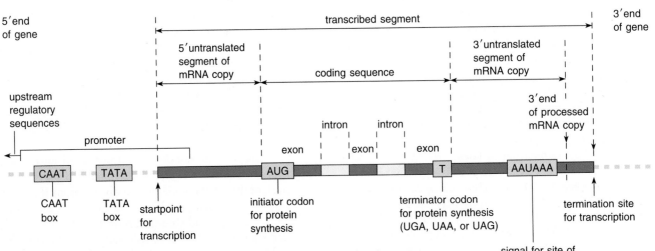

Figure 15-10 The arrangement of sequence elements in genes coding for mRNA molecules in eukaryotes (see text). By convention sequence elements in the DNA are given in their RNA equivalents except for the CAAT and TATA boxes, which are given in the form in which they appear in the DNA missense chain. One or more introns may be present in the coding sequence of the gene.

segment of the finished mRNA. At some point in the transcribed portion of the gene is a coding triplet corresponding to the AUG codon, the first codon to be translated into an amino acid by a ribosome in protein synthesis. This initiator codon marks the beginning of the sequence coding for a polypeptide. At a point farther downstream is a codon corresponding to one of the three terminator codons signifying the termination point for protein synthesis. This codon marks the end of the coding sequence of the gene. One or more introns (shown in light blue in Fig. 15-10) may be inserted in the coding sequence. The final sequences of the gene, downstream of the segment copied into pre-mRNA, are known collectively as the *3'-flanking sequences.*

Almost all eukaryotic mRNA genes contain the code for a single protein or polypeptide. However, because exons may be spliced together in different combinations during pre-mRNA processing, several different mRNA types can be produced by a single gene.

Individual sequences within any segment of a protein-encoding gene are written in the 5' → 3' direction either as an RNA equivalent or as the DNA sequence of the missense DNA chain. The missense chain is used because its sequence is identical to that of the RNA copied from the gene except for the substitution of T for U. Individual triplets or the coded message can therefore be read directly from the missense chain in the same direction as in the mRNA. For example, the

initiator codon AUG is identified in the missense DNA chain directly as AUG or as the missense sequence ATG. Because the sense chain is antiparallel to the missense chain (see p. 527), the 5' → 3' direction in the sense chain is opposite to the direction of transcription. The AUG codon, for example, corresponds to the sequence TGC read in the standard 5' → 3' direction in the sense chain of the gene.

This convention allows any DNA or RNA coding sequence to be written in the standard 5' → 3' direction used for all nucleotide sequences, which is the same direction RNA is transcribed and mRNA translated. Any DNA sequence in a nontranscribed region of the gene, such as the promoter, is also written by convention as the DNA sequence of the missense chain.

The Promoter The promoter contains several important sequence elements that appear in most eukaryotic protein-encoding genes. Similar sequences also occur in the promoters of the genes of viruses that infect eukaryotic cells. Often, these common sequence elements appear in more or less altered form in different genes. A representative sequence that describes the most frequently repeated or average form of the sequence element is accepted as a *consensus sequence.*

Footprinting (see Information Box 15-1) has been used extensively to isolate and identify promoters. Another common method uses *S1 nuclease,* an enzyme that

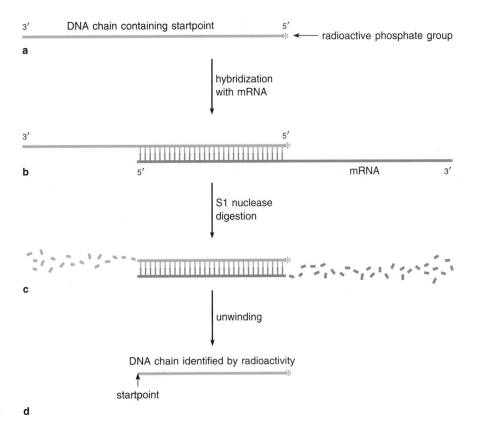

Figure 15-11 Determination of the startpoint of a gene by S1 nuclease digestion. A sample of single-stranded DNA suspected to include the startpoint, labeled with a radioactive phosphate group at its 5' end **(a)**, is hybridized with the mRNA encoded by the gene **(b)**. The hybrid is digested with S1 nuclease to hydrolyze unpaired regions **(c)** and then heated to unwind the remaining nucleotide chains. The 3' end of the undigested DNA chain **(d)** is the startpoint of the gene. (The DNA can be identified by the label at its 5' end.)

attacks only single-stranded nucleotide chains, to locate the startpoint of the gene. Once the startpoint is identified, sequences immediately upstream of this site can be assigned to the promoter. In this technique a sample of single-stranded DNA suspected to include the startpoint of a gene is labeled at its 5' end with a radioactive phosphate group (Fig. 15-11a). The sample is hybridized with the mRNA encoded by the gene (Fig. 15-11b), and the hybrid is digested with S1 nuclease to hydrolyze unpaired regions (Fig. 15-11c). Heating unwinds the undigested nucleotide chains. The 3' end of the remaining labeled DNA chain is the startpoint of the gene (Fig. 15-11d).

Substitution-deletion analysis has revealed that one sequence element, the *TATA box*, is common to almost all eukaryotic mRNA promoters. The TATA box, actually the consensus sequence TATA^A_TA in animals and lower eukaryotes, appears at a site some 20 to 30 bases upstream of the startpoint (Fig. 15-12; the position marked ^A_T indicates that either A or T may appear at this point). Substitution-deletion analysis and the observation of naturally occurring mutants demonstrate that the primary functions of the TATA box are initiation of transcription, including binding the RNA polymerase enzyme, and indication of the startpoint for transcription. The TATA box is recognized and bound by a general initiation factor, TFIID (see below), which promotes transcription of almost all mRNA genes.

The fact that the TATA box indicates the startpoint has been demonstrated by several experiments. For example, S. Hirose and his coworkers showed that substituting T for A at the second or fourth positions of the TATA box in the gene coding for the silk protein of silkworms greatly reduces the precision of startpoint location, yielding RNA copies that start randomly at or near the normal startpoint. Naturally occurring mu-

tations involving substitutions in the TATA box, one of which is responsible for a form of the human disease *thalassemia* (see below), have the same effect.

The very few genes without a TATA box include those encoding deoxynucleotidyl transferase, dihydrofolate reductase, and some genes of the SV40 virus that are transcribed at later stages of infection. Some of these genes have multiple, random startpoints, as expected from the mutant studies indicating that the TATA box fixes the startpoint. Undoubtedly, as yet unidentified sequences in these genes indicate the startpoint and provide recognition sites for initiation factors.

Other short sequence elements occur less frequently in mRNA gene promoters. In general these sequence elements bind regulatory proteins that modify the rate at which the TATA box is used by RNA polymerase II to initiate transcription. Some of these gene-specific elements are common to many genes, and some occur in a few or even only in single genes. One of the more common of these modifying elements is the CAAT box, actually the consensus sequence GC^T_CCAATCT in animals and lower eukaryotes, centered roughly at about −80 in the promoter (Fig. 15-12). As expected from its modifying function, substitution or deletion of nucleotides in the CAAT box primarily reduces or increases the rate at which RNA polymerase II binds to the promoter region and initiates transcription.

Other sequence elements with effects on promoter functions have been identified at points from about 85 to as many as 200 or more bases upstream of the startpoint. In some cases sequences affecting promoter function are even located downstream of a gene. These more distant sequence elements almost uniformly affect the frequency at which the gene under study is transcribed. Thus, like the CAAT box, their primary func-

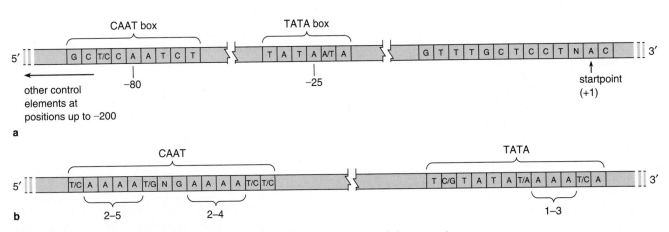

Figure 15-12 Eukaryotic promoter sequences. **(a)** Sequence elements of the promoter in animals and lower eukaryotes (see text). The CAAT and TATA boxes of these organisms are shown as consensus sequences. **(b)** Consensus sequences of the CAAT and TATA boxes of plant mRNA genes. N indicates that any of the four nucleotides may occur with approximately equal frequency at that site.

tion is to control the rate of initiation. This control is exerted through the activity of regulatory proteins that recognize and bind the sequences.

Enhancers Most striking of the distant sequences with important effects on gene function are elements some 50 to 100 nucleotides long called *enhancers* (also called *activators*, *augmentors*, or *potentiators*). Many enhancers, because they lie a hundred nucleotides or so upstream of the startpoint, might be considered part of the promoter except for their curious ability to operate when inserted experimentally at practically any location near a gene—upstream, downstream, or even within introns. Enhancers appear to work even when inserted in reverse order near their target genes. The only requirements for enhancer function seem to be reasonable proximity, within a thousand base pairs or so of the gene, and location on the same DNA molecule as the gene being transcribed.

Enhancers have been found in association with many eukaryotic mRNA genes, including those coding for antibodies, myosins, hemoglobin, ovalbumin, and an H2A histone of chickens. Few sequence features appear to be common among enhancers except that many in higher eukaryotes contain an eight-nucleotide *octamer* sequence that is recognized and bound by regulatory proteins. Mutations or substitutions altering or eliminating enhancer sequences greatly reduce the rate of transcription of their target genes, in most cases by a factor of 100 or more.

Enhancers have been detected in all eukaryotes and also in bacteria (see Supplement 15-1). How they work presents one of the more intriguing and as yet unanswered questions of molecular biology. All known enhancers act as sites that, when recognized and bound by regulatory proteins, significantly increase the rate of transcription at the nearby promoter.

Three major hypotheses have been proposed to explain how enhancers function at a distance. One maintains that the regulatory protein, after binding to the enhancer, slides or scans along the DNA until it encounters the promoter, where it stimulates tight binding of RNA polymerase. In this scanning model the primary function of the enhancer is to serve as a strong binding site for initial attachment of the regulatory protein. The conformational transmission model suggests that binding of the regulatory protein induces a conformational change in the DNA, which is transmitted along the double helix to the promoter. At the promoter the change, which could involve partial unwinding of the DNA, favors tight binding of RNA polymerase. The third hypothesis, the loop model, which is most favored by current evidence, maintains that the regulatory protein bound to the enhancer remains in place (Fig. 15-13a) and interacts directly with other proteins bound at the promoter (Fig. 15-13b); the combined regulatory proteins, according to this model,

take on a conformation that binds RNA polymerase II more avidly than the promoter complex alone (Fig. 15-13c). As part of this mechanism, the DNA segment between the enhancer and promoter is thrown into a loop. Among the several lines of evidence supporting the loop model, which was originally proposed by H. Echols, R. Tjian, M. Ptashne, and their colleagues, are electron micrographs directly showing loop formation by interaction of proteins at the enhancer and promoter sites (Fig. 15-13d and e).

Several lines of evidence make the scanning and conformational transmission models seem unlikely. H.-P. Muller and W. Schaffner, for example, tested the three models by connecting an enhancer to a promoter by a protein bridge. The enhancer still increased the rate of transcription when combined with its regulatory protein, even though no direct DNA connection existed between the enhancer and promoter. This finding appears to rule out both scanning of a regulatory protein along the DNA and transmission of conformational changes along the DNA as mechanisms increasing promoter activity. Loop formation, however, could still occur even if the enhancer and promoter were connected by a protein link. Other evidence comes from experiments testing the ability of enhancers to work if placed on different DNA molecules from their promoter. A. Wedel, for example, found that the NtrC regulatory protein shown in Figure 15-13d and e could still increase the rate of transcription of the bacterial *glnA* operon when its target enhancer sequence was placed on a different DNA molecule from the promoter and operon. The enhancer would not work unless held in close proximity to the promoter; to accomplish this, the enhancer and promoter were placed on separate but intertwined DNA circles.

Enhancers have in some instances been linked to the development of cancer. When inserted by viral infections near genes regulating cell division, enhancers sometimes overactivate the regulatory genes, triggering uncontrolled cell division and leading to tumor growth in the infected tissue (see Ch. 22). Further details of enhancers and their functions in regulating mRNA transcription are presented in Chapter 17.

Sequence Elements in the Transcribed Region: Exons and Introns Other than the initiator and terminator codons for protein synthesis, the sequence elements of primary interest in the coding segment of protein-encoding genes are those defining exons and introns. These sequences have been identified and studied by several means. One involves the S1 nuclease technique. The DNA of a gene is hybridized with its mRNA product and the RNA-DNA hybrid exposed to S1 nuclease. The introns extend from the hybrid as unpaired loops, as in the hybridized gene shown in Figure 15-9. The unpaired loops are hydrolyzed by S1 nuclease, leaving the exons intact (Fig. 15-14). Comparison

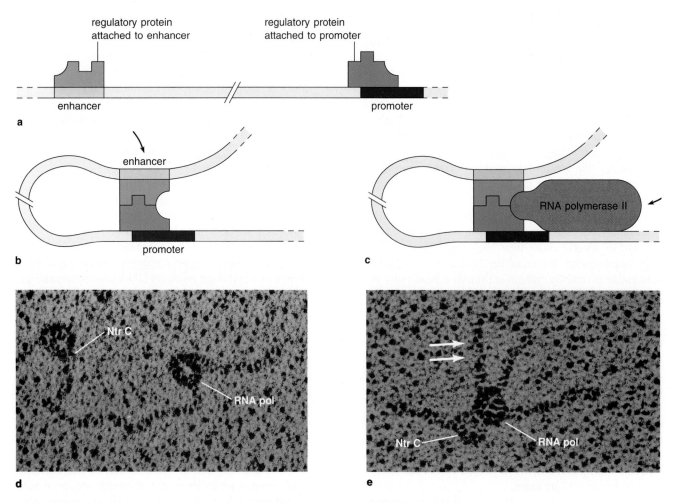

Figure 15-13 The loop model for the action of enhancers in increasing the transcription rate of a gene. **(a)** The regulatory protein bound to the enhancer. **(b)** Interaction between the regulatory protein and one or more proteins bound to the promoter creates a loop in the DNA. **(c)** The combination sets up a conformation with high binding affinity for RNA polymerase II, which adds to the complex and initiates transcription. **(d)** The enhancer-promoter region of the *glnA* operon of *E. coli*, showing the regulatory protein NtrC bound to the enhancer and RNA polymerase (RNA pol) presumably bound loosely to the promoter. The two proteins can be distinguished separately because NtrC is more lightly coated by the tungsten metal used to shadow the preparation (see p. 120). **(e)** Formation of a loop through direct interaction of NtrC and RNA polymerase. Presumably, the interaction leads to tight binding by the polymerase and initiation of transcription. **(d** and **e)** courtesy of H. Echols from *J. Biolog. Chem.* 265:14699 (1990).

of the sequences in intact, undigested copies of the gene and the DNA segments left after S1 nuclease digestion of the hybrids allows precise identification of exons and introns. In another frequently employed method, sequences of the transcribed segment of a gene are compared directly with sequences of the mature mRNA of that gene. The sequence comparison allows any introns and their boundaries to be precisely located.

Experiments of this type confirm that much of the length of many mRNA genes is occupied by introns. A gene coding for the yolk protein vitellogenin in chickens, for example, is some 23,000 nucleotides long. More than 16,000 of these nucleotides are included in

introns; the mature mRNA of this gene contains only 6700 nucleotides. In different genes introns may run from fewer than 50 to 30,000 or more nucleotides in length; some of the longer introns may contain other genes, complete with their own regulatory sequences and introns. Average intron lengths range from about 200 to 400 nucleotides. (Fig. 15-15 summarizes the exon-intron structure of several eukaryotic genes.)

Analysis of the boundaries between introns and exons reveals that almost all the introns interrupting mRNA genes, with very few exceptions, begin at their 5′ ends with the two nucleotides GU and end with the two nucleotides AG, a characteristic known as the *GU/*

DNA sense
chain of gene

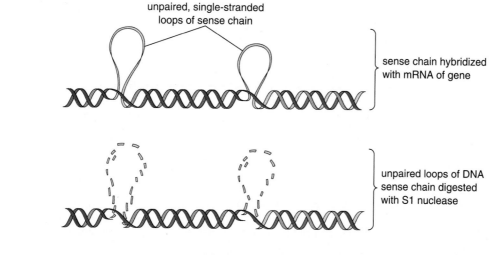

unpaired, single-stranded
loops of sense chain

sense chain hybridized
with mRNA of gene

unpaired loops of DNA
sense chain digested
with S1 nuclease

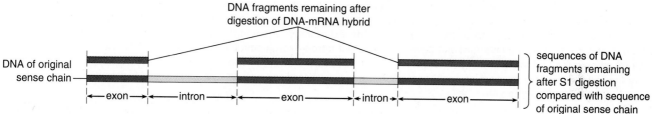

DNA fragments remaining after
digestion of DNA-mRNA hybrid

DNA of original
sense chain

←exon→ ←——intron——→ ←——exon——→ ←intron→←——exon——→

sequences of DNA
fragments remaining
after S1 digestion
compared with sequence
of original sense chain

Figure 15-14 Use of S1 nuclease digestion to identify introns, exons, and their boundaries in the DNA of a gene.

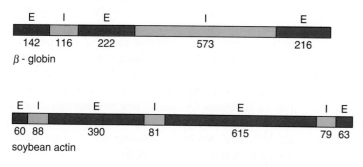

E   I   E   I   E
142 116 222   573   216
β - globin

Figure 15-15 The intron-exon structure of three eukaryotic genes, with the numbers of base pairs in introns (I) and exons (E).

E I   E   I   E   I E
60 88   390   81   615   79 63
soybean actin

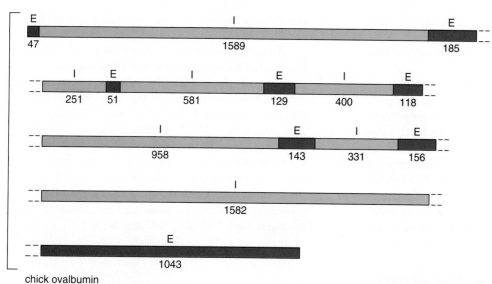

E   I   E
47   1589   185

    I   E   I   E   I   E
    251 51   581   129   400   118

              I   E   I   E
              958   143   331   156

                          I
                          1582

          E
          1043
chick ovalbumin

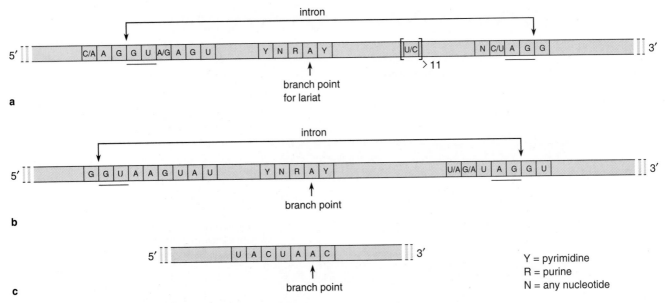

a

b

c

Y = pyrimidine
R = purine
N = any nucleotide

Figure 15-16 Consensus sequences of intron-exon boundaries in eukaryotic pre-mRNAs. **(a)** Animal pre-mRNAs; **(b)** plant pre-mRNAs. Plant introns also contain a majority of U-A base pairs distributed throughout the intron. The GU at the 5′ end and the AG at the 3′ end are essentially invariant and occur in almost all intron-exon junctions. The branch point is the site at which the 5′ end of an intron, once clipped free, joins to form the lariat during the processing reactions. **(c)** The invariant sequence around the branch point in yeast genes.

AG rule, or, when considered in DNA, as the *GT/AG rule*. Other sequences surrounding the terminal GU and AG and lying within introns are repeated frequently enough to allow construction of the consensus sequences shown in Figure 15-16. The GU/AG boundaries and the internal sequences are recognized as signposts by the enzymatic mechanisms removing introns (see below).

The majority of known eukaryotic genes and the genes of viruses infecting eukaryotic cells contain at least one intron. Exceptions include histone genes that are active during the period in which DNA is replicated; in most species these histone genes occur entirely in uninterrupted form. A relatively few other genes, such as the zein seed storage protein gene of maize, interferon genes, and some globin genes of insects, are also intronless.

Why so much of the DNA of eukaryotic genes is occupied by introns remains uncertain. In some genes, such as those coding for hemoglobins, immunoglobins, insulin, and the crystallin proteins of the vertebrate eye lens, intron-exon junctions fall at divisions between major structural or functional domains in the encoded proteins. Correspondences with intron-exon boundaries are also noted at points in which the amino acid chains of proteins make loops at protein surfaces. In other genes, such as those coding for actins, there is no

apparent relationship between the positions of intron-exon boundaries and domains in the proteins.

The correspondences between the positions of introns and protein domains have given rise to the hypothesis that many proteins may have appeared in evolution through the exchange of exons between genes, a process called *exon shuffling*. This mechanism would allow the evolution of proteins with new but related functions by assembling new combinations of existing domains already selected and perfected by the evolutionary process. Evolution by this mechanism would be much more efficient than one-by-one changes in amino acids at random points. (For a further discussion of exon shuffling as a basis for evolutionary change, see Ch. 27.)

The introns of genes in which no correlations can be detected between intron-exon structure and protein domains may have arisen by insertions of introns at random points within genes, possibly through the activity of viruses that regularly insert DNA segments into genes. Any DNA segment, once acquiring the necessary 5′ and 3′ consensus sequences, could become a permanent intron of an mRNA gene. Perfect removal of the intron from the mRNA copied from the gene would ensure that the intron has no effect on the protein encoded by the gene and the organism containing it. Such "selfish DNA" inserts, as F. H. C. Crick

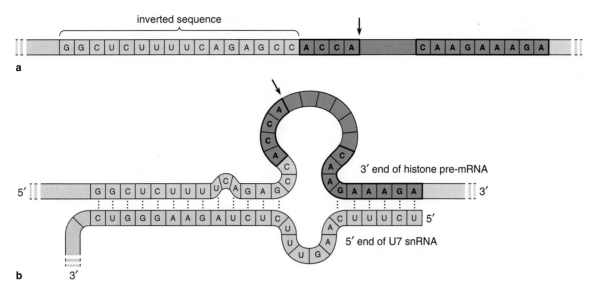

Figure 15-17 Sequence elements at the 3′ end of histone genes and histone pre-mRNA processing. **(a)** Consensus sequences in the 3′ region of histone genes. A reverse repeat is followed by the ACCA and CAAGAAAGA consensus sequences; the arrow indicates the point at which the pre-mRNA is cleaved to form the 3′ end of the finished histone mRNA. **(b)** Pairing between the 3′ end of histone pre-mRNAs and U7 snRNA. The pairing may hold the 3′ end of the mRNA in the correct conformation for clipping, which occurs at the site indicated by the arrow.

termed them, would thus be protected from evolutionary selection and maintained permanently in the DNA from generation to generation.

Whether selfish or functional, introns are maintained in all eukaryotes in spite of the considerable phosphate-bond energy required for their replication and transcription. Similar introns also occur in genes coding for tRNA and rRNA in eukaryotes (see below).

Sequences at the 3′ End of mRNA Genes Sequences following the coding segment of eukaryotic mRNA genes contain signals for processing the 3′ end of the transcript and terminating transcription. The segment of mRNA genes corresponding to the 3′ untranslated segment contains the AATAAA sequence (transcribed as AAUAAA in the pre-mRNA), which marks the site at which the 3′ end of the pre-mRNA is to be clipped off. The clipping site lies 10 to 20 nucleotides downstream of the AAUAAA signal. Most single base substitutions in the AAUAAA sequence, such as changes to AACAAA, AAUACA, AAUUAA, or AAUGAA, interfere with the clipping reaction, yielding mRNAs with greatly extended 3′ ends. Substitutions in the signal also reduce the efficiency of the reaction adding the poly(A) tail. The only natural variation of the sequence, AUUAAA, operates at about 75% of the efficiency of the usual AAUAAA signal (see Fig. 15-19).

M. Wickens and his colleagues have studied the functions of the AAUAAA signal by means of synthetic RNAs that include the signal. These studies have shown that the synthetic RNAs work, and indicate the site of clipping and addition of the poly(A) tail with virtually maximum efficiency, if they include as little as the AAUAAA signal and an additional eight downstream nucleotides. Although there is no requirement for a specific sequence downstream of the AAUAAA, a U- or UG-rich series of bases is necessary for efficient operation of the signal.

Transcription of pre-mRNAs proceeds to regions as far as a thousand nucleotides or more downstream of the AAUAAA signal. At some point in these downstream regions, transcription stops and the pre-mRNA transcript and the RNA polymerase II enzyme are released. Rather than terminating at specific sequences, transcription may be stopped by reactions associated with mRNA processing in the vicinity of the AAUAAA signal (see below).

Research by N. J. Proudfoot, M. L. Birnstiel, C. Birchmeier, and others showed that, instead of the AAUAAA signal, most histone genes have an inverted sequence followed by two highly conserved sequences, ACCA and CAAGAAAGA (Fig. 15-17a), in the 3′ untranslated segment. Nucleotides in the conserved ACCA and CAAGAAAGA sequences evidently serve as processing signals regulating the site at which the 3′ end of a histone pre-mRNA is clipped.

At least one human disease, a form of thalassemia, is caused by a mutation of the AAUAAA signal to AAUAAG. This mutation causes faulty processing of an α-hemoglobin protein necessary for normal function of hemoglobin in erythrocytes. The result is anemia ranging from mild to severe, depending on whether the individual is homozygous or heterozygous for the mutation.

Transcription of Pre-mRNAs

The initiation of pre-mRNA transcription requires an interaction of regulatory proteins and transcription factors with the promoter, enhancer sequences if present, and RNA polymerase II. Investigations by P. A. Sharp and his colleagues and others indicate that at least five *transcription factors*—*TFIIA, TFIIB, TFIID, TFIIE*, and *TFIIF*—interact with the promoter in the region of the TATA box and with RNA polymerase II to initiate tight enzyme binding, unwinding of the DNA, and initiation of transcription at the startpoint. ATP is required for initiation to progress most efficiently; indications are that TFIIE acts as an ATPase and hydrolyzes ATP during the initiation process.

The process of initiation (Fig. 15-18) begins as TFIID recognizes and binds the TATA box of an mRNA gene, in a reaction accelerated by TFIIA (Fig. 15-18a). After TFIID binds, TFIIB, TFIIE, TFIIF, and RNA polymerase II add to the complex. Although the sequence of events is not yet certain, TFIIB probably binds first to TFIID (Fig. 15-18b), setting up a conformation that binds RNA polymerase II (Fig. 15-18c). TFIIE and TFIIF then add as a closely associated TFIIE/F complex (Fig. 15-18d). Apparently, TFIIB and TFIIE have binding affinities for both TFIID and RNA polymerase II but do not recognize or bind DNA sequences in the promoter. This assembly then binds and hydrolyzes ATP, the DNA unwinds, and transcription begins. Nonhydrolyzable analogs of ATP (see p. 421) bind to the complex, but transcription does not begin, indicating that energy is required for initiation, perhaps for unwinding the DNA. The initiation complex remains somewhat unstable until two or three phosphodiester bonds have been formed in the RNA copy.

TFIID is considered most central to the initiation process because it is the first to recognize a sequence in the promoter and to bind strongly to the DNA. R. G. Roeder, R. Tjian, M. Horikoshi, and their colleagues have cloned the gene encoding TFIID in organisms as diverse as humans and yeast. The protein encoded by the gene, which contains about 240 amino acids, has a 180-residue DNA-binding core domain that is highly similar in sequence in yeasts and humans. Other segments of the molecule react with regulatory proteins.

A variety of regulatory proteins modify the initiation process. Some of these proteins bind directly to TFIID and modify its affinity for the TATA box. Others bind to other initiation factors or to DNA sequences located elsewhere in the promoter or upstream sequences. Presumably these proteins, which serve as gene-specific regulatory factors, modify the conformation of the initiation complex or the DNA to enhance or inhibit initiation. (Further details of these proteins and their activity are presented in Ch. 17.)

Once initiation has been accomplished, one or two elongation factors, one identified as *TFIIS*, add to the complex. The enzyme then proceeds along the DNA template, adding bases one at a time until regions well beyond the AAUAAA signal are reached. In genes in which termination has been investigated in detail by Proudfoot and others, the point at which termination takes place does not appear to involve interactions with specific sequences in the termination region. Instead, the enzyme appears simply to run on after passing the AAUAAA signal and to continue transcribing until the pre-mRNA is clipped at a point just beyond the AAUAAA signal. Presumably, clipping the transcript releases a factor necessary for termination or induces a conformational change in the DNA or the segment of the transcript remaining with the enzyme that stops transcription. In support of this conclusion is the fact that an intact AAUAAA signal is required for termination. As transcription stops, the enzyme and the portion of the RNA transcript remaining with the enzyme are released from the DNA.

Processing Pre-mRNAs

The reactions processing pre-mRNAs take place entirely in the cell nucleus. Addition of the 5' cap probably occurs while transcription is still in progress. Soon after transcription is complete, surplus nucleotides are cleaved from the 3' end, and the poly(A) tail is added. Not long after the 5' cap is added, the pre-mRNA enters the series of reactions that remove introns.

Adding the Cap and Tail The 5' cap is synthesized by a group of processing enzymes that are evidently loosely associated with the pre-mRNA ribonucleoprotein particle. The capping base is added directly to the first nucleotide copied from the startpoint, with no clipping reactions to shorten the 5' end of the pre-mRNA. The capping nucleotide and one or two of the succeeding nucleotides are then methylated to complete the cap.

At the opposite end of the pre-mRNA, the transcript is cleaved at a point 10 to 20 nucleotides past the AAUAAA signal. Two or more proteins may promote the clipping reaction. One is a *specificity factor (SF)*, which may recognize the AAUAAA signal. This factor is suspected to include an snRNA, *U11*, as part of its structure. The enzyme that adds adenines to the tail, *poly(A) polymerase*, binds to the complex, and the tail is cleaved, probably by an additional factor that acts as an RNA endonuclease. The poly(A) polymerase then adds As to the tail one at a time, using ATP as the source for the adenines. Removal of SF or substi-

Figure 15-18 The probable interactions among transcription factors, RNA polymerase II, and the promoter during the initiation of transcription. **(a)** The process begins as the transcription factor TFIID recognizes and binds the TATA box of an mRNA gene. The binding is accelerated by TFIIA, which may bind to TFIID before it attaches to the promoter. **(b)** After TFIID binds, TFIIB adds to the complex. **(c)** This sets up a conformation that binds RNA polymerase II and aligns it at the startpoint. **(d)** The assembly binds TFIIE/F, which hydrolyzes ATP. At this point the DNA unwinds and transcription begins.

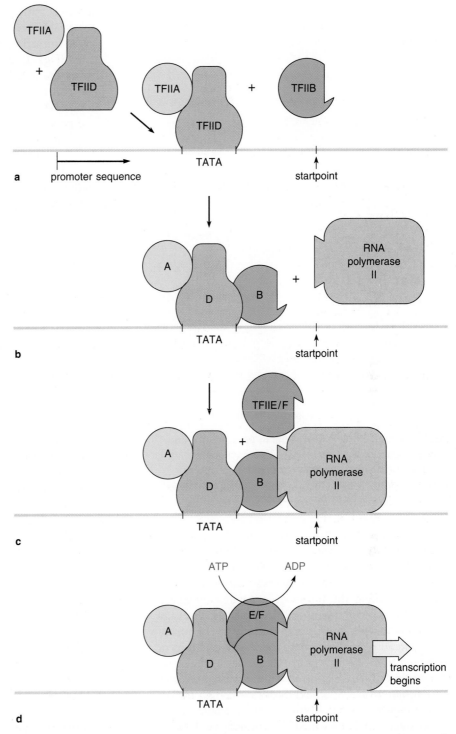

tutions or deletions in the AAUAAA signal significantly reduce the efficiency of the reaction adding the poly(A) tail (Fig. 15-19). After the poly(A) tail is about ten nucleotides long, polymerization becomes independent of the signal, SF is released, and adenines are added until the tail is complete. The conditions terminating polymerization and fixing the length of the poly(A) tail are unknown.

The tails of histone pre-mRNAs, which contain the unique ACCA and CAAGAAAGA sequences as processing signals, are clipped by a reaction that involves another snRNA, *U7*. Pairing between U7 snRNA and the CAAGAAAGA sequence in a histone pre-mRNA is a critical part of the process. M. L. Birnsteil and his colleagues found that mutations in the CAAGAAAGA sequence destroyed cleavage of the tail unless compensating mutations that preserved the pairing were introduced into the U7 snRNA. The pairing probably holds the 3′ end of a histone pre-mRNA in a conformation that serves as a recognition site for cleavage at a point just past the ACCA sequence (see Fig. 15-17b).

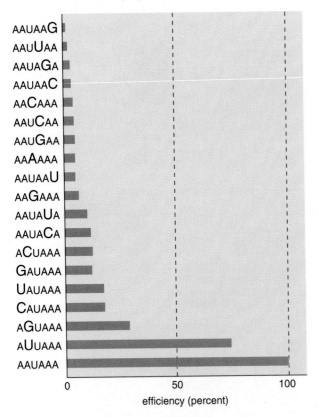

polyadenylation activity

AAUAAG
AAUUAA
AAUAGA
AAUAAC
AACAAA
AAUCAA
AAUGAA
AAAAAA
AAUAAU
AAGAAA
AAUAUa
AAUACa
ACUAAA
GAUAAA
UAUAAA
CAUAAA
AGUAAA
AUUAAA
AAUAAA

0   50   100
efficiency (percent)

Figure 15-19 Effects of variations in the AAUAAA consensus signal on the efficiency of the reactions adding the poly(A) tail. The only common natural variant is AAUUAA, which occurs in 12% of pre-mRNAs. Courtesy of M. P. Wickens, from *Trends Biochem. Sci.* 15:277 (1990).

Removing Introns The splicing reactions removing introns include two major steps: (1) precise breakage of sugar-phosphate bonds at the boundaries of introns, and (2) rejoining the free ends generated by intron removal into a continuous mRNA molecule. Breakage of the pre-mRNAs must occur precisely at the intron-exon junctions, so that no nucleotides are added to or removed from the ends of the exons at either side of the introns. Otherwise, intron removal would produce a *frameshift*; that is, it would throw the coding triplets out of register, so that all triplets past the faulty point of removal or addition would be incorrectly read during protein synthesis.

Intron removal proceeds by a number of steps that involve participation of five snRNAs—*U1, U2, U4, U5,* and *U6.* Each snRNA is associated with one or more proteins; the entire complex formed by the snRNAs is known as a *spliceosome.* Research by J. A. Steitz, Sharp, and their coworkers and others indicates that the snRNA-protein complexes pair with specific segments of the intron consensus sequences and interact with each other to fold the intron into a conformation that triggers its removal. (Fig. 15-20 shows the sequence

in which snRNAs are believed to interact in intron removal.) The biochemical steps in intron removal include formation of a branched RNA structure called a *lariat* (Figs. 15-20 and 15-21; the caption to Fig. 15-21 gives the details of lariat formation). Intron removal, with all its precision, can proceed in the test tube as long as ATP and snRNAs in the form of a spliceosome are present.

Once the splicing reaction is complete, the lariat is opened by a "debranchase" enzyme that can hydrolyze the 2' → 5' linkage holding the branch point together. After the branch is removed the intron chain is rapidly degraded into individual nucleotides.

The lariat structure is also produced by reactions that remove introns from pre-rRNAs in some lower eukaryotes (see p. 613). The reactions removing the rRNA introns in these lower eukaryotes are catalyzed by the RNA itself acting as a ribozyme (see p. 101). Because the lariat structure appears during intron removal from pre-mRNAs, it is considered possible that one or more of the steps clipping and splicing introns in pre-mRNAs are also self-catalyzed, perhaps by one of the snRNAs, and do not require the participation of enzymatic proteins.

Involvement of snRNAs in intron removal is supported by a battery of experimental results, some developed in the Steitz laboratory through an interesting relationship to the human autoimmune disease *lupus erythematosus.* Lupus patients develop antibodies that attack proteins in their own bodies. One group of these autoantibodies completely inhibits the splicing reaction removing introns from pre-mRNAs in either living cells or the test tube. These antibodies react specifically with proteins associated with snRNAs.

The snRNAs taking part in intron removal have been detected in vertebrate and invertebrate animals, in plants, and in the yeast *Saccharomyces cerevisiae.* Some differences have been noted in intron sequences required for efficient splicing among vertebrates, yeast, and higher plants. For example, in vertebrates and higher plants the loosely conserved sequence YNRAY is required for formation of the lariat (see Fig.15-16). A series of pyrimidines is also required on the 3' side of the splice point for efficient intron removal in vertebrates. In plants, the intron must have significantly more U-A than G-C base pairs throughout its structure instead. In yeasts the sequence UACUAAC is absolutely required at the branch point. In these sequences the underlined A indicates the branch point for the lariat.

As a consequence of these sequence preferences, introns in vertebrate pre-mRNAs are not spliced if introduced into yeast cells, and only a few yeast pre-mRNAs can be spliced in vertebrate cells. Plant cells do not splice vertebrate pre-mRNAs efficiently unless they happen to have a high A-U content. For example, in the pre-mRNA transcribed from the human β-globin

Figure 15-20 A possible sequence of interactions among U1, U2, U4, U5, and U6 snRNAs in intron removal by spliceosomes. Each snRNA is complexed with several proteins. **(a)** U1 and U2 bind to the intron by pairing interactions between the snRNAs and the consensus sequences of the intron. U1 binds at the 5′ end of the intron; U2 binds at the site where the lariat will form, with the circled A marking the branch point. **(b)** An interaction between U1 and U2 brings the 5′ end of the intron to the branch site. The structure is stabilized by an interaction of U1 and U5 snRNAs with U4 and U6. U4 and U6 are associated in the same ribonucleoprotein as U4/U6 snRNA. **(c)** The folded conformation of the intron promotes breakage of the 5′ end of the intron, which joins at branch point A to form the lariat. The 3′ end of the intron is then broken, freeing the lariat with the snRNAs. At the same time, the free ends of the exons on either side of the intron join into a continuous, unbroken RNA chain.

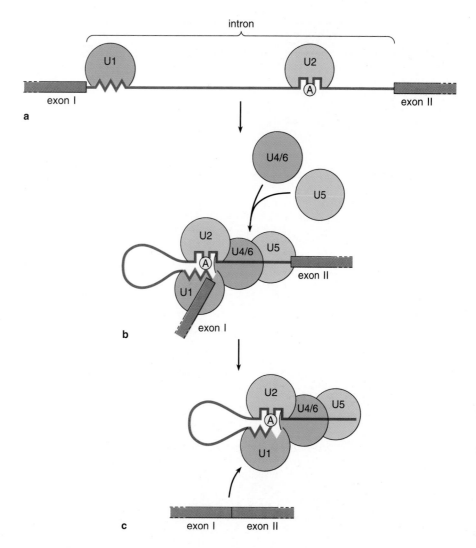

gene, intron 1, which contains 55% A-U base pairs, is not spliced, but intron 2, with 71% A-U base pairs, is removed correctly in higher plant cells. These differences indicate that although the snRNAs involved in intron removal are generally conserved, their activity is modified by differences in nucleotide sequences or associated proteins in major taxonomic groups.

A few examples have been observed of splicing reactions in which the exons brought together and linked are transcribed on separate locations in the DNA, in one case on different chromosomes. This pattern of processing, called *trans-splicing*, has so far been found to involve only the addition of a 5′ untranslated segment encoded in one part of the DNA to the coding segment of an mRNA encoded elsewhere. In the processing reactions producing a mature actin mRNA in the nematode *Caenorhabditis*, for example, a 5′ untranslated segment transcribed on one chromosome is trans-spliced to two exons of a coding sequence and a 3′ untranslated segment transcribed on a different chromosome.

N. Ringertz and his coworkers found that when mammalian nuclei are stained with fluorescent anti-

bodies against spliceosome proteins, fluorescence is concentrated in 20 to 50 discrete spots rather than diffused throughout the nuclei. This staining pattern indicates that spliceosomes are localized in aggregates, perhaps by attachment to the nuclear matrix, and not freely suspended in solution in the nucleus.

Alternative Splicing Pathways The pre-mRNAs of many genes can follow *alternative splicing* pathways that vary with the particular introns to be removed. This allows the same gene to give rise to several different but related mRNAs and, through these mRNAs, several different but related proteins. Because of alternative splicing, a DNA segment can be either an exon or an intron depending on whether it is retained or removed during processing. Alternate splicing can also act as an on-off switch for mRNA activity by producing functional or nonfunctional mRNAs from the same pre-mRNA through the inclusion or exclusion of an incapacitating intron.

For example, the pre-mRNA of a gene active in the thyroid gland and hypothalamus of higher vertebrates has six potential exons (labeled A to F in Fig.

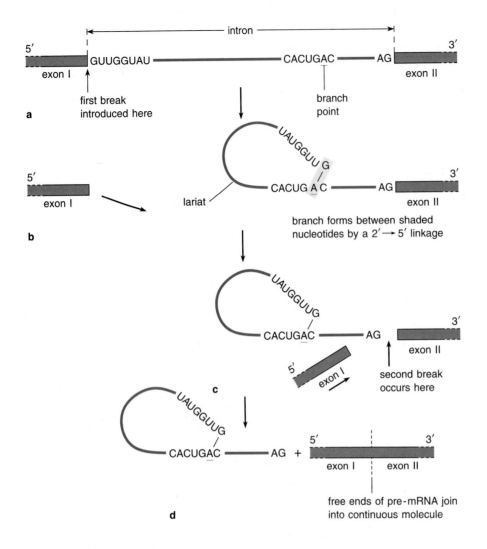

Figure 15-21 Biochemical steps in intron removal. The reactions, worked out by Green, Maniatis, and their colleagues, first introduce a break just to the left of the GU at the 5' end of the intron **(a)**. The free end of the intron then joins to the underlined adenine residue in the consensus sequence within the intron **(b)**. The link, which attaches the 5' end of the intron to the 2'-carbon of the adenine, produces a branch at this point and closes the loop of the lariat. The 3' end of the intron is still intact at this step. **(c)** The second break opens just to the right of the AG at the 3' end of the intron, and the intron is released from the pre-mRNA. **(d)** As the intron is released, the free ends of the pre-mRNA are joined into a continuous chain. The . . . CACUGAC . . . sequence is a version of the consensus sequence that occurs at the branch point of the lariat.

15-22*a*) separated by five introns. In the thyroid gland exons A or B and C, D, and E are linked together, and all other introns and exons are removed by the splicing reaction, producing a finished mRNA (Fig. 15-22*b*) that codes for the thyroid hormone *calcitonin*. In the hypothalamus exons A or B and C, D, and F are linked together, and other introns and exons removed, to produce an mRNA coding for a distinct polypeptide, the *calcitonin gene-related polypeptide* (*CGRP*; Fig. 15-22*c*), which may be involved in the function of taste receptors. In these alternative splicing reactions, either the A or B exon may be retained or eliminated; segment E is an exon in the calcitonin pathway and an intron in the CGRP pathway. Segment F is an intron for calcitonin and an exon for CGRP. Thus, a given segment, depending on whether it is removed or retained during alternative splicing, can act as either an exon or an intron.

The list of pre-mRNAs processed differentially by alternative splicing lengthens constantly as more examples of this mechanism are discovered. (Fig. 15-23 shows the variety of alternative splicing reactions observed in pre-mRNA processing.) Among others the list includes pre-mRNAs for amylase, lamins, collagens,

growth hormone, prolactin, fibrinogen, lens crystallin, myosin light and heavy chains, troponin T, tropomyosin, protein kinases C, fibronectin, vimentin, antibody polypeptides, and, in higher plants, a subunit of RuBP carboxylase. The distinct but related proteins produced by alternative splicing reactions have different functions or are routed to different final locations. The length of the list indicates that alternative splicing has often been used in evolution to increase the variety of proteins produced by the cells of an organism.

As yet little is known about the molecular basis for the mechanisms regulating which pathways are used in different cell types. However, the sequences chosen for retention or removal are thought to depend on an interaction between regulatory proteins and differences in nucleotide sequence or secondary structure in the pre-mRNA. Mutations in several *Drosophila* genes, including *sex-lethal* and *transformer*, are known to cause differences in alternative splicing pathways. These genes may encode proteins regulating the reactions. There is some evidence that the protein encoded by the *sex-lethal* gene, for example, regulates alternative splicing by binding and blocking one of two alternate splicing sites.

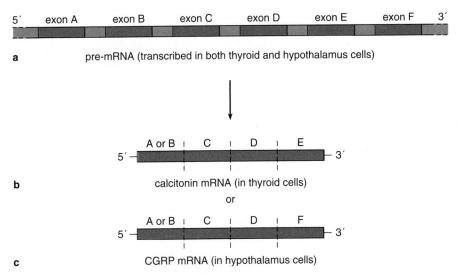

a

pre-mRNA (transcribed in both thyroid and hypothalamus cells)

b

calcitonin mRNA (in thyroid cells)

or

c

CGRP mRNA (in hypothalamus cells)

Figure 15-22 Alternative splicing of the same pre-mRNA to produce two different mRNAs: calcitonin mRNA in cells of the thyroid gland and CGRP mRNA in cells of the hypothalamus. **(a)** The pre-mRNA, which contains six potential exons (A to F), separated by five introns. **(b)** In the thyroid gland exons A or B, C, D, and E are linked together, and other introns and exons removed, producing an mRNA that codes for the hormone calcitonin. **(c)** In the hypothalamus, exons A or B, C, D, and F are linked together, and other introns and exons removed, to produce an mRNA coding for the calcitonin gene related polypeptide (CGRP), which is suspected to be involved in the function of taste receptors. Either exon A or exon B, which occur in the 5′ untranslated segment, may be present in either protein. Segment E is an exon for calcitonin mRNA and an intron for CGRP mRNA, while segment F is an intron for calcitonin and an exon for CGRP.

Other Processing Reactions Many pre-mRNAs are methylated at a few internal sites, usually at adenines in the 3′ untranslated region, as part of the processing reactions. At least some methylated adenines are retained in the finished mRNA. The functional significance of methylated sites within finished mRNAs is unknown.

A few pre-mRNAs undergo an unusual and enigmatic processing reaction called *RNA editing*, in which sequences within the coding portion are altered after transcription so that they do not match the DNA from which they were copied. Almost all known examples of RNA editing, which were discovered through comparisons of gene sequences with those of mature mRNAs, involve the addition of one or more Us at specific sites in the coding sequence, deletion of Us, or chemical conversion of C to U. The evolutionary origins and functional significance of RNA editing remain uncertain.

The most extreme cases of RNA editing have been observed in mitochondrial mRNAs, especially in trypanosome protozoans. In the mitochondrial mRNA copied from a gene encoding subunit II of cytochrome oxidase in trypanosomes, for example, four U-containing bases are added at closely spaced but not adjacent sites during RNA editing. In the mRNA copied from the mitochondrial gene encoding subunit III of cytochrome oxidase, the final mRNA becomes twice the length of the DNA code by the addition of several

hundred Us at sites throughout the molecule. Several original Us are also deleted during editing of this mRNA. Recently, R. Mahendran, M. R. Spottswood, and D. L. Miller discovered an exceptional example of RNA editing in which 54 Cs, rather than Us, are inserted into a mitochondrial mRNA in the slime mold *Physarum*.

One example of RNA editing has been observed by S.-H. Chen, L. M. Powell, and their coworkers in an mRNA encoded in the cell nucleus in mammals rather than in mitochondria. This occurs in the mRNA encoding apolipoprotein-B, in which a C is converted to U in intestinal but not in liver cells. The change from C to U converts a codon specifying an amino acid into a terminator codon. The termination produces a shortened version of the protein in intestinal cells.

In trypanosomes RNA editing may possibly follow "guide" RNAs discovered by B. Blum, N. Bakalara, and L. Simpson. The guides, which are precisely complementary to the final, fully edited version of the mRNAs, are transcribed from sites that are distinct from those encoding the mRNAs. After initial transcription the mRNA supposedly pairs with a guide RNA; Us are then entered or deleted by enzymatic mechanisms to make the mRNA complementary to the guide RNA. In the mammalian example, surrounding sequences or secondary structures in the mRNA may indicate the C to be converted to U. (See Ch. 21 for more information on RNA editing in mitochondria.)

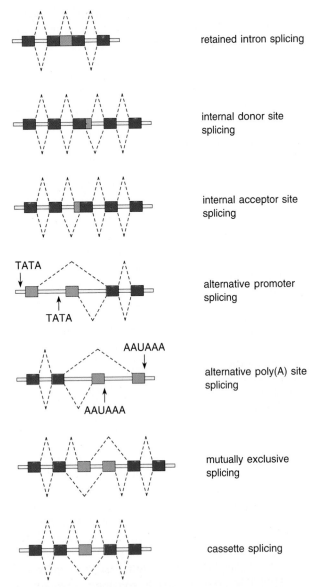

retained intron splicing

internal donor site splicing

internal acceptor site splicing

TATA

alternative promoter splicing

TATA

AAUAAA

alternative poly(A) site splicing

AAUAAA

mutually exclusive splicing

cassette splicing

Figure 15-23 Various pathways of alternate splicing. Exons are shown as boxes. The dashed lines connect exons that are incorporated into the finished mRNA. Note that some exons are spliced alternatively from internal sites so that part or all of the exon may be included in a finished mRNA. The internal sites contain the consensus sequences for 5′ or 3′ clipping. Courtesy of B. Nadal-Ginard. Reproduced, with permission, from *Ann. Rev. Genet.* 23:527 © 1989 by Annual Reviews, Inc.

TRANSCRIPTION AND PROCESSING OF RIBOSOMAL RNA

Ribosomal RNAs are transcribed in greater quantities than any other RNA type. This intensive rRNA synthesis supports the assembly of ribosomes, which are required literally in the millions per cell for protein synthesis to proceed at adequate levels.

The rRNAs

Ribosomal RNAs make up roughly half of the total mass of ribosomes in both prokaryotes and eukaryotes. Individual ribosomes are assembled from two subunits, one large and one small (Fig. 15-24). Each subunit consists of one or more rRNA molecules in combination with 30 to 40 different proteins, for a total of 70 to 80 proteins per eukaryotic ribosome (for details of ribosome structure, see p. 665).

Four distinct kinds of rRNA can be extracted from eukaryotic ribosomes, one from the small subunit and three from the large subunit. These four rRNAs are usually identified in *Svedberg*, or *S*, units (see p. 124), which reflect the relative rates at which molecules descend in a centrifuge under standard conditions. The faster the molecules descend, the larger the S number and, generally, the higher the molecular weight. According to this nomenclature, the single rRNA of the small ribosomal subunit is identified as *18S* rRNA, with a molecular weight of approximately 700,000. The three rRNAs of the large subunit are identified as *28S*, *5.8S*, and *5S* rRNA, with molecular weights of approximately 1,400,000, 50,000, and 41,000, respectively (Table 15-2). These S values are averages; actual values vary depending on the species and methods used to prepare the rRNAs for centrifugation.

In eukaryotes 28S, 18S, and 5.8S rRNA are transcribed by RNA polymerase I in the form of a large pre-rRNA. This large pre-rRNA ranges in S value over limits from about 37S to 45S depending on the species. In yeast and *Drosophila* the large pre-rRNA has a value of 37S, in *Xenopus* 40S, and in mammals 45S. Each large pre-rRNA contains an 18S, 28S, and 5.8S sequence, separated from one another by transcribed spacers. Processing reactions release the 18S, 28S, and 5.8S rRNAs as individual molecules. Transcription and processing of large pre-rRNAs take place in the nucleolus.

The remaining 5S rRNA type, located on genes outside the nucleolus, is separately transcribed as a pre-5S rRNA by RNA polymerase III. The pre-5S rRNA, slightly larger than a finished 5S molecule, subsequently enters the nucleolus for processing and assembly with 28S and 5.8S rRNA into large ribosomal subunits. 18S rRNA combines with a distinct group of ribosomal proteins to form small ribosomal subunits. In this chapter the precursor containing the 18S, 28S, and 5.8S rRNAs is referred to as "large pre-rRNA," the one containing 5S rRNA as "pre-5S rRNA."

rRNA Structure Each mature rRNA type has been completely sequenced in many eukaryotic species. The sequences have revealed many interesting and significant structural features of these molecules. The sequence of a given rRNA type, such as 18S rRNA, is very similar in closely related species. As more distantly related species are compared, the divergence in sequence becomes greater. Among distantly related or-

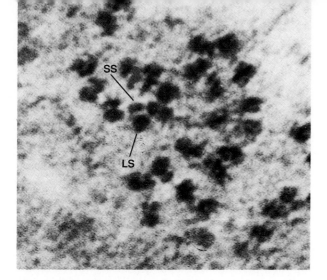

Figure 15-24 A group of ribosomes in a thin-sectioned rat liver cell. A cleft marking the division between the large and small subunits is visible in many of the ribosomes. SS, small ribosomal subunit; LS, large subunit. Courtesy of N. T. Florendo.

ganisms, relatively few sequence similarities can be detected. The degree to which the primary structure of a given rRNA type is similar between two different species, as a result, reflects the evolutionary history and "relatedness" of the organisms. The dependence of rRNA sequences on evolutionary relationships is sufficient, in fact, to allow the construction of evolutionary trees based on rRNA sequences alone. The lineages constructed in this way correspond well with evolutionary trees based on other lines of evidence such as the fossil record. In some cases the rRNA-based trees have helped to resolve the evolutionary relationships and origins of species and groups that were previously in doubt.

The sequences of the various rRNAs reveal that each type contains inverted sequences, which can form a wide assortment of hairpins and other secondary structures (see p. 575 and Fig. 15-25). The paired regions typically contain nonstandard base pairs in addition to the usual A-U and G-C pairs.

The inverted sequences of the largest rRNA molecules are so extensive that the molecules could fold into many possible patterns. For example, 18S rRNA contains enough inverted sequences to fold into more than 10,000 possible secondary structures. Choosing which of the possible structures is correct has been approached by a battery of techniques, including computer analysis, calculations of lowest energy states, chemical probes, enzymatic digestion, and comparisons among different organisms. (Information Box 15-2 explains how these methods are applied.)

Application of these methods has revealed that each rRNA can fold into a secondary structure that with certain variations is common to all organisms and can be extended even to the equivalent rRNA types of prokaryotes, mitochondria, and chloroplasts. This is true in spite of the fact that the nucleotide sequences of a given rRNA type in distantly related organisms may be quite different. The studies thus demonstrate that, with certain exceptions, it is the secondary structure, not the primary structure, of rRNA molecules that has been highly conserved in evolution. This conclusion indicates that in most regions of rRNA molecules secondary structure, rather than nucleotide sequence, is the most significant feature in the assembly and function of ribosomes.

There are important exceptions in which nucleotide sequence as well as secondary structure is conserved in rRNAs. In bacteria, for example, a sequence in the small-subunit rRNA pairs with a complementary sequence in the 5' untranslated region of bacterial mRNAs during initiation of protein synthesis. The rRNA sequence and its complement in mRNAs, known as the *Shine–Dalgarno sequence* (see p. 689), are highly conserved in bacteria.

Variations in the secondary structures of individual rRNA types among different species are limited primarily to certain regions of the molecules. Insertions and deletions of sequences in these regions, which are evidently less critical to function, also account for most of the differences in size between rRNAs of different species.

Table 15-2	Ribosomes, Ribosomal Subunits, and rRNAs in Prokaryotes and Eukaryotes (with approximate molecular weights in parentheses)			
Group	Intact Ribosomes	Ribosomal Subunits	rRNAs	Number of Nucleotides
Prokaryotes	70S (2,520,000)	30S (930,000)	16S (550,000)	~1480–1540
		50S (1,590,000)	23S (1,100,000)	~2900–2930
			5S (41,000)	116–120
Eukaryotes	80S (4,420,000)	40S (1,400,000)	18S (700,000)	~1790–1900
		60S (2,820,000)	28S (1,400,000)	~3400–4700
			5.8S (50,000)	158–163
			5S (41,000)	116–121

Determining Secondary Structures in RNAs

Four primary methods—chemical probes, enzymatic digestion, estimates of minimum energy states, and comparisons of RNA sequences in different organisms—have been used to work out the arrangement of hairpins, loops, and other secondary structures in RNA molecules. Chemical probes are used as "tags" to combine with and mark specific segments of RNA molecules in solution. Some of the probes, such as *kethoxal*, attach only to nucleotides in single-chain regions; others attach and mark only double-helical regions. The probes are often used in conjunction with crosslinkers, which tie together helical segments and preserve their arrangement. By noting the segments attached by chemical probes after stabilization of RNAs by crosslinkers, the probable positions of many single-chain and double-helical segments can be worked out.

Enzymatic digestion similarly allows detection of single-chain and double-helical segments in RNA molecules suspended in solution. Some enzymes, such as S1 and T2 nucleases, digest only single-chain segments and leave double-helical segments intact. Other enzymes, such as a nuclease from cobra venom, specifically attack double-helical segments. Sequencing the fragments remaining after either type of digestion often allows the positions of single-chain and double-helical regions to be identified.

Minimum energy states are predicted by computer programs that calculate the free energy associated with all possible secondary structures. From these possibilities the structure is chosen that allows the lowest energy state. Sequence comparisons based on computers have provided the most powerful method for determining RNA secondary structure. The basis for the approach is to compare the RNA sequences of different species and to determine which of the many potential folding patterns can actually be shared by RNAs from different species. As the relationships between species used for comparisons become more distant and sequences become more divergent, the number of possible folding patterns that all RNAs of a given type may share becomes restricted to a very few. When taken in combination with the chemical and enzymatic techniques and the minimum energy estimates, the comparative approach allows each RNA type to be assigned a most probable overall folding pattern as a "consensus" secondary structure.

For example, the bacterial small-subunit rRNA, *16S rRNA*, can fold into a pattern of hairpins and other paired structures equivalent to that of eukaryotic 18S rRNA (Fig. 15-25). The patterns are equivalent even though bacterial 16S rRNA is significantly smaller and almost completely different in nucleotide sequence from eukaryotic 18S rRNA. Most of the variations in size and secondary structure between prokaryotic 16S and eukaryotic 18S rRNAs are confined to six variable regions. The small-subunit rRNA of mammalian mitochondria can also fold into a secondary structure equivalent to those shown in Figure 15-25. This organelle rRNA, which contains only about half as many nucleotides as prokaryotic 16S rRNA, shows extensive deletion of nucleotides in the same six variable regions. (Prokaryotic rRNAs are discussed further in Supplement 15-1; the rRNAs of mitochondria and chloroplasts are described in Ch. 21.)

All rRNAs undoubtedly fold further into tertiary, three-dimensional structures as they are processed and assembled into ribosomal subunits. Investigations of these tertiary rRNA structures are just beginning to yield results. (Some details of this research are presented in the description of ribosome structure in Ch. 16.)

The functional significance of secondary and tertiary structures in rRNA remains largely unknown.

However, the folding patterns probably serve as molecular "signposts" for rRNA processing, recognition of rRNA segments by proteins during assembly of the ribosomal subunits, and maintenance of ribosome structure. The folding patterns probably also provide recognition sites for the associations formed among ribosomes, mRNAs, tRNAs, and the various factors linked to ribosomes during protein synthesis. Some secondary structures are likely to change between two or more forms during interactions between the rRNA types and during interaction with proteins. Such changes may underlie some of the individual reactions of polypeptide assembly (for details, see Ch. 16).

Chemical modifications are limited to 5% or less of individual nucleotides in rRNAs. The modifications primarily involve addition of methyl groups to any of the four bases or to the ribose sugar units, or alteration of uridine to pseudouridine (see Fig. 15-4). The modified nucleotides are restricted primarily to unpaired loops at the tips of hairpins and to the boundaries between double-helical and single-chain segments.

The base or sugar modifications in rRNAs may protect segments of the molecules against nuclease activity, particularly in unpaired regions, or may serve as processing signals indicating sites for clipping, splicing, or attachment of proteins. The modified bases

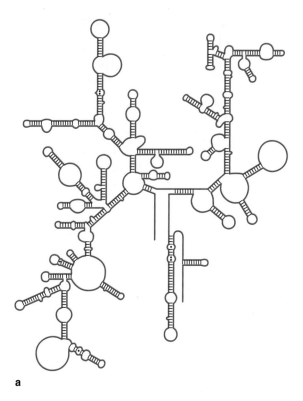

a

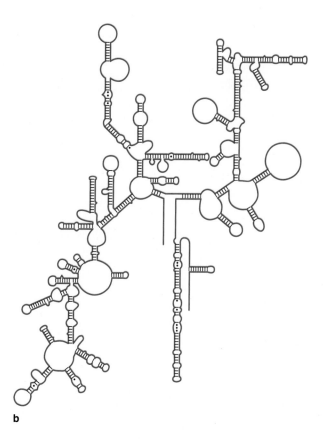

b

Figure 15-25 Comparison of the secondary structures of an *E. coli* 16S rRNA **(a)** and a yeast 18S rRNA **(b)**. Courtesy of H. F. Noller. Reproduced, with permission, from *Ann. Rev. Biochem.* 53:119, © 1984 by Annual Reviews, Inc.

might also underlie some of the recognition reactions and conformational changes that take place during protein synthesis.

Structures of the Individual Eukaryotic rRNA Types
The 18S rRNA of small ribosomal subunits folds into a structure containing 60 to 70 paired elements (see Fig. 15-25). On the order of 40 ribose sugars and about the same number of bases are methylated, amounting in total to about 2% modified nucleotides.

The major rRNA of the large ribosomal subunit, 28S rRNA, folds into a complex secondary structure containing more than a hundred paired structures. The 28S rRNAs of all eukaryotes, and the equivalent rRNA type of bacterial and organelle ribosomes, fold into essentially the same six-domain structure. (Fig. 15-26 shows the 23S rRNA of *E. coli*, which is similar in secondary structure to a eukaryotic 28S rRNA.) Variations in length are restricted primarily to insertions or deletions in seven variable regions. Chemical modifications of either the base or sugar unit occur at scattered locations in the 28S structure.

The 5.8S rRNA of large ribosomal subunits folds into a secondary structure with five double helices (Fig. 15-27). Much of the 5.8S rRNA sequence is complementary to the 5′ end of eukaryotic 28S rRNA. This allows a paired structure to form between the two rRNAs in this region (Fig. 15-28*a*). The 5′ end of prokaryotic *23S rRNA*, which is equivalent to eukaryotic 28S rRNA, can fold into essentially the same paired structure (Fig. 15-28*b*). Evidently, during the evolution of eukaryotes from their prokaryotic ancestors, a split occurred at the 5′ end of the prokaryotic 23S rRNA, liberating a separate molecule that has persisted as the 5.8S rRNA of eukaryotes.

Other splits in 28S rRNAs have apparently occurred during the evolution of some eukaryotic groups. Some insects have a 2S rRNA corresponding to the terminal 25 nucleotides of 5.8S rRNA. In *Crithidia*, a trypanosome protozoan, 28S rRNA is broken into multiple fragments that, nevertheless, assemble into functional, large ribosomal subunits. The evolutionary persistence of these fragments indicates that, at least at some points, the functions of the major rRNA type of the large ribosomal subunit are not greatly disturbed by cutting the molecule into separate pieces.

The smallest rRNA of the large ribosomal subunit, 5S rRNA, folds into a secondary structure containing four double-helical segments (Fig. 15-29). 5S rRNA is the most highly conserved of the various rRNA types. Of the approximately 120 nucleotides in this rRNA type, 21 are invariant or nearly so in all species, both eukaryotic and prokaryotic. Among eukaryotes only one example of a modified 5S rRNA base has been detected, a pseudouridine that occurs in a yeast 5S rRNA. Because eukaryotic 5S rRNAs are transcribed from genes with internal promoters (see below), a part

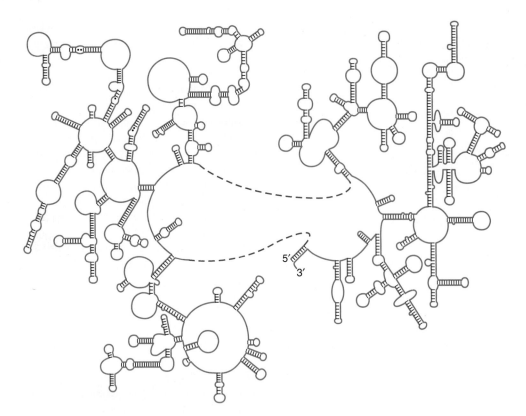

Figure 15-26 A secondary structure model proposed for the 23S rRNA molecule of *E. coli*. The dashed lines connect parts of the molecule that are continuous. Courtesy of H. F. Noller. Reproduced, with permission, from *Ann. Rev. Biochem.* 53:119, © 1984 by Annual Reviews, Inc.

Figure 15-27 A secondary structure model proposed for 5.8S rRNA. Asterisks mark the positions of modified bases. Courtesy of R. N. Nazar, from *Canad. J. Biochem.* 62:311 (1984).

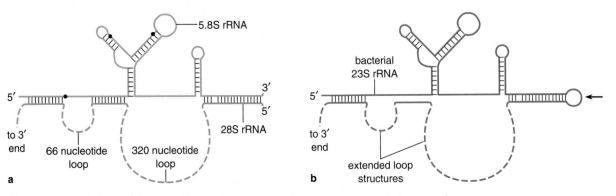

Figure 15-28 The proposed relationship of eukaryotic 5.8S rRNA to the 5′ end of prokaryotic 23S rRNA. **(a)** Pairing between 5.8S rRNA and the 5′ end of 28S rRNA in eukaryotes. **(b)** The equivalent paired structure formed by fold-back pairing at the 5′ end of a prokaryotic 23S rRNA molecule. A single break at the arrow would convert this paired structure to the eukaryotic form, with a separate 5.8S rRNA molecule rather than the continuous structure shown. The dots mark the positions of modified bases.

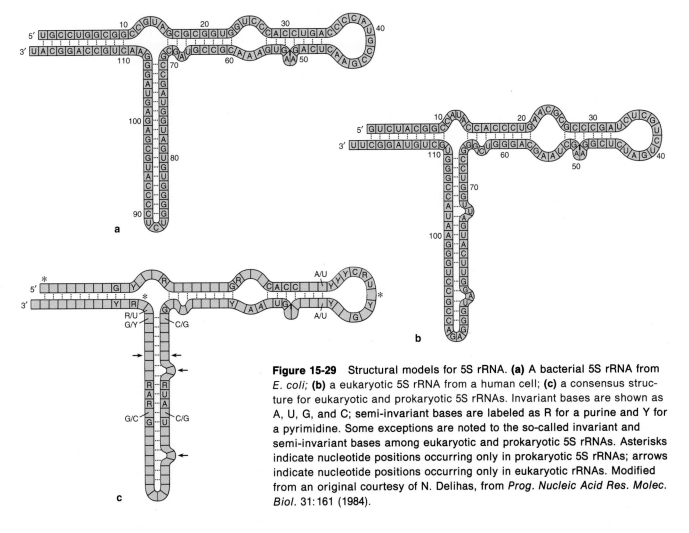

Figure 15-29 Structural models for 5S rRNA. **(a)** A bacterial 5S rRNA from *E. coli;* **(b)** a eukaryotic 5S rRNA from a human cell; **(c)** a consensus structure for eukaryotic and prokaryotic 5S rRNAs. Invariant bases are shown as A, U, G, and C; semi-invariant bases are labeled as R for a purine and Y for a pyrimidine. Some exceptions are noted to the so-called invariant and semi-invariant bases among eukaryotic and prokaryotic 5S rRNAs. Asterisks indicate nucleotide positions occurring only in prokaryotic 5S rRNAs; arrows indicate nucleotide positions occurring only in eukaryotic rRNAs. Modified from an original courtesy of N. Delihas, from *Prog. Nucleic Acid Res. Molec. Biol.* 31:161 (1984).

of the sequence, lying from about positions 55 to 83 in the finished molecule, does double duty as a promoter in the DNA sequence and part of the 5S molecule in the corresponding RNA sequence. Mutations in this region affect promoter recognition during transcription as well as 5S rRNA function in ribosomes. This may explain why this segment of the 5S rRNA sequence is especially highly conserved among different eukaryotes.

Despite its highly conserved nature, 5S rRNA variants, with small differences in sequence, occur within individual higher vertebrate species. In the amphibian *Xenopus*, in which the variants have been best studied, differing in about 5% of their sequences has been detected. Two of the variants make up the so-called *oocyte-type* 5S rRNA, transcribed only in oocytes and early embryos. The remaining 5S rRNA, the *somatic-type,* is transcribed in all *Xenopus* cells. Multiple 5S variants have been detected in at least one other vertebrate species, the chicken. The functional significance of the 5S rRNA variants is unknown.

More 5S molecules have been sequenced in different species than any other RNA type. Information from these sequencing studies has provided the best

and most complete of the RNA-based evolutionary trees.

rRNA Gene Structure, Transcription, and Processing

The genes encoding large pre-rRNAs occur in multiple copies in all eukaryotes, concentrated in clusters at one or more locations in the chromosomes of each species. This arrangement is in distinct contrast to that of mRNA genes, which occur largely in single copies scattered throughout the chromosomes. Each repeat in the large pre-rRNA gene clusters consists of a coding segment and a long *intergenic spacer* that contains sequences signaling the initiation and termination of transcription (Fig. 15-30a).

Large Pre-rRNA Genes A large pre-rRNA gene (Fig. 15-30b) contains one copy of the 28S, 18S, and 5.8S coding sequence. The sequences are arranged in the order 18S—5.8S—28S, with the 18S sequence closest to the 5' end of the gene. The three coding sequences are separated by *intragenic spacers* that are transcribed but subsequently eliminated during rRNA processing.

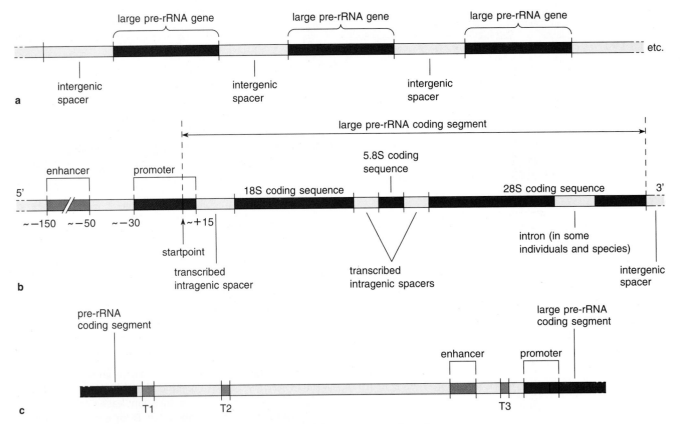

Figure 15-30 Large pre-rRNA genes and their spacers. **(a)** The pattern of tandem repetition of large pre-rRNA genes and nontranscribed intergenic spacers in large pre-rRNA gene clusters. Tens, hundreds, or thousands of repeats may occur in the clusters in different species. **(b)** The arrangement of coding sequences and transcribed intragenic spacers in large pre-rRNA genes. **(c)** The intergenic spacer of a large pre-rRNA gene.

A transcribed spacer lies between the first nucleotide copied during transcription and the beginning of the 18S coding sequence. Two other short intragenic spacers separate the central 5.8S coding sequence from the 18S and 28S coding sequences. In most species transcription apparently continues for some distance past the 3' end of the 28S coding sequence. However, the transcript of the region beyond the 3' end is degraded so rapidly that it cannot be detected unless the processing reactions are experimentally inhibited.

Experiments employing substitution-deletion analysis revealed that the promoters of most large pre-rRNA genes lie within the region bound by about 35 nucleotides upstream to about 15 nucleotides downstream of the startpoint, or from about −35 to +15. An additional sequence that acts as an enhancer lies farther upstream in many species, in most cases in the region between about −150 and −50. Wide variations occur between different organisms in the importance of individual sequences and sites in the promoter and enhancer regions.

In general no consensus sequences analogous to the TATA and CAAT boxes of mRNA genes are noted in large pre-rRNA promoters. Although closely related species may show sequence homologies in large pre-rRNA promoters or enhancers, few or no similarities are seen in these regions among different major groups such as fungi, insects, amphibians, and mammals. In some groups, as in different members of the genus *Drosophila*, large pre-rRNA promoters differ extensively or completely even among closely related species. The recognition of large pre-rRNA promoters therefore probably depends on similarities in secondary structure rather than on particular nucleotide sequences. Species-specific control proteins that recognize sequence elements unique to species or taxonomic groups may also be important. These conclusions are supported by the conditions required for initiation of transcription in cell-free systems. For example, among cell-free systems developed from mammalian species, large pre-rRNA genes isolated from the mouse can be successfully transcribed by RNA polymerase only in mouse cell extracts; attempts to transcribe mouse pre-rRNA genes with human cell extracts are unsuccessful. This is true even though some degree of sequence similarity in the promoter region of large pre-mRNA genes is noted among different mammals.

The intergenic spacer separating large pre-rRNA

genes (Fig. 15-30c) contains one or more termination signals for the gene repeat preceding it. In many species with multiple terminators, the final terminator of the series is located near the promoter for the following pre-rRNA gene. Although multiple terminators are distributed in this way in the intergenic spacer in many species, transcription in most cases appears to end at the terminators lying farthest upstream. As a consequence, most of the intergenic spacer usually remains untranscribed. The spaces between the terminators in many species contain complex combinations of short repeated sequences. As in the promoter and enhancer regions, few or no sequence homologies occur in large pre-rRNA terminators between different species and groups except for the presence of inverted sequences near three or more A-T base pairs. The inverted sequences and A-T base pairs may act in termination in a manner similar to one class of bacterial terminators (see pp. 642 and below).

The arrangement of coding segments and transcribed spacers within large pre-rRNA genes has been visualized directly by combining RNA-DNA hybridization with electron microscopy. An experiment by M. Pellegrini, J. Manning, and N. Davidson located these segments in *Drosophila* large pre-rRNA genes with unusual clarity. In their experiment, samples of 18S, 5.8S, and 28S rRNA purified from *Drosophila* were added to unwound large pre-rRNA genes from the same species under conditions that promoted hybridization. The RNA-DNA hybrids were prepared for electron microscopy by reacting them with a viral protein, T4 gene 32 protein, that binds single-stranded DNA. The protein reacts strongly with any segments left unhybridized in the preparations and makes them thick enough to be readily visible under the electron microscope.

The results of the technique (Fig. 15-31) directly show the arrangement of coding sequences in the order 18S → 5.8S → 28S, and the intragenic spacers separating them. A long intergenic spacer containing from 6000 to 10,000 nucleotides separates one large pre-rRNA coding sequence from the next in *Drosophila*. The preparations revealed an intron-like intervening sequence within the segment coding for 28S rRNA in many of the large pre-rRNA genes in *Drosophila*. Similar unspliceable intervening sequences occur in the 28S coding segment of other flies (see below).

With few exceptions the same arrangement of cod-

Figure 15-31 Large pre-rRNA gene structure as revealed by RNA-DNA hybridization. **(a)** A *Drosophila* large pre-rRNA gene hybridized with 18S, 5.8S, and 28S rRNA. **(b)** Tracing of the rRNA-rDNA hybridization pattern. In the micrograph unhybridized DNA appears thicker because it has complexed with T4 gene 32 protein. (The unpaired regions are shown as solid black lines in the tracing.) One of the unpaired segments appears in the middle of the 28S sequence (arrow), indicating that the 28S gene is interrupted by an intron. Spacers occur between the coding segments for the three rRNA species. The bar shows the length occupied by 1000 base pairs. Courtesy of M. Pellegrini, from *Cell* 10:213 (1977). Copyright Massachusetts Institute of Technology.

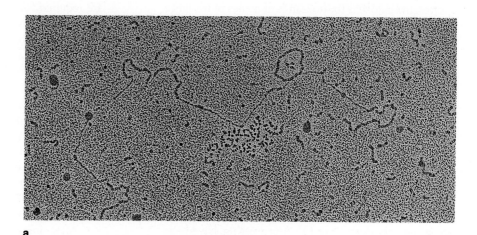

a

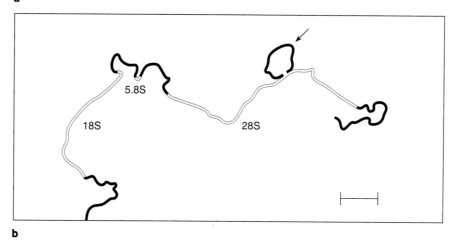

b

ing sequences and spacers (but without the intron) occurs in the large pre-rRNA genes of all eukaryotes. The coding sequences show a greater or lesser degree of sequence conservation between different species depending on the evolutionary relatedness of the organisms. Few or no sequence homologies, however, occur in the spacer regions within the gene, even between closely related species. In some organisms, such as the amphibian *Xenopus*, sequence variations in the intragenic spacers are noted even among individuals of the same species.

Wide variations are also noted in the total length of the large pre-rRNA genes among different eukaryotes. Some of this difference arises in the variable regions of the 18S and 28S rRNAs, which, as noted, may contain extensive insertions or deletions of nucleotides. In addition, the length of intragenic spacers varies widely among different species and may even vary among individuals of the same species. In general the total length of large pre-rRNA genes increases as organisms climb the evolutionary tree. Among vertebrates, for example, large pre-rRNA genes include about 8000 nucleotides in *Xenopus*, 10,500 in birds, and 13,000 in mammals.

Repeats of the large pre-rRNA genes separated by intergenic spacers are concentrated in segments of the chromatin that form the nucleolus. Depending on the species, from dozens to thousands of copies of the large pre-rRNA genes may occur in the nucleolar chromatin segment. *Drosophila* has about 400 to 450 repeats of the DNA segment coding for large pre-rRNA; humans have about a thousand repeats of this gene. These figures represent a haploid chromosome set; a diploid body cell of an individual contains twice as many large pre-rRNA genes.

Introns in Large Pre-rRNA Genes   All known introns appearing in large pre-rRNA genes occur about two-thirds of the way along the 28S coding sequence, in a region corresponding to one of the variable segments of the 28S secondary structure. These introns, which are spliced out during pre-rRNA processing, have been discovered in a number of eukaryotic organisms, including nuclear genes in the protozoans *Tetrahymena* and *Chlamydomonas* and the fungus *Physarum*, and in rRNA and mRNA genes in mitochondria and chloroplasts of lower eukaryotes. Although no consensus sequences occur at the boundaries of these introns, all contain inverted sequences that can form similar patterns of secondary structures. The paired structures evidently arrange the introns into configurations that promote self-splicing without intervention of enzymatic proteins (see Fig. 15-34).

Another type of intervening sequence occurs in the rRNA genes of *Drosophila*, other dipteran flies, and possibly also the roundworm *Ascaris*. This intervening sequence, which is not a spliceable intron, renders the large pre-rRNA gene inoperative by interfering with either the transcription or processing reactions. About two-thirds of the large pre-rRNA genes in all *Drosophila melanogaster* individuals carry the incapacitating intervening sequence, which may include as many as 5000 base pairs. The intervening sequence, which occurs in locations that are similar, but not identical, to the positions taken by introns that can be spliced in large pre-rRNA genes, typically contains short repeated sequences at both ends, in a pattern similar to movable DNA segments called *transposable elements.* These mobile genetic elements, which occur commonly in both prokaryotes and eukaryotes, may show up as insertions at many points in the DNA and may, at times, move from place to place. In many cases transposable elements destroy the function of a gene when inserted into its coding segment. (Details of transposable elements and their origins are presented in Ch. 18.) Insertion of a transposable element in the 28S coding sequence during the evolutionary history of *Drosophila* may have been the origin of the intervening sequence.

Large Pre-rRNA Transcription and Processing Transcription of large pre-rRNA genes begins as the RNA polymerase I enzyme binds to the 5'-flanking promoter sequences. One or more transcription factors are required for recognition of a large pre-rRNA gene promoter by the polymerase I enzyme. These factors, like their counterparts in mRNA transcription, bind to the promoters and set up a DNA-protein complex that is recognized and bound by the enzyme. Some of these factors are common to several species or larger taxonomic groups, and some are species-specific. A factor named *UBF*, for example, seems to be common to vertebrate animals; it appears to bind to both enhancers and the promoter of vertebrate large pre-rRNA genes. In humans a second protein, *SL1*, acts as a species-specific factor increasing the rate of transcription initiation. When UBF is present, SL1 binds to the promoters and initiation takes place.

After initiation, transcription continues entirely through a large pre-rRNA gene, copying without interruption the initial spacer following the startpoint, the 18S, 5.8S, and 28S coding sequences, and the spacers separating them. After transcribing the coding sequences, transcription terminates at one of the termination signals in the intergenic spacer. Protein factors recognizing and binding the termination sequences in the intergenic spacer may promote release of RNA polymerase I and the large pre-rRNA transcript. Several proteins considered likely to have this activity have been identified, including *TTFI* in the mouse and *Rib2* in the frog.

Processing begins while transcription of the large pre-rRNA is still in progress. Chemical modifications,

Figure 15-32 The probable sequence of events in processing the large pre-rRNA in eukaryotes. The small vertical arrows mark sites of enzymatic cleavage. In most organisms initial enzymatic cuts occur near or at the 5' and 3' ends of the 18S sequence, liberating this molecule as an almost completely processed entity. These cuts leave a segment containing the 5.8S and 28S sequences, which are released by further enzymatic cuts.

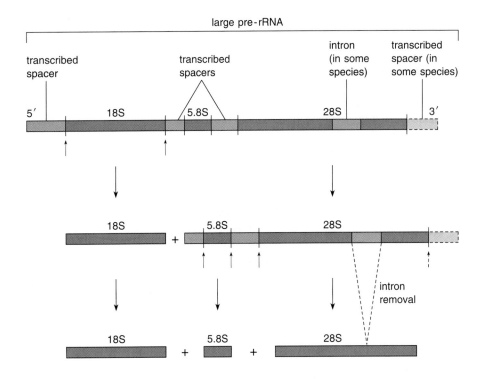

primarily methylations of either ribose sugar residues or the four RNA bases, proceed very rapidly. Methylation of ribose units forms the modified sugar *2-0-methylribose* (see Fig. 15-4; the same sugar methylation occurs in the 5' cap of mRNAs). Most methylations and other modifications occur in segments of the large pre-rRNA transcript that are preserved in the finished 18S, 5.8S, or 28S rRNA molecules.

Transcription is followed almost immediately by cutting reactions that separate the 18S, 5.8S, and 28S rRNA molecules from the large pre-rRNA. Although the cutting sites vary somewhat in different species, in most organisms initial cleavages by RNA endonucleases occur near or at the 5' and 3' ends of the 18S sequence, liberating this molecule as an almost completely processed entity (Fig. 15-32). These cuts, in most species, leave a segment containing the 5.8S and 28S sequences, which are released by further cuts.

Although both RNA endonuclease and exonuclease activity can be detected in the nucleus, the specific enzymes cutting the 18S, 28S, and 5.8S molecules from the large pre-mRNA are still unknown. The sequence features recognized as signals by the enzymes to make the cuts are similarly unclear. No doubt some of the paired structures formed by inverted sequences within the large pre-rRNA and conformations set up by the combination of processing proteins with the pre-rRNA are recognition signals for the processing enzymes.

While the clipping reactions are in progress, ribosomal proteins are added to the maturing rRNA molecules. The proteins added to the maturing rRNA molecules include more than 70 to 80 different types in eukaryotes (see p. 668 for details).

The cleavage reactions releasing 18S, 5.8S, and 28S rRNAs from the large pre-rRNA transcript were studied by a series of now-classic experiments by R. P. Perry and others. The experiments involved exposure of cells to tritiated uridine, a radioactive precursor used by the cells to make RNA, and actinomycin D, a drug that halts RNA transcription. After a brief exposure of living cells to tritiated uridine, during which RNAs undergoing transcription were labeled, further transcription was blocked by adding actinomycin D. After the block, cells could process the rRNAs transcribed during exposure to the label but could not make any more rRNA copies. If cell extracts were made a few minutes after exposure to the label and actinomycin D block, the large pre-rRNA, which had a value of 40S to 45S in the species used for the studies, was labeled. If cells were allowed to process rRNAs for about 10 to 15 minutes before extracts were made, the label was distributed between 18S rRNA and a segment centrifuging with a value of 32S. After longer periods between exposure to label and preparation of cell extracts, the labeled 32S molecules disappeared, and labeled 5.8S and 28S molecules could be found in the extracts. If the period between exposure, blockage, and preparation of cell extracts was extended to several hours, radioactivity could be detected in cytoplasmic ribosomes. Besides revealing the sequence of cleavages producing the final 18S, 5.8S, and 28S rRNA molecules, the labeling experiments were the first to demonstrate that the large pre-rRNA

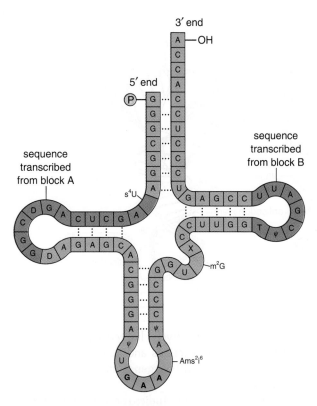

Figure 15-43 Segments of a mature tRNA molecule corresponding to blocks A and B of the tRNA gene promotor (in dark red).

of tRNA genes do double duty as sequences promoting initiation and as codes for parts of mature tRNA molecules (Fig. 15-43).

Sequences in the promoter blocks are highly conserved among eukaryotes. Sequence conservation in these regions could reflect either requirements of transcription initiation or the structure and function of finished tRNA molecules. Of the two possibilities promoter function is probably more significant because the block A consensus sequence of tRNA genes, RRYN-NARYGG, is shared with the block A of 5S rRNA genes.

Experiments by G. Ciliberto and his colleagues demonstrated the functional equivalence of the block A promoter regions of tRNA and 5S rRNA genes. These investigators constructed artificial genes by replacing the block A promoter of tRNA genes with the block A segment of a 5S rRNA promoter. The hybrid genes were fully functional in transcription when supplied with RNA polymerase III and the initiation factors necessary for tRNA gene transcription.

The 3' end of tRNA genes is marked by a series of four or more Ts that serves as a termination signal. No sequence at the 3' end of eukaryotic tRNA genes corresponds to the terminal CCA sequence found universally at the 3' end of mature tRNA molecules. This short sequence is added during processing (see below).

Some eukaryotic tRNA genes contain an intron in their coding sequences. In all cases where introns have been detected in eukaryotic tRNA genes, which include yeast, mammals, and higher plants, the intron always occurs in exactly the same position—immediately following the first nucleotide on the 3' side of the sequence corresponding to the anticodon in mature tRNAs (Fig. 15-44).

No consensus sequences are noted in the introns of different tRNA gene families. Instead, the intron and the exons of pre-tRNAs form secondary structures that provide the recognition signal for the reactions removing the intron during processing (see below). In most tRNAs the intron contains a sequence that is complementary to the anticodon. Pairing between the anticodon and its complement in the intron probably contributes to the secondary structures triggering intron removal.

The total number of tRNA genes varies widely among different eukaryotic species. Yeast cells have about 400 tRNA genes per haploid genome; humans have about 1400 per haploid genome. The champions in this regard, once again, are the amphibians: species such as *Xenopus* have some 8000 tRNA genes per haploid genome. In organisms such as yeast, with smaller totals, each tRNA type is encoded in about 18 to 20 copies; organisms at the other extreme, such as *Xenopus*, average nearly 400 copies of each tRNA gene.

tRNA Transcription and Processing

Transcription of tRNA genes requires the participation of at least two factors. These two factors, apparently identical to the TFIIIB and TFIIIC factors active in 5S gene transcription, combine with the split promoter of tRNA genes before the RNA polymerase III binds to the promoter sequences. TFIIIC binds first to the A and B blocks of the promoter, followed by TFIIIB. TFIIIB probably interacts both with TFIIIC and with DNA sequences lying in the 5'-flanking region of tRNA genes. Combination of the factors enables the RNA polymerase III enzyme to bind, and transcription begins at the startpoint. The transcription factors remain linked to the promoter as long as the gene is active, inducing successive rounds of enzyme binding and transcription.

It is uncertain whether the initiation of tRNA transcription requires any additional transcription factors specific only for tRNA genes. The TFIIIB and TFIIIC factors alone seem to be sufficient for initiation in the tRNA genes studied to date. Therefore, initiation of tRNA genes may not involve gene-specific factors equivalent to TFIIIA, which is required only for 5S genes.

The completed transcript of a tRNA gene is a pre-tRNA, longer than its finished tRNA counterpart at both its 5' and 3' ends. Pre-tRNAs in some species also

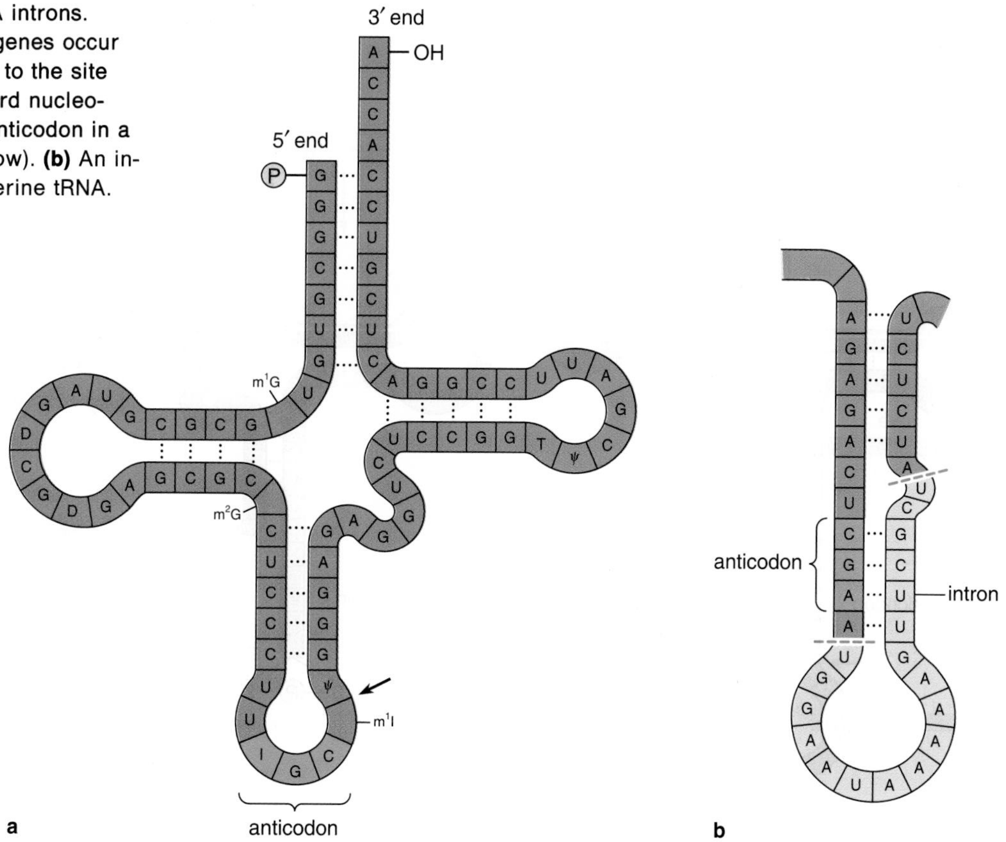

Figure 15-44 Transfer RNA introns. **(a)** The introns of all tRNA genes occur in a position corresponding to the site between the second and third nucleotides on the 3' side of the anticodon in a mature tRNA molecule (arrow). **(b)** An intron in the precursor of a serine tRNA.

include an intron that must be removed by splicing reactions.

The surplus segments are removed from the ends of pre-tRNAs by endonuclease enzymes. Little is known about the reactions processing the 3' ends of tRNAs. Cleavage of the surplus segment from the 5' end of pre-tRNAs is catalyzed in eukaryotes by *RNAse P*, an unusual enzyme that contains an snRNA, *M1 snRNA*, as part of its structure. Indications from studies of the reaction cleaving the 5' end of pre-tRNAs in bacteria, which is also catalyzed by RNAse P, are that the RNA rather than the protein of the endonuclease is active in catalysis (see Supplement 15-1). The 5' cleavage in tRNA processing is thus another reaction in which RNA molecules act as biological catalysts.

A recent experiment by H. van Tol, H. Gross, and H. Beier showed that under certain conditions 5' cleavage of pre-tRNAs can proceed in the test tube without the M1 snRNA. On this basis these investigators proposed that tRNAs are themselves capable of catalyzing removal of the 5' surplus segment. The role of the M1 snRNA, according to their proposal, may be in positioning the tRNA in a conformation that increases the efficiency and accuracy of the self-catalyzed reaction.

Following removal of surplus segments from the 5' and 3' ends of the maturing tRNA, a single processing enzyme, *tRNA nucleotidyl transferase*, adds the terminal CCA sequence in a series of three steps:

$$tRNA + CTP \longrightarrow tRNA\text{-}C + PP_i$$

$$tRNA\text{-}C + CTP \longrightarrow tRNA\text{-}C\text{-}C + PP_i$$

$$tRNA\text{-}C\text{-}C + ATP \longrightarrow tRNA\text{-}C\text{-}C\text{-}A + PP_i$$

Base modifications take place while the other processing steps are in progress. The modifications, carried out by a highly complex and varied group of enzymes active primarily in the nucleus, may include methylations of bases and sugar residues, acetylations, deaminations, addition of sulfhydryl and other groups, and rearrangements of the four original bases. Among the bases produced by these reactions is methylation of uracil to produce thymine, which occurs in tRNA molecules as well as in DNA. Few of the individual enzymes carrying out the modifying reactions have been identified in eukaryotes.

Introns are processed from pre-tRNAs in a two-step reaction that removes the intron and splices the free exon ends into a continuous molecule (Fig. 15-45). Intron removal is catalyzed by an RNA endonuclease that evidently relies for its signals primarily on the exons of the precursor rather than on sequences within the intron itself. Evidence for this conclusion was derived by M. I. Baldi and his coworkers from the effects of substitutions or deletions of nucleotides within the introns. Experimental changes in the intron sequence generally have no effect on intron removal, as long as

intron is removed from the U6 pre-snRNA transcript by a spliceosome containing the U1, U2, U5, and U4/U6 snRNAs, evidently by the same mechanism removing introns from pre-mRNAs.

U snRNA genes occurs in multiple copies in eukaryotes. For example, the U1 gene occurs in about 30 copies per haploid chromosome set in humans; *Drosophila* has fewer copies of the U1 gene, probably less than 10. *Xenopus*, as might be expected, has many U1 genes—about 500 repeats of two slightly different U1 genes that are active only in oocytes and early embryos and some 50 to 100 repeats of another U1 variant that is active later in embryonic cells and in adult *Xenopus*. Many U snRNA genes occur in tandemly repeated clusters.

scRNAs

Only one small cytoplasmic RNA type, the 7S scRNA occurring in the signal recognition particle (*SRP*), has been extensively characterized. Analysis of this particle by G. Blobel and his colleagues revealed that it contains the SRP 7S scRNA molecule in combination with six polypeptides. Sequencing shows that SRP 7S scRNA has curious relationships to the *Alu family*, a class of apparently nonfunctional sequences that occur in many repeats in humans and other mammals. (Humans have about 300,000 copies of the *Alu* sequence per haploid genome.) Part of the *Alu* sequence is the same as the first 100 and last 40 nucleotides of an SRP 7S scRNA. This suggests that *Alu* sequences may be SRP 7S scRNA *pseudogenes*, that is, faulty, nonfunctional copies of the SRP 7S scRNA gene that have duplicated and persist in the genome. (For further details of pseudogenes in general, and the *Alu* pseudogene class in particular, see Ch. 18.)

Genes coding for the SRP 7S scRNA, transcribed by RNA polymerase III, have an internal promoter. Sequences upstream of the startpoint of SRP 7S scRNA genes also influence transcription initiation, in a pattern similar to tRNA genes. Although many repeats of the *Alu* pseudogenes occur in higher eukaryotes, the number of functional repeats of the SRP 7S scRNA gene in most species is probably less than ten. The RNA product of these genes, once transcribed, combines with the six SRP proteins to form a mature SRP particle.

ALTERATIONS IN CHROMATIN STRUCTURE DURING TRANSCRIPTION

Chromatin undergoes extensive structural changes in regions that are active in transcription. Some alterations probably reflect an opening of chromatin structure required for DNA unwinding and accommodation of the RNA polymerase enzymes. Other alterations, in regions within or near promoters, probably result from interactions of the DNA with regulatory proteins.

Two primary methods have been employed to detect changes in chromatin structure as genes undergo transcription. One uses DNA endonucleases, the enzymes that gave some of the first clues leading to the discovery of nucleosomes (see p. 547 and Information Box 14-4). The second method employs direct examination and comparison of nucleosomes and higher-order structures in inactive and active genes viewed under the electron microscope.

Alterations in Regions Containing mRNA Genes

The rate at which chromatin is digested by endonucleases increases in characteristic patterns as mRNA genes become active in transcription. Genes that are about to be transcribed are typically digested two to five times more rapidly by DNA endonucleases than genes in chromatin that is to remain inactive. This increased sensitivity to endonucleases, which extends over the entire length of the gene including both its 5'- and 3'-flanking regions, probably reflects unwinding from the solenoid to a more extended chain of nucleosomes (see p. 546 for details of these structures).

As transcription begins, mRNA-encoding genes become even more sensitive to nuclease digestion, some five to ten times more than completely inactive chromatin. The increased sensitivity associated with transcription extends from about the startpoint through several hundred nucleotides into the 3'-flanking region. In addition to the increased sensitivity to endonucleases, some or all of the DNA lengths generated by endonuclease digestion of chromatin lose their regular periodicities (see p. 547) and become random. The increased sensitivity to digestion and loss of regular DNA lengths in the digest are considered to result from a change in nucleosome structure from the compact to a more open or unfolded state in the DNA being transcribed.

Scattered segments of the chromatin in regions of active mRNA genes become even more sensitive to endonuclease digestion. These regions, called *hypersensitive sites* (see also p. 551), may be more than a hundred times more sensitive to one particular endonuclease, DNAse I, than the chromatin of totally inactive genes. The hypersensitive sites, which may include lengths of DNA ranging from about 30 to as many as 300 to 400 base pairs, are concentrated in the 5'-flanking region of mRNA genes and, in some genes, scattered at other locations within the transcribed segment and 3'-flanking region. Enhancer sequences are among the sites made hypersensitive in some active genes. The hypersensitive sites are believed to be DNA lengths from which nucleosomes have been completely dislodged by regulatory proteins.

Hemoglobin genes, for example, occur in groups that include types encoding both embryonic and adult forms of the hemoglobin polypeptides. In chick embryos all the hemoglobin genes are digested by DNA endonucleases at rates characteristic of completely inactive genes up to about 20 hours after fertilization. At that time, just before transcription actually begins, embryonic hemoglobin genes and their flanking regions become more sensitive to endonuclease digestion. Shortly thereafter, transcription of these genes is initiated, and the DNA of their transcribed and 3'-flanking regions becomes still more sensitive to DNAse I digestion. At the same time, or just before transcription begins, a series of hypersensitive sites appears at various points, mostly within the 5'-flanking region of the embryonic hemoglobin genes.

Between 5 and 14 days after fertilization, transcription switches from embryonic to adult hemoglobins in chick embryos. The embryonic genes lose their increased sensitivity to endonuclease digestion as they become inactive, except for some of the hypersensitive sites in the 5'-flanking region. Transcription of the adult hemoglobin types, which reaches full production by about 14 days, is accompanied by the same series of changes in DNAse I sensitivity: (1) moderately increased sensitivity, extending over relatively long stretches within and on both sides of the genes before transcription begins, (2) the appearance of hypersensitive sites just before transcription begins, and (3) more highly increased sensitivity in the transcribed and 3'-flanking region as transcription is initiated.

The series of chromatin changes associated with hemoglobin transcription is not observed in cell types, such as those of the brain, in which hemoglobin is not produced. In brain cells endonuclease sensitivity of the hemoglobin genes remains at the levels characteristic of completely inactive chromatin in both embryonic and adult cells.

Not all genes develop hypersensitive sites before or during transcription. One gene of the hemoglobin family and the few plant genes analyzed so far, for example, show the generally increased sensitivity to DNAse I associated with activation and transcription but no hypersensitive sites. Although hypersensitive sites make their appearance in coordination with transcription in the genes in which they occur, the presence of hypersensitive regions does not obligate genes for transcription; many examples have been noted, as in embryonic hemoglobin genes, in which hypersensitive sites persist after transcription of their associated gene ceases. Thus, the appearance of hypersensitive sites is a prerequisite for transcription of some genes, but their presence alone does not necessarily obligate genes to become active. Hypersensitive sites have also been

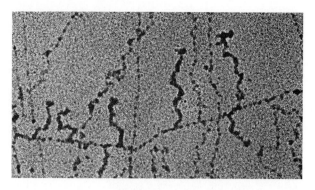

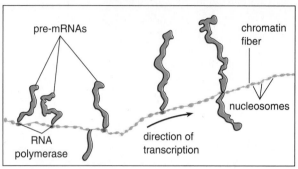

Figure 15-49 Transcribing chromatin isolated from the milkweed bug *Oncopeltus*. × 66,000. Micrograph courtesy of C. D. Laird and V. E. Foe, from *Cell* 9:131(1976). Copyright Massachusetts Institute of Technology.

detected at or near initiation points for DNA replication. (The role of hypersensitive sites in genetic regulation is discussed further in Ch. 17.)

Genes actively transcribing mRNA have been isolated and examined directly under the electron microscope. Figure 15-49 shows an active gene; in the figure the pre-mRNAs are visible as thick fibers branching off from the DNA chain. RNA polymerase molecules are frequently visible in such preparations as dense particles at the bases of the transcripts. Nucleosomes generally are not visible at the sites occupied by RNA polymerase, suggesting that nucleosomes may unfold or slide to expose the DNA. (The Experimental Process essay by M.-S. Lee and W. T. Garrard on p. 630 presents compelling evidence that nucleosomes unfold.)

Alterations in Regions Containing rRNA Genes

Generally, active large pre-rRNA genes are digested more rapidly by DNA endonucleases than inactive genes, in patterns similar to active mRNA genes. The DNA of transcribed rRNA segments is also more susceptible than inactive segments to binding by chemical crosslinkers. Both of these lines of evidence support the conclusion that chromatin unfolds in active large pre-rRNA genes. Sites hypersensitive to endonuclease digestion also appear in regions upstream of active large

The Experimental Process

What Happens When RNA Polymerase Encounters a Nucleosome?

Myeong-Sok Lee and William T. Garrard

MYEONG-SOK LEE and WILLIAM T. GARRARD have collaborated on studies of the mechanism of generating chromatin structures characteristic of actively transcribing genes. Garrard (left) has been a Professor of Biochemistry at the University of Texas Southwestern Medical Center at Dallas since 1974. He received his B.S. degree in microbiology from the University of Washington, Ph.D. in microbiology from the University of California at Los Angeles, and did postdoctoral work with Professor James Bonner at the California Institute of Technology. Lee (right) is a postdoctoral fellow and received his B.S. degree in biology from Seoul National University, Korea, and Ph.D. in biochemistry for his work with Garrard.

A variety of experimental approaches indicate that transcriptionally active genes are associated with nucleosomes, both upstream and downstream of traversing RNA polymerase II molecules. Since the topologically constrained DNA associated with histone octamers within nucleosomes would impede RNA polymerase movement, the chromatin fiber itself would be expected to undergo conformational alterations to facilitate transcription. In fact, it has been known for many years that transcriptionally active or potentially active genes have an increased DNase I sensitivity in chromatin, suggesting that their corresponding chromatin fibers are conformationally more accessible to nuclease than those of inactive genes. The nature of these alterations at the level of nucleosome structure is not understood, however. Furthermore, it is unclear whether such changes are a cause or an effect of transcription. These long-standing problems led us to address the role of transcription in generating DNase I sensitivity and nucleosome alterations in chromatin.

We used the yeast *Saccharomyces cerevisiae* for our studies because this organism possesses several advantages over other eukaryotes. Its genome size is about $200 \times$ smaller than that of a mammalian cell (thus facilitating detection of specific nucleotide sequences by hybridization); it can be grown in the haploid state (thus facilitating mutant gene constructions); and its doubling time is less than 2 hours (for ease of laboratory manipula-

tions). In addition, one can insert a mutation into the yeast genome by site-directed gene replacement,[1] allowing assessment of base changes on gene expression and chromatin structure in the normal chromosomal environment.

As a model system to study the effects of transcription on chromatin structure, we chose the heat-shock-inducible gene *HSP82*, in which transcriptional activity of the wild-type gene can be induced 20-fold simply by shifting cells from 30°C to 39°C for 10 minutes. Since the relatively high basal level transcription of the gene in control (non-heat-shocked) cells impeded our study, we created a promoter mutant by the site-directed gene replacement technique, which eliminated basal level transcription without markedly affecting heat-induced transcription. This led to a situation where heat-shock induction was now over 200-fold.[2] This mutant strain proved to be extremely valuable for our resulting studies, since by heat-shock treatment we could transcriptionally turn on the gene which was previously completely off and study the effect of transcription on the chromatin structure.

To examine the relationship between transcription and DNase I sensitivity, we prepared nuclei from both control and 10-minute-heat-shocked mutant cells and digested the nuclei lightly with DNase I. The resulting purified DNA was cleaved with a restriction endonuclease and separated by agarose gel electrophoresis; the chromatin structure surrounding the *HSP82* gene locus was analyzed via Southern blot hybridization with a radiolabeled DNA probe chosen to abut the restriction site. This technique, known as indirect end-labeling,[3] is analogous to footprinting techniques used to study various protein-DNA complexes. We were able to focus our attention to a specific region of the genome by carefully selecting the appropriate restriction enzyme(s) to generate DNA fragments about 2000 to 4000 bp long containing a region of the *HSP82* gene that we were interested in and a probe that does not cross-hybridize with the other gene family member, *HSC82*. As shown in Figure A*a*, before heat-shock (−) the 3′-region of the *HSP82* gene lacks DNase I sensitivity but possesses a hypersensitive site at the end of the gene (arrow). After heat-shock (+) the *HSP82* gene body exhibits marked sensitivity to DNase I (filled bars). Furthermore, this sensitivity exhibits a sharp boundary at the end of the gene since immediately downstream the *CIN2* gene lacks such DNase I sensitivity (a). From these results we concluded that transcription induces DNase I sensitivity in the 3′-region of the *HSP82* gene.[4]

We then became quite interested in determining the

Figure A Transcription induces DNase I sensitivity and nucleosome "splitting." The chromatin structure of the 3'-region of the *HSP82* gene and the immediate downstream *CIN2* gene: **(a)** low resolution mapping; **(b)** higher resolution (1 Kb = 1,000 base pairs). The open vertical arrows depict the *HSP82* and *CIN2* gene transcription units. Calibration on the left is absolute DNA length on the gel and on the right is the map position on a linear scale with respect to the start site of the *HSP82* gene transcription unit. Filled vertical bars and arrows mark hypersensitive regions, open circles refer to internucleosomal cleavage sites, and closed circles mark cleavage sites that exhibit a half-nucleosomal periodicity.

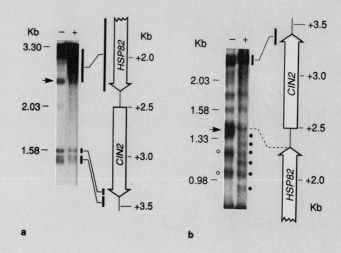

a   b

Figure B Chromatin structure of the 3'-region of the *HSP82* gene. Schematic diagram depicting the interpretation of the results of the chromatin structure analyses of Figure A.

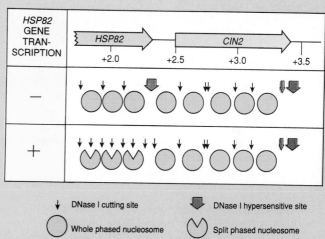

underlying structure of the DNase I-sensitive chromatin. For example, did this chromatin still possess typical nucleosomes? Therefore, we performed a higher resolution chromatin analysis simply by mapping the structure from the other direction. As shown in Figure A*b*, before heat shock (−) when the corresponding region of the *HSP82* gene lacks pronounced DNase I sensitivity it exhibits a "whole-nucleosomal" cleavage periodicity of about 160 bp (open circles). However, strikingly, after heat shock (+) the DNase I cutting interval within the same region is observed to be about 80 bp (filled circles). This novel periodicity is approximately half of the nucleosomal repeat length of yeast chromatin and thus corresponds to a "half-nucleosomal" cleavage periodicity. We have operationally defined the structures associated with these regions as "split" nucleosomes.[4]

As an important control experiment to demonstrate that such cleavage sites in the 3'-region of the gene reflect a specific chromatin structure, we performed a similar experiment with naked genomic DNA. As expected, the naked DNA was digested randomly, and exhibited a smear along the corresponding DNA sequences.

Figure B schematically summarizes the results of the

above experiments. We concluded that nucleosomes split within the 3'-region of the *HSP82* gene by observing a transcription-associated change from the DNase I cleavage periodicity for whole nucleosomes to the one for half-nucleosomes. These results led us to suggest that a split nucleosomal structure represents one of the underlying structures of DNase I-sensitive chromatin.[4]

Our definition for nucleosome splitting is operational, and the actual structure of split nucleosomes remains to be determined. However, previous studies by other researchers have revealed that major structural changes occur within nucleosomes associated with active genes, and half-nucleosomes have been directly observed by electron microscopy under specialized conditions. Moreover, the path of DNA about the histone octamer exhibits a dyad axis of symmetry, and it has been previously suggested that conformational alterations might generate DNase I cleavage sites near this dyad axis.

While our data convincingly demonstrate that transcription induces very specialized alterations in chromatin, these observations raise yet another important question. What is the precise mechanism for transcription-induced DNase I sensitivity in chromatin and nucleosome splitting?

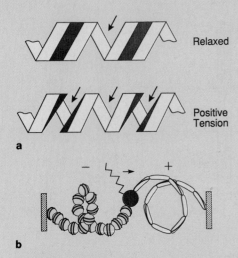

Relaxed

Positive
Tension

a

− +

b

Figure C Positive supercoils downstream of the transcription complex split nucleosomes and clear the path for RNA polymerase II. **(a)** The diagram illustrates how positive supercoiling might split nucleosomes near their centers and expose new DNase I cleavage sites at the dyad axes. **(b)** A chromosomal loop with anchored ends, composed of a polynucleosomal array that is being traversed from left to right by RNA polymerase II. The upstream nucleosomes become tightly packed and the downstream nucleosomes split as a consequence of the twin-domain model of DNA supercoiling.[5] Topoisomerases offset these processes.

Physical disruption of nucleosomes during transcription by RNA polymerase might induce DNase I sensitivity in chromatin and nucleosome splitting. This idea is not favored since studies from many laboratories have revealed that DNase I-sensitive chromatin often extends far upstream and downstream of the boundaries of transcription

units. Alternatively, such changes in chromatin might be induced by signals that transiently spread down the chromatin fiber ahead of the transcription complex, such as waves of supercoiling.

During the time we were facing this question, it was proposed[5] and later proved by other researchers that the movement of RNA polymerase induces positive DNA supercoiling in front of the transcription complex and negative DNA supercoiling behind it. Putting these facts together with our observations led us to propose that transient positive supercoiling downstream of RNA polymerase movement splits nucleosomes at their dyad axes of symmetry, leading to new DNase I cutting sites near the center of nucleosomes (see Figure C*a*). We believe that this conformational transition serves a crucial function in paving the way for RNA polymerase passage (as outlined in Figure C*b*). Such positive supercoils may also unwind the negative supercoils associated with the chromatin superhelix, leading to extensive chromatin decondensation along with nucleosome splitting. This model is consistent with experimental results suggesting that torsional stress is necessary for transcription to occur on nucleosomal templates,[6] and is amenable to experimental testing in the immediate future.

References

[1] Boeke, et al. *Methods Enzymol.* 154:161 (1987).

[2] McDaniel, et al. *Mol. Cell. Biol.* 9:4789 (1989).

[3] Wu, C. *Nature* 286:854 (1980).

[4] Lee, M.-S., and Garrard, W. T. *EMBO J.* 10:607 (1991).

[5] Liu, L. F., and Wang J. C. *Proc. Natl. Acad. Sci. USA* 84:7024 (1987).

[6] Harland, R., et al. *Nature* 302:38 (1983).

pre-rRNA genes in many species, in a pattern similar to the hypersensitive sites in active mRNA genes.

In general the DNA lengths obtained after endonuclease digestion of active large pre-rRNA genes indicate that some of the DNA remains wound into nucleosomes, and some is more or less unfolded or unwound. That is, some of the DNA released by digestion appears in the regular 200 or 145 base-pair lengths characteristic of nucleosomes (see p. 547), and some appears in altered lengths. Although the sources of the regular and altered lengths are uncertain, the regular periodicities probably represent DNA released from the intergenic spacer regions, in which direct electron microscopy shows some nucleosomes are retained. The altered lengths are probably DNA segments released from the large pre-rRNA coding segments, in which the nucleosomes are presumably displaced or unfolded by the RNA polymerases.

Active 5S rRNA genes are also more sensitive to endonuclease digestion than inactive genes, indicating that chromatin unwinds and disassembles to a greater or lesser extent in these regions during RNA transcription. Hypersensitive sites also appear in active 5S rRNA genes, at points near the 5' end of the internal promoter and just upstream of the startpoint for transcription. 5S genes are so small that an entire gene and its nontranscribed spacer contain only enough DNA to wrap around two nucleosomes.

Actively transcribed large pre-rRNA genes have been observed directly under the electron microscope by O. L. Miller, D. L. Beatty, and others. The large genes in these preparations (Fig. 15-50) appear as long axial fibers coated at intervals with a matrix of shorter fibers. Work with enzymes and other techniques (see the caption to Fig. 15-50 for details) demonstrated that the long axial fibers are DNA molecules of large pre-

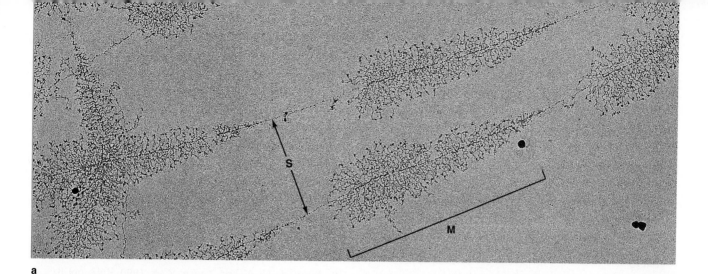

a

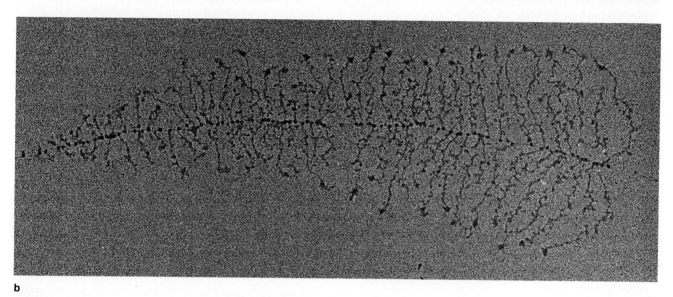

b

Figure 15-50 Transcribing large pre-rRNA genes isolated from the newt *Triturus*.
(a) The long axial fibers (S) are DNA molecules of the nucleolus. The matrix regions (M) are segments active in rRNA transcription. Large pre-rRNA transcripts extend at right angles from the DNA. The long axial fibers in such preparations are broken by DNAse, indicating that they are DNA molecules of rRNA genes; the fibers of the matrix regions are broken by RNAse, indicating that they are the rRNA product of the genes. ×26,000.
(b) An active, large pre-rRNA gene at higher magnification. The beginning, or 5′ end, of the gene is located at the end where the matrix fibers are shortest. Longer, nearly complete large pre-rRNA molecules extend from the rDNA axis at the opposite, 3′ end of the gene. The granules at the base of each ribonucleoprotein fiber are probably RNA polymerase molecules. ×46,000. Courtesy of O. L. Miller, Jr., and Barbara R. Beatty, Oak Ridge National Laboratory.

rRNA genes. The short, brushlike fibers of the matrices are pre-rRNA molecules in the process of transcription. The granule at the base of each pre-rRNA transcript is probably an RNA polymerase molecule transcribing the gene. The RNA polymerases and transcripts are crowded along the gene, indicating that transcription of large pre-rRNA genes proceeds very rapidly, with enzymes following each other closely as they transcribe the DNA. This appearance coincides with biochemical

information, which indicates that the large pre-rRNA sequences are among the most actively transcribed genes in the nucleus.

Each gene, represented by the length coated by transcripts, is separated from the next by a segment that carries no transcripts. These regions are the intergenic spacers separating the large pre-rRNA genes. Visible nucleosomes appear to be limited to the intergenic spacers.

For Further Information

DNA structure, *Chs. 2 and 14*
Endoplasmic reticulum in protein synthesis and secretion, *Ch. 20*
Genetic code, *Ch. 16*
Histone proteins, *Ch. 14*
Nonhistone regulatory proteins, *Chs. 14 and 17*
Nucleolus and nucleolar organizer in cell division, *Ch. 24*
Nucleosomes and replication, *Ch. 23*
Nucleotides and nucleic acids, *Ch. 2*
Protein synthesis, *Ch. 16*
    in mitochondria and chloroplasts, *Ch. 21*
Pseudogenes, *Ch. 18*
Regulation, transcriptional, *Ch. 17*
    posttranscriptional, *Ch. 17*
Ribosome structure, *Ch. 16*
RNA editing, *Ch. 21*
Transcription in mitochondria and chloroplasts, *Ch. 21*
Transposable genetic elements, *Ch. 18*

Suggestions for Further Reading

General Books and Articles

Darnell, J. E. 1983. The processing of RNA. *Sci. Amer.* 249:90–100 (October).

Lewin, B. 1990. *Genes IV*. New York: Wiley.

Saenger, W. 1984. *Principles of nucleic acid structure.* New York: Springer-Verlag.

Wyatt, J. R.; Puglisi, J. D.; and Tinoco, I. 1989. RNA folding: pseudoknots, loops, and bulges. *Bioess.* 11:100–106.

General Features of RNA Transcription and RNA Polymerases in Eukaryotes

Echols, H. 1990. Nucleoprotein structures initiating DNA replication, transcription, and site-specific recombination. *J. Biolog. Chem.* 265:14697–700.

Gabrielsen, O. S., and Sentenac, A. 1991. RNA polymerase III(c) and its transcription factors. *Trends Biochem. Sci.* 16:412–16.

Geiduschek, E. P., and Tocchini-Valentini, G. P. 1988. Transcription by RNA polymerase III. *Ann. Rev. Bioch.* 57:873–914.

Greenblatt, J. 1991. RNA polymerase-associated transcription factors. *Trends Biochem. Sci.* 16:408–11.

Kerpolla, T. K., and Kane, C. M. 1991. RNA polymerase: regulation of transcription elongation and termination. *FASEB J.* 5:2833–42.

McClure, W. R. 1985. Mechanism and control of transcription initiation in eukaryotes. *Ann. Rev. Bioch.* 54:171–204.

Polyarovsky, O. I., and Stepchenko, A. G. 1990. Eukaryotic transcription factors. *Bioess.* 12:205–10.

Sentenac, A. 1985. Eukaryotic RNA polymerases. *CRC Crit. Rev. Biochem.* 18:31–90.

Sollner-Webb, B. 1988. Surprises in pol III transcription. *Cell* 52:153–54.

Woychik, N. A., and Young, R. A. 1990. RNA pol II: subunit structure and function. *Trends Biochem. Sci.* 15:347–51.

Young, R. A. 1991. RNA polymerase II. *Ann. Rev. Biochem.* 60:689–715.

mRNA Gene Structure

Atchison, M. L. 1988. Enhancers: mechanisms of action and cell specificity. *Ann. Rev. Cell Biol.* 4:127–53.

Blake, C. C. F. 1985. Exons and the evolution of proteins. *Internat. Rev. Cytol.* 93:149–85.

Dynan, W. S. 1989. Modularity in promoters and enhancers. *Cell* 58:1–4.

Guarente, L. 1988. UASs and enhancers: common mechanisms of transcriptional activation in yeast and mammals. *Cell* 52:303–5.

Heidecker, G., and Messing, J. 1986. Structural analysis of plant genes. *Ann. Rev. Plant Physiol.* 37:439–66.

Jeang, K.-T., and Khoury, G. 1988. The mechanistic role of enhancer elements in eukaryotic transcription. *Bioess.* 8:104–7.

Nussinov, R. 1990. Signal sequences in eukaryotic upstream regions. *Crit. Rev. Biochem. Molec. Biol.* 25:185–224.

Wang, J. C., and Giaever, G. N. 1988. Action at a distance along a DNA. *Science* 240:300–304.

Transcription of mRNA Genes in Eukaryotes

Conaway, J. W., and Conaway, R. C. 1991. Initiation of eukaryotic mRNA synthesis. *J. Biolog. Chem.* 266:17721–24.

Greenblatt, J. 1991. Roles of TFIID in transcription initiation by RNA polymerase II. *Cell* 66:1067–70.

Lewin, B. 1990. Commitment and activation at polymerase II promoters: a tail of protein-protein interactions. *Cell* 61:1161–64.

Martin, K. J. 1991. The interaction of transcription factors and their adapters, coactivators, and accessory proteins. *Bioess.* 13:499–503.

Mermelstein, F. H.; Flores, O.; and Reinberg, D. 1989. Initiation of transcription by RNA polymerase II. *Biochim. Biophys. Acta* 1009:1–10.

Nevins, J. R. 1983. The pathway of eukaryotic mRNA formation. *Ann. Rev. Bioch.* 52:441–66.

Proudfoot, N. J. 1989. How RNA polymerase II terminates transcription in higher eukaryotes. *Trends Biochem. Sci.* 14:105–10.

Roeder, R. G. 1991. The complexities of eukaryotic transcription initiation: regulation of preinitiation complex assembly. *Trends Biochem. Sci.* 16:402–8.

Saltzman, A. G., and Weinmann, R. 1989. Promoter specificity and modulation of RNA polymerase II transcription. *FASEB J.* 3:1723–33.

Sawadogo, M., and Sentenac, A. 1990. RNA polymerase B (II) and general transcription factors. *Ann. Rev. Biochem.* 9:711–54.

Wasylyk, B. 1988. Transcription elements and factors of RNA polymerase B promoters of higher eukaryotes. *CRC Crit. Rev. Biochem.* 23:77–120.

Pre-mRNA Structure and Processing

Aebi, M., and Weissmann, C. 1987. Precision and orderliness in splicing. *Trends Genet.* 3:102–7.

Agabian, N. 1990. *Trans* splicing of nuclear pre-mRNAs. *Cell* 61:1157–60.

Andreadis, A.; Gallego, M. E.; and Nadal-Ginard, B. 1987. Generation of protein isoform diversity by alternative splicing. *Ann. Rev. Cell Biol.* 3:207–42.

Bernstein, P., and Ross, J. 1989. Poly(A), poly(A) binding proteins and the regulation of mRNA stability. *Trends Biochem. Sci.* 14:373–77.

Breitland, R. E.; Andreadis, A.; and Nadal-Ginard, B. 1987. Alternative splicing: a ubiquitous mechanism for the generation of multiple protein isoforms from single genes. *Ann. Rev. Biochem.* 56:467–95.

Cattaneo, R. 1991. Different types of mRNA editing. *Ann. Rev. Genet.* 25:71–88.

Green, M. R. 1991. Biochemical mechanisms of constitutive and regulated pre-mRNA splicing. *Ann. Rev. Cell Biol.* 7:559–99.

Guthrie, C., and Patterson, B. 1988. Spliceosomal snRNAs. *Ann. Rev. Genet.* 22:387–419.

Hoffman, M. 1991. RNA editing: what's in a mechanism?. *Science* 253:136–38.

Jackson, R. J., and Standart, N. 1990. Do the poly(A) tail and 3' untranslated region control mRNA translation? *Cell* 62:15–24.

Luhrmann, R.; Kastner, B.; and Bach, M. 1990. Structure of spliceosomal snRNPs and their role in pre-mRNA processing. *Biochim. Biophys. Acta* 1087:265–92.

Maniatis, T. 1991. Mechanisms of alternative pre-mRNA splicing. *Science* 251:33–34.

Manley, J. L. 1988. Polyadenylation of mRNA precursors. *Biochim. Biophys. Acta* 950:1–12.

Mulligan, R. M. 1991. RNA editing: when transcription sequences change. *Plant Cell* 3:327–30.

Patthy, L. 1991. Exons—original building blocks of proteins? *Bioess.* 13:187–92.

Razin, A., and Szyf, M. 1984. DNA methylation patterns. Formation and function. *Biochim. Biophys. Acta* 782:331–42.

Shapiro, M. B., and Senapathy, P. 1987. RNA splice junctions of different classes of eukaryotes: sequence statistics and functional implications in gene expression. *Nucleic Acids Res.* 15:7155–74.

Sharp, P. 1987. Splicing of messenger RNA precursors. *Science* 235:766–71.

Simpson, L. 1990. RNA editing—a novel genetic phenomenon? *Science* 250:512–13.

Smith, C. W. J.; Patton, J. G.; and Nadal-Ginard, B. 1989. Alternative splicing in the control of gene expression. *Ann. Rev. Genet.* 23:527–77.

Stuart, K. 1991. RNA editing in tropanosomatoid mitochondria. *Ann. Rev. Microbiol.* 45:327–44.

Weiner, A. M., and Maizels, N. 1990. RNA editing: guided but not templated? *Cell* 61:917–20.

Wickens, M. 1990. How the messenger got its tail: addition of poly(A) in the nucleus. *Trends Biochem. Sci.* 15:277–81.

Wahle, E. 1992. The end of the message: 3'-end processing leading to polyadenylated messenger RNA. *Bioess.* 14:113–18.

RNA Catalysts in Processing Reactions

Altman, S. 1990. Enzymatic cleavage of RNA by RNA. *Biosci. Rep.* 10:317–37.

Cech, T. R. 1987. The chemistry of self-splicing RNA and RNA enzymes. *Science* 236:1532–39.

———. 1990. Self-splicing of group I introns. *Ann. Rev. Biochem.* 59:543–68.

Cedergren, R. 1990. RNA—the catalyst. *Biochem. Cell Biol.* 68:903–6.

Jacquier, A. 1990. Self-splicing group II and nuclear pre-mRNA introns: how similar are they? *Trends Biochem. Sci.* 15:351–54.

Sharp, P. A., and Eisenberg, D. 1987. The evolution of catalytic function. *Science* 238:729–30, 807.

Symons, R. H. 1992. Small catalytic RNAs. *Ann. Rev. Biochem.* 61:641–71.

Ribosomes and rRNA Structure

Brimacombe, R. 1984. Conservation of structure in ribosomal RNA. *Trends Biochem. Sci.* 9:273–77.

Brimacombe, R.; Maly, P.; and Zweib, C. 1983. The structure of ribosomal RNA and its organization relative to ribosomal protein. *Prog. Nucleic Acid Res. Molec. Biol.* 28:2–48.

Delihas, N.; Anderson, J.; and Singhal, R. P. 1984. Structure, function, and evolution of 5S ribosomal RNAs. *Prog. Nucleic Acid Res. Molec. Biol.* 31:161–90.

Hardesty, B., and Kramer, G. 1986. *Structure, function, and genetics of ribosomes.* New York: Springer-Verlag.

Maden, B. E. H. 1990. The numerous modified nucleotides in eukaryotic ribosomal RNA. *Prog. Nucleic Acids Res. Molec. Biol.* 39:242–303.

Noller, H. F. 1984. Structure of ribosomal RNA. *Ann. Rev. Biochem.* 53:119–62.

Walker, T. A., and Pace, N. R. 1983. 5.8S ribosomal RNA. *Cell* 31:320–22.

Zuker, M. 1989. On finding all suboptimal foldings of an RNA molecule. *Science* 244:48–52.

rRNA: Genes, Transcription, and Processing

Chastain, M., and Tinoco, I., Jr. 1991. Structured elements in RNA. *Prog. Nucleic Acid Res. Molec. Biol.* 41:131–77.

Mandal, R. K. 1984. The organization and transcription of eukaryotic ribosomal RNA genes. *Prog. Nucleic Acid Res. Molec. Biol.* 31:115–60.

Reeder, R. H.; Labhart, P.; and McStay, B. 1987. Processing and termination of RNA polymerase I transcripts. *Bioess.* 6:108–12.

Schlessinger, D.; Bolla, R. I.; Sirdeshmukh, R.; and Thomas, J. R. 1985. Spacers and processing of large ribosomal RNAs in *Escherichia Coli* and mouse cells. *Bioess.* 3:14–18.

Wolffe, A. P., and Brown, D. D. 1988. Developmental regulation of two 5S rRNA genes. *Science* 241:1626–32.

rRNA and the Nucleolus

Goessens, G. 1984. Nucleolar structure. *Internat. Rev. Cytol.* 87:107–58.

Hadjilov, A. A. 1984. *The nucleolus and ribosome biogenesis.* New York: Springer-Verlag.

Hernandez-Verdun, D. 1991. The nucleolus today. *J. Cell Sci.* 99:465–71.

Reeder, R. H. 1990. rRNA synthesis in the nucleolus. *Trends Genet.* 6:390–95.

Scheer, U., and Benavente, R. 1990. Functional and dynamic aspects of the mammalian nucleolus. *Bioess.* 12:14–21.

Somerville, J. 1986. Nucleolar structure and ribosomal biogenesis. *Trends Biochem. Sci.* 11:438–42.

tRNA: Structure, Genes, Transcription, and Processing

Baldi, M. I.; Mattoccia, E.; Bufardeci, E.; Fabbri, S.; and Tocchini-Valentini, G. P. 1992. Participation of the intron in the reaction catalyzed by the *Xenopus* tRNA splicing endonuclease. *Science* 255:1404–8.

Bjork, G. R.; Ericson, J. U.; Gustafsson, C. E. D.; Hagervall, T. G.; Jonsson, Y. H.; and Wikstrom, P. M. 1987. tRNA modification. *Ann. Rev. Biochem.* 56:263–87.

Deutscher, M. P. 1984. Processing of tRNA in prokaryotes and eukaryotes. *CRC Crit. Rev. Bioch.* 17:45–71.

———. 1990. Ribonucleases, tRNA nucleotidyltransferase, and the 3′ processing of tRNA. *Prog. Nucleic Acid Res. Molec. Biol.* 39:209–40.

Pace, N. R., and Smith, D. 1990. Ribonuclease P: function and variation. *J. Biolog. Chem.* 265:3587–90.

Sharp, S. J.; Schaak, J.; Cooley, L.; Burke, D. J.; and Soll, D. 1985. Structure and transcription of eukaryotic tRNA genes. *Crit. Rev. Biochem.* 19:107–44.

snRNA and scRNA: Structure, Genes, Transcription, and Processing

Guthrie, C., and Patterson, B. 1988. Spliceosomal snRNAs. *Ann. Rev. Genet.* 22:387–419.

Kunkel, G. R. 1991. RNA pol III transcription of genes that lack internal control regions. *Biochim. Biophys. Acta* 1088:1–9.

Lerner, M. R., and Steitz, J. A. 1981. Snurps and scyrps. *Cell* 25:298–300.

Mount, S. M., and Steitz, J. A. 1984. RNA splicing and the involvement of small ribonucleoproteins. *Modern Cell Biol.* 3:249–97.

Parry, H. D.; Scherly, D.; and Mattaj, I. W. 1989. "Snurpogenesis": the transcription and assembly of U snRNA components. *Trends Biochem. Sci.* 14:15–19.

Steitz, J. A. 1988. Snurps. *Sci. Amer.* 258:56–63 (June).

Zieve, G. W., and Sauterer, R. A. 1990. Cell biology of the snRNP particles. *Crit. Rev. Biochem. Molec. Biol.* 25:1–46.

Transcription and Chromatin Structure; Hypersensitive Sites

Ausio, J. 1992. Structure and dynamics of transcriptionally active chromatin. *J. Cell. Sci.* 102:1–5.

Collins, F. S., and Weissman, S. M. 1984. The molecular genetics of human hemoglobin. *Prog. Nucleic Acid Res. Molec. Biol.* 31:315–462.

Gross, D. S., and Garrard, W. 1988. Nuclease hypersensitive sites in chromatin. *Ann. Rev. Biochem.* 57:159–97.

Grunstein, M. 1990. Histone function in transcription. *Ann. Rev. Cell Biol.* 6:643–78.

———. 1990. Nucleosomes: regulators of transcription. *Trends Genet.* 6:395–400.

van Holde, K. E.; Lohr, D. E.; and Robert, C. 1992. What happens to nucleosomes during transcription? *J. Biolog. Chem.* 267:2837–40.

Transcription and Processing in Prokaryotes

Altman, S. 1990. Ribonuclease P postscript. *J. Biolog. Chem.* 265:20053–56.

Bear, D. G., and Peabody, D. S. 1988. The *E. coli rho* protein: an ATPase that terminates transcription. *Trends Biochem. Sci.* 13:343–47.

Fournier, M. J., and Ozeki, H. 1985. Structure and organization of the tRNA genes of *Escherichia coli* K-12. *Microbiol. Rev.* 49:379–97.

Friedman, D. I.; Imperiale, M. J.; and Adhya, S. L. 1987. RNA 3′ end formation in the control of gene expression. *Ann. Rev. Genet.* 21:453–88.

Helman, J. D., and Chamberlin, M. J. 1988. Structure and function of bacterial sigma factors. *Ann. Rev. Biochem.* 57:839–72.

Reinhold-Hurek, B., and Shub, D. A. 1992. Self-splicing introns in tRNA genes of widely divergent bacteria. *Nature* 357:173–76.

Horowitz, M. S. Z. 1990. Structure-function relationships in *E. coli* promoter DNA. *Prog. Nucleic Acid Res. Molec. Biol.* 38:137–64.

Inouye, M., and Delihas, N. 1988. Small RNAs in the prokaryotes: a long list of diverse roles. *Cell* 53:5–7.

Kustu, S.; North, A. K.; and Weiss, D. S, 1991. Prokaryotic transcriptional enhancers and enhancer-binding proteins. *Trends Biochem. Sci.* 16:397–402.

Lendahl, L., and Zengel, J. M. 1986. Ribosomal genes in *E. coli*. *Ann. Rev. Genet.* 20:297–326.

Liu, X.-Q. 1991. An ancient intron in eubacteria: new light on intron origins. *Bioess.* 13:185–86.

Richardson, J. P. 1990. Rho-dependent transcription termination. *Biochim. Biophys. Acta* 1048:127–38.

Singhal, R. P., and Shaw, J. K. 1983. Prokaryotic and eukaryotic 5S RNAs: primary sequences and proposed secondary structures. *Prog. Nucleic Acid Res. Molec. Biol.* 28:177–252.

Srivastava, A. K., and Schlessinger, D. 1990. Mechanism and regulation of bacterial ribosomal RNA processing. *Ann. Rev. Microbiol.* 44:105–29.

Stragier, P. 1991. Dances with sigmas. *EMBO J.* 10:3559–66.

Travers, A. A. 1987. Structure and function of *Escherichia coli* promoter DNA. *CRC Crit. Rev. Biochem.* 22:181–219.

Review Questions

1. Outline the structure of RNA. What are inverted sequences? Hairpins? What is the conformation of RNA double helices? Give two examples of the importance of secondary structure to RNA functions.

2. Describe the overall steps in transcription. What is a phosphodiester linkage? What is the 5′ end of an RNA transcript? The 3′ end? What structural features mark the 5′ and 3′ ends of a newly synthesized RNA molecule?

3. Summarize the properties and functions of eukaryotic RNA polymerases I, II, and III. What major RNA types are transcribed by each of the eukaryotic RNA polymerases? What types of promoters are recognized by each of the polymerase enzymes? What structural features do these enzymes have in common and in what ways do they differ? What features of RNA nucleotides are important in the reactions assembling an RNA molecule on a DNA template?

4. Outline the overall steps of initiation, elongation, and termination of transcription. What factors are important in these mechanisms? What is footprinting, and how is this technique used in the study of RNA transcription?

5. What is the sense chain of a DNA molecule? The missense chain? In what direction are DNA and RNA sequences read? Write an RNA coding sequence containing four codons (include some Us) and indicate how the sequence reads in the sense and missense chains of the DNA.

6. What major steps occur in RNA processing?

7. Diagram the structure of an mRNA molecule and indicate the probable or suspected functions of each major sequence segment.

8. Diagram the structure of a pre-mRNA molecule. How were pre-mRNAs identified as the precursors of mRNAs? What are introns? Exons? How were introns discovered?

9. Diagram the gene for the pre-mRNA you outlined in the answer to question 8.

10. What is the promoter? What are the major functions of the CAAT and TATA boxes? What is a consensus sequence? What is the startpoint? What is an enhancer? How do enhancers differ from promoters? How are enhancers believed to work? What is the meaning of "upstream" and "downstream" with reference to mRNA genes?

11. Diagram the major sequence features of introns. What mechanisms may account for the existence of introns? What is "selfish DNA"?

12. Outline the steps occurring in the initiation, elongation, and termination of mRNA transcription.

13. What processing steps occur in addition of the 5′ cap? In processing the 3′ end of pre-mRNAs with the AAUAAA signal? How might processing of the 3′ end be related to the termination of mRNA transcription?

14. What processing steps occur in intron removal? Why is intron removal thought possibly to be catalyzed by RNA acting as a ribozyme? What are snRNAs? Spliceosomes? How are snRNAs thought to participate in intron removal? What is a lariat?

15. What is alternative splicing? What is the significance of alternative splicing to protein structure and function?

16. What rRNA types are found in each ribosomal subunit in eukaryotes? In prokaryotes? What are the relationships in sequence and secondary structure among the rRNAs of different groups? What RNA polymerase enzymes transcribe each rRNA type?

17. How are probable secondary structures determined for rRNA molecules? What evidence indicates that secondary structure is more important than primary sequence in many regions of rRNA molecules? What functional roles might secondary structures play in rRNA molecules?

18. Diagram the major sequence features of a large pre-rRNA gene. List two experimental procedures that identified these features.

19. What processing steps release the individual rRNA types from the large pre-rRNA? What further processing steps are involved in rRNA maturation?

20. What types of introns occur in rRNA genes? How are these introns removed? What features regulate their removal? What are group I and group II introns? How do the reactions removing them differ?

21. Diagram the structure of a 5S pre-rRNA gene, the transcript copied from the gene, and a mature 5S rRNA molecule. What evidence indicates that the promoter of 5S rRNA genes is internal? How are the promoters of 5S rRNA genes structured? What transcription factors participate in 5S transcription? How do transcription factors interact with the promoter and RNA polymerase III in transcription?

22. How are large and 5S pre-rRNA genes arranged in eukaryotes? What is the relationship between this arrangement and the nucleolus? What is a nucleolar organizer? What structures are visible inside nucleoli? What is the probable relationship between these structures and rRNA transcription and processing? What experiments and evidence established that large pre-rRNA genes are located in the nucleolus? That 5S pre-rRNA genes are located on chromosomal sites outside the nucleolus? Outline the major steps occurring in ribosome synthesis.

23. Outline the structure of tRNA molecules. What are invariant bases? Semi-invariant bases? Where does an amino acid attach to a tRNA molecule? Where is the anticodon located?

24. Diagram the structure of a tRNA gene. How are tRNA promoters structured? Compare the promoters of tRNA and 5S rRNA genes. Where do introns typically occur in tRNA genes? What transcription factors participate in tRNA transcription?

25. Diagram a pre-tRNA transcript. What processing steps convert the transcript into a mature tRNA? What is RNAse P? M1 snRNA? What evidence indicates that the RNA component of RNAse P is catalytic?

26. What structural features of pre-tRNAs regulate intron removal? Compare the mechanisms determining intron removal in mRNA, rRNA, and tRNA genes. In which of these mechanisms are RNA molecules believed to act as catalysts?

27. What are U snRNAs? What functions do these molecules carry out in eukaryotic cells? What is the relationship between U snRNAs and the human disease lupus erythematosus? Diagram a U snRNA gene. What processing steps occur in the transcripts of U snRNA genes?

28. What is SRP 7S scRNA? What are *Alu* sequences? What is the relationship between this scRNA and the *Alu* sequences?

29. What alterations occur in the sensitivity of DNA to endonuclease digestion as genes go from an inactive to an active state? What are hypersensitive sites? What is the significance of these alterations to chromatin structure? Compare the appearance of active mRNA and rRNA genes under the electron microscope.

RNA Transcription and Processing in Prokaryotes

Prokaryotes exhibit many differences from eukaryotes in the details of gene structure, transcription, and processing. A single RNA polymerase of prokaryotes transcribes all the major RNA classes—mRNA, rRNA, and tRNA.

Prokaryotic mRNA genes usually contain codes for more than one polypeptide instead of one as in eukaryotes. No introns occur in prokaryotic mRNA genes, except in one bacterial group, the archaebacteria, and no clipping and splicing reactions are necessary to convert the transcripts of protein-encoding genes to finished form. There is therefore almost no mRNA processing in most bacteria, except for a few internal methylations and the addition of a short poly(A) tail to some. The fact that prokaryotic mRNAs are transcribed in essentially finished form reflects the fact that transcription and translation occur simultaneously in prokaryotes—translation of the 5′ end of mRNAs begins before transcription of the 3′ end is complete. This simultaneous transcription and translation depend in part on the fact that prokaryotes have no nuclear envelope. The genes being transcribed and the mRNA transcripts are suspended directly in the cytoplasm, which contains ribosomes and all the factors required for protein synthesis.

In contrast to bacterial mRNAs, both rRNA and tRNA precursors contain extra nucleotides that are removed in processing reactions. As a consequence, the genes coding for rRNA and tRNA in bacteria include transcribed spacer segments that, like their counterparts in eukaryotes, enlarge the DNA segments occupied by the genes. The genes coding for rRNA and tRNA in bacteria also occur in multiple, repeated copies.

Bacterial gene structure, transcription, and processing are generally considered representative of cyanobacterial systems as well. Although this is probably a safe assumption, it should be kept in mind that most of our knowledge in these topics has been developed primarily from one bacterial species, *Escherichia coli;* in reality, comparatively little is known as yet about transcription and processing in cyanobacteria.

Bacterial RNA Polymerase

The single RNA polymerase enzyme of bacteria contains in its core structure three different polypeptide subunits called α, β, and β', with molecular weights of 36,600, 150,600, and 155,600, respectively. Two copies of the α-polypeptide are present in the core, making the structure $\alpha_2\beta\beta'$, with a total molecular weight of about 380,000 and dimensions of about 10×15 nm. The $\alpha_2\beta\beta'$ complex is known as the *core enzyme.* The

β'- and β-subunits of the bacterial core enzyme are related in sequence to the L′ and L polypeptides of eukaryotic RNA polymerases (see p. 580).

The core enzyme is potentially capable of copying DNA from any source into an RNA transcript. However, in order to initiate transcription with high efficiency on bacterial promoters, the enzyme must combine with another polypeptide subunit called a *sigma factor.* Sigma factors have binding sites that recognize the RNA polymerase and sequences within bacterial promoters (the -10 and -35 regions; see below). Without a sigma factor RNA polymerase binds weakly and unselectively at many sites in addition to promoters in bacterial DNA. None of these associations lead to the strong polymerase-DNA linkage necessary for the initiation of transcription. With a sigma factor complexed to the enzyme, binding is limited essentially to promoter sequences and leads directly to the initiation of transcription. The core enzyme plus a sigma factor is known as the *holoenzyme.* Each *E. coli* cell contains about 3000 copies of the holoenzyme.

Sigma factors increase the selectivity of bacterial RNA polymerase for bacterial promoters by about a million times. A single sigma factor combines with an RNA polymerase enzyme before initiation takes place and is released as initiation is completed and the elongation phase of transcription proceeds.

Distinct sigma factors enable RNA polymerases to distinguish between different bacterial gene groups and thus form part of the mechanism regulating transcription. At least five sigma factors have been identified in *E. coli.* The primary sigma factor recognizing most promoter sequences in this bacterium has a molecular weight of 70,000 and is identified as σ^{70}. Another sigma factor, σ^{32}, increases in concentration during heat shock and other forms of stress and enables the RNA polymerase to bind to the promoters of genes encoding proteins that increase tolerance to stress. A third sigma factor, σ^{54}, appears during growth under conditions of limited nitrogen. A fourth, σ^{23}, appears during infection with bacteriophage T4. The interactions of the fifth factor, σ^{28}, remain unknown. All these factors contain helix-turn-helix motifs (see p. 543) that recognize and bind DNA sequences in bacterial promoters.

Other bacteria have a greater diversity of sigma factors; *Bacillus subtilis,* another much studied bacterium, has at least eight. The promoters of bacterial viruses are typically recognized by sigma factors specific to the virus; some bacteriophages also direct the synthesis of their own RNA polymerases. The viral sigmas cause RNA polymerase to bind most avidly to viral genes and dominate transcription in infected cells.

Bacterial Genes and Their Promoters

Bacterial genes are organized in transcriptional units called *operons*, which usually contain more than one coding sequence (Fig. 15-51). The coding sequences of bacteriophages are organized in the same pattern. Most bacterial operons are controlled by a single promoter lying in the region in advance of the transcribed segment. Some, however, have two or more promoters arranged tandemly in the 5'-flanking region. When two or more promoters are present, each is controlled by different regulatory factors, and each may independently initiate transcription of the operon.

Bacterial promoters, like their counterparts in eukaryotes, contain consensus sequences that provide recognition and binding sites for RNA polymerase, regulate the rate of initiation of transcription, and indicate the startpoint. All bacterial genes, including those encoding mRNAs, rRNAs, and tRNAs, are transcribed from upstream promoters with more or less similar consensus sequences. This similarity reflects the fact that no specialized RNA polymerases exist in eukaryotes; a single RNA polymerase makes RNA copies of all gene types. A few bacterial genes also have control sequences upstream of the promoter region that act as enhancers.

Two consensus sequences occur in *E. coli* promoters that are recognized by RNA polymerase in combination with sigma factor σ^{70} (Fig. 15-52). One, the short sequence TATATT, called the *– 10 sequence* or *Pribnow box*, is usually centered about 10 nucleotides upstream of the startpoint. The second sequence, TTGACA, called the *– 35 sequence*, usually begins about 35 nucleotides upstream of the startpoint. The startpoint for transcription usually lies about 7 to 8 nucleotides downstream of the – 10 sequence. Other sequences with regulatory effects on RNA polymerase binding may lie upstream of the – 35 sequence. The consensus se-

Table 15-3 Promoter Consensus Sequences in *E. coli*

Sigma Factor	Consensus Sequences	
σ^{70}	TTGACA------------------TATAAT −35 −10	
σ^{54}	GTGGC-------------TTGCA −26 −14	
σ^{32}	CCCCC--------------------TATAAATA −39 −16	
σ^{28}	TAAA------------GCCGATAA −35 −10	
σ^{23}	TATAATA −15	

NOTE: Modified from an original courtesy of M. S. Z. Horwitz and L. A. Loeb, from *Prog. Nucleic Acids Res. Mol. Biol.* 38:137 (1990).

quences of promoters recognized by RNA polymerase in combination with other sigma factors are shown in Table 15-3. Because promoters with the – 10 and – 35 consensus sequences recognized by the core enzyme in combination with σ^{70} make up the great majority of genes in *E. coli*, the following discussion pertains primarily to this promoter type.

Promoters vary greatly in the strength with which they bind RNA polymerase, depending on the closeness of their – 10 and – 35 sequences to the consensus sequence. Those with perfect copies of the consensus sequences bind RNA polymerase most strongly; those with base substitutions altering the consensus sequence bind RNA polymerase more weakly. (Fig. 15-52 shows the effects of base substitutions at various points in the – 10 and – 35 sequences.)

The particular sequences present at the – 10 and – 35 positions, by binding the RNA polymerase more

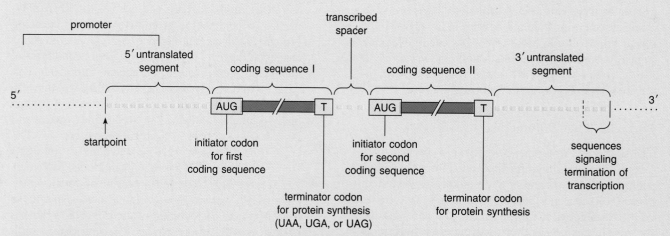

Figure 15-51 Structure of a prokaryotic mRNA gene or operon. Operons typically contain coding information for one to as many as eight or more polypeptides or proteins.

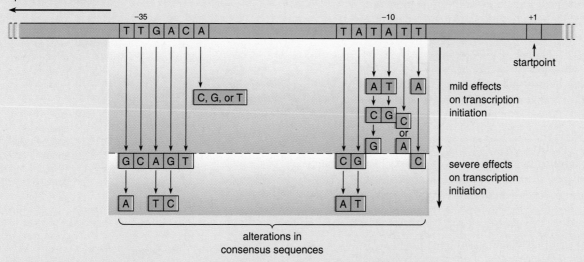

other sequences with regulatory effects on transcription located upstream as far as −200

startpoint

mild effects on transcription initiation

severe effects on transcription initiation

alterations in consensus sequences

Figure 15-52 Sequence structure of the bacterial promoters transcribed by RNA polymerase linked to sigma factor σ^{70}. Promoters with "perfect" TTGACA and TATATT consensus sequences at the −35 and −10 positions bind RNA polymerase most strongly. Alterations in the consensus sequences weaken RNA polymerase binding to the extent indicated by the lengths of the arrows extending downward from the sequences. From data presented in P. H. von Hippel et al., *Ann. Rev. Biochem.* 53:389 (1984).

or less strongly, set a base-level rate for the initiation of transcription. The base level is adjusted by two overall types of regulatory proteins called *repressors* or *activators*—repressors reduce the level of transcription below the base level, and activators increase the rate above the base level. Promoters governed by repressors are by far the most common in bacterial systems studied to date.

In general, promoters regulated by repressors match the consensus sequences for the −10 and −35 boxes most closely and have a relatively high base-level rate of transcription. Repressors evidently exert their negative regulatory effects by binding to the promoter and either directly inhibit access by the RNA polymerase holoenzyme or induce conformational changes that have the same effect. The site bound by the repressor, called the *operator*, may lie upstream of the −35 sequence or downstream of the −10 sequence. The operator may overlap these sequences or be completely separate.

The promoters of operons regulated by activators show greater departures from the −10 and −35 consensus sequences, so great that either the −10 or −35 sequence, but not both, may be entirely absent. Usually, the −35 sequence is most greatly altered. As a consequence, these promoters, without intervention of their activators, bind RNA polymerase relatively weakly and are characteristically initiated and tran-

scribed at very low base levels. Activators adjust the rate of transcription upward to a greater or lesser extent.

Activators carry out this role by binding to recognition and control sequences overlapping or upstream of the −35 position. Once bound to the promoter, they increase the strength of RNA polymerase binding either by reacting directly with RNA polymerase or by inducing conformational changes in the promoter that increase its ability to bind RNA polymerase. The promoters of the viruses infecting bacteria are usually of the activator-controlled type.

Sequence Elements Indicating Termination of Bacterial Transcription

Sequences signaling the termination of transcription of bacterial operons occur in two types. One, the *type I* or *rho⁻ terminator*, contains a short inverted sequence centered about 16 to 20 nucleotides upstream of the termination point. Also present in type I terminators is a series of four to eight As, corresponding to a series of Us in the RNA transcript, lying just upstream of or including the termination point. The other bacterial terminator, *type II* or *rho⁺*, lacks the series of As and may or may not contain an inverted sequence. Type II terminators, however, have sequences recognized by *rho*, a factor promoting termination (see below).

Bacterial Transcription

Prokaryotic transcription proceeds through the same phases of initiation, elongation, and termination as in eukaryotes. In the initiation phase the core enzyme, in combination with a sigma factor, collides randomly with the DNA. Collision with a promoter leads to strong binding between the holoenzyme and the DNA. Strong binding is very quickly followed by DNA unwinding and initiation of transcription at the startpoint.

The sigma factor is released after about 8 to 9 nucleotides have been placed in the RNA transcript. The core enzyme then continues transcribing the operon until a termination signal is encountered. Transcription proceeds at an average rate of about 30 to 50 nucleotides per second during the elongation phase.

Termination takes place in type I terminators when the enzyme reaches the inverted sequence followed by a series of As in the template chain (Fig. 15-53a). The inverted sequence, once transcribed, probably induces formation of a hairpin in the transcript, stabilized by a high proportion of G-C bonds. (G-C base pairs are held together by three hydrogen bonds as compared to two for A-U base pairs; see Fig. 14-2.) The hairpin is believed to interrupt movement of the RNA polymerase just as it completes copying the series of As in the template chain. At this point the RNA transcript would be held to the template chain only at its 3' end, in a region containing primarily A-U base pairs (Fig. 15-53b). The relative instability of the A-U base-paired region, along with interference by the hairpin, is believed to cause release of the transcript and RNA polymerase (Fig. 15-53c).

In termination by type II terminators, the *rho* factor is first activated by combination with ATP. After activation *rho* binds to its recognition sequences within the transcript of the type II terminator. Binding of *rho* triggers ATP hydrolysis and, according to current ideas, unwinds the RNA transcript from the DNA template. The unwinding stops transcription and releases the transcript and the polymerase enzyme from the DNA. The *rho* factor is a protein with 419 amino acid residues and distinct domains that interact with RNA and ATP. The molecule apparently combines into a hexamer containing eight *rho* units arranged in a circle; presumably, the RNA transcript wraps around the hexamer as part of termination.

mRNA Genes, Transcription, and Processing in Bacteria

Bacterial mRNA Genes and Their Transcription

Bacterial mRNA genes, like all the genes in these organisms, are supplied with an upstream promoter. Within the promoter is the startpoint, the first DNA base to be copied into the mRNA transcript. Following the startpoint is a 5' untranslated segment that falls

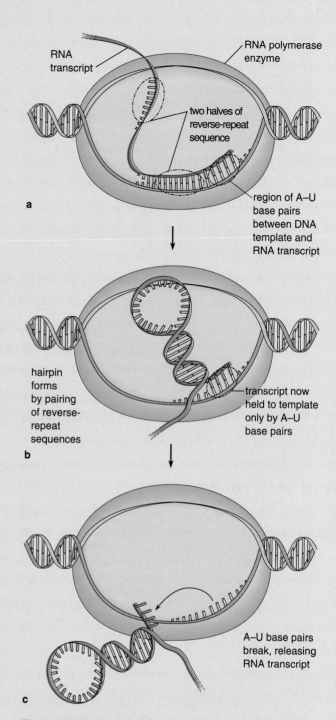

Figure 15-53 Termination by a type I terminator. Termination takes place in type I terminators when the enzyme reaches the inverted sequence followed by a series of As in the template chain **(a)**. The inverted sequence, once transcribed, probably induces formation of a hairpin in the transcript. The hairpin is believed to interrupt movement of RNA polymerase just as it completes copying the series of As in the template chain **(b)**. At this point the RNA transcript is held to the template chain only at its 3' end, in a region containing primarily A-U base pairs, which are relatively unstable. This instability plus interference by the hairpin are believed to cause release of the transcript and RNA polymerase at this point **(c)**. Adapted from an original courtesy of T. Platt, from *Cell* 20:739 (1980). Copyright Cell Press.

between the startpoint and the beginning of the first coding sequence of the operon. The coding sequence for the first polypeptide encoded in the gene begins with the initiator codon for protein synthesis, usually AUG as in eukaryotes. Also as in eukaryotes, the first coding sequence ends with one of three terminator codons, UGA, UAA, or UAG.

In most mRNA operons the second coding sequence is separated from the first by a short spacer sequence. The spacer, although copied into the mRNA transcript, contains no protein code and is not translated. The spacer between coding sequences may contain from 1 to about 100 nucleotides in different bacterial operons. In some operons no spacers occur between coding sequences; in a few the coding sequences overlap by one nucleotide, so that the last base of a UAA or UGA terminator codon of the previous coding sequence shares the first base of the AUG start codon of the next.

Each successive coding sequence in an mRNA operon follows the same pattern: each begins with the start codon for protein synthesis, ends with one of three terminator codons, and in most operons is separated from the next coding sequence by a spacer. Although most operons contain from three to four coding sequences, as few as one to as many as eight or more may be included. The individual coding sequences of operons are frequently termed *cistrons.*

Following the last coding sequence of the operon is the 3′ untranslated segment. At the end of this sequence are the signals for termination of transcription. Operons are usually transcribed into a single mRNA by bacterial RNA polymerase, beginning at the startpoint, reading through all the coding sequences, and ending at the termination site. Some operons contain sequences capable of terminating transcription in the spacers between coding sequences. These internal sequences may be switched on or off as a regulatory mechanism to shorten the mRNA transcript and to omit one or more of the downstream coding sequences. (For details of the mechanisms switching the internal sequences on or off, see p. 733.)

Bacterial mRNA Processing Because no cap structure is added to prokaryotic mRNAs, the 5′ end of an mRNA transcript is characteristically marked by the three phosphates of the first nucleotide in the mRNA chain (see Fig. 15-2). A few base pairs within the mRNA transcript may be methylated while transcription is in progress. A short poly(A) tail is added to the 3′ end of some mRNAs after transcription is complete. Translation of a bacterial mRNA into polypeptides typically begins as soon as the leader and a few codons of the first coding sequence have been copied and continues while transcription of the remainder of the mRNA is in progress (see Fig. 16-28).

Bacterial rRNAs and Their Genes, Transcription, and Processing

Bacterial rRNAs Bacterial rRNAs are significantly smaller than their eukaryotic counterparts and exist in fewer types (see Table 15-2). The small subunits of bacterial ribosomes contain 16S rRNA (see Fig. 15-25a), equivalent in secondary structure and function to the 18S rRNA of eukaryotes. Large ribosomal subunits in bacteria contain 23S rRNA (see Fig. 15-26) and a 5S rRNA (see Fig. 15-29a) equivalent to eukaryotic 28S and 5S rRNA. No separate rRNA corresponding to the 5.8S rRNA of eukaryotes exists in the large ribosomal subunits of bacteria. The sequence equivalent to this eukaryotic rRNA type is included in the 5′ end of prokaryotic 23S rRNA (see Fig. 15-28).

The generally smaller S values for the major rRNAs of prokaryotes are reflected in the overall dimensions and structure of prokaryotic ribosomes, which are significantly smaller and contain fewer proteins than eukaryotic ribosomes. In general, mitochondrial and chloroplast rRNAs and ribosomes resemble those of prokaryotes in structure and function. (Mitochondrial and chloroplast ribosomes and their activities in protein synthesis are described in Ch. 21.)

With the exception of bacterial 5S rRNA, no sequence homologies occur between bacterial and eukaryotic rRNA types. However, as noted on p. 603, each bacterial RNA type can fold into secondary structures, including hairpins and other paired structures, that are similar to eukaryotic forms. (Bacterial 16S and eukaryotic 18S rRNAs are compared in Fig. 15-25.) Bacterial 5S rRNA shares some sequence relationships as well as secondary structures with eukaryotic 5S. Of the approximately 120 nucleotides of 5S rRNAs, 21 positions are the same in eukaryotes and prokaryotes (see Fig. 15-29c).

Although no sequence homologies are noted between the prokaryotic 16S and 23S rRNAs and their eukaryotic equivalents, many similarities are noted between 16S or 23S rRNA in different prokaryotes, especially among closely related species. Even cyanobacterial rRNA types show clear similarities to their bacterial 16S, 23S, and 5S rRNA counterparts—the 16S rRNA of the cyanobacterium *Anacystis*, for example, shows 74% sequence homology with *E. coli* 16S rRNA. Mitochondrial and chloroplast rRNAs also share sequence similarities with prokaryotic rRNAs.

Bacterial rRNA Genes and Their Transcription Different bacterial species contain from about 5 to 15 repeats of the rRNA genes. In *E. coli*, the best-studied species, seven copies of the rRNA gene or operon occur in each cell, distributed at scattered points around the DNA circle. Individual rRNA operons in *E. coli* include one coding sequence for each of the three rRNA types, distributed in the gene in the order 16S → 23S → 5S

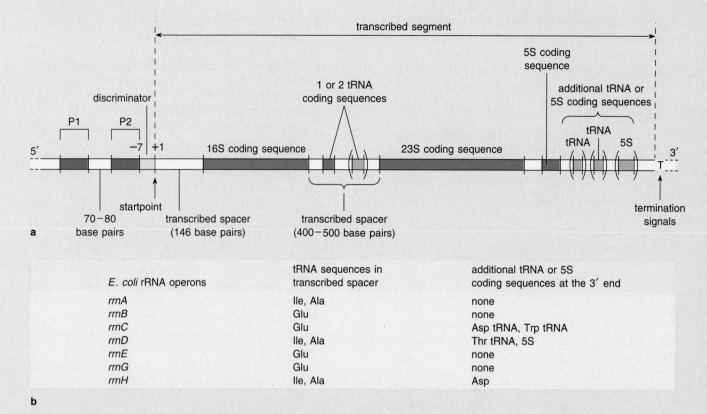

Figure 15-54 Arrangement of coding and other sequences in prokaryotic rRNA operons. **(a)** The sequences coding for 16S, 23S, and 5S rRNAs, and at least one tRNA sequence, shown in dark blue, are present in all prokaryotic rRNA genes in the indicated order. Additional tRNA or 5S rRNA sequences present in some genes are enclosed in parentheses. Each of the two promoters, P1 and P2, has equivalents of the −10 and −35 consensus sequences of bacterial mRNA promoters. **(b)** The seven rRNA operons of *E. coli* and their included coding sequences.

E. coli rRNA operons	tRNA sequences in transcribed spacer	additional tRNA or 5S coding sequences at the 3' end
rrnA	Ile, Ala	none
rrnB	Glu	none
rrnC	Glu	Asp tRNA, Trp tRNA
rrnD	Ile, Ala	Thr tRNA, 5S
rrnE	Glu	none
rrnG	Glu	none
rrnH	Ile, Ala	Asp

(Fig. 15-54). Each coding sequence is separated from the next by a transcribed spacer. Transcribed spacers also occur at the 5' and 3' ends of the gene.

The rRNA genes of *E. coli* are *mixed operons* that contain tRNA codes as well as the 16S, 23S, and 5S coding sequences. One or two tRNA coding sequences are placed in the transcribed spacer separating the 16S and 23S coding sequences; three of the seven rRNA genes of *E. coli* contain an additional one or two tRNA coding sequences following the 5S coding sequence. One of the rRNA genes of *E. coli* contains an extra 5S coding sequence following a tRNA sequence at the 3' end of the operon.

Each rRNA gene of *E. coli* has two promoters, one following the other in the 5'-flanking region (see Fig. 15-54). The first promoter, *P1*, is centered about 300 nucleotides upstream of the startpoint. The second, *P2*, is centered about 110 nucleotides downstream of the first, with its 3' end separated from the startpoint by a series of seven nucleotides that is highly conserved among the different rRNA operons. Within each of the two promoters are sequences equivalent to the −10

and −35 consensus sequences of bacterial mRNA promoters. Sequences with positive effects on transcription initiation also occur at sites upstream of the two promoters.

The two promoters of *E. coli* rRNA operons apparently respond to different regulatory mechanisms—P1 possibly to regulatory proteins active in exponential growth, in which cellular synthesis and division proceed at the most rapid rate, and P2 to regulatory proteins active during periods of more restricted growth. Presumably, the control proteins active at these different periods are tailored to match sequences in one or the other of the two promoters. The conserved seven-base sequence between the second promoter and the startpoint, called the *discriminator*, also figures importantly in the mechanisms regulating the rRNA genes in *E. coli*.

Following the startpoint is the transcribed spacer that precedes the 16S coding sequence in the rRNA operon. At the other end of the gene, at the 3' end of the transcribed spacer following the last coding sequence of the operon, lie the sequence elements sig-

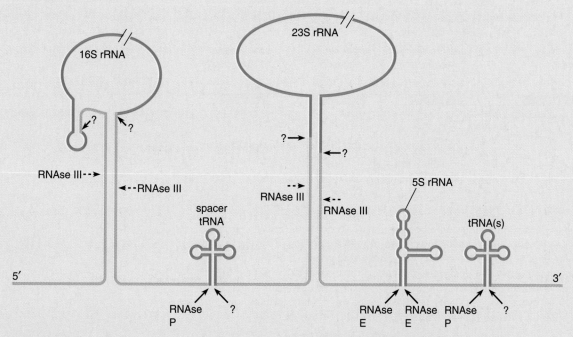

Figure 15-55 Processing reactions cutting 16S, 23S, and 5S rRNAs and tRNAs from their precursor in *E. coli*. Complementary sequences preceding and following the 16S and 23S rRNAs allow these segments to fold into two large hairpins, with the rRNA sequences forming the loops. The first cuts are made by an RNA endonuclease, RNAse III, in the stems of the hairpins (dashed arrows) to free the 16S and 23S rRNAs, still with surplus segments at their 5' and 3' ends. These initial cuts also release the tRNA in the spacer between the 16S and 23S segments with its surplus segments and another segment containing the 5S rRNA and tRNA sequences at the 3' end of the pre-rRNA. The rRNA and tRNA sequences are then cut to their mature lengths (in darkest red) by other enzymes (solid arrows). The enzymes removing surplus segments from 16S and 23S rRNA remain unknown, as do the enzymes removing surplus nucleotides from the 3' ends of the tRNAs; the 5' ends of tRNAs are cut by RNAse P. The 5S rRNA is processed to its mature length by RNAse E, an endonuclease that cuts precisely at the 5' and 3' ends of the molecule. The known enzymes were identified in mutants and through the activities of the isolated and purified enzymes in cell-free systems. Redrawn from an original courtesy of D. Schlessinger. Reproduced, with permission, from *Ann. Rev. Microbiol.* 44:105. © 1990 by Annual Reviews, Inc.

naling the termination of transcription. In bacterial rRNA genes these sequences include an inverted sequence followed by a series of Ts, in a pattern typical of type I bacterial terminators.

An entire rRNA operon is transcribed into a single precursor RNA that includes the transcribed spacers and all the individual rRNA and tRNA sequences of the operon. The rRNAs and tRNAs are released from the precursor as separate molecules by processing reactions that begin before transcription is complete (Fig. 15-55 outlines these reactions).

Bacterial tRNA Genes, Transcription, and Processing

Bacterial tRNAs and Their Genes   Bacterial tRNA molecules closely resemble their eukaryotic counter-

parts in structure and function. Invariant, semi-invariant, and many of the modified nucleotides fall at the same positions in bacterial and eukaryotic tRNAs (see Fig. 15-40). All bacterial tRNAs can also fold into the same cloverleaf in two dimensions and the same L-shaped structure in three dimensions as eukaryotic tRNAs.

Bacterial tRNA genes are dispersed throughout the DNA circle. Some occur in "pure" tRNA operons containing no other type of gene. In *E. coli* these pure tRNA operons contain from one to seven individual tRNA coding sequences. Other tRNA genes occur in mixed operons with rRNA sequences or with sequences coding for proteins. *E. coli* has 77 tRNA sequences, coding for some 60 different tRNA types, distributed throughout the genome in mixed and pure operons.

The tRNA coding sequences in both the mixed and pure operons are surrounded by transcribed spacer sequences. In pure operons containing a single tRNA,

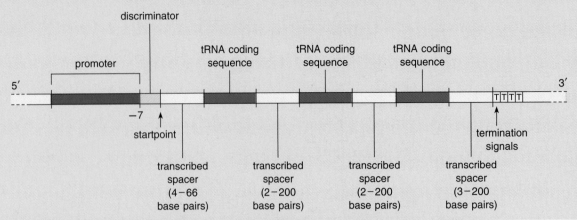

Figure 15-56 Arrangement of coding and other sequences in bacterial operons encoding only tRNA genes. The promoter resembles those of mRNA operons in structure and contains representatives of the −10 and −35 consensus sequences. A seven-nucleotide discriminator sequence occurs just upstream of the startpoint, as in rRNA genes. Sequences with positive effects on transcription initiation also occur upstream of the promoter between about −40 and −98. Bacterial tRNA genes have type I termination sequences and contain a reverse-repeat sequence followed by a series of Ts.

transcribed spacers occur at both the 5′ and 3′ ends of the coding sequence. Transcribed spacers also occur between each coding sequence in pure operons containing multiple tRNA sequences (Fig. 15-56). In *E. coli* the CCA sequence forming the 3′ end of mature tRNA molecules is encoded in the DNA in all genes sequenced to date. This is in distinct contrast to eukaryotic tRNA genes, in which the terminal CCA is added during processing and not encoded in the genes. In other bacteria, such as *Bacillus subtilis*, the CCA sequence element is included in some genes and absent in others. All bacteria apparently possess the enzymes necessary to add the CCA if it is incomplete or missing in the transcript.

Promoters for bacterial tRNA genes occur upstream of the transcribed segments. Only single promoters have been identified in the tRNA operons sequenced to date. The tRNA promoters resemble those of mRNA operons in structure and contain representatives of the −10 and −35 consensus sequences. A highly conserved, seven-nucleotide discriminator occurs just in advance of the startpoint as in rRNA genes. Sequences that enhance transcription initiation also occur upstream of the promoter in tRNA operons, in the region from about −40 to −98. The termination sequences of bacterial tRNA genes are type I and contain a reverse-repeat sequence followed by a series of Ts.

Recently, introns were discovered in several tRNA genes of purple photosynthetic bacteria, in archaebacteria (a bacterial group that also possesses eukaryotelike characteristics—see p. 1141), and in cyanobacteria. All these introns have internal sequences typical of group

I introns, and are therefore probably capable of self-splicing (see Fig. 15-33). The tRNA introns in archaebacteria were the first to be detected in bacterial genes of any type. The cyanobacterial intron, which occurs in a leucine tRNA gene, also appears in the same position in the equivalent leucine tRNA gene of chloroplasts.

Bacterial tRNA Processing All bacterial tRNAs are transcribed in the form of precursors with extra nucleotides at both their 5′ and 3′ ends. The pre-tRNAs transcribed from tRNA genes containing multiple tRNAs include several coding sequences separated by transcribed spacers. (Fig. 15-57 shows a pre-tRNA transcript with two included tRNAs.) The tRNAs are released from the precursors by a series of cuts and nibblings that involve several enzymes (Fig. 15-58). Attachment of the 3′-terminal CCA, if required, is carried out by the same series of reactions attaching the CCA to tRNAs in eukaryotes (see p. 623).

The processing steps converting the tRNAs in mixed rRNA-tRNA or polypeptide-tRNA operons to the finished state depend on the form in which they are released from the mixed precursor. Most of these tRNAs are released with fully clipped 5′ ends and require removal of surplus nucleotides only at the 3′ end.

While the cutting and trimming reactions are in progress, a battery of enzymes chemically modifies individual bases to other forms at precise locations in the tRNA sequence. The various enzymes trimming tRNAs to their mature size, adding the terminal CCA, and modifying individual bases apparently recognize signals in the sequences or secondary structure of re-

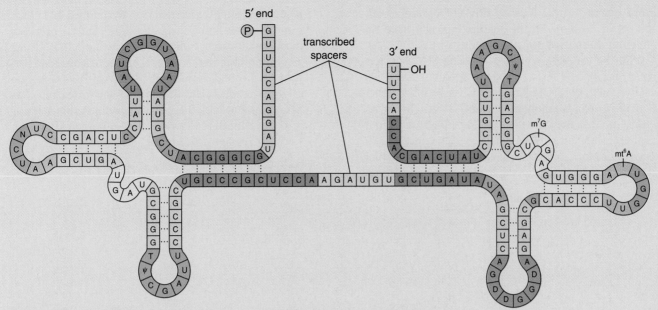

Figure 15-57 A pre-tRNA from *E. coli* containing the sequences for glycine and threonine tRNA. The nucleotides in the spacers are removed during processing. Modified from an original courtesy of J. A. Carbon.

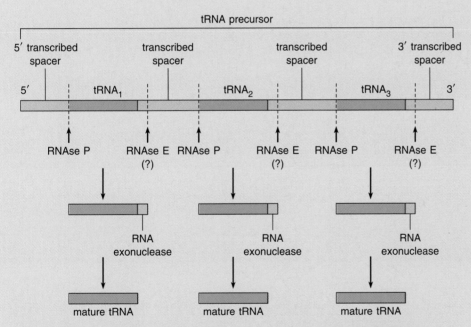

Figure 15-58 Processing of tRNA precursors transcribed from pure tRNA operons. The small vertical arrows indicate sites at which cuts are made. Initial breaks (not shown) may be made by RNAse III to release the tRNAs; the cuts leave the tRNAs with extra sequences still at their 5′ and 3′ ends. RNAse P processes the 5′ end of the tRNAs to mature length in a single reaction. The reactions processing the 3′ ends take place in at least two steps: an endonuclease, possibly RNAse E, cuts the 3′ end to a length including a few more nucleotides than the mature rRNA; these extra 3′ nucleotides are then nibbled away one at a time by an as yet unknown exonuclease. The enzyme stops its 3′ nibbling at the terminal CAA, if present; if no CCA is included in the precursor, the exonuclease trims to the nucleotide to which the 3′-terminal CCA will be attached.

gions retained in the finished tRNAs, because substitutions or deletions in the spacers typically have no effect on these processing reactions.

Of the various enzymes processing tRNA precursors, one, *RNAse P*, is of special interest because its catalytic activity appears to depend on its RNA rather than its protein component. Bacterial RNAse P, like its eukaryotic counterpart (see p. 623), is built up from a protein combined with M1 RNA, which contains 377 nucleotides in *E. coli*. Experiments in the laboratory of S. Altman and S. Pace showed that 5' cleavage of tRNA precursors could be carried out by purified M1 RNA in protein-free solutions. Although catalysis proceeds most rapidly when both the protein and M1 RNA are associated together in an intact ribonucleoprotein particle, the protein component of RNAse P alone has no catalytic activity. This makes RNAse P a ribozyme, an RNA molecule with the ability to catalyze a biochemical reaction. Altman received the Nobel Prize in 1989 with Cech for his work with ribozymes.

Suggestions for Further Reading: See p. 636.

16

THE INTERACTION OF mRNA, rRNA, AND tRNA IN PROTEIN SYNTHESIS

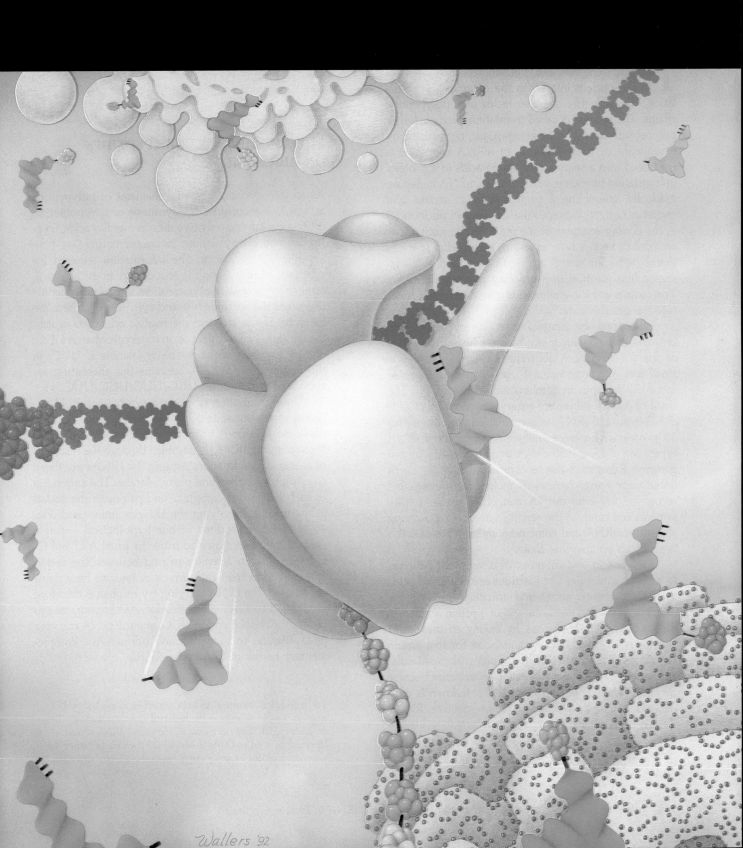

Wallers '92

The mRNA, rRNA, and tRNA molecules transcribed in the nucleus interact in the cytoplasm in translation, the assembly of amino acids into polypeptide chains. The process is called translation because during polypeptide assembly the genetic code, represented by a sequence of nucleotides in an mRNA molecule, is translated into a sequence of amino acids in a protein. In translation ribosomes read along an mRNA molecule, gradually assembling a protein with an amino acid sequence corresponding to the sequence of nucleotides in the coding segment of the mRNA. The nucleotides are read three at a time as *codons;* each codon specifies that a particular amino acid is to be added at a corresponding position in a growing polypeptide chain. The codons and the amino acids they designate constitute the genetic code.

Ribosomes are complex assemblies of rRNA and proteins that contain many binding sites and centers of catalytic activity. A ribosome is assembled from one small and one large subunit, each containing one or more rRNA types in combination with proteins. The small ribosomal subunit of eukaryotes contains a single rRNA type, 18S rRNA, in combination with more than 30 proteins. The large subunit contains three rRNA types, 28S, 5.8S, and 5S rRNA, and at least 40 different proteins. Ribosomes can be considered as immense ribonucleoprotein enzymes with multiple active sites that recognize and bind mRNA and other required components and catalyze the reactions involved in moving along the mRNA and combining amino acids one at a time into a polypeptide chain.

Each amino acid enters protein synthesis attached to a tRNA molecule. The reactions linking amino acids to tRNAs, termed *amino acid activation*, provide the ultimate basis for the accuracy of protein synthesis. This is because the reactions, by linking an amino acid with a tRNA containing an anticodon for that amino acid, directly associate a coding triplet with a particular amino acid. The reactions are called an *activation* because chemical energy released by ATP breakdown is conserved as amino acids attach to their respective tRNAs. This conserved energy, released as amino acids are transferred from tRNAs to a growing polypeptide chain, provides a major driving force for protein synthesis. The complex between an amino acid and its tRNA is an *aminoacyl-tRNA.*

The steps and elements in protein synthesis, including the assembly of amino acids into polypeptides, the structure of ribosomes, and the attachment of amino acids to tRNA, are described in this chapter. The discovery and nature of the genetic code directing the synthesis of proteins are also discussed. This chapter concentrates on protein synthesis as it occurs in eukaryotes; the equivalent mechanisms of prokaryotes are described in Supplement 16-1. A pattern of protein synthesis taking place without the participation of mRNA and ribosomes in some prokaryotes and lower eukaryotes is described in Supplement 16-2.

THE ASSEMBLY OF POLYPEPTIDES ON RIBOSOMES

As an introduction to the mechanisms of polypeptide assembly, we consider the synthesis of a hypothetical protein containing only two different amino acids, tryptophan and phenylalanine, in an alternating Trp-Phe-Trp-Phe . . . sequence. The information required for synthesis of this protein is encoded in an mRNA molecule in the sequence of nucleotides containing adenine (A), uracil (U), guanine (G), and cytosine (C). These are taken by threes to form the nucleic acid code words, the codons. The mRNA codon for tryptophan is UGG; one of several codons for phenylalanine is UUC. In this example the mRNA contains the alternating sequence of two codons, UGG/UUC/UGG/UUC . . . .

During initiation of protein synthesis, mRNA attaches to the subunits of a ribosome. Surrounding the ribosome is a pool of amino acids attached to their corresponding tRNAs, assembled through the reactions of amino acid activation. Among the tRNAs are those carrying tryptophan and phenylalanine. The anticodon of each tRNA can recognize and pair with the codon in the mRNA specifying the tRNA's amino acid (Fig. 16-1a). The anticodons in our hypothetical example contain the bases expected from the usual A-U and G-C base pairs. (In actuality pairing between the codon and anticodon frequently involves unusual base pairs; see below.) The tRNA carrying tryptophan is therefore considered to carry the anticodon ACC,[1] complementary to the mRNA codon UGG, and the tRNA carrying phenylalanine to carry the anticodon AAG, complementary to the mRNA codon UUC.

---

[1] To be technically correct, an anticodon, like all nucleic acid sequences, should be read in the standard $5' \rightarrow 3'$ direction. The anticodon ACC, corresponding to the codon UGG, should therefore be read as CCA. However, for the sake of clarity the anticodons in this example are given in the $3' \rightarrow 5'$ direction.

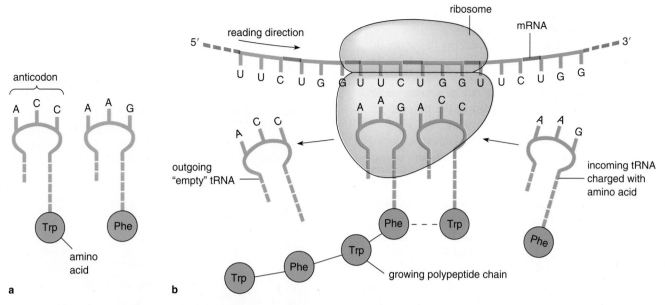

Figure 16-1 The overall reactions of polypeptide assembly. **(a)** The tRNAs for trypto-phan (Trp) and phenylalanine (Phe) and the hypothetical anticodons ACC and AAG used in the text example. The tRNA diagrams show only a small segment of the tRNA clover-leaf. **(b)** At the ribosome anticodons of the tRNAs carrying either phenylalanine (Phe) or tryptophan (Trp) form complementary base pairs with the corresponding codons of the mRNA. If the sequence UUC is present at the ribosomal pairing site, a tRNA carrying the anticodon AAG pairs with the mRNA, carrying with it the amino acid Phe. At the next triplet on the messenger, the mRNA has the codon UGG; a tRNA carrying Trp pairs at this point. Through the expenditure of energy originally derived from ATP during amino acid activation, a peptide bond forms between Phe and Trp. The ribosome then moves to the next codon on the mRNA and the process repeats, this time attaching Phe to the growing peptide chain. As each amino acid is added to the polypeptide, its now-empty tRNA is released. Translation continues in this way until the ribosome reaches the end of the mRNA code, producing, in this example, a polypeptide chain of alternating Trp and Phe amino acids.

At the ribosome the anticodons of tRNAs carrying either phenylalanine or tryptophan form complementary base pairs with the corresponding codons of the mRNA (Fig. 16-1b). If the sequence UUC is present at the ribosomal pairing site, a tRNA carrying the anti-codon AAG attaches to the mRNA and ribosome, carrying with it the amino acid phenylalanine. At the next triplet on the messenger, the ribosome encounters the sequence UGG; a tRNA carrying tryptophan attaches at this point.

Phenylalanine and tryptophan are brought into close proximity by this pairing interaction. Through the expenditure of energy originally derived from ATP during amino acid activation, a peptide bond forms between the two amino acids. The enzymatic activity catalyzing formation of the bond is part of the ribosome. The ribosome then moves to the next codon on the mRNA and the process repeats, this time attaching a phenylalanine to the growing peptide chain. As each successive amino acid is added to the polypeptide, its now-empty tRNA is released. Translation continues in this way until the ribosome reaches the end of the

mRNA code, producing in this example a long poly-peptide chain containing tryptophan and phenylalanine in an alternating sequence.

The reactions of polypeptide assembly on ribo-somes fall into three overall phases. In the first, called *initiation*, the large and small ribosomal subunits assem-ble with mRNA and a specialized aminoacyl-tRNA called an *initiator tRNA*. A series of *initiation factors* controls and speeds the process of initiation. Once the ribosomal subunits are assembled with mRNA and the initiator tRNA, the second phase of polypeptide as-sembly, *elongation*, begins. In this phase, which is speeded by a group of *elongation factors*, aminoacyl-tRNA complexes bind to the ribosome in sequence according to the mRNA code. As the sequential pairing proceeds, the amino acids are transferred from the tRNAs to a gradually lengthening polypeptide chain. The elongation cycle continues until the ribosome reaches the end of the coded message. At this point protein synthesis enters its final phase, *termination*. Dur-ing termination mRNA and the completed polypeptide are released, and the ribosomal subunits separate. Rapid progress through the termination phase requires the presence of *termination factors*.

All three phases require an energy input in the form of ATP or GTP hydrolysis. Both nucleotides are hydrolyzed as part of initiation; steps in the elongation phase also hydrolyze GTP at the rate of two molecules of GTP for each amino acid added to the polypeptide chain. Termination similarly requires the hydrolysis of GTP. The ATP and GTP molecules hydrolyzed during polypeptide assembly represent a significant energy input into protein synthesis in addition to the ATP hydrolyzed as part of amino acid activation.

The reactions of polypeptide assembly were pieced together primarily in cell-free systems (see p. 124) developed by P. C. Zamecnik, W. V. Zucker, H. M. Schulman, D. Nathans, and F. Lipmann, and their coworkers and others. These cell-free systems—extracted from sources such as *E. coli*, rat liver cells, mammalian reticulocytes (immature blood cells), wheat germ, and yeast—include ribosomes, aminoacyl-tRNAs, and the initiation, elongation, and termination factors required for the most efficient progress of protein synthesis.

Initiation

Initiation in eukaryotes takes place in a series of steps worked out by W. C. Merrick, J. W. B. Hershey, W. F. Anderson, R. E. Thach, A. J. Shatkin, and their colleagues and others. The reactions of initiation involve the interaction of the large and small ribosomal subunits, mRNA, an initiator aminoacyl-tRNA, GTP, and a large group of initiation factors (Table 16-1). The reactions unite a small ribosomal subunit with an mRNA molecule, bind the initiator tRNA, and then add the large ribosomal subunit. The complex produced by initiation is ready to enter the elongation phase of polypeptide assembly by adding a second aminoacyl-tRNA and forming the first peptide linkage.

The initiator tRNA is a specialized aminoacyl-tRNA that pairs only with the AUG appearing as the initiator codon at the beginning of both eukaryotic and prokaryotic protein codes. A different aminoacyl-tRNA pairs with an AUG codon located internally in the message. Because AUG specifies the amino acid methionine, newly synthesized proteins begin with methionine in both eukaryotes and prokaryotes. The eukaryotic initiator tRNA, with its attached methionine, is identified as *Met-tRNA$_i$*, in which *Met* designates methionine and the subscript *i* indicates that the tRNA is the specialized initiator type.

The Steps in Eukaryotic Initiation In the first reaction of initiation (step 1 in Fig. 16-2), a ribosome separates into large and small subunits. Met-tRNA$_i$, prepared for its interaction by combination with GTP in a side reaction involving an initiation factor (step 2), then adds to the small subunit (step 3). The small subunit is now

Table 16-1	Some Eukaryotic Initiation Factors
Factor	Possible Functions in Initiation
eIF1	May stabilize binding of initiator tRNA to small ribosomal subunit.
eIF2	Forms complex with GTP and initiator tRNA; promotes binding of initiator tRNA to small subunit.
eIF2B	Promotes exchange of GTP for GDP associated with eIF2 to regenerate eIF2·GTP complex.
eIF3	Promotes dissociation of ribosomal subunits; stabilizes ribosomal subunits in dissociated form.
eIF4A	Binds to 5' untranslated segment of mRNA and promotes its melting.
eIF4B	May promote unwinding function of eIF4A.
eIF4C	Promotes dissociation of ribosomal subunits; stabilizes small subunit in dissociated form.
eIF4E	Recognizes and binds 5' cap of mRNA; promotes binding of mRNA to small ribosomal subunit.
eIF4F	Composite factor consisting of eIF4A, eIF4E, and p220.
eIF5	Promotes binding of large ribosomal subunit to complete initiation.
eIF6	May stabilize large ribosomal subunit in dissociated form.
p220	May align eIF4E on 5' cap of mRNA during binding.

ready to add the mRNA in a complex reaction (step 4) driven by ATP hydrolysis. Attachment takes place at the 5' cap of the mRNA; once attached, the small subunit moves or "scans" along the mRNA until it reaches the AUG initiator codon. The movement prepares the small subunit for the final step in initiation, addition of the large ribosomal subunit. This reaction (step 5), which is driven by hydrolysis of the GTP brought to the complex with the initiator tRNA, completes initiation. The completed ribosome, with mRNA and Met-tRNA$_i$ attached, is now ready to enter the elongation phase.

Each reaction of initiation is speeded by initiation factors. (Fig. 16-3 outlines one of several models for the roles of these factors in initiation.) Many of the factors add to the growing complex as initiation proceeds; with addition of the large subunit in the final step, all are released. The most impressive things about these factors are their number and the complexity of their interactions with the ribosomal subunits, mRNA, and the initiator tRNA. In prokaryotes, in comparison,

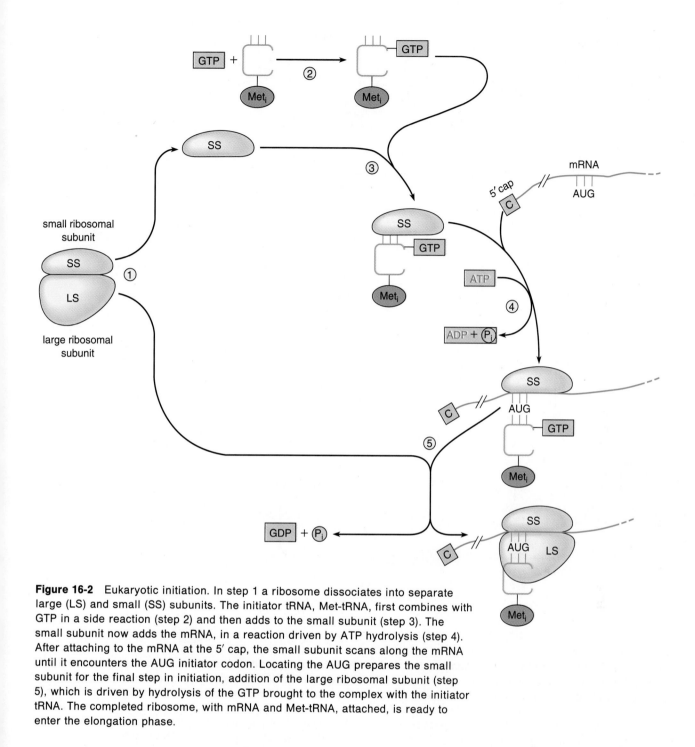

Figure 16-2 Eukaryotic initiation. In step 1 a ribosome dissociates into separate large (LS) and small (SS) subunits. The initiator tRNA, Met-tRNA, first combines with GTP in a side reaction (step 2) and then adds to the small subunit (step 3). The small subunit now adds the mRNA, in a reaction driven by ATP hydrolysis (step 4). After attaching to the mRNA at the 5' cap, the small subunit scans along the mRNA until it encounters the AUG initiator codon. Locating the AUG prepares the small subunit for the final step in initiation, addition of the large ribosomal subunit (step 5), which is driven by hydrolysis of the GTP brought to the complex with the initiator tRNA. The completed ribosome, with mRNA and Met-tRNA, attached, is ready to enter the elongation phase.

only three factors are involved in initiation (see Supplement 16-1). The greater number of eukaryotic factors reflects much finer control of the rate of individual substeps within the process of initiation, as part of the generally more complex regulatory and differentiation pathways of eukaryotic cells. (Details of translational regulation in prokaryotes and eukaryotes are presented in Ch. 17.)

The initiation and other factors speeding steps in polypeptide assembly were discovered through re-

search with cell-free systems. Investigators noted almost immediately that the purest systems, those consisting of little more than ribosomes, mRNA, aminoacyl-tRNAs, and an energy source, assembled polypeptides very slowly or not at all. In order to proceed with any efficiency, the system had to retain much of the cytoplasmic solution surrounding ribosomes in the cell. The cytoplasmic substances necessary for polypeptide assembly were later identified as soluble proteins with properties similar to enzymes—the initiation,

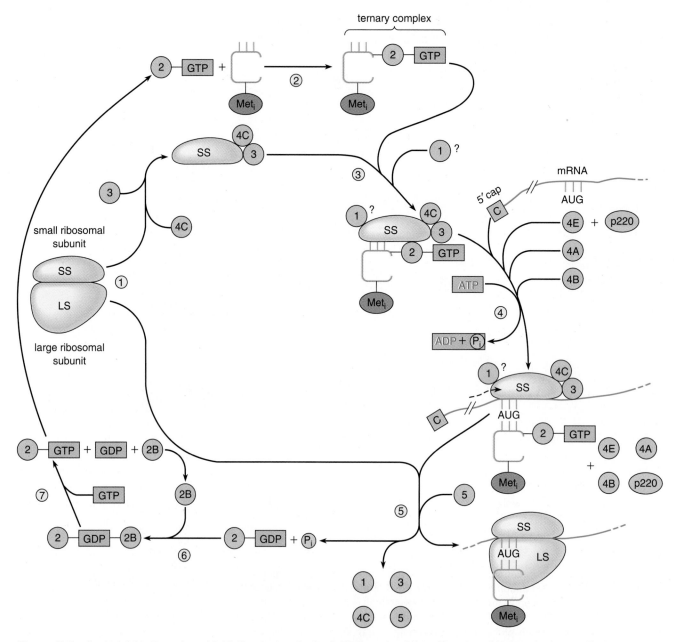

Figure 16-3 A model for the roles of initiation factors in the initiation of polypeptide assembly in eukaryotes. Each eukaryotic factor is identified as *eIF* followed by a number (see Table 16-1); the small *e* indicates *eukaryotic*. In the diagram the eIF prefixes have been omitted for each factor; each is shown as a number in an orange circle. Dissociation of the ribosomal subunits in step 1 is promoted by the activity of at least two initiation factors, eIF3 and eIF4C. The two initiation factors remain bound to the small subunit. In the side reaction preparing the initiator tRNA for addition to the complex (step 2), another initiation factor, eIF2, already bound to GTP by a side cycle (described below), adds the initiator tRNA to form the three-part, or ternary, complex eIF2·GTP·Met-tRNA$_i$. The ternary complex binds to the small subunit (step 3). Another initiation factor, eIF1, may join at this point to stabilize the complex. The small subunit is now ready to add the mRNA in step 4, which is accelerated by at least four initiation factors, eIF4A, eIF4B, eIF4E, and a polypeptide known as p220. The eIF4E factor is believed to recognize and bind the 5' cap of the mRNA; p220 may align eIF4E during this binding. The attachment of eIF4E sets up conditions favoring the addition of eIF4A and eIF4B to the complex. (The combination of p220, eIF4A, and eIF4E is often named as the composite initiation factor *eIF4F*.) Once the complex is fully assembled, the combination of factors "melts" or unwinds any secondary structures in the 5' untranslated portion of the mRNA between the 5' cap and the initiator codon. The melting is believed to be promoted primarily by eIF4A, which hydrolyzes ATP as an energy source for the unwinding; eIF4B may enhance the unwinding process. The unwinding opens out the mRNA, greatly facilitating attachment and movement of the small ribosomal subunit to the AUG initiator codon. Evidently, p220 is released during the scanning process. The final step in initiation, addition of the large ribosomal subunit (step 5), is also facilitated by an initiation factor, eIF5. As the large subunit adds, the GTP brought in with the initiator tRNA is hydrolyzed, and the remaining initiation factors associated with the small subunit are released, along with eIF5, GDP, and inorganic phosphate (P$_i$). The GDP remains bound to eIF2 at this point. The eIF2·GTP complex used in step 2 is regenerated in a side cycle that employs still another initiation factor, eIF2B (steps 6 and 7). These reactions, in effect, exchange a fresh GTP from the medium for the GDP linked to eIF2.

elongation, and termination factors. The roles of these factors in protein synthesis were determined in experiments that involved purification of individual factors and observation of the effects of their addition or removal in cell-free systems. Research continues on an intensive level today, particularly on initiation. The results of this research are highly significant because translational regulation, now recognized as nearly as important to cell differentiation development as transcriptional regulation, involves controls acting primarily on initiation factors. (The Experimental Process essay by A. C. Lopo on p. 656 describes an experiment identifying phosphorylation of an initiation factor as a critical translational control during fertilization.)

The AUG Initiator Codon and Ribosome Scanning In 90% to 95% of eukaryotic mRNAs the initiator codon is the first AUG downstream of the 5' cap. This fact led to an initial hypothesis that the small subunit simply scans along the mRNA until it reaches the first AUG. However, additional research by M. Kozak and others revealed that a few eukaryotic mRNAs contain AUG sequences upstream of the AUG initiator codon. In these unusual mRNAs the small subunit usually ignores the upstream AUG sequences and uses only the AUG at the beginning of the coding sequence to initiate for protein synthesis. The scanning hypothesis was modified to include the idea that sequences near the correct AUG initiator codon must contribute to its identification. Substitution-deletion studies by Kozak and Shatkin later identified the sequence indicating the correct AUG in higher eukaryotes as GCCGCCPuCCAUGG (see p. 583; AUG is underlined in the sequence; Pu = a purine). A few eukaryotic mRNAs, possibly by means of specialized sequence information or secondary structure in the 5' untranslated segment, are able to promote direct attachment of the small ribosomal subunit to the initiator codon without initial binding to the 5' cap and scanning.

A few mRNAs have two initiator codons that may be used alternatively. The mRNA encoded by the *myc* gene, for example (*myc* is an oncogene; see p. 934), has a CUG codon located some distance upstream of an AUG initiator codon. During initiation a ribosome either may use the upstream CUG as the initiator codon or may scan through this codon and use the downstream AUG as the initiator. (CUG regularly substitutes for AUG as an initiator codon in a few mRNAs; see p. 583.) The alternate initiation points produce two proteins that differ in the presence or absence of an amino acid segment at the N-terminal, or "front," end of the protein. This mechanism is one of several that permit two different proteins to be assembled from the same mRNA.

Special Features and Significance of the Initiator tRNA The initiator tRNA allows the protein synthesis mechanism to circumvent an operational problem with initiation. Ribosomes have two major binding sites for tRNAs. One, the *P site*, is structured to bind tRNAs linked to a polypeptide chain, termed *peptidyl-tRNAs*. The second site, the *A site*, is structured to bind aminoacyl-tRNAs, that is, tRNAs with a single linked amino acid. However, during initiation the initiator tRNA attaches to the P site, even though it carries only a single amino acid. The second aminoacyl-tRNA to bind to the ribosome, which is a noninitiator tRNA, attaches to the A site. Juxtaposition of the two aminoacyl-tRNAs at the A and P sites sets up the conditions necessary for formation of the first peptide linkage.

The basis for recognition of the initiator tRNA by the P site remains uncertain. Met-tRNA$_i$ has several unique structural features that may cause the ribosome to recognize it as a tRNA bearing a peptide chain rather than a single amino acid (see Fig. 16-4a). One or more initiation factors may also contribute to the critical attachment of the initiator tRNA to the P site.

mRNA Structure and Initiation In eukaryotic mRNAs three structural features of the 5' end of the molecules seem to be of paramount importance in initiation: (1) the 5' cap, (2) the degree of secondary structure in the 5' untranslated segment, and (3) the consensus sequence GCCGCCPuCCAUGG contributing to recognition of the AUG initiator codon.

The 5' cap is a prerequisite for initiation by almost all eukaryotic mRNAs. The only known exceptions to this rule are the capless mRNAs of several viruses infecting eukaryotic cells, including the poliovirus and the rhinoviruses responsible for the common cold, and the mRNA encoding a single cellular protein, the *immunoglobulin binding protein (BiP)*. The viruses activate an enzyme that breaks down some of the cap-binding initiation factors necessary for initiation by the capped mRNAs of the host cells. The resulting inhibition of the cap-binding complex shuts down translation of the host cell mRNAs. Initiation of the capless viral mRNAs is stimulated by a viral protein, *VPg*, which evidently attaches to specific sequences in the 5' untranslated segment of the viral mRNA and promotes its recognition and attachment to small ribosomal subunits. The mechanism effectively limits infected cells to synthesis of viral proteins and, as far as is known, BiP.

The continued synthesis of BiP, a protein that promotes association and folding of immunoglobulin polypeptides in the ER, suggests that BiP mRNA contains sequences in the 5' untranslated segment that can initiate translation. This conclusion is supported by the recent experiments of D. J. Macejak and P. Sarnow,

When Sperm Meets Egg: Initiation Factors and Translational Control at Fertilization

Alina C. Lopo

ALINA C. LOPO's interests have focused on the cellular and molecular detail of early development for most of her research career. After obtaining B.S. and M.S. degrees from the University of Miami, she completed doctoral research at the University of California at Davis and San Diego, then postdoctoral work at the U.C. San Francisco and U.C. Davis Schools of Medicine. In 1985 she joined the faculty of the Division of Biomedical Sciences at U.C. Riverside, where most of the work described here was carried out. She is now pursuing an M.D. at the UCLA School of Medicine. In her spare time, she is a wife and mother.

A key consequence of fertilization is the dramatic activation of the metabolic machinery of the dormant egg. One of the many cellular processes that is rapidly and efficiently turned on at fertilization is protein synthesis. This translational activation is a fundamental step in the activation of development, since without the full activation of protein synthesis, development cannot be completed successfully. The mechanisms regulating translational activation at fertilization have been the subject of extensive investigation by cell, molecular, and developmental biologists since egg activation was first described by Tore Hultin in 1950. A puzzling aspect of this phenomenon is that the unfertilized egg contains all the necessary molecules to carry out protein synthesis. Although progress had been made in understanding how the ionic changes at fertilization influence translational activation, until relatively recently the macromolecular changes were poorly understood.

In the early 1980s work by several laboratories demonstrated that the regulation of protein synthesis at fertilization was, as in most other eukaryotic cells, primarily by regulation of the initiation phase. Working with John Hershey, we showed that possibly the earliest change in initiation was somehow related to the activity of initiation factor eIF4F.[1] In the preceding few years several groups, including Hershey's, had shown that, in rabbit reticulocyte lysate and in mammalian cells in culture, turning protein synthesis on or off correlated well with the phosphorylation or dephosphorylation of some initiation factors. This led us to ask whether similar changes took place in the egg following fertilization.

We focused our initial studies on the smallest subunit of eIF4F, the so-called cap-binding protein, or eIF4E. The reasoning behind this centered on the knowledge that the presence of the 5' or m[7]G cap on mRNA is an absolute requirement for efficient protein synthesis in eukaryotes. Previous studies suggested that the bulk of mRNAs stored in the egg during oogenesis were capped, so we thought the key to translational activation may reside in the level of activity of the cap-binding protein. If this protein were somehow activated at fertilization, then perhaps mRNA could enter the initiation pathway. At approximately the same time we were working on this problem, Roger Duncan and John Hershey obtained evidence that, at least in HeLa cells, eIF4E appeared to be dephosphorylated when protein synthesis was repressed following heat-shock treatment. This reinforced our belief that we were on the right track, that the change was biologically likely.

We purified the eIF4E protein from unfertilized eggs of the sea urchin using m[7]G affinity chromatography.[2] We used sea urchin eggs because it is easy and inexpensive to obtain very large quantities of pure unfertilized eggs. The large quantities were necessary because, based on evidence from mammalian cells in culture, eIF4E is of relatively low abundance in cells. The sea urchin egg eIF4E turned out to be quite similar in molecular weight and amino acid composition to the mammalian factor, which was being characterized by Bob Rhoads' group at the University of Kentucky and Nahum Sonenberg at McGill University. This was not surprising, because protein synthesis is a highly conserved cellular process.

We had improved our yields of pure sea urchin egg eIF4E approximately ten-fold, and we were soon ready to start making rabbit antibodies against the factor. Our plan was this: first, raise rabbit antibodies against sea urchin eIF4E. Then, prepare two-dimensional polyacrylamide gels of homogenates from unfertilized eggs and from eggs fertilized and homogenized at 1-minute intervals up to 15 minutes after fertilization. Use each of these 2D gels to prepare Western blots, and use our antibody to probe the blots and ask how many forms of the eIF4E were present. Since phosphate groups are charged, phosphorylation or dephosphorylation of a protein will change its charge. This would be observable on the horizontal dimension of the 2D gel.

Although this plan is quite straightforward conceptually, it took six months of very long hours and hard work to perfect the procedure and obtain results. Finally, we consistently obtained the following results: In the unfertilized egg, all of the eIF4E detectable by our method was in one spot on the gel. However, no earlier than 4 and no later than 5 minutes after fertilization, a large fraction of the detectable eIF4E was converted to a more acidic form, as might be expected if it is phosphorylated.

While these results indicated we were on the right track, they did not conclusively prove that the protein was phosphorylated. There were other possible explanations. For example, acetylation of the protein would produce a change of similar magnitude and direction. There certainly is precedent for the covalent modification of proteins by acetylation, although it is less common than phosphorylation. Another even less likely possibility was that the protein was glycosylated. This, too, could have yielded a comparable change in mobility. Thus, we needed to establish the exact nature of the change before we could draw any meaningful conclusions from our data.

There were two ways we could go about demonstrating that the covalent modification was due to addition of a phosphate group. One way was to reverse the change in vitro, using an enzyme known to be specfic for the removal of phosphate groups from proteins. Examples of such enzymes are alkaline phosphatase and acid phosphatase. While this works well in some systems, in our system we encountered enough problems early on to lead us to look for a more workable alternative. This was to give the unfertilized egg radioactive phosphate, then fertilize the eggs and show that, 4 to 5 minutes after fertilization, the eIF4E contained radioactivity. While relatively straightforward in concept, this experiment proved problematic in its execution for one reason: Unfertilized eggs are exceedingly impermeable to many substances, especially phosphate. However, it was essential that we get the radioactive phosphate into the egg prior to fertilization, since we had to show that the unfertilized egg eIF4E had no phosphate. To circumvent the problem, we preloaded the unfertilized eggs with a high level of radioactive phosphate by incubating a tiny amount of eggs for six hours with a very large amount of radioactive phosphate. By increasing the gradient in this way, we were able to load the phosphate pools in the eggs with radioactive material.

What we did next required mostly timing and speed. A labeled, unfertilized egg sample was homogenized and passed over the m^7G affinity column. The eIF4E would attach to the beads, while all other proteins would go through. We removed the eIF4E bound to the column by adding a large excess of m^7GTP to the column. An identical batch of radioactively preloaded unfertilized eggs was then fertilized, and homogenized at 5 minutes after fertilization. This fertilized sample was treated in a similar fashion, by passing over the m^7G affinity column. The two samples containing eIF4E were then analyzed by slab gel electrophoresis, and autoradiograms were prepared by exposing the gels to X-ray film. Much to our delight, when we examined the autoradiograms, we saw a radioactively labeled band right where eIF4E should be in the fertilized egg sample, but not in the unfertilized sample.

This showed that the mobility change we had observed was due to phosphate, but it raised another question: was eIF4E unlabeled in the fertilized egg because we were not getting radioactive phosphate into the ATP pools of the egg? The enzymes that catalyze the reaction of protein phosphorylation use the gamma-phosphate on ATP, not free phosphate. Since the unfertilized egg is metabolically quiescent, it was possible that we were getting hot phosphate into the egg but not into the ATP.

The question, an important one to answer, turned out to be easy to address. We simply ran a slab gel of the material that had not bound to the m^7G affinity column and prepared an autoradiogram from it. This material contained all the proteins in the unfertilized egg except eIF4E. The autoradiogram showed that many proteins were radioactively labeled in the unfertilized egg. This established that we were getting the radioactive phosphate into the ATP pools in the egg, and that protein kinases were capable of using this phosphate, even in the relatively quiescent egg, to phosphorylate proteins.

A final issue that concerned us was whether the eIF4E phosphorylation was unique to *Strongylocentrotus purpuratus*, the species we used in our work, or whether this was a more general phenomenon. We examined two other species of sea urchin that were evolutionarily distant from *S. purpuratus*, *Lytechinus pictus* and *Arbacia punctulata*. In all cases we were able to demonstrate phosphorylation of the eIF4E concomitant with the activation of protein synthesis.

Additional controls further strengthened our conclusions and provided more information on the possible triggers for phosphorylation of eIF4E. Looking back, it is interesting to note that about 75% of the time spent on this project was devoted to carrying out the experimental controls necessary to establish that we were not dealing with an artifact.

What did our results tell us? At the time we carried out this work, it was the first report of a specific biochemical change in the translational machinery that correlated with the activation of protein synthesis. However, it is important to keep in mind that like all scientific endeavor, our results raised as many, if not more questions than were answered. For example, what enzyme phosphorylates the eIF4E? How is it activated at fertilization? And, most important, does phosphorylation of eIF4E allow it to participate more efficiently in binding of the 5' cap, and if so, how?

References

[1] Lopo, A. C.; MacMillan, S.; and Hershey, J. W. B. *Biochemistry* 27:351 (1988).

[2] Lopo, A. C., and Hershey, J. W. B. *J. Cell Biol.* 101:221a (1985).

Schultz, A. M.; Henderson, L. F.; and Oroszlan, S. 1988. Fatty acylation of proteins. *Ann. Rev. Cell Biol.* 4:611–47.

Yan, S. C. B.; Grinnell, B. W.; and Wold, F. 1989. Post-translational modifications of proteins: some problems left to solve. *Trends Biochem. Sci.* 14:264–68.

Protein Folding

Baldwin, R. L. 1989. How does protein folding get started? *Trends Biochem. Sci.* 14:291–94.

Bowie, J. U.; Reidhaar-Olson, J. F.; Lim, W. A.; and Sauer, R. T. 1990. Deciphering the message in protein sequences: tolerance to amino acid substitutions. *Science* 247:1306–10.

Chothia, C., and Finkelstein, A. U. 1990. The classification and origins of protein folding patterns. *Ann. Rev. Biochem.* 59:1007–39.

Creighton, T. E. 1992. Protein folding pathways determined using disulfide bonds. *Bioess.* 14:195–99.

Ellis, R. J. 1990. Molecular chaperones: the plant connection. *Science* 250:954–59.

Ellis, R. J., and Hemmingsen, S. M. 1989. Molecular chaperones: proteins essential for the biogenesis of some macromolecular structures. *Trends Biochem. Sci.* 14:339–42.

Gatenby, A. A., and Ellis, R. J. 1990. Chaperone function: the assembly of ribulose biphosphate carboxylase-oxygenase. *Ann Rev. Cell Biol.* 6:125–49.

Gethway, H.-J., and Sambrook, J. 1992. Protein folding in the cell. *Nature* 355:33–45.

Horwich, A. L.; Neupert, W.; and Hartl, F.-U. 1990. Protein-catalyzed protein folding. *Trends Biotechnol.* 8:126–31.

Kim, P. S., and Baldwin, R. L. 1990. Intermediates in the folding reactions of small proteins. *Ann. Rev. Biochem.* 59:631–60.

Matthews, C. R., and Hurle, M. R. 1987. Mutant sequences as probes of protein folding mechanisms. *Bioess.* 6:254–57.

Nilsson, B., and Anderson, S. 1991. Proper and improper folding of proteins in the cellular environment. *Ann Rev. Microbiol.* 45:607–35.

Pain, R. 1987. Protein folding for pleasure and for profit. *Trends Biochem. Sci.* 12:309–12.

Richards, F. M. 1991. The protein folding problem. *Sci. Amer.* 264:54–63 (January).

Rothman, J. E. 1989. Polypeptide chain binding proteins: catalysts of protein folding and related processes in cells. *Cell* 59:591–601.

Shortle, D. 1989. Probing the determinants of protein folding and stability with amino acid substitutions. *J. Biolog. Chem.* 264:5325–28.

Szulmajster, J. 1988. Protein folding. *Biosci. Rep.* 8:645–51.

Nonribosomal Polypeptide Assembly

Kleinkauf, H., and von Döhren, H. 1983. Non-ribosomal peptide formation on multifunctional proteins. *Trends Biochem. Sci.* 8:281–83.

Review Questions

1. Define the initiation, elongation, and termination stages of polypeptide assembly.

2. Outline the major steps in initiation of protein synthesis. What are initiation factors, and how are these factors believed to operate in initiation? How do initiator tRNAs differ from tRNAs active in elongation?

3. What is the function of the 5′ cap of mRNAs in eukaryotic initiation? How do eukaryotic initiation factors interact with the 5′ cap in initiation?

4. What does scanning mean with respect to eukaryotic initiation? What role do the initiation factors play in scanning? What structural features of mRNAs are important in the scanning mechanism?

5. What are the A, P, R, and E sites of ribosomes? How are these sites believed to operate in polypeptide assembly? What is translocation? What is the relationship between translocation and the genetic code?

6. Outline the functions of elongation factors in polypeptide assembly.

7. Diagram the reaction taking place during formation of a peptide linkage.

8. Outline the steps taking place in the termination of polypeptide assembly. What does readthrough mean with reference to termination? What is the relationship between tRNA activity and readthrough?

9. List the reactions in which phosphate-bond energy is utilized in protein synthesis. How many phosphate groups would be hydrolyzed in the synthesis of a protein 100 amino acids long?

10. Outline the structure of ribosomes. What structural features are common to prokaryotic and eukaryotic ribosomes? What features are unique to eukaryotic ribosomes? What rRNAs occur in each ribosomal subunit? How are ribosomes isolated? Once isolated, how are they disassembled into individual proteins? What findings indicate that the rRNA molecules of ribosomes play functional roles during polypeptide assembly?

11. What information established that the codons of the genetic code are nucleotide triplets? What findings indicated that the nucleotides in mRNAs are simply read three at a time with no "punctuation marks" indicating the first and last nucleotides of each codon? What experiments identified the coding assignments of the codons?

12. What does degeneracy mean with respect to the genetic code? What patterns are noted in degeneracy of the code? What is the wobble hypothesis? How does the mechanism proposed in the wobble hypothesis prevent mistakes during polypeptide assembly? What evidence indicates that wobble is more extensive than predicted by the wobble hypothesis? What is the two-out-of-three rule?

13. What does universality mean with respect to the genetic code? What is the possible relationship between tRNA function and the exceptions noted to the universality of the code?

14. What is preference in codon usage? How is codon preference related to mutation? How is codon preference used by bacteria as a regulatory mechanism?

15. Outline the steps in amino acid activation. What two major outcomes result from the reactions? Where do the reactions take place in cells?

16. Describe the major characteristics of aminoacyl-tRNA synthetases. In what form are these enzymes believed to exist in eukaryotes? In prokaryotes? What is proofreading with respect to synthetase enzymes?

17. What features of tRNA structure are recognized by synthetases? What is the relationship of these features to the three-dimensional structure of tRNA molecules? What are isoaccepting or cognate tRNAs? How does the accu-racy of protein synthesis depend on the functions of tRNAs? Of the synthetase enzymes?

18. What major modifications take place during the processing of newly synthesized proteins? What kinds of organic and inorganic groups are added to proteins during processing?

19. What overall steps are believed to take place during protein folding? What groups contribute to the stability of fully folded proteins?

20. On what steps in amino acid activation and polypeptide assembly does the accuracy of protein synthesis depend? Which are more important to the accuracy by which amino acids are entered into proteins, nucleic acids or enzymatic proteins?

Supplement 16-1
Protein Synthesis in Prokaryotes

Polypeptide assembly proceeds in bacteria through the same phases of initiation, elongation, and termination as in eukaryotes. Although there are differences in detail, particularly in the factors taking part in the reactions, these stages proceed by similar overall mechanisms. Amino acid activation also proceeds by essentially the same reaction sequence as the equivalent pathway in eukaryotes. However, the aminoacyl-tRNA synthetase enzymes evidently occur in free suspension in the bacterial cytoplasm, with no tendency to associate in large aggregates as in eukaryotes. It is generally assumed that in cyanobacteria, polypeptide assembly and amino acid activation proceed by essentially the same pathways and involve the same factors as in bacteria.

Polypeptide Assembly in Bacteria

Initiation Bacterial initiation takes place in a series of steps involving the interaction of mRNA, the large and small ribosomal subunits, an initiator tRNA, GTP, and three initiation factors, *IF1*, *IF2*, and *IF3* (Table 16-5). In the first reaction of initiation (step 1 in Fig. 16-26), ribosomes split into separate large and small subunits. Although intact bacterial ribosomes tend to dissociate spontaneously, the dissociation rate is increased by initiation factor IF1, which combines directly with small ribosomal subunits. Once in dissociated form, the small subunits are stabilized by combination with IF3 (step 2). IF3 acts as an *anti-association factor*, preventing the small subunits from reassembling with large subunits into intact ribosomes.

The small subunit is now ready for the next step, in which mRNA and the initiator tRNA, in combination with its amino acid, are added to the complex. This step is preceded by a preliminary reaction, in which the initiator tRNA combines with IF2 to form the *binary complex* of bacterial protein synthesis (step 3). IF2 evidently promotes binding of the initiator tRNA to the P site facing the AUG codon of the mRNA; without IF2 this binding site has little or no affinity for the initiator tRNA. The IF2–initiator-tRNA complex and the mRNA then add to the small subunit (step 4). The mRNA associates with the small subunit in such a way that its AUG initiator codon, indicating the startpoint for polypeptide assembly, is centered at the P site on the small subunit.

The first amino acid of a prokaryotic polypeptide chain is methionine, as it is in eukaryotes. In prokaryotes, however, the methionine carried by the initiator tRNA is modified into *formylmethionine* by the addition of a formyl (—C—H) group that blocks the amino
$$\overset{\parallel}{\underset{O}{}}$$
end of the amino acid (Fig. 16-27). The initiator tRNA of bacteria, with its attached modified methionine, is identified as *fMet-tRNA$_i$*. Like its eukaryotic counterpart the bacterial initiator tRNA differs in structure from the tRNAs participating in elongation (see Fig. 16-4b). These structural differences undoubtedly contribute to the ability of the bacterial initiator to combine with the P site of ribosomes during initiation.

The final reaction in bacterial initiation (step 5) adds the large subunit to complete the ribosome. As the large subunit binds, all three initiation factors are released from the ribosome. Release of IF2 is accompanied by GTP hydrolysis, which evidently powers a conformational change in IF2 or the small subunit, required to release this factor. IF1 may also promote IF2 release. The completed initiation complex now consists

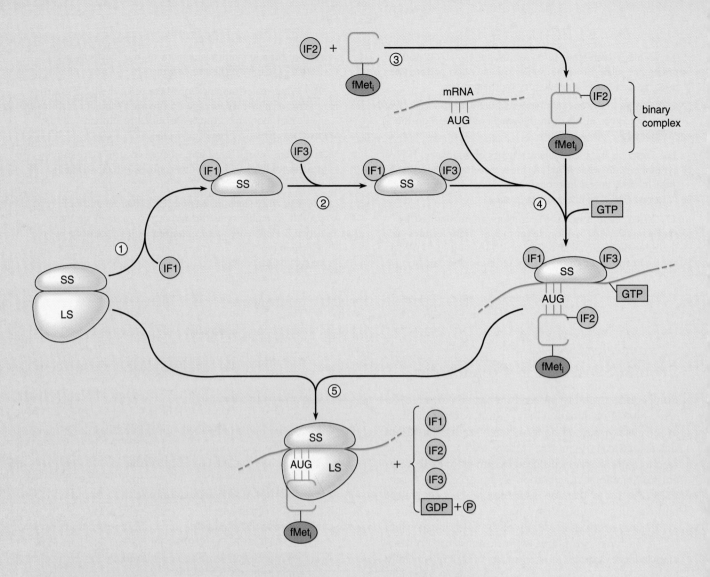

Figure 16-26 The steps in prokaryotic initiation. In step 1, a ribosome dissociates into separate small (SS) and large (LS) subunits. Dissociation is promoted by initiation factor IF1, which combines directly with small ribosomal subunits. Once in dissociated form, the small subunit is stabilized by combination with IF3 (step 2), which prevents small subunits from reassembling with large subunits. Then mRNA and the initiator tRNA in combination with its amino acid (fMet$_i$) add to the complex. This step is preceded by a preliminary reaction (step 3) in which the initiator tRNA interacts with IF2 to form the binary complex (compare with the ternary complex of eukaryotic initiation in Fig. 16-3). The IF2–initiator-tRNA complex and the mRNA then add to the small subunit (step 4) in such a way that the AUG initiator codon of the mRNA is centered at the P site on the small subunit. Step 5 adds the large subunit to complete the ribosome. As the large subunit binds, all three initiation factors are released from the ribosome. Release of IF2 is accompanied by GTP hydrolysis.

of the ribosome, mRNA, and fMet-tRNA, with the initiator tRNA linked at the P site of the ribosome opposite the AUG initiator codon.

Binding of mRNA to the small ribosomal subunit and positioning of the AUG codon at the P site evidently depend on complementary base pairing between the mRNA and the 16S rRNA molecule of the small ribosomal subunit. The mRNA sequences that associate with the small subunit were determined from footprint-

ing studies (see Information Box 15-1) by J. A. Steitz, who found that the portion of the mRNA protected by the association, some 30 to 40 nucleotides, was approximately centered on the initiator codon. Short segments of both the 5′ untranslated segment and the coding sequence following the AUG were included in the footprint. Subsequently, J. Shine and L. Dalgarno noted that a sequence near the 3′ end of 16S rRNA is complementary to a segment of the footprint.

Table 16-5	Prokaryotic Initiation Factors
Factor	Function in Initiation
IF1	Promotes dissociation of ribosomes into separate subunits.
IF2	Binds to fMet-tRNA$_i$; enhances fMet-tRNA$_i$ binding to small subunit; hydrolyzes GTP.
IF3	Stabilizes small ribosomal subunits in dissociated form; prevents reassociation of small and large subunits.

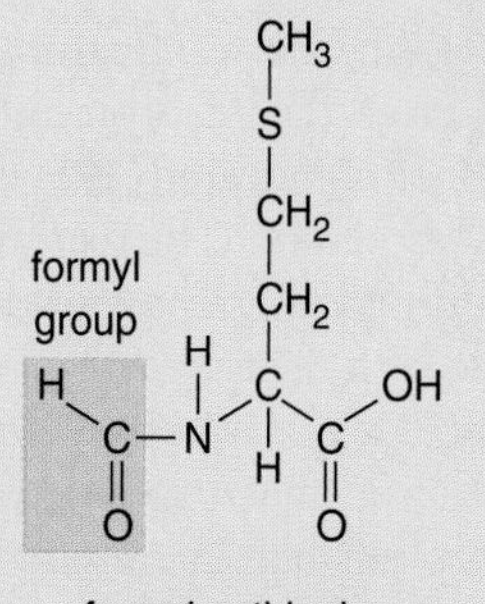

Figure 16-27 The formylmethionine (fMet) placed as the first amino acid in prokaryotic protein synthesis. The formyl group is shaded in blue.

Shine and Dalgarno proposed accordingly that binding and positioning of mRNA on the small ribosomal subunit depend on complementary base pairing between the mRNA and the 3′ end of 16S rRNA. Their hypothesis later received direct support from the experiments of Steitz and K. Jakes, who treated initiation complexes with a ribonuclease enzyme that splits 16S rRNA near the 3′ end. The 3′-end fragment of the rRNA, after separating from the small subunits, was found in a complex with mRNA, held together by hydrogen bonds as predicted by the Shine–Dalgarno hypothesis. Further confirmation was obtained from studies in which initiation was inhibited by alterations in the sequence of the segment of 16S rRNA proposed to pair with the mRNA 5′ untranslated segment. At some positions, alteration of a single base in the 16S pairing segment can completely inhibit mRNA binding to the small ribosomal subunit. The mRNA sequence pairing with the 16S rRNA during bacterial initiation, the consensus sequence AGGAGGU, is now known as the Shine–Dalgarno sequence.

The mechanism pairing mRNA with 16S rRNA in the small subunit seems to be peculiar to prokaryotes. As noted in this chapter, an entirely different mechanism, involving interaction of initiation factors with the 5′ cap of eukaryotic mRNAs followed by scanning to locate the AUG start codon, appears to be responsible for mRNA binding and positioning in eukaryotes.

Elongation and Termination The reactions of bacterial elongation follow the same pathway as in eukaryotes (see Fig. 16-7). The bacterial elongation factors are identified differently, however (see Table 16-2). The factor that interacts with GTP and an aminoacyl-tRNA and promotes binding of the aminoacyl-tRNA to the A site of ribosomes (steps 1 and 2 in Fig. 16-7) is identified as *EF-Tu* in bacteria (equivalent to eukaryotic EF1α; the *u* in *Tu* indicates that the factor is unstable when heated). The factor that promotes release of the empty tRNA and translocation of the ribosome to the

next codon after formation of the peptide bond (steps 4 and 5 in Fig. 16-7) is identified as *EF-G* (equivalent to eukaryotic EF2). Finally, the factor that allows EF-Tu to exchange its GDP for GTP after the release of EF-Tu from the ribosome (steps 6 and 7 in Fig. 16-7) is identified as *EF-Ts* (equivalent to eukaryotic EF1βγ; the *s* in *Ts* indicates that the factor is stable when heated). The elongation cycle turns more rapidly in bacteria—about 15 to 20 times per second as compared to 1 to 3 times per second in eukaryotes.

Three factors, *RF1*, *RF2*, and *RF3*, participate in bacterial termination. Appearance of a terminator codon at the A site causes either RF1 or RF2 to bind to the A site (equivalent to step 1 in Fig. 16-9). The two factors are closely related in amino acid sequence; RF1 binds if the terminator codon is UAG and RF2 if the codon is UGA. Either may bind if the terminator is UAA. The two termination factors promote hydrolysis of the bond linking the newly synthesized polypeptide chain to the tRNA held at the P site, in a mechanism equivalent to that of the single RF of eukaryotes. After the polypeptide is released, ejection of RF1 or RF2 from the A site is promoted by the third bacterial termination factor, RF3, possibly in combination with GTP. Although its role in bacterial termination is not completely certain, GTP may be hydrolyzed to provide energy for RF1 or RF2 ejection.

Termination by the interaction of the terminator codons and factors is imperfect in bacteria, as it is in eukaryotes. The readthrough occurring when terminators fail is employed by at least one bacterial virus, the *Qβ bacteriophage* infecting *E. coli*, to increase the variety of polypeptides assembled during protein synthesis. The Qβ bacteriophage requires 160 copies of a 14,000-dalton protein to assemble its coat and 1 or more copies of another 38,000-dalton protein to be fully infective. The two proteins are identical for some distance into their amino acid sequences, starting from their N-terminal ends, and are encoded in the same gene. However, the heavier protein has an extra 195 amino acids tacked onto its C-terminal end, placed there by readthrough of a UGA terminator codon of the mRNA. Termination of the readthrough polypeptide occurs at a UAG codon located 195 codons downstream of the UGA terminator. The readthrough, caused by pairing of a tryptophan aminoacyl-tRNA with the UGA codon instead of a release factor, occurs in about 3% of the polypeptides synthesized, producing the 38,000-dalton protein in much smaller quantities than the smaller coat protein but in numbers sufficient to assemble fully infective viral particles.

Posttranslational processing in bacteria primarily involves removal of one or more amino acids from either end of the newly synthesized polypeptide chain and modification of individual amino acid residues to other forms. The entire formylmethionine is removed from the N-terminal end of some bacterial proteins during processing; in others only the formyl group is removed.

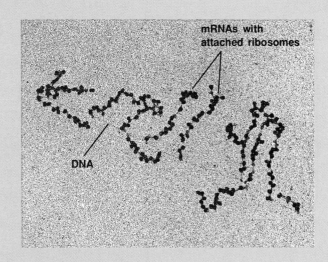

Figure 16-28 Simultaneous transcription and translation in *E. coli* (see text). × 57,000. Courtesy of O. L. Miller, Jr., Barbara A. Hamkalo, and C. A. Thomas, Jr.

Bacterial Ribosomal Proteins

The small ribosomal subunits of bacteria, after dissociation and disassembly, prove to contain a single rRNA type, 16S rRNA, in combination with 21 proteins. The prokaryotic large ribosomal subunit contains two rRNA types, 5S and 23S, in combination with 31 different proteins.

Using a numbering system based on the rate at which the isolated proteins migrate on electrophoretic gels, the small-subunit proteins are identified as S1 to S21, roughly in order of molecular weight, with the largest protein first. The 31 large-subunit proteins are numbered from L1 to L34, also with the largest protein first. Two numbers in the large-subunit sequence, L8 and L26, are unused because of misassignments early in the research identifying the proteins; another large-subunit protein, L7/12, originally thought to be two distinct proteins, later proved to be the same polypeptide except for the presence of a modifying acetyl group in L7 that is absent in L12. All the large- and small-subunit proteins occur in single copies in prokaryotic ribosomes except for L7/L12, which is present in four copies in the large subunit. The tentative positions of these proteins in the small and large ribosomal subunits of bacteria, determined using the methods described in Information Box 16-1, are shown in Figure 16-16.

The entire protein complement of *E. coli* ribosomes, as well as the *E. coli* rRNAs, has been completely sequenced. Most ribosomal proteins are small, basic, and insoluble molecules ranging in molecular weight from 6000 to 32,000. Only one, S1, with a molecular weight of 61,000, falls outside this range. The assembly of these proteins with the bacterial rRNAs into functional ribosomes, and the function of sites on bacterial ribosomes in protein synthesis are covered in the main body of this chapter.

The events in both bacterial transcription and translation can be seen in progress in Figure 16-28, a remarkable electron micrograph made by O. L. Miller and his colleagues. The thin strand running from left to right in the figure is part of the DNA of a bacterial cell. Attached to the DNA at intervals are side branches containing rows of spherical particles. The branches are individual mRNA molecules, already attached to ribosomes and engaged in translation before their transcription is complete. The small, dense particles where the side branches join the DNA are RNA polymerase molecules engaged in transcribing the gene. Because the mRNA-ribosome complexes become longer toward the right of the micrograph, the direction of transcription runs from left to right in the picture. Simultaneous transcription and translation, in which translation of the 5′ end of the mRNA molecules is initiated before transcription is complete, occur regularly in bacteria.

Actively growing *E. coli* cells are estimated to contain as many as 20,000 ribosomes, which completely pack the cytoplasm. Most of these ribosomes are engaged in protein synthesis in assemblies resembling Figure 16-28. Because the cell interior is packed with these assemblies, which include DNA and mRNA as well as ribosomes, much of the region recognized as cytoplasm actually contains extensions of the nucleoid. The nucleoid therefore probably extends throughout most of an active bacterial cell; the region recognized as the nucleoid likely contains only the tightly packed, inactive portion of the bacterial DNA.

Suggestions for Further Reading: See p. 685.

Nonribosomal Mechanisms of Protein Synthesis

A limited number of very small polypeptides, including some antibiotics and membrane components, are assembled in prokaryotes and some lower eukaryotes by enzymatic mechanisms that do not involve mRNA, tRNAs, or ribosomes. Besides being nonribosomal, the unusual mechanisms synthesizing the polypeptides of this group frequently use D-amino acids (see p. 51); in contrast, ribosomally synthesized proteins are assembled solely from L-amino acids. The presence of D-amino acids in a small polypeptide, in fact, is a defini-tive indication that the polypeptide is nonribosomally synthesized.

Nonribosomal protein synthesis takes place on the surface of large enzymes containing a series of active sites that can arrange and bind amino acids in sequence. The amino acids are bound to the enzymes in the form of aminoacyl-adenylates (AA-AMPs), equivalent to the product of the first reaction of amino acid activation (reaction 16-1 on p. 677). During binding to the enzyme, the amino acid is transferred from AMP to an

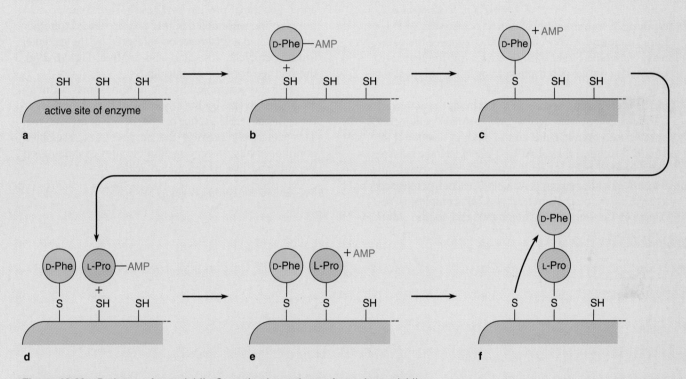

Figure 16-29 Pathway of gramicidin S synthesis on the surface of gramicidin S synthetase 2. **(a)** The active site of the enzyme contains a series of reactive —SH groups, oriented to fit the sequence of amino acids to be bound to the enzyme. In the reaction binding the first amino acid (**b** and **c**), the activated D-phenylalanine (D-Phe) is transferred from AMP to the first —SH group. Much of the energy released in hydrolysis of the aminoacyl-AMP complex is stored in the linkage between the amino acid and the sulfur atom. Next the enzyme binds the second aminoacyl-AMP, which contains L-proline (L-Pro), and the L-proline is transferred to a second —SH group (**d** and **e**). Addition of L-proline triggers formation of the first peptide linkage (**f**), between the carboxyl group of D-phenylalanine and the amino group of the L-proline. The resulting dipeptide remains linked to the surface of the enzyme by an —S— bond. The process now repeats, adding three more L-amino acids—valine, ornithine, and leucine—in sequence at other —SH groups lined up on the enzyme surface. **(g)** Two 5-amino-acid polypeptides assembled in this way are linked together head to tail in a final reaction catalyzed by the enzyme, forming a cyclic 10-amino-acid peptide, the finished form of gramicidin S.

—SH group at the active site of the enzyme. The transfer, like the transfer of amino acids from AMP to tRNA during amino acid activation, retains much of the energy released by hydrolysis of the AA-AMP complex.

Several —SH groups capable of binding amino acids are lined up on the surface of the protein-synthesizing enzyme, in positions allowing amino acids to be attached in order and in arrangements allowing formation of consecutive peptide linkages. These —SH groups are part of complex organic prosthetic groups resembling the prosthetic group of coenzyme A, a molecule also engaged in "high energy" transfers (see Fig. 9-11).

For example, in the synthesis of *gramicidin S*, a bacterial membrane polypeptide assembled by the nonribosomal pathway, two enzymes cooperate in the reactions. The first enzyme, *gramicidin S synthetase 1*, catalyzes the interaction of amino acids with ATP to form aminoacyl-AMP. The second enzyme of the pathway, *gramicidin S synthetase 2*, carries out the reactions assembling the activated amino acids into a polypeptide (Fig. 16-29).

In these reactions enzyme 2 first binds the activated D-form of phenylalanine, the only D-amino acid incorporated into gramicidin S. In the binding reaction (Fig. 16-29a through c) the activated D-phenylalanine residue is transferred from AMP to the first —SH group on the surface of the enzyme. Much of the energy released in hydrolysis of the aminoacyl-AMP complex is retained in the linkage of the amino acid to the sulfur atom of the enzyme. The enzyme now binds the second aminoacyl-AMP, which in gramicidin synthesis is L-proline. In the reaction L-proline is transferred from AMP to a second —SH group on the enzyme (Fig. 16-29d and e). Addition of L-proline triggers formation of the first peptide linkage, which extends between the carboxyl group of D-phenylalanine and the amino group of L-proline (Fig. 16-29f). Formation of the peptide bond is powered by free energy released by hydrolysis of the —S— bond linking phenylalanine to the enzyme.

The resulting dipeptide remains attached to the surface of the enzyme by its linkage to the second sulfur atom. The process now repeats, adding three more L-amino acids, valine, ornithine, and leucine, in sequence at additional —SH groups lined up on the enzyme surface. (Ornithine is a nonprotein amino acid—one that occurs naturally in cells but does not enter ribosomal protein synthesis.) Two 5-amino-acid polypeptides assembled in this way are linked together head to tail in a final reaction catalyzed by the enzyme, forming a cyclic 10-amino-acid peptide (Fig. 16-29g), the finished form of gramicidin S.

Proteins assembled by the nonribosomal pathway are limited in size by the requirement that they fit over the series of —SH groups on the surface of the enzyme. This effectively restricts the polypeptides assembled nonribosomally to very small molecules containing on the order of five to ten amino acids.

Suggestions for Further Reading: See p. 686.

17

TRANSCRIPTIONAL AND
TRANSLATIONAL REGULATION

The genes encoding mRNAs, rRNAs, and tRNAs may number as many as 30,000 or more in eukaryotic cells. At any given time, however, only a fraction of these genes, estimated to be about 5000 in most cells, is active. This is true even though almost all cells in eukaryotes retain a complete gene complement and have the capacity to synthesize any of the RNAs and proteins of the species to which they belong.

The difference between the synthetic potential of eukaryotic cells and the number of RNAs and proteins actually assembled reflects controls at two primary levels. *Transcriptional regulation* determines which of the many genes in the nucleus are copied into RNAs and which genes are maintained in an inactive state. *Translational regulation* controls the rate at which mRNAs are used by ribosomes in protein synthesis.

These controls are supplemented by two other levels of regulation. *Posttranscriptional controls* regulate events between transcription and translation, primarily RNA processing and transport of finished RNAs to the cytoplasm. *Posttranslational controls* adjust the levels of cellular proteins by regulating the rate at which newly assembled polypeptides are processed into finished form and the rate at which finished proteins are degraded.

Enzymes are among the proteins regulated in kinds and quantity at all these levels. By regulating enzyme synthesis, the mechanisms controlling transcription and translation extend indirectly to the reactions catalyzed by the enzymes, including those assembling all the remaining molecules of the cell—lipids, carbohydrates, and nucleic acids. The entire spectrum of transcriptional and translational controls thus provides cells with an exquisitely sensitive mechanism for regulating the kinds and numbers of all cellular molecules produced and the times and locations of their synthesis. The total mechanism underlies the development and maintenance of complex organisms, in which individual cells carry out specific, coordinated tasks. The mechanism also allows cells to respond and adjust to environmental changes and stimuli.

Although the molecular basis of transcriptional regulation is incompletely understood, it has become clear that cells are controlled at the most fundamental level by regulatory proteins with the capacity to recognize individual genes and turn their activity in transcription on or off. The specific transcriptional controls imposed by these proteins are the primary regulatory mechanisms underlying cell differentiation and development. Because regulatory proteins and their activities are so important to cell growth, division, and development, the research identifying and tracing the functions of these proteins is among the most exciting and fundamentally significant areas of investigation in cell and molecular biology.

This chapter describes the mechanisms controlling transcription and translation in eukaryotes and their coordination into integrated pathways of regulation. Comparisons with the regulatory mechanisms of prokaryotes are made in Supplement 17-1. The structure and transcriptional activity of polytene chromosomes, an unusual eukaryotic chromosome type in which gene transcription and regulation can be observed directly under the light and electron microscopes, are covered in Supplement 17-2.

TRANSCRIPTIONAL REGULATION IN EUKARYOTES

Chromosomal proteins of the nonhistone group are now known to recognize and control individual genes. Although these proteins were suspected to be the primary elements controlling gene activity for many years, little progress was initially made in isolating and identifying individual proteins able to regulate specific genes. The task proved difficult for several reasons. One is that regulatory proteins occur in relatively small numbers inside cells. As a result, they were difficult to detect and identify among other proteins occurring in relatively large quantities in the nonhistone fraction of chromatin (see p. 541 for details). Another source of difficulty stemmed from the fact that many regulatory proteins act cooperatively rather than singly in transcriptional control. This made the activity of individual proteins difficult to trace.

The first transcriptional regulatory proteins to be definitely identified were those responding to stimulation of cells by steroid hormones. Because these regulatory proteins directly bind the hormones, labeling the hormones with radioactivity provided a convenient and effective way to tag the proteins for isolation and purification.

The regulatory proteins responding to steroid hormones were identified during the 1970s. During the 1980s technical breakthroughs greatly expanded the detectability of proteins regulating transcription. The most productive of these techniques uses genes themselves as a "probe" to detect and trap regulatory proteins. The DNA of a gene or of the 5'-flanking control regions of a gene is increased in quantity by cloning or the polymerase chain reaction (these techniques are described in Ch. 4). The DNA is immobilized by binding to nitrocellulose filter paper or to gel beads in a separatory column (Fig. 17-1a). A bulk preparation of cellular proteins is then passed over the immobilized copies of the gene (Fig. 17-1b); those with the ability to recognize and bind the control sequences of the gene are trapped in the paper or column as they bind the DNA (Fig. 17-1c). After nonbinding proteins are

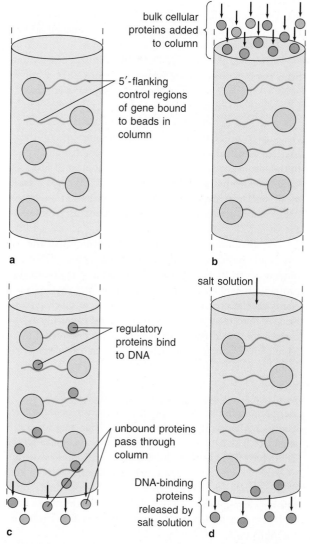

Figure 17-1 A method using the 5'-flanking control region of a gene, increased in quantity by cloning, to trap a regulatory protein recognizing sequences in the DNA. **(a)** The DNA is immobilized by binding to gel beads in a column. **(b)** A bulk preparation of cellular proteins is poured through the column. **(c)** Proteins that can recognize and bind the control sequences of the gene are trapped in the column. **(d)** After nonbinding proteins are washed away, the regulatory proteins recognizing the gene are released in purified form by pouring a salt solution through the column.

washed away, the regulatory proteins recognizing the gene are released in purified form by adding a salt solution (Fig. 17-1d).

Once individual regulatory proteins were isolated, their characteristics could be studied by molecular techniques that permit individual amino acids to be changed or deleted and by grafting techniques that split off segments or domains of a regulatory protein and attach them to other proteins. These studies allowed regions within the regulatory proteins to be identified with such activities as recognition of DNA sequences and promotion of transcription by RNA polymerase enzymes. These studies revealed functional motifs common to subclasses of regulatory proteins, such as the *zinc finger*, *helix-turn-helix*, and *leucine zipper* (see p. 543). The distinctive motifs, once characterized, provided structural features that allowed other regulatory proteins carrying the motifs to be isolated and identified. The Experimental Process essay by G. C. Prendergast on p. 696 describes his elegant experiments applying this and other approaches to the study of the Myc regulatory protein. The Myc protein also illustrates a regulatory protein assembled from two different polypeptides as a heterodimer (see below).

More than a hundred types of eukaryotic regulatory proteins have been identified by these approaches, and the list grows constantly. Among them are examples from fungi and protozoa, insects, vertebrates, higher plants, and viruses infecting these forms. (Table 17-1 lists some of these proteins and the genes they regulate.)

One of the most striking features of the proteins isolated to date is that they control transcription by mechanisms that are universally shared among eukaryotic organisms. Many regulatory proteins can be interchanged among organisms as distantly related as yeast cells and humans and still operate successfully. This indicates that the regulatory mechanisms probably evolved with the earliest eukaryotes.

For genes encoding mRNAs, footprinting studies (see Information Box 15-1) have shown that regulatory proteins bind primarily to short control sequences in 5'-flanking regions of the genes. The control sequences, which vary between about 8 and 20 nucleotides in length, form parts of promoters and enhancers (see p. 589) and also occupy sites between these elements. Because they are part of the same DNA molecule as the gene, the control sequences are called the *cis* elements of the regulatory system (*cis* = on the same side). Some *cis* elements recognized by specific regulatory proteins are listed in Table 17-1.

The *cis* control elements have been studied by substitution-deletion experiments (see p. 583) in which individual bases are changed or eliminated and the effects on transcription noted. The control sequences have also been studied by recombinant DNA techniques in which a *cis* element, either derived from an

Sequential Clues to the DNA-Binding Specificity of Myc

George C. Prendergast

GEORGE C. PRENDERGAST is senior research biochemist in Cancer Research at the Merck Sharp and Dohme Research Laboratories in West Point, Pennsylvania. He received his Ph.D. in molecular biology from Princeton University in 1989, providing the first evidence for gene regulation by the Myc protein. As an American Cancer Society Postdoctoral Fellow at the Howard Hughes Medical Institute of the New York University Medical Center, Prendergast identified the DNA binding specificity of Myc and isolated the murine form of Max, a DNA-binding partner protein for Myc. He is currently studying the biochemistry of the Myc and Ras proteins.

The chief characteristic of a cancer cell is its ability to grow uncontrollably. One of the most important reasons that cancer cells escape normal growth controls is because they have sustained extensive damage to the signaling machinery that regulates cell division. A number of the proteins that comprise this machinery are encoded by the cellular oncogenes, which are mutant in tumor cells. One of the first oncogenes to be identified, called *ras*, was cloned from tumor cell DNA that when introduced into normal cells could transfer the abnormal growth characteristics of the tumor cells. Since this first experiment in the late 1970s, many other cellular oncogenes have been identified and cloned. But the rapid progress in discovery of new oncogenes has outpaced work to determine how they normally work and why their function is altered in tumor cells.

The *myc* oncogene was identified a short time after *ras* and has since been found to be involved in many human cancers. Initial studies performed in Robert Weinberg's laboratory demonstrated that when introduced with *ras* into normal rat fibroblasts by DNA transfection, plasmids that continuously expressed *myc* from a strong promoter could force uncontrolled cell growth.[1] Later, it was shown that the Myc protein was localized in the cell nucleus and could nonspecifically adhere to DNA. These observations suggested the hypothesis that Myc may recognize specific DNA sequences and perhaps regulate transcription, DNA replication, or other nuclear processes important for cell growth.

Recent tests of this hypothesis greatly benefited from several key clues, all of which came from the comparison of amino acid sequences of Myc with other DNA-binding proteins. The first clue was the identification by Steven Mcknight's group of an amino acid sequence pattern, or motif, called the "leucine zipper" (LZ) in the carboxy-terminus of Myc.[2] The LZ motif, named for the characteristic heptad repeat of leucines it contained, was originally discovered in the transcription factor C/EBP where it was shown to be necessary for dimerization and DNA binding. Because C/EBP could bind DNA specifically, this suggested that Myc might too, since it also contained an LZ motif.

A second clue to Myc DNA binding specificity came from a comparison I made while in Edward Ziff's group between the amino acid sequences of Myc and another LZ oncogene protein called Fos. Many LZ proteins contain a section rich in basic amino acids located immediately upstream of the LZ motif. When dimerized by the LZ, this so-called basic motif could be demonstrated to form a sequence-specific DNA-binding domain in Fos, C/EBP, and other proteins. I noticed that Myc had a basic motif similar in sequence to Fos's but located in a different position, approximately 40 amino acids upstream of the Myc LZ.[3]

At this time I resolved to test the notion that the Myc basic motif could mediate sequence-specific DNA binding, even though it appeared to be in a region of the protein different from other LZ proteins. But other clues clarifying the problem continued to come in quick succession. A third clue to Myc's DNA-binding capabilities came from David Baltimore's group, who discovered in Myc another amino acid sequence pattern, called the helix-loop-helix (HLH) motif.[4] Like the LZ, the HLH was originally identified in another protein, called E12, that was isolated through its abililty to bind to a DNA element implicated in transcription of immunoglobulin genes. The E12 HLH motif was shown to be important for the ability of E12 to dimerize and specifically recognize DNA. The Myc HLH identified by Baltimore's group fit exactly into the 40-amino-acid space between the LZ and the Fos-like basic motif I had noticed in Myc. The proximity of this new dimerization motif to what I thought might be a DNA-binding region in Myc made me feel that I was on the right track.

First, I had to find a way to functionally dimerize the Myc basic motif. Studies of LZ proteins showed that a basic motif had to be dimerized to work, but analyses of the Myc carboxy-terminal region containing the HLH and LZ motifs suggested that it dimerized poorly. Therefore, to force dimerization of the Myc basic motif, I generated by recombinant DNA techniques a chimeric plasmid that would fuse the Myc basic motif to the E12 HLH (which had been shown in Baltimore's group to dimerize efficiently). The chimeric E12/Myc protein encoded on the plasmid could be produced by *in vitro* transcription and then by translation of the resultant RNA in a rabbit reticulocyte lysate. I could show that E12/Myc was capable of dimerizing through the E12 HLH by cotranslating *in vitro*-transcribed E12/Myc RNA along with that for normal E12, and then immunoprecipitating the protein mixture with anti-Myc antibodies. The presence of E12 protein in the immunoprecipitate indicated that it must have been bound to E12/Myc in the mixture, since E12 lacked the region reacting with the Myc antibody.

About this time, two more clues from the comparison of Myc amino acid sequences with newly cloned proteins suggested what DNA elements to screen for binding to the E12/Myc chimera. First, by comparing the DNA-binding sites of several HLH proteins whose specificity was known, a DNA sequence consensus pattern for recognition, CANNTG, was identified (called an E box, for enhancer box). Second, the basic regions of Myc and a novel HLH/LZ transcription factor called TFE-3, cloned by Holger Beckman in Tom Kadesch's group, were observed to be very similar.[5] TFE-3 could bind two DNA sequences, $E_{\mu E3}$ from the immunoglobulin heavy chain gene (containing the E box consensus CATGTG), and E_{USF} from the adenovirus major late promoter (containing the E box consensus CACGTG). Because of the similarity of the TFE-3 and Myc basic regions, I tested E12/Myc for specific binding to a number of DNA sequences that contained these E box consensus sequences.

To do this, I used a technique called the gel mobility shift assay. In different trials, the E12/Myc chimeric protein was incubated with various radiolabeled DNA oligonucleotides containing a particular E box sequence. Also included was a >1000-fold molar excess of nonspecific unlabeled DNA to prevent nonspecific binding by E12/Myc to the radiolabeled oligonucleotide probe. The mixtures were then separated on nondenaturing polyacrylamide gels and processed for autoradiography. A positive result would be indicated by the presence of a radioactive band toward the top of the gel, where the oligonucleotide would be bound to the slowly migrating protein, rather than at the bottom of the gel, where the unbound DNA would run. To verify the assay was working, positive controls were performed for specific DNA binding of TFE-3 to E_{USF} and $E_{\mu E3}$. Following tests with several oligonucleotides, I found that E12/Myc could specifically bind E_{USF}, suggesting that the Myc basic motif recognized the E box sequence CACGTG.[6]

To be certain of this result, two important control experiments had to be performed. First, it was important to show that the Myc basic motif was responsible for the activity, and not an E12-derived part of the chimera. Second, I had to verify that E12/Myc was recognizing the E box sequence and not another part of the oligonucleotide. For the first control, I constructed and expressed by recombinant DNA methods three mutant E12/Myc proteins with different missense mutations in the basic motif.

When tested as before, all three mutants lost DNA-binding activity compared to normal E12/Myc. This argued that E12/Myc was recognizing DNA through the Myc basic motif. For the second control, I made mutant oligonucleotides containing mutations in and around the CACGTG E box sequence and tested all for binding to E12/Myc. Only those with mutations in or adjacent to the E box sequence affected DNA binding. This demonstrated that, as predicted, E12/Myc was recognizing the E_{USF} element in the oligonucleotide.

Did the DNA binding specificity of E12/Myc reflect that of Myc *in vivo*? If it did, one would predict that E12/Myc would interfere with the action of Myc when they were co-expressed in cells, through competition for specific DNA-binding sites. When it was introduced along with Myc and Ras into normal rat fibroblasts by DNA transfection, I found that E12/Myc could suppress the uncontrolled cell growth induced in its absence. In contrast, an E12/Myc mutant that lost the ability to specifically bind DNA because it had a mutation in the Myc basic motif also lost the ability to suppress Myc/Ras-induced cell growth. This control experiment demonstrated that suppression by E12/Myc required the DNA-binding activity of the Myc basic motif. Taken together, the data argued that the specificity of the E12/Myc chimera for DNA was the same as Myc itself.

What was the conclusion of my work? I gathered the first evidence for a specific Myc DNA-binding activity *in vitro* that could be argued to be relevant to Myc *in vivo*, thereby identifying the first known target for Myc activity in the cell. The set of experiments that led to this conclusion were prompted in part from numerous clues turned up by sequence comparisons. With further work, it should be possible to test Myc's potential function as a regulator of transcription, DNA replication, or other nuclear processes important for normal and tumorigenic cell growth.

References

[1] Land, H.; Parada, L. F.; and Weinberg, R. A. Tumorigenic conversion of primary embryo fibroblasts requires at least two co-operating oncogenes. *Nature* 304:596–602 (1983).

[2] Landschultz, W. H.; Johnson, P. F.; and McKnight, S. L. The leucine zipper: a hypothetical structure common to a new class of DNA binding proteins. *Science* 240:1759–64 (1987).

[3] Prendergast, G. C., and Ziff, E. B. DNA binding motif. *Nature* 341:392 (1989).

[4] Murre, C.; McCaw, P. S.; and Baltimore, D. A new DNA-binding and dimerization motif in immunoglobulin enhancer binding, daughterless, MyoD, and Myc proteins. *Cell* 56:777–83 (1989).

[5] Beckmann, H.; Su, L.-K.; and Kadesch, T. TFE3: A helix-loop-helix protein that activates transcription through the immunoglobulin enhancer μE3 motif: *Genes Dev.* 4:167–79 (1990).

[6] Prendergast, G. C., and Ziff, E. B. Methylation-sensitive sequence-specific DNA binding by the c-Myc basic region. *Science* 251:186–89 (1991).

Table 17-1 Some Proteins Regulating Specific Genes

Regulatory Protein (*trans* element)	Sequence Recognized (*cis* element)	Source	Motif*	Genes Regulated
ACF		Mammals		Serum albumin
Antennapedia		*Drosophila*	HTH	Homeotic gene
AP-2	CCCCAGGC	Mammals		MHC class I, proenkephalin, metallo-thionein, others
AP-3		Mammals		SV40 viral enhancer
AP-4		Mammals		Proenkephalin, others
APF		Mammals		Serum albumin
CDF		Sea urchin		Histone H2B
C/EBP	TGTGGAAAG	Mammals	LZ	Serum albumin, α-globin
CREB (ATF)	TGACGTCA	Mammals	LZ	Somatostatin, proenkephalin
Estrogen receptor	GGTCANNNTGACC	Birds, mammals	ZnF	Vitellogenin, ovalbumin, others
Fos/Jun (AP-1)	TGACTCA	Mammals	LZ	Growth regulation
GAL4		Yeast	ZnF	Enzymes of galactose metabolism
GCN4		Yeast	LZ	Genes encoding enzymes synthesizing amino acids
Glucocorticoid receptor	GAACANNNTCTTC	Mammals	ZnF	Prolactin, others
GT-1		Plants		RuBP carboxylase, chlorophyll-binding protein
HSF		Mammals, others		Heat shock proteins
ITF		Mammals		β-interferon
MAT proteins		Yeast	HTH	Mating type genes
NF-1 (CTF)	GCCAAT	Mammals		α-globin, β-globin, serum albumin, heat shock proteins
OCT-1 (OTF1)	ATTTGCAT	Mammals	HTH	snRNA, histone genes, others
OCT-2 (OTF2)	ATTTGCAT	Mammals	HTH	Antibody genes
Pit-1		Mammals	HTH	Growth hormone
SP1	GGGCGG	Vertebrates	ZnF	Wide variety
TUF		Yeast		Ribosomal proteins
Ultrabithorax		*Drosophila*	HTH	Homeotic gene

Adapted from data presented in P. F. Johnson and S. L. McKnight, *Ann. Rev. Biochem.* 58:799 (1989); C. Murre et al., *Cell* 56:777 (1989); and P. J. Mitchell and R. Tjian, *Science* 245:371 (1989).
* HTH = helix-turn-helix; LZ = leucine zipper; ZnF = zinc finger.

organism or artificially synthesized, is placed in the 5′-flanking region of a gene encoding an enzyme that can be easily identified. The effects of the sequence on regulation of the gene are noted after the gene is introduced into living cells. (Fig. 17-2 outlines a typical technique used in this type of experiment.)

The regulatory proteins binding to *cis* elements, called the *trans* elements of the regulatory system (*trans* = on the opposite side), interact with DNA and with other proteins to promote or inhibit binding of an RNA polymerase and initiation of transcription. Each

protein binding to the *cis* elements of a gene usually recognizes a distinct sequence in the 5′-flanking region.

Stimulation of transcription by proteins acting as *trans* elements is called *positive regulation;* inhibition is termed *negative regulation.* Although *trans* elements usually act as either positive or negative regulators, a few may work in either capacity depending on the positions and functions of *cis* elements in the DNA and the effects of other regulatory proteins binding the control regions of the same gene.

Regulatory proteins typically control genes co-

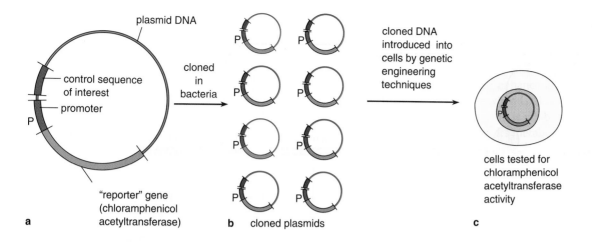

a   b   cloned plasmids   c

Figure 17-2 A general technique for testing a control sequence. **(a)** The sequence is introduced into a bacterial plasmid by recombinant DNA techniques (see p. 134). The control sequence is placed into the 5'-flanking region of a "reporter" gene encoding an enzyme that is easily detected by its activity. (Bacterial chloramphenicol acetyltransferase is often used for this purpose.) After cloning **(b)** the copies are placed in recipient cells by genetic engineering techniques **(c)** (see p. 782). After an incubation period of about one hour, the cells are tested for relative amounts of enzyme activity. To evaluate sites within the control sequence, individual bases may be changed one at a time in consecutive experiments and the amounts of enzyme activity compared.

operatively. Some regulatory combinations are made by assembling polypeptide subunits by twos to form distinct types of *heterodimers*.

The cooperative interaction of regulatory proteins has several highly significant consequences for gene control. By acting in different combinations, a relatively small number of proteins can specifically regulate a large number of genes. This circumvents a basic problem in gene regulation—if each gene were regulated by a single, distinct protein, the number of genes encoding regulatory proteins would have to be equal to the number of genes to be regulated. Cooperative interactions also allow gene activity to be adjusted upward or downward by degrees rather than simply being turned entirely on or off. In addition, regulation in combinations sets up control groups and *networks* (Fig. 17-3) in which multiple genes recognized by one or more regulatory proteins are linked in their regulatory responses. The networks become particularly extensive if genes encoding other regulatory proteins are among the group recognized by common factors (as in Fig. 17-3*b*).

These and additional operating principles of *cis* and *trans* elements in transcriptional regulation are illustrated by the following examples. The examples begin with a relatively simple system in the yeast *Saccharomyces* and continue with more complex systems, some involving networks that control major developmental and response pathways.

Regulation by Two Interacting Proteins: The GAL4/80 System of Saccharomyces

A pair of regulatory proteins of yeast cells, *GAL4* and *GAL80*, operate in a clear-cut system that serves as an excellent introduction to regulatory proteins and their activity. Although relatively simple, the GAL4/80 system has provided some of the most significant and useful information about transcriptional regulation.

The genes regulated by GAL4 and GAL80 encode enzymes that carry out steps in galactose metabolism. Genetic crosses established yeast cells with mutations of the GAL4 gene that were unable to synthesize several enzymes necessary for galactose breakdown. Research by M. Ptashne, S. A. Johnston, and their coworkers revealed that GAL4 recognizes and binds a DNA sequence in enhancerlike regions upstream of the promoter in several yeast genes including *GAL1*, *GAL7*, *GAL10*, and *GAL12*. Binding of the normal, nonmutant form of GAL4 increases transcription of these genes by about a thousand times and activates galactose metabolism.

The GAL4 protein contains two major domains. One contains the region recognizing and binding *cis* elements in the DNA. The other includes the region, distinguished by a preponderance of acidic amino acid residues, capable of directly or indirectly activating initiation of transcription by RNA polymerase II. The activating region may bind either RNA polymerase II

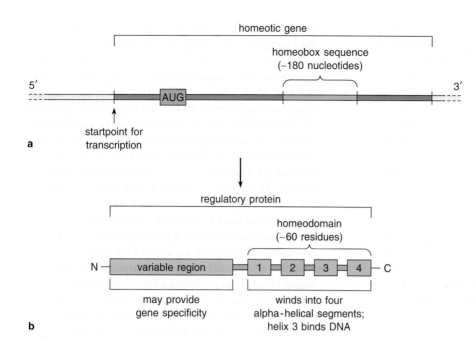

Figure 17-8 Homeotic genes and homeodomain proteins. A sequence element common to all homeotic genes, the homeobox **(a)**, encodes the homeodomain **(b)**. The homeodomain, which contains four alpha-helical segments, including a helix-turn-helix motif, binds to a recognition sequence in the 5'-flanking regions of all genes controlled as a unit by a homeotic gene.

to development in *Drosophila.* One homeotic gene of this species, the *antennapedia,* or *antp,* gene, is highly active in segments destined to form the thorax of the adult, in which legs develop. Normally, activity of the gene is limited in other segments, such as those destined to become the head of the adult fly. In Gehring's experiment the upstream control region of a heat shock gene was attached to the 5'-flanking region of the *antp* gene in developing *Drosophila* embryos. At normal temperatures development progressed without change and produced a head with antennae in adults (Fig. 17-9*a*). At elevated temperatures the *antp* gene, activated by the heat-shock control regions in the head segments, converted the cell groups that would normally form antennae to the development of legs (Fig. 17-9*b*).

Tests in other organisms with homeobox or homeodomain sequences revealed that homeotic genes occur in all major animal phyla except sponges and coelenterates. The homeobox sequences in these genes are highly conserved and encode homeodomains that differ only in conservative amino acid substitutions, that is, the replacement of one amino acid by another of similar chemical properties. Only 5 conservatively substituted amino acids out of 60, for example, are different in the homeodomains of 2 homeotic regulatory proteins of *Drosophila* and *Xenopus,* an amphibian. More than 20 different homeodomain-containing proteins have been detected in humans.

Recently, S. Hake and her colleagues discovered a gene controlling leaf development in maize that encodes a protein with a homeodomain. In maize plants with a mutant form of the gene called *knotted-1,* leaf veins are twisted into protrusions rather than lying in

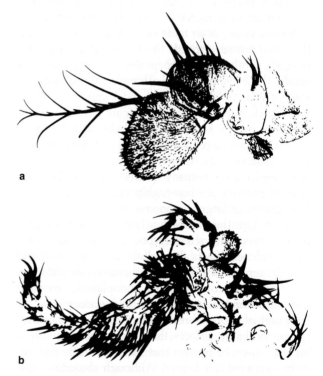

Figure 17-9 The effect of a homeotic gene on development in *Drosophila.* **(a)** Normal development of an antenna **(b)** Conversion of an antenna to a leg in a fly in which a homeotic gene, *antp,* is overactive in head segments. Courtesy of W. J. Gehring. Reprinted by permission from *Nature* 325:816; copyright © 1987 Macmillan Magazines Ltd.

flat planes as in normal plants. Using the gene as a probe for DNA of similar sequence, these investigators found two other genes in maize encoding proteins with homeodomains. Thus, homeotic genes may be common in plants as well as animals and may have appeared in evolution before the split producing the two major kingdoms of organisms.

The homeodomain folds into four alpha-helical segments (see Fig. 17-8b). The third segment, held with helix 2 of the homeodomain in a helix-turn-helix motif, recognizes and binds an eight base-pair or *octamer* sequence in the 5'-flanking region of the genes controlled as a group by a homeotic gene. The third helix fits into the large groove of the DNA molecule and recognizes nucleotide sequences exposed in the groove.

Homeodomains are identical or closely similar in different regulatory proteins encoded by homeotic genes. For many, the amino acid residue at position 9 in the third helix is critical to the ability of the homeodomain to recognize and bind specific DNA sequences. For example, J. Treisman and her coworkers found that the homeodomain of the *bicoid* regulatory protein of *Drosophila* has a lysine at this position; the homeodomain of another regulatory protein encoded by the *paired* gene has a serine at this position. If the serine of the paired protein is changed experimentally to lysine, this protein switches to binding the DNA sequence recognized by the bicoid protein. When the homeodomains of different regulatory proteins are identical, the ability of the proteins to recognize and bind different DNA sequences evidently depends on differences in amino acid sequences outside the homeodomains or on interactions with other regulatory proteins.

In an important subgroup of these regulatory proteins the homeodomain forms part of a larger DNA-binding element called the *POU domain* (pronounced "pow"). The name is taken from four major regulatory proteins containing the POU domain—Pit-1, OCT-1, and OCT-2 of mammals, and a regulatory protein encoded in the *unc-86* gene of the nematode worm *Caenorhabditis*. The POU domain in these proteins is 150 to 160 amino acids in length. The first 75 to 82 amino acids form the *POU-specific domain*; this domain is separated from the following homeodomain by a short spacer of variable length. In these proteins the entire POU domain binds to the DNA and determines both the specificity and strength of binding. All POU proteins carry a cysteine in the critical ninth position of the third helix. The existence of the POU domain shows that the regulatory proteins encoded by homeotic genes are highly conserved at structural levels above the homeodomain.

Although the functions of most homeotic genes in vertebrates are unknown, at least some appear to control major developmental pathways, particularly in the nervous and neuroendocrine systems. For example, E. M. DeRobertis and his coworkers tested the developmental effects of one homeodomain-containing protein by injecting antibodies against the protein into *Xenopus* embryos. In injected embryos regions of the central nervous system that would normally form parts of the spinal cord near the brain developed into more anterior structures of the hindbrain. The tested homeodomain protein therefore appears to direct development of major parts of the central nervous system along the anterior-posterior axis. Rather than controlling major developmental pathways, however, some homeotic genes of vertebrates appear to regulate more general mechanisms that affect both embryos and adults as a whole. OCT-1, for example, one of the regulatory proteins encoded in a homeotic gene of mammals, activates a wide variety of genes that are expressed in almost all mammalian body cells, among them snRNA and histone genes; OCT-2 activates antibody synthesis in cells of the immune system.

The homeotic regulatory proteins of *Drosophila* are capable of activating genes when injected into mammalian cells. This and the general conservation of homeotic genes among animals and higher plants confirm that this system, like many mechanisms regulating genetic activity in eukaryotes, has ancient evolutionary origins and has been maintained in largely intact form for many millions of years. The genes controlled by the conserved systems, however, often carry out distinct and unrelated functions in different major taxonomic groups.

The preceding examples clearly illustrate several principles that are common to many of the systems regulating specific genes in animals, plants, and lower eukaryotes:

1. Specific regulation depends on an interaction between control sequences in 5'-flanking regions of genes (*cis* elements) and regulatory proteins (*trans* elements). The control sequences may form parts of promoters or enhancers or may lie outside these regions; the regulatory proteins, which may have positive or negative effects, may combine directly with the DNA, other regulatory proteins, transcription factors, or RNA polymerase II.

2. Most regulatory proteins combining directly with DNA have at least two functional domains—one recognizing and binding *cis* elements in the DNA and the other directly or indirectly activating transcription. In many regulatory proteins the domain activating transcription contains a preponderance of acidic amino acid residues. Acidic activating domains are not found in all regulatory proteins; for example, some are rich in proline or glutamine residues in this domain. The DNA-binding or other domains of regulatory proteins may be activated or deactivated by modifications such as reversible phosphorylation, combination with other proteins, or linkage to regulatory molecules such as steroid hormones.

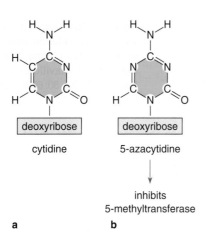

cytidine   5-azacytidine

inhibits
5-methyltransferase

a   b

Figure 17-11 Cytidine **(a)** and 5-azacytidine **(b)**. 5-aza-cytidine irreversibly inhibits 5-methyltransferase, the enzyme adding methyl groups to cytosine in DNA.

exceptions illustrate that methylation is only one of several possible modifying controls used in genetic regulation in vertebrates.

The use of methylation as a genetic regulatory mechanism appears to be limited to vertebrates. In *Drosophila* and other dipteran flies, brine shrimp, nematode worms, ciliate protozoans, and some fungi, methylation of cytosines can be detected in neither active nor inactive genes. In higher plants cytosines in DNA sequences apparently remain highly methylated at all times, whether genes are active or not. The use of methylation as a modifying gene control only among the vertebrates suggests that this mechanism appeared relatively late in the evolution of the animal kingdom as a device for transcriptional regulation.

Molecular Effects of DNA Methylation The molecular effects of methylation in transcriptional regulation remain uncertain. However, a methyl group added to the 5′-carbon of cytosine projects into the major groove of DNA (see Fig. 14-8 and p. 536), in the region where the recognition segments of regulatory proteins are believed to interact with the edges of the DNA bases. The projecting methyl group may block insertion of regulatory proteins into the major groove. Frequently, CpG doublets occur in the center of *cis* control elements of genes, in a position in which their interference with recognition and binding by regulatory proteins is expected to be greatest.

There is some evidence that methyl groups actually interfere with the binding of regulatory proteins. G. Schutz and his coworkers, for example, showed that the addition of a methyl group to a cytosine in a *cis* element of the tyrosine aminotransferase gene of the rat prevents a regulatory protein from binding to the element. In another experiment G. C. Prendergast and E. B. Ziff found that addition of a methyl group to a CpG doublet in the *cis* element ... GGCCA<u>CG</u>T-GACC ... blocked binding of the regulatory protein

encoded in the *myc* gene to this sequence. Part of this sequence including CpG forms the CACGTG element common to genes controlled by Myc and related proteins (see Prendergast's essay on p. 696).

Several lines of evidence have shown that methylation patterns are passed from one cell generation to the next. For example, the pattern of methylation of cloned DNA molecules injected into *Xenopus* eggs has been shown to be faithfully copied as the DNA is replicated in the eggs. Similarly, patterns of methylation in DNA introduced by genetic engineering techniques into cultured mouse cells have been demonstrated to be precisely copied and passed on for at least 200 generations. The duplication of methylation patterns shows that regulatory programs as well as genetic information can be passed on in cell division. (A mechanism suggested by R. Holliday and A. D. Riggs for the duplication of methylation patterns during DNA replication is shown in Fig. 17-12.)

Histone Modifications and Genetic Regulation

The histones organize chromatin at several levels in eukaryotic cells (see Ch. 14 for details). At the most fundamental level histones wind DNA into nucleosomes, the primary structural units of chromatin. Interactions between H1 molecules of different nucleosomes are believed to wind nucleosome chains into solenoids, the higher-order coils considered to be equivalent to individual chromatin fibers. Interactions between H1 molecules are also believed to be responsible for condensation of chromatin into heterochromatin, a still higher level of organization. Changes in chromatin structure at each of these levels—nucleosomes, solenoids, and heterochromatin—have been correlated with transcriptional regulation in general rather than specific patterns, that is, the shift of genes between unavailable and ready states.

Chemical modification and substitutions or loss of histone types and variants have both been implicated in chromatin alterations linked to transcriptional regulation. Of these, chemical modification of the histones has been most thoroughly supported as a control by experimental evidence.

Chemical Modification of the Histones The histone modifications correlated with transcriptional regulation, addition of either acetyl or phosphate groups to amino acid side chains, are fully reversible. Acetyl groups are added to the core histones (H2A, H2B, H3, and H4), and phosphate groups to both core histones and H1. The core histones wind DNA into nucleosome core particles; H1 closes the DNA loops around nucleosomes and sets up H1-H1 linkages between nucleosomes (see p. 546 for details). Each core histone has from 2 to 4 lysine side chains available for the addition of acetyl

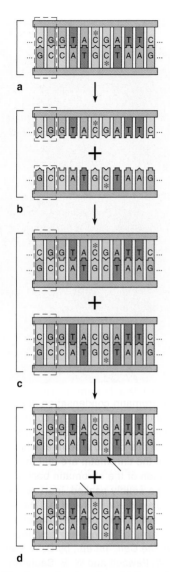

Figure 17-12 A mechanism by which methylation patterns might be passed on during DNA replication. The dashed box encloses an unmethylated doublet; a methylated doublet is shown in blue (asterisks indicate methylated Cs). **(a** and **b)** The DNA unwinds for replication. **(c)** Replication produces two DNA molecules that are hemimethylated at the blue CpG doublet. **(d)** This doublet is then fully methylated (arrows) by a putative *hemimethylase* that recognizes the hemimethylated sites. The result duplicates the methylation pattern of the parent DNA.

groups; in total an individual nucleosome has 26 sites that can be modified by reversible addition of an acetyl group.

Evidence implicating acetylation of core histones in transcriptional regulation, developed through the investigations of V. G. Allfrey, E. M. Bradbury, M. A. Gorovsky, and others, has accumulated over many years. This evidence is largely circumstantial, primarily showing a positive correlation between the addition of acetyl groups to core histones, particularly H3 and H4, and transcriptional activity. For example, in chromatin

identified as active by greater susceptibility to digestion by DNA endonucleases (see p. 547), core histones are generally more highly acetylated than those of inactive chromatin. In developing embryos histone acetylation increases with transcriptional activity. In viruses in which the viral DNA is wound into nucleosomes by host-cell histones, such as the SV40 virus (see p. 549), histones of active viral genes are about four times more highly acetylated than those of inactive genes.

Another example comes from studies with ciliated protozoans, in which cells contain two different nuclei, the *micronucleus* and *macronucleus*. In the micronucleus, which contains a complete gene complement but functions primarily in cell reproduction rather than mRNA transcription, genes are inactive and core histones are unacetylated during cytoplasmic growth. In the macronucleus, which contains only part of the genetic complement, most or all genes are active and core histones are acetylated. Many other observations support the correlation between transcriptional activity and increased acetylation of core histones.

The molecular effects of acetylation of core histones on chromatin structure are uncertain. Acetylation is expected to reduce the net positive charge of histones (for details, see p. 542). As such, it may diminish the attraction of core histones for DNA or the strength of interaction among core histones within nucleosomes. These changes might cause alterations in chromatin structure at any level, including the unfolding of nucleosomes, unwinding of solenoids, or decondensation of heterochromatin. One recent experiment by C. L. F. Woodcock and his colleagues, in fact, demonstrates that the experimental addition of acetyl groups to core histones inhibits the winding of nucleosome chains into solenoids. Any of the chromatin transitions would be expected to increase the availability of gene sequences to RNA polymerases or regulatory proteins, thereby providing a nonspecific mechanism that enhances transcription. At the same time, removal of acetyl groups might be expected to increase DNA folding into nucleosomes, solenoids, or heterochromatin, which would make genes in these structures less available for transcription.

Not all experiments demonstrate a clear correlation between histone acetylation and transcriptional activity. Some cells show no difference in acetylation levels between the core histones of active and inactive genes. Like DNA methylation, these contradictory results indicate that histone acetylation might be used as a device for transcriptional regulation in some cell types and species but not others.

Chemical Modification and Chromatin Condensation Phosphorylation of H1, and to a lesser extent of one or more of the core histones, has been correlated with chromatin condensation in many cell types and species. Chromatin condensation, in turn, accompanies

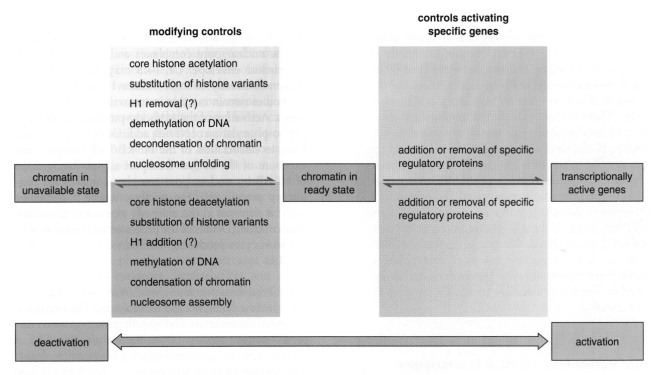

Figure 17-14 Summary of some of the possible levels of transcriptional regulation in eukaryotes. A potentially large number of modifying controls may convert individual genes or blocks of genes between an unavailable state in which they cannot be activated and a ready state in which they can be activated. Conversion between the ready state and the transcriptionally active state is moderated by specific regulatory proteins that can recognize the promoter, enhancer, and other control sequences of genes.

SPECIALIZED MECHANISMS REGULATING rRNA GENES

Genes encoding rRNAs are among the most intensively transcribed sequences in eukaryotic cells. Although these genes are controlled by specific regulatory proteins, the pattern of regulation is fundamentally different from the mechanisms regulating mRNA genes. Most mRNA genes occur in single copies and are individually regulated by controls that adjust their transcription rates upward or downward. In contrast rRNA genes occur in many copies; transcriptional regulation is accomplished primarily by activating or deactivating greater or fewer numbers of genes.

Transcription of rRNA is most active in developing oocytes. In these cells all the multiple rRNA gene copies typically become active. In many animal oocytes rRNA transcription is further increased by *amplification* of large pre-rRNA genes (see p. 1086)—the oocytes make extra copies of the genes, which supplement and greatly increase the oocyte's capacity to synthesize rRNA (see p. 1056 for details). Once oocyte development is complete, the entire set of rRNA genes is turned off and remains inactive until fertilization; extra copies of rRNA genes are degraded. The rRNA genes are activated

soon after fertilization and remain active in most cells throughout embryonic development and adult life. Typically, the number of rRNA genes activated and the level of rRNA transcription in developing and mature cells reflect the rate of cell activity in protein synthesis and the need for ribosomes.

Regulation of rRNA synthesis by control of the number of genes activated has been nicely demonstrated in mutants lacking a portion of their large pre-rRNA genes. In *Xenopus*, for example, D. D. Brown and J. B. Gurdon showed that embryos with half the usual quantity of large pre-rRNA genes still synthesize large pre-rRNAs at the same rate as normal embryos. The slack is taken up in the mutants by the activation of a larger proportion of the remaining rRNA gene copies than would normally be transcribed at this stage of development.

Ribosomal RNA genes show some of the same alterations as mRNA genes in coordination with transcriptional regulation. Reduction in the level of DNA methylation, particularly at sites near the promoter regions, is frequently noted in the active rRNA genes of vertebrates. Methylation of rRNA genes remains high in sperm cells and mature nucleated erythrocytes, in which rRNA synthesis is completely inactive. Alterations of this type, as in the equivalent regulatory changes in mRNA genes, are expected to act as general rather than specific controls.

Regulation of rRNA genes appears to depend on alterations in the number of transcription factors (see p. 610) available for the initiation process. In this sense the transcription factors act as specialized regulatory proteins as well as factors promoting initiation by RNA polymerase I and III.

The connection between transcription factors and rRNA gene regulation was elegantly demonstrated by Brown and his colleagues in their study of 5S rRNA synthesis in *Xenopus*. In developing *Xenopus* oocytes all 5S genes are active, including both oocyte-type and somatic-type 5S genes (see p. 615). Activity of both types shuts down as eggs mature. After fertilization both types of 5S genes again become active. As embryonic development proceeds, activity of oocyte 5S genes gradually diminishes, while transcription by somatic 5S genes continues. By gastrulation, transcription by oocyte 5S genes is essentially shut off, and almost all the 5S rRNA of embryonic cells is transcribed by somatic-type 5S genes. This condition continues in somatic cells throughout the remainder of embryonic development and adult life.

Control of 5S rRNA transcription appears to depend on interactions among several factors in *Xenopus*. Brown and his colleagues and others observed that the level of 5S gene transcription is paralleled by changes in concentration of the specific 5S transcription factor, TFIIIA (see p. 615). This factor exists in very high concentrations in developing *Xenopus* oocytes and early embryos, reaching a level of ten copies for every 5S gene. At this level supplies of the factor are more than sufficient to activate all 5S genes, both somatic and oocyte. The concentration of TFIIIA gradually falls as development proceeds. By gastrulation, when 5S rRNA synthesis is concentrated primarily in somatic-type 5S genes, TFIIIA concentrations have dropped to about one copy per 5S gene.

Brown and his colleagues also noted a difference in the stability of transcription complexes formed between TFIIIA and other transcription factors (see p. 615), RNA polymerase III, and the promoter regions of the two types of 5S genes. The complex formed between these elements is significantly less stable on oocyte-type than on somatic-type 5S rRNA genes. This difference appears to depend on substitutions of only six nucleotides in the two 5S rRNA genes.

Brown proposed that these characteristics contribute to 5S gene regulation in the following way. Early in embryonic development TFIIIA levels are high enough to allow maximum transcription of all 5S genes and are thus not rate limiting. As TFIIIA concentration falls later in development, somatic 5S genes form more stable complexes than oocyte 5S genes with the available TFIIIA molecules genes. As a result, initiation of transcription on oocyte 5S genes gradually diminishes as TFIIIA concentration falls. By gastrulation, when TFIIIA concentration has dropped to about one molecule per 5S gene, the remaining TFIIIA molecules are bound almost entirely to somatic 5S genes, and transcription on oocyte 5S genes essentially halts.

POSTTRANSCRIPTIONAL REGULATION

Before the mRNAs copied from active genes participate in protein synthesis, they must be processed, transported to the cytoplasm, and made available to ribosomes. Regulatory mechanisms may be imposed during each of these steps to adjust the numbers and kinds of mature mRNAs that bind to ribosomes and direct polypeptide assembly. The controls acting between transcription and the initiation of translation accomplish posttranscriptional regulation.

Posttranscriptional Regulation During mRNA Processing

Two processing steps have been established as posttranscriptional controls in eukaryotes—the reactions splicing introns from pre-mRNAs and the addition of poly(A) tails. In thyroid cells, for example, the pre-mRNA transcript of the calcitonin gene is processed by splicing together some exons but not others to produce *calcitonin*, a peptide hormone. In cells of the hypothalamus, the pre-mRNA product of the same gene is processed differently, by splicing together another combination of exons, to produce a distinct mRNA coding for a different polypeptide, *CGRP* (CGRP = Calcitonin Gene-Related Polypeptide). These observations indicate that the splicing reactions are regulated in different patterns in thyroid and hypothalamus cells as a mechanism of posttranscriptional control.

Another example of regulated alternative splicing occurs as part of sex determination in *Drosophila*, in which the *sex-lethal (Sxl)* gene encodes a regulatory protein necessary for female development. In males an exon near the 5' end of the coding segment of the pre-mRNA is included during alternative splicing; in females this exon is spliced out. The exon includes a termination signal for protein synthesis, so that a short, nonfunctional copy of the protein is produced in males. In females a fully functional protein is produced. The functional Sxl protein, interestingly enough, induces alternative splicing of the pre-mRNA copy of a second gene, *transformer (tra)*. The pattern of alternative splicing of the *tra* gene product is similar; in the presence of a functional Sxl protein, alternative splicing of the *tra* pre-mRNA produces an mRNA encoding a fully functional protein. If the Sxl protein is absent, as it is in males, alternative splicing produces an mRNA encoding an abortive, nonfunctional protein.

The Sxl protein therefore stands as an example of a regulatory protein that influences the pathway of alternative splicing. Identification of a specific protein influencing alternative splicing is as yet a rare event; the factors influencing this mechanism remain unknown in most systems. However, regulation of alternative splicing could conceivably result from differences in mRNA sequence or secondary structure, or differences in the enzymes, accessory factors, structural proteins, or snRNAs associated with the pre-mRNA product as well as specific regulatory proteins.

Some mRNAs are stored in inactive form in the nucleus or cytoplasm as incompletely spliced precursors. In the eggs of some animals, for example, large quantities of inactive mRNAs are stored in the cytoplasm. At least part of their inactivity depends on incomplete splicing. Several *Drosophila* mRNAs, including those for the L1 ribosomal protein, are also regulated posttranscriptionally by delayed splicing. Posttranscriptional regulation of *Drosophila* L1 mRNA may occur by negative feedback—the L1 ribosomal protein, if overabundant, may combine with either its pre-mRNA or spliceosomes to inhibit the splicing reactions. The L32 ribosomal mRNA of yeast is similarly regulated posttranscriptionally by delayed splicing, possibly involving a negative-feedback mechanism by the L32 protein.

Variations in the rate of poly(A) tail addition are employed as a regulatory mechanism by a virus infecting eukaryotic cells, the *adenovirus*. In this virus several genes, *L1, L2,* and *L3,* are transcribed simultaneously and at the same rates late in the infective process. (These genes code for coat proteins of the virus.) Initially, poly(A) tails are added to the L1 pre-mRNA at two or three times the rate of the L2 and L3 pre-mRNAs. Later, when viral coat protein production is nearly complete, poly(A) addition to L1 pre-mRNAs decreases, and addition of poly(A) to L2 pre-mRNA increases, so that the rate of L2 mRNA processing exceeds L1 processing by several times. These differences in poly(A) tail addition regulate the rate at which the mRNA products of the L genes complete processing and, ultimately, the rate at which the finished mRNAs reach the cytoplasm.

Posttranscriptional Regulation of mRNA Transport and Stability

Experiments with adenovirus also suggest that posttranscriptional controls may be exerted in some cells at the level of mRNA transport through nuclear pores. As the cycle of adenovirus infection is completed, the L series mRNAs cease to enter the cytoplasm even though transcription and processing continue in the nucleus. Although the mechanisms responsible for this type of posttranscriptional regulation are unknown, they may depend on the presence or absence of protein factors necessary for passage through the pore complexes of the nuclear envelope (see p. 557). Alternatively, delays in transport may reflect temporary attachment of mRNA-protein complexes to cytoskeletal elements of the nuclear matrix.

Posttranscriptional control of mRNA stability occurs widely in eukaryotes and may, in fact, be universally employed as a regulatory mechanism. Typically, mRNAs have a finite life expectancy in the cytoplasm, and differences are frequently noted in the relative stability of individual mRNA types (Table 17-2). Most significantly, the life expectancy of individual mRNAs can be observed to increase or decrease in coordination with cell activities or differentiation and development. When chicken liver cells are exposed to estrogens, for example, vitellogenin mRNA has a half-life of nearly 500 hours. If estrogens are withdrawn, or in cells never exposed to estrogens, the half-life of vitellogenin mRNA drops to 16 hours—about 30 times shorter.

In many cases experimental transfer of the 3' untranslated segment from one mRNA to another can transfer the stability characteristics of the donor mRNA, indicating that information regulating breakdown resides in the sequence or secondary structure of this segment. One structural feature often observed in the 3' untranslated segment of unstable mRNAs is the repetition of U-A sequences rich in U, such as AUUUA. Removing the U-A sequences from mRNAs originally containing them stabilizes these mRNAs; conversely, adding the sequences to the 3' untranslated segments of mRNAs originally without them destabilizes these

Table 17-2 Half-Lives of Some Mammalian mRNAs

Protein Encoded	Half-Life in Hours
β-actin (nonmuscle)	60
Cytochrome P$_{450}$	36
Dihydrofolate reductase	97
Fatty acid synthetase	48–96
Glucocorticoid receptor	2.3
Glucose 6-phosphate dehydrogenase	15
Glyceraldehyde 3-phosphate dehydrogenase	75–130
Heat shock protein 70 (hsp70)	2.0
Insulin receptor	9
Ornithine aminotransferase	19
Ornithine decarboxylase	0.5
Pyruvate kinase	30
Thymidine kinase	2.6
Tubulin	4–12
Tyrosine aminotransferase	2.0

Adapted from J. L. Hargrove and F. H. Schmidt, *FASEB J.* 3:2360 (1989).

mRNAs. For mRNAs with poly(A) tails, removal of the tail, which seems to protect the 3' end from enzymatic attack (see p. 584), precedes degradation.

Stability of histone mRNAs is coordinated with DNA replication. During replication histone mRNAs are relatively stable in the cytoplasm, with half-lives of about 60 minutes. At completion of DNA replication, or if replication is experimentally inhibited, histone mRNA half-life drops to a few minutes (see Ch. 22 for details). Experiments by W. F. Marzluf showed that the last 20 nucleotides of histone mRNAs, including an inverted sequence that forms a hairpin, figure importantly in both the stability and breakdown of these mRNAs (histone mRNAs, with few exceptions, lack poly(A) tails). Removal of this 3'-end structure makes histone mRNAs unstable under any cellular conditions and leads to their rapid breakdown. Adding the last 20 to 30 nucleotides of histone mRNAs to other, non-histone mRNAs gives the nonhistone types stability characteristics typical of histone mRNAs. These hybrid non-histone mRNAs are stable during DNA replication and are degraded rapidly when replication stops. Apparently, the 3'-end hairpin of histone mRNAs protects them from degradation by the usual cytoplasmic enzymes hydrolyzing other mRNA types; after DNA synthesis is complete, a ribonuclease capable of specifically recognizing the hairpin evidently appears and degrades the histone mRNAs.

Breakdown of histone mRNAs is coupled to protein synthesis and proceeds only if the mRNA is translated nearly to completion. Evidently, the ribonuclease responsible for the degradation recognizes the histone mRNA only when the mRNA is attached to a ribosome near its 3' end. Other mRNAs, such as the β-tubulin mRNA, are also degraded only while engaged in protein synthesis. In the case of tubulin mRNA, research by D. W. Cleveland and his associates showed that breakdown occurs immediately after protein synthesis begins, when a short polypeptide containing the first four amino acids emerges from the ribosome.

The effects of transcriptional and posttranscriptional regulation combine to control both the kinds and numbers of mRNA molecules reaching the cytoplasm. Although there is no completely exclusive separation of kinds and numbers, regulation of the *kinds* of mRNA reaching the cytoplasm is primarily the result of transcriptional controls. The distinction is not complete because some posttranscriptional regulatory mechanisms, such as alternative splicing, also regulate the kinds of mRNA produced. The *numbers* of mRNAs reaching the cytoplasm are controlled by both transcriptional and posttranscriptional regulation—transcriptionally, through differences in the rate of initiation of transcription, and posttranscriptionally, through changes in the rates of mRNA processing, transport, and degradation in the cytoplasm.

TRANSLATIONAL REGULATION

Translational regulation controls the rate and efficiency with which mRNAs are used as templates for polypeptide assembly. This pattern of regulation is particularly important in animal eggs and early embryos. It is also important in cells in which rapid, short-term adjustments in levels of protein synthesis are critical to survival because changes at this level, in some cases, can take effect within seconds or minutes after a stimulus.

The effects of translational regulation may be quantitative, adjusting the numbers of polypeptides synthesized upward or downward, or qualitative, completely turning on or off synthesis of a given polypeptide or group of polypeptides. Both quantitative and qualitative translational regulation usually proceed without changing the number of mRNA molecules in the cytoplasm.

The processes of translation, from initiation through termination, include a multitude of steps potentially subject to regulatory controls. However, most known eukaryotic translational controls operate during the initiation of protein synthesis. The regulation of protein synthesis at initiation rather than later stages represents an economy in the use of ribosomes, since controls inhibiting elongation or termination would trap ribosomes in inactive complexes.

Translational Controls Regulating Steps in Initiation

The large number of initiation factors and the complexity of the eukaryotic initiation process (see p. 654) offer many opportunities for translational controls. There is ample evidence that such controls are actually widely used in translational regulation in eukaryotes.

Techniques for Detecting Controls Several methods have been used to detect regulatory pathways based on control of individual initiation factors. One frequently used technique looks for modifications to initiation factors, such as phosphorylation or dephosphorylation, that occur in synchrony with the inhibition or activation of initiation. The effects of experimentally adding or removing the phosphate groups are then noted. Another method follows the effects of adding or removing individual initiation factors in either active or inactive form. It has been possible, for example, to identify factors inhibiting initiation by adding each factor in unmodified, active form and noting the effects on cells in which protein synthesis is downregulated. If translation increases significantly on addition of a particular factor, it can be assumed that adjustment in the supply of the factor is employed as a translational control.

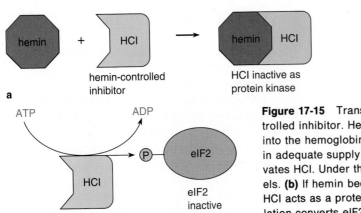

a

b

Figure 17-15 Translational regulation by hemin and HCI, the hemin-controlled inhibitor. Hemin is the iron-containing porphyrin ring incorporated into the hemoglobin molecule to complete its synthesis. **(a)** When hemin is in adequate supply in developing red bloods, it combines with and inactivates HCI. Under these conditions protein synthesis proceeds at high levels. **(b)** If hemin becomes unavailable, HCI is activated. The active form of HCI acts as a protein kinase, adding phosphate groups to eIF2. Phosphorylation converts eIF2 to inactive form and halts protein synthesis.

Controls Regulating eIF2 Activity   Among the various eukaryotic initiation factors, eIF2, which binds an initiator tRNA to the small ribosomal subunit during initiation, is a frequent target of regulatory controls. Inhibition of eIF2 is imposed by addition of a phosphate group to one of the three polypeptide subunits of the factor. The phosphorylated form of eIF2 is inactive in the side cycle that exchanges GDP for GTP after the initiation factor has attached a molecule of mRNA to the small subunit (step 7 in Fig. 16-3). The eIF2 remains as an inactive eIF2·GDP complex, unable to enter further cycles of mRNA attachment to the ribosome. Cessation of the cycles effectively stops protein synthesis.

The best-documented control system based on eIF2 inhibition regulates hemoglobin synthesis in developing red blood cells (Fig. 17-15). In these cells translation of the globin proteins depends on the supply of *hemin,* an iron-containing porphyrin ring incorporated into the hemoglobin molecule to complete its synthesis. When hemin is in adequate supply, it combines with and inactivates a regulatory protein, *hemin-controlled inhibitor* (HCI; also called *hemin-controlled repressor,* or *HCR;* Fig. 17-15a). Under these conditions protein synthesis proceeds at high levels in developing red blood cells. If hemin becomes unavailable, HCI is released from its complex with hemin and is activated. In active form the HCI protein acts as a protein kinase that specifically adds phosphate groups to eIF2 (Fig. 17-15b). Addition of the phosphates inhibits the factor's activity in initiation and halts protein synthesis. If hemin becomes available, HCI is again complexed, inhibiting its phosphorylating activity and releasing the eIF2 blockage.

As a result of the mechanism, translation runs in the presence of hemin and is halted in its absence. The regulatory control is effective in developing erythrocytes because protein synthesis becomes concentrated almost entirely in the assembly of hemoglobin molecules in these cells. Interruption of the assembly of other proteins when hemin supplies are low is not critical. The mechanism works rapidly enough to shut down protein synthesis within five minutes after hemin becomes unavailable.

The system inhibiting globin mRNA translation works in cell-free preparations as well as in intact cells. Addition of unphosphorylated eIF2 to inhibited cell-free systems extracted from red blood cells releases the blockage, indicating that the initiation factor is actually the target of the regulatory mechanism.

Similar regulatory mechanisms based on eIF2 phosphorylation have been discovered in a variety of cell types. These mechanisms, which employ distinct protein kinases, are triggered by various stimuli including elevated temperature, ATP depletion, virus infection, salt shock, general starvation, or deprivation of glucose, amino acids, or serum. Exposure to double-helical RNA also induces the activity of a kinase phosphorylating eIF2. All these stimuli, through their effect in activating enzymes phosphorylating eIF2, effectively slow or shut down protein synthesis.

The eIF2 phosphorylation triggered by virus infection provides a natural defense against viruses. The defense is effective because, in the general inhibition of protein synthesis induced by eIF2 phosphorylation, assembly of viral proteins is inhibited as well as normal host proteins. The eIF2 phosphorylation triggered by double-stranded RNA may also be part of a defense against viruses, because the genetic material of some viruses appears in this form during cycles of growth and reproduction.

The defenses against viral infection provided by translational controls based on eIF2 are promoted by *interferon,* a protein synthesized and secreted by certain cells in response to viral infection. Interferon, when taken up by other cells, greatly increases the activity of the protein kinases stimulated by viral infection or double-stranded RNA. The activity of an mRNA-digesting enzyme stimulated by double-stranded RNA is also intensified by interferon.

A few viruses, such as the influenza virus and the adenovirus, have evolved countermeasures against eIF2-based translational controls. The influenza virus somehow suppresses the protein kinase dependent on double-stranded RNA that phosphorylates eIF2; the adenovirus produces an RNA molecule with the same effect. These countermeasures allow synthesis of the viral proteins to proceed even in the presence of interferon.

Controls Regulating Other Initiation Factors Initiation factors other than eIF2 are used as control points in some systems. In sea urchin eggs, for example, A. C. Lopo and her colleagues found that two of the elements inhibiting protein synthesis before fertilization are eIF4B and eIF4E. During initiation of protein synthesis, eIF4B may promote the unwinding of secondary structure in the mRNA; eIF4E binds the 5' cap of mRNA and speeds association of mRNA with the small ribosomal subunit (see Fig. 16-3 and Table 16-1). In unfertilized eggs the two factors exist in an inactive, unphosphorylated form. At fertilization a protein kinase is activated that rapidly phosphorylates the factors and releases protein synthesis from inhibition. (Lopo's Experimental Process essay on p. 656 describes her experiments identifying phosphorylation of eIF4E as a translational control in sea urchin eggs.)

Phosphorylation of eIF4E and other initiation factors, including eIF2B, eIF3, eIF4B, and eIF5 (see Table 16-1), has also been implicated as a mechanism regulating translation in other systems. Among the protective responses of cells exposed to heat shock and other stresses is a general repression of protein synthesis. During translational inhibition as a result of heat shock, eIF4B and eIF4E are dephosphorylated; eIF4B phosphorylation levels also change during the inhibition occurring when cultured HeLa cells are deprived of fresh serum. The heat shock inhibition of another cultured cell type, Erlich ascites cells, can be reversed by addition of phosphorylated eIF4E.

One group of viruses, including the poliovirus and the rhinoviruses responsible for the common cold, employs a mechanism that favors translation of their own mRNAs by destroying the eukaryotic initiation factors recognizing and binding the 5' cap of mRNAs during initiation (see Fig. 16-3). The factors eIF4E and, in some cases, p220 are degraded by proteinases activated by the viruses. Destruction of the initiation factors prevents infected cells from translating their own mRNAs, but the viral mRNAs, which lack 5' caps, are unaffected by loss of the initiation factors. A protein encoded by the viruses specifically recognizes and binds the viral mRNAs and promotes their attachment to ribosomes and entry into polypeptide assembly.

Controls Regulating the Availability of mRNAs for Initiation During oocyte development mRNAs are made in large quantity and stored in the cytoplasm. Some of these mRNAs become active at different stages of oogenesis, and some are stored and remain inactive until the egg has been fertilized and embryonic development is under way. The availability of many of these mRNAs for initiation of protein synthesis depends on sequences in the 5' or 3' untranslated segments of the mRNAs and the combination of regulatory proteins with these sequences.

For example, the mRNA for the ribosomal protein S19 is synthesized in amphibian oocytes and stored in the cytoplasm in inactive form. The mRNA does not become active until the tailbud stage of the embryo. P. Mariottini and F. Amaldi found that if the 5' untranslated segment of the S19 mRNA was transferred experimentally to an unrelated mRNA, that for *chloramphenicol acetyltransferase (CAT)*, and introduced into early embryos, the CAT mRNA also remained untranslated until the tailbud stage. With its own 5' untranslated segment, the CAT mRNA entered protein synthesis immediately after introduction. Presumably, the S19 mRNA is regulated through a control protein that binds to a sequence element in the 5' untranslated segment and blocks entry of the mRNA into protein synthesis. At the tailbud stage other elements, possibly other binding proteins, inhibit activity of the blocking protein and allow the S19 mRNA to initiate protein synthesis.

Regulatory sequences in the 5' or 3' untranslated segments of mRNAs also control initiation in adult organisms. In mammals, for example, the relative quantities of two proteins involved in the entry and maintenance of iron in body cells, *ferritin* and the *transferrin receptor*, depend on the combination of a regulatory protein with control sequences in the untranslated regions of the mRNAs for these proteins. Iron atoms are carried throughout the body in combination with a protein known as *transferrin*. Transferrin receptors located on cell surfaces combine with transferrin and enter the cytoplasm by endocytosis. In the cytoplasm the iron atoms are released from transferrin and either enter directly into cytoplasmic reactions requiring iron or are stored by combination with ferritin. When iron is in limited supply, more transferrin receptors are synthesized, and synthesis of ferritin decreases. When iron is abundant, the opposite pattern is noted: synthesis of transferrin receptors decreases, and ferritin assembly rises.

R. D. Klausner and J. B. Harford discovered that the mRNAs for both the transferrin receptor and ferritin have a common sequence element, which they called *IRE*, for *iron-responsive element.* The IRE is recognized and bound by a control protein termed the *IRE-binding protein.* Ferritin mRNA has one IRE in its 5' untranslated segment; the transferrin receptor mRNA has five IREs in its 3' untranslated segment. When iron is in limited supply, the IRE-binding protein takes a form that binds strongly to IRE sequences. According to a model advanced by Klausner and Harford, attachment of the binding protein to the IRE element in the 5' untranslated segment of ferritin mRNA blocks its entry into protein synthesis, thereby reducing the quantity of ferritin molecules made in the cell. Attachment of the binding protein to IRE elements in the 3' translated segment of the transferrin receptor mRNA prevents its breakdown in the cytoplasm and leads to increased synthesis of the receptor.

If iron supplies are abundant, the IRE-binding protein takes a low-affinity form in which its binding to IRE elements is inhibited. In this case the IRE element in the 5' untranslated segment of the ferritin mRNA remains unbound. As a result, ferritin mRNA is readily translated, and ferritin levels rise in the cytoplasm. At the same time, the unbound IRE elements in the transferrin receptor mRNA lead to more rapid breakdown of this mRNA, reducing the quantities of the receptor assembled in the cytoplasm.

The AIDS virus (HIV) employs a translational control that greatly favors initiation of its own mRNAs. The HIV DNA encodes a protein that combines with the 5' untranslated segment of the viral mRNA, greatly speeding its translation. Addition of the HIV 3' untranslated segment to any other mRNA, viral or not, speeds translation of that mRNA by a factor of some 1000 to 2000 times if the protein is present. This translational control offers a potentially effective method for controlling the HIV virus by antibodies against the activating protein, which could possibly inhibit translation of the HIV mRNA in host cells. The AIDS system might also provide a genetic engineering technique for enhancing the synthesis of any desired protein. Attaching the HIV 3' untranslated sequence to the mRNA for a desired protein such as interferon or growth hormone, along with supplying the activating protein, would allow synthesis of the protein encoded in these mRNAs in large quantity. (For details of the AIDS virus and its mechanisms of infection, see Ch. 19.)

Translational Controls at Later Stages of Protein Synthesis

Relatively few translational controls have been detected at stages later than initiation. However, there are indications that controls are exerted during elongation in some eukaryotic systems. The activity of EF1α, an elongation factor that catalyzes addition of an aminoacyl-tRNA complex to a ribosome during each elongation cycle (step 2 in Fig. 16-7), decreases significantly in some cells when polypeptide assembly is inhibited. Such decreases are noted, for example, in aging rats, in rat liver cells following removal of the thyroid gland, and during conditions of high cell density in tissue culture, all situations in which protein synthesis is drastically reduced. Changes in the opposite direction, toward a rise in EF1α activity in coordination with

increased protein synthesis, are also noted in some systems. EF1α becomes more active in mice, for example, in coordination with the burst of protein synthesis accompanying a massive immune response; in plants similar increases in EF1α activity take place when protein synthesis is activated in wheat seedlings by the plant hormone *kinetin*. The molecular basis for these changes in EF1α remains unknown.

At least one pathogenic bacterium takes advantage of regulatory opportunities at elongation to destroy polypeptide assembly in the host. The potent toxin of the bacterium causing diphtheria catalyzes a reaction adding an ADP-ribosyl group (see the figure in Information Box 14-3) to the elongation factor EF2. This factor normally speeds the transfer of peptidyl-tRNAs from the A to the P site after peptide bond formation (step 5 in Fig. 16-7). ADP ribosylation blocks the activity of EF2 in translocation and completely stops protein synthesis in infected cells. The toxin arrests protein synthesis so effectively that a single molecule can kill a cell.

Translation is thus regulated by many different mechanisms. All indications from these observations are that translational controls are as varied as transcriptional controls and are used in the same wide variety of combinations in different species and cell types to achieve precise control of polypeptide assembly in the cytoplasm.

POSTTRANSLATIONAL REGULATION

Once their synthesis is complete, the numbers and kinds of proteins in eukaryotic cells are controlled by variations in the rate of protein breakdown. Some proteins persist for months or even years; others have half-lives on the order of hours. In some cases the rate of breakdown is altered in response to changes in environmental conditions or the developmental stage of the cell.

Proteins in the lens of the eye in higher vertebrates normally persist without breakdown for the life of the individual. For humans these proteins may last 90 years or more without degradation. Another relatively long-lived protein is hemoglobin, which persists as long as red blood cells last—about three months in humans, for example. The relatively stable proteins, with long half-lives, are generally degraded in lysosomes when the organelles or structures containing them are taken in and digested (see p. 864).

Proteins with relatively short half-lives include steroid hormone receptors, the milk protein casein, homeotic regulatory proteins, and heat shock proteins. M. Rechsteiner and his coworkers noted the curious fact that these and many other proteins with short half-lives contain four amino acids in relatively large proportions in their sequences—proline, glutamic acid, serine, and threonine. The Rechsteiner laboratory calls these proteins the *PEST* group, after the single-letter abbreviations for the amino acids (P = proline, E = glutamic acid, S = serine, and T = threonine).

During a heat shock response many proteins that are stable under unstressed conditions are rapidly degraded. The process is selective, breaking down only certain proteins. Presumably, these are proteins that have been denatured by exposure to heat or other stressful conditions or that increase the sensitivity of cells to these conditions.

Research by Rechsteiner and others, including A. Hershko and A. Varshavsky, demonstrated that relatively short-lived proteins, rather than being degraded in lysosomes, are hydrolyzed in the cytoplasmic solution. Their breakdown is catalyzed by enzymes that require ATP, even though the hydrolysis reaction yields energy.

The proteins destined for cytoplasmic hydrolysis are identified for breakdown by the attachment of ubiquitin (see also p. 681). As many as ten or more ubiquitin units may be added to lysine residues within the marked proteins, some of them in chains. During protein breakdown the ubiquitin units are split off intact and recycled.

Addition of ubiquitin depends on alterations at the N-terminal end of the proteins destined for degradation. The acetyl group added as a posttranslational modification to the N-terminal end of many proteins (see p. 681) is removed. In addition, if an acidic residue is present at the N-terminal end, it is replaced by a basic amino acid, arginine, before ubiquitins are attached. The arginine is transferred directly from an arginine-tRNA complex. These N-terminal-end alterations are evidently required for recognition of the protein destined for breakdown by the enzymes attaching ubiquitin. The relationship of the PEST amino acids to this recognition, if any, is unknown.

Ubiquitin is among the most highly conserved proteins of eukaryotes. The sequence of this protein is identical in animals from insects to mammals and differs in only four amino acids between organisms as distantly related as yeast, animals, and higher plants. Conservation at this level indicates that the function of the protein is highly critical and that almost any alteration in its sequence is likely to be lethal.

Besides marking proteins for degradation, ubiquitin has other functions. It is added to proteins such as histones H2A and H2B and to some cell-surface receptors as a posttranslational modification (see p. 542); as far as is known, the proteins modified in this way remain functional. The structural and functional significance of these modifications is unknown. Ubiquitin also apparently facilitates the folding of newly synthesized polypeptides into their final three-dimensional form.

The total spectrum of transcriptional, posttranscriptional, translational, and posttranslational controls determines the assemblage of proteins in cells and regulates the rates at which they are degraded and replaced. Which of the many controls operating at these levels is most significant in determining the numbers and kinds of proteins depends on the cell type and the protein. For most proteins transcriptional controls are most important in determining numbers and kinds. For some, however, regulatory controls acting after transcription is complete are paramount. Many of the proteins working as "housekeepers" in the cell—proteins required for standard, nonspecialized maintenance of cell reactions—fall into this category. Many of these proteins are transcribed constitutively in constant, steady numbers; their final concentrations in the cytoplasm are adjusted primarily or exclusively by posttranscriptional regulation of the number of mRNAs made available for protein synthesis or by regulation of their rates of translation or breakdown.

For Further Information

AIDS virus and the immune system, *Ch. 19*
Chromatin structure, *Ch. 14*
DNA-binding proteins and their structural motifs, *Ch. 14*
DNA structure, conformations, and modifications, *Ch. 14*
Genetic code, *Ch. 16*
Genetic rearrangements, *Ch. 18*
Histone proteins, variants, and modifications, *Ch. 14*
mRNA transcription and processing, *Ch. 15*
Nonhistone proteins, *Ch. 14*

Protein synthesis, *Ch. 16*
  in mitochondria and chloroplasts, *Ch. 21*
rRNA transcription and processing, *Ch. 15*
Steroid hormones, *Ch. 7*

Suggestions for Further Reading

General Books and Articles

Gehring, W. J. 1985. The molecular basis of development. *Sci. Amer.* 253:153–62 (October).

Holliday, R. 1989. A different kind of inheritance. *Sci. Amer.* 260:60–73 (June).

Ross, J. 1989. The turnover of mRNA. *Sci. Amer.* 260:48–55 (April).

Structural Motifs in DNA-Binding Proteins

Berg, J. M. 1990. Zinc finger domains: hypotheses and current knowledge. *Ann. Rev. Biophys. Biophys. Chem.* 19:405–21.

———. 1990. Zinc fingers and other metal-binding domains. Elements for interactions between macromolecules. *J. Biolog. Chem.* 265:6513–16.

Brennan, R. G., and Matthews, B. W. 1989. The helix-turn-helix DNA binding motif. *J. Biolog. Chem.* 264:1903–6.

Busch, S. J., and Sassone-Corsi, P. 1990. Dimers, leucine zippers, and DNA-binding proteins. *Trends Genet.* 6:36–45.

Gruissem, W. 1990. Of fingers, zippers, and boxes. *Plant Cell* 2:827–28.

Harrison, S. G. 1991. A structural taxonomy of DNA-binding proteins. *Nature* 353:715–19.

Landshulz, W. H.; Johnson, P. F.; and McKnight, S. L. 1988. The leucine zipper: a new DNA-binding structure. *Science* 240:1759–64.

McKnight, S. L. 1991. Molecular zippers in gene regulation. *Sci. Amer.* 264:54–64 (April).

Schleif, R. 1988. DNA binding by proteins. *Science* 241:1182–87.

Struhl, K. 1989. Helix-turn-helix, zinc-finger, and leucine-zipper motifs for eukaryotic transcriptional regulatory proteins. *Trends Biochem. Sci.* 14:137–40.

Vinson, C. R.; Sigler, P. R.; and McKnight, S. L. 1989. Scissors-grip model for DNA recognition by a family of leucine zipper proteins. *Science* 246:911–19.

Transcriptional Control by Regulatory Proteins, General

Adhya, S., and Garges, S. 1990. Positive control. *J. Biolog. Chem.* 265:10797–800.

Atchison, M. L. 1988. Enhancers: mechanism of action and cell specificity. *Ann. Rev. Cell Biol.* 4:127–53.

Beardsley, T. 1991. Smart genes. *Sci. Amer.* 265:86–95 (August).

DeRobertis, E. M.; Oliver, G.; and Wright, C. V. E. 1990. Homeobox genes and the vertebrate body plan. *Sci. Amer.* 263:46–52 (July).

Goodbourn, S. 1990. Negative regulation of transcriptional activation in eukaryotes. *Biochim. Biophys. Acta* 1032:53–77.

Guarente, L. 1988. UASs and enhancers: common mechanisms of transcriptional activation in yeast and mammals. *Cell* 52:303–5.

Herskowitz, I. 1989. A regulatory hierarchy for cell specialization in yeast. *Nature* 342:749–57.

Hinnebusch, A. G. 1990. Transcriptional and translational regulation of gene expression in the general control of amino acid biosynthesis in *Saccharomyces cerevisiae*. *Prog. Nucleic Acid Res. Molec. Biol.* 38:195–240.

Johnson, P. F., and McKnight, S. L. 1989. Eukaryotic transcriptional regulatory proteins. *Ann. Rev. Biochem.* 58:799–839.

Jones, N. 1990. Transcriptional regulation by dimerization: two sides to an incestuous relationship. *Cell* 61:9–11.

Lamb, P., and McKnight, S. L. 1991. Diversity and specificity in transcriptional regulation: the benefits of heterotypic dimerization. *Trends Biochem. Sci.* 16:417–22.

Levine, M., and Manley, J. L. 1989. Transcriptional repression of eukaryotic promoters. *Cell* 59:405–8.

Lobell, R. B., and Schleif, R. F. 1990. DNA looping and unlooping by Ara C protein. *Science* 250:528–32.

Mitchell, P. J., and Tjian, R. 1989. Transcriptional regulation in mammalian cells by sequence-specific DNA binding proteins. *Science* 245:371–78.

Müller, H.-P., and Schaffner, W. 1990. Transcriptional enhancers can act in *trans*. *Trends Genet.* 6:300–304.

Ptashne, M. 1988. How eukaryotic transcriptional activators work. *Nature* 335:683–89.

———. 1989. How gene activators work. *Sci. Amer.* 260:40–47 (January).

———. 1990. Activators and targets. *Nature* 346:329–31.

Renkawitz, R. 1990. Transcription repression in eukaryotes. *Trends Genet.* 6:192–97.

Roeder, R. G. 1991. The complexities of eukaryotic transcriptional initiation: regulation of preinitiation complex assembly. *Trends Biochem. Sci.* 16:402–8.

Struhl, K. 1989. Molecular mechanisms of transcriptional regulation in yeast. *Ann. Rev. Biochem.* 58:1051–77.

Wasylyk, B. 1988. Enhancers and transcription factors in the control of gene expression. *Biochim. Biophys. Acta* 951:17–35.

Whitelaw, E. 1989. The role of DNA-binding proteins in differentiation and transformation. *J. Cell Sci.* 94:169–73.

Steroid Hormone Receptors

Beato, M. 1989. Gene regulation by steroid hormones. *Cell* 56:335–44.

Berg, J. M. 1989. DNA binding specificity of steroid receptors. *Cell* 57:1065–68.

Brent, G. A.; Moore, D. D.; and Larsen, P. R. 1991. Thyroid hormone regulation of gene expression. *Ann. Rev. Physiol.* 53:17–35.

Burnstein, K. L., and Cidlowski, J. A. 1989. Regulation of gene expression by glucocorticoids. *Ann. Rev. Physiol.* 51:683–99.

Evans, R. M. 1988. On the steroid and thyroid hormone receptor superfamily. *Science* 240:889–95.

Gehring, U. 1987. Steroid hormone receptors: biochemistry, genetics, and molecular biology. *Trends Biochem. Sci.* 12:399–402.

Glass, C. K., and Holloway, J. M. 1990. Regulation of gene expression by the thyroid hormone receptor. *Biochim. Biophys. Acta* 1032:157–76.

Gronemeyer, H. 1991. Transcriptional activation by estrogen and progesterone. *Ann. Rev. Genet.* 25:89–123.

Miesfeld, R. L. 1989. The structure and function of steroid receptor proteins. *Crit. Rev. Biochem. Molec. Biol.* 24:101–17.

Rories, C., and Spelsberg, T. C. 1989. Ovarian steroid action on gene expression. *Ann. Rev. Physiol.* 51:653–81.

Samuels, H. H.; Forman, B. M.; Horowitz, Z. D.; and Ye, Z.-S. 1989. Regulation of gene expression by thyroid hormone. *Ann. Rev. Physiol.* 51:623–39.

Wahli, W., and Martinez, E. 1991. Superfamily of steroid nuclear receptors: positive and negative regulators of gene expression. *FASEB J.* 5:2243–49.

Weinberger, C., and Bradley, D. J. 1990. Gene regulation by receptors binding lipid-soluble substances. *Ann. Rev. Physiol.* 52:823–40.

Heat Shock Genes and Proteins

Ang, D.; Kiberek, K.; Skowra, D.; Zylicz, M.; and Georgopoulos, C. 1991. Biological role and regulation of the universally conserved heat shock proteins. *J. Biolog. Chem.* 266:24233–36.

Gething, M.-J., and Sambrook, J. 1992. Protein folding in the cell. *Nature* 355:33–45.

Lindquist, S., and Craig, E. A. 1988. The heat shock proteins. *Ann. Rev. Genet.* 22:631–77.

Schlesinger, M. J. 1990. Heat shock proteins. *J. Biolog. Chem.* 265:12111–14.

Vierling, E. 1991. The roles of heat shock proteins in plants. *Ann. Rev. Plant Physiol. Plant Molec. Biol.* 42:579–620.

Homeotic Genes and Homeodomain Regulatory Proteins

Gehring, W. J. 1987. Homeoboxes in the study of development. *Science* 236:1245–52.

Gehring, W. J.; Muller, M.; Affoeter, M.; Percival-Smith, A.; Billeter, M.; Qian, Y. Q.; Otting, G.; and Wuthrich, K. 1990. The structure of the homeodomain and its functional implications. *Trends Genet.* 6:323–29.

Hayashi, S., and Scott, M. P. 1990. What determines the specific action of *Drosophila* homeodomain proteins? *Cell* 63:883–94.

Levine, M., and Hoey, T. 1988. Homeobox proteins as sequence-specific transcription factors. *Cell* 55:537–40.

Ruvkun, G., and Finney, M. 1991. Regulation of transcription and cell identity by POU domain proteins. *Cell* 64:475–78.

Scott, M. P.; Tamken, J. W.; and Hartzell, G. W. 1989. The structure and function of the homeodomain. *Biochim. Biophys. Acta* 989:25–48.

Vollbrecht, E.; Veit, B.; Sinha, N.; and Hake, S. 1991. The developmental gene *Knotted-1* is a member of a maize homeobox gene family. *Science* 350:241–43.

Transcriptional Regulation in the Higher Plants

Dean, C.; Pichersky, E.; and Dunsmuir, P. 1989. Structure, evolution, and regulation of *RbcS* genes in higher plants. *Ann. Rev. Plant Physiol.* 40:415–39.

Gilmartin, P. M.; Sarokin, L.; Memelink, J.; and Chua, N.-H. 1990. Molecular light switches for plant cells. *Plant Cell* 2:369–78.

Kuhlemeier, C.; Green, P. J.; and Chua, N.-H. 1987. Regulation of gene expression in higher plants. *Ann. Rev. Plant Physiol.* 38:221–57.

Moses, P. B., and Chua, N.-H. 1988. Light switches for plant genes. *Sci. Amer.* 258:88–93 (April).

Quail, P. 1991. Phytochrome: a light-activated molecular switch that regulates plant gene expression. *Ann. Rev. Genet.* 25:389–409.

Schwarz-Sommer, Z.; Huijser, P.; Nacken, W.; Saedler, H.; and Sommer, H. 1990. Genetic control of flower development by homeotic genes in *Antirrhinum majus*. *Science* 250:931–36.

DNA Methylation and Conformational Changes in Transcriptional Regulation

Cedar, H. 1988. DNA methylation and gene activity. *Cell* 53:3–4.

Cedar, H., and Razin, A. 1990. DNA methylation and development. *Biochim. Biophys. Acta* 1049:1–8.

Doerfler, W.; Hoeveler, A.; Weisshaar, B.; Dobranski, P.; Knebel, D.; Langner, K.-D.; Achten, S.; and Müller, U. 1989. Promoter inactivation by sequence-specific methylation and mechanisms of reactivation. *Cell Biophys.* 15:21–27.

Dynan, W. S. 1989. Understanding the molecular mechanism by which methylation influences gene regulation. *Trends Genet.* 5:35–36.

Hoffman, R. M. 1990. Unbalanced transmethylation and the perturbation of differentiated state leading to cancer. *Bioess.* 12:163–66.

Holliday, R. 1989. DNA methylation and epigenetic methods. *Cell Biophys.* 15:15–20.

Lancillotti, F.; Lopez, M. C.; Arias, P.; and Alonso, C. 1987. Z-DNA in transcriptionally active chromosomes. *Proc. Nat. Acad. Sci.* 84:1560–64.

Monk, M. 1987. Methylation and the X chromosome. *Bioess.* 4:204–8.

Prendergast, G. C., and Ziff, E. B. 1991. Methylation-sensitive sequence-specific DNA binding by the c-Myc basic region. *Science* 251:186–89.

15. How does the regulation of rRNA genes differ from that of mRNA genes? What evidence shows that rRNA genes are actually regulated? How are 5S rRNA genes regulated in *Xenopus* embryos?

16. What is posttranscriptional regulation? At what levels might posttranscriptional regulation take place? What evidence indicates that regulation of this type actually occurs?

17. Contrast the mechanisms controlling the stability of histone and nonhistone mRNAs. What is the relationship between the stability of some mRNAs and protein synthesis?

18. What is translational regulation? At what steps of polypeptide assembly does translational regulation commonly occur? How is this pattern of regulation related to fertilization? To viral infections?

19. What is posttranslational regulation? What is ubiquitin? What is the relationship of this protein to posttransla-tional regulation? What other functions does ubiquitin carry out in eukaryotic cells? What protein modifications precede ubiquitin addition during protein degradation? What are PEST proteins?

20. Compare and contrast the overall mechanisms of transcriptional regulation in prokaryotes and eukaryotes. What are operons? Operators? Repressors? Activators? Inducers? Contrast transcriptional regulation by repressors synthesized in active and inactive forms. What is autoregulation?

21. How are sigma factors involved in regulation in bacteria? In cell differentiation?

22. What is attenuation? Antitermination?

23. What forms of posttranscriptional regulation occur in bacteria? What is antisense RNA? How might antisense RNA be used to treat viral or bacterial infections?

Supplement 17-1

Regulation in Prokaryotes

Although bacteria regulate both transcription and translation, their regulatory mechanisms differ in many details from those of eukaryotes. The opportunities that bacteria offer for genetic and biochemical analysis have made it possible to unravel many of these details and have made it obvious that bacteria, although relatively simple genetically, have transcriptional and translational pathways that are easily as versatile and varied as those of eukaryotes.

Because bacteria normally live in labile, continually changing environments, their regulatory processes are in a constant state of adjustment. Most bacterial regulatory systems are easily reversible, allowing rapid, short-term switches from one biochemical pathway to another. Very few bacterial regulatory pathways lead to permanent cellular changes resembling the development and differentiation of eukaryotic cells. Typically, the regulatory pathways of bacteria are highly automated, tying gene expression in a direct, mechanical way to changes in the environment.

Transcriptional Regulation in Bacteria

Bacterial genes are organized in *operons*, transcriptional units that usually contain more than one coding sequence (see Supplement 15-1 and Fig. 15-51). The coding sequences of the viruses infecting bacteria, *bacteriophages*, are organized in the same pattern. Each operon is controlled as a unit by one or more promoters and, in a few genes, by enhancerlike sequence elements located upstream of the promoters.

The *cis* elements in 5'-flanking regions of bacterial operons are bound by regulatory proteins called *repressors* and *activators*. Repressors adjust the rate of initiation by their promoters downward from a base level, which is usually relatively high. Thus, repressor-regulated operons are characterized by a relatively high transcription rate when free of their repressors and a reduced rate when their promoters are complexed with a repressor. Activators work in the opposite way, by adjusting their promoters upward from a base level that is usually low. Activator-regulated operons are therefore characterized by a very low or essentially non-existent transcription rate when free of their activators and a higher rate when their promoters are complexed with an activator. Some operons in *E. coli* are regulated by both mechanisms and are adjusted downward by repressors or upward by activators. In the relatively few bacterial operons with enhancerlike sequences, these elements are recognized and bound by regulatory proteins with activities similar to the enhancer-binding proteins of eukaryotes.

Operons are also regulated by the presence or absence of *sigma factors*, which bind to RNA polymerase and enable it to recognize promoters with specific sequence elements (see p. 639 and Table 15-3). In these systems sigma factors have the same effect as activators—presence of the factors allows initiation and transcription of their specific operons; when the factors are absent, transcription of the operon is shut off.

Various other controls regulate transcription in bacteria. One of these, *attenuation*, allows transcription to begin but when active as a regulatory mechanism cuts it off almost immediately, before the first coding sequence of the operon is reached. Another system, *antitermination*, works in long, complex operons that contain internal signals for termination of transcription in addition to their final terminators. When active, the antitermination mechanism prevents termination at one or more of the internal terminators, causing the RNA polymerase to read farther downstream in the operon and include more of its coding sequences.

Many operons are controlled by more than one of these regulatory mechanisms, and many of the regulatory elements, such as repressors, activators, or sigma factors, can repress or activate more than one operon. The result is a complex network of superimposed controls that provides total regulation of transcription and allows almost instantaneous changes to be made when environmental conditions are altered.

Regulation by Repressors   Bacterial repressors were the first proteins identified with specific gene regulation in any type of organism. They were discovered in the 1950s by F. Jacob and J. Monod of the Pasteur Institute in Paris, who were interested in the inheritance of genes controlling the metabolism of glucose and other sugars in *E. coli*. These investigators found that three enzymes involved in lactose metabolism are encoded in the same transcriptional unit in *E. coli*, which they termed the *lac operon* (Fig. 17-16). Jacob and Monod were the first to call bacterial transcriptional units operons, and introduced many terms and concepts still widely employed in studies of bacterial gene structure, transcription, and regulation.

The *lac* operon was found to be controlled by a regulatory protein, which was termed the *lac repressor*. The *lac* repressor, encoded in a gene separate from the *lac* operon, is synthesized in active form. In this form it binds to a site within the promoter and inhibits transcription of the *lac* operon (Fig. 17-17a; the sequence segment binding the repressor was termed the *operator*). RNA polymerase can still bind when the *lac* repressor is linked to the operator, even though the

Regulation in Prokaryotes   **727**

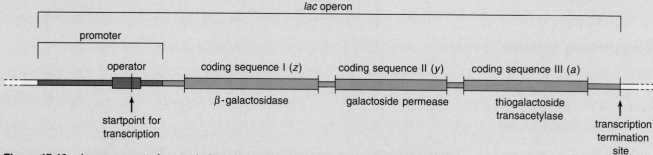

Figure 17-16 Arrangement of control elements and coding sequences in the *lac* operon of *E. coli*. The operator is the DNA site recognized and bound by the repressor. The first coding sequence (z) encodes the enzyme *β-galactosidase*, which hydrolyzes lactose into glucose and galactose. The second coding sequence (y) encodes *galactoside permease*, a protein that promotes the transport of lactose and related sugars into the cell. The third coding sequence *(a)* encodes *thiogalactoside transacetylase*, an enzyme whose function in lactose metabolism is still unclear.

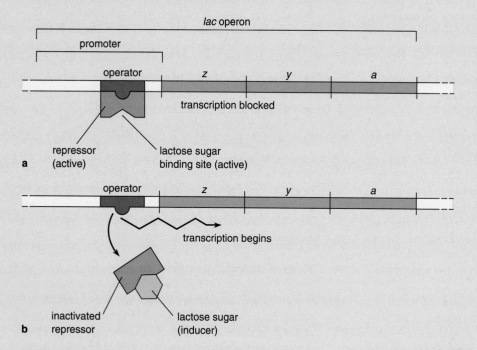

Figure 17-17 Regulatory activity of the *lac* repressor. **(a)** The active repressor binds the operator when no derivatives of lactose metabolism are available, blocking transcription. **(b)** If lactose becomes available, a derivative of the sugar binds the repressor and converts it to a form that no longer binds the operator. Conversion of the repressor to inactive form releases the gene from inhibition and allows tight binding of RNA polymerase and transcription. *z*, *y*, and *a*, the coding sequences of the *lac* operon.

operator and the region bound by RNA polymerase overlap to some degree. However, the presence of the repressor prevents the tight binding of RNA polymerase necessary for initiation of transcription. This is the situation when no lactose is present in the medium.

The *lac* repressor also has a site that can recognize and bind a chemical derivative of lactose. If lactose becomes available, some of the derivative binds to the repressor (Fig. 17-17*b*). The binding induces a conformational change in the repressor that greatly reduces its affinity for the *lac* promoter, and the repressor is released. RNA polymerase can now bind tightly to the promoter, and the operon is transcribed at its base level, which, for the *lac* operon, is high.

The repressor mechanism automatically adjusts transcription of the *lac* operator to the availability of lactose. When lactose is present, the operon is turned

on and the enzymes necessary to run the pathway are synthesized; when no lactose is available, the operon is turned off and none of the enzymes are synthesized. The mechanism ensures that the enzymes of the pathway are not made unless they are required.

The repressor-based mechanism can also operate to reduce the synthesis of the enzymes of a pathway when a product of the pathway appears in the medium. The *trp* operon encoding enzymes of the biochemical pathway synthesizing the amino acid tryptophan works this way (Fig. 17-18). The *trp* repressor has two binding sites, one for the promoter of the *trp* operon and one for tryptophan. However, in contrast to the *lac* repressor, the *trp* repressor is synthesized in inactive form. In this form the repressor has no affinity for the *trp* promoter, and transcription of the operon takes place at the high basal rate. This is the situation when the

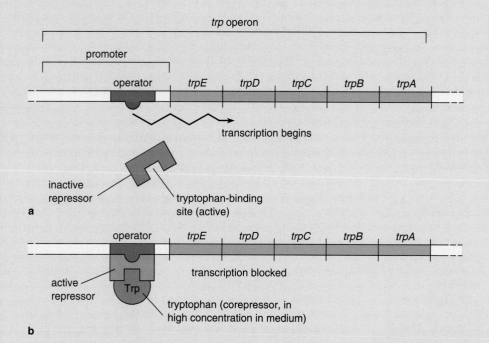

Figure 17-18 Arrangement of control elements and coding sequences in the *trp* operon and regulatory activity of the *trp* repressor. **(a)** When no tryptophan is available in the medium, the repressor is inactive in binding the operator, and transcription can proceed. **(b)** If tryptophan becomes available, the amino acid binds and activates the repressor. The activated repressor binds the operator, which blocks transcription. *trp A, B, C, D,* and *E,* coding sequences of the *trp* operon.

supply of tryptophan is low in the cell or its surrounding medium (Fig. 17-18*a*).

If tryptophan becomes available, or cellular levels of the amino acid are high because of synthesis by the tryptophan pathway, excess tryptophan combines with the tryptophan-binding site on the repressor (Fig. 17-18*b*). Binding of tryptophan to the repressor induces a conformational change that activates the promoter-binding site of the repressor. As a result, the repressor binds to the *trp* promoter, which blocks tight binding of the promoter by RNA polymerase, and stops transcription. The mechanism thus works to adjust synthesis of the enzymes of the tryptophan pathway to the availability of tryptophan. When tryptophan is available, the pathway is shut off, and when tryptophan is required, the pathway is turned on.

Figure 17-19 summarizes the two variations of the repressor mechanism. Note that in the form of the pathway regulating the *lac* operon, in which combination with a substance inactivates the repressor and activates the operon, the substance combining with the repressor is called the *inducer.* In the pathway regulating the *trp* operon, in which combination with a substance activates the repressor and shuts off the operon, the substance combining with the repressor is called the *corepressor.*

lac form: active repressor + inducer →
                         inactive repressor;
                         operon transcribed
                         at basal rate

trp form: inactive repressor + corepressor →
                         active repressor; operon
                         transcribed at levels below
                         basal rate

Jacob and Monod received the Nobel Prize in 1965 for their discovery and explanation of the operon and its regulation.

The *lac* repressor was successfully isolated and purified in 1967 by W. Gilbert and B. Müller-Hill. Since then, it has been fully sequenced and its probable three-dimensional structure worked out. The protein has 360 amino acids; the fully active repressor consists of four repressor proteins forming a tetramer with a total molecular weight of 152,000. The tetramer contains short alpha-helical segments positioned as a helix-turn-helix motif to fit into the DNA major groove.

Several other repressors, including the *trp* repressor, have also been sequenced and biochemically characterized. In general, repressors vary considerably in overall structure and in their effects on their target operons—some reduce transcription by their target operons to a greater or lesser extent; others shut down transcription entirely.

The genes coding for repressors are also operons, most of them containing only a single coding sequence. The operons encoding most repressors are *autoregulated,* that is, negatively controlled by their own protein products. As the quantity of a repressor protein increases, some repressor molecules bind to the promoter of their own operon and reduce its transcriptional activity. Thus, most repressors actually have at least three binding sites: one for the promoter of the target operon, one for the inducer or corepressor, and one for the promoter of the operon encoding the repressor. Autoregulation adjusts repressor synthesis so that there are always some repressor molecules of a given type in the cell.

The promoter sequence recognized and bound by an active repressor may lie downstream of the −10

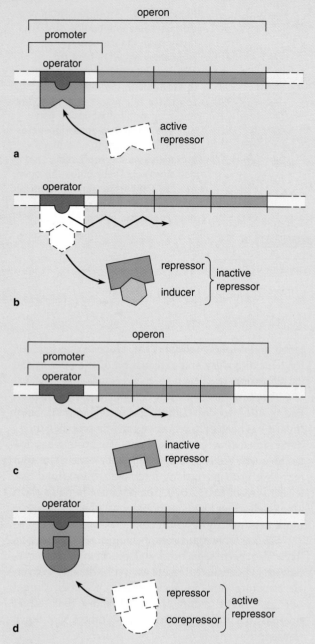

Figure 17-19 Summary of the two forms of the operon mechanism. (**a** and **b**) *lac* form, in which the repressor is synthesized in active form and inactivated by the inducer; (**c** and **d**) *trp* form, in which the repressor is synthesized in inactive form and activated by the corepressor.

sequence, as it does in the *lac* and *trp* operons, may overlap this sequence, or may overlap or lie upstream of the −35 sequence. (See p. 640 for an explanation of the −10 and −35 sequences and their activity in bacterial promoters.) Some operons are supplied with two or more sequences recognized and bound by the same or different repressors. The *lac* operon, for example, has two other sites that can be bound by the *lac* repressor in addition to the primary operator sequence lying in the region from −3 to +21. Additional

repressor molecules bound to these sites increase the degree of repression. Multiple sequences bound by different repressors place an operon under the control of distinct regulatory pathways. This multiple control increases the extent of the automated networks regulating bacterial transcription.

Regulation by Activators Activators were originally discovered through investigation of mutations that affect several operons simultaneously. One of the first activators detected in these mutants was the *cAMP receptor protein* (*CRP*; also called the *catabolite activator protein*, or *CAP*). It is a positive regulator of the *lac* and two other operons, the *gal* and *ara* operons, which encode enzymes metabolizing galactose and arabinose sugars.

CRP is activated by cAMP through a pathway linked to the concentration of glucose in the cell or the surrounding medium. CRP is synthesized in inactive form, in which it has no affinity for the promoters of the *lac*, *gal*, and *ara* operons. When glucose concentrations are low, the enzyme converting ATP to cAMP is active, elevating cAMP levels in the cell. Combination with cAMP induces a conformational change that converts CRP to active form. In this form CRP binds to its target operons and increases their frequency of initiation above base level. Production of the enzymes encoded in the operons positively controlled by CRP allows sugars other than glucose to be metabolized as an energy source.

If glucose becomes available, a sensory pathway that is not as yet understood leads to inhibition of the adenylate cyclase enzyme. As a result, cAMP concentration falls, and CRP is converted to inactive form. This shuts down the operons synthesizing enzymes that metabolize other sugars and allows the cell to concentrate on glucose, the preferred metabolite.

CRP is a dimer consisting of two identical polypeptides. Like the *lac* repressor CRP folds into a helix-turn-helix motif that holds two alpha-helical segments in the correct position to bind in the large groove of the DNA. Binding of CRP induces a sharp bend in the DNA that in some way facilitates the initiation of transcription. The bend presumably induces tight binding by RNA polymerase, possibly by inducing the DNA to unwind.

A gradually lengthening list of activators has been discovered in bacteria. Typically, activators control groups of operons encoding enzymes that carry out major functions, such as transport, sulfate assimilation, anaerobic electron transport, responses to stress or DNA damage, and phosphate metabolism. Groups of operons concerned with major functions of bacterial viruses are also controlled by activators. Like the genes encoding repressors many of those encoding activators are autoregulated—the regulator proteins act as repressors of their own operons.

In each system examined so far, activity of the positive regulator depends on its combination with nucleotide-derived substances such as cAMP. Note that the role of cAMP in activating CRP is quite different from its function as a second messenger in eukaryotic cells (see Ch. 7 for details). There is no known instance in eukaryotes in which cAMP directly activates a transcriptional regulator as it does in bacteria.

Regulatory Protein Acting as Both Repressor and Activator: AraC Investigations by R. B. Lobell and R. F. Schlief showed that at least one bacterial regulatory protein, AraC, can act as either a repressor or activator depending on the presence of an inducer. The AraC protein controls transcription of the *BAD* operon in *E. coli*, which encodes genes involved in arabinose metabolism. When no arabinose is present, two AraC molecules bind as a dimer to two distinct operator sites separated by 210 base pairs in the promoter of the *BAD* operon. The binding draws these intervening base pairs into a loop that evidently prevents access to the promoter by RNA polymerase, and the *BAD* operon is transcribed well below the base level. If arabinose becomes available, the sugar combines with the two molecules of AraC and induces a conformational change that blocks binding of the protein dimer to the operator lying farthest upstream. This opens the loop; the dimer, which remains bound to the downstream operator, now acts as an activator that enhances binding of RNA polymerase and raises transcription well above the base level.

Sigma Factors as Positive Regulators The bacterial RNA polymerase does not form the tight binding to promoter sequences necessary for the initiation unless it is bound to a sigma factor (see p. 639). In *E. coli* most operons are transcribed by RNA polymerase in combination with the "standard" sigma factor, a 70,000-dalton protein identified as σ^{70}. At elevated temperatures a second sigma factor, σ^{32}, is synthesized in quantity. This sigma factor, in combination with the RNA polymerase, specifically recognizes promoters of operons encoding the heat shock proteins in *E. coli*. Another sigma factor, σ^{54}, promotes synthesis of a specialized group of operons in *E. coli* when nitrogen supplies are limited. Controls switching the activity of bacteriophage DNA from "early" operons concerned with infection and DNA replication to "late" operons encoding proteins necessary for production of coat proteins and host cell lysis are frequently also regulated by sigma factors specific for the promoters of the two operon groups.

Positive regulation by sigma factors accounts for one of the rare examples of cell differentiation in bacteria. Under certain conditions the bacterium *Bacillus subtilis* undergoes *sporulation*, in which it differentiates into a quiescent, highly resistant form that can survive extended desiccation and lack of nutrients. During the process an initial cell, the *sporangium*, divides to produce two unlike products. One, the *forespore*, develops into the spore. The second, the *mother cell*, synthesizes a protective protein coat around the developing spore and then breaks down.

P. Stanier and his colleagues discovered that the entire mechanism is controlled by a series of five sigma factors that are synthesized only during sporulation. The first sporulation sigma, σ^H, is induced when *B. subtilis* encounters nutrient starvation. Transcription by RNA polymerase associated with σ^H activates a further series of genes encoding additional sigmas leading to the fourth sigma in the cascade, σ^G, which appears in the forespore and activates operons encoding proteins involved in production of the spore. Among the proteins synthesized in response to σ^G activity is the final sigma in the cascade, σ^K, which appears in the spore mother cell and activates operons concerned with formation of the protein coat. The operon encoding σ^K is assembled from two sequence segments that are widely separated in *B. subtilis* DNA. During formation of the spore mother cell, these sequence segments are joined and the intervening DNA eliminated by a process superficially resembling gene splicing. (Ch. 18 outlines the basis for genetic rearrangements of this type in bacteria.) Under normal growth conditions sporulation sigmas are not produced, and the operons concerned with sporulation remain inactive.

Regulation by Attenuation The attenuation mechanism is an unusual regulatory pathway in which transcription is initiated but interrupted before RNA polymerase progresses as far as the first coding sequence in an operon. Attenuation is particularly important as a regulatory mechanism of many operons encoding enzymes that synthesize amino acids. The pathway depends for its operation on the fact that translation and transcription proceed simultaneously in bacteria.

A major hypothesis explaining attenuation was developed by C. Yanovsky and his colleagues from several observed characteristics of the pathway (Fig. 17-20). The mRNA of each operon regulated by attenuation contains an AUG start codon and the code for translation of a short polypeptide upstream of the primary coding sequence (Fig. 17-20a). Just downstream of the segment coding for the short polypeptide are several inverted sequences capable of forming alternate secondary structures (Fig. 17-20b and c). Note from the figure that sequence segments 1 and 2 are capable of winding into a 1-2 helix that excludes segment 3 and that segments 2 and 3 can wind into a 2-3 helix that excludes segment 1. The 2-3 helix mimics a bacterial type 1 transcription terminator in that it is followed by a sequence of Us in the RNA transcript (see p. 641 and Fig. 15-53). Typically, several codons for the amino

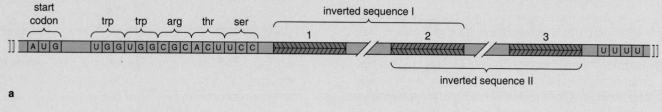

a

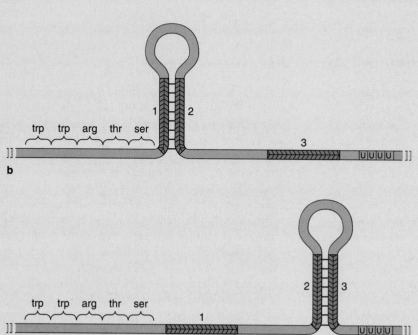

b

c

Figure 17-20 An explanation of attenuation via the formation of fold-back double helices. (a) Structure of the short coding sequence upstream of the primary coding segment of the *trp* operon (given in RNA equivalents). The short sequence contains an AUG start codon followed by a series of codons for amino acids, including tryptophan. The codons are followed by a series of inverted sequences (1, 2, and 3 in the figure) capable of forming the alternate fold-back double helices shown in (b and c). The double helix formed in (c), followed by a series of Us, mimics a signal terminating transcription in bacteria.

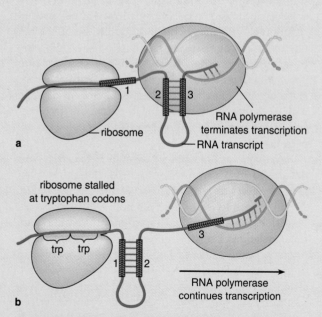

a

b

Figure 17-21 A simplified version of Yanovsky's attenuation hypothesis (see text). (a) Termination of transcription when the amino acid is in plentiful supply; (b) continuation of transcription when supplies of the amino acid are limited.

acid synthesized by the enzymes of the operon occur near the possible points of double-helix formation.

According to Yanovsky's model, RNA polymerase initiates transcription and is followed closely by a ribosome that attaches to the growing mRNA transcript. The ribosome initiates translation as soon as RNA polymerase has completed transcription of the short coding segment. The ribosome begins translation at the AUG codon of the short coding segment and continues until it reaches the sequential codons calling for the amino acid synthesized by the enzymes encoded in the operon. When the amino acid is in plentiful supply in the cell, the ribosome moves at a normal rate through these codons (Fig. 17-21a), following closely enough behind the RNA polymerase to occupy inverted sequence segment 1 before the 1-2 hairpin can form in the transcript. This permits the 2-3 hairpin to form as the RNA polymerase completes transcription of sequence segment 3 and its downstream Us. The 2-3 hairpin and the downstream Us act as a terminator, causing release of RNA polymerase and halting transcription at this point. Thus, when the amino acid encoded by the operon is available in the medium, transcription of its operon is halted by the attenuation mechanism before the first regular coding sequence of the operon can be transcribed and the *trp* mRNA does not appear in the cytoplasm.

When supplies of the amino acid are limited (Fig. 17-21b), all the steps proceed as before until the ribosome reaches the sequential codons specifying the amino acid synthesized by the enzymes encoded in the operon. Because the amino acid is in limited supply, the ribosome "stalls" at these codons. The halt in translation keeps sequence segment 1 unoccupied by the ribosome long enough for the 1-2 hairpin to form as soon as RNA polymerase completes transcription of sequence segment 2. Because segment 2 is now wound with segment 1 into the 1-2 hairpin, the 2-3 termination hairpin cannot form. The RNA polymerase therefore encounters no termination signal and continues into the primary coding sequences of the operon. The entire operon is transcribed, and the enzymes necessary for synthesis of the amino acid appear in the cytoplasm. Thus, when the amino acid is in short supply, attenuation does not occur, and the operon encoding the enzymes necessary for its synthesis is fully transcribed.

Attenuation has been detected in operons encoding the enzymes synthesizing tryptophan, histidine, isoleucine, phenylalanine, and threonine, and also several ribosomal proteins. Most of these operons are regulated by repressors, and some are controlled by activators as well, adding more links to the network of superimposed mechanisms regulating their transcription.

Regulation by Antitermination   Antitermination allows RNA polymerase to read through termination signals lying in the spacers between coding sequences in some operons (see p. 643). When the antitermination pathway is active, termination does not take place at these internal terminators, and transcription of downstream coding sequences proceeds. When the pathway is inactive, an internal terminator is recognized by RNA polymerase, and transcription stops at this point, eliminating the downstream coding sequences from the transcript of the operon.

Whether antitermination is active or inactive depends on the presence of termination factors with activity similar to the Rho factor (see p. 642). When the termination factor required for recognition of an internal terminator is present, the factor binds as the internal terminator sequence is transcribed and induces a conformational change stripping the enzyme from the transcript and DNA template. When the termination factor is absent, the enzyme is unable to respond to the internal terminator and reads through without pausing.

The antitermination pathway operates, for example, in the mechanism that determines whether bacteriophage lambda enters the *lysogenic* or *lytic phase* of its life cycle. In the lysogenic phase the viral DNA is integrated into the host cell DNA and passed on passively during cell division with little or no damage to the host cell. Under certain conditions—for example,

when the host cell encounters unfavorable growth conditions—the bacteriophage switches to the lytic phase, in which the viral DNA is liberated from the host cell DNA, replicates, directs the synthesis of coat proteins, and causes lysis of the host cell with release of mature, infective viral particles.

Sequences coding for proteins that induce the lytic phase lie within an operon of the viral DNA downstream of an internal termination sequence. When conditions favor the lysogenic phase of the life cycle, a termination factor called *N* is produced that binds to RNA polymerase, permitting the enzyme to recognize the internal terminator and preventing transcription of the downstream coding sequences. When conditions favor lysis, the operon coding for the N protein is shut down. The absence of N prevents recognition of the internal terminator. In this situation the RNA polymerase reads through the internal terminator and translates the downstream sequences of the operon encoding proteins characteristic of the lytic phase. Among these are coat proteins of the virus and enzymes releasing the viral DNA from the host DNA.

Other Mechanisms Regulating Transcription in Bacteria   Bacterial genes are also regulated by mechanisms not directly involving positive or negative regulatory factors, attenuation, or antitermination. The degree of DNA supercoiling, for example, regulates some bacterial operons. Among them are operons encoding topoisomerases, enzymes that add or remove supercoils from the bacterial DNA circle (see p. 537). One topoisomerase controlled by supercoiling adds negative supercoils to the DNA, and another removes supercoils. When negative supercoiling falls below the typical number of about one supercoil for each 200 base pairs of DNA, transcription of the operon encoding the topoisomerase adding negative supercoils is stimulated. At the same time the operon encoding the topoisomerase removing supercoils is inhibited. The opposite effects on transcription of the two genes are observed if the degree of supercoiling rises above the 1:200 base-pair level. These alterations in activity probably depend on conformational changes in the promoters of the operons when DNA supercoiling varies from the optimum. The degree of negative supercoiling also affects the transcriptional rate of unrelated operons, such as the *lac* and *gal* operons; the more negatively supercoiled the DNA, the more actively these operons are transcribed.

K. Okamoto and M. Freundlich found that transcription of at least one *E. coli* gene, encoding the cAMP receptor protein (CRP), is inhibited by an RNA molecule instead of regulatory proteins. The RNA, called an *antisense RNA*, is complementary to sequences in the 5′ region of the mRNA copied from the gene. The cAMP receptor protein, when present in quantity, stimulates transcription of the antisense RNA, which slows or stops transcription of the CRP gene. Presumably, the antisense RNA pairs with the 5′ end of the CRP mRNA, forming a secondary structure that resembles a type I transcription terminator (see p. 642). The terminator stops transcription at this point.

The mechanism inhibiting transcription of the CRP gene is the only known example of a transcriptional control based on an antisense RNA in either prokaryotes or eukaryotes. Several mechanisms are known, however, in which antisense RNAs regulate bacterial translation.

The multiplicity of mechanisms controlling bacterial operons ties the regulatory mechanisms into networks that provide a balanced, precise, and sensitive adjustment of the entire cell to suit environmental conditions and allow maximum efficiency in the use of available nutrients. The entire mechanism is aided by the fact that bacterial mRNAs are very short-lived, about 3 minutes on the average. This permits the cytoplasm to be cleared quickly of the mRNAs produced by one set of operons as they are replaced in activity by another set, and reduces the lag time for a cellular response to environmental change. In rapidly dividing bacteria, which may pass from one generation to the next in as little as 20 minutes, enzymes and other proteins are also quickly diluted in the cytoplasm of offspring if their synthesis is halted. This is in distinct contrast to the regulatory processes of relatively long-lived eukaryotic cells, particularly those imposing development and differentiation in higher eukaryotes, which are relatively slow, preprogrammed, and often irreversible under normal conditions. Although relatively little is known about regulatory pathways in cyanobacteria, the mechanisms controlling transcription in these organisms are believed to resemble those of bacteria.

Translational Regulation in Bacteria

Translational controls in bacteria primarily involve regulation of the availability of mRNAs. One of the best-documented prokaryotic systems, investigated by M. Nomura and his colleagues, involves regulation of the translation of ribosomal proteins in *E. coli* by a feedback mechanism.

The ribosomal proteins of *E. coli* are encoded in seven operons. Six of these operons encode several ribosomal proteins and some unrelated proteins as well (Fig. 17-22). Each of the mRNAs transcribed from these operons can be recognized and bound by a ribosomal protein encoded in the operon. As any of these ribosomal proteins accumulate in excess in the *E. coli* cytoplasm, they bind to their own mRNAs and block further translation. Binding occurs in the 5′ untranslated segment and the initial coding segments of the mRNAs, including the region containing the Shine–Dalgarno sequence (see p. 688). The binding evidently depends on the ability of the mRNAs to fold into secondary structures resembling those of the rRNA segments to which the ribosomal proteins bind in ribosomes (Fig. 17-23).

Translational regulation of this type, through blockage of an mRNA by a translation product of the mRNA, appears frequently in prokaryotic systems and in the viruses infecting bacteria. Each protein encoded in these mRNAs is capable of binding and blocking its own mRNA, providing a typically automated mecha-

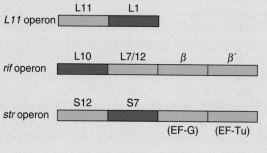

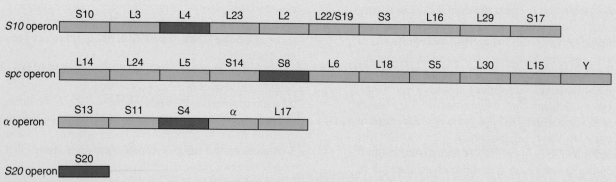

Figure 17-22 Organization of the operons coding for ribosomal proteins. The encoded ribosomal proteins that inhibit translation of their own operon mRNA (see text for explanation) are indicated by darker blue. Codes for proteins unrelated to ribosomes are also included in all the operons except *S20*. These proteins include the bacterial elongation factors EF-G and EF-Tu; RNA polymerase subunits α, β, and β'; and Y, a protein active in secretion. Modified from an original courtesy of M. Nomura. Reproduced, with permission, from *Ann. Rev. Biochem.* 53:75. © 1984 by Annual Reviews, Inc.

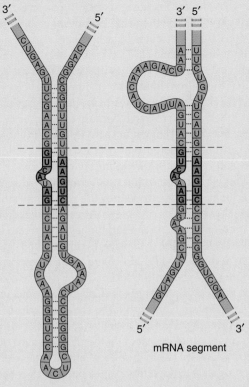

rRNA segment

mRNA segment

Figure 17-23 Translational control by a bacterial feedback mechanism in which a translational product of an mRNA can block further translation of the mRNA. The dashed lines enclose similar structures in a segment of bacterial 16S rRNA (left) bound by ribosomal protein S8, and in the mRNA for protein S8 (right). When present in excess, S8 binds the mRNA segment and blocks further translation of the mRNA. From data presented in D. P. Cerretti *et al.*, *J. Molec. Biol.* 204:309 (1988).

nism that adjusts translation downward when quantities of the protein appear in excess in the cell. As yet no feedback mechanisms regulating the availability of mRNAs in this manner have been reported in eukaryotes or the viruses infecting eukaryotic cells.

The availability of mRNAs in prokaryotes is also regulated by antisense RNAs with sequences complementary to segments in the 5′ untranslated segment of their target mRNAs. Pairing of antisense RNAs with the 5′ untranslated segments, usually in the Shine–Dalgarno region, blocks initiation of protein synthesis and effectively reduces availability of the mRNAs to ribosomes.

In *E. coli*, for example, the relative amounts of two outer membrane proteins are controlled by an antisense RNA called micF. The coding sequence for the antisense RNA is supplied with its own promoter and is located near the *ompC* operon encoding one of the two membrane proteins. Conditions that increase transcription of the *ompC* operon also promote micF transcription and cause this RNA to appear in the cytoplasm as well as the mRNA of this gene. The micF antisense RNA is complementary to the 5′ end of the mRNA for the other outer membrane protein, transcribed from the *ompF* operon. Pairing of the micF antisense RNA with the OmpF mRNA covers the Shine–Dalgarno sequence and effectively blocks synthesis of the OmpF protein, ensuring that as OmpC synthesis increases in the cytoplasm, synthesis of the OmpF protein will decrease by similar amounts.

Regulation of transcription or translation by antisense RNAs appears to be limited to prokaryotes. However, the existence of such RNAs offers a potentially powerful method for controlling transcription or translation of specific mRNAs in eukaryotic cells, particularly those of infecting viruses or bacteria. The method has been used successfully in experiments to block mRNAs of the herpes simplex virus and several cellular genes, including those encoding β-actin, a heat shock protein, and several proteins implicated in the conversion of normal cells into cancer types. Through this means it may eventually be possible to defeat infections by synthesizing antisense RNAs that can combine with and inactivate the mRNAs of the pathogen or block the growth of tumor cells.

An artificially synthesized antisense RNA has also been used by the Yanovsky group to test the attenuation model. The antisense RNA was constructed to be complementary to the region of an attenuated mRNA forming the proposed 2-3 termination hairpin and to prevent its formation. Bacterial attenuation systems exposed to the antisense RNA failed to undergo early termination, even when amino acid supplies were limiting, in support of the hypothesis.

Suggestions for Further Reading: See p. 725

Supplement 17-2
mRNA Transcription and Regulation in Polytene Chromosomes

The differential gene activity resulting from transcriptional regulation of specific genes cannot be visualized directly under either the light or electron microscopes in most cells. However, visual identification of active genes is possible in eukaryotic cells containing so-called *polytene chromosomes* (Figs. 17-24 and 17-25). These chromosomes, which appear as roughly cylindrical, crossbanded structures, are more than a hundred times larger than the chromosomes appearing during division in ordinary cells of the same organisms. At these dimensions they can even be seen at interphase under the light microscope. It is possible also to observe directly in these chromosomes that some sites are active and that the pattern of activity changes as development proceeds. The chromosomes have been of special interest in regulation research because they occur in *Drosophila*, in which differential activity of chromosome sites can be correlated with the extensive catalog of genes and mutations known in this organism.

The giant chromosomes are often termed *salivary gland chromosomes*, because they are commonly studied in the salivary glands of dipteran larvae. However, they also occur in other organs and tissues of dipteran larva such as cells of the Malpighian tubules and gastric pouch. Some adult flies also have giant chromosomes in single large cells in the footpads. At least two other insect groups, the springtails (Collembola) and grasshoppers, contain cells with large chromosomes of this type. Chromosomes with a similar appearance also occur as an intermediate stage in the development of the macronucleus in ciliate protozoa and in the embryo suspensor stalk in flowering plants. Because they appear in a wide variety of species and cell types, they are generally termed *polytene*, meaning "many threaded," rather than salivary gland chromosomes.

Development and Structure of Polytene Chromosomes

In *Drosophila*, polytene chromosomes first appear in salivary gland cells early in larval development. Within the nuclei of these cells, the DNA replicates repeatedly without division of the chromosomes or nuclei. As replication proceeds, the chromosomes gradually become visible in the nuclei as thick, elongated structures with characteristic transverse bands. The number of visible chromosomes is apparently one-half the usual diploid somatic number because homologous chromosomes *synapse*, or pair, during development of polyteny in *Drosophila*.

Measurements of DNA quantity in *Drosophila* salivary gland nuclei revealed that the DNA in these nuclei doubles successively in ten well-defined, synchronized steps. This indicates that mature salivary gland nuclei contain about a thousand times ($2^{10} = 1024$) as much DNA as ordinary somatic nuclei and that each polytene chromosome must be made up of as many strands of DNA. Not all the DNA replicates uniformly during polytene chromosome development; some remains at the ordinary diploid level. J. G. Gall and his associates and others showed that highly repetitive, apparently nonfunctional sequences are concentrated in the underreplicated chromosome regions. A few functional

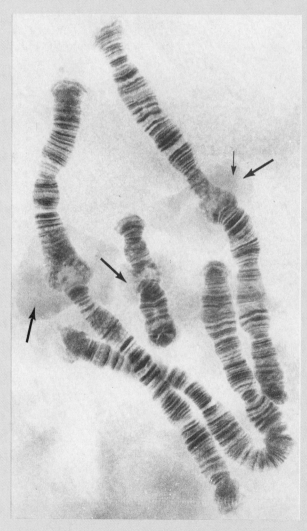

Figure 17-24 Light micrograph of polytene chromosomes from the salivary gland of the gall midge *Chironomus tentans*. The longer chromosomes in this micrograph are between 200 and 250 μm in length. The arrows point to large puffs identifiable as sites of intense mRNA transcription. ×500. Courtesy of W. Beerman and Springer-Verlag.

Figure 17-25 Electron micrograph of sectioned polytene chromosomes of *Chironomus*. ×4,500. Courtesy of J. Derksen and Springer-Verlag, from *Chromosoma* 85:643 (1982).

genes, such as those encoding the histones, are also underreplicated to some degree.

The degree of polytenization is evidently under precise metabolic control, since nuclei in other *Drosophila* tissues regularly replicate to reach different levels of polyteny. Malpighian tubule cells regularly replicate six times and gastric pouch cells nine times during polytenization, for example, as compared to the tenfold replication observed in salivary gland nuclei.

Polytene chromosomes sectioned for electron microscopy show masses of chromatin fibers running lengthwise through the chromosomes. The individual chromatin fibers are thought to extend from one end of the chromosome to the other and to be tightly coiled in the bands and extended in interband regions (Fig. 17-26a).

Active Sites on Polytene Chromosomes: Bands and Puffs

Within a given species the number and pattern of bands are repeated in all cell types containing polytene chromosomes. This is true in tissues as diverse as intestinal and salivary gland cells. Thus, differential activity of the cells in the various tissues is not reflected in variations in the number or types of bands in polytene chromosomes. However, some bands do differ in appearance in the cells of different tissues or at different times in the same tissue. In cells in these different locations, one or more bands uncoil into diffuse, brushlike structures called *puffs* (Fig. 17-26b; arrows in Fig. 17-24).

Evidence that the puffs are genes active in RNA

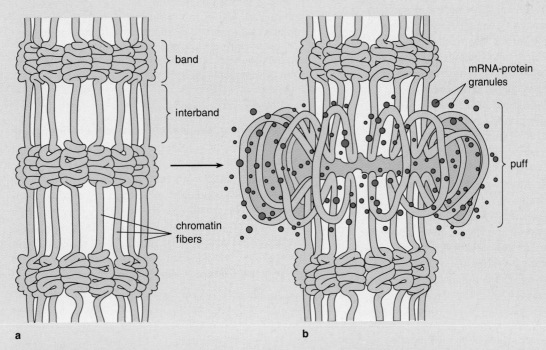

Figure 17-26 Probable structural relationships among bands, interbands, and puffs in polytene chromosomes. **(a)** Unpuffed, **(b)** puffed.

transcription comes from a number of sources. Radioactive uridine, used by cells in RNA transcription, is incorporated in large quantities into RNA in puffs. By means of fluorescent antibodies, RNA polymerase II and the various protein factors associated with mRNA transcription, processing enzymes, and mRNA-associated proteins have been shown to be associated with puffs. Also, cDNA probes copied from mRNA molecules (see p. 134) hybridize readily with the RNA transcribed in most puffs, establishing that the mRNA is synthesized in these regions. The latter method has also allowed many puffs to be identified with specific genes. Other work has shown that cytoplasmic 5S rRNA and various tRNAs originate from single puffed bands. Certain puffs can therefore be identified as active in the synthesis of rRNA and tRNA as well as mRNA.

Usually at least five to ten bands uncoil into puffs large enough to be easily recognized in salivary gland nuclei. The puffing pattern changes regularly during development as some previously active genes are repressed, and new ones become active (Fig. 17-27). The polytene chromosomes thus directly demonstrate that genes are turned on and off in definite sequences as embryonic development proceeds. The first demonstration that steroid hormones can affect gene activity, in fact, was made by U. Clever with polytene chromosomes. Clever injected ecdysone, an insect hormone that induces molting, into *Drosophila* larvae at a stage of development before the hormone would normally appear. Several puffs characteristic of the molting stage appeared prematurely in the polytene chromosomes of the larvae, indicating that the hormone was directly or indirectly responsible for this gene activity.

Although there is general agreement that puffs are sites of transcriptional activity, the relationship of bands to genes is not entirely certain. Part of the uncertainty comes from the relatively small number of countable bands in polytene chromosomes. In *Drosophila* only about 5000 bands are visible in an entire nucleus. Estimates of the number of genes in *Drosophila*, however, place the total much higher, at least 20,000 to 30,000. Some regions have been entirely sequenced or examined by other molecular techniques; in some of these regions the bands prove to contain more than one gene. Thus a significant proportion of the bands probably contains multiple genes.

The interbands, formed by segments of uncondensed chromatin between the bands, may represent regulatory sequences or nontranscribed spacers between genes. Or, as some investigators have proposed, the interbands may be an uncoiled part of a gene contained in an adjacent band. One of the more interesting

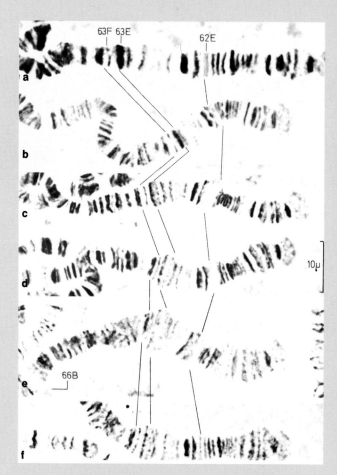

Figure 17-27 The sequence of appearance and disappearance of puffs at different points in a polytene chromosome as development proceeds in *Drosophila* larvae. Band 63F forms no puff during developmental stages (**a** through **f**); band 63E enlarges and forms a puff in (**c, d,** and **e**) and retracts slightly in (**f**); band 62E forms a puff in (**b** and **c**) and retracts in (**d** through **f**). The lines connect the same bands in (**a** through **f**). × 1300. Courtesy of M. Ashburner and Springer-Verlag, from *Chromosoma* 38:255 (1972).

characteristics of bands and interbands is their possible association with Z-DNA (see p. 530). Depending on the preparation methods, antibodies developed against Z-DNA may react with only the interbands, only the bands, or both bands and interbands in polytene chromosomes. The possibility that Z-DNA may occur in interbands prompted the hypothesis that interbands contain reservoirs of left-handed DNA turns that can be used to unwind right-handed turns in the bands when genes become active in transcription.

Suggestions for Further Reading: See p. 725.

prokaryotic genomes and some of the processes responsible for genetic rearrangements in these organisms are described in Supplement 18-1. The experimentally induced rearrangements used in genetic engineering are discussed in Supplement 18-2. Chapter 19 takes up the specialized rearrangements responsible for the generation of antibodies in higher vertebrates.

ORGANIZATION OF EUKARYOTIC GENOMES: AN OVERVIEW

The genome of a eukaryotic cell is divided among several to many linear, individual DNA molecules. Each DNA molecule, held in association with histone and nonhistone proteins, is a *chromosome* of the organism (Fig. 18-1). The ends of a chromosome are the *telomeres*; telomeres function in replication and make the chromosome tips inert to chemical interactions and enzymatic attack. Internally, a chromosome is divided into two *arms* by the *centromere*, the position at which spindle microtubules attach to the chromosome during cell division. Depending on the location of the centromere, the arms may be of equal or unequal length. Chromosomes also have sites serving as *replication origins*—sites at which enzymes bind to initiate replication. Artificial or natural DNA molecules with these three features—telomeres, a centromere, and replication origins—can function effectively during replication and cell division and can become a permanent part of the genome.

Information from a number of sources supports the conclusion that each chromosome consists of a single DNA molecule. Perhaps the best evidence comes from the yeast *Saccharomyces cerevisiae*, in which the number of large DNA molecules separated by gel electrophoresis is equal to the number of chromosomes and linkage groups.

A *haploid* or *monoploid cell* possesses one copy of each chromosome. Usually, each chromosome of a haploid cell contains a distinct group of genes and non-

---

The DNA of genes and all other functional and nonfunctional sequence elements makes up the *genome* of an organism. Among the genes are those encoding mRNAs, rRNAs, and tRNAs; other functional sequences occur as regulatory, spacing, or recognition elements, or as origins of replication. Nonfunctional sequences also occur in great numbers in eukaryotic genomes, sometimes taking up more of the available DNA than functional sequences. Much of this nonfunctional DNA consists of *repetitive* sequences—relatively short elements repeated thousands or even millions of times. Repetitive sequences inflate the genomes of many eukaryotes well beyond the amount of DNA needed for coding, regulation, and replication.

The arrangement of functional and nonfunctional sequences is not necessarily fixed in the genome. Existing sequences have been observed to move from one location to another or to be internally rearranged. Insertion of DNA sequences from outside the genome is also frequently observed. Although many of these changes, known collectively as *genetic rearrangements*, take place on the evolutionary time scale, some are relatively rapid and can be observed within the lifetime of single individuals.

The organization of genes and other sequence elements in eukaryotic genomes and genetic rearrangements are the subjects of this chapter. The structure of

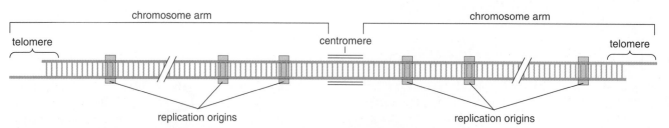

Figure 18-1 The essential structural and functional features of a eukaryotic chromosome. Each chromosome is a single, linear DNA molecule. At its tips a chromosome contains specialized sequences known as telomeres. Internally, a chromosome is divided into two arms by the centromere, the position at which spindle microtubules attach to the chromosome during cell division. Replication begins from origins spaced along the chromosome length.

coding sequences. The nuclei of a *diploid cell* contain two copies of most or all the chromosomes. Cells with three *(triploids)*, four *(tetraploids)*, or more copies of each chromosome also occur with some frequency in nature.

The genes coding for the various RNA types probably number from as few as 5000 or less in lower eukaryotes to perhaps as many as 100,000 in the most complex species. In most eukaryotic species genes encoding mRNAs occur primarily in single copies; that is, in one copy per haploid genome equivalent. Some mRNA genes, however, such as the genes encoding histones, tubulins, and actins, are regularly repeated. Repeats of mRNA genes, when present, may be identical or show individual variation, and they may be clustered or scattered in the genome. When variations occur in coding sequences among a group of repeated mRNA genes, the group is known as a *gene family*. Individual members of a gene family encode distinct but related polypeptides. Repeated genes coding for immunoglobulin, tubulin, actin, histone, and hemoglobin proteins, among others, make up gene families of this type. Some gene families contain so many different representatives with varying degrees of relatedness that they are known as *superfamilies*. The large group of genes encoding proteins related to the immunoglobulins is easily the most extensive vertebrate superfamily.

In addition to fully functional sequences, many gene families contain faulty copies that either remain untranscribed or, if transcribed and translated, result in the synthesis of nonfunctional proteins. Faulty gene copies of this type are called *pseudogenes*. In some gene families pseudogenes take up much more space in the genome than functional copies.

In contrast to mRNA genes, all rRNA and tRNA genes are repeated in eukaryotic genomes. In most eukaryotes the rRNA gene repeats, which typically number in the hundreds to thousands, are usually identical or nearly so, and are clustered in a localized region on one or a few chromosomes. The tRNA genes, repeated from tens to hundreds of times for each tRNA type, are usually scattered in small groups throughout the genome. The sn/scRNA genes, in single or repeated copies, are also distributed throughout the genome. Nonfunctional rRNA, tRNA, and sn/scRNA genes also occur as pseudogenes in the genome.

Repetitive sequences are classified as *moderately* repetitive if they occur in a few to a hundred thousand copies and *highly* repetitive if they occur in hundreds of thousands to millions of copies. Moderately repetitive sequences, which include both functional and nonfunctional elements, are distributed in clusters and also dispersed singly or in small groups in locations between mRNA, rRNA, tRNA, and sn/scRNA genes. Highly repetitive sequences, which are primarily or exclusively nonfunctional, are usually clustered, often in groups near the tips of the chromosomes or surrounding the centromeres.

THE ORGANIZATION OF CODING SEQUENCES

mRNA Genes

Most of the coding sequences of eukaryotes, 80% or more in most species, are mRNAs of different kinds. Each mRNA gene has the sequence elements diagrammed in Figure 15-10. The transcribed portion of an mRNA gene contains segments corresponding to the 5' untranslated region, coding segment, and 3' untranslated region of its mRNA copy. Introns may interrupt the coding segment of the gene. Upstream of the gene, in the 5'-flanking region, are promoter, enhancer, and other regulatory sequences.

Single-Copy mRNA Genes Messenger RNA genes occurring in single copies are positioned one after another by the thousands in the genome. Between the 3' end of one gene and the 5'-flanking control elements of the next gene are apparently nonfunctional spacers usually containing moderately repetitive sequences.

Exceptions are noted in some species to this purely tandem arrangement of mRNA-encoding genes. In some cases one gene, complete with its own 5'-flanking elements and internal introns, is located entirely within an intron of another gene. More rarely, two genes overlap in such a way that their sense chains occupy opposite strands of the double helix in the same DNA segment. For example, J. P. Adelman and his colleagues found two rat genes, one encoding the gonadotropin-releasing hormone and the other an unidentified protein, that occupy opposite nucleotide chains of the same DNA segment over a distance of some 500 nucleotides. Similar examples of gene overlap have been noted in *Drosophila*.

mRNA Gene Families Typically, the multiple mRNA copies in gene families code for the most abundant proteins of the cell, such as myosins, tubulins, histones, collagens, hemoglobins, and immunoglobulins. The individual mRNA genes of these families may be clustered or dispersed, identical or nearly so in sequence, or highly diverse.

Members of the actin and tubulin gene families, for example, vary in sequence to a limited extent and are dispersed throughout the genome without clustering in most species. Members of the histone family, which for individual histone types range from almost invariant to moderately different, are clustered in some species and dispersed in others. The hemoglobin gene family, in which individual members may vary extensively in sequence, occurs in clusters in vertebrates.

mRNA Pseudogenes Many mRNA gene families include one or more pseudogenes. These faulty copies evidently result from the operation of two completely independent mechanisms that produce recognizably different classes called *nonprocessed* and *processed pseudogenes*. Nonprocessed pseudogenes appear through mutations introducing changes in gene copies that originated as duplicates of a single ancestral gene. The mutations, including substitutions, deletions, and rearrangements of nucleotides, may occur in both transcribed segments and flanking regions. Some pseudogenes of this class are so altered in their 5'-flanking regions that they cannot form an initiation complex with the RNA polymerase II enzyme and remain untranscribed. Others retain a functional promoter and can be transcribed, but have deletions or substitutions within the transcribed segments. The sequence alterations may interfere with translation, alter coding segments to produce codon frameshifts (see p. 663), or produce internal terminator codons. If translation can be initiated, the frameshifts or internal terminators result in proteins that contain stretches of "nonsense" amino acids or are significantly shortened. In either case the polypeptides translated from mRNAs of these pseudogenes are usually nonfunctional.

Nonprocessed pseudogenes came as no real surprise when they were discovered in the late 1970s. Classical geneticists such as J. B. S. Haldane predicted their eventual discovery in the 1930s, when duplicate copies of some genes were first discovered by genetic crosses in *Drosophila*. Because duplicate copies were known to exist, it was assumed that one or more of the copies could mutate away from functional forms without detrimental effects as long as one or a few remained fully functional. Nonprocessed pseudogenes have been detected, for example, in the hemoglobin, immunoglobulin, interferon, and histocompatibility (MHC) gene families of mammals, the leghemoglobin genes of the soybean, and the cuticle genes of *Drosophila*.

In contrast, the discovery of processed pseudogenes in the late 1970s by P. Leder and his coworkers and others came as a complete surprise. These pseudogenes are called *processed* because they resemble a processed mRNA rather than an intact gene. In particular, 5'- and 3'-flanking sequences and any introns occurring in functional members of the gene family to which processed pseudogenes belong are missing. Processed pseudogenes usually even contain a series of A-T base pairs corresponding to the poly(A) tail of a processed mRNA. Because none of the normal 5'-flanking sequences of the promoter are present, processed pseudogenes of mRNA families are usually not transcribed. Pseudogenes of this class have been detected in several mammalian gene families, including tubulin, actin, and immunoglobulin families, and also as faulty copies of single-copy mRNA genes such as the tropomyosin, cytochrome *c*, and β-lipoprotein genes.

Processed pseudogenes are believed to originate from DNA copies of fully processed mRNAs. In some way the DNA copies are made in the nucleus or cytoplasm and subsequently inserted into the genome. Figure 18-2 shows how such DNA copies might be made through the activity of *reverse transcriptase*, an enzyme that makes complementary DNA (cDNA) copies from an RNA template. The probable origin of processed pseudogenes by insertion is suggested by the fact that they are characteristically flanked at either end by a directly repeated 9 to 14 base-pair sequence. Flanking direct sequence repeats of this type are typical of DNA segments that have been inserted in the genome, often through the activity of enzymes encoded in infecting viruses (see below). As far as is known, the reverse transcriptase required to make the inserted cDNA copies does not normally occur in eukaryotic cells. However, many viruses and transposable elements common in eukaryotic cells (see below) encode reverse transcriptases that could copy mRNAs as well as their usual templates.

Although nonprocessed pseudogenes are widely distributed among eukaryotes, processed pseudogenes are restricted almost exclusively to mammals. Some investigators have estimated that as much as 20% of mammalian genomes may represent processed pseudogenes. The prevalence of processed pseudogenes in mammals may reflect the high frequency of retrovirus infection of mammalian cells. During their cycle of infection, these viruses (see below) synthesize a cDNA copy of their genetic information, which is stored in the free viral particle as an RNA molecule. Besides the reverse transcriptase necessary to make this copy, retroviruses encode proteins capable of inserting a cDNA copy into the host cell DNA. Retrovirus-infected cells therefore possess all the enzymatic machinery necessary to produce and insert processed pseudogenes.

The chromosomal locations of nonprocessed and processed pseudogenes reflect their probable origins. Nonprocessed pseudogenes are distributed in patterns resembling the normal members of their gene families. If the gene family of a nonprocessed pseudogene is clustered, for example, pseudogenes occur within the cluster as one or more of the repeats. Such locations are fully compatible with the idea that nonprocessed pseudogenes represent originally intact gene copies that mutated away from functional form. Processed pseudogenes, on the other hand, may occur anywhere in the genome, usually with no tendency to be placed near their normal counterparts. This is the expected arrangement if processed pseudogenes represent DNA copies of processed mRNAs that are inserted at random points in the chromosomes.

The DNA changes leading to the formation of nonprocessed pseudogenes extend over evolutionary time, that is, over millions of years. The time involved can be estimated by comparing the numbers of accumulated mutations in nonprocessed pseudogenes with known or estimated mutation rates. (Pseudogenes are

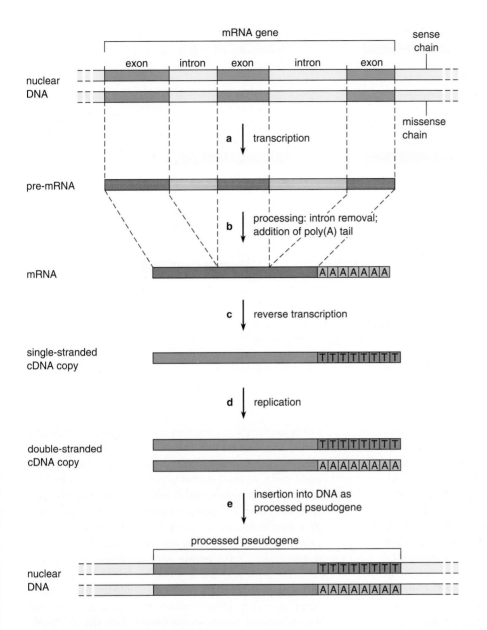

Figure 18-2 Formation of a processed pseudogene through the activity of reverse transcriptase. **(a)** An mRNA-encoding gene is transcribed by RNA polymerase, producing a pre-mRNA transcript. After removal of introns and addition of the poly(A) tail during processing **(b)**, the finished mRNA is copied into cDNA by reverse transcriptase **(c)**. The cDNA is then used as a template for replication of a complementary nucleotide chain, forming a double-stranded cDNA copy **(d)**. This reaction is also catalyzed by reverse transcriptase. Finally, the cDNA copy is inserted by an unknown mechanism into the DNA of the genome **(e)**.

generally free to mutate, since they are unexpressed and thus not subject to selective forces.) Such comparisons show that it takes from 1 to 2 million years for a duplicate copy in a gene family to accumulate enough mutations to become a nonfunctional, nonprocessed pseudogene. Therefore, most of the nonprocessed pseudogenes now identified appeared millions of years ago and are generally present in all members of a contemporary species. Often, they are present in all members of major taxonomic groups, meaning that they date at least to the common evolutionary ancestor of the groups sharing them.

In contrast to the extended time required for evolution of a nonprocessed pseudogene, processed pseudogenes can appear instantly when a DNA copy of an mRNA molecule is inserted into the genome. As with the nonprocessed type, the age of processed pseudogenes can be estimated by noting the number of mutations altering their internal sequences. Processed

pseudogenes of the metallothionein and cytochrome *c* genes in humans, for example, are perfect copies of their respective mRNAs. Their insertions into the genome are therefore evolutionarily relatively recent. Their origins are not in the immediate past, however, because the same pseudogenes are also present in rats and mice, indicating that they arose at or somewhat before the evolution of the common ancestor of these species. Other processed pseudogenes, such as three β-tubulin pseudogenes of higher vertebrates, are highly mutated, indicating that they arose much earlier in evolution. From their degree of mutation these three pseudogenes are estimated to have been inserted some 4, 11.5, and 13 million years ago.

Some Examples: The Globin, Tubulin, and Histone Gene Families Globin genes are widely distributed among animals and also occur in plants. In animals globin genes encode the polypeptide portions of hemoglobin and a

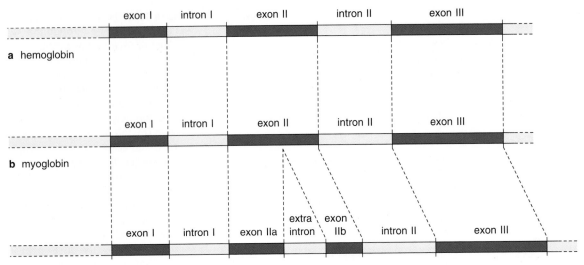

Figure 18-3 Intron-exon structure of globin genes. Almost all animal globin genes coding for hemoglobin **(a)** and the muscle globin, myoglobin **(b)**, have the same distribution of introns and exons. Although the introns may be of different lengths, they fall at the same points in the coding sequence. Leghemoglobin **(c)**, typical of the globulins found in plants, has an extra intron splitting the region corresponding to exon II of the animal globins.

closely related protein, myoglobin. The family illustrates many significant characteristics of eukaryotic genome structure, including functional genes, both nonprocessed and processed pseudogenes, and spacer regions with a variety of apparently nonfunctional repeated sequences.

The globin and myoglobin genes of animals have a similar internal structure, in which three exons are separated by two introns (Fig. 18-3a and b). Almost all known genes of this family have the same introns at exactly the same points in the sequence. The globin genes of plants also have a third intron, which does not appear in animal globin genes (Fig. 18-3c).

In mammals the globin gene family is split into two clusters. One cluster codes for α-globin polypeptides and one for β-globins. (A hemoglobin molecule is a tetramer containing two α- and two β-polypeptides.) The α-globin cluster occurs on chromosome 18 in humans (see p. 1044), occupying a total DNA length of about 30,000 base pairs (Fig. 18-4a). The β-globin cluster occupies about 70,000 base pairs on chromosome 11 (Fig. 18-4b). The α-globin cluster contains four functional genes and three nonprocessed pseudogenes; the β-globin cluster contains five functional genes and one processed pseudogene. A processed pseudogene of the α-globin subfamily has also been found at a completely separate site on chromosome 11. (A processed pseudogene that is widely separated from its parent gene family is sometimes called an *orphon*.)

Comparisons between processed pseudogenes and their normal counterparts provide insights into the origins of the globin pseudogenes. In the α-subfamily,

for example, the ζ gene (for an embryonic globin) and its ψζ pseudogene are closely similar in sequence, indicating recent evolutionary divergence. Six nucleotides differ in the two sequences. Of these, only two are in coding regions and result in amino acid changes. However, one of the two converts an amino acid codon to a terminator codon early in the sequence, causing protein synthesis to stop after production of a short, nonfunctional polypeptide only six amino acids long.

The ψα-1 and ψα-2 pseudogenes of the α-cluster show only about 70% sequence similarity to the functional α₂ and α₁ genes following them, indicating that they diverged earlier in evolution than the ψζ pseudogene. Among the mutations in the ψα pseudogenes are many base changes and deletions. Together, the deletions in these pseudogenes cause extensive frameshifts and codon changes, among them the production of several terminator codons within the coding sequences. The accumulated mutations indicate that the ψα-1 and ψα-2 pseudogenes began to diverge from functional forms about 45 million years ago. The same 20-nucleotide deletion occurs in the ψα pseudogenes in both chimpanzees and humans, indicating that this mutation took place before the evolutionary split producing the human and chimpanzee species about 25 million years ago.

Both the α- and β-globin gene clusters contain long stretches of nonfunctional sequences. The regions between the genes of the α-globin cluster contain a variety of repeated elements (see Fig. 18-4a), including several representatives of the *Alu* family. The *Alu* repeats are of special interest because they are probably pseudo-

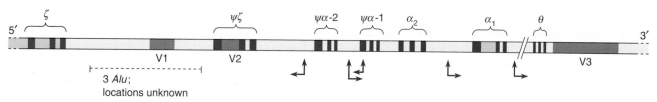

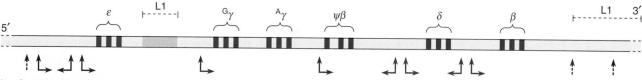

a α-globin gene cluster

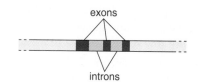

b β-globin gene cluster

Figure 18-4 Partial maps of the globin gene family of humans. **(a)** α-globin gene cluster. Closest to the 5′ end of the cluster is the gene for an embryonic globin, ζ, followed by a nonprocessed ζ pseudogene, ψζ. Two nonprocessed pseudogenes of the adult α-globin genes, ψα-1 and ψα-2, are next. Two functional adult α-globin genes, α2 and α1, and another embryonic gene, θ, lie toward the 3′ end of the α-globin cluster. *Alu* sequences are indicated by arrows; the vertical part of each arrow shows the location and the horizontal part shows the direction of the sequence. The *Alu* sequences in the α-globin cluster show two patterns of accumulated mutations, indicating that they inserted into this cluster at two separate times in evolutionary history. Three variable regions, V1, V2, and V3, contain variable numbers of repeated sequences. V1, for example, contains a 36 base-pair sequence that is repeated 32 times in some individuals and 58 times in others. Note that one variable region lies within the first intron of the ψζ pseudogene. **(b)** The β-globin gene cluster. Three embryonic β-globin genes, ε, ᴳγ, and ᴬγ, occur in sequence at the 5′ end of the cluster, followed by a nonprocessed adult pseudogene, ψβ. After the ψβ pseudogene are two adult hemoglobin genes, δ and β. Each gene in the α- and β-globin clusters has the intron-exon structure shown in the small diagram at the lower right. Some *Alu* sites are indicated by dotted arrows because exact locations and direction of insertion are unknown. Several L1 elements, members of another repeated sequence family in humans and other primates, occur within the β-globin gene cluster. (The L elements are *LINE* sequences; see Information Box 18-2.)

genes of a completely different gene, one coding for an scRNA (see Information Box 18-2). A variety of repeated sequence elements, including *Alu* sequences, also occurs in the spacers between the β-globin genes (see Fig. 18-4*b*).

Repetitive sequences make up about 84% of the DNA in the α- and β-globin clusters, leaving about 16% of the DNA occupied by the genes. Considering that half of the genes consists of introns, only 8% of the 100,000 DNA base pairs in the α- and β-globin gene clusters encodes information copied and retained in mature mRNA molecules. The unequal balance between coding and noncoding sequences in the globin genes is more or less typical of eukaryotic genomes. Only about 5% of the entire human genome, for example, is estimated to be occupied by coding sequences.

The globin genes of plants were originally thought to be limited to leguminous plants such as pea and soybean. In recent years, however, many higher plants, including nearly 200 species distributed among 20 gen-

era and 8 orders, have been found to possess globin genes. Globins, in fact, may be generally distributed among plants. If so, the globin genes of plants and animals probably stem from a common ancestral gene dating back to the time of the split between the two kingdoms some 1500 million years ago.

The globin genes of legumes, known as *leghemoglobins*, encode hemoglobinlike proteins that bind oxygen in root cells of these plants. Binding the oxygen protects the reactions of nitrogen fixation, carried out by symbiotic bacteria in root cells of the legume, from the potentially damaging effects of oxygen. The globin gene family of the soybean includes four functional leghemoglobin genes and four pseudogenes, located at four sites in the genome. Two sites are clusters containing both functional copies and pseudogenes; the other two sites contain one pseudogene each.

Other gene families share some features of the globin family and have distinctive characteristics. The tubulin gene family is generally dispersed in single copies rather than clustered. Some 15 to 20 α-tubulin genes, and about the same number of β-tubulin genes, are distributed singly at sites throughout the chromosomes in mammals. Most of these copies are pseudogenes. For example, of the 10 β-tubulin genes that have been detected in humans, 5 are processed pseudogenes and 2 are nonprocessed, leaving 3 as functional copies. Mutations within the processed β-tubulin pseudogenes are extensive, indicating that their insertion into the genome occurred many millions of years ago.

The histone gene family is distinguished by high numbers of repeats compared to other mRNA genes, as many as several hundred copies of each histone type in some species. Depending on the species, histone genes may be clustered, scattered singly or in small groups, or distributed in both arrangements. In many species histone genes are organized in a repeat unit with one representative of each histone type arranged in a fixed order (Fig. 18-5). In other species, such as those of mammals and birds, histone genes are organized in clusters with the genes for each major histone type in random order.

Although there are many copies of individual histone genes, pseudogenes of either the processed or nonprocessed type are very rare. In fact one of the most interesting characteristics of the histone gene families is the essentially perfect fidelity with which some histone gene repeats are preserved within a species. This is particularly evident in H3 and H4 genes, which

may be identical in sequence in all of the tens or hundreds of copies in the genome. Sequence preservation extends even to the spacer sequences between histone genes in some species; in sea urchins, for example, intergenic spacers are almost perfectly preserved from one repeat unit to the next within a single species. (Intergenic spacers differ in separate species, however.)

Histone genes are not unique in this regard. Other coding sequences, such as rRNA and tRNA genes, and even some nonfunctional repeated sequences are also maintained as perfect or near-perfect repeats even though they are present in multiple copies in the genome. The fidelity of the many copies suggests that a mechanism may work to correct mutations as they appear, perhaps in cell lines giving rise to gametes. The molecular basis for a correction mechanism, if it indeed exists, remains unknown and presently without even a plausible hypothesis.

rRNA, tRNA, and sn/scRNA Genes

The genes encoding rRNAs, tRNAs, and sn/scRNAs are the most highly repeated coding sequences in eukaryotes. Pseudogenes of these RNAs tend to be as repetitive as their normal counterparts, in some cases outnumbering the functional copies by many times. This is particularly evident among the genes transcribed by RNA polymerase III (see p. 580), including 5S rRNA and sn/scRNA genes.

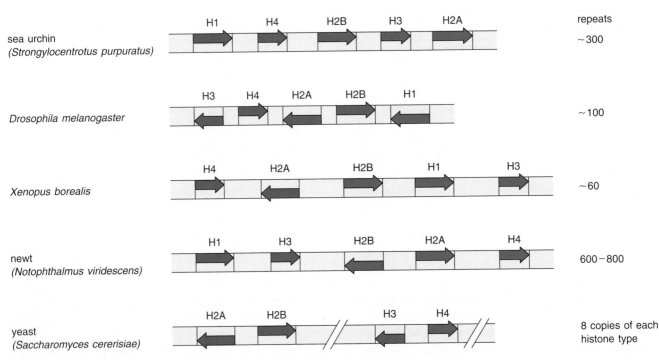

Figure 18-5 The ordered arrangement of individual histone genes within repeat units of some species. The arrows within each unit indicate the sense chain of the DNA double helix and the direction in which the genes are read during transcription. The arrangement shown for the newt is one of several variations occurring in this species.

rRNA Genes The hundreds or thousands of rRNA gene repeats are clustered at sites on one or several chromosomes in eukaryotic genomes (Table 18-1). A cluster of genes encoding the large pre-rRNA that includes 18S, 5.8S, and 28S rRNA forms a site known as a *nucleolar organizer (NOR)* in eukaryotes. This site is so called because the nucleolus forms around it when rRNA transcription and the assembly of ribosomal subunits are in progress (see pp. 616 and 1030). Although the large pre-rRNA repeats occur in large numbers in NORs, the total DNA length occupied by these genes represents a relatively small fraction of the total coding DNA, only about 1% to 2% of the genome in most species.

Within NORs individual genes are organized as shown in Figure 15-30b, with coding sequences for 18S, 5.8S, and 28S rRNAs separated by intragenic spacers. A long intergenic spacer separates one large pre-rRNA gene from the next in the clusters. Little or no variation occurs in the coding sequences of these genes from one repeat to the next within a species. Although the intergenic spacers vary somewhat, these sequence elements are also relatively uniform. Much of the variation that does occur is limited to differences in the number of repeats of short sequence elements. In most species the repeated sequences of the intergenic spacer are found only in this location in the genome. In mammals, however, some repeats of the intergenic spacers are also distributed in other locations.

Although sequence variation is limited in the inter- and intragenic spacers of large pre-rRNA genes in single species, few similarities are noted in these sequence elements between different species, frequently even between closely related ones. The spacer sequences are thus generally highly conserved within but not between species.

The large pre-rRNA gene families sometimes contain pseudogenes. The most conspicuous example occurs in *Drosophila* and some other dipteran flies. The large pre-rRNA pseudogenes of *Drosophila*, which are nonprocessed, each contain a long insert that occurs in approximately the same position as introns in the functional pre-rRNA genes of other species such as *Tetrahymena*. The *Drosophila* insert renders the genes containing it nonfunctional—only a partial transcript, terminating at the 5' end of the insert, is copied from the pseudogene. The transcript is degraded and does not contribute to the processed rRNA synthesized in *Drosophila* (see p. 610 for details). Small variations occur in the location and sequence of the *Drosophila* insert, producing several classes of closely related pseudogenes in this species. The insert is characterized by the presence of short direct repeats at the 5' and 3' ends, indicating that it may have originated as a mobile element (see p. 763). As many as one-half to two-thirds of the large pre-rRNA genes of *Drosophila* individuals may carry the incapacitating insert.

The 5S pre-rRNA genes are clustered in separate

Table 18-1 Organization of rRNA and tRNA Sequences in Eukaryotic Genomes

Organism	Large pre-rRNA Gene Repeats	Large pre-rRNA Gene Organization	5S rRNA Gene Repeats	5S rRNA Gene Organization	tRNA Gene Repeats
Tetrahymena	1–2 copies in micronucleus; thousands of copies in macronucleus	One chromosome site in micronucleus; macronuclear copies extrachromosomal	325	Unknown	800
Yeast (*Saccharomyces*)	100–120	Clustered on single chromosome	140	Intermingled with large pre-rRNA genes in same cluster	320–400
Higher plants	1000–30,000	Clusters on one or several chromosomes of haploid set	Unknown	Unknown	Unknown
Drosophila melanogaster	130–250	Single clusters on two chromosomes (X and Y)	500	Single cluster	750
Sea urchin	~175	One cluster	~175	Unknown	Unknown
Xenopus laevis	400–600	One cluster	24,000	Clusters on most or all chromosomes	7800
Human	300	Single clusters on five chromosomes of haploid set	160	One major cluster	1300

locations from the large pre-rRNA clusters in almost all eukaryotic species. There are exceptions to this arrangement; in yeast and some other fungi, 5S and large pre-rRNA genes are intermingled in the same clusters, with the 5S genes located in the intergenic spacers of the large pre-rRNA genes.

Each repeated unit in a 5S gene cluster includes a coding sequence and an intergenic spacer. (Fig. 15-35 shows the internal structure of a typical 5S pre-rRNA gene.) Repeats in the clusters vary in different species from a hundred to thousands of copies (see Table 18-1). Like the intergenic spacer of large pre-rRNA genes, 5S intergenic spacers contain combinations of repeated sequence elements that vary somewhat in number and sequence within the clusters. Sequence differences in the coding regions of 5S rRNA genes, producing 5S rRNA variants are also noted in some species.

The genomes of both *Xenopus* and humans contain 5S pseudogenes. In *Xenopus* one class of 5S pseudogenes is located in tandem with functional 5S gene copies, with one pseudogene in each intergenic spacer (Fig. 18-6). The pseudogenes, evidently of the nonprocessed type, do not function in transcription. Other 5S pseudogenes occur singly or in small clusters at scattered locations in the *Xenopus* genome; at least some of these pseudogenes may be processed. One human 5S pseudogene, located near the functional 5S gene cluster, contains an insert with similarities to the *Alu* family of repeated elements.

tRNA Genes In most species tRNA genes are dispersed on several or most chromosomes in clusters containing either single or multiple tRNA types. (Table 18-1 lists tRNA repeat numbers for several eukaryotes; the internal structure of tRNA genes is shown in Fig. 15-42.) Particularly among higher eukaryotes, tRNA pseudogenes are so common that several of the highly

repeated, nonfunctional sequences of mammalian and higher plant genomes have evolutionary origins as processed tRNA pseudogenes.

The *C family* of repeated sequence elements in cows and goats, for example, occurs in a consensus sequence that can be folded into a secondary structure equivalent to the stems and loops of a tRNA molecule except for part of the acceptor stem (Fig. 18-7). The anticodon sequence of the C-family elements of cows and goats corresponds to a cysteine tRNA. Similar tRNA-derived repeated sequences, such as the *R.dre.1* family of rats, are found in other mammalian species. The *R.dre.1* family is 73% homologous to an alanine tRNA sequence and includes complete D and TψC stems. Some of the repetitive, nonfunctional sequence elements derived from tRNA genes occur in mammals in as many as 100,000 copies per haploid genome, dispersed both singly and in tandem clusters throughout most of the chromosomes.

The well-preserved A and B promoter blocks (see p. 621 and Fig. 15-42) of these tRNA-like pseudogenes are believed to have contributed to their wide dispersion and prevalence in mammalian genomes. Many tRNA pseudogenes are actively transcribed by RNA polymerase III enzymes, providing the chance that some of the transcripts will be converted into cDNA by reverse transcriptase and reinserted at other locations in the genome. Because processed tRNA pseudogenes formed in this way retain the internal RNA polymerase III promoter, each round of transcription and reinsertion increases the chance that more cDNA copies will be made and inserted.

sn/scRNA Genes The genes encoding sn/scRNAs (see p. 628) occur in both clusters and as single genes in different eukaryotic species. The snRNA genes occur in clusters of tandem repeats in sea urchins, for example; in yeast snRNA genes occur in single copies. In humans

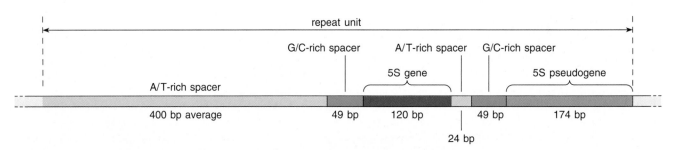

Figure 18-6 The arrangement of 5S pseudogenes in *Xenopus*. Each repeat unit contains a functional 5S gene followed by a nonfunctional pseudogene. The pseudogene, evidently of the nonprocessed type, includes 73 base pairs (bp) of the sequence flanking the 5' side of a normal *Xenopus* 5S gene and the first 101 nucleotides of the 120 base-pair coding sequence. Although the 101 base-pair sequence retained in the pseudogene includes the promoter, it is evidently not transcribed. Transcriptional inactivity may result from one or more of nine sequence substitutions in the coding region of the pseudogene.

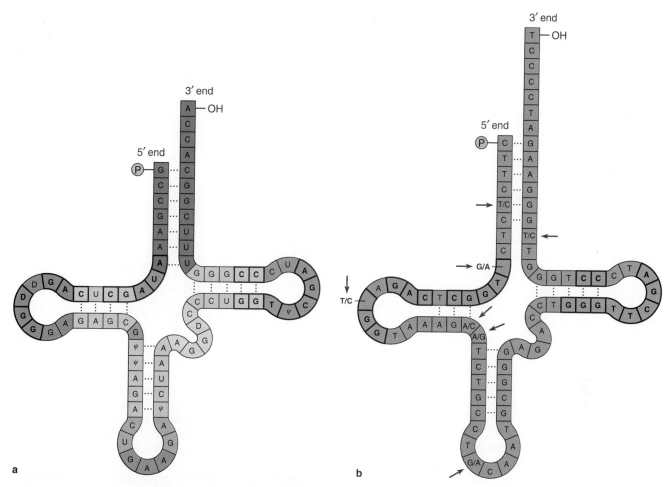

Figure 18-7 A pseudogene derived from a tRNA. **(a)** A tRNA carrying the amino acid phenylalanine. **(b)** The consensus sequence of the C family of repeated sequences in cows and goats, folded in the form of a tRNA. The bases in boldface are conserved parts of the A and B promoter blocks of a tRNA gene. The arrows indicate alternate bases in the consensus sequence. The T bases in the pseudogene correspond to modified bases U, D, and ψ in a processed tRNA molecule. Courtesy of J. H. Rogers, from *Internat. Rev. Cytol.* 93:187 (1985).

snRNA genes of the U type are clustered, often with more than one type of U snRNA gene in a cluster. The repeat numbers of functional sn/scRNA genes are relatively low, on the order of 10 to about 30 copies. However, sn/scRNA pseudogenes may occur in much greater numbers, on the order of hundreds to thousands. The U1 snRNA gene of humans, for example, occurs in about 30 functional copies and 500 to 1000 pseudogenes.

The only scRNA gene characterized in detail, the SRP 7S scRNA gene (see p. 628), encodes a small cytoplasmic RNA forming part of the signal recognition particle (SRP). The SRP participates in the reactions attaching ribosomes to the ER during the synthesis of membrane and secreted proteins (see p. 837 for details). Although functional copies of the SRP 7S scRNA gene

probably number fewer than ten in most eukaryotic species, nonfunctional copies may occur in hundreds to many thousands. The *Alu* family of repeated sequences in humans and other primates is apparently a collection of SRP 7S scRNA pseudogenes.

Most sn/scRNA pseudogenes are flanked by direct sequence repeats at their 5′ and 3′ ends, suggesting that they were inserted in genomes as cDNA copies of processed RNA transcripts. Some snRNA pseudogenes, however, contain the 5′- and 3′-flanking sequences typical of functional genes, indicating that at least some are nonprocessed. Mixed snRNA pseudogenes also occur, containing both nonprocessed and processed elements; evidently, this class originated through insertions of one pseudogene type into another.

Coding sequences occur in single and multiple copies, distributed individually or in clusters in the chromosomes of eukaryotic genomes. Genes encoding mRNAs are present in single copies, or in multiple copies as families including members that vary to a greater or lesser extent in sequence. The members of mRNA gene families may be clustered in one or a few locations, or distributed singly throughout the genome. The genes encoding rRNAs, tRNAs, and sn/scRNAs typically occur in multiple numbers. Large pre-rRNA genes are clustered in one or a few locations in the genomes of most eukaryotes; in these locations the clusters constitute the nucleolar organizer regions around which the nucleolus forms. The genes for 5S rRNA, tRNAs, and sn/scRNAs are distributed in small clusters throughout the genomes of most eukaryotes. All the major gene types also appear in the form of pseudogenes that, in some cases, outnumber functional copies. Depending on the particular gene, pseudogenes may be of the processed or nonprocessed type or both.

REPETITIVE SEQUENCES IN EUKARYOTIC GENOMES

The existence of repetitive noncoding sequences in eukaryotic genomes first came to light during the 1960s when two investigators, R. J. Britten and D. E. Kohne, developed a method, now known as *reassociation kinetics*, for detecting them in bulk DNA samples. In this technique the DNA of an organism is isolated and purified, broken into pieces about 1000 base pairs in length, and heated to unwind the DNA into separate chains (see Information Box 14-1). The preparation is then cooled slowly, and the rate at which the chains rewind into double helices is noted.

Nonrepeated sequences rewind at a basal rate determined from experiments with the DNA of organisms such as bacteria in which most sequence elements are nonrepetitive. The basal rate is relatively slow because each nonrepeated DNA segment has only one pairing partner in the solution, which it must encounter by random collisions in order to rewind (see Fig. 4-28a and b). The size and total number of different sequences (called the *complexity*) of the genome also affect the outcome; as these factors increase, the chance that a DNA segment will find its pairing partner decreases.

These relationships are described by *Cot* curves (Fig. 18-8), in which the fraction of DNA that has reassociated is plotted against C_0t, the concentration of the DNA at the beginning of the experiment (C_0) multiplied by the time (t) required for a given fraction to reassociate. The time required for half the DNA of a sample to reassociate ($C_0t_{1/2}$) allows a comparison of the relative size and complexity of the genomes under

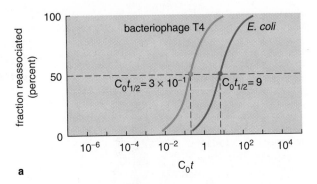

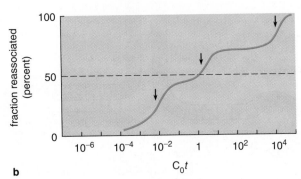

Figure 18-8 Cot curves tracing DNA reassociation. **(a)** Typical Cot curves produced by reassociation of DNA from genomes that consist primarily or exclusively of unique, nonrepeated sequences. For DNA of this type, the position occupied by the $C_0t_{1/2}$ (the point at which 50% of the DNA has reassociated) reflects the sequence complexity of the genome—the more complex the genome, the farther the $C_0t_{1/2}$ is displaced to the right. **(b)** A Cot curve obtained from the reassociation of DNA from eukaryotic genomes. The curve typically extends over more of the horizontal axis than a Cot curve from purely unique DNA, and consists of three segments. The segment farthest to the left is produced by highly repeated sequences, which reassociate most rapidly. The middle segment reflects reassociation of moderately repeated segments. The segment farthest to the right is produced by the relatively slow reassociation of unique sequences; the position taken by this segment primarily reflects the sequence complexity of the unique DNA. The arrows mark the inflection points for the individual sigmoid curves making up the three segments; the distance between these points on the horizontal axis gives a measure of the degree of sequence repetition in the moderately and highly repeated fractions.

study (Fig. 18-8a). The farther the $C_0t_{1/2}$ is displaced to the right, the larger and more complex the genome. Note from the figure that the $C_0t_{1/2}$ for *E. coli* DNA (about 10^1), is approximately 30 times larger than the $C_0t_{1/2}$ for bacteriophage T4 DNA (about 3×10^{-1}), indicating that the *E. coli* genome is about 30 times more complex than the bacteriophage genome. These relationships assume that there are no repeated sequences.

The degree of repetition of sequences has an effect on the Cot curve opposite to that of the complexity. Because they have more possible pairing partners, repeated sequences reassociate more rapidly as DNA preparations are cooled (see Fig. 4-28c and d). This pushes the value for $C_0t_{1/2}$ to the left in the plot.

The prokaryotic genomes examined by Britten and Kohne typically produced Cot curves that occupied a relatively restricted length on the scale. However, when eukaryotic DNA was tested, the length occupied on the C_0t scale was greatly expanded (Fig. 18-8b). Eukaryotic Cot curves typically have three steps, as if three consecutive sigmoid curves are combined. Britten and Kohne concluded from this that the sigmoid curve in the step farthest to the right is produced by the reassociation of nonrepeated sequences in the eukaryotic DNA. The steps to the left reflect two classes of repeated sequences—a highly repeated class, producing the step farthest to the left, and a moderately repeated class, producing the intermediate step. If this is indeed the case, a comparison of the inflection points for the three steps (arrows in Fig. 18-8b) would give a direct measure of the degree of repetition in the two repeated classes. For the typical eukaryotic Cot curve in Figure 18-8b, a comparison of the values for the inflection points for the first step ($C_0t = 10^4$) and last step ($C_0t = 10^{-2}$) shows a difference of about a million times, indicating that the first step contains sequences repeated by this amount. From their results Britten and Kohne concluded that all eukaryotes have three classes of DNA sequence elements—*unique sequences* occurring in only one copy, *moderately repetitive sequences* in copies from a few to 100,000, and *highly repetitive sequences* in hundreds of thousands to millions of copies (Table 18-2; Britten's Experimental Process essay on p. 756 describes experiments leading to their discovery and important conclusions derived from the research).

Britten and Kohne's results indicating that some sequence elements occur in millions of copies in eukaryotic genomes were at first discounted by most cell and molecular biologists. However, their results were later confirmed by several unrelated experimental approaches.

The most direct confirmation came from DNA sequencing. Some sequencing studies investigated DNA fractions obtained when the genome of a species is broken into relatively short segments and centrifuged (see p. 125). When prepared by this approach, the DNA of some species separates into a main band that contains the bulk of the genome and one or more separate, small *satellite bands*. The satellite bands separate from the main band because they contain higher proportions of adenine-thymine or guanine-cytosine pairs than the bulk DNA, giving them a markedly different density. Sequencing of the satellite bands showed that each consists of a single sequence element, in many cases less than ten base pairs in length. One of the satellites from *Drosophila melanogaster*, for example, contains nothing more than tandem repeats of the sequence AAGAG. Although the sequence is very short, hybridization studies showed that it could pair with a surprisingly large fraction of the DNA in the genome, so large that the AAGAG element must occur in segments containing more than a million copies.

Repetitive sequences have also been detected by means of restriction endonucleases, bacterial enzymes that recognize short sequence elements and cut DNA at sites containing the sequences (see p. 135 and Table 4-2). Many restriction endonucleases have been identified and purified, each specific in its recognition and cutting activity for a different DNA sequence. The enzymes cut DNA at each site occupied by the short sequence element recognized by the enzyme. Use of the enzymes allows repeated sequences to be detected and cut from the genome as fragments of uniform length. Once separated from the remainder of the DNA by techniques such as gel electrophoresis (see p. 126), the uniform fragments can readily be sequenced. (Fig. 18-9 explains the technique in further detail.) The technique also confirmed that the genomes of all eukaryotes

Table 18-2 Distribution of Unique and Repeated Sequences in Different Eukaryotes

Organism	Unique (%)	Moderately Repetitive (%)	Highly Repetitive (%)	Reference
Dinoflagellate	40	60*		*Cell* 6:161 (1975)
Polytoma (an alga)	70	10	20	*Chromosoma* 49:19 (1974)
Wheat	25	50–68	10	*Heredity* 37:231 (1976)
Rana clamitans (green frog)	22	67	9	*Molec. Genet.* 5:3 (1976)
Chick	70	24	6	*Eur. J. Biochem.* 45:25 (1974)
Rat	65	19	9	*Biochemistry* 13:841 (1974)
Human	64	25	10	*J. Mol. Biol.* 63:323 (1972)

* Moderately and highly repetitive combined.

DNA Fingerprinting

A new forensic technique detecting individual differences in DNA sequence elements, developed in 1985 by A. J. Jeffreys, V. Wilson, and S. V. Thien, has proved useful for identifying persons responsible for crimes such as murder and rape. The technique, *DNA fingerprinting*, has also been employed to determine paternity. DNA fingerprinting depends on groups of repeated sequences called *hypervariable elements*, which occur in a distinct form in each individual. Although different in each individual, hypervariable elements have *core* sequences that are similar enough from one person to the next to be represented by an average or consensus sequence. The repeated DNA sequences are frequently preserved well enough in semen or bloodstains, even in dried form, to allow their extraction and comparison with those of a suspect or victim.

In the technique (see part *a* of the figure in this box), a DNA sample obtained from hair, semen, or blood is digested with a restriction endonuclease to obtain fragment classes of different sizes. The fragments are then run on an electrophoretic gel and extracted from the gel by means of a Southern blot (see p. 129). The bands containing hypervariable sequences of interest are identified by hybridizing them with a radioactive probe consisting of DNA with the consensus sequence for the hypervariable elements.

These samples are compared with samples from the victim or suspect. In a rape case, for example, a DNA fingerprint made from the semen or blood of a suspect could be compared with the fingerprint made from a semen sample obtained from the vagina or clothing of the victim. In a murder case a DNA fingerprint made from bloodstains found on a suspect's person, clothing, or other articles could be compared with the DNA fingerprint of the victim. If the DNA fingerprints from these sources are the same or similar, the results can be taken as valid evidence.

DNA fingerprints can also be used as an indicator of paternity because the DNA fingerprints of a parent and offspring are much more similar than those of totally unrelated persons. Although the test is not definitive, similar DNA fingerprints, when used in conjunction with other tests such as blood typing, provide a good indication whether a child has been fathered or mothered by a given person. (Part *b* of the figure in this box shows a fingerprint made to determine paternity.)

To be dependable, DNA fingerprints must be made and compared with great care. The technique requires great expertise, and comparisons of bands from different preparations are inherently difficult. Degradation of DNA in the samples or contamination by bacteria can produce extra bands that invite errors in identification. The polymerase chain reaction can also produce extra bands (as in the figure in this box). Because of these difficulties, many scientists feel that laboratories and individual technicians carrying out DNA fingerprint comparisons should be tested regularly and licensed only if competence is demonstrated.

Courts have experienced some operational problems with admission of DNA fingerprinting evidence. Some experts estimate that the chance of an innocent person producing an incriminating DNA fingerprint is as small as one in hundreds of millions; others maintain that the chance may be as great as 1 in 50. Some civil libertarians have also expressed concern that DNA fingerprinting constitutes an invasion of privacy. In spite of these difficulties, further experience with the method, agreement on standards to be used for comparison of fingerprints, establishment of rules governing submission of individuals to DNA fingerprinting, and licensing of technicians and laboratories promise to make the technique useful and dependable as a source of definitive legal evidence.

Suggestions for Further Reading: See p. 772.

but distinct sequences in each species. These subfamilies alternate in twos, threes, or fours to produce clusters with regularly arranged sequence patterns repeated thousands of times. Different chromosomes within a species usually contain clusters with different repeat patterns. These clusters are generally located at either side of the centromere and, in some species, near the telomeres.

One of the most striking characteristics of highly repetitive sequences is the near-perfect fidelity with which they are conserved within a species. Separate species, however, no matter how closely related, have different highly repeated sequences arranged in distinct numbers, patterns and combinations. The degree of preservation within a species suggests that, as with some highly preserved functional sequences such as multiple histone gene copies, a correction mechanism must somehow recognize new mutations as they appear anywhere in the thousands or millions of repeats, and remove or replace nucleotides to return the mutated

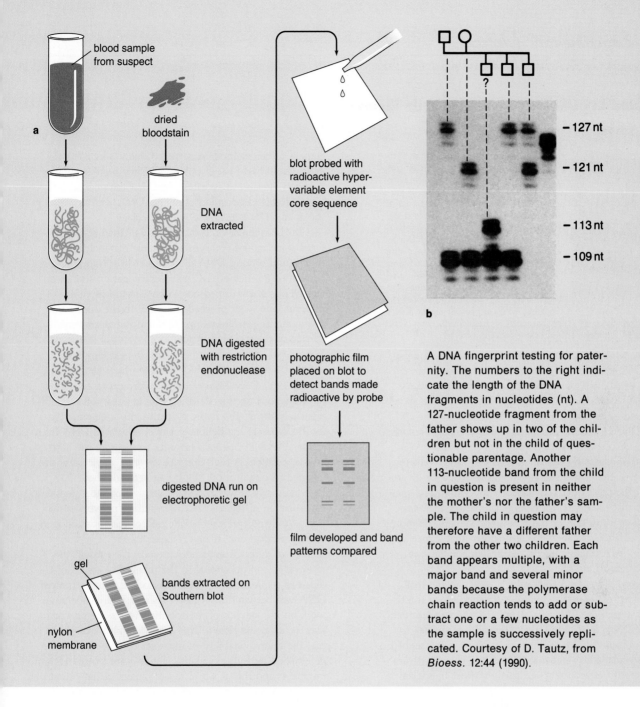

a

blood sample from suspect

dried bloodstain

DNA extracted

DNA digested with restriction endonuclease

digested DNA run on electrophoretic gel

gel

nylon membrane

bands extracted on Southern blot

blot probed with radioactive hyper-variable element core sequence

photographic film placed on blot to detect bands made radioactive by probe

film developed and band patterns compared

b

— 127 nt

— 121 nt

— 113 nt

— 109 nt

?

A DNA fingerprint testing for paternity. The numbers to the right indicate the length of the DNA fragments in nucleotides (nt). A 127-nucleotide fragment from the father shows up in two of the children but not in the child of questionable parentage. Another 113-nucleotide band from the child in question is present in neither the mother's nor the father's sample. The child in question may therefore have a different father from the other two children. Each band appears multiple, with a major band and several minor bands because the polymerase chain reaction tends to add or subtract one or a few nucleotides as the sample is successively replicated. Courtesy of D. Tautz, from *Bioess.* 12:44 (1990).

sequences to match a "standard" type. Curiously, the sequence preservation of highly repeated elements is usually not perfect; a small percentage can be detected to vary from the standard. This indicates that the correction mechanism is "leaky" and allows a small number of incorrect copies to persist.

Variations in the number of highly repeated sequences may be responsible for wide variations in total DNA quantity between some closely related species. Some salamanders, such as *Triturus*, for example,

have seven times as much DNA per nucleus as their relatives among the frogs; in plants one member of the genus *Vicia*, *V. faba* (the broadbean), has about five times more DNA per nucleus than another member of the same genus, *V. sativa.* In many cases, as in the *Vicia* example, the nuclei of the closely related species with widely different DNA quantities contain the same number of chromosomes.

Highly repeated sequences also account for the unexpectedly high DNA content of genomes in some

Repeated DNA Sequences and Mobile Elements

Roy J. Britten

After working on the Manhattan Project during the war, Roy J. Britten got a Ph.D. in nuclear physics working with a cyclotron. Because the future of particle physics appeared to require large teams, he switched to biophysics at the Department of Terrestrial Magnetism of the Carnegie Institution of Washington. During 20 years there, he worked on a variety of subjects ranging from bacterial studies (amino acid pools, feedback control of amino acid synthesis, ribosomes and their role in protein synthesis) to eukaryotic DNA organization. Since 1971 he has been at Caltech studying gene regulation in development and the evolution of DNA.

The hybridization of short DNA fragments to long denatured DNA embedded in agar was used for bacterial DNA in 1961.[1,2,3] This approach also worked for eukaryotic DNA. The rates for eukaryotes, however, were about a thousand times higher than expected for unique sequence DNA. I became increasingly disturbed by the fact that rate calculations indicated this was possible.

Measurements by Martin and Hoyer[16] and Walker[17] showed that there was a large amount of sequence divergence of the DNA that was reassociating within one species, and that the degree of divergence between species increased as a function of the time since the last common ancestor existed. Walker[17] had also shown that a specific subset of DNA fragments was responsible for the measured hybridization.

Bolton, Hoyer, and McCarthy[2,15] also showed that the same set of sequences hybridized regardless of the tissue from which the DNA was extracted, as expected for genomic DNA. In addition, they showed that the sequences responsible for the hybridization were transcribed in total cellular RNA and that RNA from different tissues showed partly overlapping and partly distinct hybridization.

In the spring of 1964 I decided to attack the question of why the DNA-agar hybridization rates were so much faster than expected in eukaryotes. The first tests (by following optical reassociation in slabs of agar) showed that agar did not have a catalytic effect on the process. Reassociation was also examined between labeled short (sheared) DNA fragments and large fragments using molecular sieve agar gel columns to separate the long and short fragments and thus to assay hybridization. This procedure convincingly showed solution reassociation at expected rates and established that the basic rate calculations were correct. However, it did not work well with animal DNA due to large losses by filtration on the gel columns. As it turned out later, this resulted from network formation in the DNA deriving from interspersed repeat reasso-

ciation. Other methods superseded this work and these early results were never published. They were significant, however, because they established that the unexpectedly high rate of reassociation of animal DNA was intrinsic in the animal DNA itself and not an artifact.

In the summer of 1964 Bill Hoyer (then at NIH) offered some mouse embryos as suitable material for the purification of long DNA molecules. Mike Waring and I performed a series of CsCl ultracentrifugation studies on this DNA, which turned out to be a fortunate choice of material. The work proceeded rapidly to the purification of the mouse satellite [see p. 751] and the measurement of its rate of reassociation in solution by optical techniques.[18] The rate of reassociation indicated a repeat length of about 300 nucleotides, which has been fully confirmed.

In 1965 Dave Kohne started hydroxyapatite studies of the general occurrence of repeats in eukaryotes. He immediately developed hydroxyapatite into a quantitative tool to separate the double-strand reassociated DNA from single strands in one pass in order to measure the kinetics of reassociation. A high-pressure shearing device was developed to obtain short DNA fragments of about 500 base pairs, which was necessary because of short period interspersion of repeats and the resulting network formation. We demonstrated that the rate of reassociation is inversely proportional to the complexity—that is, the number of different sequences present. As a result, the rate was also proportional to the number of copies of a repeated sequence that would reassociate with each other. DNA was prepared from many species, and comparative work on the kinetics of reassociation was done on what was for us at the time a large scale. We also established that the thermal stability of duplexes could accurately be determined by hydroxyapatite elution with labeled DNA fragments and that it was a reliable index of the extent of sequence divergence between reassociated sequences. The quantitative measurements were very important to this work.

We concluded[8] that a large fraction of the DNA of all of the eukaryotes we investigated was made up of repeated sequences ranging from a few copies up to millions. We also identified and assayed the single copy DNA sequences. That work and more recent studies show that the majority of DNA appears to be single copy, except in species with very large genomes such as salamanders where the repeats dominate. It is not known yet how much of the "single copy" DNA is made up of low frequency (2 to 10 copy) repeats with large divergence.

The work on sequence arrangement stopped but was taken up again after several years with quantitative measurements of interspersion repeats in calf[9] and sea urchin

DNA.[4,7] Later, DNA sequence arrangement patterns were examined in more detail.[12,13,14]

The DNA-agar measurements initiated events leading to the proposal that repeated sequence families exist.[10,12] Most important were measurements of interspecies relationships, which made possible the early recognition that the repeats were undergoing evolutionary change.

Since many DNA sequences were found to exist in repetitive families composed of many members, new repetitions must be continually created during the evolution of a species. This became evident when we realized that each of two species that have diverged significantly maintains large amounts of internal homology or repetition, but has lost some interspecies homology. In at least one of the species (and presumably both) new sets of multiple repeated sequences have arisen. Thus, the pattern of repetition in the genome of a particular animal would result from a balance of processes such as duplication, translocation, point mutation, and deletion of nucleotide sequences.

The observation that repeated sequences are both interspersed and change in frequency over evolutionary time implies a process of insertion (and deletion) of relatively short sequences in many locations in the genome. The result is formally equivalent to duplication and transposition but we know almost nothing about the mechanism of these processes in eukaryotes.

A few of the potential classes of mobile elements have been identified, but except in *Drosophila* the processes generating these sequences have yet to be described, although the insertion of reverse transcribed copies of various RNA sequences is probably a major process.

In 1969 Eric Davidson and I found we were both thinking that repeated sequences were related in some way to the mechanisms of gene regulation. In one rather long weekend we wrote a paper[5] formally describing a model in which 5' DNA regions adjacent to genes contained sets of sequences to which molecules bound that controlled transcription. The underlying idea is that embryonic development and differentiation depend on the control of transcription and the activation or repression of appropriate batteries of genes. At the time the binding sequences were considered as a subset of the repeated DNA sequences and RNA sequences were considered to be possible transactivators due to the simplicity of complementary sequence recognition, although all along Eric and I thought that protein transactivators were likely.

Further speculation on the possible evolutionary role of repeats came in 1971.[6] Then we noted that the gene regulation model we had proposed provided a possible molecular-level interpretation of the programming of gene activity in development.[11] It indicated how particular cell lines in development could be specified in the processes of localization and induction. A basic property of the model is its potential for evolutionary modification of the regulatory systems. Important parts of the regulative relationships depend on the location of sequences in the genome, and a variety of changes can be caused by sequence translocation. As a result, many individual cellular properties could be affected by genomic alterations that may occur relatively frequently, namely, translocations and chromosomal rearrangements. It is quite plausible that selectively favorable changes of the former type are much more frequent. Furthermore, the model regulatory system is receptive to the incorporation of new sequences and sets of sequences, and through these additions can achieve new functional interrelationships. It can grow as well as become rearranged. Whether or not the particular gene regulation system we proposed turns out to occur in living systems, the properties of growth and change by genomic rearrangement are likely to be necessary attributes of evolving regulatory systems.

It appears that mobile DNA elements may be responsible for many of the repeated sequences, and some of the repeats are surely mobile themselves. The confirmation of this 20-year-old prediction remains out of reach, but it certainly has not been contradicted.

References

[1] Bolton, E. T.; Britten, R. J.; Byers, T. J.; Cowie, D. B.; Hoyer, B.; Kato, Y.; McCarthy, B. J.; Miranda, M.; and Roberts, R. B. In *Carnegie Institution of Washington Yearbook* 63:366–97 (1964).

[2] Bolton, E. T.; Britten, R. J.; Byers, T. J.; Cowie, D. B.; Hoyer, B.; McCarthy, B. J.; McQuillen, K.; and Roberts, R. B. In *Carnegie Institution of Washington Yearbook* 62:303–30 (1963).

[3] Bolton, E. T.; Britten, R. J.; Cowie, D. B.; McCarthy, B. J.; Midgley, J. E.; and Roberts, R. B. In *Carnegie Institution of Washington Yearbook* 61:244–93 (1962).

[4] Britten, R. J. In *Evolution of Genetic Systems*. H. H. Smith ed. New York: Gordon and Breach, pp. 80–94 (1972).

[5] Britten, R. J., and Davidson, E. H. *Science* 165:349–58 (1969).

[6] Britten, R. J., and Davidson, E. H. *Quart. Rev. Biol.* 46:111–38 (1971).

[7] Britten, R. J., and Davidson, E. H. In *Molecular Genetics and Development Biology*. M. Sussman ed. Englewood Cliffs: Prentice-Hall, pp. 5–27 (1972).

[8] Britten, R. J., and Kohne, D. E. *Science* 161:529–40 (1968).

[9] Britten, R. J., and Smith, J. In *Carnegie Institution of Washington Yearbook* 69:378–88 (1970).

[10] Britten, R. J., and Waring, M. J. In *Carnegie Institution of Washington Yearbook* 64:316–33 (1965).

[11] Davidson, E. H.; Britten, R. J. *J. Theoret. Biol.* 32:123–30 (1971).

[12] Davidson, E. H.; Hough, B. R.; Amenson, C. S.; and Britten, R. J. *J. Mol. Biol.* 77:1–23 (1973).

[13] Goldberg, R. B.; Crain, W. R.; Ruderman, J. V.; Moore, G. P.; Barnett, T. R.; Higgins, R. C.; Gelfand, R. A.; Galau, G. A.; Britten, R. J.; and Davidson, E. H. *Chromosoma* 51:225–51 (1975).

[14] Graham, D. E.; Neufeld, B. R.; Davidson, E. H.; and Britten, R. J. *Cell:* 127–37 (1974).

[15] McCarthy, B. J., and Bolton, E. T. *Proc. Nat'l Acad. Sci. USA* 50:156–64 (1963).

[16] Martin, M., and Hoyer, B. *Biochemistry* 5:2706–13 (1966).

[17] Walker, P. M. B., and McLaren, A. *J. Mol. Biol.* 12:394–409 (1965).

[18] Waring, M., and Britten, R. J. *Science* 154:791–94 (1966).

Alu and LINE Sequences

In humans and other primates, two well-characterized examples of nonfunctional moderately repeated sequences are the *Alu* family of short interspersed elements and the LINE family of Long INterspersed Elements. The *Alu* family (see part *a* of the figure in this box) is named for the restriction endonuclease *AluI*, which recognizes and cuts a short sequence element that occurs within most *Alu* repeats. A complete primate *Alu* element is about 300 base pairs (bp) in length, and contains two similar but not identical repeats of a sequence closely resembling portions of the coding region of an SRP 7S scRNA. The two repeats are separated by an A-rich insert and followed by a short poly(A) segment. Both deletions and insertions occur in the two repeats of the SRP 7S sequence, including a 31 base-pair insertion in the second repeat. The short poly(A) segment at the 3′ end of an *Alu* element indicates that *Alu* sequences may have originated as processed pseudogenes of SRP 7S scRNA genes. In humans there is considerable variation among *Alu* sequences produced by mutations, deletions, or rearrangements. Many truncated copies missing major portions are also present.

A short direct sequence repeat flanks both ends of the *Alu* unit in about 80% of the members of this family in humans. Such flanking direct repeats are hallmarks of mobile elements that can insert in new locations in the genome.

Alu sequences occur singly, in pairs, and in small clusters in the human genome. The two globin gene clusters of humans, for example, contain *Alu* sequences singly and in inverted pairs in the spacers between individual genes. *Alu* elements also occur as inserts in the introns of as many as 25% of other mRNA genes. In this location they are transcribed and appear as RNA copies in the pre-mRNA transcripts of these genes. The *Alu* copies are cut from the pre-mRNA transcripts as the introns are removed during processing, so that they have no effect on the proteins translated from the mRNAs.

The more complete *Alu* sequences contain an RNA polymerase III promoter, making them potentially active in transcription. Relatively few *Alu* sequences of humans are actually transcribed, however, possibly because necessary flanking sequences are missing. Those that are transcribed may be used for production of cDNA copies, leading to further rounds of insertion in the genome.

Alu repeats frequently occur in higher numbers than those typical for moderately repeated sequences. In different primate species *Alu* repeats occur in 100,000 to 500,000 copies and make up from 3% to 8% of the total DNA of the genome. They occur in 300,000 to 500,000 copies in the human genome, where they constitute 6% to 8% of the total DNA. At this level there is one *Alu* sequence for every 5000 to 9000 DNA base pairs in the genome.

Differences between individual *Alu* sequences because of mutations, rearrangements, and deletions provide a measure of the time since their insertion in the genome. The *Alu* repeats within the globin gene clusters, for example, differ in sequence from a "standard" *Alu* by about 14%, indicating that their insertion predates the evolutionary split between apes and humans. Other *Alu* sequences have been inserted more recently. Indications are that *Alu*s are continually inserted into the genomes of humans and other mammals at a rate of about one insertion every 100 years. *Alu* sequences occur in all primates investigated to date and also in rodents such as the rat and mouse.

Recently two human patients were found to suffer from genetic disabilities due to insertion of an *Alu* sequence into the coding regions of mRNA genes. One of these insertions, discovered by M. R. Wallace and her coworkers, was responsible for development of neurofibromatosis, characterized by tumors of the nervous system and other defects. The *Alu* insertion responsible for the gene mutation in this patient did not appear in his parents, indicating that it occurred in his lifetime.

The LINE repeats of human, primate, and other mammalian genomes (see part *b* of the figure in this box) contain between 6000 and 7000 base pairs. Within a complete LINE are several sequences encoding proteins. Among these are a highly mutated, probably nonfunc-

eukaryotic groups (Fig. 18-10). Many relatively simple organisms, such as snails, have more DNA in their genomes than genetically more complex animals, such as mammals. The differences are most extreme among amphibians—some salamanders have as much as 25 times more DNA per nucleus than humans. Most of these differences in DNA content result from differing proportions of highly repetitive sequences rather than

from differences in the numbers of coding and other functional sequences.

Many hypotheses have been advanced to account for the existence of highly repetitive sequences in eukaryotes. One proposal maintains that these sequences are "evolutionary junk," the products of continued mutation and duplication of once-functional sequences. Another possibility sometimes advanced is the opposite

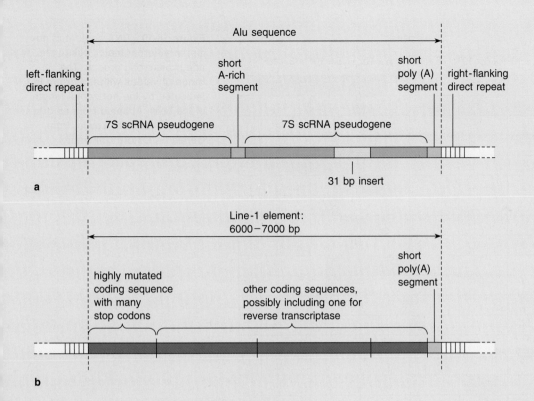

a

b

tional sequence and several apparently intact sequences that may include a reverse transcriptase. The coding sequences are followed by a short poly(A) segment at the 3' end. Few LINE repeats are complete, however. About 95% show deletions at the 5' end of greater or shorter length; in some only 30 to 40 nucleotides of the original sequence remain. Curiously, the 3' end is usually complete. Other internal substitutions, deletions, and rearrangements are common.

The short poly(A) segment at the 3' end of LINE repeats indicates that they may represent or include a processed pseudogene. Many of these elements are also bounded by short direct repeats, indicating that they can move and insert in new locations in the genome. RNA copies of LINE repeats have been detected in human cells, particularly in aberrant embryos, which implies that the coding sequences may be transcribed under some circumstances, possibly during early development.

LINE repeats occur in humans in 50,000 to 100,000 copies, enough to make up about 5% of the genome. They are widely dispersed in human chromosomes, frequently occurring between mRNA genes and within introns. They are also found as inserts within clusters of highly repeated sequences.

Two hemophiliac children were found recently by H. H. Kazazian and his coworkers to have insertions of a LINE sequence into a gene encoding a factor necessary for normal blood clotting. The insertion was not present in the parents of the children, indicating that the LINE moved into the clotting-factor gene at some time within the lifespan of these children. Thus both *Alu* and *LINE* insertions have been detected within living individuals, and are obviously ongoing processes of change in human genomes.

idea, that these sequences are maintained as evolutionary raw material, as blocks of DNA that may eventually evolve into functional genes. A third idea is that highly repetitive clusters, which appear at localized sites on almost all the chromosomes of the set in many eukaryotes, may serve as recognition sites for such processes as chromosome pairing and crossing over during meiotic cell division. Related to this idea is still another

proposal, that blocks of highly repetitive sequences act as passive spacers, positioning regions of the chromosomes where they can function most efficiently in pairing and recombination. Although these proposals stand as more or less remote possibilities, there is no experimental evidence directly supporting any of them. The only information even remotely suggesting that the sequences might be functional is the fidelity with

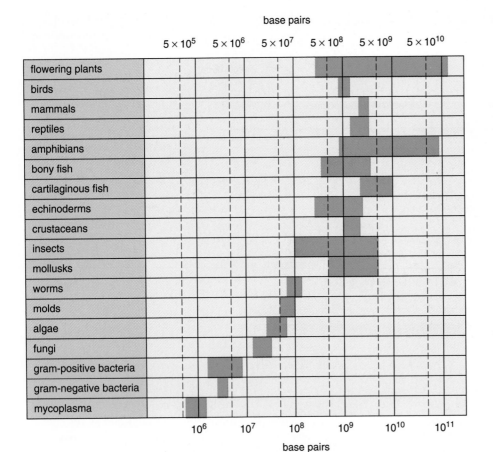

base pairs

5×10^5 5×10^6 5×10^7 5×10^8 5×10^9 5×10^{10}

flowering plants
birds
mammals
reptiles
amphibians
bony fish
cartilaginous fish
echinoderms
crustaceans
insects
mollusks
worms
molds
algae
fungi
gram-positive bacteria
gram-negative bacteria
mycoplasma

10^6 10^7 10^8 10^9 10^{10} 10^{11}

base pairs

Figure 18-10 DNA content of nuclei in different major eukaryotic and prokaryotic groups. The range of values within a group is indicated by the horizontal length of the bars. Adapted from an original courtesy of B. Lewin, from *Genes IV*, Oxford University Press, 1990.

which they are preserved within a species. The biological bases for the origin, existence, and precise maintenance of so much apparently nonfunctional DNA in eukaryotes therefore remain unknown. (Table 18-4 summarizes the characteristics of moderately and highly repeated sequences.)

The hypothetical chromosome shown in Figure 18-11 reviews the arrangement of the different sequence elements in eukaryotic genomes. Long stretches are taken up by genes encoding mRNAs. Although most mRNA genes occur in single copies, some, such as the genes coding for histones, are repeated. The genes encoding rRNAs and tRNAs are regularly repeated in eukaryotic genomes. A cluster of large pre-rRNA genes makes up the nucleolar organizer, the site on the chromosome around which a nucleolus forms. The genes for 5S rRNA are usually located in clusters at some distance from the nucleolar organizer, on the same or a different chromosome. Repeats of tRNA genes are clustered at other points. The sn/scRNA genes may be located singly or clustered. These relationships apply to a haploid equivalent. An mRNA gene occurring in a single copy in a haploid occurs in two copies in a diploid, three copies in a triploid, and so on.

Moderately repeated sequences may occur in the spacer regions between the mRNA genes and in introns within

the genes. Some of these moderately repeated elements are functional sequences, such as those of the promoter and other elements regulating transcription; others are nonfunctional sequences such as pseudogenes. Nonfunctional moderately repeated sequences may be dispersed throughout the chromosomes, placed singly or in small repeated groups between functional sequences, or clustered in groups at localized sites on the chromosomes. Highly repetitive sequences are generally clustered rather than dispersed. Frequently, clusters of highly repeated sequences are located near the centromeres and telomeres of the chromosomes. (Information Box 18-3 describes efforts under way to completely sequence and locate all elements in the human and other genomes.)

GENETIC REARRANGEMENTS

Originally, the sequences making up the genome of a species were thought to remain in place, with only very rare exceptions. However, the techniques of molecular biology have revealed that some sequence elements move from place to place or are added or deleted with much greater frequency than previously expected. Sequence elements have even been found to move be-

Table 18-4   Characteristics of Moderately Repeated and Highly Repeated Sequences

Characteristic	Moderately Repeated	Highly Repeated
Number	$\sim 10^3 - 10^5$	$10^6 - 10^7$
Transcription	Some sequences transcribed	Rarely or never transcribed
Sequence structure	Complex	Usually simple
Arrangement in genome	Dispersed and clustered	Clustered, located primarily at centromeres and telomeres
Distribution in chromatin	Both heterochromatin and euchromatin	Condensed heterochromatin
Time of replication	With coding sequences	Late
Evolutionary origin	Many as pseudogenes or transposable elements	Unknown
Function	Some functional as mRNA or accessory RNA genes; others as regulatory, transcription, or processing signals	Most or all no known function

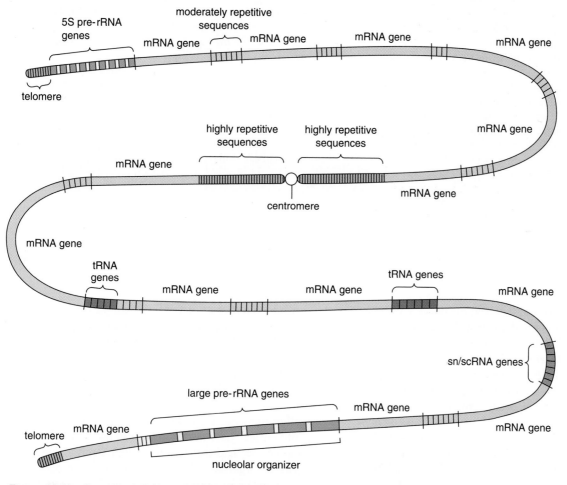

Figure 18-11   Summary of the arrangement of coding, repetitive, and other DNA sequences in a hypothetical eukaryotic chromosome. Genes encoding mRNAs occur in tandem over long stretches in the chromosome, mostly in single copies. Genes encoding rRNAs and tRNAs are repeated and occur in clusters in eukaryotic genomes. A cluster of large pre-rRNA genes forms the nucleolar organizer, around which a nucleolus forms. The genes for 5S rRNA usually cluster at some distance from the nucleolar organizer, on the same or a different chromosome. The sn/scRNA genes may be located singly or clustered. Moderately repetitive sequences may be dispersed singly throughout the chromosome, placed in small repeated groups between functional sequences, or clustered in larger groups at localized sites. Highly repetitive sequences are usually clustered around the centromeres and, in some species, near the telomeres.

Sequencing Human and Other Genomes

As yet no eukaryotic genome has been entirely sequenced. The longest genome sequenced so far, that of the cytomegalovirus, barely exceeds 200,000 base pairs. Recently completed in the laboratory of B. Barrel and his coworkers, this sequence represents only a tiny fraction of the DNA in eukaryotic genomes. However, an effort is now under way to sequence the human genome, which includes some 3.2×10^9 base pairs. Establishing the sequence will be very expensive—current estimates range from \$1.00 to \$10.00 per base pair or about \$3 billion to more than \$30 billion. If ever obtained, the complete sequence would fill the equivalent of 200 Manhattan telephone directories.

Scientists are presently divided over the advisability of sequencing the human genome. Those in favor of the project promise that obtaining the sequence would inevitably speed the search for mutations responsible for hereditary diseases and provide means for their diagnosis and treatment. The sequence could also answer fundamental questions about genome organization and the mechanisms controlling genes in development and cell differentiation. Those opposed claim that the money and effort required to sequence the human genome would be better spent in research designed to target specific problems, restricted to functional segments of the genome.

These opponents point out that, since only 3% to 5% of the human genome is functional (the remaining 95% to 97% consists of introns, repeated elements, and other nonfunctional sequences), the complete sequence would represent little more than small islands of useful information "floating in a vast sea of genetic gibberish." Whatever the pros and cons, the project is now going ahead.

A more limited project, developed by J. C. Venter and S. Brenner, is designed to avoid sequencing nonfunctional DNA in the human genome. In this technique, mRNAs are isolated from active cells and copied into cDNAs. Only these cDNAs are sequenced, so that the sequences obtained are restricted to functional ones. Hundreds of previously unknown genes have been sequenced in Venter's laboratory by this approach, using brain tissue as the source of mRNAs. (Brain tissue is ideal because as many as 60% of the mRNA-encoding genes of the genome are active in some brain cells.)

Other projects of lesser scope, including efforts to obtain the complete genome sequences of the bacteria *Escherichia coli* (4.7 million base pairs) and *Mycoplasma* (742,000 base pairs), the yeast *Saccharomyces cerevisiae* (20 million base pairs), the nematode worm *Caenorhabditis elegans* (100 million base pairs), and the plant *Arabidopsis thaliana* (70 million base pairs), are also under way.

---

tween different species or between the nucleus and the DNA circles of mitochondria and chloroplasts.

Many of these changes, known collectively as *genetic rearrangements*, are actually very rare events that appear and spread slowly among populations as part of the evolutionary developments giving rise to new species. However, some occur rapidly enough to be observed within the lifetimes of individuals. As a consequence, the genome, once considered a relatively unchanging, genetic "Rock of Gibraltar" for each species, is now known to be in a dynamic state of flux and change.

Several mechanisms provide opportunities for rapid genetic rearrangements. One source of relatively rapid change is in movable sequences known as *transposable elements*. These sequences can leave one location in the genome and insert in another, or generate copies that insert in new locations while leaving the original copy

intact. Another source of change is certain viruses, which carry sequence elements relatively rapidly to new locations in the same genome and between the genomes of different individuals. DNA transfer from individual to individual may sometimes take place among totally unrelated organisms such as bacteria, higher plants, and animals. At times DNA fragments originating from other individuals are directly taken up and integrated into nuclear genomes. Transfer also takes place between the nucleus and the two major DNA-containing cytoplasmic organelles, mitochondria and chloroplasts.

Genetic rearrangements can also arise from faulty operation of the recombination mechanism that mixes chromosome segments in new combinations during meiosis. The faulty operation, called *unequal crossing over*, produces some chromosome copies with duplicated sequences and others with deletions. Another important route of genetic rearrangement occurs through the gross breakage and random rejoining of chromosome segments. The rearrangements produced by this process, called *translocations*, have figured significantly in the evolution of some eukaryotic groups, particularly insects and higher plants. Tracing the translocation patterns in these groups has in some cases allowed evolutionary lineages to be reconstructed.

Genetic rearrangements are detected by several methods, including direct sequencing, changes in the DNA fragment classes produced by restriction endonucleases (see p. 135), the appearance of spontaneous mutations in genetic lines, and direct observation of alterations in chromosomes under the light or electron microscopes. Some genetic rearrangements, including movement of transposable elements and viral infections, take place commonly in prokaryotes as well as eukaryotes (see Supplement 18-1 for details).

The genetic rearrangements resulting from these mechanisms are of more than academic interest. Translocations have been linked to the transformation of normal cells into malignant tumors in humans; the movement of transposable elements has also been identified as a step in the development of some forms of cancer. The harnessing of rearrangement mechanisms, particularly those of transposable elements, shows promise as a means to change the genome of organisms at will. Purposeful rearrangements of this type, called *genetic engineering*, may improve domestic plants and animals and correct human genetic disabilities. (Genetic engineering is discussed in Supplement 18-2.)

Genetic Rearrangements by Movement of Transposable Elements

The first indication that individual sequence elements can move from place to place in either prokaryotic or eukaryotic genomes dates back to research with maize conducted in the 1940s by Barbara McClintock. McClintock noted that some mutations affecting kernel and leaf color in maize appeared and disappeared spontaneously and rapidly under certain conditions, and that "controlling elements" affecting these mutations appeared to move from place to place in the genome. Some of the mutations and movements of controlling elements occurred so rapidly that changes in their effects could be noticed at different developmental times in the same individual or in a single developing kernel (Fig. 18-12). McClintock reported her findings in the late 1940s and early 1950s, but her conclusions were not widely accepted because most geneticists and molecular biologists were not prepared to accept the idea that sequence elements could move from place to place in DNA.

These attitudes began to change in the 1960s, when transposable elements were detected by molecular bi-

Figure 18-12 Maize kernels showing different color patterns as a result of mutations arising from the movement of transposable elements.

ologists working with bacteria. In the 1970s, examples were discovered in yeast and higher eukaryotes including mammals. McClintock was belatedly awarded the Nobel Prize in 1983 for her pioneering work with transposable elements, after these further discoveries confirmed her early findings and showed that movable genetic elements are probably universally distributed among prokaryotes and eukaryotes.

Classes of Eukaryotic Transposable Elements The transposable elements of eukaryotes, although highly variable in internal structure, fall roughly into two classes with characteristics that depend on the mechanism by which they move from place to place in the genome. Members of the first class, called *transposons*, move by a mechanism that cuts or excises the element from its original location in genomic DNA and inserts it in a new one. Most transposons are delineated by a short sequence that is repeated in inverted form at either end of the element. A central nonrepeated sequence bounded by the terminal repeats usually includes the code for at least one protein, a *transposase* involved in the reaction cutting the transposon from one location in the genome and inserting it in a new one. Most transposases have the ability to insert the transposon at essentially any location in the genome. Some, however, are restricted to recognition and insertion at certain sequence elements.

The second class, known as *retrotransposons* or *retroposons*, moves to new locations via an RNA intermediate. The retroposon is transcribed into an RNA copy that is subsequently used to produce a cDNA copy by reverse transcriptase, along the lines shown in Figure 18-2. The cDNA copy is then inserted into the genome at a new location, leaving the original copy undisturbed and in place. Each round of movement increases the number of retroposons in the genome.

Retroposons occur in three forms—*viral, virallike,* and *nonviral.* Viral forms are completely functional viruses capable of producing free, infective viral particles; virallike forms are closely similar to viral forms in structure in the genome but are not known to form infective particles. The viral and virallike retroposons are bounded by sequence elements that contain both direct and inverted segments. These terminal repeats enclose a central nonrepeated segment that includes coding sequences for viral coat proteins and for reverse transcriptase. Host cell genes in more or less complete form may also become trapped in the central region of these retroposons. Nonviral retroposons lack ter-

minal repeats but often include at least an internal sequence encoding a reverse transcriptase. Many non-viral retroposons have a short poly(A) sequence at their 3′ ends, indicating their possible origins as processed pseudogenes.

Insertion of Transposable Elements into the Genome Most transposons and retroposons insert into genomes by creating a *staggered break* in the host DNA (Fig. 18-13). The host DNA is opened in such a way that a short, single-chain segment extends from both free ends (Fig. 18-13a). During insertion of the transposable element, the gaps opposite the single chains are filled in by DNA replication (Fig. 18-13b and c). As a result, the base pairs originally included within the staggered break are duplicated as a short direct

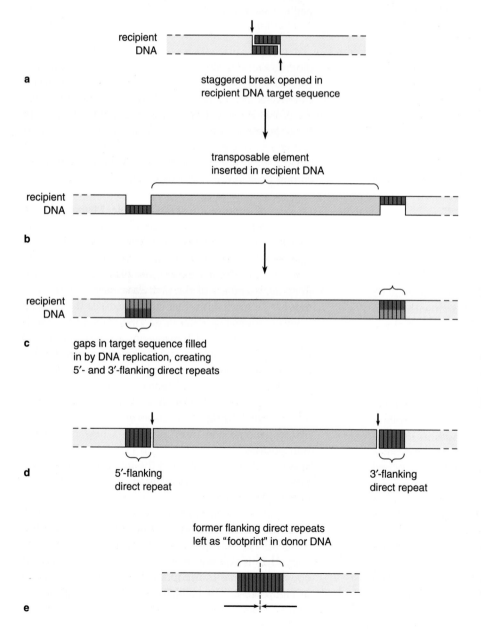

Figure 18-13 Insertion of transposable elements at staggered breaks in the recipient DNA. **(a)** A staggered break is opened in the recipient DNA molecule. **(b)** The transposable element is inserted at the staggered break, and **(c)** the single-stranded gaps in the receiving DNA are filled in. The filling-in creates 5′- and 3′-flanking direct repeats at the boundaries of the transposable element in its new location. **(d)** Removal of a transposable element occurs by cuts made by a transposase at either end of the element (arrows). **(e)** A "footprint" consisting of the former 5′- and 3′-flanking direct repeats is left at the site from which the transposable element has been removed.

repeat at both ends of the inserted element. This mechanism accounts for the fact that transposable elements are often bracketed by short direct repeats. Because most transposable elements insert at essentially any site in the DNA, the short direct repeats derived from the host differ in sequence at each insertion site. Transposable elements that are clipped from one site and moved to another usually leave the direct repeat as a "scar" or "footprint" marking their former location (Fig. 18-13d and e).

Both transposons and retroposons are detected primarily through DNA sequencing, or through their effects on genes at or near their sites of insertion. When inserted within the coding segment of a gene, a transposable element usually destroys the functional activity of the protein encoded by the gene. Insertion within the 5′-flanking regions of a gene frequently destroys or at least inhibits transcription of the gene. At times, however, insertion of a transposable element has the opposite effect—transcriptional activity of the gene is increased, particularly if the movable element blocks inhibitory sequence elements or happens to have picked up the control sequences of a highly active gene during one of its previous insertions. At times excision of a transposon leaves unsealed openings in the DNA, producing chromosome breaks that may lead to translocations of chromosome segments.

Even when a transposon is excised without leaving breaks in the DNA, the tandemly repeated sequence left as a footprint may interfere with genetic activity through the interruption of 5′-flanking control sequences or the introduction of frameshift mutations (see p. 663) in the coding segment of genes. Entire genes of the host cell DNA may become trapped within the central region of a transposable element and may move with the element to a new location during transposition. Such a trapped host gene may become continuously active through the effects of the promoter sequences in the element. Host cell genes may also become abnormally active if moved in a transposable element to the vicinity of an enhancer or promoter of an intensely transcribed gene.

Once inserted in the genome, transposable elements become more or less permanent residents, duplicated and passed on during cell division along with the rest of the DNA. Transposable elements that insert into reproductive cells may be inherited, thereby becoming a permanent part of the genetic makeup of the species.

Long-standing transposable elements are subject to mutation along with other sequences in the DNA. Such mutations may accumulate in the element, gradually altering it into a nonmobile, residual sequence in the DNA. The genomes of many eukaryotes are littered with the evolutionary junk of nonfunctional transposable elements created in this way.

Transposable elements are currently being exploited as a means to introduce desired genes into organisms or to correct genetic defects. Some successes have already been obtained with this technique, and there are indications that transposable elements will provide the key to the genetic engineering of many species (see Supplement 18-2).

Examples of Eukaryotic Transposable Elements

Ty Elements in Yeast The primary transposable elements of yeast cells, called *Ty elements*, are virallike retroposons that include more than 6000 base pairs of DNA when complete (Fig. 18-14a). A directly repeated sequence occurs at both ends of a Ty element. Enclosed by the direct repeats is a central nonrepeated region that includes codes for at least two proteins, one of which is probably a reverse transcriptase.

Most yeast cells contain about 30 to 35 complete Ty elements and as many as 100 or more partial elements that include only the directly repeated boundary sequences. Ty elements move to new locations in yeast cells at the rate of about one insertion for every 10^8 individuals. At this rate movements of Ty elements can easily be detected in large yeast cultures over periods as short as days or weeks.

Ty elements provided one of the main lines of evidence supporting the conclusion that retroposons move to new locations via an RNA intermediate. In a definitive experiment J. D. Boeke, D. J. Garfield, and C. A. Styles introduced an mRNA intron into the DNA of a Ty element. In the new locations at which the element appeared, the intron was missing. Because the enzymes and factors removing introns work only on RNA molecules, the processing reactions splicing out the intron could have taken place only while the Ty element was in the form of an RNA transcript.

Copia* and P Elements in *Drosophila Transposable elements form a major part of the *Drosophila* genome, probably as much as 10% or more. Most fall into a group called the *copia* or *copialike elements*. (The term *copia* reflects the abundance of these transposons in *Drosophila*.) Some 30 families of *copia* elements occur in *Drosophila*.

Individual *copia* elements, which may total as many as 3000 or more per haploid nucleus, are virallike retroposons that follow the structural pattern shown in Figure 18-14b. The elements, about 5000 to 8000 base pairs in total length, are flanked at each end by a short, host-derived sequence that is directly repeated at both ends of the element. Just inside the flanking direct repeats is a combination of inverted and direct repeats that occurs at both ends of the retroposon; these repeated sequences mark the boundaries of the *copia* ele-

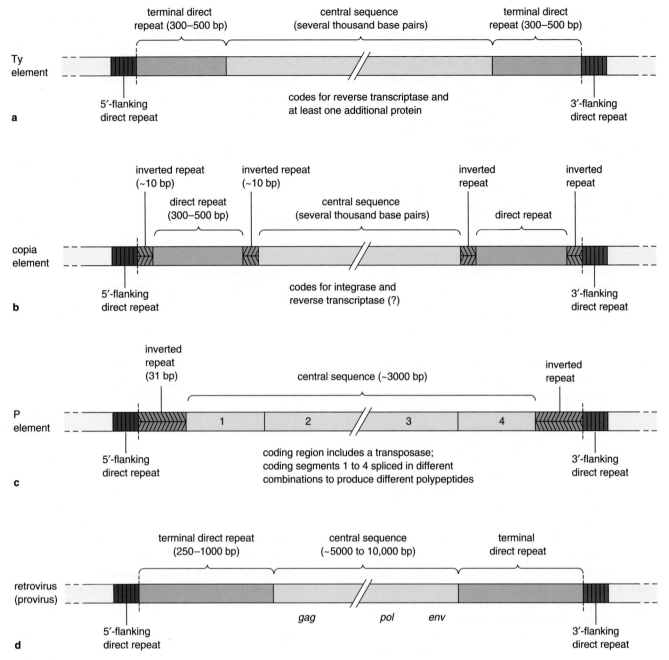

Figure 18-14 Some representative transposable elements. **(a)** Ty element of yeast. **(b)** *Copia* element of *Drosophila*. **(c)** P element of *Drosophila*. **(d)** Mammalian retrovirus in the proviral form in which it is inserted in the DNA. Vertical dashed lines mark the outer boundaries of each transposable element; the 5'- and 3'-flanking direct repeats outside the vertical dotted lines are derived from the host cell DNA.

ment. The terminal repeats enclose a central sequence that in most *copia* elements includes several thousand base pairs. In one *copia* element recently sequenced by S. M. Mount and G. M. Rubin, the 4227 base-pair central element contains codes for several polypeptides. Among these may be a reverse transcriptase and an *integrase*, an enzyme that catalyzes insertion of a newly formed cDNA copy into the genome.

Copia elements move to new locations in the genome with a frequency of about one detectable translocation in every 10^4 to 10^5 individuals. The movement is so frequent that all *Drosophila* strains examined to date have distinct *copia* families at different sites in the genome.

Drosophila also includes transposons known as P elements (Fig. 18-14c). These transposons are structured with inverted boundary repeats enclosing a central coding region. The central region includes a code for a transposase that catalyzes excision of the elements from one site and insertion in another. Some *D. melanogaster* strains include as many as 50 P elements per individual; others have none. Those with P elements are called *P strains;* those without are *M strains.*

A curious effect of P elements is that males of P strains, if mated with M-strain females, rarely produce fertile offspring. The combination of paternal chromosomes with P elements and maternal chromosomes with none in the P/M hybrids causes a burst of excisions and movements of the P elements. Many of the excisions lead to chromosome breaks and translocations that produce infertile sperm and eggs in the hybrids. The reverse cross, M males with P females, is not characterized by extensive movement of P elements. This is evidently because a factor in the cytoplasm of eggs derived from an M female is required for the burst of P element excision and movement. The factor, whose identity is unknown, is not present in the eggs of P females.

Two features of the present distribution suggest that the P elements of *D. melanogaster* may have been introduced from another *Drosophila* species, possibly within the last 50 years. One is that older laboratory stocks of *D. melanogaster* maintained in continuous isolation since before 1950 are free of P elements. The second is that the present worldwide distribution of P elements is uneven among wild *D. melanogaster* strains, as if the elements have not had enough time to permeate the entire global population. M. Houck and M. Kidwell have suggested that a possible source is *D. willistoni*, which also has P elements. Even though the two species separated millions of years ago and do not interbreed, the P elements in *D. willistoni* are essentially the same as those in *D. melanogaster*. If P elements were inherited from a common ancestor when *melanogaster* and *willistoni* separated millions of years ago, the elements would likely have mutated into significantly different forms in the two species. Houck and Kidwell suggest instead that parasitic mites feeding on both species may have transferred DNA containing P elements from *willistoni* to *melanogaster* about 50 years ago. In support of their hypothesis is the observation that one such mite, *Proctolaelaps regalis*, actually picks up and retains DNA containing P elements when feeding on the flies.

The Retroviruses of Mammals and Other Vertebrates

The primary transposable elements of mammals and other vertebrates are viral retroposons derived from a group of infecting viruses called *retroviruses*. These viruses range from harmless residents of the genome to some that can be highly detrimental.

A retrovirus, known as a *provirus* when inserted in the DNA of a host cell, consists of a set of terminal direct repeats enclosing a central nonrepeated sequence (Fig. 18-14*d*). The terminal direct repeats include sequences capable of acting as enhancer, promoter, and termination signals for transcription. The central sequence, which includes from about 5000 to 10,000 base pairs in different retroviruses, usually contains three coding regions. Located closest to the 5' end of the central sequence is the *gag* region, which encodes four proteins that associate with RNA in the core of a retrovirus in its free, infective form outside the host cell. Occupying the approximate center of the central sequence is the *pol* region, which contains codes for a reverse transcriptase and an integrase catalyzing insertion of the retrovirus in new locations. At the 3' end of the central sequence is the *env* region, which contains codes for glycoproteins forming the coat of free retroviral particles. As in other transposable elements, a retrovirus is flanked by short direct repeats derived from the host DNA when inserted into the genome.

A retroviral provirus is transcribed into an RNA copy that is subsequently processed by addition of a poly(A) tail. Transcription begins at the promoter region in the 5' terminal repeat and ends at the termination signal in the 3' terminal repeat. The promoter sequence in the 3' terminal repeat and the terminator in the 5' repeat, for unknown reasons, remain inactive and have no effect on transcription. The RNA copy of the provirus may be used as a template for a cDNA copy, which is inserted into the genome in a new location. Movement of this type is relatively rare, however. The RNA copy may also be used by ribosomes as the directions for making the viral core and coat proteins, which lead to production of viral particles that can infect new individuals. Most new retroviral insertions originate by this route.

A free retroviral particle contains two copies of the viral genome, packed into the virus in the form of two single-stranded RNA molecules. The two RNA copies contain the same genes in the same order but may carry different forms (alleles) of the retroviral genes. On infection the RNA molecules of the retrovirus are released into the host cell, along with reverse transcriptase molecules that are included in the viral coat. The reverse transcriptase makes a double-stranded cDNA copy of either of the two RNA molecules of the virus, which is subsequently inserted in the genome by the integrase.

Essentially all humans and most other mammals contain from 1 to as many as 100 or more integrated retroviruses. In mice, where the most detailed investigations have been carried out, new insertions of retroviral elements appear on the average of about 1 in every 30 generations. Most of these originate from new cycles of infection.

15. What genetic effects result from the movement of transposable elements to new locations? What is the possible relationship between transposable elements and pseudogenes?

16. Outline the life cycle of a retrovirus. What is the relationship of retroviruses to cancer? How are retroviruses used in genetic engineering?

17. What is unequal crossing over? What is the relationship of this mechanism to the development of nonprocessed pseudogenes? How might introns and repetitive sequences function in the production of unequal crossovers?

18. What are chromosome translocations? What is the relationship of this mechanism to cancer?

19. What is the relationship between the crown gall tumor and genetic engineering in plants? How is the organism responsible for crown gall tumors used to introduce genes into plants?

20. What evidence indicates that DNA sequences can move between organelles and the nuclear genome?

Supplement 18-1

Genome Structure and Genetic Rearrangements in Prokaryotes

Prokaryotic Genomes

The genomes of prokaryotes are organized in patterns that are fundamentally different from eukaryotic genomes. A prokaryotic nucleoid contains only a fraction of the DNA typical of eukaryotic cells—*E. coli* nucleoids, for example, contain only about 1500 μm of DNA, as compared to the centimeters or even meters of DNA in a typical eukaryotic nucleus. Although the nucleoids of cyanobacteria contain somewhat larger quantities of DNA, up to about two and one-half times more than bacteria, the DNA content per cyanobacterial cell is still only a fraction of the DNA complement of eukaryotic cells.

Bacterial nucleoids also differ from their eukaryotic counterparts in containing a single DNA molecule, arranged as a covalently closed, continuous circle. As a consequence, all the genes of a bacterial genome are genetically linked. In eukaryotes, in contrast, the genome is divided among several to many DNA molecules in which the genes are organized in separate linkage groups.

The diminutive size of prokaryotic genomes is reflected in a general economy of sequence usage in these organisms. Almost all the DNA directly makes up parts of genes or forms the 5' or 3' control elements regulating transcription. Several to many genes are grouped together as operons controlled by common 5'-flanking elements. Furthermore, very few nonfunctional repetitive sequences can be detected in bacteria. Introns, too, are uncommon—they have been detected only in a limited number of genes in a few bacterial groups (see p. 646). This economy in sequence usage differs fundamentally from the arrangement of sequences in most eukaryotes, in which functional elements make up a relatively minor fraction of the genome. The absence of superfluous sequences in prokaryotes may reflect the small size of bacterial and cyanobacterial cells, which have limited space for housing a genome. The time available for DNA replication is also limited in the rapid division cycles and short generation times characteristic of these organisms.

Another peculiarity of prokaryotic systems is the presence of *plasmids*, additional DNA circles suspended in the cytoplasm outside the nucleoid. Plasmids range from very small circles containing only a few hundred nucleotides to larger ones including many thousands of base pairs. Plasmid circles contain replication origins, so they are duplicated and passed on during cell division. Genes originating from the same or a different prokaryotic species, or from viruses infecting bacteria, may be included in plasmids. Bacterial plasmids frequently also contain transposons. In addition to their natural functions, plasmids are widely used to introduce DNA sequences into bacteria for cloning (see Ch. 4).

The Arrangement of Coding Sequences Genes encoding mRNAs are distributed throughout the bacterial DNA circle. The arrangement of prokaryotic mRNA genes in operons including the codes for one to several polypeptides, controlled by the same promoter and transcribed as a unit into a single mRNA transcript (see Fig. 15-51), is fundamentally different from the disposition of mRNA genes in eukaryotic genomes. Because prokaryotic mRNA genes have no introns, mRNA transcripts require little or no processing in these organisms.

The mRNA genes of prokaryotes are packed into the genome with little or no space separating the flanking control regions of one operon from the next or between individual coding segments within operons. Some coding segments within operons even overlap so that nucleotides do double duty as parts of two adjacent genes.

Ribosomal RNA operons occur in multiple copies in bacteria, ranging from 2 or 3 up to a maximum of about 15 repeats. The rRNA genes may be arranged singly or distributed among several small clusters. The seven rRNA operons of *E. coli*, structured internally as shown in Figure 15-54*b*, are distributed singly at scattered points around the genome. Each rRNA operon includes the code for 5S rRNA as well as the 16S and 23S coding sequences. The rRNA operons of bacteria also contain from one to three tRNA sequences, some placed between the 16S and 23S codes and in some rRNA operons at the 3' end of the gene (see Fig. 15-54*a*).

Other tRNA genes are scattered around the genome. In *E. coli* 46 different tRNA types are encoded in 77 coding sequences located singly, in clusters, and in mixed operons containing both tRNA and protein-encoding sequences. Some tRNA genes of *E. coli* occur in multiple copies ranging from two to a maximum of seven, arranged in the same or different clusters.

Repeated Sequences Although repetitive sequences are limited in prokaryotes, some sequences in this category have been detected in a few species. *E. coli* and *Salmonella* contain a number of different moderately repeated sequence elements dispersed around the genome, ranging from about 4 to 40 base pairs in length. Among these are functional sequences serving as signals for genetic recombination and integration or excision of DNA segments. For example, an 8 base-pair element known as the *Chi sequence* is located at more than 900 sites around the *E. coli* DNA circle. This sequence is recognized by enzymes catalyzing recombination. Another repeated element of unknown function, an inverted sequence about 25 to 30 base pairs long, occurs

at 100 to 200 sites in the *E. coli* genome. These elements have been suggested as binding sites for proteins that tie the *E. coli* DNA circle into looplike domains (see p. 568).

Repeated sequences have also been reported in archaebacteria. This interesting bacterial group is one that also contains introns, giving it two characteristics that are more typical of eukaryotes than prokaryotes. Other eukaryotic characteristics, such as some eukaryotelike sequence patterns in rRNA genes and part of the mechanism assembling proteins on ribosomes, have been noted in these bacteria. The several eukaryotelike characteristics of archaebacteria have led some investigators to propose that this group may represent the survivors of an offshoot of the prokaryotic evolutionary tree that gave rise to eukaryotes or perhaps a remnant of the most ancient cells of all, those that gave rise to both prokaryotes and eukaryotes. (For details of archaebacteria and their possible evolutionary relationships, see Ch. 27.)

Almost all the information concerning the arrangement of coding and other sequences in prokaryotes has been derived from studies in bacteria, much of it from a single bacterium, *E. coli*. Therefore, it is not certain whether the other major prokaryotic group, the cyanobacteria, shares the same patterns of sequence distribution or even whether cyanobacterial genomes are circular as in bacteria. Few mRNA genes have been mapped in cyanobacteria, and the numbers and disposition of rRNA and tRNA genes in these organisms are mostly unknown. Plasmids have been detected in cyanobacteria, however, and studies tracing the rate at which DNA rewinds after dissociation indicate that there are few repeated sequences. These and the generally prokaryotic characteristics of cyanobacteria suggest that the arrangement of genes and other sequence elements in these organisms is probably similar to that

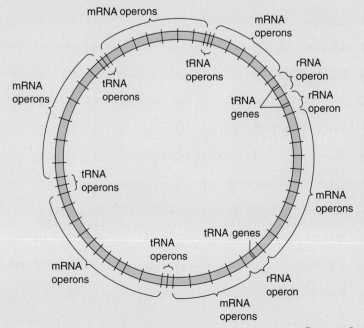

Figure 18-17 A summary of the arrangement of coding sequences in prokaryotes. Located at points around the circle are rRNA operons containing coding sequences for 16S, 23S, and 5S rRNAs and some tRNAs. Other tRNAs are in tRNA operons at other locations. Some of the tRNA operons are "pure" operons containing only tRNAs; some are mixed operons containing both tRNAs and protein-encoding sequences. Other segments are filled in with mRNA operons. The mRNA operons may be continuous, with no intervening spacers, or may be separated by small amounts of spacer DNA. Repetition of sequences is confined to rRNA and tRNA genes and a relatively small amount of repetitive DNA. Distribution of sequences in bacteria and cyanobacteria is expected to be similar.

of bacteria. Figure 18-17 summarizes the arrangement of coding sequences in the circular DNA molecules of bacterial and presumably cyanobacterial genomes.

Genetic Rearrangements in Bacteria

Prokaryotic genomes are even more subject to genetic rearrangements than those of eukaryotes. Sequences are exchanged by recombination among the nucleoid, plasmids, and infecting viruses. Transposable elements also insert at a low but constant frequency in both the nucleoid and plasmids. In addition, bacteria at times take up and integrate DNA segments from the surrounding medium that originate from the breakdown of other cells. The more or less constant movement of sequence elements to and from the nucleoid and plasmids provides an important source of variability that is vital to the survival of prokaryotic species in their constantly changing environments.

Transposable Elements   Transposable elements were first detected in bacteria by H. Saedler, P. Starlinger, and J. A. Shapiro and others in the late 1960s, a little more than a decade after McClintock discovered movable sequences in maize. Most bacterial transposable elements (see Fig. 18-18) are bounded by a repeated sequence that is inverted at the two ends of the element. The terminal repeats enclose a central nonrepeated sequence usually encoding one or more proteins. At least some of the encoded proteins are enzymes that catalyze movement of the sequence elements. Like their eukaryotic counterparts, bacterial transposable elements are bracketed by a short direct repeat, indicating that insertion takes place at a staggered break in the bacterial host DNA.

Bacterial transposable elements fall into the transposon rather than retroposon class (see p. 763). However, rather than being excised from one site and inserted in another, bacterial transposons move primarily by a process that involves replication of the original element and insertion of the replicated copy in the new location. The mechanism leaves the original element in place as well as inserting a copy in a new location. Excisions also occur but evidently by a mechanism not directly connected with movement. The precise molecular mechanism underlying the primary bacterial transposon movement, called *replicative transposition*, is unknown. However, several models, each supported to some extent by experimental evidence, have been advanced (see below).

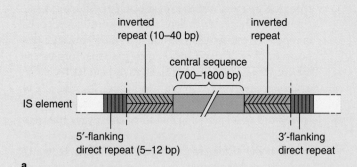

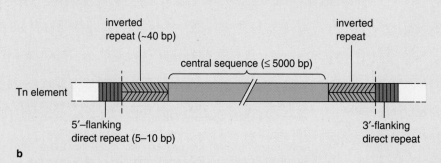

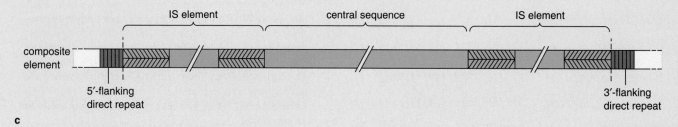

Figure 18-18   Bacterial transposable elements. **(a)** IS element; **(b)** Tn element; and **(c)** composite element, consisting of two complete IS elements enclosing a central sequence.

Bacterial transposons appear at a new location about once in 10^5 to 10^7 individuals, at rates comparable to the mutation rate in bacteria. Excisions occur much more slowly, about once in 10^6 to 10^{10} individuals. These rates would suggest that transposons are constantly on the increase in bacteria and that eventually most of the DNA of bacterial genomes would be occupied by movable sequences. However, the maximum number of transposons is limited to about 1% of the genome in most bacterial species. In at least some bacterial transposons, the limit is maintained by a repressor protein encoded in the transposon, which inhibits transcription of the genes encoding enzymes necessary for further transpositions.

Three classes of transposons, each with distinct properties, have been detected in bacteria (Fig. 18-18). One, *IS elements* (IS = *Insertion Sequence*), contains the minimum sequence structure required for movement in bacteria: terminal inverted sequence repeats enclosing a central nonrepetitive sequence, in which the central sequence includes one or more genes that encode enzymes catalyzing replicative transposition. The second class of transposons, *Tn elements* (*Tn* = *Transposition*), are structured similarly to IS elements except that the central nonrepetitive sequence is longer and often includes genes originating from the host DNA that are unrelated to transposition of the elements. The third class of bacterial transposons, *composite elements*, consists of a combination of IS and Tn sequences. The boundaries of a composite element are formed by complete IS elements, one at each end. The bracketing IS elements contain codes for the proteins necessary for transposition of composite elements. The central sequence usually includes only host cell genes. Composite elements are often designated as Tn elements, with only the identifying number distinguishing them from "pure" Tn elements. For example, the much-studied *Tn3* element is a pure Tn element, while another transposable sequence, *Tn10*, is a composite element.

The three kinds of transposable elements may insert in plasmids as well as the main genome of a bacterial cell. Their insertion in plasmids facilitates the transfer of the elements to other cells and species, since plasmids are frequently passed from cell to cell among prokaryotes, and even sometimes from prokaryotes to eukaryotes (as, for example, in the Ti plasmids transferred from *Agrobacterium* to higher plants).

IS Elements IS elements are so common in bacteria that they can almost be regarded as regular components of the genome. These elements, which usually number between 5 and 20 per cell, are relatively small, with central elements that include 700 to 1800 base pairs (see Fig. 18-18*a*). Mutations in the inverted boundary sequences of an IS element inhibit or interfere completely with the appearance of the element in a new location, indicating that the repeats contain signals recognized by the enzymes causing translocation.

The central sequence of an IS element usually encodes from one to three polypeptides concerned with insertion of the transposon in new locations. Where more than one polypeptide is encoded in the central sequence, the coding sequences characteristically overlap, so that a single DNA sequence codes for parts of two different polypeptides. In most IS elements with overlapping codes, the entire length of one of the two DNA chains in the central region is occupied by the code for one polypeptide. The opposite nucleotide chain encodes one or more smaller polypeptides within the same DNA length. Since the two chains of a DNA helix are antiparallel (see p. 527), the 5' and 3' ends are reversed and the coding sequences read in opposite directions on the two chains (Fig. 18-19).

Several models have been advanced for the mechanism by which IS elements appear in new locations in the genome or plasmids. Although the models differ in detail, almost all involve: (1) replication of the original IS element, and (2) recombination, in which breaks are

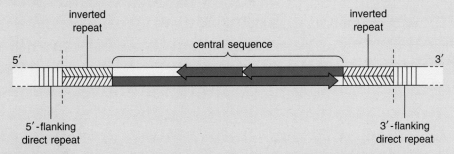

Figure 18-19 Overlapping protein codes in the central sequence of IS elements. In most IS elements one large polypeptide is encoded in one of the two nucleotide chains (bottom arrow). The opposite chain contains codes for one or more smaller polypeptides (top arrows), reading in the opposite direction within the same length of DNA. The encoded polypeptides include a transposase and other polypeptides required for movement of IS elements from one location to another in the genome.

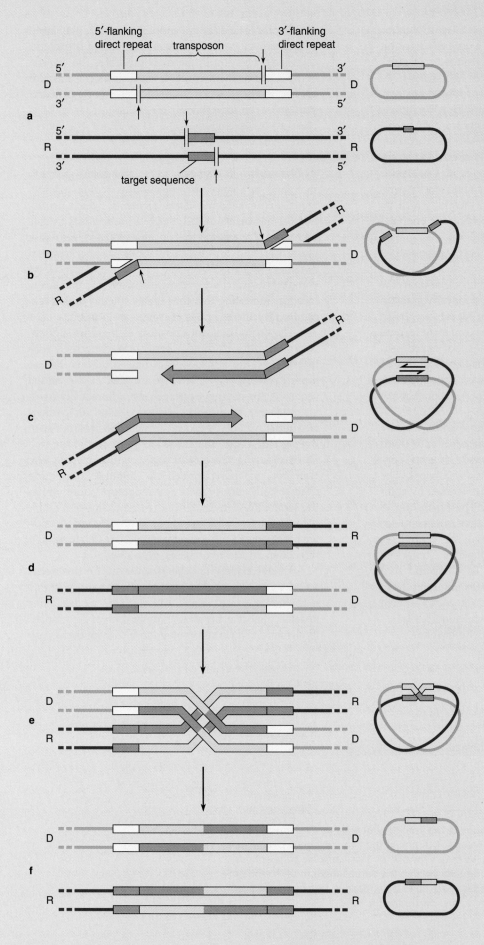

Figure 18-20 A possible mechanism for replicative transposition of transposons in bacteria. **(a)** The donor DNA molecule D containing the transposon, and the recipient DNA molecule R. Both nucleotide chains of each DNA molecule are shown. As a first step, breaks appear in the donor and recipient DNA molecules at the points indicated by arrows. **(b)** The recipient DNA molecule is linked to the donor DNA at the points indicated by arrows. **(c)** Unwinding and replication of the transposon. The large arrows indicate the replicating nucleotide chains and the direction of replication. **(d)** Completion of replication. **(e)** Crossing over or recombination between the replicated transposons. **(f)** Completion of recombination. The transposon is now present in the recipient DNA molecule as well as the donor, and both DNA molecules are completely separate. The small diagrams to the right of each step show the consequences of the replication and recombination mechanisms if the donor and recipient molecules are circular, as they would be if movement of the transposon takes place between the bacterial genome and a plasmid.

introduced and the free ends generated by the breaks are exchanged and rejoined. (Fig. 18-20 shows one of the transposition models.) Depending on the model, replication may precede or follow recombination.

Insertion of an IS element generates a 5 to 12 base-pair direct repeat originating from the host DNA. The flanking repeat varies in degree of conservation depending on the particular family of IS elements. In some the repeat is not conserved, indicating that the element inserts randomly with no preference for any particular host sequence at the point of insertion. Other IS elements are characteristically surrounded by more or less conserved direct repeats, indicating that the enzymatic mechanisms inserting them recognize particular, specific host sequences as hotspots for the insertion site.

Once inserted at two or more locations in the genome or in plasmids of an individual cell, IS elements of the same family form sites where DNA of two different regions may pair, providing further opportunities for DNA rearrangement by recombination. The breakage and exchange of DNA segments resulting from recombination between IS elements may move entire segments of the genome from one location to another, may reverse the order of genes, or may exchange sequences between the genome and plasmids. (Fig. 18-21 shows movement of sequences from a plasmid to the genome by recombination between IS elements.) Recombination between IS elements and between repeats of other transposons, in fact, is the primary mechanism by which genes move between plasmids and the main genome in bacteria.

Tn Elements   Bacterial Tn elements (Fig. 18-18b) may contain up to about 5000 base pairs of DNA, more than twice as much as the longest IS elements. The central sequence enclosed by the inverted repeats contains several nonoverlapping genes that code for enzymes required for movement of the elements. Among the proteins encoded in Tn elements are usually a transposase concerned with the appearance of Tn elements in new locations and a *resolvase*, an enzyme that catalyzes part of the recombination phase of transposition.

The central coding sequence of Tn elements also frequently contains one or more host cell genes, often including some that provide drug resistance. Many drugs such as penicillin, erythromycin, tetracycline, ampicillin, and streptomycin, originally highly effective against bacterial pathogens, have been greatly reduced in their effects by the spread among bacteria of resistance genes carried in Tn elements. Most often, the resistance-conferring genes encode enzymes that recognize and catalyze the breakdown or neutralization of the antibiotics. The genes have made diseases caused by more than 50 different genera of bacteria either untreatable or highly resistant to the standard antibiotics.

Tn elements transpose by mechanisms similar to those outlined in Figure 18-20 for IS elements. Like IS elements, insertion of a Tn element generates a short direct repeat, originating from the host DNA, that brackets the inserted element. The target sequences for transposition, depending on the particular Tn family, vary from random to highly specific. Because Tn ele-

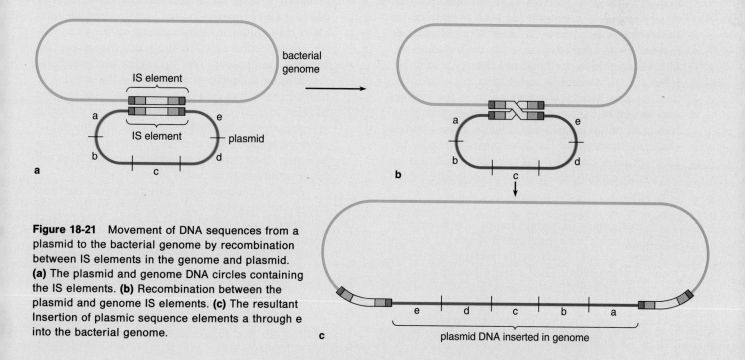

Figure 18-21 Movement of DNA sequences from a plasmid to the bacterial genome by recombination between IS elements in the genome and plasmid. **(a)** The plasmid and genome DNA circles containing the IS elements. **(b)** Recombination between the plasmid and genome IS elements. **(c)** The resultant Insertion of plasmic sequence elements a through e into the bacterial genome.

ments insert in plasmids as well as the main bacterial genome, they can move readily by recombination between plasmids and the genome, and between different individuals and species by processes such as that shown in Figure 18-21. Recombination greatly increases the spread of host cell genes carried by Tn elements, including those conferring antibiotic resistance.

Composite Elements   The structural arrangement of composite elements (see Fig. 18-18c), consisting of two complete IS elements surrounding a central sequence, makes these elements the longest and most complex bacterial transposons. One much-studied composite element, *Tn10*, for example, is formed by two IS elements, each 1400 base pairs long, enclosing a central sequence some 6500 base pairs in length. Genes necessary for transposition of composite elements are carried in the bracketing IS elements. The central sequence usually carries one or more host cell genes. Among these, as in Tn elements, may be genes conferring drug resistance.

Composite elements appear in new locations in the genome and plasmids by mechanisms similar to those of IS elements. Insertions are bracketed, as in other transposons, by a short direct repeat originating from the host DNA target sequence. The presence of the IS sequences at each end of a composite element provides additional opportunities for pairing and recombination between the two ends of the element or between composites and single IS elements elsewhere in the genome or in plasmids. Such recombination events greatly add to the spectrum of genetic rearrangements generated by composite elements.

Composite elements probably evolved through the chance inclusion of host cell sequences between two closely spaced IS elements. Supporting evidence for this evolutionary pathway is taken from observations that two closely spaced IS sequences can transpose any DNA sequences trapped between them. For example, when a Tn10 element is inserted into a plasmid (Fig. 18-22a), new insertions generated from the plasmid may consist of the usual Tn10 element, excluding the remainder of the plasmid (Fig. 18-22b), or an alternate arrangement in which the IS elements transpose the DNA of the plasmid located between them, leaving the former central sequence of the composite behind (Fig. 18-22c).

Transpositions of Bacterial Viruses   Several viruses infecting bacterial cells insert into the bacterial genome and move to new locations by patterns resembling those of transposons. Their activity greatly increases the variety of genetic disturbances and the variability generated in bacterial cells by sequence transposition. Among the most thoroughly investigated of these viruses is bacteriophage *Mu*, which infects *E. coli*. Mu is

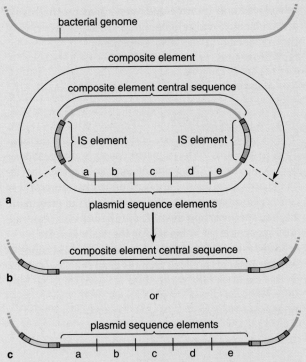

Figure 18-22   Alternate patterns of sequence transposition by a composite element. **(a)** A plasmid containing a composite element. **(b)** Movement into the bacterial genome of the central sequence contained within the element during transposition, excluding the plasmid sequence elements. **(c)** Movement instead of plasmid sequence elements a through e as the sequence enclosed between the terminal IS elements of the composite element. This alternate pattern excludes the central sequence formerly included in the composite element.

a relatively large virus that encodes more than 20 proteins. Once inserted in the *E. coli* genome, the viral DNA may appear in new locations by the same replicative mechanisms used by bacterial transposons. Like bacterial transposons, Mu may generate mutations, deletions, and insertions, and modify the activity of host cell genes at or near its points of insertion. Inserted Mu elements differ from bacterial transposons in lacking inverted repeats at their boundaries. Mu resembles other transposons, however, in generating a short, five base-pair direct repeat originating from the host cell target sequence at its points of insertion.

Another much-studied bacterial virus, bacteriophage *lambda*, enters and leaves sites in the *E. coli* genome by a nonreplicative form of recombination that does not leave a copy at the original site. The genome of the free, infective form of the lambda virus consists of a large, 47,000 base-pair piece of double-stranded DNA. Although no inverted repeats occur at the ends of the viral DNA molecule, the viral DNA has single-stranded, "sticky" ends with complementary sequences allowing the two ends to pair and form a closed circle.

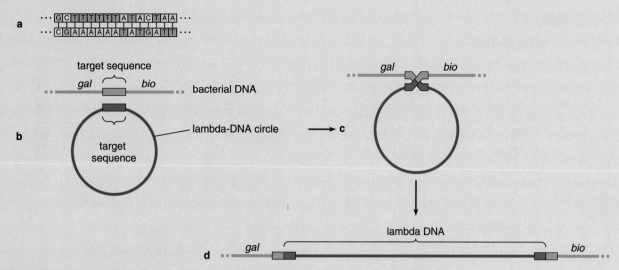

Figure 18-23 Insertion of bacteriophage lambda into the *E. coli* genome. **(a)** The sequence forming the target site for insertion of the lambda DNA circle into the *E. coli* genome. The sequence appears in exactly the same form in the lambda DNA circle and at a point in the *E. coli* genome between the *gal* and *bio* operons. **(b)** The target sequence in the *E. coli* and lambda DNA molecules. **(c)** Pairing and recombination between the *E. coli* and lambda target sequences. **(d)** The resulting insertion of the lambda DNA into the *E. coli* genome.

One site within the lambda DNA molecule contains a 15 base-pair sequence that is precisely homologous to a sequence located between the *gal* and *bio* operons in the *E. coli* genome (Fig. 18-23a and b). The viral DNA inserts at this site through pairing and recombination between the homologous viral and target site sequences (Fig. 18-23b to d). Insertion is catalyzed by an integrase encoded in the viral DNA. Excision occurs by reversal of the recombination pathway, leaving both the viral DNA and the host cell genome complete and separate. Excision is catalyzed by another enzyme encoded in the virus, *excisionase*, which coordinates with integrase in cutting the viral DNA free from the host genome. The genes coding for the integrase and excisionase of bacteriophage lambda provide another example of gene overlap—the coding sequences of the two genes share 23 nucleotides in common, in this case in the same nucleotide chain of the DNA double helix.

Once inserted into the *E. coli* genome, the lambda DNA is replicated along with the host cell DNA and transmitted from parent to offspring during cell division. In this form, the *lysogenic phase* (see p. 733), the virus has little effect on its host cell. At some point, however, the sequence encoding the integrase and excisionase is transcribed, and the two enzymes necessary for excision of the virus are produced in the host cell. (For details of the mechanism switching the lambda genome to transcription of the excision genes, see p. 733.) Excision triggers the *lytic phase* of the viral life cycle, in which the viral DNA is replicated and infective

viral particles are produced and released by lysis of the cell. Several other bacteriophages are inserted and excised from bacterial genomes by similar pathways.

Bacterial transposons cause the same spectrum of genetic disturbances as eukaryotic elements—mutations, deletions, substitutions, inversions of DNA sequences, and modification of the transcriptional activity of nearby genes—when they appear in new locations in the genome. Although many of the genetic disturbances are deleterious, the process of transposon insertion is an important, beneficial source of variability.

Evidence that the balance between deleterious and beneficial effects of transposons to bacteria leans toward benefit comes from several sources. One is the development of drug resistance brought about by genes carried in transposons. Another line of evidence comes from experiments testing the survival of bacterial strains with and without certain transposons. In these experiments bacterial cells possessing either IS or Tn elements consistently become the dominant cell type when placed together with cells lacking the elements. The advantage conferred by transposons in these experiments presumably results from the variability generated by mutations, gene rearrangements, or modifications of gene activity when the elements move to new locations. The benefits of the mechanism are marked in prokaryotes, in which offspring are produced in large numbers in generations that quickly come and go.

Suggestions for Further Reading: See pp. 772 and 773.

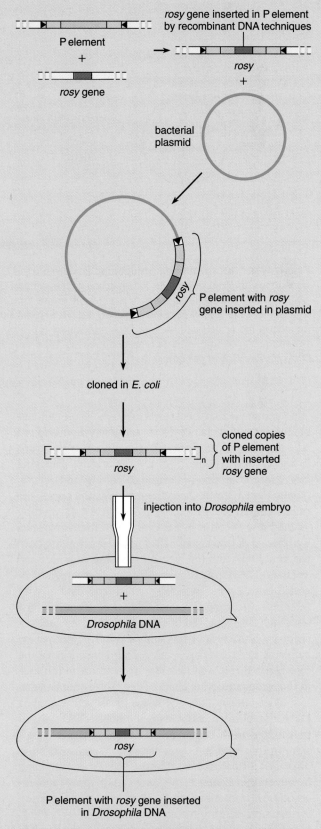

Supplement 18-2
Genetic Engineering

The discovery of genetic rearrangements opened the possibility that the mechanisms underlying some of the alterations could be used for genetic "engineering"— the experimental introduction of new genes or genetic information into living organisms. The introduced genes might correct hereditary defects, improve crop plants, provide a means for the study of developmental programs and gene regulation, or allow the production of needed proteins in microorganisms. The techniques of molecular biology have already made some of these possibilities a reality; genes have been successfully introduced in functional form into plants, animals, and prokaryotes, and bacteria have been used to manufacture proteins for medicine and research.

Genetic engineering takes advantage of several natural sources of genetic rearrangement, including plasmids, transposable elements, viruses, and the ability of both prokaryotic and eukaryotic cells to absorb and integrate pieces of foreign DNA. To track the uptake, incorporation, and function of introduced genes, easily detectable genetic markers, such as genes imparting drug resistance, are usually included in the DNA segments being introduced. If the marker gene can be detected in stable form in the recipient organism, the introduced genes may be assumed also to have been incorporated into the DNA of the recipient cells. Once incorporated, the introduced genes remain potentially functional, and are replicated and passed on intact to cell descendants.

Genetic Engineering in Animals

Experiments with animal genetic engineering have employed a variety of techniques, including transposable elements, viruses, and direct DNA uptake to introduce genes. Each of these technical approaches has produced at least partial successes in several animal species, including *Drosophila* and mice.

Successes with *Drosophila* Some genes introduced into *Drosophila* via transposable elements have retained full functional capabilities and corrected genetic defects in individuals and their offspring. The most productive *Drosophila* experiments have employed P elements (see p. 765). In one particularly successful experiment (Fig. 18-24) G. M. Rubin and A. G. Spradling introduced the gene for *rosy* eye color to a line of flies lacking a normal form of the gene. In its normal form the *rosy* gene encodes a fully functional form of *xanthine dehydrogenase*, an enzyme that catalyzes a critical step in the pathway producing normal eye color in *Drosophila*. Rubin and Spradling added the normal *rosy* gene to a P element by recombinant DNA techniques (see

Figure 18-24 The techniques used in the successful introduction of a normal form of the *rosy* gene into a *Drosophila* mutant with a defective gene. The *rosy* gene codes for xanthine dehydrogenase, which is required for development of normal eye color in *Drosophila*.

Ch. 4) and cloned the altered transposon in bacteria to obtain enough copies for introduction into the mutant strain. The P elements carrying *rosy* were injected into a region of *Drosophila* embryos destined to form the germ line of the animal. The P elements inserted into the DNA of the germ line cells in some of the injected embryos, producing gametes with the normal form of the *rosy* gene. The descendants of these individuals produced functional xanthine dehydrogenase enzymes and had normal eye color.

A number of other genes have been successfully introduced into *Drosophila* by the same approach. In all cases the introduced genes functioned normally, even though they were inserted into the genome at random locations by their P-element vector.

Genetic Engineering in Mice Mice have been genetically engineered, with varying degrees of success, by several techniques. One of the most successful experiments, carried out by R. E. Hammer, R. D. Palmiter, and R. L. Brinster, corrected a genetic defect producing dwarf mice. The condition, called *little*, results from deficiencies in production of growth hormone.

In the experiment cloned DNA containing a normal rat growth hormone gene was injected directly into fertilized mouse eggs. If the cloned DNA successfully inserted in the DNA of the fertilized egg, all the cells of the resulting individual, since they were produced by division of the fertilized egg, would contain the inserted growth hormone gene. After implantation in the uterus of an adult female mouse, some of the eggs developed normally to form individuals with ample supplies of growth hormone—a bit too ample, in fact. Mice in which the rat gene was successfully inserted grew to about one and one-half times normal size (Fig. 18-25). Because the gene also inserted in germ line cells, it could be passed on to all descendants of the altered individuals, producing a line of giant mice.

The mice receiving the growth hormone gene grew to giant size because of efforts to circumvent a problem with gene activity encountered generally in attempts to insert genes in mammals. When introduced into mammals, DNA combines with chromosomal proteins and becomes heavily methylated. These and possibly other alterations usually make genes carried in the introduced DNA inactive in transcription (see p. 708). In the experiment with *little* mice, the promoter of the mammalian *metallothionein* gene was grafted to the 5' end of the rat growth hormone gene. The metallothionein protein binds heavy metal ions, effectively removing them from solution in the body and cancelling their toxic effects. The metallothionein gene is converted to active form by the presence of heavy metals. When combined with the metallothionein promoter, the growth hormone gene could be made active by exposing mice receiving the gene to heavy metals—the same pathway activating the regular metallothi-

Figure 18-25 A giant mouse (right) produced by the experimental introduction of a rat growth hormone gene. A mouse of normal size is on the left. Courtesy of R. L. Brinster and R. E. Hammer, School of Veterinary Medicine, University of Pennsylvania.

onein gene of the mice also activated the implanted growth hormone gene. The method worked a little too well, activating the inserted gene to overproduce growth hormone, so that the genetically altered individuals grew to giant size.

Several other genes have been inserted in mice by the same pathway, including those coding for interferon, insulin, and β-globin. For example, F. Constantini and his coworkers injected a normal β-globin gene into fertilized eggs of a mouse line with the hereditary disease *thalassemia*, in which β-globin production is deficient. Introduction of the normal β-gene corrected the deficiency.

Several methods have been employed to circumvent the technically difficult process of directly injecting DNA into cells with a microneedle. Some take advantage of the fact that cells in culture take up DNA directly. A small fraction of the DNA becomes inserted by unknown mechanisms into the genome of the cultured cells. DNA uptake in this way is called *transfection*. The rate at which cells take up DNA by transfection is frequently increased by adding calcium phosphate to the surrounding medium. The Ca^{2+} precipitates the DNA into particulate masses that enter cells rapidly by endocytosis. Uptake is also increased by a technique called *electroporation*, in which cells are exposed to an electric current. The current creates transient, microscopic pores in the plasma membranes of the cultured

One of the major successes in this field of genetic engineering is the production of insulin, a peptide hormone needed in quantity by persons with diabetes. Formerly, insulin was obtained almost entirely from pork and beef pancreas, the supply of which was unable to meet the steadily rising demand for the hormone. The fact that the porcine and bovine insulin molecules have different amino acid sequences from human insulin also tends to induce an immune or allergic reaction in long-term users. The problem was solved in the late 1970s by implanting human insulin genes in *E. coli*, in one of the first successful commercial applications of genetic engineering.

The insulin molecule contains two polypeptide subunits, A and B. Because bacteria do not possess the processing enzymes necessary to remove introns, codes for the insulin polypeptides had to be inserted in *E. coli* in processed form, with introns removed. To accomplish this, synthetic genes for the human A and B peptides, which contain 20 and 30 amino acids, respectively, were assembled in the test tube by linking individual nucleotides one by one into the required sequences. The artificial genes were then attached to the *lac* promoter, a "strong" bacterial promoter that initiates RNA transcription at a high rate. Once inserted into a plasmid and introduced into *E. coli*, the synthetic A and B genes were actively transcribed.

In commercial applications of the technique, bacteria containing the implanted genes are grown in large batches. Isolation of the insulin polypeptides is facilitated by a further genetic engineering step in which the code for a short polypeptide sequence acting as a signal for secretion is added to the genes for the polypeptides. The secretion signal causes the bacteria to release the insulin polypeptides into the surrounding medium.

Several other proteins, including human growth hormone, interferon, and a vaccine against hoof and mouth disease, have been developed for commercial production by similar methods. Many other polypeptides are under development for production in bacteria, including other hormones and serum albumin for use as a supplement for blood transfusions.

There is always the danger that genetically engineered bacteria may be accidentally released, with unknown consequences to the environment. This danger is minimized by growing the bacteria in laboratories that are carefully sealed and protected by several levels of barriers to escape. In addition, genetically engineered bacteria are usually altered with "failsafe" genes that are lethal outside the laboratory environment. For example, the *hok* gene, which encodes a protein that causes a lethal loss of membrane potential, is often added in combination with the *lac* promoter (see p. 728). Exposure to lactose sugars, which are common in the environment, activates the *hok* gene and kills the cells.

The Ethics of Genetic Engineering

The techniques, results, and possible future accomplishments of genetic engineering have raised ethical questions that are in many respects more difficult than the scientific ones. Are scientists playing God by attempting to modify the genetic makeup of humans and other organisms? Should the work of genetic engineering be limited to the correction of genetic defects in existing individuals or should defects be corrected in germ lines, so that descendants will be free of the disability? Will genetic engineering release "monsters," impossible to control, into the environment? Might unscrupulous governments use the techniques of genetic engineering to develop weapons of biological warfare or to modify human abilities or behavior? In short, are the benefits of genetic engineering worth the risk?

Consideration and discussion of these questions have spread far beyond the scientific community. Public fears have led to legal efforts to stop scientific investigation or applications of genetic engineering. Some activists have even suggested, because the germ plasm of plants and domestic animals is a natural resource, that tampering with this resource by genetic engineering should be considered a federal offense punishable by fine or imprisonment!

Providing the answers to questions about the ethics and legality of genetic engineering will require careful analysis by people both inside and outside the scientific community—lawmakers, governmental executives, judges, and citizens, as well as scientists. Hopefully, the final answers and decisions will be based on an informed evaluation of scientific merits and disadvantages rather than on ignorance, fear, and emotion.

Suggestions for Further Reading: See p. 773.

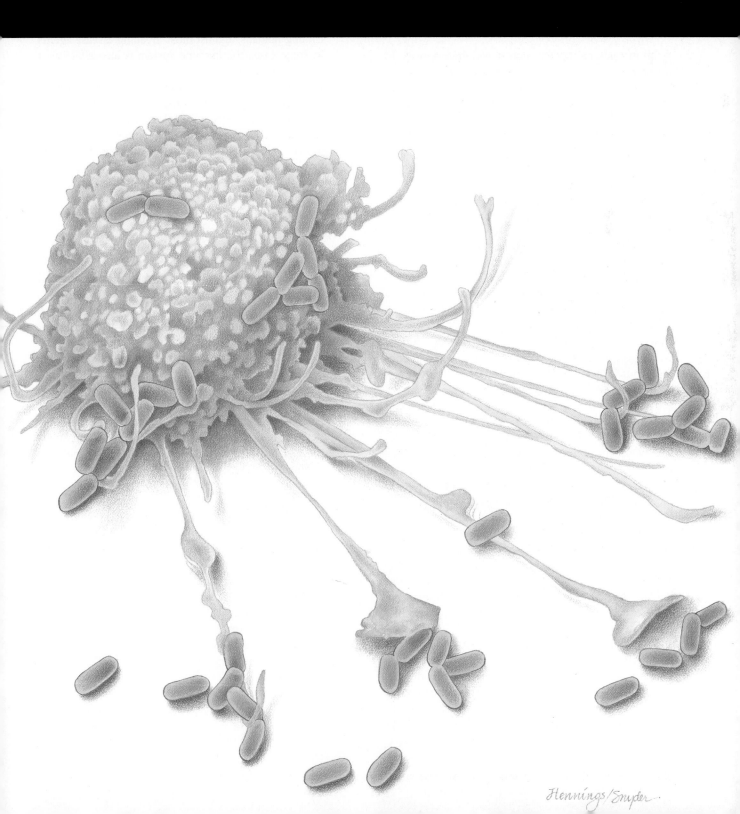

Hennings/Snyder

- *Antibody structure* - *Antibody gene groups* -
*Genetic rearrangements producing functional
antibody genes* - *Clonal selection of B cells* - *The
immune response* - *MHC* - *Killer and helper T
cells* - *Development of immune tolerance* -
Macrophages - *Interleukins* - *Autoimmune diseases*
- *Organisms avoiding or defeating the immune
response* - *AIDS*

Mammals and other higher vertebrates react to invasion by foreign organisms and substances by an immune response. Among the primary reactions in the immune response is the generation of *antibodies*, which provide the first line of defense against invading pathogens that include bacteria, viruses, toxins, fungi, and protozoa. Higher vertebrates are even capable of developing antibodies against artificial, man-made substances never encountered before by any living organism.

Substances that stimulate antibody production are called *antigens.* Most natural antigens are proteins or polysaccharides. However, certain nucleic acid types, such as Z-DNA (see p. 530), and other molecules can also stimulate antibody production. Substances acting as antigens may be freely suspended in solution or may be a surface or internal molecule of a pathogen such as an infecting bacterium or virus.

Vertebrates have the capacity to produce antibodies that specifically recognize and react with millions or even billions of different antigens. For a time it was thought that each different antibody type must be encoded in a distinct gene in the genome and, therefore, that perhaps a million or more genes might be required for antibody production. However, further research revealed that antibody diversity results from programmed genetic rearrangements assembling a relatively small number of gene subunits in different combinations to produce a very large number of functional genes. The total number of subunits used in the rearrangements is probably not more than several hundred to a thousand or so in most vertebrates. The rearrangements mixing these subunits in different combinations in the final antibody genes, along with mutations and variations in DNA and RNA splicing that form part of the mechanism, can generate distinct antibodies in numbers that may indeed amount to billions.

Antibodies are produced in *lymphocytes*, a type of leukocyte, or white blood cell. The antibody-producing lymphocytes are called *B cells* because they originate and undergo much of their maturation in bone marrow. The genetic rearrangements producing antibodies take place in these cells. Completed antibody molecules ap-

pear on the surfaces of B cells, and they are also released into the bloodstream and other body fluids and secretions.

Production of antibodies by B cells is a vital part of a larger immune response that includes additional cell types, among them *T cells* and *macrophages.* (T cells are a type of lymphocyte that matures in the thymus; macrophages directly engulf invading bacteria and foreign cells and stimulate T-cell development.) The activities of these cells are coordinated with the production of antibodies by B cells to produce a total defense that effectively protects vertebrate animals, including humans, against most invading disease agents. In some cases the immune system is also effective in killing cancer cells.

Among the many highly unusual and interesting features of the immune system is the fact that it is capable of a form of molecular learning and has memory. In the reaction to an invading antigen, the immune system selects antibodies that recognize the antigen from among the large repertory available and produces these antibodies in quantity. In this sense the immune system "learns" which of its antibodies to produce in quantity to make an effective immune response. After an immune response is initiated and the antigen is eliminated, B and T cells active against the antigen remain in elevated numbers in the circulation as a "memory" of the response, ready to mount another attack if the antigen invades again.

This chapter describes antibody structure, the structure of the gene groups encoding antibodies, the genetic rearrangements and other processes giving rise to functional antibody genes, and the coordinated activities of B cells, T cells, and macrophages in the immune response. Autoimmune diseases and pathogens that defeat or circumvent the immune system, such as the AIDS virus and the protozoan causing African sleeping sickness, are also discussed. Further information on the molecular and clinical aspects of AIDS is presented in Supplement 19-1.

THE PRODUCTION OF ANTIBODIES

Antibodies work by binding tightly to specific antigens located either on the surfaces of invading viruses or organisms or suspended as free particles in the circulation. Binding by an antibody has one or more effects. By linking to the surfaces of invading pathogens, antibodies block reactive groups that would otherwise bind the pathogen to surface receptors on body cells and promote its entry into the cell interior. The presence of antibodies on a pathogen surface also marks the pathogen for engulfment and destruction by macrophages. For antigens suspended in the body circulation, combination with antibodies causes aggregation or precipitation of the antigen, and also marks the antigen for engulfment and elimination.

Antibody Structure

Antibody molecules, also called *immunoglobulins*, are large, complex proteins built up from four polypeptide subunits (Fig. 19-1a and b). The two smaller polypeptide subunits, each about 220 amino acid residues in length, are identical in structure; these are the *light chains* of an antibody molecule. The two larger chains, identical to each other but different from the light chains, contain about 440 amino acid residues each; these are the *heavy chains* of an antibody molecule. The light and heavy chains are structured from domains called *variable* and *constant regions*. As the names suggest, the variations in amino acid sequence responsible for differences in the binding specificities of antibodies are confined almost entirely to the variable regions. The constant regions are highly conserved between different antibodies.

An antibody light chain is structured from two domains, one variable and one constant. Each heavy chain has a single variable domain and a long constant region containing three to four domains.

The four-part complex formed by the polypeptides of an antibody approximates the Y-shaped structure in Figure 19-1a and b. Note from the figure that the variable regions of the light and heavy chains form the arms of the Y. The tips of the two arms have identical binding sites that recognize and bind a specific antigen. The arms are connected to the tail of the Y by flexible "hinges" that allow the arms to flex or rotate to accommodate antigen binding. The tail of the Y determines overall features of antibody function, such as location in different regions of the body, and whether the antibody is secreted or remains attached to cell surfaces. The four polypeptides are held in the Y-shaped structure by disulfide (—S—S—) linkages between the light and heavy chains (shown in Fig. 19-1a).

An antigen-binding site takes the form of a depression or cleft about 2 nm long and 3 nm wide at the tip of a Y arm. (Fig. 19-2 shows the binding site of an antibody in combination with an antigen.) With these dimensions an antibody can recognize and bind only a small portion of most antigens. Recognition and binding depend on a close fit between the antibody site and the surface of the antigen, and on opportunities for the formation of hydrophobic, ionic, and van der Waals attractions. For protein antigens, the portion recognized may include from as few as 3 or 4 to a maximum of about 16 amino acid residues. The portion of an antigen recognized and bound by an antibody is termed the *epitope*. The presence of two binding sites allows antibodies to crosslink one antigen molecule to another, forming the aggregates or precipitates that are recognized for engulfment by macrophages.

Although the variable and constant regions of the light and heavy chains may differ extensively in amino acid sequence, the domains forming these regions are remarkably similar in number of amino acid residues (about 110). The three-dimensional folding pattern of the amino acid chain within any of the domains is also similar (see Fig. 19-1c and d). These similarities suggest that the constant and variable domains originated in evolution by duplication of a single ancestral domain.

Within the variable region of either a light or heavy chain, three short segments display most of the variation in amino acid sequence between different antibodies. Relatively little variation occurs in other segments. The three short variable segments are called the *hypervariable segments (HV)* of a variable region. The relatively invariant segments are termed the *framework segments (FR)* (see Fig. 19-1a). The folding patterns taken by the four framework segments hold the hypervariable segments in the antigen-binding clefts at the tips of the Y arms.

Antibody structure and function have been most thoroughly studied in mice and humans. Two families of light chains, λ *(lambda)* and κ *(kappa)*, and one family of heavy chains occur in these species. Each family is encoded in a separate group of gene subunits: the *lambda light chain* (L_λ) gene group, the *kappa light chain* (L_κ) gene group, and the *heavy chain (H)* gene group.

A completed antibody molecule has either a pair of lambda light chains or a pair of kappa light chains in combination with a pair of heavy chains. Antibodies therefore have the structure $L_{\lambda 2}H_2$ or $L_{\kappa 2}H_2$. Although antibodies reacting with distinct antigens have different variable regions, all antibodies of the same type, formed in a single B cell to combine with a single specific antigen, are identical in their variable regions. However, the antibodies of a single type, with identical variable regions, fall into several *classes* depending on the combination of heavy chain domains used in their assembly.

Five different combinations of heavy chain domains produce the α-, γ-, δ-, ϵ-, or μ-heavy chain classes of each antibody type. For a given antibody type an antibody assembled with heavy chains of the α-class is termed an *IgA* antibody (where *Ig* = immunoglobulin); an antibody assembled with heavy chains of the γ-class is known as an *IgG* antibody. Similarly, antibodies with the δ-, ϵ-, or μ-heavy chain classes are known as *IgD*, *IgE*, and *IgM* antibodies, respectively.

The five classes of a single antibody type all react with the same antigen. However, each has a different function and is found primarily in different parts of the body. IgA antibodies, for example, are the primary type occurring in body secretions such as tears, milk, and the fluids coating the surfaces of mucous membranes. IgD antibodies are primarily anchored to the surfaces

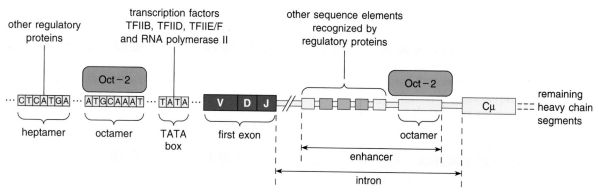

Figure 19-10 Interaction of regulatory proteins and control sequences in antibody genes. Upstream of the TATA box are a seven-nucleotide sequence recognized by a regulatory protein and an octamer sequence recognized and bound by the regulatory protein Oct-2. The general transcription factors TFIIB, TFIID, and TFIIE/F bind to the TATA box located downstream of the octamer. The enhancer in the long intron contains another octamer recognized by Oct-2 and several other sequence elements recognized by regulatory proteins. Shown is the enhancer of heavy chain genes, which contains five sequence elements in addition to the octamer. At least three of them (in darker blue) are bound by regulatory proteins specific for antibody genes. Interactions between the regulatory proteins and transcription factors bound to the promoter and enhancer regions lead to tight binding of RNA polymerase II and the initiation of transcription.

type, and somatic mutation, which serves primarily to make fine adjustments in binding affinities of the antibody. The total number of B cells is so great—10 million are produced each day in the mouse—that all possible antibodies are likely to be made.

In maturing B cells a functional antibody gene is assembled in only one of the two chromosomes in pairs carrying the antibody gene groups. The gene groups in the other chromosome of the pairs may undergo some rearrangement but do not produce functional genes that are transcribed into antibody mRNAs. The basis of the mechanism limiting effective gene rearrangement to only one of the two chromosomes in a pair, called *allelic exclusion*, remains unknown. However, there are indications that some of the proteins produced through transcription of the functional genes may inhibit rearrangement of the excluded ones. Some support for this hypothesis is taken from the results of transfection experiments in which a fully rearranged antibody gene is introduced into embryonic B cells. In the transfected cells rearrangement of antibody genes of the same type (lambda, for example) as the introduced gene is inhibited.

Promoter and Enhancer Activity in Rearranged Antibody Genes Recall that, in the rearrangements producing functional antibody genes, the promoters forming part of V elements are activated by the enhancer located in the long intron separating J and constant region elements. Several of the regulatory proteins binding as *trans* elements (see p. 698) to the promoter

and enhancer regions have been identified (Fig. 19-10). Upstream of the TATA box of the promoter is an eight-nucleotide, or *octamer*, sequence recognized by the regulatory protein Oct-2; the same regulator binds to an octamer sequence in the enhancer region in the long intron between the J and constant region elements. A seven-nucleotide sequence recognized by other regulatory proteins lies upstream of the octamer in the promoter. The general transcription factors TFIIB, TFIID, and TFIIE/F bind to the TATA box. The enhancer in the long intron contains several 6 to 12 base-pair sequence elements in addition to the octamer recognized by Oct-2. (The enhancer of heavy chain genes, which contains five sequence elements in addition to the octamer, is shown in Fig. 19-10.) At least three of these are bound by regulatory proteins specific for antibody genes, including one identified as *NF-κB*. Interactions between the regulatory proteins and transcription factors bound to the promoter and enhancer regions lead to tight binding of RNA polymerase II and the initiation of transcription (for details of these interactions, see Ch. 17).

The enhancer of antibody genes may promote transcription of any gene placed nearby by malfunction of the rearrangement mechanism. This occurs, for example, when faulty rearrangements place a potential oncogene next to an antibody gene enhancer (see p. 937). The faulty placement activates or greatly increases transcription of the potential oncogene, converting it to a full-fledged oncogene and triggering the uncontrolled and rapid cell division characteristic of cancer cells (for further details, see Ch. 22).

Stimulation of B Cells Producing an Antibody Type: Clonal Selection

B cells continue to mature through the lifetime of the individual, yielding individual cells with the capability of synthesizing a specific antibody type. In early B cells the initial IgM or IgD antibodies, rather than being released to the medium, are inserted as recognition molecules at the cell surface.

Unless stimulated by contact with an antigen that can combine with their exposed IgM or IgD antibodies, the early B cells remain inactive in cell division. If the B cell encounters an antigen capable of combining with the variable region of the antibody exposed at its surface and interacts with a T cell (see below), cell division is triggered, leading to production of a rapidly dividing clone of B cells. Because they are descendants of the original stimulated cell, all the cells of the clone synthesize antibodies interacting with the same antigen.

Production of a clone through stimulation by an antigen is termed *clonal selection*. The process was proposed by Macfarlane Burnet some 30 years ago, long before the cellular and molecular mechanisms underlying the process were discovered. Burnet, who based his hypothesis partly on ideas advanced by D. W. Talmage and N. K. Jerne, proposed clonal selection as a form of cellular natural selection: antigens select B cells recognizing them to reproduce and become dominant in the population. (Burnet received the Nobel Prize in 1960, and Jerne in 1984.) The Experimental Process essay by G. Ada on p. 804 describes his experiments supporting clonal selection, which were critical to the acceptance of the theory.

As division of a B-cell clone proceeds, somatic mutations, which increase greatly in frequency after the cell line is stimulated by antigen, lead to small differences in the variable regions of the antibodies. During further rounds of interaction with T cells, the B cells most likely to be stimulated are those with variations of the antibody binding most strongly to the antigen. As a consequence, the antibody type produced by the clone, becomes progressively stronger and more specific in binding capacity.

Class switching within the clone as individual B cells mature confronts the antigen with antibodies of the five different classes. After stimulation most of the IgM and IgD antibodies, along with the remaining classes, are secreted from the maturing B cells rather than inserted into the cell surface. The change from membrane-bound to secreted forms of IgM and IgD antibodies results from a variation in pre-mRNA processing. The variation eliminates the final exon of the transcript (the M exon in Fig. 19-5*d*), which codes for a polypeptide segment that anchors the IgM or IgD antibody molecules to the plasma membrane. Elimination of the exon removes the membrane anchor and releases the antibody molecules from the cell surface.

Secretion of the various antibody classes leads to a full-blown immune response against the antigen, which, if successful, eliminates the antigen from the body and its secretions.

Elimination of the antigen interrupts the immune response, and division of the clone is reduced to the low levels characteristic of unstimulated B cells. Most or all of the fully mature B cells gradually die and are eliminated from the bloodstream and other body fluids. (Life expectancy of mature B cells is a matter of days.) However, immature B cells of the same clone remain in the bloodstream in elevated numbers. These quiescent cells, with IgM and IgD antibodies exposed as cell surface receptors, remain ready as *memory B cells* that can provide a quick response if the antigen is again encountered by the individual. (Other elements of the immune response and some of the more physiological aspects of immunity are described later in this chapter.)

The ability of a clone of B cells to produce antibodies against a single antigen is widely used as a means to produce *monoclonal antibodies* for scientific research and medicine (Fig. 19-11). In the technique, first developed by G. Kohler and C. Milstein, an animal such as a mouse or rat is injected with an antigen of interest. Antibody-producing B cells are then extracted from the spleen of the animal; because the B cells do not divide readily by themselves, they are placed in a culture vessel with myeloma cells, a cancerous lymphocyte type. The B cells and myeloma cells are cultured under conditions that cause them to fuse into large, composite cells called *hybridomas*. Hybridomas combine the antibody production of B cells with the rapid, continuous division of myeloma cells. The B cells are said to be "immortalized" by fusion with the myeloma cells into hybridomas.

Single hybridoma cells are then removed from the fusion culture and used to start clones. The clones can be grown in immense numbers, producing large quantities of the desired antibody. Because the clones are derived from single hybridoma cells, the antibody produced is of a single, highly specific type.

The monoclonal antibody technique circumvents a problem inherent in methods simply using an animal, such as a horse or rabbit, as a source of antibodies. When an animal is injected with an antigen, a wide spectrum of antibodies is produced that reacts to different parts of the antigen with varying degrees of specificity. Frequently, some of the antibodies produced also cross react with other molecules sharing structural characteristics with the antigen. The monoclonal technique produces a single antibody type that can be selected to be highly specific for the antigen of interest.

Monoclonal antibodies have some disadvantages. Cell fusion is a relatively rare event so many of the B cells are left in an unfused, nondividing state. As a result, much of the potential for specific antibody pro-

Instruction or Selection: The Role of Antigen in Antibody Production

Gordon Ada

GORDON ADA worked first with Macfarlane Burnet and then with Gus Nossal at the Walter and Eliza Hall Institute in Melbourne, Australia (1948–68), and received his D.Sc. at the University of Sydney in 1959. He was Professor of Microbiology at the John Curtin School of Medical Research in Canberra until his retirement in 1987. He left Australia in 1988 to work at the Johns Hopkins School of Public Health in Baltimore Maryland, becoming Director of their Center for AIDS Research. He has recently returned to Canberra.

Antibodies (antitoxins) were discovered, isolated, and shown to be proteins in the late 19th century. They occur in response to an infection, and extensive studies showed them to have a remarkable specificity in their interaction with antigens, approaching that of enzyme-substrate reactions. Until the 1960s, there was no understanding of the mechanisms of protein synthesis within cells.

The first attempt to explain antibody production by cells was by a remarkable German chemist, Paul Ehrlich. He proposed in 1901 that white blood cells possessed receptors at their surface. These receptors had "sidechains" which bound foreign substances (antigens) derived from infectious agents. This binding event stimulated the cell, which responded by producing these receptors in excess, and many receptors—the antibodies—were shed into the blood. This concept implied that these receptors occurred *naturally* in a sufficient range of specificities to bind components from many infectious agents.

It was soon found, however, that proteins from many other sources, such as milk, could induce formation of antibodies and bind to them. Karl Landsteiner's finding in the 1930s that antibodies could be formed to and would bind with exquisite specificity to *completely synthetic* compounds seemed to discredit Ehrlich's concept. How could the body recognize substances that had not existed previously? This led to the notion that antigen instructed a cell to make antibody of a complementary specificity. In one popular form of the Instructive Theory of antibody formation, the Template Theory, it was proposed that antigen entered the lymphocyte and acted as a template inside the cell; flexible precursor antibody molecules moulded themselves on this antigen template to assume a conformation with the correct antigen-binding specificity.

In the 1940s, Macfarlane Burnet pointed out that the Template Theory did not explain several facts about antibody formation, such as a second (memory) response to an antigen, in which antibody production is greater than in the primary response; and immunological tolerance, in

which antigen administration failed to induce an antibody response. Building upon ideas recently reformulated first by Niels Jerne and particularly by David Talmage which were reminiscent of Ehrlich's concept that antibodies existed normally, Burnet in 1957 proposed the Clonal Selection Theory of antibody formation. The crux of this theory was that in the animal host there existed populations of lymphocytes, each possessing immunoglobulin receptors of a single specificity. Antigen *selected* those cells with receptors that recognized polypeptide segments called epitopes on that antigen, causing the selected cells to differentiate and proliferate to become clones of antibody-secreting cells (ASCs). Allowing for increases in specificity due to somatic mutation, the secreted antibody from the progeny of a single precursor cell would have a specificity similar to the precursor's Ig receptors. Tolerance occurred due to the deletion of "forbidden clones," i.e., cells reacting against self components.

The Template and Clonal Selection theories were so different as to be irreconcilable. In the first, the cell produced a molecule whose specificity could be modified by a foreign agent, the antigen; in the second, antigen selected cells that were already genetically programmed to produce a molecule of a particular specificity. Which theory was correct? This essay describes a series of experiments that established the essential correctness of the Clonal Selection Theory.

A colleague, Gus Nossal, working with Joshua Lederberg, argued that individual lymphocytes exposed to two different antigens would make antibodies either of both specificities (the Template Theory) or of only one specificity (Clonal Selection). They immunized rats with two chemically related but immunologically distinct proteins, the flagellar proteins of two different *Salmonella* bacteria. Individual ASCs isolated some time later from each rat were exposed in microdroplets first to one and then to the other antigen in the form of motile bacteria; when viewed in a microscope, each of 62 separate ASCs examined were found to immobilize only those *Salmonella* bacteria with flagella of a single specificity. This result certainly favored the Clonal Selection Theory. There was the possibility, however, that a precursor cell might have the capability of making antibody of many specificities but that the first contact with an antigen channeled the cell's response into a more restricted if not a single specificity.

Nossal and I then decided to try a different tack. The Template Theory predicted that an individual ASC must contain many thousands of molecules of intact antigen in order for the latter to act as a template in a cell for the entire time antibody was being synthesized and secreted.

In fact, in earlier experiments when large amounts (milligrams) of a particulate antigen (visible by electron microscopy) had been administered, one group reported that ASCs seen in thin sections seemed to contain particles of antigen. We changed the methodology in two ways. As the immunogen, we used flagella because they are highly immunogenic, nanogram quantities inducing a strong antibody response; and we labeled it with the recently available carrier-free preparations of the isotope iodide-125 (^{125}I), in order that tiny amounts of the antigen could be traced in the body. Individual specific ASCs isolated from the draining lymph nodes of rats immunized with the labeled antigen were fixed on glass slides, dipped in photographic emulsion, exposed in the dark for 60 days, and then developed. We had calculated that a single grain (above background levels) over an ASC would correspond to about 10 molecules of antigen. In fact, the great majority of more than 200 cells examined were found to have no associated grains. A group in London led by John Humphrey later obtained similar results. Again, this result argued against the template theory. (As we now know, 25 years later, antigen is in fact taken into lymphocytes but it is degraded and does not persist to become a template.)

However, we now needed an approach that allowed direct examination of the interaction of antigen with precursor immunocompetent lymphocytes before they became ASCs. Two Israelis, Dov Sulitzeanu and David Naor pointed the way.

Sulitzeanu and Naor showed that if a suspension of cells including lymphocytes from lymphoid tissues of unimmunized animals was exposed at 0°C to a labeled antigen (bovine serum albumin) for a short time, the unbound antigen washed away, and the cells autoradiographed, distinct labeling patterns were seen. Rather than all or most lymphocyte-like cells being labeled (a Template Theory prediction), only a very small minority (<1%) were labeled, and of these, some cells were labeled more than others. Pauline Byrt and I repeated and extended these findings with our antigens. Cells from the mouse thoracic duct, a source of almost pure lymphocytes, showed the clearest labeling pattern of all (Ref. 2, Fig. 9.2). In addition, we were able to show that if the cells were exposed to anti-Ig sera before exposure to labeled antigen, this prevented the subsequent binding of the antigen. We were seeking evidence that the antigen was binding to the cells via a specific Ig receptor, but it might be that the labeled cells were coated nonspecifically with cytophilic antibody. What was needed was a *functional* test to show this binding pattern was more than a coincidence.

Lymphocytes were known to be very sensitive to ionizing radiation, and I recalled that at a recent Cold Spring Harbor Symposium (1967), Richard Dutton had described how pulsing with tritiated thymidine had selectively killed dividing lymphocytes. It was not feasible for us to use tritiated antigen as the level of radioactivity of the labeled antigen would be far too low. However, we knew that the path length of emitted β rays from ^{125}I was about 10 μm, about the mean diameter of resting lymphocytes. Would the radiation from ^{125}I-labeled flagella

adsorbed to a lymphocyte surface selectively damage that cell so that it would not proliferate and mature to become an ASC when later exposed to the same (but unlabeled) antigen? The experiment was worth a try!

Splenocytes from normal, unimmunized mice were exposed to ^{125}I-labeled flagella of one of two different specificities. After standing for 1 hour at 4°C, unadsorbed antigen was washed away, and the cells immediately adoptively transferred to mice with the same gene complement whose own immune lymphoid system had been destroyed by previous x-irradiation. Both groups of mice were then immunized with unlabeled flagella of both specificities, and the quantities of antibody reacting with both antigens measured some days later. It was hoped that mice which had received cells exposed to hot antigen would respond poorly if at all after immunization with antigen of the same immunological specificity, but give a normal response to flagella of the other specificity. Such a result would have demonstrated that cells to which the labeled antigen bound were necessary for the production of antibody of that specificity.

In the first experiment, however, no such differences were found. What was the reason for this result? Were the cells that had bound the antigen unimportant in an immune response? Or was the design of our experiment inadequate in some way? One possibility was that the short exposure of the labeled cell to radiation was insufficient to cause enough cellular damage.

To test this possibility, the experiment was repeated but after exposure to the antigen for 1 hr, followed by removal of unbound antigen. The cells were let stand in the test tube at 4°C for 16–20 hr before adoptive transfer and challenge. This time the expected results were obtained; exposure to labeled antigen of one specificity significantly reduced the ability of the lymphocyte population to respond to that antigen, but not to the immunologically distinct antigen.

This result was difficult to reconcile with any theory other than one requiring individual cells to have a very restricted potential for antigenic stimulation, such as the Clonal Selection Theory. The results were completely contrary to the prediction of the Template Theory. Together with amino acid sequence data on different immunoglobulins, which also were becoming available at that time, these findings sounded the death knell for any form of an Instruction Theory of Antibody Formation.

References

Burnet, F. M. A modification of Jerne's theory of antibody production using the concept of clonal selection. *Aust. J. Sci.* 20:67–69 (1957).

Nossal, G. J. V., and Ada, G. L. In *Antigens, Lymphoid Cells, and the Immune Response.* New York: Academic Press (1971).

Ada, G. L., and Nossal, G. J. V. The clonal selection theory. *Scientific American* 257:62–69 (1987).

Silverstein, A. M. The history of immunology. In *Fundamental Immunology.* Ed. Paul, W. E. New York: Raven Press, 21–38 (1989).

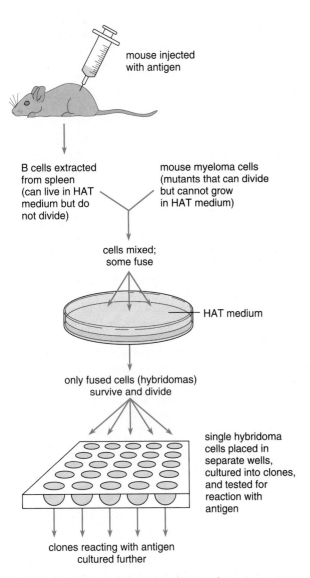

mouse injected
with antigen

B cells extracted
from spleen
(can live in HAT
medium but do
not divide)

mouse myeloma cells
(mutants that can divide
but cannot grow
in HAT medium)

cells mixed;
some fuse

HAT medium

only fused cells (hybridomas)
survive and divide

single hybridoma
cells placed in
separate wells,
cultured into clones,
and tested for
reaction with
antigen

clones reacting with antigen
cultured further

Figure 19-11 A technique used to produce mono-clonal antibodies. *HAT* medium (*HAT* = *H*ypoxanthine + *A*minopterin + *T*hymidine) contains aminopterin, which blocks direct synthesis of purine and pyrimidines but contains thymidine and hypoxanthine. Normal cells can use thymidine and hypoxanthine to make purines and pyrimidines. The B cells removed from the mouse cannot survive in cultures for extended periods because they do not divide in the absence of T-cell stimulation and other interactions of the immune system. The myeloma cells can divide and grow indefinitely in cultures but lack the enzymes necessary to make purines and pyrimidines from thymidine and hypoxanthine in HAT medium. The fused cells—hybridomas—contain genetic information from both the B and myeloma cells and can grow, divide, and survive indefinitely in culture.

duction remains untapped. Moreover, at present monoclonal antibodies can be produced readily only with cells from the mouse; human B cells, for unknown reasons, fuse with myeloma cells much less frequently than mouse B cells do. Using mouse-derived mono-

clonals as treatments for human disease frequently induces an immune response against the mouse antibodies in the patient, greatly reducing the effectiveness and persistence of the monoclonal antibodies.

The experiments revealing antibody structure and the genetic rearrangements producing antibody genes have been among the most challenging and difficult of molecular biology. The complexity of structure and function revealed by these experiments only scratches the surface and, as is often the case in biological investigation, poses many more questions than have been answered so far. Much mapping and sequencing remains to be done to establish the exact numbers of V elements and the details of their arrangement in the genome, particularly in humans. The enzymes and molecular mechanisms producing antibody gene arrangements remain largely unknown. Many more questions exist at the level of interaction between B and T cells and other components of the immune response (see below).

The antibody gene families are members of a large and diverse *immunoglobulin (Ig)* superfamily encoding cell surface molecules of various kinds (Fig. 19-12). Many of the receptors and other cell surface molecules taking part in the immune response, such as the T-cell receptors and the MHC surface molecules of lymphocytes and body cells (see below), are members of the Ig superfamily. The family also includes cell surface molecules that have no apparent functional relationship to immunity, such as the cell adhesion molecules N-CAM and I-CAM (see p. 254). The many members of the Ig superfamily share at least one domain that resembles the variable and constant domains of antibodies. For this reason the entire Ig superfamily is thought to have evolved, through many branches, from a single ancestral gene encoding the original Ig-type domain.

THE IMMUNE RESPONSE

An immune response involves the activities of macrophages and T cells as well as B cells. All these cells are different types of lymphocytes, which in total amount to some 10^{10} cells in humans and form a significant part of body mass. Macrophages take up invading bacteria and foreign cells by endocytosis and promote T-cell development. T cells occur in two major types, *killer T cells* and *helper T cells*. Killer, or *cytotoxic*, T cells can recognize and kill body cells containing infective viruses. Killer T cells can also kill some kinds of tumor cells and cells originating from outside the body. Helper T cells stimulate the activity of B cells in division and antibody secretion. Without helper T cells, little or no B-cell response occurs.

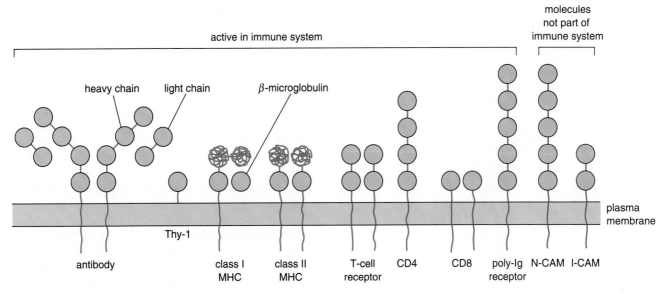

Figure 19-12 Some members of the immunoglobulin superfamily. The circles represent domains with sequence similarities to the basic immunoglobulin domain. Thy-1 is a surface marker on T cells. Many members of the immunoglobulin superfamily serve cell surface functions not directly related to the immune system; two of these, the cell adhesion molecules N-CAM and I-CAM, are shown.

The interactions of macrophages, B cells, and helper and killer T cells enable humans and other vertebrates to survive in a world full of toxic substances and potentially lethal viruses, bacteria, and other parasites. Interactions of the immune system are also responsible for some unfavorable outcomes for humans, including allergic reactions, rejection of transplanted tissues and organs, and autoimmune diseases, in which an immune response is actually mounted against self instead of foreign cells and substances. A number of infective microorganisms and viruses have developed ways to circumvent the immune system and survive in the body without detection by lymphocytes (see below).

MHC and the Immune Response

The various cell types engaged in the immune response recognize each other and body cells of the same individual through cell surface molecules. Primary among these are proteins encoded in the gene group known as the *major histocompatibility complex* (*MHC*; see p. 234). Research by M. L. Gefter, H. M. Grey, J. Smith, and their coworkers and others has shown that MHC molecules participate directly in the immune response by binding and displaying fragments of antigens on cell surfaces. This display plays a vital role in the cellular interactions leading to macrophage, B-cell, and T-cell activation.

MHC molecules also play a role in the recognition of foreign cells. Cells from different individuals of the

same or different species have different MHC molecules on their surfaces, enabling lymphocytes to recognize a cell introduced from a different individual as foreign and to destroy it. As such, MHC molecules are the primary barrier to organ and tissue transplantation among humans. For this reason MHC molecules are sometimes called *transplantation antigens*.

The MHC gene family, which occupies more than 25,000 base pairs of DNA, includes three classes of genes. Two of these, called *class I* and *class II genes*, encode cell surface molecules that act in antigen presentation. Class I genes encode *class I MHC* molecules, which occur on all body cells of an individual except those of the immune system. Class II genes encode *class II MHC* molecules, which occur on lymphocytes of various kinds. Class I and II MHC show enough homologies to antibodies to indicate that they are members of the Ig superfamily (see Fig. 19-12). *Class III genes* code for a group of proteins that form part of the *complement system*. These proteins, which circulate in the bloodstream, participate in reactions destroying foreign and infected cells during an immune response.

Class I MHC genes encode the larger of the two polypeptides making up an MHC molecule of this class (the green segment in Fig. 19-13). An analysis by D. C. Wiley and his coworkers showed that this polypeptide has three domains, which extend outward from the surface of the plasma membrane. The domain nearest the plasma membrane (domain III in Fig. 19-13) includes a transmembrane segment anchoring the MHC

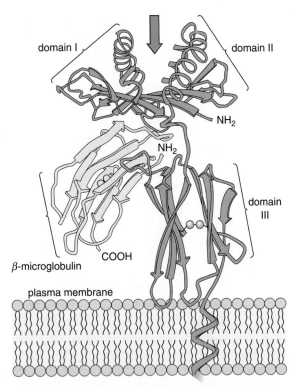

domain I

domain II

NH₂

NH₂

domain III

COOH

β-microglobulin

plasma membrane

Figure 19-13 A class I MHC. The β-microglobulin polypeptide is shown in yellow. The arrow marks the site binding and displaying a processed antigen. Redrawn from an original courtesy of D. C. Wiley. Reprinted by permission from *Nature* 329:506; copyright © 1987 Macmillan Magazines Ltd.

to the outer side of the cell. This domain demonstrates clear sequence similarities to the constant domains of antibody molecules. The other two domains (domains I and II in Fig. 19-13) form a cleft that binds an antigen fragment (arrow in the figure). The small polypeptide subunit of class I MHC is a relatively small protein, *β-microglobulin*. Although encoded in a separate gene on a different chromosome, the β-microglobulin polypeptide contains a domain with clear sequence similarities to the constant region of antibody genes, indicating that its gene is yet another member of the Ig superfamily.

A class II MHC, which also shares structural homologies with immunoglobulin molecules, consists of two polypeptides, each with two domains extending from the plasma membrane (see Fig. 19-12). The domain that ties each polypeptide to the cell surface shares sequence similarities to the constant regions of antibody molecules. The outer domains form a cleft binding an antigen fragment.

Both class I and class II MHC have variable and constant domains. However, unlike antibody genes, each version of an MHC is encoded in its own gene. Thus differences between the MHC molecules of different individuals depend on different genes and alleles rather than genetic rearrangements. The number of genes and alleles is so extensive that no two individuals except identical twins are likely to have the same class I and II MHC. The sequence differences distinguishing

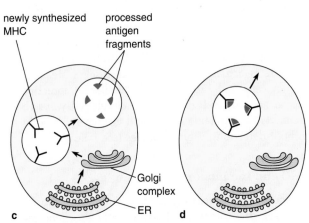

Figure 19-14 Endocytosis, processing, and display of antigen fragments at the cell surface by class II MHC. **(a)** Antigens originating as foreign proteins from outside the cell combine with surface receptors and are taken inside by endocytosis. **(b)** Once inside vesicles in the cytoplasm, the antigens are processed into fragments about 10 to 20 residues in length. **(c)** Newly synthesized class II MHC are subsequently introduced into vesicle membranes by fusion with vesicles from the ER and Golgi complex. **(d)** Some of the antigen fragments produced by the processing reactions fit the binding sites of the MHC. **(e and f)** The antigen fragments bound to MHC are brought to the cell surface for display by fusion of the vesicles with the plasma membrane.

the class I and II MHC of different individuals are concentrated primarily in the variable domains.

The antigen fragments displayed by MHC arise from cellular *processing reactions* within the cell (Fig. 19-14). The peptide fragments arise from different sources for class I and class II MHC. For class II MHC, the best understood of the two types, the fragments originate from foreign proteins taken into leukocytes by endocytosis. Once inside, the antigens are processed in vesicles by proteolytic and other enzymes into fragments about 8 to 20 residues in length (Fig. 19-14*a* and *b*). Later, newly synthesized class II MHC are introduced into the vesicle membranes by fusion with vesicles from the ER and Golgi complex (Fig. 19-14*c*). Of the several to many antigen fragments produced by the processing reactions, some fit the binding sites of one or more of the class II MHC (Fig. 19-14*d*). The selected antigen fragments, bound to the class II MHC, are then brought to the cell surface for display by fusion of the vesicle with the plasma membrane (Fig. 19-14*e* and *f*).

The peptide fragments displayed by class I MHC follow a similar pathway except for their intracellular origin. The proteins from which these fragments are derived originate from infecting viruses or other parasites, or from the cell itself. The proteins are digested in the cytoplasm, possibly by a multienzyme complex

known as a *proteasome,* and then carried through ER membranes. In the ER, the fragments join with class I MHC undergoing assembly. From this point the fragments are delivered to the cell surface by the usual ER-Golgi complex-secretory vesicle route. For either class I or class II MHC, the display of antigen fragments at the cell surface is termed *antigen presentation.* The two MHC pathways are not entirely exclusive; for example, class II MHC have been demonstrated to display polypeptide fragments of cellular origin in some cases.

MHC Binding by T Cells

T cells possess surface receptors that, like the antibodies synthesized by B cells, are tailored to recognize and bind specific antigens. Unlike an antibody, however, a T-cell receptor can bind an antigen only when it is displayed at a cell surface as a fragment linked to a class I or II MHC. The binding site of an MHC folds around the antigen fragment, forming a composite region that includes segments of both the antigen and MHC. This region is recognized by the T-cell receptor (Fig. 19-15). Killer T cells can bind an antigen fragment only when it is displayed by a class I MHC, the type occurring on body cells. Helper T cells can bind an antigen fragment only when it is displayed by a class

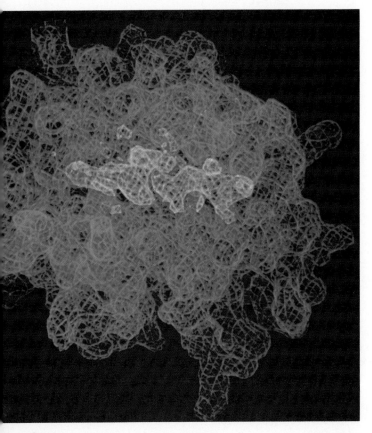

a

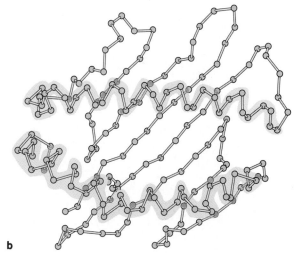

b

Figure 19-15 The MHC–antigen complex, deduced from X-ray diffraction by Wiley and his colleagues. **(a)** A computer reconstruction of an antigen fragment (red) bound to the cleft of a class I MHC for display to a killer T cell, looking directly into the cleft. **(b)** A diagram of the cleft. Courtesy of D. C. Wiley.

II MHC, the type occurring on cells of the immune system.

Although each T-cell clone can recognize only a single antigen, the entire population of T cells in an individual, like the antibodies of B cells, is collectively able to recognize and bind essentially any antigen, whether natural or artificial.

The T-Cell Receptor and Its Genes A T-cell receptor is encoded in genes related to but distinct from antibody genes. Variability in the receptor polypeptides is generated through genetic rearrangements by a mechanism closely similar to the process assembling the variable and constant regions of antibody polypeptides.

A T-cell receptor consists of two dissimilar polypeptides, each with a variable and a constant region (Fig. 19-16a). The variable region of both polypeptides extends away from the surface of T cells; the constant domain includes a transmembrane segment that anchors the receptor to the plasma membrane. The variable regions of the two polypeptides combine to form a binding site that recognizes an antigen fragment presented by an MHC (Fig. 19-16b).

Two forms of T-cell receptor have been identified. The most common form, which occurs on the surfaces of almost all T cells circulating in the body, has the two polypeptides α and β. The alternate form, which has the two polypeptides γ and δ, appears to be confined to T cells associated with body epithelia, especially in the skin, lungs, and intestines.

The gene groups encoding T-cell receptor polypeptides, like the antibody heavy chain genes, consist of clusters of V, D, J, and C elements separated by extensive spacers. One each of the V, D, J, and C elements is selected at random and linked together into a continuous, single α-, β-, γ-, or δ-polypeptide gene. Complementary sequences similar to the R sequences of antibody genes lie in the same arrangement as in antibody genes (as in Fig. 19-6b), indicating that the mechanism combining V, D, and J elements in T-cell receptor genes probably follows the same pathway as the rearrangements generating antibody genes. The random assembly of V, D, and J elements produces an essentially endless variety of T-cell receptors, one for each T cell and its descendants.

As in antibody genes, imprecision in V-D-J joining contributes to the variability of the T-cell receptor.

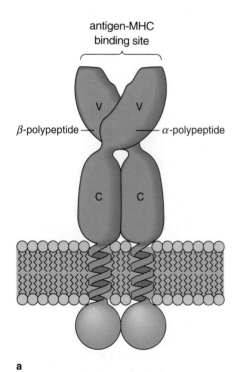

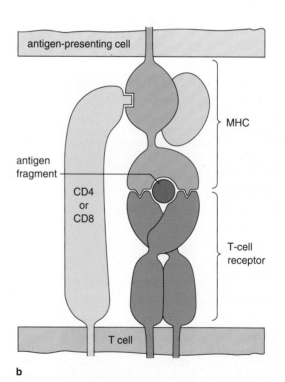

a   b

Figure 19-16 The T-cell receptor. **(a)** The receptor consists of two polypeptides, each with a constant (C) and a variable (V) domain. The variable domains of the T-cell receptor, like their counterparts in immunoglobulin molecules, are encoded in genes assembled by genetic rearrangements. A receptor assembled from polypeptides α and β, the type occurring on circulating T cells, is shown. **(b)** A T-cell receptor binding to an MHC in combination with an antigen fragment. The combination is stabilized by a CD4/CD8 receptor also carried on T cells, which recognizes and binds the constant regions of MHC. The diagram shows a class I MHC, which would be recognized by CD8.

However, there is as yet no evidence that somatic mutation, another source of variability in antibody genes, operates in T-cell receptor genes. The lack of somatic mutation may act as a failsafe to keep the receptor from changing to a type that could respond to a protein of the individual, which would activate a B cell capable of producing autoantibodies.

T-Cell Differentiation and the Development of Immune Tolerance   T cells originate, as do all lymphocytes, from the division of cells in the bone marrow called *hematopoetic stem cells.* T cells subsequently develop in the thymus, a glandlike organ just beneath the breastbone in humans. Early in human development embryonic T cells migrate from the bone marrow to the thymus, where the genetic rearrangements that produce T-cell receptors take place.

During T-cell development in the thymus, T cells with receptors that can recognize molecules of the individual itself are eliminated or made nonfunctional, so that they cannot contribute to autoimmune reactions in the body. T cells that fail to react with self-antigens are preserved and released to the body. The elimination of self-reacting T cells, called *clonal deletion,* and the retention of those that fail to react with self, produce *immune tolerance.* Although the molecular processes underlying the development of immune tolerance are not completely understood, they are considered to depend on interactions between antigens, T-cell receptors, MHC molecules, and another receptor type, the *CD4/CD8 receptors,* in the thymus. The CD4/CD8 receptors recognize and bind constant regions on MHC molecules, CD4 binding to the constant region of class II MHC, and CD8 to the constant region of class I MHC (see Fig. 19-16b).

One of the major hypotheses explaining the interaction of these factors in the development of immune tolerance is shown in Figure 19-17. When they arrive

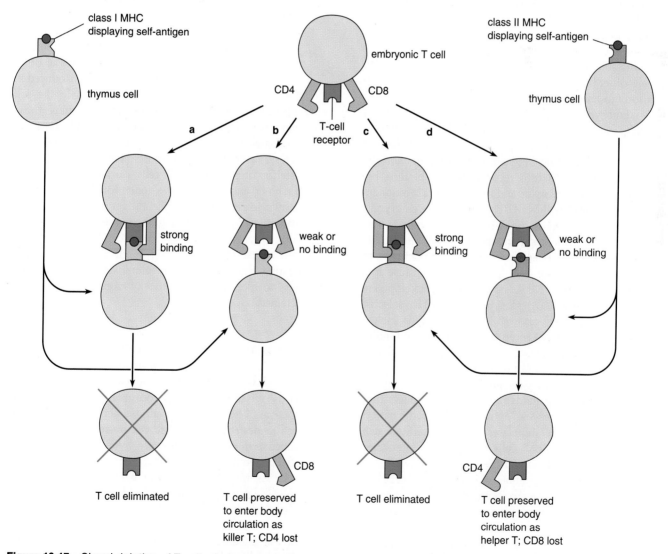

Figure 19-17   Clonal deletion of T cells that can recognize self-antigens in the development of immune tolerance (see text).

in the thymus, embryonic T cells produce both CD4 and CD8 receptors in addition to their individual T-cell receptor type. Within the thymus are cells bearing class I or class II MHC in combination with peptide fragments from the embryo itself. These fragments are derived from proteins forming parts of the thymus or from intact polypeptides or fragments carried into the thymus from other regions of the developing embryo.

Embryonic T cells with receptors that bind class I MHC link to thymus cells that carry this MHC class in combination with a self-polypeptide fragment; the linkage is presumably stabilized by binding between the CD8 receptors carried on the embryonic T cells and the class I MHC. If there is recognition and binding between the T-cell receptor and the peptide fragment carried by the class I MHC (Fig. 19-17a), the embryonic T cell is destroyed. If the T-cell receptor fails to recognize the peptide fragment and does not bind tightly (Fig. 19-17b), the T cell is preserved. Because it links to the class I MHC, it is shunted into the pathway leading to a mature killer T cell. As part of its development, the CD4 receptor is lost. At this point the selected killer T cell is released into regions of the embryo outside the thymus; it carries the CD8 receptor and is able to bind to a class I MHC (carried by body cells other than those of the immune system) displaying a foreign antigen fragment.

Embryonic T cells with receptors that bind class II MHC follow a similar pathway. These T cells link to thymus cells that carry this MHC class; the linkage is stabilized by binding between CD4 receptors and the class II MHC. If there is recognition and binding between the T-cell receptor and the self-peptide fragment carried by the class II MHC (Fig. 19-17c), the embryonic T cell is destroyed. If the T-cell receptor fails to recognize the peptide fragment and does not bind tightly (Fig. 19-17d), the T cell is preserved and continues to develop, in this case as a helper T cell because it is linked to the class II MHC. As part of this developmental pathway, the CD8 receptor is lost. The selected helper T cell is then released from the thymus. Helper T cells carry the CD4 receptor and are able to bind to a class II MHC (carried by B cells or macrophages) displaying a foreign antigen fragment.

The T cells that pass surveillance in the thymus and enter the body fluids remain inactive unless they encounter their specific antigen-MHC combination. Such an encounter stimulates division of the T cell, establishing a clone of T cells, all capable of interacting with the same antigen and the appropriate MHC.

It is likely that a number of embryonic T cells that can recognize self antigens escape the screening process in the thymus. Among these may be some that react too weakly with self antigens to be recognized as self-reactive. Others may escape surveillance because not all polypeptides of the embryo are likely to circulate through the thymus to be presented by MHC. These escaped cells differentiate into helper or killer T cells according to the type of T-cell receptor they carry but somehow normally remain inactive (or *anergic* in the language of immunology) in the body. Although normally inactive, these T cells, which are potentially able to react with self antigens, may become active under unusual conditions. This activation may underlie development of autoimmune disease (see below).

Clonal deletion as the basis of immune tolerance was first proposed by Burnet, the same investigator who advanced the clonal selection hypothesis for the activation of B cells. Many fundamental additions to the model were supplied later by J. Lederberg. Several lines of evidence support the development of immune tolerance through clonal deletion. One is that only 1% to 5% of developing T cells that enter the thymus survive to enter the body circulation. Another line comes from experiments showing that foreign antigens are tolerated if they are injected into mice soon after birth, during the period in which T-cell selection occurs. For example, R. E. Billingham and his coworkers injected cells from one strain of mice (identified as strain B) into a second strain during embryonic development (strain A). Adults from the injected strain A mice tolerated tissue grafts from strain B but not from mice of other strains. Presumably, the foreign antigens introduced in these experiments are presented along with the self antigens to developing T cells in the thymus; those that react with the foreign antigen at this time are eliminated along with those that react with self antigens. In a series of experiments more directly supporting the deletion of self-reactive T cells, carried out by H. von Boehmer and P. Kisielow, genes for T-cell receptors known to recognize specific antigens were introduced into mice by genetic engineering techniques. If the antigen recognized by the receptor type was inserted into mouse embryos during T-cell development, cells bearing the receptor were eliminated; if the antigen was not present, the genetically engineered cells were retained.

The Responses of Killer T Cells The combination bound by a killer T cell—a processed antigen displayed by a class I MHC—usually appears on the surfaces of body cells infected by a microorganism or virus. Binding the class I MHC with its antigen activates the killer T cell to destroy the infected cell by releasing a protein called *perforin*. Perforins have hydrophobic segments that penetrate into the plasma membrane of the infected cell. As they penetrate, the perforins assemble into tubelike structures that open wide pores in the plasma membrane (Fig. 19-18a). The pores permit free diffusion of ions and small molecules, leading to lysis and death of the infected cell (Fig. 19-18b). How killer T cells avoid killing themselves at the same time remains a mystery.

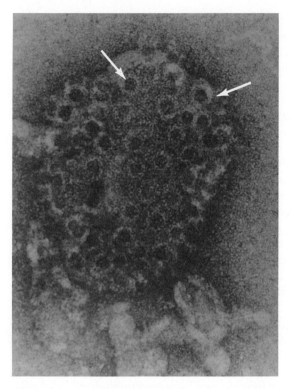

a

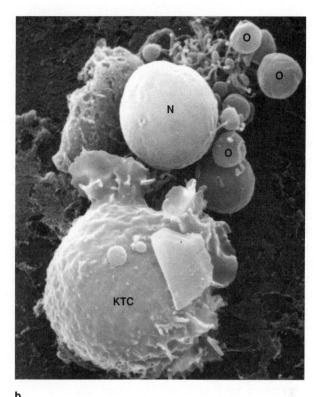

b

Figure 19-18 The action of killer T cells in killing body cells displaying foreign antigens. **(a)** Tubular pores (arrows) assembled by polymerization of perforin molecules released by an active killer T cell. Courtesy of E. R. Podack, from *Proc. Nat. Acad. Sci.* 82:8629 (1985). **(b)** Lysis of a target cell by a killer cell. The lysed cell is toward the top of the figure. N, nucleus of the lysed cell; O, cytoplasmic organelles; KTC, killer T cell. Courtesy of G. Kaplan, from S. Joag et al. *J. Cellular Biochem.* 39:239 (1989).

Killer T cells respond similarly when they encounter a class I MHC on a cell from a different individual. Evidently, the foreign class I MHC is recognized as a class I MHC of the same individual would be if it carried a foreign antigen fragment. Tumor cells displaying altered surface molecules may also be recognized as foreign and lysed by killer Ts.

The Responses of Helper T Cells The combination recognized by the receptors of helper T cells, a processed antigen bound to a class II MHC, appears primarily on B cells. In B cells the antigen is first encountered and bound by IgM or IgA antibodies extending from the B-cell surface. The antibodies with the bound antigen are then taken into the cell by endocytosis, where the antigen is released, processed, and attached to a class II MHC. The MHC with the processed antigen is subsequently introduced into the plasma membrane as shown in Figure 19-14*e* and *f*.

When a helper T cell binds an antigen displayed by a class II MHC on a B cell, the cellular response is very different from that of a killer T. Instead of inducing lysis, the helper T secretes several growth factors known as *interleukins* (*IL*). *IL-4*, *IL-5*, and *IL-6*, the in-

terleukins released by a T cell that encounters a B cell displaying an antigen, stimulate the B cell to enter rapid division and switch to secreted forms of the antibody. All members of the B-cell clone produced by the division make antibodies against the antigen recognized by both the original B cell and the helper T cell responding to it.

The interaction between helper T and B cells explains why many potentially self-reactive B cells remain inactive in the body. Many B cells capable of producing antibodies that can recognize and combine with body proteins are produced and retained. However, these B cells normally remain inactive because they do not encounter a helper T cell with a receptor that can recognize their displayed antigen—such T cells are eliminated in the thymus or are converted to the anergic state. It is also considered possible that an as yet undiscovered mechanism may screen and eliminate some self-reactive B cells.

Several drugs used to suppress activity of the immune system have helper T cells as their targets. *Cyclosporin A*, for example, is used routinely after organ transplants to reduce the chance that the immune system will reject the transplant. The drug evidently works by blocking a Ca^{2+}-dependent signal pathway that triggers T-cell activation. A major problem with cyclosporin A and other immunosuppressive drugs is that, while they lower the chance of transplant rejection, they leave the treated individual significantly more susceptible to bacterial, fungal, and viral infections.

B Cells and the Secretion of Antibodies to the Body Exterior

We have noted that after interacting with helper T cells, B cells switch to antibody secretion. One of the secreted antibody classes, IgA, deserves special attention because it is released to the outside of the body. It is also of interest because yet another cell surface receptor encoded in genes of the Ig superfamily, the *poly Ig receptor*, is involved in IgA secretion. This receptor consists of a cytoplasmic tail, a transmembrane domain, a constant domain, and four variable surface domains with sequence similarities to antibody proteins (Fig. 19-19a). The poly Ig receptor can recognize and bind the constant regions of both IgA and IgM antibodies.

The IgA secreted by B cells into the blood and plasma fluids is bound to poly Ig receptors on the surfaces of epithelial cells that line body cavities. The binding occurs on the sides of the epithelial cells that face the body interior (Fig. 19-19b). Once bound to the inner surfaces of epithelial cells by the poly Ig receptor,

the IgA antibodies enter the epithelial cell cytoplasm by endocytosis. The endocytotic vesicle containing the IgA antibodies is transported through the epithelial cell cytoplasm until it reaches the opposite side of the cell. This side forms the surfaces of mucous membranes or lines a body cavity communicating with the exterior, such as the digestive tract or the ducts of salivary, mammary, and tear glands. On this side of the epithelial cell, the vesicle fuses with the plasma membrane, placing the poly Ig receptors and their attached IgA antibodies on the cell surface. At this point the poly Ig receptor is cleaved at its base, releasing the IgA antibodies with a segment of the receptor still attached.

A recent investigation by J. E. Casanova and his colleagues indicates that phosphorylation of the poly Ig receptor is an important signal routing the receptors and their bound antibodies through the cytoplasm. The receptor is phosphorylated at a serine residue as it enters the cytoplasm; substitution of an alanine, which cannot be phosphorylated, for the serine results in very slow progress of the receptor through the cytoplasm. Substitution of an aspartic acid for the serine (aspartic acid mimics the negative charge of a phosphate group) results in typically rapid transport through epithelial cells.

The released IgA antibodies bind antigens in the body cavities or in secretions such as saliva, tears, and milk. On mucous membranes, such as those of the vagina in females, the released IgA antibodies form a sort of "antiseptic paint" that helps eliminate antigens

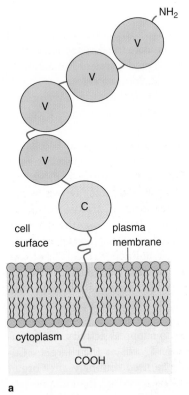

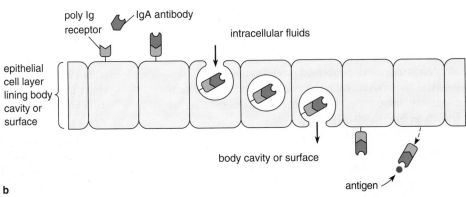

Figure 19-19 Secretion of antibodies to the cell exterior. **(a)** The poly Ig receptor, a protein containing constant (C) and variable (V) domains with amino acid sequences related to antibody domains. **(b)** Movement of IgA antibodies through epithelial cells to reach body cavities or surfaces (see text).

and infective agents. A similar mechanism, involving a different receptor carried on placental cells, transports IgG antibodies from the maternal circulation into the bloodstream of the developing fetus. (A blood barrier formed by cells of the placenta prevents the maternal and fetal blood from directly mixing.)

MALFUNCTIONS OF THE IMMUNE SYSTEM

Although the complex mechanisms of the immune system defend vertebrates remarkably well against invading cells and molecules, the system is not foolproof. Some malfunctions of the immune system cause reactions against proteins or cells of the body itself, producing autoimmune diseases. In addition, some viruses and other pathogens have evolved means to escape surveillance and destruction by the immune system. A number of these pathogens, including the virus causing AIDS, even use elements of the immune response to promote infection.

Autoimmune Diseases

One of the most remarkable features of the normal immune system is that foreign molecules but not molecules forming parts of the body are recognized and eliminated. As we have seen, a major part of this self-tolerance depends on elimination by the thymus of T cells capable of recognizing body proteins as antigens. The absence of these T cells inhibits the activation of B cells capable of making antibodies against body proteins. However, the mechanisms of self-tolerance are slightly "leaky," and almost all individuals probably produce antibodies against their own cells or proteins at some time. In most cases the effects of such anti-self antibodies are not serious enough to produce recognizable disease. However, in some individuals, about 5% to 10% of the population, the attack on self becomes a serious problem.

Several autoimmune diseases are not uncommon. *Insulin-dependent juvenile onset diabetes* results from an autoimmune attack on a specific protein on the surfaces of the β-cells producing insulin in the pancreas. The attack gradually eliminates the β-cells over a period of a few years until the individual is incapable of producing insulin. In *acquired hemolytic anemia* an autoimmune response destroys circulating red blood cells. *Lupus erythematosus* is caused by the production of a wide variety of autoantibodies active against blood cells, blood clotting factors, and internal cell molecules and structures such as mitochondria and the polypeptides of snRNAs (see p. 626). The disease is characterized by anemia, uncontrolled bleeding, and circulatory and renal difficulties produced by the deposition of antibody-antigen complexes in capillaries and kidney glo-

meruli. *Rheumatoid arthritis* results from an autoimmune attack on connective tissue, particularly in the joints, causing pain and inflammation. *Myasthenia gravis* is associated with the production of a variety of autoantibodies, frequently including one against the receptor for acetylcholine, one of the primary elements in the transmission of nerve impulses across synapses (see p. 223). The anti-acetylcholine receptor antibodies may account for the extreme muscular weakness of patients with this disease. In *multiple sclerosis* antibodies produced against myelin basic protein, a component of the myelin sheaths (see p. 219) insulating the surfaces of nerve cells, seriously disrupt the nervous system.

Almost all autoimmune diseases are related to the genetic alleles determining the structure of MHC molecules, particularly class II MHC. Individuals possessing certain alleles are much more susceptible to development of autoimmunity. For example, J. A. Todd and H. O. McDevitt discovered that individuals with alleles placing alanine, valine, or serine at position 57 in the amino acid sequence of the β-polypeptide of class II MHC are much more susceptible to insulin-dependent juvenile onset diabetes (Fig. 19-20); the presence of aspartic acid at this position provides resistance to the disease.

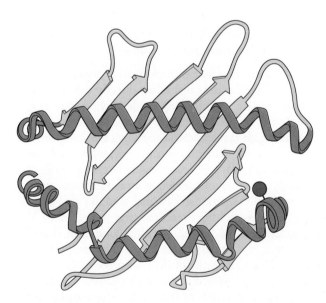

Figure 19-20 The site (red sphere) in the region binding antigen fragments in class II MHC that confers susceptibility or resistance to insulin-dependent juvenile onset diabetes. Individuals with class II MHC alleles placing alanine, valine, or serine at this position (57 in the amino acid sequence of the β-polypeptide) are much more susceptible to the disease; aspartic acid at this position provides resistance. Redrawn from an original, courtesy of D. C. Wiley. Reprinted by permission from *Nature* 329:506; copyright © 1987 Macmillan Magazines Ltd.

The molecular events initiating these and other autoimmune diseases are unknown in most cases. In some instances the onset of an autoimmune disease may be traced to environmental stress or to an injury that exposes body cells not normally associated with lymphocytes to cells of the immune system. In other cases autoimmune disease is triggered by a massive infection in which the body is exposed to an antigen that stimulates production of antibodies that by chance can cross react with some body proteins. For example, antibodies formed against infecting streptococcus bacteria sometimes also react with human heart tissue, producing a form of rheumatic fever; antibodies against the Epstein–Barr and hepatitis B viruses are also active against myelin basic protein, the body protein attacked in multiple sclerosis. Antibodies produced through injection of a vaccine sometimes also cross react with body cells or molecules.

Many of the more serious autoimmune diseases appear early in life, during adolescent or early adult years. Another peak in onset occurs during the forties or fifties. A general failure of immune tolerance, along with a generalized reduction in the effectiveness of the immune system, often occurs in old age. For some unknown reason most autoimmune diseases are more common—some as much as three or four times—in females than in males.

Infective Agents That Defeat or Sidestep the Immune Response

A number of pathogenic viruses or organisms have evolved mechanisms that either directly defeat the immune system by inhibiting or destroying part of the response, or circumvent it by one means or another. Several pathogenic bacteria, including those responsible for rheumatic fever, gonorrhea, meningococcal meningitis, and relapsing fever, regularly change their surface groups to avoid detection by the immune system. In many viruses the "spikes" that bind the viral coat to cell surfaces during infection (see p. 24) have the segments bound by cell receptors located in clefts or pits that are too narrow for entrance by antibody proteins. Other segments of the coat proteins exposed to antibodies are altered so rapidly by constant mutations that the virus avoids or reduces the effectiveness of an immune response. These systems are used by many viruses causing human disabilities, including those responsible for polio, influenza, the common cold, and AIDS.

Many viruses take advantage of features of the immune system to get a free ride to the cell interior. For example, HIV (for *Human Immunodeficiency Virus*), which is responsible for AIDS (*Acquired Immune Deficiency Syndrome*) has a surface molecule that is recognized and bound by the CD4 receptor on helper T cells. Binding to the receptor facilitates entry of the virus into T cells, their primary site of infection.

HIV causes catastrophic failure of the immune response, primarily through the gradual destruction of helper T cells. Elimination of these vital links in the activation of B cells eventually disables the immune system. Although incapacitation of the immune system does not directly kill a person infected with HIV, a host of secondary infections such as fungal diseases and pneumonia, unopposed by an immune response, usually does. (Further details of HIV and the characteristics of AIDS are presented in Supplement 19-1.)

The trypanosome causing African sleeping sickness, *Trypanosoma bruceii*, employs a unique system of genetic rearrangements and switches in gene activity to stay ahead of the immune response (see Information Box 19-1). The rearrangements and activity changes continually generate new cell surface markers, altering them frequently so that individuals escape destruction long enough to initiate further rounds of cell division and infection.

T. bruceii is transmitted by the bite of the tsetse fly, a common insect in equatorial Africa. The trypanosomes enter the bloodstream during a bite and then multiply until the body mounts an effective immune response, a reaction that takes about seven days. During this time a small number of the trypanosomes change their surface markers. These trypanosomes are not recognized by antibodies produced against the initial population of infecting cells and survive the initial immune response. The surface changes continue, constantly producing a few offspring that avoid each successive immune response and act as founders of the next population. There is no cure for the disease, which eventually leads to extreme lethargy and death as the trypanosomes invade the central nervous system.

ANTIBODIES IN RESEARCH

The ability of higher vertebrates to produce antibodies specifically recognizing and binding antigens of an almost unbelievable variety has been used in many lines of scientific research, particularly in the identification and isolation of cellular molecules. By adding markers such as heavy metals or fluorescent dyes to antibodies, the cellular locations of molecules of various kinds can be highlighted for electron or light microscopy. Antibodies attached to gels in separatory columns (see p. 131) have been used to extract and purify single proteins from among mixtures obtained from whole cells. Antibodies have also been widely used to interfere with the activity of specific molecules inside cells, thereby providing clues to their normal cellular functions. In addition, antibodies are being developed as a possible

The Production of Variable Surface Glycoprotein in *Trypanosoma Bruceii*

The surface marker responsible for the ability of *Trypanosoma bruceii* to avoid destruction by the immune system, the *VSG marker* (*VSG* = *V*ariable *S*urface *G*lycoprotein), is a membrane glycoprotein containing a variable region. This region of the glycoprotein differs in the new markers produced as the trypanosomes change from one surface marker to the next. VSG surface markers are so closely packed on the trypanosome cell surface that no other surface antigens are exposed for recognition by the immune system.

The genes encoding VSG markers occur in thousands of copies, amounting in total to about 10% of the trypanosome genome. The copies are distributed over many of the chromosomes of the protozoan, often near the telomeres. Each gene copy codes for a distinct version of the VSG protein. All of the many gene copies lack functional promoters and remain inactive except for those that move by genetic rearrangements to chromosomal regions called *expression sites* (see the figure in this box). *T. bruceii* has 20 expression sites, each containing a promoter capable of activating a VSG gene, and each carrying a different version of the gene. At any time only one of the genes at the expression site is active; the activity of one expression site apparently inhibits the others. Activity regularly switches from one expression site to another, however, leading to the appearance of a different VSG protein on the trypanosome surface. The variation of VSG groups thus results from a combination of genetic rearrangements and switches in activity from one expression site to another.

Most movements of VSG genes to expression sites occur without removal of the original copy from the VSG gene bank. In this way the movement resembles the migration of some types of transposable elements (see p. 763). Change from one VSG gene to another is effected by excision of the existing VSG gene copy from an ex-

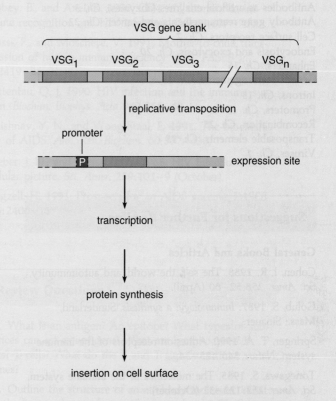

pression site and insertion of a copy of a different VSG gene. Repeated sequences in the 5'- and 3'-flanking ends of each VSG gene copy may be involved in the mechanism accomplishing the movements. Changes resulting from movements of VSG genes and switches in expression site occur at a low but regular frequency of about 1 in every 10,000 to 100,000 cells, producing the new VSG types that are missed by a mounting immune response.

means for delivering toxic substances to specific cell types, such as those of tumors. (Details of these *immunotoxins*, as they are called, are presented in Ch. 20.) For all of these purposes, monoclonal antibodies provide exceptional and exclusive specificity for the molecules or systems under study.

Antibodies have also provided critical insights into the mechanisms by which enzymes operate. By using molecules with structural similarities to the transition state for a chemical reaction as antigens, investigators have induced test animals to develop antibodies capable of catalyzing chemical reactions. The resulting antibody enzymes, or *abzymes*, have demonstrated the importance of the transition state in enzymatic catalysis and also have provided a way to produce "designer enzymes"—enzymes capable of catalyzing a desired reaction. (Details of abzymes and their importance to studies of enzymatic catalysis are presented in Ch. 3.)

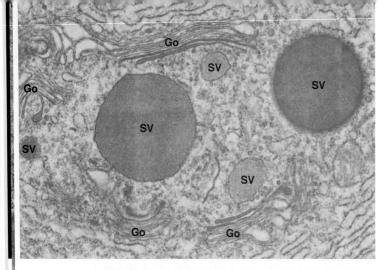

Figure 20-8 A group of Golgi complexes (Go) and secretory vesicles (SV) in a gastrodermal cell of a coelenterate. A moderately dense, granular substance, probably a secreted protein, is visible in the *trans* elements of the Golgi complexes, in small vesicles of intermediate size, and in the large secretory vesicles. The smaller vesicles fuse to form secretory vesicles after budding from the Golgi complex. × 16,000. Courtesy of H. W. Beams and R. G. Kessel.

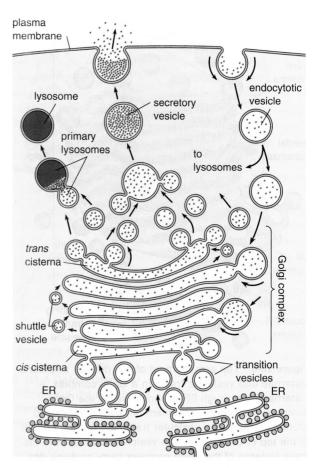

Figure 20-9 Vesicle traffic to and from the Golgi complex. Transition vesicles carry newly synthesized proteins and glycoproteins with their core carbohydrate structures from the ER to the *cis* cisterna of the Golgi complex. Proteins and glycoproteins then move through the Golgi complex from the *cis* to the *trans* face in shuttle vesicles that bud from one cisterna and fuse with the next in line (from the bottom to the top of the Golgi complex in the diagram). At the *trans* cisterna vesicles bud off and fuse to form lysosomes and secretory vesicles. Some material also enters the Golgi complex from outside the cell via endocytotic vesicles.

extensive assembly of enzymes. Among these are enzymes adding or removing sugar groups to complete the processing of glycoproteins and glycolipids. Other Golgi enzymes add sulfate, acetyl, and phosphate groups to glycoproteins and glycolipids, add fatty acids to proteins, and remove segments from the interior or ends of the amino acid chains of proteins. (Further details of processing reactions taking place in the Golgi complex are given below and in Supplement 20-2.)

Also present are enzymes or complexes of more generalized function, such as an H^+-ATPase pump that lowers the internal pH of cisternae at the *trans* face of the Golgi complex. Golgi membranes probably also contain proteins concerned with sorting and routing molecules through the complex.

Some Golgi enzymes are restricted to animals or plants. Enzymes that add sugar groups to proteoglycans, the complex molecular assemblies that form parts of the extracellular matrix, occur only in animal Golgi complexes (see p. 279). Plant Golgi membranes contain a complex collection of enzymes that assemble sugars into precursors of cell wall molecules.

Many of the enzymes and other molecules associated with the Golgi complex are unequally distributed among the Golgi cisternae, indicating that different cisternae have distinct and specialized functions. The enzymes adding sugar groups to glycoproteins and glycolipids, for example, are distributed in Golgi cisternae roughly in order of the reactions they catalyze— enzymes catalyzing early steps are concentrated in *cis* cisternae and those catalyzing later and final steps in

medial and *trans* cisternae. The distribution of these and other enzymes in the cisternae of the Golgi complex, in fact, may ensure that the reaction sequences take place in the correct order.

There are other chemical and morphological differences among cisternae of the Golgi stacks. Golgi membranes tend to be noticeably thinner on the *cis* side of the complex, approximating the dimensions of ER membranes. On the *trans* side the membranes are thicker, similar in dimensions to the plasma membrane. Cholesterol, a lipid typical of plasma membranes, occurs in higher concentrations in *trans* than in *cis* cisternae. (Table 20-1 summarizes the distribution of enzymes and biochemical properties in the Golgi cisternae.)

Table 20-1 Compartmentation of Golgi Components and Conditions

Component or Condition	Cis Cisternae	Medial Cisternae	Trans Cisternae
Membrane thickness	Thinner, similar to ER	Intermediate	Thicker; similar to plasma membrane
pH	Near neutral	More acid	More acid
Fatty acylase	+		
Mannosidase I	+		
Acetylglucosamine transferase I		+	
Mannosidase II		+	
NADPase		+	
Phosphatases		+	
Adenylate cyclase	+	+	+
5' nucleosidase	+	+	+
Acid phosphatase			+
Galactosyl transferase			+
Nucleoside diphosphatase			+
Sialyl transferase			+
Thiamine pyrophosphatase			+

Much of the work localizing enzymes within *cis*, medial, or *trans* cisternae has been accomplished by reacting Golgi complexes with antibodies developed against individual enzymes. The antibodies, tagged with metal atoms to make them visible under the electron microscope, mark the compartments in which the enzymes are concentrated. The *cis*, medial, and *trans* cisternae can also be separately purified by centrifugation in sucrose density gradients.

The number of Golgi complexes per cell varies greatly in different tissues or species. Although the average number is about 20, some cell types may have many more. For example, corn root tip cells contain several hundred complexes; more than 25,000 have been counted in cells of *Chara*, a green alga. In plants Golgi complexes are more or less evenly scattered through the cytoplasm. In animal cells Golgi complexes are frequently concentrated in masses near the ER or just outside the nucleus, in a region near the centrioles or cell center. In motile animal cells the region containing the cell center and Golgi complexes typically takes up a position on the side of the nucleus toward the direction of movement.

Microtubules have been implicated as cytoskeletal supports holding both the ER and Golgi complexes in position in animal cells. Destroying the microtubule network with agents such as colchicine causes both the ER and Golgi complex to contract and lose their typical orientation in the cytoplasm. Elements of the ER and Golgi complex, as well as vesicle traffic moving between the Golgi complex and plasma membrane, may also travel along microtubules toward either the cell center or periphery. These movements are probably powered by "motors" such as dynein, kinesin, and dynamin (see p. 429) that use energy released by ATP hydrolysis to push the ER and Golgi cisternae and vesicles along the microtubules.

Evidence That Proteins Move Sequentially Through the System

The movement of proteins through the ER and Golgi complex has been followed by a variety of techniques. The first evidence was developed by L. G. Caro and G. E. Palade, who gave pancreatic cells a pulse of radioactive amino acids and followed the progress of label through the cells by autoradiography (see p. 118). (Palade received the Nobel Prize in 1975 for his pioneering work tracing cellular distribution pathways.) In this and many confirming experiments by later investigators, label first appeared in rough ER membranes and then, within several minutes, in rough ER vesicles and cisternae (Fig. 20-10a). Soon after this, within 10 to 30 minutes, label appeared in transition vesicles and *cis* cisternae of the Golgi complex (Fig. 20-10b). Subsequently, label appeared in medial and *trans* cisternae of the Golgi complex and in secretory vesicles between the Golgi complex and the plasma membrane (Fig. 20-10c). Finally, within one to four hours after initial exposure, radioactivity appeared in the extracellular space, showing that the secretory vesicles eventually discharge their contents to the cell exterior. Similar experiments using labeled amino acids or sugar units, or antibodies developed against individual proteins, have traced the routes followed by proteins destined to enter lysosomes or storage vesicles or to become components of the plasma membrane. Cell fractionation and centrifugation (see p. 123) have also been widely employed to trace protein movement and distribution through the ER, Golgi complex, and later destinations.

In addition, the movement of proteins and other molecules through the ER and Golgi complex has been followed by tagging them with markers making them visible under the light or electron microscopes. Heavy metals are frequently used as markers for electron mi-

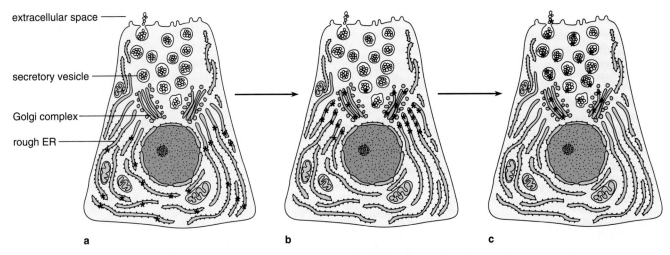

extracellular space

secretory vesicle

Golgi complex

rough ER

a   b   c

Figure 20-10 Results of a typical experiment tracking the progress of newly synthesized proteins through the ER and Golgi complex by autoradiography. The proteins have been made radioactive by exposing cells of the guinea pig to a three-minute pulse of tritiated leucine. **(a)** Three minutes after exposure, only the rough ER is labeled. **(b)** Three-minute exposure to the label, followed by a 17-minute "chase" by exposure to unlabeled medium. Label is distributed in the rough ER, Golgi complex, and secretory vesicles. **(c)** Three-minute exposure followed by 117-minute chase. Label is concentrated in the secretory vesicles. Redrawn from an original courtesy of P. Favard. From *Handbook of Molecular Cytology*, North-Holland Publishing Company, Amsterdam, 1969.

croscopy; for light microscopy fluorescent dyes are commonly used. The marking approach is readily combined with antibody studies by attaching the dyes or heavy metals to antibodies developed against proteins under study. Some of these experiments showed that proteins synthesized by ribosomes attached to the outer membrane of the nuclear envelope make their way to the ER and other elements of the distribution system

(Fig. 20-11). The experiments using these and other techniques also traced the routes of proteins traveling in the opposite direction, from the plasma membrane or exterior to the cell interior, by endocytosis.

One of the many interesting questions about the distribution system concerns travel within the Golgi complex itself. Cisternae were once thought to move through the Golgi complex by a mechanism known as

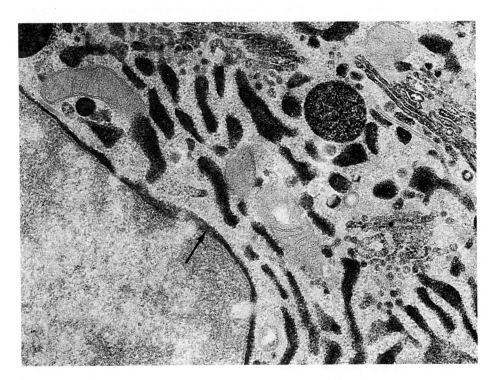

Figure 20-11 Location of a newly synthesized enzyme, peroxidase, in the perinuclear compartment (arrow) as well as cisternae of the ER in a cell of the uterine lining of the rat. The locations of the enzyme have been traced by means of a reaction that precipitates a heavy metal deposit wherever the enzyme is present. ×32,000. Courtesy of A. R. Hand, from *J. Cell Biol.* 74:399 (1977), by copyright permission of the Rockefeller University Press.

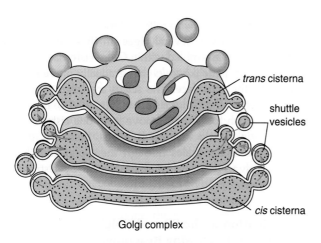

Figure 20-12 Movement of materials from *cis* to *trans* cisternae of the Golgi complex via shuttle vesicles.

cisternal progression. They were believed to assemble at the *cis* face through the fusion of transition vesicles, move progressively through the complex as an intact unit, and disassemble at the *trans* face by budding into secretory vesicles. Proteins were considered to be carried from the *cis* to the *trans* face by this progressive cisternal movement.

More recently, cell biologists have become convinced that individual Golgi cisternae remain more or less fixed in position, and that proteins and glycoproteins move from one cisterna to the next via *shuttle vesicles* (Fig. 20-12). In this mechanism, first proposed by M. G. Farquhar, the shuttle vesicles are thought to bud from one cisterna, travel the intervening distance, and fuse with the next cisterna in line. Eventually, proteins and glycoproteins reach the cisterna at the *trans* face. At this point budding forms secretory vesicles, lysosomes, or storage vesicles.

A major line of evidence supporting the stationary cisternae model comes from a group of experiments carried out by J. E. Rothman and his colleagues with two Chinese hamster cell lines. In one line galactose units were added normally to complete assembly of glycoprotein carbohydrate groups in the Golgi complex. In the other, mutant line the enzyme adding the terminal galactose units was missing from the Golgi complex. The mutant line was infected with *vesicular stomatitis virus (VSV)*, which induces infected cells to make large quantities of a viral glycoprotein that requires the terminal glycosylation. Because of the missing enzyme, the mutant cells were unable to complete processing of the viral protein. The mutant cells were then fused with uninfected cells of the wild-type line possessing the required enzyme. As a result of the cell fusion, the wild type and mutant cells shared the same cytoplasm. Within minutes, too quickly for viral proteins to be newly synthesized, completely processed viral glycoproteins appeared in the fused cells. Presum-

ably, shuttle vesicles carried incompletely processed viral glycoproteins from *cis* cisternae of mutant Golgi complexes to *trans* cisternae of normal Golgi complexes, where the enzymes missing in the mutants added the required galactose units.

More recent work in the Rothman laboratory indicates that the shuttle mechanism requires several proteins held in solution in the cytoplasm (Fig. 20-13). One of these proteins forms a coat around the vesicles as they develop. The coat is removed before the vesicles fuse with the next cisterna, in a step that involves GTP

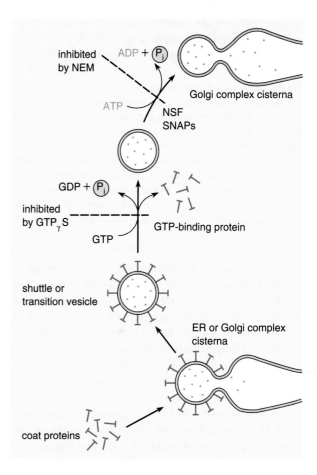

Figure 20-13 Factors involved in movement of shuttle vesicles in the Golgi complex, and in movement of transition vesicles form the ER to the *cis* cisterna of the Golgi complex. The vesicles pick up a protein coat as they bud from either an ER or Golgi complex cisterna and move through the distance separating them from the target cisterna. Just before fusing with the target cisterna, the vesicle sheds its protein coat in a step that involves a GTP-binding protein and requires the binding and hydrolysis of GTP. The uncoating step, which is required for vesicle fusion, is blocked by GTP$_\gamma$S, a nonhydrolyzable GTP analog. After uncoating, fusion takes place; this step requires a fusion protein, NSF, and is blocked by NEM (*N*-methyl maleimide). SNAPs bind NSF to a membrane receptor as fusion proceeds.

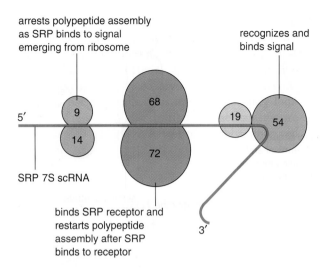

arrests polypeptide assembly
as SRP binds to signal
emerging from ribosome

recognizes and
binds signal

5′

SRP 7S scRNA

binds SRP receptor and
restarts polypeptide
assembly after SRP
binds to receptor

3′

Figure 20-16 The signal recognition particle (SRP) and the functions of some of its polypeptides. The numbers indicate the molecular weights of the SRP polypeptides in thousands.

hydrolyzed as the ribosome attaches to an ER membrane.

Another protein was found to be involved in the attachment of the SRP to ER membranes. This protein, the *SRP receptor* (also called the *docking protein*), which forms a part of ER membranes, binds strongly to an SRP linked to a signal sequence.

The SRP and SRP receptor are believed to participate in the attachment of ribosomes to the ER by the process shown in Figure 20-17. When the signal sequence extends from the ribosome (Fig. 20-17*a*), the SRP binds to the signal and the ribosome along with GTP (Fig. 20-17*b*). Protein synthesis is arrested as the SRP binds. The ribosome-SRP complex is recognized and bound by the SRP receptor in the ER membrane (Fig. 20-17*c*). Binding by the receptor attaches the ribosome firmly to the membrane, and the hydrophobic signal sequence buries itself in the membrane. As the ribosome binds, GTP is hydrolyzed, the SRP is released, and protein synthesis resumes. Extension of the polypeptide into or through the ER membrane then proceeds (as in Fig. 20-15*c* through *f*).

The RNA molecule forming part of the signal recognition particle is now known as *SRP 7S scRNA*. Investigations by Walter and others showed that if this RNA is removed or broken by enzymatic attack, the

preparations. Further investigation revealed that the SRP recognizes and binds the N-terminal signal as it emerges from a ribosome and binds the ribosome as well. GTP is bound as the SRP binds the signal and is

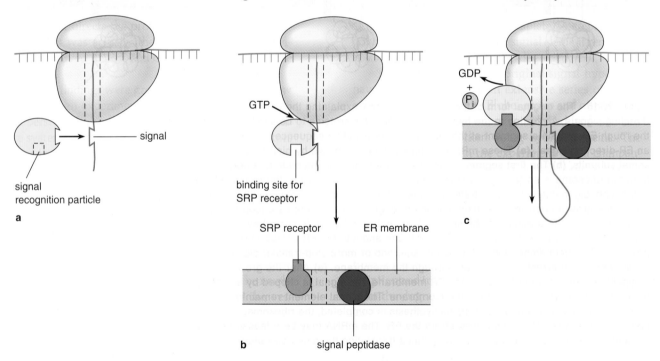

Figure 20-17 Operation of the signal recognition particle and SRP receptor in the ER-directed signal mechanism. **(a)** The mechanism proceeds as in Figure 20-15*a* and *b* until the signal extends from the ribosome. **(b)** As the signal emerges, the SRP binds to the signal and ribosome along with GTP. Polypeptide assembly is arrested as the SRP binds. Binding the signal and ribosome alters the SRP to a conformation in which it is recognized and bound by the SRP receptor in the ER membrane **(c)**. As the ribosome binds, GTP is hydrolyzed, the SRP is released, and protein synthesis resumes. Extension of the polypeptide through the ER membrane then proceeds as in Figure 20-15*c* through *f*.

SRP disassembles. Thus the scRNA probably acts as a backbone holding the SRP together. This conclusion is supported by the ability of SRP particles to self-assemble in fully functional form if the separate polypeptides are added to intact SRP 7S scRNA molecules. Although the RNA evidently holds the structure together, the functions of the SRP in binding the ribosome, signal, and docking protein appear to depend on the protein component of the particle. (Fig. 20-16 shows functions identified with the SRP proteins.)

Definite identification of the SRP and its receptor is presently limited to mammalian systems, but scRNAs with sequences homologous to mammalian SRP 7S scRNA have been detected in amphibians, insects, and higher plants. When assembled with mammalian SRP proteins, amphibian and insect scRNAs can form functional signal recognition particles. In the yeast *Saccharomyces*, Walter and his colleagues found a 54,000-dalton protein that is similar in sequence to the 54,000-dalton protein in the mammalian SRP (see Fig. 20-16). RNAs similar to the SRP 7S scRNA can also be detected in yeast. These observations indicate that an ER-directing signal mechanism similar to that of mammals may exist in these organisms and may indeed be universal among eukaryotes. Even bacteria have RNAs with sequence similarities to SRP 7S scRNA, and a protein similar to the 54,000-dalton protein of the SRP (see p. 871).

Several components besides those shown in Figures 20-15 and 20-17 may be involved in the ER-directing signal mechanism. After the ribosomes attach to the ER, two membrane proteins are found tied to the ribosomes if protein crosslinkers (see p. 547) are added as probes. These proteins, called *ribophorins I* and *II*, may stabilize ER-ribosome binding. Protein crosslinkers also reveal that a 35,000-dalton membrane glycoprotein is located in close proximity to the N-terminal signal after ribosomes attach to the ER. This protein may be a *signal receptor* that binds and stabilizes the signal in the membrane after the SRP delivers it to the ER.

The original signal hypothesis proposed that a protein or group of proteins forms a hydrophilic channel or pore through the membrane through which the newly synthesized protein extends. It is still not clear whether protein channels actually exist. However, a recent experiment by S. M. Simon and Blobel gives strong indications that such channels are actually opened during protein synthesis. Simon and Blobel fused patches of ER membranes containing ribosomes active in protein synthesis with a bilayer separating two compartments containing salt solutions (see p. 163). Adding puromycin to the side of the patches containing ribosomes caused a large increase in conduction of ions through ER membranes. Presumably, by removing growing proteins from ribosomes attached to the ER, the puromycin exposed protein-conducting, hydrophobic channels in the ER membranes. The channels closed when ribosomes were removed from the ER membranes, indicating that the ribosomes interact with the membranes to open the channels.

Cotranslational vs. Posttranslational Protein Insertion The original signal hypothesis proposed that newly synthesized polypeptide chains were pushed out of ribosomes with enough force to drive them through the ER membrane. One essential requirement for such a mechanism is that protein synthesis and passage across the membrane must be simultaneous. In other words, transfer across the membrane must be *cotranslational* for the mechanism to work.

This idea had to be revised when it was discovered that many proteins can enter ER membranes *posttranslationally*, that is, after much or all of their synthesis is complete. Posttranslational entry is particularly common in bacteria, in which the signal mechanism attaches ribosomes to the cytoplasmic surface of the plasma membrane. (The plasma membrane acts as the counterpart of the ER in prokaryotes; see Supplement 20-1.) Some eukaryotic proteins that normally pass through ER membranes cotranslationally have been discovered experimentally to be capable of passing through ER membranes posttranslationally. For example, when synthesized in a cell-free system, the human glucose transporter protein can pass through ER membranes readily if the membranes are added after synthesis of the protein is complete. Some proteins penetrating other membrane systems, such as those of mitochondria, also enter posttranslationally under normal or experimental conditions (see below).

Proteins that enter posttranslationally are usually in an intermediate folding state. ATP is frequently required for posttranslational entry, along with two or more proteins forming part of the soluble cytoplasm. These characteristics have led to the proposition that the cytoplasmic proteins either act as *chaperones* (see p. 684) that stabilize entering proteins in the unfolded state or work as "unfoldases" that use energy from the hydrolysis of ATP or GTP to open proteins to a conformational state allowing their entry into membranes. (Part of SRP function, in fact, is probably to maintain the signal in an unfolded form in which it can penetrate ER membranes.) As the chaperoned proteins pass into or through the membranes, they fold into more compact states. The free energy released during this compaction may supply a component of the energy required to push the proteins through the membrane.

Characteristics of ER-Directing Signals

Compilation of hundreds of ER-directing signals, also called *transit peptides*, *leader sequences*, or *presequences*, has revealed that most contain combinations of amino

The Experimental Process

Redirecting Protein Traffic from Cytoplasm to the Secretory Pathway: A Test of the Signal Hypothesis

Vishwanath R. Lingappa

VISHWANATH R. LINGAPPA received the B.A. degree from Swarthmore College in 1975, after which he pursued graduate research with Gunter Blobel at The Rockefeller University. He received the Ph.D. degree in 1979 and the M.D. degree from Cornell University Medical College in 1980. After residency in internal medicine at the University of California Hospitals in San Francisco, he joined the UCSF faculty where he is Professor of Physiology and Medicine. A Board Certified internist, Dr. Lingappa is engaged in medical student teaching, research in molecular mechanisms of protein trafficking, and general internal medicine practice in San Francisco.

For several years after it was first proposed, the signal hypothesis[1] (see p. 836) was met with skepticism from some quarters. Some critics doubted that a small, discrete sequence could be the sole determinant of chain targeting to, and translocation across, the ER membrane, especially since considerable variation in amino acid sequence occurs between signal sequences even within a single species. Rather, they favored notions such as (1) that global features of protein folding distinguished secretory from cytosolic proteins, or (2) that other specific regions of secretory proteins in addition to the signal sequence might be involved in translocation, or (3) that a specific sequence in cytosolic proteins prevented their translocation. Others wondered if cell-free systems, from which most of the data in favor of the signal hypothesis had been derived, were an adequate reflection of events occurring *in vivo*.

We thought of a simple experiment that would provide a dramatic test of the predictions of the signal hypothesis, and thereby resolve some of the controversy. Our idea was to see if a signal sequence was sufficient to direct a normally cytosolic protein across the ER membrane, both in cell-free systems and in living cells. To do such an experiment we would have to take the DNA encoding the signal sequence from one protein and attach it, using molecular genetic techniques, in frame at the 5'-end of the DNA for the coding region of a normally cytoplasmic protein. The resultant hybrid DNA molecule would encode a protein composed of the signal sequence from the secretory protein followed by the "passenger" coding sequence of the cytosolic polypeptide. If mRNA encoding such a *chimeric protein* was transcribed from this DNA, and then translated into protein either in a cell-free

system or in a living cell, the location of the protein would provide a powerful test of the signal hypothesis.

In planning experiments it is often useful to consider the possible outcomes in a preliminary "thought experiment" before actually investing a major amount of time and effort. We considered the following possibilities:

1. The signal hypothesis is incorrect or overly simplistic. Hence a signal sequence is either unnecessary or, if necessary, may not be sufficient for translocation of a normally cytosolic protein across the ER membrane. In either case, the chimeric protein would remain in the cytoplasm.

2. The signal hypothesis is correct. Hence a signal sequence, being both necessary and sufficient for translocation across ER membranes, would redirect an otherwise cytosolic protein into the secretory pathway.

3. Finally, regardless of whether the hypothesis is correct or incorrect, the experimental approach might be flawed, for either systematic or fortuitous reasons. For example, juxtaposition of the specific sequences from the two proteins composing the chimera might cause the chain to fold into an insoluble aggregate and hence be scored as nontranslocating. A different chimera might have given a different result, and the one we chose to study may have been an (unfortunate) exception to the general rule.

Experimental science is an inductive rather than a deductive process. Thus, a negative result (e.g., lack of translocation of a particular chimeric protein) is relatively uninformative because it does not distinguish between possibilities (1) and (3). On the other hand, a positive result (e.g., translocation of the chimera) suggests that the signal hypothesis is correct for at least a subset of proteins. It does not, however, rule out the existence of other proteins that are exceptions to these rules. A crucial feature of well-planned experiments is the inclusion of controls that allow the results to be properly interpreted. For example, it is necessary to confirm that the chosen cytoplasmic protein *without* the signal sequence is expressed exclusively in the cytoplasm.

Globin, the protein component of hemoglobin, was chosen as the cytoplasmic protein for these experiments. This is a small, highly conserved protein whose tertiary structure is known. A colleague provided a full-length cDNA clone of chimpanzee alpha globin for these studies.

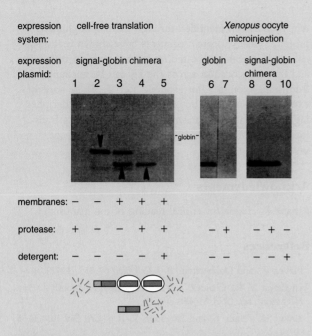

expression system:	cell-free translation					*Xenopus* oocyte microinjection				
expression plasmid:	signal-globin chimera					globin		signal-globin chimera		
	1	2	3	4	5	6	7	8	9	10
membranes:	−	−	+	+	+					
protease:	+	−	−	+	+		−	+	−	+	−
detergent:	−	−	−	−	+		−	−		−	−	+

Figure A Signal globin expression and translocation in a cell-free translation system (with cartoon interpretation) and in *Xenopus* oocytes.

This full-length cDNA had been engineered into a site within the *E. coli* beta lactamase gene. Beta lactamase, being a secretory protein of *E. coli*, has a signal sequence. Moreover, a convenient restriction endonuclease site (see p. 135) at the start of the globin coding region made deletions of the upstream lactamase codons beyond the signal sequence technically quite easy. Thus, for the initial experiments we chose to use the beta lactamase signal sequence. These DNA molecules were constructed behind an active promoter (see p. 589). This allowed us to make mRNA from the cloned DNA in a test tube and then translate this mRNA in a cell-free translation system.

In the absence of added ER membranes, the full chimera was synthesized, which was larger than authentic globin by the size of the signal sequence (Fig. A, lane 2, downward pointing arrowhead). When proteases were added to this translation product it was totally digested, as expected (lane 1). However if ER-derived membrane vesicles were present *during* translation, a different pattern was observed. A new product the size of authentic globin was seen (lane 3, upward pointing arrowhead), in addition to some residual full-length chains. When these products were digested by added proteases, the authentic globin-sized product was protected from digestion while the residual full-length chains were degraded (lane 4). Protection of authentic globin-sized chains from digestion by added proteases was lost if the vesicle bilayer was solubilized with a mild detergent, strongly suggesting that protection from proteases was due to translocation into the vesicle lumen (analogous to translocation into the lumen of the ER *in vivo*) and not due to intrinsic protease resistance of the protein itself. Moreover, if membrane vesicles were added after completion of protein synthesis, no authentic globin-sized products were observed and the protein remained completely protease sensitive.[2] This demonstrated that translocation of globin occurred only while the signal sequence-bearing chain was being synthesized. As a necessary control, globin without a signal sequence was shown not to be translocated even when membrane vesicles were present during its synthesis.[2]

To test whether these results applied to events occurring *in vivo*, we used microinjection to introduce the mRNA for authentic globin or the signal-globin chimera into the cytoplasm of *Xenopus* oocytes, large living cells with an active secretory pathway. mRNA so introduced is translated by ribosomes in the oocyte cytoplasm. Radiolabeled amino acids included in the medium which bathes the oocytes are taken up and incorporated into the newly synthesized proteins. Homogenization of the oocytes converts the ER into vesicles, and analysis of localization can be carried out either by fractionation of the homogenate into cytosol and vesicles or by protease digestion, as previously described. Authentic-sized globin was synthesized from both mRNAs (lanes 6 and 8). However, globin synthesized from the chimeric mRNA was protected from added protease (lane 9) unless detergent was added (lane 10), while that synthesized from globin mRNA not encoding a signal sequence was not protected from added proteases (lane 7). Moreover, fractionation demonstrated that authentic globin synthesized from globin mRNA was entirely in the cytosol while authentic-sized globin synthesized from the chimeric mRNA was entirely in the vesicles[3] (not shown). These data suggested that, in the case of the chimera, authentic globin was generated as a result of cleavage of the signal sequence from the growing chain as it translocated to the ER lumen. Thus, both *in vitro* and *in vivo*, the beta lactamase signal sequence was sufficient to direct chimpanzee alpha globin out of the cytoplasm and into the secretory pathway.[2,3]

An ironic twist to this story was that, some time later, an insect globin gene was cloned and found to have a remarkable structure. Globin is found in the cytoplasm of red blood cells in most animal species. However insects lack red blood cells. Instead, they secrete their globin into a fluid, called hemolymph, which bathes the cells. Thus insect globin, whose mature sequence is quite conserved with other animal globins, is naturally a secreted rather than a cytoplasmic protein. When the mRNA for insect globin was examined, it was found to encode a signal sequence at its amino terminus. Unbeknownst to us, we had been "scooped" by evolution 800 million years ago (when insects branched off from the lineage leading to vertebrates) on the neat idea of redirecting cytoplasmic globin into the secretory pathway!

Subsequent work over the years has shown that these findings are true for a wide range of signal and passenger sequences, although passenger sequences are not always passive in the translocation process. Presumably due to features of chain folding, the choice of the passenger can significantly influence the efficiency of every known step in the process of translocation. Furthermore, at least some passenger domains fold in such a way that they cannot be translocated by a given signal sequence. Nevertheless, the simple experiments described above established that, in principle, a signal sequence would redirect a cytoplasmic protein into the secretory pathway.

Although experiments are usually carried out to answer a specific question, they often are most valuable for the subsequent issues they raise and the new ways of looking at a problem that they foster. The simple experiment described above not only provided a test that many critics of the signal hypothesis found compelling, but also contributed to the realization that protein chimeras were a powerful approach to the study of specific sequences involved in protein trafficking. Subsequent work on trafficking of protein chimeras has allowed the study of protein sequences that target proteins to other intracellular organelles besides those of the secretory pathway (e.g., chloroplasts and mitochondria). Similarly, sequences that cause proteins to stop on their way across particular membranes (a crucial step in the assembly of integral membrane proteins) have been identified through the use of chimeric proteins. Finally, protein chimeras have shown that a signal sequence can be located far from the amino terminus and that translocation across the ER can be uncoupled from translation (see, for example, Ref. 4). These sorts of experimental manipulations have contributed to deciphering the intricate mechanisms of protein targeting, translocation, and trafficking and remain a fruitful line of investigation today.[5]

Acknowledgments

I thank R. Hegde for critical reading of the manuscript.

References

[1] Blobel, G., and Dobberstein, B. *J. Cell Biol.* 67:852 (1975).

[2] Lingappa, V. R.; Chaidez, J.; Yost, C. S.; and Hedgpeth, J. *Proc. Nat'l Acad. Sci. USA* 81:456 (1984).

[3] Simon, K.; Perara, E.; and Lingappa, V. R. *J. Cell Biol.* 104:1165 (1987).

[4] Perara, E.; Rothman R. E.; and Lingappa V. R. *Science* 232:348–52 (1986).

[5] Chuck, S. L., and Lingappa, V. R. *Cell* 68:1–13 (1992).

acids forming three subregions (Fig. 20-18a). At the N-terminal end of the signal, from one to seven amino acids form a polar region that usually includes from one to three positively charged residues such as lysine. This positively charged segment may promote initial attachment of the signal to the negatively charged ER membrane surface. (Many phospholipid molecules are negatively charged at their polar ends; see p. 156.)

Following the positively charged subregion is a hydrophobic *core* that includes from at least 6 to 12 or more amino acids. The amino acids in this region are capable of forming an alpha helix or beta strand long enough to span the hydrophobic interior of a membrane. The ability of the core to assume a hydrophobic membrane-spanning structure is considered an essential feature that promotes insertion of the signal in the membrane interior. Following this insertion, a hydrophilic pore may open in the membrane to allow passage of the remaining polypeptide chain.

Following the core is the third region of the signal, including the point at which the signal is cleaved after insertion of a protein into the membrane. In most signals this region contains amino acids with small, uncharged side chains at the first and third positions upstream of the cleavage site, and often an amino acid with a bulky, polar or charged side group at the second position upstream of the cleavage site (Fig. 20-18b). These structural features presumably take on a configuration recognized by the enzyme removing the signal.

The Signal Peptidase

The activity of the signal peptidase in removing the signal is a requirement for successful transfer of most ER-directed proteins. If the signal peptidase is inhibited, transfer is usually incomplete. In particular, proteins that would normally pass entirely through the membrane, such as secreted proteins, become "stuck" in the membrane if the signal peptidase activity is faulty.

Presence of the signal peptidase is readily detected in isolated ER membranes by its specific activity in clipping off the signal. Although the enzyme has not as yet been precisely located in ER membranes, it is thought to face the inner compartment of ER cisternae or, in view of its extreme hydrophobic nature, to be buried entirely in the membrane interior. The signal peptidase has been isolated and partially purified in the Blobel laboratory.

Both the eukaryotic signal peptidase and its bacterial equivalent, *signal peptidase I* (see Supplement 20-1), are universal in their activity and can recognize and correctly cleave N-terminal signals from any source, either eukaryotic or prokaryotic.

Stop-Transfer Signals

Once a protein is directed to the ER by an N-terminal signal, it may either remain inserted in the membrane or pass through the membrane to enter the solution enclosed within the ER cisternae. Retention in the ER

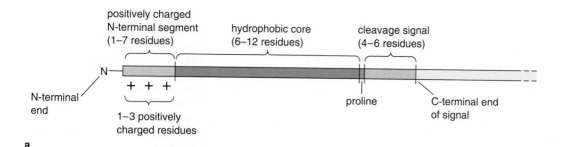

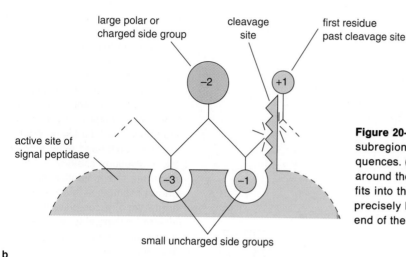

Figure 20-18 ER-directing signals. **(a)** Functional subregions common to most ER-directing signal sequences. **(b)** Arrangement of amino acid residues around the cleavage site. This arrangement probably fits into the active site of the signal peptidase enzyme, precisely locating the cleavage site at the C-terminal end of the signal.

membrane depends on stop-transfer signals, which arrest progress of one or more segments of a new amino acid chain through the membrane.

For example, glycophorin A, the nonsecreted forms of IgD and IgM antibodies, and many viral proteins are anchored to the plasma membrane by a stretch of hydrophobic or neutral amino acids some 20 to 24 residues long, lying just inside the C-terminal end of the protein. The segments anchoring these proteins are stop-transfer signals, which arrest progress of the proteins through ER membranes during their initial synthesis.

Once inserted into the membrane by the N-terminal signal, proteins continue to extend through the membrane until a stop-transfer signal is encountered. At that point the hydrophobic sequence becomes locked in the membrane. Further segments of the protein may pass through the membrane, followed by additional stop-transfer sequences. Presumably, the stop-transfer sequence removes its segment of the amino acid chain from the hydrophilic channel, and anchors it in the hydrophobic membrane interior. This arrangement produces segments that loop back and forth across the membrane. (Fig. 20-19 shows one pattern by which the consecutive loops might form.) In general, the number of stop-transfer signals determines the number of membrane-spanning segments in a pro-

tein and the number of times the amino acid chain crisscrosses the membrane.

Like the ER-directing signal, stop-transfer signals have been identified and evaluated in hybrid proteins. For example, C. S. Yost, J. Hedgpeth, and Lingappa constructed a hybrid protein with no stop-transfer sequences by linking 182 amino acids from the N-terminal end of bacterial β-lactamase to 142 amino acids from the C-terminal end of chimpanzee α-globin. When translated in a cell-free system complete with dog ER cisternae, the hybrid protein passed completely through ER membranes and wound up in the ER compartment. Adding a segment that contained the stop-transfer sequence of an antibody to the middle of the hybrid, between the β-lactamase and α-globin segments, produced a protein that remained anchored in the ER membranes. In the anchored hybrid the α-globin segment extended from the cytoplasmic side of the ER cisternae and the β-lactamase segment extended into the ER compartment. If the stop-transfer segment was placed at the N-terminal end of the hybrid, the protein failed to anchor in the membrane, indicating that stop-transfer sequences are influenced to some extent by surrounding sequence elements or domains. Most identified stop-transfer signals consist of a hydrophobic segment 8 to 20 amino acids in length, flanked at either end by hydrophilic regions.

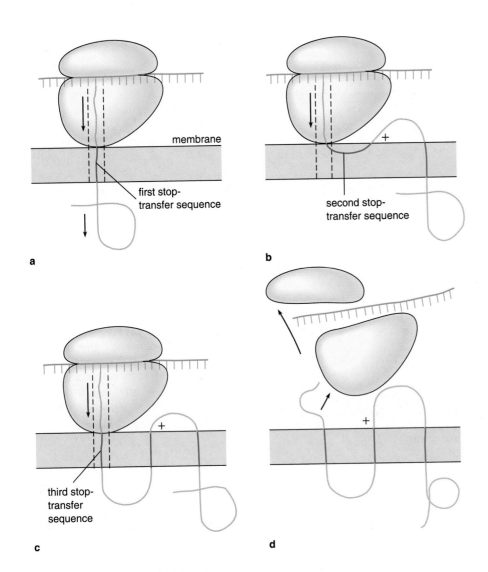

Figure 20-19 Insertion of proteins that loop back and forth across a membrane by consecutive stop-transfer sequences. **(a)** The first stop-transfer sequence emerges from the ribosome, locking the amino acid chain in the membrane at this point. **(b)** A second stop-transfer sequence emerges from the ribosome. A charged or polar segment of the amino acid chain just in front of the second stop-transfer sequence holds this end of the sequence at the membrane surface. **(c)** Continued emergence of the protein from the ribosome buries the second stop-transfer sequence in the membrane. The charged segment is still at the membrane surface facing the ribosome. Continued emergence of the polypeptide chain from the ribosome brings another stop-transfer sequence into the membrane. **(d)** The completely inserted polypeptide, with segments looping back and forth across the membrane.

membrane

first stop-
transfer sequence

second stop-
transfer sequence

third stop-
transfer
sequence

a

b

c

d

Postinsertion Signals and Further Destinations for ER-Directed Proteins

After being retained in the ER membrane, or passing into the ER compartment if no stop-transfer signal is present, a newly synthesized protein may become a permanent resident in these locations or travel to further stations along the ER–Golgi complex–plasma membrane pathway. Direction to any of these locations depends on the presence or absence of *postinsertion signals* that target proteins to final destinations in the ER or Golgi complex, lysosomes, storage vesicles, the plasma membrane, or the central vacuole of plants and yeast cells.

In most cases the sequences or structures forming postinsertion signals for ER-directed proteins have proved difficult to work out because surrounding sequences greatly modify the activity of the internal signal. As a result, most postinsertion signals have been identified only as an entity of unknown location and sequence within a long stretch of amino acids in a

protein. However, it has been possible definitely to identify two such signals and their probable mechanism of operation—the sequence indicating retention of proteins in the ER and the modified carbohydrate unit that directs newly synthesized glycoproteins to lysosomes.

The ER Retention Signal Among the several proteins residing permanently in the ER compartment is *BiP* (for *Binding Protein*), a soluble protein that promotes protein folding and the assembly of antibody and other polypeptides into multisubunit proteins. Another is *PDI* (for *Protein Disulfide Isomerase*), an enzyme catalyzing rearrangement of disulfide bonds in newly synthesized proteins. These and other proteins retained in the ER were found by S. Munro and H. R. B. Pelham to share a sequence of four amino acids at their C-terminal ends, most commonly Lys-Asp-Glu-Leu, or *KDEL* in the one-letter code. Some natural substitutions are noted in the ER retention signal in a few proteins, such as arginine for the first amino acid (giving RDEL) or glutamic acid

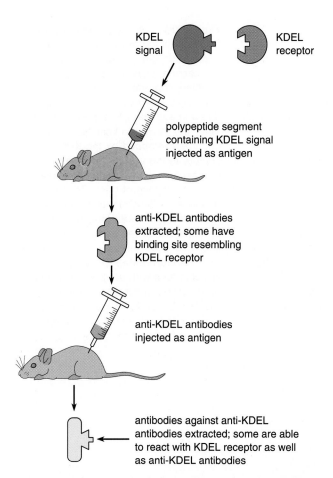

KDEL
signal

KDEL
receptor

polypeptide segment
containing KDEL signal
injected as antigen

anti-KDEL antibodies
extracted; some have
binding site resembling
KDEL receptor

anti-KDEL antibodies
injected as antigen

antibodies against anti-KDEL
antibodies extracted; some are able
to react with KDEL receptor as well
as anti-KDEL antibodies

Figure 20-20 The second-generation or anti-ideotype technique used to generate antibodies against signal receptors. For the KDEL receptor, a peptide with the KDEL signal is used as an antigen to develop the first generation of antibodies. One or more of these anti-KDEL antibodies might have an active site resembling that of the KDEL receptor. The anti-KDEL antibodies are used as antigens to develop a second generation of antibodies that can recognize the KDEL-binding site of the first antibody. The second-generation antibodies are used as a probe for the KDEL receptor in ER or nearby membranes.

for the second (giving KEEL). In yeast the standard form of the ER retention signal uses histidine as the first amino acid, giving HDEL. Monro and Pelham found that removal of the ER retention signal causes the proteins to be secreted rather than retained in the ER. Conversely, addition of the KDEL signal to the C-terminal end of a non-ER protein causes retention in the ER.

D. Vaux and his colleagues used "second generation" or *anti-ideotype* antibodies to identify a receptor for the KDEL signal and the probable mechanism by which the receptor operates (Fig. 20-20). Antibodies were first developed against the KDEL signal. If a membrane receptor recognizes and binds the KDEL signal

as part of the ER retention mechanism, it was considered possible that one or more of the anti-KDEL antibodies might have active sites resembling that of the KDEL receptor. Next the anti-KDEL antibodies were used as antigens to develop second-generation antibodies that could recognize the KDEL-binding site. These antibodies were then used as a probe for the KDEL receptor in ER or nearby membranes. The antibodies proved to bind to an integral membrane protein located primarily in vesicles adjacent to the ER or in the *cis* face of the Golgi complex.

These findings clearly supported a hypothesis for salvage and retention of resident ER proteins advanced earlier by Munro and Pelham. According to their idea (Fig. 20-21), resident ER proteins are not fixed rigidly in the ER. Instead, these proteins are carried away in the transition vesicles budding from the ER and fusing with the Golgi complex, along with proteins addressed to sites in the ER-directed pathway. As the ER proteins enter the transition vesicles, they are bound and trapped by the KDEL receptor in the vesicle membranes (Fig. 20-21a and b). These vesicles then fuse with the *cis* face of the Golgi complex (Fig. 20-21c). After reaching the *cis* cisterna, the KDEL receptors with their bound ER proteins are sorted and placed in vesicles that bud from the Golgi complex and return to the ER (Fig. 20-21d and e). As the vesicles fuse with the ER membranes (Fig. 20-21f), the KDEL receptors release the ER proteins, and the cycle of release and salvage of ER proteins is ready to begin again.

Several lines of evidence had supported the Munro and Pelham hypothesis prior to the work identifying the KDEL receptor. Among the most definitive was the effect of temperature on retention of ER proteins. If mammalian cell temperature is lowered to 16°C, ER proteins, marked by combination with fluorescent antibodies (see p. 119), gradually accumulate in vesicles near the ER or in the Golgi complex. If the temperature is raised to normal levels, the accumulated proteins are released from the vesicles and Golgi complex and return rapidly to the ER.

Thus retention of proteins in the ER depends on a C-terminal KDEL signal, a membrane-bound receptor recognizing and binding the KDEL signal, and a *salvage compartment* consisting of vesicles in which ER proteins bound by the KDEL receptor are sorted out and returned to the ER. If no KDEL signal is present, the proteins leaving the ER continue on their pathway through the system.

The Signal Directing Enzymatic Proteins to Lysosomes Lysosomes are membrane-bound sacs containing a combination of hydrolytic enzymes capable of breaking down most biological substances. Many enzymes destined to enter lysosomes are marked by a postinsertion sorting signal that consists of a mannose

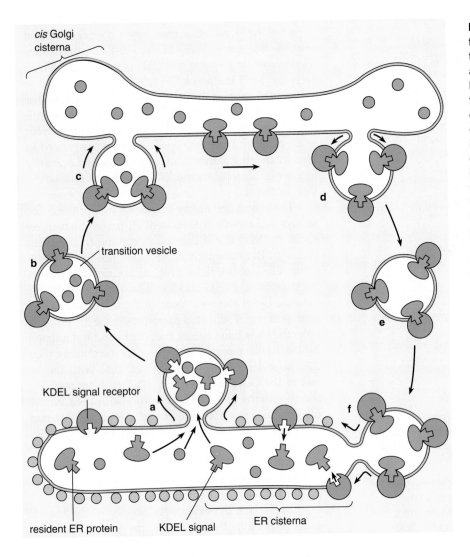

cis Golgi
cisterna

c

transition vesicle

b

d

KDEL signal receptor

a

e

f

resident ER protein   KDEL signal   ER cisterna

Figure 20-21 A mechanism proposed for the salvage of resident ER proteins. (**a** and **b**) Resident ER proteins are carried away in transition vesicles budding from the ER, along with proteins addressed to sites in the ER-directed pathway. As the ER proteins enter the transition vesicles, they are bound and trapped by the KDEL receptors in the vesicle membranes. (**c**) The transition vesicles then fuse with the *cis* face of the Golgi complex. (**d** and **e**) In the Golgi cisterna the KDEL receptors with their bound ER proteins are sorted and placed in vesicles that bud from the Golgi complex (**d**) and return to the ER (**e**). As the vesicles fuse with the ER membranes (**f**), the KDEL receptors release the resident ER proteins, and the cycle of release and salvage of ER proteins is ready to begin again.

sugar with a phosphate group added to its 6-carbon (Fig. 20-22; lysosome structure and function are described in detail later in this chapter). This *mannose 6-phosphate signal*, identified through the research of S. Kornfeld and others, is attached in the Golgi complex.

Tests by W. J. Brown and Farquhar detected a protein that acts as the receptor for the lysosome signal. The receptor, which is present in *trans* cisternae of the Golgi complex, in vesicles near the complex, and in lysosomal membranes, is maximally active in binding the mannose 6-phosphate signal near pH 7. Binding weakens as the pH falls below 6; at about pH 5, the pH characteristic of the lysosomal interior, binding affinity is completely lost.

These characteristics suggested that the mechanism sorting lysosomal proteins operates as shown in Figure 20-23. The internal pH of the Golgi complex is only slightly acid in *trans* cisternae. As a consequence, hydrolytic enzymes with the mannose 6-phosphate signal are bound tightly to the inner surfaces of the membranes containing the receptor (Fig. 20-23*a*). These membranes pinch off as vesicles destined to form lysosomes (Fig.

20-23*b* and *c*). As the internal pH of these vesicles falls because of the activity of H$^+$-ATPase pumps in the vesicle membranes, the receptors lose their affinity for the lysosomal enzymes and release them into the interior of the vesicles (Fig. 20-23*d*). After the enzymes are released, the receptors return to the Golgi complex, presumably by means of shuttle vesicles that pinch off from the maturing lysosomes and fuse with the Golgi (Fig. 20-23*e* and *f*).

Brown and Farquhar found that treating cells with ammonium ions (NH$_4^+$) interfered with the lysosomal sorting mechanism. The ammonium ions became concentrated in lysosomes, raising their internal pH. Under these conditions the mannose 6-phosphate receptors failed to release the hydrolytic enzymes and remained linked to them in lysosomes. The linkage jammed the mechanism so that the receptors could not recycle to the Golgi complex. As a result, the supply of mannose 6-phosphate receptors in the Golgi complex became exhausted, and the lysosomal enzymes passing through the Golgi complex were secreted to the exterior of the cell instead of entering lysosomes. Freeing the receptors

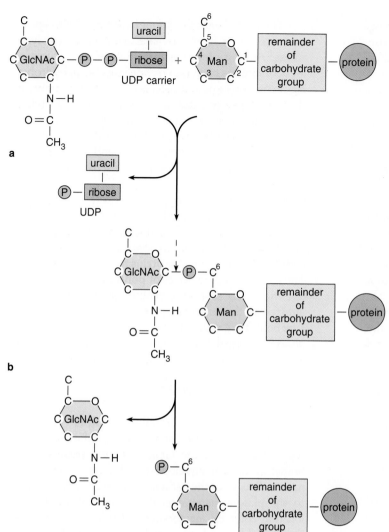

Figure 20-22 The two-step reaction forming the mannose 6-phosphate signal, which directs proteins to lysosomes. In the initial step **(a)** an acetylglucosamine unit is transferred from a UDP carrier to the 6-carbon of a mannose in the carbohydrate group of the lysosomal enzyme. At the close of this reaction the two sugars are linked by a phosphate group (P) from UDP, which extends between the 1-carbon of the acetylglucosamine and the 6-carbon of the mannose. In this structure the acetylglucosamine covers the mannose 6-phosphate signal. The second step **(b)** uncovers the signal by cleaving the bond between the acetylglucosamine and the phosphate group (at the dashed arrow), releasing the acetylglucosamine and leaving the phosphate attached to the 6-carbon of the mannose. This step, catalyzed by an "uncovering" enzyme, takes place in *trans* cisternae of the Golgi complex. GlcNAc, *N*-acetylglucosamine; Man, mannose; UMP, uridine monophosphate.

Figure 20-23 The mechanism sorting lysosomal enzymes by means of receptors and the mannose 6-phosphate signal. The internal pH of the Golgi complex is only slightly acid in *trans* cisternae. As a consequence, hydrolytic enzymes possessing the mannose 6-phosphate signal are bound tightly to the inner surfaces of the membranes containing the receptor **(a)**. These membranes are incorporated into vesicles destined to form lysosomes (**b** and **c**). As the H^+-ATPase pumps in the vesicle membranes lower the internal pH of the vesicles, the receptors lose their affinity for the lysosomal enzymes and release them into the interior of the vesicles **(d)**. After the enzymes are released, the receptors return to the Golgi complex, presumably via shuttle vesicles that pinch off from the maturing lysosomes and fuse with the Golgi complex (**e** and **f**).

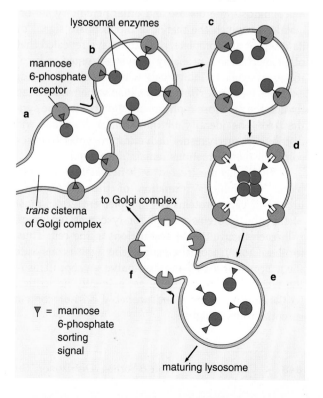

by removing NH_4^+ from the cells restored the sorting mechanism, switching the enzymes with the mannose 6-phosphate signal from secretion to their normal destination in lysosomes.

Mannose 6-phosphate receptors can also be detected in quantity in the plasma membranes of many animal cell types. Their presence in this location may provide a fail-safe mechanism retrieving lysosomal enzymes that are incorrectly secreted to the cell exterior. At the neutral pH characteristic of the cell surface, any secreted lysosomal enzymes are tightly bound by the mannose 6-phosphate receptors and returned to the cell in endocytotic vesicles. Most of these vesicles fuse with lysosomes or the Golgi complex, automatically returning the errant lysosomal enzymes to their normal distribution pathways. About 5% to 10% of the enzymes in lysosomes are estimated to have been retrieved from the cell exterior by this mechanism.

Permanent secretion of lysosomal enzymes occurs as an abnormality in individuals with mutations in the genes coding for parts of the lysosome sorting and distribution network. For example, persons suffering from the hereditary disease *mucolipidosis* lack the transferase enzymes carrying out the initial step in production of the mannose 6-phosphate signal (the step shown in Fig. 20-22a and b). As a consequence, the signal is not completed and none of the lysosomal enzymes are bound by the mannose 6-phosphate receptor. Instead, they enter secretion vesicles and are released in large quantities to the cell exterior, where they attack the extracellular matrix. Persons with the disease suffer from a variety of skeletal and mental abnormalities, caused in large part by destruction of the extracellular matrix by the lysosomal enzymes. The disease, which is presently incurable, is eventually fatal.

While tragic in its consequence for persons with mucolipidosis, the disease and other hereditary abnormalities in which the mannose 6-phosphate signal or its receptor are completely inactive have revealed that other mechanisms probably deliver enzymes to lysosomes as well. In individuals with mucolipidosis, for example, lysosomes form normally in some tissues even though the mannose 6-phosphate signal cannot be synthesized. The identity of the additional sorting and distribution mechanisms delivering enzymes to lysosomes in these situations remain unknown.

When cells are treated with *tunicamycin*, an antibiotic that inhibits formation of the carbohydrate groups of glycoproteins (see Supplement 20-2), most of the glycoproteins, except for lysosomal enzymes, still reach their ultimate destinations in the cell. Thus, for these nonlysosomal glycoproteins, which constitute the majority of sorted and distributed glycoproteins in the cell, carbohydrate groups are evidently not critical to the sorting process and probably do not contain essential routing signals.

The central vacuole of fungi and plant cells appears in part to be a functional equivalent of animal lysosomes. Among other elements the central vacuole of these organisms contains enzymes capable of hydrolyzing many biological molecules, particularly polysaccharides. These enzymes and other proteins, including proteins to be stored in seeds, tubers, or roots, enter the central vacuole after synthesis in the ER and processing in the Golgi complex, evidently by a sorting mechanism that recognizes a postinsertional, vacuole-directing signal. B. W. Tague and M. J. Chrispeels showed that phytohemagglutinin, a protein entering the vacuole of plants, is routed to the vacuole of yeast cells when introduced into these organisms. Yeast cells and higher plants thus evidently share a sorting mechanism and receptors that direct proteins to the central vacuole. The experimental removal of polysaccharide segments does not affect routing to central vacuoles, so proteins directed to this organelle probably are not marked by a carbohydrate segment equivalent to the mannose 6-phosphate signal. Instead, grafting experiments indicate that the vacuole-directing signal resides in the polypeptide, usually within the first 50 amino acids following the ER-directing signal.

Routing ER-Directed Proteins to Other Locations
Some proteins entering the ER ultimately become integral elements of the Golgi complex, plasma membrane, or nuclear envelope membranes. Longest and most extensive of the pathways followed is that leading to the plasma membrane. Proteins addressed to this location enter ER membranes and subsequently travel through the membranes of transition vesicles, Golgi complex, and secretory vesicles to reach their final destination. Among the molecules taking this route are active and passive transport proteins and cell surface receptors. In some membranes, such as the epithelial cells lining body cavities, the sorting and distribution mechanism places different proteins in distinct subparts of the plasma membrane. Proteins transporting sugars and amino acids in intestinal epithelial cells, for example, are distributed to the plasma membrane region facing the intestinal cavity. MHC molecules and the poly-Ig receptors binding antibodies (see p. 814) wind up in segments of the plasma membrane on the opposite side of the epithelial cells facing the body interior. The signals and mechanisms sorting proteins to different membrane segments in such cells remain unknown.

Proteins with final destinations in the Golgi complex follow the same route as membrane proteins initially but are arrested in their progress as they reach the Golgi cisternae. These proteins, which include a wide variety of sugar transferases, proteinases, phosphorylases, sulfatases, and other processing enzymes, remain locked in the Golgi membranes, possibly through interactions with receptors, when their own processing is complete. Some ER proteins probably travel the same route as Golgi membrane proteins but reverse course after modification in the Golgi and return to permanent locations in the ER.

Although a postinsertion signal retaining proteins in the Golgi complex has not been identified, an experiment by A. M. Swift and C. E. Machamer indicates that such signals actually exist. Swift and Machamer studied the E1 protein of avian corona virus, which is targeted to *cis* Golgi cisternae in infected cells. They tested the effects of grafting segments of the E1 protein to two proteins, the stomatitis virus G protein and human chorionic gonadotropin, that normally pass through the Golgi complex without delay. The first transmembrane segment of the E1 protein, when grafted to the test proteins, caused them to remain in the Golgi complex. Thus some or all of the E1 transmembrane segment must constitute a Golgi-targeting signal. This conclusion was supported by the effects of single amino acid substitutions, which could destroy the segment's ability to target proteins to the Golgi complex. A similar transmembrane segment able to direct retention in the Golgi complex has been detected by others in a galactosyltransferase enzyme.

Proteins forming parts of the nuclear envelope may be synthesized on ribosomes attached either to the ER or to the outer membrane of the nuclear envelope. If synthesized on the outer nuclear membrane, the proteins may simply remain in this location or travel through the connections made with the ER (see Fig. 20-4) for modification in the ER or Golgi before returning to take up permanent residence in the nuclear envelope. If synthesized on the ER, they may enter the nuclear envelope through the connections, either directly or after modification in the Golgi complex. It is likely that postinsertion signals and receptors are responsible for retention or direction of proteins to the nuclear envelope membranes.

Proteins entering the ER-based distribution system have a variety of possible cellular destinations ranging from the ER itself to the nuclear envelope, Golgi complex, lysosomes, plasma membrane, and cell exterior. It is obvious from the way in which the system operates that secretion is the fate of proteins that have only an ER-directing signal at the N-terminal end and no further routing information in the form of stop-transfer signals or postinsertion signals. For example, lysosomal enzymes as well as ER proteins are secreted if their routing signals are eliminated or when segments of the receptor mechanism distributing these proteins go awry. Secretion is thus the "default" pathway for ER-directed proteins. If one or more stop-transfer sequences follow the ER-directing signal, so that no postinsertion signals are present, placement in the plasma membrane is evidently the default pathway for the system.

DISTRIBUTION OF PROTEINS TO LOCATIONS OUTSIDE THE ER-GOLGI PATHWAY

Distribution to Mitochondria and Chloroplasts

Both mitochondria and chloroplasts contain DNA, ribosomes, and all the components necessary for synthesizing proteins. However, the proteins encoded and assembled inside the two organelles comprise only a small part of their total complement. Most of their proteins are encoded in the cell nucleus and synthesized on cytoplasmic ribosomes.

Proteins originating in the cytoplasm outside the organelles may be directed to one of several locations. For mitochondria these locations include the outer boundary membrane, the intermembrane compartment, the inner boundary membranes or cristae, and the innermost compartment, the matrix (see Fig. 9-7 and p. 321). Chloroplasts have no less than six possible locations for proteins entering from the outside: (1) the outer boundary membrane, (2) the intermembrane compartment, (3) the inner boundary membrane, (4) the stroma, the interior compartment analogous to the mitochondrial matrix, (5) the thylakoid membranes, and (6) the compartment enclosed by the thylakoids (see Fig. 10-3 and p. 369). Recent research has shown that the information sorting and distributing proteins to locations in the organelles is spelled out by a signal mechanism with overall properties similar to ER-directed pathways.

Experiments linking organelle-directing signals to nonorganelle proteins demonstrated their capabilities. This approach was used by G. van den Broeck and his coworkers to examine the signal directing the small subunit of RuBP carboxylase (see p. 388) to chloroplasts. They produced a hybrid protein by grafting the RuBP carboxylase signal onto the N-terminal end of a bacterial marker enzyme, *neomycin phosphatase*, which normally remains in solution in the bacterial cytoplasm. When supplied with the RuBP carboxylase signal and introduced into tobacco plants, the bacterial enzyme was routed correctly into tobacco chloroplasts. Similar retargeting experiments have been carried out by grafting mitochondria-directing signals onto bacterial cytoplasmic proteins; in these experiments the mitochondrial signal successfully directs the protein to the mitochondrial interior.

All cytoplasmically synthesized mitochondrial and chloroplast proteins studied to date are capable of inserting into the organelle membranes posttranslationally. In spite of this capability, many of these proteins are evidently assembled on ribosomes attached to the outer boundary membrane of the organelles and inserted cotranslationally. Ribosomes are frequently ob-

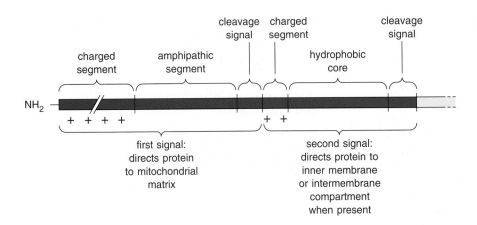

Figure 20-24 Structure of N-terminal signals directing proteins to mitochondria. Chloroplast-directing signals are similar except that the core of the first signal is neutral or hydrophobic rather than amphipathic.

served in this position under the electron microscope. In mitochondria, where the most extensive studies have been carried out, ribosomes also prove to be attached to outer boundary membranes isolated from cells for analysis. These ribosomes can be removed only by agents, such as the drug puromycin (see p. 658), that can separate ribosomes from growing polypeptide chains. Because insertion is potentially posttranslational, however, attachment of ribosomes to the outer membrane probably simply reflects the penetration of the N-terminal segments of organelle proteins before assembly of their C-terminal portions is complete.

Initially proteins are apparently attached to the outer surface of the organelles by an interaction between the signal and the outer boundary membrane. This interaction is promoted by soluble cytoplasmic factors and by receptor proteins in the organelle membranes. The soluble cytoplasmic factors may act primarily as chaperones (see p. 684) maintaining the organelle proteins in the partially unfolded condition necessary for membrane penetration. There are also indications that ATP is hydrolyzed as an energy source to maintain the unfolded state.

Mitochondrial Sorting and Distribution Signals The N-terminal signal sequences of proteins directed to mitochondria are typically longer than ER-directing signals, up to 70 amino acids in length. Comparisons of the signals directing different proteins to mitochondria reveal similarities in the distribution of amino acids with certain chemical properties (Fig. 20-24). All the mitochondrial signals examined to date, with only one exception, contain a long region with positively charged residues near the N-terminal end. These charged residues, including lysine and arginine, are located singly, separated by stretches of neutral amino acids some four to eight residues in length. In a central region following the initial charged segment, many mitochondrial signals contain a stretch of about 20 to 25 polar and nonpolar amino acid residues, long enough to span a phospholipid bilayer. The amino acids of this segment are spaced so that the structure is polar on

one side and nonpolar on the other when wound into an alpha helix or beta strand (see Fig. 5-11). Structures of this type, termed *amphipathic*, are water soluble but can take up conformations allowing them to penetrate spontaneously into membranes. Frequently, the amphipathic sequence is flanked at either end by positively charged residues. Following the amphipathic sequence is a cleavage segment, which is necessary for removal of the signal once insertion of the protein is complete. As yet, no sequence or structural homologies that might serve as the cleavage indicator are apparent in this segment of the signal, except, in some, for a cluster of positively charged residues just upstream of the cleavage site. In many proteins directed to mitochondria, a second signal region immediately follows the cleavage indicator for the first signal. The second signal consists of a hydrophobic stretch of amino acids bounded by charged residues and followed by its own cleavage signal; this signal, when present, determines routing within the mitochondrial interior.

Experiments with hybrid proteins show that the first signal, the one with the amphipathic segment, directs proteins entirely through the outer and inner boundary membranes and into the mitochondrial matrix. A signal peptidase removes the signal as the protein reaches this location. If no second signal is present, the protein remains in solution in the matrix. If a second signal does follow, removal of the first signal places the second at the N-terminal end of the protein. The second signal then directs the protein to insert into the inner boundary membrane. Depending on other sequence information, the protein may remain in the inner boundary membrane or pass through it into the intermembrane compartment. A second signal peptidase associated with the inner mitochondrial membrane removes the second signal as these movements become complete.

N. Pfanner, W. Neupert, G. Schatz, and their colleagues and others have found many proteins that act as chaperones, receptors, and other elements in the mitochondrial import mechanism (outlined in Fig. 20-25). Much of their research detecting these proteins

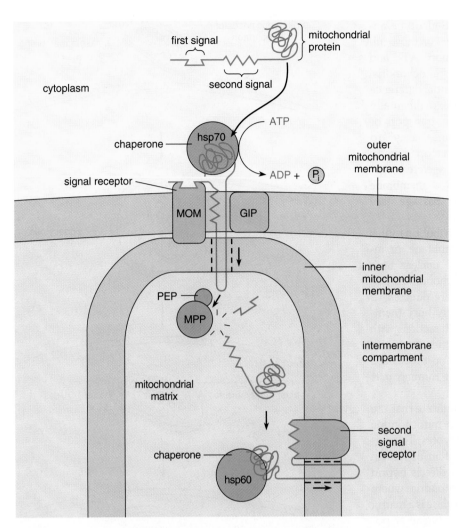

Figure 20-25 The mechanism directing proteins to mitochondria, deduced primarily from experiments with *Neurospora*. While still in the cytoplasm, a protein with an N-terminal signal directing it to mitochondria combines with a chaperone that maintains the protein in a partially unfolded state. The unfolded state evidently holds the N-terminal signal and the remainder of the protein in a configuration that permits insertion in the mitochondrial membrane. Binding to the chaperone, which is a member of the hsp70 family of heat shock proteins (see p. 705), requires ATP. As the protein collides with a mitochondrion, it is transferred from the chaperone to an outer membrane receptor forming part of a group known as MOM (for *Mitochondrial Outer Membrane*) proteins. ATP is hydrolyzed as the transfer takes place. The protein next moves through the membrane, in a process promoted by a second outer membrane protein, GIP (for *General Insertion Protein*). Insertion is thought to occur at sites in which the inner membrane makes contact with the outer membrane, allowing the protein to pass through both membranes simultaneously. Once the protein is in the matrix, its N-terminal signal is removed by the signal peptidase in the matrix, MPP (for *Mitochondrial Processing Peptidase*), which is activated by a second protein, PEP (for *Processing Enhancing Protein*). If a second signal follows the first, the protein combines with another chaperone in the mitochondrial matrix. This chaperone, a protein of the hsp60 (*heat shock protein*) family, holds the protein in a conformation in which the second signal, now at the N-terminal end, can insert in the inner mitochondrial membrane. Presumably, a receptor associated with the inner membrane, not identified, recognizes and binds the second signal. After insertion, the signal is removed and the protein reaches its final destination in the inner membrane or intermembrane compartment.

employed the second-generation or anti-ideotype antibody technique outlined in Fig. 20-20. For most mitochondrial proteins insertion requires both ATP and the potential difference resulting from mitochondrial electron transport (see p. 335). If the potential difference across the inner membrane is destroyed experimentally by agents such as uncouplers of electron transport or *ionophores* that make the inner membrane "leaky" to ions (see p. 344), sorting and distribution stop immediately. Bacterial plasma membrane or secreted proteins similarly require both ATP and a membrane potential for membrane insertion or passage (see Supplement 20-1).

The mitochondrial signaling and distribution routes direct proteins to the matrix, inner membrane, or intermembrane compartment. It remains unclear how proteins that form part of the outer mitochondrial membrane, such as the porins (see p. 349), are directed to this location. These proteins may directly bury themselves in the outer membrane through interactions with membrane lipids or interact in some other way with outer membrane proteins. In any event no specialized signals appear to be present in proteins forming part of the outer mitochondrial membrane.

Experiments with hybrid proteins indicate that the N-terminal signals directing proteins to mitochondria are universally recognized among eukaryotes. The signal from the ornithine carboxylase protein of human mitochondria, for example, successfully directs hybrid proteins into the matrix of yeast mitochondria, where it is correctly cleaved. (Ornithine carboxylase is a matrix protein.) A mutation from arginine to glycine at position 23 in the ornithine carboxylase signal blocks insertion in both yeast and human mitochondria. A yeast N-terminal sequence directs insertion of a protein into plant mitochondria, and the signal of a plant mitochondrial protein, a malate dehydrogenase enzyme of the watermelon, inserts hybrid proteins into oocyte mitochondria of *Xenopus*, an amphibian. The universal effectiveness of these signals, and probably also of the receptors binding them, indicates that the mitochondrial sorting and distribution mechanism has very ancient origins among eukaryotic organisms and has changed relatively little in the course of evolution.

The mitochondrial distribution system has interesting relationships to bacterial protein secretion. Mitochondria are widely considered to have evolved from symbiotic, bacterialike ancestors (see Ch. 27 for details). Their route of entry is believed to have been through enclosure in an endocytotic vesicle (Fig. 20-26). The inner mitochondrial membrane, according to this hypothesis, is the evolutionary remainder of the plasma membrane of the original bacterial symbionts; the outer membrane is derived from the membrane of the endocytotic vesicle. As mitochondria evolved from the bacterial symbionts, most genes encoding the original bacterial proteins were transferred to the cell nucleus.

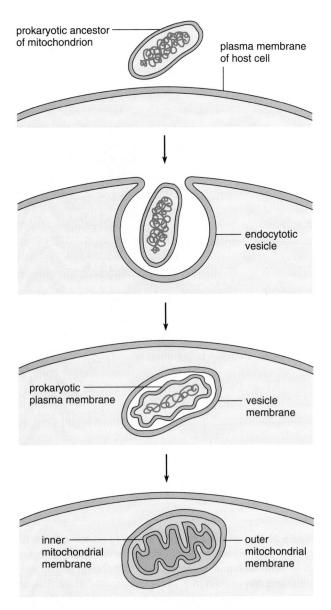

Figure 20-26 The probable evolutionary origins of the two mitochondrial boundary membranes. After entering a host cell by endocytosis, the prokaryotic mitochondrial ancestor was enclosed in a vesicle membrane. The outer mitochondrial membrane is considered to have been derived from the vesicle membrane and the inner mitochondrial membrane from the plasma membrane of the prokaryote. As mitochondria evolved from bacterial symbionts, most genes encoding the original bacterial proteins were transferred to the cell nucleus. The proteins encoded in these genes, which are synthesized in the cytoplasm, enter the organelle by means of an N-terminal signal developed during mitochondrial evolution.

The two parts of mitochondrial signals are believed to reflect these origins. The signal directing proteins to the mitochondrial matrix, which contains the amphipathic segment, is a eukaryotic evolutionary development necessary to get proteins inside the organelle. These proteins are encoded in genes that were transferred from the organelle to the nucleus during its evolutionary development. Once inside the organelle, the proteins remain in the matrix if no second signal is present. These proteins, supposedly, had no membrane-directing signal in the bacterial ancestors of mitochondria and remained in solution in the prokaryotic cytoplasm. The second signal, if present, is a remnant from the prokaryotic ancestor. In the ancestor this signal determined whether the protein would become part of the plasma membrane or be released to the outside of the membrane. In mitochondria the signal directs the protein to the inner membrane (the equivalent of the ancestral plasma membrane) or the intermembrane compartment (equivalent to the space outside the ancestral plasma membrane). In keeping with this proposed relationship, the second signal has structural features closely resembling the present-day bacterial signal directing insertion into the plasma membrane or release to the outside. In addition, the hsp60 protein acting as a chaperone in the mitochondrial matrix of yeast cells is 60% identical in sequence to *GroEL*, a protein acting as a chaperone for bacterial secretion (see Supplement 20-1).

Not all proteins enter inner mitochondrial membranes and compartments by the signal mechanism. Cytochrome *c*, for example, which ends up as a peripheral protein associated with the inner membrane, enters mitochondria with no identifiable signal sequences and no known interactions with receptors or other membrane elements. These proteins may spontaneously fold into forms that automatically direct them to their final locations. In a sense much or all of the amino acid sequence and folding conformation of these proteins acts as a signal sequence.

Chloroplast-Directing Signals   The N-terminal signals directing proteins to chloroplasts range from about 30 to nearly 100 amino acids in length; most contain more than 50 amino acids. Although relatively few chloroplast signals have been sequenced, most resemble their mitochondrial counterparts in overall structure. Investigations by K. Keegstra, S. Smeekens, and P. Weisbeck and others revealed that proteins directed through the two boundary membranes and into the stroma contain an N-terminal signal that has few evident structural features except an enrichment in serine and threonine residues and few or no acidic residues. This signal is equivalent in function to the amphipathic mitochondrial signal directing proteins to the mitochondrial matrix. However, most chloroplast-directing signals are not amphipathic. This signal ends in a region that somehow indicates a precise site for cleavage by a signal peptidase in the stroma.

In some chloroplast proteins a second signal with properties closely similar to prokaryotic signals follows the first, stroma-directing signal. The second signal begins with a few charged residues followed by a hy-

drophobic core sequence and ends with a cleavage signal closely resembling that of prokaryotic signals. The second signal, exposed at the N-terminal end when the chloroplast-directing signal is removed, targets the protein to a thylakoid membrane. Depending on further signal information, the protein may remain embedded in the thylakoid membrane or pass into the compartment enclosed by the membrane. A second signal peptidase associated with the thylakoid membrane removes the second signal as these movements take place. The second signal is thus the equivalent of the second signal in mitochondria directing proteins to the inner membrane and intermembrane compartment.

Smeekens and Weisbeck proposed that the second signal directing proteins to thylakoids dates back to the prokaryotic ancestors of chloroplasts. In these ancestral prokaryotes the signal directed proteins to thylakoids located in the cytoplasm. As the forerunners of chloroplasts became established in the cytoplasm of cell lines destined to found eukaryotic plants, many of the genes encoding thylakoid proteins moved to the cell nucleus (see pp. 770 and 1140). During this transfer the chloroplast-directing signal was added as an extra amino acid sequence at the N-terminal region of these proteins.

Chloroplasts also contain proteins that are encoded and synthesized inside the stroma and targeted to thylakoid membranes or compartments. These proteins include an N-terminal signal with structure and properties closely similar to the second signals of cytoplasmically synthesized proteins that are targeted to thylakoids.

Research identifying receptors and other factors promoting import of chloroplast proteins has not progressed as far as the equivalent studies in mitochondria. The entry of proteins into chloroplasts requires ATP and cytoplasmic factors probably acting as chaperones. Experimentally added crosslinkers tie proteins entering chloroplasts to a membrane protein that is believed to be a signal receptor. The receptor has also been identified by workers in the Blobel laboratory using the anti-ideotype antibody technique. As in mitochondria insertion of proteins is believed to occur at sites at which the inner and outer boundary membranes come in contact, so that movement across both membranes takes place simultaneously (see Fig. 20-25). Soluble factors in the stroma are necessary for entry of proteins into thylakoids; at least one of these proteins, a molecule of about 65,000 daltons, is believed to act as a chaperone.

The signals directing proteins to mitochondria and chloroplasts in plant cells differ sufficiently in structure to prevent misdirection of proteins from one organelle to the other. Hybrid proteins made by grafting a chloroplast signal onto a marker protein, for example, regularly enter chloroplasts; the same protein with a mitochondrial signal enters mitochondria. The differentiation undoubtedly involves specific recognition of the correct signal by receptors in the outer membranes of the organelles.

The work accomplished to date with chloroplast-directed proteins indicates that the signal mechanisms of this organelle, like the mitochondrial pathway, are ancient and highly conserved. Chloroplast-directed proteins of algae such as *Chlamydomonas*, for example, can be inserted and correctly cleaved by chloroplasts of a higher plant such as tobacco, or vice versa.

The mechanisms targeting proteins to the outer chloroplast boundary membrane as a final destination remain unknown. As in the equivalent membrane in mitochondria, these proteins, which have no identifiable signals, may directly bury themselves in the outer membrane through interactions with membrane lipids or outer membrane proteins.

Signals Directing Proteins to the Nuclear Interior

The signals directing proteins to the nuclear interior differ fundamentally from ER- and organelle-directing signals. A nuclear signal, rather than appearing at the N-terminal end, can occur at any site within a protein. The only restriction appears to be that the signal must be exposed at the surface. Unlike the ER- and organelle-directing signals, the nuclear signal is generally retained when a protein reaches its destination in the nucleus. Nuclear signals are relatively short, strongly basic elements consisting of clusters of lysines or arginines broken by interspersed amino acids of other types, often neutral or hydrophobic (Table 20-2). A proline or an amino acid with a bulky side chain usually lies at or near one end of the signal.

In some instances alteration of single amino acids in the signal can destroy its nucleus-directing activity. In the SV40 T antigen (a protein involved in replication of viral DNA in the nucleus), for example, a single mutation from lysine to asparagine at the third position

alters the signal enough to keep the proteins in the cytoplasm.

Retention of the nuclear signal after proteins arrive in the nucleus may reflect the fact that in many cell types, particularly in higher eukaryotes, the nucleus breaks down during division and releases its contents to the cytoplasm. Retention of the nucleus-directing signal ensures that nuclear proteins return to the re-forming nucleus as division becomes complete. The process can be imitated by injecting bulk nuclear proteins into the cytoplasm of essentially any cell; the injected proteins quickly enter and remain inside the nucleus. Several experiments with hybrid proteins by D. Kalderon and his associates and others demonstrated that addition of the nuclear signals can send cytoplasmic proteins to the nucleus.

Nuclear proteins are apparently assembled entirely on freely suspended ribosomes in the cytoplasm and make their way into the nucleus posttranslationally via the nuclear pore complexes. The passage is a two-step reaction involving: (1) attachment of the protein to a pore complex, probably by a receptor recognizing and binding the nuclear signal, and (2) penetration of the protein through the pore. The second step requires ATP to proceed. (For details of the reactions transporting proteins through nuclear pore complexes, see p. 561.)

Distribution to Microbodies

Microbodies (see p. 351) are small, cytoplasmic structures in which a single boundary membrane encloses an interior matrix. These organelles, which include peroxisomes, glyoxisomes, and glycosomes, contain batteries of enzymes catalyzing a variety of primarily oxidative pathways.

S. J. Gould, G. A. Keller, S. Subramani, and others discovered that a major signal element sending proteins to microbodies consists of a three-amino acid sequence. The sequence, which usually appears at or near the C-

Table 20-2 Some Nucleus-Directing Signals	
Protein	Signal
Adenovirus E1A	K R P R P
Histone H2B	C P P G K K R S K A
Influenza virus nuclear protein	P K K A R E P
Nucleoplasmin*	R P A A T K K A G E A K K K K L D K E D E
Polyoma virus large T antigen	V S R K R P R P A
SV40 T antigen	P K K K R K V
SV40 UP1	A P T K R K G S
Yeast ribosomal protein L3	P R K R
Yeast ribosomal protein L29	K T R K H R G and K H R K H P G

* The nucleoplasmin sequence contains two consecutive nuclear signals.

terminal end, has serine, alanine, or cysteine at the first position, a positively charged amino acid (either lysine, arginine, or histidine) at the second position, and a hydrophobic amino acid (leucine) at the third position. In the majority of microbody-targeted proteins examined so far, the signal takes the form Ser-Lys-Leu, or SKL. The signal remains part of the microbody proteins after entry into the organelle. Like the entry of proteins into many other cellular membranes, the process in microbodies requires ATP to proceed.

Addition of the signal is enough to send proteins normally located in the soluble cytoplasm to microbodies. Gould and his colleagues found that a gene encoding *luciferase,* a microbody enzyme of fireflies, was expressed correctly, with insertion of the protein encoded in the gene into peroxisomes, in yeast and plant cells; the same investigators also discovered that the C-terminal signal of a yeast peroxisomal protein, PMP20, could direct proteins to peroxisomes in mammalian cells. The microbody signal mechanism has evidently also been preserved intact since very ancient times.

Eukaryotic cells employ elaborate systems of signals, receptors, and other factors such as chaperones to deliver proteins to both internal and external locations. These mechanisms evidently arose very early in the evolution of eukaryotes as a necessary part of the development of separate internal compartments, bounded and set apart by membranes, as specialized regions of the cell. These ancient mechanisms were highly conserved and passed on essentially intact as the various kingdoms of eukaryotic organisms evolved.

EXOCYTOSIS AND ENDOCYTOSIS

The default pathway for ER-directed proteins lacking further distribution signals is secretion via exocytosis. This pathway provides the route by which proteins synthesized inside cells are exported to the cell exterior. Some of the final steps of the pathway are reversed in endocytosis to provide a route of entry for selected materials from outside the cell. Endocytosis is more than a simple reversal of exocytosis, however; it employs receptors and other elements that do not operate in exocytosis.

Substances entering cells by endocytosis are enclosed in pockets that invaginate inward from the plasma membrane. The pockets pinch off from the plasma membrane as *endocytotic vesicles* (also called *endosomes* or *receptosomes*) and move into the cytoplasm. Once inside, the vesicles have various fates depending on the types of proteins or other substances enclosed within them. In some pathways the vesicles fuse with lysosomes, a process that exposes the proteins in the

vesicles to hydrolysis by lysosomal enzymes. Other pathways deliver endocytotic vesicles to the Golgi complex or to other vesicles in the cytoplasm (see Figs. 20-9 and 20-28).

Segments of the plasma membrane are constantly cycled between the cell interior and the cell surface by the combined workings of exocytosis and endocytosis. As exocytosis proceeds, vesicle membranes are introduced into the plasma membrane. This input of membrane material is counterbalanced by the removal of membrane segments from the plasma membrane by endocytosis. The balance of the two mechanisms maintains the surface area of the plasma membrane at controlled levels.

In some epithelial cells endocytosis is coupled to exocytosis, so that proteins carried by endocytotic vesicles entering one side of the epithelium are sorted into secretory vesicles that travel through the cytoplasm and fuse with the plasma membrane on the opposite side. The sorting in this mechanism, called *transcytosis,* evidently takes place soon after endocytotic vesicles enter the cytoplasm, without participation of the Golgi complex. For example, some antibody types are secreted from the internal circulation to body cavities and surfaces by this pathway (see p. 814).

The vesicle movements of both exocytosis and endocytosis are apparently powered by microtubule-based mechanisms. Microtubules lie in tracks extending in the directions followed by vesicles in their travels between the Golgi complex and the plasma membrane, with the plus ends of the microtubules (see p. 421) placed toward the plasma membrane. J. Tooze and his coworkers and others observed vesicles sliding in both directions over the surfaces of the microtubules, as if their movements are powered by "motors" working on the microtubule surfaces. As expected if microtubules are involved in this motion, vesicle movements cease in cells exposed to colchicine, a drug that promotes microtubule disassembly.

Exocytosis

The experiments by Palade and others revealing the pathways of protein secretion (see p. 831) led to the conclusion that all eukaryotic cells are capable of releasing proteins to the cell exterior by exocytosis. Different animal cells secrete a variety of proteins by exocytosis, such as digestive enzymes, mucus, peptide hormones, milk proteins, and hormones. Epithelial cells lining the digestive tract, for example, secrete mucus and digestive enzymes; pancreatic cells secrete the peptide hormones insulin, glucagon, and somatostatin, and the precursors of several digestive enzymes, including amylase and trypsin. Plant cells secrete cell wall enzymes such as peroxidase, acid phosphatase, and α-amylase, and structural proteins of the cell wall. A number of plant cell types actively secrete proteins that leave the cell wall. Cells of the stigma in flowers, for example, secrete to the outside large quantities of the *S-allele glycoproteins,* which determine self-incompatibility during pollination.

Many cells secrete more than one protein type. In such cells secretory vesicles contain a mixture of the secreted proteins. In liver cells, for example, two commonly secreted but unrelated proteins, albumin and transferrin, are packed into the same secretory vesicles at the Golgi complex and released together to the extracellular medium.

Constitutive vs. Regulated Secretion Although many proteins are secreted without segregation as to type, some are segregated according to the *mode* of secretion—either continuously or at intervals in response to a stimulus. Continuous vesicle release is termed *constitutive secretion;* release at intervals is *regulated secretion.* Many cell types secrete proteins constitutively, including activated B cells of the immune system, mucus-secreting cells of the intestinal lining, and fibroblasts secreting collagen. Others, such as neurons, pituitary cells, and mast cells, secrete some proteins regulatively. Mast cells, for example, store about 1000 secretory vesicles densely packed in their cytoplasm. On receiving an activating signal from outside, as much as 80% of these vesicles are released by exocytosis in a matter of minutes. Most cells secreting proteins regulatively also secrete others constitutively. Cells of the pituitary gland, for example, secrete the peptide hormone ACTH regulatively and laminin (a protein of the extracellular matrix) constitutively.

In cells secreting proteins in both modes, constitutively secreted proteins are sorted into vesicles that move directly to the plasma membrane. Regulated proteins are stored in vesicles for varying lengths of time before fusion with the plasma membrane. Regulated vesicles are distinguished by their larger size and more densely concentrated contents. The concentration of proteins in regulated vesicles is often hundreds of times greater than their concentration in *trans* cisternae of the Golgi. In constitutively secreted vesicles concentrations are rarely more than twice the levels in Golgi cisternae.

Proteins probably have a signal that directs them to one of the two pathways. Research by R. B. Kelley with pituitary cells clearly supports this conclusion. Kelley studied the effects of adding genes for foreign proteins to the pituitary cells. Foreign proteins such as insulin and growth hormone, secreted regulatively in other cell types, were also sorted into storage vesicles to be secreted by the regulated pathway in pituitary cells. Foreign proteins secreted constitutively in their normal locations were correctly sorted into vesicles destined for immediate secretion in the pituitary cells.

The signal determining constitutive or regulated secretion is likely to be limited to the one directing regulated secretion. These proteins probably possess a signal diverting them from the constitutive pathway into storage vesicles. Proteins to be secreted constitutively probably contain no additional diverting signal and are simply routed into vesicles for immediate secretion. That is, constitutive secretion is presumably the "default" pathway. If this is the case, the signal indicating the regulated pathway would act similarly to the mannose 6-phosphate signal of lysosomal enzymes, which diverts proteins from the constitutive secretory pathway into lysosomes. Removal of the mannose 6-phosphate signal, as noted, relegates lysosomal enzymes to constitutive secretion.

Regulated secretion appears to take place primarily or exclusively in response to signals from outside the cell, received by receptors forming parts of intracellular signaling pathways such as those employing $InsP_3$/DAG, Ca^{2+}, or cAMP as second messengers (see pp. 245 and 247). An increase in Ca^{2+} concentration is noted, for example, in pancreatic acinar cells immediately before the release of secretory vesicles containing insulin. Increases in cytoplasmic Ca^{2+} concentration are also noted immediately before secretion in neurons, mast cells, and pancreatic cells, and in plant cells such as those of the pea stem or growing pollen tubes. In many cell types in which Ca^{2+} forms part of the pathway regulating secretion, experimentally added Ca^{2+} will also induce secretion. For example, adding Ca^{2+} to chromaffin cells of the adrenal medulla, made permeable to the ion by the addition of a Ca^{2+} ionophore, induces the immediate secretion of chromaffin granules containing adrenalin and noradrenalin. In other cells increases in cAMP or GTP rather than Ca^{2+} concentration appear to be involved in regulated secretory vesicle release.

The molecular events leading to regulated fusion of secretory vesicles with the plasma membrane remain unknown. Presumably, Ca^{2+} or cAMP activate protein kinases that directly or indirectly promote fusion of secretory vesicles with the plasma membrane by phosphorylating proteins in the vesicle membranes or surrounding cytoplasm. The phosphorylations may cause conformational changes in vesicle membrane proteins that expose hydrophobic regions, thereby inducing the vesicles to fuse with the plasma membrane.

There is also evidence that Ca^{2+} may directly activate some proteins promoting the fusion of secretory vesicles with the plasma membrane. For example, cytoplasmic proteins known as *annexins* are activated by direct combination with Ca^{2+}; the activated annexins bind with the membranes of secretory vesicles and promote vesicle fusion. Ca^{2+} has even been noted to promote the fusion of pure phospholipid vesicles, suggesting that the ion may have direct effects on the lipid as well as the protein component of membranes. G proteins binding and hydrolyzing GTP have also been implicated in the fusion mechanism. (For a discussion of molecular mechanisms involved in membrane fusion, see Supplement 5-1.)

Polarized Secretion   Exocytosis takes place at specialized locations in some cells. Epithelial cells of the small intestine, for example, secrete digestive enzymes from the surface of the cell that faces the intestinal cavity. On the side facing the interior of the body, a totally different spectrum of proteins, including those of the extracellular matrix, are secreted. Perhaps the most extreme example of *polarized secretion*, as this phenomenon is called, is observed in neurons and neurosecretory cells, in which vesicle release is restricted to axon terminals. The existence of polarized secretion suggests that certain proteins, such as the digestive and extracellular matrix proteins released by intestinal cells, probably contain signals recognized by receptors sorting them into vesicles destined for release at different points on the cell surface.

Endocytosis

Most eukaryotic cell types are able to take up materials from the surrounding medium by endocytosis. These materials may follow one of three distinct but closely related pathways. In one mechanism, *receptor-mediated endocytosis*, the substances taken in are recognized and bound to the cell surface by receptors. The receptors, which are glycoproteins forming part of the plasma membrane, recognize and bind only certain molecules from the medium. Typically, the molecules entering the cell by receptor-mediated endocytosis become significantly more concentrated as they are tied to the cell surface and packed into vesicles.

A second pathway, *bulk-phase endocytosis*, simply takes in bulk fluid from the surrounding medium. No binding by cell surface receptors occurs, and any molecules that happen to be in solution in the extracellular fluid are taken in with no increase in concentration. Bulk-phase endocytosis apparently proceeds at a fairly constant rate in all eukaryotic cells. The endocytosis carried out by higher plant cells is primarily or exclusively of the bulk-phase type, and evidently functions primarily to remove segments of the plasma membrane introduced by exocytosis.

The third endocytotic pathway, *phagocytosis* (meaning "cell eating"), differs from both receptor-mediated and bulk-phase endocytosis in that the materials taken in are large insoluble aggregates of molecules, cell parts, or even whole cells. The process resembles receptor-mediated endocytosis to the extent that the materials taken in are bound by specific cell-surface receptors. Both receptor-mediated and bulk-phase endocytosis are

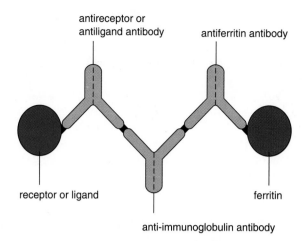

Figure 20-27   The three-way complex used to link ferritin, an iron-containing protein, to ligands or receptor molecules. The complex uses an antiligand or antireceptor antibody, an anti-immunoglobulin antibody, which reacts with antibodies themselves, and an antiferritin antibody.

widely distributed among eukaryotic cells, but phagocytosis is more limited. In mammals, for example, only white blood cells (leukocytes) can take in materials by phagocytosis.

Substances entering cells by endocytosis can be followed in their travels by one or more of several experimental techniques. For light microscopy, molecules entering cells are frequently linked to a fluorescent dye such as fluorescein. For electron microscopy, the substances are tagged by linking them to heavy metals.

Antibodies are also widely used in endocytosis research. Both a receptor and the molecule it recognizes and binds (called the *ligand*) in receptor-mediated endocytosis can be specifically tagged by linking them to antibodies. In one of the most frequently employed variations of this approach, different antibodies are developed against the receptor or its ligand, against the antibody proteins themselves, and against *ferritin*, an iron-containing protein that is visible in the electron microscope. Reacting the antibodies sets up a three-way complex that links the ferritin protein specifically to a receptor or its ligand (Fig. 20-27).

Receptor-Mediated Endocytosis   Of the three endocytotic mechanisms, receptor-mediated endocytosis has been studied experimentally in greatest detail, particularly in mammalian cells. Mammalian cells take up a wide variety of substances by receptor-mediated endocytosis. The list in different mammalian cells involves growth factors and peptide hormones; blood serum proteins, including molecules that carry metabolites such as cholesterol and other lipids, iron, and vitamins; lysosomal enzymes; antibodies; and defective serum

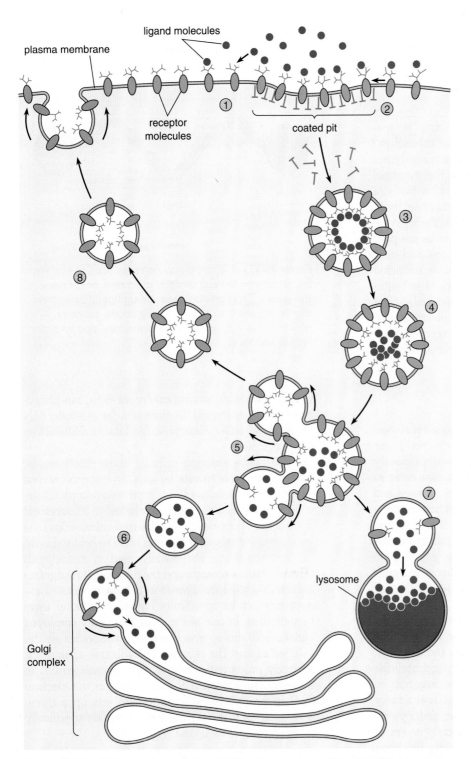

Figure 20-28 Pathways in receptor-mediated endocytosis. After binding to a receptor specifically recognizing them (step 1), ligands entering cells by endocytosis are concentrated in coated pits (step 2). The pits invaginate and sink into the underlying cytoplasm as endocytotic vesicles (step 3). As a vesicle forms, it loses its surface coat. During vesicle movement and fusion the ligands within the vesicles are sorted and, in many cases, released from their receptors (steps 4 and 5). The sorted molecules may proceed to the Golgi complex (step 6) or to lysosomes (step 7). Substances that travel to the Golgi complex may remain there or may be distributed to the ER or perinuclear space. They may also be resorted and secondarily routed in vesicles to the plasma membrane or to lysosomes. After removal from their ligand, some receptors are recycled to the plasma membrane (step 8); others are degraded in the vesicles that fuse with lysosomes.

proteins and glycoproteins. Some of these substances, such as blood serum proteins carrying iron and vitamins, are recognized and taken in by essentially all mammalian cell types; others, such as defective blood proteins and glycoproteins, are taken in by only a relatively few cells, such as those of the liver. Some pathogens or toxins, including viruses and the diphtheria and cholera toxins, also enter cells by the receptor-mediated endocytotic pathway.

Receptor-mediated endocytosis proceeds in several steps (Fig. 20-28). After binding to a receptor that specifically recognizes them (step 1 in Fig. 20-28), substances entering cells are concentrated in segments of the plasma membrane that will invaginate to form endocytotic vesicles (step 2). These membrane segments, called *coated pits*, are characterized by a thick layer of dense material that covers their cytoplasmic surface (see Fig. 20-29). Invagination of the pits, which may be driven by formation of the surface coat, proceeds rapidly until the pits pinch free from the plasma membrane and sink into the underlying cytoplasm as endocytotic vesicles (step 3). As the vesicle forms, it loses its surface coat. At this stage the vesicles may be spherical or extended into elongated and variously branched tubules. As they travel into the underlying cytoplasm along microtubules, vesicles may fuse into larger structures.

During vesicle movement and fusion substances in the vesicles are sorted and, in many cases, released from their receptors (step 4). Although the basis for this sorting is presently unknown, it presumably involves an interaction between recognition sites on the endocytosed molecules and receptor sites on the endocytotic vesicles. The sorted molecules may proceed to the Golgi complex or lysosomes within their original endocytotic vesicle or in smaller vesicles that pinch off from the endocytotic vesicle (step 5).

The substances sorted into vesicles that travel to the Golgi complex (step 6) may remain there or be distributed further to the ER or perinuclear space. They may also be resorted and secondarily routed in vesicles to the plasma membrane or to lysosomes. Fusion of vesicles with lysosomes (step 7) exposes their contents to degradation by the lysosomal enzymes. The material in vesicles returning to the plasma membrane (step 8), primarily unbound receptors, is reintroduced into the plasma membrane by exocytosis. This accomplishes recycling of the receptors. Not all receptors are recycled, however; some are degraded in the vesicles that fuse with lysosomes.

Membrane Receptors and Formation of Coated Pits
Receptors for substances entering cells by endocytosis are usually present in the plasma membrane in many thousands of copies. Liver cells, for example, contain as many as 150,000 receptors for *transferrin*, a blood plasma protein that transports iron ions; another

receptor, one for *low-density lipoprotein* (*LDL*, a carrier for cholesterol), occurs in about 20,000 copies in the plasma membrane of a typical mammalian cell such as a fibroblast. The receptors, which are free to move laterally in the fluid plasma membrane, collide randomly with segments of the membrane forming coated pits. From the numbers of receptors and coated pits, which form constantly at many sites in the plasma membrane, the receptors are calculated to collide with a forming pit at least once every three to four seconds. At this rate it is unlikely that any receptor binding its ligand will remain for a significant time in the plasma membrane without encountering a coated pit and undergoing endocytosis.

Some receptors, such as those for transferrin and *epidermal growth factor* (*EGF*), are trapped in coated pits only if they are bound to their ligand. In this case the receptor evidently undergoes a conformational change on binding the ligand that exposes a site recognized and bound by the coated pit. Other receptors, such as LDL and lysosomal protein receptors, are trapped in coated pits whether bound to their ligands or not. These receptors apparently possess binding sites for coated pits that are always active. At the same time, nonreceptor membrane proteins, such as transport carriers, which have no sites recognized by coated pits, do not bind to forming pits and are not taken into the cytoplasm by endocytosis.

Receptors are constantly removed from the plasma membrane and packed into the forming pits. The mechanism packing ligand-bound receptors into coated pits may concentrate the ligands to 1,000 to 10,000 times their concentrations in the surrounding medium.

The Proteins Forming Coated Pits The coats of coated pits appear in sections as a thick layer of "bristles" extending at right angles from the membrane surface (Fig. 20-29). When viewed face-on in isolated preparations, the bristles prove to be parts of a basketlike assembly that covers the cytoplasmic side of the pits (Fig. 20-30). The basketlike appearance results from an interlocking network of polypeptides that trace out pentagons, hexagons, and other regular geometric patterns on the membrane surface. The coat persists until the pocket pinches off as a free vesicle below the plasma membrane; then it usually disassembles and disappears.

The network forming the coat is stable and has proved relatively easy to isolate from plasma membrane preparations. First accomplished by B. M. F. Pearse and her colleagues in 1975, the isolated network of animal cells proved on analysis to contain a 180,000-dalton polypeptide as its major constituent. Termed *clathrin* by Pearse (from *clathratus* = lattice), the polypeptide is capable of assembling reversibly into baskets in the test tube under various conditions. For example, P. P. van Jaarsveld and his coworkers showed that purified clathrin remains in monomer form at about pH 8; as

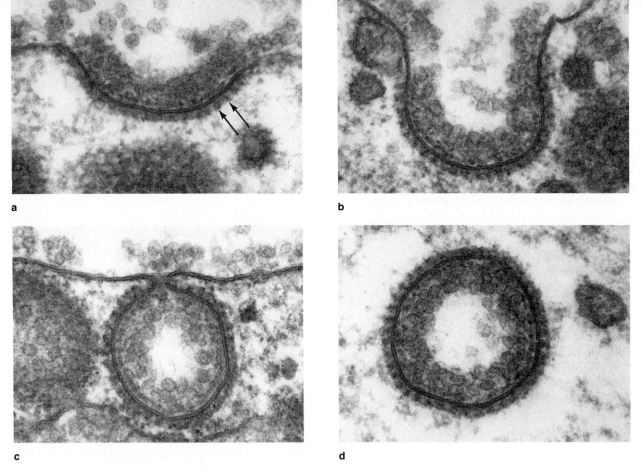

Figure 20-29 Formation of coated pits and vesicles in a developing chick oocyte. In thin sections the coat appears as a covering of short, knobbed bristles on the cytoplasmic surface of the pit or vesicle membrane (arrows). × 135,000. Courtesy of M. M. Perry.

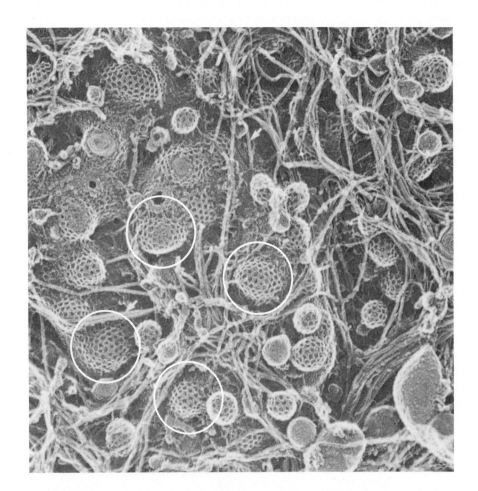

Figure 20-30 Coated pits (circles) viewed face-on from the inside surface of a plasma membrane isolated from a mouse liver cell. The fibers (arrows) are elements of the cytoskeleton. × 80,000. Courtesy of J. Heuser, from *Cell* 30:395 (1982). Copyright Cell Press.

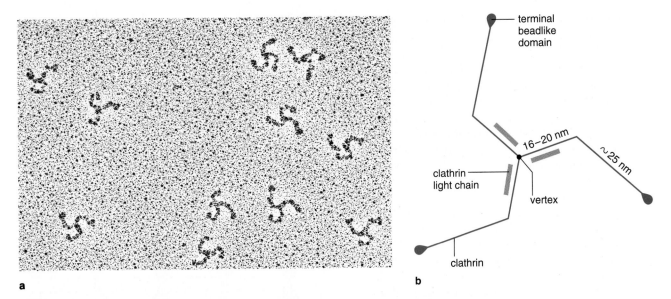

Figure 20-31 Clathrin structure. **(a)** Isolated clathrin triskelions prepared for electron microscopy by shadowing. Courtesy of D. Branton. **(b)** An individual triskelion, consisting of three clathrin molecules joined at the vertex; each clathrin molecule is linked to a clathrin light chain.

the pH is lowered to 6 to 6.8, the monomers self-assemble into typical basketlike structures.

An isolated clathrin subunit appears in the electron microscope as a three-legged structure called a *triskelion* (Fig. 20-31). Each leg is bent to a shallow angle at a position roughly halfway along its length. The molecular weight of intact triskelions lies near 600,000, indicating that the three-legged structure probably contains three clathrin polypeptides, linked together at the center of the triskelion. Much of the mass of each clathrin polypeptide is concentrated in a domain that forms a beadlike swelling at the tip of each triskelion arm.

Triskelions are believed to assemble in an overlapping network to form the lattice lining a coated pit (Fig. 20-32a). Only slight changes in the angles separating the legs at the center of triskelions are necessary to accommodate pentagonal, hexagonal, or other lattice patterns (Fig. 20-32b). The variations in lattice patterns may be required to accommodate the different surface curvatures of coated pits and vesicles.

Several other polypeptides occur in coated pits and vesicles, including one group that is closely associated with clathrin triskelions. This group, *clathrin light chains*, has molecular weights ranging from 30,000 to 36,000 in different species and occurs in quantities approxi-

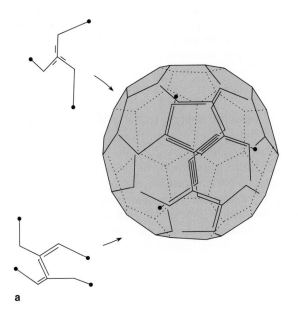

Figure 20-32 Clathrin coat structure. **(a)** Packing of triskelions to form a clathrin coat. Modified from an original courtesy of T. Kirchhausen, from *Cell* 33:650 (1983). Copyright Cell Press. **(b)** Packing of triskelions to form either pentagons or hexagons. Switches between the polygon types require only a slight adjustment of the angles traced by the triskelions. Modified from an original courtesy of B. M. F. Pearse. Reproduced from *J. Cell Biol.* 91:790 (1981) by copyright permission of the Rockefeller University Press.

mately equal to the clathrin polypeptide. Each triskelion, therefore, probably contains three clathrin light chains, one associated with each clathrin molecule of the triskelion. Antibodies against the clathrin light chains react with triskelions at a position on the legs near the center of the structure, indicating that light chains associate with the first segment of each leg in advance of the bend. Investigations by F. M. Brodsky and his colleagues detected at least five different kinds of clathrin light chains in higher animals; cells in different tissues contain distinct combinations of light chain types.

Four additional polypeptides also occur in clathrin coats as subunits of *adaptors*, proteins able to bind the globular tips of clathrin triskelions as well as the cytoplasmic extensions of membrane receptors. These binding capabilities indicate that the primary function of the adaptors, which line the surface of the coat facing the plasma membrane, is to bind surface receptors to the coated pits (Fig. 20-33). They may also speed polymerization of clathrin into assembled baskets. Clathrin and the adaptor polypeptides have been detected in animals, plants, fungi, algae, and protozoa. The amino acid sequence of the clathrin polypeptide in the yeast *Saccharomyces cerevisiae*, deduced from the gene sequence by S. K. Lemmon and her coworkers, is 50% identical to mammalian clathrin.

Functions, Assembly, and Disassembly of the Clathrin Coat The clathrin coat may serve to reinforce the plasma membrane on its cytoplasmic side during coated pit formation. Assembly of the coat may also provide energy for membrane invagination during endocytosis. Self-assembly of the basket under the plasma membrane may depress the membrane into an invagination. As more clathrin subunits add to the forming pit, the invagination deepens until it pinches off from the plasma membrane. Once the vesicle is fully formed, the coat disassembles from its surface and joins a cytoplasmic pool of other disassembled subunits. At any time about half the total clathrin complement of eukaryotic cells is estimated to be in the pool of unassembled subunits.

Assembly of the clathrin coat evidently proceeds in cells without an energy input. However, disassembly of the coats from fully formed vesicles is an energy-requiring process that involves an "uncoating protein" with ATPase activity. The activity of the uncoating protein, which is a member of the hsp70 group of heat shock proteins, depends on the presence of assembled baskets containing both clathrin and clathrin light chains. When activated, the ATPase hydrolyzes ATP, on the order of about three molecules of ATP per triskelion, and promotes rapid disassembly of clathrin baskets.

The uncoating protein binds to sites on the clathrin light chains. According to a current model for the action of the uncoating protein, these binding sites are exposed when a coated pit finishes its invagination and rounds

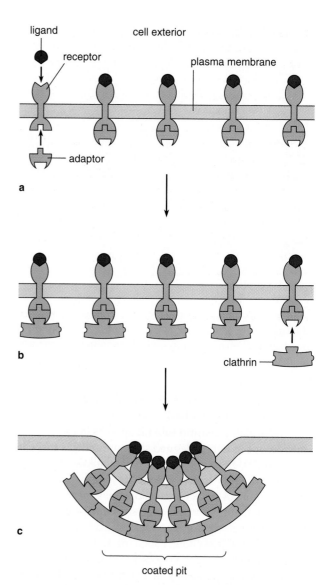

Figure 20-33 The possible interactions of clathrin, adaptors, and surface receptors in coated pit formation. The adaptors recognize and bind both the cytoplasmic extensions of surface receptors (a) and clathrin (b). The interaction between the adaptors and clathrin promotes the assembly of the clathrin lattice and a coated pit (c).

up into a vesicle. Binding to the clathrin light chains activates the ATP-dependent "unfoldase" function of the uncoating protein, which induces conformational changes in the triskelions that cause the vesicle coat to disassemble. Removing the coat converts the vesicle into a smooth-walled form that travels further along the endocytotic pathway. As soon as the clathrin triskelions disassemble from the vesicles, the sites on clathrin light chains binding the uncoating protein are presumably blocked, releasing the uncoating protein and converting the triskelion to the conformation in which it is ready to enter another round of coated vesicle formation.

Coated Vesicles Associated with the Golgi Complex

Coats are also frequently seen around vesicles budding from the Golgi complex (Fig. 20-34). Antibodies against clathrin react with the coats of budding vesicles on *trans* cisternae, indicating that clathrin is present in these coats. These vesicles give rise to secretory vesicles or lysosomes. Coats forming around vesicles at other sites, including the shuttle vesicles moving between the Golgi cisternae, do not react with antibodies against clathrin. Although these vesicle coats resemble those of coated pits to some extent, they are evidently assembled from a different group of proteins. At least one of the nonclathrin coated vesicle proteins, however, has been found by R. Duden and his colleagues to have sequence similarities to one of the adaptor polypeptides of endocytotic coated pits. This similarity opens the possibility that the cell-surface and Golgi-associated coated pits and vesicles are related structures with a single evolutionary origin. Coats that fail to react with anticlathrin antibodies are also sometimes observed around transition vesicles.

The coats around vesicles in the Golgi complex probably also function to reinforce, shape, and provide an energy source for membrane segments forming pockets that pinch off in vesicular form. They may also bind the cytoplasmic ends of receptors involved in sorting proteins passing through the Golgi complex. The coats around Golgi vesicles disassemble as the vesicles fuse with their target membranes—shuttle vesicles, for example, lose their coats as they fuse with the next Golgi cisterna in line. The research by Rothman and his coworkers showed that uncoating of shuttle vesicles requires GTP hydrolysis (see Fig. 20-13). If GTP is replaced by the nonhydrolyzable analog GTP$_\gamma$S, the shuttle vesicles fail to uncoat and accumulate around the Golgi complex in sufficient quantity to be isolated by cell fractionation and sucrose gradient centrifugation (see p. 125). Analysis of the coated shuttle vesicle fraction by the Rothman group confirmed that clathrin is absent from the coats.

Not all plasma membrane segments involved in endocytosis form coated pits. Membrane coats are rare in cells taking in material by phagocytosis; the pockets forming in the plasma membrane in this process are usually smooth walled. Significantly, membrane invagination in phagocytosis is frequently observed to be ATP dependent, in contrast to coated pit formation, which is ATP independent. This observation suggests that another energy-requiring motile system, such as microtubules or microfilaments, may be active in the reactions forcing segments of the plasma membrane to form pockets in phagocytosis.

Endocytotic Vesicles, pH, and Receptor-Ligand Sorting The vesicles that pinch off from coated pits during receptor-mediated endocytosis initially contain a mixture of different receptors, each still linked to its ligand.

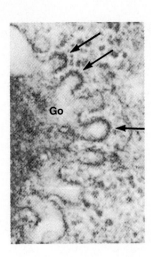

Figure 20-34 Coated vesicles (arrows) forming at the margins of a Golgi complex (Go) of a carrot cell. ×81,000. Courtesy of John G. Robertson and The Company of Biologists Ltd., from *J. Cell Sci.* 58:63 (1982).

Very shortly after a newly formed endocytotic vesicle separates from the plasma membrane, many receptors release their ligand into the vesicle interior.

Ligand release appears to depend on both pH changes in endocytotic vesicles and the sensitivity of the receptor-ligand complexes to the changes. The pH of the medium immediately surrounding the cell, which is trapped and carried into endocytotic vesicles as they form, is initially near the neutral value of pH 7. Immediately after an endocytotic vesicle releases from the plasma membrane, its internal pH drops rapidly to 5 to 5.5. The pH drop can be measured experimentally by attaching pH-sensitive dyes such as fluorescein to ligands. (The dyes absorb light differently at different pHs.) The reduction in pH probably results from the activity of an H$^+$-ATPase pump in the vesicle membranes. At the lower pH characteristic of the endocytotic vesicle, some receptors, such as the LDL receptor, lose their affinity for their ligand and release it into the vesicle interior. Other receptors, such as the receptor binding EGF, are insensitive to the pH change and remain tightly bound to their ligand.

In many but not all cases, the sensitivity of receptors to the pH change in endocytotic vesicles determines whether the receptors are degraded in lysosomes or recycled to the plasma membrane. Most receptors remaining bound to their ligands as endocytotic vesicle pH falls travel to lysosomes and are degraded with their ligand. Those releasing their ligand are sorted into membrane segments that pinch off from endocytotic vesicles and eventually return to the plasma membrane. As expected from the operation of this mechanism, agents that raise the pH of endocytotic vesicles, such as NH$_4$Cl, block ligand release by pH-sensitive receptors. At the artificially elevated pH the receptors are trapped with their ligands into the lysosomal pathway. As a consequence, they are degraded and fail to reappear at the cell surface.

In some mammalian cells, vesicles in regions where receptors separate from their ligands can be distinguished from the surrounding vesicular traffic by their

characteristics, and comparisons with the eukaryotic process. *Microbiol. Rev.* 53:333–66.

Schatz, P. J., and Beckwith, J. 1990. Genetic analysis of protein export in *E. coli. Ann. Rev. Genet.* 24:215–48.

Wandersman, C. 1989. Secretion, processing, and activation of bacterial extracellular proteases. *Molec. Microbiol.* 3:1825–31.

Wickner, W.; Driessen, A. J. M.; and Hartl, F. V. 1991. The enzymology of protein translocation across the *Escherichia coli* plasma membrane. *Ann. Rev. Biochem.* 60:101–24.

Lipid Synthesis and Distribution

Alfsen, A. 1989. Membrane dynamics and molecular traffic and sorting in mammalian cells. *Prog. Biophys. Molec. Biol.* 54:145–57.

Bishop, W. R., and Bell, R. M. 1988. Assembly of phospholipids into cellular membranes: biosynthesis, membrane movement, and intracellular translocation. *Ann. Rev. Cell Biol.* 4:579–610.

Dawidowicz, E. A. 1987. Lipid exchange: transmembrane movement, spontaneous movement, and protein-mediated transfer of lipids and cholesterol. *Curr. Top. Membrane Transport* 29:175–202.

van Meer, G. 1989. Lipid traffic in animal cells. *Ann. Rev. Cell. Biol.* 5:247–75.

Voelker, D. R. 1991. Organelle biogenesis and intracellular lipid transfer in eukaryotes. *Microbiol. Rev.* 55:543–60.

Wickner, W. 1989. Secretion and membrane assembly. *Trends Biochem. Sci.* 14:280–83.

Wirtz, K. W. A. 1991. Phospholipid transfer proteins. *Ann. Rev. Biochem.* 60:73–99

Glycosylation in the ER and Golgi Complex

Abeijon, C., and Hirschberg, C. B. 1992. Topography of glycosylation reactions in the ER. *Trends Biochem. Sci.* 17:32–36.

Kornfeld, R., and Kornfeld, S. 1985. Assembly of asparagine-linked oligosaccharides. *Ann. Rev. Biochem.* 54:631–64.

Moore, P. J.; Swords, K. M. M.; Lynch, M. A.; and Staehelin, L. A. 1991. Spatial organization of the assembly pathways of glycoproteins and complex polysaccharides in the Golgi apparatus of plants. *J. Cell Biol.* 112:589–602.

Paulson, J. C., and Colley, K. J. 1989. Glycosyltransferases. *J. Biolog. Chem.* 264:17615–18.

Roth, J. 1987. Subcellular organization of glycosylation in mammalian cells. *Biochim. Biophys. Acta* 906:405–36.

Yamamoto, F. I.; Clausen, H.; White, T.; Marken, J.; and Hakamori, S. 1990. Molecular genetic basis of the histo-blood group ABO system. *Nature* 345:229–33.

Review Questions

1. Outline the structure of the ER and Golgi complex. In what ways are the two membrane systems similar? Different? In what ways is the nuclear envelope related to the ER? What are transition vesicles? Shuttle vesicles? Secretory vesicles? Storage vesicles? Endocytotic vesicles? What molecular events are involved in the fusion of these vesicles?

2. Outline the major functions of the ER and Golgi complex. What major groups of proteins occur in the ER and Golgi complex? What major functions are associated with *cis* and *trans* cisternae of the Golgi complex?

3. Outline the major steps in assembly of the carbohydrate groups of glycoproteins in the ER and Golgi complex.

4. What major lines of evidence indicate that proteins move from the ER, through the Golgi complex, and to the cell exterior?

5. What experimental results led to formulation of the signal hypothesis? Outline the events taking place when a ribosome initiates protein synthesis on an mRNA for a secretory protein.

6. What is the signal recognition particle? How does it function in ER-directed distribution mechanisms? What other elements and factors are believed to operate in direction and insertion of proteins into the ER, and how are they believed to interact? What structural characteristics are common to most ER-directing signals? How are these elements believed to act in directing proteins to the ER?

7. What evidence indicates that the insertion of ER-directed proteins is potentially posttranslational? What are the possible energy sources for the insertion of ER-directed proteins?

8. What are stop-transfer signals? Postinsertion signals? How are signals in these categories believed to operate in the insertion, sorting, and distribution of ER-directed proteins? What is the salvage compartment, and how is it believed to interact with the KDEL signal to return resident proteins to the ER?

9. Outline the steps followed in the assembly, modification, sorting, and distribution of a lysosomal protein, beginning with an mRNA entering protein synthesis. What mechanisms guard against unscheduled secretion of lysosomal enzymes? What is the role of pH in the distribution of proteins to lysosomes?

10. Outline the sources of membrane proteins for the ER, Golgi complex, plasma membrane, nuclear envelope, mitochondria, and chloroplasts.

11. What signals route proteins to mitochondria and chloroplasts? What separate membranes and compartments may proteins be assigned to in the two organelles? Compare the signals of ER- and organelle-directed proteins.

12. Outline the mechanism directing proteins to the mitochondrial matrix and chloroplast stroma. What mechanisms distribute proteins further from these locations? In what ways are these distribution mechanisms related to prokaryotic secretion? Compare the energy sources for mitochondrial and chloroplast protein import.

13. What signals send proteins to the nucleus? What is the role of nuclear pore complexes in the distribution of proteins to the nucleus? Compare the routes followed by a nuclear envelope protein and a chromosomal protein.

14. What is the difference between constitutive and regulated secretion? What evidence suggests that constitutive secretion is the "default" pathway? What is polarized secretion?

15. Trace the possible routes followed by a protein taken in by endocytosis. Compare receptor-mediated endocytosis, bulk-phase endocytosis, and phagocytosis. What cellular pathways may be followed by the receptors active in receptor-mediated endocytosis? What is the relation of pH changes to these pathways?

16. What are coated pits? How do coated pits function in endocytosis? What major protein groups occur in coated pits? What overall functions are associated with each protein group? What is a triskelion? What is the apparent role of ATP in coated pit formation and disassembly?

17. What types of enzymes occur in lysosomes? How do lysosomes function in endocytosis?

18. What is the relationship among endocytosis, exocytosis, and the surface area of the plasma membrane? What lines of evidence support this relationship?

19. Compare protein secretion in bacteria and eukaryotes. What are the characteristics of membrane-directing signals in bacteria? What are the apparent sources of energy for the insertion of proteins into the bacterial plasma membrane?

Supplement 20-1
Protein Sorting, Distribution, and Secretion in Prokaryotes

Although prokaryotes are more simply organized than eukaryotes, proteins are also routed to various locations in these organisms. In gram-positive bacteria (see p. 302) possible destinations include the cell interior, the plasma membrane, and the cell exterior including the cell wall. In the more complex gram-negative bacteria and cyanobacteria, different destinations include the cell interior, the plasma membrane, the periplasmic compartment between the plasma membrane and outer membrane, the outer membrane itself, and the cell exterior (see Fig. 8-31).

In bacteria, proteins destined to remain inside the cell are assembled on ribosomes suspended in the cytoplasmic solution. Proteins addressed to the plasma membrane, cell wall or exterior may be assembled on ribosomes that are freely suspended or attached to the cytoplasmic surface of the plasma membrane. In either case insertion and passage of proteins through the plasma membrane is actually or potentially posttranslational.

With some exceptions the assignment of proteins to the plasma membrane in bacteria depends on an N-terminal signal similar to the ER-directing signal of eukaryotes. The bacterial signal includes an initial positively charged segment, a hydrophobic middle segment capable of spanning a membrane as an alpha helix or beta strand, and a terminal region specifying cleavage of the signal. These have properties similar to the ER-directing signals of eukaryotes except that the initial segment is frequently more positively charged and the core often less hydrophobic.

Insertion of newly synthesized proteins into the plasma membrane in *E. coli* requires a membrane potential and ATP. Several proteins have been implicated in the process through the effects of mutations that slow or stop movement of proteins through the plasma membrane (Fig. 20-39). Two of these, *SecB* and *GroEL*, are cytoplasmic proteins that probably act as chaperones by binding to newly synthesized proteins with an N-terminal signal and holding them in loosely folded conformations in which they can penetrate the plasma membrane. Two membrane proteins, *SecA* and *SecY*, may act in concert to promote passage of proteins through the membrane. SecA is a peripheral membrane protein with ATPase activity, and SecY is an integral transmembrane protein. SecA is stimulated as an ATPase when it binds a protein with an N-terminal signal from the cytoplasm; ATP hydrolysis may be associated with transfer of the protein from SecB or GroEL to SecY, which may be the protein primarily involved in insertion and movement of the protein across the plasma membrane. Mutant forms of several other integral membrane proteins—*SecD*, *SecE*, and *SecF*—also interrupt movement into or across the plasma membrane. These proteins may join with SecY as parts of a translocation complex in the *E. coli* plasma membrane.

Recently A. Dobberstein, P. Walter, and their associates discovered a particle in *E. coli* that has some of the properties of the eukaryotic signal recognition particle. The possible prokaryotic counterpart to the SRP contains a 4.5S RNA that can be substituted for SRP 7S scRNA in eukaryotic SRP, and a 48,000-dalton

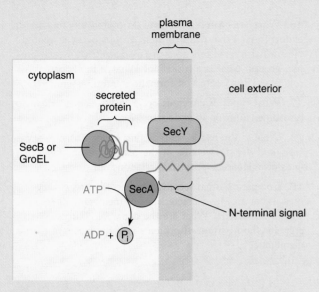

Figure 20-39 Elements implicated in bacterial protein secretion. Two cytoplasmic proteins, SecB and GroEL, probably act as chaperones binding to newly synthesized proteins with an N-terminal signal and holding them in conformations that can penetrate the plasma membrane. Passage through the membrane is promoted by two membrane proteins: SecA, a peripheral membrane protein with ATPase activity, and SecY, an integral transmembrane protein. SecA is stimulated as an ATPase when it binds a protein with an N-terminal signal from the cytoplasm; ATP hydrolysis may be associated with transfer of the protein from SecB or GroEL to SecY, which may be the protein primarily involved in insertion and movement of the protein across the plasma membrane. Insertion of newly synthesized proteins into the plasma membrane in *E. coli* also requires a membrane potential.

polypeptide that can take the place of a critical 54,000-dalton polypeptide in eukaryotic SRP. The prokaryotic 4.5S RNA and 48,000-dalton protein also have sequence similarities to their apparent eukaryotic counterparts. These findings indicate that *E. coli* and other bacteria may have an SRP that recognizes N-terminal signals and assists binding of proteins to the bacterial plasma membrane as part of membrane insertion or secretion.

If proteins are to remain part of the plasma membrane, stop-transfer sequences or conformations hold the protein in the membrane. Proteins that are to be placed in the cell wall or secreted in gram-positive bacteria, or released into the periplasm in gram-negative bacteria, pass entirely through the plasma membrane.

Once its function in directing insertion of proteins in the plasma membrane is complete, the N-terminal signal is removed from almost all prokaryotic proteins. Signals are cleaved in *E. coli* by one of two different signal peptidases. *Signal peptidase I,* a 37,000-dalton protein, occurs in both the plasma membrane and outer membrane. This enzyme removes the N-terminal signal from all membrane-directed proteins except lipoproteins, the unusual protein-lipid complexes that form part of the outer membrane of gram-negative bacteria (see p. 305 and Fig. 8-37). The other enzyme, *signal peptidase II,* located exclusively in the plasma membrane, is specialized for removal of the N-terminal signal from lipoproteins.

Relatively little is known about the signals and mechanisms directing placement of proteins in the outer membrane of gram-negative bacteria or indicating that they are to be secreted by passing entirely through the outer membrane. Many of these proteins travel to the periplasmic space by means of an N-terminal signal that directs their insertion and passage through the plasma membrane. Once in the periplasm, further information, probably involving folding states promoted by their amino acid sequences, directs their insertion in the outer membrane. Passage entirely through the outer membrane, which is highly impermeable to most substances, may involve receptor and translocator proteins in the membrane that are unique to each secreted protein.

Some proteins, such as OmpA are distributed to the outer membrane of *E. coli* with no evident N-terminal signal. Evidently, the conformation of this protein promotes its passage through the plasma membrane; on reaching the periplasmic space, it refolds into a different conformation that promotes its insertion in the outer membrane. It is possible that chaperones in the cytoplasm and periplasmic space may maintain OmpA in the correct states to penetrate the two membranes. Some proteins with no N-terminal signals are also secreted through the outer membrane in *E. coli.*

Although the bacterial pathways distributing newly synthesized proteins are simpler than eukaryotic routes, the interchangeability of signals demonstrates the close relationship of the two systems. As noted in the main part of this chapter (see p. 835), eukaryotic secretory and membrane proteins, when introduced by recombinant DNA techniques into bacterial cells, are correctly routed to the plasma membrane or secreted to the cell exterior. Similarly, bacterial secreted proteins such as β-lactamase are correctly sorted into secretory vesicles and released to the exterior when their genes are introduced into yeast cells or *Xenopus* oocytes.

Suggestions for Further Reading: See p. 869.

Reactions Adding Sugar Units to Proteins in the ER and Golgi Complex

Several sugar units are added in the glycosylation of proteins, primarily glucose, galactose, mannose, fucose, acetylglucosamine, and acetylgalactosamine (see Fig. 5-5 and p. 162). Before being added, the sugar units are activated in the soluble cytoplasm surrounding the ER and Golgi complex by reaction with one of the nucleoside triphosphates UTP, CTP, or GTP. When mannose is activated, for example, UTP reacts with the sugar to produce a UDP-mannose derivative:

$$UTP + mannose \rightarrow UDP\text{-}mannose + P_i \quad (20\text{-}1)$$

Some activated sugars are transported across ER and Golgi membranes in this form by proteins that exchange a nucleotide-sugar complex from outside for an unbound, or "empty," nucleotide on the inside. The sugar is then transferred directly from the nucleotide residue to the growing carbohydrate group by enzymes called *glycosyl transferases*, which are active within the organelle compartment. Energy for the additions is derived from hydrolysis of the nucleoside diphosphate–sugar complex, which retains much of the energy released by removal of the phosphate in Reaction 20-1. The glycosyl transferase enzymes, themselves glycoproteins, are integral membrane proteins of the ER or Golgi complex, with their active sites directed toward either the cytoplasmic surfaces or the cisternae of the two organelles.

Some sugars are transported into the ER and added to glycoproteins via a *carrier lipid* that acts as an intermediate in the reactions. In this pathway one or more activated sugars interact with the carrier lipid on the cytoplasmic side of the ER membranes. The sugar is transferred from a nucleoside diphosphate to the carrier lipid. For a UDP-linked sugar, for example:

$$UDP\text{-}sugar\ unit\ +\ carrier\ lipid \longrightarrow$$
$$carrier\ lipid\text{-}sugar\ +\ UDP \quad (20\text{-}2)$$

Several sugars may be linked together on the carrier lipid, producing a partly assembled carbohydrate tree on the cytoplasmic side of the ER. The sugar or carbohydrate is then transported across the ER membranes while linked to the carrier lipid and transferred to the glycoprotein being modified. Primary among these carrier lipids in both plants and animals are *dolichols*, long-chain alcohols containing an internal series of five-carbon *isoprenoid* subunits (Fig. 20-40).

Acetylglucosamine, mannose, and glucose are among the sugars commonly linked into the core units of carbohydrate groups which are added in the ER. Acetylgalactosamine, fucose, galactose, and, in animals, sialic acid, are added to the cores as terminal or capping sugars in the Golgi complex. These sugars are transferred from their nucleotide carriers to the carbohydrate groups under assembly by glycosyl transferases named for the sugar they add.

The reactions linking sugars to glycoproteins take two primary forms depending on the amino acid side chains to which carbohydrates are linked in the proteins (see also p. 162). Carbohydrates that attach to an amino ($-NH_2$) group in the side chain of asparagine residues in the proteins are called *N-linked carbohydrates*. Carbohydrates that attach to a hydroxyl ($-OH$) group on the side chains of serine or threonine (or, more rarely, to hydroxyproline or hydroxylysine) are called *O-linked carbohydrates*. Both *N-* and *O*-linked carbohydrate groups may occur on the same protein. Glycophorin, a major plasma membrane protein of mammalian red blood cells, for example, contains 1 *N*-linked and no less than 15 *O*-linked carbohydrate groups bound to different points in its amino acid sequences (see Fig. 5-7).

The interactions producing *N*-linked carbohydrates, which are presently the best understood of the two pathways, give an idea of the complexity of the glycosylation reactions (Fig. 20-41). The *N*-linked process begins on the cytoplasmic side of the ER, with transfer of acetylglucosamine from a nucleoside diphosphate carrier to dolichol (Fig. 20-41*a*). A second acetylglucosamine residue is then added to the first, forming a short (acetylglucosamine)$_2$ chain attached to dolichol (Fig. 20-41*b*). Five mannose units are next transferred from their nucleotide carriers to the carbohydrate group, forming the branched (acetylglucosamine)$_2$(mannose)$_5$ structure shown in Figure 20-41*c*. All these reactions occur on the cytoplasmic side of

$$H_3C-\underset{\underset{CH_3}{|}}{C}=CH-CH_2(CH_2-\underset{\underset{CH_3}{|}}{C}=CH-CH_2)_n-CH_2-\underset{\underset{H}{|}}{\overset{\overset{CH_3}{|}}{C}}-CH_2-CH_2-CH_2OH$$

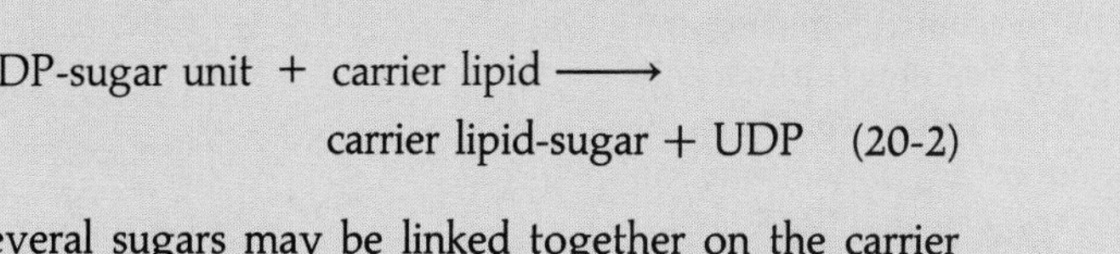

dolichol

Figure 20-40 Dolichol, a membrane carrier lipid that transfers sugar groups from the cytoplasm into the ER lumen. The number of isoprenoid subunits (in red) varies from 15 to 18.

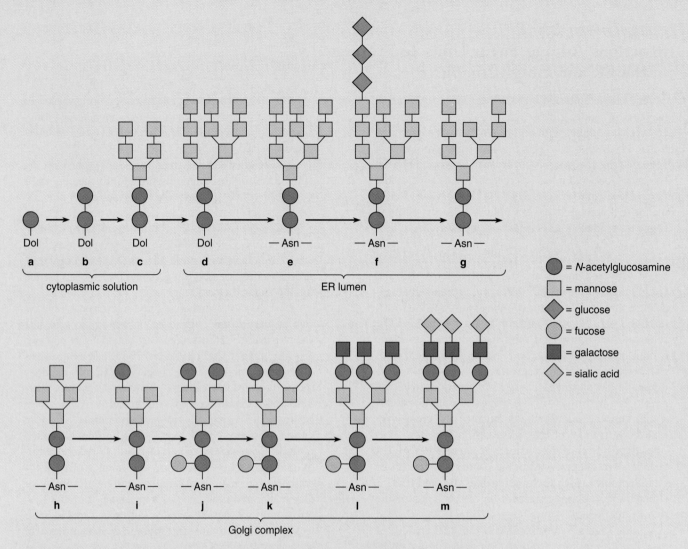

Figure 20-41 The assembly of *N*-linked carbohydrate groups in the ER and Golgi complex. Dol, dolichol; Asn, asparagine. The process begins on the cytoplasmic side of the ER, with transfer of acetylglucosamine from a nucleoside diphosphate carrier to dolichol **(a)**. A second acetylglucosamine residue is added to the first, forming a short chain attached to the dolichol **(b)**. Five mannose units are then transferred from their nucleotide carriers to the dolichol complex, forming a branched structure **(c)**. The dolichol-carbohydrate complex is next transported across the ER membrane by the dolichol carrier. In the ER lumen four more mannose residues are added **(d)** and the resulting structure is transferred as a unit from dolichol to an asparagine residue **(e)**. In some organisms this step completes the core structure; in others three glucose residues are added from nucleotide carriers **(f)**. In most cases the three glucose residues and one mannose unit are subsequently removed while the glycoprotein is still in the ER **(g)**. The glycoprotein is then transported to the Golgi complex, where three more mannose residues are usually removed, leaving the core structure shown in **(h)**. Steps **(i** through **m)** outline a typical mammalian pathway removing further mannose residues and adding acetylglucosamine, fucose, galactose, and sialic acid to complete the carbohydrate group.

the ER membranes; once assembled to this extent, the complex is transported across the ER membrane by the dolichol carrier. In the ER lumen four more mannose residues are added to the structure (Fig. 20-41*d*). This (acetylglucosamine)$_2$(mannose)$_9$ structure is now trans-

ferred as a unit from dolichol to an asparagine residue in the protein being glycosylated (Fig. 20-41*e*). In some organisms this step completes the core structure; in others three glucose residues are transferred from their nucleotide carriers on the lumen side of the ER to form

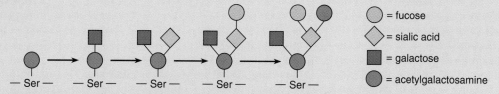

Figure 20-42 A typical pathway assembling a small *O*-linked carbohydrate group. Ser, serine. A single mannose or acetylgalactosamine unit is transferred from a nucleotide carrier directly to the side-chain —OH group of a serine or threonine residue or, more rarely, to an —OH group of hydroxyproline or hydroxylysine. This initial core glycosylation evidently occurs on the lumen side of ER membranes or while the glycoprotein is in transit from ER to the Golgi complex. To this, or to other relatively simple core structures, additional capping sugars, including the same spectrum used in *N*-linked glycosylation, are transferred from nucleotide carriers in the Golgi complex to complete the carbohydrate group. The pathway shown is one of many possibilities.

the core shown in Figure 20-41*f*. These glucose units are transported individually across the membrane by protein-based carrier systems rather than dolichol. The same *N*-linked core structure, with or without the glucose residues, is assembled by a closely similar pathway in the ER of plants, mammals, and yeast.

Later steps vary to some degree depending on the glycoprotein or organism. In most cases the three glucose residues and one mannose unit are removed while the glycoprotein is still in the ER (Fig. 20-41*g*). After this initial trimming the glycoprotein is transported in transition vesicles to the Golgi complex, where three more mannose residues are usually removed, leaving the trimmed (acetylglucosamine)$_2$(mannose)$_5$ core structure (Fig. 20-41*h*). Why the transient glucose and mannose residues are first added and then usually removed almost immediately is unknown; perhaps they serve as processing signals for later steps in the pathway or are necessary for movement of the glycoproteins to the Golgi complex.

In some plant glycoproteins the core structure undergoes no further change and remains as the carbohydrate group of the mature glycoprotein. In other plant and in all mammalian glycoproteins, further glycosylations add capping sugars to assemble a more complex structure. The steps diagramed in Figure 20-41*i* through *m* outline a typical mammalian pathway removing mannose residues and adding acetylglucosamine, fucose, galactose, and sialic acid to complete the carbohydrate group.

Information on the pathways of *O*-linked glycosylation is more limited. In the most simple *O*-linked core structures (Fig. 20-42) either a single mannose or acetylgalactosamine unit is transferred from a nucleotide carrier directly to the side-chain —OH groups of serine or threonine residues or, more rarely, to an —OH group of hydroxyproline or hydroxylysine in the recipient protein. This initial core glycosylation evidently

occurs on the lumen side of ER membranes or while the glycoprotein is in transit from the ER to the Golgi complex. To this or other relatively simple core structures (at least four different cores have been detected in *O*-linked glycosylation) additional capping sugars, including the same spectrum used in *N*-linked glycosylation, are transferred from nucleotide carriers in the Golgi complex to complete the carbohydrate group.

Although the number of different sugar residues added to core structures to complete *N*- and *O*-linked glycosylation is relatively small, an almost infinite variety of carbohydrate trees can be assembled by variations in the sequence, type, and number of sugar residues added in the capping reactions. These variations account for the highly varied glycoprotein types noted among plant, animal, and other eukaryotic glycoproteins.

The human A, B, and O blood antigens (see p. 236) have an interesting relationship to the reactions adding sugar groups to glycoproteins. Recent investigations by F. Yamamoto and his coworkers revealed that the three antigens depend on mutant forms of a glycosyl transferase active in the Golgi complex. The A and B antigens result from base substitutions at seven different positions in the gene encoding the glycosyl transferase. The version of the enzyme producing the A antigen adds a terminal acetylgalactosamine unit at one site in the carbohydrate group of a membrane glycoprotein; the version producing the B antigen adds a galactose at this position. The O antigen depends on a single base deletion in the gene. The deletion causes a shift in reading frame that produces a completely inactive enzyme. As a result, the position glycosylated in the A and B forms of the glycoprotein remains vacant.

Suggestions for Further Reading: See p. 870.

21

CYTOPLASMIC GENETIC SYSTEMS

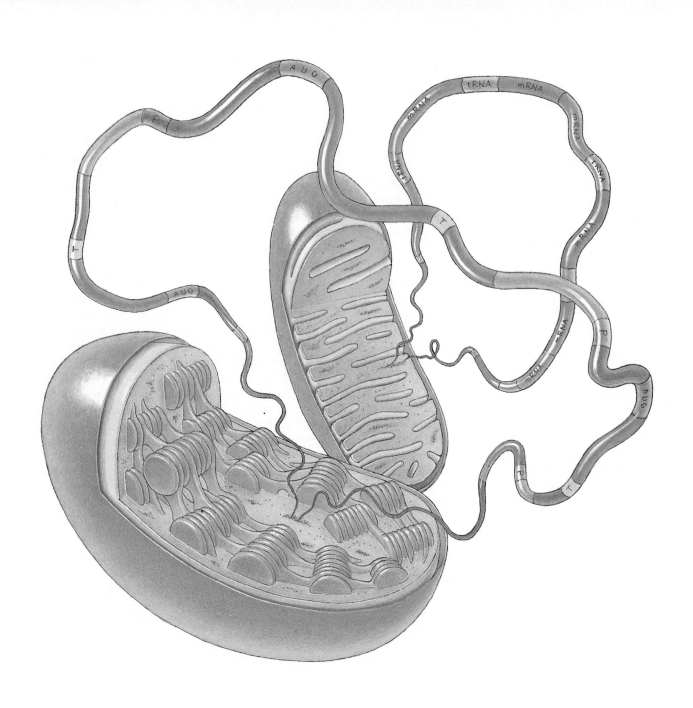

■ *Cytoplasmic inheritance* ■ *Mitochondrial and chloroplast DNAs* ■ *Organelle genes* ■ *Genome arrangement* ■ *Transcription and translation* ■ *Organelle genetic codes* ■ *Regulation* ■ *Origins of organelle genomes* ■ *Genetic information in centrioles*

G enetic information is stored and transcribed in the cytoplasm as well as in the cell nucleus in eukaryotic cells. Two cytoplasmic organelles, mitochondria and chloroplasts, contain their own genetic systems, complete with information encoded in DNA molecules, the enzymatic mechanism for transcribing the information into mRNA, tRNA, and rRNA, and ribosomes and other factors required to translate mRNAs into polypeptides. Although mitochondria and chloroplasts are residents in the cytoplasm of eukaryotic cells, their genomes and their transcription and translation mechanisms are primarily prokaryotic in nature. The genetic mechanisms of the two organelles make them semi-independent structures within eukaryotic cells.

The mitochondrial or chloroplast DNA molecules of several species have been completely sequenced. The sequences reveal that although polypeptides are encoded in the organelle DNAs, the total includes only a fraction of the total protein complement of either organelle. Most organelle polypeptides are encoded in the cell nucleus and synthesized in the cytoplasm outside mitochondria and chloroplasts. These polypeptides make their way into the organelles by the distribution mechanisms discussed in Chapter 20.

The study of mitochondrial and chloroplast genomes and the transcription and translation mechanisms of the two organelles has proved to be a fascinating exercise in its own right. Though obviously prokaryotic in origins, the genetic systems of the organelles have evolved many features that are unique and without parallels in other systems. In addition, the organelles have been an invaluable source of new information about the workings of genes in general and of important exceptions tempering the conclusions of molecular biology. Several of the most fundamental dogmas of the field have been revised as a result of discoveries made in mitochondria and chloroplasts, including one so basic as the supposed universality of the genetic code. In a very real sense the unusual and versatile genetic systems of the organelles illustrate what is possible within the boundaries of molecular rules.

DNA has also been detected by some investigators in centrioles, the microtubule-based structures that generate cilia and flagella. Two known mutations in this DNA interfere with the generation of flagella in the unicellular alga *Chlamydomonas*, suggesting that cen-

triolar DNA may contain information that is necessary for centriole structure and function.

This chapter discusses the genetic systems of mitochondria and chloroplasts, beginning with the lines of evidence originally indicating that the two organelles are partially independent in heredity. The DNA of each organelle, its organization into the genome, and its transcription and translation are then described. The chapter ends with a discussion of the possibility of a genetic system in centrioles, for which supporting evidence is relatively limited.

THE DISCOVERY OF DNA AND GENES IN MITOCHONDRIA AND CHLOROPLASTS

Cytoplasmic Inheritance of Mutations Affecting the Organelles

The first indications that mitochondria and chloroplasts have their own genetic systems appeared quite early in the history of genetics, from crosses carried out in the early 1900s by C. Correns and E. Bauer. Correns noted that in some individuals of the plant *Mirabilis* (the "four-o'clock"), segments in leaves and other structures fail to develop normal green color. The color differences were found to depend on color in chloroplasts—green segments of the plants contain apparently normal chloroplasts, and colorless segments contain colorless or yellow chloroplasts. Crosses between flowers on normal and colorless segments revealed that inheritance of the color differences is determined entirely by the segment giving rise to the egg (female gametophyte) and is not affected by the source of the pollen.

Correns noted that rather than following the expected Mendelian patterns, leaf color in *Mirabilis* parallels inheritance of the cytoplasm in angiosperms. The pollen donates nuclei but little or no cytoplasm to the fertilized egg in this group of higher plants (see Ch. 26 for details). In most fertilizations chloroplasts and other cytoplasmic components of the zygote come exclusively from the egg. On this basis Correns proposed that the inheritance of leaf color in *Mirabilis* is due to genes carried in the maternal cytoplasm, perhaps in the chloroplasts.

More examples of color inheritance resembling the *Mirabilis* pattern were soon discovered in other plants and algae. In all these examples, chloroplast color was inherited primarily through the maternal line, with no linkage evident to nuclear genes.

Research beginning in the 1950s showed that some mitochondrial traits are also controlled by cytoplasmic rather than nuclear genes. Among yeast cells maintained on agar plates, colonies that grow more slowly than normal strains frequently appear. These *petite* mutants were studied in detail by B. Ephrussi in the early 1950s. Crosses between petite yeast revealed two separate patterns of inheritance. One group, called *segregational petites*, behaves in typical Mendelian fashion. Genes controlling this pattern of inheritance are located on the chromosomes and show linkage to other nuclear genes. A second group proved to be independent of the nucleus. These mutants, called *suppressive petites*, follow a strictly cytoplasmic pattern of inheritance. That is, formation of suppressive petite colonies depends on the source of the mitochondria and not on genes carried in gamete nuclei. (Fig. 21-1 outlines an experiment

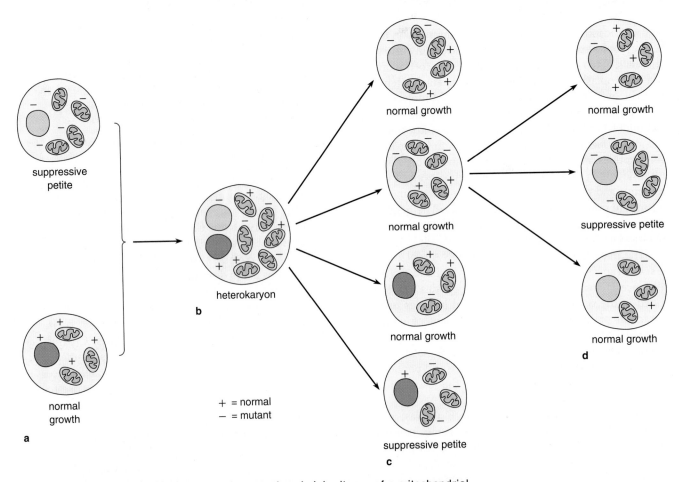

Figure 21-1 An experiment demonstrating cytoplasmic inheritance of a mitochondrial trait. In yeast sexual reproduction takes place by fusion of two haploid cells. Normally, the nuclei of the cells fuse to produce a diploid zygote nucleus, but in some strains, only the cytoplasm fuses, leaving separate haploid nuclei in a common cytoplasm. A cell of this type, with a fused cytoplasm but separate nuclei, is termed a heterokaryon. When the heterokaryon divides, the haploid nuclei are placed in separate daughter cells. In the experiment, a yeast cell carrying suppressive petite mitochondria along with a nuclear mutation **(a)** was fused with a cell that has normal mitochondria and the wild-type form of the nuclear gene to form a heterokaryon **(b)**. If the suppressive trait were carried on a chromosome in the nucleus, normal cells and cells with suppressive petite mitochondria would be expected with equal probability among the offspring. (The nuclear mutation is shown in the expected 1:1 ratio in **c**.) However, depending on the random combinations of normal and suppressive petite mitochondria received by daughter cells, the offspring may be normal or suppressive petites in any ratio, with no relationship to presence or absence of the nuclear mutation. (As long as normal mitochondria are present, a cell will be normal with respect to the mitochondrial trait.) Further, as long as suppressive petite mitochondria are present in a normal cell, the trait will appear in any offspring that happen to receive only suppressive petite mitochondria **(d)**. The random appearance of the trait among offspring and the lack of relationship to the nuclear mutation indicate that the suppressive petite trait is carried in the cytoplasm.

demonstrating cytoplasmic inheritance of a suppressive petite trait.)

From biochemical analyses Ephrussi and his colleagues found that cellular oxidations in suppressive petites proceed only by the glycolytic pathway. The suppressive petites lack elements of the mitochondrial electron transport system (see p. 327) such as cytochrome *b* and are unable to carry out oxidative phosphorylation. The inner mitochondrial membranes are transformed from the usual pattern of cristae folds into irregular whorls.

A similar group of mutants was studied by M. B. Mitchell and H. K. Mitchell in the fungus *Neurospora*. These *poky* mutants are also characterized by slow growth, absence of cytochrome *b* or other elements of the mitochondrial electron transport system, and alterations in the morphology of inner mitochondrial membranes. Offspring of crosses between poky and normal *Neurospora* are stable and follow a strictly maternal line of inheritance. No linkage of the poky trait to any chromosomal genes can be established.

More recent genetic work has revealed additional patterns of cytoplasmic inheritance in higher plants and animals. In higher plants, for example, male sterility frequently results from mitochondrial mutations that are inherited cytoplasmically, with no linkage to nuclear genes. In humans several diseases have been identified with mutations in mitochondrial genes (Table 21-1). The diseases are characterized by alterations in mitochondrial structure and enzyme activity and by damage to the organ systems most dependent on mitochondrial reactions for energy: the central nervous system, skeletal and cardiac muscle, the liver, and the kidneys. Within families the defects are transmitted through the mother in non-Mendelian patterns. (In mammals mitochondria of the offspring are usually derived entirely from the egg cytoplasm; mitochondria carried in the sperm cell are degraded soon after fertilization; see Ch. 26).

The Search for DNA in Organelles

Indications that nucleic acids are present in mitochondria and chloroplasts came as early as 1924. At that time E. Bresslau and L. Scremin first detected DNA inside mitochondria by use of the Feulgen stain for light microscopy, a procedure that gives DNA a red-magenta color. In the 1960s H. Ris and W. Plaut observed thin fibers with the dimensions expected for DNA in electron micrographs of the chloroplasts of a green alga, *Chlamydomonas*. Light microscope preparations of the *Chlamydomonas* chloroplasts, stained by the Feulgen reaction, showed that the filamentous regions contain DNA. Previous digestion with deoxyribonuclease eliminated the thin fibers and the DNA stain from the chloroplasts. Shortly after this discovery, Ris detected fibers with the same dimensions and structure in mitochondria from another alga, *Micromonas*.

Similar fibers were found by other investigators in mitochondria and chloroplasts from many different organisms. Although the fibers are more evident in organelles from embryonic or rapidly growing cells (Figs. 21-2 and 21-3), it is generally accepted that DNA fibers are a regular constituent of mitochondria and chloroplasts at all stages. Typically, the DNA fibers visible in the organelles are much thinner than nuclear chromatin fibers and approach the dimensions of the fibers in bacterial nucleoids.

DNA was first isolated from mitochondria by G. Schatz, M. M. K. Nass, and their colleagues; soon after, R. Sager, A. Rich, and J. T. O. Kirk isolated DNA from chloroplasts. Analysis of the organelle DNAs revealed that although the genetic material of the two organelles is primarily prokaryotic in structure and sequence organization, it does have some features resembling those of eukaryotes. Organelle DNAs also have unique features, that is, features unlike the DNA of either prokaryotes or eukaryotes.

Observations that mitochondria and chloroplasts

Table 21-1 Some Human Diseases Caused by Mutations in Mitochondrial Genes

Disease	Symptoms
Kearns–Sayre syndrome	May include muscle weakness, mental deficiencies, abnormal heartbeat, short stature.
Lebers hereditary optic neuropathy*	Vision loss from degeneration of the optic nerve, abnormal heartbeat.
Mitochondrial myopathy and encephalomyopathy	May include seizures, strokelike episodes, hearing loss, progressive dementia, abnormal heartbeat, short stature.
Myoclonic epilepsy	Vision and hearing loss, uncoordinated movement, jerking of limbs, progressive dementia, heart defects.

* Found by D. C. Wallace and his coworkers to result from a single substitution of guanine for adenine in mitochondrial DNA, which changes arginine to histidine at one position in subunit 4 of NADH-coenzyme Q oxidoreductase.

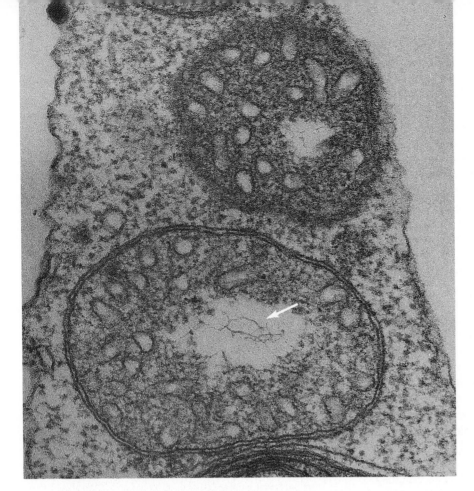

Figure 21-2 DNA fibers (arrow) within a mitochondrion of the brown alga *Egregia*. ×79,000. Courtesy of T. Bisalputra and A. A. Bisalputra, from *J. Cell Biol.* 33:511 (1967), by permission of the Rockefeller University Press.

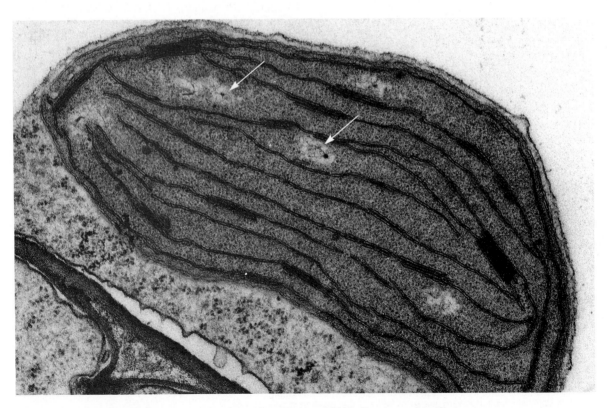

Figure 21-3 DNA fibers (arrows) in a maize chloroplast. ×37,000. Courtesy of L. K. Shumway and T. E. Weier.

contain DNA and are at least partially independent genetically are paralleled by the behavior of the structures during the cell cycle. The DNA of the organelles replicates, and both mitochondria and chloroplasts can be observed to grow and divide by a process resembling bacterial cell division (for details, see Chs. 22 and 24). In general the behavior of the two organelles during the cell cycle is entirely consistent with the conclusion that both mitochondria and chloroplasts arise only from preexisting organelles.

THE ORGANELLE DNAs

Isolation of the DNA

DNA is isolated from mitochondria and chloroplasts by several techniques. In one of the most-used procedures, cells are broken and the mitochondria or chloroplasts are separated and concentrated by centrifugation. As part of the isolation procedure, the organelles are exposed to deoxyribonuclease while their boundary membranes are still intact. This step destroys any DNA of nuclear origin that might be present as a contaminant in the suspending solution. After successive centrifugations and washes to remove deoxyribonuclease and degraded DNA from the suspending medium, the boundary membranes of the organelles are disrupted, usually by exposure to detergents. Breakage of the membranes releases the organelle DNA, which is purified and concentrated by centrifugation.

In some species the physical density of mitochondrial or chloroplast DNA differs significantly from that of nuclear DNA, allowing the organelle DNA to be separated and purified by density-gradient centrifugation (see p. 125). In these preparations the organelle DNA separates into distinct "satellite" bands at some distance from the nuclear DNA. The density differences depend on distinct proportions of A-T and G-C pairs in the organelle and nuclear DNA.

The DNA isolated from mitochondria and chloroplasts by these techniques, with very few exceptions, exists as closed, supercoiled circles (Figs. 21-4 and 21-5). The DNA is largely unassociated with structural proteins except for small quantities of a histonelike protein similar to the HU proteins of *E. coli* and other bacteria (see p. 568).

The DNA Circles of Mitochondria

Mitochondrial DNAs vary considerably in size among major taxonomic groups. In animals mitochondrial

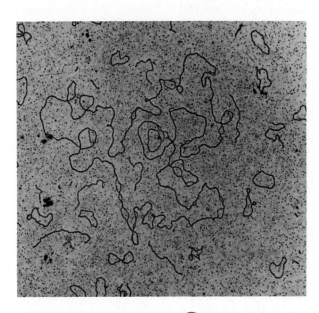

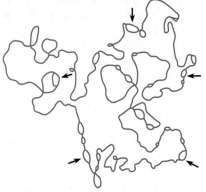

Figure 21-5 DNA circle from a pea chloroplast. The "eyes" in the circle (arrows on the tracing) are regions where the nucleotide chains of the DNA have unwound. The small circles in the background are ϕX 174 viral DNA, included for size comparisons. × 15,000. Courtesy of R. Kolodner and K. K. Tewari.

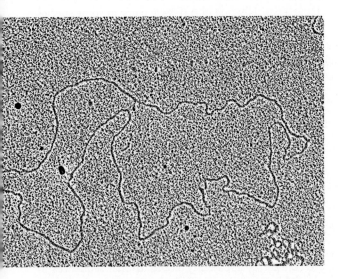

Figure 21-4 DNA circle isolated from a mitochondrion of mouse fibroblast cells. × 20,000. Courtesy of M. M. K. Nass, from *Science* 165:25 (1989). Copyright © 1969 by the American Association for the Advancement of Science.

DNA circles are relatively small and fairly uniform in size, ranging from 4.5 to 5.5 μm of included DNA (about 16,000 to 20,000 base pairs) in most species. The human mitochondrial DNA circle, for example, contains 16,596 base pairs. The mitochondrial DNA of several animal species has been nearly or completely sequenced, among them two *Drosophila* species, a locust, the amphibian *Xenopus*, two nematodes, a starfish, two sea urchins, the chicken, the mouse, the rat, the cow, and *Homo sapiens.* These sequences reveal a remarkable similarity in sequence structure and gene complement among animal mitochondrial DNAs.

The mitochondrial DNA circles of plants are relatively large and vary over extreme limits from a minimum of about 35 μm to more than 700 μm of included DNA. Wide variations in size are observed even among closely related plant species. For example, in the cucurbit family, which includes cucumbers, zucchini, and watermelon, the perimeter of mitochondrial DNA circles measures from 103 to 754 μm. These dimensions are equivalent to a range of 218,000 to 2.5 million base pairs in the cucurbit family; the largest cucurbit mitochondrial DNA circles are about half the size of an *E. coli* genome.

Fungal and protozoan mitochondrial DNA circles, with approximately 10 to 25 μm of included DNA, lie between the animal and plant extremes. The DNA circle of the yeast *Saccharomyces cerevisiae,* which is more than 90% sequenced, includes about 78,000 base pairs. The mitochondrial DNAs of two protozoans, *Tetrahymena* and *Paramecium,* are linear molecules about 14 μm long. These are the only known examples of linear mitochondrial DNA.

Despite their varied lengths, the mitochondrial genomes of plants, fungi, and protozoa contain approximately the same number of genes as the relatively small animal mitochondrial DNAs—about 40 to 50 genes, of which 13 to 20 encode polypeptides and the rest rRNAs and tRNAs. The DNA segments accounting for most of the differences in genome size among the mitochondria of different taxonomic groups contain sequences that are largely nonfunctional. In some, as in higher plants, much of the nonfunctional DNA consists of repeated sequences.

Mitochondrial DNA is characteristically rich in A-T base pairs, ranging from slightly more than 50% in higher vertebrates such as the chicken to 74% in *Drosophila.* In some species there is a significant difference in density between the two nucleotide chains of the mitochondrial DNA circle, so that one (identified as the *heavy,* or *H, chain*) is characteristically more dense than the other (the *light,* or *L, chain;* see also p. 997).

Mitochondrial DNAs also frequently contain small quantities (0.1% to 0.2%) of nucleotides with ribose instead of deoxyribose as the sugar and uridine as a possible base. The significance of these ribonucleotides in mitochondrial DNA remains unknown. They may result from mistakes during replication or may be remainders of *primers* (RNA primers are used to start DNA replication; see p. 960). It is also possible that they may have an as yet unknown function.

In most cells the total complement of mitochondrial DNA circles is less than 1% of total cellular DNA. Higher percentages occur in a few cell types. In rapidly dividing yeast cells mitochondrial DNA may make up as much as 18% of the cellular total; in large animal eggs, such as those of amphibians, mitochondrial DNA may represent as much as 99% of total cellular content. The large proportion of mitochondrial DNA in these eggs reflects the fact that most of their volume is cytoplasm densely packed with mitochondria. Dividing the mitochondrial DNA total by the number of mitochondria gives enough DNA in most cells for about five to ten circles per mitochondrion.

Chloroplast DNA

Chloroplast DNA is also circular except in one known species, the green alga *Acetabularia.* Considerable differences in the size of chloroplast DNA circles are noted among different species, especially in green algae. In these algae chloroplast DNA circles range from a minimum of 27 to as much as 200 μm of included DNA or about 80,000 to 600,000 base pairs. Among most land plants, including bryophytes, ferns, and seed plants, variations are more limited and range from about 40 to 50 μm of included DNA or about 120,000 to 160,000 base pairs. As in mitochondria, these size differences result primarily from variations in the content of apparently nonfunctional sequences; the gene complement of chloroplast DNA in different taxonomic groups is remarkably similar. Chloroplast DNA is also characteristically rich in A-T base pairs; in some species the fraction of A-T pairs is more than 80%.

The DNA circles of chloroplasts contain several times as many genes as those of mitochondria. The two chloroplast DNA circles sequenced to date, from *Marchantia* (a liverwort) and the angiosperm *Nicotiana* (tobacco), contain a closely similar complement of about 120 genes. The genes are arranged in the same order and pattern, except for one reversed segment. The similarity of the genes and their arrangement in the two species is remarkable in view of the evolutionary divergence of the bryophytes and the angiosperms, estimated to have taken place some 350 to 400 million years ago.

Most of the difference in length between the *Marchantia* and tobacco chloroplast DNA circles reflects the content of apparently nonfunctional sequences in a long inverted repeat, which contains 10,058 base pairs in *Marchantia* and 25,339 base pairs in tobacco. Only a few genes, such as those encoding the rRNAs, are distributed among the nonfunctional sequences in the inverted repeat (Fig. 21-6). The same inverted repeat,

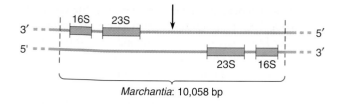

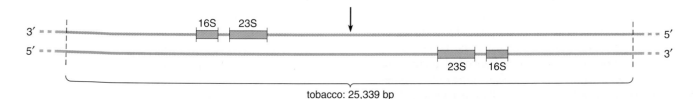

Figure 21-6 The inverted repeat sequence regions of *Marchantia* and tobacco. The arrow above each inverted sequence marks the center of the repeated region; to the left of the arrow the top chain is the sense chain of the DNA; to the right of the arrow the top chain is the missense chain. The rRNA genes are not drawn to scale.

with differences in the amount of nonfunctional DNA, is present in the chloroplasts of most plants except for a few species, including the pea, broad bean, and pine.

Chloroplast DNA also contains a small quantity of ribonucleotides of unknown function, scattered at several locations around the circles. Pea and spinach chloroplast DNA circles, for example, contain 18 ± 2 ribonucleotides; those of lettuce contain 12 ± 2. Although chloroplast DNA exists primarily in unbound form, like mitochondrial DNA, small quantities of an HU-like protein resembling bacterial HU have also been detected in some plants.

Chloroplast DNA may make up as much as 6% of the total DNA of a plant cell. At these levels there is enough DNA for about 50 to 100 circles per chloroplast. Chloroplasts thus typically contain more DNA circles than mitochondria.

GENES AND GENOME STRUCTURE IN THE ORGANELLES

Mitochondrial Genes

The research identifying mitochondrial genes was initially carried out by genetic crosses and biochemical analysis. Genes for several mitochondrial proteins, such as cytochrome *b*, were located inside the organelle in this way. Later, RNA-DNA hybridization (see p. 997) was used as a test for the presence of these and other genes in mitochondria. In this technique the RNAs isolated from mitochondria are added to mitochondrial DNA; the regions in the organelle DNA forming hybrid pairs with the added RNAs can be assumed to code for the RNA types. This work established that all mRNAs, rRNAs, and tRNAs inside the organelles are encoded there; none enter from the surrounding cytoplasm. Direct sequencing, carried out by G. Attardi and his coworkers and others, confirmed the presence of the genes identified in mitochondria by the other techniques and revealed the presence of additional, un-

identified genes that were at first designated as *ORFs* or *URFs* (for Open or Unidentified Reading Frames). Most of the ORFs have since been identified by comparing their encoded amino acid sequences with those of mitochondrial proteins.

These experimental approaches established that the mitochondrial DNA circles of all protozoan, fungal, plant, and animal species encode a relatively small number of proteins (Table 21-2). In higher vertebrates, in which the most complete studies have been carried out,

Table 21-2 Genes of Yeast, Mammalian, and Plant Mitochondria*

Gene	Yeast	Mammalian	Plant
Cytochrome oxidase			
subunit I	+	+	+
subunit II	+	+	+
subunit III	+	+	+
Cytochrome *b*	+	+	+
ATPase complex			
F_1 subunit α	−	−	+
F_0 subunit 6	+	+	+
F_0 subunit 8	+	+	+
F_0 subunit 9	+	−	+
NADH-coenzyme Q oxidoreductase subunits	0 (other fungi 6)	7	6?
Ribosomal proteins	1	−	+
16S rRNA equivalent	+	+	+
23S rRNA equivalent	+	+	+
5S rRNA equivalent	−	−	+
tRNAs	24	22	~30
RNA component of RNAse P	+	−	?

* Genes for *maturase* or *transposase* polypeptides are also present within the introns of some fungal mitochondria.

creases, the rate of transcription increases in cells containing chloroplasts. In addition, in some cell types and tissues, such as the relatively rare epidermal cells containing chloroplasts, the overall rate of chloroplast transmission is set at very low levels; in other cells, such as the palisade cells of leaves, chloroplast transcription is very high. Within the chloroplasts regulated at high or low levels, X.-W. Deng and W. Gruissem and others found that the relative rates at which individual genes are transcribed remains approximately the same.

Some exceptions have been noted in chloroplasts undergoing transformation to other plastid types (see p. 404). H. Kobayashi and his coworkers found that as chloroplasts change into chromoplasts in ripening tomatoes or into amyloplasts in sycamore cells, transcription of some genes is selectively turned down or off. Most of the individually regulated genes encode polypeptides taking part in photosynthesis, such as subunits of photosystems, the ATPase complex, or RuBP carboxylase. Kobayashi and his colleagues discovered that reduction in transcription was correlated with methylation of the regulated genes, and proposed that the methylation may interfere with tight binding of RNA polymerase to the promoters of these genes. Thus methylation may be used as a regulatory mechanism in chloroplasts undergoing transformation to other plastid types as well as in vertebrate nuclear genes (see p. 708).

Posttranscriptional and Translational Regulation

Although transcriptional regulation in mitochondria and chloroplasts appears to be accomplished primarily by adjusting the overall rate of transcription upward or downward, with relatively few adjustments of individual genes, the amounts of individual mature mRNAs maintained inside the organelles, and the rates at which they are translated, vary widely. In yeast mitochondria, for example, the genes encoding cytochrome b and ATPase subunit 9 are transcribed at equal rates. The amount of mature mRNAs for the two polypeptides, however, varies widely under different conditions.

These variations appear to depend primarily on posttranscriptional controls at two levels: (1) adjustments in the rates at which specific individual pre-mRNAs are processed, and (2) alterations in the rates at which finished mRNAs are degraded. For example, P. Klaff and W. Gruissem found that in mature spinach leaves, the half life of the mRNA for the D1 protein of photosystem II was more than twice as long as that of young spinach leaves. In contrast, the half life of the large RuBP carboxylase subunit mRNA was the same in young and old leaves. The basis for these controls remains obscure. However, the controls are effective in adjusting the levels of different mRNA types to meet the needs of mitochondria and chloroplasts. The post-transcriptional controls also help to maintain the balance between synthesis of protein subunits in the organelles and in the surrounding cytoplasm.

Translational controls include adjustments in both the overall rate of polypeptide assembly and the synthesis of individual types. The overall rate of protein synthesis appears to be adjusted upward or downward, like the rates of transcription, by proteins encoded in the cell nucleus. Although the identity of the proteins is unknown, they are probably primarily factors promoting the initiation of protein synthesis, as in the cytoplasmic systems of both eukaryotic and prokaryotic cells. Superimposed on the overall regulation of protein synthesis are controls that adjust the translation of individual polypeptides upward or downward. For example, T. Fox and his coworkers detected a nuclear-encoded factor, a product of the *CBP1* gene, that binds to the 5′ end of cytochrome b mRNA and promotes its translation in yeast mitochondria (see Fig. 21-15). Evidently, there are also posttranslational controls that regulate the rate at which finished proteins are degraded. In chloroplasts, for example, an overabundance of the large RuBP carboxylase subunit, which is encoded and synthesized inside the organelle, is compensated for by an increase in the rate at which the polypeptide is broken down.

Thus the overall rates of transcription and translation within the organelles are balanced with the levels at which organelle polypeptides are assembled in the surrounding cytoplasm by factors encoded in the cell nucleus. This overall balance is fine-tuned by posttranscriptional, translational, and posttranslational controls that adjust the levels at which individual polypeptides are produced within the organelles.

ORIGINS OF THE ORGANELLE GENOMES

There are many unanswered questions about the origins of the organelle genetic systems and their relationships to nuclear genomes. One of the primary questions involves the mechanism of gene transfer. If mitochondria and chloroplasts originated from ancient, symbiotic prokaryotes, as most investigators agree, how were genes transferred from the developing organelles to the nucleus? As P. E. Thorsness and Fox have shown, such movements can actually take place—they were able to detect gene transfer from yeast mitochondria to the nucleus at the rate of two migrations for every 100,000 cells per cell division (see the Experimental Process essay by Thorsness on p. 902).

Observations of "promiscuous DNA" also leave little doubt that genes can move between organelles and the cell nucleus. Promiscuous DNA refers to DNA sequences that occur in more than one of the three DNA-containing membrane-bound compartments of cells—the nucleus, mitochondria, and chloroplasts. For example, fragments of ribosomal RNA and cytochrome *b* genes, along with parts of the replication origin of yeast mitochondria, can be detected in the nuclei of yeast cells. The degree of sequence divergence between the copies in the mitochondria and nucleus indicates that the transfer took place about 25 million years ago. Mitochondrial sequences have been discovered in the nuclei of other species, including sea urchins, locusts, and humans. Transfer of chloroplast genes to the nucleus has also been detected. In spinach sequence fragments sufficient to make up about six complete chloroplast genomes occur at various locations in the nucleus. Chloroplast genes also appear in quantity in plant mitochondria.

The mechanisms that might account for these transfers are unknown. However, the fact that such transfers have taken place repeatedly supports the idea that the genes encoding organelle proteins probably moved to the nucleus from the evolutionary progenitors of mitochondria and chloroplasts.

Other questions surround the complement of genes still existing inside the organelles. Why are any genes still inside the organelles, and why are the groups remaining inside so similar in different taxonomic groups? The 13 or so proteins encoded inside mitochondria probably make up less than 5% of the total in the organelle, which may easily amount to 300 or more different polypeptides. Although chloroplasts encode many more proteins than mitochondria, the polypeptides encoded inside probably still represent only about 15% to 20% of the total. With very few exceptions the polypeptides encoded and synthesized in either mitochondria or chloroplasts are incomplete—they are subunits of more complex proteins including polypeptides encoded in the cell nucleus. Maintenance of the relatively few genes remaining inside the organelles requires more than 100 polypeptides, including enzymes, transcription and translation factors, and ribosomal proteins, almost all encoded in the cell nucleus and synthesized in the cytoplasm outside mitochondria and chloroplasts.

Many investigators have considered that the complement remaining inside represents polypeptides that for one reason or another cannot be distributed to the organelle in functional form if synthesized in the surrounding cytoplasm. A polypeptide encoded in such a gene, for example, might be too hydrophobic to be synthesized on cytoplasmic ribosomes and distributed in soluble form from its site of synthesis to the organelle. Standing against this line of argument is the fact that ATPase subunit 9, one of the most hydrophobic polypeptides of mitochondria, is encoded in the nucleus in most species. Further, a great many genes still present in chloroplasts, such as the RuBP carboxylase gene, code for soluble rather than hydrophobic proteins.

Also standing against this line of argument are the results of recent experiments in which proteins now encoded in the organelles have been successfully transferred to the nucleus. In these experiments, conducted in mitochondria by P. Nagley and R. J. Devenish, and in chloroplasts by A. Y. Cheung, L. Bogorad, M. van Montagu, and J. Schell, the genes for organelle-encoded polypeptides were isolated and provided with eukaryotic promoters and the N-terminal signals necessary for routing the polypeptides to the organelles. These genes were introduced into the nucleus by techniques that promote their incorporation into the nuclear genome (see p. 782). Once in the nucleus, the genes were transcribed and the encoded polypeptides, after synthesis on cytoplasmic ribosomes, entered the organelles in fully functional form. Genes for a photosystem II polypeptide and the large RuBP carboxylase subunit of chloroplasts and for ATPase subunit 8 of mitochondria were successfully transferred to the cell nucleus in this way.

Others have proposed that gene transfer is still in progress, and that the genes remaining inside the organelles represent a transitory step on the way to eventual transfer of all organelle genes to the nucleus. Supporting this conclusion is the fact that, against the background of general similarity in gene complement, some differences are noted in the locations of a few genes in the organelles. The gene for mitochondrial ATPase subunit 9 is located in the nucleus in animals and in at least one fungus, *Aspergillus;* in yeast this gene is in mitochondria. In *Neurospora* copies of the gene are present in both the nucleus and mitochondria, but only the nuclear copy is active. Similarly, genes for both subunits of RuBP carboxylase are present in red algal chloroplasts; in green algae and higher plants, only the gene for the larger of the two subunits remains in the organelle.

If gene transfer is still in progress, it is likely that movement from the organelles is probably much less likely today than during the very ancient period when the organelles first took up residence in the cytoplasm of cells destined to form eukaryotes. Initially, the developing organelles and their hosts presumably had similar, prokaryotelike genetic systems. During this period transfer of genes in functional form would have been relatively likely. However, as eukaryotic features appeared in cell nuclei, including distinct promoter and regulatory sequences, the chance that an organelle gene could be transferred in functional form was greatly reduced. The present collection of organelle genes may, therefore, represent a sort of evolutionary cul-de-sac from which functional organelle genes may now move to the nucleus only at very low rates.

GENETIC INFORMATION IN CENTRIOLES

Centrioles are small, barrel-shaped structures assembled from a circle of nine triplet microtubules (see Fig. 11-28). Their primary function seems to be to give rise

Escape of DNA from Mitochondria to the Nucleus During Evolution and in Real Time

Peter E. Thorsness

PETER THORSNESS attended The Colorado College and received a B.A. in chemistry in 1982. He pursued graduate studies in biochemistry at the University of California, Berkeley. He investigated principles of metabolic regulation and regulation of enzymes via covalent modification in the laboratory of Daniel E. Koshland, Jr., receiving his Ph.D. in 1987. Dr. Thorsness then accepted a postdoctoral fellowship from the American Cancer Society and joined Thomas D. Fox's research group at Cornell University. In 1991, he joined the faculty of the Department of Molecular Biology at the University of Wyoming.

The concepts and principles of evolution have been the subject of much controversy. Perhaps the most troubling aspect of evolution for a molecular biologist is the inability to experimentally determine the true path of evolution for a given protein or gene—scientists are often left to "best guess" the process of evolution. On the cellular level, it has long been proposed that eukaryotic cells arose by an endosymbiotic process in which one organism (the pro-eukaryote) was colonized by a second (the promitochondrion in the case of the mitochondrial compartment).[1]

Present day mitochondria contain only about a half-dozen protein coding genes plus the tRNAs and rRNAs necessary for translation; the nucleus of the cell contains hundreds of additional genes that are necessary for functional mitochondria. Since pro-mitochondria were presumed to have had a full complement of genetic material, mitochondrial evolution is proposed to have followed a path of loss or transfer of genetic material to the nucleus. In addition to this circumstantial case for gene transfer, there is evidence for the presence of pseudogenes of mitochondrial origin in the nuclei of a variety of organisms. To test for the presence of a pathway for transfer of genetic information between intracellular compartments, a researcher would ideally be able to place a DNA of specific sequence in the mitochondrial compartment and have a method to detect its presence in the nucleus if it ever made the move. However, for a number of years it was impossible to do this test as it was not feasible to genetically transform mitochondria with standard techniques.

A major technical advance was made in 1988 when a method was developed to introduce DNA into mitochondria of the budding yeast *Saccharomyces cerevisiae*. The method, biolistic transformation, was amazingly simple in design, but the obvious is often overlooked. Soon after this technique was developed, I joined the laboratory of Thomas D. Fox at Cornell University in Ithaca, New York, who had developed a particularly useful version of the biolistic transformation process.[2] Dr. Fox's scheme took advantage of several useful genetic tricks that are largely available only when working with *S. cerevisiae*. This yeast can grow well with no mitochondrial DNA whatsoever as long as it has a fermentable carbon source. Yeast that lack mitochondrial DNA make a particularly good target for mitochondrial transformation since plasmid DNAs introduced into their mitochondria are not subject to the rapid recombination processes that can occur between different mitochondrial DNAs. To transform the yeast we used a plasmid DNA that contained both the nuclear gene *URA3*, to mark the presence of the DNA in the nucleus by complementation of the recessive mutation *ura3-52*, and a mitochondrial gene, *COX2*, to mark the presence of the DNA in mitochondria.[3]

The transformation procedure involved precipitation of the plasmid DNA onto 0.5 μm tungsten particles. An explosive charge accelerated the DNA-coated particles at a lawn of yeast cells on an agar plate in an evacuated chamber. Yeast that were pierced by a particle, survived, and received plasmid DNA were identified by complementation of the nuclear mutation *ura3-52*. A small subset of those cells also had DNA delivered to the mitochondrial compartment. These mitochondrial transformants were identified by a marker rescue technique. Nuclear transformants from the biolistic transformation were replica-plated to a lawn of haploid yeast cells of opposite mating type whose mitochondrial genome lacked the *COX2* gene. Neither the transformants nor the tester strain were able to grow on nonfermentable carbon sources (such as glycerol or ethanol) due to their incomplete mitochondrial genomes. However, a diploid formed from one haploid that contained the mitochondrial gene *COX2* on the plasmid and a second haploid that contained the rest of the mitochondrial genome would be able to utilize nonfermentable carbon sources and grow. Using this screen, we were able to isolate the 1 transformant in 2000 that had DNA delivered to both the nucleus and mitochondria.

At this point in our experiment we had a yeast that was a uracil prototroph (cell that can make uracil) because it carried the plasmid in the nucleus that complemented the *ura3-52* mutation, and also had the *COX2* gene in mitochondria. As our goal was to detect the movement of DNA from one compartment in the cell to another, it was necessary to isolate versions of this transformed yeast that contained the plasmid only in the nucleus or only in mitochondria. Subsequently, migration of DNA from mitochondria to nucleus would be revealed by the appearance of uracil prototrophs within a pool of what was formerly entirely uracil auxotrophs (cells that cannot make uracil). Likewise, the movement of DNA from the nucleus

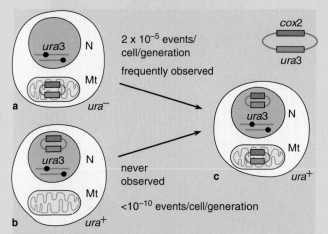

Figure A Experimental scheme. A plasmid carrying the nuclear gene *URA3* and the mitochondrial gene *COX2* could be maintained in either the nucleus (N), the mitochondrion (Mt), or both compartments. **(a)** Cells containing the plasmid in their mitochondria frequently gave rise to cells containing the plasmid in both the nucleus and mitochondria **(c)**. **(b)** Cells containing the plasmid in the nucleus failed to give rise to cells with the plasmid in both compartments C at detectable frequency. All strains carried a nonreverting chromosomal *ura3* mutation.

to mitochondria could be followed by the appearance of *COX2* DNA in mitochondria that formerly was devoid of any mitochondrial DNA. The DNA in mitochondria would be scored by rescuing a *COX2* mutation in a tester strain, as previously described. It is important to note that we assumed the *URA3* gene, when present in the cell only in mitochondria, would not be expressed because of the altered genetic code of mitochondria and the lack of protein synthesis in these mitochondria. Likewise, the *COX2* gene is not functional when located in the nucleus because of differences in genetic codes. It was possible to isolate both a strain that contained the plasmid DNA only in mitochondria and a strain that contained the plasmid DNA only in the nucleus by taking advantage of the mitotic instability inherent with plasmids propagated in these compartments.

The next question was straightforward: do cells that contain the plasmid DNA only in their mitochondria ever move it to their nucleus and consequently become uracil prototrophs? Not knowing what to expect, we grew up the uracil auxotroph strain and plated about 10^8 cells on an agar plate lacking uracil. After several days of incubation the plate had given rise to hundreds of colonies, indicating complementation of the *ura3-52* mutation and thus the presence of DNA with this information in the nucleus. When we measured the rate at which these uracil prototrophs arose, we found it exceeded the normal mutation rate for a nuclear gene in yeast. The rate of DNA escape and migration from mitochondria to the nucleus was on the order of 2×10^{-5} events/cell/cell division (Fig. A). Frankly, we were somewhat astonished that, at least for yeast and for an evolutionary time scale, DNA was leaking out of mitochondria all the time. As for any other experiment directed at questions of evolution, our results

only provided evidence for a pathway for transfer of genetic information to the nucleus from mitochondria, and did not prove that evolution actually happened in that fashion.

Of course we conducted several control experiments to convince ourselves that DNA had been transferred from mitochondria to the nucleus. In the first, we showed that the nuclear mutation *ura3-52* never reverts; yeast cells lacking the plasmid containing the wild-type copy of the *URA3* gene never became uracil prototrophs. A related but even more important observation was that yeast which previously had plasmid DNA in mitochondria or in the nucleus but had lost it during nonselective outgrowth never gave rise to uracil prototrophs. In fact, only those yeast containing the plasmid DNA in mitochondria were capable of generating uracil prototrophs from a clonal population of cells that were originally exclusively uracil auxotrophs.

What about the migration of DNA in the opposite direction—from nucleus to mitochondria? We ran essentially the same trials as for detecting the migration of DNA from mitochondria to nucleus: growth of a large number of yeast cells containing the plasmid DNA only in the nucleus, followed by genetic analysis of what DNA was in mitochondria. Despite examining over 10^{11} yeast, we never detected DNA in mitochondria that had formerly resided in the nucleus (see Fig. 1). That is not to say that DNA cannot migrate in that direction, but we estimated that the rate of migration from nucleus to mitochondria cannot be any more frequent than 1×10^{-10} events/cell/cell division.

The concepts of evolution stimulated our research approach, but they were not the only factors. We also were interested in the prospects for using this novel genetic assay to analyze the mitochondrial compartment. We found that the escape of DNA from mitochondria was increased by changing the temperature at which the yeast were grown; yeast grown at 37°C or 16°C rather than the optimal 30°C had a 2-fold higher rate of escape. Other environmental stresses such as freezing cells in 15% glycerol at −70°C or incubation of cells with $1M$ sorbitol also induced the escape and migration of DNA from mitochondria to the nucleus without cell growth. Likewise, a different nuclear genetic background also influenced the rate of DNA escape from mitochondria. Different strains of yeast had rates of DNA escape that varied 5-fold. To further extend our genetic analysis of the mitochondrial compartment, we have isolated mutant strains of yeast with an increased rate of DNA escape from mitochondria. We hope that the isolation and analysis of nuclear genes that complement mutations affecting the rate of DNA escape from mitochondria will generate new insights into the biogenesis and metabolism of mitochondria.

References

[1] Margulis, L. *Symbiosis in cell evolution: Life and its environment on the early earth,* San Francisco: Freeman (1981).

[2] Fox, T. D.; Sanford, J. C.; and McMullin, T. W. *Proc. Nat'l, Acad. Sci. USA* 85:7288 (1988).

[3] Thorsness, P. E., and Fox, T. D. *Nature* 346:376 (1990).

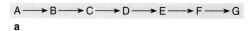

a

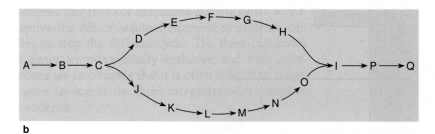

b

Figure 22-12 Gene cascades in cell cycle control. **(a)** A straight-line pathway, in which each gene stimulates activity of the next gene in line. **(b)** A branched pathway, in which one or more branches produce parallel pathways of gene control at some parts of the cell cycle. When the pathways are separated, the activity of genes in one pathway does not affect the genes in the other pathway.

Equivalents of the *cdc2/28* and *cdc13* Genes in Higher Eukaryotes

Genes encoding polypeptides closely related to the yeast *cdc2/28* and *cdc13* gene products have also been identified in a wide range of higher eukaryotes including *Drosophila*, starfish, *Xenopus*, and humans. The human *cdc2/28* gene is identified as *cdc2* as it is in *S. pombe*; the protein encoded in human *cdc2* is 62% identical in sequence to the yeast *cdc2/28* polypeptides. P. C. L. John and his colleagues recently found that antibodies against the *cdc2* gene product react with algal cells and higher plants such as wheat, carrot, and oats, indicating that an equivalent gene is also present in these organisms.

The *cdc2/28* Gene Product and MPF Research identifying the *cdc2/28* gene product revealed that it forms part of MPF, the protein complex that triggers passage from G2 to mitosis in a variety of higher eukaryotes. MPF was originally discovered through investigations of early *Xenopus* embryos. Individual cells removed from the embryos during the cleavage divisions continue to divide for a time in synchrony with those still in the embryo, indicating that an internal molecular "clock" keeps the divisions on schedule. Some parts of the *Xenopus* division schedule remain intact even if the nucleus is removed. For example, foreign DNA injected into enucleated cells removed from the embryo is still replicated according to the time schedule followed by the cells in the embryo.

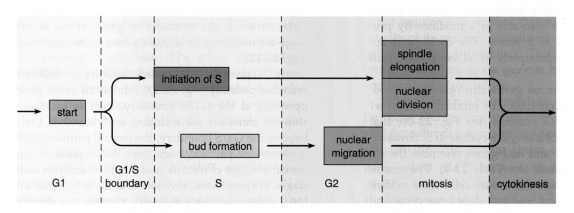

Figure 22-13 The branched pathway of cell cycle regulation in *Saccharomyces cerevisiae*. At late G1 mutations in the genes initiating "start" can stop or inhibit the entire cycle. During S there are two separate pathways of cell cycle control, in which mutations on a pathway affect only that pathway. Just before cytoplasmic division is completed (cytoplasmic division occurs by bud formation in *S. cerevisiae*), the pathways rejoin, so that mutations can once again stop or inhibit the entire progress of cell division.

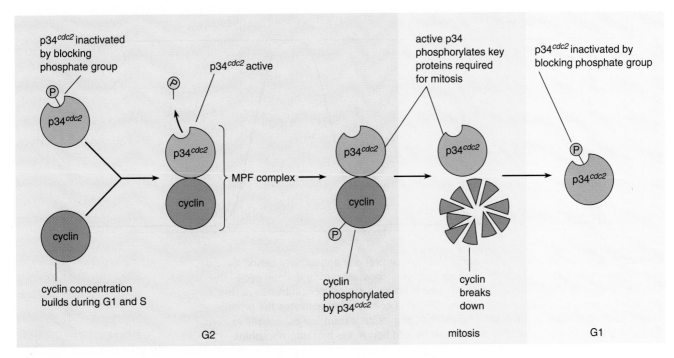

Figure 22-14 The interactions of cyclin and the p34^{cdc2} protein kinase in the MPF complex during the G2/mitosis transition. Late in G2 cyclin, which gradually increases in concentration during interphase, combines with p34^{cdc2} to form the MPF complex. At this point the protein kinase is inactivated by a phosphate group linked to a tyrosine residue in the site hydrolyzing ATP. Formation of the MPF complex leads to removal of the blocking phosphate and activation of the kinase. Among the first substrates for p34^{cdc2} is probably cyclin, which becomes phosphorylated soon after MPF forms. Among the other targets of p34^{cdc2} are histone H1, lamins of the nuclear envelope, and microtubule proteins; phosphorylation of these target proteins initiates mitosis. Soon after mitosis begins, cyclin breaks down, probably through activation of a specific proteinase by the MPF complex. Cyclin breakdown promotes rephosphorylation of the active site in p34^{cdc2}, inactivating the protein kinase by the beginning of the following G1.

The first insights into the nature of the molecular clock were obtained by M. W. Kirschner, who found a protein originally known as *maturation promoting factor* whose activity cycled in coordination with the divisions. During the S stage of a cleavage division (no G1 is discernible in *Xenopus* during the cleavage divisions), MPF activity is low or nonexistent. Activity of the factor rises following S and peaks as mitosis begins. By the end of mitosis MPF activity is again at low levels. The coordination is more than coincidental, because injection of active MPF into S cells can induce some of the events of mitosis, including breakdown of the nuclear envelope and condensation of the chromosomes. Because later work showed that MPF is present as a mitotic inducer in most, if not all, eukaryotic cell types, MPF is now taken to mean *mitosis or M-phase promoting factor*.

M. J. Lohka and his coworkers extracted and purified MPF and found that it consists of two polypeptides with molecular weights of 34,000 and 45,000. The smaller polypeptide of the MPF complex is a serine/threonine protein kinase that was subsequently identified as the product of the *cdc2* gene, p34^{cdc2}. The larger polypeptide turned out to be identical to cyclin, a protein previously found by T. Hunt and J. Ruderman to cycle in concentration in coordination with cell division in cleaving starfish eggs. Cyclin, equivalent to the *cdc13* gene product of *S. pombe*, was directly implicated in cell cycle control by an extended series of experiments in sea urchins and other organisms. (The Experimental Process essay by Hunt on p. 930 tells the story of his discovery of cyclin and subsequent developments.)

Action of p34^{cdc2} and Cyclin Further work by Kirschner, Hunt, and others, including A. W. Murray, P. Beach, Nurse, W. D. Dunphy, J. Newport, and their colleagues provided insights into the activity of p34^{cdc2} and cyclin in the MPF clock of higher eukaryotes (Figs. 22-14 and 22-15). The p34^{cdc2} polypeptide is assembled almost continuously and remains in approximately constant

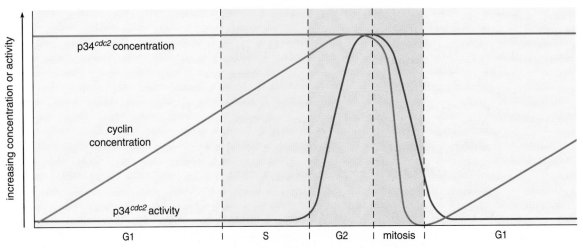

Figure 22-15 A summary of cyclin and p34^{cdc2} concentrations and activity during the cell cycle. The concentration of p34^{cdc2} remains relatively constant during the cell cycle. Cyclin concentration builds during G1 and S, and reaches a critical concentration during G2, when it combines with p34^{cdc2} to form MPF. This combination activates the protein kinase. During mitosis cyclin breaks down, reducing its concentration essentially to zero. At about the same time p34^{cdc2} is deactivated by addition of the blocking phosphate group.

concentrations in the cytoplasm throughout the cell cycle. Until combination with cyclin, however, the protein kinase is inactive because a phosphate group attached to a tyrosine blocks its active site. Cyclin concentration is at a minimum following a previous mitosis and gradually builds as the protein is assembled at a more or less constant rate in the cytoplasm. By late G2 cyclin levels reach a critical concentration, and p34^{cdc2} combines with cyclin to produce the MPF particle. Subsequently, MPF becomes fully active when p34^{cdc2} is dephosphorylated, a reaction that may be triggered by a vertebrate equivalent of the *cdc25* gene active in this role in yeast.

In active form MPF catalyzes a series of phosphorylations that initiate mitotic division. The addition of phosphate groups to histone H1, for example, which is thought to trigger chromatin condensation, is catalyzed by p34^{cdc2} at early mitotic stages; other phosphorylations by the protein kinase cause nuclear envelope breakdown through phosphorylation of lamins (see pp. 560 and 1017) and promote spindle assembly and activity.

Full activation of the MPF particle leads to degradation of the cyclin subunit and inhibition of MPF activity just before anaphase begins. Supposedly, degradation occurs through activation by MPF of a proteinase with high specificity for cyclin. The breakdown of cyclin is necessary for cells to exit mitosis and enter the subsequent G2; if the breakdown is inhibited, or if

intact cyclin is added during mitosis, cells arrest their progress and fail to complete mitosis. As cyclin breaks down in the normal course of events, p34^{cdc2} is rephosphorylated and remains in the cytoplasm in inactive form until cyclin again reaches high enough levels to produce fully active MPF particles at the next G2/mitosis transition.

The *cdc2* gene product and cyclin are thus strongly implicated as central controls triggering the transition from G2 to mitosis. The presence of *cdc2* and *cdc13* gene products in highly conserved form in a wide range of eukaryotes, including algae, fungi, higher plants, and animals, indicates that cell cycle controls based on these polypeptides were established very early in evolution and have remained as critical regulatory elements through a billion years of evolutionary history. It is interesting that although the MPF clock offers ample opportunities for interruption of the cell cycle—as, for example, by inhibition of cyclin synthesis or breakdown—arrests at the G2/mitosis transition or during mitotic division are relatively rare in nature. This may reflect a cellular economy, because cells arrested during G2 would have paid the price in energy necessary to replicate their DNA and assemble other molecules required for mitosis. Arrest at G1 saves this expenditure.

Although the MPF clock is known primarily as a control regulating the G2/mitosis transition, there are indications that p34^{cdc2} and cyclin, or closely related proteins, may also be involved in the G1/S transition.

In *S. cerevisiae*, cells with mutations in the *cdc28* gene fail to enter S at restrictive temperatures. Other evidence that $p34^{cdc2}$ and cyclinlike proteins may be involved in the initiation of S comes from recent research by G. D'Urso and his colleagues, who worked with a cell-free system extracted from rapidly growing human cells. D'Urso and his coworkers found a large, 250,000-dalton protein complex in the extract that was required for the system to enter replication of added DNA. On analysis the protein complex was found to contain $p34^{cdc2}$ and cyclin or closely similar proteins. The researchers proposed that the $p34^{cdc2}$ protein kinase in the complex is activated by combination with a cyclin protein in late G1. The protein kinase then phosphorylates a group of proteins necessary for the initiation of DNA replication. Phosphorylation activates the proteins, according to this proposal, and the S phase begins. All indications now point to families of cyclin and $p34^{cdc2}$ genes, with some members of the families active at the G1/S transition and others at G2/mitosis. The cyclin type active at the G1/mitosis transition, which remains the only fully characterized member of its family, is now known as *cyclin B*. Another cyclin, *A*, reaches peak activity at the G1/S transition and is believed to regulate this step. Recent experiments by F. Girard and his coworkers, for example, show that cyclin A is required for cultured mammalian fibroblasts to enter DNA replication; elimination of cyclin A by injection of anticyclin A antibodies prevented the cells from entering S. Equivalents of cyclin A and B have been identified in higher plants as well as animals. Other cyclins, identified as *C, D,* and *E,* have been identified in animals and shown to peak during G1, suggesting a multiple and interlocking system of cyclin-based controls.

With the exception of the MPF system, the study of cell cycle genes and their effects in other eukaryotes has not as yet progressed to the levels attained in yeast. An equivalent of the *cdc25* gene has been identified through a mutation called *string* in *Drosophila*. Other mutants identified in *Drosophila* include 16 affecting the G1/S transition, 17 causing faulty chromosome condensation, and 13 altering or stopping the anaphase movement of chromosomes by the spindle. Most of these mutations are known only by the stage at which they interrupt the cell cycle; little or nothing is known about products of the mutant genes or their roles in cell division. The exceptions are primarily genes in the maintenance class encoding enzymatic proteins necessary for functions such as DNA replication but not directly or indirectly regulating the cell cycle.

A number of mutants have been found in both lower and higher eukaryotes in which detailed steps in mitosis are affected. Among these are several that interfere with the complex mechanisms assembling the spindle and carrying out the anaphase movement. These mutations, which primarily affect gene products with maintenance rather than regulatory functions, are described in Chapter 24.

Cell Cycle Genes Identified through the Study of Oncogenes

As noted, most oncogenes are altered forms of genes that also occur in unaltered form as proto-oncogenes in normal body cells. As such, they are specialized examples of mutants, in this case associated with a particular phenotype—one characterized by the uncontrolled growth typical of cancer cells.

Identifying Oncogenes The most rapid progress in this area has come from the study of oncogenes carried by the retroviruses (see p. 767). Although the majority of retroviral infections appear to be harmless or to cause other types of disabilities, some have been linked to tumor formation in humans and other mammals. In many of these cases the retroviruses carry oncogenes or cause disruptions in the genome by altering a proto-oncogene to an oncogene. (The mechanisms by which retroviruses carry or alter host genes are outlined in Ch. 18.) In either case identification of the oncogene has frequently led to discovery of the corresponding proto-oncogene and its normal role in the cell cycle.

Among the most productive of the investigators active in this effort were J. M. Bishop and H. E. Varmus, who received the Nobel Prize in 1989 for their research establishing the relationship between retroviral oncogenes and their normal proto-oncogene counterparts. Bishop, Varmus, and their colleagues studied the *src* oncogene carried by the retrovirus associated with a cancer known as Rous sarcoma. They showed that the DNA sequence of the *src* oncogene was also present in a closely related form in normal cells.

Many more oncogenes have been identified by *transfection* (see p. 783), a genetic engineering technique in which the DNA of human cancer cells is introduced into cultured mouse cells. Frequently, the mouse cells are transformed into types that grow much more rapidly than normal cells. The human oncogene responsible for the transformation is then isolated from the mouse cells by cloning. (Fig. 22-16 outlines a typical method used in this approach.) Among the pioneers using transfection to isolate human oncogenes were J. Cooper and R. A. Weinberg.

Not all the cellular functions affected by oncogenes are directly or indirectly related to regulation of cell division. Besides growing through uncontrolled cell division, cancer cells *metastasize*: that is, they break loose readily from their tissues of origin and lodge in other body locations where they grow and divide to form additional tumors (see below). A few oncogenes

Finding Cyclins and After

Tim Hunt

TIM HUNT did his Ph.D. on hemoglobin synthesis with the late Asher Korner at Cambridge University. From 1968 through 1970 he worked on the control of globin synthesis with Irving London at the Albert Einstein College of Medicine in New York. Between 1971 and 1990 he shared a laboratory with Richard Jackson at Cambridge, studying rabbit reticulocyte lysates and teaching biochemistry. Since then, he has been at ICRF Clare Hall Laboratories, doing single-minded research on cyclins, cdc2s, and the control of the cell cycle.

The discovery of cyclin was a lucky accident. It came about through teaching summer courses at the Marine Biological Laboratory, Woods Hole. I first went there in 1977 to teach protein synthesis in the embryology course; the *quid pro quo*, apart from the beautiful surroundings, was the supply of sea urchin eggs in the same place as well-equipped laboratories. For the first two summers we studied the increase in the rate of protein synthesis which occurs in these eggs after fertilization. We never did find out, really, what turned on protein synthesis in sea urchin eggs, and it slowly dawned that what looked like a big stimulation was really only a small relief from overwhelming inhibition. Going from 99.9% inhibition to 98% inhibition is equivalent to going from 0.1% activity to 2% activity, a 20-fold increase in rate.

One of the instructors in the 1979 embryology course was Joan Ruderman. Her graduate student, Eric Rosenthal, had recently started to investigate a different kind of change in protein synthesis in fertilized eggs of the clam, *Spisula*. His elegant experiments showed that the patterns of protein synthesis changed soon after fertilization in these cells; new mRNAs were recruited onto ribosomes, and some of the previously translated mRNAs stopped being used.[1] This made sense, because changing an egg into an embryo presumably needs new proteins. But we had no idea what the new proteins were, and called the main ones *A*, *B*, and *C* in descending order of molecular weight. Sequencing of cDNA by Nancy Standart identified protein *C* as the small subunit of ribonucleotide reductase.

It was much more difficult to understand the situation in sea urchin eggs, because of a series of careful experiments performed by Bruce Brandhorst, another Woods Hole habitué. Unlike what happened in clams, Brandhorst could find no change in the pattern of protein synthesis after fertilization of sea urchin eggs.[2] His data were very convincing.

Thus the paradox. It had long been known that if protein synthesis was inhibited after fertilization of sea ur-

chin eggs, DNA replication was permitted but entry into mitosis was blocked. How could this be, if the eggs (1) made so little protein after fertilization, compared to what they already possessed, and (2) there was no difference in the kind of proteins synthesized?

By 1982, our studies of protein synthesis and its control in sea urchin eggs had explored many alleys without great success. On July 22, in a quiet moment, I did a simple experiment, purely out of curiosity, to compare the patterns and rates of protein synthesis in fertilized and parthenogenetically activated sea urchin eggs. I added ^{35}S-methionine to the eggs, and took samples at regular intervals for running on an SDS-polyacrylamide gel. The result was intriguing and completely unexpected. The most strongly labeled band at early times faded away after about an hour, whereas most of the other bands got stronger. Repetition of the experiment with carefully controlled batches of fertilized eggs put the matter beyond doubt, for in the next experiment, the protein appearing as the strongly labeled band—which we termed *cyclin*—came and went twice. Its disappearance preceded cell division by about 10 minutes. When division was inhibited by colchicine, or by inhibition of DNA synthesis, the destruction of cyclin was delayed or failed to occur. We could see cyclin continuing to cycle several hours after fertilization by labeling older embryos for 30 minutes, and then adding emetine to block further synthesis. The cyclin band went away, while all the others stayed.

These results were written up in a state of high euphoria.[3] People had been looking for so-called "periodic proteins" for several years, for it was widely suspected that if cells engaged in different activities at different stages of their cycle, they would contain different sets of enzymes. I do not recall having seen anyone point out what seems plain in retrospect, namely that for such a mechanism of restricting certain activities to particular phases of the cycle to work, the proteins in question would have either to be very unstable, or to be eliminated by some special mechanism when they were no longer required. Such appeared to be the case with cyclin.

Two years later, Joan Ruderman and I made very detailed studies of the stability of cyclins during the cell cycle, and found that they were stable proteins during interphase, and very rapidly destroyed just before the metaphase→anaphase transition. Clams contained two cyclins, corresponding to the two translationally regulated proteins *A* and *B*.[4,5]

John Gerhart had given a seminar on MPF in 1979, which had introduced me to this fascinating and important entity, and he turned up on the very day that cyclin was discovered. He told me about experiments that he, Marc Kirschner, and Mike Wu had been doing to measure

MPF levels during meiosis in *Xenopus* embryos.[6] I was particularly excited by the news that after its initial, auto-catalytic appearance, new protein synthesis was required for MPF to reappear after meiosis II. Could cyclin be a component of MPF? Or was cyclin perhaps the enzyme that catalysed the activation of MPF? If the latter, did it act directly and enzymatically, or by titrating out a hypothetical "anti-MPF"?

The descriptive phase of cyclin studies came to an end in 1986, when Katherine Swenson discovered that clam cyclin *A* mRNA promoted maturation in frog oocytes. This was a very important development, because it clearly demonstrated that cyclins could promote entry into M-phase. The question then arose whether frogs (or indeed, any organism apart from marine invertebrates) had cyclins. One day in Berkeley, Mike Wu offered to show me a real MPF assay. I thought it would be fun to try in parallel if maternal mRNA would score in this assay, as Swenson's experiments implied. Fortunately, Eric Rosenthal was in Berkeley working on oogenesis in *Urechis caupo*, an invertebrate worm that lives in the mudflats of Bodega Bay. He gave me some oocyte and embryo RNA, and Wu injected it into oocytes. The ones injected with two-cell embryo mRNA duly matured, whereas the oocyte mRNA-injected cells just sat there. We decided to see if *Xenopus* mRNA could do the same, and were thrilled to find that mRNA from mature or fertilized eggs was active. All the controls looked good, and we guessed this meant that *Xenopus* has cyclins.

Back in England, my graduate students, John Pines and Jeremy Minshull, were making good progress cloning and sequencing cyclin mRNA from *Arbacia*. Minshull decided to make a cDNA library from *Xenopus* and screen it with an oligonucleotide based on a consensus sequence developed from clam cyclin *A* and sea urchin cyclin *B*. Since they were, evolutionarily speaking, so far apart, it was likely this sequence would find any cyclin (indeed, flies, cows, and soybeans all eventually yielded up cyclins to oligonucleotide 40). In early 1987, we saw our first vertebrate *B*-type cyclin.

By now, cell-cycle studies had taken an enormous boost from the work of Paul Nurse and Melanie Lee at the ICRF in London, who found the human version of the yeast cell cycle control gene, cdc2.[7] In 1988, Joan Ruderman and David Beach showed that clam cyclin immuno-precipitates possessed protein kinase activity, and contained a polypeptide that probably corresponded to clam cdc2.[8] That same summer, various consortia of frog and yeast workers showed that MPF contained cdc2,[9,10] and when antibodies to frog cyclins became available, the large subunit of MPF proved to be a *B*-type cyclin.[11]

Meanwhile, how could we directly show that cyclin was really essential for driving cell cycles? Two approaches put the matter beyond doubt, both relying on the cell-free system from frog eggs that Fred Lohka had originally developed in Yoshio Masui's laboratory.[12] This preparation could replicate the DNA of added nuclei and then enter mitosis, which, like mitosis in intact sea urchin eggs, required protein synthesis. Minshull, Blow, and I used complementary oligonucleotides capable of pairing with cyclin mRNA to eliminate this mRNA. We found that without cyclin mRNA, DNA replication was OK, but mitosis failed.[13]

At about this time, Andrew Murray did a much more daring experiment in Kirschner's laboratory. He removed all the endogenous mRNA from the frog egg extract using ribonuclease, and then added back Pines's *Arbacia* cyclin B mRNA and a ribonuclease inhibitor. The only protein made by the extract was cyclin, and amazingly, the nuclei replicated and entered mitosis.[14] In fact, they entered mitosis not once, but two or three times, and at each mitosis the cyclin was destroyed. A simple construct lacking the first 90 codons of the cyclin mRNA gave rise to a stable protein, because the "destruction box" had been removed. When this mRNA was added to the extracts, they entered mitosis but got stuck there![15] This convinced even the most hardened sceptics that cyclin synthesis was needed to get into mitosis, and its destruction was required for the cell to return to interphase.

1988 was also the year that a nest of new cyclins turned up in yeast. It was especially exciting that some of them, discovered independently by Fred Cross, Bruce Futcher, and Steve Reed, appeared to be involved in control of the G1-S transition. Thus it emerged that the periodic protein of sea urchin eggs was a representative of a family of cell cycle regulators, whose function now emerges as the control, in time and in space, of the cdc2 family of protein kinases. The size and variety of these families, their richness for the optimist or complexity for the pessimist, presents a considerable challenge. The pressing question now, it seems to me, is to identify the key substrates whose phosphorylation can bring about cell cycle transitions. Life was simpler in 1982.

References

[1] Rosenthal, E. T.; Hunt, T.; and Ruderman, J. V. *Cell* 20:487–94 (1980).

[2] Brandhorst, B. P. *Dev. Biol.* 52:310–17 (1976).

[3] Evans, T.; Rosenthal, E. T.; Youngblom, J.; Distel, D.; and Hunt, T. *Cell* 33:389–96 (1983).

[4] Westendorf, J. M.; Swenson, K. I.; and Ruderman, J. V. *J. Cell Biol.* 108:1431–44 (1989).

[5] Hunt, T.; Luca, F. C.; and Ruderman, J. V. *J. Cell Biol.* 116: 707–24 (1991).

[6] Gerhart, J.; Wu, M.; and Kirschner, M. *J. Cell Biol.* 98:1247–55 (1984).

[7] Lee, M. G. and Nurse, P. *Nature* 327:31–35 (1987).

[8] Draetta, G., et al. *Cell* 56:829–38 (1989).

[9] Dunphy, W. G.; Brizuela, L.; Beach, D.; and Newport, J. *Cell* 54:423–31 (1988).

[10] Gautier, J.; Norbury, C.; Lohka, M.; Nurse, P.; and Maller, J. *Cell* 54:433–39 (1988).

[11] Gautier, J., et al. *Cell* 60:487–94 (1990).

[12] Lohka, M. J., and Masui, Y. *Science* 220:719–21 (1983).

[13] Minshull, J.; Blow, J. J.; and Hunt, T. *Cell* 56:947–56 (1989).

[14] Murray, A. W., and Kirschner, M. W. *Nature* 339:275–80 (1989).

[15] Murray, A. W.; Solomon, M. J.; and Kirschner, M. W. *Nature* 339:280–86 (1989).

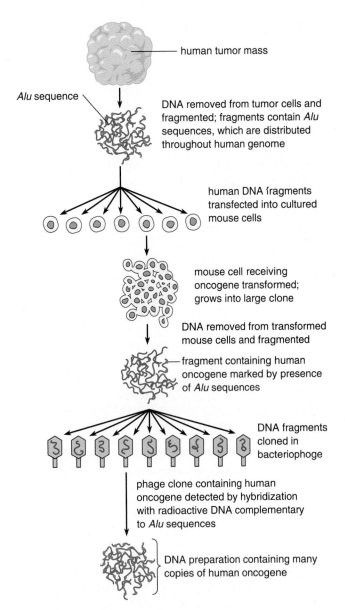

human tumor mass

Alu sequence

DNA removed from tumor cells and fragmented; fragments contain *Alu* sequences, which are distributed throughout human genome

human DNA fragments transfected into cultured mouse cells

mouse cell receiving oncogene transformed; grows into large clone

DNA removed from transformed mouse cells and fragmented

fragment containing human oncogene marked by presence of *Alu* sequences

DNA fragments cloned in bacteriophage

phage clone containing human oncogene detected by hybridization with radioactive DNA complementary to *Alu* sequences

DNA preparation containing many copies of human oncogene

Figure 22-16 A technique used for identifying human oncogenes by transfection of cultured mouse cells. (In transfection, cells take up DNA fragments from the surrounding medium and integrate them into their own DNA; see p. 783 for a description of this genetic engineering technique.) The human DNA can be identified among the mouse DNA sequences by the presence of *Alu* repeats (see p. 758), which are distributed widely and take a unique form in the human genome.

promote metastasis rather than, or in addition to, cell cycle controls.

Oncogene Types About 100 oncogenes have been discovered through the study of retroviruses, transfection experiments, and other techniques. (Table 22-4 lists some oncogenes commonly found in tumors of humans and other animals.) Most of these oncogenes

prove to be restricted to a relatively small number of activities such as: (1) protein kinases, (2) polypeptide growth factors, (3) surface receptors for hormones or growth factors, (4) the G proteins active in reaction cascades activated by surface receptors, (5) steroid or thyroid hormone receptors, or (6) nuclear proteins probably active as gene regulators.

In addition to the oncogenes in these classes, several recently discovered genes known as *tumor suppressor genes, recessive oncogenes,* or *anti-oncogenes* have the unusual effect of inhibiting tumor growth promoted by other oncogenes. The genes have this beneficial effect if they are present in normal form in at least one of the two chromosomes of the set. Genes in the tumor suppressor group, including *rb* and *p53*, have generated considerable interest because of their potential use as a means to control cancer.

The oncogenes encoding these proteins are named according to several conventions. Most are named for an associated cancer type—for example, *myb*, an oncogene encoding a nuclear protein, is named for its association with *myeloblastosis*, a retrovirus-induced cancer of chickens. Some of the derivations are rather obscure. The oncogene *jun* is named for the Japanese word *ju-nena*, meaning 17 and referring to the *avian sarcoma virus 17* in which it was first identified. An oncogene is not necessarily restricted to the type of cancer or species from which it is named. Some, such as *ras* and *myc*, are found in many different kinds of tumor cells, although they are named for a single cancer type. Frequently, oncogenes carried by retroviruses are identified as v- oncogenes, as in v-*ras;* the normal proto-oncogenes are often designated as the c-equivalents, as in c-*ras.*

Oncogenes Encoding Protein Kinases Among oncogenes those encoding protein kinases are most frequently identified in tumors. Of these, the majority, including *src, abl, fps, yes, fgr, met,* and *ros,* encode protein kinases that add phosphate groups to tyrosine residues in their target proteins. Other protein kinases, products of the *mos, raf,* and *mil* oncogenes, phosphorylate serine or threonine residues in their target proteins.

Some of the target proteins phosphorylated by these protein kinases undoubtedly take part in pathways that directly or indirectly regulate the cell cycle or affect functions that go awry in cancer cells. For example, the kinases encoded in two of the genes in this group, *src* and *ros,* add phosphate groups to membrane phospholipids taking part in the InsP₃/DAG pathway of cellular activation (see p. 247). The preponderance of protein kinases among the direct or indirect oncogene products reflects the fact that the addition of phosphate groups to cellular proteins is a primary mechanism of cell cycle regulation in eukaryotes. In general the oncogene forms of protein kinases are produced in a state of activity

Table 22-4   Some Oncogenes and Their Original Sources

Oncogene	Original Cancer Type	Original Source	Activity or Product	Cellular Location
abl	Abelson leukemia	Mouse	Tyr protein kinase	Cytoplasm
erbA	Erythroblastosis	Chicken	Thyroid hormone receptor	Nucleus
erbB	Erythroblastosis	Chicken	EGF receptor	Plasma membrane
ets	Myeloblastosis	Chicken	Regulatory protein	Nucleus
fes*	Feline sarcoma	Cat	Tyr protein kinase	Cytoplasm
fgr	Feline sarcoma	Cat	Tyr protein kinase	Plasma membrane
fms	Feline sarcoma	Cat	CSF receptor	Plasma membrane
fos	Osteosarcoma	Mouse	Regulatory protein	Nucleus
fps*	Sarcoma	Chicken	Tyr protein kinase	Cytoplasm
jun	Sarcoma	Chicken	Regulatory protein	Nucleus
kit	Sarcoma	Cat	Tyr protein kinase; probable GF receptor	Plasma membrane
met	Osteosarcoma	Mouse	Tyr protein kinase	Cytoplasm
mil†	Sarcoma	Chicken	Ser/Thr protein kinase	Cytoplasm
mos	Sarcoma	Mouse	Ser/Thr protein kinase	Cytoplasm
myb	Myeloblastosis	Chicken	Regulatory protein	Nucleus
myc	Myelocystosis	Chicken	Regulatory protein	Nucleus
neu	Neuroblastoma	Rat	EGF receptor-like protein	Plasma membrane
p53	Many cancers		Tumor suppressor gene	Nucleus
raf†	Sarcoma	Mouse	Ser/Thr protein kinase	Cytoplasm
ras	Sarcoma	Rat	G protein	Plasma membrane
rb	Retinoblastoma	Human	Tumor suppressor gene	Nucleus
rel	Reticuloendotheliosis	Turkey	Regulatory protein	Nucleus
ros	Sarcoma	Chicken	Tyr protein kinase; probable GF receptor	Plasma membrane
sis	Sarcoma	Monkey	PDGF	Secreted
ski	Carcinoma	Chicken	Regulatory protein	Nucleus
src	Sarcoma	Chicken	Tyr protein kinase	Cytoplasm
trk	Carcinoma	Human	Tyr protein kinase	Cytoplasm
yes	Sarcoma	Chicken	Tyr protein kinase	Cytoplasm

* Same oncogene initially identified in different organisms.
† Same oncogene initially identified in different organisms.
  EGF, epidermal growth factor; CSF, colony stimulating factor; GF, growth factor; PDGF, platelet-derived growth factor.

that promotes cell division, in situations in which their normal counterparts would be either less active or inactive.

At least some of the oncogenes encoding protein kinases promote malignancy by increasing the tendency of cancer cells to metastasize. Among the target proteins phosphorylated by *src*, for example, is the surface receptor for fibronectin, one of the major proteins of the extracellular matrix (see p. 286). The phosphorylation, which takes place on the cytoplasmic extension of the fibronectin receptor, is believed to inhibit binding between the receptor and fibronectin, thereby loosening attachment of the cell to the extracellular matrix and promoting its tendency to break loose and migrate.

Oncogenes Encoding Growth Factors or Surface Receptors   Only one known oncogene, *sis*, encodes a polypeptide growth factor. The protein encoded in *sis*, a segment of PDGF, is able to induce division in cells with a PDGF receptor. We have noted that fibroblasts, smooth muscle cells, and the glial cells of the central nervous system are among the cell types with PDGF

receptors. As such, they are susceptible to tumor formation by exposure to the product of the *sis* oncogene.

Several oncogenes, including *erbB* and *fms* encode growth factor receptors. The *erbB* oncogene encodes a receptor for EGF; *fms* encodes a receptor for *colony stimulating factor (CSF)*, a growth factor that normally triggers growth in macrophages, the white blood cells that scavenge foreign cells and substances from the body. Two oncogenes, *kit* and *ros*, probably also encode receptors for growth factors.

Oncogenes Encoding G Proteins   The altered forms of the G proteins encoded in *ras* oncogenes probably participate in the cascade of reactions transmitting a signal to the underlying cytoplasm that a surface receptor has bound to its target protein (see Fig. 22-6*b* and *c* and p. 243). Three *ras* genes, H-*ras*, N-*ras*, and K-*ras*, have been identified in mammals; each has a mutant form implicated in the generation of cancer. Although the effects of the oncogene *ras* genes are uncertain, one of the internal enzymes suspected to be stimulated by pathways including the *ras* G proteins is phospholipase C, the enzyme that breaks down a membrane phospholipid to release the InsP₃ and DAG second messengers.

The faulty forms of the G proteins encoded in *ras* oncogenes generally have reduced ability to break down GTP. This limits the self-deactivation function of these proteins (see p. 243). In this form the G proteins remain active in stimulating the regulatory cascades in which they take part. The result, if other conditions are met, is uncontrolled cell division. The *ras* oncogenes are found in a very large percentage of tumors of certain kinds, as in 97% of pancreatic carcinomas and 20% to 40% of cancers of the colon and rectum. They are also common in cancers of epithelial tissues and the blood. The *ras* oncogenes occur in some noncancerous growths, such as benign forms of growths called *polyps*, which often form in the intestine. (Further conversions of other proto-oncogenes to oncogenes can alter intestinal polyps to malignant form; see below.)

An Oncogene Encoding a Steroid or Thyroid Hormone Receptor   Only a single known oncogene, *erbA*, encodes a receptor for steroid or thyroid hormones. The DNA sequence of this oncogene closely resembles that of genes encoding normal receptors for thyroxin and a particular class of steroid hormones, glucocorticoids (see p. 704). As noted, these and other normal forms of the steroid or thyroid hormone receptors are activated by binding the hormone; in active form the receptor recognizes and binds the control regions of specific genes and regulates their transcription. The *erbA* oncogene encodes a faulty receptor that is continuously active in gene regulation, whether the steroid or thyroid hormone is bound or not.

Oncogenes Encoding Nuclear Proteins   The nuclear proteins encoded in *ets, fos, jun, myb, myc, rel,* and *ski* are known or suspected to be regulatory proteins that control genes fundamentally important to cell cycle regulation. The normal forms of these oncogenes are stimulated very early during the G1/S transition, as expected if they are fundamental controls. Recent research in the laboratories of R. Tjian, T. Curran, R. Franza, and P. Vogt has revealed that the *fos* and *jun* proto-oncogenes encode subunits of a family of regulatory proteins activating transcription, known as the *AP-1* family in higher eukaryotes and *GCN4* in yeast. The products of the *fos* and *jun* genes combine through a leucine-zipper motif (see p. 544) to form an active regulatory protein. In some systems a regulatory protein encoded in the *ets* proto-oncogene cooperates with the *fos-jun* complex in gene regulation. Among the many genes activated by these regulatory proteins are some known to promote cell division, such as other oncogenes and genes encoding growth factors. The *myc* gene product can bind directly to DNA and probably acts to regulate genes controlling the G1/S

transition. (The Experimental Process essay by G. C. Prendergast on p. 696 describes his research identifying the Myc DNA-binding activity.)

Oncogenes, Proto-Oncogenes, and Evolution   The normal proto-oncogene forms of the vertebrate oncogenes are descendants of gene families that have been conserved throughout eukaryotic evolution. The many protein kinases encoded in different oncogenes share enough sequence homologies to suggest that those encoding tyrosine kinases, for example, probably stem from a single, common ancestral gene. Similar relationships to ancient ancestral genes are noted in other oncogene families, such as *ras*, which occurs in normal and mutant forms in yeast as well as vertebrates. As noted, the similarities in the *ras* family are so close that the human *ras* gene in its normal form can substitute for mutant *ras* in *S. cerevisiae*. The proteins encoded in the yeast *ras* genes, in fact, are as much as 90% homologous to human *ras* in sequence regions near their N-terminal ends; extensive sequence differences are noted only at the C-terminal ends. Other oncogenes with obviously ancient roots are the *myc-fos-myb* group encoding nuclear proteins, which have been detected in eukaryotes ranging from yeast to humans. The wide distribution of these genes reinforces the conclusion that some fundamental mechanisms controlling cell division are similar in all eukaryotic cells.

Tumor Suppressor Genes   The normal counterparts of tumor suppressor genes encode products that inhibit cell division. A mutation eliminating the activity of one of the two copies of a gene in this group usually has no noticeable effect, because the remaining copy is still active in suppressing cell division. A mutation eliminating the activity of the second copy of the gene, however, may lead to uncontrolled cell growth. The genes are thus *recessive* in the sense that both copies of the gene must be in mutant form to produce an effect. The recessive nature of tumor suppressor genes is in stark contrast to dominant oncogenes such as *ras* or *myc*, in which a single copy of the oncogene often produces uncontrolled growth.

In some human families, for example, individuals inherit one normal copy of the *rb* gene, located at a site on chromosome 13 of the human set. The gene encodes the RB protein, which has been shown by P. D. Robbins and his coworkers to bind to DNA and repress transcription of *fos* and other regulatory proteins activating genes related to cell division. As long as one normal copy of the gene is present, individuals are not unusually susceptible to developing cancer. However, the chromosome region containing the normal copy is unusually subject to deletions; a deletion impairing the normal copy leads to development of a form of retinoblastoma, a cancer of the eye most commonly found in children. The development of two other cancers of children, Wilms' tumor (a cancer of kidney cells) and neuroblastoma, follows a similar pattern. Neuroblastoma, a cancer of sympathetic neurons, is the most common solid tumor of children. Loss of normal *rb* genes in both chromosomes has been demonstrated in adult bone, lung, and prostate cancers in addition to the retinoblastoma occurring in children.

The RB protein is bound by proteins encoded in a number of DNA tumor viruses (see p. 941), including SV40, papillomavirus, and adenovirus. Presumably, binding by the viral proteins inactivates RB and leads to uncontrolled cell division. Because of its activity in suppressing cell division, there is now much interest in the normal RB protein. J. Nevins, D. Livingston, and L. Bandara and their colleagues found that one binding target of normal RB protein is *E2F*, a regulatory protein that promotes transcription of several cell cycle genes, including *fos*. RB binds and inactivates E2F, thereby blocking transcription of *fos*. There is some evidence that RB directly inactivates another cell division gene, *myc*, by binding and blocking one of its 5′ control sequences. In cells entering division by normal processes, RB is phosphorylated at the G1/S boundary; the phosphorylation inhibits activity of the RB protein and allows entry into DNA replication and subsequent stages of the cell cycle.

Other tumor suppressor genes have been detected by fusion of normal cells with tumor cells. Often the tumor cells are converted to normal growth patterns by the fusion, as if a factor supplied by the normal cells suppresses uncontrolled cell division. The chromosomal locations of the genes responsible for tumor suppression have been worked out by a technique known as *microcell transfer.* In a microcell all the chromosomes have been eliminated except for one or two. Microcells carrying different chromosomes of the set from normal cells are fused with tumor cells until a chromosome is located that suppresses uncontrolled division of the tumor cells. This chromosome, suspected to carry a tumor suppressor gene in normal cells that is deleted in the tumor cells, is then analyzed by restriction endonucleases to pinpoint the site of the deletion. Once the site is located, the gene and its product can be sequenced and identified. This technique led to the discovery of *p53*, a tumor suppressor gene implicated in generation of bladder, blood, brain, breast, colorectal, esophageal, liver, and lung cancers when doubly mutant. The p53 protein encoded in the gene has been identified as a DNA-binding protein. More than 10 additional tumor suppressor genes have been found.

Introduction of normal copies of the *rb* and *p53* genes has suppressed growth of some cultured tumor cells; presumably, these are cells in which these genes are doubly mutated into oncogenic form. The successes with cultured cells gives promise that introduction of normal tumor suppressor genes may provide a means to control at least some forms of cancer.

Alterations Changing Proto-Oncogenes into Oncogenes

Sequencing the known oncogenes has revealed that many are altered in some way as compared to their normal cellular counterparts. The characteristics of these alterations have provided clues to the factors giving rise to cancer. In addition, they have afforded valuable insights into the patterns by which their normal counterparts, the proto-oncogenes, may work to regulate cell division.

The alterations distinguishing oncogenes from proto-oncogenes fall into several clearly defined classes involving: (1) sequence substitutions in the gene or its control regions; (2) deletion of sequences from the gene or its control regions; (3) movement or *translocation* of part or all of a gene to a different location in the chromosome; or (4) *amplification* of the gene to produce extra copies. Occasionally, more than one of these changes is noted in the same oncogene, producing a copy that is extensively altered from its normal form. In most cases the alterations activate the gene or its products, leading directly or indirectly to changes associated with the transformation of normal cells to cancer cells.

Most of the gene alterations involved in the development of cancer occur in body cells, not in reproductive cells. Therefore, although the primary causes of these tumors are genetic, they are not hereditary. They die with the individual and are not passed on to offspring. A relatively few gene alterations producing oncogenes, however, are inherited and give the persons receiving them an increased likelihood of developing cancer.

Sequence Substitutions in Coding or Control Regions Mutations altering coding segments, promoters, or enhancers are among the most common alterations converting proto-oncogenes to oncogenes. The oncogenic forms of the *ras* gene detected in cancer cells, for example, usually differ from their normal counterparts by single base substitutions in one of three positions in the coding sequence. The alterations cause amino acid substitutions in corresponding positions in the G protein encoded in the gene. In the *ras* oncogene associated with human bladder carcinoma, for example, a mutation leads to substitution of valine for glycine at position 12 in the amino acid sequence of the encoded G protein. In one form of human lung cancer, the *ras* oncogene contains a mutation substituting leucine for glutamine at position 61 in the encoded G protein.

Oncogenes Created by Deletions The second type of alteration producing oncogenes, deletion of sequences from either control or coding regions, is also frequently noted. The *myc* gene in its normal form, for example,

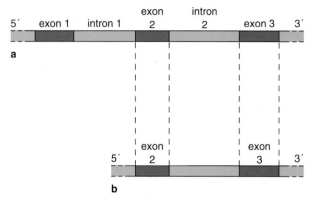

Figure 22-17 The *myc* gene in normal and oncogene forms. **(a)** The normal form, which has three exons and two introns. **(b)** A common oncogenic form of *myc*, in which the first exon and intron are missing. Promoter-like sequences forming part of the second exon allow transcription of the oncogene.

consists of three exons separated by two introns (Fig. 22-17a). In many of its oncogenic forms, the first exon and intron and the control sequences in advance of the gene are missing (Fig. 22-17b). Although the effects of the deletions are uncertain, removal of the first exon,

which is normally copied into the *myc* mRNA but not translated into a protein sequence, may alter RNA processing to increase the amounts of *myc* mRNA available to ribosomes in the cytoplasm. Removal of the promoter region of the *myc* gene does not prevent transcription of the gene, because additional signals capable of acting as promoters form parts of the second exon.

Several other oncogenes follow the same patterns in their conversion from normal forms: deletions alter the function of normal promoters, RNA processing, or the protein encoded in the gene. The deletions are often coupled with placement of the altered gene near strong promoters or enhancers of other genes, which increase its rate of transcription. In many of its oncogenic forms, for example, the *myc* gene is placed in locations that increase its transcription above normal levels (see below). Deletions are also common among the alterations converting tumor suppressor genes to oncogenic form. In this case the deletion destroys activity of the gene product in inhibiting cell division.

Translocations and Oncogene Formation Translocation of a gene to a new position in the chromosomes often places the gene in a control environment that is

Table 22-5   Translocations and Other Chromosome Defects Noted in Cancer Cells

Cancer Type	Chromosome Defect
Translocations	
carcinomas	
ovarian papillary cystadenocarcinoma	Between chromosomes 6 and 14
leukemias	
acute lymphocytic leukemia	
type L1, L2	Between chromosomes 9 and 22
type L2	Between chromosomes 4 and 11
type L3	Between chromosomes 8 and 14
acute nonlymphocytic leukemia	
type M1	Between chromosomes 9 and 22
type M2, myelogenous	Between chromosomes 8 and 21
type M3, promyelocytic	Between chromosomes 15 and 17
type M4, myelomonocytic	Between chromosomes 9 and 11
chronic lymphocytic leukemia	Between chromosomes 11 and 14
chronic myelogenous leukemia	Between chromosomes 9 and 22
lymphomas	
Burkitt's	Between chromosomes 8 and 14
follicular	Between chromosomes 14 and 18
small cell lymphocytic	Between chromosomes 11 and 14
Other chromosomal defects	
acute nonlymphocytic leukemia	Deletions in chromosomes 5 and 7; extra chromosome 8
chronic lymphocytic leukemia	Extra chromosome 12
chronic lymphocytic lymphoma	Extra chromosome 12
neuroblastoma	Deletion in chromosome 1
retinoblastoma*	Deletion in chromosome 13
small cell lung carcinoma	Deletion in chromosome 3
small cell lymphocytic lymphoma	Extra chromosome 12
Wilms' tumor*	Deletion in chromosome 11

From tables and data presented in J. J. Yunis, *Science* 221:227 (1983). See this paper for original references.
* Can be inherited.

Figure 22-18 A chromosome translocation between chromosomes 8 and 14 commonly found in Burkitt's lymphoma. **(a)** The translocation; the solid arrows indicate the points of breakage; the dashed arrows show how the broken chromosome segments exchange places in the translocation. The light and dark crossbands are produced by a staining technique that identifies individual human chromosomes for light microscopy (see Fig. 24-43). **(b)** The region of chromosome 14 at which the fragment from chromosome 8 joins. The translocation places the second and third exons of the *myc* gene next to exons encoding heavy chain segments in an antibody gene.

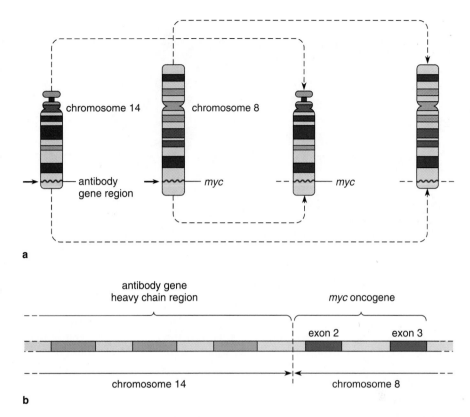

distinctly different from its previous location. Controls that limited activity of the gene in its previous location may be absent in the new environment, causing the gene to be more active. Or, the gene in its new location may come under the influence of a highly active promoter or enhancer element belonging to another gene, also making the gene more active. If the gene is related to cell cycle control, the translocation may result in rapid and uncontrolled cell division.

Many translocations responsible for gene movements are the result of gross exchanges in which whole chromosome segments separate and attach to sites in other chromosomes. In some translocations breaks occur in two chromosomes simultaneously, setting up a reciprocal exchange of segments between the chromosomes involved (see p. 769).

Chromosome translocations have been observed in association with many forms of human cancer, including several types of leukemias and Burkitt's lymphoma (Table 22-5). In one line of Burkitt's lymphoma cells, for example, a reciprocal exchange has taken place between segments of chromosomes 8 and 14 of the human set (Fig. 22-18). The original break in chromosome 8 in this line occurred at the *myc* proto-oncogene, separating *myc* between its first and second exons. In chromosome 14 the break occurred within the segment coding for the immunoglobulin heavy chain constant region. The broken segments then exchanged places, placing the shortened *myc* gene next to the heavy chain coding region of chromosome 14. The translocation increases the concentration of the

protein encoded in the *myc* gene, either through an increase in its rate of transcription or, if the first exon is lost, through an increase in processing of its mRNA. One possibility for the increase in transcription is that the *myc* gene is brought under the influence of enhancer elements in the immunoglobulin gene. Another possibility is that movement to the new location separates the *myc* gene from as yet unidentified sequence elements or other genes that regulate or suppress its activity in its normal location on chromosome 8.

Other translocations converting proto-oncogenes to oncogenes are noted. For example, in the translocation responsible for *chronic myelogenous leukemia, abl* has been moved from its normal location as a proto-oncogene at one end of chromosome 9 to a site on chromosome 22 containing the normal form of the gene *bcl*. The combination of the two genes produces an *abl-bcl* hybrid that is highly active as an oncogene. The translocation produces a characteristically shortened chromosome 22 known as the *Philadelphia chromosome*, which is a hallmark of chronic myelogenous leukemia.

Leukocytes of the B-cell and T-cell types (see p. 790) are frequently involved in leukemias developing as a result of chromosome translocations. Their involvement probably results from a failure of the breakage and rearrangement mechanisms that normally assemble genes of the immunoglobulin family into antibody genes in B cells and receptors in T cells (see p. 850). Rarely, a breakage occurring as part of these rearrangements remains open, providing an opportunity for segments broken from chromosomes at other

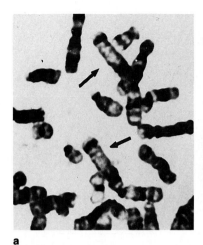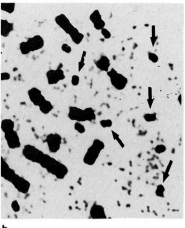

a   b

Figure 22-19 Chromosomal abnormalities produced by DNA amplification. **(a)** Homogeneously staining regions, or HSRs (arrows), produced when amplified segments remain covalently linked to the remainder of the DNA. **(b)** Double minutes (arrows) produced when amplified DNA is released from the chromosomes. Courtesy of J. K. Cowell. Reproduced with permission, from *Ann. Rev. Genet.* 16:21. © 1982 by Annual Reviews, Inc.

sites to mistakenly join. Why the site on chromosome 8 containing the *myc* gene often shares in translocations with the immunoglobulin genes remains unknown. Perhaps an arrangement of chromosomal proteins in the *myc* region creates an exposed site that is easily broken.

Amplification in Oncogene Production Gene amplification results from overreplication of restricted segments of the DNA (see p. 979). The long chromosomes of eukaryotic cells contain multiple sites from which DNA replication proceeds (see p. 975). Each site triggers replication of a subpart of the DNA molecule known as a *replicon*. Normally, each replicon is activated only once during an S period, so that the entire DNA molecule of the chromosome is duplicated only once. Rarely, one or more replicons may be activated more than once during an S phase, causing a segment of the DNA molecule to be overreplicated, or amplified. DNA amplification produces extra copies of genes or other DNA sequences, increasing the "dosage" of these elements in the cell.

Amplification in oncogene formation is probably a random mechanism that may involve any region of the DNA. If an amplified segment happens to contain a gene related to cell division, the extra copies of the gene may lead to overproduction of the gene's encoded protein and stimulation of division. The amplified DNA may remain covalently linked to other DNA sequences in its chromosome (as in the arrangement shown in Fig. 23-24), or it may be released to the surrounding nucleoplasm as small, unattached fragments of chromatin. If extensive enough, amplified regions remaining linked to the chromosome appear under the light microscope as highly characteristic, evenly stained segments known as *homogeneously staining regions*, or *HSRs* (Fig. 22-19a). Unattached amplified fragments often take a form known as *double minutes*, which resemble replicated but undivided chromosomes of very small size (Fig. 22-19b).

Gene amplification is frequently observed in association with human and other animal cancers. Amplified copies of the *myc* gene, for example, appear in several human tumors, including neuroblastomas and some breast and colon cancers. In cells of mouse adrenocortical tumors, the *ras* gene is amplified from 30 to 60 times. Amplified segments containing the gene are retained in the chromosomes as HSRs and also released as double minutes. Amplification of the oncogenes in each of these examples results in increased transcription of the gene, contributing toward conversion of the affected cells to malignant types. The oncogenes *erbB* and *neu* are also commonly found in amplified form.

Amplification also frequently accompanies other forms of chromosome alterations. For example, in leukemias involving translocations between the *myc* and immunoglobulin gene regions, the *myc* oncogene is often amplified in its new location.

Oncogene Formation without Alterations in Gene Sequences In some instances normal genes are converted to oncogenes without changes in the coding or control sequences of the gene itself. This takes place, for example, in the process termed *insertional activation*, in which a retrovirus inserts into the DNA near a gene related to cell division. Insertion of the retroviral DNA brings the proto-oncogene under the influence of strong viral promoters or enhancers. Stimulation by the nearby viral elements activates the gene and promotes cell division. For reasons that are not understood, a hotspot for retroviral insertion lies near the *myc* gene, close enough for the gene to become activated to oncogenic form when the viral DNA inserts. Insertional activation of the *myc* gene in this pattern takes place, for example, in a cancer of birds known as avian leukosis.

Conversion of a single proto-oncogene to an oncogene is usually not sufficient in itself to produce cancer. Instead, cancer cells are commonly characterized

Table 22-6 Malignant and Benign Tumor Types	
Tumor	Tissue
Malignant	
adenocarcinoma	Glandular
carcinoma	Epithelial
glioma	Glial tissues of central nervous system
hepatoma	Liver
leukemia	White blood cells (leukocytes)
lymphoma	Lymphocytes
melanoma	Pigment cells
myeloma	Plasma cells, usually in bone marrow
nephroblastoma	Kidney
neuroblastoma	Nerve cells
retinoblastoma	Eye (retina)
sarcoma	Connective tissue
seminoma	Reproductive cells
squamous	Epidermal
teratoma	Reproductive cells
Usually benign	
adenoma	Glandular
chondroma	Cartilage
fibroma	Fibroblasts
osteoma	Bone
papilloma	Surface epithelia

by several active oncogenes, sometimes as many as three or more. In one study of human malignancies by D. J. Slamon and his coworkers, for example, tumors from 54 patients with 20 different kinds of cancer were examined. More than one oncogene was identified in each of the 54 tumors; in most, three oncogenes—*fos*, *myc*, and *ras*—were active. These observations support the conclusion that the genes controlling cell division form parts of complex networks in which the balance of activity among many genes, rather than control by single genes, is responsible for activation or deactivation of the division process. They also add support to the generally held conclusion that cancer develops through successive changes involving the additive effects of several mutations or other changes.

CANCER: CHARACTERISTICS AND CAUSES

Cancer is among the most dreaded of human diseases. Recognized as a major threat to health since the earliest days of recorded history, cancer still counts as one of the most frequent causes of human fatality, particularly in technically advanced countries. In these countries it accounts for about 15% to 20% of deaths each year. In 1988 over 450,000 persons died from cancer in the United States, more than the total number of Americans killed in World War II and the Vietnam War combined.

The Characteristics of Cancer

More than 200 types of cancer have been cataloged. (Some of the most common forms are listed in Table 22-6.) All are characterized by uncontrolled cell division and metastasis, in which tumor cells break loose and spread from their sites of origin to lodge and grow in other locations in the body. Of the types listed in Table 22-6, *carcinomas*—including tumors of the skin, colon and rectum, lung, breast, and prostate—are by far the most common cancers of humans.

The detrimental effects of cancerous growths result from interference with the activity of other normal cells, tissues, and organs or from loss of vital activities due to conversion of essential cells from functional to nonfunctional forms within tumors. Solid tumors destroy surrounding normal tissues by compression and interference with blood supply and nerve function. They may also break through barriers such as the outer skin, internal membranes, epithelia, or gut wall. The breakthroughs cause bleeding and infection and destroy the separation of body compartments necessary for normal function. Both compression and breakthroughs cause pain that, in advanced cases, may become extreme.

Tumors of blood cells convert dividing cell lines from functional types to nonfunctional forms that crowd the bloodstream but are unable to carry out required activities such as oxygen transport or the immune response. Some tumors of glandular tissue upset bodily functions through excessive production and secretion of hormones. When the total mass of tumor tissue becomes large, the demands of the actively growing and dividing cancer cells for nutrients may deprive normal cells of their needed supplies, leading to generally impaired body functions, muscular weakness, fatigue, and weight loss.

Tumors with these destructive characteristics are described as *malignant*. Not all tumors are malignant; some types of unprogrammed tissue growth, such as common skin warts, usually cause no serious problems to their hosts and are therefore classified as benign (see Table 22-6). Benign tumors, in contrast to the malignant types, grow relatively slowly and do not invade surrounding tissues or metastasize. Often, benign tumors are surrounded by a fibrous capsule of connective tissue that prevents or retards expansion or breakup. Some initially benign tumors, including even common warts, however, may change with time to take on malignant characteristics.

Individual cells of a malignant tumor are set apart from cells in normal tissues by differences in biochemistry, physiology, and structure (Fig. 22-20). First and foremost is the characteristic of uncontrolled division. Cancer cells typically cycle through DNA replication and mitosis much more rapidly than their normal counterparts. Cancer cells also fail to differentiate to fully

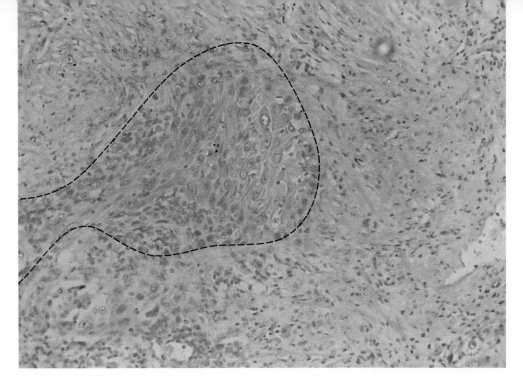

Figure 22-20 A mass of tumor cells (dashed line) embedded in normal tissue. Tumor cells typically have less cytoplasm than normal cells, making the tumor cells appear more densely colored in preparations stained for light microscopy. ×270. Courtesy of P. Chambon. Reprinted by permission from *Nature* 348:699. Copyright © 1990 Macmillan Magazines, Ltd.

mature and functional form and typically lose tight attachments to their neighbors and become mobile. Within a tumor cell disruptions of the cytoskeleton are frequently visible, and biochemical changes characteristic of dividing cells become established: high metabolic rates, increases in protein phosphorylation, raised cytoplasmic concentrations of Na^+, K^+, and Ca^{2+}, and elevated pH. Often chromosomal abnormalities are present, including extra chromosomes, missing chromosomes, exchanges of segments between chromosomes, and breakage.

Much has been learned about the biochemical and structural characteristics of cancer cells by the study of cells in culture. Normal body cells placed in culture eventually become quiescent and stop dividing. In contrast tumor cells, or cells that have undergone some of the genetic and other alterations leading to development of the cancerous condition, continue to divide indefinitely. Such cells are said to be transformed. Among the changes noted in transformed cultured cells, in addition to continued division, are increased metabolic rates and transport of substances across the plasma membrane; loss of cytoskeletal structures and reduced attachment to other cells, leading to a more rounded shape; alterations in structure and density of surface carbohydrate groups; partial or complete loss of differentiation; loss of contact inhibition of movement; and often the ability to grow and divide without a requirement for added growth factors. (Nontransformed cells usually require added growth factors to divide in culture.) Some transformed cells become self-sustaining by secreting their own growth factors. Transformed cells may show from a few to all of these characteristics.

Cultured cells are frequently used as test objects evaluating the ability of chemicals, radiation, or viral infections to induce transformation. The degree of transformation by the exposure can be evaluated by injecting the cells into a test animal; highly transformed cells usually divide rapidly to form malignant tumors.

Establishing the Genetic Basis of Cancer

Hints that cancer might have a basis in altered gene activity began to appear more than 170 years ago. In 1820 a British physician, Sir William Norris, noted that *melanoma*, a cancer involving pigmented skin cells, was especially prevalent in one family under study. More than 40 kinds of cancer, including common types such as cancer of the breast and colon, have since been noticed to occur more frequently in some families than in others.

Other indications that cancer has a genetic basis were taken from the fact that the chromosomes of many tumor cells show gross rearrangements that are readily visible under the light microscope. These observations suggest that although cancers themselves are rarely directly inherited, a tendency to develop cancer could be hereditary. The tendency might be expressed in terms of unstable sites in genes or other segments of the genome that lead to chromosome breakage or rearrangements.

These indications were put on a firmer foundation by research with tumors caused by viruses infecting animal cells, most notably those caused by retroviruses. Cancer-causing retroviruses were found to carry genes capable of transforming normal cells to the malignant state. The transforming genes were at first thought to be purely viral in origin. However, sequencing and other molecular approaches revealed that the viral oncogenes, including v-*abl*, v-*erbA*, v-*erbB*, v-*ets*, v-*jun*, v-*mos*, v-*myb*, v-*myc*, v-*ras*, and v-*sis*, were derived from genes normally present in cells—the proto-oncogenes first discovered by Bishop and Varmus.

The discovery of altered cellular genes in cancer-inducing retroviruses prompted a search for similar genes in nonviral cancers. Much of this work was accomplished by transfection experiments. Many of the oncogenes identified by transfection turned out to have counterparts in the oncogenes carried by retroviruses, confirming by a different route that these genes are capable of contributing to the development of cancer.

Viruses and Cancer

Viral infections contribute to many forms of cancer by introducing oncogenes or other genes promoting division, or by suppressing genes that inhibit cell division. Two viral groups—retroviruses and DNA tumor viruses—are linked directly to the generation of cancer. The tumors caused or promoted by viruses are common enough to account for as much as 20% of cancer worldwide.

Retroviruses The retroviral genome, which is encoded in RNA molecules in free viral particles, is integrated into host cell genomes in the form of a DNA copy (see p. 767). All mammalian individuals, including humans, are likely to contain several to many infecting retroviruses integrated into their DNA. Once established in the DNA of the reproductive cells, the retroviral DNA is passed from generation to generation as a more or less permanent part of the DNA complement of each cell (for details, see p. 767). Rarely, retroviruses carry genes capable of inducing uncontrolled cell division and many of the remaining characteristics of malignant cells. The division-inducing genes provide an evolutionary advantage to the retroviruses carrying them because the cell lines they infect grow in large numbers and thereby increase the capacity of the virus for reproduction and further infections. The included genes are particularly advantageous to a retrovirus if the growth they induce is slow enough to permit extended survival of the host as an agent causing further infections.

DNA Tumor Viruses The *DNA tumor viruses* contain DNA as their hereditary material. Although these viruses may or may not integrate into the host cell genome as part of their cycles of infection, integration usually occurs when they are associated with cancer. Unlike retroviruses, none of the DNA tumor viruses contain oncogenes that are recognizable as abnormal forms of host cell genes. Instead, the genes causing uncontrolled growth in these viruses appear to be of purely viral origin.

Many DNA tumor viruses are members of the most common viral groups infecting humans and other mam-

Table 22-7 Viral Groups in Which DNA Tumor Viruses Occur

Viral Group	Tumor Type
Adenoviruses	Tumors usually induced only in species other than natural host
Hepadnaviruses animal hepatitis (woodchuck, duck, ground squirrel)	Liver cancer (?)
human hepatitis B	Liver cancer (?)
Herpesviruses cytomegalovirus Epstein–Barr virus	Burkitt's lymphoma, nasopharyngeal carcinoma
herpes simplex I herpes simplex II	Cervical cancer (?)
Papillomaviruses	Skin warts, venereal warts, cervical carcinoma in humans, alimentary carcinoma in cattle
Papoviruses BK JC polyoma SV40	Various cancers in host and other species

mals (Table 22-7). Some of the more than 50 known *papillomaviruses* are relatively harmless and cause benign tumors such as skin and venereal warts in humans. Other papillomaviruses, however, have been implicated in cervical carcinomas, which are a leading cause of worldwide cancer deaths among women. About 90% of cervical carcinoma cells are found to be infected by one or both of the papillomaviruses *HPV-16* and *HPV-18*. Some of the normally benign tumors induced by papillomaviruses can also become malignant if exposed to other factors promoting cancer, such as ultraviolet light or X-rays.

Several members of another group, the *herpesviruses*, are suspected but not as yet proved to be responsible for human cancers. Two of these viruses, *herpes simplex I* and *II*, are universally spread among human populations. Herpes simplex I, in particular, may infect more than 90% of the people in some regions. In most persons herpes simplex I causes lesions (cold sores) in skin or mucous membranes that are painful but otherwise not dangerous. Herpes simplex II, which causes genital lesions, is frequently present in cervical tumor cells, however. It is not presently known whether herpes simplex II contributes to cancer of the cervix or is present in most cervical tumors simply because it infects so much of the population.

Links to cancer are stronger for another herpesvirus, the Epstein–Barr virus. Most people become infected with this virus very early in life and show no symptoms. However, some infected individuals develop cancer. The Epstein–Barr virus has been implicated as a cause of one form of Burkitt's lymphoma that is common in parts of Africa; a nasopharyngeal carcinoma linked to infections by the virus is common in China. Cancer is apparently more likely to be associated with Epstein–Barr viral infections if the immune system of the affected individual is suppressed by factors such as extreme stress or other infections.

Some DNA tumor viruses induce malignant tumors primarily or exclusively when they infect organisms other than their normal hosts. The *adenoviruses* fall into this category. Most humans experience several adenovirus infections during their lifetimes. Typically, the infections cause respiratory and intestinal upsets but not cancer. The human adenovirus, however, can induce malignant tumors when injected into newborn hamsters. The *hepadnaviruses*, which are responsible for human diseases such as hepatitis B, can also cause tumors in nonhuman hosts. In this case, however, there are indications that the viruses in this group may be responsible for human cancers as well. For example, persons with lifelong, residual hepatitis B infections are several hundred times more likely to contract liver cancer than uninfected individuals, particularly if they develop cirrhosis of the liver from alcoholism or other causes. More than 250 million people worldwide are estimated to harbor chronic hepatitis B infections. The appearance of cancer related to the infection in these people may be delayed by as much as 20 to 30 years after initial exposure to the virus. The *papovaviruses*, another DNA virus group, includes the SV40 virus of monkeys. SV40 is not associated with cancer in its natural monkey hosts, but it is capable of transforming cultured mouse and human cells to malignant types, especially if oncogenes such as *ras* are also present.

The growth-inducing genes carried by DNA tumor viruses, as noted, have no obvious counterparts among the gene complement of their host cells. SV40, for example, possesses a gene encoding a series of proteins known as *T antigens*, which promote DNA replication and cell division. This gene is capable of inducing DNA replication when included in DNA transfected into a suitable cell type. At least some effects of the SV40 T antigens are exerted through their ability to combine with and inactivate the products of tumor suppressor genes, including *rb* and *p53*. Similarly, the adenovirus *E1A* gene is capable of inducing cell division in transfection experiments because the protein encoded in this gene combines with and inactivates the *rb* gene product. When combined with the division-inducing effects of the E1A protein, the product of a second adenovirus gene, *E1B*, is able to transform cultured cells into fully malignant types by binding and inactivating p53.

Why many DNA tumor viruses cause cancer only in species not acting as their normal hosts is unknown. It is possible that in their normal hosts, the effects of the viral genes are balanced and kept in check by host cell genes that have evolved this function. In nonhost species these genes may enter a cellular environment in which the controlling genes are missing or have activities that differ sufficiently from the normal host types to release the viral genes from control. In this situation the resulting uncontrolled cell division is a significant step toward the development of a malignant tumor. Often, immune suppression is a contributing factor in cancers induced by DNA tumor viruses.

Multistep Progression from Initiation to Malignancy

Fully malignant cancer cells do not usually develop from a single cause. Multiple alterations, sometimes requiring many years to appear, are necessary for full establishment of malignancy. This feature is known as the *multistep progression* of cancer. In most cases the complete sequence of steps leading from an initiating alteration to full malignancy is unknown.

An initiating genetic alteration may be induced by a long list of factors, including exposure to radiation or certain chemicals (see below), insertion of a retrovirus, generation of random mutations during replication, random DNA amplification, or random loss of a DNA segment.

The initial changes may also be inherited. A small percentage of human cancers, about 5%, show a strong genetic predisposition. Depending on the gene and cancer type, as many as 100% of the persons receiving the oncogenes or faulty tumor suppressor genes involved are likely to develop cancer. Among these strongly predisposed cancers are familial retinoblastoma, familial adenomatous polyps of the colon (FAP), and multiple endocrine neoplasia, in which tumors develop in the thyroid, adrenal medulla, and parathyroid glands. In addition to the strongly predisposed cancers, some cancers, including breast, ovarian, and colon cancers other than FAP, show a disposition in some family lines—members of these families show a greater tendency to develop the cancer than individuals in other families.

Subsequent steps from the initiating change to a fully malignant state usually include conversion of additional genes to oncogenic form. Also important during intermediate stages are further alterations to the initial and succeeding genes that increase their activation. The initial conversion of a proto-oncogene to an oncogene by translocation, for example, may be compounded at successive steps by sequence changes and amplification. These steps are usually driven by the same sources of change responsible for the initiating step. Because the first change may have induced an increase in the rate of DNA replication and cell division, opportunities for additional changes are likely to become more frequent as progression advances. Progression is also advanced by alterations that reduce the activity of tumor suppressor genes. These changes may be as important, or even more important, to the full development of cancer as activation of oncogenes.

At any stage along the way, a change required to advance progression toward the fully malignant state may happen soon after a previous change or only after a considerable passage of time. Or, it may not occur at all, leaving the transformation at an intermediate stage without development of full malignancy within the lifetime of the individual. During the intermediate stages avoidance of factors that promote genetic change, including exposure to tobacco smoke or radiation sources such as X-rays, may delay or prevent progression to full malignancy.

The last stage in progression to malignancy is often metastasis. The separation and movement may result from the development of motility by the cancer cells or may occur through penetration into elements of the circulatory system, which carry the malignant cells throughout the body. Movement to secondary locations may be followed by growth of the displaced cells into additional tumors at these sites.

Metastasis is promoted by changes during transformation that alter cell surface molecules in tumor cells and break connections to other cells, or to molecules of the extracellular matrix such as collagen, fibronectin, laminin, and glycosaminoglycans (see Ch. 7). Often, changes in cytoskeletal elements linked to the plasma membrane, such as microfilaments and microtubules, are associated with metastasis. We have noted, for example, that phosphorylations catalyzed by the *src* protein kinase reduce the affinity of the fibronectin receptor for this extracellular matrix protein; cells with *src* and some other oncogenes also secrete proteinases that break connections to the extracellular matrix and aid their movement into and through surrounding tissues.

Relatively few of the cells breaking loose from a tumor survive their passage through the body to lodge and grow in new locations. Most metastasizing cells are destroyed by various factors, including deformation by passage through narrow capillaries and destruction by blood turbulence around structures such as the heart valves and vessel junctions. Tumor cells often develop changes in their MHC's that permit their detection and elimination by the immune system, particularly in the bloodstream. Unfortunately, the rigors of travel may act as a sort of natural selection for the cells that are most malignant, that is, those most able to grow uncontrollably and spread by metastasis. Also, the surviving cells are often those most resistant to methods such as chemotherapy employed to treat cancer.

Some of the genetic alterations taking place in the multistep progression of specific cancers have been traced. For example, B. Vogelstein and his coworkers reconstructed a sequence of alterations leading to ma-

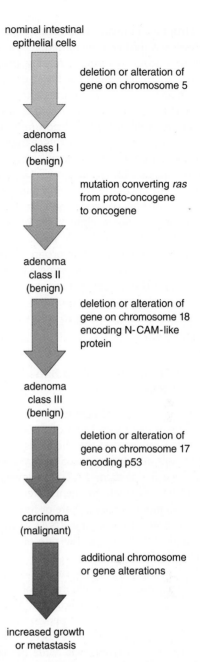

Figure 22-21 One of several possible sequences of events in the multistep progression of colorectal cancer. The genes altered on chromosomes 5 and 17 are tumor suppressor genes.

lignancy in the development of colorectal cancer (Fig. 22-21). The proposed sequence, one of several possible pathways for progression of this cancer, begins with a deletion leading to loss of a gene on chromosome 5 of the human set. This gene is missing on one of the two chromosomes in pair 5 as a hereditary deficiency in some persons; these individuals have a high incidence of colorectal cancer. Loss of the gene leads to limited proliferation of intestinal epithelial cells, producing a benign tumor known as an *adenoma class I* that grows as a polyp from the intestinal epithelium. At some point the progression is furthered by mutation of a *ras* proto-oncogene to oncogene form, increasing proliferation of the tumor to a still benign but faster-growing form known as an *adenoma class II.* Progression continues as a gene on chromosome 18 encoding a cell surface glycoprotein resembling cell adhesion molecules such as N-CAMs (see p. 254) is deleted; this change advances the adenoma to *class III,* which is still benign. Finally, an alteration appears on chromosome 17 mutating or deleting the gene encoding the tumor suppressor protein p53. Loss of p53 converts the class III adenoma into a malignant carcinoma; eight to ten further genetic changes increase the capability of the carcinoma cells to break loose and migrate. Other progressions leading to colorectal cancer involve the same alterations in a different order or distinct gene changes. The alteration in the gene carried on chromosome 5, however, is apparently the first step leading to colorectal cancer in most progressions.

The Role of Chemicals and Radiation in Multistep Progression

Many natural and artificial agents initiate or promote the development of cancer. Most of these agents, collectively called *carcinogens,* are chemicals or forms of radiation capable of inducing chemical changes in DNA (for details, see Supplement 23-2). However, carcinogens are not limited solely to agents altering DNA; in some cases carcinogens may initiate or further the progression of cancer by modifying RNAs or proteins or may act simply by increasing the rate of DNA replication and cell division.

The first evidence that exposure to certain chemicals could promote the development of cancer was recognized more than 200 years ago. In 1761 an English physician, John Hall, noted that persons taking snuff through the nose were especially susceptible to nasal cancer. A similar association was made between clay pipe smoking and lip cancer in 1795. During the same period a connection was also noted between the tarry chemicals encountered by chimney sweeps and scrotal cancer. Various industrial operations, such as asbestos mining, the preparation of asbestos products, and the extraction and purification of heavy metals, were also found to place workers at high risk of cancer. (Table 22-8 lists some of the industrial chemicals linked to cancer.) The first clue that some forms of radiation could induce cancer was discovered at about the turn of the century, when X-ray technicians were noted to develop skin and other cancers with unusually high frequency.

Experiments with test animals directly supported the idea that these and other agents could cause cancer. The first of these efforts was carried out in China in 1915. In this experiment coal tars suspected to produce cancer were rubbed on the ears of rabbits. The rabbits

Table 22-8 Industrial Chemicals Linked to Cancer	
Industry or Chemical	Type of Cancer
Arsenic and arsenic compounds	Lung, skin
Asbestos mining and manufacturing	Membranes lining chest and body cavities, lungs
Benzene production	Leukemia
Chromium and chromium compounds	Lung
Ether production or exposure	Lung
Furniture manufacturing solvents	Nasal cavity
Mining with radon exposure	Lung
Nickel refining	Nasal cavity, lung
Radioactive isotopes	Leukemias
Shoe manufacturing solvents	Nasal cavity, urinary bladder
Solvents used in making tires and other rubber products	Urinary bladder
Soots, tars, oils, and smoke exposure	Skin, lung
Vinyl chloride production and products	Liver, mesenchyme

Adapted from D. H. Phillips, in *The Molecular Basis of Cancer* (P. B. Farmer and J. M. Walker, eds.), p. 133. Copyright © 1985 John Wiley and Sons, Inc.

developed tumors of the same types noted in persons exposed to the coal tars. These observations were followed by many experiments along the same lines. In general this approach, which continues today, confirms that many chemicals and forms of radiation can induce cancer in humans and other animals.

The Characteristics of Chemical Carcinogens   The most directly potent chemical carcinogens (Figs. 22-22 and 22-23) are *alkylating agents*—substances capable of substituting a monovalent organic group such as a methyl or ethyl group for a hydrogen atom. Other molecules, such as many of the *polycyclic hydrocarbons,* are not directly carcinogenic in themselves. Instead, chemical derivatives of these substances, produced by natural biochemical processes in the body, convert them to cancer-inducing molecules. Ironically, one of the primary routes for conversion of substances such as the polycyclic hydrocarbons into cancer-causing derivatives is detoxification, carried out primarily in the liver (see p. 827). Detoxification reactions normally oxidize poisons into nontoxic, soluble substances that can readily be cleared from cells and excreted. In the case of polycyclic hydrocarbons, detoxification converts these

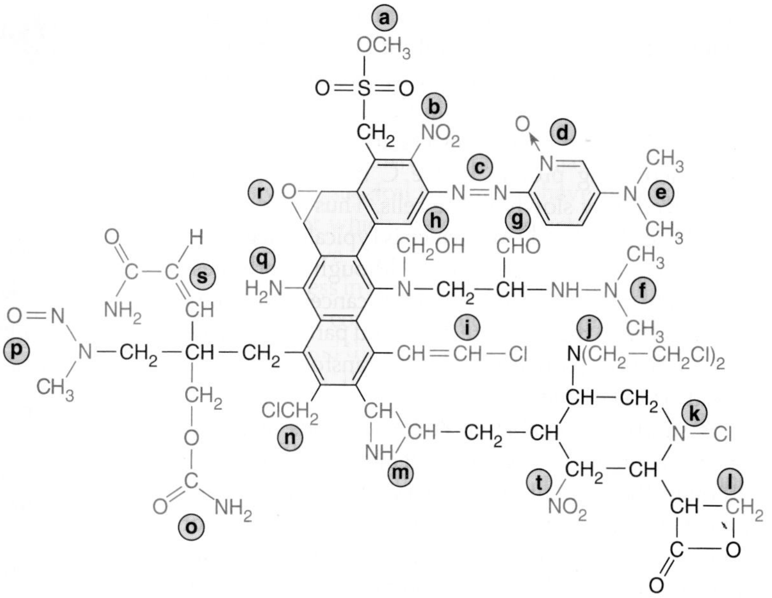

Figure 22-22 A hypothetical molecule showing chemical groups known or suspected to be carcinogenic (in red). **(a)** Alkyl esters of phosphonic or sulfonic acids; **(b)** aromatic nitro groups; **(c)** aromatic azo groups; **(d)** aromatic ring N-oxides; **(e)** aromatic mono- and dialkylamino groups; **(f)** alkyl hydrazines; **(g)** alkyl aldehydes; **(h)** N-methylol derivatives; **(i)** monohaloalkenes; **(j)** N and S mustards; **(k)** N-chloramines; **(l)** propiolactones and propiosultones; **(m)** aromatic and aliphatic aziridinyl derivatives; **(n)** aromatic and aliphatic substituted primary alkyl halides; **(o)** urethane derivatives (carbamates); **(p)** alkyl N-nitrosamines; **(q)** aromatic amines and their derivatives; **(r)** aliphatic and aromatic epoxides; **(s)** α, β-unsaturated carbonyl structure; **(t)** aliphatic nitro group. Adapted from an original courtesy of R. W. Tennant, from *Mutat. Res.* 257: 209 (1991).

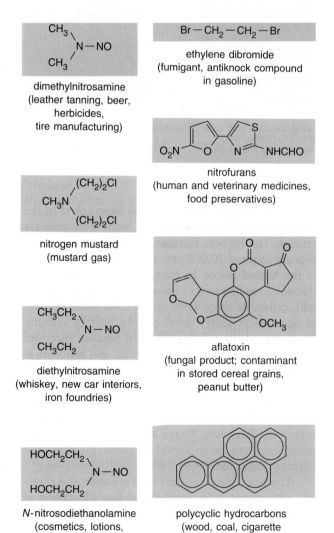

Figure 22-23 Some common chemical carcinogens and their sources.

dimethylnitrosamine (leather tanning, beer, herbicides, tire manufacturing)

nitrogen mustard (mustard gas)

diethylnitrosamine (whiskey, new car interiors, iron foundries)

N-nitrosodiethanolamine (cosmetics, lotions, shampoos)

ethylene dibromide (fumigant, antiknock compound in gasoline)

nitrofurans (human and veterinary medicines, food preservatives)

aflatoxin (fungal product; contaminant in stored cereal grains, peanut butter)

polycyclic hydrocarbons (wood, coal, cigarette smoke)

usually insoluble molecules into soluble derivatives that are highly reactive inside cells. Polycyclic hydrocarbons, incidentally, are the primary carcinogens of tobacco and other smokes produced by the incomplete combustion of organic substances or fossil fuels. Tobacco smoke, the most frequent cause of cancer, is estimated to be responsible for one-third of cancer cases in the United States.

The *phorbol esters* provide an interesting example of a group of chemicals that promotes the growth of cancer cells rather than inducing mutations. These molecules resemble the DAG second messenger generated by the InsP$_3$/DAG pathway (Fig. 22-24; see also Figs.

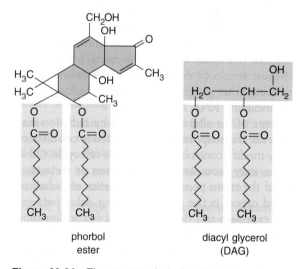

phorbol ester

diacyl glycerol (DAG)

Figure 22-24 The structural similarities between a phorbol ester and DAG, a second messenger of the InsP$_3$/DAG pathway. The similarities make the phorbol esters effective as stimulants of the InsP$_3$/DAG pathway, leading to promotion of division in some cell types.

Cancer: Characteristics and Causes **945**

Liotta, J. 1992. Cancer cell invasion and metastasis. *Sci. Amer.* 266:54–63.

Liotta, J.; Steeg, P. S.; and Stetler-Stevenson, W. G. 1991. Cancer metastasis and angiogenesis: an imbalance of positive and negative regulation. *Cell* 64:327–36.

Carcinogens

Ames, B. N. 1983. Dietary carcinogens and anticarcinogens. *Science* 221:1256–64.

Cohen, S. M., and Ellwein, L. B. 1990. Cell proliferation in carcinogenesis. *Science* 249:1007–11.

George, A. M., and Cramp, W. A. 1987. The effects of ionizing radiation on structure and function of DNA. *Prog. Biophys. Molec. Biol.* 50:121–69.

Henderson, B. E.; Ross, R. K.; and Pike, M. C. 1991. Toward the primary prevention of cancer. *Science* 254:1131–38.

Pitot, H. C., and Dragan, Y. P. 1991. Facts and theories concerning the mechanisms of carcinogenesis. *FASEB J.* 5:2280–86.

Singer, B., and Kusmierek, J. T. 1982. Chemical mutagenesis. *Ann. Rev. Biochem.* 51:655–93.

Review Questions

1. Outline the major stages of the cell cycle. What major events take place in G1? In S? G2? Compare the total DNA quantity per cell in G1 and G2 with that of a gamete in a diploid organism. What stage is most variable in duration?

2. What is start or the restriction point of the cell cycle? At what stage are genes likely to be activated that trigger entry into S?

3. Outline the major events taking place in prophase, metaphase, anaphase, and telophase of mitosis. What events mark the beginning and end of each mitotic stage? What features of mitosis ensure that the products of division receive exactly the same complement of chromosomes?

4. What are chromosomes and chromatids? At what stages do chromatids exist in the cell cycle? When are chromatids converted to chromosomes? Diagram the steps in mitotic division of a haploid and a diploid cell.

5. Compare the major events of cytoplasmic division in plants and animals. What role does the spindle play in cytoplasmic division?

6. What is the G0 stage of the cell cycle? At what other stages may the cell cycle arrest? What evidence indicates that replication, mitosis, and cytoplasmic division are potentially separable?

7. What major systems regulate cell division through receptors on the cell surface? Outline the operation of the systems. What are G proteins? How do they operate in the reaction cascades triggered by cell surface receptors? What are second messengers? How are second messengers generated in the reaction systems triggered by cell surface receptors? What are protein kinases? What is their role in the pathways linked to surface receptors?

8. What is density-dependent inhibition of cell division? How is this mechanism related to receptors on the cell surface?

9. Outline the mechanism by which steroid hormone receptors may regulate cell division.

10. Outline the changes observed in concentrations of the cyclic nucleotides during the cell cycle. How might these changes be related to cell surface receptors?

11. Outline the changes noted in ion transport and pH during the cell cycle. How might these changes be related to cell surface receptors?

12. What alterations are noted in histone and nonhistone chromosomal proteins during the cell cycle? How might these alterations be related to cell surface receptors?

13. What are polyamines? How are polyamines synthesized? Why are polyamines believed to be involved in modification or regulation of the cell cycle?

14. What evidence relates cell volume to cell cycle regulation? What natural systems bypass regulation by cell volume?

15. How are temperature-sensitive mutants used in the study of cell cycle genes? What major functions are carried out by genes affecting the cell cycle? List one yeast gene operating in each of the functional categories.

16. What is MPF? $p34^{cdc2}$? Cyclin? How do $p34^{cdc2}$ and cyclin interact in regulation of the G2/mitosis transition? What role do phosphorylations play in this interaction? What evidence indicates that MPF is widely distributed among eukaryotes?

17. What primary methods have been used to identify oncogenes? What types of cellular activities are carried out by oncogenes? Give examples of oncogenes carrying out these activities.

18. What are proto-oncogenes? What is the relationship between oncogenes and proto-oncogenes? What alterations convert proto-oncogenes into oncogenes? What are recessive oncogenes or tumor suppressor genes? What characteristics distinguish the operation of recessive and other oncogenes in cancer cells?

19. How are chromosome translocations related to the development of cancer? Why are translocations significant in the development of blood cancers?

20. What major characteristics of malignant tumors are responsible for their destructive effects? What is the relationship of viral infections to human cancer?

21. What is multistep progression? What factors are likely to initiate progression of cells from a normal to malignant state? What is the relationship of tumor suppressor genes to the development of cancer? What is metastasis? What factors promote metastasis?

22. What are carcinogens? How are carcinogens believed to promote the development of cancer? What agents in the environment may have a carcinogenic effect?

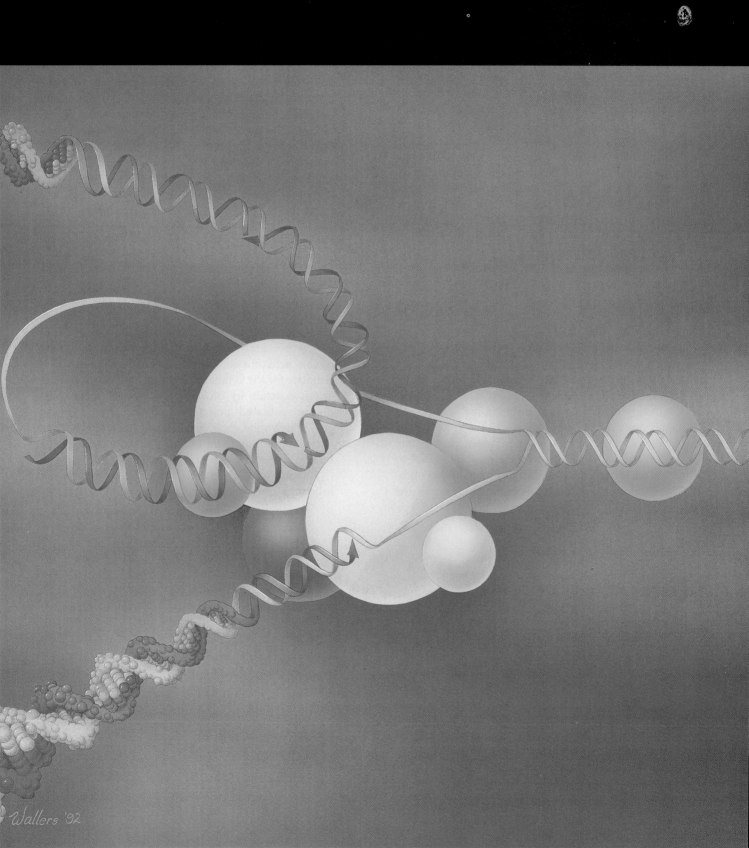

Wallers '92

- *DNA replication* ▪ *Unwinding and priming*
▪ *Unidirectional replication* ▪ *DNA polymerases and other enzymes of replication* ▪ *Proofreading*
▪ *Telomere replication* ▪ *Replication origins and replicons* ▪ *DNA amplification* ▪ *Mismatch repair*
▪ *Other types of DNA repair* ▪ *Duplication of chromosomal proteins* ▪ *Nucleosomes and replication* ▪ *Replication in mitochondria and chloroplasts*

During the S phase of interphase, cells replicate their DNA in preparation for mitotic and meiotic division. The replication, in which the parental DNA molecule is duplicated into two exact, sequence-by-sequence copies, is remarkable for its almost perfect fidelity. Once replication is complete, many of the few mistakes that do occur are corrected by repair mechanisms that scan the DNA to detect base mismatches and other irregularities. The result is a nearly perfect duplication of the genetic information of the parental cell, ready to be divided and parceled into daughter cells by the division mechanisms.

Mistakes do slip by the DNA replication and repair mechanisms in very small numbers. These mistakes persist as *mutations*, which constitute any difference in sequence from the parental template that appears and persists in the replicated copies. The few mistakes remaining after DNA replication and repair are highly important to the evolutionary process because they are the ultimate source of the variability acted upon by natural selection.

During the S phase the chromosomal histone and nonhistone proteins are duplicated as well as the parental DNA. The histone and nonhistone proteins are synthesized and combined with the replicated DNA to reproduce the quantities and arrangement of these proteins in the G1 chromosomal parent. Synthesis of the chromosomal proteins is almost as significant to cell reproduction as DNA replication because this duplication preserves the patterns of genetic regulation and cell differentiation imposed by these proteins.

Of the enzymatic mechanisms catalyzing chromosome duplication, only those carrying out DNA replication and repair are understood in any detail. Many uncertainties surround the synthesis of chromosomal proteins and preservation of their arrangement with the DNA.

This chapter describes the enzymatic mechanisms replicating DNA and repairing errors that arise during replication. Other sources of errors and structural deficiencies in the DNA are described, along with the mechanisms detecting and correcting these lesions and the characteristics of the few errors that persist as mutations. The problems of duplication of the chromosomal proteins and their solution in eukaryotes are also considered. DNA replication in mitochondria and chloroplasts is outlined in Supplement 23-3.

SEMICONSERVATIVE DNA REPLICATION

When J. D. Watson and F. H. C. Crick discovered the molecular structure of DNA in 1953, they pointed out that a possible replication mechanism is inherent in its structure (Fig. 23-1; see Ch. 14 for details of Watson and Crick's discovery and the molecular structure of DNA). Because the two nucleotide chains of a DNA double helix are complementary, each can serve as a template for the synthesis of the missing half when the chains separate by unwinding. The mechanism, producing replicated molecules that each consist of one "old" nucleotide chain used as the template and one "new" nucleotide chain assembled on the template, is termed *semiconservative replication* (Fig. 23-2a).

Soon after Watson and Crick made their discovery, it became apparent that the semiconservative pathway

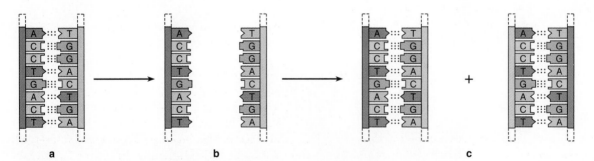

Figure 23-1 Semiconservative DNA replication. The two nucleotide chains of a DNA double helix **(a)** unwind **(b)**; each half of the original molecule serves as a template for a complementary chain assembled according to the base pairing rules. The result is two DNA molecules that are exact duplicates of the original molecule entering replication **(c)**.

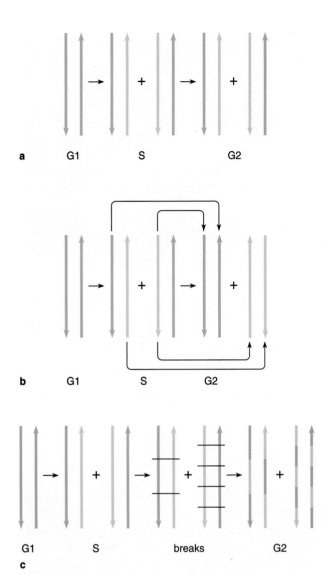

a G1   S   G2

b G1   S   G2

G1   S   breaks   G2
c

Figure 23-2 Three hypothetical pathways for DNA replication. Template chains are shown in dark blue and copies in light blue. **(a)** Semiconservative replication, in which each copy chain remains wound with its template after replication, producing two daughter molecules that are one-half "old" and one-half "new." **(b)** Conservative replication, in which the parental chains rewind into an all "old" helix after replication; the copy chains wind together into an all "new" DNA chain. **(c)** One possibility for dispersive replication, in which neither "old" nor "new" chains remain intact. All living organisms replicate DNA by the semiconservative pathway.

was only one of three possible mechanisms for the replication process. In one of the alternate pathways, *conservative replication* (Fig. 23-2*b*), the two nucleotide chains of the original molecule, after unwinding and serving as templates, would rewind into an all "old" molecule. The two new nucleotide chains would sep-

arate from their templates and wind together into an all "new" molecule. The terms *semiconservative* and *conservative* imply that in both pathways the original nucleotide chains, although separating and acting as templates, remain intact

In the third possible pathway, *dispersive replication* (Fig. 23-2*c*), the continuity of the original DNA chains would be broken and the parts distributed randomly between the product molecules. One mechanism by which this might occur is shown in Figure 23-2*c*. In this case numerous breaks appear in the original nucleotide chains before they unwind and separate. Distribution of the broken chains between the replicated molecules is random.

The definitive experiment demonstrating that replication is semiconservative in prokaryotes was conducted by W. Meselson and F. W. Stahl in 1958 with the bacterium *E. coli*. (Supplement 23-1 outlines an equivalent experiment with a eukaryote, the plant *Vicia*, by J. H. Taylor.) Meselson and Stahl grew *E. coli* in a medium containing a heavy isotope of nitrogen (^{15}N) for several generations, long enough for the DNA to become completely labeled. Then the bacteria were transferred to a ^{14}N medium. At the time of transfer, and at intervals afterward, DNA extracts were made and analyzed by buoyant density centrifugation (see p. 125). DNA extracted at the time of transfer from ^{15}N to ^{14}N medium centrifuged into a single band characteristic of ^{15}N DNA (generation 0 in Fig. 23-3). After an interval sufficient for most of the cells to replicate once (generations 0.3 to 1.0 in Fig. 23-3), the ^{15}N band disappeared, and a new band appeared that was intermediate in density between ^{15}N and ^{14}N DNA. After longer periods of growth (generations 1.1 to 4.1 in Fig. 23-3), a new, less dense band characteristic of pure ^{14}N DNA began to appear. The intermediate ^{15}N-^{14}N band did not disappear, however.

These results are consistent with semiconservative but not conservative replication (Fig. 23-4). If replication were conservative, as in Figure 23-2*b*, two distinct bands, one characteristic of ^{14}N and one of ^{15}N DNA, would appear in the centrifuge after one cell generation. In this case the two ^{15}N nucleotide chains of the replicating DNA would separate and serve as templates for copies made with ^{14}N DNA. After replication the original, old ^{15}N nucleotide chains would reassociate, as would the new ^{14}N chains. Therefore, after conservative replication the two DNA products would be identifiable after one generation as separate ^{15}N and ^{14}N bands. The production of a single band of intermediate density after one generation, actually observed by Meselson and Stahl, showed that all the DNA molecules were hybrids containing one ^{15}N and one ^{14}N chain. This result eliminated conservative replication as a possibility and was the expected result if replication is semiconservative, as in Figure 23-2*a*.

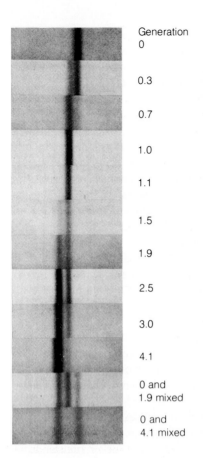

Generation
0

0.3

0.7

1.0

1.1

1.5

1.9

2.5

3.0

4.1

0 and
1.9 mixed

0 and
4.1 mixed

Figure 23-3 The results of centrifugation of *E. coli* DNA extracted at various times after actively dividing cells were transferred from a growth medium containing ¹⁵N to one containing ¹⁴N. Density increases toward the right. Bands displaced toward the left therefore represent lighter DNA. The three bands in the next-to-lowest frame correspond, reading from left to right, to ¹⁴N-¹⁴N DNA, ¹⁴N-¹⁵N hybrid DNA, and ¹⁵N-¹⁵N DNA. Generation 0 represents DNA at the time of transfer from ¹⁵N to ¹⁴N medium. After one round of replication (generation 1.0), all the DNA is of intermediate density, consistent with molecules containing one ¹⁴N and one ¹⁵N chain. This result ruled out conservative replication, which would have produced two bands, one at the ¹⁴N level and one at the ¹⁵N level. After two rounds of replication (generation 1.9 to 2.5) there are two distinct bands, one hybrid ¹⁴N-¹⁵N helix and one all new ¹⁴N-¹⁴N helix. This result is consistent with semiconservative replication but rules out dispersive replication. Semiconservative replication is therefore the only pathway consistent with the results. Courtesy of J. Meselson and F. W. Stahl.

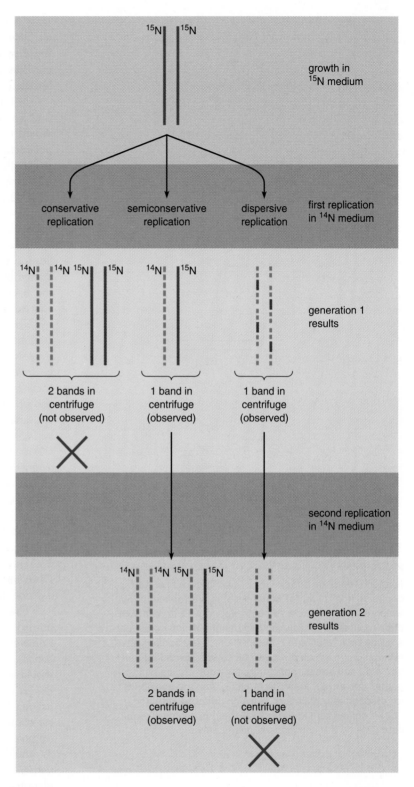

Figure 23-4 The combinations of ¹⁴N and ¹⁵N DNA and the number of bands expected in the centrifuge if replication is conservative, semiconservative, or dispersive.

The same outcome could result from dispersive replication. However, the distribution of density observed in the centrifuge after two generations ruled out this possibility. If replication were dispersive, the DNA would be expected to remain intermediate in density after two generations, because little pure ^{14}N and ^{15}N DNA would be retained or produced. But after two generations, Meselson and Stahl actually observed a pure ^{14}N band, in addition to the intermediate ^{14}N-^{15}N hybrid band. This was exactly the result expected if replication is semiconservative. All the DNA molecules after the first replication, according to the semiconservative mechanism, should contain one ^{15}N and one ^{14}N chain. At the second replication these separate and serve as templates for the synthesis of complementary ^{14}N chains. Because the templates remain paired with the new chains, one set of product molecules should consist of ^{14}N-^{15}N DNA and one of ^{14}N-^{14}N DNA. This explains the appearance of the new ^{14}N band and the persistence of the ^{14}N-^{15}N hybrid band after two generations of growth. The results obtained by Meselson and Stahl in bacteria were therefore compatible only with the semiconservative pathway of DNA replication. Later experiments demonstrated that all prokaryotes and eukaryotes replicate their DNA by the semiconservative pathway.

THE REACTIONS OF DNA REPLICATION

An Overview of Replication

The process by which a template DNA molecule is copied in replication resembles RNA transcription, but with several important exceptions. The assembly of individual DNA nucleotides into a chain is catalyzed by a group of enzymes known as *DNA polymerases.* The enzymes use as raw materials the four DNA nucleotides in the form of nucleoside triphosphates: *deoxyadenosine triphosphate (dATP), deoxyguanosine triphosphate (dGTP), deoxycytidine triphosphate (dCTP),* and *thymidine triphosphate (TTP).* These nucleotides differ from their counterparts in RNA synthesis by the presence of deoxyribose rather than ribose as their five-carbon sugar (indicated by the *d* in front of the triphosphate; see Fig. 2-31) and the use of thymidine triphosphate instead of uridine triphosphate in the assembly of nucleotide chains.

DNA polymerases differ fundamentally from RNA polymerases in their requirement for a *primer*—they can initiate synthesis only by adding nucleotides to the end of an existing nucleotide chain that acts as a primer for the reaction. RNA polymerases, in contrast, can put the first nucleotide of a new chain in place with no primer requirement. Although either DNA or RNA chains can act as primers for DNA replication, RNA is generally used in nature as the primer. At some point after DNA synthesis is initiated, the RNA primer is removed and replaced by DNA.

Polymerization of a DNA chain proceeds from the primer as shown in Figure 23-5. DNA polymerase binds to the exposed 3'-OH group of the primer (Fig. 23-5a) and recognizes the first base to be copied from the template. In Figure 23-5a the first template base is a guanine. According to the base-pairing rules, presence of a guanine at this site causes the enzyme to bind dCTP from the nucleoside triphosphates in the surrounding medium (Fig. 23-5b). All four nucleoside triphosphates collide and bind weakly to the DNA polymerase, but normally only the nucleotide forming the correct base pair with the template base, in this case dCTP, will proceed from the initial weak binding to tight binding. The tight binding holds the dCTP opposite the guanine of the template, in a position that favors formation of a covalent linkage between the 3'-OH group at the end of the primer and the innermost phosphate group bound to the 5'-carbon of the dCTP (Fig. 23-5c). The last two phosphates of the dCTP are split off, and the remaining phosphate is bound to the oxygen of the 3'-OH, creating a $3' \rightarrow 5'$ *phosphodiester linkage* (see p. 72) between the primer and the added nucleotide.

In response to formation of the first phosphodiester linkage, the enzyme moves to the next base on the DNA template, shown as a thymine in Figure 23-5. The enzyme then binds a dATP nucleotide from the medium and catalyzes formation of the second phosphodiester linkage. As before, the linking reaction splits off two phosphate groups from the nucleoside triphosphate most recently bound by the polymerase. The process then repeats, adding complementary nucleotides in succession into the growing DNA chain.

Each nucleotide added to an exposed 3'-OH group provides the 3'-OH group for the next assembly reaction. As a result, a 3'-OH group is always present at the newest end of the growing chain, and synthesis is said to proceed in a $5' \rightarrow 3'$ direction. All known DNA polymerases add nucleotides only in this direction.

Because the two chains of a DNA double helix are *antiparallel* (see p. 527), the newly synthesized DNA chain in Figure 23-5 runs in a direction opposite to the template chain. In the figure the 5' end of the newly synthesized chain is at the bottom, and its 3' end is at the top. The template chain is oriented with its 5' end at the top of the figure and its 3' end at the bottom.

The reactions adding nucleotides to a growing DNA chain thus proceed by the same pathway as RNA transcription except for the primer requirement. The entire reaction of DNA replication, however, differs fundamentally from RNA transcription in that the template DNA double helix must unwind and separate completely for replication to be semiconservative.

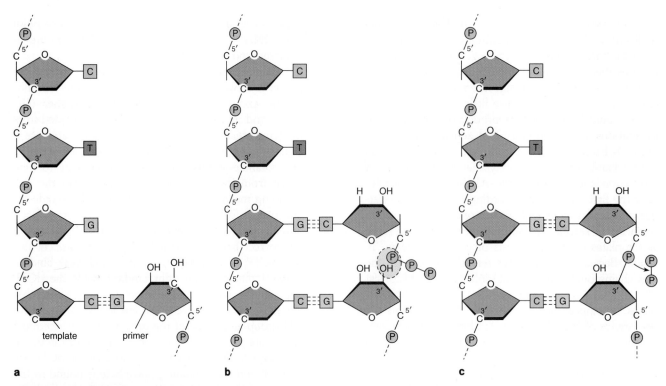

Figure 23-5 The polymerization reaction linking DNA nucleotides into a copy chain during replication. **(a)** DNA polymerase binds at the 3' end of the primer and recognizes the first base to be copied from the template, shown as a guanine. **(b)** The presence of a guanine at this site causes the enzyme to bind dCTP from the nucleoside triphosphates in the surrounding medium. The guanine is held in a position that favors formation of a covalent linkage between the 3'-OH group at the end of the primer and the innermost phosphate group bound to the 5'-carbon of the dCTP. **(c)** In this reaction the last two phosphates of the dCTP are split off, and the remaining phosphate is bound to the oxygen of the 3'-OH of the primer, creating a 3' → 5' phosphodiester linkage between the primer and the added nucleotide. The enzyme then moves to the next base on the DNA template, and the cycle of reactions shown in **(a** through **c)** repeats. Each successive nucleotide added to the chain provides the 3'-OH group required for addition of the next nucleotide to be added.

The problems in unwinding and separating the two template DNA chains during replication are nicely illustrated by a classic experiment carried out by J. Cairns, who followed the replication of *E. coli* DNA molecules by autoradiography (see p. 118). In Cairns's experiment the bacterial DNA was labeled by growing cells in the presence of radioactive thymidine, which is used only in the assembly of DNA. The DNA was then isolated and purified. When placed under a photographic emulsion, the circular DNA molecules, made radioactive by the thymidine label, exposed a pattern of grains that could be traced under the light microscope (Fig. 23-6). The sites of replication could be identified with forks at two points around the DNA circles (arrows in Fig. 23-6).

By comparing the total length of DNA in the circles with the time required for *E. coli* to replicate its DNA, Cairns estimated that the two replication forks must move along the molecule at a combined rate of 30 μm per minute. At this rate the DNA would be required

to unwind at approximately 13,000 rpm for replication to proceed! How could the DNA in a covalently closed circle unwind at this rate? Cairns originally proposed that a "swivel" exists at the point at which synthesis terminates, around which the DNA rotates to accomplish unwinding. More recently, it has become apparent

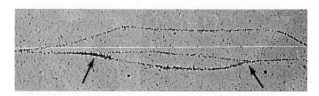

Figure 23-6 Pattern of grains traced in a photographic emulsion by a replicating *E. coli* DNA molecule. The DNA has partly replicated. The two forks (arrows) are the advancing sites of replication. ×260. From "The Bacterial Chromosome" by J. Cairns, *Scientific American* 214 (1966). Copyright © 1966 by Scientific American, Inc. All rights reserved.

that the swivel is supplied by an enzymatic mechanism that introduces a break in the DNA just in advance of a replication fork, allows one chain to revolve around the other, and closes the break. The total swiveling reaction evidently takes place about 6500 times per fork per minute in *E. coli.*

A second fundamental difference between transcription and replication is related to unwinding. This difference results from the fact that DNA unwinding and replication proceed at a fork that apparently moves *unidirectionally* along the template DNA. To produce a unidirectional fork, the two nucleotide chains of the template DNA must unwind and replicate simultaneously in the same overall direction. However, because the two chains of a DNA double helix are antiparallel, only one is presented to the replication mechanism in the required $3' \rightarrow 5'$ direction. How is the other, "wrong way" chain replicated in the same overall direction followed by the moving fork? This problem is solved by a mechanism that replicates the wrong-way

chain in short bursts that actually run in the direction opposite to fork movement (Fig. 23-7). The short lengths produced by this *discontinuous replication*, as it is called, are then covalently linked into a continuous nucleotide chain.

The enzymes and factors required to solve the special problems of DNA replication—unwinding, priming, and unidirectional replication—have been pieced together through research with a variety of prokaryotic and eukaryotic systems, and with viruses infecting both prokaryotic and eukaryotic cells. The most successful research efforts in DNA replication have been those using *E. coli,* particularly in the laboratory of A. Kornberg. (Kornberg received the Nobel Prize in 1959 for his work in DNA replication.) Much of Kornberg's research was conducted with temperature-sensitive mutants (see p. 924), in which cells could be grown in large quantities at permissive temperatures and then analyzed for the effects of the mutations at restrictive temperatures. By this means a particular mu-

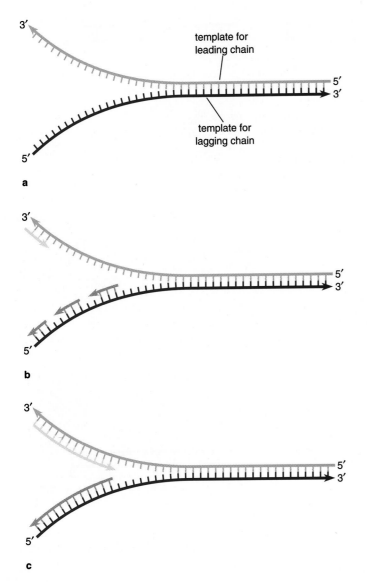

Figure 23-7 Solution to the problem of unidirectional replication at a fork. The template chain presented to the replication complex in the "wrong" $5' \rightarrow 3'$ direction—the chain on the bottom in **(a)**—is copied in short bursts that run counter to the direction of fork movement **(b)**. The short lengths are then linked into a continuous chain **(c)**. The overall effect is unidirectional synthesis in the direction of fork movement. The copy chain synthesized in the direction of fork movement is called the leading chain, and the chain synthesized in short bursts opposite to the direction of fork movement is called the lagging chain.

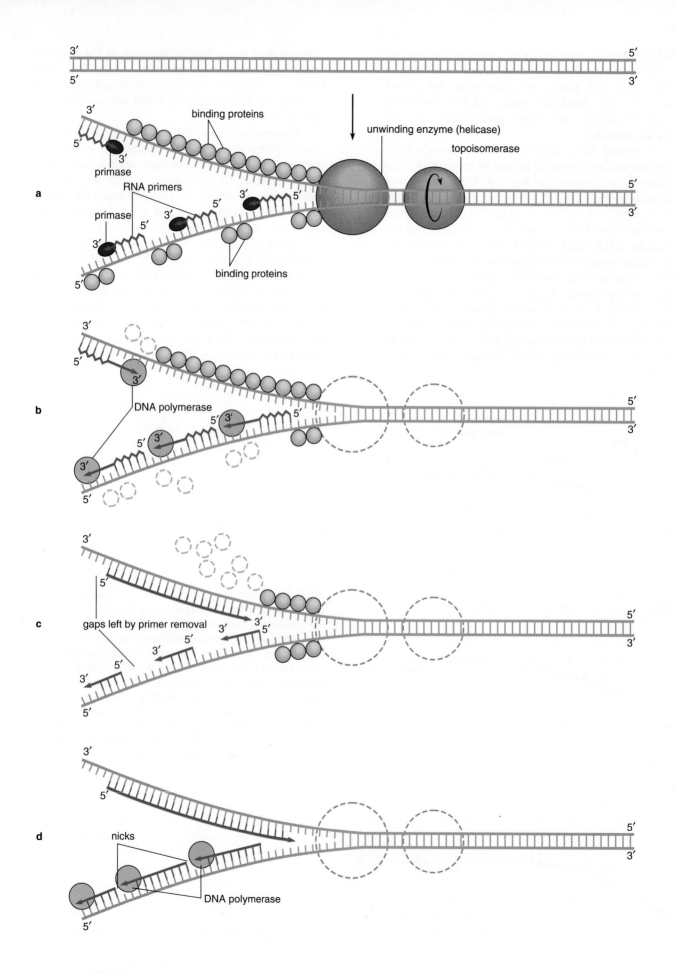

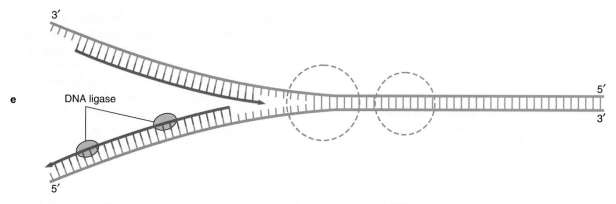

Figure 23-8 The enzymes and factors catalyzing DNA replication in diagrammatic form. In living cells the enzymes are organized into complexes that carry out several of the steps in close coordination. **(a)** Unwinding and primer synthesis. The template helix is unwound by a helicase that hydrolyzes one molecule of ATP for each turn unwound. Just in front of the unwinding site, a topoisomerase relieves supercoils generated by the unwinding. As the chains unwind, DNA-binding proteins stabilize them in single-chain form and prevent them from rewinding. These proteins are displaced by short RNA chains, which are laid down as primers by a primase. **(b)** After primers are laid down, they are extended as new DNA chains by a DNA polymerase. **(c)** The primers are then removed, leaving gaps at their former sites. **(d)** The gaps are filled in by another DNA polymerase. Gap filling leaves a single-chain nick because the DNA polymerase cannot join the 3' end of the chain it has just completed to the 5' end of the next chain in line. **(e)** The single-chain nicks are closed by DNA ligase in a reaction that seals the copies into continuous nucleotide chains. Primer synthesis, removal, gap filling, and sealing occur primarily in the lagging chain.

tant could be identified with a missing function in DNA replication at the restrictive temperature, allowing the protein encoded in the gene in many cases to be identified with a single reaction in the replication process. Also highly successful has been research with several viral systems, primarily viruses infecting *E. coli* and a few other bacteria.

In more recent years the work with prokaryotic systems has been complemented by investigations in eukaryotes. The yeast *Saccharomyces* and several viruses infecting eukaryotic cells, including adenovirus and SV40, have been particularly useful in this research. The SV40 system, developed by T. J. Kelly and his coworkers and others, has been especially valuable because the virus encodes only one protein, the *T antigen*, that acts in the initiation of replication. The remaining enzymes and factors of replication are supplied by the eukaryotic host cell and can be identified as such. SV40 also allows study of the roles and fates of nucleosomes in replication because the viral DNA circle is complexed with histones originating from host cells. Recently, T. Tsurimoto, T. Melendy, and B. Stillman constructed the first completely defined eukaryotic cell-free replicating system including the SV40 DNA circle, the SV40 T antigen, and eight purified eukaryotic proteins. (The functions of these proteins are outlined below.)

The Specialized Mechanisms of DNA Replication

Research with prokaryotic and eukaryotic systems has allowed a generalized mechanism to be pieced together that, with some differences in detail, applies to both groups of organisms and also to many of the viruses infecting prokaryotes and eukaryotes. Each of the three mechanisms peculiar to DNA replication—unwinding, priming, and unidirectional fork movement—involves the coordinated activity of specialized enzymes and factors (Fig. 23-8).

Unwinding Unwinding depends on an *unwinding enzyme*, or *helicase*, that moves along the template double helix just in front of the DNA polymerase, separating the template chains as it goes (Fig. 23-8a). The helicase uses at least one molecule of ATP per turn of helix unwound to drive the unwinding reaction. The unwound nucleotide chains are stabilized by *DNA binding proteins* (also called *helix destabilizing proteins* or *single-strand binding proteins*), which bind avidly to the single chains and prevent them from rewinding in the region just behind the fork.

As the helicase proceeds, the unwinding twists the template double helix in front of the fork in the same right-handed direction as the double helix turns. The long helix in front of the fork cannot rotate readily to compensate for this twisting force; nor can turns be compressed more tightly in advance of the fork because the number of turns per unit length is fixed at an almost constant value (see p. 529). Instead, the DNA in advance of the fork is forced into one positive supercoil (see p. 537) for each full turn unwound.

Rather than accumulating in front of the fork, the supercoils are relieved by one or both of two groups of enzymes called *DNA topoisomerases*. These enzymes break the DNA in front of the fork, allow the DNA to rotate around the breaks, and reseal the DNA. A single cycle of breakage, rotation, and resealing by a DNA topoisomerase eliminates one positive supercoil. Either or both types of topoisomerases relieve supercoils during DNA replication in different organisms: *DNA topoisomerase I* introduces a break in one of the two DNA chains, and *DNA topoisomerase II* introduces breaks in both chains (see below).

Priming and Polymerization As the DNA unwinds, primers are synthesized by a specialized RNA polymerase called a *primase*. The primase attaches to a template chain and catalyzes synthesis of an RNA primer about five to ten nucleotides long (see Fig. 23-8a). RNA primers are assembled in the $5' \rightarrow 3'$ direction on both sides of the fork: in the direction of unwinding on one chain and in the opposite direction on the other. Primer assembly evidently displaces the binding proteins from the region, freeing them to cycle ahead for reuse at the advancing unwinding site.

DNA polymerases add DNA nucleotides sequentially from the primers (Fig. 23-8b). As in primer synthesis, polymerization proceeds in the $5' \rightarrow 3'$ direction on both sides of the fork, so that nucleotides are added in the unwinding direction on one chain and in the opposite direction on the other.

The DNA polymerases are highly accurate and usually enter only the correct complementary base opposite a base on the template chain. However, base mismatches do occur at a regular but low frequency, primarily as a result of the formation of *tautomers*— alternate configurations of the four DNA nucleotides that allow unusual base pairs to form. (Information Box 23-1 describes tautomers and their effects on base pairing.) Most mismatches arising during replication are corrected either by a proofreading mechanism carried out by the DNA polymerases themselves or by reactions that repair mismatched bases after replication is complete (see below).

Once the primers have served their functions, they are removed, either by the DNA polymerase molecules or by an RNAase specialized for this function (Fig. 23-8c). The gaps created by primer removal are filled in by another DNA polymerase, which adds nucleotides until the last remaining bases in the gaps are paired

and filled in (Fig. 23-8d). Since the DNA polymerase cannot link the last nucleotide added to the 5' end of the next segment, a single-chain nick is left open at the 3' end of the filled gap. The nicks are closed by the final major enzyme acting in DNA replication, *DNA ligase*. This enzyme, which uses ATP or NAD as an energy source, forms the final phosphate linkages required to seal the copies into continuous DNA nucleotide chains (Fig. 23-8e). Table 23-1 summarizes the activities of the major enzyme classes and factors in DNA replication.

Unidirectional Fork Movement Replication in the same direction on both sides of a replication fork, as shown in Figure 23-7, is more apparent than real. On the template chain exposed in the "correct" $3' \rightarrow 5'$ direction a single primer or at most a very few primers are laid down in the direction of unwinding. The DNA copy assembled on this template chain is called the *leading chain*. On the template chain exposed in the "incorrect" $5' \rightarrow 3'$ direction, primer synthesis proceeds in the direction opposite to unwinding. The DNA copy assembled on this template chain is the *lagging* chain. The primers are laid down in the lagging chain at intervals of about 1000 to 2000 nucleotides in prokaryotes and 100 to 200 nucleotides in eukaryotes. These discontinuous lengths, after primer removal and gap filling, are covalently sealed into a continuous nucleotide chain (see Fig. 23-7c).

Thus the apparent assembly of both new chains in the direction of unwinding, as seen at the microscopic level in preparations such as Cairns's micrographs, is actually accomplished at the molecular level by continuous synthesis on the leading chain template in the direction of unwinding and short bursts of discontinuous synthesis on the lagging chain in the direction opposite to unwinding. The combination produces an overall polymerization that proceeds, in effect, in the

Table 23-1	Major Enzymes and Factors of DNA Replication	
Enzyme or Factor	Activity
Helicase	Unwinds DNA helix
Binding proteins	Stabilize DNA in single-chain form
DNA topoisomerases	Relieve supercoils in front of replication fork
Primase	Synthesizes RNA primers
DNA polymerases	Assemble DNA chains on primers; fill gaps left after primer removal
DNA ligase	Seals nicks left after DNA primases fill gaps left by primer removal

Tautomers and Base Mismatches During Replication

The standard arrangement of atoms in DNA bases actually represents the most probable among several possible arrangements. Alternate arrangements of the atoms of each base, called tautomers, occur with low but regular frequency. Although the tautomers allowing alternate base pairing are rare (about 1 in every 10,000 to 100,000 bases may undergo a shift to a different tautomer at any instant), and persist only briefly when formed, they may lead to a base mismatch if a template or copy base undergoes a tautomeric shift at the instant it enters the replication process.

For example, a shift of adenine from its most common form, in which it pairs with thymine (part *a* of the first figure in this box), to one of its tautomers would allow it

to pair with cytosine (part *b*). Similarly, a tautomeric shift of guanine would allow a change in its pairing partner from C to T (parts *c* and *d*).

If left unrepaired, the incorrect base insertion resulting from a tautomeric shift would be faithfully copied at the next round of replication (see the second figure in this box), barring another error at the same site. Once copied, the incorrectly paired base is preserved in further replications of the cell line receiving the altered DNA molecule.

Tautomers are not the only possible source of misrepairs. Evidently the DNA double helix is flexible enough to accommodate considerable distortion, allowing even purine–purine or pyrimidine–pyrimidine base pairs to form as rare events.

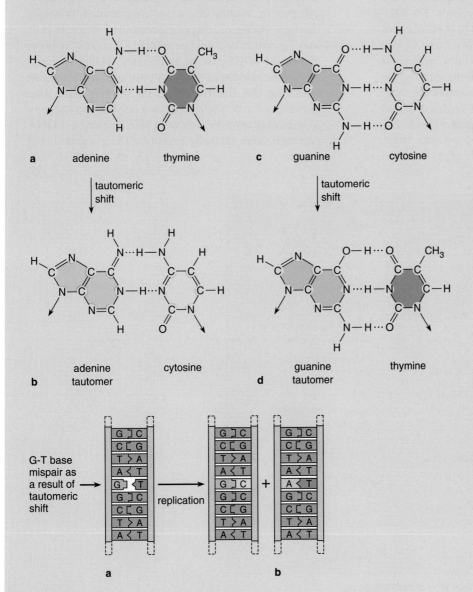

direction of fork movement in both the leading and lagging chains.

Replication advances at a rate of about 500 to 1000 nucleotides per second in prokaryotes and about one-tenth this rate, some 50 to 100 per second, in eukaryotes. The entire process is so rapid that the RNA primers and gaps left by discontinuous synthesis persist for only seconds or fractions of a second. As a consequence, the replication enzymes and factors operate only in the vicinity of the fork. A few micrometers behind the fork the new DNA chains are fully continuous and wound with their template chains into complete DNA double helices.

The Enzymes and Factors of DNA Replication

Although the overall pattern of steps in DNA replication is basically the same for prokaryotes and eukaryotes, details of the reaction sequence vary significantly in the two groups of organisms. The variations depend primarily on differences in the properties and types of the enzymes and factors carrying out replication, and in their organization into replication complexes in the two groups of organisms. Among the enzymes and factors, the DNA polymerases are most varied and distinct (Table 23-2). The central reaction of DNA replication, the addition of nucleotides to the end of an RNA primer, is catalyzed by one DNA

polymerase in prokaryotes, and two different types in eukaryotes. Other distinct DNA polymerases catalyze subsidiary reactions such as gap filling after primer removal and repair of DNA lesions resulting from replication. The enzymes and factors catalyzing the other reactions of DNA replication—unwinding, priming, primer removal, and DNA ligation—also differ to a greater or lesser extent in properties and organization between prokaryotes and eukaryotes.

Prokaryotic DNA Polymerases Three DNA polymerases, designated *I, II,* and *III,* have been identified in *E. coli* through research in the laboratories of Kornberg, Cairns, M. L. Gefter, and others (see Table 23-2). Each of these enzymes can work as a DNA polymerase assembling nucleotides into a chain or in the opposite sense as a DNA exonuclease removing nucleotides one at a time from the end of a nucleotide chain.

The roles of the three prokaryotic DNA polymerases in replication were unscrambled primarily through studies of temperature-sensitive mutants in *E. coli.* Temperature-sensitive mutants deficient in polymerase III activity were completely unable to polymerize nucleotides into DNA at restricted temperatures, indicating that this enzyme is the primary replication enzyme of *E. coli.* Mutants lacking polymerase I activity at restricted temperatures were able to carry out DNA polymerization normally except for short segments left

---

Table 23-2 Prokaryotic and Eukaryotic DNA Polymerases

Enzyme	Direction of Synthesis	Exonuclease Activity	Probable Functions
Prokaryotic			
polymerase I	$5' \rightarrow 3'$	$5' \rightarrow 3'$; $3' \rightarrow 5'$	Gap filling after primer removal; DNA repair
polymerase II	$5' \rightarrow 3'$	$3' \rightarrow 5'$	Gap filling after primer removal; DNA repair
polymerase III	$5' \rightarrow 3'$	$5' \rightarrow 3'$; $3' \rightarrow 5'$	Primary replication enzyme
Eukaryotic			
polymerase α	$5' \rightarrow 3'$	None*	Primary replication enzyme (with polymerase δ); DNA repair
polymerase β	$5' \rightarrow 3'$	None*	DNA repair
polymerase γ	$5' \rightarrow 3'$	$3' \rightarrow 5'$	Primary replication enzyme of mitochondria
polymerase δ	$5' \rightarrow 3'$	$3' \rightarrow 5'$	Primary replication enzyme (with polymerase α)
polymerase ϵ	$5' \rightarrow 3'$	$3' \rightarrow 5'$	DNA repair

* May be associated with other proteins that have $3' \rightarrow 5'$ exonuclease activity.

open in the newly assembled DNA chains. This indicates that polymerase I fills the gaps left by primer removal.

Mutants lacking polymerase II activity at restricted temperatures were able to replicate their DNA apparently normally, leaving the role of this enzyme in doubt. However, mutants lacking both polymerase I and polymerase II activity showed even greater deficiencies in gap filling at restricted temperatures than mutants lacking polymerase I alone. This suggests that polymerase II may act in an auxiliary role, facilitating gap filling by DNA polymerase I.

Taken together, these findings indicate that polymerization of the new DNA chains on templates, starting from primers, is catalyzed by prokaryotic DNA polymerase III. DNA polymerase I functions to fill the gaps left after removal of the primers, with DNA polymerase II working in coordination or possibly functioning as a backup for DNA polymerase I.

E. coli DNA polymerase III is a complex enzyme that contains at least 20 polypeptide subunits (Fig. 23-9a). Polypeptide α, a major subunit present in two copies, is a "core" polypeptide that carries out the fundamental polymerization of nucleotides into a chain. Many other polypeptides of the complex, including the one responsible for the $3' \rightarrow 5'$ exonuclease activity (ϵ in Fig. 23-9a), are also present in two copies. This finding led to a proposal by C. S. McHenry and K. D. Johansen that DNA polymerase III is an *asymmetric dimer* whose structure may enable it to carry out replication of the leading and lagging chains simultaneously (see below). The other *E. coli* enzyme directly implicated in replication, DNA polymerase I, is a relatively simple molecule consisting of a single polypeptide subunit.

The $3' \rightarrow 5'$ exonuclease activity of DNA polymerase III (see Table 23-2) evidently provides the primary replication enzyme with the ability to correct most of its own errors in base pairing during DNA replication. According to a hypothesis advanced by D. Brutlag and Kornberg, polymerase III adds additional nucleotides to a growing chain only if the most recently added base is correctly paired (Fig. 23-10a). If the most recently added nucleotide is incorrectly paired (Fig. 23-10b) and unstable because of the mispairing, the reverse $3' \rightarrow 5'$ exonuclease activity of polymerase III rather than its forward $5' \rightarrow 3'$ polymerase activity is favored. As a result, the enzyme goes backward and removes the mispaired base or bases until a correct pair, forming a stable association resistant to disruption, is reached (Fig 23-10c). At this point forward activity of the polymerase again becomes favored, and the enzyme returns to the polymerization reaction (Fig. 23-10d).

The effect of the combined polymerase-exonuclease activity, according to this model, is to *proofread* the new chain as it is assembled and to eliminate pairing

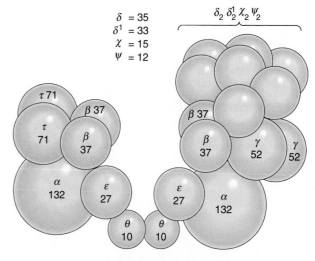

$\delta = 35$
$\delta^1 = 33$
$\chi = 15$
$\psi = 12$

$\delta_2 \delta_2^1 \chi_2 \psi_2$

a   *E. coli* DNA polymerase III

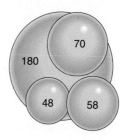

b   eukaryotic DNA polymerase α

Figure 23-9 DNA polymerase structure. **(a)** Polypeptide subunits of *E. coli* DNA polymerase III, the primary replication enzyme complex. The numbers give the molecular weights of the subunits in thousands. The unequal distribution of the subunits, which occur in two or four copies, gives the polymerase its structure as an asymmetric dimer. Polymerase activity is associated with the two α-subunits; the ϵ-subunits are the sites of $3' \rightarrow 5'$ exonuclease activity implicated in proofreading. Redrawn from an original courtesy of A. Kornberg. **(b)** Polypeptide subunits of eukaryotic DNA polymerase α. Polymerase activity is associated with the 180,000-dalton subunit; primase activity is associated with the 48,000- and 58,000-dalton subunits.

errors as they occur. Several lines of evidence support the proofreading model. For example, the error rate of bacterial DNA polymerase with functional exonuclease activity is about 10^{-6} in cell-free systems. If the exonuclease activity is inhibited, the error rate increases to 10^{-3} to 10^{-4}. The frequency of mispairing varies considerably, depending on the particular template base and its surrounding sequence and the relative concentrations of the four nucleoside triphosphates in the surrounding medium.

Another factor reducing the error rate is a DNA repair mechanism that detects and corrects mismatched bases that escape proofreading (see below). The com-

The Reactions of DNA Replication   **963**

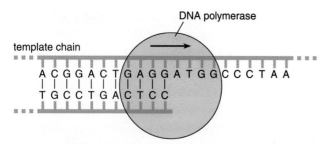

enzyme continues activity in the forward direction as a $5' \rightarrow 3'$ polymerase as long as the most recently added nucleotide is correctly paired

a

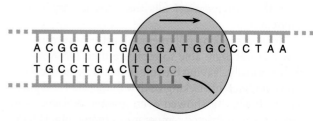

enzyme adds mispaired nucleotide

b

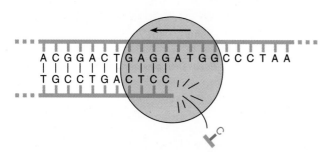

enzyme reverses, acting as a $3' \rightarrow 5'$ exonuclease to remove mispaired nucleotide

c

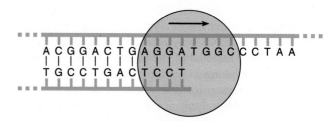

enzyme resumes forward activity as a $5' \rightarrow 3'$ polymerase

d

Figure 23-10 Proofreading by bacterial DNA polymerase III, according to Brutlag and Kornberg.

bination of the two processes in living cells achieves a final error rate as low as 10^{-9} to 10^{-10}. This is equivalent to only one pairing mistake in a billion or more bases linked into growing DNA chains.

The characteristics of the proofreading mechanism may explain why the replicating enzymes must proceed from a primer and why RNA rather than DNA primers are used in DNA replication. If the DNA polymerase were to start synthesis of a new DNA chain with no existing nucleotides in place, the first nucleotide would more than likely be mispaired. This is because the bases in very short double helices have greater freedom of movement, allowing them to take up alternate configurations that permit "nonstandard" base pairs to form (see p. 961). As a consequence, the enzyme would turn around and remove the mispaired base. The enzyme would thus be likely to cycle back and forth, adding and removing bases without being able to proceed to stable elongation of the copy chain. An RNA primer, in contrast, is laid down by primases with no proofreading ability. The primases ignore mispairs and assemble an RNA chain that is long enough to restrict base pairs primarily to the correct ones toward its 3'-OH end, from which the DNA polymerase begins polymerization. The frequent mispairs toward the 5' end of the primers are unimportant to the final outcome because they are removed and replaced later in the replication process.

Eukaryotic DNA Polymerases Work with eukaryotes has established that higher organisms possess at least five DNA polymerases, designated α, β, γ, δ, and ϵ (see Table 23-2). All five eukaryotic enzymes, like their bacterial counterparts, polymerize DNA chains only in the $5' \rightarrow 3'$ direction.

Several lines of evidence initially implicated DNA polymerase α as the primary replication enzyme of eukaryotes. Among this evidence were observations by H. J. Edenberg, S. Anderson, M. L. DePamphilis, and others that eukaryotic replication slows significantly or stops when the system is exposed to inhibitors or antibodies against DNA polymerase α. Temperature-sensitive mutations also implicated polymerase α as a primary replication enzyme. There are problems with DNA polymerase α as the primary replication enzyme, however. One is that the *processivity* of the enzyme is relatively low. Processivity describes the number of nucleotides a polymerase assembles before it stops catalysis and releases from the template. In the case of polymerase α, processivity is only moderate, so that the enzyme is expected to polymerize only 100 to 200 nucleotides before releasing. Another problem is that the DNA polymerase α of most eukaryotic species lacks the $3' \rightarrow 5'$ exonuclease activity necessary for proofreading. Two of its subunits can work as a primase, however.

Later work by J. J. Byrnes and others implicated

DNA polymerase δ as another enzyme directly involved in primary DNA replication. Agents inhibiting DNA polymerase δ also greatly slow or inhibit DNA replication in eukaryotes. Further, a factor called *PCNA* (for *Proliferating Cell Nuclear Antigen*), previously found necessary for DNA replication to progress in eukaryotes, was discovered by G. Prelich, R. Bravo, and others to exert its effects by stimulating the activity and increasing the processivity of DNA polymerase δ. In addition, DNA polymerase δ has the $3' \rightarrow 5'$ exonuclease activity required for proofreading and is characterized by very high processivity. Polymerase δ also acts as a helicase.

The activities of the two primary eukaryotic DNA polymerases make it seem likely that they cooperate in DNA replication, through a mechanism in which polymerase δ catalyzes synthesis of the leading chain and polymerase α assembles the lagging chain (see Fig. 23-16). The high processivity of polymerase δ is well suited to the extended assembly characteristic of the leading chain, and the moderate processivity of polymerase α to the short lengths characteristic of the discontinuously assembled lagging chain. The primase activity of polymerase α is also consistent with its involvement in lagging-chain synthesis, in which frequent primers must be laid down. In support of this hypothesis is the finding that in the completely purified SV40 cell-free system, only the lagging chain is assembled if polymerase δ is omitted from the reaction mixture. The "asymmetric dimer" of the eukaryotic system may thus consist of a combination of the DNA polymerase α- and δ-enzymes.

Although the combination of enzymes and factors that accomplish proofreading remains uncertain, there is good evidence from experiments by T. A. Kunkel and others that proofreading actually takes place in eukaryotes. Measurements of mispairing when exonuclease activity is inhibited show that DNA polymerases α and δ place the wrong base in DNA copies once in several thousand times. The proofreading function of polymerase δ, when active, improves the fidelity by about 100 times. Therefore, polymerase δ is believed to supply the proofreading function for the coordinated synthesis. The combination of proofreading and mismatch repair (see below) sets the accuracy of replication at very high levels, equivalent to one mispairing for every 10^9 to 10^{12} nucleotides placed in growing eukaryotic DNA chains.

Because DNA polymerase γ is found primarily in association with mitochondria in eukaryotic cells, this enzyme is thought to be specialized for DNA replication in these organelles (see Supplement 23-3). Polymerase ϵ, which has structure and properties similar to polymerase δ, has been implicated in DNA repair. DNA polymerase β is apparently absent from protozoa, fungi, and at least some higher plants, indicating that its functions are probably not central to the replication mechanism. Other research indicates that DNA polymerase β may work in coordination with polymerases α and ϵ in DNA repair (see below).

Eukaryotic DNA polymerases have proved to be highly difficult to isolate in intact form. For this reason the subunit composition of the enzymes remains uncertain. However, DNA polymerase α appears to consist of at least four different subunits (see Fig. 23-9b). The largest of these subunits is the core unit catalyzing polymerization of a nucleotide chain, and the two smallest subunits have primase activity. DNA polymerase δ has at least two subunits plus the PCNA factor, which is closely associated with the enzyme. Another factor, *RF-C*, may set up a link between the two primary polymerases in the replication complex (see Fig. 23-16). Polymerase ϵ, as noted, is similar to polymerase δ, so similar that for a time δ and ϵ were thought to be alternate forms of the same enzyme. DNA polymerase β appears to consist of a single 40,000-dalton subunit. Most of the various polypeptides of the primary replication complex, including those of polymerases α and δ and the PCNA factor, are highly conserved among eukaryotes from yeasts to humans.

The Priming Enzymes   The primases working in DNA replication have the capacity to incorporate DNA as well as RNA nucleotides during their polymerization reactions. However, the primers synthesized for DNA replication usually consist only of RNA because of the greater affinity of the primases for RNA than DNA nucleotides as reactants for primer assembly.

The primase of prokaryotes appears to be closely associated with the helicase unwinding the DNA for replication, forming a so-called *primosome* that catalyzes unwinding and priming as a coordinated reaction. The primase activity of eukaryotes, as noted, is evidently associated with the DNA polymerase α-enzyme, forming a complex that carries out priming and DNA polymerization as a closely coordinated sequence.

Both prokaryotic and eukaryotic primases require ATP or GTP as an energy source for initiating the priming reaction. Frequently, the nucleoside triphosphate used as the energy source for primer initiation is placed as the first nucleotide of the primer, without regard to the base opposite this position in the template chain. Further mispairing may occur as the primer is assembled because of the ability of the bases in short nucleotide chains to take up alternate pairing arrangements. As noted, these mispairs do not lead to errors in the final product because they are eliminated during primer removal at later steps of DNA replication.

Although RNA primases assemble the primers for prokaryotic and eukaryotic replication, other priming mechanisms are used by some viruses. In various bacterial viruses replication proceeds simply by the addition of nucleotides to a 3'-OH end exposed by a single-chain nick in the template double helix; no RNA primer of any kind is inserted. In the *fd* virus infecting *E. coli*, priming is carried out by the RNA polymerase normally carrying out transcription. In the retroviruses infecting mammalian cells, a tRNA molecule is used as primer; by pairing with a complementary sequence at the 3' end of the template, the tRNA provides the required 3'-OH end for addition of DNA nucleotides by DNA polymerase. In the adenovirus, and in bacteriophage φ 29, a protein binds to the end of the template chain and covalently links a DNA nucleotide, positioning it so that its 3'-OH group is exposed to act as a primer (Fig. 23-11).

Evidence that primers are laid down at short intervals in synthesis of the lagging chain was first obtained in a now-classical experiment carried out in 1968 by R. Okazaki and his coworkers. In their experiment replicating bacteria were exposed to a "pulse" of tritiated thymidine lasting a few seconds. If DNA was isolated from the bacteria immediately after the pulse, radioactivity was found to be associated almost entirely with short pieces of DNA 1000 to 2000 nucleotides in length. Short RNA primers could be detected at the 5' ends of these fragments. If longer periods were allowed between the pulse of label and DNA extraction, radioactivity was associated with DNA of very high molecular weight. This finding indicated that the DNA was initially synthesized in short pieces that were later covalently linked into long, continuous strands, as expected if replication of the lagging chain is discontinuous.

Initial synthesis of these short DNA lengths, now called *Okazaki fragments*, has since been detected in a wide variety of prokaryotes, eukaryotes, and viruses. In eukaryotes Okazaki fragments are some 100 to 200 nucleotides long, approximately the length of DNA wrapped around a nucleosome, suggesting that eukaryotic DNA may be released one nucleosome at a time for priming of the lagging chain.

Not all systems replicate discontinuously and release Okazaki fragments. In adenovirus replication one chain of the linear DNA replicates in the 5' → 3' direction from one end of the molecule and the other chain from the opposite end (see Fig. 23-11). As a result, both DNA chains act as templates for leading chains. The

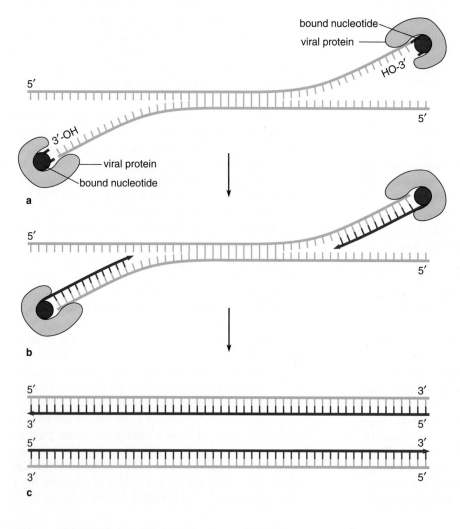

Figure 23-11 The adenovirus priming mechanism, in which a protein linked to a nucleotide provides the primer for replication. **(a)** The protein binds at the 3' end of each template chain. Bound to the protein is a nucleotide that supplies a 3'-OH group as a primer for DNA replication. **(b)** Replication in progress. Note that both copy chains are synthesized continuously in the 5' → 3' direction so that there is no lagging chain. **(c)** Completion of replication.

unusual pattern of replication observed in mitochondria also synthesizes DNA by a mechanism in which both DNA chains act as templates for leading chains (see Supplement 23-3).

Enzymes Removing Primers Primer removal in prokaryotes appears to depend on the ability of DNA polymerase I to work simultaneously as a $5' \rightarrow 3'$ polymerase and a $5' \rightarrow 3'$ exonuclease (Fig. 23-12). In its role as a $5' \rightarrow 3'$ exonuclease, the enzyme nibbles RNA bases one at a time from the primer, starting at the single-chain nick between a newly synthesized DNA segment and the RNA primer (Fig. 23-12a). As it removes the RNA bases, it works simultaneously as a $5' \rightarrow 3'$ polymerase, adding a new DNA nucleotide as each RNA nucleotide is removed (Fig. 23-12b). This *displacement synthesis*, as it is called, continues until the enzyme reaches the end of the primer. The reaction leaves a single-chain nick at the $3'$ end of the segment formerly occupied by the primer (Fig. 23-12c), which is sealed by DNA ligase.

Some experiments indicate that another enzyme, *ribonuclease H*, may also be involved in primer removal in prokaryotes, possibly in a secondary or backup role. The enzyme has the capacity to recognize and remove the RNA chain of hybrid RNA-DNA helices. Primer removal, although not completely inhibited, is delayed in *E. coli* mutants with defective forms of ribonuclease H. The enzyme can also be detected in eukaryotes, suggesting that it may play some role in primer removal in these organisms.

The Enzymes and Factors Unwinding DNA The activity of the DNA binding proteins in stabilizing DNA in the single-stranded form is well documented in *E. coli* and several prokaryotic and eukaryotic viruses. In *E. coli*, temperature-sensitive mutants with defects in the DNA-binding protein *SSB* stop DNA replication immediately at restrictive temperatures. The ability of SSB to stimulate DNA replication in isolated systems also supports its direct involvement in DNA replication. Binding by the SSB proteins is cooperative, so that several molecules attach in rows to the unwinding chains just behind the fork, probably by polymerizing into chains. C. R. Wobbe and his coworkers and others found that *RF-A*, a eukaryotic DNA binding protein, carries out a similar function in the SV40 system.

The helicase unwinding DNA in prokaryotes, like so many of the enzymes and factors of replication, was first implicated by studies of temperature-sensitive *E. coli* mutants in Kornberg's laboratory. These studies implicated a protein called *DNA B* as the *E. coli* helicase. As noted, the bacterial helicase is complexed with the primase in the primosome, in a combination associated with the lagging chain.

The helicase unwinding DNA for eukaryotic replication remains uncertain. However, the helicase ac-

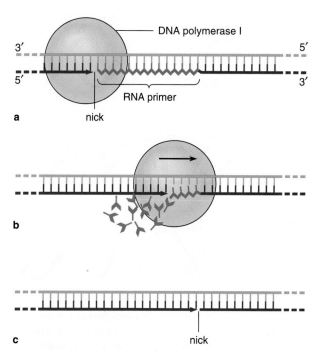

Figure 23-12 Displacement synthesis by bacterial DNA polymerase I. **(a)** The enzyme starts at a single-chain nick between a segment of newly synthesized DNA and the next primer in line. **(b)** The enzyme then displaces the primer one nucleotide at a time, using its $5' \rightarrow 3'$ exonuclease activity; new DNA nucleotides are added at the same time in the $5' \rightarrow 3'$ direction by the polymerase activity of the enzyme. **(c)** After the primer has been replaced by displacement synthesis, a single-chain nick remains that is sealed by DNA ligase.

tivity associated with polymerase δ may carry out this function in eukaryotes. If so, the helicase function in eukaryotes is associated with the leading rather than the lagging chain.

The DNA topoisomerases relieving the positive supercoils generated by DNA unwinding have been identified in both prokaryotes and eukaryotes. As noted, DNA topoisomerase I, discovered by J. C. Wang and others, introduces a break in one of the two chains of a DNA double helix (Fig. 23-13a). The enzyme allows the unbroken chain to pass through the opening created by the break (Fig. 23-13b) and then reseals the break (Fig. 23-13c). DNA topoisomerase I works without an input of phosphate-bond energy. DNA topoisomerase II, discovered by M. Gellert, introduces breaks into both nucleotide chains of a DNA helix (Fig. 23-14a). The enzyme then allows an intact region of the double helix to pass through the break (Fig. 23-14b). Following the passage, both chains are resealed, leaving the DNA molecule intact. In contrast to DNA topoisomerase I, a DNA topoisomerase II requires energy from ATP or GTP to catalyze its reaction. Both enzymes form covalent linkages with the broken DNA ends during their catalytic cycles.

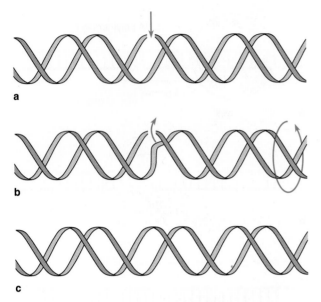

a

b

c

Figure 23-13 Action of DNA topoisomerase I in relieving supercoils in front of the replication fork. The enzyme makes a break in one of the two nucleotide chains of the DNA double helix (arrow in **a**) and passes the unbroken chain through the break **(b)**. This movement causes one end of the molecule to rotate in the direction opposite to the supercoil (arrow at the right end of the molecule in **b**). Passage of the chain through the break relieves the supercoil, and the molecule is resealed **(c)**.

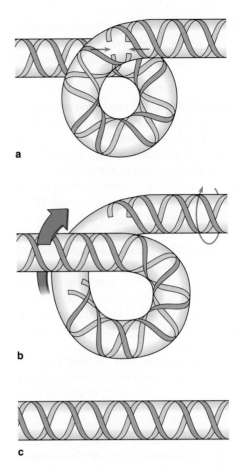

a

b

c

Figure 23-14 Action of DNA topoisomerase II in relieving supercoils in front of the replication fork. The supercoil is shown as a loop in this diagram. The enzyme makes a break in both chains of the DNA double helix (arrows in **a**) and passes an unbroken region of the double helix through the breaks **(b)**. The movement has the effect of rotating one end of the molecule in the direction opposite to the supercoil (arrow at the right end of the molecule in **b**). The passage relieves the supercoil, and the molecule is resealed **(c)**.

The enzyme unwinding positive supercoils in *E. coli* is a DNA topoisomerase II also known as *DNA gyrase*. Enzymes that inhibit DNA gyrase, such as the drug novobiocin, strongly inhibit DNA replication in *E. coli*. In yeast or cell-free systems replicating the DNA of the eukaryotic virus SV40, both DNA topoisomerases I and II seem to be active in DNA replication. Of the two, however, only DNA topoisomerase I can be eliminated without stopping the process. DNA topoisomerase II may, therefore, be the primary enzyme that unwinds positive supercoils in eukaryotes as well as in prokaryotes.

DNA topoisomerases also carry out steps in other vital cell processes such as DNA repair, recombination, and the release of tangles created during replication and recombination. The release of tangles, carried out by DNA topoisomerase II through its ability to pass one double helix through another, is highly critical to mitosis because tangles frequently appear during replication of the long DNA molecules typical of eukaryotes. Without release of the tangles by the enzyme, the chromosomes would be unable to separate at anaphase (for details, see p. 1020).

DNA Ligase DNA ligase, which closes the single-chain nicks remaining in newly synthesized chains after the gaps left by primer removal are filled, catalyzes the last major step in DNA replication. The activity of DNA ligase in replication was first identified in bacteria and viruses with mutant forms of the enzyme. In the mutants most of the DNA appeared intact if the molecules were isolated by procedures that maintained them in double-helical form. However, if the DNA was unwound by heating (see p. 528), the newly synthesized nucleotide chains separated into short pieces, indicating that single-chain nicks were left open during replication. The activity of the enzyme in sealing nicks was subsequently characterized by I. R. Lehman and his coworkers. (Fig. 23-15 outlines the mechanism.) Since its discovery in prokaryotes and viruses, DNA ligase has been isolated from a variety of eukaryotes as well.

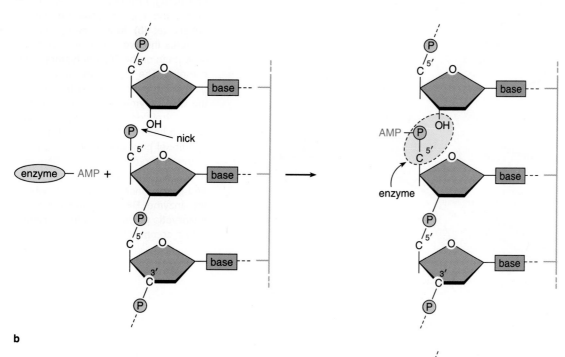

enzyme + NAD ⟶ enzyme – AMP + NMN or enzyme + ATP ⟶ enzyme – AMP + PP

a

b

Figure 23-15 Activity of DNA ligase in sealing single-chain nicks left in new DNA nucleotide chains by the DNA polymerases. In the first step in bacteria **(a)**, the ligase hydrolyzes NAD into AMP, which remains attached to the enzyme by a covalent bond, and *nicotinamide mononucleotide (NMN)*, which is released. In eukaryotes the ligase interacts with ATP in the first step, hydrolyzing the two terminal phosphates from the molecule and retaining the AMP residue through a covalent linkage. In either case the ligase-AMP product retains much of the free energy released by the hydrolysis. In the second step **(b)** the AMP group is transferred from the enzyme to the 5'-phosphate exposed at a nick in a newly replicated DNA chain. Reaction of this activated group with the 3'-OH exposed at the other side of the nick is then catalyzed by the ligase **(c)**; the 5'-phosphate group is split from AMP and attached to the adjacent 3'-oxygen, closing the nick by a phosphodiester linkage. The AMP split off by the reaction is released to the medium.

c

Organization of the Enzyme Systems Replicating DNA The distinct properties of prokaryotic and eukaryotic replication enzymes are reflected in differences in their organization at the replication fork (Fig. 23-16). The asymmetric dimer formed by bacterial DNA polymerase III is considered to sit astride the replication fork, with one-half of the dimer catalyzing assembly of the leading chain and the other half assembling the lagging chain (Fig. 23-16a). The primosome, consisting

of the combined helicase-primase activities, is thought to work with the DNA binding proteins on the lagging chain just in advance of the polymerase III complex. Immediately in front of the fork is DNA gyrase, the topoisomerase II unwinding positive supercoils in the bacterial system.

Eukaryotic systems are believed to be organized in a similar pattern with several exceptions (Fig. 23-16b). DNA polymerases α and δ have been proposed

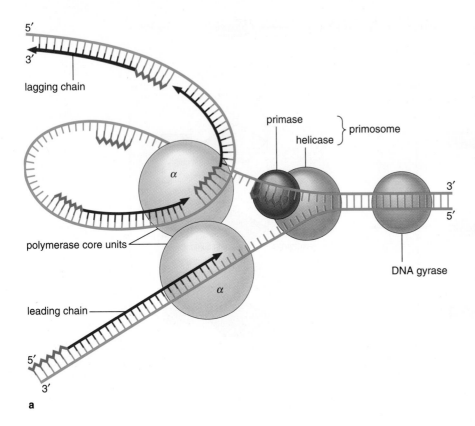

5′
3′
lagging chain

primase
helicase } primosome

α

polymerase core units

α

leading chain

3′
5′

DNA gyrase

5′
3′

a

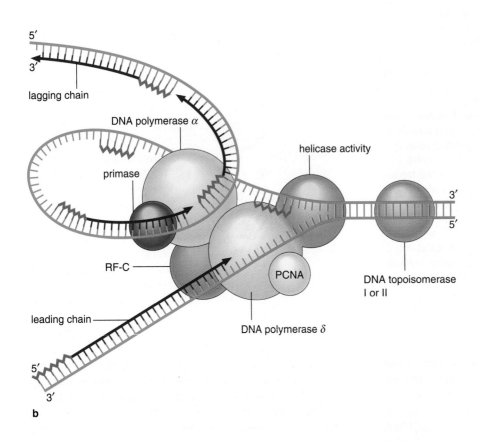

5′
3′
lagging chain

DNA polymerase α

helicase activity

primase

RF-C

PCNA

DNA polymerase δ

DNA topoisomerase I or II

3′
5′

leading chain

5′
3′

b

Figure 23-16 Proposed arrangements of the enzymes carrying out unwinding, priming, and DNA polymerization in prokaryotes **(a)** and eukaryotes **(b)**. In the prokaryotic complex the two α-subunits are the core units of the DNA polymerase III complex. The loop shown in the lagging chain in both (**a** and **b**) allows the DNA polymerases to move in the same direction on both lagging and leading chains. PCNA and RF-C are additional polypeptides forming part of the eukaryotic complex; PCNA is necessary for polymerase-δ activity and greatly increases processivity of the enzyme; RF-C may link the two eukaryotic polymerases in the replication complex.

to be coordinated in replication, with polymerase δ catalyzing assembly of the leading chain and polymerase α the lagging chain. The primase activity associated with polymerase α assembles the short primers required for discontinuous synthesis of the lagging chain. Note from Figure 23-16b that the helicase unwinding the fork is considered part of the polymerase-δ complex and that supercoils can be released by either topoisomerase I or II.

In Figure 23-16a and b, the lagging chain is shown looped around the enzyme. The loop, first proposed by B. M. Alberts and his colleagues, allows the two halves of bacterial polymerase III, or the combined polymerase-α–polymerase-δ complex of eukaryotes, to assemble both leading and lagging chains in the same overall direction and progress as a unit in the direction of fork movement.

In summary the complexes carrying out the primary replication reactions in prokaryotes and eukaryotes are similar except for several details. Prokaryotes have a single primary polymerase, III; eukaryotes have two polymerases, α and δ, acting in coordination. In prokaryotes the primase is associated with the helicase in the primosome complex; in eukaryotes the primase forms part of polymerase α. In eukaryotes helicase activity may be associated with polymerase δ.

Telomere Replication: Replacing the Primer of the Leading Chain

The linear DNA molecules of eukaryotes present special problems to the replication mechanism at the end of the leading chain. Removal of the primer laid down to begin replication of the leading chain leaves an unfilled gap opposite the 3′ end of the template; no 3′-OH group is available to act as primer for the enzyme filling this gap (Fig. 23-17).

If left unfilled, the gap would make the leading chain shorter than its template by a length equivalent to the primer. Because the same problem would exist at every successive replication, the DNA would become progressively shorter. Eventually, the DNA molecule would become too short to replicate and would disappear from the genome. Obviously, this does not happen, at least in reproductive cells; in these cells the total length of replicating eukaryotic chromosomes is maintained through successive rounds of divisions and preserved from one generation to the next.

Elegant work by E. W. Blackburn and her colleagues showed that maintenance of chromosome length depends on a group of repeated sequences that caps the ends, or telomeres (from telo = end and mere = segment), of all eukaryotic chromosomes. These terminal repeats typically consist of hundreds of copies of short sequences that are rich in G and T bases in the nucleotide chain that has its 3′ end at the telomere

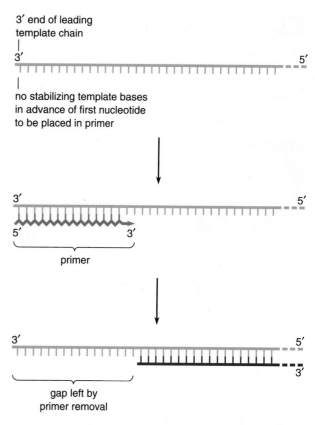

Figure 23-17 The gap left by primer removal at the 5′ end of the leading chain. There is no available 3′-OH group for a DNA polymerase to use as the starting point for filling the gap.

(see Fig. 23-18; see also Table 18-3 and p. 752). For example, the protozoan *Tetrahymena* has repeats of the sequence TTGGGG at the telomere in this nucleotide chain; *Saccharomyces* has variations of the sequence TGTGGG in this position; humans and other vertebrates have variations of the sequence TTAGGG.

Research by Blackburn, C. W. Greider, and J. W. Szostak revealed that chromosome lengths are probably maintained by *telomerase*, a specialized polymerase that can add telomere sequence repeats to chromosome ends. The telomerase does not directly fill the gap left at the end of the leading chain by primer removal. Instead, it adds telomere repeats to the end of the template chain (Fig. 23-18a and b). This would allow the gap at the end of the leading chain to be filled by the usual process of primer synthesis and extension of a copy chain by DNA polymerase (Fig. 23-18c). Although the subsequent primer removal would still leave an unfilled gap at the end of the leading chain (Fig. 23-18d), the 5′ end of the chain would be extended to maintain the average length of the chromosome.

Blackburn and Greider found that the *Tetrahymena* telomerase contains an RNA molecule with a sequence complementary to the telomere repeats of this species. They proposed that, rather than copying a DNA tem-

The Experimental Process

Telomerase: The Search for a Hypothetical Enzyme

Carol W. Greider

CAROL GREIDER was raised and educated in Davis, California. She received her bachelor's degree from the University of California at Santa Barbara in 1983. She did her graduate work with Dr. Elizabeth Blackburn at U.C. Berkeley where she initially discovered the telomerase enzyme. After receiving her Ph.D., in 1988 she took a position as a Cold Spring Harbor Fellow at Cold Spring Harbor Laboratory, on Long Island, New York. Dr. Greider now has an independent laboratory working on telomerase at Cold Spring Harbor where she is a Senior Staff Investigator.

Replication of the molecular end of a chromosome presents a unique problem. It was first pointed out by Watson[1] that DNA polymerases would not be able to completely replicate the ends because they polymerize DNA only in the 5' to 3' direction and they require a primer. One would imagine that at each round of division a region at the end of the chromosome would not be fully replicated [see Fig. 23–17], and after many rounds of division chromosomes would be significantly shorter. A number of theoretical mechanisms were proposed for how chromosome ends might replicate. However, until the structure of telomeres was characterized, it was not possible to test the numerous models that had been proposed.

Telomeres were first cloned and characterized in ciliated protozoa; they consist of tandem repeats of very simple sequences, e.g., (TTGGGG)n in *Tetrahymena*. The number of repeats on each chromosome end is variable; thus telomeres characteristically appear as 'fuzzy' bands on Southern blots. In 1983 Szostak and Blackburn showed that *Tetrahymena* telomeres would function as telomeres in the yeast *Saccharomyces cerevisiae*. This cross-kingdom conservation of telomere function not only allowed yeast telomeres to be cloned, but it also provided evidence for how telomeres might be replicated.[2] When the *Tetrahymena* (TTGGGG)n telomeres were cloned back out of yeast, they had yeast telomeric $G_{1-3}T$ sequences added onto them.[3] This result—along with the fact that telomeres normally have a variable number of (TTGGGG)n

repeats, and evidence that under some circumstances trypanosome and *Tetrahymena* telomeres increase in length—suggested that there must be a mechanism, maybe an enzyme, that adds telomeric sequences *de novo* onto chromosome ends. We imagined that this mechanism could overcome the end-replication problem indirectly: sequence loss due to incomplete replication might be restored by *de novo* addition of telomere repeats.[3]

In Liz Blackburn's laboratory at U.C. Berkeley in April 1984, I set out to look for such a hypothetical enzyme activity that would add telomeric sequences onto chromosome ends. The lesson I learned from looking for an enzyme activity that had never been reported before was this: The assay you choose is most important; try as many different assays as possible. At the time, we did not have detailed information about the molecular structure of the chromosome end. Liz Blackburn and I developed an assay in which we hoped to get specific addition of DNA nucleotides with labeled phosphate (^{32}P-dNTPs) onto a restriction fragment having telomeric sequences at the end to compare with ends having nontelomeric sequences. Initially, I tried using a large DNA fragment that had (TTGGGG)n repeats at only one end. This fragment was incubated in a buffer with *Tetrahymena* cell extracts and ^{32}P-dNTPs. After incubation, the fragment was digested at two internal sites with a restriction enzyme to generate three pieces, one with the (TTGGGG)n end, one with the nontelomeric sequence end, and one internal piece as a control. We then compared the label incorporated into the different fragments to determine if there was some specificity for label incorporation into telomeric ends. Because we did not know what the molecular end structure was, we first treated the fragment with either of two different single-strand exonucleases to generate either 5' or 3' overhangs, or left the fragment blunt-ended. We tried many different extract preparations and buffer conditions as well as different combinations of ^{32}P- and unlabeled nucleotide. In all the experiments both end fragments incorporated labeled nucleotide and, depending on the extract used, the internal band often was also labeled. The fragments with a recessed 3'-end labeled best, as they could be filled in by endogenous DNA polymerases. Having a large number of hot bands in an agarose gel was not very informative.

To look more closely at the labeled products, I began running the fragments out on sequencing gels after incubation with the *Tetrahymena* extracts and ^{32}P-dNTPs. In retrospect this was a very good move; it brought us one small step closer to the 'right' assay. We hoped not only to see labeling but to visualize any increase in size, which would be indicative of *de novo* addition. Such an increase of a few nucleotides would not be seen on an agarose gel but could be resolved on a high-resolution gel of the kind used for DNA sequencing. A 'fill in' reaction by DNA polymerase should stop at the length of the original input DNA, while *de novo* addition might generate products that are larger than the input DNA fragments. After a number of these experiments, I got a result suggesting (if you were being optimistic) that a fragment with a (TTGGGG)$_n$ 3' overhang was getting longer. Liz suggested in the next experiment that I try using as a substrate (TTGGGG)$_4$ oligonucleotide, which a postdoc in the lab, Eric Henderson, was characterizing. In the very first experiment where I used this oligonucleotide, an elongation pattern with a six-base periodicity was apparent. That was Christmas 1984, over nine months after we started looking for activity. Subsequently, the six-base periodicity seen in that first successful experiment turned out to be the result of the novel enzyme *telomere terminal transferase* or *telomerase*, which we had set out to look for.

The next year was spent trying to determine whether this activity was in fact a novel enzyme and not some side reaction of DNA polymerase that was fooling us.[4] However, the initial breakthrough, in my mind, was having an activity to work on and to characterize in the first place. Using synthetic DNA oligonucleotides and sequencing gels in the assay made the difference. Using oligonucleotides, we could obtain a 100-fold molar excess of primer over what we had been adding when we were using DNA restriction fragments. In retrospect, we approached the 'right' assay slowly but continuously; we just kept making new and reasonable changes to the assay.

Having identified the activity and shown that (TTGGGG)$_n$ repeat synthesis was independent of added template, the next big question was: How does an enzyme know how to synthesize

TTGGGGTTGGGGTTGGGG . . .

over and over without a template? A plausible and testable (!) model came to mind: the enzyme carries its own template with it. Half the answer came quickly. I treated active telomerase fractions with micrococcal nuclease or RNase A and found both would destroy enzyme activity. This suggested that RNA was somehow required for telomerase activity. The other half of the answer, that the essential RNA component does provide the template for (TTGGGG)n repeat synthesis, took several more years to prove.

First I had to purify the enzyme biochemically and identify the RNA species that copurified (the lability of telomerase activity made this a large task).[5] I then sequenced the 159-nucleotide copurifying RNA directly and used the sequence information to clone the gene. The RNA contained the sequence CAACCCCAA, exactly what you would expect for a template. But having an RNase-sensitive enzyme and an RNA that copurified with a CCCCAA sequence did not prove that this RNA had anything to do with telomerase. It could have been a cruel joke (on me) that an unrelated RNA copurified with telomerase activity and had that sequence. To prove that the 159-RNA species was required for telomerase I used a trick that was popular in the field of RNA splicing, where most factors (snRNPs) have essential RNA components. DNA oligonucleotides complementary to the RNA were incubated with telomerase and RNase H. RNase H will cleave the RNA of a DNA/RNA duplex. With specific oligonucleotides, RNase H cleaved the 159-RNA and inactivated telomerase. Thus that particular RNA was essential for telomerase activity.[6]

The final proof for the template mechanism came a year later when workers in Liz Blackburn's lab showed that mutations in the CAACCCCAA sequence of the telomerase RNA gene would template the synthesis of mutant telomeres *in vivo* in *Tetrahymena*.[7] Today telomerase activity has been identified in several ciliates as well as human cells. A telomerase-mediated dynamic length equilibrium is widely expected to be the mechanism used by most eukaryotes to overcome the end-replication problem.

References

[1] Watson, J. D. *Nature New Biology* 239:197–201 (1972).

[2] Szostak, J. W., and Blackburn, E. H. *Cell* 29:245–55 (1982).

[3] Shampay, J.; Szostak, J. W.; and Blackburn, E. H. *Nature* 310: 154–57 (1984).

[4] Greider, C. W., and Blackburn, E. H. *Cell* 43:405–13 (1985).

[5] Greider, C. W., and Blackburn, E. H. *Cell* 51:887–98 (1987).

[6] Greider, C. W., and Blackburn, E. H. *Nature* 337:331–37 (1989).

[7] Yu, G.-L.; Bradley, J. D.; Attardi, L. D.; and Blackburn, E. H. *Nature* 344:126–32 (1990).

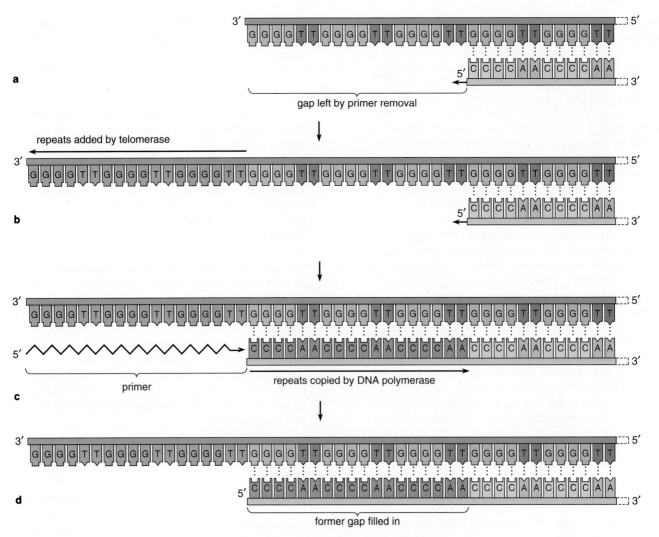

Figure 23-18 The indirect mechanism by which telomerases are believed to maintain the lengths of eukaryotic chromosomes. **(a)** The gap in the leading chain left by primer removal. **(b)** The telomerase extends the template chain by adding telomere repeats. **(c)** The extended chain is used as template for DNA synthesis by insertion of a primer and copying by DNA polymerase. **(d)** After primer removal, the entire mechanism has filled in the original gap left by primer removal. The mechanism is shown as it would occur in *Tetrahymena*, which has the telomere sequence TTGGGG when read in the conventional 5′ → 3′ direction. Adapted from an original courtesy of C. W. Greider. Reprinted by permission from *Nature* 337:331. Copyright © 1989 by Macmillan Magazines, Ltd.

plate, the enzyme adds telomere repeats by making copies from its own RNA template (Fig. 23-19).

Their proposal is directly supported by experiments in which the telomerase of one species is used to fill telomere sequences in another. The telomerase of *Saccharomyces*, for example, can add telomere sequences at the ends of *Tetrahymena* DNA molecules. The yeast telomerases, however, add the typical yeast telomere repeats to the ends of *Tetrahymena* telomeres. In another experiment Blackburn and her colleagues carried out site-directed mutagenesis to alter the RNA molecule associated with the telomerase. When the RNA molecule was altered in sequence, the sequence inserted by the telomerase was also altered to remain

complementary to the telomerase RNA. Both experiments demonstrate that telomere repeats are copied from the RNA molecule in the enzyme rather than the DNA template. In general the telomerases of all eukaryotes appear to be completely interchangeable. (Greider's Experimental Process essay on p. 972 describes the experiments leading to the discovery of telomerase and its activity in telomere replication.)

Although telomere sequences are maintained by telomerases in mammalian reproductive cells, the telomere repeats of somatic cells gradually decrease in number as cells undergo repeated divisions and as individuals age. For this reason some investigators have suggested that telomerase activity is reduced in somatic

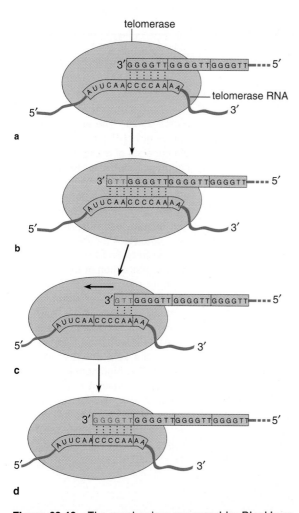

Figure 23-19 The mechanism proposed by Blackburn and Greider by which telomerase adds telomere sequences to the template chain, using the *Tetrahymena* system as model. **(a)** The RNA of the telomerase pairs with the terminal telomere repeat. **(b)** The telomere is elongated by addition of the first three nucleotides of the repeat; synthesis occurs in the usual 5′ → 3′ direction using the RNA of the telomerase as template. **(c)** The enzyme translocates, so that the 3′ end of the RNA template now pairs with the three nucleotides just added to the telomere. **(d)** The remainder of the repeat is filled in, using the RNA of the telomerase as template. This completes the telomere repeat; more repeats may be added by additional cycles from (**a** to **d**). Modified from an original courtesy of C. Greider, from *Bioess.* 12:363 (1990).

cells, leading to eventual loss of telomeres and development of chromosome instability. This instability may contribute to the debilitating effects of old age in humans and other mammals.

Replication of the circular DNA molecules of prokaryotes, mitochondria, chloroplasts, and the viral genomes with circular DNA molecules presents no equivalent problems to the mechanism because there are no free ends. As a result, a nick opened at any point around the circle provides a 3′-OH group from which primer insertion or gap filling may proceed.

INITIATION OF REPLICATION: REPLICATION ORIGINS AND REPLICONS

Replication Origins

The initiation of DNA synthesis is associated with sequence elements that act as *replication origins.* In many respects replication origins function in a manner similar to promoters of transcription. The origins are recognized by proteins that bind and destabilize the DNA. The destabilization opens or "melts" the DNA in the region of the origin and arranges it into a conformation that is recognized and bound by the replication enzymes.

The *E. coli* DNA circle has a single replication origin, *oriC*, that consists of a 245 base-pair segment containing several repeats of 9 and 13 base-pair sequences known as *9-mers* and *13-mers* (Fig. 23-20). Initiation leads to the formation of two replication forks, which proceed in opposite directions from the origin (Fig. 23-21*a*). The two forks meet and complete replication at a point approximately 180° from the origin. The termination region contains several sequence elements that are essential for release of the replication complex and cessation of DNA replication (Fig. 23-22).

Replication origins with 9-mers and 13-mers have been isolated and identified in several other bacteria. Each of these origins, when inserted into *E. coli*, successfully initiates replication.

Experiments by A. Kornberg and his associates and others implicated several proteins in initiation of replication at the *oriC* element in *E. coli*. One of these, *DnaA*, is an ATP-binding protein that recognizes and links in multiple copies to the four 9-mers within the *oriC* origin of *E. coli*. Kornberg and his colleagues propose that the binding wraps the DNA into a loop and forces the DNA to unwind in the adjacent 13-mers, which are rich in A-T base pairs (see Fig. 23-20*c*). The high concentration of A-T base pairs, held together by only two hydrogen bonds as compared to the three holding G-C base pairs together, makes the double helix of the 13-mers easier to unwind. After the DNA unwinds, the helicase binds and expands the open region, in an interaction that requires ATP and is enhanced by another initiation protein, *DnaC*. Binding and brief transcription by RNA polymerase from a promoter near the *oriC* element also enhances initiation, probably by facilitating unwinding of the DNA in the origin. The short RNA chain assembled by the RNA polymerase does not serve as a primer and apparently has no role in DNA replication other than promoting initial unwinding at the replication origin. Once the

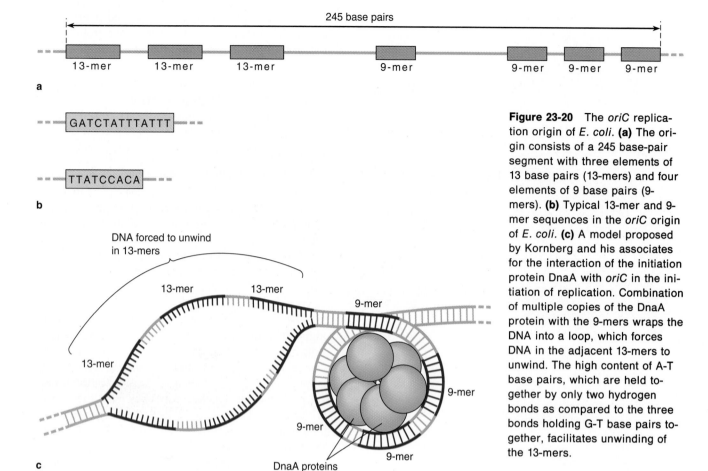

helicase is bound, the primase adds to the complex and assembles primers on the unwound DNA templates, and replication proceeds.

Replication origins have as yet been identified in only one eukaryotic species, the yeast *Saccharomyces.* The origins were identified by adding various *Saccharomyces* DNA sequences to DNA circles and testing which could promote the initiation of DNA replication in living cells. The sequences capable of serving as origins, termed *ARS* in yeast (for *Autonomously Replicating Sequence*), consist of an element about 100 to 120 base pairs in length, including a central group of several repeats of an 11 base-pair core element rich in A-T base pairs. The central group of core repeats is bounded on either side by longer flanking sequences. The A-T base pairs in the core sequences, as in the *E. coli* 13-mers, probably facilitate unwinding of the DNA. The flanking sequences contain a complex combination of multiple, overlapping sequence elements that include variations of the core. Both core and flanking sequences are required for successful initiation of replication.

M. Budd and J. L. Campbell and others identified a group of proteins that recognize and bind sequences in the flanking regions of the *ARS* element. Among these proteins may be some that act as initiators for DNA replication in yeast cells. As in bacteria there are

indications that an RNA polymerase recognizes promoterlike elements in the flanking sequences and that RNA transcription may contribute to the initiation of replication.

About 400 copies of the *ARS* sequences are distributed through the yeast genome. Within the 400 copies are subgroups with highly conserved copies of the core sequence but different variations of the flanking sequences. W. L. Fangman and his coworkers and others found that the different *ARS* subgroups are activated at distinct times in S, probably through initiator proteins that specifically recognize the sequence subtypes. The controlled activation of *ARS* subgroups replicates subparts of the DNA in a definite order, with some regions replicating early in S and others later. Replication of the genome in a definite order appears to be typical of eukaryotes (see below).

Replication usually starts in the same locations from one cell to the next within higher eukaryotes, indicating that specific origins are probably generally present in eukaryotes. Although it has not yet been possible definitely to identify higher eukaryotic replication origins, evidence from several sources indicates that they lie very near or share sequence elements with transcriptional promoters. For example, R. Tjian and his coworkers and others found that one group of proteins

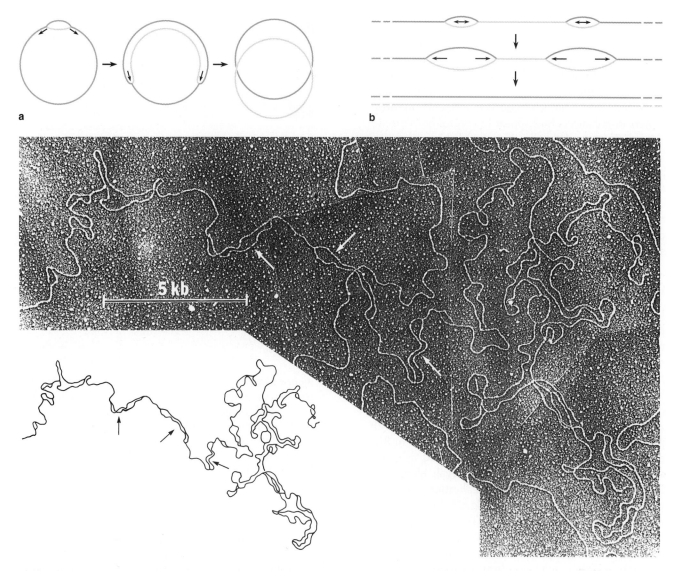

Figure 23-21 Replication from a single replication origin in the circular DNA molecule of a prokaryote **(a)** and from multiple origins in the linear molecules of eukaryotes **(b)**. **(c)** Replication "bubbles" (arrows) in replicating DNA isolated from *Drosophila*. Each bubble results from bidirectional movement of two replication forks from an initiation site, as in **(b)**. The bar shows the length occupied by 5000 base pairs in the DNA. Micrograph courtesy of D. S. Hogness.

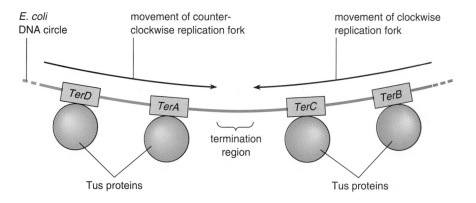

TerA = AATTAGTATGTTGTAACTAAAGT
TerB = AATAAGTATGTTGTAACTAAAGT
TerC = ATATAGGATGTTGTAACTAATAT
TerD = CATTAGTATGTTGTTAACTAAATG
consensus = AATTAGTATGTTGTAACTAAANT

Figure 23-22 The region signaling termination of replication in *E. coli*, which lies 180° around the DNA circle from the replication origin. Four *Ter* sequence elements are necessary for termination, two on each side of the termination site. The *Ter* sequences are recognized and bound by Tus proteins, which evidently inhibit the helicase unwinding DNA for replication. When the *Ter* sequences are bound by the Tus proteins, replication forks moving clockwise or counterclockwise pass through the sequences and terminate as they meet in the termination region. From data presented in P. L. Kuempel et al., *Cell* 59:581 (1989).

binding the CAAT box of transcriptional promoters (the CTF/NF-I group; see p. 702) can initiate replication as well as transcription. Other investigators, including P. J. Rosenfeld and T. J. Kelly, discovered additional proteins that apparently regulate both transcription and replication, or identified sequence elements in promoters or enhancers that appear to control both mechanisms. The regulatory proteins probably open the DNA and place it in a conformation that can be bound by the enzymes and factors carrying out eukaryotic replication. A brief period of RNA transcription, carried out by an RNA polymerase recognizing elements in the promoter sequences associated with replication origins, may promote DNA unwinding in the origins as in yeast and prokaryotes.

Replicons

Higher eukaryotes, in which genomes are considerably larger than in *Saccharomyces*, typically have many more replication origins. Estimates of the total have been made by a technique originally worked out in 1968 by J. A. Huberman and A. D. Riggs. These investigators labeled replicating DNA by briefly exposing actively dividing Chinese hamster cells to radioactive thymidine. After labeling, the replicated DNA was extracted and prepared for light microscope autoradiography. In the preparations radioactivity could be detected in short segments spaced at 15 to 60 μm intervals along the DNA molecules, as expected if replication proceeds simultaneously at multiple sites. (Fig. 23-23 shows the results of a similar experiment with *Triturus* DNA.) This spacing indicates that each hamster chromosome, which contains several centimeters of DNA on the average, probably initiates replication from thousands of sites.

Subsequent work established that replication in all eukaryotic chromosomes proceeds from multiple origins. A length of DNA whose replication is initiated from one of these origins is termed a *replicon*. A replication origin lies near the center of each replicon. Activation of an origin produces two replication forks that move in opposite directions (see Fig. 23-21*b* and *c*) until they meet forks from other replicons. Replicons vary in length from as short as 4 μm, or 13,000 base pairs, to as long as 280 μm, or 900,000 base pairs, in different species. In all there may be from 20,000 to as many as 100,000 or more replicons per haploid genome in higher eukaryotes.

Rather than activating all replicons simultaneously, higher eukaryotic cells, like yeast cells, initiate replication in specific subgroups or clusters of replicons at certain times during S. The patterns in which replicons are activated have been traced by exposing cells in S to a pulse of 5-bromodeoxyuridine (BrdU), which is incorporated into replicating DNA. Segments of the chromosomes undergoing replication during the pulse can be identified at the next metaphase by a fluorescent

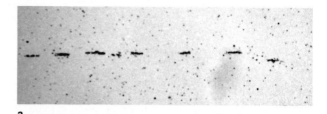

a

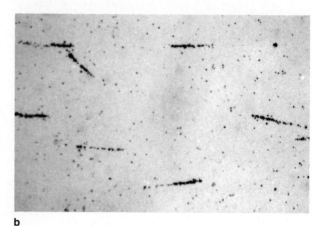

b

Figure 23-23 Replication origins detected by autoradiography in **(a)** embryonic and **(b)** premeiotic cells of the newt *Triturus*. The origins are more widely spaced in the premeiotic cells, indicating that fewer origins per unit length are activated in premeiotic replication (see text). × 450. Courtesy of H. G. Callan, from *Biol. Zbl.* 95:531 (1976).

stain that binds specifically to BrdU. Experiments of this type show that subregions of the chromosomes replicate at more or less fixed times within the S phase.

Typically, genes in active segments of the chromosomes replicate early in S, and genes in inactive or heterochromatin segments replicate later. In some species entire chromosomes may enter replication early or late. In mammalian females, for example, the X chromosome that is active in transcription replicates early in S and the inactive X later in somatic cells.

The pattern of replicon activation in eukaryotic cells changes with development. H. G. Callan found that in amphibian *(Triturus)* embryos, the S phase runs its course in only 1 hour at the blastula stage. Later, at the neurula stage, S lasts 4 to 6 hours. In adult *Triturus* body cells, S lasts 48 hours, and in cells about to enter meiosis it extends to 200 hours. By preparing autoradiographs after cells were exposed to radioactive precursors at these stages, Callan observed that the variations are correlated with differences in the distances that separate replicating segments in the DNA. During the neurula stage, for example, the segments taking up label are spaced about 40 μm apart (see Fig. 23-23*a*). In premeiotic cells, spacing between the replicating segments extends to 100 μm or more (see Fig.

23-23b). Therefore, differences in the overall rate of replication in *Triturus* appear to be correlated with the number of replicons activated per unit length of chromosome.

Equivalent correlations between the duration of S and the number of replicons activated have been observed in a variety of eukaryotes. In *Drosophila*, for example, 20,000 replicons, separated by an average DNA length of 7900 base pairs, are activated simultaneously during the S phase of early embryonic cells, which requires only about 3.5 minutes for completion. In later embryos, when the S phase has lengthened to 20 minutes, replicons are activated in subgroups rather than simultaneously. In larvae, in which S requires 10 to 12 hours for completion, the average distance separating activated replicons is 40,000 to 100,000 base pairs. Not all organisms regulate the duration of S by variations in the number of replicons activated; some, such as mouse cells in the interphase before meiosis, increase the duration of S by slowing the rate of DNA unwinding and replication.

Callan and others suggested that control of the time and place of replicon activation is exerted through regulatory proteins that recognize different groups of origins. Binding of the control proteins to these origins promotes DNA unwinding and initiation of the replication mechanism at these sites. Presumably, origins activated simultaneously share the same DNA sequences and are activated by the same control proteins, in a pattern similar to the *ARS* subgroups of yeast.

Normally, each replicon is activated only once during a given S period so that no segments of the genome are overreplicated. Although the basis for this control is unknown, it is believed to depend on one or more proteins that combine with and block the origin of each replicon once its replication is complete. Support for this interpretation comes from the results of fusion of cells (see p. 186) in the S stage of interphase with cells in G1 or G2. In fusions between S and G1 cells, factors carried by the S cells can induce G1 nuclei to enter replication. Fusion of an S cell with a G2 cell, however, induces no further replication in the G2 nucleus, as if the replication origins in the G2 cell are indeed blocked and unable to initiate. Nuclei in mitosis also fail to enter replication when fused with an S cell, indicating that the block to replication persists until the next G1. Presumably, the blocking proteins are removed during the mitosis/G1 transition.

Replicons and DNA Amplification

Though replicons in most eukaryotic cell types are normally activated only once during S, in certain cells some replicons are activated repeatedly, so that portions of the genome are copied more than once. Local increases in the number of DNA sequences by multiple activations of replicons are one source of *DNA amplification*, a process by which extra copies of genes are generated (see also p. 938).

DNA amplification may be programmed as a normal part of development or may be a random, unscheduled event leading to alterations in cell function. Programmed DNA amplification takes place in cells in which rRNAs or proteins are required in greater quantities than can be produced through maximum activity of the normal gene complement. Unscheduled DNA amplification underlies the development of drug resistance in some cell types and is implicated in the development of some types of cancer.

Programmed DNA Amplification One of the best-documented examples of programmed DNA amplification occurs in the development of polytene chromosomes in *Drosophila* larvae (see Supplement 17-2). In salivary gland nuclei of *Drosophila* larvae, for example, replicons in some areas of the genome are activated repeatedly during G1, producing a thousand or more extra copies of some DNA segments. The replicons of other segments, such as those containing highly repetitive sequences, are activated only once or even remain inactive during development of the polytene nuclei, so that these portions of the genome remain at the usual G1 or G2 level. The result is a branched structure in which replicated copies of the original DNA molecule occur in different numbers in different regions of a chromosome (Fig. 23-24).

Drosophila also provides a second example of programmed DNA amplification. This mechanism amplifies two groups of genes encoding proteins of the outer coat (the *chorion*; see p. 1105) of the eggs in this species. Within follicle cells secreting the chorion, genes encoding about 20 different egg-coat proteins occur in clusters on separate chromosomes. In follicle cells, but not other *Drosophila* body cells, the clusters are amplified about 20 times in one cluster and 80 times in the other. Isolated chromatin containing the amplified chorion genes appears multibranched (Fig. 23-25), as expected if amplification occurs by repeated activation of replicons.

Not all examples of programmed DNA amplification occur by repeated activation of replicons. In amphibian oocytes, extra copies of sequences encoding rRNAs are made by a process called the *rolling circle mechanism*. In this mechanism a circular copy of the rRNA genes is replicated continuously to produce many copies (for details, see Supplement 25-1).

Unprogrammed DNA Amplification Unscheduled amplification of DNA segments, presumably by random reactivation of one or more replicons, evidently occurs at a low but significant rate in many eukaryotic cells. The mechanisms producing randomly generated amplification are unknown. However, the extra replications may reflect a failure of blocking proteins to bind to

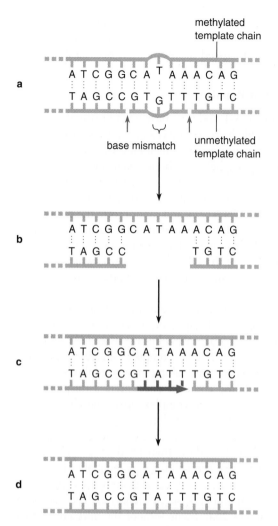

Figure 23-26 Excision repair of mismatched bases in *E. coli.* **(a)** Repair enzymes recognize a mispaired region, break one chain of the DNA, and remove several to many bases including the mismatched base. **(b)** The excision leaves a gap in the DNA, which is filled in **(c)** by a DNA polymerase using the intact template chain as a guide. **(d)** The single-chain nick left after gap filling is sealed by DNA ligase to complete the repair.

chains as compared to bacterial DNA, the degree of methylation might also provide one signpost distinguishing between the template and copy chains, as it does in bacteria. Detection of single-chain nicks may also contribute to recognition of newly synthesized chains in eukaryotes.

Repair of DNA Alterations Arising from Chemicals or Radiation

Most alterations caused by chemical modifications or radiation are breaks in the sugar-phosphate backbone chain, missing bases, or alteration of bases or sugars to other forms. In some cases chemical alterations pro-

duce covalent linkages between adjacent bases on the same nucleotide chain. (Supplement 23-2 gives details of some of these alterations.) Some enzymatic systems repairing these defects recognize local distortions created in the DNA double helix by the change; others are able to recognize individual altered nucleotides.

In general the pathways repairing alterations are varied and *redundant*—the same repair may be carried out by several different pathways acting in coordination. If one system becomes inoperative or ineffective, the repair may be accomplished by another route using different enzymes.

Although complex and highly varied, mechanisms repairing DNA damage fall into a relatively few general types. One type involves what is known as *direct repair*. The enzymes carrying out this repair recognize chemically altered bases and convert them to their original, correctly paired form. Most enzymes in this category recognize and remove specific modifying chemical groups added to bases, such as methyl, ethyl, or larger organic groups.

The second type of damage repair involves *base removal* rather than conversion of altered bases to their original form. The enzymes catalyzing this mechanism, the *DNA glycosylases*, recognize an altered base and break the bond linking it to the deoxyribose sugar. There are several types of DNA glycosylases, each able to recognize a specific class or group of altered bases. One DNA glycosylase, for example, recognizes and removes bases from which an amino ($-NH_2$) group has been removed (see Supplement 23-2).

The third repair mechanism carries out excision repair by a mechanism similar to that detecting and repairing mismatches. Many of the enzymes of this mechanism, like those operating in mismatch repair, are stimulated by distortions or kinks in the DNA helix rather than by specific altered bases. Frequently, DNA glycosylases and the enzymes catalyzing excision repair work in coordination to cut mutated bases from DNA. For example, in one pathway a site lacking a base because of the activity of a DNA glycosylase subsequently serves as a distorted site recognized for excision by a DNA endonuclease.

The gaps left in DNA nucleotide chains by the repair mechanisms remain unfilled in bacteria only when all three DNA polymerases are defective in mutant cells. Mutations in any one of the three enzymes cause a drop in the rate of gap filling after excision repair but do not completely halt the process. This suggests that, rather than being the exclusive function of any one enzyme, gap filling associated with DNA repair may potentially be carried out by all three prokaryotic enzymes working in coordination or as substitutes for each other. Because the most drastic reductions in gap-filling efficiency are noted in bacterial mutants deficient in DNA polymerase I, this enzyme is thought to be the primary repair enzyme.

Mutations and Their Effects

The astonishing fidelity of DNA replication achieved by proofreading and DNA repair produces copies that are highly faithful to the parental molecules entering the replication process. However, even with the safeguards built into the system, a small but significant number of errors escape detection and persist in the replicated copies. These persistent errors are the primary source of mutations in all living organisms.

Though the final error rates are very low, the large amounts of DNA replicated and the frequency of replication magnifies the numbers that persist. In a bacterium such as *E. coli*, for example, with about 4 million base pairs, an error rate of 1×10^{-10} means that an incorrectly matched base pair is likely to leak past the proofreading and repair mechanisms about once every two or three replications. Although the total per cell is small, the rapidity of bacterial replication and cell division and the large number of cells involved can produce millions of mistakes per generation in large bacterial populations. In eukaryotes such as humans, with about 3×10^{10} base pairs per nucleus, incorrectly paired nucleotides are likely to appear and persist somewhere in the genome several times each replication. With literally billions of replications taking place each day in the human body (the divisions maintaining red blood cells alone proceed at the rate of 2 million per second!), the total number of persisting mispaired bases becomes potentially huge—on the order of at least millions or more per day per individual.

Any change in a DNA molecule that alters its sequence so that it no longer represents an exact copy of its parental DNA molecule constitutes a mutation. Mutations in functional sequences of the genome have varying effects depending on their positions. Mutations in the promoter or other control regions of a gene may alter transcriptional activity but have no effect on the amino acid sequence of the protein encoded in the gene. Similarly,

mutations in the 5' or 3' untranslated regions may alter the activity of the mRNA encoded by the gene in protein synthesis without affecting the amino acid sequence of the encoded protein.

Mutations within the coding sequence of a gene, however, may change a codon for one amino acid to a codon for another. For example, a single substitution of C for A at the third position of the codon GAA changes the amino acid specified at this position from glutamine to asparagine (parts *a* and *b* of the figure). The effects of such a change in a single amino acid may range from negligible to severe depending on the position of the change and the degree of difference in size and chemical properties between the original and substituted amino acid.

Base substitutions may also change a codon for an amino acid to a terminator codon. This type of change is likely to have drastic effects on the activity of a protein because the entire amino acid sequence after the position of the new terminator codon will be lost.

Not all mutations in coding regions cause amino acid substitutions. For example, mutation of the GAA codon for glutamine to GAG causes no change in the amino acid sequence of the encoded protein because both GAA and GAG code for the same amino acid (parts *a* and *c* of the figure). Alterations that cause no change in the amino acid sequence of the encoded protein are called *silent mutations.*

Most mutations in the coding regions of mRNA genes are detrimental. They affect only the individual if they occur in somatic cells. However, those persisting in reproductive cells are passed on to offspring and may become the source of hereditary defects or disease; a very few are beneficial. Whether detrimental or beneficial, they are the ultimate source of variability for the evolutionary process.

Suggestions for Further Reading: See p. 990.

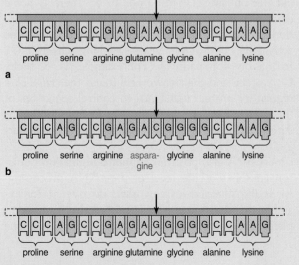

a

b

c

gene

protein encoded in gene

The effects of mutations on proteins encoded in a gene. **(a)** The gene sequence and the corresponding sequence of amino acids in a protein. A mutation changing adenine to cytosine at the position marked by the arrow would cause substitution of asparagine for glutamine in the encoded protein **(b)**. A change from adenine to glycine at the same position **(c)** would not change the amino acid sequence because the codons GAA and GAG both stand for glutamine.

Several eukaryotic DNA polymerases, including α, β, and ε, appear to be associated with gap filling in connection with DNA repair. The eukaryotic enzymes appear to be specialized for greatest efficiency in coordination with different repair pathways. Inhibitors of polymerase α, for example, have the most pronounced effects on the repair of DNA damage resulting from ultraviolet light or agents adding straight-chain or cyclic organic groups to DNA bases (*alkylating agents;* see Supplement 23-2). Repair of damage from bleomycin, an antibiotic that produces lesions in DNA similar to the effects of ionizing radiation (see Supplement 23-2), is more greatly affected by inhibitors of DNA polymerase β.

Under normal circumstances the enzymes carrying out DNA repair of all kinds operate so rapidly that neither base mismatches arising from replication errors nor alterations from mutagens persist long enough to cause problems for the individual. In human cells, for example, alterations caused by exposure to ultraviolet light are normally repaired within 24 hours. Active genes are repaired before inactive ones; although the basis for this preference is not understood, it may depend on structural differences between active and inactive chromatin.

Deficiencies in DNA repair are suspected to underlie several human hereditary diseases (Table 23-3). Affected persons are highly susceptible to DNA damage from agents such as radiation or mutagens and show an unusually high incidence of cancer. Breaks and gaps are frequent in their DNA, and recovery from DNA breaks introduced by exposure to mutagens is slow or nonexistent. Individuals with *xeroderma pig-*

mentosum, for example, are deficient in enzymes carrying out excision repair. These persons are so susceptible to damage by ultraviolet light that pigmented spots, lesions, and tumors, frequently malignant, appear rapidly in the skin in any body regions exposed to sunlight. Mutations in at least seven different genes may contribute to xeroderma pigmentosum, which occurs in 1 out of every 250,000 persons.

The characteristics of individuals with deficiencies in DNA repair underscore its importance to survival. The serious skin lesions of persons with xeroderma pigmentosum also show how injurious ultraviolet light would be without our DNA repair mechanisms—without them, none of us could venture outside except at night!

DUPLICATION OF CHROMOSOMAL PROTEINS IN EUKARYOTES

Although DNA replication is the most critical event of the S phase for the distribution of hereditary information to daughter cells, duplication of the chromosomes in preparation for cell division also involves the doubling of structural, regulatory, and other proteins forming regular parts of the chromatin. The five histone proteins, in particular, exist in chromatin in precise ratios with DNA that must be maintained to preserve chromatin structure as the DNA quantity doubles during S. The doubling of histone and nonhistone proteins assures that the products of chromosome duplication are structurally and functionally the same as the chromosomes entering the process and are fully prepared for mitotic division.

Synthesis of Chromosomal Proteins

The bulk of the five histones associated with DNA in chromatin—H1, H2A, H2B, H3, and H4—is synthesized in close coordination with DNA replication. Transcription of the genes encoding the five histones is initiated in late G1, just prior to S. There is some evidence that regulatory proteins stimulating transcription of the histone genes are produced or become activated as part of the events switching cells from G1 to the division pathway. For example, G. S. Stein, G. L. Stein, and their coworkers found that *HiNFD,* a regulatory protein that binds to control sequences of the histone H4 gene and promotes its transcription, is barely detectable at G1 but becomes highly active during S.

There is also posttranscriptional regulation of histone mRNAs at the level of pre-mRNA processing. M. L. Birnstiel and his coworkers discovered, for example, that the segment of U7 snRNA that forms complementary base pairs with the 3' end of histone

Table 23-3	Human Diseases Known or Suspected to Be Caused by Deficiencies in DNA Repair
Disease	Characteristics
Ataxia-telangiectasia	Neurological disorders; immunological deficiencies; high susceptibility to lymphomas
Cockayne syndrome	Skin sensitive to sunlight; dwarfism; mental retardation; premature aging
Hereditary retinoblastoma	High susceptibility to eye cancer
Santis–Caccione syndrome	Xeroderma pigmentosum with severe neurological involvement and mental retardation
Xeroderma pigmentosum	Marked susceptibility to skin, eye, and tongue cancer from exposure to sunlight

mRNAs is masked in nondividing cells. In cells at S the segment is exposed, allowing it to pair with histone pre-mRNAs and participate in the processing reactions that clip surplus sequences from their 3' ends. (For details of the role of U7 in histone pre-mRNA processing, see p. 597).

Fully processed mRNAs for all five histone types begin to appear in the cytoplasm as soon as S begins. These mRNAs are immediately translated to form the histone proteins. The newly assembled histones are subsequently routed into the nucleus for assembly with DNA into nucleosomes (see p. 546). Because this nucleosome assembly follows closely behind replication forks, the time during which replicated DNA remains in the naked, uncomplexed form is very limited (Fig. 23-27).

Transcription of the histone genes peaks early in S but continues at elevated levels as long as the DNA is replicating. The accumulated histone mRNAs, which have a half-life during S in higher eukaryotes of about 45 to 60 minutes, drive histone synthesis at high levels throughout S. As DNA replication becomes complete, transcription of the histone genes shuts down, and newly synthesized and processed histone mRNAs cease to enter the cytoplasm. Cessation of histone gene transcription is coupled with a rapid drop in the half-life of histone mRNAs to about 10 to 15 minutes. This halt in histone gene transcription and greatly increased mRNA breakdown rapidly clears the cytoplasm of histone mRNAs. As a result, further synthesis of histone proteins essentially stops.

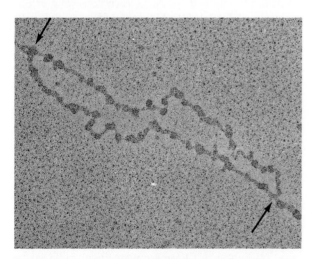

Figure 23-27 Replicating chromatin isolated from a *Drosophila* embryo, showing two replication forks (arrows). Nucleosomes are visible in close proximity to the replication forks in both replicated and prereplicative segments, indicating that DNA is free of nucleosomes only in the immediate vicinity of the replication fork, and that nucleosomes are added to the replicated segments almost immediately after the fork passes. × 96,000. Courtesy of S. L. McKnight, from *Cell* 12:795 (1977). Copyright Massachusetts Institute of Technology.

Thus histone synthesis begins through transcriptional regulation of the histone genes and posttranscriptional control of histone pre-mRNA processing at the beginning of S, and ends through a combination of transcriptional and posttranscriptional controls at the close of S. The signals turning off histone transcription probably involve destruction or deactivation of regulatory proteins recognizing and binding control sequences in the promoters of histone genes. N. B. Pardey and W. F. Marzluff showed that breakdown of histone mRNAs at the end of S is regulated by an interaction between hydrolytic enzymes and the unusual stem-loop structures occurring near the end of the 3' untranslated region (see p. 595). The stem-loop structure, which occurs at the 3' end of all histone mRNAs degraded at the end of S, is necessary and sufficient for the breakdown to occur. Addition of the histone stem-loop structure gives unrelated mRNAs the stability characteristics of histone mRNAs. A globin mRNA to which the histone stem-loop structure has been added, for example, is degraded rapidly at the close of S along with the histone mRNAs.

Histone mRNAs are degraded only when they are being translated on ribosomes, suggesting that the translation apparatus causes alterations in the stem-loop structure that promote its recognition by the ribonuclease catalyzing the breakdown. Experiments by S. W. Peltz and J. Ross demonstrated that histone proteins are also necessary for the breakdown, suggesting that the excess histones accumulating when DNA replication ceases may work in a feedback mechanism to activate the breakdown process.

Not all histone synthesis is confined to the S phase. In many cell types histones are synthesized during G1 and G2 at about 5% to 8% of their level during S. This steady histone assembly is probably a "housekeeping" function that compensates for histones lost from the chromosomes as a result of random breakdown during G1 and G2. Usually, the histone variants (see p. 540) synthesized at low levels during G1 and G2 differ markedly from the histones assembled in bulk at S. In mouse liver cells, for example, the H2A.1, H2B.2, H3.1, and H3.2 sequence variants are synthesized exclusively during S; the H2A.3 and H3.3 variants are assembled at low but steady levels throughout interphase.

The mRNAs encoding histones synthesized at steady levels lack the 3' stem-loop structure characteristic of S-phase histones and, unlike the S-phase histones, have poly(A) tails. The genes encoding the continuously synthesized histones are also frequently interrupted by introns, in contrast to those active only during S, which are intronless. The continuously transcribed histone genes differ further in being scattered singly rather than clustered in groups as are the histone genes transcribed in close coordination with DNA replication. Although the differences in structures at the 3' ends of the mRNAs for constitutive and S-phase histones are undoubtedly related to the relative stability of the mRNAs, the significance of the differences in gene structure and distribution remains unknown.

Even histones normally synthesized only during DNA replication may be uncoupled from S in some cell types, particularly in developing oocytes and early embryos. In *Xenopus* and other amphibians, for example, the cytoplasm of developing oocytes becomes packed with both histone mRNAs and proteins, synthesized at a time when DNA replication has totally stopped. After fertilization DNA replication and cell division proceed rapidly with no detectable histone synthesis until the blastula stage, when the *Xenopus* embryo reaches 30,000 cells. All the histones added to the rapidly replicating DNA up to this stage are derived from the large quantities stored in the egg cytoplasm. Thus in organisms such as *Xenopus*, histone synthesis without DNA replication takes place during oocyte development, and DNA replication without histone synthesis occurs during early embryogenesis. At the late blastula/early gastrula stage, the rate of cell division slows, and histone gene transcription and synthesis of the histone proteins begin in the *Xenopus* embryo. This new histone synthesis is tightly coupled to DNA replication.

In general, storage of large quantities of histones in oocytes is characteristic of species in which cell division in early embryos takes place at rates so high that cycles of DNA replication and division proceed without detectable pauses in G1. Insects, amphibians, reptiles, and birds are among the animals showing this developmental pattern.

In contrast to the marked increases in histone synthesis during DNA replication, the synthesis of bulk nonhistone proteins does not change appreciably during S. This is not too surprising, since many nonhistone proteins are enzymes associated with the more or less constant chromosomal functions of transcription, or proteins combining with the RNA products of transcription (see p. 541). It is likely, however, that against this background of relatively constant nonhistone protein synthesis, at least some regulatory proteins and any nonhistones forming part of the structural framework of chromosomes are duplicated during S in patterns similar to the histones.

Nucleosomes and DNA Replication

The histone proteins combine with DNA to form nucleosomes, the structural subunits of chromatin. In nucleosomes two molecules each of the core histones—H2A, H2B, H3, and H4—form an octamer around which approximately two turns of DNA are wrapped. The wrapped DNA is stabilized by a single H1 molecule positioned outside the core structure (see Fig. 14-17 and pp. 546 to 550).

For replication to proceed nucleosomes probably unfold or disassemble temporarily from the DNA during the brief period when a replication fork passes. This conclusion is supported by electron micrographs of replicating chromatin, which show that the forks are free of nucleosomes. Measurements of the nucleosome-free region around replication forks reveal that only some 200 to 300 base pairs of DNA are exposed, roughly one to one and one-half times the length of DNA wrapped around a nucleosome. This suggests that the short DNA lengths exposed at the fork reassociate with the histones or that nucleosomes refold as soon as enough DNA is replicated in either copy to wrap completely around a nucleosome core. It is still uncertain whether nucleosomes dissociate completely from the DNA as a replication fork passes or merely unfold or open out in some way without detaching to accommodate passage of the replication enzymes.

Electron micrographs also show that nucleosomes are present in their usual numbers and separation in both DNA chains behind the fork. This means that new nucleosomes must be added following the fork to bring the total number up to twice the G1 amount. In what form are the new nucleosomes added?

Experiments designed to answer this question are complicated by the fact that there are several possible patterns by which the addition might occur (Fig. 23-28). It is possible that (1) the octamers of the old core particles are completely conserved as a unit, so that the nucleosomes on the daughter chains are constructed entirely from either new or old octamers (Fig. 23-28a); (2) only half-octamers, each consisting of one molecule of H2A, H2B, H3, and H4, are conserved, with the result that the octamers on daughter chains may contain old-old, old-new, or new-new halves (Fig. 23-28b); or (3) there is no strict conservation of core structure, so that the octamers on daughter chains contain mixtures of old and new histones (Fig. 23-28c). Attempts to sort out these possibilities have produced conflicting results.

One series of experiments, conducted in the laboratories of H. Weintraub and A. Worcel and others, gave indications that octamers are conserved as a unit in the pattern shown in Figure 23-28a. I. M. Leffak, R. Granger, and Weintraub, for example, labeled newly synthesized histones by exposing replicating cells to amino acids containing heavy carbon and nitrogen (^{13}C and ^{15}N) during S. Octamers were then stabilized by exposing them to a protein crosslinker. The stabilized octamers were isolated and centrifuged by the buoyant density technique (see p. 125), which separated the octamers according to density.

If octamers are completely conserved, as in Figure 23-28a, the crosslinked cores should contain either all light (old) or all heavy (new) histones (made heavy by the incorporation of ^{13}C and ^{15}N amino acids). In this case the two types would migrate to different levels

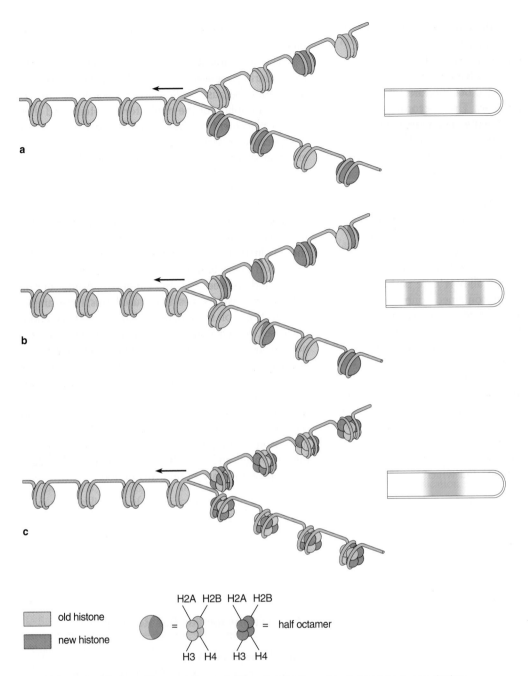

old histone

new histone

H2A H2B H2A H2B

= half octamer

H3 H4 H3 H4

Figure 23-28 Three possible patterns for the distribution of individual histones during replication. Shown to the right of each possible outcome are the results expected from centrifugation if newly synthesized histones are assembled from density-labeled amino acids (see text). **(a)** Conservation of octamers; **(b)** conservation of half-octamers; **(c)** mixed or hybrid distribution of histones in nucleosomes after replication. The arrows show the direction of fork movement.

in the centrifuge tube, producing two distinct bands. This was the result obtained by Weintraub and his coworkers, indicating that old octamers are conserved as a unit, and new core histones are added in the form of exclusively new octamers.

Very different results were obtained in a similarly designed experiment by V. Jackson. Jackson, using a different crosslinker, formaldehyde, found many of the newly synthesized core particles to be hybrid or intermediate in density, in agreement with the outcome shown in Figure 23-28c. After obtaining the hybrid core particles by centrifugation, Jackson disassembled them by reversing the crosslinker. Gel electrophoresis of the disassembled core particles clearly supported their hybrid nature. Some of the core particles contained all old H3 and H4 histones, evidently combined into

Campbell, J. 1988. Eukaryotic DNA replication: yeast bares its ARSs. *Trends Biochem. Sci.* 13:212–17.

Diffley, J. F. X., and Stillman, B. 1990. The initiation of chromosomal DNA replication in eukaryotes. *Trends Genet.* 6:427–32.

Fangman, W. L., and Brewer, B. J. 1991. Activation of replication origins in yeast chromosomes. *Ann. Rev. Cell Biol.* 7:375–402.

Hurwitz, J.; Dean, F. B.; Kwong, A. D.; and Lee, S.-H. 1990. The *in vitro* replication of DNA containing the SV40 origin. *J. Biolog. Chem.* 265:18403–6.

Linskens, M. H. K., and Huberman, J. A. 1990. Two faces of higher eukaryotic DNA replication origins. *Cell* 62:845–47.

Marahrens, Y., and Stillman, B. 1992. A yeast chromosomal origin of DNA replication defined by multiple functional elements. *Science* 255:817–23.

Spradling, A., and Orr-Weaver, T. 1987. Regulation of DNA replication during *Drosophila* development. *Ann. Rev. Genet.* 21:373–403.

Stillman, B. 1989. Initiation of eukaryotic DNA replication *in vitro. Ann. Rev. Cell Biol.* 5:197–245.

Umek, R. M.; Linskens, M. H. K.; Kowalski, D.; and Huberman, J. A. 1989. New beginnings in studies of eukaryotic DNA replication origins. *Biochim. Biophys. Acta* 1007:1–14.

DNA Amplification

Hamlin, J. L.; Leu, T.-H.; Vaughn, J. P.; Ma, C.; and Dijkwel, P. A. 1991. Amplification of DNA sequences in mammalian cells. *Prog. Nucleic Acids Res. Molec. Biol.* 41:203–39.

Kafatos, F. C.; Orr, W.; and Delidakis, C. 1985. Developmentally regulated gene amplification. *Trends Genet.* 1:301–6.

Orr-Weaver, T. L. 1991. *Drosophila* chorion genes: cracking the eggshell's secrets. *Bioess.* 13:97–105.

Schimke, R. T. 1988. Gene amplification in cultured cells. *J. Biolog. Chem.* 263:5989–92.

Spradling, A., and Orr-Weaver, T. 1987. Regulation of DNA replication during *Drosophila* development. *Ann. Rev. Genet.* 21:373–403.

Stark, G. R.; Debatisse, M.; Guilooto, E.; and Wahl, G. M. 1989. Recent progress in understanding mechanisms of DNA amplification. *Cell* 57:901–8.

DNA Repair

Bohr, V. A., and Wassermann, K. 1988. DNA repair at the level of the gene. *Trends Biochem. Sci.* 13:429–33.

Cleaver, J. E., and Karentz, D. 1987. DNA repair in man: regulation by a multigene family and association with human disease. *Bioess.* 6:123–27.

Downes, C. S., and Johnson, R. T. 1988. DNA topoisomerases and DNA repair. *Bioess.* 8:179–84.

Grossman, L.; Caron, P. R.; Mazur, S. J.; and Oh, E. Y. 1988. Repair of DNA-containing pyrimidine dimers. *FASEB J.* 2:2696–2701.

Heywood, L. A., and Burke, J. F. 1990. Mismatch repair in mammalian cells. *Bioess.* 12:473–77.

Mitchell, D. L., and Hartmann, P. S. 1990. The regulation of DNA repair during development. *Bioess.* 12:74–79.

Modrich, P. 1991. Mechanisms and biological effects of mismatch repair. *Ann. Rev. Genet.* 25:229–53.

Sancar, A., and Sancar, G. 1988. DNA repair enzymes. *Ann. Rev. Biochem.* 57:29–67.

Wood, R. D., and Coverly, D. 1991. DNA excision repair in mammalian cell extracts. *Bioess.* 13:447–53.

Duplication of Chromosomal Proteins

Dilworth, S. M., and Dingwall, C. 1988. Chromatin assembly *in vitro* and *in vivo. Bioess.* 9:44–49.

Fairman, M. P. 1990. Nucleosome segregation—divided opinions? *Bioess.* 12:237–39.

Gruss, C., and Sogo, J. M. 1992. Chromatin replication. *Bioess.* 14:1–8.

Heintz, N. 1991. The regulation of histone gene expression during the cell cycle. *Biochim. Biophys. Acta* 1088:327–39.

Marzluff, W. F., and Pandey, N. B. 1988. Multiple regulatory steps control histone mRNA concentration. *Trends Biochem. Sci.* 13:49–52.

Osley, M. A. 1991. The regulation of histone synthesis in the cell cycle. *Ann. Rev. Biochem.* 60:827–61.

Svaren, J., and Chalkley, R. 1990. The structure and assembly of active chromatin. *Trends Genet.* 6:52–56.

Mutagens and Mutations

Drake, J. W. 1991. Spontaneous mutation. *Ann. Rev. Genet.* 25:125–46.

Lett, J. T. 1990. Damage to DNA and chromatin structure from ionizing radiation and the radiation sensitivities of mammalian cells. *Prog. Nucleic Acid Res. Mol. Biol.* 39:305–52.

Singer, B., and Kusmierek, J. T. 1983. Chemical mutagenesis. *Ann. Rev. Biochem.* 51:655–93.

Replication in Mitochondria and Chloroplasts

Cantatore, P., and Saccone, C. 1987. Organization, structure, and evolution of mammalian mitochondrial genomes. *Internat. Rev. Cytol.* 108:149–208.

Clayton, D. A. 1991. Replication and transcription of vertebrate mitochondrial DNA. *Ann. Rev. Cell Biol.* 7:453–78.

Nelson, B. D. 1987. Biogenesis of mammalian mitochondria. *Curr. Top. Bioenerget.* 15:221–72.

Schinkel, A. H., and Tabak, H. F. 1989. Mitochondrial RNA polymerase: dual role in transcription and replication. *Trends Genet.* 5:149–54.

Review Questions

1. Define semiconservative, conservative, and dispersive replication. Outline two experiments indicating that replication proceeds by the semiconservative pathway.

2. Outline the steps taking place in the polymerization of a DNA chain on a DNA template. How do these steps differ from the assembly of an RNA chain on a DNA template? What is a primer? Why are primers required in replication?

3. How are the problems of template unwinding and unidirectional replication solved in replicating systems? Define the roles of DNA-binding proteins, helicases, unwinding enzymes, DNA polymerases, topoisomerases, and DNA ligases in replication. What is the difference in activity between topoisomerases I and II?

4. What are the leading and lagging chains in replication? What are Okazaki fragments, and how do they function in replication? What is the possible relationship between the lengths of Okazaki fragments in eukaryotes and nucleosomes?

5. Compare the DNA polymerases of prokaryotes and eukaryotes. What evidence indicates that polymerase III of prokaryotes and polymerases α and δ of eukaryotes are the primary replication enzymes? What is meant by the processivity of a DNA polymerase?

6. What is an asymmetric dimer? What is the possible equivalent of the prokaryotic asymmetric dimer in eukaryotes? Compare the association of helicase and primase activity with the DNA polymerases of prokaryotes and eukaryotes.

7. What is proofreading? What evidence indicates that proofreading actually takes place? What combinations of enzymes or activities are believed to accomplish proofreading in prokaryotes and eukaryotes? Why is proofreading necessary? What are tautomers, and how are they related to the requirement for proofreading?

8. What are telomeres? What special problems occur at the telomere in replication of the leading chain? How are these problems apparently solved in eukaryotes?

9. What are replication origins? Replicons? What evidence indicates that replicons are activated in a definite sequence in eukaryotes? How are replicons related to differences in duration of the S stage? To DNA amplification? To the development of drug and herbicide resistance?

10. What is mismatch repair? How is it believed to take place? What mechanism appears to ensure that mismatches are repaired in the newly synthesized DNA copy instead of the template chain in bacteria?

11. What are mutagens? Carcinogens? What chemical alterations are induced in DNA by mutagens? How does radiation damage DNA? What major pathways of DNA repair correct damage caused by chemical mutagens and radiation?

12. What regulatory processes control the synthesis of histones during the cell cycle?

13. In what patterns might nucleosones be distributed on replicating DNA? What experimental evidence supports these possibilities?

14. Compare the initiation and progress of replication in *E. coli*, a eukaryotic chromosome, a mitochondrion, and a chloroplast.

The Taylor Experiment Demonstrating That DNA Replication Is Semiconservative in Eukaryotes

An experiment conducted by J. H. Taylor with a eukaryote, the plant *Vicia* (the broad bean), complemented the Meselson and Stahl experiment demonstrating that replication is semiconservative in *E. coli*. To demonstrate semiconservative replication in this species, Taylor grew *Vicia* roots for one generation in a medium containing radioactive thymidine and followed the distribution of label in the chromosomes through subsequent divisions by autoradiography.

After exposure of the root tips to labeled thymidine during S, Taylor noted that both chromatids of each chromosome were labeled during the following mitosis (Fig. 23-29a). This finding, equivalent to the observation of ^{14}N-^{15}N hybrids after one generation in the Meselson and Stahl experiment, ruled out the conservative pathway and established that replication is either semiconservative or dispersive.

To understand this conclusion, consider that each G1 chromosome entering replication consists of one long DNA double helix. The two nucleotide chains of each G1 double helix unwind and serve as templates for replication during S. Because this replication occurs in the presence of radioactive thymidine, the newly synthesized nucleotide chains are labeled. If replication followed the conservative pathway, the two new nucleotide chains would unwind from their templates and associate in an all-new double helix. The two old (unlabeled) nucleotide chains would likewise reassociate

into an entirely old DNA molecule. During mitosis these DNA molecules, one completely old and one new, would condense to form the two chromatids of each chromosome. In autoradiographs made at this mitosis, only the new chromatid of each chromosome would be expected to show the presence of label. However, Taylor noted that both chromatids of each chromosome are labeled at the first mitosis following DNA synthesis in the presence of labeled thymidine, eliminating conservative replication as a possibility.

Tracing the cells through one more cycle of division eliminated dispersive replication and confirmed that replication in this eukaryote is semiconservative. In this extension of the experiment, cells were allowed to replicate once more, in a medium from which the radioactive precursor was removed. Autoradiographs were made at the next mitosis. During this mitotic division only one of the two chromatids of each chromosome was labeled (Fig. 23-29b).

This result is consistent only with semiconservative replication. After the first replication, if replication is semiconservative, each daughter double helix would contain one labeled and one unlabeled nucleotide chain. This explains the observation that both chromatids of each chromosome were labeled during the first mitotic division following a replication in the presence of label (as in Fig. 23-29a). During the first mitotic division following exposure to label, the two chromatids of each

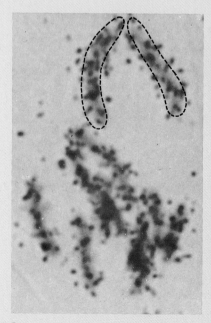
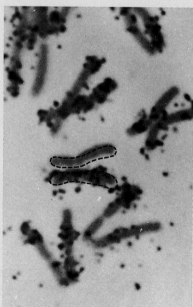

a   b

Figure 23-29 Distribution of label in *Vicia* chromosomes after exposure to a pulse of radioactive thymidine during S. **(a)** During the first mitotic division after exposure, both chromatids of each chromosome are labeled. (One chromosome, showing both chromatids labeled, is outlined in dashed lines.) The presence of label is indicated by the dark grains of photographic emulsion exposed over radioactive regions. **(b)** During the second mitotic division after exposure, only one of the two chromatids of each chromosome is labeled. (One chromosome, showing one chromatid labeled and one unlabeled, is outlined in dashed lines.) At some points the label switches between the arms, indicating that the chromatids exchanged segments at these points following replication. ×2,700. From *Molecular Genetics*, ed. J. H. Taylor, 1983. Courtesy of J. H. Taylor and Academic Press, Inc.

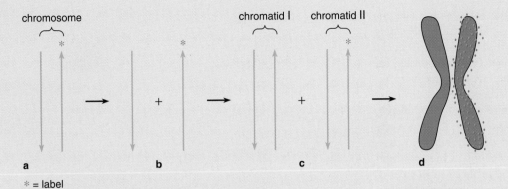

chromosome   chromatid I   chromatid II

a   b   c   d

✳ = label

Figure 23-30 The basis for production of one labeled chromatid and one unlabeled chromatid in each chromosome during the second mitotic division in Taylor's experiments. **(a)** A chromosome delivered to a nucleus at the close of the first division, in which, according to semiconservative replication, one DNA chain of the double helix is labeled, and one is unlabeled. **(b)** In the interphase following the first division, the two nucleotide chains of each chromosome separate to act as templates for replication. **(c)** The result of replication in unlabeled medium. The newly synthesized chains, which are unlabeled, remain wound with their template chains. Because one template chain is labeled and one is unlabeled, one of the two daughter DNA molecules consists of entirely unlabeled DNA. The other daughter molecule contains one labeled nucleotide chain and one unlabeled chain. **(d)** During the subsequent mitotic division, the daughter molecules condense to form the two chromatids of a metaphase chromosome, one of which is labeled and one unlabeled.

chromosome were separated and pulled to opposite ends of the cell. One of these chromatids is followed diagrammatically through the subsequent replication and mitosis in Figure 23-30. During interphase, the two nucleotide chains of this chromatid separate to act as templates for replication (Fig. 23-30a and b). Because this second replication takes place in unlabeled medium, the newly synthesized nucleotide chains are unlabeled. If replication is semiconservative, the unlabeled chains remain wound with their template chains. Since only one of the two template chains entering the second replication is labeled, after semiconservative replication one of the two daughter DNA molecules consists of

entirely unlabeled DNA, and in the other molecule one nucleotide chain is labeled and one is unlabeled (Fig. 23-30c). During the mitotic division following this replication, one of the two chromatids of each chromosome will be labeled, and one unlabeled (as in Fig. 23-29b).

If replication were dispersive, a mixture of old and new DNA segments would likely be present in all chromatids. Both chromatids would therefore be expected to be labeled during the second mitosis as well as the first. Thus, the results of the first and second Taylor experiments are compatible only with semiconservative replication.

Supplement 23-2

Alteration of DNA by Chemicals and Radiation

The four nucleotide bases of DNA molecules are highly reactive groups capable of interacting chemically with a variety of substances. Some of these substances are natural products of biochemical reactions breaking down metabolites inside cells. Others are taken up by cells from their surroundings. The energy absorbed from radiation has the effect of increasing the reactivity of DNA or of the substances occurring naturally in cells or taken up from the surroundings. This reactivity frequently results in chemical modification of the DNA,

including the introduction of breaks and the alteration or deletion of bases or other components of DNA nucleotides.

Alteration of DNA by Chemicals

Although chemical modifications of the DNA bases are potentially almost endlessly varied, the most likely possibilities fall into three categories. One involves the substitution of other chemical groups for the amino groups carried by three of the four bases (adenine, cytosine, and guanine). The second includes *alkylating reactions*, in which a monovalent straight-chain or cyclic organic group—methyl, ethyl, etc.—is substituted for a hydrogen atom. Most alkylations add groups to nitrogens or oxygens forming parts of the ring structure of the bases. Reactions in the third category completely remove bases from the template chain, leaving a site with an intact sugar-phosphate backbone but no base.

Chemicals carrying out reactions in the first category, including such natural and artificial substances as hydroxylamine, methoxyamine, hydrazine, bisulfite, and nitrous acid, are frequently termed *nonalkylating* agents because they substitute simple inorganic chemical groups for the amino groups of the A, C, and G bases of DNA. Bisulfite, for example, found as a natural product of the breakdown of sulfur-containing amino acids and as the aqueous form of sulfur dioxide, a common air pollutant, substitutes a carbonyl oxygen (=O) for an amino group of the DNA bases (Fig. 23-31a). Many nonalkylating substances have the same effect, converting cytosine to uracil (as in Fig. 23-31a), adenine to the modified base hypoxanthine (Fig. 23-31b), or guanine to xanthine (Fig. 23-31c). Removal of the amino group from 5-methylcytosine (or m⁵C; see p. 536), a common modified base in DNA, converts the methylated cytosine to thymine (Fig. 23-31d).

Conversion of C to U in the template chain by a nonalkylating agent results in entry of A rather than G in the copy during replication; conversion of m^5C to T also results in placement of A rather than G in the copy. Similar copying errors result from conversion of A to hypoxanthine, which pairs with C instead of T, and conversion of G to xanthine, which pairs with T as well as C.

Substances in the second category, the alkylating agents, include relatively simple molecules such as formaldehyde, nitrosamines, and nitrosoureas and more complex or cyclic compounds such as the N and S mustards (mustard gas), epoxides, lactones, and benzopyrene. Many of the more complex alkylating agents occur in materials such as coal tar, wood smoke, and cigarette smoke.

The more simple alkylating substances primarily add methyl or ethyl groups to DNA bases, converting them to modified forms that may pair differently from the original A, T, C, and G types. Others cause changes that block normal base pairing, leading to nonspecific, random entry of nucleotides into the copy chain at positions opposite an alkylated template base. The cyclic groups added by more complex alkylating substances may be so large that passage of the replicating enzymes is blocked. More complex alkylating agents may also introduce crosslinks between bases of the same or opposite DNA chains, providing another alteration that blocks passage of replicating enzymes or

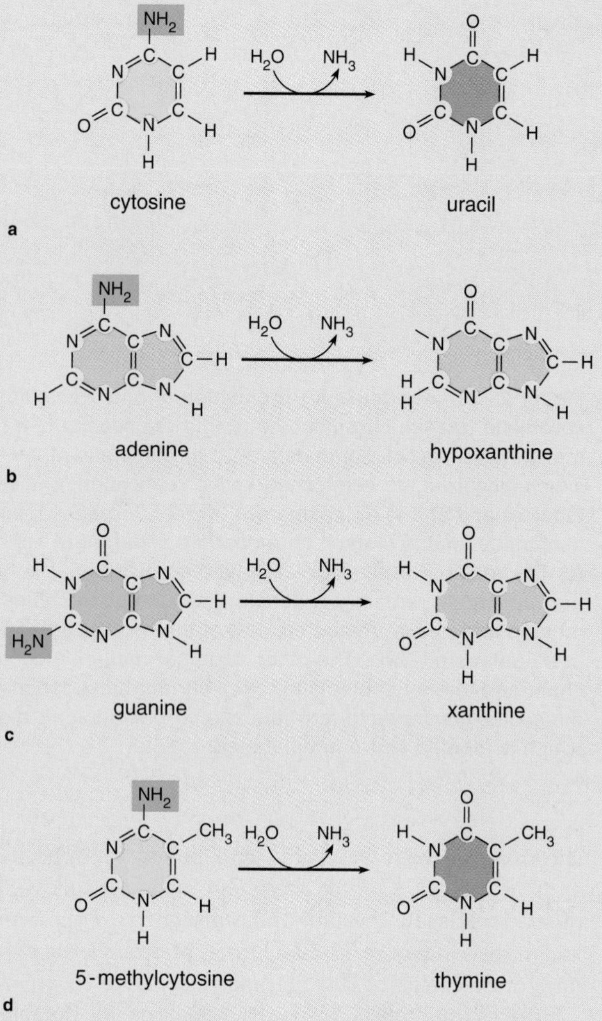

Figure 23-31 Conversion of bases to other forms through removal of amino groups (in green) by non-alkylating agents.

prevents DNA from unwinding for replication or transcription. Blockage of replication leads to "skips" in which the copy chain has an open break of greater or shorter length.

The third category of chemical modification, removal of entire bases, may result from the action of natural or manufactured chemicals. Removal of a base from the template chain leaves a blank; during replication, an unpaired base is placed opposite the blank in the copy chain. The purine bases A or G are most commonly lost in such reactions. Such *depurinations*, caused by the activity of substances occurring naturally in cells, may amount to many thousands per day in each mammalian cell.

Alterations Resulting from Radiation

Radiation-induced alterations may result from direct absorption of energy by parts of the DNA molecule

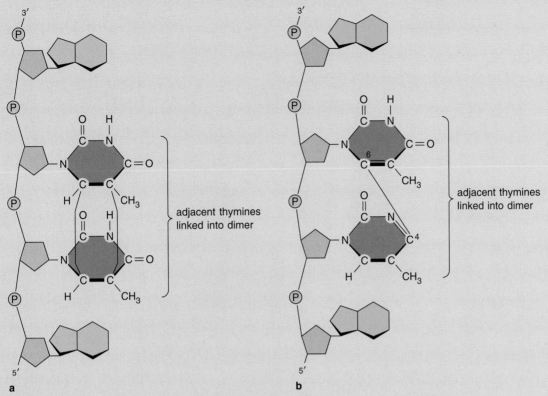

Figure 23-32 Thymine dimers. (**a** and **b**) Two types of thymine dimers formed by the action of ultraviolet light on DNA. Dimers of the type shown in (**a**) are produced most frequently.

itself or by chemical activation of substances absorbing radiation in the vicinity of DNA. Direct absorption of radiant energy raises electrons within DNA from ground to excited states (see p. 370). In the excited state, electrons may promote interactions with atoms both inside and outside the DNA bases. Primary among the radiation sources directly inducing alterations in DNA is ultraviolet light, which is strongly absorbed by DNA at wavelengths at or near 260 μm.

Although direct absorption of ultraviolet light can lead to a variety of chemical modifications in DNA, its most common result is formation of covalent crosslinks between adjacent pyrimidine bases on the same DNA chain. Adjacent C-C and C-T pairs may link together as a result of ultraviolet light absorption, but the most frequent linkage is of adjacent thymines into what is known as a *thymine dimer* (Fig. 23-32). The various dimers and other base modifications caused by direct absorption of ultraviolet light may lead to base substitutions, difficulties in unwinding, or other faults in the replication mechanism.

Radiation absorbed by molecules in the vicinity of DNA excites these substances to more reactive forms in which they may cause chemical modifications in DNA or split the sugar-phosphate backbone chain to produce single- or double-chain breaks. Both electromagnetic radiation in the form of X-rays or gamma rays, and particulate radiation such as electrons, neutrons, and protons may be responsible for excitation of molecules in the vicinity of DNA. This effect of radiation, since it frequently produces reactive ions or chemical radicals, is often termed *ionizing.*

Ionizing radiation creates reactive substances that interact with atoms of the DNA bases in various ways. The number of substances ionized by absorbing radiation is extensive, and the base modifications induced by the ionized groups highly varied, because essentially any organic or inorganic substance may be activated by ionizing radiation. However, the high content of water in biological systems makes it the most likely target. Therefore, the activated products produced when water absorbs the energy of radiation—hydrogen atoms, hydrogen peroxide, and hydroxyl radicals—are the primary reactive substances produced by ionizing radiation. The ionized products may convert the four DNA bases to modified forms (Fig. 23-33a shows some of the modified bases produced from thymine in irradiated water), change purines to modified pyrimidines (Fig. 23-33b), or cut the sugar-phosphate backbone.

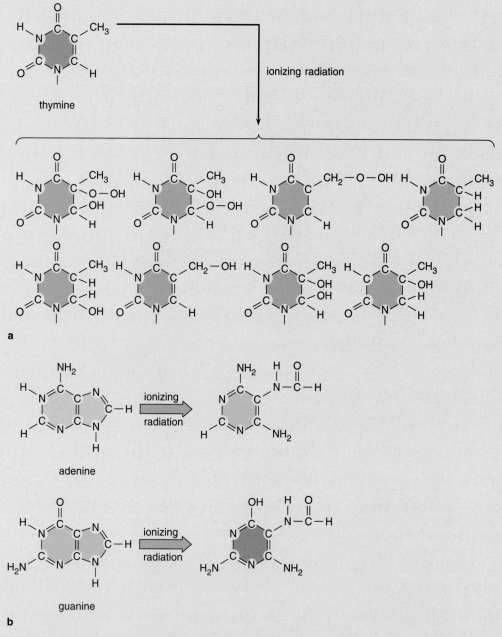

Figure 23-33 Effects of ionizing radiation on DNA bases. The most common substance ionized by radiation, and interacting with bases to produce modified forms, is water. **(a)** Modified forms of thymine produced by the action of ionizing radiation. **(b)** Conversion of the purines adenine and guanine to modified forms by ionizing radiation.

Many, if not most, of the alterations produced in DNA by chemicals and radiation are corrected by the DNA repair mechanisms described in this chapter. The mechanisms correcting these deficiencies are so vital to survival that they probably evolved along with DNA-based informational systems at the earliest times in the evolution of life.

Suggestions for Further Reading: See p. 990

DNA Replication in Mitochondria and Chloroplasts

Mitochondria and chloroplasts contain all the enzymes and factors necessary to replicate their DNA. This has been amply demonstrated in isolated mitochondria and chloroplasts, which are able to incorporate the four nucleoside triphosphates into newly synthesized DNA.

Both organelles contain DNA polymerases with properties distinct from those active in the cell nucleus. The DNA polymerase of mitochondria, as demonstrated by W. J. Adams and G. Kalf, and A. Weissbach and his coworkers, is DNA polymerase γ (see Table 23-2), which can act in the forward direction as a $5' \rightarrow 3'$ polymerase, and has the $3' \rightarrow 5'$ exonuclease activity required for proofreading. A DNA topoisomerase with similarities to bacterial DNA topoisomerase II has also been detected in mitochondria, along with a primase. The DNA polymerase enzymes of mitochondria and chloroplasts, and evidently all the other enzymes and factors concerned with DNA replication, are encoded in the cell nucleus rather than in the organelles.

The pattern of DNA replication inside both mitochondria and chloroplasts differs significantly from the replication pathways observed in the nucleus. Much of this difference rests on the fact that the DNA molecules of the two organelles, like the DNA of bacteria, are circular rather than linear. In mitochondria, for example, replication begins from an origin at one point on the circle, as it does in bacteria. In contrast to bacterial replication, however, replication from this origin is unidirectional—only one replication fork progresses around the circle from the origin.

D. D. Chang and Clayton showed that mitochondrial replication in animals begins in the nontranscribed region, at a point that coincides with the transcription promoter for the L chain (Fig. 23-34; see p. 882). Replication is primed by an RNA transcript that is initiated at the L-chain promoter, evidently by an RNA-containing primase identified by the Clayton group. The RNA polymerase associated with transcription may promote initiation, possibly by opening the DNA at

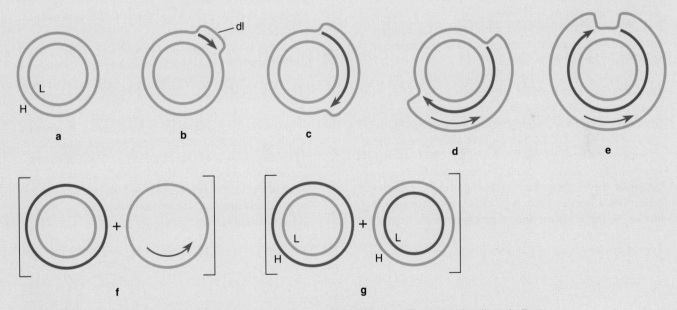

Figure 23-34 The pattern of DNA replication in mammalian mitochondria. The light blue lines indicate template chains, the dark blue lines newly synthesized copies. **(a)** An unreplicated circle, showing the light (L) and heavy (H) nucleotide chains. **(b)** Replication begins at a point that coincides with the transcription promoter for the L chain and proceeds around this chain **(b** and **c)**. Replication of the L chain is continuous because it is presented to the enzyme in the correct $3' \rightarrow 5'$ orientation. The H chain is displaced as a single-stranded loop (dl) at this stage in the process. **(d)** Displacement of the H chain continues until about two-thirds of the L chain has replicated. At this point, corresponding to a site between genes for cysteine and asparagine tRNAs on the L chain, replication begins in the opposite direction on the H chain. **(e)** Replication continues on both chains; because replication of the H chain proceeds in the opposite direction to L-chain replication, it also takes place continuously. Both copy chains are, therefore, replicated as leading rather than lagging chains. **(f)** As replication of the L chain is completed, the L and H chains are released as free circles. **(g)** The H chain then completes its replication. Modified from an original courtesy of D. L. Robberson.

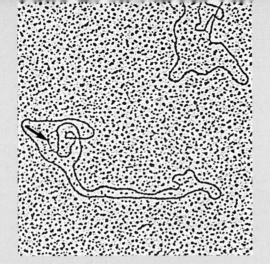

Figure 23-35 A single-stranded displacement loop (arrow) in a replicating mitochondrial DNA circle from a mouse cell. The loop is the H chain, displaced by replication of the L chain. Courtesy of H. Kasamatsu.

the replication origin. At a site about 50 to 190 nucleotides from the promoter, assembly of the primer ceases, and DNA replication begins at the 3' end of the primer. Replication then proceeds around the L chain. Because the L chain is presented to the enzyme in the correct 3' → 5' orientation, replication of the copy proceeds continuously in the required 5' → 3' direction. The opposite H chain, rather than acting as a template for discontinuous synthesis of the lagging chain, is initially displaced simply as a single-stranded loop (Figs. 23-34*b* and *c* and 23-35).

In mammalian mitochondria displacement of the H chain continues until about two-thirds of the L chain has replicated. At this point, corresponding to a site between genes for cysteine and asparagine tRNAs on the L chain (see Fig. 21-11), replication begins in the opposite direction on the H chain (Fig. 23-34*d*). Research in the Clayton laboratory indicates that replication of the H chain may be initiated by a primase distinct from the one initiating L-chain replication. Replication of the H chain, since it proceeds in a direction opposite to L-chain replication, also occurs continuously in the 5' → 3' direction. As a consequence, both chains act as leading chains for continuous replication, and no discontinuous Okazaki fragments are generated. As replication of the L chain is completed, the H chain is released as a free circle that finishes its replication separately (Fig. 23-34*e* to *g*).

Replication in *Drosophila* mitochondria proceeds by a similar pattern, except that the point at which H-chain replication begins is more variable. Little is known about patterns of replication in plant mitochondria.

Chloroplast replication initially resembles the mitochondrial pattern except that replication proceeds simultaneously from two origins, forming two displacement sites that travel toward each other (Fig. 23-36*a* through *c*). As the loops meet, they form two regular replication forks that proceed in opposite directions around the circle as in bacteria (Fig. 23-36*d* and *e*). The forks advance until replication of the circle is complete (Fig. 23-36*f*). The origins for replication of the chloroplast DNA circle have not been determined except in *Chlamydomonas*, in which replication of the two chains begins at separate, but closely adjacent points near one of the two rRNA genes, and in the garden pea, in which one origin lies in the spacer between the 16S and 23S rRNA genes and the other lies downstream of the 23S gene.

Replication of mitochondrial DNA is not confined to the S phase. Instead, mitochondria appear to replicate their DNA almost continuously during interphase. Chloroplasts also replicate their DNA independently of S.

At some time during interphase, usually in advance of mitosis and cytokinesis, mitochondria and chloroplasts divide by a mechanism that parcels out the replicated DNA circles and other organelle structures. The organelle division mechanism, which resembles prokaryotic cell division, is described in Chapter 24.

Suggestions for Further Reading See p. 990.

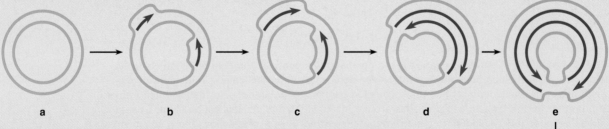

Figure 23-36 Mechanism of DNA replication in chloroplasts. In this pattern, replication of both template circles begins simultaneously, producing two displacement loops that expand toward each other (**a** through **c**). As the loops meet (**d**), they form two regular replication forks that proceed in opposite directions around the circle as in bacteria (**e**). The forks advance until replication of the circle is complete (**f**). Redrawn from an original courtesy of R. D. Kolodner and K. K. Tewari. Reprinted by permission from *Nature* 256:708. Copyright 1975 by Macmillan Magazines, Ltd.

NUCLEAR AND CYTOPLASMIC DIVISION

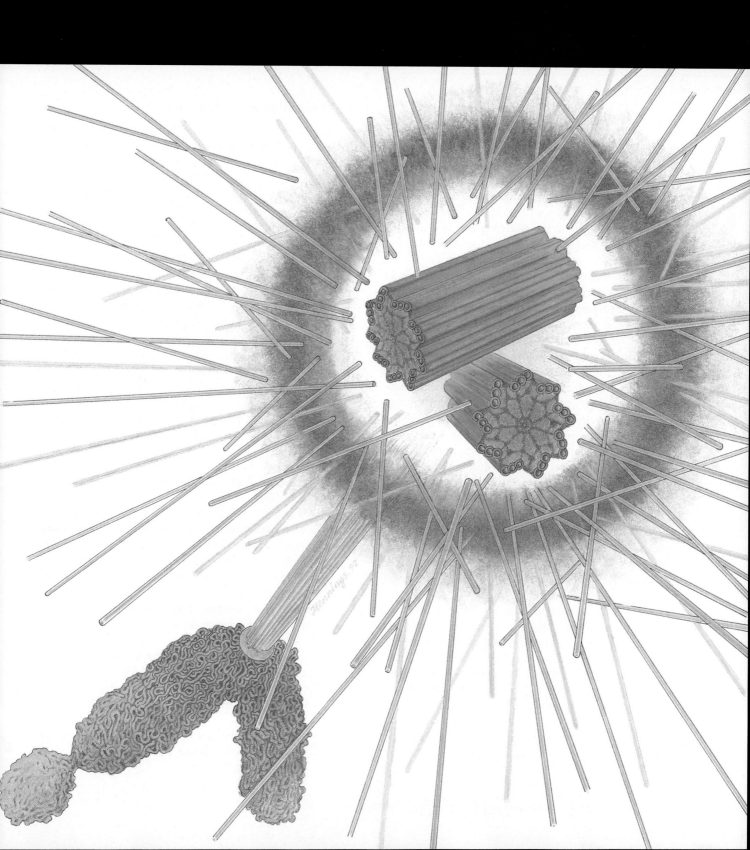

The major steps and outcome of mitosis and cytokinesis (cytoplasmic division), as outlined in the overview to Chapter 22 (pp. 910–916), have been understood in overall terms since the late 1800s. However, these processes include many details that have not been explained completely at the molecular and biochemical level, including mechanisms as basic to cell division as the movement of chromosomes by the spindle.

Interest and research in mitosis and cytokinesis suffered a near-total eclipse when molecular biology revolutionized the study of cells. In more recent years, however, realization that the cell cycle is fundamentally important to development and differentiation has resulted in a resurgence of interest in nuclear and cytoplasmic division. Interest has also been spurred by research into the genetics of the cell cycle, which has revealed that many mutations capable of interrupting the cycle affect steps during mitotic and cytoplasmic division. A further source of attention has come from a renewed appreciation of the many fundamental cellular systems that come into play during mitosis and cytokinesis, including the microtubule- and microfilament-based motile systems that figure so prominently in division of the nucleus and cytoplasm.

Current interest in mitosis and cytokinesis centers in several areas, which are the subjects of this chapter. These include the factors accomplishing chromosome condensation, the structure of chromosomes during mitosis, the role of centrioles and the cell center in mitosis, the mechanisms moving chromosomes during anaphase, breakdown and re-formation of the nuclear envelope and the nucleolus, the relationships of microtubules and microfilaments to cytoplasmic division, and the division of mitochondria and chloroplasts. In some of these areas recent research has provided answers to questions posed for nearly a century. In others the present state of knowledge amounts to not much more than the conclusions reached by classical cytologists more than a hundred years ago.

EVENTS OF PROPHASE

Prophase is considered to begin when chromosomes start to condense. Condensation continues until the chromosomes have been converted from their extended interphase forms to highly compact rodlets. In addition, the nucleolus breaks down and disappears during prophase in most species, centrioles and asters divide and move to opposite ends of the cell, and the spindle forms. In higher eukaryotes breakdown of the nuclear envelope marks the end of prophase and the transition to metaphase.

Chromosome Condensation

Condensation reduces the duplicated chromosomes from their interphase state, in which they may be as much as several centimeters in length, to compact rods a few microns long. Condensation continues throughout prophase and may extend well into metaphase. Chromosomes reach their most compact state by the end of metaphase, when they are short enough to be divided by the spindle without tangling or breaking.

Condensation seems to be controlled by specific factors that appear or become active at prophase and metaphase. This was nicely demonstrated in cell fusion experiments (see p. 123) conducted by R. T. Johnson and P. N. Rao. These investigators found that fusion of a metaphase cell with cells in G1 or S causes chromatin fibers in the G1 or S cells to condense prematurely into threads thick enough to be visible under the light microscope. Johnson and Rao suggested that condensation of the G1 and S chromatin is induced in the fused cells by a soluble *condensation factor* that diffuses from the metaphase nucleus or cytoplasm into the interphase nucleus. The condensation factor was later identified as a protein that, when injected into interphase cells, can induce premature chromosome condensation.

There are indications that the condensation factor has been highly conserved in evolution. Human metaphase cells can induce premature chromatin condensation, for example, when fused with avian, amphibian, or insect cells in G1 or S. A recent study by B. Von Der Haar, K. Sperling, and D. Gregor showed that mitotic cells of *Xenopus*, an amphibian, can even induce chromosome condensation when fused with interphase cells of a plant.

It now seems likely that the condensation factor identified by Johnson and Rao is *MPF* (see p. 926 and below), a regulatory element that is fully activated at the G2/prophase transition. When fully active, MPF is able to induce chromosome condensation, formation of the spindle, and breakdown of the nuclear envelope, all events that occur during mitotic prophase.

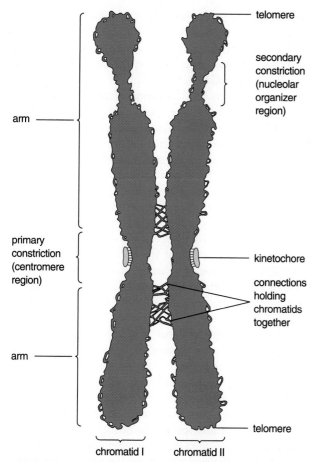

arm

telomere

secondary constriction (nucleolar organizer region)

primary constriction (centromere region)

kinetochore

connections holding chromatids together

arm

telomere

chromatid I · chromatid II

Figure 24-1 Major structures and regions of a fully condensed metaphase chromosome. The chromosome is divided into two identical sister chromatids, which are the result of DNA replication and duplication of chromosomal proteins during the previous interphase. Each chromatid consists of a single linear DNA molecule with its associated histone and nonhistone proteins. The primary constriction or centromere carries the kinetochore, the structure that links to spindle microtubules during mitosis. On either side of the primary constriction are the arms of the chromosome, each terminating in a telomere. In at least one chromosome or chromosome pair of the set a secondary constriction marks the nucleolar organizer region (NOR), a site containing many repeats of rRNA genes. The nucleolus reforms at this site at the close of mitosis.

MPF consists of two major components, the $p34^{cdc2}$ protein kinase and cyclin. As cyclin concentrations reach maximum levels during G2, $p34^{cdc2}$ is activated (see p. 927 for details). The activated kinase phosphorylates a number of proteins critical to entry into mitosis, including histone H1. This phosphorylation coincides with chromosome condensation and may be a step that promotes H1-H1 interactions that lead to the folding and packing of chromatin fibers. Increases in phosphorylation of histone H3 are also noted in coordination with chromosome condensation (see p. 921 and Fig. 22-8).

Condensation apparently proceeds without major alterations in the molecular structure of individual chromatin fibers. Many observations of condensed chromosomes have demonstrated that nucleosomes are retained. Further, DNA is released in the usual 200 and 140 base-pair lengths characteristic of nucleosomes (see p. 547) if condensed chromosomes are digested with DNA endonucleases.

One question that has intrigued cell biologists for many years is how the very long chromatin fibers of chromosomes, which are intimately intertwined during G1, S, and G2, are able to condense without tangling into separate, distinct bodies during prophase. It now appears that DNA topoisomerase II, the enzyme that can pass one DNA helix entirely through another (see Fig. 23-14), relieves interchromosome tangles set up during condensation. T. Uemara and his colleagues found in the yeast *Schizosaccharomyces pombe* that DNA topoisomerase II activity is absolutely required for condensation to proceed normally; tangles do indeed persist, leading to failure of chromosomes to separate normally at metaphase, if this topoisomerase is inactive.

Major Substructures of Fully Condensed Chromosomes Several characteristic substructures are visible in the fully condensed chromosomes of most eukaryotes (Figs. 24-1 and 24-2). Each condensed chromosome clearly consists of two duplicate halves lying side by side. These two subparts, the *sister chromatids* of a chromosome, result from DNA replication and duplication of chromosomal proteins during the S phase of interphase. Each chromatid consists of a single long, linear DNA molecule. At some point along the chromosome, both sister chromatids show a differentiated region called the *centromere* or *primary constriction*. This region, which is usually smaller in diameter than other parts of the chromosome, contains the *kinetochores*, sites at which spindle microtubules attach to the chromatids during metaphase. A kinetochore appears in many species as a disc- or ball-like structure on the surface of the chromatid (see Figs. 24-6 and 24-7). Each chromatid has its own kinetochore, so there are two kinetochores per chromosome in the centromere region.[1]

---

[1] There are some unresolved differences among cell biologists in usage of the terms *primary constriction, centromere,* and *kinetochore.* Some use centromere interchangeably with primary constriction to describe the narrowed region in condensed chromosomes, as is done here. In this usage kinetochore designates the structure, superimposed on the chromatin in the centromere region, to which spindle microtubules attach. Others use only primary constriction for the narrowed region of the chromosome, and centromere and kinetochore interchangeably for the microtubule attachment site.

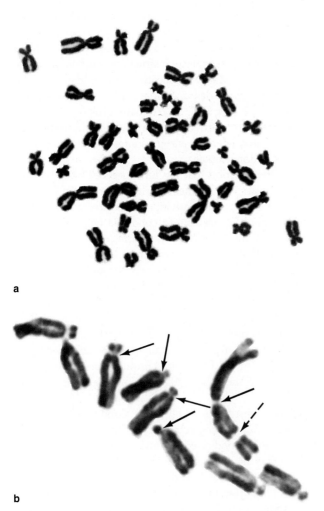

a

b

Figure 24-2 Karyotypes. **(a)** Light micrograph of chromosomes from a human cell at metaphase. × 1,500. Courtesy of S. Brecher. **(b)** Light micrograph of metaphase chromosomes of the plant *Vicia*. Solid arrows indicate primary constrictions, the sites at which chromosomes connect to spindle microtubules; the dashed arrow indicates the secondary constriction containing the nucleolar organizer. × 1,700. Courtesy of S. Wolff.

The position of the centromere, which is fixed for a given chromosome of a set, may be in the approximate middle of the chromosome or at a point closer to either end. The chromatin on either side of the primary constriction forms the *arms* of the chromosome. At the tips of the arms are the *telomeres* of the chromosome. Often the telomeres are slightly enlarged, giving the tips of the chromosomes a knoblike appearance.

Other structures are frequently visible on one or more of the fully condensed chromosomes of a species. Usually at least one chromosome or chromosome pair of a set has a narrowed region in addition to the primary constriction called the *secondary constriction* or *nucleolar organizer region* (*NOR*; see Figs. 24-1 and 24-2). The NOR contains a large cluster of rRNA gene repeats

and marks the site where the nucleolus forms during interphase (see pp. 616–618). The nucleolus persists at this site until it breaks down at late prophase and reappears at the same location during telophase and the subsequent G1.

The patterns and locations of primary and secondary constrictions, the relative lengths of the arms, and the structure of the telomeres are constant for each condensed chromosome in a eukaryotic species. Collectively, the morphology and number of the entire set of chromosomes form what is known as the *karyotype* of the species. In many cases the karyotype is so distinctive that a species can be identified from this characteristic alone.

Although the total karyotype is highly distinctive in many species, some of the chromosomes within a karyotype are often so similar in structure that they cannot be separately distinguished. This is the case in the human karyotype (see Fig. 24-2*a*). The problem is circumvented in mammals and other higher vertebrates by staining techniques that produce a unique pattern of crossbands in each chromosome (see Figs. 24-42 and 24-43). The crossbands, which reflect underlying differences in molecular structure among the chromosomes, have made it possible to identify unambiguously each condensed chromosome of humans and many other higher animal species. In turn this unambiguous identification has allowed mutant genes to be identified with individual chromosomes of the set and the chromosome rearrangements taking place in association with disorders such as cancer to be traced (see p. 936; further details of the banding technique and its applications are given in Supplement 24-1).

An overall coiling of whole chromatid arms into regular gyres or spirals often becomes visible in chromosomes as they condense during mitotic prophase (Fig. 24-3). Ordinarily, regular coiling does not extend below the level of a whole chromatid arm; no regular coils or gyres can be seen at the level of individual chromatin fibers. However, the folding pattern of chromatin fibers within the major gyres, although irregular, is probably ordered and precisely determined by the condensation mechanism. Otherwise, condensation would not be expected to produce the characteristic sizes and shapes for individual chromosomes.

A number of investigators have proposed that ordered folding during condensation depends on an arrangement of chromatin fibers into loops radiating from a central core that runs the length of each chromosome. The primary evidence for this hypothesis comes from studies of condensed chromosomes by U. K. Laemmli and his coworkers. These investigators treated chromosomes with heparin and dextran sulfate, which are negatively charged molecules that extract most of the proteins from chromatin. After the extraction a residual "scaffold" becomes visible that traces an apparent core

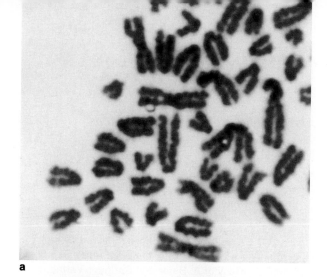

a

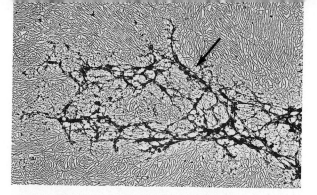

Figure 24-4   The residual scaffold (arrow) remaining after treatment of a human metaphase chromosome to remove chromosomal proteins. The DNA extends in loops from the scaffold framework. ×14,000. Courtesy of U. K. Laemmli, from *Cell* 12:817 (1977). Copyright Massachusetts Institute of Technology.

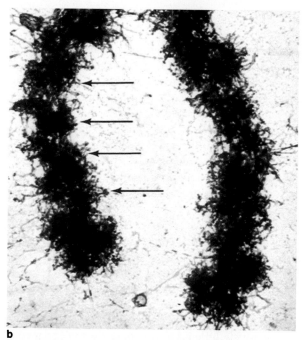

b

Figure 24-3   Gross coiling in metaphase chromosomes. **(a)** Light micrograph showing gross coiling in metaphase chromosomes of *Marmosa*, a marsupial. Courtesy of K. M. Fisher. **(b)** Gross coiling (arrows) visible in an electron micrograph of an isolated bovine chromosome. ×22,000. Courtesy of R. Jahn.

or axis of a chromosome (Fig. 24-4; see also p. 554). Surrounding the scaffold are whorls of DNA fibers, which can be traced as loops extending from the scaffold. The individual loops contain from 50,000 to 150,000 base pairs of DNA.

Scaffolds remain somewhat controversial because they become visible only when chromosomes are prepared for examination by the Laemmli technique. No structures resembling scaffolds can be seen in sectioned chromosomes or in chromosomes isolated for electron microscopy by other techniques. However, the scaffolds

produced by the Laemmli technique may reflect an underlying organization of the chromatin fibers of condensed chromosomes as loops extending from a central axis (Fig. 24-5). The loops may represent the *domains* of the genome, constituting segments of genetic material that operate as a unit in replication and other functions, often proposed as part of eukaryotic chromosome structure (see p. 555). During condensation a protein spaced at sites between the domains may act as a "folder" by setting up linkages that pull the domains together as loops (see Fig. 24-5). These proteins, arranged along the central axis of fully condensed chro-

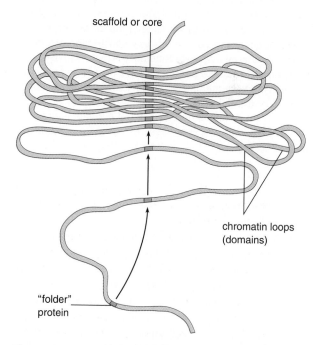

Figure 24-5   A possible arrangement of chromatin fibers in metaphase chromosomes in which the fibers are held as loops extending from a central core. The core is formed by the association of "folder" proteins along the central axis of a chromatid.

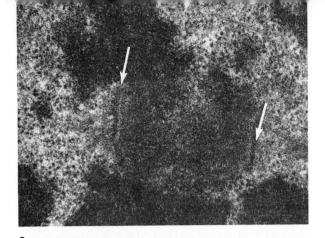

a

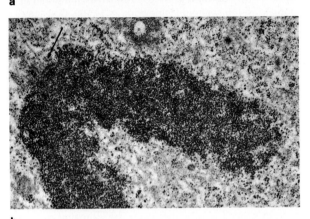

b

Figure 24-6 Kinetochores in chromosomes at metaphase and anaphase. **(a)** The two kinetochores (arrows) at the primary constriction of a Chinese hamster metaphase chromosome. Three layers are clearly visible in the kinetochore on the left. Because of the thinness of the section, only short segments of the chromosome arms on either side of the primary constriction are included in the micrograph. Kinetochores of this type range from 0.4 to 0.9 μm in diameter in different species (about 2.2 μm in mammals), and from 50 μm to as much as 0.1 μm in total thickness. × 49,000. Courtesy of L. I. Journey. **(b)** A kinetochore (arrow) of a Tasmanian wallaby chromosome at anaphase. × 21,000. Courtesy of B. R. Brinkley.

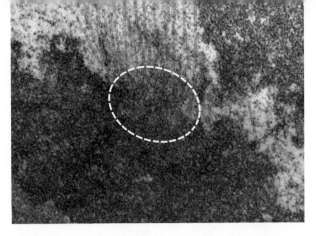

Figure 24-7 A ball-like kinetochore (outlined by the dashed line) on a chromatid of the plant *Haemanthus*. × 29,000. Courtesy of A. Bajer and Springer-Verlag.

mosomes, may be precipitated into a residual scaffold by the Laemmli technique.

Analysis of the scaffolds revealed two major proteins and several others in lesser concentrations. One of the two major proteins has been shown by Laemmli and his coworkers to be topoisomerase II. The other major protein, with a molecular weight of 135,000, remains to be identified. This protein may be the folder that brings the bases of chromatin loops together along the cores of the chromatid arms.

Kinetochores Kinetochores appear disclike in many groups, including mammals, some insects and protozoa, slime molds, and green algae (Fig. 24-6). A disclike kinetochore is usually structured from three more or

less distinct layers—an outermost dense layer; a middle, less dense, faintly fibrous zone; and an innermost, coarsely granular layer closely applied to the underlying chromatin. In some species the outermost surface of the kinetochore is coated with fibers that extend outward from its surface. This outer fiber layer, which is most prominent in kinetochores that have not as yet attached to the spindle, is called the *corona*.

Kinetochores appear as ball-like, uniformly dense masses in higher plants (Fig. 24-7). In other organisms, such as some fungi, insects, and protozoa, no differentiated structure can be seen where spindle microtubules attach to chromosomes. In these organisms the molecules forming the kinetochore may be distributed too sparsely to be visible in the electron microscope. Or they may simply be insufficiently preserved by the techniques used to prepare cells for electron microscopy.

Relatively little is known about the molecular constituents of kinetochores. Some of the best evidence showing that proteins form part of their structure comes from the use of antibodies produced in patients with *scleroderma*, an autoimmune disease in which antibodies are developed against proteins in the centromere and kinetochore region. The antibodies react with a group of proteins that seems to be common to the kinetochore regions of many organisms, including those of both animals and plants (Table 24-1).

Kinetochores also react positively with stains for RNA. The RNA stain is reduced or eliminated by techniques removing RNA, such as ribonuclease digestion or cold perchloric acid extraction. The positive stain for RNA suggests that some of the proteins associated with kinetochores may be ribonucleoproteins.

When supplied with tubulin under conditions promoting microtubule assembly, kinetochores can serve as *microtubule-organizing centers* (MTOCs; see p. 426 and below) in the test tube. Under such conditions microtubules initiate growth at kinetochores and extend to considerable length. Kinetochores can also attach in the

Table 24-1	Proteins Identified in Centromeres and Kinetochores	
Protein	Molecular Weight	Tentative Location
CENP-A	18,000	In chromatin under kinetochore; probably a centromere-specific histone H3 variant
CENP-B	80,000	In centromere chromatin; binds to alphoid sequences in centromere
CENP-C	140,000	In kinetochore
CENP-D	50,000	In kinetochore

Adapted from H. F. Willard, *Trends Genet.* 6:410 (1990).

test tube to the ends of previously existing microtubules. Most investigators are convinced that the latter process accounts for kinetochore activity under normal conditions—during attachment to the spindle, kinetochores "capture" the ends of previously existing microtubules rather than acting as MTOCs (see below). The ability of kinetochores to act as MTOCs, according to this point of view, may possibly result from the unusual conditions used in the test tube to promote microtubule assembly, among them very high tubulin concentrations.

The microtubules attaching to a kinetochore collectively form a bundle called a *spindle fiber.* In different species a spindle fiber may contain as few as one (as in the yeast *Saccharomyces* and many other fungi) to as many as a hundred or more microtubules (as in the higher plant *Haemanthus*). In most higher eukaryotes spindle fibers usually contain from 15 to 35 or so microtubules; human kinetochores, for example, link to about 15 microtubules. Recent research into the anaphase movement of chromosomes indicates that kinetochores, besides serving as attachment sites for spindle microtubules, take part in generation of the force for chromosome movement (see below).

DNA Structure in the Centromere Region   The DNA sequences in the centromeres of the yeast *Saccharomyces cerevisiae* have been identified by L. Clarke and J. A. Carbon and others. Among other techniques, centromere sequences have been isolated by identifying the minimum sequences capable of functioning as centromeres, that is, as sequences allowing DNA molecules containing them to attach and be separated correctly by the spindle (see below). Sequencing revealed that centromeres of this species, which contain from about 120 to 140 base pairs of DNA, include three characteristic sequence elements identified as *CDEI, CDEII,* and *CDEIII* (Fig. 24-8a).

Several proteins bind to the centromere sequences of *S. cerevisiae.* These proteins protect a region about 220 to 250 base pairs in length from endonuclease digestion (see p. 547), indicating that they bind tightly to the centromere region. The protected region is flanked on either side by a short DNA length that is hypersensitive to endonuclease digestion; outside the hypersensitive sites the digestion patterns indicate that nucleosomes are held in highly ordered positions on either side of the centromere (Fig. 24-8b). These findings suggest that kinetochore proteins bind to the centromere sequences and overlap about 50 base pairs on either side. Short lengths of "naked" DNA that is unbound or only very loosely associated with proteins separate the centromere from the surrounding chromatin, which is wound with histones into nucleosomes. The highly ordered combination of proteins with the centromere elements places the nucleosomes in phase (see p. 550) for some distance on either side of the centromeres.

Although *S. cerevisiae* centromere sequences are presently the best characterized of any species, there are no indications that they are shared by other organisms. Even the related species *Schizosaccharomyces pombe* has radically different centromeres. In *S. pombe* the centromere DNA sequences are much longer elements containing 40,000 to 80,000 base pairs that include complex combinations of several repeated sequence elements (Fig. 24-8c). The *S. cerevisiae* centromere sequences are not recognized by the human kinetochore proteins identified by scleroderma antibodies. Further, the *S. cerevisiae* sequences have no ability to induce microtubule attachment when introduced into the cells of higher eukaryotes or even into *S. pombe.*

The structure of *S. pombe* centromeres—relatively long centromere elements containing complex combinations of several repeated elements—may be more characteristic of higher eukaryotes than the *S. cerevisiae* pattern. Human centromeres, for example, appear to contain several repeated sequence elements, including one class of 170 base-pair elements known as the *alphoid* sequences. H. Matsumoto and his coworkers found, incidentally, that the 80,000-dalton kinetochore protein identified with scleroderma antibodies binds to the alphoid sequence elements of human centromere DNA.

In most higher eukaryotes long tracts of repetitive sequences occur outside the relatively limited segment of the centromere region to which the kinetochore attaches. These flanking repetitive sequences vary from simple to complex in different species, vary between different chromosomes of the set within a species, and may even vary in a given chromosome between different individuals of the same species. In some species the flanking repeated sequences occur in other regions of the chromosomes as well, frequently in regions just inside the telomeres (see below). Although generally known as *centromere DNA* or *C-DNA,* these flanking

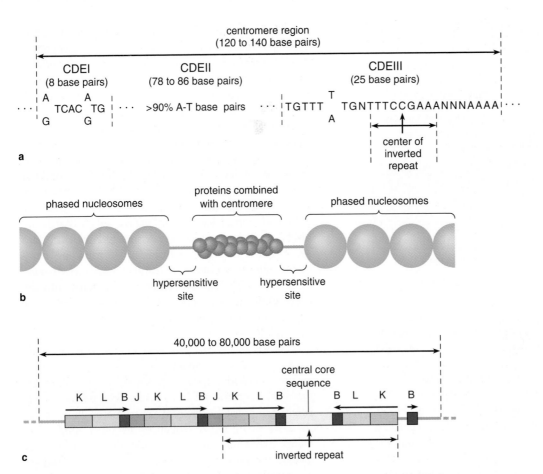

Figure 24-8 Centromere structure in *Saccharomyces cerevisiae* and *Schizosaccharomyces pombe*. **(a)** Sequence structure of the centromere region in *S. cerevisiae*. CDEI is an eight-nucleotide sequence. Although conserved in this form in *S. cerevisiae* centromere DNA, elimination of most or all of this sequence has little effect on attachment to the spindle but disturbs separation of chromatids at anaphase. The CDEII sequence, which varies from 78 to 86 base pairs, is nonconserved except for a preponderance of A-T base pairs, which make up more than 90% of this element. Experimental changes in CDEII have little effect as long as the A-T content is not reduced significantly below 90%. Changes in CDEIII amounting to as little as one base substitution, especially in the CCG element at the center of the inverted repeat, can completely eliminate the ability of chromosomes to attach to the spindle. **(b)** Organization by proteins of the *S. cerevisiae* centromere and surrounding DNA, as indicated by sensitivity to endonuclease digestion. The centromere sequence is protected from digestion, indicating it is tightly complexed with proteins. On either side of the centromere sequence are short lengths of DNA that are hypersensitive to endonuclease digestion, indicating that these regions are uncombined or only loosely associated with proteins. The DNA on either side of the hypersensitive sites is wound into nucleosomes that take up fixed, highly ordered (phased) positions for some distance on either side of the centromere. **(c)** Sequence structure of the centromere region in *S. pombe*. The elements B, J, K, and L are different types of repeated sequences. Essentially the entire segment shown is necessary for full centromere activity. **(a, b,** and **c)** are not drawn to relative scale; **(c)** adapted from an original courtesy of L. Clarke, from *Trends Genet.* 6:150 (1990).

sequences probably have little or no relationship to the function of the centromere in binding kinetochore proteins. They may simply be functionless "filler" DNA that frequently flanks centromeres or may be one of the elements holding sister chromatids together until anaphase begins (see below).

Telomeres Specialized sequences also define telomeres in all eukaryotic species. These sequences consist of short repeated elements that are rich in G and T bases in the DNA chain that has its 3′ end at the telomere (see pp. 740 and 973). Recent research by E. W. Blackburn and her coworkers (detailed in Ch. 23) showed that these sequences are recognized by telomerases, a group of RNA-containing enzymes that solve a problem inherent in replicating the 3′ end of DNA nucleotide chains.

There are also repeated sequence elements that lie

inboard of the short telomere repeats in many species. These *subtelomeric sequences,* as they are called, are longer elements that repeat in complex and varying patterns in many species. In some higher eukaryotes, as noted, the same repeated sequence elements occur in the subtelomeric region and in the C-DNA sequences flanking the centromeres.

Besides taking part in DNA replication, telomeres are essential for the functional and biochemical stability of chromosomes. Chromosomes lacking telomeres are highly sensitive to enzymatic breakdown and readily interact chemically with other cellular structures including other chromosomes. Frequently, as a result of such interactions, chromosomes fuse into multiple structures that cannot be distributed correctly by the spindle. This stabilizing effect appears to depend on "capping" of DNA molecules by telomere repeats. Several recent experiments have shown that the ends of unstable chromosomes lacking telomeres can be stabilized functionally and biochemically through the addition of telomere repeats by telomerase.

Telomeres may also supply what is known as a *topological restraint* for the DNA of linear chromosomes by preventing DNA supercoils from unwinding. This restraint may contribute to the supercoil density characteristic of eukaryotic DNA, which averages about one supercoil for each 200 base pairs (see p. 537).

Telomeres enter into a variety of interactions that form parts of the mitotic and meiotic mechanism in different species, including attractions or attachments between telomeres and the nuclear envelope, centrioles, or the cell center. The molecular basis for these interactions, which probably contribute to the organization of chromosomes before and during cell division, remains unknown.

Centromeres, Telomeres, Replication Origins, and Artificial Chromosomes

The basic elements necessary for DNA molecules to function stably as chromosomes in cell division have been nicely worked out by a series of recombinant DNA experiments with plasmids (see p. 567) of yeast cells. Plasmids do not normally connect to the spindle. Instead, they are distributed to daughter cells simply by random distribution during cytoplasmic division. J. W. Szostak, Blackburn, and A. W. Murray found that inclusion of an *S. cerevisiae* centromere sequence allows plasmid circles to connect to the spindle and be distributed with the normal yeast chromosomes during mitosis. (Fig. 24-9 shows how the experiments were carried out.) If the plasmid circles are opened, creating linear DNA molecules unprotected by telomere sequences at their tips, the circles become unstable and are lost. However, addition of yeast telomere sequences to the opened tips make the linear plasmids stable and

able to maintain their capacity for normal distribution during mitosis (see Fig. 24-9).

Szostak and Murray found that addition of a yeast replication origin (an ARS element; see p. 976) to a linear plasmid sequence containing a centromere and telomeres completes their division function. These *artificial chromosomes* replicate during interphase and are distributed correctly by the spindle to daughter cells during mitosis in *S. cerevisiae.* These experiments defined the minimum elements necessary for the distribution of linear genetic information to daughter cells via replication and mitosis: a replication origin, a centromere, and telomeres at both ends. A minimum length of 100,000 to 150,000 base pairs is necessary for artificial chromosomes to function without significant error in the yeast experiments. Besides demonstrating the minimum structural elements necessary for a linear DNA molecule to act as a chromosome, the artificial chromosomes have proved to be highly useful as a means to clone DNA in large quantities in yeast cells for sequencing.

The condensed chromosomes produced by the folding and packing of chromatin fibers are characterized by several essential structural and functional features. Each is divided lengthwise into two sister chromatids, which are the products of DNA replication and the duplication of chromosomal proteins during the S phase preceding division. At some point the chromosome contains a centromere region that carries the kinetochores, one for each sister chromatid, attaching the chromosomes to the spindle. The tips of the chromosomes are defined by the telomeres, structures that figure in replication and stabilize the chromosomes against chemical interactions. (Fig. 24-10 summarizes the structure of fully condensed chromosomes.)

Spindle Formation

In higher eukaryotes the spindle, assembled primarily from large numbers of microtubules, organizes during prophase according to one of two basic patterns known as *astral* or *anastral formation.* Which pattern is followed depends on whether an *aster* is present during division. The aster develops from the *cell center* (see p. 514), which is a system of microtubules radiating from a region located near one side of the nucleus (Fig. 24-11). In interphase cells the microtubules radiating from the cell center act as cytoskeletal elements that organize cytoplasmic systems such as the endoplasmic reticulum and Golgi complex and hold them in position. Inside the cell center is a pair of centrioles usually arranged at right angles to each other (as in Figs. 24-11*b* and 24-17; centrioles are barrel-shaped structures containing nine sets of microtubule triplets arranged in a circle—see p. 437 and Fig. 11-28). Late in G2 the microtubules radiating from the cell center become shorter and more numerous, and lose their association with organized

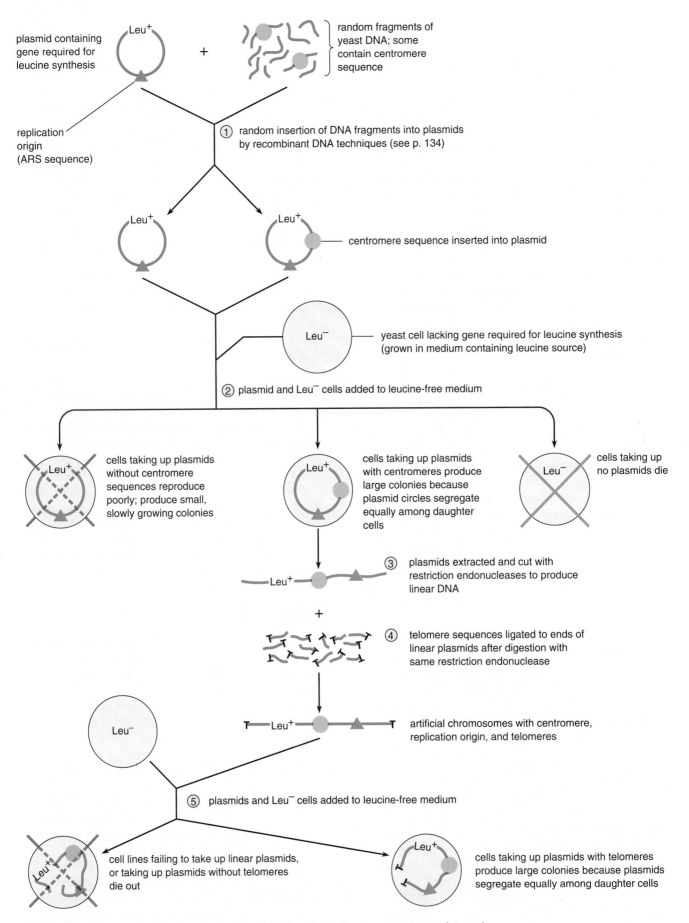

Figure 24-9 An experimental procedure used to identify centromere sequences (steps 1 and 2) and create artificial chromosomes (steps 3, 4, and 5).

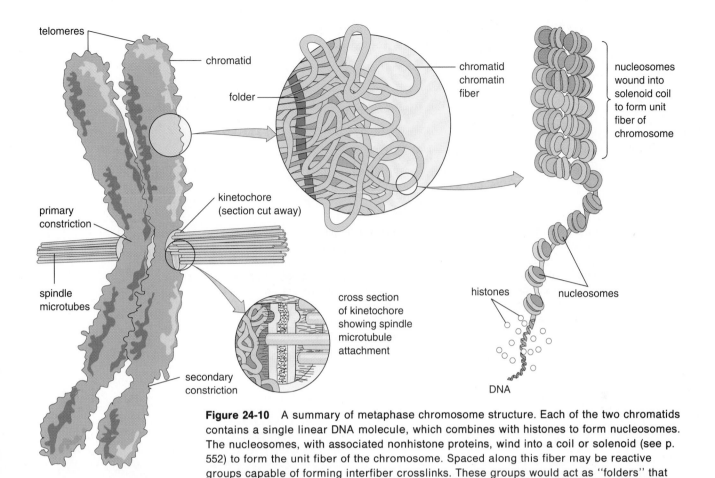

Figure 24-10 A summary of metaphase chromosome structure. Each of the two chromatids contains a single linear DNA molecule, which combines with histones to form nucleosomes. The nucleosomes, with associated nonhistone proteins, wind into a coil or solenoid (see p. 552) to form the unit fiber of the chromosome. Spaced along this fiber may be reactive groups capable of forming interfiber crosslinks. These groups would act as ''folders'' that condense the chromosome from interphase to metaphase state during mitosis. The folders may become aligned along the axis of the chromosome, forming a core or scaffold from which the condensed chromatin fibers extend as looped domains. The final folding pattern produces the gross features recognizable as chromosome substructures at metaphase: a primary constriction, arms of characteristic length and thickness, and telomeres. Secondary constrictions associated with the NOR may be located on one or more chromosomes or pairs of the set. Two kinetochores, one for each chromatid of the chromosome, form the attachment sites for spindle microtubules within the primary constriction.

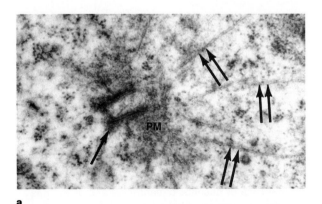

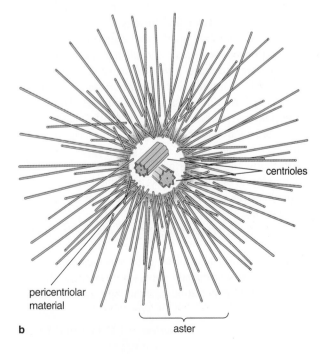

Figure 24-11 Centrioles and asters. **(a)** One of a pair of centrioles (single arrow) at the center of an aster from chick spleen. Microtubules (double arrows) of the cell center or aster are anchored in dense pericentriolar material (PM) near the centriole but do not contact the centriole. ×41,000. Courtesy of J. André. **(b)** The arrangement of microtubules, centrioles, and pericentriolar material in an aster.

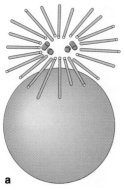

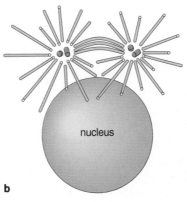

nucleus

a

b

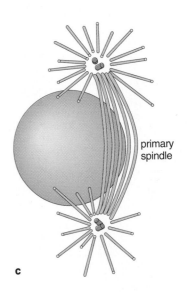

primary
spindle

c

Figure 24-12 Formation of the primary spindle in organisms with asters.
(a) The original pair of centrioles duplicates, producing two centriole pairs.
(b) While the centrioles are duplicating, the aster divides into two parts, with
each part containing a parent and a daughter centriole. Microtubules lengthen
between the asters as they separate. **(c)** Separation continues until the asters
lie at opposite sides of the nucleus. The mass of microtubules extending between them forms the primary spindle. After the nucleus breaks down at the
close of prophase, the primary spindle moves into the location formerly occupied by the nucleus. The microtubules and centrioles are not drawn to scale.

elements of the ER and Golgi complex. At the same
time several proteins not identifiable in interphase take
up residence in the cell center. In this form the cell
center is known as the aster.

Asters figure in spindle formation in most animal
cells and in the cells of some lower plants. At the
beginning of S the centrioles within the cell center
separate slightly and begin to duplicate. The duplication, which continues throughout S, G2, and into mitotic prophase in some species, produces two pairs of
centrioles. As prophase begins, the cell center, now
converted into an aster, divides in half, each half containing one centriole of the original pair and one of
the new copies (Fig. 24-12a). The asters with their
included centriole pairs continue to separate until they
lie at opposite ends of the nucleus (Figs. 24-12b and
c). As the asters move apart, microtubules lengthen in
the direction of movement and fill the space between
them. By late prophase, when the asters are fully separated, the microtubules extending between them form
a large mass around one side of the nucleus. This mass
is the *primary spindle* (Fig. 24-12c). As the nuclear envelope breaks down at the close of prophase, the developing spindle moves into the region formerly
occupied by the nucleus and takes on a symmetrical
shape. This type of spindle is identified as an *astral
spindle* because of the asters at its two ends or *poles*
(Fig. 24-13a).

Anastral spindles have no asters or centrioles at their
poles (Fig. 24-13b), and no changes associated with
spindle formation take place until early prophase. At
this time light microscopy shows a clear space developing in a narrow zone around the nucleus (Fig.

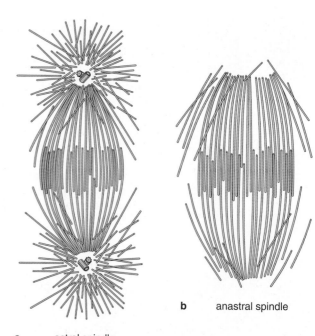

b   anastral spindle

a   astral spindle

Figure 24-13 Astral and anastral spindles. **(a)** Astral
spindles, with centrioles and asters at their tips, are
typically more pointed at the poles. **(b)** The microtubules of anastral spindles are less convergent at the
poles, producing a structure that is broader at the tips.
The presence or absence of centrioles and asters has
no apparent effect on spindle function in division of
chromatids.

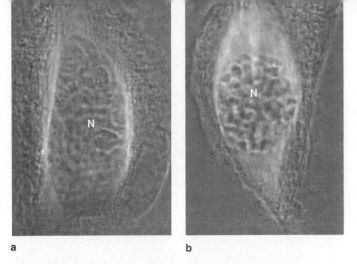

a

b

Figure 24-14 Formation of an anastral spindle in a living cell of the plant *Haemanthus*, as observed under polarized light. The developing spindle appears bright because of the effect of the parallel arrangement of the microtubules on polarized light. **(a)** At early stages the spindle appears as a clear zone surrounding the nucleus. **(b)** Extension of the clear zone toward the poles. N, nucleus. ×700. Courtesy of S. Inoué and A. Bajer and Springer-Verlag.

24-14*a*). The clear zone grows until it takes on a spindle shape and completely surrounds the nucleus (Fig. 24-14*b*). Electron microscopy reveals that the clear zone is filled initially with masses of randomly oriented microtubules. As development proceeds, the microtubule bundles become packed into more or less parallel arrays that converge at opposite sides of the nucleus and form the poles of the spindle. Spindle formation follows this pathway in higher plants lacking asters or centrioles, including all angiosperms and most gymnosperms. Some animal cells, such as certain protozoa and the developing eggs of many higher animals, also lack asters or centrioles. The presence or absence of asters and centrioles has no apparent effect on spindle function in metaphase and anaphase.

A different pattern of spindle formation occurs in lower eukaryotes including some algal and fungal species. In these organisms the spindle forms between *spindle pole bodies* (*SPBs*; Fig. 24-15*a*), which function in

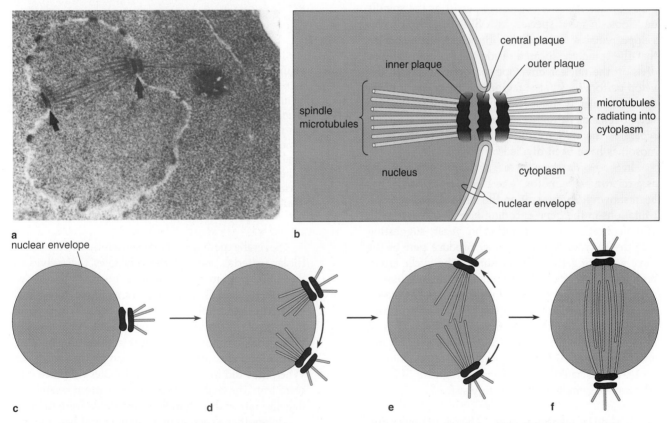

Figure 24-15 Spindle pole bodies (SPBs). **(a)** SPBs (arrows) at opposite sides of a *Saccharomyces cerevisiae* nucleus. Spindle microtubules extend from the SPBs into the nucleus on the nuclear side of the nuclear envelope and from the SPBs into the cytoplasm on the opposite side. ×42,000. Courtesy of A. E. M. Adams and J. R. Pringle. Reproduced from *J. Cell Biol.* 107:1409 (1988) by copyright permission of the Rockefeller University Press. **(b)** Major structural elements of an *S. cerevisiae* SPB. Parts **c** through **f** show the function of SPBs in generating the spindle in an organism such as *S. cerevisiae*. **(c)** An SPB at interphase. Late in interphase the SPB divides into two parts **(d)**. As the two SPBs move apart, microtubules grow from their inner surfaces into the nucleoplasm. These microtubules elongate and interdigitate as the SPBs continue to separate **(e)**. By the time the SPBs have reached opposite sides of the nucleus, the spindle has fully formed between them **(f)**.

a manner similar to the asters of higher organisms. SPBs consist of a series of layers of dense plates or plaques separated by less dense layers, usually embedded partially or completely in the membranes of the nuclear envelope. In *S. cerevisiae*, for example, the SPB consists of an outer dense plaque lying just outside the nuclear envelope, a central plaque embedded in the nuclear envelope, and an inner plaque just inside the nuclear envelope. Microtubules extend from the outer dense plaque into the surrounding cytoplasm during interphase (Fig. 24-15*b*).

During interphase a single SPB is located at one side of the nucleus (Fig. 24-15*c*). Late in interphase the SPB divides into two parts, and microtubules elongate from the inner surfaces of the SPBs into the nucleoplasm (Fig. 24-15*d*). As the SPBs separate, the microtubules extending from the inner plaques elongate and interdigitate (Fig. 24-15*e*). The two SPBs continue to migrate around the nucleus until they take up final positions at opposite sides of the nucleus. As the SPBs separate, the interdigitating microtubules elongate between them, forming a spindle that stretches across the nucleus (Fig. 24-15*f*). Microtubules extending from the outer dense plaque into the cytoplasm persist as the spindle develops. In most species with SPBs the nuclear envelope remains intact during these developments so that the spindle is located inside the nucleus. Later in division the nuclear envelope may remain intact, develop perforations in the regions of the poles, or break down completely. In species in which the nuclear envelope remains partially or completely intact, a furrow separates the nucleus into two parts after the chromosomes have been divided by the spindle.

In *S. cerevisiae*, kinesin, one of the proteins that acts as a microtubule "motor" (see below and p. 441), has been shown to be present on the inner plaque of SPBs. Mutations in the gene encoding kinesin interfere with SPB duplication and separation in yeast, suggesting that this microtubule motor may produce part of the force separating the SPBs in fungi. (The cyclic cross-bridging action of kinesin and other microtubule motors such as dynein that produce microtubule-based movements is described in Ch. 11.) The p34^{cdc2} protein kinase has also been detected by fluorescent antibodies in association with the SPBs of *S. pombe* by J. S. Hyams and his colleagues, indicating that phosphorylation of proteins associated with SPBs may regulate SPB duplication or spindle formation in yeasts.

MTOCs and Spindle Formation The microtubules giving rise to astral spindles grow from a region surrounding but not directly connected to the centrioles. In the electron microscope the origins of microtubule growth appear as deposits of dense material from which microtubules radiate. In astral spindles these dense deposits are known as the *pericentriolar material* or *satellite bodies* (Fig. 24-16; see also Fig. 24-11).

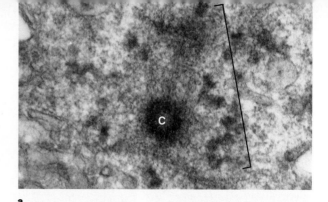

a

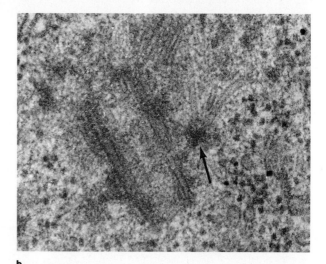

b

Figure 24-16 Pericentriolar material. **(a)** Pericentriolar material (bracket) surrounding a centriole (C) in a cultured mammalian cell. ×30,000. Courtesy of C. L. Rieder, from *Electron Micr. Rev.* 3:269. Copyright 1990, Pergamon Press Ltd. **(b)** Portion of the astral region of a mouse cell, showing microtubules originating from a region of the pericentriolar material (arrow) near a centriole. ×100,000. From *The nucleus*, ed. A. J. Dalton and F. Haguenau, 1968. Courtesy of E. de Harven and Academic Press, Inc.

Several experiments have shown that the pericentriolar material, and not the centrioles themselves, is the MTOC from which spindle microtubules grow. R. R. Gould and G. G. Borisy, for example, found that isolated pericentriolar material can act as a center for microtubule assembly if supplied with tubulin under polymerizing conditions. In most isolated preparations, no microtubules formed directly from the centrioles. When they did form, these microtubules extended outward from the centriole triplets, in a pattern resembling flagellar rather than spindle microtubule formation.

In another experiment S. Brenner and her coworkers discovered a mutant line of rat kangaroo cells containing twice the normal number of chromosomes. These cells entered division without replicating either their DNA or centrioles and formed a spindle with centrioles at one end but not the other. The end without centrioles also functioned normally in mitosis as a center of microtubule organization. Examination of these spindles in the electron microscope revealed that although centrioles were present at only one pole, the dense pericentriolar material was present at both poles.

Recent experiments by L. Clayton, C. M. Black, and C. W. Lloyd used antibodies to test plant cells for the presence of MTOCs with properties similar to the pericentriolar material of astral spindles. The antibodies were obtained from scleroderma patients, who make autoantibodies against pericentriolar material as well as kinetochores. The antibodies, when linked to a fluorescent dye and reacted with onion root tip cells, stained the zone around the nucleus in which spindle microtubules first appear at early prophase. As mitosis progressed to anaphase, the stain became more concentrated at the poles. Thus microtubule-organizing structures similar to the pericentriolar material of animal cells may also generate the anastral spindles of higher plants. The fact that antipericentriolar antibodies produced in humans react with the spindle-forming structures of plants also indicates that the molecular components of the centers organizing spindle microtubules have been highly conserved in evolution.

The identity of most components of the pericentriolar material remains unknown. However, recent research by M. Kirschner and Y. Zheng and their co-workers showed that a newly discovered tubulin type, γ-tubulin (see p. 419), is localized in the pericentriolar material. These investigators found that antibodies against γ-tubulin react positively with human, mouse, *Drosophila*, and *Xenopus* pericentriolar material and with the spindle pole bodies of the fungi *Aspergillus* and *Schizosaccharomyces*. The γ-tubulin, which makes up only about 1% of total cellular tubulin, appears to be restricted entirely to the pericentriolar material in animal cells. Because the antibody reaction is most intense during prophase through metaphase, when spindle microtubule assembly is most active, γ-tubulin is thought to take part in the microtubule-organizing function of the pericentriolar material.

This information makes it clear that centrioles do not directly give rise to spindle microtubules. Instead, spindles form through the activity of MTOCs that are recognizable as pericentriolar material.

Centriole Duplication During Spindle Formation
Centrioles duplicate in an interesting pattern, first described by J. G. Gall, in which a parental chromosome apparently directly gives rise to a copy. During this duplication the centrioles move apart slightly, and each produces a small, budlike extension, the *procentriole* (Fig. 24-17). The procentriole grows at right angles from one end of the parent centriole and remains separated from the parent by a narrow space. The procentriole gradually increases in length during S, G2, and prophase. By the end of prophase, duplication is usually complete, and the copies are indistinguishable from the parents.

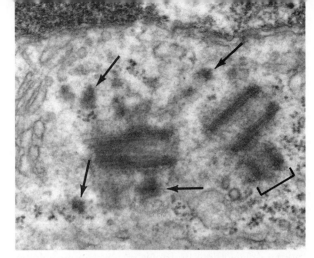

a

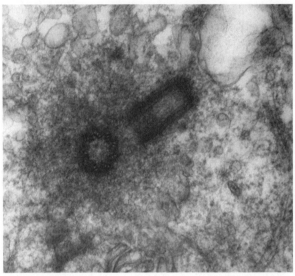

b

Figure 24-17 Centriole duplication. (**a**) Centrioles duplicating in a rat cell. The plane of section has caught one of the procentrioles (bracket) arising as a bud from one end of a parent centriole. Several deposits of pericentriolar material (arrows) are also present. ×60,000. Courtesy of R. G. Murray, A. S. Murray, and A. Pizzo, from *J. Cell Biol.* 26:601 (1965), by permission of the Rockefeller University Press. (**b**) The duplicated centrioles and surrounding pericentriolar material in a cultured mammalian cell. ×55,000. Courtesy of C. L. Rieder.

By the time the spindle is fully assembled, separation of the asters has divided the centrioles into two pairs, each consisting of a parent and copy centriole. In this sense the duplication and separation of the centrioles resembles semiconservative replication of DNA. Depending on the species, the centrioles within the asters may remain oriented at right angles or may take up other positions including parallel or tandem arrays.

The pattern of centriole duplication has been interpreted by many investigators to mean that these structures have an independent line of inheritance similar to mitochondria and chloroplasts. Although highly controversial, there is some evidence that DNA may be present in centrioles and that these structures may indeed be partially autonomous (see p. 901 for details).

The Role of Centrioles in Spindle Formation   To investigators at the turn of the century, the movements of the centrioles and asters suggested that centrioles give rise to the spindle. The accumulated evidence now makes this unlikely.

We have noted that spindle microtubules actually form from pericentriolar material rather than from centrioles in astral spindles. Completely functional spindles form in cells without centrioles such as those of higher plants; in these cells the microtubules of anastral spindles form from deposits of dense material similar to the dense pericentriolar material in astral spindles. Fully functional spindles also form in the many animal oocytes that lose their centrioles during the last few divisions before maturation.

A series of experiments by R. Dietz indicated that spindles can also form without centrioles in cells that normally possess them. By flattening crane fly cells under a microscope slide, Dietz prevented migration of the asters and centrioles around the nucleus during early stages of cell division. A spindle still formed in these cells; in the spindle one pole was anastral, and the other contained two asters and centriole pairs. The following mitosis and cytoplasmic division produced two cells, one with two asters and centriole pairs and one with neither centrioles nor asters. In the next division the cell without asters or centrioles formed a spindle similar in appearance to astral spindles except for a slight broadening at the poles. The division stages proceeded normally in the anastral cell.

These observations and experiments indicate that centrioles are simply passengers during astral spindle formation, and are passively separated and moved to the spindle poles by dividing asters. In these locations they are incorporated into separate daughter cells at the following mitosis and cytoplasmic division. Thus, the developing spindle microtubules divide the centriole pairs and place them at opposite ends of the cell, just as the spindle divides and moves the chromosomes later in mitosis.

Rather than forming the spindle, centrioles appear to function directly only in the formation of flagella or cilia. During development of a flagellum, a centriole gives rise to the system of microtubules that forms the axis of the flagellar shaft. This centriole usually persists as the *basal body* of the flagellum (for details, see p. 437). In keeping with this function, the presence of centrioles in an organism appears to be correlated with the existence, at some time in its life cycle, of cells that possess cilia or flagella. In plants, for example, the species containing centrioles reproduce by means of flagellated, motile gametes. Because of the importance

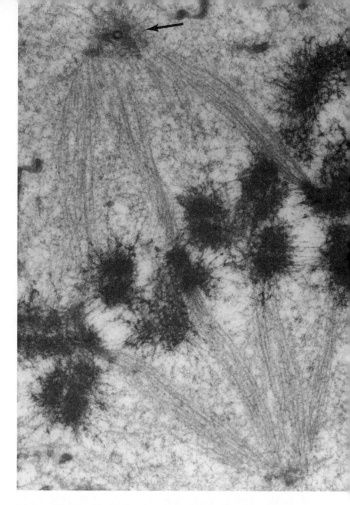

Figure 24-18   The fully developed spindle of a mammalian cell. Only kinetochore microtubules have been caught in the plane of section. One of the centrioles is visible in cross section in the aster at the top of the micrograph (arrow). × 14,000. Courtesy of C. L. Rieder, from *Electron Micr. Rev.* 3:269. Copyright 1990, Pergamon Press Ltd.

of flagella to motility, reproduction, and other functions in animals and some protozoa, algae, and plants, it seems likely that the movements of the centrioles during astral spindle formation ensure that these structures are distributed without loss or error to daughter cells during cell division.

The Fully Organized Spindle: Microtubules   Depending on the species, the completed spindle (Fig. 24-18) may contain from tens to many thousands of microtubules. The fungus *Phycomyces*, for example, has only 10 microtubules in its spindle; the rat kangaroo has about 1500. The spindle of *Haemanthus*, a flowering plant, is estimated to contain 10,000 microtubules.

The attachment of chromosomes to the spindle at metaphase (see below) differentiates two types of spindle microtubules. The spindle microtubules that attach to chromosomes are known as *kinetochore microtubules* (*KMTs*). These microtubules extend without breaks or overlaps from a kinetochore to either pole of the spindle. Other microtubules, which extend from the poles

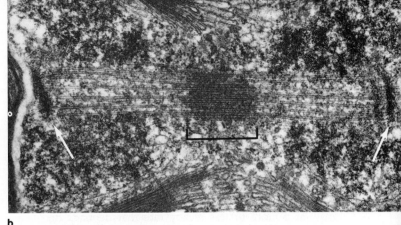

a

b

Figure 24-19 Two variations in the arrangement of interpolar microtubules in spindles. **(a)** The spindle of the fungus *Phycomyces*, in which interpolar microtubules extend from pole to pole without breaks. Even though no breaks are visible, the microtubules in this type of spindle probably interdigitate, with one microtubule originating from one spindle pole body (SPB) overlapping over its entire length with one or more microtubules originating from the opposite pole. SPBs (arrows) are visible at the two ends of the spindle. × 37,000. Courtesy of W. W. Franke. Copyright Academic Press, from *Internat. Rev. Cytol.*, suppl. 4:71 (1974). **(b)** The spindle of the diatom *Fragilaria* showing a zone of overlap (bracket) between interpolar microtubules at the spindle midpoint. SPBs (arrows) are visible at both poles. × 36,000. Courtesy of D. H. Tippit, from *J. Cell Biol.* 79:737 (1978), by permission of the Rockefeller University Press.

but do not make connections to chromosomes or kinetochores, are known as *interpolar microtubules* (IMTs).

In the spindles of most and perhaps all species, IMTs originating from one pole overlap and interdigitate with those from the opposite pole. In some primitive eukaryotes the overlap is complete and extends the length of the spindle (Fig. 24-19*a*). In other lower eukaryotes and all higher eukaryotes, the IMTs from one pole overlap with the ends of IMTs originating from the opposite pole only in the central region of the spindle (Fig. 24-19*b*). In either case the overlapping arrangement creates two *half spindles*, in which the microtubules of either half spindle, including both KMTs and IMTs, are oriented with their minus ends

toward the spindle poles and their plus ends toward the spindle midpoint. (The plus ends are the ends on which tubulin subunits assemble most rapidly; see p. 421.) Because the IMTs originating from either pole have their plus ends directed toward the spindle midpoint, the interdigitated microtubules in the zone of overlap run in opposite directions, or are *antipolar* (Fig. 24-20). Both the overlapping arrangement of IMTs and the antipolar arrangement of the two half spindles are significant for the mechanisms generating the anaphase movement (see below).

Spindles may also contain so-called *free microtubules* that neither run from pole to pole nor make connections to kinetochores. These microtubules often lie at various

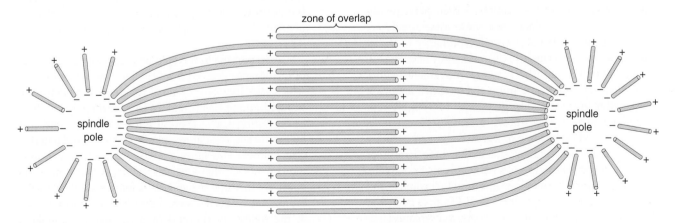

Figure 24-20 The arrangement of interpolar microtubules in the zone of overlap. Microtubules originating from either pole extend past each other and end beyond the spindle midpoint, forming two overlapping half spindles. Note that the IMTs are arranged uniformly with their minus ends at the spindle poles and their plus ends directed toward the midpoint.

angles to KMTs and IMTs, disturbing, to some extent, the otherwise parallel arrangement of elements in the spindle.

Fully organized spindles are in a state of equilibrium with a pool of unassembled tubulin molecules. Microtubules continually assemble and disassemble within the structure, so there is a constant turnover of microtubule subunits. The characteristics of assembly and disassembly follow the pattern known as *dynamic instability*—at any instant some microtubules in a population grow at a more or less steady rate that depends on factors such as tubulin and Ca^{2+} concentration; others undergo a rapid, "catastrophic," and complete breakdown. (For details of the dynamic instability of microtubules at equilibrium, see p. 422.)

Many treatments, including exposure to colchicine, cold, hydrostatic pressure, and ultraviolet radiation, can shift the equilibrium in the direction of unpolymerized subunits and cause almost instantaneous disassembly of the spindle. During prophase, removal of the dispersing agent allows rapid spindle reassembly, often within seconds or minutes after return to normal conditions. Other treatments, including exposure to heavy water and the drug taxol, can shift the equilibrium toward greater polymerization and thereby increase the number and length of spindle microtubules. (see p. 427 and below for details). This ready conversion between assembled and disassembled microtubules also figures importantly in the anaphase movement of chromosomes. (For details of microtubule assembly, including the relative rates of plus and minus ends, see p. 422.)

The rate at which spindle microtubules assemble and disassemble varies according to the spindle microtubule type and the stage of mitosis. Before any spindle microtubules make contact with kinetochores, all are dynamically unstable and undergo constant assembly and disassembly. (The half-life of spindle microtubules at prophase is estimated to be about 0.5 seconds.) The spindle microtubules attaching to chromosomes are evidently stabilized by the fixation of their plus ends to the kinetochores, so that once these attachments are made, dynamic instability is limited primarily to IMTs. This situation, in which KMTs are more stable than IMTs, continues until the anaphase movement begins, when IMTs also become more stable. Once anaphase begins, dynamic instability gives way to a highly regulated pattern of microtubule assembly/disassembly, in which KMTs become progressively and steadily shorter, and IMTs either remain the same length or become longer (see below). After anaphase is complete, all spindle microtubules become highly unstable and disappear except for segments of IMTs that persist at the spindle midpoint and participate in cytokinesis.

Other Components of the Fully Formed Spindle

Mature spindles contain a variety of components in addition to microtubules. Among the molecular elements is an Mg^{2+}-dependent ATPase with properties similar to flagellar dynein (see p. 429). Antibodies developed against flagellar dynein also interact with the spindle, adding evidence that dynein or a very similar microtubule motor is present. Tests with fluorescent antibodies by J. R. McIntosh, M. P. Sheetz, and their colleagues showed that the dyneinlike motor is present in the zone of overlap of IMTs at the spindle midpoint, at the poles, and in or near kinetochores. Another microtubule motor, kinesin (see p. 441), can also be detected in the spindle, particularly in the vicinity of kinetochores. Creatine phosphate, a substance capable of taking up and releasing energy in the form of phosphate groups (see p. 478), and calmodulin, the control protein activated by Ca^{2+}, are also present. Other molecular components are a Ca^{2+}-dependent ATPase associated with pumping calcium across membranes; actin in the form of microfilaments; myosin, the ATPase associated with microfilament-generated motility (see p. 463); and a variety of microtubule-associated proteins (MAPs). Among the latter are MAP-1, tau (see p. 425 for the functions of these MAPs), and STOP (for *Stable Tubule Only Polypeptide*). STOP, a MAP that stabilizes microtubules in assembled form, is associated preferentially with KMTs and may be partially responsible for the greater stability of these microtubules during prophase and metaphase. Like STOP, many of the other MAPs associated with the spindle are concentrated in specific locations and may be responsible for differences in structure, function, or stability of microtubules in these regions. B. M. Alberts and his coworkers, for example, developed antibodies to more than 50 *Drosophila* proteins that bind to microtubules. When linked to a fluorescent dye and added to cells, some of the antibodies bound primarily to asters, others to pericentriolar material or kinetochores, and some to the entire spindle. Another as yet unnamed MAP that was recently discovered by McIntosh and his colleagues seems to be concentrated in the zone of IMT overlap. This MAP may stabilize and link antipolar IMTs in this region.

Also present in the spindle is an extensive system of membranes derived from the ER. These membranes, which take the form of small tubules and vesicles distributed among KMTs and IMTs, are probably involved in regulating Ca^{2+} concentration in the spindle and, by this means, regulating spindle assembly and function (see below). The kinesin in the spindle may power movements of these vesicles along spindle microtubules. By the time the spindle is fully formed, the small tubules and vesicles have become most concentrated at the poles. By the end of prophase the spindle is primed and set for the central events of mitosis— the attachment of chromosomes and the anaphase movement.

Breakdown of the Nucleolus and Nuclear Envelope

Nucleolar Breakdown In most organisms the nucleolus breaks down and disappears during prophase and reappears during telophase. In these organisms nucleolar breakdown is first detected as a loosening of nucleolar structure. The nucleolus becomes progressively smaller and finally disappears toward the end of prophase. As the nucleolus disappears, rRNA synthesis halts.

Some of the molecules of the disaggregated nucleolus probably simply diffuse into the nucleoplasm and later, after the nuclear envelope breaks down at the close of prophase, intermix freely with the cytoplasmic solution. Several experiments indicate, however, that a significant portion of the disaggregated nucleolar material collects on the surfaces of the condensing chromosomes and is carried with the chromosomes to daughter nuclei during anaphase. This material evidently contributes to re-formation of the nucleolus during telophase (see p. 1030).

Breakdown of the Nuclear Envelope Disassembly of the nuclear envelope marks the transition from prophase to metaphase. Depending on the species, the breakdown may occur after migration of the asters is complete or while they are still separating. The disassembly begins as perforations appear in the envelope at scattered sites. The inner and outer membranes fuse at the perforated sites, producing separate flattened, closed vesicles resembling ER cisternae. As these changes take place, the vesicles drift from the nuclear margin into the cytoplasm. For a time pore complexes persist in the vesicles, but soon these disappear, and the remaining sacs become indistinguishable from the ER.

Disassembly of the nuclear envelope and its re-formation during telophase are paralleled by the behavior of the lamins, the cytoskeletal proteins that line and support the inner membrane of the nuclear envelope (see p. 560). Work with fluorescent antibodies by L. Gerace and his associates showed that just before the nuclear envelope begins to break down at late prophase in mammalian cells, the lamins disassemble. The lamin most closely associated with the membranes of the nuclear envelope, lamin B (see p. 560), remains bound to the membranous vesicles released from the nuclear envelope to the cytoplasm. R. Stick and his coworkers detected this lamin in association with all the larger cisternae of the ER visible during mitosis, indicating that the vesicles derived from nuclear envelope breakdown probably become fully integrated into the ER. The other two lamins of mammals, A and C, are released from the nuclear membranes as they disassemble; these lamins either enter the cytoplasmic solution or bind to the surfaces of the metaphase chromosomes, or are distributed between the two locations. The lamins reassemble on the inner surface of the nuclear envelope as it re-forms during telophase (see p. 1031).

Disassembly and reassembly of the lamins appear to be controlled by phosphorylation. Just before the lamins are released from the disintegrating nuclear envelope in late prophase, their phosphorylation increases by at least four times. The lamins remain in a highly phosphorylated state until telophase, when they are dephosphorylated as the nuclear envelope reassembles.

In general, experimental work and observations indicate that removal of the lamins from the nuclear envelope is a necessary step for nuclear envelope breakdown. Removal of the lamins is only part of the process, however, because nuclear envelopes can persist in apparently intact form after the lamins are removed. Lamin removal is, therefore, necessary but not in itself sufficient for nuclear envelope disassembly.

The regulatory mechanisms controlling breakdown and reconstruction of the nuclear envelope are uncertain. However, experiments with fused cells suggest that one or more soluble factors capable of inducing envelope breakdown and re-formation are directly involved in the process. The factors were detected in cell fusion experiments carried out by Johnson and Rao (see p. 1000). When interphase cells were fused with cells in metaphase, either the nuclear envelope of the interphase nucleus broke down, or a new nuclear envelope formed prematurely around the metaphase chromosomes. Whether breakdown or assembly of the nuclear envelope occurs may depend on the relative concentrations of the factors or the degree of their chemical alteration by modifications such as phosphorylation.

One likely possibility for a control factor is MPF (see p. 926 and below). The $p34^{cdc2}$ protein kinase forming part of MPF has been shown by M. Peter and his coworkers to be capable of adding phosphate groups to the same sites on the lamins that are phosphorylated during nuclear envelope breakdown in mitotic cells. G. Dessev and his colleagues found that the $p34^{cdc2}$ kinase can also directly induce the lamins to disassemble when added to isolated nuclei. These findings indicate that as the concentration of cyclin, the other component of the MPF particle, increases during prophase, the $p34^{cdc2}$ protein kinase becomes active and adds phosphate groups to the lamins.

Research by R. A. Steinhardt and his colleagues also implicated Ca^{2+}, calmodulin, and a Ca^{2+}/calmodulin-dependent protein kinase in the control of nuclear envelope breakdown. Several groups of investigators, including the Steinhardt laboratory, found that a transient increase in calcium concentration occurs just before the nuclear envelope breaks down. Further, experimental increases in Ca^{2+} concentration during prophase can cause premature breakdown of the nuclear envelope. For example, Steinhardt and his coworkers injected into living cells *nitr-5*, a substance that releases Ca^{2+} when exposed to light. After nitr-5 injection, nuclear envelope breakdown could be triggered by a bright flash of light. On the other hand substances that take up Ca^{2+} can block nuclear envelope breakdown when injected into mitotic cells. The Steinhardt group found that antibodies developed against a Ca^{2+}-dependent protein kinase active during mitosis block nuclear envelope breakdown. This finding implicates the protein kinase in the process, in a mechanism that involves: (1) a regulated release of Ca^{2+} from vesicles in the vicinity of the nuclear envelope, (2) activation of calmodulin by combination with Ca^{2+}, (3) activation of the protein kinase by the Ca^{2+}/calmodulin complex, and (4) phosphorylation by the activated protein kinase of proteins directly or indirectly triggering nuclear envelope breakdown.

Because the lamins do not appear to be among the proteins directly phosphorylated by the Ca^{2+}/calmodulin-dependent protein kinase, this kinase may be involved in pathways directing disassembly of the nuclear envelope membranes rather than the lamins. MPF and the Ca^{2+}/calmodulin-dependent protein kinase may, therefore, work in coordination to induce lamin breakdown and nuclear envelope disassembly.

EVENTS OF METAPHASE

During metaphase three events take place that are of major significance to the mitotic mechanism. One is movement of the spindle to the region previously occupied by the nucleus and the completion of spindle assembly. The second is attachment of chromosomes to spindle microtubules through linkage of kinetochores to previously existing microtubules. In organisms in which the nuclear envelope remains partially or completely intact during mitosis, attachment to the spindle, which is located inside the nucleus, proceeds by the same mechanism. The third event is movement of the chromosomes to the spindle midpoint. This movement, called *congression*, evidently depends on the connections made by kinetochores; it does not take place if the connections are broken. Some authorities call the stage during which these three steps are in progress *prometaphase* and reserve *metaphase* for the brief period during which the chromosomes are fully attached to the spindle and awaiting initiation of the anaphase movement.

The three events place each chromosome at the spindle midpoint, with the kinetochores of its sister chromatids attached to microtubules leading to opposite spindle poles. As metaphase gives way to anaphase, these connections ensure that the two chromatids are separated and moved to opposite spindle poles.

Because each chromatid is an exact duplicate of its sister, the kinetochore connections and precise division place exactly the same complement of chromosomes and the same genetic information in each daughter nucleus.

Chromosomes may make complicated excursions and oscillate between the poles during their movement to the spindle midpoint. Research by R. B. Nicklas showed that the movements of congression are absolutely dependent on spindle microtubules. Nicklas detached chromosomes from the spindle of living cells with a microneedle and moved them to distant locations in the cytoplasm. The detached chromosomes remained motionless for a time and then returned rapidly to the spindle midpoint. Electron micrographs made from cells at various stages in this sequence showed that the return movement did not begin until the chromosomes reattached to spindle microtubules.

Nicklas's experiments also showed that if kinetochore connections are scrambled by disturbing the chromosomes with a microneedle, the kinetochores reattach so that sister kinetochores are again connected only to microtubules leading to opposite poles. Nicklas proposed that the correct connections depend on tension developed through the attachments. If connections are made to opposite poles, the development of force pulls the two kinetochores of a chromosome in opposite directions, creating tension at the sites where the microtubules connect. This tension stabilizes the attachments, and the connections persist. If the two kinetochores of a chromosome attach to the same pole, they will both be pulled in the same direction, and no tension will develop at the attachment points. This condition, according to Nicklas, makes the connections unstable and leads to their release.

Early tension developed by spindle microtubules may also account for oscillations made by the chromosomes during congression. As the connections are made, the number of microtubules attached to either kinetochore may be temporarily unequal. The inequality may pull the chromosomes toward the pole with more connections; as additional connections are made, the inequality may alternate from side to side, causing the chromosome to move back and forth. The chromosomes stabilize at the spindle midpoint as the number of microtubules attached to the two kinetochores reaches an equal maximum. Several observations and experiments support this view. For example, T. S. Hays and E. D. Salmon showed that if some of the KMTs leading to one sister chromatid are cut at metaphase by a laser microbeam, the chromosome moves toward the opposite pole. The greater the dose of laser light, the closer the chromosome moves to the opposite pole. The tension responsible for these movements may depend on a microtubule motor forming part of the kinetochore (see below).

Another factor contributing to the back-and-forth movements of congression may be reversals of the kinetochore motors. Recently, A. A. Hyman and T. J. Mitchison found that isolated chromosomes move along microtubules in the poleward direction at low ATP concentrations, but can reverse and move in the opposite direction at higher ATP concentrations. The switch to movement away from the pole could also be induced if the isolated chromosomes were exposed to the ATP analog ATP-γ-S, which causes irreversible phosphorylation of target proteins when used as a phosphate source by protein kinases. This effect suggests that phosphorylation of kinetochore proteins may regulate the reversals.

The movements of chromosome congression, which take place at a measured rate of 0.05 to 1 μm per second, are characteristically much faster than the approximate 1 to 2 μm per minute rate observed for the anaphase movement (see below). The rate of congression is within the range observed for movements powered by microtubule motors such as dynein and kinesin, indicating that a cyclic crossbridge produces congression without being limited by other factors such as the rate of microtubule assembly or disassembly. Both dynein and kinesin, as noted, have been detected by fluorescent antibodies in kinetochores; the two motors may coordinate their activities in congression.

EVENTS OF ANAPHASE

Analysis of the chromosome movements taking place during anaphase revealed that separation of chromatids and their movement to the poles depend on several distinct processes. Chromatid separation is an autonomous process that can take place independently of the spindle. Subsequent movement of chromatids to the poles, the anaphase movement proper, is completely dependent on spindle microtubules. This movement results from a combination of separate but coordinated movements associated with interpolar and kinetochore microtubules.

Chromatid Separation

If the spindle is disassembled experimentally by exposing metaphase cells to microtubule "poisons" such as colchicine (see p. 427), the chromosomes are left without microtubule attachments at the former spindle midpoint. Under these conditions the chromatids still separate more or less in unison at the time at which anaphase would normally begin. Subsequently, the chromosomes decondense and form a single nucleus or a group of several nuclei of varying size. A single nucleus, if produced, is functional and apparently normal except that it contains twice the usual complement of chromosomes. (Viable polyploids are commonly produced in plants by the use of colchicine in this manner.)

In addition, the sister chromatids of chromosomes cut from the spindle by a microneedle still separate at the same time as those of chromosomes remaining attached to the spindle. These observations make it clear that separation of sister chromatids depends on a mechanism that is independent of the spindle or spindle microtubules.

It was once thought that chromatid separation depended on delayed replication of the centromere region or duplication of kinetochores. The centromere region or kinetochores was believed to remain single and unduplicated until the anaphase movement begins. Replication or duplication of either structure was thought to free the chromatids so that they could separate.

The idea that delayed centromere replication is responsible for chromatid separation was abandoned when experiments measuring the uptake of DNA precursors such as tritiated thymidine showed that no significant fraction of the DNA remains unreplicated after the S phase preceding mitosis. Further, no replication can be detected as the anaphase movement begins. In yeast R. M. McCarroll and W. L. Fangman demonstrated directly that centromere sequences replicate during S, along with the rest of the DNA.

Delayed kinetochore duplication has also been made unlikely by the accumulated evidence. Direct examination in the light or electron microscopes makes it clear that kinetochores are fully duplicated and separate when chromosomes become visible at early prophase. For example, B. R. Brinkley and his colleagues showed that reaction of cells with fluorescent antikinetochore antibodies clearly produces two distinct spots of fluorescence on each chromosome, one for each kinetochore, by the G2 stage preceding mitosis. (Fig. 24-21 shows double kinetochores on prophase chromosomes.)

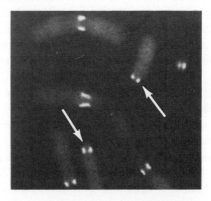

Figure 24-21 Kinetochores in prophase chromosomes reacted with fluorescent antibodies against kinetochore proteins. Each chromosome clearly has two kinetochores (arrows). Courtesy of B. R. Brinkley and R. P. Zinkowski. Reproduced from *J. Cell Biol.* 113:1091 (1991), by copyright permission of the Rockefeller University Press.

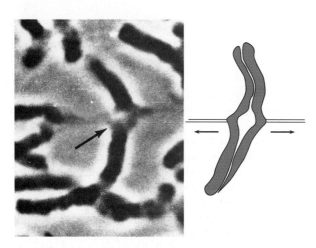

Figure 24-22 A metaphase chromosome of the plant *Haemanthus* photographed in a living cell. Tension developed by the spindle microtubules has pulled the kinetochores toward the poles. The chromatids are held together by attachments above and below the primary constriction. The centromere region at the site marked by the arrow lies out of the plane of focus. × 1,600. Courtesy of A. Bajer and Springer-Verlag.

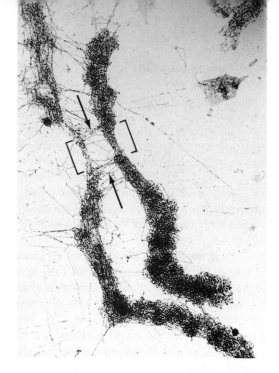

Figure 24-23 Fibrous connections (arrows) that appear to anchor the chromatids together on either side of the centromere region (brackets) of an isolated human chromosome. × 12,000. Courtesy of J. G. Abuelo and D. E. Moore, from *J. Cell Biol.* 41:73 (1969), by permission of the Rockefeller University Press.

A number of morphological observations suggest that the connections holding chromatids together before anaphase lie entirely outside the centromere region. At metaphase early tension developed by spindle microtubules frequently pulls the centromere region apart. Chromatids are still held together, however, at sites above and below the centromere region (Fig. 24-22). Fibers making connections between the chromatids at points above and below the centromere region are also frequently seen in metaphase chromosomes isolated for electron microscopy (Fig. 24-23).

Experiments by B. A. Hamkalo and her coworkers implicated the repetitive C-DNA sequences flanking the centromere region in the connections holding the chromatids together. Hamkalo noted that in mice C-DNA was present in all the regions at which chromatids were held together before anaphase. Digestion of the C-DNA by restriction endonucleases (see p. 135) disrupted the connections and released the chromatids. This DNA may be involved in holding chromatids together through chromatin loops that remain entangled until the anaphase movement is ready to begin. Supporting this hypothesis are observations by T. Uemara, who found that DNA topoisomerase II, which allows one DNA helix to pass through another, is required for separation of chromatids at anaphase as well as for condensation in *Schizosaccharomyces pombe.* A DNA topoisomerase is also among the proteins detected in the centromere region by the antikinetochore antibodies of scleroderma patients.

J. B. Rattner and his coworkers have identified a protein which they term *CLiP* (for *C*hromatid *Li*nking *P*rotein) which antibodies have shown to be located in the fibers holding sister chromatids together in or near the centromeres. After chromatid separation CLiPs can no longer be detected. Rattner suggests that CLiPs may contribute to the structures holding chromatids together, and may be released or degraded as part of chromatid separation.

Thus, the structures holding chromatids together until anaphase begins may possibly be tangled chromatin loops spaced on either side of the centromere, stabilized by CLiPs. Digestion by topoisomerase II of the tangles in chromatin loops persisting from DNA replication and degradation of CLiPs may free the chromatids for their anaphase movement.

Components of the Anaphase Movement

The movement of chromatids to the poles is clearly dependent on spindle microtubules. The movement stops immediately if the spindle is disrupted by anti-microtubule agents such as colchicine. If the disrupting agent is removed, the anaphase movement does not resume until the spindle has reassembled. Connections between chromatids and spindle microtubules are also necessary for the movement to proceed. If the connections are experimentally broken, the chromatids remain stationary until attachments to spindle microtubules are remade.

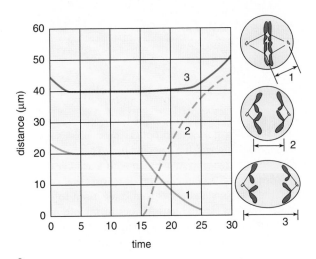

Figure 24-24 The two components of the anaphase movement. **(a)** Measurements showing the relative contributions of kineto-chore and interpolar microtubules to the anaphase movement. 1, distance separating chromosomes from the poles; 2, distance between sister chromatids; 3, total distance between the poles. At 15 minutes anaphase separation begins (curve 2). During the period from 15 to 30 minutes, as the anaphase movement takes place, the distance between the kinetochores and the poles becomes shorter and the separation between the poles increases. The total distance traveled by the chromatids is the sum of the two components. Redrawn from ''How Cells Divide'' by D. Mazia. *Sci. Amer.* 205:100 (1961). Copyright © 1961 by Scientific American, Inc. All rights reserved. **(b)** The two components in diagrammatic form. Anaphase A (curve 1 in **a**) is the component of the total movement contributed by the decrease in distance between the kinetochores and the poles. Anaphase B (curve 3 in **a**) is the component of the total movement contributed by a lengthening of the entire spindle.

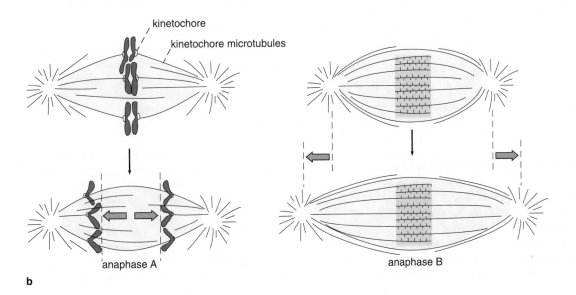

Careful measurements by H. Ris revealed some time ago that in higher organisms the total distance moved by chromatids during the anaphase movement has two components. One results from a reduction in the distance between the kinetochores and the poles. The second component results from an increase in the distance separating the spindle poles, so that the entire spindle becomes longer (Fig. 24-24). The component of the total anaphase movement caused by a reduction of the kinetochore-to-pole distance is termed *anaphase A;* that caused by lengthening of the entire spindle is *anaphase B* (see Fig. 24-24*b*). The individual contribution of each component to the total distance traveled by the chromatids varies in different cell types and species. In most higher organisms the total movement is derived from a fairly equal combination of the two. In some algae and fungi, the kinetochore-to-pole distance remains the same during anaphase, and separation of the chromosomes is entirely the result of the spindle elongation producing anaphase B.

The rate at which both components of the anaphase movement take place, about 1 to 2 μm per minute, is very slow compared to other cellular motions such as cytoplasmic streaming, flagellar beating, or muscle contraction. The anaphase movement, in fact, occurs at approximately the rate at which microtubules lengthen or shorten through the addition or release of microtubule subunits. This observation suggests that, although part or all of the anaphase movement may be powered by microtubule motors such as dynein or kinesin, microtubule assembly/disassembly is the rate-limiting process.

The Experimental Process

Spindle Elongation *in vitro:* The Relative Roles of Microtubule Sliding and Tubulin Polymerization

W. Zacheus Cande

W. Z. CANDE studies the mechanism of mitosis and meiosis in diatoms, yeast, and maize. He is a Professor of Cell and Developmental Biology in the Department of Molecular and Cell Biology at the University of California, Berkeley, and has been a member of the Berkeley faculty since 1976.

Although substantial progress has been made in describing spindle ultrastructure, relatively little is known about how the spindle operates at a biochemical level. *In vitro* model systems have been important tools for analyzing complex motility events such as muscle contraction and flagellar beat. Muscle cell models have provided the bridge that allowed us to understand the roles of individual proteins, such as myosin, during muscle contraction. Until recently, the mitosis field lacked this type of system for understanding how the spindle generates force for moving chromosomes. In part this is due to the number of mechanochemical events involved. Anaphase chromosome movement consists of at least three distinct events, and the conditions required for the reactivation of one phase of movement may not be compatible with maintenance of the other phases of chromosome separation *in vitro*. In most spindles, chromosomes move poleward (anaphase A) as the kinetochore microtubules shorten. Then, or at the same time, spindles elongate (anaphase B) and the polar or nonkinetochore-attached microtubules [IMTs in this book] grow and are rearranged as the spindle increases in length. Finally, in some cells, there may be interactions between elements of the surrounding cytoplasm and the poles, mediated by astral microtubules, which help move the reforming nuclei further apart. This last phase may not be present in all cells; however, it certainly plays a major role in fungi where spindles are intranuclear and cytoplasmic microtubule arrays are attached to the nuclear envelope and persist throughout mitosis.

The major challenge in isolating functional spindles from dividing cells is developing a medium that will stabilize the spindle microtubules against the rigors of isolation yet still preserve spindle function. This is a difficult task since the distinction between unnatural fixation and physiologically useful stabilization may only be a matter of degree. As was first demonstrated by Shinya Inoué in the

late 1950s using polarization optics, the spindle is a very dynamic structure and the spindle fibers, i.e., the microtubules, are in a dynamic equilibrium with a subunit pool.[1]

With respect to organization of cytoskeletal components, the diatom spindle stands in comparison to other spindles much as striated muscle does to smooth muscle and other nonmuscle motility systems; it is a representative, yet much more highly ordered system.[2] In the diatom spindle, the machinery responsible for anaphase A and B are spatially separated, thus allowing the possibility of studying one process independently of the other. Second, the central spindle, the structure responsible for spindle elongation, is constructed of two interdigitating sets of microtubules that originate from platelike pole complexes. These are arranged in paracrystalline arrays and antiparallel microtubules display specific near-neighbor interactions in the zone of microtubule overlap. Since the zone of microtubule overlap is visible even by light microscopy, it is possible to monitor changes in overlap zone organization during spindle elongation. This degree of order is also found in the mammalian spindle, but only after midbody formation during telophase. My colleague, Kent McDonald, and I reasoned that the unique structural attributes of the diatom spindle that make it so useful for morphological studies would also be advantageous for developing an *in vitro* system for studying spindle elongation. For example, we thought that the highly ordered and crossbridged microtubules of the diatom central spindle would more easily withstand the rigors of isolation than the microtubules of the more loosely organized mammalian spindle.

McDonald and I used the large spindles (300 microtubules per half spindle, spindle length 12 μm) isolated from the diatom *Stephanopyxis turris* to describe the behavior of microtubules in the zone of microtubule overlap *in vitro*.[3] After screening numerous species from several collections, we selected this diatom because of its natural mitotic synchrony in culture and the presence of a spindle that was large enough for light microscopy yet not so large as to prohibit quantitative analysis in the electron microscope. The homogenization medium we used was designed to stabilize microtubules and the isolation procedure was kept as simple as possible. Populations of mitotic cells were collected by filtration, cell walls disrupted by mechanical shear, and spindles and nuclei were then collected on coverslips by low-speed centrifugation. Each coverslip contained hundreds of spindles, embedded in chromatin. After ATP addition, the spindles increased in length and a prominent gap developed between the two half-spindles, which was bridged by only a few microtubules. As visualized on-line with video microscopy and polarization optics or with populations of spindles using indirect immunofluorescence, the two half-spindles slid completely apart with a concurrent decrease in the extent of the zone of microtubule overlap.[3]

These results can best be interpreted by imagining that the microtubules of one half-spindle push off against the microtubules of the other half-spindle generating sliding forces between them. Our observations are consistent with models that postulate mechanochemical interactions in the zone of microtubule overlap that generate the forces necessary for anaphase B. Movement apart of the two half-spindles is not likely to be due to the autonomous swimming of each half-spindle, since it is limited by the extent of microtubule overlap.

The changes in distribution of newly incorporated tubulin in the spindle midzone give further support to our interpretation of the mechanism of force generation.[4] When spindles are incubated in biotin-labeled tubulin before ATP addition, labeled tubulin is incorporated as new microtubules nucleated at the poles and into the spindle midzone as two bands that flank the original zone of microtubule overlap [see Fig. 24–27b]. Ultrastructural studies demonstrate that the new tubulin adds onto the ends of the original half-spindle microtubules in the midzone and does not form new microtubules next to the old ones. Thus, this new tubulin is a marker for the free ends of the original half-spindle microtubules. After ATP addition the spindle elongates, and the two bands in the midzone become one [see Fig. 24–27d]. This change can be explained if the microtubules containing the newly incorporated tubulin in each half-spindle slide over each other toward the center of the overlap zone. We have also used this method of labeling the ends of microtubules in the overlap zone to demonstrate that microtubule sliding occurs during spindle elongation in the small spindles of the diatom *Cylindrotheca fusiformis* and in the fission yeast *Schizosaccharomyces pombe*, where it is difficult to measure changes in spindle length.

Our observations eliminate several possible mechanisms of force generation. First, the poles are not pulled apart *in vitro*. There are no cytoplasmic structures attached to the isolated spindles that could generate this force. Second, microtubule polymerization cannot be responsible for generating the forces required to move the spindle poles apart because spindle elongation *in vitro* can occur in the absence of tubulin polymerization. Finally, it has been suggested that tension generated during spindle formation, stored in chromatin and kinetochore microtubules, and released during chromosome-to-pole movement may be responsible for spindle elongation. However, spindles elongate in the absence of chromatin and with kinetics inconsistent with this model.

When tubulin derived from brain microtubules is included in the reactivation medium, the extent of spindle elongation is no longer limited to the size of the original overlap zone, and some spindles elongate by the equivalent of 3 to 5 times their length. However, it is possible to uncouple tubulin polymerization from spindle elongation.[4] In the absence of ATP, tubulin is incorporated into the spindle and the zone of microtubule overlap increases in extent. After removal of tubulin and subsequent addition of ATP, the spindle elongates to an extent determined by the size of the new overlap zone, not the original overlap zone. Spindle elongation under these conditions is sensitive to the same inhibitors in the presence or absence of added tubulin, suggesting that microtubule polymerization does not contribute to the generation of forces required for anaphase B. Thus, the process of tubulin polymerization is distinct from that of force generation and its role is to define the extent of polymer that can slide through the zone of microtubule overlap. Depending on the relative rates of microtubule assembly and sliding as defined experimentally by the relative tubulin and ATP concentrations, it is possible *in vitro* for the overlap zone to increase, decrease, or remain constant as the spindle elongates.

The pharmacology of spindle elongation is similar in fission yeasts, diatoms, and mammalian cells, suggesting that a similar force-generating enzyme is used in all eukaryotic cells. It has several novel features: like kinesin it is inhibited strongly by the nonhydrolyzable ATP analog AMPPNP; however, unlike kinesin it is sensitive to low concentrations of N-ethyl maleimide and vanadate. Recently, in collaboration with Takashi Shimizu, we have described the substrate specificity of the anaphase B motor using 20 different ATP analogues and other nucleotides, and have shown that the motor is more kinesinlike than dyneinlike in the range of nucleotides that will support spindle elongation. We are now using biochemical and molecular approaches to directly identify and characterize the anaphase B motor in diatoms.

References

[1] Inoue, S., and Sato, H. Cell motility by labile association of molecules: the nature of mitotic spindle fibers and their role in chromosome movement. *J. Gen. Physiol.* 50:259–92 (1967).

[2] Pickett-Heaps, J. D., and Tippit, D. H. The diatom spindle in perspective. *Cell* 14:455–67 (1978).

[3] Candle, W. Z., and McDonald, K. L. *In vitro* reactivation of anaphase spindle elongation using isolated diatom spindles. *Nature* (Lond.) 316:168–70 (1985).

[4] Masuda, H.; McDonald, K. L.; and Cande, W. Z. The mechanism of anaphase spindle elongation: uncoupling of tubulin incorporation and microtubule sliding during *in vitro* spindle reactivation. *J. Cell Biol.* 107:623–33 (1988).

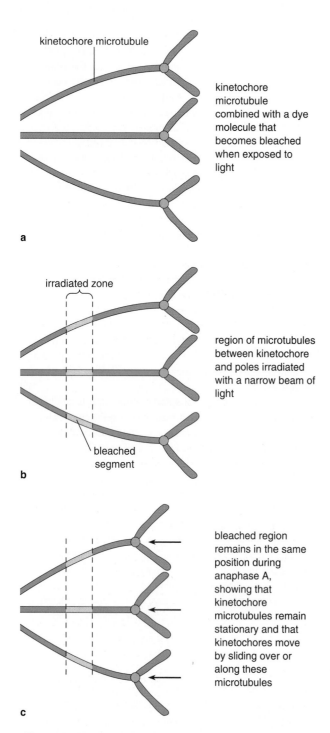

kinetochore microtubule

kinetochore microtubule combined with a dye molecule that becomes bleached when exposed to light

a

irradiated zone

region of microtubules between kinetochore and poles irradiated with a narrow beam of light

bleached segment

b

bleached region remains in the same position during anaphase A, showing that kinetochore microtubules remain stationary and that kinetochores move by sliding over or along these microtubules

c

Figure 24-25 An experiment by Mitchison and his colleagues demonstrating that kinetochore microtubules remain stationary during the anaphase A movement and that the kinetochores must move by sliding over or along the kinetochore microtubules.

Mechanisms of the Anaphase Movement

Many hypotheses have been advanced to explain the anaphase movement (almost as many as there are microtubules in the spindle). For many years there was little definitive evidence favoring one hypothesis over another. Recent investigations, however, have nar-

rowed the possibilities and clearly support distinct mechanisms for the anaphase A and B movements.

Anaphase A   Much has been learned about anaphase A from experiments by T. Mitchison, M. W. Kirschner, D. E. Koshland, and G. J. Gorbsky in which kinetochore microtubules were given reference marks that allowed their movement to be traced during anaphase. In one group of these experiments, spindle microtubules were combined with a dye molecule that bleaches when it is exposed to light (Fig. 24-25a). The marked spindles were exposed to a narrow beam of light that bleached a stripe across the kinetochore microtubules perpendicular to the spindle axis (Fig. 24-25b). The bleached segment of the kinetochore microtubules remained in the same position, at the same distance from the pole, as anaphase A progressed (Fig. 24-25c). This finding indicates that kinetochore microtubules remain stationary during anaphase A and that the kinetochores move by sliding along or over these microtubules. Similar results were obtained in experiments in which a segment of the kinetochore microtubules was labeled with biotin, a marker molecule that can be detected under the light microscope by antibiotin antibodies combined with a fluorescent dye (see p. 119). The segments of the kinetochore microtubules labeled with biotin also remained stationary as the chromosomes moved to the poles.

Because kinetochore microtubules become shorter as anaphase A progresses, they must disassemble at the end nearest the chromosome as the kinetochore passes over them. Evidence directly supporting this conclusion was obtained in the labeling experiments by Mitchison, Kirschner, and their colleagues. During metaphase labeled tubulin was incorporated at the kinetochore end of the KMTs, indicating that new subunits are continually added at this end before anaphase begins. At anaphase label disappeared from the KMTs, as expected if microtubules disassemble at the chromosome end as the kinetochores move over them.

The movement of kinetochores is evidently therefore much like a locomotive moving along a railroad track, except that the track is disassembled as the locomotive passes over it. The mechanism generating the motile force is presently uncertain. Either or both of the dynein and kinesin motors detected in association with kinetochores may be involved in production of the anaphase A movement. In living cells ATP appears to be necessary for the anaphase A movement, in support of the hypothesis that an ATP-dependent microtubule motor such as dynein or kinesin produces the movement. Of the two, dynein seems more likely as the motor because it produces motion toward the minus ends of microtubules, in the direction anaphase A progresses. (Fig. 24-26 shows one possibility for the action of a microtubule motor in producing kinetochore movement.)

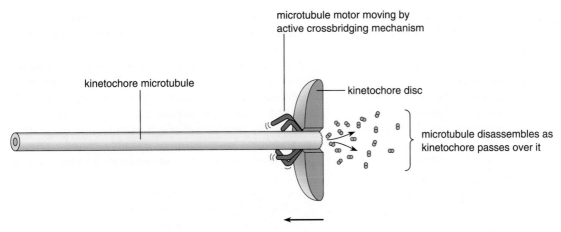

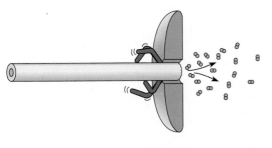

Figure 24-26 One possibility for the mechanism by which kinetochores move over kinetochore microtubules. Active crossbridging by a microtubule motor such as dynein or kinesin slides the kinetochore over the microtubule. ATP hydrolysis supplies the energy for the crossbridging cycle. As the kinetochore passes, the microtubule disassembles.

Observations by W. Z. Cande and his associates indicate that microtubule disassembly may also produce part of the force responsible for anaphase A. These investigators found that anaphase A can proceed in isolated spindles without a supply of ATP if the preparations are held under conditions promoting microtubule disassembly.

The idea that microtubule assembly or disassembly might produce the force for part of the anaphase movement was originally advanced by S. Inoué and his coworkers. The Inoué model, called the *dynamic equilibrium hypothesis,* was developed through studies of the effects of various treatments on spindle assembly and disassembly. Inoué and his colleagues began their studies before spindles were known to contain microtubules. As a result of their early studies, in which spindles were observed to disassemble when exposed to agents such as colchicine and cold temperatures and to reassemble rapidly when the agent is removed, Inoué proposed for the first time that the spindle exists in equilibrium with a pool of disassembled spindle subunits. When spindles were discovered to be microtubule-based structures, Inoué proposed that adjustments in the tubulin-microtubule equilibrium might provide part of the force for the anaphase movement.

Support for Inoué's hypothesis was taken from observations of the effects of small adjustments in the tubulin-microtubule equilibrium. Exposing cells to colchicine at higher concentrations completely destroys the spindle and stops the anaphase movement. How-

ever, colchicine at very dilute concentrations leaves the spindle largely undisturbed and actually increases the rate of anaphase A. Hydrostatic pressure has the same effect. As the pressure is gradually increased, chromosome movement first speeds up, reaching velocities between two and three times the normal values. These results indicated that the increased rates of microtubule disassembly are reflected in a greater rate of shortening of the KMTs. As pressures reach several thousand pounds per square inch, the spindle disassembles, and all movement ceases. Adding taxol, a substance that increases microtubule polymerization, has the opposite effect and reverses anaphase A by pushing the chromosomes farther away from the poles. These observations support the idea that microtubule disassembly may be involved in the mechanism generating the force for anaphase A. However, they do not determine whether microtubule assembly/disassembly provides a portion of the force for chromosome movement or merely serves as a rate-limiting step for other mechanisms of force generation.

Anaphase B The accumulated evidence of more than two decades of investigation clearly indicates that anaphase B—the increase in separation of the spindle poles—occurs by an active sliding mechanism. The sliding, probably powered by a microtubule motor, occurs between IMTs in the zone of overlap at the spindle midpoint.

The first indications that IMTs slide at the spindle

midpoint, a mechanism originally proposed by Mc-Intosh, D. H. Tippet, and J. D. Pickett-Heaps and their colleagues, were obtained by the laborious method of sectioning entirely through spindles from one pole to the other and tracing the path of each microtubule. This approach showed that in spindles in which IMTs overlap at the midpoint the interdigitation in the zone of overlap is near perfect. The nearest neighbor of any microtubule in the overlap zone is likely to be a microtubule originating from the opposite pole. The sectioning also showed that as anaphase B proceeds in some organisms, the degree of overlap decreases by an amount equivalent to the increase in distance between the poles.

Further evidence that microtubule sliding occurs at the spindle midpoint was obtained by H. Masuda, K. L. McDonald, and Cande using labeling techniques. These investigators labeled tubulin subunits with biotin and added the labeled subunits to isolated spindles under conditions that favored microtubule polymerization. Under these conditions the labeled tubulin subunits added to the plus ends of microtubules overlapping at the spindle midpoint (Fig. 24-27a and b). The labeled tubulin was then removed from the medium, and further polymerization was carried out with unlabeled tubulin (Fig. 24-27c). Although the zone of overlap lengthened, the distance between the poles did not increase as either the labeled or unlabeled tubulin was added, indicating that the force for anaphase B is unlikely to be produced by microtubule assembly.

After incorporation of labeled and unlabeled tubulin, ATP was added to the isolated spindles. The addition caused the poles to move apart; at the same time the segments labeled by biotin moved over each other in a pattern indicating that progressive sliding occurred between IMTs in the zone of overlap at the midpoint (Fig. 24-27d and e). The Experimental Process essay by Cande on p. 1022 describes his experiment showing that microtubule sliding drives anaphase B.

It remains unclear whether a dyneinlike or kinesinlike microtubule motor, or both motors working in coordination, power anaphase B. Vanadate, an inhibitor of flagellar dynein, and antidynein antibodies stop anaphase B. Antidynein antibodies also react with the spindle, especially in the zone of overlap. Kinesin can also be detected in the spindle; Cande and his coworkers have found that, like kinesin, the anaphase B movement is strongly inhibited by AMPPNP. In general, Cande and his colleagues find that the anaphase B motor seems more kinesinlike than dyneinlike in isolated spindles under conditions that support spindle elongation (see Cande's essay).

One problem with dynein as an anaphase B motor is that dynein typically produces movement toward the minus ends of microtubules, opposite to the direction of anaphase B sliding. However, a few examples of dyneinlike motors producing movement toward mi-crotubule plus ends have been observed, and it is possible that one of these unusual dyneins is involved in anaphase B.

Though microtubule sliding appears to generate the force for anaphase B, microtubule assembly also is probably involved indirectly in this movement. As noted, anaphase B, like anaphase A, takes place at the relatively slow rate at which microtubules lengthen or shorten by assembly or disassembly. In most spindles IMTs become longer by some two to three times during the anaphase movement, through a process in which tubulin subunits are continually added to the plus ends of the IMTs as they slide over each other. The rate of this addition may be a major factor limiting the rate of anaphase B.

Anaphase A probably occurs through activity of kinetochores moving actively along KMTs. Although the mechanism producing the force for this movement is still uncertain, it is probably produced by a microtubule motor. Microtubule disassembly, which occurs as kinetochores pass over KMTs, appears to limit the rate of anaphase A. Anaphase B takes place through a forceful sliding of microtubules that decreases the zone of overlap at the spindle midpoint. The motor for this movement is a dynein- or kinesinlike ATPase that produces force by a cyclic crossbridging motion. Microtubule assembly probably limits the rate of anaphase B.

Other Models for Anaphase Movement

A number of additional hypotheses have been advanced for the anaphase movement. Most of these models regard spindle microtubules as passive cytoskeletal elements that are not directly involved in producing movement. Although the weight of evidence now favors kinetochores as the source for anaphase A and microtubule sliding for anaphase B, these other models have persisted, some only because there is no absolutely definitive evidence directly ruling them out. It is also possible that a few groups of organisms have evolved unique anaphase mechanisms or that in some primitive groups such as dinoflagellates (see below) distinct mechanisms have persisted as relics from transitional stages in the first evolution of mitotic cell division.

Among the more important of these additional proposals is the hypothesis that part or all of the anaphase movement is powered by microfilaments rather than microtubules. Another model proposes that anaphase B results from tension applied to the asters or spindle poles by molecular mechanisms originating from outside the spindle. According to this model, tension lengthens the spindle by pulling it apart at the poles. A third alternative to be considered proposes that in some primitive organisms in which the nuclear envelope persists during mitosis, movement of chromatids is provided by growth or expansion of segments

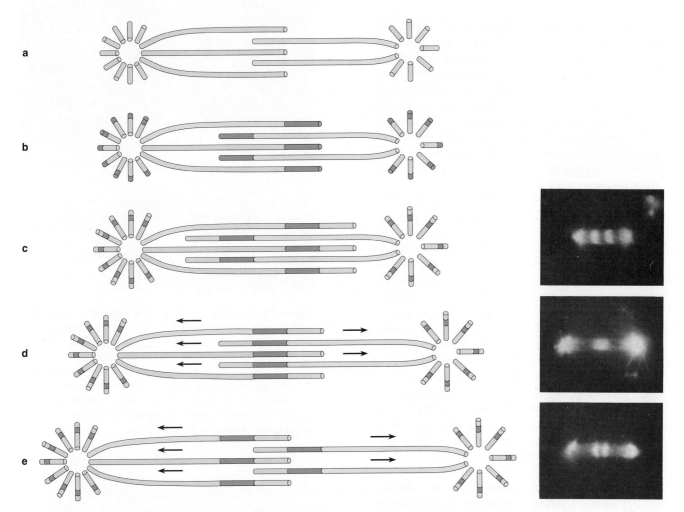

Figure 24-27 The experiment conducted by Matsuda, McDonald, and Cande demonstrating microtubule sliding in the zone of overlap at the spindle midpoint. Spindles were exposed to biotin-labeled tubulin under conditions that favored polymerization of the tubulin into spindle microtubules. **(a)** An unlabeled spindle; **(b)** a spindle after exposure to label under polymerizing conditions. The labeled tubulin subunits have added to the plus ends of the microtubules overlapping at the spindle midpoint. **(c)** The labeled tubulin was removed from the medium and further polymerization was carried out with unlabeled tubulin. The unlabeled tubulin subunits also added to the plus ends of the spindle microtubules. After the labeled and unlabeled tubulins were incorporated, ATP was added to the isolated spindles. **(d and e)** The addition caused the poles to move apart; at the same time the segments labeled by biotin moved over each other in a pattern indicating that the IMTs originating from opposite poles slide over each other in the zone of overlap. The labeled zones are shown in purple in the diagram and are visible as regions of bright fluorescence in the micrographs. Micrographs, ×2,400. Courtesy of W. Z. Cande. Reproduced from *J. Cell Biol.* 107:623 (1988), by copyright permission of the Rockefeller University Press.

of the nuclear envelope connected to the chromatids rather than by microtubules.

The Sliding Microfilament Model The proposal that microfilaments are responsible for part or all of anaphase movement, put forward most forcefully by A. Forer and his coworkers, is based primarily on the observation that microfilaments can be detected in the spindles of some species, in orientations consistent with movement

of chromosomes to the poles. Spindles in some species also stain strongly with fluorescent antibodies developed against either actin or myosin, particularly near kinetochores. According to the Forer model, spindle microtubules act primarily or exclusively as passive cytoskeletal supports that limit the speed of chromosome movement to the rate at which the supporting microtubules disassemble, but do not directly produce the force for motion.

Several experiments produced results that argue strongly against the Forer model. Experimental addition of antimicrofilament agents such as cytochalasin B (see p. 459) has no effect on chromosome movement. Similarly, antibodies against actin or myosin fail to interrupt either anaphase A or B. One of the most telling experiments of this type was carried out by P. Kiehart, Inoué, and I. Mabuchi, who injected antiactin and antimyosin antibodies into dividing cells. The antibodies stopped movements known to be based on microfilaments, such as the furrowing responsible for cytoplasmic division, but had no effect on the anaphase movement. The results of the antibody experiments are especially significant because furrowing begins while anaphase is still in progress in most animal cells. Thus, it is possible to show in the same cell that furrowing is stopped by antiactin or antimyosin antibodies, while the anaphase movement continues without interruption or delay.

A series of genetic experiments carried out by J. A. Spudich and his colleagues and others also appears to rule out the involvement of actin-based motility in the anaphase movement. Spudich and his colleagues produced cells of the fungus *Dictyostelium* in which the single gene encoding the heavy chain component of myosin (see p. 460) in this species was disrupted and made completely inactive. In spite of this inactivation, which interrupted cytoplasmic division, anaphase proceeded without hindrance.

The Polar Tension Model According to the polar tension hypothesis, a pulling force applied to the poles or asters from outside the spindle contributes to anaphase B. Evidence for the model is limited to a few observations involving micromanipulation or application of drugs to spindles in living cells. One of these experiments was conducted by J. R. Aist and M. W. Berns with the fungus *Fusarium*. In this fungus the two spindle halves sometimes pop completely apart at the close of anaphase under normal conditions, as if a pulling force is exerted on the poles from outside the spindle. Destruction of the central spindle region in *Fusarium* by a laser microbeam increased the anaphase B movement by about three times, as if the central spindle retards rather than promotes anaphase B in this species. In the same experiment irradiation in the region between the poles and the plasma membrane retarded anaphase B, as might be expected if a pulling force is generated in this region. Microtubules radiating from the poles to the cell periphery may thus be responsible for generating polar tension in *Fusarium*. Similar results were obtained by Aist, Berns, and their colleagues in experiments with spindle forces in another fungus, *Nectria*.

Other experiments failed to detect polar tension. For example, diatom spindles, isolated in such a way that the poles are freely suspended in the isolating medium, with no attachments to other structures that could produce a pulling force, still slide completely apart

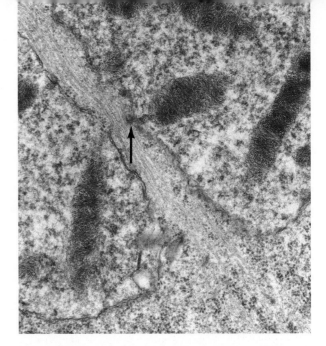

Figure 24-28 Longitudinal section through a cytoplasmic channel in a dividing nucleus of the dinoflagellate *Gyrodinium*. Microtubules run in parallel arrays through the channel. Chromosomes make contact (arrow) with the nuclear envelope on the nuclear side of the channels but not directly with microtubules. × 30,000. Courtesy of D. F. Kubal and H. Ris, from *J. Cell Biol.* 40:508 (1969), by permission of the Rockefeller University Press.

and fall into separate spindle halves. In other organisms, such as the algae *Vaucheria* and *Acetabularia*, anaphase B proceeds even though the spindles move freely in the cytoplasm, as if unconnected to any structures that could pull on the poles. These observations suggest that polar tension, if actually present as a pulling force operating from outside the spindle, may contribute to anaphase B only in limited groups of organisms.

The Membrane Growth Model The possible involvement of nuclear membranes in the anaphase movement depends on observations made in some dinoflagellates. In the dinoflagellate *Gyrodinium*, investigated by D. Kubai and Ris, the nuclear envelope remains intact throughout mitosis. During prophase cytoplasmic channels extend through the nucleus by invagination of the nuclear envelope. No organized spindle forms, and microtubular structures are limited to bundles of parallel microtubules that fill the cytoplasmic channels (Fig. 24-28). Chromosomes attach to the inside surface of the nuclear envelope along these channels (arrow in Fig. 24-28). During anaphase the chromosomes separate and are pulled apart at the attachment points. No direct connections can be detected between the microtubules in the channels and either the chromosomes or the nuclear envelope. Further, microtubules in the channels do not change in length during division of the chromosomes. Kubai and Ris proposed from these observations that the primary function of the microtubule bundles in this species is cytoskeletal and that the force for chromatid movement is supplied by the nuclear envelope. The microtubules, according to their proposal, support and provide rigidity to the nuclear channels and establish two ends, or poles, toward which the chromosomes are moved. Their proposal is supported by the observation that colchicine has no effect on chromosome movement in *Gyrodinium*.

The system observed in *Gyrodinium* shows interesting parallels to a division mechanism proposed for bacterial nuclei, in which the replicated nucleoids are considered to be separated by the growth of plasma membrane segments lying between them (see Supplement 24-2). The chromosomes of dinoflagellates also show similarities to bacterial nucleoids in both structure and molecular composition (see p. 570). The system in *Gyrodinium* may therefore be a remnant of an early step in the evolution of eukaryotes in which membranes still retained the primary responsibility for division of the replicated genetic material, and microtubules played a secondary role. Membranes may also contribute to a greater or lesser extent to anaphase in other primitive eukaryotes, such as fungi, in which the nuclear envelope remains intact during mitosis.

Regulation of the Anaphase Movement

Several lines of experimental evidence have linked changes in Ca^{2+} concentration to regulation of the anaphase movement. Among the most significant are the results of experiments by M. Poenie, C. H. Keith, and M. L. Shelanski and their colleagues using indicators such as *aequorin* and *quin-2*, which fluoresce in the presence of Ca^{2+}. Injection of these molecules during prophase or until late metaphase produces only very low levels of fluorescence, indicating that Ca^{2+} is present at low concentrations in the spindle up to this time. However, just before anaphase, one or more "spikes" of fluorescence lasting some 20 seconds are emitted by the indicators. The spikes indicate that Ca^{2+} is released into the cytoplasmic solution in the vicinity of the spindle microtubules as the chromosomes begin their movement to the poles. In some experiments a steady, prolonged rise in Ca^{2+} concentration rather than a spike is observed to begin just before anaphase and to continue through the movement.

The results of injection also suggest that Ca^{2+} contributes to anaphase regulation. If injected at metaphase, Ca^{2+} speeds entry of cells into anaphase; if injected at anaphase, Ca^{2+} increases the rate of anaphase movement. Conversely, reductions in Ca^{2+} concentration at metaphase by exposing cells to a calcium chelator such as EGTA, which binds Ca^{2+} and removes it from solution, or a Ca^{2+} buffer, which also removes

excess Ca^{2+} from solutions, slow the entry of cells into anaphase.

For example, P. Hepler and his coworkers followed the effects of injected Ca^{2+} on the anaphase movement in living stamen hair cells in the plant *Tradescantia*. They found that Ca^{2+} concentration normally increases just after the sister chromatids separate and peaks as the chromatids reach the poles. Injecting Ca^{2+} to levels of about 1 μM caused a twofold increase in the rate of the anaphase movement. Injected Ca^{2+} buffers or the calcium chelator EGTA stopped the anaphase movement. W. Z. Cande obtained essentially the same results with isolated spindles: adding Ca^{2+} to levels of about 0.5 μM speeded the anaphase movement briefly.

Electron microscopy reveals that spindles are packed with vesicles originating from the ER (Fig. 24-29). Through the use of antimonate ions, which form an insoluble, electron-dense precipitate when complexed with calcium, S. M. Wick and Hepler demonstrated that the vesicles contain stored calcium ions. The spindle is stained by fluorescent antibodies against the Ca^{2+}-ATPase pump of muscle cell ER membranes; injection of the same antibody into living cells blocks mitosis at metaphase. It has also been possible to detect the polypeptides of the calcium pump among the proteins isolated from spindles.

These observations suggest that the Ca^{2+} concentration of resting spindles is maintained at low con-

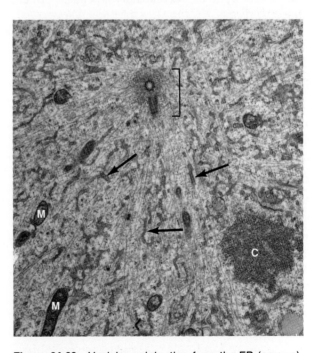

Figure 24-29 Vesicles originating from the ER (arrows) in a rat kangaroo cell at early metaphase. Note the centriole pair at the spindle pole (bracket); the two centrioles are oriented at 90° to each other. C, chromosome; M, mitochondrion. ×8,000. Courtesy of K. L. McDonald, from *Internat. Rev. Cytol.* 90:169 (1984).

centrations by Ca^{2+}-ATPase pumps, which continually transport Ca^{2+} from the cytoplasmic solution into the spindle vesicles. At the onset of anaphase, gated Ca^{2+} channels open briefly in the vesicle membranes, releasing the ion into the surrounding cytoplasm and setting off a Ca^{2+} spike in the vicinity of spindle microtubules. This increase in Ca^{2+} concentration is part of the pathway triggering anaphase.

The triggering pathway may involve activation of calmodulin (see p. 251), which can be detected in quantity in the spindle. In activated form calmodulin may induce microtubule sliding. (There are indications that the activated Ca^{2+}/calmodulin complex has this capability in flagella.) The activated calmodulin complex may also be involved in triggering the disassembly of microtubules associated with anaphase A. Another effect of the Ca^{2+}/calmodulin complex may be activation of protein kinases that add phosphate groups to proteins directly involved in triggering or producing anaphase. L. Wordeman and Cande, for example, found that activation of anaphase in isolated spindles is correlated with phosphorylation of a 250,000-dalton protein.

In addition, increases in Ca^{2+} concentration are associated with spindle disassembly after anaphase is complete. At this time injection of Ca^{2+} or activated Ca^{2+}/calmodulin complex greatly speeds spindle breakdown. J. H. Dinsmore and R. D. Sloboda showed that breakdown of isolated spindles is triggered by addition of a phosphate group to a 62,000-dalton spindle protein by a Ca^{2+}/calmodulin-dependent protein kinase.

EVENTS OF TELOPHASE

As chromatids reach the poles of the spindle and the anaphase movement ceases, anaphase changes to telophase. During telophase the spindle disassembles except for short microtubules that persist at the former spindle midpoint and take part in cytoplasmic division. The chromosomes gradually decondense and reach the extended interphase state by the end of telophase. During decondensation the nucleolus reappears at NOR sites on the chromosomes, and the nuclear envelope re-forms. Completion of these changes marks the transition from telophase to the G1 stage of the following interphase. Of these alterations some progress has been made in working out the molecular steps underlying reorganization of the nucleolus and re-formation of the nuclear envelope.

Reorganization of the Nucleolus

During telophase the nucleolus reappears on one or more chromosomes as a small, spherical body that gradually increases in size until the typical interphase structure is established. B. McClintock was the first to demonstrate that a specific chromosomal site is required for reappearance of the nucleolus at telophase. Working with maize, McClintock succeeded in obtaining mutant individuals in which the nucleolus failed to reappear at telophase. Instead, numerous small, nucleoluslike bodies were distributed among the chromosomes. McClintock proposed the term *nucleolar organizer* (*NOR*) for the site necessary for normal reappearance of the nucleolus at telophase. We now know that the NOR contains many repeats of large pre-rRNA genes (see p. 607).

Classical cytologists believed that the nucleolar substance, after breakdown of the nucleolus at prophase, became distributed over the condensing chromosomes as a surface matrix. This matrix supposedly adhered to the chromosomes and was carried to the poles at anaphase. During reconstitution of the nucleus in telophase, this material, called at this stage the *prenucleolar material*, was thought to flow back to the NOR to be incorporated into the reorganizing nucleolus. This interpretation was supported by two observations. First, the apparent prenucleolar material, visible between the chromosomes at telophase, showed the presence of RNA when stained by cytochemical techniques. Second, as the nucleolus reorganized, the prenucleolar material gradually disappeared from between the telophase chromosomes as if it was indeed reincorporated into the developing nucleolus.

More recent evidence confirms that such material can be identified on or near the chromosomes at metaphase, anaphase, and telophase, and that this substance disappears as the nucleolus is reorganized. In one of these experiments, N. Paweletz and M. C. Risueno used a silver-staining technique that specifically tags nucleolar proteins with dense deposits, making them visible in the electron microscope. During interphase the silver technique stains only the nucleolus. As the nucleolus breaks down during mitotic prophase, the densely staining material becomes distributed over and within the condensing chromosomes. At telophase the dense material once again becomes associated with the re-forming nucleolus at the NORs. At least one nucleolus-associated protein, ribosomal protein S1 (see p. 668), has been specifically identified on the surfaces of mitotic chromosomes by the use of fluorescent antibodies (Fig. 24-30).

Other techniques confirmed that the prenucleolar material associated with chromosomes contains RNA. The prenucleolar material stains positively for RNA and becomes labeled if cells are exposed to tritiated uridine, a radioactive RNA precursor, before mitosis. Moreover, rRNA can be detected in association with isolated metaphase chromosomes. Analysis shows that this rRNA includes complete and partially processed pre-rRNAs.

These experiments indicate that during telophase the NOR may actually collect and organize nucleolar

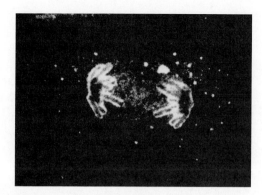

Figure 24-30 Distribution of S1, a ribosomal protein derived from the nucleolus, on the surfaces of the separating chromatids at anaphase. Labeled material can also be seen in the cytoplasm. The protein has been labeled by combination with a fluorescent anti-S1 antibody. × 1,250. Courtesy of W. W. Franke. Reproduced from *J. Cell Biol.* 100:873 (1985), by copyright permission of the Rockefeller University Press.

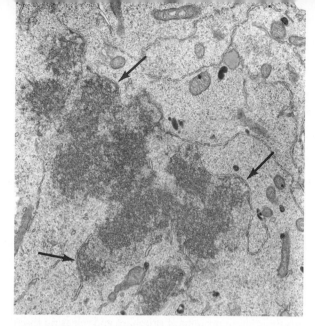

Figure 24-31 Segments of the nuclear envelope (arrows) reappearing at the margins of the decondensing chromatids in a human cell at telophase. The membrane segments extend and fuse until the chromosomes are surrounded by a complete and continuous nuclear envelope. × 44,000. Courtesy of G. G. Maul and Academic Press, Inc., from *J. Ultrastr. Res.* 31:375 (1970).

material released from the chromosomes during prophase. Experiments with actinomycin D, a drug that blocks RNA transcription, support this idea more directly. D. M. Phillips and S. G. Phillips found that cells blocked with actinomycin D during the mitotic division sequence organized their nucleoli successfully during telophase, even though rRNA synthesis was blocked at this time.

These observations suggest that the cycle of nucleolar disappearance and reappearance during mitosis reflects the following molecular changes. At the beginning of prophase, the nucleolus contains rRNA precursors in association with ribosomal proteins, all at various stages in processing into ribosomal subunits. As prophase progresses, rRNA transcription and processing stop, and the partially processed ribosomal precursors are released from the nucleolus. These collect on the surfaces of the condensed chromosomes as the prenucleolar material and possibly also disperse into the cytoplasm along with the processing enzymes associated with rRNA maturation. At telophase the partially processed ribosomal subunits and the processing enzymes return to the NOR and assemble at this site. Processing of the preexisting material then resumes. Transcription of new rRNA also begins at this time, so that, as telophase proceeds into interphase, a progressively greater part of the nucleolus represents newly synthesized material.

Presumably, dispersal of rRNA, ribosomal proteins, and processing enzymes during metaphase and anaphase is an adaptation that reduces the mass of the chromosomes bearing NORs. Dispersal of the nucleolus during anaphase would also reduce the chance of tangling and nonseparation of the chromatids containing NORs.

Re-Formation of the Nuclear Envelope

The nuclear envelope begins to reappear when chromatids reach the spindle poles. At first, flattened membranous vesicles apparently identical to ER cisternae become closely applied to the surfaces of the decondensing chromatids (Fig. 24-31). Whether these are the same vesicles that disperse from the nuclear envelope during breakdown remains unknown. These vesicles increase in number and gradually fuse until the chromatids are enclosed by an envelope consisting of two concentric membranes separated by a narrow space. Pore complexes develop as the membranes fuse into a continuous layer. By the end of telophase, the daughter nuclei are completely separated from the cytoplasm by a fully differentiated nuclear envelope. The envelope develops in such close association with the surfaces of the chromatids that the cytoplasm appears to be largely excluded from the newly formed nuclei.

We have noted that the lamins are dephosphorylated as the nuclear envelope re-forms (see p. 1017). These proteins reassemble into a fibrous framework on the surface of the inner nuclear membrane facing the nucleoplasm. Lamins A and C are released to the cytoplasmic solution or bind to the chromosome surfaces during metaphase and anaphase; lamin B remains bound to membranous vesicles after nuclear envelope breakdown. Recent work by Gerace and his colleagues indicates that during telophase in mammals, lamins A and C, if not already bound, link tightly to the surfaces of the chromosomes. The lamin B molecules carried on ER vesicles contact lamin A and C proteins on the surfaces of the chromosomes. This contact brings the vesicles into close proximity on the chromosome surfaces, and the nuclear envelope, with its supporting lamin network, reassembles. As the envelope reassembles, pore complexes appear and connections are presumably set up between the lamin network and proteins of the nuclear matrix—the framework of fibrous material proposed as a support for the nucleoplasm and chromatin.

The proteins forming parts of the nuclear matrix and other nonchromosomal constituents of the nucleoplasm return from the cytoplasm to the nucleus by two major routes. Some attach to the surfaces of the chromosomes as they condense and remain attached until the newly formed nuclear envelope separates the nucleus from the cytoplasm at telophase. The factors attaching these proteins at prophase and releasing them at telophase remain unknown. Others return by virtue of their nucleus-directing signals (see p. 854). These proteins are left in the cytoplasm as the nuclear envelope reorganizes but return to the nucleus via cellular sorting and distribution mechanisms that direct them through the nuclear pore complexes during or soon after telophase.

DIVISION OF THE CYTOPLASM: CYTOKINESIS

Cytoplasmic division in most cells begins during late anaphase and continues into telophase. By the end of telophase, when the daughter nuclei reach G1, division of the cytoplasm is usually complete. The two primary mechanisms by which cytoplasmic division takes place, *furrowing* in animals and *cell plate formation* in plants (see p. 914), share one major feature. In both systems a band of overlapping microtubules persisting at the former spindle midpoint is intimately involved in the early stages of cytoplasmic division.

Furrowing

As anaphase and telophase progress in animal cells, dense material begins to appear in a layer surrounding the zone of overlapping microtubules at the spindle midpoint. By late telophase the dense material has increased in quantity until a solid layer extends across the spindle midpoint (Fig. 24-32). At this time the spindle breaks down rapidly except for microtubules that persist at the midpoint. The persisting microtubules are the remnants of the zone of overlap at the former spindle midpoint and retain the interdigitated, antipolar arrangement characteristic of this region. Other structures, such as vesicles, may also move to the equatorial layer. The fully developed dense layer, with associated

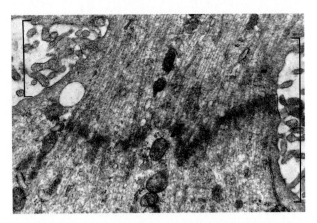

Figure 24-32 An early stage of cytokinesis in a human cell. Dense material surrounding microtubule remnants at the former spindle midpoint forms a continuous layer across the cell (see text). The deepening furrow is visible on either side of the cell opposite the dense layer (brackets). × 35,000. Courtesy of A. Krishan and R. C. Buck, from *J. Cell Biol.* 24:443 (1965), by permission of the Rockefeller University Press.

microtubules and vesicles, is termed the *midbody*. By late anaphase this structure is large enough to be visible under the light microscope.

As the midbody develops, a furrow appears in the plasma membrane, extending around the cell at the level of the midbody. The furrow gradually deepens and compresses the midbody until it is only a narrow, fibrous connection between the daughter cells (Fig. 24-33). The midbody may persist in this form for a time, but in most cells it breaks and disappears soon after furrowing is complete. In some cells the residual midbody is engulfed by one of the daughter cells. The furrow, once induced, may form in as little as 20 sec-

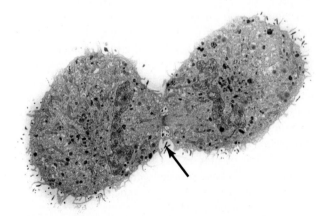

Figure 24-33 A human tissue culture cell in an advanced stage of cytoplasmic division. The furrow, advancing in the region of the midbody (arrow), has cut the cytoplasm almost in two. × 2,800. Courtesy of G. G. Maul and Academic Press, Inc., from *J. Ultrastr. Res.* 31:375 (1970).

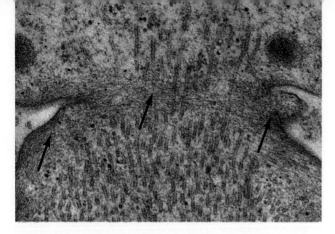

Figure 24-34 Microfilaments (arrows) under the plasma membrane in the region of an advancing furrow in a rat cell. ×31,000. Courtesy of D. Szollosi, from *J. Cell Biol.* 44:192 (1970), by permission of the Rockefeller University Press.

onds; cytoplasmic division by furrowing in some cells is complete in 10 minutes.

A series of experiments showed that development and inward extension of the furrow depends on microfilaments. Fine filaments about 5 to 7 nm in diameter are visible at the apex of the furrow, as in Figure 24-34, and run in a complete band around the dividing cell. The HMM decoration technique (see p. 453) produces "arrowheads" on these fibers, indicating that they are actin microfilaments. Fluorescent staining using both anti-actin and antimyosin antibodies produces bright fluorescence just under the furrow, demonstrating that both components of the actin-myosin contractile system are present in the band. Further evidence that furrow formation and contraction depend on microfilament activity comes from observations that antiactin and antimyosin antibodies and microfilament poisons such as cytochalasin B stop contraction and reverse furrow formation in most animal cells.

Although the developing furrow follows the plane of the former spindle midpoint, a series of elegant experiments by R. Rappaport with dividing sand dollar eggs showed that the initial stimulus for furrow formation involves an interaction between astral microtubules and the cell cortex (Fig. 24-35). By anaphase astral microtubules have extended until they make contact with the cell cortex just under the plasma membrane (Fig. 24-35a). The densest region of contact is the narrow band around the cell in which microtubules from both asters overlap (bracket in Fig 24-35a). Rappaport proposed that the densely packed microtubule tips stimulate formation of a furrow in this region (Fig. 24-35b). His experiments showed that the spindle midbody is not directly required for furrow formation; furrows will form in any region of the cell cortex in which astral microtubules overlap, in the pattern shown in Figure

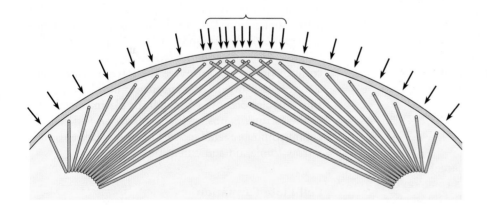

a

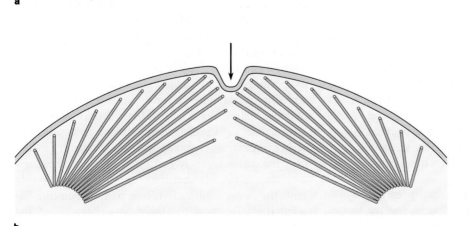

b

Figure 24-35 Stimulation of the cell cortex to form a furrow by astral microtubules. **(a)** During anaphase the astral microtubules lengthen and make contact with the cell cortex just under the plasma membrane. The density of contact between the microtubules and the cortex is greatest in a narrow region in which microtubules from both asters overlap (bracket). This region of overlap extends in a band around the cell. **(b)** The microtubule tips stimulate formation of a furrow in the densely packed region.

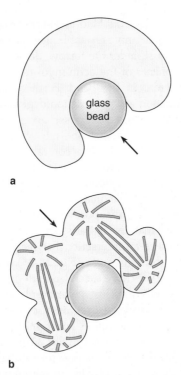

a

b

Figure 24-36 An experiment by Rappaport showing that a furrow will form between the asters of adjacent spindles. **(a)** A glass bead was pushed into the side of a fertilized sand dollar egg, distorting the cell into a horseshoe shape. **(b)** At the next division an extra furrow (arrow) formed between the asters of adjacent spindles brought together by the distorting glass bead, even though no spindle structure extended between the asters. The result of this experiment supports the conclusion that an interaction between asters and the cell cortex induces furrow formation and shows that a midbody remaining from a spindle equator is not necessary for formation of a furrow. Redrawn from an original courtesy of R. Rappaport, from *J. Exptl. Zool.* 148:81. Copyright © 1961 by Wiley-Liss, a division of John Wiley and Sons, Inc.

24-35a. For example, if the asters of adjacent spindles are brought together experimentally in a common cytoplasm, a furrow will develop between the asters even though no spindle lies between them (Fig. 24-36).

Adjustments in Ca^{2+} concentration have been implicated in the regulation of furrowing. T. G. Hollinger and A. W. Schuetz, for example, induced formation of cleavage furrows in frog eggs by injecting Ca^{2+} just below the cell surface. Injection of fluorescent Ca^{2+} indicators also shows that a brief rise in cytoplasmic Ca^{2+} concentration occurs in coordination with the onset of furrowing. Although Ca^{2+} is thus implicated in regulating furrow formation, the molecular steps in the regulation and the connection between the calcium signal and stimulation of the cortex by astral microtubules remain unknown. Calmodulin has not been detected in furrows, indicating that Ca^{2+} acts directly rather than by activating this control protein.

In most animal cells the asters and spindle are located symmetrically in the center of the cell. As a result, cytoplasmic division cuts the dividing cell into two more or less equal parts, following a plane that extends through the former spindle midpoint. In a few animal cell types, such as the cells of some early animal embryos, the asters and spindle are located asymmetrically, leading to unequal division of the cytoplasm.

Moving the spindle before anaphase, which can be accomplished by centrifugation or by pushing the spindle with a microneedle, causes a corresponding shift in the plane of cytoplasmic division. In most animal cells movement of the spindle after anaphase does not disturb the site of division, which then passes through a plane corresponding to the position of the spindle midpoint before displacement. This shows that once the furrow is stimulated in the cell cortex by the asters, the stimulating alteration persists in the original location even though the asters and spindle are later displaced. The asters and spindle can even be destroyed by agents such as colchicine without affecting the position and progress of cytoplasmic division as long as the destruction occurs after mid-to-late anaphase. If applied before anaphase, such treatments inhibit rather than displace furrow formation.

Nonmicrofilament systems may act in cytoplasmic division in some animal cells. In animal cells that are closely packed in tissues, for example, division of the cytoplasm seems to proceed by the addition of cytoplasmic vesicles to the furrow margin rather than by active contraction. The vesicles increase in number and coalesce until the daughter cells are separated. In the fungus *Thraustochytrium* microtubules rather than microfilaments seem to provide the force for deepening a furrow that divides the cytoplasm in this organism. Microtubules occur just beneath the plasma membrane in the furrow. Exposing the *Thraustochytrium* cells to antimicrotubule agents such as colchicine interferes with furrow formation and extension.

Cell Plate Formation

Spindle remnants take part in plant cytokinesis in a pattern that is remarkably similar to midbody formation in animals. As in animal cells, the spindle disassembles after anaphase except for short lengths of overlapping, antipolar microtubules that persist at the former spindle midpoint. Newly assembled microtubules may also be added in overlapping patterns in this region. Deposits of dense material form around these short microtubules as in animal cells undergoing cytokinesis; in addition, actin microfilaments also accumulate in quantity in and near the dense layer. Myosin decoration of these microfilaments reveals that they are oriented uniformly with the arrowhead pattern (see p. 454) pointing toward the interior of the layer on either side of the former spindle equator.

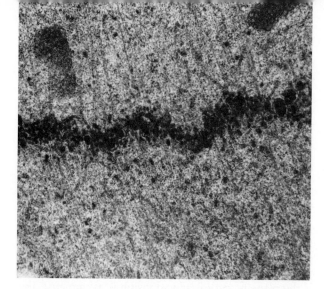

Figure 24-37 Phragmoplast formation in *Haemanthus*. A layer of dense material enclosed in vesicles has collected around microtubule remnants at the former spindle midpoint. × 18,000. Courtesy of A. Bajer and Springer-Verlag.

As the dense material accumulates, vesicles originating from the ER and Golgi complex move into the developing layer at the former spindle equator. The vesicles apparently move along both microtubule remnants and oriented microfilaments to reach the dense layer. Experiments using radioactive tracers have shown that the vesicles contain precursors of cell wall materials such as cellulose and pectins.

The dense material and the vesicles increase in quantity until a continuous layer forms (Fig. 24-37). This layer is termed the *phragmoplast*. At first the phragmoplast is confined to the area of the cell formerly occupied by the spindle midpoint. At the earliest stages the phragmoplast may be incomplete in the center, so that when viewed from the poles it appears hollow. The phragmoplast then grows both inward and outward by the accumulation of additional microtubules, microfilaments, and vesicles, until the layer spreads entirely across the dividing cell. As the phragmoplast expands, vesicle fusion begins in its older regions (Fig. 24-38a and b; see also Fig. 22-5). The fusion, which amounts to a specialized form of exocytosis, dumps cell wall material into a gradually expanding layer between the daughter cells. As in exocytosis in other cell types (see p. 856), a rise in Ca^{2+} concentration can be detected in the regions of vesicle fusion. Fusion continues until the entire phragmoplast is replaced by a continuous cell wall (Fig. 24-38c). The new cell wall is lined on either side by plasma membranes derived from the vesicle membranes. When the plasma membranes and new cell wall are fully formed, the partition at the midregion is termed the *cell plate*.

The cell plate does not entirely cut off cytoplasmic communication between daughter cells. Numerous narrow openings, lined by plasma membrane, persist in the cell plate and remain as direct cytoplasmic con-

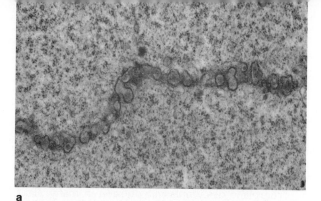

a

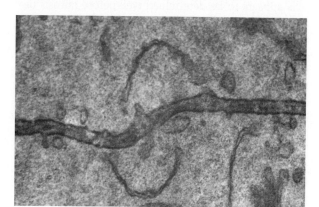

b

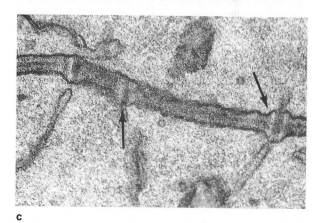

c

Figure 24-38 Successive stages in the fusion of phragmoplast vesicles to form the cell plate. **(a)** An early stage in the blood lily *Haemanthus*, in which vesicle fusion is under way. × 25,000. Courtesy of W. G. Whaley, M. Dauwalder, J. E. Kephart, and Academic Press, Inc. from *J. Ultrastr. Res.* 15:169 (1966). **(b)** A later stage in the grass *Phalaris*, in which vesicle fusion is nearly complete. × 59,000. **(c)** A fully formed cell plate in *Phalaris;* several plasmodesmata (arrows) remain as direct connections between the cytoplasm of daughter cells. × 70,000. Courtesy of A. Frey-Wyssing, J. R. Lopez-Saez, K. Muhlethaler, and Academic Press, Inc., from *J. Ultrastr. Res.* 10:422 (1964).

nections between the daughter cells (arrows in Fig. 24-38c; see also Fig. 8-30). These connections, called *plasmodesmata* (see p. 298), evidently form where tubular extensions of the ER are caught in the developing cell plate.

Cytoplasmic division by phragmoplast and cell plate formation occurs in almost all plant cells. Exceptions among plants are noted in only a few cell types, such as the microspores giving rise to pollen (see p. 1115). In these cells cytoplasmic division occurs by a process resembling furrowing. Although most algal cells also divide by a process resembling furrowing, a few, such as *Spirogyra* and *Oodegonium*, form cell plates.

The plane of cytoplasmic division in many plant cells appears to be determined well before mitosis begins. In these cells the nucleus takes up a position, usually in the center of the cell, some 8 to 20 hours prior to mitosis. The nucleus then becomes surrounded by a band of radiating microtubules and microfilaments that extends outward as a disclike layer and contacts the cell wall. In the region of contact, a belt of microtubules and microfilaments, called the *preprophase band* (*PPB*), forms around the cell periphery (see Fig. 13-22). The PPB persists for a time but usually disappears before the onset of mitosis. In wheat seedlings, for example, B. E. S. Gunning and M. Sammut found that the preprophase band appears just after S and disassembles at the G2/mitosis transition. The position taken by the PPB marks the site at which the cell plate will contact the old cell wall during cytokinesis.

Once the PPB has marked the future site of cytoplasmic division, displacement of the nucleus or spindle usually has no effect on the site at which the cell plate attaches to the side walls of the dividing cell. If the spindle and developing phragmoplast are displaced experimentally, the developing cell plate curves at its edges so that it still meets the original wall at the site previously marked by the PPB. Some investigators have proposed that a layer of microfilaments persisting in the region of the former PPB directs the margins of the developing cell plate to the cell wall.

PPBs appear before mitosis in plants from bryophytes to angiosperms. The PPBs are characteristic of ordered tissues in which the position of cells is critical for development of plant form. Less organized tissues such as callus and liquid endosperm enter division without forming PPBs. In most ordered tissues the position taken by the nucleus and subsequent formation of the PPB are symmetrical, so that the plane of division cuts the dividing cell into two equal parts. However, in some cells, such as those giving rise to the guard cells of stomata, the nucleus and PPB take up an eccentric position, so that the future plane of cytoplasmic division produces daughter cells of unequal size.

Formation of the PPB, spindle, and phragmoplast involves an essentially complete reorganization of the cytoskeleton. Microtubules and microfilaments, which are located primarily in the cortex of interphase plant cells, become concentrated in the PPB as this structure forms. As dividing cells enter prophase, the PPB disassembles, and microtubules reassemble into the spindle. During anaphase the spindle disassembles, and microtubules reorganize into the phragmoplast. After mitosis is complete, the arrangement typical of interphase cells reappears (see also p. 519).

Distribution of Cytoplasmic Organelles by Cytokinesis

The relative quantities of cytoplasmic organelles distributed to daughter cells depend in most systems simply on the volume of cytoplasm received. If the plane of cytoplasmic division cuts approximately through the middle of the cell, the division products receive approximately equal numbers of cytoplasmic organelles. If the plane of cytoplasmic division is asymmetric, the numbers of cytoplasmic organelles distributed to the daughter cells vary proportionately. This relative distribution reflects the fact that there are no processes equivalent to the spindle mechanism that precisely organize and divide cytoplasmic organelles between the products of division. Only the fact that daughter cells both receive cytoplasm, usually in equivalent amounts, ensures that cytoplasmic organelles are apportioned approximately equally.

Exceptions to this passive distribution pattern are seen in a few cell types in which mitochondria or chloroplasts separate before cytoplasmic division into two approximately equal groups concentrated at the poles of the spindle or collect around the equator of the spindle in such a way that the plane of division cuts the collection into two approximately equal groups. The forces or mechanisms arranging mitochondria or chloroplasts into groups before cytoplasmic division in these cases remain unknown.

Cytoplasmic division in animal and plant cells completes the process of cell division. The new plasma membranes cut the cytoplasm of the parent cell into two portions and distribute the various cytoplasmic organelles, including mitochondria, chloroplasts if present, and vesicles of the ER and Golgi complex, between daughter cells.

Although cytoplasmic division is closely linked to mitosis in most cells, the two processes are potentially separable and independent (see p. 916). In insect eggs, for example, mitosis proceeds for some time after fertilization without any accompanying cytoplasmic division. After hundreds to thousands of nuclei are formed, depending on the species, furrowing then divides the egg cytoplasm and surrounds each nucleus with its own cytoplasm and plasma membrane. In this case a furrow forms between the single asters of any two adjacent interphase nuclei until all have been enclosed in separate cells. Subsequent divisions proceed by the usual pattern, in which furrowing is triggered at anaphase, and each mitosis is accompanied by a cytoplasmic division.

Division of Mitochondria and Chloroplasts

The distribution of mitochondria and chloroplasts during mitosis and cytoplasmic division is complicated because both organelles contain their own DNA and are partly autonomous (see Ch. 21 for details). Both organelles replicate their DNA during interphase. At some point following replication, both organelles divide by a furrowing process that resembles cytoplasmic division in miniature (Figs. 24-39 to 24-41). In advance of the furrow, the inner boundary membrane grows inward from one or both sides until it extends as a complete wall or septum cutting the mitochondrial matrix or chloroplast stroma into two parts. The outer membrane then invaginates into the septum, cutting the organelle in two. As part of the division, the internal structures of both organelles, including soluble and membrane-bound enzymatic systems, DNA circles, and

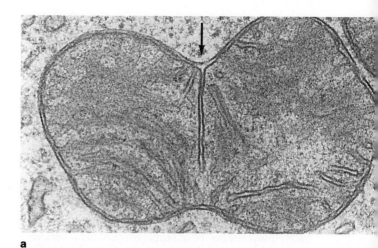

a

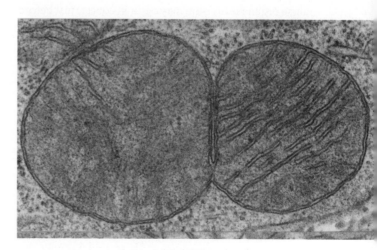

b

Figure 24-40 Electron micrographs of dividing organelles. **(a)** A dividing mitochondrion from the gastric mucosa of the mole. The division is at a stage comparable to Figure 24-39*b*. Micrograph by T. Kanaseki, from D. W. Fawcett, *The Cell*, W. B. Saunders Co., 1987. **(b)** An equivalent stage in a dividing fern chloroplast. × 10,000. Courtesy of E. Gantt and H. J. Arnott, from *J. Cell Biol.* 19:446 (1963), by permission of the Rockefeller University Press.

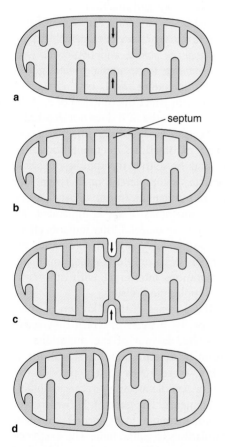

Figure 24-39 Steps in division of a mitochondrion; chloroplast division follows a similar pattern. **(a)** The inner boundary membrane grows inward; **(b)** growth continues until the inner membrane cuts the matrix into two parts. **(c)** The outer membrane then invaginates into the septum; **(d)** invagination continues until the organelle is separated into two parts. As part of the division, the internal structures, including soluble and membrane-bound enzymatic systems, DNA circles, and ribosomes, are distributed in approximately equal portions between the two division products.

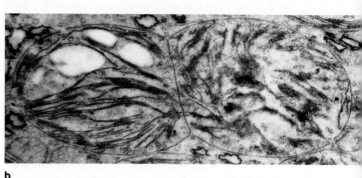

Figure 24-41 A later stage in the division of a rat liver mitochondrion, equivalent to Figure 24-39*c*. Micrograph by T. Kanaseki, from D. W. Fawcett, *The Cell*, W. B. Saunders Co., 1987.

ribosomes, are distributed in approximately equal portions between the two division products.

The molecular mechanisms accomplishing organelle division are unknown. Distribution of DNA circles to the division products may be moderated by the organelle boundary membranes, in a process analogous to prokaryotic cell division—by attachment of the circles to the inner boundary membrane, followed by membrane growth that separates the circles. If this is the case, division of the organelle DNA would count as another of the many organelle mechanisms resembling those of prokaryotes (see Ch. 21 for details). Alternatively, there may be no specialized mechanism dividing the DNA circles of the organelles; the multiple circles may simply be distributed randomly between the division products as the septum forms. In this case the apportionment of DNA circles in mitochondria and chloroplasts would follow the same pattern as other molecules and internal structures of the organelles.

In most cells organelle DNA replication and division proceed throughout much or all of the cell cycle. Organelle numbers are smallest just after cytoplasmic division is complete. From the initiation of G1 onward, division increases the number of the organelles until a maximum number is reached. In rapidly dividing cells the maximum organelle count is reached late in G2, just before the next cytoplasmic division takes place.

Coordinated organelle and cell division is observed in only a few cell types. In some single-celled algae, such as *Micromonas*, only one mitochondrion and chloroplast are present at interphase. Each organelle divides in coordinated fashion just before cell division. Cytoplasmic division separates the doubled mitochondria and chloroplasts equally between the daughter cells, so that the interphase number of one each per cell is maintained. Organelle division is also coordinated to some extent in other single-celled algae with greater numbers of mitochondria and chloroplasts. In these algae organelle division may occur in concert just before or just after cell division. Mitochondria have also been observed to divide synchronously in the slime mold *Physarum*. In this organism mitochondria in the *plasmodium*, a generalized cytoplasm shared by all nuclei, undergo several rounds of coordinated division while DNA replication is in progress.

GENETIC REGULATION OF MITOSIS AND CYTOPLASMIC DIVISION

Division of the chromosomes and cytoplasm is an incredibly complex process with many subparts and events that must be closely coordinated for a successful outcome. As a consequence, there undoubtedly are many control points that regulate the process to ensure that all the events occur in the correct manner and sequence. These control points have been investigated through some of the same methods described in Chapter 22 for study of the entire cell cycle: identification of genetic mutants and detection of factors that modify portions of mitosis or cytoplasmic division.

A number of genes have been identified in both yeasts and higher organisms encoding factors that control, modify, or are necessary for parts of mitosis or cytoplasmic division to proceed (Table 24-2). One of the most significant discoveries is the *cdc2* gene, which has been detected in organisms from yeast to mammals. This gene encodes the $p34^{cdc2}$ polypeptide forming part of MPF. When fully activated, MPF triggers the primary events of mitotic prophase: condensation of the chromosomes, formation of the spindle, and nuclear envelope breakdown. The protein kinase activity of $p34^{cdc2}$ has been directly implicated in the H1 phosphorylation associated with condensation of the chromosomes and the lamin phosphorylation accompanying nuclear envelope breakdown, and may carry out phosphorylations regulating spindle assembly and function.

Many additional genes affecting segments of mitosis and cytoplasmic division have been detected, particularly in *S. cerevisiae*, *S. pombe*, and *Drosophila* (see Table 24-2). In *S. cerevisiae*, for example, several genes have been identified, including *cdc31*, *esp1*, *kar1*, and *spa1*, that in mutant form affect the function of spindle pole bodies. In the mutants the SPB does not duplicate; nor does it form the chromosome attachments necessary for separation of chromatids. Other mutations alter or block spindle function. Either the spindle does not assemble or assembles incompletely; or if assembled, there is no anaphase movement. In one unusual set of mutants certain chromosomes of the complement fail to be included in daughter nuclei. Other mutants affect chromatid separation generally, so that chromatids remain locked together. With only one exception (*top2*; see below), the mutations interfering with mitosis in yeast also block cytoplasmic division.

For most of these genes, little is known of their encoded products or the manner in which they exert their effects. However, several of the mutations affecting spindle form or function involve alterations in tubulin. Presumably, these genes merely provide the tubulin necessary for spindle assembly rather than acting as regulatory elements. The gene *cdc31*, which prevents SPB duplication in mutant form, encodes a protein with sequence similarities to calmodulin. The existence of this gene suggests that Ca^{2+} may be involved in some way in regulating or triggering SPB duplication. One of the mutants blocking chromatid separation, *top2*, encodes DNA topoisomerase II. The failure of chromatids to separate in *top2* mutants indicates that the enzyme eliminates intertwined loops produced during replication or tangles formed as chromosomes condense.

The mitosis and cytoplasmic division mutants detected in higher eukaryotes in addition to *cdc2* have

Table 24-2 Some Mutations Affecting Mitosis or Cytokinesis

Mutation	Encoded Protein	Effect of Mutation
Saccharomyces cerevisiae		
cdc31	Similar to Ca²⁺-binding proteins	SPBs do not duplicate; spindle does not form.
esp1	Unknown	Extra SPBs form.
kar1	53kD	Defects in SPB duplication and chromatid separation.
ndc1	Unknown	Chromatids do not separate; go to same pole.
spa1	56kD	Extra SPBs form.
top2	DNA topoisomerase II	Chromatids do not separate.
Schizosaccharomyces pombe		
cdc2	p34^{cdc2} protein kinase	Failure to enter mitosis; no chromosome condensation or nuclear envelope breakdown.
cut7	Similar to kinesin	Spindle formation faulty.
dis1, 2, 3	Unknown	Chromatids do not separate.
nda2	α1-tubulin	Spindle does not assemble.
nda3	β-tubulin	Spindle does not assemble.
nuc2	Unknown	Spindle does not elongate.
Drosophila		
asp	Unknown	Abnormal spindle; mitosis arrests at metaphase.
gnu	Unknown	Asters divide, but spindle does not form.
mrg	Unknown	Asters do not duplicate; forms spindle with single pole.
polo	Unknown	Multiple asters form.
Mammals		
MS1-1	Unknown	Cleavage furrow does not form.
RCC1	45kD, binds DNA	Faulty chromosome condensation.
Tax18	Unknown	Spindle assembly blocked; taxol can reverse effects.
ts2	Unknown	Chromosomes do not attach to spindle.
ts111	Unknown	Cleavage furrow does not form.
ts546	Unknown	Similar to *ts2*.
ts654, 655	Unknown	Faulty chromosome condensation.
ts687	Unknown	No anaphase movement.
ts745	Unknown	Asters do not divide; forms spindle with single pole; no chromosome division.

similar effects. Several of the mutants identified in mammals inhibit spindle formation, either partially or completely; others form spindles that cannot produce the anaphase movement. The effects of one of the latter mutants, *Tax18*, can be reversed by the drug taxol, which promotes microtubule assembly. This suggests that in the *Tax18* mutant either tubulin or a MAP necessary for correct adjustment of the tubulin-microtubule equilibrium may be faulty. One mutation, *ts745*, produces only a half spindle; in this mutant the chromosomes are not divided. Some mutations, such as *RCC1*, cause faulty condensation; others interfere with attachment of chromosomes to the spindle. Another mutant, *ts111*, appears to involve faults in cytoplasmic division; mitosis proceeds normally, but the cleavage furrow fails to form. All these mutations have been detected in mammalian cells, many of them in cultured cells.

One general characteristic made clear by these studies is the potential independence of many mitotic processes. Although the spindle may not form or chromatids may not separate, the chromosomes eventually decondense and the nuclear envelope still forms at approximately the usual times. This indicates that some segments of the complex mitotic pathway are controlled by separate banks of genes that do not depend for their operation on the completion of earlier steps. However, as in yeast, cytoplasmic division is usually dependent on earlier steps. If spindle formation and the anaphase movement do not proceed, cytoplasmic division is usually inhibited.

Although it has not as yet been possible to identify the proteins encoded in most of these genes or their mode of action, genetic studies, at the very least, confirm that individual steps of mitosis and cytoplasmic division are under genetic control and that the operation of a series of genes, working in defined and regulated cascades, is responsible for the complex events of cell division. Other indications of specific genetic activity and regulation follow from the appearance of proteins that cannot be detected at other times in the cell cycle, and from the general reduction in RNA transcription noted during mitosis.

The entire cycle of replication, mitosis, and cytoplasmic division, in all of its elegant complexity, occurs countless billions of times in the growth, development, and maintenance of structure in many-celled eukaryotes. In humans mitotic divisions following fertilization of the egg produce an adult individual with an estimated total of 65 trillion cells. D. M. Prescott estimated that to maintain the supply of red blood cells in humans, more than 2 million mitotic divisions per second occur in each individual. The mechanism is so close to perfection that these repeated division cycles occur almost without error throughout the lifetime of the organism. This order and precision depend ultimately on the activity of specific genes, activated at the right time and place before and during cell division.

For Further Information

Suggestions for Further Reading

General Books and Articles

Brooks, R.; Fantes, P.; Hunt, T.; and Wheatley, D. 1989. The cell cycle. *J. Cell Sci.* suppl. 12.

Dustin, P. 1984. *Microtubules.* New York: Springer-Verlag.

Mazia, D. 1987. The chromosome cycle and the centrosome cycle in the mitotic cycle. *Internat. Rev. Cytol.* 100:49–92.

McIntosh, J. R., and Koonce, M. P. 1989. Mitosis. *Science* 246:622–28.

Wise, D. 1988. The diversity of mitosis: the value of evolutionary expression. *Biochem. Cell Biol.* 66:515–29.

Chromosome Condensation and Structure

Balczon, R. D., and Brinkley, B. R. 1990. The kinetochore and its roles during cell division. In *Chromosomes, eukaryotic, prokaryotic, and viral* (Adolph, K. W., ed.). Boca Raton, Florida: CRC Press, pp. 167–89.

Blackburn, E. H. 1990. Telomeres: structure and synthesis. *J. Biolog. Chem.* 265:5919–21.

Bloom, K.; Hill, A.; Kenna, M.; and Saunders, M. 1989. The structure of a primitive kinetochore. *Trends Biochem. Sci.* 14:223–27.

Brinkley, B. R. 1990. Toward a structural and molecular definition of the kinetochore. *Cell Motil. Cytoskel.* 16:104–9.

———. 1991. Chromosomes, kinetochores, and the microtubule connection. *Bioess.* 13:675–81.

Clarke, L. 1990. Centromeres of budding and fission yeasts. *Trends Genet.* 6:150–54.

Earnshaw, W. C. 1988. Mitotic chromosome structure. *Bioess.* 9:147–50.

———. 1991. When is a centromere not a kinetochore? *J. Cell Sci.* 99:1–4.

Godward, M. B. E. 1985. The kinetochore. *Internat. Rev. Cytol.* 94:77–105.

Murphy, M., and Hayes-Fitzgerald, M. 1990. *Cis*- and *trans*-acting factors involved in centromere function in *Saccharomyces cerevisiae. Molec. Microbiol.* 4:329–36.

Murray, A. W., and Szostak, J. W. 1987. Artificial chromosomes. *Sci. Amer.* 257:62–68 (November).

Pluto, A. F.; Cooke, C. A.; and Earnshaw, W. C. 1990. Structure of human centromere at metaphase. *Trends Biochem. Sci.* 15:181–85.

Rao, P. N. 1990. The discovery (or rediscovery?) of the phenomenon of premature chromosome condensation. *Bioess.* 12:193–97.

Rattner, J. B. 1991. The structure of the mammalian centromere. *Bioess.* 13:51–56.

Schulman, I., and Bloom, K. S. 1991. Centromeres: an integrated protein/DNA complex required for chromosome movement. *Ann. Rev. Cell Biol.* 7:311–36.

Willard, H. F. 1990. Centromeres of mammalian chromosomes. *Trends Genet.* 6:410–16.

Zakian, V. A. 1989. Structure and function of telomeres. *Ann. Rev. Genet.* 23:579–604.

Artificial Chromosomes

Blackburn, E. H. 1985. Artificial chromosomes in yeast. *Trends Genet.* 1:8–12.

Schlessinger, D. 1990. Yeast artificial chromosomes: tools for mapping and analysis of complex genomes. *Trends Genet.* 6:248–58.

Spindle Formation and Structure

Bajer, A. S. 1990. The elusive organization of the spindle and kinetochore fiber: a conceptual retrospect. *Adv. Cell Biol.* 3:65–93.

Baskin, T. I., and Cande, W. Z. 1990. The structure and function of the mitotic spindle in flowering plants. *Ann. Rev. Plant Physiol. Plant Molec. Biol.* 41:277–315.

Kuriama, R., and Nislow, C. 1992. Molecular components of the mitotic spindle. *Bioess.* 14:81–88.

Leslie, R. J. 1990. Recruitment: the ins and outs of spindle pole formation. *Cell Motil. Cytoskel.* 16:225–28.

Lloyd, C. 1988. Actin in plants. *J. Cell Sci.* 90:185–88.

Rieder, C. L. 1990. Formation of the astral mitotic spindle: ultrastructural basis for the centrosome-kinetochore interaction. *Electron Micr. Rev.* 3:269–300.

Stearns, T.; Evans, L.; and Kirschner, M. 1991. γ-tubulin is a highly conserved component of the centrosome. *Cell* 65:825–36.

Vorobjev, I. A., and Nadezhdina, E. S. 1987. The centrosome and its role in the organization of microtubules. *Internat. Rev. Cytol.* 106:227–93.

Zheng, Y.; Jung, M. K.; and Oakley, B. R. 1991. γ-tubulin is present in *Drosophila melanogaster* and *Homo sapiens* and is associated with the centrosome. *Cell* 65:817–23.

The Anaphase Movement

Aist, J. R.; Bayles, C. J.; Tao, W.; and Berns, M. W. 1991. Direct experimental evidence for the existence, structural basis, and function of astral forces during anaphase B *in viro. J. Cell Sci.* 100:279–88.

Cande, W. A., and Hogan, C. J. 1989. The mechanism of anaphase spindle elongation. *Bioess.* 11:5–9.

Hogan, C. J., and Cande, W. Z. 1990. Antiparallel microtubule interactions: spindle formation and anaphase B. *Cell Motil. Cytoskel.* 16:99–103.

McIntosh, J. R., and Hering, G. E. 1991. Spindle fiber action and chromosome movement. *Ann. Rev. Cell. Biol.* 7:403–26.

McIntosh, J. R., and Koonce, M. P. 1989. Mitosis. *Science* 246:622–28.

McIntosh, J. R., and McDonald, K. L. 1989. The mitotic spindle. *Sci. Amer.* 261:48–56 (October).

Mitchison, T. J. 1988. Microtubule dynamics and kinetochore function in mitosis. *Ann. Rev. Cell Biol.* 4:527–49.

Nicklas, R. B. 1988. The forces that move chromosomes in mitosis. *Ann. Rev. Biophys. Biophys. Chem.* 17:431–49.

Pickett-Heaps, J. 1991. Cell division in diatoms. *Internat. Rev. Cytol.* 128:63–108.

Vale, R. D., and Goldstein, L. S. B. 1990. One motor, many tails: an expanding repertoire of force-generating motors. *Cell* 60:883–85.

Warner, F. D., and McIntosh, R., eds. 1989. *Cell movement.* Vol. II: *Kinesin, dynein, and microtubule dynamics.* New York: Alan R. Liss.

Warner, F. D.; Satir, P.; and Gibbons, I. R., eds. 1989. *Cell movement.* Vol. I: *The dynein ATPases.* New York: Alan R. Liss.

The Nuclear Envelope Cycle

Gerace, L., and Burke, B. 1988. Functional organization of the nuclear envelope. *Ann. Rev. Cell Biol.* 4:335–74.

Scheer, U., and Benavente, R. 1990. Functional and dynamic aspects of the mammalian nucleolus. *Bioess.* 12:14–21.

Regulation of Mitosis

Glover, D. M. 1989. Mitosis in *Drosophila. J. Cell Sci.* 92:137–46.

Hayles, J., and Nurse, P. 1989. A review of mitosis in the fission yeast *Schizosaccharomyces pombe. Exptl. Cell Res.* 184:273–86.

Hepler, P. K. 1989. Ca^{2+} transients during mitosis: observations in flux. *J. Cell Biol.* 109:2567–73.

Huffaker, T. C.; Hoyt, M. A.; and Botstein, D. 1987. Genetic analysis of the yeast cytoskeleton. *Ann. Rev. Genet.* 21:259–84.

Newlon, C. S. 1988. Yeast chromosome replication and segregation. *Microbiol. Rev.* 52:568–601.

Nurse, P. 1990. Universal control mechanism regulating onset of M-phase. *Nature* 344:503–7.

Pelech, S. L.; Sanghera, J. S.; and Daya-Makin, M. 1990. Protein kinase cascades in meiotic and mitotic cell cycle control. *Biochem. Cell Biol.* 68:1297–1330.

Schlegel, R. A.; Halleck, M. S.; and Rao, P. N. 1987. *Molecular regulation of nuclear events in mitosis and meiosis.* New York: Academic Press.

Wolniak, S. M. 1988. The regulation of mitotic spindle function. *Biochem. Cell Biol.* 66:490–514.

Zhang, D. H.; Callahan, D. A.; and Hepler, P. K. 1990. Regulation of anaphase chromosome motion in *Tradescantia* stamen hair cells by calcium and related signalling agents. *J. Cell Biol.* 111:171–82.

Cytoplasmic Division (Cytokinesis)

Birky, C. W. 1983. The partitioning of cytoplasmic organelles at cell division. *Internat. Rev. Cytol.*, suppl. 15:49–89.

Devore, J. J.; Conrad, G. W.; and Rappaport, R. 1989. A model for astral stimulation of cytokinesis in animal cells. *J. Cell Biol.* 109:2225–32.

Gunning, B. E. S., and Sammut, M. 1990. Rearrangement of microtubules involved in establishing cell division planes start immediately after DNA synthesis and are completed just before mitosis. *Plant Cell* 2:1273–82.

Warn, R. M. 1990. Cytokinesis genetics takes wing. *Trends Genet.* 6:309–10.

Chromosome Banding

Bickmore, W. A., and Sumner, A. T. 1989. Mammalian chromosome banding—an expression of genome organization. *Trends Genet.* 5:144–48.

Holmquist, G. P. 1989. DNA sequences in G- and R-bands: evolution and molecular ecology. *Chromosomes Today* 10:21–32.

Manuelidis, L. 1990. A view of interphase chromosomes. *Science* 250:1533–40.

Schweizer, D.; Strehl, S.; and Hagemann, S. 1991. Plant repetitive DNA elements and chromosome structure. *Chromosomes Today* 10:33–43.

Sumner, A. T. 1990. *Chromosome banding.* London: Unwin Hyman.

Cell Division in Bacteria

Cook, W. R.; deBoer, P. A. J.; and Rothfield, L. I. 1989. Differentiation of the bacterial cell division site. *Internat. Rev. Cytol.* 118:1–31.

Cooper, S. 1991. Synthesis of the cell surface during the division cycle of rod-shaped, gram-negative becteria. *Microbiol. Rev.* 55:649–74.

D'Ari, R., and Bouloc, P. 1990. Logic of the *E. coli* cell cycle. *Trends Biochem. Sci.* 15:191–94.

de Boer, P. A. J.; Cook, W. R.; and Rothfield, L. I. 1990. Bacterial cell division. *Ann. Rev. Genet.* 24:249–74.

Firshein, W. 1989. Role of the DNA/membrane complex in prokaryotic DNA replication. *Ann. Rev. Microbiol.* 43:89–120.

Hiroga, S. 1992. Chromosome and plasmid partition in *E. coli. Ann. Rev. Biochem.* 61:283–306.

Luthenhaus, J. 1990. Regulation of cell division in *E. coli. Trends Genet.* 6:22–25.

Ogden, G., and Schaechter, M. 1986. The association of the *E. coli* chromosome with the cell membrane. In *Bacterial chromatin*, ed. G. O. Gualerzi and C. L. Pons, pp. 45–49. Berlin: Springer-Verlag.

Review Questions

1. What is condensation? When does it begin? What characteristics of the condensation mechanism suggest that it may be built into the chromosomes themselves? What is the relationship of phosphorylation to condensation? Of DNA topoisomerase II activity to condensation? What is the significance of condensation to mitotic division?

2. Define chromosome, chromatid, centromere, kinetochore, chromosome arm, primary and secondary constriction, and telomere. What functions are telomeres known or believed to carry out? What is a karyotype? What is the chromosome scaffold or core? What is C-DNA? What is an "artificial chromosome"? What are the minimum components of an artificial chromosome?

3. Outline the two major patterns of spindle formation. What are centrioles and asters? What are KMTs? IMTs? SPBs? What characteristics indicate that spindles are in equilibrium with a pool of microtubule subunits? List the major components of spindles other than microtubules.

4. What is congression? What is the relationship of congression to spindle microtubules?

5. What experimental results indicate that separation of the chromatids is a distinct process from the anaphase movement? What experimental results rule out the possibility that late centromere replication or kinetochore duplication is responsible for separation of the chromatids? What is the possible role of DNA topoisomerase II in chromatid separation?

6. What is anaphase A? Anaphase B? What evidence indicates that anaphase A and B proceed by separate mechanisms? Outline one experiment indicating that during anaphase A kinetochores move along KMTs. What is the dynamic equilibrium hypothesis, and what is its possible relationship to anaphase A? Outline two experiments indicating that anaphase B occurs by microtubule sliding at the spindle midpoint. What evidence indicates that a dynein- or kinesinlike ATPase is active in anaphase B?

7. Outline the sliding microfilament, polar tension, and membrane growth models for the anaphase movement. What evidence covered in the text supports these models or makes them seem unlikely?

8. What evidence implicates Ca^{2+} as a major factor in regulation of the anaphase movement? What effects might increases in Ca^{2+} concentration have on the spindle?

9. Outline the structure of centrioles. What is the probable relationship of centrioles to spindle formation and function? What pattern is followed in centriole duplication? What is the major cellular function of centrioles? What is the pericentriolar material? What is the possible relationship of this material to spindle formation?

10. Outline the nucleolar cycle during mitosis. What is the NOR? What is the relationship of rRNA genes to this region?

11. Outline the cycle of nuclear envelope breakdown and re-formation during mitosis in higher organisms. What is the relationship of the nuclear envelope to the ER? What are the lamins? What is the relationship of the lamins to nuclear envelope breakdown and re-formation? What role does phosphorylation play in the breakdown and reassembly of the lamins during mitosis? What happens to the lamins after nuclear envelope breakdown? How do nuclear proteins return to the nucleus at the close of mitosis?

12. Outline the events taking place in cytoplasmic division by furrowing. What is the midbody? What evidence indicates that microfilaments provide the force for furrowing? What apparently fixes the plane of cytoplasmic division in animals?

13. Outline the events taking place in cytoplasmic division in plants. What is the phragmoplast? Cell plate? Preprophase band? What events apparently fix the plane of cytoplasmic division in plants? What is the relationship of cytoplasmic division in plants to exocytosis? In what ways is cytoplasmic division in plants similar to the process in animals? In what ways is it different?

14. What evidence indicates that mitosis and cytoplasmic division are under genetic control? What is the possible role of MPF in the regulation of mitosis?

Supplement 24-1
Q, G, C, and R Bands in Metaphase Chromosomes

In 1969 T. Caspersson and his coworkers discovered that the metaphase chromosomes of higher animals reveal a series of alternating bright and dark transverse bands when stained with the fluorescent dye *quinacrine.* The bands, called *Q bands,* were subsequently found to correspond to regions of the chromosomes with slightly higher concentrations of adenine-thymine (A-T) base pairs.

Not long after Caspersson's discovery, a series of experiments revealed that essentially the same banding pattern can be produced in metaphase chromosomes by special applications of a stain much used in light microscopy, the *giemsa stain.* (The giemsa stain contains a mixture of several acidic and basic dyes, including *azure A, azure B,* and *eosin.*) When produced by the giemsa stain, the bands are termed *G bands* (Fig. 24-42).

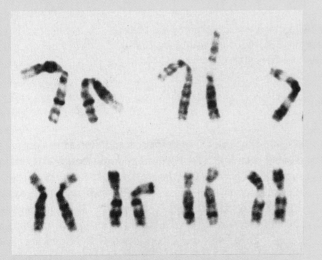

Figure 24-42 G banding in human metaphase chromosomes. Courtesy of W. Schnedl. © Academic Press, Inc., from *Internat. Rev. Cytol.,* suppl. 4:237 (1974).

The use of this stain in banding studies originated as an unexpected side development from the in situ hybridization method developed by M. L. Pardue and J. G. Gall (see p. 617). As part of their technique, mouse chromosomes were treated with dilute alkali, followed by exposure to a saline solution at 66°C. After this treatment, Pardue and Gall noted that giemsa staining was limited to regions of the chromosomes around the centromeres. Their application of the in situ technique revealed that these intensely stained bands contain highly repetitive DNA sequences. These strongly staining bands around the centromeres were later identified as *C bands.* Other investigators found that the giemsa mixture would stain bands at other points, producing G bands, if the preliminary exposure to alkali and saline was modified by reducing the duration of treatment or concentration of the reagents. With further experimentation it was found that G bands could be reliably produced by exposing metaphase chromosomes to a warm solution of salts (0.15 M NaCl and 0.015 M sodium citrate) or the proteinase trypsin, followed by giemsa staining. More than 2000 bands are revealed in the human chromosome set by the various G-banding techniques. (Fig. 24-43 shows the more prominent bands.)

The DNA segments responsible for G banding were identified by G. Holmquist and others in experiments using a label incorporated into DNA during replication. If cells were exposed to label late in the S phase, the G bands became labeled. The DNA replicating at this time contains relatively few genes and consists primarily of repetitive sequences. The molecular basis for production of G bands by the repetitive sequences is not understood. D. E. Comings and others found that the various techniques producing G bands have little effect on the chromosomes other than protein denaturation. In some way the denaturation aggregates segments containing repetitive DNA into masses that

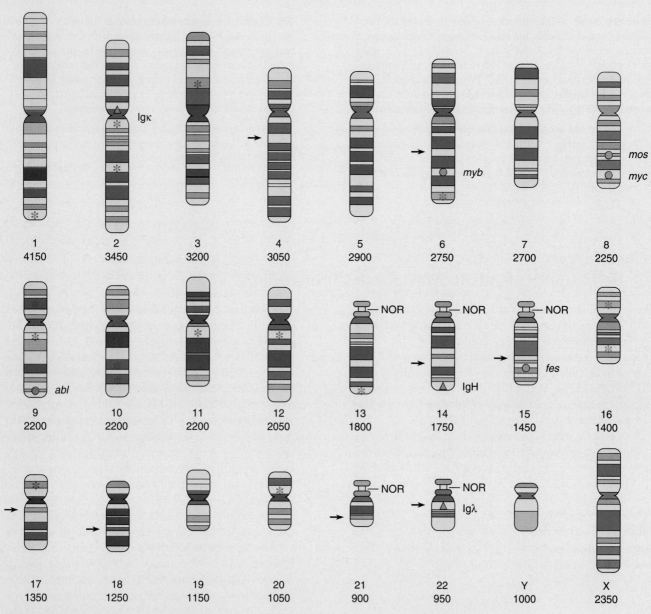

Figure 24-43 The pattern of G bands in human chromosomes, showing the sites occupied by antibody genes (triangles) and several oncogenes (dots), nucleolar organizers (NOR), and sites prone to breakage (asterisks). *abl, fes, mos, myb,* and *myc,* oncogenes; IgH, antibody heavy chain genes; 1gκ, kappa light chain genes; Igλ, lambda light chain genes. Sites where translocations frequently occur are marked with an arrow. The numbers give the gene content estimated for each chromosome. Modified from an original courtesy of J. J. Yunis, from *Science* 221:227. Copyright 1983 by the AAAS.

are colored intensely as crossbands with the giemsa stain. The more drastic techniques used to produce C bands actually remove most of the DNA from the chromosomes except in centromere regions. Why this group of repetitive sequences, which replicates very late during S, is more resistant to breakdown is unknown. However, it is possible that one or more proteins held in association with C-DNA make this fraction more stable.

The G-banding pattern can be reversed into a negative image by treating metaphase chromosomes with

hot buffer solutions or restriction endonucleases before giemsa staining. (Restriction endonucleases are enzymes that break DNA internally at specific short sequences; see p. 135.) The reverse bands produced by this method, called *R bands,* are rich in G-C base pairs and contain a relatively higher proportion of mRNA genes, particularly those concerned with "housekeeping" functions. This segment of the DNA replicates early during S. Evidently the restriction endonucleases produce R bands by attacking sequences that occur frequently in repetitive segments, extracting this DNA,

Table 24-3	Metaphase Chromosome Bands
Band Type	Characteristics of DNA in Bands
Q bands	A-T rich; mostly repetitive sequences with relatively few mRNA genes; replicate late in S.
G bands	Same as Q bands.
C bands	Highly repetitive DNA flanking centromere region; replicate very late in S.
R bands	Reverse of G and Q bands; G-C rich; contain relatively more mRNA genes; nonrepetitive; replicate early in S.

and leaving the nonrepetitive regions intact to react with the giemsa stain. (Table 24-3 summarizes the various banding patterns and the DNA segments contained within them.)

Chromosome banding would be little more than a scientific curiosity except for the fact that the bands are characteristic and different for each chromosome of the complement in higher animals. As a result, they allow each chromosome to be unambiguously identified. This was previously impossible in many species, including humans, because many chromosomes have the same overall size and shape. The banding patterns in mammals have made it possible to assign genes to specific chromosomes and to locations with respect to

G bands in the chromosomes. In this work the DNA of a gene of interest is cloned and made radioactive by growth of the cloning organism in tritiated thymidine. The DNA is then added to a preparation of isolated chromosomes under conditions that allow the DNA to hybridize with complementary sequences in the chromosome. After staining to reveal G bands and coating of the preparation with a photographic emulsion (see p. 118), the site occupied by the gene is marked by exposed grains. The banding patterns have also permitted identification of diseases such as several forms of leukemia and Down's syndrome with human chromosome rearrangements such as translocations, deletions, and insertions (see p. 769). Figure 24-43 diagrams the locations of antibody genes, several oncogenes, and sites that are prone to breakage.

G bands can be produced in reptiles, birds, mammals, and a few species of bony fishes. Lower vertebrates, invertebrates, and plants show relatively few bands when stained by the technique. In plants C- and G-band techniques produce essentially the same result—bands usually appear only in regions surrounding the centromeres. Evidently, the patterns producing Q, G, and R bands reflect arrangements of repetitive sequences in chromosomes that evolved only in higher vertebrates.

Suggestions for Further Reading: See p. 1042.

Supplement 24-2
Cell Division in Bacteria

Although there is little evidence to support the proposal, division of the nuclear material in bacteria has long been believed to involve either the plasma membrane or the cell wall. According to a model first proposed by F. Jacob (Fig. 24-44), the DNA circle of bacterial cells is attached to the plasma membrane or, by means of connections made through the plasma membrane, to the cell wall (Fig. 24-44a). The connections were proposed to be in the region of the bacterial DNA circle containing the replication origin. The attachment site is replicated along with the DNA, producing two complete DNA circles that are both attached to the plasma membrane (Fig. 24-44b and c). After replication is complete, the membrane or cell wall grows between the attachment points, separating the DNA circles and placing them in opposite ends of the cell (Fig. 24-44d and e). Cytoplasmic division (Fig. 24-44e and f) encloses the DNA molecules in separate daughter cells (Fig. 24-44g).

The model is supported by observations that the DNA of bacterial cells, when isolated, is frequently

attached to segments of the plasma membrane. Various experiments tracing the insertion of new wall material, however, indicated that new molecules are added throughout the cell wall rather than in a growth region restricted to a site between the separating DNA circles. Thus, if membrane attachments do indeed separate the replicated DNA circles of bacteria, the localized growth powering the separation must take place in the membrane itself rather than the cell wall.

Cytoplasmic division in bacteria occurs through a process that superficially resembles furrowing in animal cells. A *septum* of wall material gradually extends into the cytoplasm from the inner surface of the cell wall until the cell is cut in two (Fig. 24-45). The septum is lined by an inward extension of the plasma membrane, so that the cell products each retain a complete and unbroken surface membrane when division is complete.

Genetic studies using *E. coli* revealed several genes that control steps in cytoplasmic division in this bacterium. In *E. coli* the septum forms in a region in which the plasma membrane, peptidoglycan layer, and outer

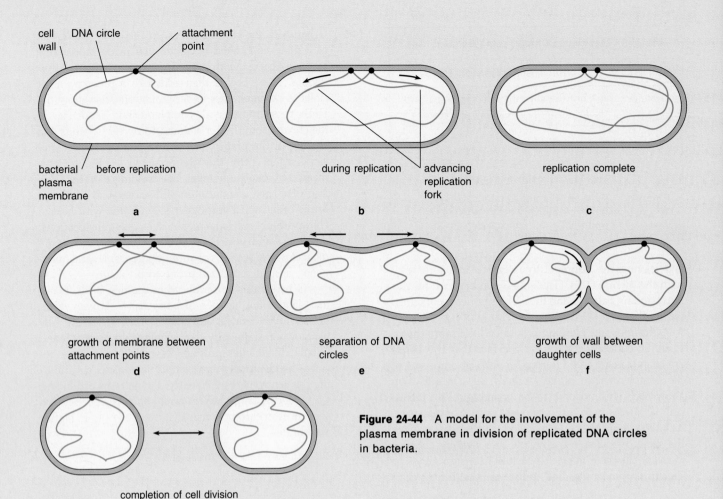

Figure 24-44 A model for the involvement of the plasma membrane in division of replicated DNA circles in bacteria.

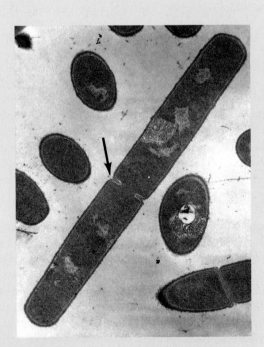

Figure 24-45 Growth of the septum (arrow) dividing the cytoplasm in the bacterium *Bacillus cereus.* × 13,000. Courtesy of L. Santo.

membrane become attached in two closely spaced, ring-like belts around the cell (Fig. 24-46a). The septum later forms in the narrow region between the two rings. The two rings, called *periseptal annuli*, have been proposed by L. I. Rothfield and his colleagues as a pair of "gas-kets" that seal off the region of septum formation from the rest of the wall and mark future sites of cytoplasmic division.

Periseptal annuli are laid down in the cell cycle before the one in which septum formation will take place at the marked site. Thus, the double rings form and move to locations one-quarter and three-quarters of the way along a cell during one cycle (Fig. 24-46b). As that cycle proceeds through cytoplasmic division, the sites occupied by the rings now lie in the middle of the two cell products (Fig. 24-46c). When the products replicate their DNA and enter the next cytoplasmic division, septums form at the locations marked by the rings in the previous cycle (Fig. 24-46d). By this time the locations for septum formation in the next cycle have already been marked.

A number of genes have been found through the isolation of temperature-sensitive mutants to affect the progress of septum formation in *E. coli* (Table 24-4).

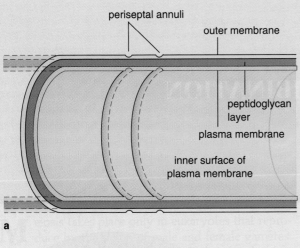

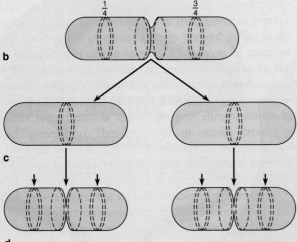

a

b

c

d

Figure 24-46 The periseptal annuli marking the site of future septum formation in gram-negative bacteria. **(a)** Structure of periseptal annuli, which consist of two closely spaced belts around the cell in which the plasma membrane, peptidoglycan layer, and outer membrane become tightly attached. The septum later forms between the two rings. **(b)** The appearance of periseptal annuli in the cell cycle before the one in which septum formation will take place at the sites marked by the annuli. Annuli form at locations one-quarter and three-quarters of the way along a cell during one cycle. **(c)** As that cycle proceeds through cytoplasmic division, the sites marked by annuli are placed in the middle of the two cell products. **(d)** As the products enter the next cytoplasmic division, septums form at the locations marked by the annuli in the previous cycle. By this time the locations for septum formation in the next cycle have already been marked (arrows).

Three of these genes, *min*C, *min*D, and *min*E, appear to affect the site at which septum formation takes place, perhaps by regulating which of the regions marked by periseptal annuli become active in septum formation. If *min*C and *min*D are in mutant form and *min*E is normal, *minicells* are formed through septum formation at all

Table 24-4 Genes Affecting Cytoplasmic Division in *E. coli*

Gene	Function
*fts*A	Necessary for final steps in septum formation; component of septum that may react with PBP3 (encoded in *fts*I).
*fts*I	Essential for septum formation; gene encodes PBP3, a protein with transglycolase and tran-speptidase activity.
*fts*Y	Encodes protein with sequence similarities to mammalian signal recognition protein; may be involved in secretion of proteins necessary for cytoplasmic division.
*fts*Z	Necessary for early step in septum formation.
*mif*1 *mif*2	Necessary for incorporation of chromosomes in daughter nuclei.
*min*B	Inhibits cell division, perhaps inhibiting *fts*Z or its encoded protein affects septum formation.
*min*C	Same as *min*B; no cell division if overactive.
*min*D	Same as *min*C; no cell division if overactive.
*min*E	May stimulate *fts*Z or its encoded protein; multiple sites of septum formation if overactive.
*sfi*C	Probably inhibits *fts*Z or its encoded protein.
*sul*A	Probably inhibits *fts*Z or its encoded protein; becomes active during stress response.

sites marked by periseptal annuli, including those positions that normally would not become active until the next cell cycle. If *min*C and *min*D are active and *min*E is defective, septa do not form at any position. Thus correct septum formation at the one-half position appears to depend on a balance of factors encoded in the *min*C, *min*D, and *min*E genes: *min*E in some manner limits inhibition of septum formation by *min*C and *min*D to the one-quarter and three-quarters positions, and allows formation at the one-half position.

Another gene, *fts*Z, appears to control a very early and essential step in initiation of the septum. A number of additional genes, including *sul*A, *sfi*C, *min*B, and possibly *min*C and *min*D, regulate septum formation by inhibiting either transcription of the *fts*Z gene or activation of its product (see Table 24-4).

Another gene, *fts*I, encodes an enzyme with transglycolase and transpeptidase activity. Although this gene does not regulate cell division, normal activity of its product is essential for the reactions that insert new molecules into the peptidoglycan layer. The *fts*I product binds penicillin and is one of the targets of this antibiotic. For this reason the *fts*I-encoded enzyme is identified as *PBP3*, for *Penicillin Binding Protein*.

Suggestions for Further Reading: See p. 1042.

products of meiosis and the union of gametes in fertilization are discussed in Chapter 26.

AN OVERVIEW OF MEIOTIC CELL DIVISION

Meiosis is preceded by a premeiotic interphase during which DNA replicates, chromosomal proteins are duplicated, and many of the proteins and other molecules required for cell division are assembled. Premeiotic interphase is followed by two sequential meiotic divisions. During the first of these divisions, which is highly specialized and differs significantly from a mitotic division, the chromosome number is reduced, recombination takes place, and, in animals in particular, RNAs and proteins are synthesized in quantity. The second meiotic division proceeds much like a mitotic division in the same species.

Premeiotic Interphase

The interphase preceding meiosis has G1, S, and G2 stages that resemble their counterparts in a premitotic interphase (see p. 910). However, there are some differences. The most conspicuous is in DNA replication, which is characteristically more extended in premeiotic S than in a premitotic S of the same species. For example, in the newt *Triturus*, premeiotic S occupies 10 days, as compared to 12 hours for premitotic S in the same species. An equivalent difference is also observed in other animals, as well as in plants, protists, and many fungi (Table 25-1). In some organisms, such as the newt *Triturus*, the greater length of premeiotic S appears to reflect the activation of fewer replicons (see p. 978) per unit time rather than a reduction in the rate at which the DNA unwinds and replicates. In others, such as the yeast *Saccharomyces* and the mouse, the opposite seems to be true: replicons are activated at approximately the same rate in premeiotic and premitotic S, but the DNA unwinds and replicates more slowly.

Other, more subtle differences are noted in the pattern of DNA synthesis during premeiotic S. The sex chromosomes, which replicate toward the end of premitotic S in most organisms, replicate at the beginning of S in premeiotic cells in many of the same species. In addition a small part of the DNA, about 0.1% to 0.2%, remains unreplicated until meiotic prophase. The unreplicated fraction, first detected by Y. Hotta and H. Stern in premeiotic cells of the plant *Lilium*, consists in this species of about 5,000 to 10,000 DNA segments, ranging from 1,000 to 5,000 base pairs in length, distributed throughout the genome. According to Hotta and Stern, a protein, which they term the *L protein*, binds to these sequence segments during premeiotic interphase and blocks their replication. A similar DNA fraction is delayed in replication in the mouse and probably in other sexually reproducing eukaryotes as well. Most investigators now consider the extended period of DNA replication during premeiotic interphase and the delay in replication of a DNA fraction until meiosis to be related in some way to chromosome pairing or genetic recombination (see below).

During premeiotic interphase in most species, the nucleus grows to a greater volume than in somatic cells. Often, the patterns in which the chromatin condenses into masses of heterochromatin (see p. 553) also differ in premeiotic nuclei. Although the significance of these structural changes is unknown, they also are believed to be related in some manner to pairing and recombination.

Depending on the species, the G2 following premeiotic S ranges from essentially nonexistent to intervals approximating a premitotic G2. In a few species, such as the plant *Trillium*, cells enter a cell cycle arrest in premeiotic G2, which persists until the G2 cells are exposed briefly to cold temperatures. The cold shock breaks the arrest, and the cells enter and complete meiosis.

Meiosis

Each of the two meiotic divisions can be separated for convenience into the same four stages as mitosis: prophase, metaphase, anaphase, and telophase (see p. 912). The stages are identified as part of the first or second meiotic division by a I or II after the stage name, as in prophase I or prophase II. An abortive interphase of greater or shorter duration, called *interkinesis*, may separate the two meiotic divisions. No S phase and no additional DNA replication take place between the two divisions.

There are many variations in the details of meiosis in different eukaryotes. In general the spindle forms and functions along the same lines as in a mitotic division in the same species. Depending on the organism, spindle formation may involve centrioles or spindle pole bodies or may proceed without the presence of obvious polar structures. This chapter concentrates on meiotic division as it takes place in most higher plants and animals.

Table 25-1	Duration of Premeiotic and Premitotic DNA Replication	
Species	Premeiotic S	Premitotic S
Triturus (newt)	10 days	12 h
Mouse	14 h	5–6 h
Triticum (a higher plant)	12–15 h	3.8 h
Saccharomyces (yeast)	1.0 h	~0.5 h

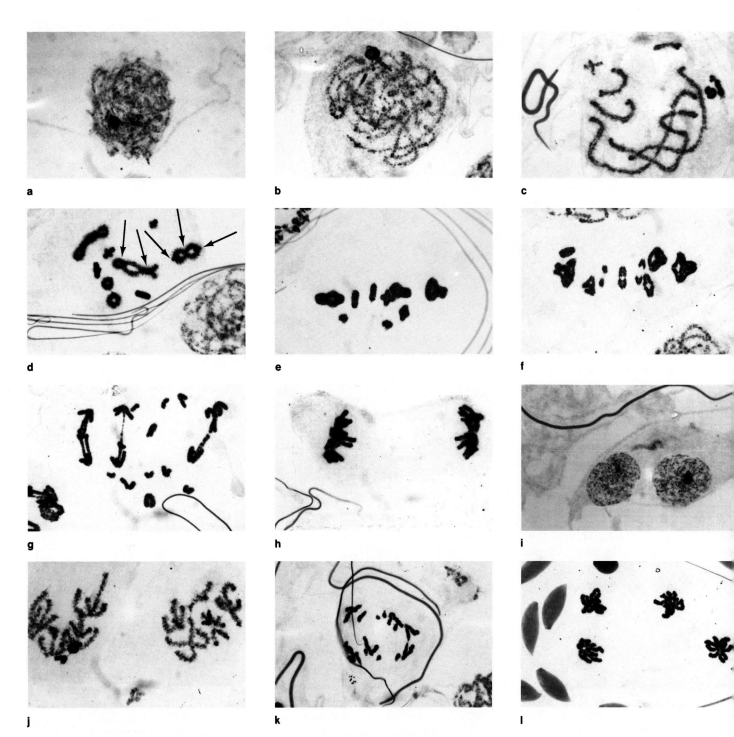

Figure 25-2 Stages of meiosis in the grasshopper *Chorthippus*. **(a)** Leptotene; **(b)** zygotene; **(c)** pachytene; **(d)** diplotene, showing several crossover points (arrows); **(e)** diakinesis; **(f)** metaphase I; **(g)** anaphase I; **(h)** telophase I; **(i)** interkinesis; **(j)** prophase II; **(k)** anaphase II, showing only one of the two spindles; **(l)** telophase II, showing both spindles. × 2,200. Courtesy of J. L. Walters.

Meiotic Prophase I

In prophase I of meiosis the chromosomes pair, recombination takes place, and greater or lesser amounts of RNA and proteins are synthesized. The relative complexity of prophase I is reflected in the length of time

meiotic cells spend in the stage, which may extend for weeks, months, or even years depending on the species. Although continuous, prophase I is divided for convenience into five sequential substages according to major structural characteristics and activities of the chromosomes: (1) *leptotene*, in which initial condensation

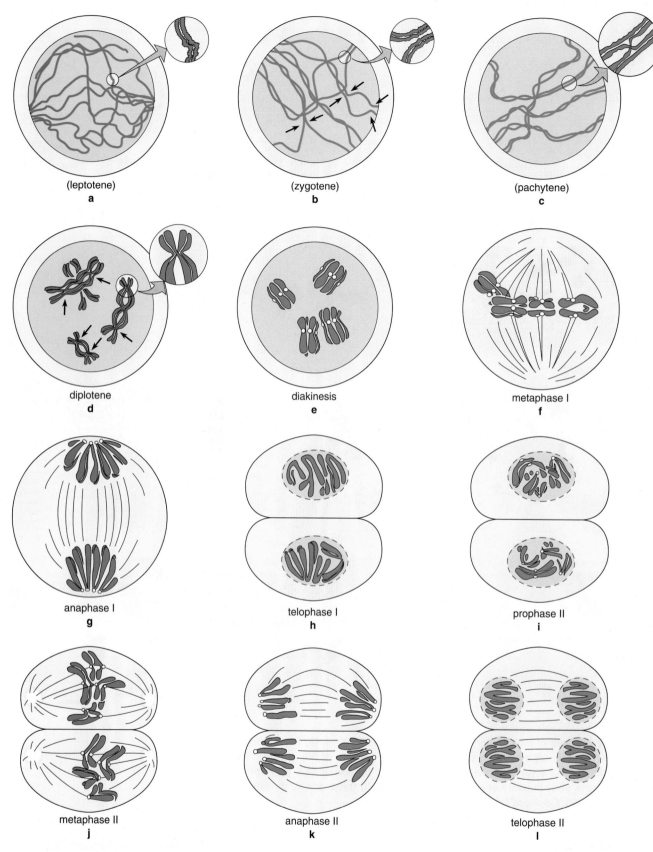

Figure 25-3 Meiosis in diagrammatic form. **(a)** Leptotene. Although both chromatids are shown for diagrammatic purposes in the magnified circles for (**a** through **c**) of this figure, the split between the two chromatids of a chromosome is not actually visible during these stages. **(b)** Zygotene. Pairing is in progress at several points (arrows). **(c)** Pachytene. **(d)** Diplotene. The chromosomes are held in pairs only by crossovers (arrows) remaining from recombination during pachytene. In most organisms all four chromatids usually become visible in tetrads at this time. **(e)** Diakinesis. **(f)** Metaphase I. **(g)** Anaphase I. **(h)** Telophase I. **(i)** Prophase II. **(j)** Metaphase II. **(k)** Anaphase II. **(l)** Telophase II.

takes place; (2) *zygotene*, in which the chromosomes pair; (3) *pachytene*, in which recombination takes place; (4) *diplotene*, during which extensive RNA transcription occurs in some species; and (5) *diakinesis*, in which condensation becomes complete and final alterations prepare the chromosomes for division.

Leptotene   The onset of leptotene (from *leptos* = fine or thin; Figs. 25-2*a* and 25-3*a*) is marked by the first appearance of chromosomes as extended threads in the nucleus as a result of the initiation of chromosome condensation. As in mitosis, condensation involves an apparently irregular folding of chromatin fibers from the extended interphase state into shorter, thicker structures (see p. 1000).

Although leptotene is superficially similar to the beginning of mitotic prophase, several differences can be seen in the structure and arrangement of the chromosomes. One is that sister chromatids are so closely associated that they cannot be separately distinguished as they condense. As a result, chromosomes appear as single rather than double threads. One factor holding chromatids in close association at this time may be the DNA segments left unreplicated in premeiotic S. These segments might form short regions in which the DNA molecules of the two chromatids branch from a single, unreplicated helix (Fig. 25-4).

A second difference, which also becomes obvious as soon as the chromosomes begin to condense, is that the telomeres, or tips of the chromosomes, are attached to the nuclear envelope (diagramed in Figs. 25-3*a* through 25-3*c*). The chromosome arms extend from the connections in loops or folds into the nucleoplasm. The telomere attachments may contribute to the mechanisms aligning and pairing the chromosomes (see below).

Another difference is seen in the pattern of condensation in the chromosome arms. The chromatin in the arms packs into a series of spherical masses sepa-

rated by more constricted regions (faintly visible in Fig. 25-2*a* and *b*), giving the arms the appearance of a series of irregularly sized beads on a string. The spherical masses of chromatin, called *chromomeres*, persist through the diplotene stage of meiotic prophase I. The functional significance of chromomeres, other than representing units of condensation, remains unknown.

Zygotene   Leptotene ends and zygotene (from *zygon* = yoke or Y) begins as the chromosomes begin to pair (Figs. 25-2*b* and 25-3*b*). In diploid organisms each chromosome, except for one type of sex chromosome (see p. 1059), is present in two copies as a *homologous pair*. One member or *homolog* of each pair is derived from the male parent of the organism and the other from the female parent. The two homologs of a pair contain the same genes aligned in the same order. However, different forms of the genes, called *alleles*, may be present in either homolog. The name given to the pairing stage refers to the Y- or fork-shaped appearance of the chromosomes at the sites where homologs come together.

The pairing, called *synapsis*, proceeds until the two chromosomes of each homologous pair are aligned along their entire length. Synapsis places the chromomeres of opposite homologs in register, so that the paired chromosomes appear as a double string of beads. The time required for complete pairing may extend from hours to several days.

Synapsis takes place in several stages. Initially, the homologs move from their original, often widely separated locations in the nucleus and begin pairing at one or more scattered points. The initial pairing frequently, but not always, begins where homologs are attached by their telomeres to the nuclear envelope—the attachment sites move together, and pairing of chromosomes proceeds inward from the telomeres. The molecular mechanisms accomplishing this first stage of pairing, in which homologous chromosomes may ap-

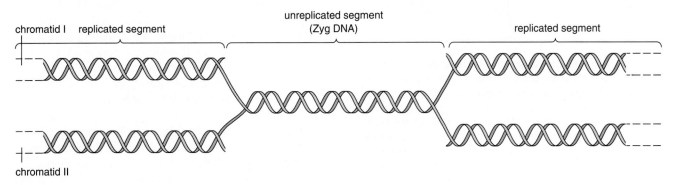

Figure 25-4   A segment of DNA (zygDNA) that remains unreplicated during premeiotic interphase. Unreplicated segments of this type may hold the two chromatids of each chromosome in close register during the early stages of meiotic prophase.

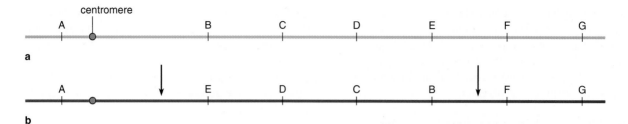

a

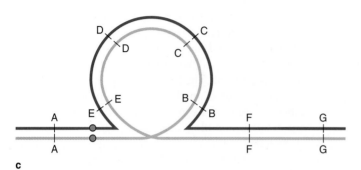

b

c

Figure 25-5 Chromosome pairing in a region containing an inversion. (The subdivision of chromosomes into chromatids is not shown.) **(a)** The normal form of the chromosome, present in one chromosome of a homologous pair. **(b)** The other chromosome of the homologous pair, in which the segment between the arrows has become inverted. **(c)** The pattern of pairing between the chromosomes, in which the normal chromosome forms a loop to allow pairing with the inverted segment. Such loops, which are commonly observed during leptotene and pachytene in nuclei with an inversion, show that synapsis occurs only between homologous segments, and does not proceed as a nonspecific "zippering" of homologs from the tips inward.

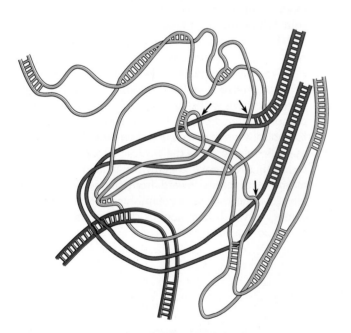

Figure 25-6 Diagram of tangles (arrows) between three different chromosome pairs during leptotene in a meiotic cell of rye (*Secale*). By pachytene the pairs untangle and all interlocks are released, presumably by breakage and rejoining of DNA helices through the activity of DNA topoisomerase II. The two chromosomes of one homologous pair are shown in light blue, those of the second in dark blue, and those of the third in gray. Paired segments are shown as regions with cross lines between homologs. Redrawn from an original courtesy of M. Abirached-Darmency and Springer-Verlag, from *Chromosoma* 88:299 (1983).

proach from distances amounting to many micrometers, are unknown.

Synapsis is more than a simple, nonspecific "zippering" of homologs once initial contacts are made. This is evident from the pattern of pairing noted between chromosomes in which a segment in one of the two homologs contains an inversion—a segment that has been broken out and reinserted in reverse order (Fig. 25-5*a* and *b*). If the inversion is long enough, one of the two chromosomes will form a loop to accommodate pairing between homologous sites in the normal and inverted segments (Fig. 25-5*c*).

Frequently, synapsis produces tangles between different chromosome pairs (Fig. 25-6). The tangles reflect the extended and intertwined arrangement of chromatin fibers during interphase. By the time recombination begins, essentially all interlocking tangles are resolved and eliminated. The tangles are evidently released by DNA topoisomerase II, which can pass one DNA helix entirely through another (see p. 968 and Fig. 23-14).

The experiments of Hotta and Stern indicate that one part of the DNA left unreplicated in premeiotic interphase in *Lilium*, consisting of segments 3000 to 5000 base pairs in length, completes its replication as the chromosomes pair. Hotta and Stern term these segments *zygDNA* because they replicate during zygotene. Pairing does not take place in *Lilium* if this replication is inhibited, suggesting that zygDNA functions in some as yet unknown manner in the pairing mechanism. Hotta and Stern's research showed that

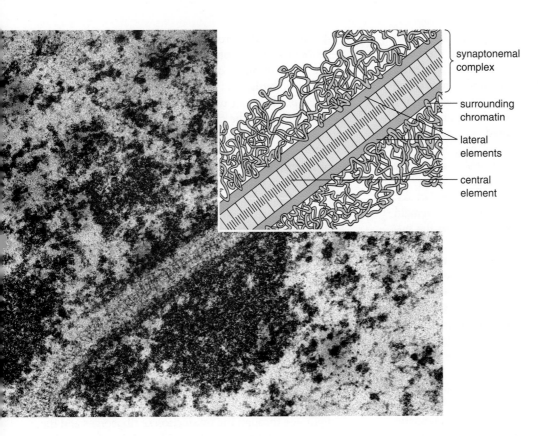

Figure 25-7 The synaptonemal complex, set up at pachytene between a pair of chromosomes in a grasshopper. The inset shows the relationship of the synaptonemal complex to the chromatin fibers of the paired chromosomes. ×74,000. Micrograph courtesy of P. B. Moens, from *J. Cell Biol.* 40:542 (1969), by permission of the Rockefeller University Press.

synaptonemal complex

surrounding chromatin

lateral elements

central element

replication of the zygDNA is complete except for single-chain nicks that are left unsealed until the end of prophase.

Some zygDNA elements are intensively transcribed during zygotene. The transcription produces a *zygRNA*, which makes up a large fraction of the total RNA of lily meiotic cells at this stage. The zygRNA, which is processed by the addition of a poly(A) tail (see p. 584), persists only until recombination begins. When pairing is inhibited, zygRNA is not transcribed in lily plants, and neither unreplicated zygDNA nor its zygRNA product can be detected in mitotic cells. These observations seem compatible with the conclusion that zygDNA and the zygRNA product of some of its sequences are involved somehow in chromosome pairing during meiotic prophase. A similar group of zygDNA sequences and zygRNA can also be detected in the mouse, suggesting that zygDNA may be generally distributed and active in synapsis.

Although the pairing is very close, synapsed homologs appear under the light microscope to remain separated by a space about 1.5 to 2 μm wide. Electron microscopy shows that this space is filled by an elaborate structure called the *synaptonemal complex* (Fig. 25-7). This structure, which assembles as homologs come together, is probably intimately involved in the pairing and recombination mechanisms. In synapsis it may function to bring homologs together or to reinforce and stabilize the chromosomes after they have paired.

Pachytene Pachytene (from *pachus* = thick; Figs. 25-2*c* and 25-3*c*) begins as the pairing of homologous chromosomes is completed. As the name suggests, condensation has proceeded until the chromosome arms are much thicker than in earlier stages. The fully paired homologs are called either *tetrads*, referring to the fact that each consists of four closely associated chromatids, or *bivalents*, referring to the fact that each consists of two chromosomes.

The primary event of pachytene is genetic recombination. In recombination the chromatids of opposite homologs exchange segments and generate new combinations of alleles (see p. 1053). The second fraction of the DNA left unreplicated during premeiotic S in the lily, which Hotta and Stern term *P-DNA*, replicates during pachytene. The P-DNA, made up of segments ranging from 1000 to 3000 base pairs, includes moderately repetitive sequences and genes encoding enzymes active in DNA nicking and repair. Some of these genes are transcribed during pachytene. The delayed replication of the P-DNA segments, as Hotta and Stern suggest, may be connected with recombination.

Another molecular change detected at this time is synthesis of a group of histone variants that are unique to meiosis. These *meiotic histones* replace some or all of the histone variants occurring in nucleosomes of body cells in the same organism. This replacement may also be involved in some manner in the mechanism of recombination or may reflect the alterations in chromosome structure taking place during prophase.

An Overview of Meiotic Cell Division **1055**

In the developing oocytes of many animals, another pattern of replication makes extra copies of the DNA segments encoding rRNA during pachytene. The extra copies produced by this pattern of replication, called *rDNA amplification,* are released to the nucleoplasm. (Details of the mechanism believed responsible for rDNA amplification are presented in Supplement 25-1.) During the diplotene stage of prophase I, the amplified rDNA copies form extra nucleoli that become active in rRNA synthesis.

Pachytene may last for days or even weeks. During this time homologous chromosomes remain closely paired and retain their attachments to the nuclear envelope.

R. Stick and H. Schwartz found that the lamin proteins underlying the nuclear envelope (see p. 560) disassemble during pachytene and do not reappear until diplotene begins. The significance of this disassembly is unknown. It is possible, as Stick and Schwartz suggest, that loss of the lamins is related to the fact that the telomeres are quite mobile during this stage in many species. Perhaps, loss of the lamins allows telomere attachments to move laterally without hindrance in the lipid bilayer of the inner nuclear membrane. The observations of Stick and Schwartz also indicate that the lamins are not absolutely required for the nuclear envelope to remain intact.

Diplotene   The close of pachytene and the beginning of diplotene (from *diplos* = double; Figs. 25-2*d* and 25-3*d*) are marked by pronounced changes in chromosome structure and activity. At this stage sister chromatids become individually visible, making each chromosome appear clearly double (the name given to the stage reflects this double appearance). As a consequence, all four chromatids of a tetrad can be individually distinguished. The separation between homologs also widens during diplotene. At the same time the synaptonemal complex disappears from most sites between homologs, and the chromosomes lose their attachments to the nuclear envelope.

At diplotene, homologs remain attached only at scattered points (arrows in Fig. 25-2*d*) at which two of the four chromatids cross over (shown diagrammatically in Fig. 25-3*d*). These crossing places, called *crossovers* or *chiasmata* (singular *chiasma* = crosspiece), reflect chromatid exchanges taking place during recombination. The number and position of the crossovers vary in different cells undergoing meiosis in the same species. However, there is usually at least one crossover per chromosome arm. Electron microscopy reveals that small segments of the synaptonemal complex persist at the crossovers, possibly as reinforcements holding the chromosomes together at these sites.

In many animals the chromosomes decondense to a greater or lesser extent and become active in RNA transcription during diplotene. The degree of decondensation varies from barely discernible (as in Fig. 25-2*d*) to almost complete reversion to an interphase-like state. In many organisms, particularly female amphibians, birds, and reptiles, decondensation at this stage produces an unusual *lampbrush* configuration in which much of the chromatin forms loops that extend outward from chromomeres over part or all of the chromosome arms (see Fig. 25-32). The loops are sites of intensive RNA transcription throughout diplotene. (Lampbrush chromosomes are discussed further in Supplement 25-2.)

Decondensation or the formation of lampbrush chromosomes reflects intensive activity in transcription of mRNA, rRNA, and tRNA during the diplotene stage. Structurally, the rRNA synthesis is marked by extensive growth of nucleolar material. If extra rDNA sequences have been produced by DNA amplification, these also become active in transcription and develop into extra nucleoli called *accessory nucleoli* at this time. (In at least some species with accessory nucleoli, especially those in which lampbrush chromosomes develop, most or all of the activity in rRNA transcription and ribosome assembly is confined to the accessory nuclei.) The cytoplasm of animal oocytes grows extensively during diplotene through the production of ribosomes and the synthesis of large quantities of proteins, fats, carbohydrates, and other cytoplasmic components. In birds and reptiles oocytes become truly massive during diplotene growth—the yolk of a hen's egg, for example, is a single, large cell produced by diplotene growth. Most of this growth is limited to the cytoplasm; the oocyte nucleus remains microscopic or may become just large enough to be visible to the naked eye. (Details of oocyte development in animals during diplotene are presented in Ch. 26.)

The diplotene stage may be greatly extended. In developing amphibian oocytes this part of meiotic prophase may last up to nearly one year. In humans oocytes reach this stage in unborn females at about the fifth month of fetal life and remain arrested in diplotene during the remainder of prenatal life, through birth and childhood, and until the individual reaches sexual maturity. Then, just before ovulation, one oocyte each month breaks arrest and continues the meiotic sequence. The time between the onset and completion of diplotene in human females may thus amount to as much as 50 years or more. Intensive RNA and protein synthesis, however, occurs only during the part of the stage preceding arrest in the developing fetus.

Diakinesis   At the close of diplotene, chromosomes recondense and pack further into short rodlets. The loops of lampbrush chromosomes, if present, are withdrawn, all RNA transcription ceases, and nucleoli disappear. These changes mark the transition to the final stage of meiotic prophase I, diakinesis (from *dia* = across and *kinesis* = movement; Figs. 25-2*e* and 25-3*e*).

The name refers to the fact that chromosomes become so compact at this stage that the chiasmata are forced to move laterally, toward the tips of the chromosomes. The chromosomes are now ready to enter metaphase of the first meiotic division.

The five sequential stages of meiotic prophase I produce four results of significance to the outcome of meiosis: (1) chromosome condensation, (2) synapsis, (3) recombination, and (4) synthesis of most or all of the RNA, protein, lipid, and carbohydrate molecules required for the differentiation of gametes and the early stages of embryonic development. The remaining events of meiosis are concerned primarily with the divisions that reduce the chromosome number.

The Meiotic Divisions

Near the end of prophase I the spindle for the first meiotic division forms. Centrioles, if present, complete their replication, and the two resultant centriole pairs migrate within their asters to opposite sides of the nucleus and form spindle poles as they do in mitosis. (In the oocytes of many female animals the centrioles disappear at some point during the meiotic divisions; see p. 1113.)

Metaphase I   As in mitosis, breakdown of the nuclear envelope provides a convenient reference point for the transition from prophase I to metaphase I (Figs. 25-2*f* and 25-3*f*). The developing spindle invades the position formerly occupied by the nucleus, and the tetrads, left scattered by the breakdown of the nuclear envelope, move to the spindle midpoint. Except for the pairing of homologous chromosomes into tetrads, meiosis at this point closely resembles the transition from prophase to metaphase in mitosis.

The first major divergence from the mitotic pattern—and, in fact, the distinctive event of the first meiotic division—occurs as the tetrads attach to spindle microtubules. The two sister chromatids of one homolog of a tetrad make microtubule attachments to the same spindle pole (Fig. 25-8*a*), and the sister chromatids of the other homolog make microtubule attachments to the opposite pole. This pattern differs fundamentally from mitosis, in which sister chromatids make connections to opposite spindle poles (Fig. 25-8*b*).

A series of observations by R. B. Nicklas demonstrated clearly that the metaphase I attachments are functions of the chromosomes, and not of factors in the surrounding cytoplasm. In his experiments Nicklas detached tetrads from a metaphase I spindle with a microneedle and positioned them at various angles to the poles. Although a large proportion of the disturbed tetrads initially made incorrect microtubule connections, these were quickly corrected, so that, in each case, sister chromatids again made connections to only

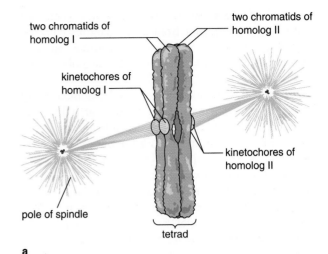

a

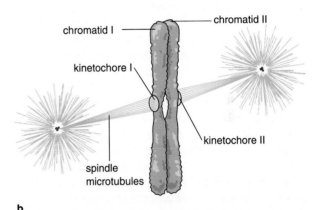

b

Figure 25-8   Kinetochore connections in meiosis and mitosis. **(a)** Spindle connections made by chromosomes of a tetrad at metaphase I of meiosis (equivalent to Fig. 25-3*f*). Both kinetochores of one chromosome make connections leading to the same spindle pole; both kinetochores of the other chromosome connect to the opposite pole. **(b)** Spindle connections made by chromosomes at metaphase of mitosis or at metaphase II of meiosis. The two kinetochores of the chromosome connect to opposite spindle poles.

one pole. Connections were also made only to one pole if a metaphase I tetrad was pushed into a metaphase II spindle. (At metaphase II the chromosomes are unpaired, and sister chromatids make connections leading to opposite poles as in mitosis; see below.) Conversely, the two chromatids of a metaphase II chromosome, if pushed into a metaphase I spindle, made connections leading to opposite spindle poles as they normally would at metaphase II. This behavior indicates that the metaphase I kinetochore connections, and the opposite connections made in metaphase II, depend on whether the chromosomes are paired as tetrads or are single. Pairing as tetrads during metaphase I, in turn, depends on chiasmata. An extended series of observations made over many years has shown that if chiasmata are disturbed or are absent from a homologous pair, the homologs usually come apart and make random, metaphase II–like attachments to the spindle at metaphase I.

Although the connections made by homologs are normally opposite, mistakes do occur at a low but constant rate. These mistakes cause both chromosomes of a homologous pair to move to the same pole at anaphase I. As a result of this *nondisjunction* of homologs, two of the nuclei resulting from meiosis receive an extra copy of one chromosome, and two are deprived entirely of a copy. The complete loss of a chromosome is usually fatal to a zygote formed by a deprived gamete. However, zygotes receiving an extra copy, which gives them three copies of one chromosome instead of two, are often viable but develop into individuals that have varying degrees of disability. In humans, for example, Down's syndrome results from nondisjunction of chromosome 21 during formation of one of the two gametes uniting to form the individual, giving the individual three copies of this chromosome.

Anaphase I   Anaphase I begins as tetrads are pulled apart, and homologs start their movement toward opposite spindle poles (see Figs. 25-2g and 25-3g). As a result of the movement, the two members of each homologous pair are separated, and the poles receive the haploid number of chromosomes. Anaphase I therefore accomplishes the reduction in chromosome number that is one of the primary outcomes of meiosis. However, because each chromosome still contains two chromatids, the $2\times$ amount of DNA is present at each pole.

Studies in the yeast *Saccharomyces cerevisiae* indicate that DNA topoisomerase II is necessary for initiation of anaphase I. In mutants with faulty forms of the enzyme, chromosomes fail to separate, and anaphase I does not proceed normally. Because this enzyme allows one DNA double helix to pass completely through another, it may allow separation of homologs at the onset of anaphase I by breaking and resealing any DNA molecules intertangled at the chiasmata.

Different homologous pairs separate independently from each other at anaphase I, so that any combination of chromosomes of maternal and paternal origin may be delivered to a spindle pole. The random combinations delivered to the poles contribute to the total genetic variability of the products of meiosis. The 23 pairs of chromosomes in humans, for example, allow 2^{23} combinations of maternal and paternal chromosomes to be delivered to the poles, producing 8,400,000 different types of gametes from this source of variability alone. Even without recombination, therefore, the possibility that two children of the same parents will receive the same combination of maternal and paternal chro-

mosomes is 1 in $(2^{23})^2$ or 1 in 70 trillion! The further variability introduced by recombination, which mixes chromatid segments randomly between maternal and paternal chromatids, makes it practically impossible for humans and most other sexually reproducing organisms to produce genetically identical gametes.

Telophase I and the Interphase Between Divisions I and II   The degree to which meiotic cells proceed through recognizable telophase I and interkinesis stages (see Figs. 25-2h, 25-2i, and 25-3h) varies widely among different species. In most animal species telophase I and the interkinesis between the two divisions are so transitory that meiotic cells move essentially directly from anaphase I to prophase II. In monocotyledonous plants, in contrast, the chromosomes decondense completely, and nuclear envelopes temporarily surround the polar masses of chromatin during telophase I and interkinesis. All possible gradations between these extremes are found in nature. Whatever the degree of chromosome decondensation and reversion to an interphaselike state between the two meiotic divisions, no DNA replication occurs at this time in any known organism.

During this period the single spindle of the first division disassembles and reorganizes into two spindles that form in the regions previously occupied by the anaphase I spindle poles. Centrioles and asters, or spindle pole bodies, if present, also divide at this time. The centrioles do not duplicate, however. As a result, centriole and aster division places a single centriole at each pole of the two spindles in cells in which centrioles are present. (This centriole divides later during sperm development in many animals.) These events complete the cellular rearrangements leading to the second meiotic division.

The Second Meiotic Division   After a brief or even nonexistent prophase II (see Figs. 25-2j and 25-3i), the chromosomes left at the two poles of the first meiotic division move to the midpoints of the two newly formed spindles. If decondensation has occurred during telophase I, chromosomes recondense into tightly coiled rodlets as they move to the metaphase II spindle midpoint. Attachment of chromosomes to the spindle takes place in the same pattern as mitosis: the two kinetochores of each chromosome attach to spindle microtubules leading to opposite poles (see Fig. 25-3j).

As a consequence of the opposite kinetochore connections, sister chromatids separate and move to opposite spindle poles at anaphase II (see Figs. 25-2k and 25-3k). Each pole therefore receives the haploid number of chromatids and the $1\times$ quantity of DNA. Sex chromosomes, when present in a species, are distributed among the products of meiosis in patterns that depend on whether they occur in a pair or singly in the diploid cells entering meiosis, and on whether they contain

Division of the Sex Chromosomes in Meiosis

In most animals and a few plants, one or more chromosomes are different in structure or number in the opposite sexes. Chromosomes that are regularly different in opposite sexes are called *sex chromosomes* (see the figure in this box). Chromosomes that are the same in both sexes are *autosomes.* In most species with sex chromosomes, one sex contains a homologous pair of sex chromosomes, and the other sex contains only one chromosome of this pair. By convention the sex chromosomes present in a homologous pair in one sex are called X chromosomes. In the opposite sex the single X may exist alone or in association with another sex chromosome, the Y *chromosome,* which does not occur in the other sex. In most animals females are XX, and males are XY or XO (the O indicates that no Y is present). This distribution is reversed in a few animal groups; in the butterflies and moths, copepods, and birds, males are XX, and females are XY.

In the sex with two X chromosomes, the two Xs pair normally and pass through the meiotic divisions in the same pattern as the autosomes. In XY individuals the X and Y chromosomes may or may not contain homologous segments capable of pairing during prophase I. (In humans a short region at the tip of one arm of the X and Y is capable of pairing.) Whether pairing occurs or not, the X and Y line up on the spindle along with the rest of the chromosomes at metaphase I of meiosis. Both the X and Y contain two chromatids at this stage, as do the rest of the chromosomes. In most species, particularly where the X and Y undergo partial pairing and recombination, the kinetochores of both chromatids of the X make connections to one pole, and the kinetochores of the Y make connections to the opposite pole, in a pattern similar to autosomes. In this case the X and Y separate and move to opposite poles of the spindle at anaphase I. Anaphase II separates the chromatids of these chromosomes, so that the haploid products of meiosis receive either an X or Y chromatid. Less frequently, the two kinetochores of each sex chromosome make opposite spindle connections at

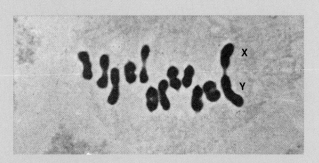

Sex chromosomes (X and Y) in the plant *Silene.* Courtesy of H. E. Warmke, from *Amer. J. Bot.* 33:648 (1946).

metaphase I, along the lines of a mitotic division, with the result that both poles receive an X and a Y chromatid at anaphase I. The X and Y make opposite spindle connections at metaphase II, so that the final haploid products still receive either an X or a Y.

In XO individuals (common in insects) both chromatids of the single X usually go to the same pole of the spindle at anaphase I. The other pole receives no X. At anaphase II the nucleus receiving the X chromosome divides again and distributes one X chromatid to each of the two division products. The other nucleus, which received no X at anaphase I, yields two O nuclei without X chromosomes. The reverse pattern is also noted, in which the two chromatids of the X make opposite spindle connections and separate at anaphase I. In the second meiotic division, the single X chromatid moves to one of the two poles, leaving the other with no X. The result of the two patterns is the same: of the four nuclei resulting from meiosis, two contain an X chromosome, and two have no X (O nuclei). At fertilization, gametes containing an X or Y (or no Y) fuse to form the diploid zygote, restoring the XX or XY (or XO) complement of sex chromosomes.

---

homologous segments that can synapse and undergo recombination. (Information Box 25-1 outlines the division of sex chromosomes during meiosis.)

At telophase II (see Figs. 25-2*l* and 25-3*l*) chromatids decondense, and nuclear envelopes form around the four division products. These four nuclei have widely divergent fates in various species of plants and animals. In the males of animal species, each nucleus is enclosed in a separate cell by cytoplasmic division, and each cell differentiates into a functional sperm cell. In

female animals only one of the four nuclei becomes functional as the egg nucleus. The other three are compartmented by unequal division of the egg cytoplasm into small, nonfunctional cells called *polar bodies* at one side of the oocyte (see Fig. 26-9). This unequal division concentrates most of the cytoplasm into a single large cell, which develops into the oocyte. Higher plants undergo a somewhat similar developmental pattern, in which all four products of meiosis give rise to functional sperm nuclei. Functional egg nuclei, however, arise from

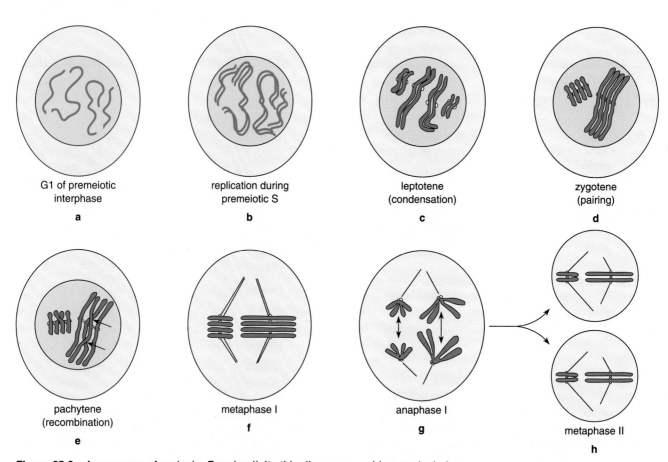

G1 of premeiotic interphase	replication during premeiotic S	leptotene (condensation)	zygotene (pairing)
a	**b**	**c**	**d**

pachytene (recombination)

e

metaphase I

f

anaphase I

g

metaphase II

h

Figure 25-9 A summary of meiosis. For simplicity this diagram considers meiosis in a hypothetical cell containing only two pairs of chromosomes, one long and one short. At premeiotic G1 **(a)**, the chromosomes of these pairs are unassociated in the nucleus. During premeiotic S **(b)**, each chromosome replicates, becoming double at all points, and now contains two chromatids. After G2, as leptotene begins **(c)**, the chromosomes condense into threads that become visible under the light microscope. Synapsis at zygotene **(d)** brings homologous chromosomes together, producing two tetrads in the nucleus, each containing two chromosomes (four chromatids). During pachytene **(e)**, recombination occurs by the exchange of segments between chromatids of the homologs (arrows). At the next stage, diplotene (not shown), at least some decondensation and RNA transcription take place in most species. The chromosomes condense again at diakinesis (not shown). At metaphase I **(f)** the two chromatids of each homologous chromosome connect to microtubules leading to the same pole of the spindle. Anaphase I **(g)** separates the two homologs of each pair and moves the haploid number of chromosomes to each pole of the spindle. Thus, one long chromosome and one short chromosome are present at each pole; tetrads no longer exist, but the chromosomes still contain two chromatids. During interphase II and prophase II two new spindles form at the metaphase I division poles. At metaphase II **(h)** the chromosomes move to the spindle midpoints and make microtubule attachments as in mitosis, in such a way that the two chromatids of each chromosome connect to microtubules leading to opposite spindle poles. Separation of the chromatids at anaphase II **(i)** delivers the haploid number of chromatids to the poles. The four division products at telophase II **(j)** each now contain the haploid quantity of DNA and possess only two chromatids, one long and one short. If, in this hypothetical organism, meiosis and gamete formation produce eggs and sperm, the end products will each contain two chromosomes, one long and one short. Fusion of an egg and sperm nucleus in fertilization **(k)** rejoins the pairs and returns the chromosome complement to the G1 level of premitotic cells.

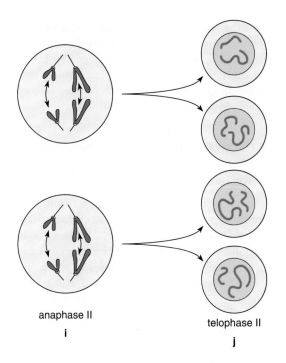

anaphase II
i

telophase II
j

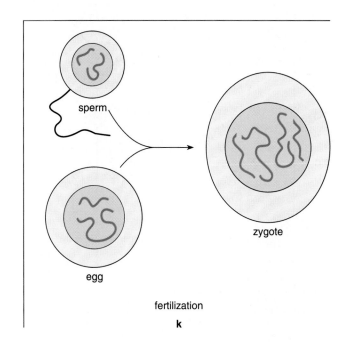

sperm

egg

zygote

fertilization
k

only one of the four meiotic products. (Fig. 25-9 summarizes the events of meiotic cell division.)

The meiotic division sequence may become arrested at various points in different animal oocytes. In female humans, as noted, an initial arrest takes place at diplotene of prophase I and persists until the time of ovulation. Meiosis then resumes and proceeds to metaphase II, when the human egg enters a second arrest that persists until fertilization. A similar pattern is followed in the mouse and rat. Mollusc eggs arrest in prophase I and remain arrested at this stage until fertilization; frog eggs arrest at metaphase II.

RNA Transcription and Protein Synthesis in Meiosis

RNA transcription can be detected during prophase I in all organisms. This has been demonstrated, for example, by autoradiographs of cells exposed to ^{3}H-uridine, a radioactive precursor of RNA. This transcription, which includes both mRNA and rRNA, peaks in many species during the diplotene stage. In some animal oocytes enormous quantities of ribosomes may be synthesized and packed into the cytoplasm at this stage. As cells reach metaphase I, RNA transcription drops to undetectable levels. A brief period of RNA transcription, primarily of mRNA, may occur again after telophase II during development of the sperm cells in male animals.

Incorporation of radioactive amino acids into proteins can be detected throughout meiosis. In many female animals, particularly those with large, yolky eggs, protein synthesis reaches its peak during the dip-

lotene stage of meiotic prophase I. Most of this protein is stored in the egg cytoplasm and remains inactive until fertilization. (Further details of RNA and protein synthesis in animal oocytes are presented in Supplement 25-2 and Ch. 26.)

The Time and Place of Meiosis

Three major variations are found in the time and place of meiosis in the life cycles of eukaryotic organisms (Fig. 25-10). The most familiar pattern takes place in animals, a few algae, and many protozoa. In this sequence, called *gametic* or *terminal meiosis*, meiosis occurs immediately before gamete formation (Fig. 25-10a).

A second pattern, called *sporic* or *intermediate meiosis* (Fig. 25-10b), takes place in higher plants and in some algae and fungi. These organisms alternate in each generation between haploid and diploid individuals. Fertilization produces the diploid *sporophyte* generation. At some point meiosis occurs in the sporophyte, producing *spores* rather than gametes. Since the spores result from meiosis, they are haploid and genetically diversified. The spores germinate and grow by mitotic divisions into haploid individuals that form the alternate *gametophyte* generation. At maturity the gametophyte generation produces eggs and sperm by differentiation of cells following an ordinary mitosis. All the eggs or sperm from a single haploid gametophytic plant, since they arise through mitosis, are genetically identical. Fusion of gametes returns the cycle to the diploid sporophyte generation.

A third major variation is observed in some fungi and algae and in a few protozoa. In this form, termed

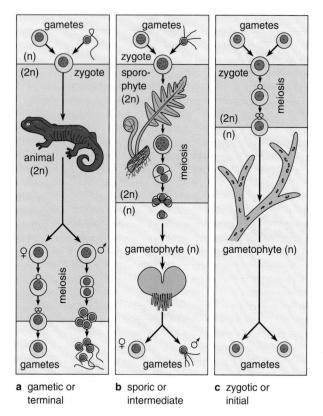

a gametic or terminal

b sporic or intermediate

c zygotic or initial

Figure 25-10 The three major patterns in the time and place of meiosis in eukaryotes (see text). The blue portions mark the diploid phase of the life cycle. n = haploid number of chromosomes; 2n = diploid number. **(a)** The gametic or terminal pattern. **(b)** The sporic or intermediate pattern. **(c)** The zygotic or initial pattern. Redrawn with permission of the Macmillan Company, Inc., from *The Cell in Development and Heredity* by E. B. Wilson. Copyright © 1925 by Macmillan Publishing Company, Inc., renewed 1953 by Anne M. K. Wilson.

zygotic or *initial meiosis* (Fig. 25-10c), meiotic divisions take place immediately after fertilization. Two haploid gamete nuclei designated as egg and sperm (or, since they are usually undifferentiated, simply as plus and minus) fuse to produce a diploid zygote. The zygote immediately enters meiosis, producing four haploid cells. These develop after one or more mitotic divisions into haploid spores that germinate to give rise to haploid individuals. Eventually, gametes are produced in these individuals as they are in the gametophytic generation of higher plants, by differentiation of cells following ordinary mitotic divisions. Several fungal species with initial meiosis supplied the critical evidence establishing that recombination occurs at prophase I of meiosis and not during premeiotic replication, as many investigators had originally proposed.

Many parts of meiosis are not completely understood and are currently under investigation. Foremost

of these are the mechanism of recombination, the role of the synaptonemal complex in this mechanism, and the patterns and functional significance of the DNA, RNA, and protein synthesis taking place during prophase I. The following sections of this chapter take up these subjects in greater detail.

THE MECHANISM OF RECOMBINATION

Classical Recombination

The experiments of classical geneticists, starting early in this century, demonstrated the overall characteristics of the recombination mechanism (Fig. 25-11). Prior to recombination the two members of a homologous pair might contain one gene with the alleles A and a at a site and another gene with the alleles B and b at a different site on the same chromosome arm (Fig. 25-11a). After replication (Fig. 25-11b), pairing during meiotic prophase I brings the homologs together and places the genes in side-by-side register (Fig. 25-11c). Recombination between these genes (Fig. 25-11d and e) produces two chromatids with the new allelic combinations A-b and a-B (Fig. 25-11f), while two of the chromatids remain unchanged, with the A-B and a-b combination. The two chromatids with the new combination of alleles are the *recombinants*, and the two that remain unchanged are the *parentals*. Because the exchange is equal and all the alleles under study are retained, this pattern of exchange is termed *reciprocal recombination*.

With the techniques available to classical geneticists, sites over most of the chromosome arms seemed equally likely to undergo recombination; therefore, the position of a recombination event was considered to be random. Exceptions were noted in a few regions where the frequency of recombination was greatly reduced. These regions, which include the centromeres and telomeres, are now known to contain highly repetitive sequences.

The probability of recombination between two genes was found to depend on the distance between them on a chromosome. As the distance increases or decreases, the space available for a random "hit" by the recombination mechanism increases or decreases in direct proportion. To classical geneticists the minimum distance for recombination seemed to be the separation between adjacent genes. The gene was therefore thought to be the unit of recombination, and it was considered unlikely that recombination could occur within the boundaries of a gene. These conclusions of classical genetics, (1) that recombination is reciprocal, (2) that it is random except at scattered locations, and (3) that it involves the gene as the unit of exchange, each required revision when the techniques of molecular genetics allowed very rare recombination events to be studied.

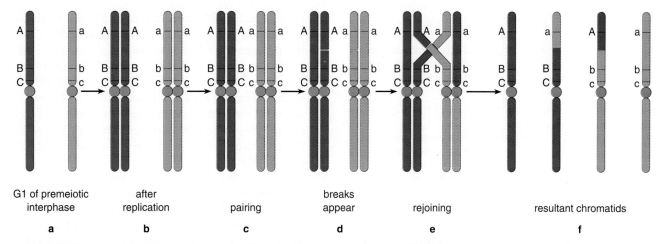

G1 of premeiotic interphase	after replication	pairing	breaks appear	rejoining	resultant chromatids
a	b	c	d	e	f

Figure 25-11 The pattern of exchanges between chromatids during recombination as determined by the techniques of classical genetics. **(a)** Before recombination the two members of a homologous pair contain a gene with the alleles A and a at one site and another gene with the alleles B and b at a different site. After replication **(b)** pairing during meiotic prophase I brings the homologs together and places the genes in side-by-side register **(c)**. Recombination by **(d)** breakage and **(e)** exchange between these genes produces two chromatids with the new allelic combinations A-b and a-B and two that retain the A-B and a-b combination **(f)**. Chromatids with the original combination of alleles are called *parentals* (P); those with new combinations of alleles are *recombinants* (R). According to the conclusions of classical genetics, an allele of the gene C at a site near the B gene will retain the parental combination with the B and b alleles unless a second crossover occurs between genes B and C.

The recombination events detectable by the techniques of classical genetics were shown, with a fair degree of certainty, to correspond to the crossovers that become visible between chromosome arms late in prophase I of meiosis. In general, the number of detectable recombination events on single chromosome arms averages about the same as the number of chiasmata, about one per arm. In addition, mutations that reduce the recombination rate cause a proportionate reduction in the number of visible chiasmata.

The physical appearance of crossovers, in which chromatids appear to switch between homologs, led many classical investigators to assume that recombination takes place by a process of physical breakage and exchange of chromatid segments. Because crossovers appear during meiotic prophase, recombination was assumed to take place at this time. These conclusions, that recombination takes place by breakage and exchange and occurs during meiotic prophase, have been fully supported by subsequent research.

Experiments Establishing the Mechanism and Timing of Recombination

Two experiments, carried out in the 1960s, ended more than a decade of debate among geneticists and molecular biologists by establishing that recombination occurs by breakage and exchange during prophase I of meiosis. Before the issue was settled, most molecular biologists felt it was unlikely that chromosomes could break and rejoin with the perfect precision required to prevent loss or addition of base pairs to either molecule entering recombination. Instead, it was proposed that recombination takes place during replication by a "copy choice" mechanism in which the replicating enzymes switch templates from one homolog to the other. Certain rare events, such as nonreciprocal recombination (see below), could also be explained more easily if recombination occurred by copy choice. The majority of geneticists and cytologists held out for breakage and exchange, which seemed most compatible with the appearance of chiasmata during meiosis and the results of classical genetic crosses.

Evidence for Breakage and Exchange The definitive experiment establishing that physical breakage and exchange of DNA molecules occur in eukaryotic recombination was carried out by J. H. Taylor in 1965, using meiotic cells of the grasshopper *Romalea*. Taylor injected grasshoppers with a radioactive DNA precursor (tritiated thymidine) and followed the distribution of label in meiotic cells by autoradiography (see p. 118). The cells of interest in the experiment were those replicating their DNA in the interphase just before the last premeiotic mitosis (Fig. 25-12). Before replication began, none of the DNA in these cells was labeled (Fig. 25-12a). After replicating in the presence of the label, the cells entered mitotic metaphase with one nucleotide

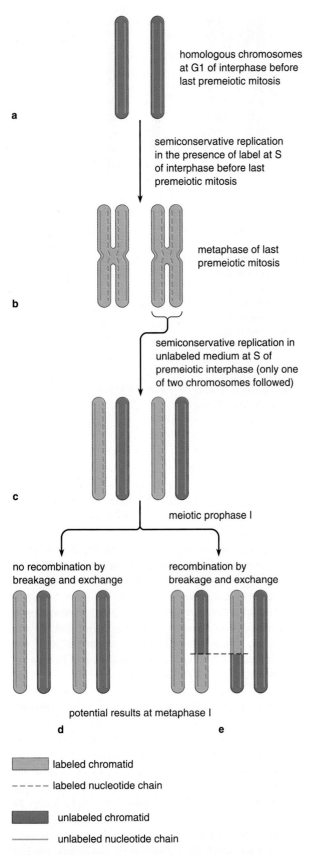

a — homologous chromosomes at G1 of interphase before last premeiotic mitosis

semiconservative replication in the presence of label at S of interphase before last premeiotic mitosis

b — metaphase of last premeiotic mitosis

semiconservative replication in unlabeled medium at S of premeiotic interphase (only one of two chromosomes followed)

c — meiotic prophase I

no recombination by breakage and exchange

recombination by breakage and exchange

potential results at metaphase I

d   e

☐ labeled chromatid

- - - - - labeled nucleotide chain

■ unlabeled chromatid

──── unlabeled nucleotide chain

Figure 25-12 Taylor's experiment demonstrating that physical breakage and exchange of chromatid segments take place during meiosis in the grasshopper *Romalea* (see text).

chain of each DNA molecule labeled and one unlabeled (Fig. 25-12b). As a consequence, all chromatids showed the presence of radioactive label. Following this last premeiotic division, the cells entered the interphase before meiosis. At this point the excess label had been washed from the tissues, and replication took place in unlabeled medium. Semiconservative replication of each chromosome produced two chromatids, one showing the presence of label and one completely unlabeled (Fig. 25-12c).

Up to this point the results were the same as those obtained in Taylor's demonstration of semiconservative replication by the same techniques (see Supplement 23-1). If no physical exchanges occurred between chromatids as part of recombination during meiotic prophase I, any single chromatid at the subsequent metaphase would be either completely labeled or unlabeled (as in Fig. 25-12d). If breakage and exchange did take place, chromatids with labeled and unlabeled segments would be detectable at the subsequent meiotic divisions (Fig. 25-12e). Reciprocal recombination, in fact, would produce a reciprocal pattern of exchange between labeled and unlabeled chromatids.

Taylor actually observed chromatids with labeled and unlabeled segments distributed in reciprocal patterns (Fig. 25-13). Because these could arise only by physical breakage and exchange between labeled and unlabeled chromatids, Taylor's experiment established that this mechanism occurs during meiosis in higher organisms. Later experiments using equivalent methods showed that recombination also occurs by breakage and exchange in prokaryotes.

Evidence That Recombination Occurs During Prophase I Demonstration that recombination takes place during prophase I of meiosis rather than replication came from the research of J. M. Rossen and M. Westergaard in 1966. Rossen and Westergaard used the fungus *Neottiella*, in which meiosis is zygotic or initial (see Fig. 25-10c), for their experiments. In this fungus both body and gamete cells are haploid. The gametes, derived from body cells after an ordinary mitosis, fuse to form the zygote. Meiosis then takes place immediately, so that the diploid condition persists for only the brief period between gamete fusion and anaphase I. Rossen and Westergaard found by measuring the quantity of DNA per nucleus that replication occurs while the gamete nuclei are still haploid, before they fuse to form the zygote. Since the two members of homologous pairs are in separate nuclei during replication, there is no possibility that the replication mechanisms could switch from one homolog to the other. Recombination, therefore, could not take place by copy choice at this time. After fusion of the gametes, when the homologs are placed in the same nucleus, meiosis occurs immediately, with typical pairing of homologous chromosomes during meiotic prophase I. Since the only

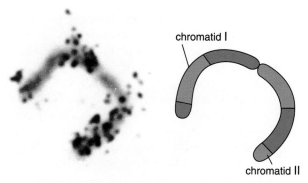

chromatid I

chromatid II

Figure 25-13 Two chromatids recovered at metaphase II of meiosis in Taylor's *Romalea* experiment. The darker color in the tracing shows labeled segments. Reciprocal exchanges of labeled segments in this manner could take place only by physical breakage and exchange. × 4,000. Courtesy of J. H. Taylor, from *J. Cell Biol.* 25:67 (1965), by permission of the Rockefeller University Press.

time both chromosomes of each homologous pair are present in the same nucleus is during prophase I of meiosis, this is the only time recombination could take place. Equivalent experiments were later carried out by others in the fungus *Schizophyllum*, the alga *Spirogyra*, and the mushroom *Coprinus*, which also follow the initial pathway for meiosis and replicate their DNA while still in the haploid state.

Intragenic Recombination

Geneticists began to discover exceptions to the conclusions of classical genetics—that the gene is the unit of recombination and that recombination is generally random and reciprocal—when they took up the study of inheritance in fungi, bacteria, and viruses. In these forms the numbers of offspring are potentially very large, and generation times are on the order of hours or minutes, allowing very rare recombinational events to be studied. This is in contrast to the two species used for classical genetic crosses, *Drosophila* and maize, in which generation times are relatively long and comparatively few offspring are produced.

Although meiosis does not occur in viruses and bacteria, recombination can be readily detected. In viruses, recombination takes place between DNA molecules originating from different viruses infecting the same cell. Although bacteria are haploid, recombination occurs between the main DNA circle and plasmids or between the main circle and segments of DNA originating from outside the cell. Many of the molecular mechanisms accomplishing recombination in bacteria and viruses appear to be the same as or similar to those of eukaryotes.

There are three primary routes by which DNA enters bacterial cells from outside. In bacterial *conjugation* a cytoplasmic bridge forms between two cells of the same species, and part or all of the DNA circle or plasmids of one cell pass into the other (Fig. 25-14). Bacteria also readily take up DNA fragments from the surrounding medium, which may enter into recombination with the main DNA circle or plasmids. This route of entry, *transformation*, is the primary route by which DNA is introduced into bacterial cells in cloning techniques (see p. 134). In the third route, *transduction* (Fig. 25-15), fragments of DNA from a bacterial cell infected by a virus are incorporated into viral particles and introduced into another cell during a subsequent infection. If the DNA fragments introduced by conjugation, transformation, or transduction are sufficiently homologous with the recipient cell DNA, a diploid condition is set up over the region of homology. Recombination may occur within the diploid region.

The short generation times and numerous offspring obtainable in bacteria and viruses, and to some extent in fungi, allow very rare recombination events to be detected with relative ease. Detection of rare events is aided by the fact that recombination occurs at a much higher frequency in bacteria and viruses than in eukaryotes. Genetic crosses with these forms revealed very quickly that recombination can occur within the boundaries of a gene or even within a codon. This type of genetic exchange, called *intragenic recombination*, may appear in only one out of hundreds of thousands or even millions of offspring. However, since billions of offspring can easily be produced, these rare events can be readily detected.

Once intragenic recombination was established as a fact in viruses, prokaryotes, and some lower eukaryotes, laborious observations in maize and *Drosophila*, requiring many years to complete, demonstrated that intragenic recombination also occurs in higher eukaryotes. This work made it clear that the unit of recombination in both prokaryotes and eukaryotes is potentially as small as a single base pair.

The study of recombination at the molecular level also revealed that the process is frequently *nonreciprocal*. Instead of appearing in perfectly reciprocal 2:2 ratios, as in classical genetic crosses, recombination between very closely spaced points sometimes produces a 1:3 ratio of alleles in offspring, as if one of the alleles is converted into the opposite type. In terms of A and a alleles of a gene undergoing recombination, for example, this would mean that the alleles might appear among offspring in a ratio of 1A:3a instead of the 2A:2a ratio shown in Figure 25-11.

Nonreciprocal recombination usually occurs without deletions or additions of nucleotides in either recombinant. In general nonreciprocal recombination in which an expected allele disappears and is replaced by the opposite allele is termed *gene conversion*.

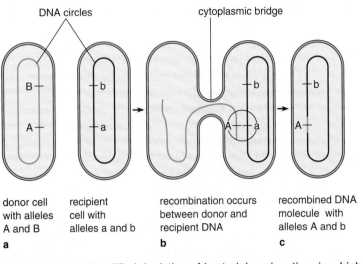

DNA circles   cytoplasmic bridge

donor cell	recipient	recombination occurs	recombined DNA
with alleles	cell with	between donor and	molecule with
A and B	alleles a and b	recipient DNA	alleles A and b
a		b	c

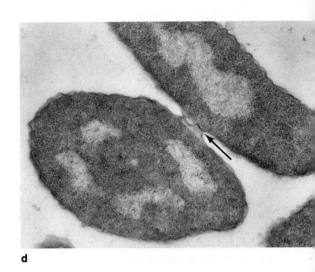

d

Figure 25-14 A simplified depiction of bacterial conjugation, in which part or all of a DNA molecule is transferred from a donor to a recipient cell. **(a)** The donor and recipient cells. The donor cell DNA has the two alleles A and B at different sites, and the recipient cell has the alleles a and b at the equivalent sites. **(b)** Conjugation, in which a cytoplasmic bridge forms between donor and recipient cells. A segment of the donor cell DNA containing the A allele has entered and paired with the homologous region of the recipient cell DNA circle. Recombination occurs within the paired region (circle), probably according to the single-chain mechanism shown in Figure 25-19. **(c)** A cell product of the conjugation, containing a recombined DNA molecule with the new combination of alleles A and b. In bacterial conjugation the donor cell dies; for the single recombination event shown in this figure the a allele would be lost and, barring other recombination events or mutations, only the A-b combination would appear in the descendants of the recipient cell. However, if a large number of bacterial cells with the alleles under study enter conjugation, all the parental (A-B, a-b) and recombinant (A-b and a-B) types would be expected among the descendants. If the sites under study are within the boundaries of a single gene, nonreciprocal allelic combinations typical of intragenic recombination would also appear among the descendants. **(d)** A pair of conjugating *E. coli* cells. A cytoplasmic bridge (arrow) has formed between the cells in the region of contact. × 44,000. Micrograph courtesy of L. G. Caro and Academic Press, Inc. (London) Ltd., from *J. Molec. Biol.* 16:269 (1966).

Other unexpected distributions of alleles are noted in intragenic exchanges. Some of the most interesting occur in the ascomycete fungi, in which the cellular products of meiosis are enclosed in a sac called the *ascus* (Fig. 25-16a). These four cells undergo two mitotic divisions following meiosis, producing a mature ascus containing 16 haploid spore cells. Because the walls of the ascus are too tight for spores to slip past one another in *Neurospora* and some other fungi, the nucleus of each spore can be traced to a particular cellular product of meiosis. This allows each of the four chromatids of a tetrad to be followed and assigned to an individual spore nucleus.

In *Neurospora*, genetic analysis is usually carried out after the first postmeiotic mitosis, using asci containing eight spores (as in Fig. 25-16a and b). At this eight-spore stage, a perfect 2:2 reciprocal exchange produces four spores containing one allele and four containing a different allele (see asci marked with an arrow in Fig. 25-16b). A nonreciprocal recombination giving a 3:1 segregation after meiosis can be directly observed as

a 6:2 combination at the eight-spore stage (see asci marked with an asterisk in Fig. 25-16b).

Asci are also found with a 5:3 combination of alleles (as diagramed in Fig. 25-16a), indicating that one of the nuclei resulting from meiosis, even though haploid, contained both alleles at the four-cell stage. For this to happen, a DNA molecule in the region containing the alleles must be a hybrid or *heteroduplex molecule*, that is, it must contain sense and missense chains originating from opposite homologs (Fig. 25-17a). During the first mitotic interphase following meiosis, these noncomplementary chains would separate (Fig. 25-17b) and serve as templates for replication. Because the templates contain noncomplementary sequences over short lengths, the two DNA molecules resulting from replication would have different sequences in these regions (Fig. 25-17c). If the noncomplementary regions contain a genetic marker included in a cross, these differences would appear as opposite alleles following the first postmeiotic mitosis, producing the 5:3 distributions noted at the eight-spore stage in

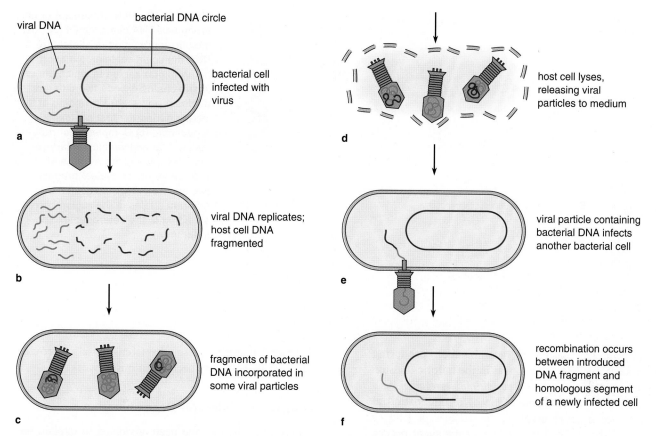

Figure 25-15 Introduction of DNA into bacterial cells by viral transduction of DNA fragments from one host cell to another. **(a)** Viral DNA is injected into a bacterial cell by an infecting virus. **(b)** During the infection cycle the viral DNA replicates and the host cell DNA breaks into fragments. **(c)** Some of the host cell DNA fragments are incorporated into viral particles assembled in the infected cell. **(d)** Lysis of the infected cell releases the newly assembled viral particles containing host cell DNA fragments to the medium. **(e)** Infection of another cell by the viral particles introduces the original host cell DNA fragments into the receiving cell. If the fragments are homologous to segments of the receiving cell DNA, pairing and recombination may occur **(f)**.

The labels in the upper figure read:

viral DNA — bacterial DNA circle

bacterial cell infected with virus (a)

viral DNA replicates; host cell DNA fragmented (b)

fragments of bacterial DNA incorporated in some viral particles (c)

host cell lyses, releasing viral particles to medium (d)

viral particle containing bacterial DNA infects another bacterial cell (e)

recombination occurs between introduced DNA fragment and homologous segment of a newly infected cell (f)

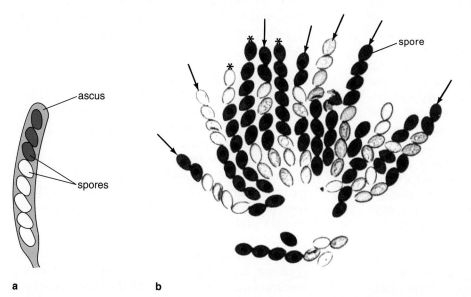

ascus

spores

spore

a   b

Figure 25-16 Asci used to evaluate genetic crosses in *Neurospora*. **(a)** An ascus containing eight spores, produced by one mitotic division following meiosis. **(b)** A cluster of asci containing spores. The transparent walls of the asci are not visible in this light micrograph. The alleles under study produce light and dark colors in the spores. Reciprocal recombination, producing a 4:4 segregation, can be seen in the asci marked by arrows; 6:2 segregation, resulting from loss of an allele by gene conversion, is visible in the asci marked by asterisks. Courtesy of H. L. K. Whitehouse, from *Genetic Recombination.* Copyright © 1982 John Wiley and Sons, Inc.

The Mechanism of Recombination   **1067**

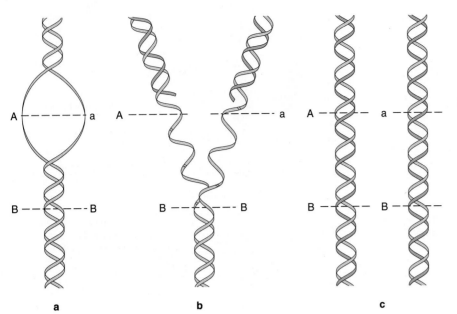

Figure 25-17 How a segment of heteroduplex DNA can account for postmeiotic segregation of alleles. **(a)** A DNA molecule containing a short heteroduplex region in which the opposite chains encode two different alleles A and a. **(b)** Unwinding of the chains during the first postmeiotic replication. **(c)** Completion of replication. The noncomplementary regions containing the A and a alleles both serve as templates, producing two separate DNA molecules with different alleles in the region corresponding to the heteroduplex segment in **(a)**.

a   b   c

Neurospora. This pattern of recombination, producing differences in alleles that appear in the mitotic division following meiosis, is called *postmeiotic segregation.*

Another novel characteristic of recombination at the molecular level involves *nonexchange of flanking markers.* When a breakage and exchange event occurs, genetic markers located close to an allele under study are expected to be transferred with it to the opposite chromatid, as long as they lie within the segment being exchanged. For example, because they lie close together, allele C in the chromosome on the left in Figure 25-11*a* would most often be transferred with allele B to the opposite chromatid by the depicted breakage and exchange (as shown in the second chromatid from the left in Fig. 25-11*f*). The only event expected to prevent this transfer is a second breakage and exchange between the B and C genes, which would return allele C to its original chromatid. Second crossovers of this type are rare, however, and become rarer as the distance between the alleles under study becomes shorter. Within the boundaries of a gene, second crossovers are expected to be very rare indeed. However, in intragenic recombination gene conversion takes place without exchange of flanking alleles at an unexpectedly high frequency, about 50% of the time, as if every other crossover event is accompanied by a second one occurring nearby.

The rare events detected by molecular genetics thus revealed three major and unexpected characteristics of the recombination mechanism at the molecular level: (1) nonreciprocal recombination involving gene conversion, (2) formation of heteroduplex DNA leading to postmeiotic segregation, and (3) nonexchange of flanking alleles at unexpectedly high frequencies. Although they present special problems to molecular biologists attempting to understand the molecular basis of recombination, these characteristics also provide clues as to how the mechanism might work.

Molecular Models for Recombination

Many hypotheses have been advanced to explain the characteristics of recombination at the molecular level. Although these models differ in detail, all assume that "nicks" are opened in the sugar-phosphate backbones of the two DNA molecules taking part in an exchange. Nucleotide chains are then considered to unwind from the nicks and "invade" the DNA of the opposite homolog by rewinding with one of its DNA chains (as in Fig. 25-18). The rewinding is commonly assumed to lead to formation of a crossed structure called a *Holliday intermediate,* named after R. Holliday, one of the foremost theorists in molecular recombination. The models fall into two major groups that differ as to whether the initial nicks leading to a Holliday intermediate are made in one or both nucleotide chains of a recombining DNA molecule.

Single-Chain Models Models proposing that one of the two chains is initially nicked are based on the first comprehensive model for molecular recombination, proposed by Holliday in 1964 (Fig. 25-18). In this hypothetical mechanism a single-chain nick is introduced by DNA endonucleases in one nucleotide chain of each of the two DNA molecules involved in the recombination (arrows, Fig. 25-18*a*). A nucleotide chain then unwinds from the nick in each molecule (Fig. 25-18*b*) and invades the opposite molecule (Fig. 25-18*c*). This initial invasion is considered to be a search by the invading chains for regions of complementary sequence. If a complementary region is encountered, the invading chains rewind with their complements in

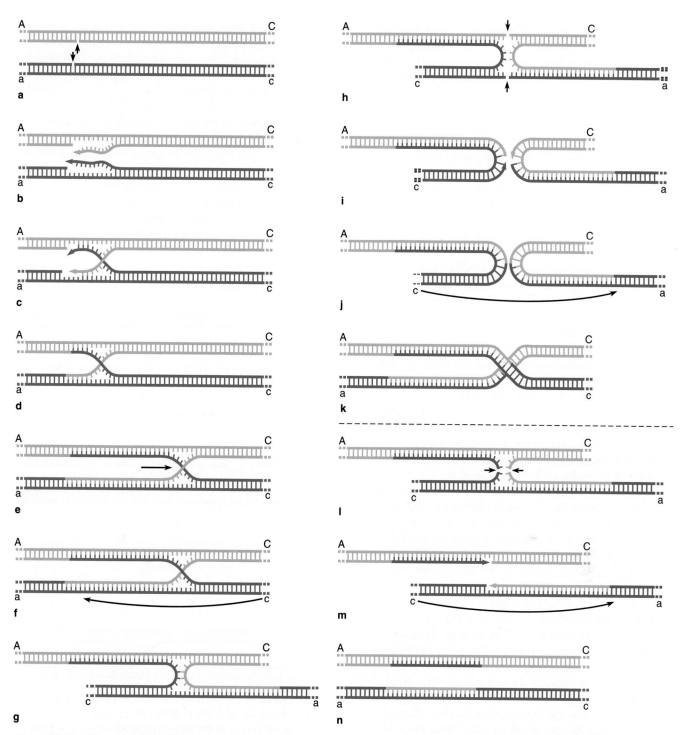

Figure 25-18 The original Holliday model for molecular recombination. **(a)** Nicks are opened (arrows) in antiparallel chains of two homologous DNA molecules. **(b)** Single chains unwind from the nicks and **(c)** rewind with complementary regions in the opposite homolog. **(d)** Gaps are filled, surplus lengths are nibbled away, and nicks are sealed, forming the Holliday intermediate. **(e)** Migration of the crossover point, called branch migration. **(f)** To follow later steps in the model more easily, the bottom double helix is considered to be rotated, so that the end containing allele c changes places with the end containing allele a. **(g)** The result of this rotation. Diagrams (**h** through **k**) show one pathway by which the structure in **(g)** may be resolved with exchange of flanking markers. **(h)** Vertical nicks (arrows) are made in the structure shown in **(g)**, any resulting gaps are filled **(i)**, and the nicks are sealed **(j)**. After rotation of the bottom double helix back to its original position, the result **(k)** is a recombination event in which flanking markers are exchanged; that is, the flanking marker allele a is now linked to C and A to c. Diagrams (**l** through **n**) show a pathway by which the structure in **(g)** may be resolved without exchange of flanking markers. **(l)** Horizontal nicks (arrows) are made in the structure shown in **(g)**, any resulting gaps are filled **(m)**, the nicks are sealed, and the bottom double helix is rotated back to its original position. The result **(n)** is a recombination event with no exchange of flanking markers; that is, the flanking allele A is still linked to C as in the top homolog originally entering the recombination event in **(a)**, and allele a is still linked to c as in the bottom double helix in **(a)**.

the opposite homolog, producing the crossed structure shown in Figure 25-18c. Perfect complementarity is not required for rewinding; small noncomplementary differences in sequence, as might be expected if the rewinding DNA originates from different alleles of the same gene, will produce regions of heteroduplex DNA. The heteroduplex regions may persist or may be "corrected" by mismatch repair (see p. 981). If the mismatch is corrected, either of the two chains may be used as template. No matter which is used as template, an allele present in the chain that is removed in the correction is lost and is replaced by the complementary copy of the allele present in the template chain. The single-chain model can thus produce gene conversion as well as heteroduplex DNA.

At this point, gaps are filled in by DNA polymerase, and all remaining nicks are sealed by DNA ligase (as described in Ch. 23). These reactions generate the closed, crossed structure shown in Figure 25-18d; this is the Holliday intermediate. While the two molecules are in this configuration, the DNA of both may unwind and rewind simultaneously through the crossing point. The effect of this simultaneous unwinding and rewinding, depending on the direction, will be to push the crossover point to the right or left. (The movement is shown to the right in Fig. 25-18e.) Movement of the crossover point, called *branch migration*, may generate additional regions of heteroduplex DNA as nucleotide chains of opposite alleles wind together.

Subsequent steps in the model free or *resolve* the crossed region of the Holliday intermediate. The mechanism proposed for this resolution, originally advanced by N. Sigal and B. M. Alberts, is easier to understand if two legs of the crossed structure are rotated through 180° (Fig. 25-18f), producing the arrangement shown in Figure 25-18g. Single-chain cuts are now made to free the crossed structure. Vertical cuts (arrows in Fig. 25-18h) will sever the two DNA chains not involved in the original invasion and rewinding. This cutting generates free ends (Fig. 25-18i) that when reciprocally sealed (Fig. 25-18j) complete the crossover. Rotating the lower half back to its original position (broken arrow, Fig. 25-18j) shows that the cutting and rejoining produce the new combinations of flanking alleles A-c and a-C (Fig. 25-18k). Flanking alleles thus recombine if the resolving cuts are made in the pattern shown in Figure 25-18h.

If horizontal cuts follow the rotation shown in Figure 25-18f, the two DNA chains involved in the initial invasion and rewinding are severed (Fig. 25-18l). Sealing these free ends (Fig. 25-18m and n) resolves the DNA molecules in their original form, effectively eliminating the crossover so that no recombination of flanking alleles takes place. (Note that the flanking alleles are in their original A-C and a-c arrangement in Fig. 25-18m.) However, within the DNA region involved in the recombination event, heteroduplex

regions and gene conversion can still lead to nonreciprocal exchanges and postmeiotic segregation. Because vertical and horizontal cuts are equally probable, gene conversion or postmeiotic segregation without recombination of flanking alleles would be expected about half the time, as observed in nature.

The initial steps in the Holliday model were later modified by M. S. Meselson and C. M. Radding and others to conform more closely with the properties of DNA and the known activity of "recombinase" enzymes (see below). In the modification, shown in simplified form in Figure 25-19, recombination begins with a nick in one of the two chains of a donor DNA molecule (Fig. 25-19a). A free end generated by the nick invades

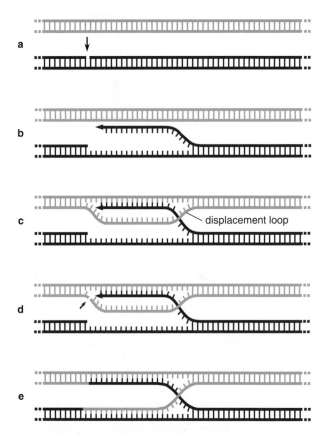

Figure 25-19 Meselson and Radding's modification to the initial steps in the Holliday model. **(a)** A nick (arrow) is made in one of the two nucleotide chains of a DNA molecule acting as donor for the exchange. **(b)** The free end generated by the nick unwinds and invades the receiving DNA molecule. If a homologous region is encountered, the invading chain winds with its complement in the receiving molecule **(c)**. The winding forces the opposite chain of the receiving molecule away from its former complement as a single-chain displacement loop. **(d)** The displacement loop is then broken, creating a free end that can invade the donor molecule to set up the Holliday intermediate **(e)** in which branch migration may occur as in Figure 25-18e. The Holliday intermediate is then resolved according to the same steps as in Figure 25-18f through k or l through n.

the receiving DNA molecule (Fig. 25-19b). If a homologous region is encountered, the invading chain winds with its complement in the receiving molecule (Fig. 25-19c). The invasion by a single chain is termed *strand transfer*. The winding forces the opposite chain of the receiving molecule away from its normal complement as a single-chain *displacement loop*. The displacement loop is then broken (Fig. 25-19d), creating a second free end that can invade the donor molecule to set up the Holliday intermediate (Fig. 25-19e). The Holliday intermediate is resolved by the same steps as in Figure 25-18e through *m*.

Double-Chain Models Double-chain models assume that the initial nicks are made in both nucleotide chains of one of the two DNA molecules entering recombination. Advanced in most complete form by J. W. Szostak and his colleagues, the double-chain models also explain nonreciprocal recombination, generation of heteroduplex DNA, and nonrecombination of flanking alleles. In addition, they explain the observation that intragenic recombination in some species is often highly *asymmetric*, proceeding as if alleles originating from one of the two DNA molecules taking part are preferentially eliminated.

The double-chain model is shown in simplified form in Figure 25-20. The mechanism begins as nicks are made in both nucleotide chains of one of two DNA molecules (Fig. 25-20a). This DNA molecule is the *donor*. After the nicks are opened, the free ends are nibbled by DNA exonucleases, opening an uneven, extended gap in the donor molecule (Fig. 25-20b). Because both the sense and missense chains are nibbled away to form the gap, any alleles in the DNA lost as the gap is opened are completely eliminated and do not contribute to the outcome. This preferential elimination of alleles in the donor molecule accounts for the asymmetry noted in some patterns of recombination.

In subsequent steps the 3′ end of a nucleotide chain extending from the gap invades the opposite homolog (Fig. 25-20c). If a complementary sequence is encountered, the invading chain rewinds with its complement in the opposite homolog (as in Fig. 25-20c). As in the modified single-chain model, the rewinding produces a displacement loop in the invaded DNA molecule; regions of heteroduplex DNA may be generated if the invading chain is not completely complementary with its mate.

Extension of the invading 3′ end eventually makes the displacement loop large enough to bridge the gap in the donor DNA (Fig. 25-20d). Repair synthesis then fills the gap in the donor molecule, using the displacement loop as template (Fig. 25-20e). Once the gap is filled, branch migration may take place (Fig. 25-20f). The migration provides an opportunity for further formation of heteroduplex DNA, as it does in the single-chain model. In addition branch migration can

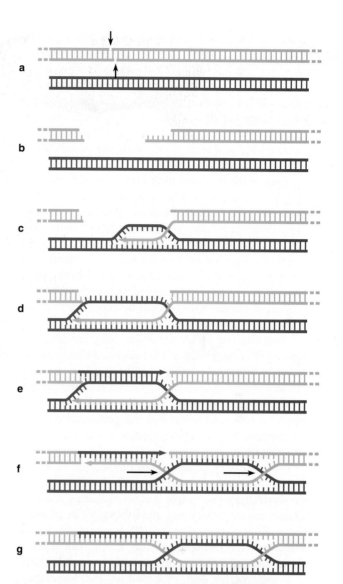

Figure 25-20 A simplified form of the double-chain model for recombination proposed by Szostak and his colleagues. **(a)** Nicks (arrows) are made in both nucleotide chains of one DNA molecule; this DNA molecule is the donor for the recombination event. **(b)** Free ends in the donor molecule are nibbled by DNA exonucleases, opening an uneven, extended gap. **(c)** The 3′ end of a nucleotide chain extending from the gap invades the opposite homolog; if a complementary sequence is encountered, the invading chain rewinds with its complement in the opposite homolog. The rewinding produces a displacement loop in the invaded DNA molecule, which bridges the gap in the donor DNA **(d)**. The displacement loop is now used as a template for repair synthesis filling the gap in the donor molecule **(e)**. Once the gap is filled, branch migration may take place **(f)**. DNA ligase seals the nicks, tying the chains into a fully closed structure with two crossover sites **(g)**. Each of the two sites is equivalent to the Holliday intermediate shown in Figure 25-18d and may be resolved in either of the patterns shown in Figure 25-18h through *k* or *l* through *n*.

result in transfer of the original invading 3′ end back to the donor molecule (shown as the crossover at the left in Fig. 25-20*f*). DNA ligase now seals the nicks, tying the chains into a fully closed structure with two crossover sites (Fig. 25-20*g*). The heteroduplex regions may remain, to produce postmeiotic segregation if they contain opposite alleles. Or they may be corrected by mismatch repair, with the possibility of gene conversion.

Each of the two crossover sites is equivalent to the Holliday intermediate shown in Figure 25-18*d*. Because this is the case, both crossovers can undergo the resolution sequence shown in either Figure 25-18*h* through *k* or *l* through *n*, leading to gene conversion and postmeiotic segregation with or without recombination of flanking alleles. As in the single-chain models, gene conversion or postmeiotic segregation without recombination of outside alleles would be expected from half of the recombination events.

Recombining DNA molecules with the structure expected for the Holliday intermediate have been detected in DNA extracted from bacterial, yeast, and human systems and photographed in the electron microscope (Fig. 25-21). These observations give strong support to the conclusion that this configuration is an intermediate step in both prokaryotic and eukaryotic recombination, as proposed in the models.

The enzymes catalyzing recombination in *E. coli* appear to operate in patterns most closely matching the single-chain model (see below), indicating that this mechanism may carry out recombination in bacteria. The *recBCD* enzyme (see below), for example, which is required for recombination in *E. coli*, can make a single-chain nick and unwind a single nucleotide chain, in a pattern like that proposed in the Meselson–Radding modification of the single-chain model. In eukaryotes a number of observations tip the balance in favor of the double-chain model. Szostak and his coworkers, for example, detected double-strand breaks in plasmids recombining with chromosomal DNA molecules of *Saccharomyces cerevisiae*. In addition, the experimental introduction of double-strand breaks, which would set up the initial step in the model, promotes recombination between plasmids and yeast chromosomes. Double-strand breaks have also been detected at "hot spots" for meiotic recombination during pachytene in *Saccharomyces*. Finally, long gaps in genes included in the plasmids can be filled in during recombination with genes in chromosomal DNA, in agreement with the steps proposed in the double-chain model.

Enzymes and Factors of Recombination

Many of the steps proposed in the recombination models can be carried out by enzymes active in DNA

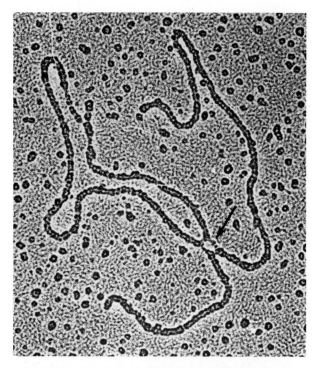

Figure 25-21 A Holliday intermediate (arrow) in DNA molecules undergoing recombination. Courtesy of C. M. Radding, from *Cell* 25:507 (1981). Copyright Cell Press.

replication, including DNA endonucleases and exonucleases, DNA binding proteins, unwinding enzymes, DNA topoisomerases, DNA polymerases, and DNA ligase (Table 25-2). A growing body of evidence indicates that many of these enzymes are actually active and required in cells undergoing recombination. A few steps in the models, including strand invasion and branch migration, do not occur in DNA replication. Enzymes with these activities have been identified in both prokaryotes and eukaryotes.

Temperature-sensitive mutants of *E. coli* (see p. 924) have been particularly valuable in the research identifying enzymes unique to recombination. One recombination enzyme identified in this way is *recBCD*, composed of three polypeptides encoded in the *recB*, *recC*, and *recD* genes. This enzyme can introduce single-chain nicks in DNA molecules and can act as an ATP-dependent helicase (see p. 967) to unwind a nucleotide chain from the nick. The single chain unwound by recBCD can be stabilized by the SSB protein, which acts in this capacity in DNA replication in *E. coli* (see p. 967).

A second enzyme necessary for recombination to proceed in *E. coli* is *recA*, which can catalyze invasion of an intact DNA helix by a single nucleotide chain (Fig. 25-19*c*). The recA enzyme, originally identified in *E. coli* by A. J. Clark and A. D. Margulies, binds to an exposed single chain and pairs it with the antiparallel

Table 25-2 Enzymes and Factors Probably Active in Recombination

Enzyme or Factor	Probable Role in Recombination
DNA endonuclease	Opens single- or double-chain breaks in DNA molecules.
Unwinding enzymes (helicases)	Promote unwinding of nucleotide chains.
DNA binding proteins	Stabilize unwound DNA in single-chain form.
DNA topoisomerase I	Promotes DNA winding or unwinding.
DNA topoisomerase II	Resolves interlocks and tangles resulting from recombination.
DNA exonuclease	Nibbles exposed ends of DNA nucleotide chains.
DNA polymerase	Extends chains and fills gaps.
DNA ligase	Seals nicks after rewinding and gap filling.
Recombination enzymes (recBCD, recA, rec1)	Promote strand invasion, displacement loop formation, and branch migration.
RuvC	Resolves Holliday intermediates.

nucleotide chain (see p. 527) of the receiving intact double helix. At first the pairing is a simple side-by-side alignment that does not require complementary sequences and does not involve winding of the donor and receiving chains. However, as soon as the alignment encounters a segment of homologous sequence, recA catalyzes rewinding of the invading chain with its complement in the receiving molecule. The opposite chain is forced away as a displacement loop, creating the configuration shown in Figure 25-19c. The enzyme is highly efficient in its search for homologous regions and can rapidly locate and rewind regions of homology. The rewinding requires ATP, which is hydrolyzed rapidly as the invading chain rewinds with the receiving molecule. The activity of recA in these steps is promoted by SSB and also enhanced by DNA topoisomerase I, which can relieve the supercoils generated by the rewinding.

Once a Holliday intermediate is formed, recA can catalyze branch migration in a $5' \rightarrow 3'$ direction with respect to the missense chain. The migration, which proceeds at about 1000 base pairs per minute, also requires ATP hydrolysis.

The complex activities of recA are especially remarkable in view of its small size—the enzyme consists of a single 352-amino acid polypeptide of only 37,842 daltons. During recombination in *E. coli*, recA polymerizes with DNA chains into nucleoprotein filaments that contain as many as thousands of individual recA molecules. The significance of recA filament formation in recombination is not understood; neither is the enzyme's unusually high rate of ATP consumption—about 100 ATPs are hydrolyzed for each base pair of rewound, heteroduplex DNA. The recA protein is required for recombination between homologous DNA molecules in *E. coli*; mutants with faulty recA are completely deficient in genetic recombination.

Holliday intermediates can be generated in the test tube by adding the recBCD, recA, and SSB proteins to homologous DNA molecules. Another enzyme, *RuvC*, recently identified in *E. coli* by S. C. West and his colleagues, can catalyze the next step in the molecular models, resolution of the Holliday intermediate when added to the test-tube system. Mutant *E. coli* cells lacking *RuvC* are unable to carry out recombination. B. Müeller and his associates found that resolution can also be catalyzed by an enzyme encoded in the DNA of the T4 bacteriophage infecting *E. coli*. The test-tube system shows that the major steps of the single-chain model can be duplicated from the initiation of recombination through resolution of the Holliday intermediate.

Enzymes with properties similar to recA have been discovered in eukaryotes. The *rec1* enzyme of the fungus *Ustilago* can catalyze most of the same steps in recombination as *E. coli* recA, but with branch migration in the opposite direction. In the lily Hotta and Stern found a 45,000-dalton protein, *m-rec*, which can substitute for recA in *E. coli* test-tube systems. (Stern's Experimental Process essay on p. 1074 describes his experiments identifying m-rec activity and its significance for the recombination mechanism in eukaryotes.)

Some of the other enzymes and factors proposed in the models have also been detected in *Lilium* by Hotta and Stern. These investigators detected a helicase with properties similar to recBCD, DNA topoisomerases I and II, and DNA endonuclease, exonuclease, ligase, and polymerase activities in lily meiotic cells during prophase I. The activities of most of these enzymes appear or increase during leptotene and zygotene, peak during pachytene, and fall to low or unmeasurable levels thereafter, in agreement with their expected roles in the recombination mechanism.

Hotta and Stern detected DNA synthesis during recombination as well as during pairing in the lily. Other investigators have also detected DNA synthesis during recombination in additional eukaryotes. In *Lilium*, DNA synthesis during recombination, in contrast to that of zygotene, is insensitive to hydroxyurea, which strongly inhibits semiconservative DNA replication. This indi-

The Experimental Process

m-rec: An Enzyme That Effects Genetic Recombination in Meiotic Cells

Herbert Stern

HERBERT STERN received his Ph.D. in Botany at McGill University, Montreal, Canada, in 1945. He spent two years as a Royal Society of Canada fellow in the Department of Biochemistry at the University of California, Berkeley, five years at the Rockefeller Institute for Medical Research (now Rockefeller University), and five years at the Research Branch of Canada Department of Agriculture. He served as Professor of Botany at the University of Illinois, Urbana, from 1960–1965. Since then, he has been Professor of Biology at the University of California, San Diego. A former president of the Society for Development Biology, his research has centered on biochemical and molecular events during meiosis.

The mechanism governing crossing-over (recombination) in meiotic cells (meiocytes) has been uninterruptedly studied in a variety of ways ever since it was recognized as a prominent feature of the process about 100 years ago. Recombination itself is not unique to meiosis. It occurs in bacteria and in mitotically dividing cells of species as diverse as baker's yeast and humans (cells of the immune system in particular). The uniqueness of meiotic recombination lies in its *regularity*, which applies to its unfailing occurrence not only in every meiotic division (rare exceptions excluded), but also in every chromosome pair. One major research challenge, still unmet, is to account for the regularity just described when it is well known that the total recombinations in any particular meiotic cell are rare events relative to the number of potential recombination sites. In general, the larger the genome size the lower the number of recombinations per DNA base pair. The mouse genome, for example, is about 20 times as large as that of *Drosophila*, while the number of crossovers per DNA unit is about 6 times higher in *Drosophila*. The issue of regularity would not exist if meiotic recombinations were extremely abundant, but such abundance is clearly unwelcome as judged by the course of evolution.

We sought an answer to this issue by asking how the enzymes relevant to recombination are organized in meiosis. Mere identification of the recombinogenic enzymes present in meiocytes would be inadequate because the process requires a still undetermined number of enzymes, and the way in which their behavior is coordinated with that of the chromosomes is an essential component of the answer. One barrier to discovering the nature of that coordination is the extremely few species in which different biochemical events can be characterized and assigned to specific stages of meiosis. A satisfactory system is one in which the cells undergo meiosis synchronously and slowly enough that a sufficient number of cells at different meiotic stages can be isolated for biochemical study. The system is even more treasured if the cells can progress through meiosis under *in vitro* conditions, thus permitting a variety of experimental manipulations. The pollen mother-cells (microsporocytes) of the Easter lily and several closely related species are a unique source of such a system. We initiated our meiotic studies using the anthers of *Lilium* and, in the course of time, it became evident that cells undergoing meiosis display a number of distinctive biochemical features that are coordinated with chromosome behavior.

We turned to the enzyme m-rec after we had found a striking change in DNA behavior in the transition from zygotene when the chromosomes are undergoing pairing, to pachytene when, as determined by cytogenetic evidence, crossing-over occurs.[1] At zygotene the chromosomes replicated select sequences of DNA that were suppressed in their replication during the S-phase. By contrast, the DNA at pachytene seemed to be in a state of turmoil. At the start of that stage a meiosis-specific endonuclease was found to introduce a large number of nicks in selected chromosomal regions that accounted for about 50% of the genome. The nicking was accompanied by intense repair synthesis. The nick-repair process did not occur in the absence of pairing and crossing-over. We were confident that recombinogenic enzymes would be active during pachytene. We selected the enzyme, later named m-rec, as one of our targets because genetic evidence unquestionably pointed to the highly purified recA protein of *E. coli* as a major component of a central recombination process. Methods have been designed for its assay although a display of its critical recombinogenic characteristics in previously untested species does not imply an identity with all other characteristics of recA protein.

The essential feature of the assay centers on the interaction between a stretch of single-stranded DNA and that of double-stranded DNA in which one strand is identical in sequence with the single-strand component. Generally, one component is circular and the other linear. If the duplex component is circular it must be supercoiled. The single strand, if circular, is necessarily intact. In an assay mixture of circular duplex and linear single strand, the enzyme catalyzes the invasion of the single strand into the duplex circle, displacing the one identical to itself by hybridizing with the complementary strand. Inasmuch as the duplex is intact, the displaced strand can neither escape the circle nor be fully displaced. The resulting structure has been named a D-loop, a product of the D-loop assay. The extent of D-loop formation is determined by using radiolabeled duplex and passing the reaction mixture through a nitrocellulose filter, which binds single-strand but not duplex DNA. Since D-loops have single-strand regions the entire complex is bound to the filter. The radioactivity bound to the filter is a measure of the number of duplexes that have undergone D-loop formation. If the substrate is a combination of linear duplex and a circular single strand, the reaction catalyzed is slightly different. In this case the invasion is effected by that strand of the duplex that is complementary to the circular DNA, and the entire strand can be transferred to form a duplex circle, a product of the strand-transfer assay. If one of the DNA components is radiolabeled, the extent of transfer can be determined by measuring the susceptibility of the reaction mixture to S1 nuclease, which does not attack double-strand DNA. If the linear duplex is radiolabeled, the transfer of one of its strands to its circular complement renders the remaining strand susceptible to nuclease digestion, so that the loss of DNA radioactivity measures the proportion of the original duplex that has been transferred to form a circular duplex. If the linear circle is labeled, the increase in DNA radioactivity measures the proportion of circular duplex formed. These recA reactions require ATP, but the requirement is not general for all recA-like proteins.

To check for the presence of a recA-like enzyme in meiotic cells, extracts of meiocytes from both lily anthers and mouse testes at unselected stages were assayed by suitable procedures. To our delight, recA-like activity was found in both groups of cells. Moreover, the requirements for D-loop and strand-transfer activities were identical to those of *E. coli* recA. We then checked vegetative tissues of various plants and somatic tissues of mice for the activity and found a 4- to 5-fold lower level in the more active nonmeiotic tissues. One distinguishing difference was found. Gel analysis of the partially purified extracts indicated the molecular weights of the rec-like proteins in meiocytes of lilies and mice to be about 43 and 45 kD (kilodaltons) respectively and those of the vegetative and somatic tissues to be about 70 and 75 kD respectively. Given the partial purity of the extracts, the significance of that finding is uncertain. A significant difference was found when the activities of the recA-like enzymes were tested at different temperatures. Lily breeders are aware that the plants are more fertile at 23–25°C than at higher temperatures at which the plants happily grow. Also, human and mouse testes prefer 33–37°C for spermatocyte fertility. We therefore compared the strand transfer activity of meiocyte and nonmeiocyte at two temperatures. Lily meiocyte recA-like enzyme was four times more active at 23°C than at 33°C. No such difference was found for the vegetative enzyme, which was somewhat lower in activity at the lower temperature. Mouse meiocyte activity was five times higher at 33°C than at 37°C, whereas that of spleen cells was slightly higher at 37°C than at 33°C. The recA-like activity of meiocytes is distinctive compared with somatic cells and the activities were designated as m-rec and s-rec respectively.

The similarities between m-rec and recA made it highly likely that m-rec functioned in meiotic recombination. The question then addressed was how its behavior during meiosis might contribute to the regularity of recombination. An answer was sought by assaying m-rec activities at different stages of meiosis, a task readily achieved with lily because of its meiotic synchrony. The result was striking. Activities were negligible during the interval of chromosome replication and remained so through the first meiotic stage, leptotene. A rise began when the cells entered the pairing stage, zygotene; activity rose sharply as the cells approached pachytene, reaching a peak value at early pachytene. A correspondingly sharp decline occurred after the cells reached midpachytene. A strong coordination exists between m-rec activity, chromosomal nick-repair activity, and cytogenetic evidence for the occurrence of crossing-over at pachytene. In a later study it was found that interference with chromosome pairing results in a much reduced rise in m-rec activity.

Two conclusions may be drawn from this study. The first of these is relevant to the largely unaccepted claim that crossing-over occurs during chromosome replication. The timing of m-rec activity validates that lack of acceptance and thus constitutes a minor contribution of the study to meiotic recombination. The second conclusion relates to the more profound consideration of the mechanism underlying the regularity of meiotic recombination. This and related studies do not fully provide a mechanism but they make possible a reasonable speculation. The latter addresses the question of how regularity might be achieved under conditions in which the frequency of recombination relative to genome size is extremely low and in which each chromosome pair is assured at least one recombination event. Clearly, one rare factor such as the recombination nodule [see p. 1080] could determine the site and frequency of recombination. If so, either of two very different mechanisms might account for regularity amidst rarity. It is conceivable that the recombination system is so tightly organized that the limiting factor contains all the biochemical ingredients essential to recombination. The alternative is to consider the limiting factor as a single essential ingredient in the process, with all other ingredients present in excess. This alternative involves the siting of a few copies of a probably complex factor whose fruitful function is assured by an overwhelming excess of all the necessary supportive mechanisms. The overwhelming excess of nick-repair and m-rec activities, as well as others not described here, all coordinated with chromosome behavior at zygotene and pachytene, make the flooding of rare sites with an excess of support mechanisms a reasonable way to achieve a regularity of rare events. The more wasteful it seems, the more efficient it is.

Reference

[1] Hotta, Y.; Tabata, S.; Bouchard, R. A.; Pinon, R.; and Stern, H. General recombination mechanisms in extracts of meiotic cells. *Chromosoma* 93:140–51 (1985).

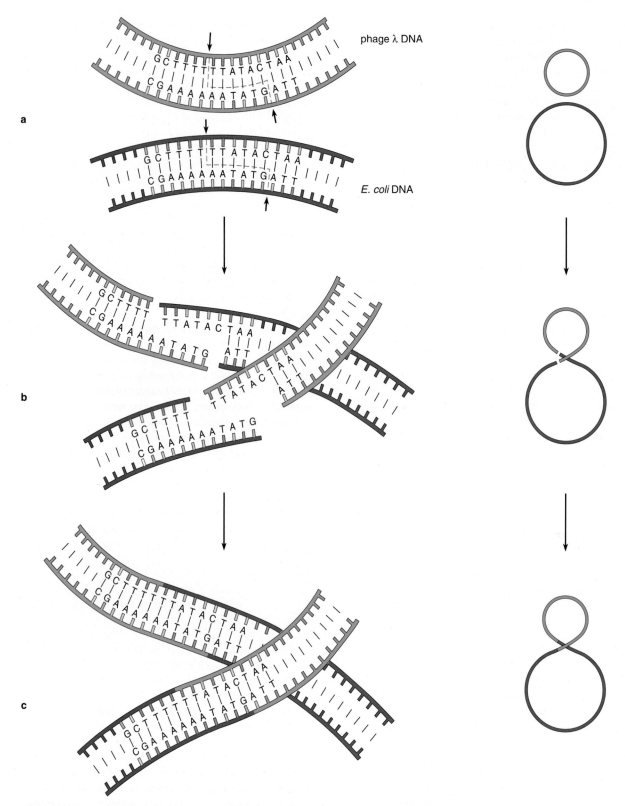

Figure 25-22 Integration of DNA from phage lambda into the *E. coli* DNA circle by site-specific recombination. A sequence serving as the recognition site for recombination occurs in both the lambda and *E. coli* DNA circles. **(a)** The sequence is nicked at corresponding locations in the lambda and *E. coli* circles (arrows). **(b)** The DNA of both molecules unwinds from the nicks, exposing single chains that are complementary to chains in the opposite DNA molecule. **(c)** The single chains wind with their complements, and the remaining nicks are ligated. Ligation seals the lambda and *E. coli* DNA into a single, continuous circle.

cates that the DNA synthesis noted during recombination is repair synthesis rather than replication, as expected if gap-filling and mismatch correction occur during recombination as proposed in the molecular models. Thus, many of the enzymes, factors, and processes predicted by molecular models have been detected in the lily, in which investigations of the molecular basis of recombination have been advanced further than any other eukaryote.

Recombination as outlined in this section, involving the exchange of segments between homologous DNA molecules, is known as *general* or *homologous recombination*. It occurs, probably by the same or similar molecular mechanisms, in all forms from viruses and bacteria to higher plants and animals. The wide distribution of general recombination among living organisms reflects its importance to survival. Through this mechanism an almost infinite variety of genetic types is presented for testing by the environment. Unless environmental change becomes too drastic, at least some of the resultant genetic types are likely to survive and reproduce successfully.

An additional major pattern, *site-specific recombination*, is also observed in viruses and bacteria. In this mechanism (Fig. 25-22) exchanges between DNA molecules are initiated at locations within the genome marked by a specific sequence recognized by the recombination enzymes. Site-specific recombination allows integration of one DNA molecule into another, as in the integration of DNA from bacteriophage lambda into the *E. coli* DNA circle (shown in Fig. 25-22). Sequences recognized by the enzymes catalyzing site-specific recombination may be limited to as few as one per genome.

THE SYNAPTONEMAL COMPLEX AND GENETIC RECOMBINATION

The synaptonemal complex almost certainly takes part in eukaryotic recombination, possibly as a framework organizing the enzymes and factors involved in breakage and exchange. Some of the evidence supporting this role is circumstantial—the synaptonemal complex is present in the right time and the right place to take part in recombination. The complex lies in the space between the homologs, where close molecular pairing and recombination probably occur. It appears between homologs as they pair, persists throughout recombination, and disappears when recombination is complete.

There is also a correlation between presence of the complex and the normal progress of recombination. In male *Drosophila* pairing occurs, but synaptonemal complexes do not form between homologs. No visible crossovers appear in chromosomes, and recombination cannot be detected in genetic crosses. In normal *Drosophila* females the synaptonemal complex is present,

chiasmata appear in chromosomes, and recombination can be detected. *Drosophila* females homozygous for a mutation eliminating recombination (the *c(3)G* mutation) have no crossovers or evidence of recombination; in these females the synaptonemal complex fails to develop between homologs. Similarly, in *S. cerevisiae*, recombination in the temperature-sensitive *cdc7* mutant is absent at elevated temperatures; examination of meiotic cells shows that the synaptonemal complex is also absent. At permissive temperatures in which the mutation is inactive, both the synaptonemal complex and recombination are present. Additional examples reinforcing the correlation between presence of the synaptonemal complex and recombination have been observed in other species, including humans. In a few species, including some insects, crossovers are restricted to certain regions of the chromosomes. The synaptonemal complex is found only in these regions.

Some organisms fail to form chiasmata or undergo recombination even though pairing occurs and apparently normal synaptonemal complexes are assembled. In females of the silkworm *Bombyx*, for example, apparently normal synaptonemal complexes are set up between homologs even though no recombination occurs. Therefore, the presence of the complex, although evidently required for recombination, does not ensure that recombination will take place. This is consistent with the role of the complex as a framework organizing the enzymes that carry out recombination.

Structure of the Synaptonemal Complex

The synaptonemal complex, first identified by M. J. Moses in 1956, is remarkably similar in appearance in different eukaryotes. It consists of a longitudinal *central element* bounded on either side by a *lateral element* (see Figs. 25-7, 25-23, and 25-24). The central element in most species appears simply as an aggregation of dense material running through the middle of the complex (as in Figs. 25-7 and 25-24). In a few species, as in some insects, the central element appears ladderlike (Fig. 25-25*a*). The two lateral elements also usually appear to be made up of dense granular material. In the ascomycete fungi and a few grasshoppers the lateral elements show a striking pattern of regular crossbands (as in Fig. 25-24). In cross section the synaptonemal complex appears as a flat, ribbonlike array (Fig. 25-25*b*). The side elements thicken where chromosomes attach to the nuclear envelope, forming dense, caplike structures closely fused to nuclear envelope membranes (Fig. 25-26). In all forms of the synaptonemal complex, very thin *transverse fibers* cross the central space and connect the lateral elements with the central element.

Chromatin fibers are densely packed around the side elements. Research by P. Moens and his coworkers indicates that chromatin extends as loops from the side elements into the surrounding nucleoplasm.

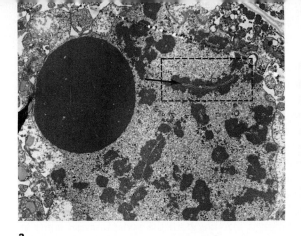

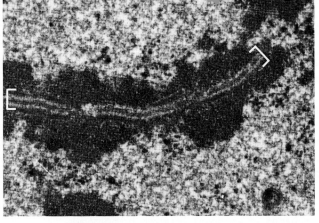

a

b

Figure 25-23 The synaptonemal complex in the lily. **(a)** The paired homologs at low magnification. The synaptonemal complex lies in the narrow, regular space separating the homologs (arrow). ×5,500. **(b)** The region enclosed by the dashed lines at higher magnification, showing the synaptonemal complex in the narrow space separating the homologs (brackets). ×41,000. Courtesy of P. B. Moens and Springer-Verlag, from *Chromosoma* 23:418 (1968).

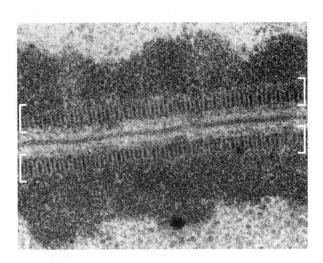

Figure 25-24 The synaptonemal complex of the fungus *Neottiella*, which has an apparently unstructured central element and striated lateral elements (brackets). ×22,000. Courtesy of D. von Wettstein.

The entire complex, including lateral elements, central elements, and transverse fibers, is digested by proteinases such as trypsin and pronase, indicating that proteins form a major part of its framework. The Moens group and others have isolated synaptonemal complexes and identified several proteins of unknown function that appear to form parts of the structure. Antibodies were developed against these proteins and combined with markers visible in the electron microscope. Using these antibodies as probes for the proteins in the synaptonemal complex showed that two, with molecular weights of 30,000 and 33,000, form part of the side elements; a third 125,000-dalton protein is located in the central element; a fourth 190,000-dalton protein occurs in both side and central elements. The

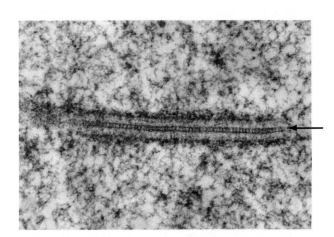

a

b

Figure 25-25 Synaptonemal complexes in *Philaenus*, a homopteran insect. **(a)** The ladderlike central element (arrow) of the synaptonemal complex. ×53,000. **(b)** Two synaptonemal complexes caught in cross section (arrows). ×20,000. Courtesy of P. L. Maillet and R. Folliot.

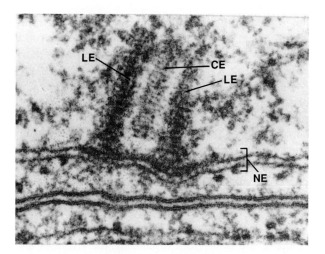

Figure 25-26 The thickened, caplike structures formed where the lateral elements (LE) of the synaptonemal complex attach to the nuclear envelope in the silkworm *Bombyx*. CE, central element; NE, nuclear envelope. Courtesy of R. C. King.

same approach showed that DNA topoisomerase II is present in side elements and in chromatin surrounding the synaptonemal complex.

The central and lateral elements stain positively with a test for RNA used in electron microscopy, indicating that these elements contain RNA. The framework of the synaptonemal complex may therefore include a ribonucleoprotein complex.

Moens and his coworkers found that short DNA segments of the chromatin are protected by the complex when isolated pachytene chromosomes are digested by DNAses. These short segments, containing from 50 to 550 base pairs, are probably the DNA of chromatin fibers bound to the surfaces or enclosed within the lateral elements. The protected sequences are random and vary from cell to cell. This indicates that no particular sequences are recognized and bound by the synaptonemal complex and that any part of the chromatin may become associated with the lateral elements. It is possible that DNA passes entirely through the lateral elements and extends into the central element, which may be the locale of recombination between homologous chromatids.

Formation and Disassembly of the Synaptonemal Complex

The synaptonemal complex first appears during leptotene, in the form of single lateral elements attached to the unpaired homologs. As the homologs pair, the lateral elements move together until they are separated by approximately 100 nm. As the lateral elements come together, the central element forms. The synaptonemal complex persists between homologs as long as they remain closely paired. As close pairing ceases at the

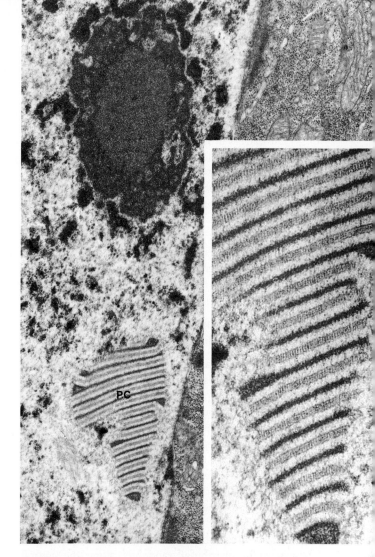

Figure 25-27 Polycomplexes (PC) in the nucleus of the mosquito *Aedes*. ×20,000. Inset, the polycomplex mass at higher magnification. ×53,000. Courtesy of T. F. Roth.

transition from pachytene to diplotene, the complex disappears except at points where homologs are held together by chiasmata.

In some organisms elements of the synaptonemal complex reassemble into masses called *polycomplexes* after disassembly from the chromosomes. These unusual structures—which have been observed in humans and in species among the insects and other arthropods, nematodes, molluscs, fungi, and higher plants—contain assemblies of regularly alternating lateral and central elements (Fig. 25-27). They may be relatively flat, with no greater thickness than an ordinary synaptonemal complex, or may extend in all directions, so that they take on the appearance of a multilayered, three-dimensional sandwich. Usually, there are no connections between polycomplexes and chromatin fibers.

In most organisms in which they occur, polycomplexes disappear by the end of prophase I. However, in some animals they can still be detected in developing sperm cells after meiosis is complete. The polycomplexes disappear in these cells before the sperm cells mature. In a few organisms, including crickets, grasshoppers, *Ascaris*, the mouse, wheat, and rye, polycomplexes appear before meiosis begins or very early in prophase I and disappear as the chromosomes pair.

The pattern in which polycomplexes appear, before or after but not during chromosome pairing and recombination, suggests that the structures serve as temporary storage sites for subunits of the synaptonemal complex. The pattern shows that chromatin is not directly required for assembly of subunits of the synaptonemal complex and suggests that the subunits may be capable of self-assembly. If so, they may exist in an equilibrium between a pool of unassembled subunits and the assembled form.

Synaptonemal Complexes and Recombination Nodules

Evidence from a variety of sources suggests that the primary element missing in organisms with synaptonemal complexes but no recombination is the *recombination nodule*, a structure first described in *Drosophila* by A. T. C. Carpenter. Recombination nodules appear in sections as dense, egg-shaped structures about 0.2 μm long, embedded in the central element of the synaptonemal complex (Fig. 25-28). The nodules, which have also been observed in protozoa, fungi, plants, and other animals, appear as the homologs pair and persist until the end of pachytene.

Initially, recombination nodules are numerous and appear to be distributed randomly along the central element of the synaptonemal complex. Later in pachy-

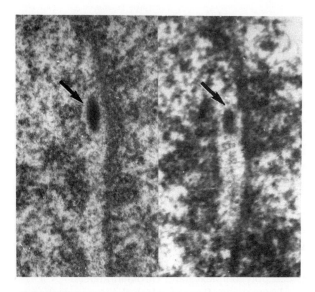

Figure 25-28 Recombination nodules (arrows) in the central elements of synaptonemal complexes of rye *(Secale)*. × 50,000. Courtesy of M. Abirached-Darmency and Springer-Verlag, from *Chromosoma* 88:299 (1983).

tene they are reduced to about one per chromosome arm—a number and distribution that correlates very closely with the crossovers that become visible later in prophase I.

Generally, recombination nodules are present in organisms or types in which recombination takes place and absent in those without recombination. In female *Bombyx*, for example, with synaptonemal complexes but no recombination, recombination nodules are absent; in several *Drosophila* mutants with reduced recombination rates, such as *mei41* and *mei218*, recombination nodules are reduced proportionately in number.

These observations indicate that the synaptonemal complex is a framework organizing the enzymes and factors carrying out recombination and suggest that recombination nodules are assemblies of these enzymes and factors. Carpenter noted that recombination nodules take up DNA precursors during recombination, as expected if DNA is being replicated or repaired as part of the recombination mechanism.

THE REGULATION OF MEIOSIS

In most organisms cells entering meiosis are products of cell lines that reproduce by mitotic division. In lower eukaryotes, including protists and fungi, the change from mitosis to meiosis is frequently triggered by environmental changes, usually toward less favorable growth conditions. In *S. cerevisiae*, for example, growth in a medium lacking carbon and nitrogen sources leads to meiosis and the formation of resistant spores. In many animals induction of meiosis involves a programmed cellular response to hormones contacting receptors at the cell surface.

Some of the genes and factors producing the response of *S. cerevisiae* to alterations in the growth medium have been worked out. In these cells the response appears to be controlled by a cAMP-based regulatory cascade (see p. 245). When internal cAMP levels are high, yeast cells divide mitotically; low cAMP levels trigger meiosis. Mutations reducing the activity of adenylate cyclase, the enzyme that converts ATP to cAMP, lead to meiosis in yeast cells even if the medium is adequate for normal growth. Conversely, mutations making the gene encoding adenylate cyclase (*CYR1*) overactive inhibit meiosis even in poor media. The cAMP appears to exert its regulatory effects through a cAMP-dependent protein kinase. Mutations in the regulatory subunit of this enzyme (encoded in the *BCY1* gene) that make the kinase activity independent of cAMP concentration prevent cells from entering meiosis even in poor growth medium. On the other hand mutations making the catalytic subunits of the protein kinase inactive (the subunits are encoded in *TPK1*, *TPK2*, or *TPK3* genes) trigger entry into meiosis.

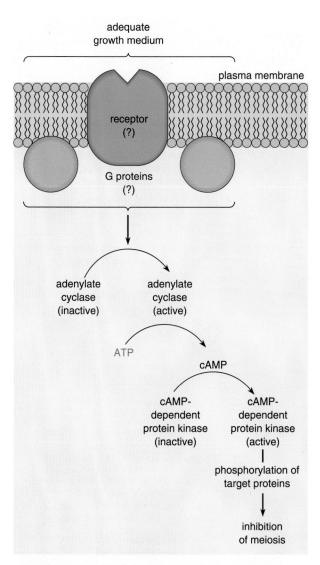

adequate
growth medium

plasma membrane

receptor
(?)

G proteins
(?)

adenylate
cyclase
(inactive)

adenylate
cyclase
(active)

ATP

cAMP

cAMP-
dependent
protein kinase
(inactive)

cAMP-
dependent
protein kinase
(active)

phosphorylation of
target proteins

inhibition
of meiosis

Figure 25-29 The cAMP-based pathway controlling meiosis in *S. cerevisiae*. A poor growth medium inactivates adenylate cyclase, which reduces cAMP concentration and turns off the pathway. This leads to dephosphorylation of target proteins by continuously active protein phosphatases, and meiosis is released from inhibition. Yeast cells then enter meiosis and form resistant spores.

These results suggest the following regulatory cascade in *S. cerevisiae* (Fig. 25-29). In adequate growth medium adenylate cyclase activity is high, maintaining cAMP at high levels. This activates the cAMP-dependent protein kinase, which phosphorylates a target protein. In phosphorylated form the protein inhibits entry of cells into meiosis. In poor growth medium a signal is transmitted that inactivates adenylate cyclase, reducing cAMP levels and inactivating the cAMP-dependent protein kinase. This leads to dephosphorylation of the target protein and triggers meiosis. The identity of the target protein and its activity remain unknown. However, the protein may directly or indi-

rectly control two genes, *IME1* and *IME2*, whose activity is required for very early steps in meiosis in *S. cerevisiae*.

The few detailed studies carried out so far in higher eukaryotes suggest that the first regulatory alterations leading toward meiosis are imposed during mitotic cell cycles well in advance of meiosis. The duration of DNA replication, for example, gradually increases through several mitotic cell cycles preceding meiosis in *Lilium* and the mouse.

Hotta and Stern's studies indicate that the early changes in the lily leading to a meiotic division can be reversed as late as the S stage preceding meiosis, so that a mitotic instead of a meiotic division occurs. Cells removed from their normal location in flower anthers and placed in a culture medium during premeiotic G1 and S revert after a few days to ordinary mitotic divisions. In this case the cells replicate the small fraction of DNA left unreplicated in premeiotic S before they divide mitotically. Cells allowed to enter early G2 before removal were unable to undergo a normal mitosis when placed in culture. Although spindles formed, the chromosomes failed to separate at anaphase. Eventually, these cells reverted to an interphaselike state without further replication or division. Cells removed from anthers later in G2 entered an abnormal form of meiosis rather than mitosis in culture. These cells exhibited irregularities in pairing during prophase, so that meiosis was never completed. The programming changes that make meiosis irreversible in these cell lines therefore appear to occur in G2 of premeiotic interphase.

These observations suggest that until G2 the changes regulating entry into a meiotic division primarily include the usual genes controlling mitosis, with activation of relatively few genetic sites unique to meiosis. Once in G2, a series of genes is activated that fundamentally changes the sequence and produces the chromosome movements and activities of meiotic prophase I. Since the second meiotic division proceeds much like a normal mitosis, the sequence at this stage is probably activated primarily by genes controlling mitosis.

The characteristics of mutations altering meiosis in *Drosophila* and *S. cerevisiae* support these conclusions (Table 25-3). Most of these mutations affect the processes of prophase I and metaphase I unique to meiosis: chromosome pairing, formation of the synaptonemal complex and recombination nodules, recombination itself, chiasma formation, and formation of the kinetochore connections leading to separation of homologous chromosomes at metaphase I. The mutations affecting recombination are especially interesting because various mutants in this category may show alterations in the rate, total amount, and location of recombination events. These effects indicate that all these characteristics are under genetic control. The surprisingly large number of mutants affecting prophase I and meta-

Table 25-3 Some Meiotic Mutants and Their Effects in Yeast and *Drosophila*

Mutation	Effect	Mutation	Effect
Saccharomyces cerevisiae		*Drosophila melanogaster*	
cdc4	Synaptonemal complex incomplete; central elements collect in nucleolus.	*c(3)G*	No synaptonemal complex; homologs fail to separate at first meiotic division.
cdc7	No synaptonemal complex.	*Df(3)sbd*[103]	Premature disassembly of synaptonemal complex; recombination reduced.
cdc9	DNA ligase faulty.		
cdc31	Chromatids fail to separate during second meiotic division.	*mei-9*	Recombination reduced; recombination nodules in normal numbers.
hop1	No synapsis; synaptonemal complex incomplete.	*mei-41*	Recombination nodules reduced in number; recombination reduced.
ndc1	Chromatids fail to separate during second meiotic division.		
rad1	No recombination.	*mei-218*	Recombination nodules reduced in number; intergenic, but nonintragenic recombination reduced.
rad50	No synaptonemal complex; no recombination.		
spo10	Synaptonemal complex incomplete; central elements collect in nucleolus.	*mei-S282*	Recombination nodules reduced in number; recombination reduced near telomeres.
spo11	Synaptonemal complex present but no recombination.	*mei-S322*	Homologs fail to separate during first meiotic division.
spo13	First meiotic division blocked; forms diploid spores.	*mei-332*	Recombination increased; effects more pronounced near centromeres.
top2	Topoisomerase II faulty; chromosomes do not separate at metaphase I.	*mei-551*	Recombination reduced by one-half; chromosome 4 and X chromosome fail to separate during first meiotic division.
		ord	Homologs fail to separate during first meiotic division.
		pal	Chromosomes of paternal origin lost during meiosis.

phase I—more than 40 in *Drosophila* and more than 50 in yeast—reflects the complexity of these events and the multiplicity of factors, enzymes, and regulatory elements required to bring them about.

Relatively few meiosis-specific mutations affect premeiotic DNA replication or the second meiotic division. Instead, most mutations identified at these stages are the same as those altering mitosis. This supports the conclusion that the major regulatory events channeling cells toward meiosis occur after premeiotic replication, during G2 or early prophase I. It also supports the conclusion that the second meiotic division proceeds by essentially the same pathway as an ordinary mitotic division.

These observations suggest that the evolutionary adaptations which established meiosis as a separate and distinct division mechanism involved the development of pairing, recombination, and separation of homologous chromosomes during the first meiotic division. These meiotic adaptations are the most complex from the genetic standpoint and, evidently, the most susceptible to disturbance through mutation.

These adaptations are not entirely unique to meiosis. Pairing of homologous chromosomes also occurs in the somatic cells of some species, such as those in the salivary glands of dipteran flies. Breakage and exchange of segments between the chromatids of replicated chromosomes also take place regularly during mitosis in many organisms including higher vertebrates. Because this exchange, called *sister strand exchange*, takes place between identical chromatids, it does not lead to genetic recombination. (The functional significance of these exchanges is unknown.) Even movement of both chromatids of a chromosome to the same pole as in metaphase I of meiosis, rather than to opposite poles as in mitosis, takes place regularly in some nonmeiotic cell types.

Meiosis has three outcomes that are vital to the process of sexual reproduction in eukaryotes. It reduces the chromosomes to the haploid level, so that chromosome number does not double at fertilization. Through re-combination and segregation of maternal and paternal chromatids, meiosis produces genetic variability in the haploid products of the division sequence. This variability is a major source of differences in the offspring of sexually reproducing organisms. Finally, through the RNA transcription and protein synthesis that occur during the diplotene stage, meiosis provides the ribosomes, enzymes, structural proteins, and raw materials needed for gamete production, fertilization, and the early stages of embryonic development.

For Further Information

Suggestions for Further Reading

General Books and Articles

Dickinson, H. G. 1989. The physiology and biochemistry of meiosis in the anther. *Internat. Rev. Cytol.* 107:79–109.

Evans, C. W., and Dickinson, H. G. 1984. *Controlling events in meiosis.* Cambridge, U.K.: Company of Biologists, Ltd.

Hennig, W., ed. 1987. *Structure and function of eukaryotic chromosomes.* New York: Springer-Verlag.

John, B. 1990. *Meiosis.* New York: Cambridge University Press.

Moens, P. B., ed. 1987. *Meiosis.* New York: Academic Press.

Murray, A. W., and Szostak, J. W. 1985. Chromosome segregation in mitosis and meiosis. *Ann. Rev. Cell Biol.* 1:289–315.

Schlegel, R. A.; Halleck, M. S.; and Rao, P. N. 1987. *Molecular regulation of nuclear events in mitosis and meiosis.* New York: Academic Press.

Stern, H. 1990. Meiosis. In *Chromosomes, Eukaryotic, Prokaryotic, and Viral,* ed. K. W. Adolph, pp. 3–37. Boca Raton, Florida: CRC Press.

Stern, H., and Hotta, Y. 1987. The biochemistry of meiosis. In *Meiosis,* ed. P. B. Moens, pp. 303–31. New York: Academic Press.

Swanson, C. P.; Merz, T.; and Young, W. J. 1981. *Cytogenetics.* 2nd ed. Englewood Cliffs, N.J.: Prentice-Hall.

Whitehouse, H. L. K. 1982. *Genetic recombination.* New York: Wiley.

The Mechanism of Recombination

Cox, M. M., and Lehman, I. R. 1987. Enzymes of general recombination. *Ann. Rev. Biochem.* 56:229–62.

Dunderdale, H. J.; Benson, F. E.; Parsons, C. A.; Sharples, G. A.; Lloyd, R. G.; and West, S. C. 1991. Formation and resolution of recombination intermediates by *E. coli* RecA and RuvC proteins. *Nature* 354:506–10.

Holliday, R. 1990. The history of the DNA heteroduplex. *Bioess.* 12:133–42.

Kowalczykowski, S. C. 1991. Biochemistry of genetic recombination: energetics and mechanism of DNA strand exchange. *Ann. Rev. Biophys. Biophys. Biochem.* 20:539–75.

Meyer, R. R., and Lane, P. S. 1990. The single-strand DNA-binding protein of *E. coli. Microbiol. Rev.* 54:342–80.

Moens, P. 1990. Unravelling meiotic chromosomes: topoisomerase II and other proteins. *J. Cell Sci.* 97:1–3.

Richardson, C., and Lehman, R., eds. 1990. *Molecular mechanisms in DNA replication and recombination.* New York: Alan R. Liss.

Roca, A. I., and Cox, M. M. 1990. The RecA protein: structure and function. *Crit. Rev. Biochem. Molec. Biol.* 25:415–56.

Roeder, G. S. 1990. Chromosome synapsis and genetic recombination: their roles in meiotic chromosome segregation. *Trends Genet.* 6:385–89.

Smith, G. R. 1987. Mechanism and control of homologous recombination in *E. coli. Ann. Rev. Genet.* 21:179–201.

———. 1988. Homologous recombination in prokaryotes. *Microbiol. Rev.* 52:1–28.

Stahl, F. W. 1987. Genetic recombination. *Sci. Amer.* 256:90–101 (February).

Sun, H.; Treco, D.; and Szostak, J. W. 1991. Extensive 3'-overhang, single-strand DNA associated with the meiosis-specific double-strand breaks at the ARG4 recombination initiation site. *Cell* 64:1155–61.

Szostak, J. W.; Orr-Weaver, T. L.; Rothstein, R. J.; and Stahl, F. W. 1983. The double-strand-break repair model for recombination. *Cell* 33:25–35.

West, S. C. 1992. Enzymes and molecular mechanisms of genetic recombination. *Ann. Rev. Biochem.* 61:603–40.

The Synaptonemal Complex in Genetic Recombination

Carpenter, A. T. C. 1987. Gene conversion, recombination nodules, and the initiation of meiotic synapsis. *Bioess.* 6:232–36.

Goldstein, P. 1987. Multiple synaptonemal complexes (polycomplexes): origin, structure, and function. *Cell Biol. Internat. Rep.* 11:759–96.

Moens, P. B., and Pearlman, R. E. 1988. Chromatin organization at meiosis. *Bioess.* 9:151–53.

von Wettstein, D.; Rasmussen, S. W.; and Holm, P. B. 1984. The synaptonemal complex in genetic segregation. *Ann. Rev. Genet.* 18:331–413.

The Regulation of Meiosis

Malone, R. E. 1990. Dual regulation of meiosis in yeast. *Cell* 61:375–78.

McLeod, M. 1989. Regulation of meiosis: from DNA binding protein to protein kinase. *Bioess.* 11:9–14.

Pelech, S. L.; Sanghera, J. S.; and Daya-Makin, M. 1990. Protein kinase cascades in meiotic and mitotic cell cycle control. *Biochem. Cell Biol.* 68:1297–1330.

Propst, F. 1988. Proto-oncogene expression in germ cell development. *Trends Genet.* 4:183–87.

Smith, G. R. 1989. Homologous recombination in *E. coli*: multiple pathways for multiple reasons. *Cell* 58:807–9.

Lampbrush Chromosomes

Callan, H. G. 1986. *Lampbrush chromosomes.* New York: Springer-Verlag.

———. 1987. Lampbrush chromosomes as seen in historical perspective. In *Structure and function of eukaryotic chromosomes*, ed. W. Hennig, pp. 5–26. New York: Springer-Verlag.

Macgregor, H. C. 1986. The lampbrush chromosomes of animal oocytes. In *Chromosome structure and function*, ed. M. S. Risley, pp. 152–86. New York: Van Nostrand Reinhold.

———. 1987. Lampbrush chromosomes. *J. Cell Sci.* 88:7–9.

Review Questions

1. Compare the chromosome number and DNA content per nucleus during G1, S, G2 and division in cells undergoing mitosis and meiosis. In what ways does premeiotic interphase differ from a premitotic interphase? What mechanisms account for the observed differences in duration of the S phase in premeiotic and premitotic S?

2. Outline the stages of meiotic prophase I. What events mark the beginning and end of leptotene? What is the major event of this stage? What differences are noted between the structure and arrangement of the chromosomes at the beginning of mitosis and the first stage of meiotic prophase? What are chromomeres?

3. What events mark the beginning and end of zygotene? What is the major event of zygotene? What is a chromosome pair? A homolog? An allele? What is zygDNA? zygRNA?

4. What events mark the beginning and end of pachytene? What major event takes place during pachytene? What is a bivalent? A tetrad? rDNA amplification?

5. What events mark the beginning and end of diplotene? What major events take place during this stage? What are chiasmata? Lampbrush chromosomes?

6. What is diakinesis? What major event takes place during this stage? What are the results of meiotic prophase?

7. Outline the patterns of spindle formation during the two meiotic divisions. What patterns are followed by centrioles, if present, during the two divisions? Compare the spindle connections made by chromosomes during the first and second meiotic divisions. What is the importance of chiasmata to these connections? How do the connections lead to a reduction in the number of chromosomes? What is nondisjunction?

8. Compare the first and second meiotic divisions with a mitotic division. What are the similarities? The differences?

9. What are maternal and paternal chromosomes? How are maternal and paternal chromosomes distributed during meiosis? During mitosis?

10. Outline the differences between gametic or terminal meiosis, sporic or intermediate meiosis, and zygotic or initial meiosis.

11. Outline the steps and outcome of classical recombination. What are recombinant chromatids? Parental chromatids? What is reciprocal recombination? Why was the gene thought to be the unit of recombination by classical geneticists?

12. What experiment established that recombination takes place by breakage and exchange in eukaryotes? What experiment demonstrated that recombination takes place during meiotic prophase rather than replication?

13. What characteristics of fungi, bacteria, and viruses made the detection of rare recombination events practical?

What is intragenic recombination? Nonreciprocal recombination? Gene conversion? Postmeiotic segregation of alleles? What is heteroduplex DNA, and how does it lead to postmeiotic segregation? What is meant by the nonexchange of flanking markers? What is the significance of this phenomenon to the mechanism of recombination?

14. Outline the major steps in the Holliday model for recombination. What is strand exchange? A displacement loop? Branch migration? What is a Holliday intermediate? In what two ways may the Holliday intermediate be resolved? What is the significance of this resolution to the nonexchange of flanking markers?

15. What are the major differences between one- and two-strand break models for recombination?

16. List the major enzymes thought to participate in recombination and describe their possible roles in the process. What steps are catalyzed by the recA enzyme of

E. coli? What evidence indicates that enzymes with similar activities occur in eukaryotes?

17. Outline the major differences between general or homologous recombination and site-specific recombination.

18. Outline the structure of the synaptonemal complex. What evidence indicates that the synaptonemal complex takes part in recombination? What are recombination nodules? What evidence indicates that recombination nodules are associated with sites of recombination? What are polycomplexes? What is the probable relationship of these structures to the synaptonemal complex?

19. At what stage of the cell cycle do cells apparently become committed to a meiotic division? What stages of meiosis are most greatly affected by mutations?

20. What major adaptations led to the appearance of meiosis? What is the significance of meiosis to sexually reproducing organisms?

Supplement 25-1
The Rolling Circle Mechanism of DNA Amplification

A specialized mechanism of DNA amplification, known as the *rolling circle mechanism*, is evidently the source of the extrachromosomal copies of rRNA genes produced during pachytene in many animal oocytes (Fig. 25-30). To start the mechanism a segment of rDNA is released from the NOR in some as yet unknown way as a covalently closed, circular molecule. Replication begins at a single-stranded nick introduced into the released circle (Fig. 25-30*a*), from which the free 5′ end begins to unwind and extend as a single-chain tail (Fig. 25-30*b*). As the tail extends, the circle rolls in the

direction of the arrow in Figure 25-30*b*, and replication begins on both single-stranded regions exposed by the unwinding (Fig. 25-30*c*). On the circle new DNA bases are added continuously to the exposed 3′ end, using the unnicked DNA chain of the circle as an endless template. On the tail new bases are polymerized into a complementary copy that lengthens progressively as the tail unwinds. These bases are added discontinuously in Okazaki fragments in the direction of the arrows. The Okazaki fragments are later closed by ligases into a continuous nucleotide chain (Fig. 25-30*d*).

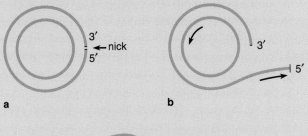

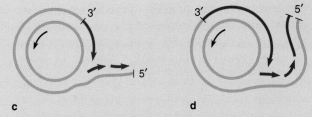

Figure 25-30 Amplification of rDNA sequences by the rolling circle mechanism (see text).

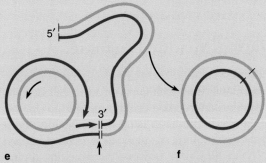

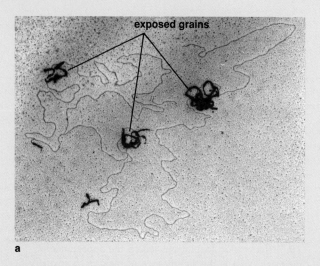
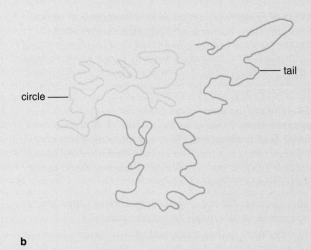

Figure 25-31 A rolling circle replicating rDNA, isolated from a *Xenopus* oocyte. **(a)** The labeled molecule, showing exposed grains over the circle and tail. **(b)** A tracing of the micrograph outlining the circle and tail. Courtesy of A. H. Bakken and the Histochemical Society, Inc.

One linear copy of the entire template DNA helix is completed each time the circle rolls through a turn. As the completed DNA copies extend into the tail, they are clipped off (Fig. 25-30*e*). The freed copies may remain linear or may roll into circles that are covalently sealed into closed, continuous DNA molecules (Fig. 25-30*f*). Amplification by the rolling circle mechanism may produce thousands of extra copies of the rRNA genes during oocyte development. In *Xenopus*, for example, this pattern of DNA amplification produces some 2500 extra rDNA copies.

J. Rochaix and his coworkers found strong evidence that amplification of rRNA genes in *Xenopus* takes place by the rolling circle mechanism. They exposed *Xenopus* oocytes to tritiated thymidine and, after allowing time for amplification to proceed, isolated the amplified rDNA from the nucleoplasm. The amplified rDNA proved to be in the form of circles, some with tails and some without. (Fig. 25-31 shows one of the tailed circles.) Autoradiographs revealed radioactivity in both

the circles and tails (as in Fig. 25-31), as expected if these configurations represent DNA replicating by the rolling circle mechanism.

Several mechanisms have been proposed as the source of the rDNA circles required as templates for the rolling circle mechanism. One maintains that an RNA copy of the rRNA genes is used as template for generation of a DNA copy by reverse transcription (see p. 742). The extrachromosomal copies produced by reverse transcription then roll into the circle used as template for rolling-circle replication. Another proposal holds that the initial circles are produced from an rDNA copy generated by extra activation of a replicon within the rRNA gene cluster. The extra activation produces a doubled segment from which one copy is released through excision and DNA repair. Still another proposed source of template circles is *replicative recombination*, a mechanism by which extra copies of transposable genetic elements are generated without loss of the original copy.

Supplement 25-2
Lampbrush Chromosomes and RNA Synthesis During Meiotic Prophase

During the diplotene stage of prophase I, in which meiotic cells in many species enter a period of growth, chromosomes partially or completely decondense, and all major types of RNA are transcribed. In most animal oocytes chromosomes become completely covered during this stage by loops extending outward from the

chromosome axes (Fig. 25-32). First noted in the 1880s, chromosomes in this configuration were called *lampbrush chromosomes* because of their resemblance to the brushes then used to clean lamp chimneys. Initially of interest because of their morphological peculiarities, and later because they were obviously a structural config-

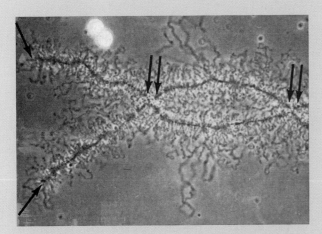

The Structure of Lampbrush Chromosomes

The loops of lampbrush chromosomes extend from the chromomeres in each chromatid of a tetrad. At any level along a tetrad there are therefore four loops, one for each chromatid. (Fig. 25-33 diagrams the loops formed by one of the two homologs.) A complete set of chromosomes has thousands of loops. Some loops in a set differ in size, thickness, or the structure of the coating that covers their surfaces. The tetrad forming a lampbrush chromosome is held together at widely spaced intervals by crossovers (double arrows in Fig. 25-32).

In the electron microscope the loops appear as a central, axial fiber covered by a matrix of smaller, projecting fibers (Fig. 25-34). The loop and matrix fibers appear very similar to transcribing ribosomal genes isolated for electron microscopy by the same techniques (see Fig. 15-50). Like the matrix fibers of active ribosomal genes, the fibers extending from lampbrush loops are conspicuously shorter at one end of the loop and become progressively longer toward the other end. Some loops contain rows of several coats of matrix fibers in tandem, each one with fibers that become progressively longer in one direction.

Cytochemical experiments by J. G. Gall, H. C. Macgregor, and H. G. Callan established that the loops of lampbrush chromosomes contain DNA, RNA, and proteins. The single fiber forming the central axis of the loops is broken only by deoxyribonucleases and is therefore evidently DNA. The matrix fibers extending from the DNA are digested by RNAses and proteinases, indicating that they are molecules of RNA in association with proteins. The RNA is labeled rapidly if nuclei containing lampbrush chromosomes are exposed to ra-

Figure 25-32 Part of a lampbrush chromosome isolated from the newt *Triturus*, as seen under the phase contrast light microscope. The arrows indicate the two chromosomes of the tetrad, held together at two points by crossovers (double arrows). Multiple paired loops extend from each chromosome; one loop of each pair extends from one chromatid of the homolog and the second loop from the other chromatid. × 400. Courtesy of J. G. Gall.

uration related to intense transcriptional activity, lampbrush chromosomes have been the subject of active research since their discovery.

Lampbrush chromosomes have been observed during diplotene in the developing oocytes of all animals except some insects and reptiles. In the males of some animal species, primarily among insects, a very limited region of the Y chromosome also extends into loops resembling those of lampbrush chromosomes.

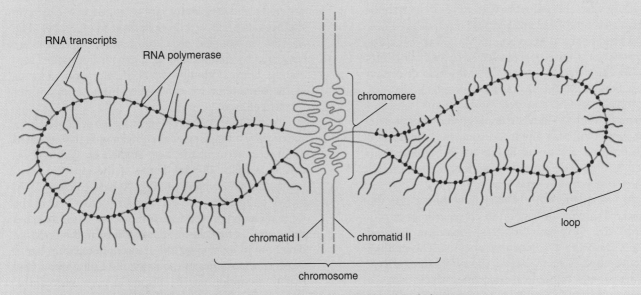

Figure 25-33 The relationships among loops, chromatids, chromomeres, and chromosomes in a lampbrush chromosome.

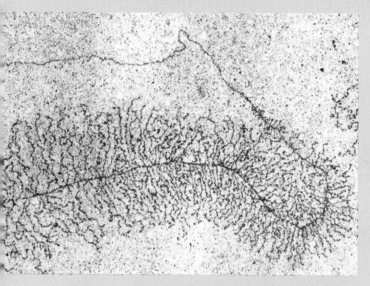

Figure 25-34 A single loop of a *Triturus* lampbrush chromosome, as seen in the electron microscope. The matrix fibers extending at right angles to the axis are probably mRNA molecules in the process of transcription. × 5,600. Courtesy of U. Sheer.

dioactive RNA precursors, indicating that the RNA is transcribed on the loops. These observations show that the loops contain genes intensely active in RNA transcription. The ends of the matrix coats in which the fibers are shortest are the sites where transcription is initiated. Loops containing several matrix coats in tandem evidently contain as many transcription units.

The Function of Lampbrush Chromosomes

Although the loops of lampbrush chromosomes are obviously engaged in RNA transcription, the functional significance of this intensive synthesis was uncertain for many years. Recent work with RNA-DNA hybridization (see p. 528) established that much of the RNA transcribed on the loops is pre-mRNA, which is quickly processed into mRNA. Transfer RNA and 5S rRNA can also be detected among the transcripts of certain loops. Some mRNAs, such as those encoding histones, ribosomal proteins, and yolk proteins, are translated into proteins in the oocyte cytoplasm during diplotene. Many other mRNAs, after processing and combination with proteins, are stored in the oocyte cytoplasm in inactive form, where they remain inactive until fertilization of the mature egg.

In one of the major studies establishing that the primary transcript of lampbrush chromosomes is pre-mRNA, M. Roshbash and his coworkers isolated mRNA from the cytoplasm of oocytes. This mRNA was then used as a template for the synthesis of cDNA by the reverse transcriptase enzyme, which makes a complementary DNA (cDNA) copy of RNA molecules (see p. 585). The cDNA synthesis was carried out in a medium containing tritiated thymidine, which made the

cDNA copy radioactive. When mixed with isolated lampbrush chromosomes under conditions promoting DNA rewinding, the cDNA preparation hybridized extensively with the loops of lampbrush chromosomes, marking them with radioactivity. The hybridization showed that the mRNAs from which the cDNA was copied are synthesized on the loops.

Although estimates vary, somewhere between about 5% and 20% of the total chromatin of the nucleus is believed to be active in transcription during the lampbrush stage. This is about the same as the amount of chromatin extended into lampbrush loops. The remaining 80% or more apparently remains condensed in inactive form in the chromosome axis. As many as 20,000 different messages can be detected in oocytes of some species, such as the amphibian *Xenopus*, during meiotic prophase. Presumably, most of these mRNAs are transcribed on the loops of lampbrush chromosomes during diplotene.

R. W. Old, Callan, and K. W. Gross directly identified the specific loops transcribing histone genes. In their experiments labeled DNA molecules containing histone genes were denatured into single-stranded form and hybridized with lampbrush chromosomes of the newt *Triturus* by the in situ technique (see p. 617). Approximately seven or eight loops hybridized with the histone DNA sequence, indicating that histone mRNAs, which are stored in high concentration in the oocyte cytoplasm, are transcribed in these locations. Specific loops responsible for 5S rRNA and tRNA transcription have also been identified by others using in situ hybridization.

Gall and his coworkers found that the histone genes of the newt *Notophthalmus* are transcribed in lampbrush chromosomes in a pattern very different from somatic transcription. Histone genes, each with its own promoter and terminator, are arranged in a cluster in the sequence H1 → H3 → H2B → H2A → H4 (Fig. 25-35a). In somatic cells the genes of a cluster are separated by spacers, and each gene is transcribed separately. Transcription begins at the promoter and ends at the terminator for each gene; the spacers are not transcribed.

Four clusters of histone genes, each separated from the next by a long segment containing highly repeated sequences, occur in the loop studied by Gall and his colleagues (Fig. 25-35b). Some of the repetitive segments contain as many as 200,000 base pairs of DNA. Genes of three of the clusters are transcribed in the same direction; in the fourth the sense chain switches sides, so that this cluster is transcribed in the opposite direction. In examining the transcripts made in lampbrush chromosomes of the newt, the Gall group found that transcription usually begins at sites upstream of any of the individual histone promoters and runs through the clusters without stopping at internal terminators. The transcripts include not only the coding

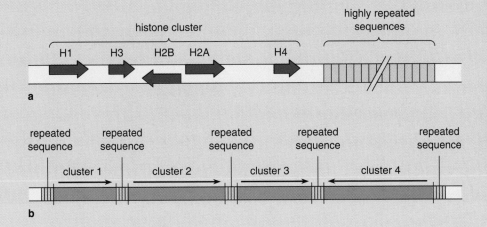

Figure 25-35 The arrangement of histone genes and gene clusters in the newt *Notophthalmus.* **(a)** The arrangement of genes in a histone cluster. Each gene is supplied with its own transcription promoter and terminator. **(b)** The arrangement of histone gene clusters in a lampbrush chromosome loop. Arrows indicate the direction of transcription for each cluster.

sequences but also any intervening spacers and even the highly repetitive sequences lying between the histone clusters. For unknown reasons, RNA polymerase II and its transcription factors evidently ignore the usual initiation and termination signals in the loop and transcribe segments that normally remain untranscribed in somatic cells. Presumably, the extremely long transcripts produced by this "readthrough" transcription are subsequently processed to release individual histone mRNAs.

Repetitive sequences have been detected among the transcripts of most other loops of lampbrush chromosomes, indicating that transcription in most or all loops probably follows the unusual readthrough pattern detected in the histone loop. This pattern of extended transcription produces extremely long RNA transcripts—in the newt studied by Gall and his colleagues, RNA transcripts up to 50 μm long, equivalent to 150,000 nucleotides of RNA, are common at the lampbrush stage. Why transcription follows this unusual pattern in lampbrush chromosomes is presently unknown.

Nucleolar Transcription During the Lampbrush Stage

The 28S, 18S, and 5.8S rRNA types transcribed in the large pre-rRNA genes of the nucleolar organizer (see p. 607) are conspicuously absent among the RNAs made in quantity on the loops of lampbrush chromosomes. This absence reflects the fact that the rDNA of nucleolar organizer segments, which encodes these rRNA types, is usually silent in transcription during the lampbrush stage. In most oocytes, as a consequence, there is little or no nucleolar material associated with the organizer sites. Instead, 28S, 18S, and 5.8S rRNA in nuclei with the lampbrush configuration are typically transcribed within accessory nucleoli (see p. 1056) that are freely suspended in the nucleoplasm at this time. As many as a thousand or more of these extra nucleoli may be present in oocyte nuclei. Each contains a core of

DNA in the form of a covalently closed circle, including repeats of the large pre-rRNA genes.

Synthesis of the extra rDNA associated with accessory nucleoli can be traced to pachytene. During this stage in *Xenopus* or *Bufo,* for example, a large quantity of rDNA is synthesized by DNA amplification (see p. 979) and released from the chromosomes. If the rDNA is isolated while its synthesis is in progress, many circles with tails (see Fig. 25-31) can be detected in the preparations. J. D. Rochaix and his coworkers found that exposing *Xenopus* oocytes to radioactive DNA precursors during rDNA amplification produces circles with labeled tails and other configurations expected if the extra replication proceeds by the *rolling circle* mechanism of DNA amplification (see Supplement 25-1).

Amplification of rDNA sequences has been detected during the early stages of meiotic prophase I in several groups in addition to amphibians, including molluscs, insects, and fish. D. J. Wolgemut and his coworkers found by in situ hybridization that limited amplification of the rDNA sequences also occurs in human oocytes, up to about four times the normal somatic rDNA level. Formation of the amplified rDNA in humans, which occurs during pachytene as in other species, is followed by the appearance of accessory nucleoli during diplotene, when transcription of the amplified rDNA takes place.

Research with lampbrush chromosomes thus establishes that this unusual configuration probably represents a specialization for the intensive transcription of mRNAs, 5S rRNA, and tRNAs required for growth of the oocyte cytoplasm during diplotene. At the close of diplotene, RNA transcription in both lampbrush chromosomes and accessory nuclei drops to unmeasurable levels. The loops on the lampbrush chromosomes retract into the chromosomes and the accessory nuclei disappear.

Suggestions for Further Reading: See p. 1084

26

GAMETOGENESIS AND FERTILIZATION

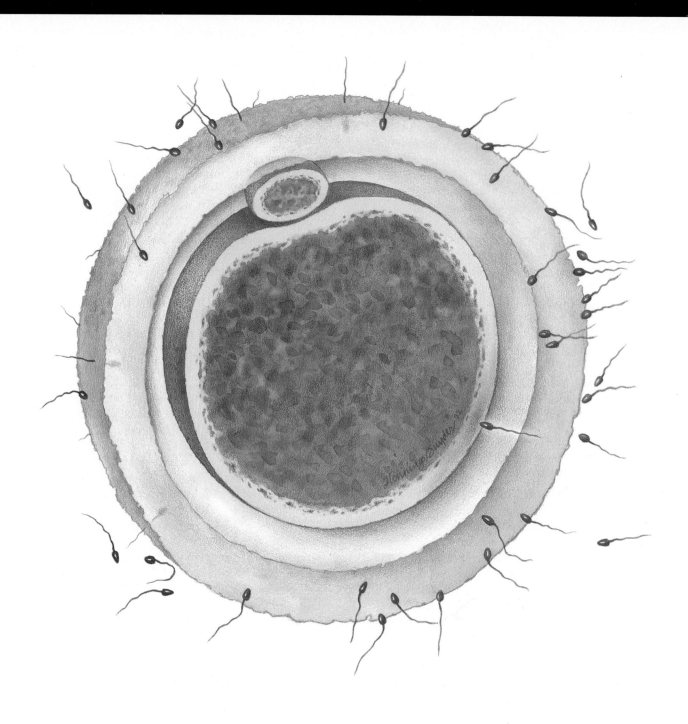

- *Spermiogenesis and sperm cell structure*
- *Oogenesis and egg structure* ▪ *RNA and protein synthesis during oogenesis* ▪ *Fertilization*
- *Egg activation and blocks to polyspermy*
- *Gametogenesis and fertilization in higher plants*
- *Histone replacement by protamines*

G ametes are specialized cells programmed for fertilization, the central event of sexual reproduction. Fertilization restores the diploid number of chromosomes and initiates embryonic development. Gametes of different major groups develop specialized structures and functions that enable them to join in fertilization. These specializations range in complexity from those of protozoa, algae, and some fungi, in which male and female gametes hardly differ from each other or from ordinary body cells, to those of animals, in which gametes are distinctive cell forms highly adapted for their roles in fertilization.

The structural and functional changes that produce gametes, whether limited or extensive, are known as *gametogenesis*. In animals gametogenesis occurs immediately after meiosis, with no intervening mitotic divisions before gamete development and maturation. In plants the haploid cells produced by meiosis grow by mitotic divisions to produce an alternate haploid generation. At some point individuals of the haploid generation produce gametes by differentiation of cells produced by mitotic divisions.

In plants and animals the male gametes are specialized for transport of the paternal set of chromosomes to the egg. In the swimming male gametes of animals and some plants, the cell nucleus is streamlined and reduced in volume, and the cytoplasm is limited to structures required for transport of the nucleus to the site of fertilization. In animal sperm cells the cytoplasm also contains structures specialized to penetrate the egg surface. The sperm cells of higher plants, contained within pollen, are similarly reduced in volume and associated with another cell that is specialized for transport. Once mature, the specialized male gametes of animals and plants remain almost totally inactive in RNA transcription and protein synthesis until the time of fertilization.

Although primarily nonmotile, the mature egg cells of animals and plants contain almost all of the cytoplasm to be contributed to the product of fertilization, the *zygote*. Packed into the cytoplasm of animal eggs are the nutrients required for some or all of embryonic development, instructions for development in the form of stored mRNAs, and the ribosomes and other factors needed to translate the messages into proteins. Also stored in the animal egg cytoplasm are large quantities of proteins, among them histones, tubulins, actins, and other molecules required for the rapid cell division characteristic of early embryos. Most animal eggs are surrounded by specialized *egg coats* that protect the egg cell and regulate fertilization. The cytoplasm of the egg cells of plants is not as complex, but the cytoplasm of the zygote, embryo, and adult plant cells is still derived largely or completely from the egg cytoplasm. Like male gametes the female gametes of animals and plants are largely inactive in RNA transcription and protein synthesis once mature and remain so until fertilization.

This chapter surveys the developmental changes producing male and female gametes in animals and plants. The discussion is followed by a description of the union of gametes in fertilization and the biochemical changes taking place as the egg is activated and progresses to the first mitotic division following fertilization.

DEVELOPMENT OF MOTILE MALE GAMETES

The motile male gametes of animals and plants consist of two major parts—the paternal set of chromosomes, packaged in the gamete nucleus, and a propulsion system consisting of one or more flagella. Most animal sperm cells also contain a packet of enzymes that are released to digest a pathway through the external coats of the egg. These specializations are produced by a series of developmental changes termed *spermiogenesis*.

Spermiogenesis in Animals

Spermiogenesis, which produces mature male gametes, begins after meiosis is complete. The changes include major alterations in both nucleus and cytoplasm. Before the onset of animal spermiogenesis, the products of meiosis, the *spermatids*, resemble ordinary body cells. Early spermatids contain all the major cytoplasmic organelles. The nucleus is surrounded by a nuclear envelope and appears similar internally to the nuclei of body cells except for the absence, in many species, of a recognizable nucleolus. As spermatids differentiate into mature sperm cells, each organelle system undergoes extensive changes.

The Structure of Mature Sperm Cells Mature sperm cells, or *spermatozoa* (singular = *spermatozoon*), of most animal species are greatly elongated cells consisting typically of a *head, middle piece,* and *tail* (Figs. 26-1 and 26-2). The nucleus forms almost all the mass of the head, which in different species may be flattened, greatly elongated, hooked, or spiraled. In the electron microscope the mature sperm nucleus appears uniformly dense in most species and shows no evidence of chromatin fibers or a nucleolus (Fig. 26-3). Although the

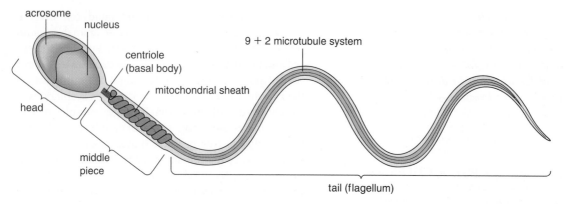

Figure 26-1 Typical structures of an animal sperm cell. Although one or two centrioles may be present at the base of the sperm tail, only one generates the microtubules of the tail.

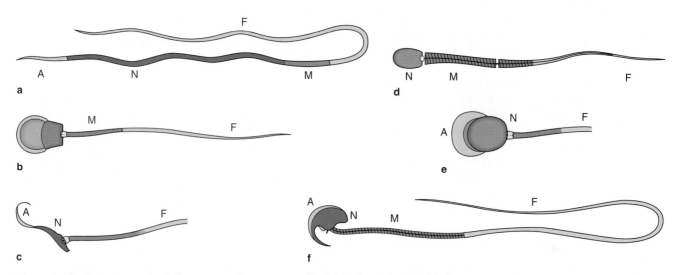

Figure 26-2 Some representative mammalian sperm cells. **(a)** Spiny anteater; **(b)** deer; **(c)** squirrel; **(d)** bat; **(e)** guinea pig; **(f)** mouse. A, acrosome; F, flagellum; M, middle piece; N, nucleus.

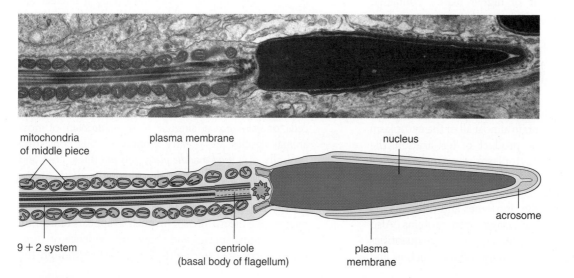

Figure 26-3 A fully condensed nucleus in a marmoset sperm cell. × 15,000. Courtesy of J. B. Rattner, B. R. Brinkley, and Academic Press, Inc., from *J. Ultrastr. Res.* 32:316 (1970).

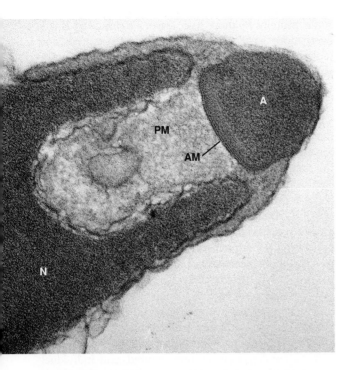

Figure 26-4 The acrosome (A) and periacrosomal material (PM) of the sea urchin *Strongylocentrotus*. The periacrosomal material, containing actin as the major constituent, polymerizes into a fiber that shoots outward as an acrosomal filament during fertilization. AM, acrosomal membrane; N, nucleus. ×92,000. Courtesy of V. D. Vacquier, from *Proc. Nat. Acad. Sci.* 74:2456 (1977).

sperm nucleus remains surrounded by a nuclear envelope, pore complexes are usually absent from the envelope except for a restricted region near the base of the tail. Lamins (see p. 560) also appear to be absent from the inner surface of the nuclear envelope in mature sperm cells.

Only a thin layer of cytoplasm remains around the nucleus in the sperm head. None of the usual cytoplasmic structures, such as mitochondria, ER, Golgi complex, and ribosomes, are present in this layer. At the anterior end of the sperm head, often covering this end of the nucleus like a cap, is the *acrosome*, a specialized secretion vesicle enclosed by a single, continuous membrane (see Figs. 26-1 and 26-3). This vesicle, which lies between the plasma membrane and the nucleus, contains the hydrolytic enzymes breaking down the surface coats of mature eggs. All animal spermatozoa have an acrosome except those of teleost fishes and a few arthropods.

Among the acrosomal enzymes in mammalian sperm are *hyaluronidase* and *zona lysin*. Hyaluronidase specifically attacks *hyaluronic acid,* a constituent of the extracellular matrix of animal cells (see p. 280). Zona lysin is specific in activity for the *zona pellucida,* the primary extracellular coat of mammalian eggs (see be-

low). Mammalian acrosomes contain other enzymes that hydrolyze components of the extracellular matrix and egg coats, including acetylglucosaminidase, sulfatase, neuraminidase, proteinases such as acrosin and collagenase, and phospholipase A. The acrosomes of nonmammalian sperm cells contain a similar spectrum of hydrolytic enzymes.

In some spermatozoa a layer of *periacrosomal material* surrounds the acrosome and separates the acrosomal membrane from the adjacent plasma and nuclear membranes (Fig. 26-4). In many marine invertebrates the periacrosomal material includes a concentrated deposit of actin in either disassembled or fibrous form. This actin deposit may be so large that it fills a deep pit in either the surface of the nucleus (as in Fig. 26-4) or the base of the acrosome. During fertilization the actin deposit in these sperm polymerizes into a long fiber that springs from the anterior end of the sperm cell and aids the attachment of sperm cells to the egg and penetration of the egg coats (see below).

The plasma membrane surrounds the nucleus and acrosome and extends posteriorly to cover the middle piece and tail. The middle piece consists primarily of a sheath of mitochondria surrounding the axis of the tail just behind the head. Frequently, the mitochondria of the middle piece are highly modified; in the spermatozoa of some organisms, such as butterflies and moths, they are so altered that they can be identified as mitochondria only by tracing their developmental origins during spermiogenesis. One or two centrioles and a variety of structures associated with the attachment of the tail to the base of the nucleus are also usually present in the middle piece. The presence or absence of centrioles, and their number, depend on whether the centriole delivered to the spermatid by the second meiotic division (see p. 1058) remains single, divides, or degenerates during spermiogenesis. Even if a pair of centrioles is present, in most animal spermatozoa a single centriole generates the microtubules forming the axis of the tail (see Fig. 26-3). After giving rise to the tail microtubules, this centriole may persist as the *basal body* of the tail or may degenerate.

The system of microtubules in the tail produces the whiplike motions that propel the sperm cell through a fluid medium. In most animal spermatozoa the microtubules of the tail are arranged in the 9 + 2 system typical of flagella (see p. 432). The microtubule system extends from the basal body, through the middle piece, to the tip of the tail. In most animals the tail microtubules are surrounded only by a sparse sheath of cytoplasm enclosed by the plasma membrane. Exceptions to the 9 + 2 pattern of microtubules, and accessory fibers in addition to the 9 + 2 system, occur in the sperm tails of some animal species, including mammals.

The accessory fibers of mammalian sperm tails consist of a thick, dense structure outside each peripheral doublet (see Fig. 11-26). The nine accessory fibers are unequal in size and shape, giving the mammalian sperm tail an asymmetrical appearance in cross section. Isolation and analysis of the accessory fibers from rat, bull, and human sperm tails show that they all contain five major types of polypeptides, with molecular weights of about 14,000, 28,000, 38,000, 44,000, and 87,000. Because none of these proteins resembles tubulin or actin, or has properties similar to microtubule- or microfilament-dependent motors such as dynein, kinesin, or myosin, the accessory fibers are thought to be nonmotile reinforcements that provide elasticity to mammalian sperm tails. This elasticity may contribute in some way to movement of mammalian spermatozoa through the relatively viscous fluids of the female genital tract.

The basic arrangement of the sperm cell into a head, middle piece, and tail is typical of almost all animals. Nonflagellate sperm occur in a few species of nematodes, some crustaceans, a few arachnids (mites), and some myriopods. Usually, absence of the tail in these sperm is correlated with absence of the organelles of the middle piece, including mitochondria and centrioles. Multiple sperm tails appear regularly in a very few animal species, such as some flatworms. Other less extensive modifications of the basic arrangement are found in copepods, a number of barnacle species, and a few insects and annelids.

Sperm Cell Development The organelles of mature sperm cells develop during a series of highly programmed rearrangements initiated at the close of meiosis. Mitochondria and the single centriole delivered to the spermatid move to the side of the nucleus that will form the posterior end of the sperm head. The Golgi complex and much of the ER move temporarily to the anterior end of the nucleus, where they become involved in formation of the acrosome (see below). After acrosome formation the ER and Golgi complex move to the posterior end of the nucleus. Later, as spermiogenesis becomes complete, they are eliminated along with most of the rest of the cytoplasm of the spermatid in a bleb that pinches off near the posterior end of the nucleus.

During spermiogenesis the nucleus in most species undergoes progressive elongation and reduction in volume. Within the nucleus chromatin coalesces into coarse granules or flattened sheets that eventually fuse into a uniformly dense mass. At the same time the extrachromosomal material of the nucleus is eliminated, often in a vesicle that buds from the nuclear envelope into the cytoplasm. Most of this material is expelled from the cell in the bleb that pinches off from the sperm cell late in its development.

The structural changes in the nucleus are accompanied in most animal species by partial or complete replacement of the histone and nonhistone chromo-

somal proteins by more basic *sperm histones* or by small, even more basic proteins called *protamines.* Histone replacement by sperm histones or protamines is probably a major factor in condensation of the sperm head, a change that reduces the energy required to propel sperm cells through a watery medium. Combination with protamines may also lock DNA in a tight linkage that protects it from enzymatic attack during transit to the egg and during early postfertilization stages, when the sperm chromatin is suspended in the egg cytoplasm (see below). In addition combination with sperm histones or protamines may be part of the changes shutting down RNA transcription in the sperm nucleus and arresting virtually all biochemical reactions except those generating tail movement. (Details of protamine structure and the replacement of histones by protamines are presented in Supplement 26-1.)

The degree of nuclear condensation varies in different major animal groups. In most, condensation produces a uniformly dense, compact nucleus such as that shown in Figure 26-3. In a few groups, including sea urchins, condensation is more limited, and individual chromatin fibers, although larger in diameter than those of early spermatids, can still be distinguished. In crustaceans, the nuclear material remains uncondensed and, in fact, appears to become even more dispersed than in body cells of the same species. No matter what structural changes take place, the nucleus becomes totally inert in RNA transcription and processing as sperm cells mature.

The acrosome first appears as a collection of vesicles arising from the Golgi complex at the anterior end of the developing sperm cell. These vesicles gradually fuse into a single, large vesicle that attaches to the anterior end of the nuclear envelope. Subsequently, the vesicle undergoes internal differentiation, often developing extensive substructure. When its development is complete, the acrosome remains in inert form at the tip of the sperm nucleus until fertilization, when it releases its enzymes to the exterior by a process closely resembling exocytosis in somatic cells (see p. 856).

As spermiogenesis proceeds, the 9 + 2 system of microtubules forming the tail axis grows outward from the distal centriole (for details of flagellar generation, see p. 437). As the microtubules elongate, mitochondria take their position near the base of the nucleus. Upon reaching this position, they align in a sheath surrounding the base of the tail to form the middle piece. Depending on the species, the mitochondria may be modified by fusion of the inner and outer mitochondrial membranes, alteration or disappearance of the inner mitochondrial membranes, or deposition of dense, crystalline material. The result in these cases, termed the *mitochondrial derivative,* may bear little resemblance to typical somatic mitochondria (Fig. 26-5). In mammalian sperm the mitochondria, which undergo relatively little internal alteration, line up end to end in a tight spiral

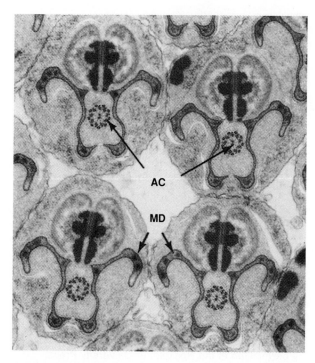

Figure 26-5 Mitochondrial derivatives (MD) in cross section in sperm cells of the plant hopper *Acanalonia*. AC, axial complex (9 + 2 system) of microtubules in the sperm tails. ×21,000. Courtesy of D. M. Phillips, from *J. Cell Biol.* 44:243 (1970), by permission of the Rockefeller University Press.

around the microtubules of the middle piece (as in Figs. 26-1 and 26-3).

During spermiogenesis in both vertebrates and invertebrates a great variety of proteins and glycoproteins is synthesized and inserted into the plasma membranes of developing sperm cells. As spermiogenesis proceeds, many of these molecules become restricted in distribution on the sperm cell surface. This localized distribution, evidently maintained by a cytoskeletal system just beneath the membrane, produces distinct subregions of the sperm cell plasma membrane that are undoubtedly specialized in function. What most of these functions might be is uncertain at present. However, the plasma membrane over the acrosome is thought to contain proteins or glycoproteins that recognize the external egg coats and trigger the release of acrosomal enzymes. The region just behind the acrosome in mammals is believed to contain molecules that promote fusion of the sperm and egg cell plasma membranes (see below).

RNA Transcription and Protein Synthesis in Animal Spermiogenesis The highly specialized structures of sperm cells are reflected in a large group of specific polypeptides that are synthesized during and before spermiogenesis. Some of these polypeptides persist and form parts of the mature sperm—acrosomal enzymes, specific proteins and glycoproteins of the plasma membrane, and protamines are examples of proteins of this type. Other proteins appear transiently during sperm cell development; these proteins evidently contribute to structural changes during spermiogenesis but are not retained in mature sperm.

In cell lines giving rise to spermatozoa, the mRNAs encoding these proteins are transcribed during two major periods: the diplotene stage of meiotic prophase I (see p. 1056) and after meiosis, during early stages of spermiogenesis. The mRNAs synthesized in these two periods depend on different genetic complements because during meiotic prophase I the nuclei undergoing transcription are still diploid and contain all the alleles typical of body cells of the same individual. During the second period early in spermiogenesis the nuclei are haploid and genetically different because of recombination and independent segregation of maternal and paternal chromatids during meiosis. As a result, the mRNAs and proteins synthesized in the two stages may be encoded in different genetic alleles, not only between the first and second peaks but also among the four haploid products of each meiotic division.

Some of the mRNAs transcribed during the first peak in meiotic prophase I are translated immediately into proteins in the cytoplasm. Others are stored in inactive form until after meiosis, when spermiogenesis is under way. In fishes, for example, protamine mRNAs, although transcribed early during meiosis, are stored in the cytoplasm in inactive form until the onset of spermiogenesis.

The transcription of rRNAs follows a similar pattern of two peaks, one during meiotic prophase I and one early in spermiogenesis. Transcription of all RNA types ceases as histones are replaced by protamines and nuclear condensation becomes complete. In mammals the shutdown in RNA transcription is correlated with increased methylation of cytosines in the DNA. This suggests that in mammals cessation of transcription as sperm cells mature may be regulated by methylation as well as by histone replacement.

By the time spermiogenesis is complete, most of the available DNA sites are fully methylated except for a few in the control regions of genes that become active in very early stages of embryonic development. These undermethylated sites remain hypersensitive to endonuclease digestion (see p. 628). Evidently, although completely arrested in RNA transcription, certain genes in sperm nuclei are held in a "ready" state for immediate activation after fertilization.

The assembly of proteins taking part in spermiogenesis or incorporated into mature sperm cells is initiated during diplotene and continues throughout most of spermiogenesis. Protein synthesis halts when ribosomes, ER, and Golgi complex are expelled from the sperm cell as sperm development nears completion.

Spermiogenesis in Plants

Many algae, bryophytes, mosses, ferns, and primitive gymnosperms produce swimming gametes with structural characteristics that are remarkably similar to those of animal spermatozoa. The nucleus of these cells is elongated and condensed, and in at least some species somatic histones are replaced by protaminelike proteins. The flagella propelling plant sperm cells, which range in number from two to hundreds in different species, contain a typical 9 + 2 system of microtubules. These flagella are generated by centrioles that are closely similar to their animal counterparts. In most plants the centrioles arise de novo in cell lines destined to give rise to gametes, apparently without descent from pre-existing centrioles (see p. 905).

The sperm cells of lower plants do not develop from the immediate products of meiosis as in animals. Instead, postmeiotic cells differentiate into haploid *spores* that, on germination, produce the gametophyte generation of the plant. The haploid gametophyte grows by mitotic divisions; at some point groups of cells differentiate into male gametes. Because this differentiation follows mitosis instead of meiosis, all sperm and eggs of an individual gametophyte are genetically identical except for chance differences that might arise from mutations.

The development of sperm in the alga *Nitella*, investigated by F. R. Turner, provides an example of the processes occurring in lower plants. In *Nitella* the sperm cell body is divided into three regions (Fig. 26-6). About a fourth of the cell at the anterior end is occupied by a mass of mitochondria; the nucleus takes up the middle half, and a group of plastids is located at the posterior end. The mature *Nitella* sperm cell has two flagella, which attach to the anterior end of the cell and extend posteriorly.

The development of *Nitella* sperm cells resembles animal spermiogenesis in many respects. The cell destined to form a sperm cell is undifferentiated at the beginning of spermiogenesis. As development proceeds, the nucleus elongates, and the chromatin gradually·condenses into a densely packed mass similar in appearance to nuclei in animal sperm cells. The mitochondria and plastids separate and move to opposite ends of the elongating nucleus. Two centrioles are present in the cell giving rise to the *Nitella* sperm; these take up a position at the anterior end of the nucleus and generate the 9 + 2 microtubule systems of the two flagella. At maturity the sperm cell becomes motile and is released through a pore in the cell wall that originally enclosed the developing spermatid.

Sperm cell development follows generally similar pathways in other plants with swimming male gametes. One interesting difference is noted in some plants producing multiflagellate sperm. In the primitive gymnosperm *Marsilea*, for example, centrioles first appear in

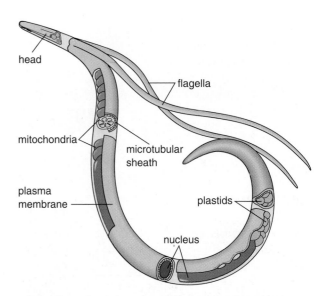

Figure 26-6 A sperm cell of the alga *Nitella.* The microtubular sheath may participate in elongation and reinforcement of the nucleus. Redrawn from an original courtesy of F. R. Turner, from *J. Cell Biol.* 37:370 (1968), by permission of the Rockefeller University Press.

two large masses in cells destined to differentiate into sperm cells, one at either side of the nucleus (see Fig. 21-22). As nuclear elongation commences, the centriole masses gradually separate, and individual centrioles, numbering between 100 and 150, take up positions around the nucleus. When nuclear elongation is complete, each centriole gives rise to a flagellum, producing at maturity a highly multiflagellate sperm cell (Fig. 26-7).

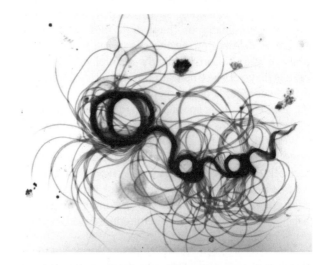

Figure 26-7 A multiflagellate, spiral-shaped sperm cell of the primitive gymnosperm *Marsilea.* ×2,400. Courtesy of I. Mizukami and J. G. Gall, from *J. Cell Biol.* 29:97 (1966), by permission of the Rockefeller University Press.

The completion of spermiogenesis in animals and in plants with motile sperm produces a cell specialized for transport of the paternal chromosomes to the egg. The mass of the sperm cell is greatly reduced, and the nuclear contents are packaged in a form that is transportable with a minimum expenditure of energy. Little remains in the cytoplasm except the propulsion mechanism and, in animal sperm, the acrosome.

DEVELOPMENT OF FEMALE GAMETES

The mature female gamete in animals and most lower plants, known as the *egg, egg cell,* or *ovum,* is nonmotile. Rather than motility, the specializations of the egg meet requirements for fertilization and early embryonic development. The developmental changes accomplishing differentiation of the egg cell are termed *oogenesis;* during development the female gamete is called an *oocyte.*

Animal Oogenesis

Mature eggs are easily the largest and most complex cells produced by animals. The many specializations of mature animal eggs include (1) a mechanism for attachment of sperm cells and entry of the sperm nucleus; (2) nutrients required for part or all of embryonic development; (3) stored directions, primarily in the form of mRNA, required for early stages in embryonic development; (4) stored proteins such as histones, tubulins, and actins required for early embryonic cell division; (5) complete machinery for synthesizing proteins, including ribosomes, ER, and Golgi complex; (6) all other cytoplasmic structures of the future zygote except, in some species, asters and centrioles; (7) egg coats, which function in fertilization, protect the egg cell, and protect the embryo after fertilization in some species; and (8) *cortical granules,* which contribute to the outer egg coats during fertilization and, in many animal species, induce changes that block additional sperm cells from fertilizing the egg.

Animal oogenesis differs fundamentally from spermiogenesis in that only one of the four haploid products of meiosis becomes a mature egg cell. The other cells produced by meiotic divisions, called *polar bodies,* eventually degenerate in most species and do not contribute to fertilization or embryonic development.

Development of the Egg Cytoplasm Most of the increase in the oocyte mass that occurs during oogenesis involves the cytoplasm and takes place during the diplotene stage of meiotic prophase I. During this period the nucleus and cytoplasm synthesize a variety of substances to be stored in the mature egg. Great quantities of RNA, including mRNA, rRNA, and tRNA, are transcribed and become concentrated in the oocyte cyto-

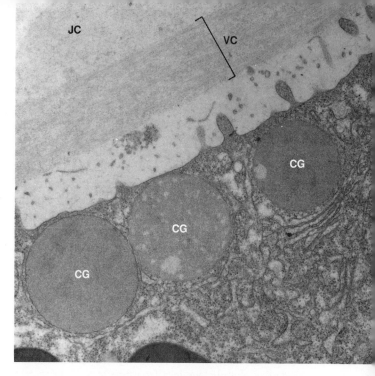

Figure 26-8 Cortical granules (CG) in a frog oocyte. VC, vitelline coat; JC, innermost layer of the jelly coat. Courtesy of R. D. Grey.

plasm. The rRNA is assembled with ribosomal proteins into large numbers of ribosomes. Some of the mRNAs are translated immediately into proteins; others are stored in inactive form in the cytoplasm. Lipids and polysaccharides are also synthesized in quantity and stored in the cytoplasm. Much of the stored lipid and protein is packed into membrane-bound structures called *yolk granules,* which fill much of the cytoplasm of all animal eggs except those of mammals.

The intense synthetic activity of developing oocytes is marked by proliferation of the various cytoplasmic organelles, including the ER, Golgi complex, and mitochondria. So many mitochondria are present in maturing oocytes, in fact, that the greatest proportion of oocyte DNA is mitochondrial rather than nuclear—in amphibian oocytes, for example, mitochondrial DNA makes up more than 90% of the total cellular DNA.

Cortical granules (Fig. 26-8), which are highly variable in both internal contents and structure among different species, are synthesized during oocyte development by pathways typical for secretory vesicles. These structures are bounded by a single membrane and contain a mixture of glycoproteins, muco- and other polysaccharides, and various enzymes, including proteinases with properties similar to trypsin. Mature cortical granules pack into a dense layer just under the plasma membrane. Sea urchin eggs, for example, contain an estimated 15,000 to 18,000 cortical granules; in mammalian eggs cortical granules number in the vicinity of 4000.

Growth of the oocyte cytoplasm in many animal species also results from the entry of proteins, lipids, and carbohydrates from outside the cell. Some of these materials are absorbed from the surrounding extracellular fluids by endocytosis. Other materials, including even intact yolk bodies and mitochondria, may be delivered to the oocyte cytoplasm from surrounding *follicle* or *nurse cells* through direct cytoplasmic connections.

The final outcome of diplotene synthesis is growth of the oocyte, which in some animals produces a cell of relatively immense size. In birds and reptiles the diameter of mature egg cells (the egg yolk) may reach several centimeters. In species with extremely large eggs, as much as 95% of the cytoplasmic volume is occupied by yolk granules. Although the nucleus also enlarges to some extent, it remains microscopic, or nearly so, even in eggs that attain very large dimensions. The eggs of humans and other mammals remain relatively small—a mature human oocyte is only about 100 μm in diameter, approximately one-quarter the diameter of the nucleus in a frog egg.

Oocyte growth during meiotic prophase may last from weeks to many months depending on the species. As the oocyte attains its final dimensions, the rate of RNA transcription, protein synthesis, and respiration falls to low levels.

Programmed Meiotic Arrest During Oocyte Development In some oocytes meiosis arrests at diplotene and does not restart until fertilization. Other oocytes arrest at metaphase I and remain at this stage until fertilization. Arrest at metaphase I is typical of many invertebrates, including arthropods, tunicates, and some annelids and molluscs. Other eggs, including those of most vertebrates, are blocked at metaphase II. In only a few groups, such as sea urchins, is meiosis completed before eggs are released. In mammals oocytes arrest initially at late diplotene within a few weeks after birth of the female and remain arrested at this stage until the female is sexually mature. At this time one or a few oocytes break arrest periodically and advance to metaphase II of meiosis. They are released from the ovary as mature eggs at this time and do not complete meiosis until fertilization.

Recent research has begun to unravel the molecular events underlying meiotic arrest in oocytes. Some years ago Y. Masui and C. L. Markert discovered a cytoplasmic activity, which they termed *cytostatic factor* (CSF). The factor appeared to be responsible for holding *Xenopus* eggs at meiotic metaphase II. Cytoplasm extracted from the arrested *Xenopus* eggs, when injected into developing frog embryos, could induce arrest of dividing cells at metaphase. CSF proved difficult to purify, however, and as yet has not been directly identified and characterized except for a notable sensitivity to Ca^{2+}; elevated Ca^{2+} concentrations lead to rapid and complete breakdown of the factor.

Recently, CSF was provisionally identified through an unexpected route, an investigation by S. Sagata and N. Watanabe, G. F. Van de Woude and Y. Ikaura, and others of the product of *mos*, a gene that in faulty form is an oncogene associated with certain forms of cancer (see p. 932). The protein encoded in *mos*, known as $p39^{mos}$, is normally synthesized only during oogenesis and spermiogenesis. In *Xenopus* oocytes it is present in significant concentration only during metaphase arrest. Immediately after fertilization, $p39^{mos}$ breaks down and disappears, evidently in response to a sudden increase in cytoplasmic Ca^{2+} concentration, which is one of the initial responses of the egg to fertilization. The investigators working with $p39^{mos}$ found that, if injected into early frog embryos, the protein duplicated the effects of CSF—it arrested dividing cells at metaphase. Further, antibodies against $p39^{mos}$, if injected into frog oocytes, destroyed CSF activity and broke meiotic arrest. These and other results make it seem likely that CSF and $p39^{mos}$ are one and the same or that $p39^{mos}$, which acts as a serine-threonine protein kinase, is a component of CSF.

Although the molecular effects of CSF/$p39^{mos}$ are unknown, it may work by inhibiting breakdown of cyclin, one of the two proteins of MPF, the factor that triggers entry of cells into division (see p. 926). In dividing cells cyclin breaks down as cells make the transition from metaphase to anaphase; this breakdown is necessary for completion of mitosis. Perhaps the CSF/$p39^{mos}$ protein inhibits the protease that hydrolyzes cyclin, thereby keeping this subunit of MPF intact and preventing the initiation of anaphase. At fertilization the sudden increase in Ca^{2+} activates an enzyme that breaks down CSF/$p39^{mos}$, releasing the inhibition of the protease attacking cyclin. This leads to cyclin breakdown and resumption of the meiotic divisions. The same or a similar mechanism may be generally responsible for meiotic arrest at metaphase I or II in animal oocytes.

The Completion of Meiosis No matter when meiosis arrests and resumes, the pattern of divisions after metaphase I is essentially the same in all animal oocytes or eggs. The nucleus breaks down, and the chromosomes and spindle take up a position near the plasma membrane. At anaphase I and telophase I, the accompanying cytoplasmic division is unequal, separating one of the two prophase II nuclei into a small, abortive cell, the first polar body (Fig. 26-9). The second cytoplasmic division at anaphase II and telophase II is also unequal, producing a second polar body. During the second division the first polar body either disintegrates, remains quiescent, or also divides, depending on the species. The result, in various animals, is a single large cell containing the egg nucleus and the majority of the cytoplasm, with one, two, or three polar bodies attached at one side. In the mature egg the nucleus is termed the *female pronucleus*.

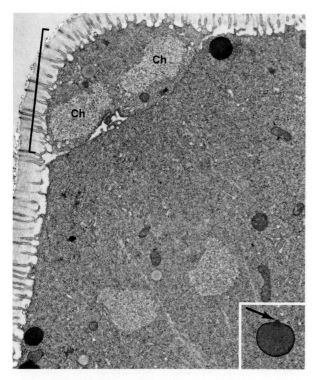

Figure 26-9 A *Spisula* (surf clam) egg with first polar body (bracket in main figure; arrow in inset). Ch, chromosomal material from the first meiotic division. Main figure ×8,800; inset ×175. Courtesy of F. J. Longo, from *J. Ultrastr. Res.* 33:495 (1970).

Polar bodies, which in most species are nonfunctional, abortive cells, vary considerably in degree of development. In mammals the first polar body fails to divide during the second meiotic division of the oocyte and soon disintegrates, leaving the egg with one polar body at maturity. The polar bodies of mammalian eggs are relatively large and may represent up to 1/20 of the volume of the mature oocyte. In a few groups, such as insects and crustaceans, no distinct polar bodies are formed. Instead, polar chromosomes are enclosed only by nuclear envelopes, forming polar nuclei that remain nonfunctional in the peripheral egg cytoplasm.

Polar bodies do function in a few insects. In some, fertilization does not occur; instead, a polar nucleus fuses with the egg cell nucleus to trigger a form of parthenogenetic development. In one rather curious instance, in coccid insects, the polar nuclei regularly become infected by a fungus and remain functional in a symbiotic role.

RNA Transcription and Protein Synthesis We have noted that the intensive transcription carried out during oocyte development in meiotic prophase I includes production of all major RNA types. Thousands of different mRNAs are assembled at this time—estimates run from as few as 1000 to 2000 in *Drosophila* to as many as 20,000 in *Xenopus*. Some types are transcribed in huge quantities. In sea urchin oocytes, for example, each of the H2A, H2B, H3, and H4 histone mRNAs is transcribed into a million copies; H1 mRNA transcripts number about 100,000. All together the histone mRNAs alone account for some 10% to 15% of the total mRNA content of mature sea urchin eggs. The total number of mRNAs of all kinds may be very large; in *Drosophila*, in which mature eggs are small compared to those of many animals, the total mRNA complement transcribed during oocyte development is estimated to be about a billion molecules.

Much of the mRNA stored in oocytes remains unprocessed. The long, unprocessed mRNA-containing transcripts resemble pre-mRNAs (see p. 585) because they contain introns that have not yet been spliced out. However, they also contain sequences that are normally never transcribed in somatic cells, such as intragenic spacers, extensive sequences lying downstream of somatic transcription terminators, and highly repeated sequences (see Supplement 25-2). The mRNAs that are translated immediately during oocyte development are fully processed.

Other mRNAs are fully processed but remain inactive until fertilization or later stages in embryonic development. One factor contributing to this inactivity is combination with "masking" proteins. T. Hunt and his coworkers, for example, found that most egg mRNAs were inactive when added to a cell-free system containing ribosomes and all factors required for protein synthesis. The mRNAs could be activated, however, by exposure to salt at concentrations high enough to remove bound proteins.

Other experiments indicate that specific sequences in 5' or 3' untranslated regions are involved in the inactivity of stored mRNAs, probably through combination with masking proteins. In one experiment implicating the untranslated regions, P. Mariottini and F. Amaldi compared two mRNAs stored in inactive form in amphibian eggs. One of the mRNAs, encoding *chloramphenicol acetyltransferase* (*CAT*), becomes active almost immediately after fertilization. The second, encoding ribosomal protein S19, remains inactive until the tailbud stage of embryonic development. Mariottini and Amaldi produced a hybrid gene in which the code for the 5' untranslated region of the S19 mRNA was grafted to the coding and 3' untranslated regions of the *CAT* gene. When introduced into oocytes, the resulting hybrid mRNA was not translated until the tailbud stage.

The mRNAs stored in the egg cytoplasm provide the future embryo with the capacity to synthesize proteins needed in quantity during the early stages of development. Other stored mRNAs represent information required for early embryonic development. Because the genes from which these mRNAs are copied during meiotic prophase constitute the diploid, maternal set, they may represent alleles not present in the mature haploid egg cell or in the embryo after fertilization.

Ribosomal RNAs in most species are synthesized in even greater quantities during oogenesis than mRNAs. In sea urchins, for example, about 85% of the total RNA transcribed during oocyte development is ribosomal. Most of this rRNA is assembled into ribosomes that remain inactive in protein synthesis—in the sea urchin less than 1% of the ribosomes become immediately active in translation in the oocyte cytoplasm. The stored ribosomes provide the fertilized egg and early embryo with the capacity for intensive protein synthesis until ribosome assembly begins in the embryo itself. Research by W. M. Worthington and E. Z. Baum and Amaldi and his colleagues, for example, revealed that mature *Xenopus* eggs contain approximately 1 trillion ribosomes, enough for 1 million embryonic cells. *Xenopus* egg cytoplasm also contains large quantities of untranslated mRNAs encoding ribosomal proteins. Like the mRNAs transcribed during oogenesis, the rRNAs of the oocyte cytoplasm are products of the maternal genome and thus may differ from equivalent molecules synthesized later in the embryo.

We have already noted that proteins required for rapid cell division, such as histones, actin, and tubulin, are synthesized in quantity and stored in the oocyte cytoplasm during meiotic prophase. Many other proteins are also made in quantity in developing oocytes. Some of these proteins, such as the enzymes catalyzing transcription and oxidative metabolism, become active in the synthetic processes of the oocyte, and others are stored in inactive form in the cytoplasm. Proteins forming parts of cortical granules, yolk granules, and egg coats are also assembled in large quantities during meiotic prophase I.

The Synthesis of Yolk and Other Stored Nutrients Yolk is a complex mixture of fats, carbohydrates, and proteins that serves as a source of energy and raw materials for part or all of embryonic development. Yolk is concentrated in the yolk granules or bodies, which may be produced by one or more of several pathways during oogenesis. In the oocytes of many animal species, particularly those of invertebrates, yolk is produced primarily through the same pathways as secretory proteins in somatic cells. The mRNAs for these yolk substances are translated on ribosomes attached to the ER; after synthesis the proteins are enclosed in small vesicles that bud off from the ER. These vesicles may contribute directly to yolk granule formation or may first be routed to the Golgi complex for modification of the enclosed proteins by addition of carbohydrate or lipid groups. In either case the vesicles fuse to form finished yolk granules (Fig. 26-10).

Yolk granules may also form through materials entering the developing oocyte from outside by endocytosis (see p. 857). The substances taken in by en-

docytosis are secreted by follicle or nurse cells or are carried by the circulatory system from more distant secretory cells, frequently located in the liver, to the site of oogenesis.

An additional route for the deposition of yolk is the unexpected and little understood formation of yolk granules within mitochondria (see Fig. 9-8d). In amphibians, in which this process has been most clearly demonstrated, the yolk material is deposited between the inner and outer mitochondrial membranes but apparently not in the matrix. As deposition proceeds, cristae gradually disappear, and the mitochondrion is transformed into a yolk granule. Among the many unanswered questions about this pattern of yolk development are the role of mitochondria in the process and the origin of the yolk material deposited in the intermembrane space.

Despite considerable variation in origin and chemical makeup of the enclosed material, the substructure of yolk granules is remarkably uniform in different animal species. The granules usually contain a dense, crystalline protein *core* that at higher magnification shows a regular periodicity (see Fig. 26-10). The core, often angular in outline, fills most of the yolk granule. A small amount of less structured *matrix* is usually present between the core and the surrounding membrane. Yolk granules of this type are observed in most animals except mammals, teleost fishes, and birds.

In egg-laying vertebrates much of the yolk is synthesized by liver cells in the form of a large, 170,000-dalton precursor, *vitellogenin*. Vitellogenin is secreted by liver cells into the bloodstream, in which it is carried to the developing oocytes. After endocytotic uptake by oocytes, vitellogenin is split and processed into three smaller proteins: *phosvitin*, a phosphorus-containing protein, and two lipid-containing proteins, *lipovitellins* I and II. These become the primary substances stored as yolk.

Mammalian oocytes synthesize relatively little yolk. For example, less than 5% of the cytoplasm of human eggs is taken up by yolk granules as compared to more than 95% in avian eggs. The small quantity of yolk in mammalian eggs reflects the fact that fertilization and embryonic development proceed entirely within the female. Most of the nutrients required for embryonic development are extracted from the maternal bloodstream and delivered to the embryonic circulation through the placenta.

Besides yolk bodies, the cytoplasm of many eggs also becomes packed with lipid droplets. These droplets, which are freely suspended in the cytoplasm without boundary membranes, usually consist of concentrations of triglycerides and phospholipids. In amphibians and birds much of this material is also synthesized in liver cells and delivered through the bloodstream to developing oocytes. Some oocytes also contain deposits of glycogen (animal starch; see p. 52) as a nutrient reserve.

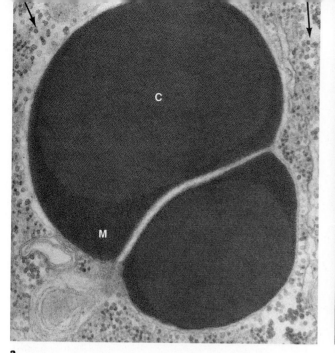
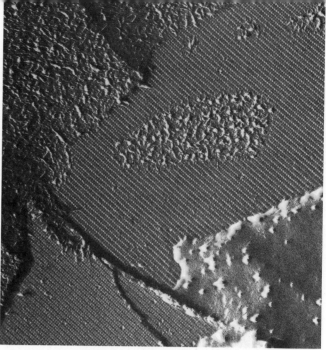

a

b

Figure 26-10   Yolk granules. **(a)** A yolk granule in a *Xenopus* oocyte. Typically, a yolk granule contains an angular crystalline core (C) embedded in a matrix substance (M). Regular periodicity is visible in the core. Numerous glycogen deposits (arrows) are visible in the cytoplasm surrounding the yolk granule. × 49,000. Courtesy of P. B. Armstrong. **(b)** Fractured surface of a *Xenopus* yolk granule prepared by freeze fracture. The regular, periodic structure of the crystalline core is clearly visible. The less structured areas are produced by ice crystals in the preparation. × 84,000. Courtesy of R. L. Leonard.

External Egg Coats   Surface coats of the egg, depending on the species, may be added during initial oocyte growth at meiotic prophase I, at later stages of meiosis, or even, in some species, after fertilization. The egg coats, which in most species contain polysaccharides, proteins, and glycoproteins as major constituents, protect the egg from attack by microorganisms, from mechanical or chemical injury and, in species such as insects, reptiles, and birds, from desiccation. The egg coats may persist after fertilization and remain as a protective layer through part or all of embryonic development.

The covering of animal eggs, called the *vitelline coat* in most animals and the *zona pellucida* in mammals (see Figs. 26-8 and 26-11), is built up from relatively simple combinations of polysaccharides, proteins, or glycoproteins. In mammals from one to four glycoproteins can be detected in the zona pellucida. P. M. Wassarman and his coworkers found that the zona pellucida of mouse and human egg cells, for example, contains three glycoproteins, which they termed *ZP1*, *ZP2*, and *ZP3*, with molecular weights of 200,000, 120,000, and 83,000, respectively. The three glycoproteins, which are interlinked by disulfide bonds, make up 90% or more of the dry weight of the zona pellucida. Some mammals, such as the pig, have a fourth zona

pellucida protein, ZP4. In mice ZP2 and ZP3 combine to form long fibers about 7 nm in diameter that comprise most of the mass of the zona pellucida. These fibers are crosslinked into a network by ZP1. ZP3 also appears to act as a sperm receptor and induces the release of acrosomal enzymes (see below). The mature zona pellucida, which ranges from 2 to 35 μm in thickness in different mammals, is a porous structure that allows passage of molecules as large as antibodies and enzymes.

The vitelline coat may be secreted by the oocyte or by cells surrounding the oocyte in ovarian tissue. In mammals antibodies against ZP glycoproteins react with the cytoplasm of both the oocyte and the surrounding follicle cells (see Fig. 26-11 and below) during deposition of the zona pellucida, suggesting that both cell types may contribute to formation of the external coat. Other observations point to the oocyte as the exclusive site of ZP glycoprotein synthesis. Radioactive sugars such as ^{3}H-fucose are taken up only by oocytes during secretion of the zona pellucida; similarly, probes for ZP glycoprotein mRNAs react positively with oocytes but not with follicle cells. In any event, whether synthesized in both oocytes and follicle cells or in oocytes alone, the zona pellucida persists in mammals until the embryo is implanted in the placenta.

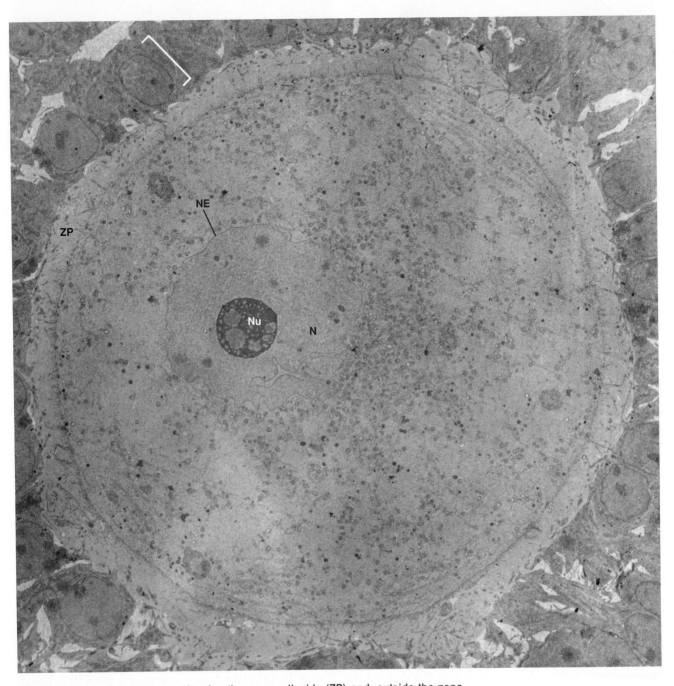

Figure 26-11 A mouse oocyte, showing the zona pellucida (ZP) and, outside the zona pellucida, a layer of follicle cells (bracket). Numerous microvilli extend from the oocyte and follicle cells into the zona pellucida. N, nucleus; Nu, nucleolus; NE, nuclear envelope. ×2,800. From *Atlas of Ultrastructure* by J. A. G. Rhodin. Courtesy of J. A. G. Rhodin and W. B. Saunders Company.

In some animals, such as insects, amphibians, sea urchins, birds, and reptiles, additional surface coats are secreted outside the vitelline coat. The additional surface coat secreted in insects, known as the *chorion*, is a hard, impermeable layer built up from different kinds of proteins. The protein complement of the chorion may be extensive; in *Drosophila* the chorion contains 6 major and 14 minor proteins. In some moths the total

approaches 200 different types. Nearest the egg surface the insect chorion has an open, columnar structure forming air spaces that promote gas exchange with the atmosphere. Nearer its outermost surface the chorion is constructed from successive, solid layers of fibers. The fibers of each layer run in different directions, giving this portion of the chorion an arrangement similar to plywood, with equivalent structural strength.

The chorion provides rigid mechanical support and protection to the egg, prevents entry by microorganisms, and allows gas exchange without loss of water. At one end the chorion is perforated by a narrow channel, the *micropyle*, the passage through which the fertilizing sperm reaches the egg surface.

In amphibians and sea urchins the vitelline coat is surrounded by a thick layer called the *jelly coat* (see Fig. 26-8). This layer, built up from highly hydrated polysaccharides and glycoproteins, provides the egg and embryo with resistance to drying and mechanical damage. In sea urchins the acrosome reaction is triggered by the jelly coat rather than the vitelline coat. In some species the jelly coat also binds eggs together in masses or to a supporting substrate.

In the eggs of birds and reptiles the egg white and shell constitute equivalent layers outside the vitelline coat. The egg white, containing the protein *ovalbumin* as a major constituent, provides additional nutrients and water for the developing embryo. The outer shell of the bird and reptile egg serves a function similar to the insect chorion in barring microorganisms, providing mechanical protection, and allowing gas exchange without extensive loss of water. Both the egg white and egg shell are secreted around the egg after fertilization by oviduct cells as the egg passes from the ovary to the exterior.

Follicle and Nurse Cells In most animals the developing oocyte is surrounded by one or more follicle or nurse cells (see Fig. 26-11). No clear distinction exists between follicle and nurse cells except for number: when present in small numbers, the cells surrounding a developing oocyte are called nurse cells; when present in relatively large numbers, they are called follicle cells.

In vertebrates, developing oocytes are surrounded by follicle cells derived from the ovary. Although vertebrate follicle cells are of mesodermal origin, it is still uncertain whether the cell lines giving rise to oocytes are derived from mesoderm or endoderm. Follicle cells completely surround the zona pellucida in mammals and make contact with the enclosed oocyte via narrow surface projections, called *microvilli*, which extend through the zona pellucida to the oocyte plasma membrane. (Microvilli are faintly visible in Fig. 26-11.) At the points of contact between follicle and egg plasma membranes, gap junctions (see p. 262) facilitate direct transfer of ions and small molecules between the two cell types.

In mammals follicle cells initially surround each oocyte in three layers containing in total about 900 cells. These cells may contribute to formation of the zona pellucida and may also secrete estrogen and progesterone, hormones partially responsible for control of the female reproductive cycle. Just before ovulation the follicle cells divide rapidly, reaching levels of more than 50,000 around the oocyte.

At ovulation most of the follicle cells of mammalian oocytes are shed except for a thin layer, the so-called *cumulus oophorus*, which remains around the zona pellucida. This layer stays with the egg until fertilization, when it too is released.

The Mature Animal Egg

The animal egg at maturity represents a developmental system ready to be triggered into intense activity. Because the information regulating much or all of early development is stored in the egg cytoplasm, these early embryonic stages are under the control of maternal genes. In some organisms, such as *Drosophila*, the effects of some maternal genes are exerted through their mRNA or protein products throughout embryonic development. In others, such as mammals, the effects of maternal genes are much reduced and limited to very early embryonic stages.

The degree to which developmental information is localized in particular regions of the egg cytoplasm varies by major animal group. In some organisms, such as ascidians and nematodes, information in the egg seems to be highly localized. In eggs of this type, called *mosaic eggs*, certain regions of the egg cytoplasm give rise only to certain parts of the embryo after fertilization. In such eggs, removing part of the egg cytoplasm causes the corresponding regions of the future embryo to be lost; other regions of the cytoplasm cannot supply the missing information. In mosaic eggs, transferring parts of the cytoplasm from one region of the egg to another causes a corresponding rearrangement of parts during embryonic development.

At the other extreme are eggs such as those of sea urchins, salamanders, and mammals, in which all parts of the egg cytoplasm seem to be equivalent in developmental capacity. In such eggs, called *regulative eggs*, removal of portions of the egg cytoplasm, up to certain limits, has little or no effect on embryonic development after fertilization. Transplantation of cytoplasm from one place to another in the egg is similarly without effect. Many animal eggs lie somewhere between the extremes represented by fully mosaic and fully regulative eggs or have a mixture of mosaic and regulative characteristics.

The developmental plasticity of the mammalian egg cytoplasm is largely responsible for the production of identical twins. In humans identical twins result from separation of the embryo into two parts very early in development. Both parts have the capacity to proceed through embryonic development in completely normal fashion, with no parts or regions defective or missing. Because the embryos originate from mitotic divisions of the fertilized egg, they are genetically identical.

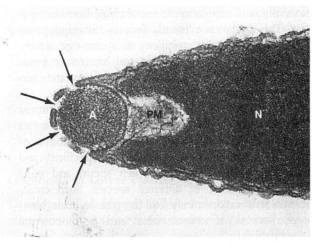

Figure 26-12 Initiation of the acrosome reaction in the sea urchin *Lytechinus*. The plasma and acrosomal membranes have fused and become continuous at several points, creating openings (arrows) that expose the acrosomal contents to the exterior. A, acrosome; N, nucleus; PM, periacrosomal material. ×14,000. Courtesy of B. L. Hylander and R. G. Summers.

tration of the coat is aided by the enzymes released by the acrosomal reaction; among these are proteinases capable of hydrolyzing the vitelline coat (for molecular details of acrosomal filament extension, see p. 486).

Similar acrosomal filaments are extended during fertilization in many marine invertebrates. In horseshoe crabs a preformed actin filament, stored in the mature sperm cell as a coil, extends by uncoiling to form the acrosomal filament (see p. 487 and Fig. 12-32).

In mammals penetration of egg coats by sperm also appears to take place by a combination of enzymatic and mechanical effects. In mammals sperm must

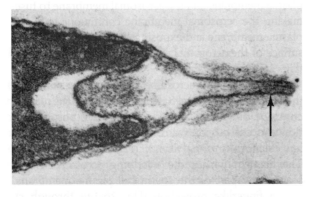

Figure 26-13 Extension of the acrosomal filament (arrow) in the sea urchin *Lytechinus* during fertilization. The material coating the surface of the filament includes bindin and enzymes released from the acrosome. ×83,000. Courtesy of D. Epel and F. Collins, from *Exptl. Cell. Res.* 106:211 (1977).

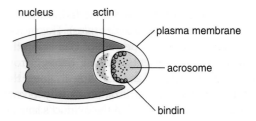

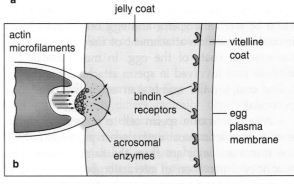

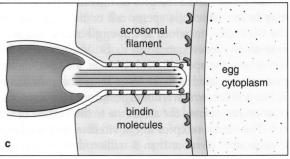

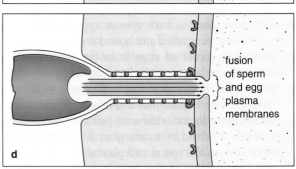

Figure 26-14 Role of the acrosomal filament and bindin in sea urchin fertilization. **(a)** The unreacted sperm cell. **(b)** Initiation of the acrosome reaction, in which the acrosomal membrane fuses with the sperm cell plasma membrane. The fusion releases acrosomal enzymes to the exterior and exposes bindin molecules as a coat on the membrane surface. Actin has begun to polymerize into microfilaments in the region between the acrosomal membrane and the nucleus. **(c)** Projection of the acrosomal filament through growth of the microfilaments. The growth has brought the tip of the acrosomal filament to the vitelline coat, where receptors link the bindin molecules and the acrosomal filament firmly to the egg surface. **(d)** Penetration of the acrosomal filament through the vitelline coat, aided by digestion of the coat by acrosomal enzymes, and fusion of the sperm and egg cell plasma membranes.

first pass through the layer of follicle cells surrounding the egg, the cumulus oophorus, to reach the zona pellucida. Although the conclusion is controversial, some investigators believe that passage of a fertilizing sperm through the follicle cell layer is aided by hyaluronidase, one of the mammalian acrosomal enzymes, released from sperm cells arriving before the fertilizing sperm. This enzyme supposedly hydrolyzes the hyaluronic acid polymer (see p. 280) forming part of the extracellular matrix of the follicle cells to facilitate passage of the fertilizing sperm between the cells of the layer. In the mouse, contact of the fertilizing sperm with the zona pellucida after passage between follicle cells triggers its acrosome reaction, as does ZP3, the ligand in sperm-egg binding.

The mammalian acrosome reaction occurs through fusions between the sperm cell plasma membrane and acrosomal membrane as in the sea urchin (Fig. 26-15). Among the enzymes released by the mammalian acro-

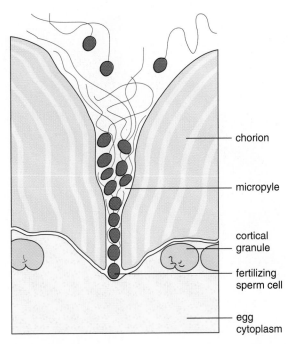

Figure 26-16 Diagram of sperm cells entering the micropyle of a trout egg during fertilization. The narrow channel through the chorion is just wide enough to admit a single fertilizing sperm cell to the egg plasma membrane. Redrawn from an original courtesy of A. S. Ginsburg.

some reaction is *acrosin,* a proteinase capable of hydrolyzing zona proteins. Passage is also driven by the sperm tail, which continues beating while the zona pellucida is penetrated.

In both vertebrates and invertebrates Ca^{2+} seems to be absolutely required for the acrosome reaction. Removal of Ca^{2+} from the fertilization medium blocks the acrosome reaction; the reaction is similarly blocked if sperm cells are exposed to agents, such as procaine, that close calcium channels in the plasma membrane. Conversely, addition of a calcium ionophore which greatly increases Ca^{2+} penetration across plasma membranes, can induce the acrosome reaction. The Ca^{2+} requirement for the acrosome reaction is typical of exocytosis in general (see p. 856). The reaction also involves activation of G proteins and a Ca^{2+}-dependent phospholipase C, suggesting that it may follow a reaction pathway similar to receptor-mediated responses in somatic cells (see Ch. 7 for details).

In animal eggs possessing a relatively impermeable exterior chorion, such as insects and fishes, the sperm passes to the immediate egg surface through the narrow channel formed by the micropyle. In teleost fishes the dimensions of the micropyle, which is wide enough to admit only a single sperm cell to the egg surface (Fig. 26-16), prevent fertilization by multiple sperm cells. The absence of an acrosome in most teleost fishes is probably related to the open channel provided by the micropyle.

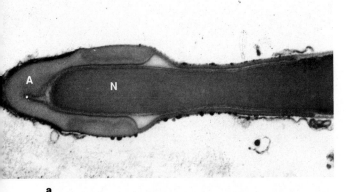

a

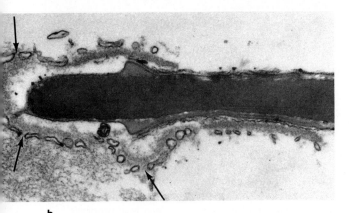

b

Figure 26-15 The acrosome reaction in mammalian sperm. **(a)** Unreacted acrosome (A) of a hamster sperm cell. N, nucleus. ×30,000. **(b)** Acrosome reaction in progress in the hamster. The acrosomal membrane and the plasma membrane of the sperm cell have fused, opening the acrosome to the exterior at numerous sites (arrows). ×31,000. Courtesy of R. Yanagimachi and Y. D. Noda, from *J. Ultrastr. Res.* 31:465 (1970), by permission of Academic Press, Inc.

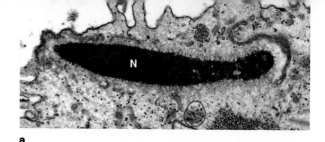

a

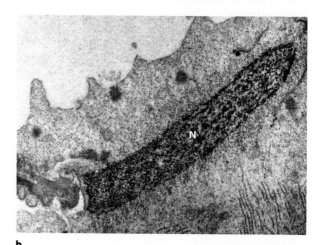

b

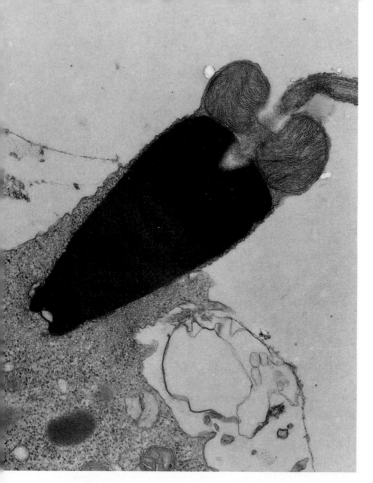

Figure 26-17 Membrane fusion and engulfment of the sperm cell in the sea urchin *Arbacia*. After initial fusion between the sperm cell plasma membrane at the tip of the acrosomal process and the egg cell plasma membrane (as in Fig. 26-14*d*), the egg cytoplasm flows into the fusing region and engulfs the sperm nucleus. ×25,500. Courtesy of E. Anderson. Reproduced from *J. Cell Biol.* 37:514 (1968), by copyright permission of the Rockefeller University Press.

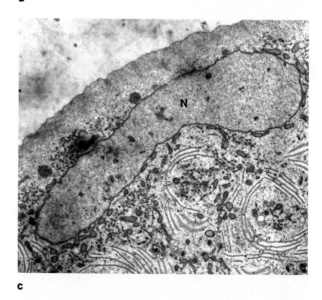

c

Figure 26-18 Stages in the decondensation and transformation of the sperm nucleus into the male pronucleus. **(a)** A sperm nucleus (N) of the hamster soon after being engulfed by the egg cytoplasm. The sperm nuclear envelope has already broken down. ×14,000. **(b)** A later stage in nuclear transformation in the rat; decondensation of the sperm nucleus is in progress. ×18,000. **(c)** A sperm nucleus of the hamster in the final stages of its transformation into the male pronucleus. Decondensation and formation of the new nuclear envelope are nearly complete. ×7,000 **(a** and **c)** Courtesy of R. Yanagimachi and Y. D. Noda, from *J. Ultrastr. Res.* 31:465 (1970), by permission of Academic Press, Inc. **(b)** Courtesy of L. Piko, from *Fertilization*, vol. II (Metz, C. B., and Monroy, A., eds.), Academic Press, Inc., 1969.

Fusion of Sperm and Egg Plasma Membranes Once the sperm cell penetrates the egg coats, contact between the sperm and egg plasma membranes causes immediate membrane fusion. Fusion makes the sperm and egg cytoplasm continuous, allowing the egg cytoplasm to surround and engulf the sperm nucleus (Fig. 26-17 and Fig. 26-18). In most vertebrate and invertebrate species the sperm tail stops beating as engulfment proceeds.

Recently, P. Primakoff, D. G. Myles, and C. P. Blobel and their colleagues discovered *PH-30*, a sperm protein that promotes fusion of mammalian egg and sperm membranes. Like many viral fusion proteins (see p. 186), PH- consists of α- and β-subunits. The α-subunit has sequence similarities to viral fusion peptides and appears to directly promote membrane fusion; the β-subunit contains a site capable of recognizing and binding integrin receptors in the egg plasma membrane. Evidently, the β-subunit binds the sperm plasma membrane tightly to the integrin receptors. The α-subunit then penetrates the egg plasma membrane and promotes fusion by a mechanism similar to that diagrammed in Figure 5-32.

Sperm engulfment in sea urchins is arrested if eggs are exposed to cytochalasin, an inhibitor of microfilament activity. Microfilaments are therefore probably responsible for the streaming movements of the egg cytoplasm engulfing the sperm in this group.

Egg Activation and the Block to Polyspermy Fusion of the sperm and egg plasma membranes initiates a series of reactions with three major effects in most animals. One is establishment of a barrier to entry by additional sperm cells, the *block to polyspermy*, which is set up by alterations in the egg plasma membrane or vitelline coat, or both. The second is a modification that makes the vitelline coat/zona pellucida less soluble and more resistant to enzymatic breakdown. The third reaction is activation of the biochemical pathways of the egg, including rapid and extensive increases in respiration and protein synthesis.

In sea urchins, where these reactions have been most completely investigated, the first measurable response, occurring within seconds after fusion of the sperm and egg membranes, is rapid depolarization of the egg plasma membrane. In this depolarization, which is apparently triggered by a rapid inward Na^+ flow, the egg shifts suddenly from a negative resting potential of -20 to -60 mV to a positive potential of $+5$ to $+20$ mV. The depolarization, which resembles an action potential in a nerve axon (see p. 213), spreads rapidly over the egg surface and persists for several minutes.

Several lines of evidence indicate that depolarization of sea urchin eggs on fertilization establishes what is known as the *fast block to polyspermy* in this animal group. L. A. Jaffe discovered that if unfertilized sea urchin eggs are held at a positive membrane potential by a voltage clamp (see p. 217), membrane fusion with sperm cells is blocked, and fertilization does not occur. Conversely, if eggs are clamped at a negative potential, the fast block to polyspermy is circumvented, and many sperm cells can fuse with the egg plasma membrane. Jaffe's results indicate that a molecule in the sperm cell plasma membrane acts as a "voltage sensor" that goes to an inactive binding state in response to the change to positive potential. (Jaffe's Experimental Process essay on p. 1110 describes her elegant experiments revealing the effects of membrane potential on fertilization.) The depolarization responsible for the fast block lasts only for a few minutes. However, changes in the vitelline coat (see below) set up a permanent block before the fast block runs its course.

A fast block to polyspermy associated with a change in membrane potential is typical of marine in-vertebrates and also occurs in at least one vertebrate group, the amphibians. In mammals the potential of the egg plasma membrane appears to oscillate regularly for a time after sperm and egg fusion. The relationship of this oscillation to a block to polyspermy is uncertain.

Not all animal eggs develop an immediate block to polyspermy. Multiple fusions of sperm cells with the egg plasma membrane, producing polyspermic eggs, occur regularly in some animal groups, including reptiles, birds, insects, and some molluscs. Although multiple sperm nuclei may enter the egg cytoplasm in these animals, only one of the nuclei eventually fuses with the egg nucleus. The remaining sperm nuclei degenerate and disappear. The mechanism limiting nuclear fusion to a single sperm nucleus in these animals is unknown.

The second reaction to sperm-egg fusion, modification of egg coats, is accomplished through cortical granule release. Cortical granules fuse with the plasma membrane, spilling their contents into the region between the plasma membrane and the vitelline coat. Among the substances released from the cortical granules are several enzymes that alter molecules in the vitelline coat. In sea urchins one of these enzymes hydrolyzes bonds linking the vitelline coat to the egg plasma membrane. Breaking the links releases the vitelline coat from the membrane and creates a space that fills with other substances released from the cortical granules, among them a polysaccharide. The substances entering the space promote the osmotic movement of water into the region between the coat and the egg surface, which forces the vitelline coat farther away from the egg plasma membrane.

Other enzymes released from cortical granules "harden" the vitelline coat by setting up covalent crosslinks between coat proteins. The alterations induced by cortical granule release are called the *zona reaction* in mammals.

Another enzyme released from cortical granules in sea urchins alters the bindin receptors of the vitelline coat so that they no longer recognize and bind the sperm ligand. This alteration, known as the *slow block to polyspermy*, is permanent and prevents any additional sperm from binding to the egg surface. The combined action of the fast and slow blocks effectively limits fertilization to a single sperm cell in sea urchins.

The relationship of the zona reaction to a block to polyspermy in mammals is uncertain. In different mammals, the block to polyspermy operates at either the zona pellucida or egg plasma membrane or at both locations. In the mouse, for example, ZP3 is altered, so that it is no longer recognized and bound by the sperm receptor. The chemical alterations hardening the zona pellucida in mice and other mammals remain unclear. The only significant alteration that can be detected is a limited hydrolysis of ZP2.

The Electrical Polyspermy Block

Laurinda A. Jaffe

LAURINDA A. JAFFE is professor of physiology at the University of Connecticut Health Center. She was an undergraduate at the University of Wisconsin and at Purdue University, where she worked in the laboratory of her father, Lionel F. Jaffe. She was a graduate student at UCLA with Susumu Hagiwara, where she did her Ph.D. thesis on the electrical polyspermy block. This work continued during post-doctoral research with Meredith Gould at the University of California in San Diego, and with Lewis Tilney at the Marine Biological Lab in Woods Hole; it is still a major interest of her research.

At fertilization, one sperm and one egg combine to form an embryo; if two or more sperm fertilize an egg, the embryo fails to develop. The problem of polyspermy prevention has been solved in diverse ways in different organisms.[1] One common mechanism is the electrical block to polyspermy.

Since the cellular events of fertilization were first observed over 100 years ago, it was clear that one component of the block to polyspermy in organisms such as sea urchins, starfish, and frogs is the elevation of a protective envelope around the egg, the fertilization envelope. However, there was controversy as to whether this relatively slow process, which takes about one minute, was fast enough to block polyspermy completely. There was speculation that since fast events in nerve and muscle were electrically mediated, an electrical mechanism might provide a fast block to polyspermy, before the fertilization envelope is fully elevated. The tools to make electrical measurements in eggs were not available until the 1950s, when an electrical change at fertilization—now called the fertilization potential —was first measured, in the egg of a starfish.[2] At this time, the fertilization potential was described, but its function was not explored further.

The discovery of the electrical polyspermy block is a story of serendipity; it began with a project having nothing to do with polyspermy. I was a graduate student in the laboratory of Dr. Susumu Hagiwara at UCLA, and at that time, there was great interest in the idea that an influx of calcium might activate the egg at fertilization. With this in mind, Dr. Hagiwara suggested that I try to block egg activation at fertilization by clamping the voltage of the sea urchin egg membrane at a very positive

potential, so that positively charged calcium ions could not enter the egg. This experiment was based on an earlier experiment[3] which showed that holding the voltage across the nerve terminal membrane at about +200 mV blocked calcium entry and therefore blocked synaptic transmission. Could fertilization also be blocked?

Using a fine microelectrode inserted through the egg membrane to pass current, I held the voltage of the egg membrane at about +200 mV. When sperm were added, the neighboring eggs fertilized, but the voltage-clamped egg did not! Was this because calcium entry was blocked? To test this, I held the voltage of the egg membrane at successively less positive potentials, to determine the threshold required to block fertilization. In the nerve terminal, potentials greater than about +100 mV are required before inhibitory effects on calcium entry are seen. However, in the sea urchin egg, potentials as small as +5 mV inhibited fertilization. The work with nerves had shown that calcium entry would not be significantly blocked at +5 mV, so the result that I was seeing was clearly unrelated to what I had set out to find originally.

At this point it occurred to me that my findings might be related to the old idea of an electrically mediated polyspermy block. +5 mV was approximately the voltage attained during the fertilization potential. Could it be that by holding the egg's membrane potential at +5 mV, I was mimicking the fertilization potential and that the natural function of the fertilization potential was to block polyspermy?

If this was so, I realized that it should be possible to induce polyspermy by suppressing the positive potential shift. After observing the rise of the fertilization potential, I applied current to bring the egg's membrane potential back to −30 mV. When such eggs were observed 2 hours later, they were seen to have cleaved into 3 cells, a clear indicator of polyspermy.

These findings[4] opened up many new questions. Did this mechanism operate in other organisms as well? How was the positive potential suppressing fertilization? In the past 15 years we have partially answered these questions.

The electrical polyspermy block functions in many but not all species; it even operates in at least one plant species, the sea weed *Fucus*.[5] In most cases, a positive shift in potential accounts for the block, but in crabs, the block occurs as a result of a negative shift in potential.[6]

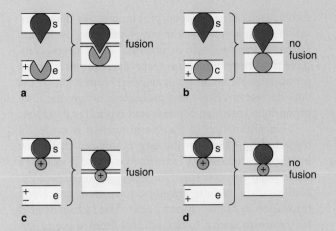

Figure A Two models of how the egg's membrane potential might regulate sperm-egg fusion. Parallel lines represent the sperm (s) and egg (e) plasma membranes; the solid figures represent membrane molecules important in fusion. **(a, b)** Potential-controlled conformational change in an egg membrane molecule. When the egg's membrane potential is negative **(a)** the successful interaction of egg and sperm molecules leads to membrane fusion. Positive potential **(b)** prevents fusion. **(c, d)** Potential-controlled insertion of a sperm membrane molecule into the egg plasma membrane. When the egg's membrane potential is negative **(c)** the positively charged portion of the sperm molecule inserts fully and fusion follows. Positive potential **(d)** inhibits full insertion and prevents fusion. Redrawn from *Ann. Rev. Physiol.* 48:191 (1986).

The evidence so far indicates that electrical polyspermy blocks do not occur in mammals; sperm access to mammalian eggs may occur slowly enough that a fast electrical mechanism for polyspermy prevention is not needed.

As for the mechanism by which membrane potential regulates the fusion of the sperm with the egg, we have made two important findings.[7] First, the block to fertilization is due to the voltage change itself and not to the accompanying ion movements. Second, and surprisingly, the "voltage sensor" appears to be in sperm rather than the egg. In principle, the voltage sensitivity of fertilization could result from a receptor molecule in the egg membrane that undergoes a conformational change from receptive to nonreceptive in response to a change in transmembrane electrical potential (Fig. A*a, b*). Ion channels in cell membranes open and close by this means. However, this does not seem to be the case for fertilization. Instead, it appears that the voltage-sensitivity of fertilization results from a voltage sensor in the sperm, perhaps a positively charged region of a sperm membrane protein that must insert in the egg membrane to initiate sperm-egg fusion (Fig. A*c, d*). The idea is that membrane insertion of such a positively charged peptide would be favored if the potential on the inside of the egg membrane is negative and opposed if it is positive.

Evidence for this model came from a series of cross-fertilization experiments between different animal species that showed differing degrees of voltage-dependence in fertilization. In some species, fertilization occurs with the same probability regardless of the voltage across the egg membrane; an example is the Japanese salamander *Cynops*. *Cynops* eggs can, in the laboratory, be fertilized with sperm from another Japanese salamander *Hynobius*. Unlike *Cynops*, *Hynobius* fertilization is blocked by a positive egg membrane potential.

During a visit to Yamaguchi University in Japan, I collaborated with Dr. Yasuhiro Iwao to find out whether the cross-fertilization of *Cynops* eggs by *Hynobius* sperm was voltage-dependent.[8] We voltage-clamped *Cynops* egg at positive potentials and then added *Hynobius* sperm; we found that positive but not negative potentials inhibited *Hynobius* sperm entry. Since fertilization of *Cynops* eggs by *Cynops* sperm is voltage-independent, we concluded that the voltage-sensitivity could only have come from the *Hynobius sperm*. Together with earlier work showing that fertilization of a voltage-sensitive egg species by a voltage-insensitive sperm species is voltage *insensitive*, this result demonstrated that the voltage sensor must be in the sperm.

The challenge now is to identify this voltage-sensitive molecule. Perhaps it will be a fusion protein like the fusion proteins that mediate virus-host cell fusion.[9] Viral fusion proteins have a hydrophobic region that inserts in the host cell membrane. If this hydrophobic region included some positively charged amino acids such as arginine or lysine, its insertion into a membrane would be voltage-dependent. The presence or absence of these charged amino acids could determine whether fertilization in a particular species was or was not voltage-dependent.

References

[1] Jaffe, L. A., and Gould, M. In *Biology of Fertilization*, C. B. Metz and A. Monroy, eds. Orlando: Academic Press, p. 223 (1985).

[2] Tyler, A.; Monroy, A.; Kao, C. Y.; and Grundfest, H. *Biol. Bull.* 111:153 (1956).

[3] Katz, B., and Miledi, R. *J. Physiol.* 192:407 (1967).

[4] Jaffe, L. A. *Nature* 261:68 (1976).

[5] Brawley, S. H. *Devel. Biol.* 144:94 (1991).

[6] Goudeau, H., and Goudeau, M. *Devel. Biol.* 133:348 (1989).

[7] Jaffe, L. A., and Cross, N. L. *Ann. Rev. Physiol.* 48:191 (1986).

[8] Iwao, Y., and Jaffe, L. A. *Devel. Biol.* 134:446 (1989).

[9] White, J. M. *Ann. Rev. Physiol.* 52:675 (1990).

In all animals a transient rise in cytoplasmic Ca^{2+} concentration coincides with cortical granule release. Release can be triggered experimentally by injections of Ca^{2+} into the cytoplasm or exposure of eggs to a calcium ionophore, indicating that the rise in Ca^{2+} concentration is an immediate trigger for cortical granule release as it is in other forms of exocytosis.

Relatively little is understood of the third reaction to sperm-egg fusion, activation of biochemical pathways in the egg. One of the first activations noted in many animal eggs at fertilization is an abrupt increase in respiration. Protein synthesis, which proceeds in many unfertilized animal eggs at very low levels, also increases rapidly in many species after fertilization.

Research by A. C. Lopo and her colleagues revealed that at least part of the activation of protein synthesis at fertilization in sea urchins depends on phosphorylation of two of the initiation factors for protein synthesis, eIF4B and eIF4E (see p. 652 and Lopo's Experimental Essay on p. 657). The initiation factors are unphosphorylated and inactive in unfertilized eggs. At fertilization addition of phosphate groups to the factors breaks their inhibition and stimulates their activity. Other investigators detected additional changes in elements and factors occurring at fertilization, including removal of proteins that mask mRNAs and prevent their translation before fertilization, and completion of processing, including the addition of poly(A) tails to previously tailless mRNAs. The combined effects of these changes lead to a steep increase in protein synthesis in most species soon after the sperm and egg membranes fuse.

In many animals egg activation can be stimulated artificially by such treatments as pricking the egg surface with a microneedle. This demonstrates that the role of the sperm cell in egg activation in these animals is only to provide an initiating stimulus. The elements and factors required for the activation reactions are parts of egg cytoplasm and evidently have no absolute requirement for sperm cell molecules or components.

Transformation of the Sperm Nucleus and Fusion of Male and Female Pronuclei   After the sperm nucleus is fully engulfed, it changes from the tightly condensed structure characteristic of the sperm cell into a form resembling an interphase nucleus. The changes transform the sperm nucleus into the *male pronucleus.*

Transformation of the sperm nucleus takes place in several more or less clearly defined steps. The sperm nuclear envelope breaks down during or shortly after engulfment by the egg cytoplasm. As soon as the sperm nuclear material is exposed to the egg cytoplasm, decondensation of the sperm chromatin begins (Fig. 26-18). At the same time, the sperm nuclear proteins are replaced by histones from the egg cytoplasm. In mammals decondensation requires breakage of the

disulfide linkages between protamines. As decondensation ends, a new nuclear envelope, lined by lamins, forms around the sperm chromatin. These changes complete transformation of the sperm nucleus into the male pronucleus. At some point between formation of the new nuclear envelope and pronuclear fusion, the male pronucleus enters an S phase and replicates its DNA.

Transformation of the sperm nucleus is evidently triggered by factors in the egg cytoplasm that show little species specificity. Extracts from the cytoplasm of unfertilized *Xenopus* eggs, for example, can induce decondensation, nuclear envelope formation, and DNA replication in human sperm nuclei. The factors needed to transform sperm nuclei are lost from the egg cytoplasm soon after fertilization. Sperm nuclei introduced experimentally into zygotes after pronuclear fusion typically remain tightly condensed.

While these changes are under way in the male pronucleus, the egg nucleus breaks its arrest and, in species in which meiosis is incomplete in mature eggs, completes the meiotic divisions. The polar bodies formed by the divisions are pushed into the space between the egg plasma membrane and the elevated vitelline coat. Once meiosis is complete, the female pronucleus also replicates its DNA.

While changes in the egg and sperm nuclei are in progress, the two nuclei move toward each other in the egg cytoplasm. In different species, microtubules or microfilaments, or both motile elements working together, are responsible for the movements bringing the male and female pronuclei together.

Fusion of the male and female pronuclei follows different pathways in different species and major groups. Only in a relatively few species, such as sea urchins, do the male and female pronuclei fuse completely before the first mitotic division following fertilization. In many species the male and female pronuclei condense into metaphase chromosomes after replicating their DNA. The condensed chromosomes from the two nuclei then attach to the midpoint of a single spindle formed for the first mitotic division. At telophase of this first division, the separated chromosomes decondense into two nuclei in which the maternal and paternal chromosomes are fully mixed. This pattern occurs, for example, in *Drosophila* and in mammals.

The Fate of Cytoplasmic Organelles of the Fertilizing Sperm   During sperm-egg fusion the egg cytoplasm engulfs the cytoplasmic organelles of the sperm cell as well as its nucleus. In general the contents of the entire sperm cell eventually enter the egg, except for the tail in some species. Sperm structures that can be traced morphologically, such as the tail, can be seen to disintegrate in the egg cytoplasm. (Mitochondria and chloroplasts are also inherited primarily from the maternal line in plants; see p. 877.)

Traditionally, the centrioles found in the asters of the first mitotic division following fertilization were thought to be derived from the sperm. During the series of mitotic divisions immediately preceding meiosis in oocytes, centrioles and asters disappear in protozoa, copepods, insects, and numerous vertebrate groups. Because the centrioles persist during spermiogenesis in most animal species and give rise to the 9 + 2 system of the sperm tail, it has been assumed that these centrioles are delivered to the egg to found the centriole line of the embryo and adult. In many animal species it is possible that centrioles do originate from the fertilizing sperm. However, in fertilized mouse eggs, no asters or centrioles can be detected until the third mitosis following fertilization; in the sperm of many insects, the sperm centrioles apparently disappear after giving rise to the microtubules of the flagellum and do not contribute to the zygote. Evidently, in these species the centrioles of the early embryo arise without derivation from previously existing centrioles. Therefore, depending on the species or major animal group, the centrioles appearing in the early embryonic divisions may be derived from the sperm or arise spontaneously in the egg. Centrioles may also arise de novo, with no participation of previously existing centrioles, in eggs stimulated to enter development artificially, as R. E. Pallazo and his coworkers and others have shown (see also p. 901).

GAMETOGENESIS AND FERTILIZATION IN HIGHER PLANTS

Gametogenesis and fertilization occupy characteristically different positions in the life cycles of higher plants from the equivalent events in animals. These differences reflect the fact that in plants the life cycle alternates between distinct haploid and diploid generations.

Fertilization in higher plants, as in animals, produces a diploid individual. These individuals, termed *sporophytes* in plants, make up the predominant phase of the life cycle in higher plants and are conspicuous as the conifers and flowering trees, shrubs, and smaller plants of the environment. The sporophyte grows to maturity through mitotic divisions. At some point meiosis occurs in certain cells of the sporophyte, giving rise to haploid products known as *spores*. As a result of the meiotic divisions, the spores have new combinations of genetic information that differ from those of the parent sporophyte. Spores grow by mitosis into a haploid phase of the life cycle termed the *gametophyte*. In higher plants the gametophyte is an essentially microscopic form dependent on the sporophyte for part or all of its existence.

Gametophytes give rise to male or female gametes, the sperm and egg cells, through differentiation of the products of mitotic divisions. Fusion of a sperm and an egg cell in fertilization restores the diploid chromosome number and establishes the sporophyte phase of the life cycle. This section considers these events as they occur in flowering plants (angiosperms).

Gametogenesis in Flowering Plants

Gametogenesis in flowering plants occurs within structures of the flower (Fig. 26-19), which differentiates from tissues of the diploid sporophyte. One segment of the flower, the *carpel*, gives rise to the female gametophyte. Male gametophytes are formed within another segment, the *stamen*. When fully formed, carpels differentiate into the *ovule*, in which the female gametophyte develops, and two accessory structures called the *style* and *stigma*. The style, a stalklike outgrowth, carries the stigma at its tip. These structures function in attachment and guidance of the male gametophyte to the ovule during fertilization. The stamens develop swellings at their tips, the *anthers*, in which the male gametophytes form. The external forms taken by the carpels, stamens, and other structures of flowers vary widely among different angiosperms.

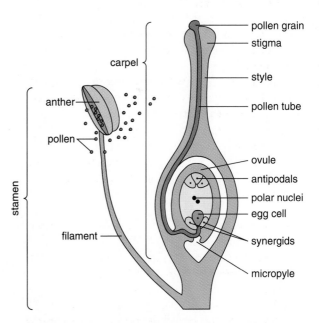

Figure 26-19 The parts of a flower. The carpel consists of the ovule, in which the female gametophyte develops, the style, and the stigma. The stamen includes the anther, in which the male gametophyte (pollen) develops, supported by the filament. Pollen grains contacting the style germinate to form a pollen tube containing the male gametophyte. The pollen tube grows through the stigma to reach the micropyle and the female gametophyte.

Development of the Female Gametophyte and Egg Cell Development of the female gametophyte begins with meiosis, which takes place in a large cell within the ovule called the *megasporocyte.* The nuclear events occurring during this meiosis closely resemble those of meiosis in animal cells. Many details of DNA synthesis and the biochemistry of recombination have, in fact, been worked out through investigations of megasporocyte meiosis in the lily (see p. 1073). Major differences between plant and animal meiosis, however, are noted in the patterns of RNA and protein synthesis during meiotic prophase I and in the degree of development of the cytoplasm during this meiotic stage. In contrast to the highly active RNA and protein synthesis and extensive cytoplasmic growth characteristic of animal oocytes, RNA and protein synthesis is reduced throughout meiosis in megasporocytes, and cytoplasmic structures actually become less, rather than more, complex.

Once meiosis is complete, three of the four haploid products degenerate and disappear. The remaining haploid product, the *megaspore,* then undergoes a series of three or more mitotic divisions to produce the female gametophyte with its included egg cell. Because the cells of the female gametophyte are all mitotic descendants of a single megasporocyte, they are genetically identical.

Figure 26-20 shows one of the more common developmental pathways followed by female gametogenesis in flowering plants. After meiosis (Fig. 26-20*a* through *e*) the surviving haploid megaspore undergoes three successive mitotic divisions to produce a total of eight nuclei (Fig. 26-20*f* through *k*). Initially, these nuclei share a common cytoplasm, with no cell walls or plasma membranes separating them. As plasma membranes and partial walls form between the nuclei through a modified process of cell plate formation (see p. 1034), two of the eight nuclei are enclosed in cells called *synergids.* These cells take up a position at the end of the ovule facing the micropyle, the opening through which the male gametophyte penetrates during fertilization. Synergids are the first cells to receive the male gametes during fertilization; they also provide nutrients to the female gametophyte in some flowering plants. Another nucleus is enclosed in a third cell that takes up a position just behind the synergids in most flowering plants. This cell becomes the *egg cell* of the female gametophyte. The three cells together, the two synergids and the egg cell, form a structure known as the *egg apparatus.* The cell walls forming around the egg cell and the synergids are incomplete, so that regions of direct contact remain between the plasma membranes of the adjacent cells. These regions probably facilitate transport of materials between the cells.

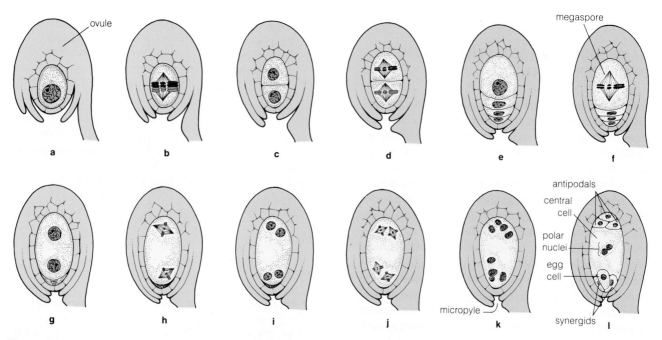

Figure 26-20 A common pathway for development of the female gametophyte in a flowering plant **(a** through **e)** Meiosis in the megasporocyte produces four cells **(e)**, of which only one survives. **(f** through **k)** Three successive mitotic divisions of the surviving cell produce eight nuclei, which become enclosed in seven cells **(l)**. Five of these cells develop into synergids (two cells) and antipodals (three cells). One of the remaining cells is the egg cell; the other cell, the central cell, encloses the two polar nuclei.

Two more of the original eight nuclei take up a position near the egg apparatus. These *polar nuclei* share a common cytoplasm in a large *central cell* that comprises most of the volume of the female gametophyte. The central cell is closely associated at one end with the synergids; in this region cell walls are incomplete, so that the central cell and synergid plasma membranes are also in close contact. Most of the volume of the central cell consists of one or more large central vacuoles; the cytoplasm is concentrated in a thin layer primarily on the side of the cell facing the synergids.

The remaining three nuclei of the gametophyte become enclosed in separate cells that take up a position at the end of the gametophyte opposite the egg apparatus. These cells, called *antipodals*, show much variation in activity, structure, and function in different flowering plants. When active, they may also supply nutrients to the remainder of the gametophyte in some flowering plants. The antipodals undergo additional mitotic divisions or become highly polyploid in some species. When its development is complete, the female gametophyte remains relatively quiescent in the ovule of the flower until fertilization.

Development of the Male Gametophyte   Male gametophytes develop from *microsporocytes* within the anther. The microsporocytes undergo meiosis, each producing four haploid *microspores*. During meiosis RNA transcription and protein synthesis are reduced as in megasporocytes. Similarly, cytoplasmic ribosomes and mRNAs are degraded, and cytoplasmic organelles dedifferentiate or are eliminated.

Each of the four microspores develops into a male gametophyte. During this development a haploid microspore undergoes one mitotic division, producing a male gametophyte that contains only two cells at maturity. The cytoplasmic division accompanying this mitosis is unequal, producing one large and one small cell. The smaller cell, called the *generative cell*, becomes surrounded by the second, larger cell, the *vegetative cell.* Eventually, the generative cell undergoes a second mitotic division to form two *sperm cells*. Although this division takes place during gametophyte development in a few species, such as maize, it occurs during growth of the pollen tube during fertilization in most species. The vegetative nucleus, which does not divide further, directs growth of the pollen tube during fertilization.

During the unequal cytoplasmic division setting off the vegetative and generative cells, most plastids and mitochondria are restricted to the vegetative cell. This preferential exclusion of organelles from the generative cell provides one basis for the maternal inheritance of chloroplasts and mitochondria in flowering plants (see below).

As its development nears completion, the male gametophyte loses water until it has only about 6% to 35% of its original water content. In this dehydrated form it is almost completely inactive biochemically. The cell wall surrounding the developing male gametophyte is replaced by a hard, nearly impermeable surface coat. The mature male gametophyte, known as a *pollen grain*, remains quiescent until it is released from the stamen to be carried by wind, water, insects, or other means to the female parts of a flower of the same species.

There is no indication that gametes in higher plants arise from separate reproductive cell lines set aside early in embryonic development as in animals. Many cell types of the sporophyte generation in flowering plants, if exposed to the correct conditions, can give rise to reproductive structures and gametes. For example, stem or leaf cuttings can be stimulated to generate complete plants that eventually form fully active and normal reproductive structures.

Fertilization in Flowering Plants

The first events in fertilization in flowering plants occur as a pollen grain contacts the stigma of a flower. In many respects the surface of the stigma interacts in patterns that resemble the outer coats of animal eggs. Although the pollen of many different plant species may contact the stigma, only pollen grains of the same species germinate to release the vegetative and generative cells. The reaction may be determined by species-specific recognition proteins at the surface of the pollen grains and receptors for these proteins on the surface of the stigma.

Binding between the stigma and a compatible pollen grain sets off a series of reactions that promote (1) penetration of water into the pollen grain, rehydrating the male gametophyte and opening escape pores in the outer coat of the grain; (2) activation of respiration, RNA transcription, and protein synthesis in the vegetative cell of the pollen grain; and (3) activation and release of enzymes and other proteins carried in the surface layers of the pollen grain. In plants in which the stigma is dry and covered with a protective cuticle, some of these enzymes are apparently capable of digesting the cutin layer.

Growth of the Pollen Tube   As the pollen grain expands, the vegetative cell extends from it and elongates into the *pollen tube*. The tube grows outward through the surface coat of the grain and penetrates downward through the stigma and into the style toward the site of fertilization (see Fig. 26-19). A thin, flexible cell wall covers the growing pollen tube. Only the advancing end of the pollen tube contains the cytoplasm of the vegetative cell and the generative cell (or sperm cells if the generative cell has already divided); at regular intervals older portions of the tube are emptied and sealed off by partitions. The pollen tube continues to grow until it reaches the micropyle of the ovule and the synergid cells of the female gametophyte.

Fertilization If the generative cell has not divided already, it divides mitotically to produce two sperm cells as the pollen tube penetrates into the style. In some species the two sperm cells are indistinguishable; in others plastids and mitochondria remaining in the generative cell are preferentially enclosed in one of the two sperm cells. In either event the nuclei of the two cells are genetically equivalent. A thin cell wall originally surrounding the generative cell is lost when the sperm cells are formed.

In most species the pollen tube first enters one of the two synergid cells at the end of the female gametophyte facing the micropyle. As the pollen tube enters the synergid, an opening appears at its tip and the two sperm cells are released. One of the sperm cells migrates through the synergid and fuses with the egg cell plasma membrane at a point where the synergid and egg cells are in direct contact. The second sperm cell fuses with the central cell at a region of contact between the synergid and this cell. In most species in which plastids and mitochondria are preferentially enclosed in one of the two sperm cells, the plastid-free cell fuses with the egg cell, and the other sperm cell fuses with the central cell.

In the relatively few species in which the sperm cell fusing with the egg cell contains plastids, these plastids, along with the mitochondria of the sperm cell, are usually degraded. As a result, the plastids and mitochondria of the embryo and adult are normally derived from the egg cell. The mechanism excluding organelles from the sperm cell is somewhat "leaky," so that mitochondria or chloroplasts may escape degradation and contribute to the zygote in a small percentage of fertilizations.

Fusion of the sperm and egg nuclei follows a pattern similar to animal species. In various plants the two nuclei may fuse completely before the first embryonic mitosis or during or after alignment of the chromosomes on the spindle of the first division. During this process DNA replication occurs, and the diploid zygote nucleus enters its first mitotic division.

The fertilized egg of plants is equivalent to the zygote of animals. All the tissues of the sporophyte embryo and mature plant arise directly from the fertilized egg by mitotic divisions. Directions for the growth of the plant are completely coded and programmed into the chromosomes of the zygote nucleus and possibly to some extent in molecules stored in the egg cell cytoplasm.

The sperm nucleus entering the central cell fuses with its two polar nuclei. Depending on the species, the polar nuclei may still be separate at this stage or may already have fused into a single diploid nucleus. In either case the triploid cell resulting from fusion of the second sperm nucleus with the polar nuclei enters a series of rapid mitotic divisions to produce a triploid tissue, the *endosperm.* The endosperm surrounds the zygote and provides nutrients to the developing embryo but does not contribute cells directly to any part of the embryo or mature plant. Because two fusions of sperm nuclei occur in fertilization in the angiosperms—one with the egg nucleus to form the zygote and one with the polar nuclei to produce the first endosperm cell—the process is called *double fertilization.* From germination of the pollen grain to nuclear fusion, the entire process requires from 1 to 48 hours in different angiosperm species.

The fertilized egg nucleus subsequently divides mitotically to produce the embryo, which becomes embedded in masses of tissue derived from the parent sporophyte. Eventually, the embryo becomes dormant at an intermediate stage of development and is embedded in hard, impervious coats to form a *seed.* Much diversity exists among different angiosperms in the relative contributions of endosperm and parent sporophyte tissue to the mass of the seed.

For Further Information

Cell cycle regulation, *Ch. 22*
Centrioles
    and the generation of flagella, *Ch. 11*
    origins, *Ch. 21*
    and the plane of cytoplasmic division, *Ch. 24*
    and spindle formation, *Ch. 24*

Cytoplasmic division (cytokinesis) in plants and animals, *Ch. 24*
Extracellular structures, *Ch. 8*
Flagellum and 9 + 2 system of microtubules, *Ch. 11*
Genetic regulation, *Ch. 17*
Meiosis, *Ch. 25*
Microfilaments and cell motility, *Ch. 12*
Microtubules and cell motility, *Ch. 11*
Mitosis, *Ch. 24*
Translational regulation, *Ch. 17*

Suggestions for Further Reading

General Books and Articles

Browder, L. W., ed. 1985. *Oogenesis*. New York: Plenum.

Hedrick, J. L., ed. 1986. *The molecular and cellular biology of fertilization: gamete interactions.* New York: Plenum.

Metz, C. B., and Monroy, A., eds. 1985. *Biology of fertilization.* New York: Cambridge University Press.

Moens, P. B. 1986. *Meiosis.* New York: Academic Press.

Spermiogenesis and Spermatozoa

Hennig, W., ed. 1987. *Spermatogenesis: genetic aspects.* Vol. 15: *Results and problems in cell differentiation.* New York: Springer-Verlag.

Metz, C. B., and Monroy, A., eds. 1985. *Biology of fertilization.* Vol. 2: *Biology of the sperm.* New York: Academic Press.

Phillips, D. M. 1974. *Spermiogenesis.* New York: Academic Press.

Risley, M. S. 1990. Chromatin organization in sperm. In *Chromosomes, eukaryotic, prokaryotic, and viral*, ed. K. W. Adolph, pp. 61–85. Boca Raton, Florida: CRC Press.

Animal Oogenesis and the Mature Egg

Bakken, A. H., and Hines, P. J. 1990. Chromosome structure and function during oogenesis and early embryogenesis. In *Chromosomes, eukaryotic, prokaryotic, and viral*, ed. K. W. Adolph, pp. 40–59. Boca Raton, Florida: CRC Press.

Browder, L. W., ed. 1985. *Oogenesis*. New York: Plenum.

Dunbar, B. S., and Wolgemuth, D. J. 1984. Structure and function of the mammalian zona pellucida, a unique extracellular matrix. *Modern Cell Biol.* 3:77–111.

Metz, C. B., and Monroy, A., eds. 1985. *Biology of fertilization.* Vol. 1: *Model systems and oogenesis.* New York: Academic Press.

Sommerville, J. 1992. RNA-binding proteins: masking proteins revealed. *Bioess.* 14:337–39.

Wasserman, P. M., and Mortillo, S. 1991. Structure of the mouse egg extracellular coat, the zona pellucida. *Internat. Rev. Cytol.* 130:85–110.

Winkler, M. 1988. Translational regulation in sea urchin eggs. A complex interaction of biochemistry and physiological regulatory mechanisms. *Bioess.* 8:157–61.

Fertilization in Animals

Blobel, C. P.; Wolfsberg, T. G.; Turck, C. W.; Myles, D. G.; Primakoff, P.; and White, J. M. 1992. A potential fusion peptide and an integrin ligand domain in a protein active in sperm-egg fusion. *Nature* 356:248–52.

Garbers, D. L. 1989. Molecular basis of fertilization. *Ann. Rev. Biochem.* 58:719–42.

Hart, N. H. 1990. Fertilization in teleost fishes: mechanisms of sperm-egg interactions. *Internat. Rev. Cytol.* 121:1–66.

Hathaway, H. J., and Shur, B. D. 1988. Novel cell surface receptors during mammalian fertilization and development. *Bioess.* 9:153–58.

Hendrick, J. L., ed. 1986. *The molecular and cellular biology of fertilization.* New York: Plenum.

Hoeh, W. R.; Blakley, K. H.; and Brown, W. M. 1991. Heteroplasmy suggests limited biparental inheritance of *Mytilus* mitochondrial DNA. *Science* 251:1488–90.

Kline, D. 1991. Activation of the egg by the sperm. *Biosci.* 41:89–95.

Longo, F. J. 1987. *Fertilization.* London: Chapman and Hall.

———. 1988. Reorganization of the egg surface at fertilization. *Internat. Rev. Cytol.* 113:233–69.

Metz, C. B., and Monroy, A., eds. 1985. *Biology of fertilization.* Vol. 3: *Fertilization: response of the egg.* New York: Academic Press.

Nuccitelli, R. 1991. How do sperm activate eggs? *Curr. Top. Dev. Biol.* 25:1–16.

Oura, C., and Toshimori, K. 1990. Ultrastructural studies on the fertilization of mammalian gametes. *Internat. Rev. Cytol.* 122:105–51.

Richter, J. D. 1991. Translational control during early development. *Bioess.* 13:179–83.

Schatten, G., and Schatten, H. 1987. Cytoskeletal alterations and nuclear architecture changes during mammalian fertilization. *Curr. Top. Devel. Biol.* 23:23–54.

Shapiro, B. M. 1987. The existential decision of a sperm. *Cell* 49:293–94.

Sidhu, K. S., and Guraya, S. S. 1991. Current concepts in gamete receptors for fertilization in mammals. *Internat. Rev. Cytol.* 127:253–88.

Trimmer, J. S., and Vacquier, V. D. 1986. Activation of sea urchin gametes. *Ann. Rev. Cell Biol.* 2:1–26.

Vacquier, V. D. 1986. Activation of sea urchin spermatozoa during fertilization. *Trends Biochem. Sci.* 11:77–81.

Wassarman, P. M. 1987. Early events in mammalian fertilization. *Ann. Rev. Cell Biol.* 3:109–42.

———. 1987. The biology and chemistry of fertilization. *Science* 235:553–60.

———. 1988. Fertilization in mammals. *Sci. Amer.* 259:78–84 (December).

Winkler, M. 1988. Translational regulation in sea urchin eggs: a complex interaction of biochemical and physiological regulatory mechanisms. *Bioess.* 8:157–61.

Yanagimachi, R. 1988. Sperm-egg fusion. *Curr. Top. Membr. Transport* 32:3–43.

Gametogenesis in Plants

Dickinson, H. G. 1989. The physiology and biochemistry of meiosis in the anther. *Internat. Rev. Cytol.* 107:79–109.

Kapil, R. N., and Bhatnagar, A. K. 1981. Ultrastructure and biology of female gametophyte in flowering plants. *Internat. Rev. Cytol.* 70:291–341.

Knox, R. B. 1984. The pollen grain. In *Embryology of angiosperms*, ed. B. M. Johri, pp. 197–271. New York: Springer-Verlag.

Mascarenhas, J. P. 1989. The male gametophyte of flowering plants. *Plant Cell* 1:657–64.

———. 1990. Gene activation during pollen development. *Ann. Rev. Plant Physiol. Plant Molec. Biol.* 41:317–38.

McCormick, S. 1991. Molecular analysis of male gametogenesis in plants. *Trends Genet.* 7:298–303.

Russell, S. D. 1991. Isolation and characterization of sperm cells in flowering plants. *Ann. Rev. Plant Physiol. Plant Molec. Biol.* 42:189–204.

Willemse, M. T. M. 1984. The female gametophyte. In *Embryology of angiosperms*, ed. B. M. Johri, pp. 159–96. New York: Springer-Verlag.

Fertilization in the Flowering Plants

Connett, M. B. 1987. Mechanisms of maternal inheritance of plastids and mitochondria: development and ultrastructure. *Plant Molec. Biol. Reporter* 4:193–205.

Cornish, E. C.; Anderson, M. A.; and Clarke, A. E. 1988. Molecular aspects of fertilization in flowering plants. *Ann. Rev. Cell Biol.* 4:209–28.

Dickinson, H. G. 1990. Self-incompatibility in flowering plants. *Bioess.* 12:155–61.

Haring, V.; Gray, J. E.; McClure, B. A.; Anderson, M. A.; and Clarke, A. E. 1990. Self-incompatibility: a self-recognition system in plants. *Science* 250:937–41.

Heslop-Harrison, J. 1987. Pollen germination and pollen-tube growth. *Internat. Rev. Cytol.* 107:1–78.

Knox, R. B., and Singh, M. B. 1987. Perspectives in pollen biology and fertilization. *Ann. Bot.* 60 (Suppl. 4):15–37.

Russell, S. D. 1991. Isolation and characterization of sperm cells in flowering plants. *Ann. Rev. Plant Physiol. Plant Molec. Biol.* 42:189–204.

Tiezzi, A. 1991. The pollen tube cytoskeleton. *Electron Micr. Rev.* 4:205–19.

van Went, J. L., and Willemse, M. T. M. 1984. Fertilization. In *Embryology of angiosperms*, ed. B. M. Johri, pp. 273–317. New York: Springer-Verlag.

Review Questions

1. Define gametogenesis, spermiogenesis, spermatid, spermatozoon, oogenesis, oocyte, and egg.

2. Outline the structures present in the sperm head of a mature sperm cell. What is the acrosome? What enzymes are present in the acrosome? Compare the acrosome to a lysosome in a somatic cell. What is periacrosomal material? How does it function in fertilization?

3. Outline the structures present in the middle piece and tail of a mature sperm cell. How do centrioles function in sperm development? What is the 9 + 2 system of sperm flagella? What variations in this system are noted among animals? What is the effect of these variations on sperm motility? What are accessory fibers?

4. Outline the changes occurring in the nucleus, nuclear envelope, ER, Golgi complex, and mitochondria during sperm development. What is a mitochondrial derivative? What are protamines and sperm histones? What biochemical events take place during the replacement of somatic histones by sperm histones or protamines? What is the functional significance of the replacement of somatic histones during sperm development?

5. Outline the patterns of RNA transcription and protein synthesis observed during sperm development.

6. Outline the major structural specializations of oocytes, giving the major functions during fertilization of each specialization. What are polar bodies?

7. What are the sources of the materials packed into the egg cytoplasm during oogenesis? What are follicle cells? Nurse cells? Cortical granules? Compare the structure and function of cortical granules and lysosomes.

8. At what meiotic stages are egg cells arrested at maturity?

9. Outline the major patterns of RNA transcription and protein synthesis during oocyte development. How do these products differ in fate and function from the RNAs and proteins synthesized during sperm development?

10. Outline the structure of the vitelline coat of invertebrate eggs and the zona pellucida of mammalian eggs. What proteins occur in the zona pellucida of mice? What are their probable structural and functional roles? What structures occur outside the vitelline coat in sea urchins? In amphibians? In insects? In birds? What is the micropyle, and how does it function in fertilization?

11. What are yolk bodies or granules? What substances typically occur in yolk granules?

12. What are the sources of the materials concentrated in yolk granules? How do these bodies function after fertilization?

13. What evidence indicates that attachment of sperm cells to egg coats involves a receptor-ligand interaction?

14. What is the acrosome reaction? What steps take place in the acrosome reaction? What is the acrosomal process or filament? How does it function in fertilization? Compare the acrosome reaction in the sea urchin and mammals.

15. What triggers egg activation? What events take place in egg activation? What is the fast block to polyspermy? What events account for the fast block in sea urchins? What events are triggered by cortical granule release in sea urchins? In mammals? What is the slow block to polyspermy? What changes occur in the male pronucleus after it enters the egg cytoplasm? How are DNA replication, fusion of the pronuclei, and the first mitotic division coordinated during fertilization?

Supplement 26-1
Replacement of Somatic Histones by Sperm Histones or Protamines

The extent of nuclear condensation during sperm development is generally correlated with the degree to which somatic histones are replaced by protamines. Protamines are very small, highly basic proteins with molecular weights ranging from about 4000 to 6000. In contrast to the strongly conserved nature of somatic histones (see p. 539), protamines are highly variable and show few similarities in size or amino acid sequence between plants and animals or among the major animal groups.

In most animals only one or at most a few protamine types are present in mature sperm nuclei. Among mammals, for example, ram and rat sperm nuclei contain only one protamine type at maturity. Mouse sperm nuclei contain two protamine types and human sperm three.

The basicity of protamines results from an unusually high content of arginine residues (Fig. 26-21). Lysines also supply basic groups in some mammalian protamines. In the most basic protamines, which occur in sperm nuclei of some fishes, arginine may make up as much as three-quarters of the total amino acid complement. These proteins, which contain only about 30 amino acids in total, are also the smallest of the animal protamines. The protamines of most other animals, depending on the species, range from an arginine content of 40% or 50% to nearly 70%. The few lower-plant sperm cells studied to date also contain protaminelike proteins with about 50% arginine content.

In mammalian protamines arginines make up about 50% to 60% of the total amino acid content. The protamines of placental mammals contain cysteine residues, which set up disulfide linkages (see p. 47) between protamine molecules as nuclear condensation becomes complete. The disulfide linkages tie the entire mammalian sperm nucleus into a highly insoluble and chemically stable structure. Only one species outside placental mammals, the dogfish shark, is known to produce protamines containing cysteine residues.

The degree of replacement of somatic histones by protamines varies greatly among different species. In some, such as the trout, virtually all somatic histones

Salmon	Ala	Arg	Arg	Arg	Arg	Ser	Ser	Ser	Arg	Pro	Ile	Arg	Arg	Arg	Arg	Pro	Arg	Arg	Arg	Thr	Thr
Bull	Ala	Arg	Tyr	Arg	Cys	Cys	Leu	Thr	His	Ser	Gly	Ser	Arg	Cys	Arg	Arg	Arg	Arg	Arg	Arg	Arg
Ram	Ala	Arg	Tyr	Arg	Cys	Cys	Leu	Thr	His	Ser	Arg	Ser	Arg	Cys	Arg	Arg	Arg	Arg	Arg	Arg	Arg
Mouse	Ala	Arg	Tyr	Arg	Cys	Cys	Arg	Ser	Lys	Ser	Arg	Ser	Arg	Cys	Arg	Arg	Arg	Arg	Arg	Arg	Cys
Human P1	Ala	Arg	Tyr	Arg	Cys	Cys	Arg	Ser	Gln	Ser	Arg	Ser	Arg	Tyr	Tyr	Arg	Gln	Arg	Gln	Arg	Ser

Salmon	Arg	Arg	Arg	Arg	Ala	Gly	Arg	Arg	Arg	Arg											
Bull	Cys	Arg	Arg	Arg	Arg	Arg	Arg	Phe	Gly	Arg	Arg	Arg	Arg	Arg	Arg	Val	Cys	Tyr	Val	Ile	Arg
Ram	Cys	Arg	Arg	Arg	Arg	Arg	Arg	Phe	Gly	Arg	Arg	Arg	Arg	Arg	Arg	Val	Cys	Cys	Arg	Arg	Tyr
Mouse	Arg	Arg	Arg	Arg	Arg	Arg	Cys	Cys	Arg	Arg	Arg	Arg	Arg	Arg	Cys	Cys	Arg	Arg	Arg	Arg	Ser
Human P1	Arg	Arg	Arg	Arg	Arg	Arg	Ser	Cys	Gln	Thr	Arg	Arg	Arg	Ala	Met	Arg	Cys	Cys	Arg	Pro	Arg

Bull	Cys	Tyr	Arg	Gln				
Ram	Thr	Val	Val	Arg	Cys	Thr	Arg	Gln
Mouse	Tyr	Thr	Ile	Arg	Cys	Lys	Lys	Tyr
Human P1	Tyr	Arg	Pro	Arg	Cys	Arg	Arg	His

Figure 26-21 Protamine sequences in different animal species. Protamines typically have long blocks of repeated arginine residues (in red) separated by other amino acids. Lysines also occur as basic residues in some mammalian protamines. Salmon sequence: T. Ando and K. Suzuki, *Biochim. Biophys. Acta* 10:375 (1967); bull: J. P. Coelingh et al., *Biochim. Biophys. Acta* 285:1 (1972); ram: P. Sautiere et al., *Eur. J. Biochem.* 144:121 (1984); mouse: K. C. Kleene et al., *Biochem.* 24:719 (1985); human P1: D. J. McKay et al., *Biosci. Rep.* 5:383 (1985).

are replaced. In mammals and some molluscs 10% to 20% of the DNA retains somatic histones; in the winter flounder somatic histones form 75% of the total protein content of mature sperm nuclei. In a limited number of species, including the frog (*Rana*), sea urchin, sea cucumber, horseshoe crab, and goldfish, sperm nuclei contain only histones; no protamines are present. The sperm nuclei in the pollen of higher plants also apparently retain somatic histones.

In some of the animal species in which sperm nuclei retain histones, such as the goldfish, sperm histones are entirely of the somatic type. In others, including the sea urchin, some somatic histones, such as H1 and H2B, are replaced by variants that appear only in sperm nuclei. In a very few species, including the crab, lobster, and other crustaceans, no proteins of any kind can be detected in association with the DNA in mature sperm nuclei.

Nonhistone proteins disappear entirely or are reduced to a small fraction of the nuclear proteins as sperm cells mature in most animal species. The identity of the small nonhistone fraction retained and its functions in sperm nuclei remain unknown.

The degree to which the sperm chromatin retains nucleosome structure (see p. 546) parallels the quantity of histones retained in the sperm nuclei at maturity. If 20% of the chromatin retains somatic histones, for example, approximately 20% of the chromatin fibers contain nucleosomes; the remainder is smooth and nucleosome free. The DNA complexed with protamines is completely protected from digestion by DNA endonucleases and reveals none of the periodicities typical of somatic chromatin when exposed to these enzymes (see p. 547).

A number of protamines, particularly in fishes and mammals, have been completely sequenced (see Fig. 26-21). The sequencing reveals several interesting features of protamine structure, including a general lack of sequence conservation between different major animal groups. The lack of conservation suggests that the overall basicity of these proteins is more important to their function than a particular amino acid sequence.

The molecular structure of the DNA-protamine complex is uncertain. X-ray diffraction studies indicate that the DNA is held in sperm nuclei in a form closely similar to the B conformation (see p. 529). Several models have been advanced that propose interaction of protamines, in extended or alpha-helical form, with either the major or minor groove of B-form DNA. Although the models differ in detail, all agree that the forces holding the structure together are strong electrostatic attractions between the numerous positively charged arginine groups of protamines and the negatively charged phosphate groups of DNA. Neutralization of the phosphate groups no doubt contributes to the compaction of sperm cell DNA brought about by the protamines.

Research by G. H. Dixon and his colleagues and others showed that somatic histones are replaced by protamines in a regular series of biochemical changes. Initially, protamines are synthesized in quantity in the cytoplasm and modified by the addition of numerous phosphate groups. The phosphorylated protamines then penetrate through the nuclear envelope and accumulate in the nucleus. Just before their replacement begins, histones become highly acetylated. Presumably, this acetylation greatly reduces the electrostatic attraction between the histones and DNA and facilitates histone removal. At the same time the protamines are dephosphorylated. Removal of phosphate groups from the protamines greatly increases their positive charge and attraction for DNA. Histone replacement then proceeds, probably simply by competition for the phosphate groups of DNA by the more strongly charged protamines.

In some animal species proteins with size and basicity ranging between histones and protamines appear as temporary replacements before the final sperm protamines combine with the DNA. In the cricket, for example, R. McMaster-Kaye and J. S. Kaye detected a series of seven proteins of intermediate size that temporarily replace the histones before a final group of four protamines appears. A similar group of proteins also appears as transitory intermediates in the development of mammalian sperm.

The biochemical and molecular changes occurring after fertilization, in broad outline, are the reverse of those adding protamines during sperm development. As the sperm nucleus penetrates the egg cytoplasm during fertilization, the protamines become highly phosphorylated and are replaced by somatic histones stored in the oocyte cytoplasm.

THE ORIGINS OF CELLULAR LIFE

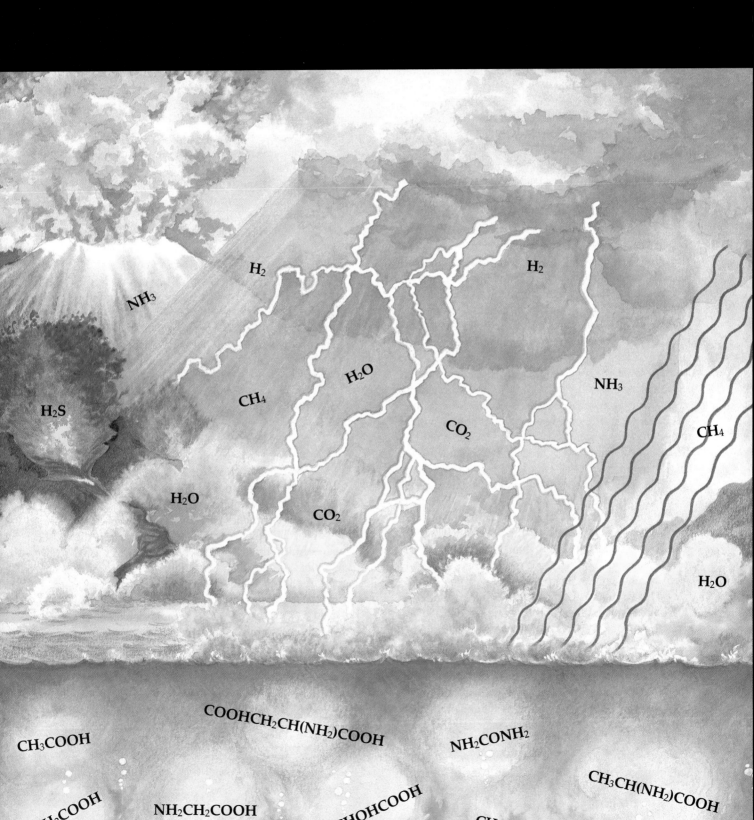

The most ancient rocky material of the solar system has been dated as approximately 4.6 billion years old. At that distant time our planet condensed out of the primordial matter and began its long transition into the environment we know today. The oldest known fossils of bacterialike cells, although controversial, have been claimed to exist in the Archean deposits near North Pole, Australia, laid down about 3.5 billion years ago. If the North Pole fossils are authentic, life must have appeared on earth between about 4.5 and 3.5 billion years ago, during the first billion years of its existence.

How did life appear on earth? A divine origin for life has been proposed by many religions. However, the possibility of a divine origin falls outside the realm of science because it cannot be tested. The same can be said for the idea that life may have originated here through contamination on a space probe launched from another planet billions of years ago. From a scientific standpoint it is necessary to hypothesize that life evolved from nonliving matter through spontaneous chemical and physical processes no different from those in operation in the universe today. Hypotheses made under these assumptions are testable to the extent that the chemical and physical processes can be duplicated in the laboratory.

The scientific view of the origin of life is a special form of a very old belief, easily as old as beliefs in divine origins, that life can arise by spontaneous generation. As late as the mid-nineteenth century, laymen and scientists alike believed that lesser forms of life could arise spontaneously and more or less constantly from rotting organic matter. A prime example is the recipe for spontaneous production of mice, reported in the 1600s by the chemist J. B. van Helmont: "If you press a piece of underwear soiled with sweat together with some wheat in an open mouth jar, after about twenty-one days the odor changes and the ferment, coming out of the underwear and penetrating through the husks of the wheat, changes the wheat into mice."

The common belief in spontaneous generation was refuted in 1862 by Louis Pasteur's famous demonstration that nutrient fluids, sterilized and sealed against contamination, could be kept indefinitely without generation of microbial or other life. In a sense Pasteur's refutation of spontaneous generation was too effective; scientists generally came to reject the idea that life could have arisen spontaneously at any time. As a result, even scientific hypotheses proposing chemical or molecular origins for life were not accepted as valid.

This generally negative attitude remained unchanged until the 1920s, when A. I. Oparin, a Russian, and J. B. S. Haldane, an Englishman, independently published a hypothesis that forced a reconsideration of spontaneous generation. Oparin and Haldane agreed that spontaneous generation of cellular life is impossible under present-day conditions. They argued, however, that the earth's surface and atmosphere during the first millions of years of its existence were radically different from today. Primordial conditions, according to Oparin and Haldane, may have favored rather than inhibited spontaneous generation of life.

These investigators proposed that earth's primitive atmosphere contained primarily reduced substances such as methane (CH_4), ammonia (NH_3), and water instead of high concentrations of oxygen as in today's atmosphere. In such an atmosphere electrons and hydrogens were readily available for the conversion of inorganic substances to organic forms. Through the absorption of solar energy and the effects of electrical discharges, great quantities of organic molecules were supposedly produced in the atmosphere and the earth's surface. The organic molecules would have accumulated because the two main routes by which such substances break down today, oxidation and decay by microorganisms, could not take place. As these organic compounds became more and more concentrated over many millions of years, they interacted spontaneously to produce still more complex organic substances such as nucleic acids and proteins.

These complex substances, according to the Oparin–Haldane hypothesis, constantly aggregated into random collections of molecules, some of which were able to carry out primitive living reactions. Presumably, these combinations were more successful than nonliving assemblies in competing for space and raw materials and therefore persisted. Eventually, the most successful of these aggregations developed the full qualities of life, including the ability to self-reproduce.

As the chemical activities of the first collections of living matter increased, the store of organic molecules used as an energy source became depleted. Life persisted, however, through the development of photosynthesis, which took advantage of sunlight as an inexhaustible energy source for the production of organic molecules from inorganic precursors.

Photosynthesis and the organisms obtaining energy by this pathway gradually became more complex until the level was achieved in which water was used as a primary source of electrons and hydrogens for photosynthetic-synthetic pathways, and oxygen was released as a by-product (see p. 367). Life at this level was definitely cellular and had advanced in complexity to organisms that were the direct ancestors of present-day bacteria and cyanobacteria.

The release of oxygen in ever greater quantities by early cyanobacteria gradually changed the atmosphere from a reducing to an oxidizing character. Once this change came about, spontaneous generation of life was no longer possible because organic molecules generated outside the confines of living cells were quickly oxidized to inorganic form. From this time on, life could arise only from preexisting life as in today's environment.

The Oparin–Haldane hypothesis was not widely accepted at first because of the general weight of opinion against spontaneous generation and the lack of an effective scientific test of the ideas. The situation remained much the same for the next 30 years, until several discoveries stimulated new interest in thinking about the origins of life. One was a successful test by S. L. Miller in 1953 of a major proposal of the Oparin–Haldane hypothesis: that an energy source acting upon a reducing atmosphere can spontaneously generate organic molecules. The second discovery came from analyses of the light transmitted or reflected by the atmospheres of other planets in the solar system or by dust clouds in interstellar space. The outermost planets, in particular Jupiter, Saturn, and Uranus, proved to have atmospheres containing some of the components proposed by Oparin and Haldane, including methane and ammonia. The interstellar dust clouds were also found to contain reduced gases and even organic molecules (see Table 27-1). Finally, some meteors arriving from outer space were found to contain a variety of organic molecules including amino acids.

These discoveries showed that key assumptions of the Oparin–Haldane hypothesis were feasible and, in fact, appeared to have taken place. The findings set off a wave of further experimentation and thinking about the conditions and reactions involved in the transition from nonliving to living matter. The contemporary conclusion from this work is that life originated through spontaneous, inanimate processes taking place under the conditions existing on the primitive earth. The molecular assemblies harboring the spark of life advanced through a series of intermediate stages that, although not yet cellular, were able to carry out successively more complex activities. Eventually, the molecular aggregates took on fully cellular characteristics.

STAGES IN THE EVOLUTION OF CELLULAR LIFE

Because conditions on the primitive earth were critical for the evolution of cellular life, the formation of the earth and its atmosphere is considered as the first of five stages leading from inanimate matter to living forms. This stage provided inorganic raw materials for the evolution of life and set up conditions for their interaction. The second stage produced organic molecules through interactions between inorganic substances, driven by energy sources such as lightning and ultraviolet radiation from the sun. In the third stage the organic molecules produced in the second stage assembled randomly into collections capable of chemical interaction with the environment. As the collections formed, interactions taking place within them produced still more complex organic substances, including polypeptides and nucleic acids. Some of these collections of molecules were capable of carrying out reactions associated with the quality of life.

There is little agreement on the form taken by the first spark of life in these primitive aggregates. Some investigators propose that in its most primitive form life consisted simply of the ability of a molecular aggregate to absorb organic molecules from the environment and use them as raw materials and as an energy source to grow in mass. A simple form of reproduction may have occurred through physical breakage of larger aggregates into two or more smaller collections retaining the organization necessary to carry out the primitive living reactions.

In the fourth stage a genetic code appeared in the primitive living aggregates. The code regulated the duplication of information required for reproduction of the molecular aggregates and established the link between nucleic acids and the ordered synthesis of proteins. With these developments providing the basis for directed synthesis and reproduction, life, although primitive and precellular, was fully established in the molecular assemblies.

The genetic code, built into the structure of the nucleic acids, provided a means for introducing and recording mutations. This development established natural selection of favorable mutations as the basis for further evolutionary change. At this stage evolution progressed from chemical to biological. The gradual accumulation of favorable mutations by natural selection accomplished the fifth and final stage in the origin of cellular life: the precellular assemblies were converted into fully organized cells, complete with a nuclear region and cytoplasm enclosed by an outer boundary membrane.

The first cells to appear on earth were probably roughly equivalent in complexity to the most primitive prokaryotes known today. Presumably, this level was reached by the time the Archean deposits were laid down in Australia some 3.5 billion years ago.

Stage One: Formation of the Earth and Its Primitive Atmosphere

According to current hypotheses, the solar system, and in fact all the stars and other bodies in space, condensed

Figure 27-1 The Horsehead nebula, a cloud of gas and dust particles some 1300 light years from earth. Courtesy of Hale Observatories.

Table 27-1	Substances and Chemical Groups Important in the Evolution of Life Detected In Cosmic Clouds or Outer Space
Atom, Molecule, or Radical	**Symbol**
Hydrogen atom	H
Hydroxyl radical	OH $^-$
Ammonia	NH_3
Water	H_2O
Formaldehyde	HCHO
Carbon monoxide	CO
Cyanogen radical	CN $^-$
Hydrogen cyanide	HCN
Cyanoacetylene	HC_2CN
Methyl alcohol	CH_3OH
Formic acid	HCOOH
Carbon monosulfide	CS
Formamide	$HCONH_2$
Silicon oxide	SiO
Carbonyl sulfide	OCS
Acetonitrile	CH_3CN
Isocyanic acid	HNCO
Hydrogen Isocyanide	HNC
Methylacetylene	CH_3C_2H
Acetaldehyde	CH_3CHO
Thioformaldehyde	HCHS
Hydrogen sulfide	H_2S
Methylene imine	H_2CNH

Adapted from S. W. Fox, *Mol. Cell. Biochem*, 3:1291 (1974); courtesy of S. W. Fox.

out of vast clouds of gas and dust particles. The matter in these clouds, which still persist in space (Fig. 27-1), has been identified by analysis of light or other forms of radiant energy passing through them. The interstellar clouds consist mostly of hydrogen gas at extremely low concentrations. Lesser amounts of helium and neon are also present. Other elements and compounds, including metallic iron and nickel, the silicates, oxides, sulfides, and carbides of these and other metals, inorganic carbon compounds, ammonia, and water, are also present. In addition, most notably for the evolution of life, organic molecules such as formic acid, methyl alcohol, formaldehyde, acetaldehyde, and cyanoacetylene can be detected in the clouds (Table 27-1).

According to the condensation hypothesis, first proposed by Immanuel Kant, stars and planetary systems are continually condensing and disintegrating in the interstellar dust and gas clouds. Some 4 to 5 billion years ago our solar system condensed from one of the clouds. Rapid condensation of most of the matter around one center caused high pressure and heat to develop in a rotating gas cloud. The intense heat and pressure set off a thermonuclear reaction, establishing the star of our solar system, the sun. The remainder of the spiraling dust and gas condensed into the planets and other bodies surrounding the sun. As the planets formed, internal pressure and heat were also generated in these bodies.

On earth internal temperatures probably rose as high as 1000° to 3000°C, causing melting and stratification of its solid matter. Metallic elements settled to form the molten core of the earth, and lighter substances, such as silicates, carbides, and sulfides of the metallic elements, floated to the surface. As the planet radiated away some of its heat, layers at the surface cooled and solidified into the rocky materials of the earth's crust. Water, initially present only as a vapor, condensed into droplets and rained down on the rocks

of the crust. Eventually, after millennia of torrential rains and further cooling, water collected into the rivers, lakes, and seas of the primitive earth.

Some of the water was retained as vapor in the atmosphere. Other components of the initial atmosphere probably included H_2, N_2, CO, and CO_2. Carbon dioxide would have been steadily lost from the atmosphere through its interaction with elements at the earth's surface to form solid carbonates. However, significant quantities would probably have returned to the atmosphere in the gases released by erupting volcanoes. H_2S may also have been among the gases released by volcanoes. Any molecular oxygen initially present would have been removed by reaction with elements of the crust and atmosphere to form oxides.

Whether ammonia and methane were present in quantity in the primordial atmosphere, as proposed by Oparin and Haldane, remains a subject of considerable controversy. These gases, according to some scientists, would have been produced by spontaneous interaction of the hydrogen, nitrogen, and carbon compounds of the primitive atmosphere and crust. This proposal is supported by spectroscopic analysis of the atmospheres of some other planets and large satellites of the solar system, in which either ammonia or methane or both are present along with water vapor, and free oxygen is absent. However, geochemists, in analyzing the oldest known sedimentary rock strata in Isua, Greenland (3.8 billion years old), find evidence of carbonates but none of the compounds expected if ammonia or methane was present in quantity on the primitive earth. According to the geochemists, the only source of hydrogen was likely to have been the H_2 condensing from the original cloud of primordial matter. In any case the early atmosphere is expected to have stabilized as moderately reducing if only hydrogen was present or strongly reducing if gases such as ammonia, methane, or H_2S were present.

All agree that free oxygen was absent from the primitive atmosphere. The reducing character of the primitive earth's atmosphere was fundamentally important to the second stage in the evolution of life, the spontaneous appearance of organic molecules characteristic of living systems. The time from the initial condensation of matter until the earth's environment favored the spontaneous formation of organic compounds may have been as much as half a billion years.

Stage Two: Spontaneous Formation of Organic Molecules

In the second stage in the evolution of life organic molecules were produced by the action of various energy sources on the components of the primitive atmosphere and crust. These energy sources are believed to be essentially the same as those acting on the earth and its atmosphere today: solar radiation, electrical discharges during storms, cosmic rays, radioactivity produced by atomic decay in surface rocks, shock waves from meteorites passing through the atmosphere, and heat released from the earth's core, particularly by volcanoes. (Table 27-2 lists the energy inputs from these sources in the present-day environment.) The organic

Table 27-2	Present-day Sources of Energy on the Earth
Source	Energy (cal/cm²/yr)
Sun (total radiation including ultraviolet)	260,000
Ultraviolet light	4,000
Electrical discharges	4
Shock waves	1.1
Radioactivity (to 1-km depth)	0.8
Volcanoes	1.13
Cosmic rays	0.0015

Adapted from S. L. Miller, H. C. Urey, and J. Oro, *J. Mol. Evol.* 9:59 (1976): courtesy of S. L. Miller and Springer-Verlag.

molecules produced in the atmosphere through these energy sources were trapped in condensing raindrops and carried down to the crust by rainstorms. Additional material was produced at the surface and possibly at sites where vents spurted hot water under the sea. Gradually, the organic molecules accumulated in lakes and seas. As their concentrations increased, interactions between organic substances and the elements and compounds of the crust and atmosphere produced more varied and complex molecules.

Evidence That Organic Molecules Can Be Synthesized Under Primitive-Earth Conditions Miller's experiment in 1953 was the first to demonstrate that organic molecules can be synthesized spontaneously under the conditions proposed for the primitive earth. Miller, then a graduate student working with H. C. Urey, developed an apparatus to test the effects of electrical discharges on a simulated primitive atmosphere (Fig. 27-2). Miller assumed that the primitive atmosphere contained both

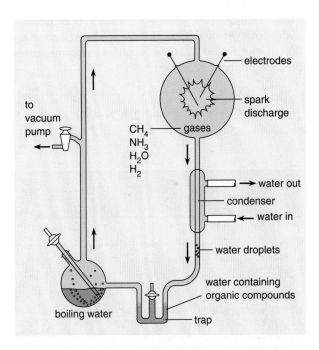

Figure 27-2 The Miller apparatus demonstrating that organic molecules can be spontaneously synthesized under conditions simulating the primitive earth. The apparatus contained a simulated atmosphere consisting of methane, ammonia, hydrogen, and water vapor. The gases were exposed to an energy source consisting of constantly sparking electrodes. Water was boiled into water vapor at one site in the apparatus and condensed back into water at another site. The water condensation removed and trapped organic molecules produced by the action of the energy source on the atmosphere. Operation of the apparatus for one week produced a surprising variety of organic compounds. Courtesy of S. L. Miller. Copyright 1955 by the American Chemical Society.

methane and ammonia, along with hydrogen and water vapor, and thus was strongly reducing. He placed these gases in a closed apparatus and exposed them to an energy source in the form of continuously sparking electrodes. Water vapor was added to the atmosphere in the chamber from water boiled at one point in the apparatus and continually removed by condensation at another point. The condensation also removed and trapped organic molecules produced by the action of the energy source on the atmosphere. Operation of the apparatus for one week produced a surprising assortment of organic compounds, including urea, biological and nonbiological amino acids, and lactic, formic, and acetic acids (Table 27-3). As much as 15% of the carbon in the simulated atmosphere was converted into organic compounds. The apparatus provided a convincing demonstration that organic compounds can actually be produced through the action of energy sources on a reducing atmosphere.

After Miller published his results, many more experiments were carried out using variations in the starting substances and energy input. An important substance included in many of these experiments was hydrogen cyanide (HCN), readily produced by the action of electrical discharges on presumed components of the primitive atmosphere:

$$CO + NH_3 \longrightarrow HCN + H_2O \qquad (27\text{-}1)$$

$$2CH_4 + N_2 \longrightarrow 2HCN + 3H_2 \qquad (27\text{-}2)$$

When HCN is present, Miller-type experiments readily produce additional amino acids, purines, pyrimidines, and porphyrins. HCN is also easily converted into a variety of *condensing agents* that promote the assembly of amino acids into polypeptides and nucleotides into nucleic acids. (Condensing agents are substances that promote condensation reactions, in which molecular subunits are assembled by removal of the elements of a water molecule from the reactants; see p. 46.) Formaldehyde, also readily produced by interactions of the primitive gases, leads to production of sugars when included in experiments simulating primitive earth conditions.

Biological amino acids are common among the organic substances produced in these experiments. Large quantities of nonbiological amino acids are also produced, including D-forms. (Only the L-forms of the 20 biological amino acids are used by ribosomes in protein synthesis; see Ch. 2.) Why only 20 of the many amino acids probably produced in this stage were later adapted for use in protein synthesis, and these only in the L-form, are among the many unanswered questions about the origin of life.

Sugars have been produced in experiments duplicating primitive conditions by reactions in which formaldehyde units interact in alkaline solutions to produce four-, five-, and six-carbon sugars such as glyceraldehyde, ribose, deoxyribose, glucose, fructose, mannose, and xylose. Fatty acids have also been produced; in

Table 27-3 Organic Compounds Formed in Sample Runs in the Miller Apparatus (mole × 10⁵)			
	Run 1	Run 3	Run 6
Glycine	63 (2.1%)*	80 (0.46%)*	14.2 (0.48%)*
Alanine	34	9	1.0
Sarcosine	5	86	1.5
β-alanine	15	4	7.0
α-aminobutyric acid	5	1	—
Methylalanine	1	12.5	—
Aspartic acid	0.4	0.2	0.3
Glutamic acid	0.6	0.5	0.5
Iminodiacetic acid	5.5	0.3	3.9
Iminoacetic-propionic acid	0.5	—	—
Formic acid	233	149	135
Acetic acid	15.2	135	41
Propionic acid	12.6	19	22
Glycolic acid	56	28	32
Lactic acid	31	4.3	1.5
α-hydroxybutyric acid	5	1	—
Succinic acid	3.8	—	2
Urea	2	—	2
Methylurea	1.5	—	0.5
Total yields of compounds listed	15%*	3%*	8%*

Reprinted with permission from S. L. Miller, in A. I. Oparin, ed., *The Origin of Life on the Earth* (I.U.B.: Sympos. Series). New York; Pergamon Press, 1959.
* Percent yield based on carbon placed in the apparatus as methane.

one experiment by W. R. Hargreaves, S. J. Muvihill, and D. W. Deamer, fatty acids and glycerol combined spontaneously to produce phospholipids when heated to dryness at 65°C, as they might have been in an evaporating tidepool along a primitive sea.

Adenine has proved to be the most easily produced of the nitrogenous bases occurring in nucleic acids. C. Ponnamperuma and his coworkers detected synthesis of this purine, along with small amounts of guanine, in mixtures of CH_4, NH_3, and H_2O exposed to a beam of electrons. Purines are also readily synthesized when experimental atmospheres containing HCN are irradiated with ultraviolet light. Spontaneous synthesis of pyrimidines requires more extreme conditions, such as exposure of urea to cyanoacetylene ($HC\equiv C-C\equiv N$) at elevated temperatures. Cyanoacetylene, produced spontaneously by the action of electrical discharges on CH_4 and N_2, acts as a condensing agent in this reaction.

Experiments with primitive atmospheres and energy sources have thus produced organic building blocks for all major biological molecules—proteins, carbohydrates, lipids, and nucleic acids. The numbers and variety of organic molecules produced in the simulation experiments depend directly on the degree to which the primitive atmosphere is reducing in character. The richest variety is obtained if methane and ammonia are among the gases being tested. If only hydrogen is present, the yield of organic material is much smaller. Even if only H_2 and CO_2 were present in the primitive atmosphere, J. Pinto and his coworkers estimate that the action of solar radiation alone on these substances would have produced 3 million tons of formaldehyde each year. At this rate formaldehyde would have reached concentrations in the oceans high enough to polymerize into more complex organic molecules within 10 million years.

One of the fundamental requirements for the synthesis and accumulation of organic molecules in any quantity in the experiments simulating primitive conditions is the absence of oxygen. If molecular oxygen is present among the reactants, converting the atmosphere from a reducing to an oxidizing character, the yields of organic molecules are vanishingly small or nonexistent.

An analysis by K. A. Kvenvolden, J. Oro, and their colleagues of meteorites that have struck the earth directly supports conclusions from the Miller-type experiments. For example, a large stony meteorite found near Murchison, Australia, contained no fewer than 77 amino acids, including many of the 20 occurring in proteins. Purines and pyrimidines are also suspected to be among the organic molecules of the meteorite. Analysis of the tail of Halley's comet by space probes and spectroscopy also indicates that the body of the comet contains a complex mixture of organic molecules. W. F. Heubner even found evidence of a formaldehyde polymer, polyoxymethane ($H_2CO)_n$, in Halley's comet.

Interstellar clouds, as noted, also contain formaldehyde, HCN, acetaldehyde, and cyanoacetylene.

Some authorities criticize the results obtained with the Murchison and other meteorites, pointing out that the amino acids and other organic compounds could have been picked up as contaminants during passage through the atmosphere and collision with the earth's surface. This is not thought likely, however, because nonbiological amino acids preponderate and because D- and L-forms occur in approximately equal quantities in the Murchison meteorite. If the amino acids were picked up as contaminants from living organisms, they should be primarily or exclusively in the L-form and mostly biological rather than nonbiological.

The arrival of organic molecules in meteorites and comets may have actually contributed significantly to the complement of organic molecules on the primitive earth during the first billion years of the solar system's existence. During this period dust and asteroids were densely distributed within the boundaries of the solar system as remnants of the dust cloud that condensed to form the sun and planets. Much of this material was swept from space by the planets and larger satellites as they moved through their orbits during the period from about 4.5 to 3.8 billion years ago. For example, most of the craters of the moon were formed by collisions with interplanetary debris during this period. The collisions are estimated by C. F. Chiya to have delivered at least 1 to 10 million kilograms of organic matter per year to the primordial earth.

Spontaneous Combination of Organic Subunits into Larger Molecules Once amino acids, purines, pyrimidines, monosaccharides, and other organic subunits formed in significant quantities, they presumably assembled spontaneously into macromolecules such as proteins and nucleic acids. A major problem presented by such assembly reactions is that they primarily involve chemical condensations. In aqueous solutions reactions tend to run spontaneously in the opposite direction, toward hydrolysis, in which covalent bonds holding molecular subunits together are broken by the addition of the elements of a molecule of water (see p. 46). Thus, it is not clear how condensations became predominant in the primitive environment.

The probability that condensations will proceed spontaneously is greatly increased if water is reduced in concentration or eliminated from the reacting systems. One means to accomplish this condition, already mentioned, is the presence of condensing agents among the reactants. Another is the anhydrous condition produced in ponds or pools evaporated to dryness by the sun or volcanic activity.

A third possibility is adsorption of reacting molecules on clays, originally proposed by J. C. Bernal and more recently by A. G. Cairns-Smith. Clays consist of stacked, extremely thin layers of negatively charged aluminosilicates separated by layers of water. Depending on the state of hydration, the water layers range from 1 to 5 nm in thickness. The aluminosilicate layers can adsorb and concentrate organic molecules in large quantities. Adsorption into the layers excludes water molecules, which sets up the anhydrous conditions promoting condensation reactions. The strongly negative charge of the aluminosilicate layers also attracts positive ions such as Na^+, K^+, Ca^{2+}, and Mg^{2+} to their surfaces. In addition metallic ions such as Fe^{2+}, Zn^{2+}, and Ni^{2+} substitute for aluminum atoms at sites in the aluminosilicate layers. These metallic ions can act as catalysts for condensation reactions, speeding the combination of organic building blocks into more complex assemblies at moderate temperatures.

All three routes for the condensation of macromolecules have been successfully tested in experiments simulating primitive conditions. The most extensive research in this area was carried out with the assembly of polypeptides and, to a lesser extent, of nucleotides and nucleic acids.

Perhaps the most spectacular results from experiments testing the spontaneous assembly of polypeptides were obtained by S. W. Fox and his colleagues. In their experiments mixtures of anhydrous amino acids were melted together at temperatures ranging from 120° to 200°C for seven days. The procedure caused the amino acids to link into long polymers with molecular weights approaching 20,000, which Fox called *proteinoids*. The polymers exhibit some of the characteristics of natural proteins when dissolved in water, including hydrolysis by proteinases such as trypsin, pepsin, and chymotrypsin, and even a limited ability to catalyze reactions such as the hydrolysis of ATP. Fox proposed that the somewhat extreme conditions required to form proteinoids might have prevailed around sites of volcanic activity, where solutions of amino acids might have been dried and heated to the required temperatures.

Polypeptides have also been produced by exposing amino acids to condensing agents under anhydrous conditions. One group of condensing agents used in these experiments, based on cyanamide (H_2N—$C\equiv N$) and its derivatives, is relatively stable in water. Mixtures of these compounds with amino acids in aqueous solution yield polypeptides at moderate temperatures.

Several laboratories demonstrated that polypeptides can also be produced from amino acids by the third major route, adsorption on clays. One of the most significant of these experiments was carried out by N. Lahav and S. Chang, who detected polypeptide formation in mixtures of clay and amino acids exposed to variations in water content and temperatures fluctuating between 25° and 94°C (the Experimental Process essay by Lahav and Chang on p. 1130 describes the rationale and results of their experiments).

Nucleotides and nucleic acids have been produced chiefly in simulated systems containing condensing agents. The condensing agents of choice have been *polyphosphates*, phosphate groups linked into chains of varying length. Besides promoting condensation reactions, polyphosphates release relatively large increments of energy when hydrolyzed, acting as an inorganic version of ATP. Ponnamperuma and R. Mack showed that nucleotides form spontaneously in mixtures of nucleosides and polyphosphates. Similarly, A. W. Schwartz and Fox induced spontaneous assembly of nucleic acids by heating nucleotides to 65°C in the presence of polyphosphates. Polyphosphates have also been used experimentally to promote spontaneous formation of AMP from adenine, ribose, and inorganic phosphate, polysaccharides from glucose, and polypeptides from amino acids at moderate temperatures. The polyphosphates required for these reactions could have been formed through the exposure of inorganic phosphorus or phosphate groups to high temperatures (200° to 350°C), possibly in regions of volcanic activity or at hot vents in the sea floor, or, as L. E. Orgel and his associates demonstrated, at more moderate temperatures through the interaction of phosphate groups with urea and ammonia.

Reactions of this type, as well as the spontaneous interactions producing simpler organic molecules, are considered to have produced significant quantities of all the major biological molecules over the millions of years following the initial formation of the earth. Accumulation of these compounds set up conditions for the third stage in the evolution of life, the formation of molecular aggregates capable of living reactions.

Stage Three: Formation of Molecular Aggregates

The formation of aggregates requires that molecules in solution must be precipitated or brought together in arrangements that are stable enough to resist disruption or return to solution. Four mechanisms have been widely considered as possibilities for spontaneous aggregation of discrete, reactive assemblies of this type: (1) *coacervation*, originally proposed by Oparin; (2) formation of *proteinoid microspheres*, proposed by Fox; (3) formation of lipid bilayers, first proposed by R. J. Goldacre; and (4) adsorption on clays, advanced as a possibility by Bernal. Each of these mechanisms has been duplicated in the test tube under conditions approximating the primitive environment.

Coacervation In coacervation, polypeptides in solution assemble into discrete, stable droplets 1 to 500 μm in diameter as a result of attractions between charged groups on the surfaces of polypeptide chains. If many polar or charged groups face the surface of a coacervate formed in this way, the surrounding water molecules tend to form an ordered film or boundary layer several molecules in thickness over the surface. The charged and polar surface groups and the film of ordered water molecules give the boundary of the droplet some of the properties of a membrane. As a result, coacervates may absorb molecules from the surrounding solution and shrink or swell in response to differences in molecular concentration inside or outside the droplets.

Oparin showed in a series of experiments that enzymes, if included in the solutions used to form coacervates, could be taken up and concentrated inside the protein droplets. Inclusion of starch phosphorylase, for example, an enzyme that catalyzes polymerization of glucose 1-phosphate, led to starch formation inside coacervates and growth of the droplets if glucose 1-phosphate was present in the solution outside the coacervates.

Although functional enzymes are not expected to be among the polypeptides included in coacervates on the primitive earth, the experiments demonstrate that reaction systems can be separately concentrated and confined in active form inside coacervate droplets. If such reactions led to extensive growth, the coacervates could have separated into smaller droplets with the same properties, thus achieving the most primitive form of growth and reproduction. By this route, according to Oparin, coacervates carrying out such "favorable" reactions could have increased in number and displaced droplets containing less favorable reaction systems. One problem with coacervates, however, is stability. The droplets are highly sensitive to conditions in the surrounding solution and break down readily if conditions stray from optimum values.

Proteinoid Microspheres Fox and his colleagues showed that if a solution of proteinoids is heated in water and allowed to cool, highly stable particles 1 to 2 μm in diameter, which he termed *microspheres*, separate out of the solution (Fig. 27-3). The microspheres are surrounded by a discrete surface layer superficially resembling a membrane and can carry out internal reactions among molecules absorbed from the surrounding solution. Some microspheres retain the weak catalytic activity of proteinoids, including the ability to increase the rate of biological reactions such as ATP hydrolysis.

Microspheres appear in electron micrographs in forms that suggest budding or fragmenting, implying that these aggregates can also carry out a primitive form of reproduction (see Fig. 27-3b). Fox proposed that microspheres formed readily in the primitive environment, producing molecular aggregates in which the first reactions of life may have appeared and reproduced.

Lipid Bilayer Assembly Phospholipids and some other lipid molecules can assemble spontaneously into bilay-

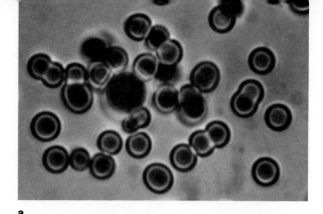

a

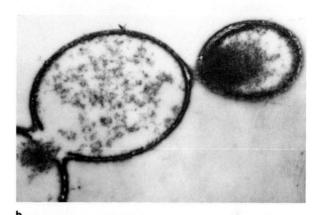

b

Figure 27-3 Fox's microspheres. **(a)** A collection of microspheres made by heating proteinoids in water, followed by cooling. Light micrograph. ×640. **(b)** Microspheres in a configuration superficially resembling cell division. Electron micrograph. ×7,800. Courtesy of S. W. Fox, from *Molecular Evolution and the Origin of Life*, by S. W. Fox and K. Dose. San Francisco: W. H. Freeman and Company, 1972.

ers when suspended in water (see p. 57). Frequently, the spontaneously formed bilayers round up into closed vesicles consisting of a saclike, continuous boundary "membrane" enclosing an inner space. Hargreaves and Deamer, for example, showed that phospholipids formed under simulated primitive earth conditions can assemble spontaneously into vesicles with an outer boundary consisting of a continuous lipid bilayer (Fig. 27-4).

Spontaneously formed bilayers are relatively stable and exhibit many properties of living membranes. Various substances, including proteins or polypeptides, can be absorbed on the bilayer surfaces or into the hydrophobic bilayer interior. Other substances can be trapped inside the vesicles as they form. The collection and concentration of organic molecules in the bilayers could have had catalytic effects and promoted a variety of biochemical reactions. In the primitive environment the vesicles could have increased in size through the further uptake of lipids and proteins. The vesicles could also have undergone a primitive form of reproduction by fragmentation caused by mechanical disturbances such as wave action.

The Experimental Process

How Were Peptides Formed in the Prebiotic Era?

Noam Lahav and Sherwood Chang

NOAM LAHAV is a professor at the Hebrew University of Jerusalem, Israel, who has been active in origin of life research since 1975. His major interests are the possible role of minerals in prebiotic reactions, especially peptide formation, and devising scenarios for the first prebiotic self-replication and translation reactions. SHERWOOD CHANG received his Ph.D. (in organic chemistry with a minor in physical chemistry) from the University of Wisconsin in 1967. Since then, he has been a research scientist at NASA's Ames Research Center, where he is now Chief of the Planetary Biology Branch. His interests are related to the origin of organic matter in the cosmos and the origin of life on Earth.

Many researchers in the early fifties were excited by the possibility of making contributions to the scientific understanding of the origin of life. Two scientific breakthroughs in 1953 stimulated their enthusiasm. The first, by James Watson and Francis Crick, was discovery of the double helix structure of deoxyribonucleic acid (DNA), which carries the information needed to replicate cells in all living organisms. With this knowledge the biochemistry of genetic processes could be studied in detail.

The second discovery resulted from an experiment outlined theoretically by Harold Urey and carried out by his graduate student, Stanley Miller, at the University of Chicago. This experiment [see p. 1125] simulating the passage of lightning through Earth's early atmosphere over its oceans, demonstrated unambiguously that some of the biochemicals necessary for all life could have been produced by a strictly nonbiological process prior to the origin of life. Although theories about how life originated had been published earlier in the century by A. I. Oparin and J. B. S. Haldane, Miller was the first researcher to provide compelling experimental evidence in support of one of the central ideas expressed in the theories, that the interaction of energy with inorganic chemicals on the primordial Earth led to synthesis of simple organic compounds.

Soon after the publication of Miller's experiment, scientists all over the world adopted his approach and studied other energy sources such as sunlight, heat, and radioactivity for their ability to promote organic syntheses. The origin of biochemical building blocks, which used to be considered a formidable barrier to understanding the origin of life, became a new area of chemical research. Moreover, it became clear that many questions about the chemical reactions leading to life's origin could be fruitfully explored through scientific research. This realization, together with the new discipline of molecular biology launched by the discovery of DNA structure, seemed to hold promise of a rapid solution to the mysteries of life's origin.

Miller's observations also paved the way for research on another key hypothesis of theories on the origin of life, namely, that simple compounds were converted by other natural processes to more complex organic compounds such as peptides and polynucleotides, the precursors of proteins and nuclei acids.

In living cells, protein synthesis occurs by chemical condensation of amino acids, a complex process catalyzed by enzymes. Enzymes also catalyze the synthesis of nucleic acids from nucleotides. Assuming that amino acids and nucleotides were readily formed on the early Earth, prebiotic processes would have been needed to convert these monomers into polymers in the absence of enzymes. Researchers who investigated experimental models for this process soon realized that the condensations were more difficult to achieve under plausible conditions than were the syntheses of the simple monomers.

As early as 1951, however, a biophysicist named J. D. Bernal suggested that minerals on the prebiotic Earth could have catalyzed the reactions of prebiotic monomers. He favored clay minerals, a family of ubiquitous aluminosilicates consisting of extremely small crystallites and possessing very chemically active surfaces. In 1966, an organic chemist, A. G. Cairns-Smith, further underscored the importance of minerals and made a case that clay crystals might have been the first life forms. Against this background in the mid-70s several scientists at NASA's Ames Research Center discussed a variety of experimental strategies to tackle the problem of peptide synthesis under prebiotic conditions. Since few experimental studies had addressed the role of clays in the origin of life, we concluded that the use of these minerals to assist the conversion of amino acids to peptides offered a promising and relatively unexplored approach.

To test our hypothesis that clay minerals could catalyze prebiotic amino acid condensation, we developed an experimental approach that also took into account natural variations in environmental conditions. Aqueous slurries of the amino acid glycine and either kaolinite or montmorillonite clay minerals were dried out by heating and held at 94°C for several days. In the second reaction cycle the mixture was cooled to room temperature, rewetted with water, and heated to dryness again. After these cycles were repeated the desired number of times, up to 27, the slurries were extracted with small amounts of ammonium hydroxide solution, and the water was analyzed for peptides. Three sets of control experiments were carried out in which the procedure was modified by lowering the temperature, omitting clays in the reaction cycles, and eliminating the rewetting step. The objectives of the control experiments were to determine how significant temperature, presence of clay, and rewetting were as independent factors in the overall results.

The rationale behind the procedures was based on the following considerations. Glycine was the major product of almost all experiments on the prebiotic synthesis of amino acids. Therefore, it seemed reasonable to use it as our starting material. Kaolinite and montmorillonite are widespread on Earth today and represent readily available model minerals known to be associated with aqueous environments. The thermal stability of glycine and its shorter polymers assure survival at the temperatures employed. Although it is known that amino acids in the solid state condense readily at temperatures above 150°C, it was considered important to carry out the reaction at temperatures below the boiling point of water. (Interestingly, microbes of ancient ancestral lineage that thrive at 105°C have been isolated recently from deep sea hydrothermal vents.) The wetting-drying procedure was intended to simulate conditions in environments where both temperature and water content fluctuate cyclically on a daily or seasonal basis. Presumably, such environments were common on the prebiotic Earth, as they are today.

Analyses of the reaction mixtures using standard chromatographic procedures showed that condensation of glycine to peptides did occur under the experimental conditions. In the absence of clays, however, negligible amounts of product were observed. The yields of peptides obtained with clays decreased with peptide length in the order diglycine > triglycine > tetraglycine > pentaglycine. The total amount of peptides increased with the number of cycles, but was always less than 1% of the initial amount of glycine. Peptide formation also took place at temperatures as low as 60°C, but at a very low rate. Heating alone was much less efficient than heating with wetting-drying cycles.

Our results showed that productivity in the condensation of glycine (and probably other amino acids) into peptides is greatly increased by the presence of clay minerals. Furthermore, the reactions are enhanced by environmental fluctuations of temperature and water content. Therefore, analogous prebiotic reactions could have occurred in common environmental settings like tidepools and seashores in the absence of implausible chemical condensing agents or enzymes.

Yet to be solved, however, are two related problems. First, how important were environmental rhythms to the origin of life? In our experiments, enhanced peptide synthesis occurred as a result of a dynamic balance between condensation and hydrolysis in a cyclically fluctuating environment. The significance of environmental fluctuations for stages of chemical evolution beyond peptide synthesis is unknown and remains to be studied. Such fluctuations might have been more important on the prebiotic Earth when day lengths were shorter and tides were more frequent and intense than now.

Second, how were peptides with non-random amino acid sequences formed and the directions for their synthesis encoded in the evolving assemblages of prebiotic polymers? Prebiotic peptides with random sequences would have limited functional capability. Eventually, a primitive mechanism would have to be "invented" by which peptides with essentially the same amino acid sequence could be produced repeatedly. A step toward such "directed" synthesis of peptides might have been provided by a process similar to that of our simulation. Given a fluctuating environment containing various amino acids, it is probable that certain peptides would be preferred over others because of the characteristics of specific monomers and short polymers. These factors include solubility, strength of adsorption to clay particles, stability to hydrolysis, and ease of incorporation into a growing peptide chain. Peptides with non-random sequences of amino acids would be of great interest. Conceivably, some could catalyze peptide bond formation itself, while others could fulfill different functions necessary in the chemical evolution of polymers into more complex systems.

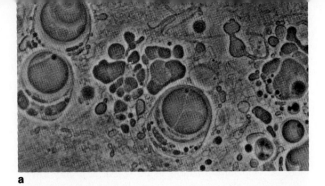

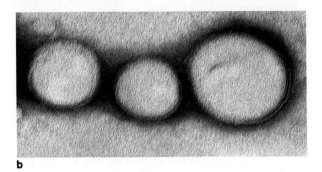

Figure 27-4 Vesicles consisting of a continuous lipid bilayer enclosing an interior space, assembled from phospholipids synthesized under primitive earth conditions. **(a)** Light micrograph. Courtesy of D. W. Deamer. **(b)** Electron micrograph of negatively stained preparation. Courtesy of W. R. Hargreaves, from *Nature* 266:78 (1977).

One problem with the spontaneous assembly of lipid bilayers from phospholipids is that the straight-chain fatty acids required for phospholipid formation have proved very difficult to produce under simulated primitive earth conditions. It is uncertain whether these molecules could have reached concentrations high enough to produce lipid bilayers in any quantity.

Adsorption on Clays Another difficulty common to coacervation, microsphere formation, and bilayer assembly is that their component molecules—proteinlike polymers or phospholipids—must be present in relatively high concentrations for spontaneous aggregation to occur. The locales in which such concentrations were reached might have been too limited for aggregation to progress to any extent by these mechanisms. Spontaneous assembly by adsorption on clays, in contrast, may have been much more common because of the extensive clay deposits expected to form in tidal flats and river estuaries. Clays, as noted, may also have been sites of extensive formation of both simple and more complex organic substances. The organic molecules formed and aggregated by the clays could have provided large, relatively stable areas of concentrated organic matter, heated by sunlight, in which a variety of interactions could have taken place. Besides the polymerization reactions demonstrated to be enhanced by clays, hydrolytic and other reactions, including some

releasing free energy, can also be promoted by clays, especially those containing metallic ions in abundance.

In addition clays have been shown to be capable of storing and transforming energy. Thus, an energy input such as heat, light, or mechanical compression can induce molecular alterations in a clay that represent a form of potential energy. On return to an original molecular arrangement, some of the energy can be released, in some cases in a form different from the form absorbed. Mechanical energy absorbed by clays, for example, can be released under changing conditions as fluorescence. Clays could thus have channeled environmental energy sources into interacting systems as well as providing centers of aggregation of organic molecules.

Of the various mechanisms aggregating macromolecules into interacting assemblies, adsorption on clays seems to be the most likely and prevalent route. All of the proposed mechanisms, however, may have contributed in regions where the special conditions required for their activity were met in the environment.

Stage Four: The Development of Life in the Primitive Aggregates

Of all the processes proposed as stages in the evolution of life, the specific reactions accomplishing the transformation of nonliving to living matter have proved to be the most difficult to imagine and test. However, there are some general ideas about the pathways by which this difficult feat may have been accomplished. These ideas fall into three areas: (1) the developments harnessing an energy source to drive metabolic pathways; (2) the appearance of enzymes, permitting reactions to take place in ordered, highly specific sequences at moderate temperatures; and (3) the development of informational systems necessary for directed synthesis and reproduction in kind.

Development of Reaction Pathways Utilizing an Energy Source Oxidation-reduction reactions, in which electrons are removed from one substance (an oxidation) and delivered to another (a reduction; see p. 312), were likely to have been among the initial energy-releasing reactions of the primitive macromolecular assemblies. At first the energy from these reactions would have been utilized in single steps, through the direct transfer of electrons from substances being oxidized to others being reduced as part of synthetic reactions. However, the advantages and efficiency of energy release in stepwise increments would have favored the development of intermediate carriers, opening the way for primitive electron transport systems.

Among the prime candidates for the first intermediate carriers are *porphyrins*. These complex molecular groups, which could have been readily synthesized in the primitive environment, gain or lose electrons

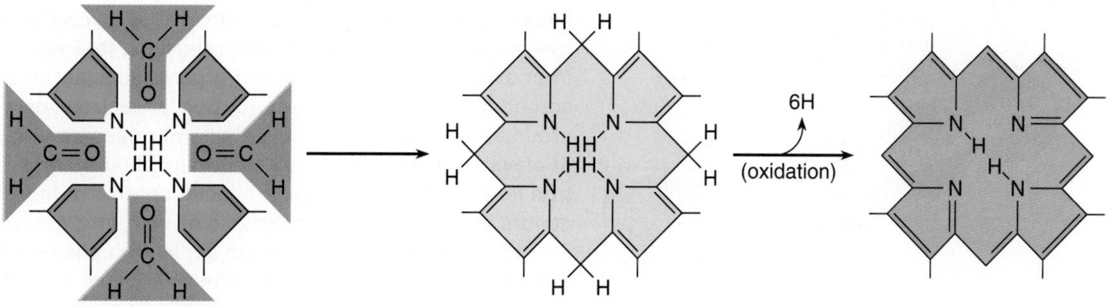

4 pyrrole rings + 4 formaldehyde molecules   porphyrin ring system

Figure 27-5 A possible route by which porphyrins may have assembled spontaneously on the primitive earth.

with relative ease. (Fig. 27-5 shows a possible synthetic pathway for porphyrins.) In this capacity they serve as the active groups transferring electrons in the *cytochromes*, carriers that occur in the electron transport systems of both respiration and photosynthesis in present-day cells.

Porphyrins are also pigments, with electrons capable of absorbing light energy and jumping to excited orbitals at elevated energy levels (see p. 370). In this capacity they form the active centers of chlorophyll molecules in all present-day photosynthetic organisms. In photosynthesis chlorophyll accepts electrons at relatively low energy levels, absorbs light energy, and transfers the excited electrons to acceptor substances. In primitive precells porphyrins could also have served in both electron transport and light absorption.

Initially, the developing oxidative and photosynthetic pathways probably used a common electron transport pathway, as they do in some contemporary photosynthetic bacteria (see p. 408). Electrons at elevated energy levels might enter the pathway after removal from highly reduced organic molecules absorbed from the environment (Fig. 27-6a). Alternatively, electrons might enter after being elevated to higher energy levels in a light-absorbing molecule such as a porphyrin (Fig. 27-6b and c). In the latter case the primitive photosynthetic pathway might operate either cyclically or noncyclically (see p. 379). In a cyclic pathway (Fig. 27-6b) electrons return to the light-absorbing molecule after passing through a chain of electron carriers. In a noncyclic pathway (Fig. 27-6c) electrons pass from the system containing the light-absorbing molecule to an acceptor outside the electron transport system. The electrons donated to acceptors in primitive systems could have been replaced by electrons removed through oxidation of an organic molecule or an inorganic substance such as H_2S. In some variations of the noncyclic pathways, electrons could pass from the joint photosynthesis-oxidation systems to organic molecules as final acceptors, accomplishing reductions in pathways synthesizing molecules required by the precells.

Development of ATP as a Coupling Agent In contemporary cells ATP is universally used as an agent that couples oxidative reactions releasing energy to synthetic reactions requiring energy (see p. 82). In this coupling activity ATP is assembled from ADP and inorganic phosphate through energy released by oxidations. The potential energy stored in ATP is released to reactions requiring energy by removal of the phosphate, which converts ATP back to ADP. (Two phosphates are added or released in some of the coupling reactions.)

ATP is synthesized in present-day cells by two primary routes. In one the energy released by a hydrolysis or oxidation is used directly to add a phosphate group to ADP, in a process termed *substrate-level phosphorylation* (see p. 311). This occurs, for example, in the direct chemical transfer of a phosphate group from phosphoenolpyruvate to ADP in glycolysis. Although relatively little ATP is produced by this mechanism in contemporary cells, it is still critical to those existing under anaerobic conditions and is considered likely to have been the initial pathway for ATP production in precells.

The second, more efficient but more complex pathway, *oxidative phosphorylation*, synthesizes ATP by the indirect use of free energy released by electron transport. The free energy released as electrons cascade through a series of electron carriers is used to set up an H^+ gradient across the membrane housing the electron transport system. ATP synthesis is then driven by the H^+ gradient, in effect by reversing an ATP-dependent H^+ pump (the F_oF_I ATPase; see p. 337 for details of this mechanism).

The earliest forms of life probably utilized hydrolysis as well as oxidation to obtain free energy. ATP may first have entered cells simply as one of many organic molecules absorbed from the environment and hydrolyzed as an energy source. It is likely that energy-requiring synthetic reactions were directly coupled to the release of free energy by the hydrolysis of ATP and other substances. Reversal of the reactions hydro-

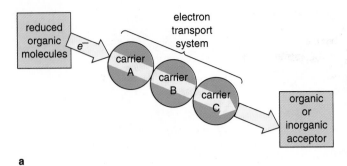

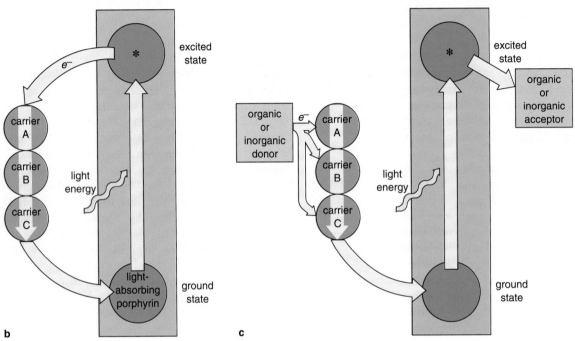

Figure 27-6 Possible electron transport pathways for primitive oxidative and photosynthetic reactions. **(a)** Electron flow from oxidation of highly reduced organic substances absorbed from the environment. **(b)** Cyclic electron flow around a porphyrin absorbing light and releasing electrons at elevated energy levels. **(c)** Noncyclic photosynthetic electron flow starting from highly reduced organic or inorganic donors and ending with an organic or inorganic acceptor. The same electron carriers could act in both photosynthetic and oxidative pathways. Porphyrins could have served as electron carriers as well as photosynthetic pigments in the primitive systems, as they do in contemporary prokaryotes and eukaryotes.

lyzing ATP, synthesizing ATP from ADP and phosphate, probably occurred to some extent in the primitive environment.

Some of the free energy released during photosynthesis or electron transport by primitive systems was probably also used in substrate-level synthesis of ATP. The likelihood of this outcome was demonstrated by W. S. Brinigar and his colleagues, who mixed a porphyrin, AMP, and inorganic phosphate with an imidazole group as a catalyst. Exposure of the mixture to ultraviolet or visible light resulted in the direct synthesis of ATP.

The development of the much more efficient pathway of oxidative phosphorylation for ATP synthesis probably appeared after substrate-level phosphorylation was widely established among precells. Because oxidative phosphorylation occurs only in association with membranes, development of this pathway for ATP production must have been linked intimately to the appearance of lipid bilayers. Oxidative phosphorylation

would also have required the evolution of electron transport systems including two or more molecules capable of accepting and releasing electrons. Establishment of a concentration gradient and membrane potential may have resulted spontaneously from electron transport in some of the membrane-bound systems. This gradient provided an energy source that could be used with greater efficiency than direct substrate-level phosphorylation to drive ATP synthesis. The ATPase activity necessary for this reaction could have been present initially in weak form in the membrane or molecular cluster enclosed by the membrane, as in Fox's microspheres. Once established, the much greater efficiency of oxidative phosphorylation assured that it would become the primary oxidative mechanism through which cells produce ATP.

The ability of precells to synthesize ATP in quantity would promote the use of this substance as a means to supply usable chemical energy to synthetic reactions. Because of the efficiency and versatility of energy transfer via phosphate bonds carried by ATP, this nucleotide gradually became established as the primary substance coupling reactions releasing and requiring energy.

Development of Metabolic Pathways Synthesizing Required Molecules

As these energy-releasing systems appeared, synthetic pathways also became more complex. N. H. Horowitz suggested a pattern by which complex, sequential synthetic pathways might have appeared. Suppose that a contemporary cellular pathway makes a required substance such as an amino acid through the sequence $A \rightarrow B \rightarrow C \rightarrow D \rightarrow E$, in which A is a simple inorganic substance and E is the final organic product. Initially, E was abundant in the environment and was absorbed directly by primitive aggregates. Later, as E became scarce because of use, chemical selection favored precells that could make E from D, a slightly less complex organic substance still found in abundance in the environment. As D became exhausted, selection favored assemblies that developed the pathway $C \rightarrow D \rightarrow E$, in which the even simpler substance C could be absorbed and used as a raw material to make D. This process continued until the entire synthetic pathway, based on an essentially inexhaustible inorganic substance, was established.

The Development of Enzymes

All these developing reaction systems yielding and utilizing energy were probably initially increased in rate by the relatively nonspecific, general catalytic activities of lipid vesicles, microspheres, coacervates, or clays. Gradually, more specialized molecules took over these catalytic functions, leading to the development of enzymes.

The first polypeptides included in the early catalytic systems may simply have provided a stable framework anchoring an inorganic catalyst such as a metallic ion to a primitive molecular cluster. However, since many amino acid side groups, particularly those with acidic or basic properties, also have catalytic activity, some of the catalyst-polypeptide complexes would have been more efficient in speeding reactions than the inorganic catalyst alone. The ability of polypeptide chains to undergo conformational changes in response to substrate binding probably also enhanced the effect of a bound catalytic group in speeding reaction rates.

As the structure of polypeptides became more ordered, some took on a folding conformation that held the catalyst in a position favoring certain substrates, which provided the beginning of specificity in catalyzed reactions. In some cases the catalytic activity of amino acid side groups in the folded active site reduced the importance of an included inorganic group, making possible the transfer of catalytic function to the protein. Selection then singled out and perpetuated the proteins providing the greatest stability, specificity, and increases in reaction rate. With the development of di-

rected protein synthesis, catalytic polypeptides became increasingly more specific and reproducible until the first fully active and specific enzymes appeared.

Origins of the Information System

Development of any of the metabolic pathways to levels in which more than a few coordinated steps were required would not have been likely without an information system to direct the synthesis and reproduction of the required components, particularly enzymes. Therefore, a mechanism for storing, reproducing, and translating information probably developed in parallel with the metabolic systems.

The developments leading to the appearance of a nucleic acid acting as a coding system present the most difficult conceptual problems for theorists interested in the evolution of life. The problem is central to understanding the origins of life, because once a system evolved that could store information and direct the synthesis of proteins of ordered sequence, the subsequent development of greater complexity and specificity through mutation and natural selection is relatively easy to visualize.

The major difficulty in understanding how a nucleic acid–based information system evolved is the total interdependence between nucleic acids and proteins in contemporary organisms. Without enzymatic proteins of essentially perfect specificity, the contemporary processes of replication, transcription, and translation of the nucleic acid code could not take place. The specificity of the enzymatic proteins involved in these processes, however, depends on their amino acid sequences, which are determined by the sequences of nucleotides in nucleic acids. Thus, in contemporary systems proteins depend on nucleic acids for their structure, and nucleic acids depend largely on proteins to catalyze their activities.

A good example of the interdependence of the two systems is the central event in translation, in which amino acids are attached to their corresponding tRNA molecules (see p. 676). Although the assembly of a protein depends on reading the genetic code in an mRNA molecule, translation of the code into the sequence of amino acids in a protein is entirely dependent on the fidelity by which amino acids are attached to tRNAs. This attachment depends on a group of proteins, the *aminoacyl-tRNA synthetases*. These proteins recognize both an amino acid and a tRNA that pairs with a codon corresponding to that amino acid, and join the two together. If the wrong amino acid is placed on a tRNA, either through failure of an aminoacyl-tRNA synthetase or through experimental manipulation, the translation system at the ribosome fails to recognize the error and places the incorrect amino acid where the correct one would have appeared (for details, see p. 676). Thus, the fidelity of protein synthesis depends on the ability of a group of enzymatic proteins to match amino acids with tRNAs. However, the properties of these enzymes depend on their amino acid sequences, which, in turn, are determined by the sequences of nucleotides in the genes encoding the proteins.

This close interdependence between nucleic acids and enzymatic proteins seemed to present an insoluble paradox until the recent discovery that some RNA molecules can act as "ribozymes" capable of self-catalysis (see p. 101). This ability may initially have allowed RNAs to act as both informational molecules and catalysts in the primitive precells, with no requirement for enzymatic proteins to catalyze their activities. Thus, a self-catalyzed "RNA world," as proposed by Orgel, F. H. C. Crick, C. Woese, W. Gilbert, and others, may have preceded the development of DNA as the hereditary material and provided an initial system free of close interdependence between nucleic acids and enzymatic proteins.

The RNA molecules developing as informational molecules may have been assembled initially from ribonucleotides taking part in oxidative and other metabolic reactions. Nucleotides containing a ribose sugar such as ATP, NAD, and coenzyme A form important parts of many metabolic pathways, including the oxidation-reduction reactions of glycolysis, respiration, and protein synthesis. Their universal presence indicates that they are likely to have been prevalent among the molecules participating in metabolic reactions in precells and thus available for assembly into polymers as RNAs. These RNAs could subsequently have developed into informational molecules and ribozymes.

Experiments by J. A. Doudna and J. W. Szostak showed that the RNA-based system that catalyzes the splicing of RNA molecules in the protozoan *Tetrahymena* (see p. 613) can be induced to splice nucleotides together into short copies of an RNA template. According to Gilbert, in the primitive RNA world self-replicative reactions of this type were the beginnings of an RNA-based informational system. Gilbert proposes that tRNA-like molecules linked to amino acids could have paired with the RNA informational molecules and influenced the assembly of polypeptides of ordered sequence. Some of the more ordered polypeptides, in turn, could have combined with the RNA informational molecules and tRNA-like molecules, stabilizing their three-dimensional form and increasing the efficiency of their interactions. Selection of the most efficient ribonucleoproteins gradually increased the specificity of the mechanism and led to a gradual transfer of most of the enzymatic activities associated with nucleic acid functions from RNA to proteins. Selection also gradually perfected and restricted the coding relationship between sequences in the RNA informational molecules and the amino acid sequences of proteins.

DNA, according to the proponents of the RNA world, entered the picture some time after RNA became established as the informational molecule. At first, DNA nucleotides may simply have been RNA nucleotides modified by removal of an oxygen from the ribose subunits of the ribonucleotides. At some point, supposedly, the DNA nucleotides were assembled into DNA copies of RNA informational molecules through the activity of an enzyme with reverse transcriptase activity. Once the copies were made, DNA supplanted RNA as the molecule storing information because it has greater chemical stability and can be assembled into much longer coding sequences than RNA. Transfer of the storage function to DNA left RNA with its functions at intermediate steps between the stored information and protein synthesis.

By the time DNA took up its activity as the informational storage molecule, interdependence between nucleic acids and proteins was fully established. The contemporary reactions self-catalyzed by RNA molecules, and the viruses with RNA genomes, according to this proposal, are molecular "fossils" persisting from the days of the RNA world.

The proposal of an RNA world solves so many conceptual problems connected with the origin of life that many thinkers in this field have accepted it almost as fact. However, the proposal is not without its problems. RNA is among the most difficult molecules to produce under conditions that simulate the primitive environment. Although ribose is among the sugars produced by self-condensation of formaldehyde, the yields are very low. Further, although the purines adenine and guanine can be made relatively easily, the pyrimidines uracil and cytosine, also required to make RNA polymers, are extremely difficult to produce under simulated conditions. Thus, it is uncertain whether RNA could have been produced in large enough quantities to develop as the primary informational molecule in the primitive environment. In fact, its appearance in quantity may have had to await the development of enzymatic proteins. Further problems are presented by the fact that ribose is synthesized by the simulated systems in both D- and L-forms (see p. 51). G. F. Joyce and his coworkers found that RNA replication stops immediately in simulated systems if mixed D- and L-forms interact.

These problems notwithstanding, there are some ideas about how the coding system may have stabilized around three-nucleotide codons specifying 20 amino acids. Systems based on one- or two-letter codons were probably eliminated because not enough amino acids could be specified. (One-letter codons can encode only 4 different amino acids and two-letter codons only 16.) Wobble (see p. 673) in two-nucleotide codon-anticodon pairing would likely have reduced the practical total considerably below 16. Codon-anticodon pairing with one- or two-nucleotide codons would also be highly unstable. Although three-letter codons allow 64 different combinations, offering the possibility of encoding this many different amino acids, the problems imposed by wobble evidently limited the mutually exclusive combinations to 21: 20 for different amino acids and 1 to specify "stop." Codons containing four or more nucleotides, which would have offered a much greater coding capacity, may have proved impractical simply because too many combinations were generated for any one group to be fixed by selection into a uniform genetic code. Thus, after considerable evolutionary hits and misses, a system based on three-letter codons, specifying only 20 amino acids and "stop," may have been the only likely outcome of the selection process.

Crick has suggested that the assignment of codons to particular amino acids was simply a historical accident. According to this idea, there are no special reasons why a particular codon stands for a given amino acid. The codons were simply established by random matching to amino acids very early in the evolution of cellular life. Once established, the codon assignments remained fixed because a mutation that switched a coding assignment could potentially alter the amino acid sequence of almost every polypeptide in an individual and would very likely be lethal.

Another possibility is that there are definite but as yet unknown chemical or physical relationships between codons or anticodons and amino acids that limit coding assignments to the present ones. The fact that substitutions in the genetic code occur in specialized and restricted systems such as mitochondria and other locations makes this possibility seem less likely. Crick's historical-accident hypothesis may therefore be correct.

Once the coding system appeared and the relationships between the codons and the amino acids became fixed, the evolutionary pathway leading to fully directed synthesis and reproduction was opened. At this stage the primitive molecular assemblies took on the full qualities of life. At the same time the coding system provided the basis for heritable changes caused by mutations and with it the change from chemical evolution to biological evolution by natural selection.

Stage Five: The Evolution of Cells from Precells

Natural selection of favorable mutations subsequently led to development of all the features of fully cellular life. One of these features was a nuclear region containing the DNA of the coding system and mechanisms for precisely replicating the DNA and transcribing it into RNA. Another feature was a cytoplasmic system, including ribosomes and the required enzymes, for translating the RNA into proteins of specified sequence. The cytoplasm also contained a photosynthetic or oxidative system supplying chemical energy for protein synthesis and the assembly of other required molecules. All these systems were enclosed by a membrane consisting of a lipid bilayer that contained transport pro-

teins controlling the molecules and ions entering and leaving the cell. Finally, the first cells included a mechanism precisely separating replicated DNA molecules and dividing the nuclear and cytoplasmic regions among daughter cells. The evolutionary pathways from precells to this level may have occupied millions of years and undoubtedly included many molecular trials that ended in failure.

LATER METABOLIC DEVELOPMENTS AND THE ORIGINS OF EUKARYOTIC CELLS

Photosynthesis, Oxygen Evolution, and Respiration

Further development of metabolic pathways in the newly emerging cells led to three sequential events of supreme importance for the appearance of eukaryotic cells. One was the development of an extended electron transport system allowing the use of water as an electron donor for photosynthesis. Splitting water molecules to release electrons for this newly developed pathway resulted in the second event critical for the evolution of more advanced cells: the release of large quantities of oxygen into the atmosphere. This changed the atmosphere from a reducing to an oxidizing character and made possible the third event, the evolution of respiratory electron pathways using oxygen as final electron acceptor.

According to R. E. Dickerson and his associates and others, these developments may have taken place by the following sequence of events. Natural selection of electron transport pathways in photosynthetic cells led to both cyclic and noncyclic systems. Among these systems were some operating in the sequences shown in Figures 27-6b and c, roughly equivalent to some of the pathways of photosynthesis in the modern green and purple sulfur bacteria (see p. 407). In these systems porphyrin-based light-absorbing pigments had evolved fully into bacteriochlorophylls, contained in a photosystem similar to that of photosystem I in modern photosynthetic cells (see p. 373). As such the systems were limited to donors, such as H_2S, releasing electrons at relatively high energy levels. Such donors were not plentiful enough in the environment for these photosynthetic cells to become dominant forms. The first evolutionary event of fundamental importance to the later evolution of eukaryotes resulted from the appearance of a second, modified form of the photosystem capable of accepting the low-energy electrons removed from water. This established the two-step pathway for light absorption and electron excitation in photosynthesis (Fig. 27-7) and allowed the photosynthetic cells possessing this pathway to use the most abundant molecule of the environment as the electron donor for the photosynthetic pathway. A critical part of this de-

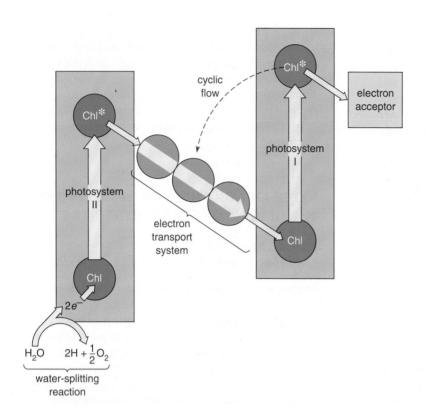

Figure 27-7 Noncyclic pathway of electron flow from water through two photosystems to an electron acceptor. The photosystems contain chlorophyll in ground (Chl) and excited (Chl*) states. The ability of the pathway to use water as an electron donor freed the photosynthetic organisms evolving this pathway from the use of less abundant substances such as H_2S as donors. In addition, the oxygen evolved as a by-product of the two-photosystem pathway altered the atmosphere from a reducing to an oxidizing character. The system may also operate cyclically by the flow of electrons from photosystem I to the electron transport system connecting the two photosystems (dashed arrow).

velopment was the evolution of a mechanism catalyzing the oxidation of water:

$$H_2O \longrightarrow 2H^+ + 2e^- + \tfrac{1}{2}O_2 \qquad (27\text{-}3)$$

As a consequence of this evolutionary development, which appeared in ancestral prokaryotes giving rise to cyanobacteria, oxygen was released to the atmosphere in ever-increasing quantities.

The gradual increase in atmospheric oxygen set the stage for the development of respiratory electron transport systems. At some point a group of cells developed an extra step in their electron transport systems involving the addition of cytochromes able to deliver electrons at very low energy levels to oxygen. This provided the advantage to these organisms of tapping the greatest possible amount of energy from the electrons before releasing them from electron transport. Just as the ability to use water as an electron donor made groups with this adaptation dominant among photosynthetic organisms, the much greater efficiency of cells able to use oxygen as final electron acceptor made these forms dominant among those living by oxidations alone.

Some indication of the time scale of these events can be taken from the geological and fossil record. The oldest prokaryotic cells yet discovered, as noted, date back about 3.5 billion years. Evolution of oxygen-evolving photosynthetic cells from these early prokaryotes may have required another 500 million years or more. This conclusion is based on discoveries of limestone-containing deposits, similar to stromatolites formed today by certain types of cyanobacteria, in rocks

Figure 27-8 Fossil and recently formed stromatolites. **(a)** Fossil stromatolites from western Australia that are more than 3 million years old. Courtesy of S. M. Awramik. **(b)** Stromatolites along a shoreline in western Australia. These stromatolites were formed relatively recently, between 1000 and 2000 years ago. Courtesy of J. W. Schopf.

laid down 1.6 to 2.7 billion years ago (Fig. 27-8). The length of time required for the atmosphere to change from a reducing to an oxidizing character has been estimated from the relative degree of oxidation of iron-containing compounds in rock layers. Rocks in relatively reduced form occur in layers deposited as recently as 1.8 billion years ago. These give way to relatively oxidized "red beds" that are about 1.5 billion years old. Therefore, the transition to an oxidizing atmosphere, begun by cyanobacteria nearly 3 billion years ago, may have occupied a period of 1.5 billion years and reached completion not much more than 1.5 billion years ago (Fig. 27-9).

The Appearance of Eukaryotic Cells

Eukaryotic cells evolved from one or more prokaryotic groups in a series of developments that may have begun as long as 3 billion years ago. The process culminated in the many adaptations that characterize present-day eukaryotic cells: separation of nucleus and cytoplasm by a nuclear envelope; structural and regulatory chromosomal proteins allowing long-term differentiation and specialization; cytoplasm containing membrane-bound compartments specializing in metabolic and synthetic functions—mitochondria, chloroplasts, ER, and the Golgi complex, among others; movement by systems based on microtubules and microfilaments; endocytosis; and reproduction by mitotic cell division.

An initial development believed to underlie the appearance of eukaryotic characteristics is the appearance of motile proteins, the forerunners of contemporary microfilament- and microtubule-based motile systems. Motile proteins enabled the plasma membrane to take in materials by endocytosis, an adaptation thought to have led to the appearance of mitochondria, chloroplasts, and other eukaryotic cell structures including the ER, Golgi complex, and nuclear envelope.

The origins of mitochondria and chloroplasts through membrane activity due to motile proteins have been proposed as the *endosymbiont hypothesis*, developed in greatest detail by L. Margulis. This hypothesis proposes that mitochondria and chloroplasts originated from prokaryotic cells that were ingested by endocytosis in cell lines destined to evolve into eukaryotes. Instead of breaking down, the ingested cells persisted as symbionts in the cytoplasm of the feeding cells. Mutations increasing the interdependence between the symbiotic prokaryotes and their host cells eventually converted the symbionts into mitochondria and chloroplasts.

Margulis proposes that in the development of mitochondria, prokaryotes evolved to a point at which complete photosynthesis was common, and oxygen was present in large quantities in the atmosphere. Among the prokaryotes were some nonphotosynthetic types that developed the capacity to engulf other cells through invaginations of the plasma membrane; these cells lived by oxidizing organic molecules ingested by this means. Some of these nonphotosynthetic cells possessed a complete respiratory system, including an electron transport system that used oxygen as final electron acceptor. These cells were therefore fully aerobic. Others were capable only of glycolytic fermentations and thus were limited to a less efficient anaerobic existence.

Interactions among these cells set the stage for the evolution of mitochondria, which began when groups of nonphotosynthetic, anaerobic cells ingested nonphotosynthetic aerobic cells in large numbers. Some of the ingested aerobes persisted in the cytoplasm of the capturing cells and continued to respire aerobically in their new location. As a result, the cytoplasm of the host anaerobe, formerly limited to the use of organic molecules as final electron acceptors, became the place of residence of an aerobe capable of carrying out the much more efficient transfer of electrons to oxygen. As the symbiotic relationship became established, many functions duplicated in the aerobe were taken over by the host cell. As part of this transfer of function, most of the genes of the aerobe moved to the cell nucleus and became integrated into the host cell DNA. At the same time the host anaerobe became dependent for survival on the respiratory capacity of the symbiotic aerobe. This gradual process of mutual adaptation culminated in the transformation of the cytoplasmic aerobes into mitochondria.

This new cell type, a former anaerobe containing a symbiotic aerobe in its cytoplasm, gave rise to the eukaryotic cell line. Other features of eukaryotic cells gradually appeared in the developing eukaryotes.

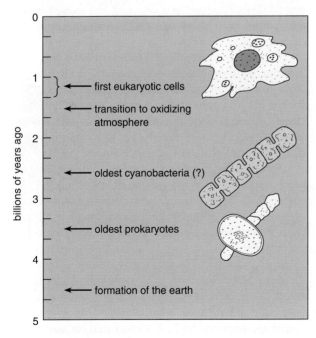

Figure 27-9 A time scale for the evolution of cellular life.

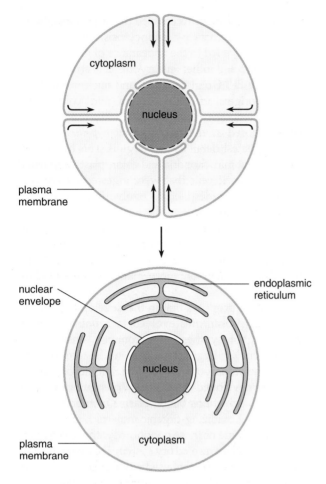

Figure 27-10 A hypothetical route for the formation of the nuclear envelope and ER through invaginations of the plasma membrane.

Development of the nuclear envelope and ER may also have been dependent on the ability of the plasma membrane to invaginate as part of endocytosis (Fig. 27-10). The invaginations carried enzymatic mechanisms associated with the plasma membrane—including those controlling the attachment of ribosomes and secretion of proteins—to the cell interior. The invaginations extended inward and surrounded the nucleus and left vesicles giving rise to the ER in the cytoplasm. The Golgi complex may also have developed by the same pathway or as a secondary development from vesicles budding off from the ER.

Some time after the ancestral eukaryotic cells developed to a level that included mitochondria, nuclear envelope, ER, and Golgi complex, a second major symbiotic relationship led to the development of chloroplasts. Members of a nonphotosynthetic eukaryotic subgroup ingested photosynthetic prokaryotes resembling present-day cyanobacteria, capable of splitting water and evolving oxygen. The ingested photosynthetic prokaryotes were gradually transformed into chloroplasts by the same evolutionary processes establishing mitochondria as permanent symbionts. These

early eukaryotic groups that contained developing chloroplasts founded cell lines giving rise to modern algae and plants.

Several major lines of evidence support the endosymbiotic hypothesis for the origins of mitochondria and chloroplasts. One is the similarity in structure and biochemistry between aerobic bacteria and mitochondria and between cyanobacteria and chloroplasts (for details, see Chs. 9, 10, and 21). Mitochondria and chloroplasts contain circular DNA molecules, with no associated histone or nonhistone proteins, that closely resemble prokaryotic DNA in structure. The organelle DNA codes for rRNA and tRNA molecules that in sequence resemble their prokaryotic counterparts more closely than the equivalent RNAs of the eukaryotic cytoplasm. The two organelles also possess ribosomes that resemble prokaryotic ribosomes in structure and carry out protein synthesis by mechanisms that most closely resemble those of prokaryotes. For example, mitochondria, chloroplasts, and contemporary prokaryotes use formylmethionine (see p. 687) as the first amino acid in a polypeptide chain instead of the unmodified methionine used in the eukaryotic cytoplasm outside the organelles. Similarities between present-day prokaryotes and the two organelles are also noted in their reactions to inhibitors of protein synthesis. For example, chloramphenicol inhibits protein synthesis in bacteria, chloroplasts, and mitochondria but has no effects on ribosomes in eukaryotic cytoplasm. Conversely, cycloheximide inhibits protein synthesis in the cytoplasm of eukaryotic cells but has no effect on bacterial or organelle protein synthesis. Another striking similarity is in the $F_o F_1$ ATPase carrying out ATP synthesis in association with electron transport. This complex enzyme occurs nowhere in living systems except in the membranes of prokaryotes and in mitochondria and chloroplasts.

Of the two organelles chloroplasts have the most pronounced biochemical and structural similarities to their prokaryotic counterparts, the cyanobacteria. Mitochondria and present-day bacteria appear to be less closely related. This difference probably reflects the more recent origins of chloroplasts from their prokaryotic ancestors.

Another line of evidence supporting the endosymbiont hypothesis for the origins of mitochondria and chloroplasts derives from the fact that contemporary organisms offer some striking examples of symbiotic relationships with a similar history. More than 150 genera distributed among 11 phyla contain eukaryotic algae or cyanobacteria as photosynthetic symbionts. Cyanobacteria, which occur as symbionts in protozoa and diatoms, can in some cases hardly be distinguished from true chloroplasts. One of the most interesting examples, described by R. K. Trench and his associates, involves a group of marine snails that utilizes chloroplasts derived from algae. Initially, the snails develop

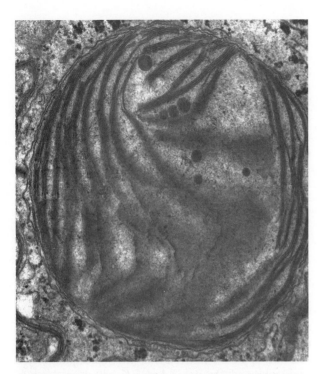

Figure 27-11 A chloroplast in a cell lining the gut of the snail *Elysia viridis.* Courtesy of R. K. Trench, R. W. Greene, B. G. Bystrom, and the Rockefeller University Press.

Reconstruction of Ancient Lineages Leading to Contemporary Prokaryotes and Eukaryotes

There is much current interest in tracing the lineages leading from the first true cells to contemporary prokaryotes and eukaryotes. Woese, G. E. Fox, and their colleagues and others approached this problem by comparing rRNA sequences among various living cells. Their work indicated that contemporary cells fall into three groups that share certain characteristics but differ enough to set them clearly apart. One group consists of present-day eukaryotes. The second consists of bacteria and cyanobacteria (called *eubacteria*, from *eu* = true or typical). The third group, the *archaebacteria*, are bacterialike organisms that include the *extreme halophiles*, which live in highly saline environments, the *extreme thermophiles*, which inhabit hot springs, and the *methanogens*, which release methane as a final metabolic product. Although the archaebacteria are bacterialike, their rRNAs are distinctly different from those of eubacteria—easily as different as they are from eukaryotic rRNAs.

Archaebacteria also have other features that are either unique, typically prokaryotic, or typically eukaryotic. For example, introns, a primarily eukaryotic characteristic, occur in tRNA genes in archaebacteria. The introns in these organisms occur in exactly the same position in the tRNA coding sequences as eukaryotic tRNA introns—at a site corresponding to one nucleotide to the 3' side of the anticodon—and include a sequence that can pair with the anticodon (see Fig. 15-44). Archaebacteria also initiate protein synthesis with unformylated methionine as do eukaryotes. On the other hand, archaebacteria lack nuclear envelopes and contain circular DNA molecules organized similarly to those of prokaryotes. Although the ribosomes of archaebacteria resemble those of eubacteria in dimensions, S. Altamura and his coworkers showed that the ribosomal subunits of archaebacteria form functional ribosomes when combined with eukaryotic subunits but not eubacterial subunits.

Comparisons among archaebacteria, eubacteria, and eukaryotes suggest that a series of very early evolutionary splits may have given rise to three separate lineages leading to these contemporary forms (Fig. 27-12). Introns, which may have been present in the common ancestor of all three lineages as a remnant of the RNA world, persisted in archaebacteria and the line leading to eukaryotes but were almost completely lost in eubacteria in response to selection pressures for maintenance of the smallest possible genome. In the line leading to eukaryotes, the advantages of the intronexon structure of genes to protein evolution by exon shuffling (see p. 594) led to the proliferation of introns in the genome. Thus, according to this proposal, advanced by Gilbert in particular, introns are very ancient sequence elements that were present in the genomes

as larvae containing no chloroplasts. As the larvae begin feeding on algae, chloroplasts from the algal cells are taken up by endocytosis in cells lining the gut. This process continues as the larvae develop into adult snails. The chloroplasts persist in the gut cells (Fig. 27-11) and continue to carry out photosynthesis in their new location. Trench showed by the use of radioactive tracers that organic molecules synthesized in the ingested chloroplasts diffuse into the cytoplasm of the host cells and are used as fuel substances by the snails. Individual chloroplasts may remain active in the gut cells for months. Thus, the route envisioned by Margulis and others for the origins of mitochondria and chloroplasts—ingestion by endocytosis and retention in the cytoplasm—can be demonstrated to take place in contemporary organisms.

There is also ample demonstration of the ability of DNA sequences to move from organelles into the nucleus. In one experiment by P. E. Thorsness and T. D. Fox, for example, genes carried on DNA initially inserted into mitochondria appeared later in the nucleus (see p. 903 and the Experimental Process essay by Thorsness on p. 902 for further details). In higher plants fragments of the entire chloroplast DNA circle can be found inserted at scattered locations in the nuclear DNA in many species. Enough of the fragments appear in spinach nuclei to make up six complete chloroplast genomes.

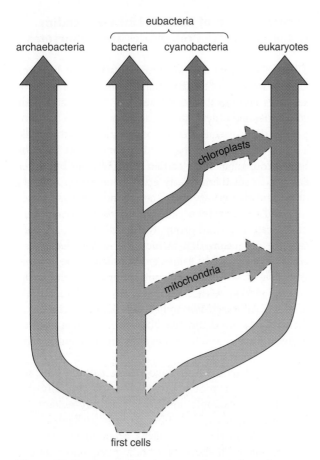

eubacteria

archaebacteria   bacteria   cyanobacteria   eukaryotes

chloroplasts

mitochondria

first cells

Figure 27-12 A possible pathway of evolution of archaebacteria, eubacteria, and eukaryotes from the first true cells. There is now considerable controversy surrounding the sequence of events leading from the first cells to the three main branches (dashed segment at the bottom of the diagram). Among the several hypotheses is one maintaining that the forerunners of archaebacteria were the most ancient line to develop from the first cells and that eubacterial and eukaryotic lines split from the archaebacterial line. Another idea maintains that the eubacterial line is most ancient and that the archaebacterial line split at a later time from the eubacterial line. Subsequently, according to this model, the archaebacterial line split to form the forerunners of the eukaryotes.

of the first cells, largely lost in the evolution of eubacteria, and retained and elaborated in eukaryotes. It should be mentioned, however, that there is no solid evidence indicating whether eukaryotic introns are evolutionary survivors from an ancient group or an innovation that appeared separately in eukaryotes and archaebacteria. Similarly, it is unknown whether the very few introns found in eubacteria are remnants from an earlier time or a sequence element introduced in some way from eukaryotes.

The present-day bacteria with features most similar to those of mitochondria appear to be a subgroup known as the *α-purple sulfur bacteria*, including *Rho-*

dospirillum and *Paracoccus.* The ancient forebearers of this group may have been the prokaryotic cells ingested and converted into the symbionts leading to present-day mitochondria. Among the cyanobacteria, a group known as the *cyanelles* exists today as chloroplastlike symbionts of protozoa. These symbiotic cyanobacteria, including *Cyanophora,* appear to have genomes most closely related to those of modern chloroplasts. In particular the DNA circle of *Cyanophora* has an inverted repeat containing two copies of the rRNA genes, a feature typical of chloroplast genomes (see p. 889). Thus, the direct ancestors of this group may have been the prokaryotic cells ingested and converted into chloroplasts.

The total time required for evolution of the first fully eukaryotic cells may have extended over a billion years or more. Fossils in rocks as old as 1.4 billion years have been claimed by J. W. Schopf and D. Z. Oehler to be eukaryotic cells on the basis of apparent nuclei and division figures resembling mitosis. These conclusions are controversial, however; others maintain that the apparent nuclei result from precipitation of prokaryotic cell structures during fossilization. According to the dissenting investigators, eukaryotic cells are not likely to have been on the earth much longer than about 1 billion years. Assuming that the first fully eukaryotic cells appeared between 1.4 and 1 billion years ago, and that the first cells appeared 3.5 billion years ago, the evolution of fully eukaryotic cells from the first cells required 2 billion years or more. If so, this interval is twice as long as the interval from the time of the earth's origins to the appearance of the first prokaryotes (see Fig. 27-9). The amount of time required undoubtedly reflects the complex adaptations of eukaryotic cells.

The events outlined in this chapter, leading from the earth's origins to the appearance of eukaryotic cells, are admittedly hypothetical and so tenuous that they may seem impossible. But given the total time span of these events, 3.5 billion years,

> the impossible becomes possible, the possible probable, and the probable virtually certain. One has only to wait; time itself performs the miracles.[1]

There are even those who say that, given the chemical conditions and environment of the primitive earth, the evolution of prokaryotic and eukaryotic cellular life, no matter how complex it has become, was inevitable—it was a predictable outcome of the spontaneous chemical reactions taking place since the earth's origin. Therefore, according to this idea, the origin of life was chemically predetermined by the initial conditions. It may well be that a similar sequence of chemical and organic evolutionary events is now under way on other planets in the universe.

[1] From G. Wald, The origin of life, *Sci. Amer.* 191:45 (1954).

For Further Information

ATP synthesis, *Chs. 9 and 10*
Chloroplasts, structure and biochemistry, *Ch. 10*
   DNA, transcription, and protein synthesis in, *Ch. 21*
Electron transport in mitochondria and bacteria, *Ch. 9*
   in chloroplasts and cyanobacteria, *Ch. 10*
Endocytosis, *Ch. 20*
Eukaryotic cell structure, *Ch. 1*
Genetic code, *Ch. 16*
Glycolytic fermentation, *Ch. 9*
Mitochondria, structure and biochemistry, *Ch. 9*
   DNA, transcription, and protein synthesis in, *Ch. 21*
Oxidative phosphorylation, *Ch. 9*
Photosynthesis, *Ch. 10*
Prokaryotic cell structure, *Ch. 1*
Protein synthesis, *Ch. 16*
Respiration, *Ch. 9*
Substrate-level phosphorylation, *Ch. 9*

Suggestions for Further Reading

General Books and Articles

Bernal, J. D. 1967. *The origin of life.* New York: World.

Cairns-Smith, A. G. 1985. *Seven clues to the origin of life.* New York: Cambridge University Press.

Calvin, M. 1969. *Chemical evolution.* New York: Oxford University Press.

Day, W. 1984. *Genesis on planet earth: the search for life's beginnings.* New Haven, Conn.: Yale University Press.

DeDuve, C. 1991. *Blueprint for a cell. The nature and origin of life.* Burlington, N.C.: Niel Patterson Publishers.

Dickerson, R. E. 1978. Chemical evolution and the origin of life. *Sci. Amer.* 239:70–86 (September).

Dyson, F. 1985. *Origins of life.* New York: Cambridge University Press.

Fox, S. W., and Dose, K. 1977. *Molecular evolution and the origin of life.* 2nd ed. New York: Marcel Dekker.

Horgan, J. 1991. In the beginning . . . . *Sci. Amer.* 264:116–25 (February).

Kenyon, D. H., and Steinman, G. 1969. *Biochemical predestination.* New York: McGraw-Hill.

Keosian, J. 1964. *The origin of life.* New York: Reinhold.

Loomis, W. J. 1988. *Four billion years.* Sunderland, Mass.: Sinauer.

Margulis, L. 1970. *Origins of eukaryotic cells.* New Haven, Conn.: Yale University Press.

Nei, M., and Koehn, R. K. 1983. *Evolution of genes and proteins.* Sunderland, Mass.: Sinauer.

Scott, A. 1987. *The creation of life: past, future, and alien.* New York: Basil Blackwell.

Formation of the Earth and Its Primitive Atmosphere

Pace, N. R. 1991. Origins of life—facing up to the physical setting. *Cell* 65:531–33.

Prebiotic Synthesis of Organic Molecules

Chyba, C. F., and Sagan, C. 1992. Endogenous production, exogenous delivery, and impact-shock synthesis of organic molecules: an inventory for the origins of life. *Nature* 355:125–32.

Chyba, C. F.; Thomas, P. J.; Brookshaw, L.; and Sagan, C. 1990. Cometary delivery of organic molecules to the early earth. *Science* 249:366–73.

Ferris, J. P. 1987. Prebiotic synthesis: problems and challenges. *Cold Spring Harbor Symp. Quant. Biol.* 52:29–35.

Laszlo, P. 1987. Chemical reactions on clays. *Science* 235:1473–77.

Mann, A. P. C., and Williams, D. A. 1980. A list of interstellar molecules. *Nature* 283:721–25.

Miller, S. L. 1987. Which organic compounds could have occurred on the prebiotic earth? *Cold Spring Harbor Symp. Quant. Biol.* 52:17–27.

Evolution of Metabolic Systems

Dickerson, R. E. 1980. Cytochrome c and the evolution of energy metabolism. *Sci. Amer.* 242:136–53 (March).

Nelson, N., and Taiz, L. 1989. The evolution of H^+-ATPases. *Trends Biochem. Sci.* 14:113–16.

Sharp, P. A., and Eisenberg, D. 1987. The evolution of catalytic function. *Science* 238:729–30, 807.

Towe, K. M. 1990. Aerobic respiration in the Archaean? *Nature* 348:54–56.

Evolution of a Coding System, Genes, and Genomes

Benner, S. A.; Ellington, A. D.; and Tauer, A. 1989. Modern metabolism as a palimpsest of the RNA world. *Proc. Nat. Acad. Sci.* 86:7054–58.

Cech, T. R. 1986. RNA as an enzyme. *Sci. Amer.* 255:64–75.

Crick, F. H. C. 1968. The origin of the genetic code. *J. Molec. Biol.* 38:367–79.

Darnell, J. E., and Doolittle, W. F. 1986. Speculations on the early course of evolution. *Proc. Nat. Acad. Sci.* 83:1271–75.

Doudna, J. A.; Couture, S.; and Szostak, J. W. 1991. A multisubunit ribozyme that is a catalyst of and template for complementary strand RNA synthesis. *Science* 251:1605–8.

Gilbert, W. 1981. The exon theory of genes. *Cold Spring Harbor Symp. Quant. Biol.* 52:901–5.

Joyce, G. F. 1989. RNA evolution and the origins of life. *Nature* 338:217–24.

Lamond, A. I., and Gibson, T. J. 1990. Catalytic RNA and the origin of genetic systems. *Trends Genet.* 6:145–49.

Orgel, L. E. 1986. RNA catalysis and the origin of life. *J. Theor. Biol.* 123:127–49.

———. 1989. Evolution of the genetic apparatus. *Cold Spring Harbor Symp. Quant. Biol.* 52:9–16.

Waldrup, M. M. 1989. Did life really start out in an RNA world? *Science* 246:1248–49.

Wintersberger, U., and Wintersberger, E. 1987. RNA makes DNA: a speculative view of the evolution of DNA replication mechanisms. *Trends Genet.* 3:198–202.

Origins of Mitochondria and Chloroplasts

Gray, M. W. 1989. Origin and evolution of mitochondrial DNA. *Ann. Rev. Cell Biol.* 5:25–50.

———. 1989. The evolutionary origins of organelles. *Trends Genet.* 5:294–99.

Gray, M. W.; Cedergren, R.; Abel, Y.; and Sankoff, D. 1989. On the evolutionary origin of the plant mitochondrion and its genome. *Proc. Nat. Acad. Sci.* 86:2267–71.

Mahler, H. R. 1983. The exon:intron structure of some mitochondrial genes and its relationship to mitochondrial evolution. *Internat. Rev. Cytol.* 82:1–98.

Margulis, L. 1981. *Symbiosis in cell evolution.* New York: W. H. Freeman.

Evolutionary Lineages of Prokaryotes and Eukaryotes

Knoll, A. H. 1991. End of the proterozoic era. *Sci. Amer.* 265:64–73 (October).

Lake, J. A. 1991. Tracing origins with molecular sequences: metazoan and eukaryotic beginnings. *Trends Biochem. Sci.* 16:46–50.

Liu, X.-Q. 1991. An ancient intron in eubacteria: new light on intron origins. *Bioess.* 13:185–86.

Wais, A. C. 1986. Archaebacteria: the road to the universal ancestor. *Bioess.* 5:75–78.

Woese, C. R. 1987. Bacterial evolution. *Microbiol. Rev.* 51:221–71.

Review Questions

1. What evidence indicates that life appeared on the earth between 4.5 and 3.5 billion years ago?

2. Why are the views that life had a divine origin or arrived on earth as contamination on a spaceship considered unscientific? Why is it necessary for scientific purposes to assume that life originated through inanimate chemical processes?

3. What is the distinction between the common and scientific attitudes about spontaneous generation? Why may spontaneous generation have been possible on the primitive earth but not now?

4. Outline the basic tenets of the Oparin–Haldane hypothesis. What gases were proposed to be present in the earth's primitive atmosphere? What is the significance of the reducing quality proposed for the primitive atmosphere? What evidence supports the idea that the primitive atmosphere was mildly or strongly reducing?

5. List the stages in the evolution of cellular life. How are the earth and its atmosphere believed to have originated? What interactions would probably have removed any gaseous oxygen from the atmosphere? What is the significance of this removal for the evolution of life?

6. Outline the structure and operation of the Miller apparatus. What biological compounds were produced in the apparatus?

7. What is the significance of HCN as a component of the early atmosphere? Of formaldehyde? What major classes of organic molecules have been produced in experiments simulating prebiotic conditions? What energy sources may have acted on the inorganic chemicals of the primitive environment to produce organic molecules? What observations of natural systems support the results of experiments indicating that organic molecules could have been synthesized on the primitive earth?

8. What is a condensation? What conditions may have promoted condensations on the primitive earth? What are clays? How may clays have figured in condensation reactions? What experiments indicate that condensations producing polypeptides and nucleic acids may actually have taken place?

9. What minimum characteristics would you consider to be required for the presence of life?

10. What are coacervates? Proteinoid microspheres? What lifelike reactions have been demonstrated in coacervates and proteinoid microspheres?

11. How may lipid bilayers have been important in molecular aggregation? What major problem is connected with the possible involvement of lipid bilayers in aggregation of molecules in the primitive earth?

12. What characteristics of clays may have made these substances ideal as sites of molecular aggregation and interaction?

13. What is the difference between substrate-level phosphorylation and oxidative phosphorylation in the synthesis of ATP? Does oxidative phosphorylation require oxygen to be present? How may substrate-level phosphorylation and oxidative phosphorylation have arisen in the primitive environment? How may porphyrins have been important in these developments? What are the possible relationships between respiration and photosynthesis in the development of electron transport pathways?

14. What advantages does RNA offer over DNA as a primitive informational molecule? What experiments and observations support the idea that RNA may have been the first informational molecule? What problems limit this possibility?

15. What characteristics may have favored the selection of three-letter codons as opposed to one-, two-, or four-letter codons? What is meant by the "historical accident" that may have fixed the codons in their present assignments? How did the appearance of a coding system alter evolution from chemical to organic?

16. What adaptations were probably important in the conversion of precells to cells? Trace the developments that may have led to the use of water as an electron donor for photosynthesis. What is the significance of photosynthesis to the evolution of cellular life? What evidence indicates that oxygen may have appeared in quantity in the earth's atmosphere some 1.8 to 1.4 billion years ago?

17. Outline the primary differences between prokaryotic and eukaryotic cells. What environmental conditions were necessary for the appearance of eukaryotic cells?

18. Outline the possible steps in the origin of mitochondria and chloroplasts according to the endosymbiont hypothesis. What evidence supports the idea that mitochondria and chloroplasts originated from ancient prokaryotes? How may the ER, Golgi complex, and nuclear envelope have originated?

19. What characteristics of archaebacteria are similar to those of eubacteria? Of eukaryotes? What relationship may ancestors of archaebacteria have had to present-day prokaryotes and eukaryotes?

INDEX